Human Biology

Test Yourself

Take a chapter quiz on the *Human Biology* ARIS site. Each quiz is specially constructed to test your comprehension of key concepts. Feedback on your responses helps you gauge your mastery of the material.

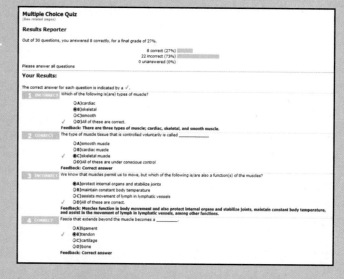

Access to Premium Learning Materials

The *Human Biology* ARIS site is your portal to exclusive study tools such as McGraw-Hill's animations, Virtual Labs, and ScienCentral videos.

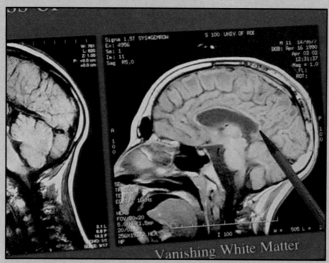

Vanishing White Matter

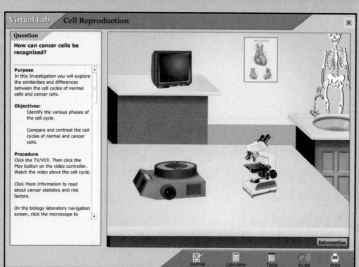

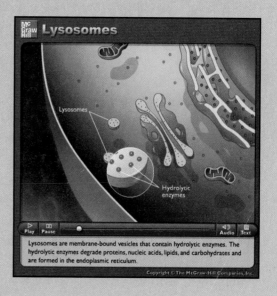

tenth **edition**

Human **Biology**

Sylvia S. **Mader**

With contributions by

Linda D. Smith-Staton
Pellissippi State Technical Community College

Susannah Nelson Longenbaker
Columbus State Community College

 **Higher Education**

Boston   Burr Ridge, IL   Dubuque, IA   New York   San Francisco   St. Louis
Bangkok   Bogotá   Caracas   Kuala Lumpur   Lisbon   London   Madrid   Mexico City
Milan   Montreal   New Delhi   Santiago   Seoul   Singapore   Sydney   Taipei   Toronto

Higher Education

HUMAN BIOLOGY, TENTH EDITION

Published by McGraw-Hill, a business unit of The McGraw-Hill Companies, Inc., 1221 Avenue of the Americas, New York, NY 10020. Copyright © 2008 by The McGraw-Hill Companies, Inc. All rights reserved. No part of this publication may be reproduced or distributed in any form or by any means, or stored in a database or retrieval system, without the prior written consent of The McGraw-Hill Companies, Inc., including, but not limited to, in any network or other electronic storage or transmission, or broadcast for distance learning.

Some ancillaries, including electronic and print components, may not be available to customers outside the United States.

 This book is printed on recycled, acid-free paper containing 10% postconsumer waste.

1 2 3 4 5 6 7 8 9 0 DOW/DOW 0 9 8 7

ISBN 978–0–07–298686–0
MHID 0–07–298686–7

Publisher: *Janice Roerig-Blong*
Sponsoring Editor: *Thomas C. Lyon*
Developmental Editor: *Darlene M. Schueller*
Marketing Manager: *Tamara Maury*
Senior Project Manager: *Jayne Klein*
Lead Production Supervisor: *Sandy Ludovissy*
Senior Media Project Manager: *Jodi K. Banowetz*
Senior Media Producer: *Eric A. Weber*
Designer: *Laurie B. Janssen*
Cover/Interior Designer: *Christopher Reese*
 (USE) Cover Image: © *GettyImages/Ross Woodhall*
Senior Photo Research Coordinator: *Lori Hancock*
Photo Research: *Evelyn Jo Hebert*
Supplement Producer: *Melissa M. Leick*
Compositor: *Electronic Publishing Services Inc., NYC*
Typeface: 10/12 *Palatino*
Printer: *R. R. Donnelley Willard, OH*

The credits section for this book begins on page C-1 and is considered an extension of the copyright page.

Library of Congress Cataloging-in-Publication Data

Mader, Sylvia S.
 Human biology / Sylvia S. Mader. -- 10th ed.
   p. cm.
 Includes index.
 ISBN 978-0-07-298686-0 --- ISBN 0-07-298686-7 (hard copy : alk. paper)
 1. Human biology. I. Title.
QP36.M2 2008
612--dc22
                         2006103270

www.mhhe.com

Brief Contents

Contents

Readings

Bioethical Focus

Health Focus

Science Focus

Preface

When I was teaching general biology, it became apparent to me that students were very interested in how their bodies worked, how to keep them healthy, and how they can occasionally malfunction. Students also recognized the importance of environmental concerns to their lives. I decided it would be possible to write a text and develop a course for nonmajors that built on these interests while still teaching biological concepts and how scientists think and carry out research.

The application of biological principles to practical human concerns is now widely accepted as a beneficial approach to the study of biology. Students should leave college with a firm grasp of how their bodies normally function and how the human population can become more fully integrated into the biosphere. We are frequently called upon to make health and environmental decisions. Wise decisions require adequate knowledge and can help ensure our continued survival as individuals and as a species.

In this edition, as in previous editions, the text presents concepts clearly, simply, and distinctly so that students will feel capable of achieving an adult level of understanding. Detailed, high-level scientific data and terminology are not included because I believe that true knowledge consists of having a working understanding of concepts rather than technical facility.

The Tenth Edition of *Human Biology*

It can truly be said that this edition of *Human Biology* is an achievement beyond all the other editions. The illustration program, the text, and the pedagogy have all been substantially improved. In short, Mader's *Human Biology* has been revitalized for this important anniversary edition.

The aim of the revision was to keep the content basically the same but to increase relevancy and student appeal. The goal was achieved through the very able assistance of two highly talented professors of nonmajors. Linda D. Smith-Staton, who teaches at Pellissippi State Technical Community College, and Susannah Nelson Longenbaker, who teaches at Columbus State Community College, are recognized for their significant contributions to this edition on the title page of the book. Many other professors also lent their talents and their names are listed in the acknowledgment section.

Brilliant New Illustration Program

A spectacular illustration program has been combined with innovative page layouts to enhance the artistic attractiveness of every page of *Human Biology*. See pages 63 to 65 for a first-hand example. Virtually every illustration in this edition is either completely new or significantly revised to convey the excitement of the concept being depicted. In addition to new artwork, almost all visuals are new. Photographs of humans, emphasizing ethnicity, and micrographs of structures have been incorporated into illustrations to increase student involvement and interest. See Figures 6.1 and 13.7, for example.

The pedagogical value of the art has been improved. For example, the systems illustrations not only label the various organs, they also state their functions. See Figure 8.1, the digestive system, and Figure 12.4, the superficial muscles of the body, for example.

Engaging New Chapter Openers

The new chapter opener page will immediately encourage student interest in the content of the chapter. Linda Smith-Staton wrote the engaging new vignettes which are accompanied by an oversized photograph perfectly selected to suit the story line of each. These real-life situations pertain to the content of the chapter in such a way that it was possible to refer to them throughout the chapter. No longer are the opening stories isolated from the rest of the chapter. For example, in chapter 3 see references to Calvin, the boy in the opening story, on pages 42, 47, and 55. Integration of the vignettes into the chapter makes them an additional way by which students can learn.

Linda also wrote "Thinking Critically About the Concepts," a feature in the chapter end matter which asks questions combining the content of the opening story and the chapter concepts, which are listed on the chapter opening page. The answers to these questions are given in Appendix A.

New Parts, Chapters, and AIDS Supplement

In this edition, parts have been reorganized and chapters have been rewritten in order to achieve our goal of more student involvement in the text.

- Chapter 1 entitled "Exploring Life and Science", is a new chapter. The characteristics of life and the role of humans in the biosphere are discussed before the chapter takes up the process of science. This portion of the chapter has been expanded in a way that will heighten student interest and increase knowledge of science. Now, the discussion, including the example of a controlled study, centers around the discovery by Dr. Barry Marshall that a bacterial infection causes ulcers. The focus reading will be of interest to students because it concerns the best spacing of births according to statistical data.

- Chapter 7, a new chapter called " Lymphatic System and Immunity" combines the chapters and sections that pertained to this topic in the previous edition. The biology of pathogens, including viruses and bacteria, is covered before the lymphatic system and then immunity are discussed.
- Part VI, Human Genetics, has been reorganized and the chapters have been rewritten to increase their applicability to students. The material previously found in Genetic Counseling has been moved into the other chapters of this part to increase student interest. For example, Chapter 18 which discusses cell division now includes chromosomal abnormalities, Chapter 20 on genetic inheritance now includes autosomal and sex-linked genetic diseases, and Chapter 21 on DNA biology and technology now includes genomics.
- An AIDS supplement is new to this edition. The supplement reviews the origin and prevalence of HIV, phases of an HIV infection, and the structure and life cycle of HIV as a precursor to reviewing the treatment of AIDS.
- With the movement of sexually transmitted diseases into the reproductive chapter and the movement of the cancer chapter into Part VI, it was possible to reduce the number of chapters in this edition from 27 to 24 and to have a total of six parts instead of seven parts. However, the book remains the same length as in the previous edition.

New Boxed Readings

All boxed readings have been revised and most are new to this edition. The topics were chosen for relevancy and interest to students.

Science Focus readings, which pertain to biological topics of interest to students, are new to this edition. For example, see Chapter 4 for a Science Focus about nerve regeneration, Chapter 11 for one about how skeletal remains are identified, and Chapter 12 for one that describes the process of rigor mortis.

Health Focus readings discuss health issues of current concern as before. As an example, see Chapter 9 for a new Health Focus which answers questions about smoking, tobacco, and health.

Bioethical Focus readings describe a modern ethical dilemma as before. For example, see Chapter 3 whose bioethical issue discusses stem cell research.

New Pedagogy

Check Your Progress boxes now appear at the end of each section in this edition. They test student understanding before the student moves on to the next section. Check Your Progress questions are answered in Appendix A.

New Summaries   An extensive chapter summary is organized according to the major sections in the chapter. Brief statements, lists, tables, and artwork help students review the important topics and concepts.

Thinking Critically About the Concepts   This feature asks conceptual questions pertinent to both the vignette (opening story) and the chapter concepts. In this way this feature integrates the chapter and the text as a whole. Answers to these questions can be found in Appendix A.

New Applications

Improvement of previously used applications and addition of new applications was a prominent aim of this edition. The new vignettes helped because they served as a way to apply conceptual ideas to real-life situations throughout the chapter. The addition of new applications in the text (aside from boxed readings) are exemplified by such new topics as

- page 31   use of nutrition labels to determine the amount of fat in a food
- page 51   malformed microtubules and infertility
- page 72   how to prevent skin cancer
- page 109   blood doping
- page 145   disorders associated with the intestinal wall
- page 151   obesity and diabetes type 2 and cardiovascular disease
- page 172   Heimlich maneuver
- page 217   manual strangulation and the hyoid bone
- page 220   glenoid cavity and the rotator cuff
- page 267   drugs of abuse which was rewritten for this edition
- page 291   cause of motion sickness
- page 330   graphs compare the effects of pregnancy and birth control pills on hormone levels and reproductive organs
- page 368   effects of pregnancy on mother
- page 413   recommendations for early detection of cancer
- page 532   practical ways to make our society more sustainable

New Emphasis on Disorders

Students will also be attracted to the discussions of disorders that accompany the systems chapters. More of these have been added because of their appeal to students. For example, see Section 9.7, Respiratory Disorders, 10.5, Kidney Disorders, and 12.4, Muscular Disorders.

Homeostasis and Working Together illustrations

Because of their popular appeal we have retained the homeostasis sections that include an illustration demonstrating how systems work together. These five sections make use of real life situations to show how homeostasis is maintained in the body. As an example see Section 6.6.

Overview of Changes to *Human Biology*, Tenth Edition

Illustration Program

A spectacular new art program combined with many new photographs and micrographs and utilizing an innovative page layout enhances the pedagogical value of the text.

New Vignettes

New opening stories bring biology to life such that the story-line is referred to throughout the chapter. Critical thinking questions (in the chapter end matter) are based on the opening story and the concepts of the chapter.

New Introductory Chapter

Chapter 1 was rewritten to give a better overview of the text and a more student-friendly explanation of the process of science and example of a controlled study.

Additional Applications

Many new and/or improved applications have been added to the text. The opening stories provide real-life applications that are mentioned throughout the chapter and the number of applications in general has increased.

New Genetics Chapters

The genetics chapters were rewritten to increase their interest. Chromosomal disorders are now in the cell division chapter; genetic disorders are now in the genetic inheritance chapter; and genomics is now in the DNA chapter.

New Boxed Readings

Science focus boxes that discuss topics of interest to students, such as how to identify skeletal remains and what happens during rigor mortis, have been added. Students will also want to know if humans can be cloned and how swallowing a camera can detect colon cancer.

New AIDS Supplement

Worldwide data is included in this up-to-date supplement. The origin and prevalence of HIV, phases of an HIV infection, and the structure and life cycle of HIV as a precursor to reviewing the treatment of AIDS are also covered.

Vivid and Engaging Illustrations

The vivid and engaging illustrations in *Human Biology* bring the study of biology to life! The figures have been rendered to convey realistic detail and close coordination with the text discussions.

Combination Art

Drawings of structures are paired with micrographs to provide students with two perspectives: the explanatory clarity of line drawings and the realism of photos.

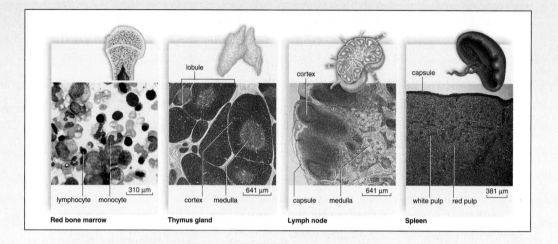

Multi-Level Perspective

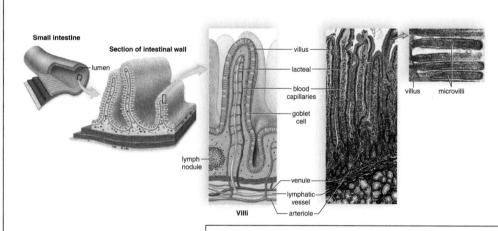

Such illustrations guide students from the more intuitive macroscopic level of learning to the functional foundations revealed through microscopic images.

Icons

Icons orient students to the whole structure or process, by providing small drawings that help students visualize how a particular structure is part of a larger one.

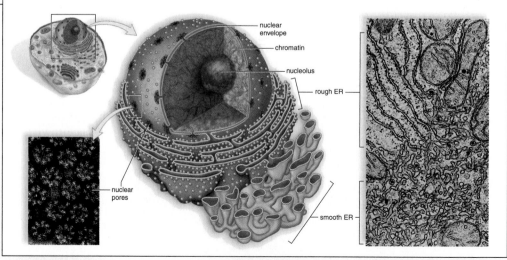

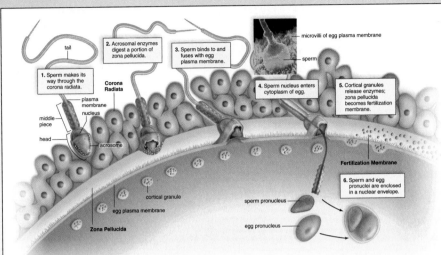

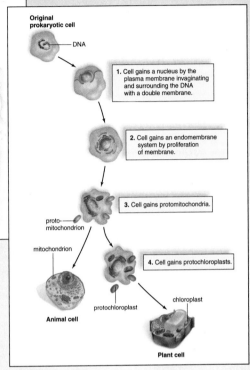

Process Figures

These figures break down processes into a series of smaller steps and organize them in an easy-to-follow format.

Human Systems Work Together

Working together illustrations use brief concise statements to tell students how various other systems help a featured system achieve homeostasis.

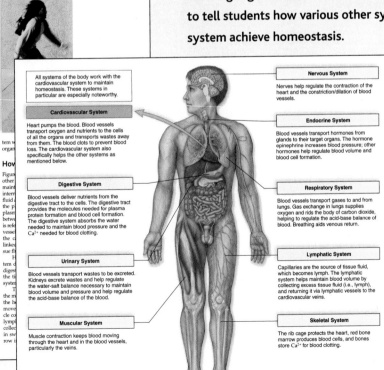

116   Part II   Maintenance of the Human Body

6.6 Homeostasis

Just think of how important it is to homeostasis for all the human systems to work together. For example, imagine what must be one of a student's worst nightmares. You slowly open your eyes in the morning, blearily gaze at your alarm clock, and realize that you've overslept. In fact, it's just a few minutes before class begins. Then, you snap wide awake as you remember why you were up so late the night before—you were studying for the final exam! Most likely, your heart starts beating faster, due to sympathetic *nervous system* stimulation just as Ashley's did when she was watching a scary movie in the Chapter 1 opening story. The resulting rise in blood pressure helps prepare the body for action. Hormones, such as epinephrine (adrenaline) from the adrenal glands, pour into the bloodstream to prolong your body's physical response to stress. The *endocrine system* is at work.

Frantic, you throw on your clothes and sprint to the bus. Your *muscular* and *respiratory systems* meet the demand. As you run, your muscles require more oxygen to support their activity. Your breathing and heart rate increase, in order to deal with the demand. The hard-working muscles generate carbon dioxide and heat. Carbon dioxide is carried via the blood to the lungs so it can be expelled. To keep you from overheating, you not only sweat, but blood flow to your skin increases so that heat radiates away.

The nightmare continues. In your mad dash up the steps to get to your exam, you fall, skinning your knee (Fig. 6.12). At first, blood flows from the wound, but a clot will soon form to seal off the injured area and allow a scab to form so that healing can begin. Any injury that breaks the skin can allow harmful bacteria into the body, but the immune system usually deals with them in short order. Later, you might notice that your knee is not only skinned, but also black and blue. The bruise is where blood leaked out of damaged vessels underneath the skin, and then clotted.

You make it to class just in time to get your exam from the instructor. As you settle down to take the exam, your heartbeat and breathing gradually slow, since the need of your muscles is not as great. Now that you are in thinking mode, blood flow increases in certain highly active parts of your brain. The brain is a very demanding organ in terms of both oxygen and glucose.

You start to feel hungry. Since there is little glucose in your bloodstream to provide fuel for your body's cells, glycogen stored in the liver is broken down to make glucose available to the blood. Finally, you hand in the exam and go to the dining hall for breakfast. When most of the nutrients from your meal are absorbed by the *digestive system*, they enter the bloodstream. However, the fats enter lacteals, which are vessels of the *lymphatic system*, for transport to the blood.

Notice in this scenario how many organ systems had to interact with the cardiovascular system. The nervous, endocrine, muscular, respiratory, digestive, and lymphatic systems helped to serve the cells' needs. Throughout all of the challenges to homeostasis described above, the urinary sys-

All systems of the body work with the cardiovascular system to maintain homeostasis. These systems in particular are especially noteworthy.

Cardiovascular System

Heart pumps the blood. Blood vessels transport oxygen and nutrients to the cells of all the organs and transports wastes away from them. The blood clots to prevent blood loss. The cardiovascular system also specifically helps the other systems as mentioned below.

Digestive System

Blood vessels deliver nutrients from the digestive tract to the cells. The digestive tract provides the molecules needed for plasma protein formation and blood cell formation. The digestive system absorbs the water needed to maintain blood pressure and the Ca^{2+} needed for blood clotting.

Urinary System

Blood vessels transport wastes to be excreted. Kidneys excrete wastes and help regulate the water-salt balance necessary to maintain blood volume and pressure and help regulate the acid-base balance of the blood.

Muscular System

Muscle contraction keeps blood moving through the heart and in the blood vessels, particularly the veins.

Nervous System

Nerves help regulate the contraction of the heart and the constriction/dilation of blood vessels.

Endocrine System

Blood vessels transport hormones from glands to their target organs. The hormone epinephrine increases blood pressure; other hormones help regulate blood volume and blood cell formation.

Respiratory System

Blood vessels transport gases to and from lungs. Gas exchange in lungs supplies oxygen and rids the body of carbon dioxide, helping to regulate the acid-base balance of blood. Breathing aids venous return.

Lymphatic System

Capillaries are the source of tissue fluid, which becomes lymph. The lymphatic system helps maintain blood volume by collecting excess tissue fluid (i.e., lymph), and returning it via lymphatic vessels to the cardiovascular veins.

Skeletal System

The rib cage protects the heart, red bone marrow produces blood cells, and bones store Ca^{2+} for blood clotting.

In this edition the working together illustrations have been integrated into homeostasis sections making a united whole. The homeostasis sections show how the systems achieve homeostasis despite real-life experiences that could alter the internal environment. For example, see page 116.

The Learning System

Proven pedagogical features that will facilitate your understanding of biology.

Chapter Concepts

The chapter outline contains a concise preview of the topics covered in each section.

Opening Vignette

A short, thought-provoking vignette applies chapter material to real-life situations. The vignette is referred to numerous times throughout the chapter.

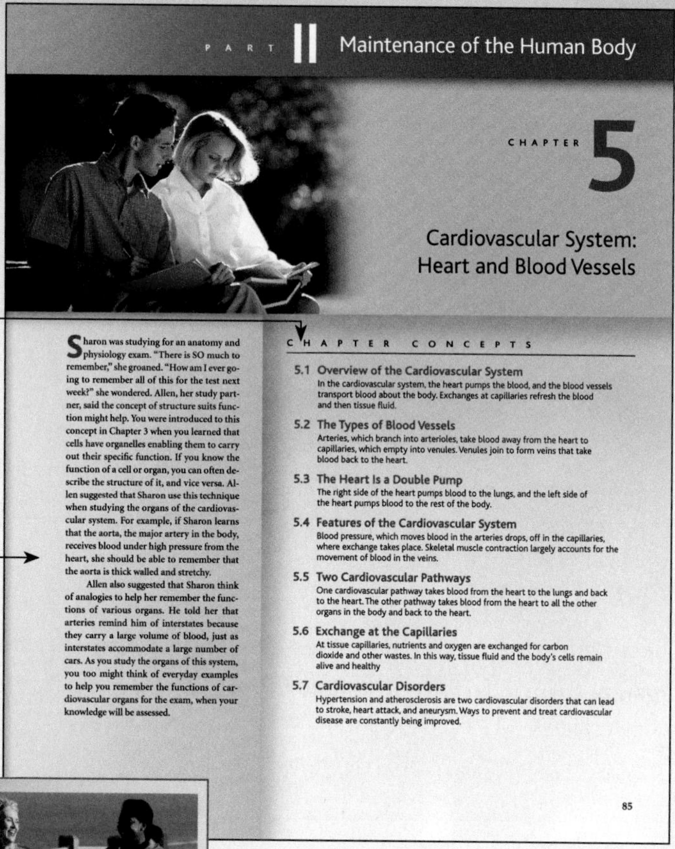

Readings

Human Biology offers three types of boxed readings that put the chapter concepts in the context of modern-day issues:

- **Health Focus** readings review procedures and technology that can contribute to our well being.

- **Science Focus** readings describe how experimentation and observations have contributed to our knowledge about the living world.

- **Bioethical Focus** readings describe modern situations that call for value judgments and challenge students to develop a point of view.

PART **II** Maintenance of the Human Body

CHAPTER **5**

Cardiovascular System: Heart and Blood Vessels

Sharon was studying for an anatomy and physiology exam. "There is SO much to remember," she groaned. "How am I ever going to remember all of this for the test next week?" she wondered. Allen, her study partner, said the concept of structure suits function might help. You were introduced to this concept in Chapter 3 when you learned that cells have organelles enabling them to carry out their specific function. If you know the function of a cell or organ, you can often describe the structure of it, and vice versa. Allen suggested that Sharon use this technique when studying the organs of the cardiovascular system. For example, if Sharon learns that the aorta, the major artery in the body, receives blood under high pressure from the heart, she should be able to remember that the aorta is thick walled and stretchy.

Allen also suggested that Sharon think of analogies to help her remember the functions of various organs. He told her that arteries remind him of interstates because they carry a large volume of blood, just as interstates accommodate a large number of cars. As you study the organs of this system, you too might think of everyday examples to help you remember the functions of cardiovascular organs for the exam, when your knowledge will be assessed.

CHAPTER CONCEPTS

5.1 Overview of the Cardiovascular System
In the cardiovascular system, the heart pumps the blood, and the blood vessels transport blood about the body. Exchanges at capillaries refresh the blood and then tissue fluid.

5.2 The Types of Blood Vessels
Arteries, which branch into arterioles, take blood away from the heart to capillaries, which empty into venules. Venules join to form veins that take blood back to the heart.

5.3 The Heart Is a Double Pump
The right side of the heart pumps blood to the lungs, and the left side of the heart pumps blood to the rest of the body.

5.4 Features of the Cardiovascular System
Blood pressure, which moves blood in the arteries, drops off in the capillaries, where exchange takes place. Skeletal muscle contraction largely accounts for the movement of blood in the veins.

5.5 Two Cardiovascular Pathways
One cardiovascular pathway takes blood from the heart to the lungs and back to the heart. The other pathway takes blood from the heart to all the other organs in the body and back to the heart.

5.6 Exchange at the Capillaries
At tissue capillaries, nutrients and oxygen are exchanged for carbon dioxide and other wastes. In this way, tissue fluid and the body's cells remain alive and healthy

5.7 Cardiovascular Disorders
Hypertension and atherosclerosis are two cardiovascular disorders that can lead to stroke, heart attack, and aneurysm. Ways to prevent and treat cardiovascular disease are constantly being improved.

85

Health Focus

Prevention of Cardiovascular Disease

Certain genetic factors predispose an individual to cardiovascular disease, such as family history of heart attack under age 55, male gender, and ethnicity (African Americans are at greater risk). Those with one or more of these risk factors need not despair, however. It means only that they should pay particular attention to the following guidelines for a heart-healthy lifestyle.

The Don'ts

Smoking
When a person smokes, the drug nicotine, present in cigarette smoke, enters the bloodstream. Nicotine causes [...] blood flow and cold hands are associated with smo[...] most people. More serious is the need for the heart [...] harder to propel the blood through the lungs at a tim[...] the blood's oxygen-carrying capacity is reduced.

Drug Abuse
Stimulants, such as cocaine and amphetamines, can c[...] irregular heartbeat and lead to heart attacks in peop[...] are using drugs even for the first time. Intravenous d[...] may also result in a cerebral blood clot and stroke.
    Too much alcohol can destroy just about every o[...] the body, the heart included. But investigators hav[...] ered that people who take an occasional drink have [...] lower risk of heart disease than do teetotalers. Two [...] drinks a week is the recommended limit for men [...] three drinks for women.

Weight Gain
Hypertension (high blood pressure) is prevalent in [...] who are more than 20% above the recommended we[...] their height. More tissues require servicing, and th[...] sends the blood out under greater pressure in tho[...] are overweight. It may be harder to lose weight o[...] gained, and therefore, it is recommended that weig[...] trol be a lifelong endeavor. Even a slight decrease i[...] can bring with it a reduction in hypertension.
    Being overweight increases the risk of diabetes [...] in which glucose damages blood vessels and mak[...] prone to the development of plaque. Diabetics es[...] need to follow the diet discussed next.

The Do's

Healthy Diet
A diet low in saturated fats and cholesterol is pr[...] against cardiovascular disease. Cholesterol is ferrie[...] blood by two types of plasma proteins, called LDL (lo[...] sity lipoprotein) and HDL (high-density lipoprotei[...] (called "bad" lipoprotein) takes cholesterol from the [...] the tissues, and HDL (called "good" lipoprotein) tr[...] cholesterol out of the tissues to the liver. When the L[...] in blood is abnormally high or the HDL level is abn[...]

Science Focus

Nerve Regeneration

In humans, axons outside the brain and spinal cord can regenerate, but not those inside these organs (Fig. 4A). After injury, axons in the human central nervous system (CNS) degenerate, resulting in permanent loss of nervous function. Not so in cold-water fishes and amphibians, where axon regeneration in the CNS does occur. So far, investigators have identified several proteins that seem to be necessary to axon regeneration in the CNS of these animals, but even so, it may be a long time before biochemistry can offer a way to bring about axon regeneration in the human CNS. Still, it's possible that these proteins might one day be used as drugs or that following gene therapy humans might produce these same proteins when CNS injuries occur.
    In the meantime, some accident victims are trying other ways to bring about a cure. In 1995, Christopher Reeve, best known for his acting role as "Superman," was thrown headfirst from his horse, crushing the spinal cord just below the neck's top two vertebrae. Immediately, his brain lost almost all communication with the portion of his body below the site of damage, and he could not move his arms and legs. Many years later, Reeve could move his left inde[...] slightly and could take tiny steps while being held u[...] a pool. He had sensation throughout his body and c[...] his wife's touch.
    Reeve's improvement was not the result of cutt[...] drugs or gene therapy—it was due to exercise (F[...] Reeve exercised as much as five hours a day, espe[...] ing a recumbent bike outfitted with electrodes that [...] leg muscles contract and relax. The bike cost him $[...] could cost less if commonly used by spinal cord p[...] tients in their own homes. Reeve, who was an activi[...] disabled, was pleased that insurance would pay for[...] about 50% of the time.
    It's possible that Reeve's advances were the [...] improved strength and bone density, which lead to [...] nerve signals. Normally, nerve cells are constantly s[...] one another, but after a spinal cord injury, the[...] Perhaps Reeve's intensive exercise brought back so[...] normal communication between nerves. Reeve's p[...]

Figure 4A   Nerve regeneration.
Outside the CNS, nerves regenerate because new neuroglia called Schwann cells form a pathway for axons to reach a muscle. In the CNS, comparable neuroglia called oligodendrocytes do not do this.

axon
site of injury

Figure 4B   Treatment today for spinal cord injuries.
a. Reeve suffered a spinal cord injury when horseback riding in 1995. **b.** He exercised many hours a day, including aqua therapy. Reeve died in 2004.

Bioethical Focus

Polar Bears: How Long Will They Last?

Polar bears are keystone species, which means that their extinction will send a ripple effect throughout the entire Arctic community. Right now the polar bear is listed as vulnerable to extinction. The chief reasons are these:

1. **Global warming.** Global warming decreases the amount of ice pack available for hunting and does not allow the bears to build the necessary layer of insulation to survive hibernation and nurse their young. Global warming is due to burning of fossil fuels, which releases carbon dioxide into the atmosphere (see page 501).
2. **Chemical pollution.** Polychlorinated biphenyls (PCBs) and organochlorine pesticides have entered the Arctic ecosystem, and they have been found in polar bears, beluga whales, seals, and seabirds. Such pollutants might negatively affect the reproductive abilities of polar bears and other animals
3. **Oil and gas exploration.** Drilling in the Arctic National Wildlife refuge in Alaska will impact about half of the onshore denning sites that are used by pregnant females. Therefore, drilling will significantly lower the survival rate of polar bear females as well as their young.

Polar bears prey on ringed seals; therefore, as polar bears decrease in number, the ringed seal population is expected to increase in size. This might sound good except that the prey of ringed seals—namely, cod, shrimp and other crustacean populations—will decrease, and the Inuit tribes living in the Arctic are dependent on these same species for survival.

Our industrialized society is dependent on the consumption of oil and gas for its way of life. Therefore, even though people know

that the burning of fossil fuels is contributing to global warming, they would most likely find it very difficult to reduce their consumption of these fuels. Also, the drilling for oil and gas in the refuge will help decrease our dependency on foreign sources of oil, as well as provide hundreds of jobs for the citizens of Alaska.

Figure 1B   Polar bear mother and cub.

The possibility of the loss of a keystone species can serve as a warning light for concern. When the light turns red, it is an indication there are serious problems within an ecosystem. The human species depends on all the other species of the biosphere for survival. As we lose species, we only increase the problems that we will have to face in the near future. Therefore, it might behoove us to take all possible steps to conserve species and to reduce pollutants, including those that lead to global warming

Decide Your Opinion
1. What selfish reasons can you give for wanting the Arctic community to continue functioning as before?
2. Is drilling for oil worth the risk of contributing to polar bear extinction?
3. What is more valuable, maintaining our lifestyle or the Arctic community?

Check Your Progress Boxes

Questions follow main sections of the text and help students assess their understanding of the material presented. Answers to these questions appear in Appendix A.

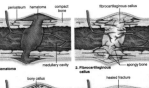

Summarizing the Concepts

A bulleted summary is organized according to the major sections in the chapter and includes art to helps students review the important topics and concepts.

Understanding Key Terms

The boldface terms in the chapter are page referenced, and a matching exercise allows students to test their knowledge of the terms.

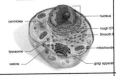

Testing Your Knowledge of the Concepts

Objective and art-based questions allow students to review material and prepare for tests. Answers to these questions appear in Appendix A.

Thinking Critically About the Concepts

This set of questions encourages students to apply what they've just learned to the real-life story that opened the chapter.

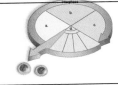

Teaching Supplements for the Instructor

Dedicated to providing high quality and effective supplements for instructors, the following instructor supplements were developed for Human Biology:

ARIS for Human Biology

aris.mhhe.com

McGraw-Hill's ARIS—Assessment, Review, and Instruction System—for *Human Biology* is a complete, online tutorial, electronic homework, and course management system. Instructors can create and share course materials and assignments with colleagues with a few clicks of the mouse. All PowerPoint lectures, assignments, quizzes, and tutorials are directly tied to text-specific materials. Instructors can edit questions, import their own content, and create announcements and due dates for assignments. ARIS has automatic grading and reporting of easy-to-assign homework, quizzing, and testing. All student activity within ARIS is automatically recorded and available to the instructor through a fully integrated grade book that can be downloaded to Excel. Contact your McGraw-Hill sales representative for more information on getting started with ARIS.

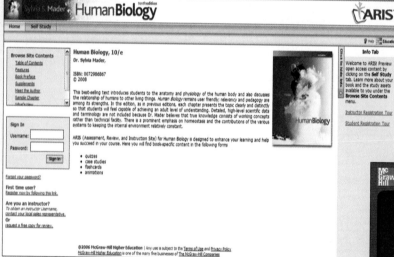

ARIS Presentation Center

Build instructional materials wherever, whenever, and however you want!

ARIS Presentation Center is an online digital library containing assets such as photos, artwork, animations, PowerPoints, and other media types that can be used to create customized lectures, visually enhanced tests and quizzes, compelling course websites, or attractive printed support materials.

Access to your book, access to all books!

The Presentation Center library includes thousands of assets from many McGraw-Hill titles. This ever-growing resource gives instructors the power to utilize assets specific to an adopted textbook as well as content from all other books in the library.

Nothing could be easier!

Accessed from the instructor side of your textbook's ARIS website, Presentation Center's dynamic search engine allows you to explore by discipline, course, textbook chapter, asset type, or keyword. Simply browse, select, and download the files you need to build engaging course materials. All assets are copyright McGraw-Hill Higher Education but can be used by instructors for classroom purposes.

Art

Full-color digital files of all illustrations in the book can be readily incorporated into lecture presentations, exams, or custom-made classroom materials.

Photos

Digital files of all photographs from the book can be reproduced for multiple classroom uses.

Tables

Tables from the text are available in electronic format.

Animations

A wealth of full-color animations support the topics discussed in the text. Instructors are able to harness the visual impact of processes in motion by importing these files into classroom presentations or online course materials.

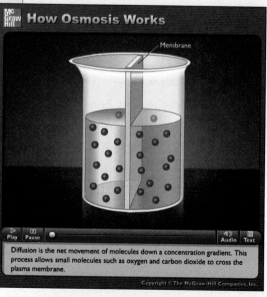

PowerPoint Lecture Outlines
Ready-made presentations that combine art and lecture notes are provided for each of the chapters in the text. These outlines can be used as they are or can be tailored to reflect your preferred lecture topics and sequences.

PowerPoint Image Slides
For instructors who prefer to create individualized lectures, illustrations, photos, and tables are preinserted by chapter into PowerPoint slides for convenience.

Human Biology Laboratory Manual

The *Human Biology Laboratory Manual*, Tenth Edition, was written by Terry Damron and Eric Rabitoy. With few exceptions, each chapter in the text has an accompanying laboratory exercise in the manual. Every laboratory has been written to help students learn the fundamental concepts of human biology and the specific content of the chapter to which the lab relates, as well as gain an understanding of the scientific method.

Instructor's Testing and Resource CD-ROM

This cross-platform CD-ROM provides these resources for instructors:

- **Instructor's Manual** This manual contains lecture outlines, learning objectives, key terms, lecture suggestions and enrichment tips, and critical-thinking questions.

- **Test Bank** The test bank provides questions that can be used for homework assignments or the preparation of exams.

- **Computerized Test Bank** The computerized test bank allows the user to quickly create customized exams. This user-friendly program allows you to search for questions by topic or format; edit existing questions or add new ones; and scramble questions and answer keys for multiple versions of the same test.

Transparencies

A set of full-color acetate transparencies is available to supplement classroom lectures. These transparencies include key figures from the textbook for classroom projection, and they are available to adopters.

eInstruction Classroom Performance System (CPS)

Wireless technology brings interactivity into the classroom or lecture hall. Instructors and students receive immediate feedback through wireless response pads that are easy to use and engage students.

This system can be used by instructors to:
- Take attendance
- Administer quizzes and exams
- Create a lecture with intermittent questions
- Manage lectures and student comprehension through use of the gradebook
- Integrate interactivity into their PowerPoint presentations

Course Delivery Systems

With help from our partners WebCT, Blackboard, eCollege, and other course management systems, professors can take complete control over their course content. Course cartridges containing content from the ARIS textbook website, online testing, and powerful student tracking features are readily available for use within these platforms.

McGraw-Hill: Biology Digitized Video Clips

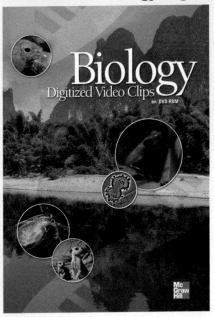

McGraw-Hill is pleased to offer adopting instructors a new presentation tool—digitized biology video clips on DVD! Licensed from some of the highest quality science video producers in the world, these brief segments range from about five seconds to just under three minutes in length and cover all areas of general biology from cells to ecosystems.

Engaging and informative, McGraw-Hill's Biology Digitized Video Clips will help capture students' interest while illustrating key biological concepts and processes such as mitosis, how cilia and flagella work, and how some plants have evolved into carnivores.

ScienCentral Videos

McGraw-Hill is pleased to announce an exciting new partnership with ScienCentral, Inc., to provide brief biology news videos for use in lecture or for student study and assessment purposes. A complete set of ScienCentral videos are located within this text's ARIS course management system, and each video includes a learning objective and quiz questions. These active learning tools enhance a biology course by engaging students in real-life issues and applications such as developing new cancer treatments and understanding how methamphetamine damages the brain. ScienCentral, Inc., funded in part by grants from the National Science Foundation, produces science and technology content for television, video, and the Web.

Virtual Labs

A complete set of over 20 online introductory biology laboratory exercises are available on this text's ARIS course management system. Each exercise is self-contained with all instructions and assessment for students. For use either stand-alone or as a supplement to a traditional lab course, this interactive learning tool will help students understand key biological concepts and processes.

Learning Supplements for the Student

Designed to help students maximize their learning experience in human biology—we offer the following options to students:

ARIS for *Human Biology*

aris.mhhe.com

McGraw-Hill's ARIS—Assessment, Review, and Instruction System—for Human Biology offers access to a vast array of premium online content to fortify the learning experience.

- **Text-Specific Study Tools** The ARIS site features quizzes, study tools, and other features tailored to coincide with each chapter of the text.
- **Course Assignments and Announcements** Students of instructors choosing to utilize McGraw-Hill's ARIS tools for course administration will receive a course code to log into their specific course for assignments.

Student Study Guide

This valuable student resource prepared by Linda Smith-Staton, contains activities and questions to help reinforce chapter concepts. The guide provides learning objectives, study tips, activites, and quizzes.

Student Interactive CD-ROM

This interactive CD-ROM is an indispensable resource for studying topics covered in the text. It includes chapter outlines, chapter-based quizzes, animations of complex processes, flashcards, PowerPoint lecture outlines, and PowerPoint slides of art and photos found in the textbook. All of the materials are organized chapter-by-chapter. Direct links of the text's ARIS website are also provided.

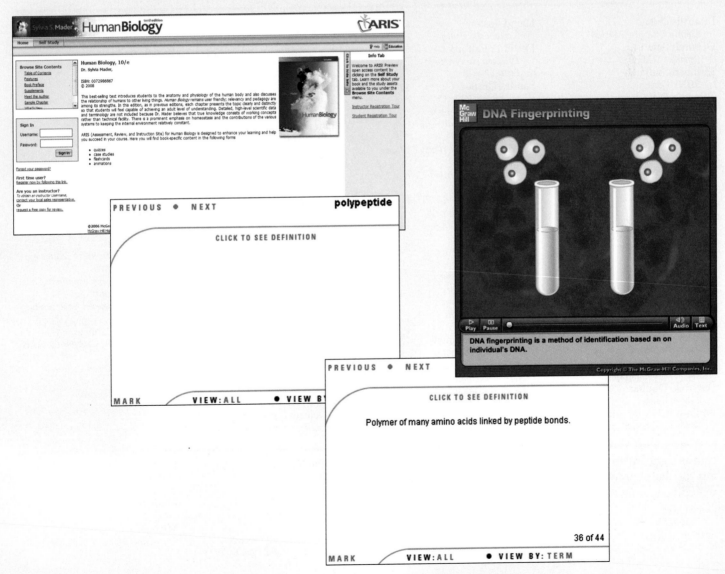

Acknowledgments

The hard work of many dedicated and talented individuals helped to vastly improve this edition of *Human Biology*. Let me begin by thanking the people who guided this revision at McGraw-Hill. I am very grateful for the help of so many professionals who were involved in bringing this book to fruition. My editor Tom Lyon, marketing manager Tamara Maury, and developmental editor Darlene Schueller who worked tirelessly to help me bring you a text and ancillaries that will serve your needs in every way. They planned well and supplied creativity, advice, and support whenever it was needed. Jayne Klein, my project manager, carefully steered the book through the production process.

Rick Noel was the designer who chose everything from the different type styles, to the colors of the opening pages, to the cover of the book. Electronic Publishing Services, Inc. created and reworked every illustration, emphasizing pedagogy, and beauty to arrive at the best presentation on the page. Evelyn Jo Hebert and Lori Hancock did a superb job of finding just the right photographs and micrographs.

In my office, Beth Butler worked faithfully on all aspects of clerical preparation and she made sure all was well before the book went to press. As always, my family was extremely patient with me as I remained determined to make every deadline on the road to publication. My husband, Arthur Cohen, is also a teacher of biology. The many discussions we have about the minutest detail to the gravest concept are invaluable to me.

The content of the tenth edition of *Human Biology* is not due to my efforts alone. I want to thank the many specialists who were willing to share their knowledge to improve *Human Biology*. Also, this edition was enriched by Linda D. Smith-Staton and Susan Nelson Longenbaker who provided invaluable assistance through all phases of this project. I am grateful to Debbie McCool for her assistance with the end-of-chapter material.

The tenth edition of *Human Biology* would not have been the same excellent quality without the suggested changes of the many reviewers who are listed here.

Rita Alisauskas
County College of Morris
Deborah Allen
Jefferson College
Elizabeth Balko
SUNY-Oswego
Tamatha R. Barbeau
Francis Marion University
Marilynn R. Bartels
Black Hawk College
Erwin A. Bautista
University of California, Davis
Robert D. Bergad
Metropolitan State University
Hessel Bouma III
Calvin College
Frank J. Conrad
Metropolitan State College of Denver
William Cushwa
Clark College
Debbie A. Zetts Dalrymple
Thomas Nelson Community College
Diane Dembicki
Dutchess Community College
Charles J. Dick
Pasco-Hernando Community College
Kristiann M. Dougherty
Valencia Community College
David A. Dunbar
Cabrini College
William E. Dunscombe
Union County College

David Foster
North Idaho College
David E. Fulford
Edinboro University of Pennsylvania
Sandra Grauer
Limestone College
Mary Louise Greeley
Salve Regina University
Esta Grossman
Washtenaw Community College
Gretel M. Guest
Alamance Community College
Martin E. Hahn
William Paterson University
Rosalind C. Haselbeck
University of San Diego
Timothy P. Hayes
Marshall University
Mark F. Hoover
Penn State Altoona
Anna K. Hull
Lincoln University
Laurie A. Johnson
Bay College
Mary King Kananen
Penn State Altoona
Patricia Klopfenstein
Edison Community College
J. Kevin Langford
Stephen F. Austin State University
Lee H. Lee
Montclair State University

Edwin Lephart
Brigham Young University
Martin A. Levin
Eastern Connecticut State University
Nardos Lijam
Columbus State Community College
William J. Mackay
Edinboro University of Pennsylvania
Terry R. Martin
Kishwaukee College
Deborah J. McCool
Penn State Altoona
V. Christine Minor
Clemson University
Nick Nagle
Metropolitan State College of Denver
Roger C. Nealeigh
Central Community College-Hastings
Polly K. Phillips
Florida International University
Shawn G. Phippen
Valdosta State University
Mason Posner
Ashland University
Donna R. Potacco
William Paterson University
Mary Celeste Reese
Mississippi State University

Jill D. Reid
Virginia Commonwealth University
Kay Rezanka
Central Lakes College
April L. Rottman
Rock Valley College
Deborah B. Schulman
Cleveland State University
Lois Sealy
Valencia Community College
Jia Shi
Skyline College
Mark Smith
Chaffey College
Alicia Steinhardt
Hartnell Community College
West Valley Community College
Lei Lani Stelle
Rochester Institute of Technology
Kenneth Thomas
Northern Essex Community College
Chad Thompson
SUNY-Westchester Community College
Jamey Thompson
Hudson Valley Community College
Doris J. Ward
Bethune-Cookman College
Susan Weinstein
Marshall University

I am also grateful to the following for their contributions to this edition of *Human Biology*:
Dave Cox
Lincoln Land Community College
Patrick Galliart
North Iowa Area Community College

Sandra Grauer
Limestone College
Sharron Jenkins
Purdue University North Central
Jill Kolodsick
Washtenaw Community College
Edwin Lephart
Brigham Young University

Susannah Nelson Longenbaker
Columbus State Community College
Debbie J. McCool
Penn State Altoona
Jodi Rymer
Christine Wildsoet Laboratory University of California, Berkeley

Linda D. Smith-Staton
Pellissippi State Technical Community College
Linda Strause
University of California–San Diego
Michael Thompson
Middle Tennessee State University

CHAPTER 1

Exploring Life and Science

As the door slowly creaked open, a young woman cautiously stepped into the room. "Hello, is someone here?" she asked tentatively. From the darkness, a huge, hairy monstrous thing leaped out, grabbed her by the throat, and proceeded to shake her fiercely from side to side. Several people in the audience screamed in horror.

"Man, did you see my friend Ashley? She jumped right up when that girl got attacked," Jesse commented once the movie came to an end. "Oh my gosh, I thought my heart would leap out of my chest, it was pounding so hard," exclaimed Ashley. "Yeah, and I thought my fingers would be crushed—she gripped my hand so tightly," interjected Jesse.

The increased heart rate experienced by Ashley was due to the activity of the sympathetic nervous system (NS). The sympathetic NS triggers what is commonly called the "fight-or-flight" response. It helps us escape danger by running away as fast as possible or stopping and putting up a good fight. The danger doesn't have to be real; in Ashley's case, the danger was only a perceived threat. Your well being is not actually threatened by watching a scary movie, yet the sympathetic NS causes us to experience the same effects as if we are in a life-threatening situation.

This text discusses human adaptations and responses, which are similar to those of other living organisms. In this chapter, you will learn about the characteristics of life and the process of science.

CHAPTER CONCEPTS

1.1 The Characteristics of Life
The process of evolution accounts for the diversity of life and why all living things share the same characteristics of life. Living things are organized; acquire materials and energy; reproduce; grow and develop; are homeostatic; respond to stimuli; and have an evolutionary history.

1.2 Humans Are Related to Other Animals
Humans are eukaryotes classified as mammals in the animal kingdom. We differ from other mammals, including apes, by our highly developed brain, upright stance, creative language, and the ability to use a wide variety of tools. Our cultural heritage makes it difficult for us to see that preservation of the biosphere is essential to our well being.

1.3 Science as a Process
Biologists use the scientific process when they study the natural world. A hypothesis is formulated and tested to arrive at a conclusion. Scientists have developed traditional methods that give them added confidence in the conclusions of studies.

1.4 Making Sense of a Scientific Study
Experimental data are widely appreciated, especially if the results are clearly presented in the form of a graph and they are accompanied by a standard error or the statistical significance.

1.5 Science and Social Responsibility
Science, technology, and society have interacted throughout human history, and scientific investigations and technology have always been affected by human values. Every member of society has a responsibility to participate in how science and technology are used for the good of all.

1.1 The Characteristics of Life

So, you've signed up to take a course in human biology? **Biology,** as you probably know, is the scientific study of life, and human biology is a specialty in this field. Before you begin, it is appropriate to take a look at who humans are and how they fit into the world at large. One of the first things for you to consider is that human beings are a part of the natural world; specifically, the world of living things that encompasses long-necked giraffes, beautiful butterflies, giant sequoia trees, low-lying mushrooms, and colorful angelfish, to name a few. Although the many different kinds of living things boggle the mind, life is not as complicated as it seems, because all living things share the characteristics of life (Fig. 1.1). Living things (i.e., organisms)

- are organized, from atoms to the biosphere.
- take materials and energy from the environment.
- reproduce; they produce offspring that resemble themselves.
- grow and develop by undergoing various stages from fertilization to death.
- are homeostatic; internal conditions stay just about the same.
- respond to stimuli; they react to external and internal changes.
- have an evolutionary history and have adapted modifications to a particular way of life.

Life Is Organized

Figure 1.2 illustrates that **atoms** join together to form the **molecules** that make up a cell (Fig. 1.2). A **cell** is the smallest structural and functional unit of an organism. Human beings are **multicellular** because they are composed of many different types of cells. A nerve cell is one of the types of cells in the human body. It has a structure suitable to conducting a nerve impulse.

A **tissue** is a group of similar cells that perform a particular function. Nervous tissue is composed of millions of nerve cells that transmit signals to all parts of the body. Several types of tissues make up an **organ,** and each organ belongs to an organ system. The organs of an **organ system** work together to accomplish a common purpose. The brain works with the spinal cord to send commands to body parts by way of nerves. This is why, as mentioned in the opening story, Ashley's heart pounded so when she was watching a scary movie. **Organisms,** such as trees and humans, are a collection of organ systems.

The levels of biological organization extend beyond the individual. All the members of one **species** (group of interbreeding organisms) in a particular area belong to a **population.** A tropical grassland may have a population of zebras, acacia trees, and humans, for example. The populations of various animals and plants in the forest make up a **community.** The community of populations interacts with the physical environment to form an **ecosystem.** Finally, all the Earth's ecosystems make up the **biosphere.**

Figure 1.1   Diversity of Life.
Biology is the study of life. Many diverse forms of life are found on planet Earth.

Masai giraffes

sulfur tuft mushrooms

queen angelfish

giant sequoia

monarch butterfly emerging from cocoon

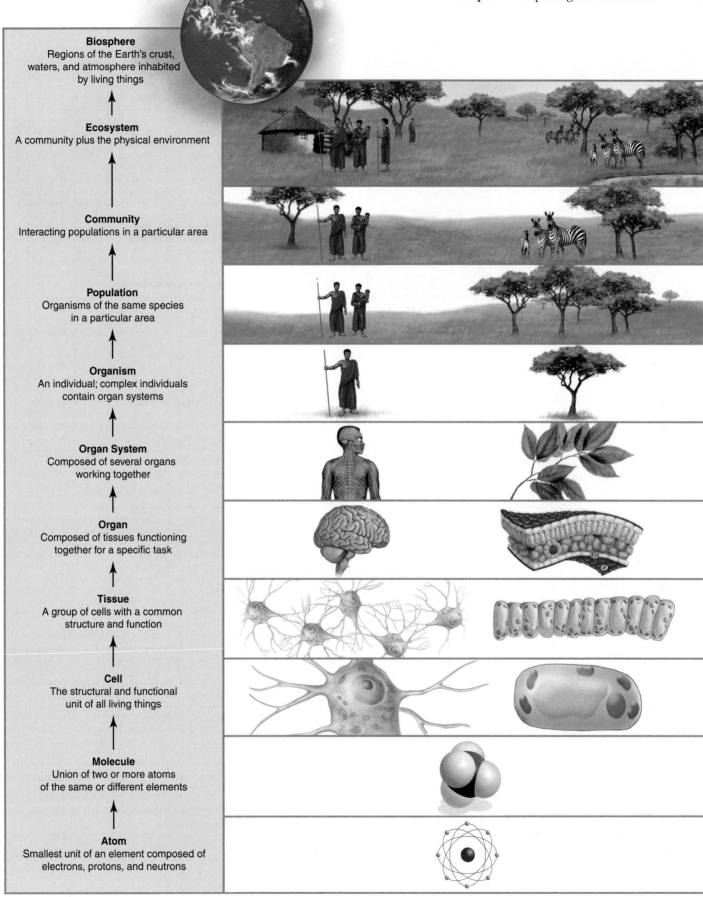

Figure 1.2   Levels of biological organization.

a.

b.

Figure 1.3 **Acquiring materials and energy.**
a. A red-tailed hawk has captured a rabbit, which it is feeding to its young.
b. Humans eat plants and animals they raise for food.

Acquiring Materials and Energy

Human beings cannot maintain their organization or carry on life's activities without an outside source of materials and energy. Human beings and other animals acquire materials and energy when they eat food (Fig. 1.3).

Food provides nutrient molecules, which are used as building blocks or for energy. It takes energy (work) to maintain the organization of the cell and of the organism. Some nutrient molecules are broken down completely to provide the necessary energy to convert other nutrient molecules into the parts and products of cells.

Most living things can also convert energy into motion. Self-directed movement, as when Ashley jumped up in the movie, is even considered by some to be one of life's characteristics.

Reproducing

Reproduction is a fundamental characteristic of life. Cells come into being only from preexisting cells, and all living things have parents. When living things **reproduce,** they create a copy of themselves and ensure the continuance of their own kind (Fig. 1.3).

The presence of genes, in the form of DNA molecules, allows cells and organisms to make more of themselves. DNA contains the hereditary information that directs the structure of the cell and its **metabolism,** all the chemical reactions in the cell. Before reproduction occurs, DNA is replicated so that exact copies of **genes** are passed on to offspring. When humans reproduce, a sperm carries genes contributed by a male into the egg, which contains genes contributed by a female. The genes direct development so that the organism resembles the parents. Red-tail hawks only produce red-tail hawks, and humans only produce humans, for example.

Growing and Developing

Growth, recognized by an increase in size and often the number of cells, is a part of development. In humans, **development** includes all the changes that occur from the time the egg is fertilized until death and, therefore, all the changes that occur during childhood, adolescence, and adulthood. Development also includes the repair that takes place following an injury.

All organisms undergo development. Figure 1.4*a* illustrates that an acorn progresses to a seedling before it becomes an adult oak tree. In humans, growth occurs as the fertilized egg develops into the newborn (Fig. 1.4*b*).

Being Homeostatic

Together, the organ systems maintain **homeostasis,** an internal environment for cells that usually varies only within certain limits. For example, human body temperature normally fluctuates slightly between 36.5 and 37.5°C during the day. In general, the lowest temperature usually occurs between 2 A.M. and 4 A.M., and the highest usually occurs between 6 P.M. and 10 P.M., but activity can cause the body temperature to rise, and inactivity can cause it to decline. Even though the body has mechanisms such as shivering when we are cold and perspiring when we are hot, the body's ability to maintain a normal internal temperature is somewhat dependent on the external temperature—we will die if the external temperature becomes overly hot or cold.

This text emphasizes how all the systems of the human body help maintain homeostasis. The digestive system takes in nutrients, and the respiratory system exchanges gases with the environment. The cardiovascular system distributes nutrients and oxygen to the cells and picks up their wastes. The metabolic waste products of cells are excreted

a.

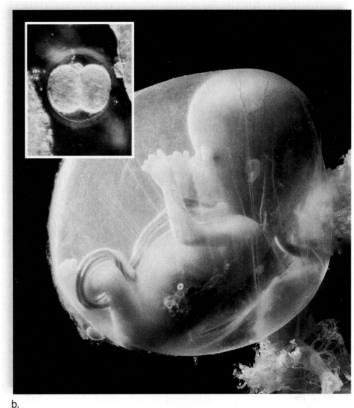

b.

Figure 1.4 Living things grow and develop.
a. A small acorn becomes a tree, and (**b**) a two-celled embryo becomes a human being by the process of growth and development.

by the urinary system. The work of the nervous and endocrine systems is critical because they coordinate the functions of the other systems.

Responding to Stimuli

Living things respond to external stimuli, often by moving toward or away from a stimulus, such as the sight of food. Living things use a variety of mechanisms in order to move, but movement in humans and other animals is dependent upon their nervous and musculoskeletal systems. The leaves of plants track the passage of the sun during the day, and when a houseplant is placed near a window, its stems bend to face the sun. The movement of an animal, whether self-directed or in response to a stimulus, constitutes a large part of its behavior. Ashley's behavior was largely directed toward minimizing a possible threat. Other behaviors help us acquire food and reproduce.

Homeostasis would be impossible without the ability of the body to respond to stimuli. Response to external stimuli is more apparent to us because it does involve movement, as when we quickly remove a hand from a hot stove. However, certain sensory receptors detect a change in the internal environment, and then the central nervous system brings about an appropriate response. When Ashley experienced a rapid and strong heartbeat, her blood pressure probably shot right up. If blood pressure rises too high, the brain directs blood vessels to dilate, helping to restore normal blood pressure.

Life Has an Evolutionary History

Evolution is the process by which a species changes through time. When a new variation arises that allows certain members of the species to capture more resources, these members tend to survive and to have more offspring than the other, unchanged members. Therefore, each successive generation will include more members with the new variation that represents an **adaptation** to the environment. Consider, for example, a red-tailed hawk (see Fig. 1.3*a*), which catches and eats rabbits. A hawk can fly, in part, because it has hollow bones to reduce its weight and flight muscles to depress and elevate its wings. When a hawk dives, its strong feet take the first shock of the landing, and its long, sharp claws reach out and hold onto the prey. All these characteristics are a hawk's adaptations to its way of life.

Evolution, which has been going on since the origin of life and will continue as long as life exists, explains both the unity and the diversity of life. All organisms share the same characteristics of life because their ancestry can be traced to the first cell or cells. Organisms are diverse because they are adapted to different ways of life.

✓ Check Your Progress 1.1

1. What are the seven characteristics of life?

2. Why would you expect all living things on Earth to exhibit these characteristics?

1.2 Humans Are Related to Other Animals

The classification of living things mirrors their evolutionary relationships. Living things are now classified into three **domains** (Fig. 1.5*a*). Of these, domain Eukarya contains four **kingdoms,** and humans are vertebrates in the kingdom Animalia. **Vertebrates** (Fig. 1.5*b*) have a nerve cord that is protected by a vertebral column, whose repeating units (the vertebrae) indicate that we and other vertebrates are segmented animals. Among the vertebrates, we are mammals (animals with hair and mammary glands) and so are apes,

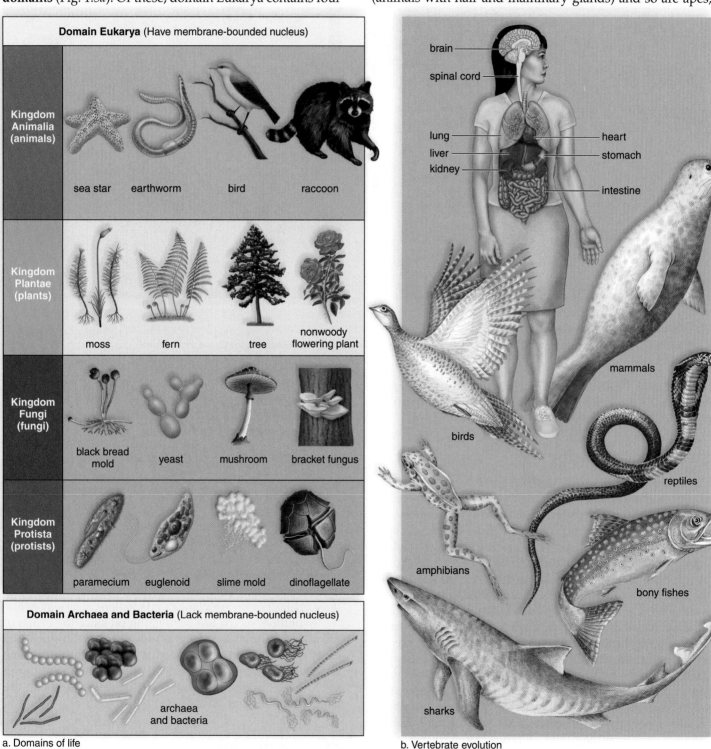

a. Domains of life

b. Vertebrate evolution

Figure 1.5 Classification of humans.
a. Living things are classified into three domains: Eukarya, Archaea, and Bacteria. Humans are in the domain Eukarya and the kingdom Animalia. **b.** Like raccoons and seals, humans are mammals, a type of vertebrate. The evolutionary tree of life has many branches; the vertebrate line of descent, indicated here in broad outline, is just one of them.

from whom we are distinguished by our (1) highly developed brains, (2) completely upright stance, (3) creative language, and (4) ability to use a wide variety of tools.

Human beings are most closely related to apes, but humans did not evolve from apes. Rather, humans and apes evolved from a common apelike ancestor. Today's apes are our evolutionary cousins, and we couldn't have evolved from our cousins because we are contemporaries—living on Earth at the same time. Our relationship to apes is analogous to you and your first cousin being descended from your grandparents.

Humans Have a Cultural Heritage

Aside from biological heritage, human beings have a cultural heritage. **Culture** encompasses human activities and products that are passed on from one generation to the next outside of direct biological inheritance. Among animals, only humans have a language that allows them to communicate information and experiences symbolically. We are born without knowledge of an accepted way to behave, but we gradually acquire this knowledge by adult instruction and imitation of role models. The previous generation passes on their beliefs, values, and skills to the next generation. Many of the skills involve tool use, which can vary from how to hunt in the wild to how to use a computer. Human skills have also produced a rich heritage in the arts and sciences. The culture of highly civilized people, in particular, gives us the impression we are separate from other animals and makes us think we are not a part of nature.

Humans Are Members of the Biosphere

All living things on Earth are part of the biosphere, a living network that spans the surface of the Earth into the atmosphere and down into the soil and seas. Although humans can grow a large portion of their food and raise farm animals, they still depend on the environment for innumerable services. If it were not for microorganisms that decompose the waste we dump into the biosphere, it would soon cover the entire surface of the Earth. The populations of natural ecosystems are also capable of breaking down and immobilizing pollutants, such as heavy metals and pesticides.

Aside from supplying us with fish as a food source, freshwater ecosystems, such as rivers and lakes, provide us with drinking water and water to irrigate crops. The water-holding capacity of forests prevents flooding, and the ability of forests and other ecosystems to retain soil prevents soil erosion. Most of our crops and prescription drugs were originally derived from plants that grew wild in an ecosystem. Some human populations around the globe still depend on wild animals as a food source. And we must not forget that almost everyone prefers to vacation in the natural beauty of an ecosystem.

Figure 1.6   Human-impacted versus a natural ecosystem.

Humans Threaten the Biosphere

The human population tends to modify existing ecosystems for its own purposes (Fig. 1.6). Humans clear forests or grasslands in order to grow crops; later, they build houses on what was once farmland; and finally, they convert small towns into cities. Almost all natural ecosystems are altered by human activities, which also reduce biodiversity. The present **biodiversity** of our planet has been estimated to be as high as 15 million species and, so far, under 2 million have been identified and named. **Extinction** is the death of a species or larger group of organisms. It is estimated that, presently, we are losing as many as 400-species per day due to human activities. Many biologists are alarmed about the present rate of extinction and believe it may eventually rival the rates of the five mass extinctions that have occurred during our planet's history. The dinosaurs became extinct during the last mass extinction 65 million years ago.

One of the major bioethical issues of our time is preservation of the biosphere and biodiversity. If we adopt a conservation ethic that preserves the biosphere and biodiversity, we are helping to ensure the continued existence of our species.

✔ Check Your Progress 1.2

1. What anatomical feature(s) tells us that humans are vertebrates and mammals?

2. How do humans differ from the other mammals, including apes?

3. Why should humans want to preserve the biosphere, where a variety of organisms live?

1.3 Science as a Process

Science is a way of knowing about the natural world. When scientists study the natural world, they aim to be objective, rather than subjective. It is very difficult to make objective observations and to come to objective conclusions because we are often influenced by our own particular prejudices. Still, scientists strive for objective observations and conclusions, while keeping in mind that scientific conclusions can change due to new findings. New findings may have been possible because of recent advances in techniques or equipment.

Importance of Scientific Theories in Biology

Science is not just a pile of facts. The ultimate goal of science is to understand the natural world in terms of scientific theories. **Scientific theories** are concepts that tell us about the order and the patterns within the natural world; in other words, how the natural world is organized. For example, these are some of the basic theories of biology.

Theory	Concept
Cell	All organisms are composed of cells, and new cells only come from pre-existing cells.
Homeostasis	The internal environment of an organism stays relatively constant.
Genes	Organisms contain coded information that dictates their form, function, and behavior.
Ecosystem	Populations of organisms interact with each other and the physical environment.
Evolution	All things have a common ancestor, but each is adapted to a particular way of life.

Evolution is the unifying concept of biology because it makes sense of what we know about living things. For example, the theory of evolution enables scientists to understand the variety of living things and the anatomy, physiology, and development of organisms—even their behavior. Because the theory of evolution has been supported by so many observations and experiments for over a hundred years, some biologists refer to the **principle** of evolution. This term is preferred terminology for theories that are generally accepted as valid by an overwhelming number of scientists.

The Scientific Method Has Steps

Unlike other types of information available to us, scientific information is acquired by a process known as the **scientific method.** The approach of individual scientists to their work is as varied as they themselves; still, for the sake of discussion, it is possible to speak of the scientific method as consisting of certain steps (Fig. 1.7).

After making initial observations, a scientist will, most likely, study any previous **data,** which are facts pertinent to

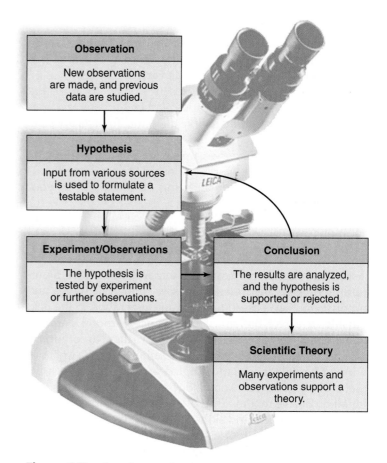

Figure 1.7   Flow diagram for the scientific method.
On the basis of new and/or previous observations, a scientist formulates a hypothesis. The hypothesis is tested by further observations and/or experiments, and new data either support or do not support the hypothesis. The return arrow indicates that a scientist often chooses to retest the same hypothesis or to test a related hypothesis. Conclusions from many different but related experiments may lead to the development of a scientific theory. For example, studies pertaining to development, anatomy, and fossil remains all support the theory of evolution.

the matter at hand. Imagination and creative thinking also help a scientist formulate a **hypothesis** that becomes the basis for more observation and/or experimentation. The new data help a scientist come to a **conclusion** that either supports or does not support the hypothesis. Because hypotheses are always subject to modification, they can never be proven true; however, they can be proven untrue. When the hypothesis is not supported by the data, it must be rejected; therefore, some think of the body of science as what is left after alternative hypotheses have been rejected.

Science is different from other ways of knowing by its use of the scientific method to examine a phenomenon. Any suggestions about the natural world that are not based on data gathered by employing the scientific method cannot be accepted as within the realm of science. Scientific theories are concepts based on a wide range of observations and experiments.

How the Cause of Ulcers Was Discovered

Let's take a look at how the cause of ulcers was discovered so we can get a better idea of how the scientific method works. In 1974, when Barry James Marshall was a young resident physician at Queen Elizabeth II Medical Center in Perth, Australia, he saw many patients who had bleeding stomach ulcers. A pathologist at the hospital, Dr. J. Robin Warren, told him about finding a particular bacterium, now called *Helicobacter pylori*, near the site of peptic ulcers (open sores in the stomach). Using the computer networks available at that time, Marshall compiled much data showing a possible correlation between the presence of *Helicobacter pylori* and the occurrence of both gastritis (inflammation of the stomach) and stomach ulcers. On the basis of these data, Marshall formulated a hypothesis: *Helicobacter pylori* is the cause of gastritis and ulcers.

Marshall decided to make use of Koch's[1] postulates, the standard criteria that must be fulfilled in order to show that a pathogen (bacteria or virus) causes a disease.

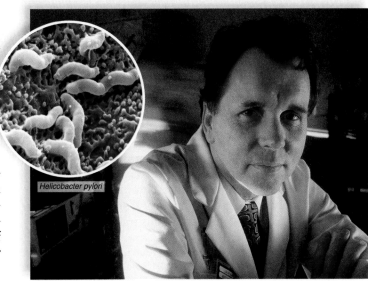

Figure 1.8   Dr. Barry Marshall.
Over a ten-year span, Dr. Marshall fulfilled Koch's postulates to show that *Helicobacter pylori* is the cause of peptic ulcers. The inset shows the presence of the bacterium in the stomach.

Koch Postulates:

- The suspected pathogen (virus or bacterium) must be present in every case of the disease;
- the pathogen must be isolated from the host and grown in a lab dish;
- the disease must be reproduced when a pure culture of the pathogen is inoculated into a healthy susceptible host; and
- the same pathogen must be recovered again from the experimentally infected host.

The First Two Criteria

By 1983, Marshall had fulfilled the first and second of Koch's criteria. He was able to isolate *Helicobacter pylori* from ulcer patients and grow it in the laboratory. (Success was achieved only after a petri dish was inadvertently left in the incubator for six, instead of two, days.) Further, he had determined that bismuth, the active ingredient in Pepto-Bismol, could destroy the bacteria in a petri dish.

Despite presentation of these findings to the scientific community, most physicians continued to believe that stomach acidity and stress were the cause of stomach ulcers. In those days, patients were usually advised to make drastic changes in their lifestyle or seek psychiatric counseling to "cure" their ulcers. Many fellow scientists believed that no bacterium would be able to survive the normal acidity of the stomach.

The Last Two Criteria

Marshall had a problem in fulfilling the third and fourth of Koch's criteria because he had been unable to infect guinea pigs and rats with the bacterium. The bacteria just didn't flourish in the intestinal tract of these animals, and our society does not believe in using humans as experimental subjects, even if it would mean saving the lives of many humans. Marshall was so determined to support his hypothesis that, in 1985, he decided to perform the experiment on himself! To the utter disbelief of those in the lab that day, he and another volunteer swallowed a foul-smelling and -tasting solution of *Helicobacter pylori*. Within the week, they felt lousy and were vomiting up their stomach contents. Examination by endoscopy showed that their stomachs were now inflamed, and biopsies of the stomach lining contained the suspected bacterium (Fig. 1.8). Their symptoms abated without need of medication, and they never developed an ulcer. The next time Marshall gave a talk, however, he challenged his audience to refute his hypothesis, and though many tried, the investigators found themselves supporting it instead.

The Conclusion

In science, many experiments that involve a considerable number of subjects are required before a conclusion can be reached. By the early 1900s, at least three independent studies involving hundreds of patients had been published showing that antibiotic therapy could eliminate *Helicobacter pylori* from the intestinal tract and cure patients of ulcers wherever they occurred in the tract.

Dr. Marshall received all sorts of prizes and awards, but he and Dr. Warren were especially gratified to receive a Nobel prize in Medicine in 2005. The Nobel committee commented, "Thanks to the pioneering discovery by Marshall and Warren, peptic ulcer disease is no longer a chronic, frequently disabling condition, but a disease that can be cured by a short regiment of antibiotics and acid secretion inhibitors."

---

[1]Robert Koch was a German microbiologist who helped verify the germ theory of disease and established the standard as to whether an organism causes a particular disease.

How to Do a Controlled Study

The work that Marshall and Warren did was largely observational. Often, scientists perform an **experiment,** a series of procedures to test a hypothesis. As an example, let's say investigators wanted to determine which of two antibiotics was best for the treatment of an ulcer. When scientists do an experiment, they try to vary just the **experimental variables,** in this case, the medications being tested. A **control group** is not given the medications, but one or more **test groups** are given the medications. If by chance, the control group shows the same results as a test group, the investigators immediately know the results of their study are invalid because it would mean the medications may have nothing to do with the results.

The study depicted in Figure 1.9 shows how investigators may study this hypothesis:

> **Hypothesis:** Newly discovered antibiotic B is a better treatment for ulcers than antibiotic A, which is in current use.

Next, the investigators might decide to use three experimental groups. It is important to reduce the number of possible variables (differences) such as sex, weight, other illnesses, and so forth between the groups. Therefore, the investigators *randomly* divide a very large group of volunteers equally into the three groups. The hope is any differences will be distributed evenly among the three groups. This is possible only if the investigators have a large number of volunteers. The three groups are to be treated like this:

> *Control group:* Subjects with ulcers are not treated with either antibiotic.
>
> *Test group 1:* Subjects with ulcers are treated with antibiotic A.
>
> *Test group 2:* Subjects with ulcers are treated with antibiotic B.

After the investigators have determined that all volunteers do suffer from ulcers, they will want the subjects to think they are all receiving the *same* treatment. This is an additional way to protect the results from any influence other than the medication. To achieve this end, the subjects in the control group can receive a **placebo,** a treatment that appears to be the same as that administered to the other two groups but actually contains no medication. In this study, the use of a placebo would help ensure the same dedication by all subjects to the study.

Figure 1.9   A controlled laboratory experiment to test the effectiveness of a medication in humans.

In this study, a large number of people were divided into three groups. The control group received a placebo and no medication. Test group 1 received antibiotic A and test group 2 received antibiotic B. The results are depicted in a graph, and it shows that antibiotic B was found to be a more effective treatment than antibiotic A for the treatment of ulcers.

State Hypothesis:
Antibiotic B is a better treatment for ulcers than antibiotic A

Large number of subjects were selected

Subjects were divided into three groups

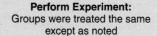

Perform Experiment:
Groups were treated the same except as noted

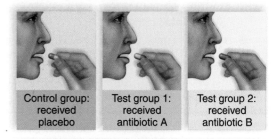

Control group: received placebo

Test group 1: received antibiotic A

Test group 2: received antibiotic B

Collect Data:
Each subject was examined for the presence of ulcers

Conclusion: Hypothesis is supported: Antibiotic B is a better treatment for ulcers than antibiotic A

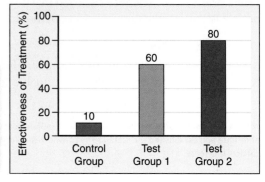

The Results

After two weeks of administering the same amount of medication (or placebo) in the same way, the intestinal tract of each subject is examined to determine if ulcers are still present. Endoscopy, a procedure depicted in the photograph, is one possible way to examine a patient for the presence of ulcers. This procedure is performed under sedation and involves inserting an endoscope—a small, flexible tube with a tiny camera on the end—down the throat and into the stomach. Now, the doctor can see the lining of the stomach, and check for possible ulcers. Tests performed during an endoscopy can also determine if *Helicobacter pylori* is present.

Because endoscopy is somewhat subjective, it is probably best if the examiner is not aware of which group the subject is in; otherwise, the prejudice of the examiner may influence the examination. When neither the patient nor the examiner is aware of the specific treatment, it is called a *double-blind* study.

In this study, the investigators may decide to determine effectiveness of the medication by the percentage of people who no longer have ulcers. So, if 20 people out of 100 still have ulcers, the medication is 80% effective. The difference in effectiveness is easily read in the graph portion of Figure 1.9.

> **Conclusion:** On the basis of their data, the investigators conclude that their hypothesis has been supported.

Publication of Scientific Studies

Scientific studies are customarily published in a scientific journal, so that all aspects of a study are available to the scientific community. Before information is published in scientific journals, it is typically reviewed by experts who ensure the research is credible, accurate, unbiased, and well executed.

Another scientist should be able to read about an experiment in a scientific journal, repeat the experiment in a different location, and get the same (or very similar) results. Some articles are rejected for publication by reviewers when they believe there is something questionable about the design or manner in which an experiment was conducted.

Scientific journals often discuss important studies in an issue of the journal. Also, each study begins with a short synopsis of the study. In this way, a scientist may be clued into which studies in an issue are of particular interest, and which should be studied thoroughly.

Further Study

As mentioned previously, the conclusion of one experiment often leads to another experiment. Scientists reading the study described in Figure 1.9 may decide that it would be important to test the difference in the ability of antibiotic A and B to kill *Helicobacter pylori* in a petri dish. Or, they may want to test which medication is more effective in women than men, and so forth. The need for scientists to expand on findings explains why science changes and the findings of yesterday may be improved upon tomorrow.

Scientific Journals Versus Other Sources of Information

The information in scientific journals is highly regarded by most scientists because of the review process and the fact that it is "straight from the horse's mouth," so to speak. The investigator who actually did the research is generally the primary author of a published study. Reading the actual results tends to prevent the possibility of misinformation and/or bias. Remember playing "pass the message" when you were young? If someone started a message and it was passed around a circle of people, the last person to hear the message rarely receives the original message. As each person passes along the message, information may be added or deleted from the original. That same thing happens to scientific information when it is published in magazines, books, or reported by someone other than the original investigator.

Unfortunately, the studies in scientific journals may be very technical and difficult for a layperson to read and understand. Therefore, most likely, the general public does rely on secondary sources of information for their science news. The information may have been taken out of context or misunderstood by the reporter, and the result is transmission of misinformation. Ideally, a reference to the original source (scientific journal article) will be provided so that the second-hand information can be verified. Remember also that it often takes years to do enough experiments for the scientific community to accept findings as well founded. So, be wary of claims that have only limited data to support them and any information that may not be supported by repeated experimentation.

People should be especially careful about scientific information that is available on the Internet, which is not well regulated. Reliable, credible scientific information can be found at websites with URLs (Web addresses or uniform resource locators) containing .edu (for educational institution), .gov (for government sites such as the National Institutes for Health or Centers for Disease Control), and .org (for non-profit organizations such as the American Lung Association or the National Multiple Sclerosis Society). Unfortunately, quite a bit of scientific information on the Internet is intended to entice people into purchasing some sort of product for weight loss, prevention of hair loss, or similar types of maladies. These websites have URLs ending with .com or .net. It pays to question and verify the information from these websites with another source (primary, if possible).

☑ Check Your Progress 1.3

1. In general, what steps do investigators use when utilizing the scientific method?

2. What is the standard experimental design for a controlled study?

3. Why might information in scientific journals be more reliable than that found in magazines or on the Internet?

1.4 Making Sense of a Scientific Study

When evaluating scientific information, it is important to consider the type of data given to support it. Anecdotal data, which consists of testimonials by individuals rather than results from a controlled, clinical study, are never considered reliable data. An example of anecdotal data would be someone who claims that a particular diet helped them lose weight. Obviously, this doesn't mean the diet will work for everyone. Testimonial data are suspect because the effect of whatever is under discussion has not been studied in a large number of subjects.

We also have to keep in mind that just because two events occur at the same time, one factor may not be the cause of the other. Dr. Marshall had this problem when his data largely depended on finding *Helicobacter pylori* at the site of ulcers. More data was needed before the scientific community could conclude that *Helicobacter pylori* was the cause of an ulcer. Similarly, the fact that a human papillomavirus (HPV) infection usually precedes cervical cancer could be viewed only as limited evidence that HPV causes cervical cancer. In this instance, HPV did turn out to be a cause of cervical cancer, but not all correlations (relationships) turn out to be causations. For example, scientific studies do not support the well-entrenched belief that exposure to cold results in colds. Instead, we now know that viruses cause colds.

What To Look For

Although most everyone who examines a study is tempted to first read the abstract (synopsis) at the beginning and then skip to the conclusion at the end of a study, we really should also examine the investigators' methodology and results before going to the conclusion. The methodology tells us how they conducted their study, and the results tell us what facts (data) they discovered. Always keep in mind that the conclusion is not the same as the data. The conclusion is an interpretation of the data. It is up to us to decide if the conclusion is justified by the data.

Graphs

Data are often depicted in the form of a bar graph (see Fig. 1.9) or a line graph (Fig. 1.10). A graph shows the relationship between two quantities, such as the taking of an antibiotic and the disappearance of ulcer. As in Figure 1.9, the experimental variable (study groups) is plotted on the horizontal (x-axis), and the result (effectiveness) is plotted along the vertical (y-axis). Graphs are useful tools to summarize data in a clear and simplified manner. For example, Figure 1.9 immediately shows that antibiotic B produced the best results.

The title and labels can assist you in reading a graph; therefore, when looking at a graph, first check the two axes to determine what the graph pertains to. For example, in Figure 1.10, we can see that the investigators were studying tree trunk diameters at four sites (places). By looking at this graph, we know that trees with the greatest diameter are found at

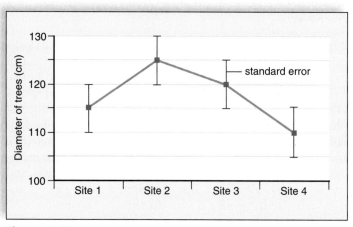

Figure 1.10   Line graph.
Line graph showing that the diameter of tree trunks varies at four different sites (places).

site 2, and we can also see to what degree the tree trunk diameters differed between the sites.

Statistical Data

Most authors who publish research articles use statistics to help them evaluate their experimental data. In statistics, the **standard error** tells us how uncertain a particular value is. Suppose you predict how many hurricanes Florida will have next year by calculating the average number during the past ten years. If the number of hurricanes per year varied widely, your standard error would be larger than if the number per year is usually about the same. In other words, the standard error tells you how far off the average could be. If the average number of hurricanes is four and the standard error is ± 2, then your prediction of four hurricanes is actually between two and six hurricanes.

Statistical Significance When scientists conduct an experiment, there is always the possibility that the results are due to chance or due to some factor other than the experimental variable. Investigators take into account several factors when they calculate the probability value that their results were due to chance alone. If the probability value is low, researchers describe the results as statistically significant. A probability value of less than 5% (usually written as p<.05) is acceptable, but, even so, keep in mind that the lower the p value, the *less* likely that results are due to chance. Therefore, the lower the p value, the *greater* the confidence the investigators and you can have in the results. Before you take a medication for ulcers, you probably want a significance as low as p<.001.

✅ Check Your Progress 1.4

1. What is wrong with **(a)** anecdotal data (testimonial data) and **(b)** with correlation data?

2. **a.** What is the difference between the data and the conclusion of a study? **b.** Which is more likely to be questionable?

3. How do scientific studies benefit from the use of graphs and statistics?

Science **Focus**

The Benefits and Limitations of Statistical Studies

Many of the studies in scientific journals and reported in the news are statistical studies, so it behooves us to be aware of their benefits and limitations. At the start, you should know that a statistical study will gather numerical information from various sources and then try to make sense out of it, for the purpose of coming to a conclusion.

Example of a Statistical Study

Let's take a look at a study that allows us to conclude that babies conceived 18 months to five years after a previous birth are healthier than those conceived at shorter or longer intervals. In other words, spacing children about two to five years apart is a good idea (Fig. 1A). Here is how the authors collected their data and the results they published in the *Journal of the American Medical Association.**

Objective To determine if there is an association between birth spacing and a healthy baby when data are corrected for maternal characteristics or socioeconomic status.

Data The authors collected data from studies performed around the world in 1966 through January 2006. The studies were published in various journals, reported on at professional meetings, or were known to the authors by personal contact. The authors gathered a very large pool of data that included over 11 million pregnancies from 67 individual studies. Twenty of the studies were from the United States, with the remaining 47 coming from 61 different countries. The authors attempted to adjust the data (by elimination of certain data) for factors such as mother's age, wealth, access to prenatal care, and breast-feeding. These adjustments allow the findings to be applied to both developed and developing countries.

Conclusion

1. A pregnancy that begins less than six months after a previous birth has a 77% higher chance of being preterm and a 39% higher chance of lower birth weight.
2. For up to 18 months between pregnancies, the chance of a preterm birth decreases by 2% per month, and the chance of a low-weight birth decreases by 3% per month as the 18-month time period is approached.
3. Babies conceived after 59 months have the same risk as those conceived in the less-than-six-months group.
4. The optimum spacing between pregnancies appears to be 18 months to five years after a previous birth.

The study leader, Agustin Conde-Agudelo, said, "Health officials should counsel women who have just given birth to delay their next conception by 18 to 59 months."

Limitations of Experimental Studies

The expression "statistical study" is a bit of a misnomer because most scientists collect quantitative data and use it

Figure 1A   Does spacing pregnancies lead to healthier children? A recent statistical study suggests that it does. If so, which mother, left or right, may have a healthier younger child?

to come to a conclusion. However, if we compare this study to the experimental study described in this chapter, we can see that the experimental study randomly divided a large number of qualified subjects into a control group and test groups. The groups were treated the same except for the experimental variable. Obviously, you wouldn't be able to divide women of the same child-bearing age into various groups and tell each group when they will conceive their children for the purpose of deciding the best interval between pregnancies for the health of the newborn. So, what is the next best thing? Do a statistical study utilizing data already available about women who became pregnant at different intervals.

A statistical study is really a correlation study. In our example, the authors studied the correlation between birth spacing and the health of a newborn. The more data collected from more varied sources makes a correlation study more reliable. When Marshall and an assistant drank a solution of *Helicobacter pylori,* the data was based on a sample size of two people—not very convincing. However, the study by Conde-Agudelo has a very large sample size, which goes a long way to validating the results. Even so, we know very well that a correlation does not necessarily translate to causation. So, it is not surprising that Dr. Mark A. Klebanoff, director of the National Institute of Child Health and Human Development, commented that many factors will affect birth spacing and that the study is not detailed enough to take all factors into consideration. Is any statistical study detailed enough? Most likely not.

Benefits of Statistical Studies

Before we give up on statistical studies, let's consider that they do provide us with information not attainable otherwise. Regardless of whether we understand the intricacies of statistical analysis, statistical studies do allow science to gain information and insights into many problems. True, further study is needed to find out if a correlation does mean causation, but science is always a work in progress, with additional findings being published every day.

* Agustin Conde-Agudelo, MD, MPH, Anyeli Rosas-Bermudez, MPH, Ana Cecilia Kafury-Goeta, MD. "Birth Spacing and Risk of Adverse Perinatal Outcomes." JAMA, 2006;295:1809 –1823. Abstract.

1.5 Science and Social Responsibility

As we have learned in this chapter, science is a systematic way of acquiring knowledge about the natural world. Biologists are scientists that study living things, from the tiniest microbes to the tallest trees (Fig. 1.11). Religion, aesthetics, and ethics are other ways in which human beings seek order in the natural world. Science differs from these other ways of knowing and learning by its process, which is based on the scientific method. Science considers hypotheses that can be tested only by experimentation and observation. Only after an immense amount of data has been gathered do scientists arrive at a scientific theory, a well-found concept about the natural world. Knowing this should help you realize that all scientific theories have merit.

Science is a slightly different endeavor from technology, but science is the driving force behind technology. **Technology** is the application of scientific knowledge to the interests of humans. Western civilization has always believed that science and technology offer us ways to improve our lives. Our ability to build houses, pave roads, grow crops, and cure illnesses all depend on technology. Just think of how many things you use each day that are made of plastic and you will know how much technology means to your daily life. Even the field of nuclear physics is of direct benefit to human beings. The findings of nuclear physics play a role in cancer therapy, medical imaging, and homeland security, for example. It has also given us nuclear power and the nuclear bomb. In other words, science and technology are not risk free; and uncontrolled technology can result in unanticipated side effects, such as the possible loss of polar bears, as described in the Bioethical Focus on page 15.

Science and Technology, Benefits Versus Risks

Investigations into cell structure and genes led to the biotechnology revolution of current times. We now know how to manipulate genes. If you are a diabetic and are using insulin, it was produced by genetically modified (GM) bacteria. Despite its benefits, there are risks associated with biotechnology. For example, ecologists are concerned that GM crops could endanger the biosphere. To take an example, many farmers are now planting GM cotton that produces a toxin that kills insect pests, but the natural predators that feed on these pests may also be killed by the toxin. It may not be wise to kill off friendly insects. Or closer to home, people are now eating GM foods and/or foods that have been manufactured by using GM products. Some people are concerned about the possible effects of GM foods on human health.

In medicine, gene technology raises even more difficult ethical issues, such as whether humans should be cloned or whether gene therapy should be used to modify the inheritance of people, even before they are born. A current debate centers around the use of stem cells to cure human illnesses, such as the spinal cord injury suffered by the actor Christopher Reeve or Alzheimer disease, which afflicts so many seniors. Most likely, you will not appreciate this debate until you know that early human embryos are composed only of cells, and these cells are called embryonic stem cells. Embryonic stem cells are genetically capable of becoming any type of tissue needed to cure a human illness. Should human embryos be dismantled and used for this purpose? It means they will never have the opportunity to become a human being.

a.

b.

Figure 1.11   Biologists at work.
Biologists can be found collecting data (**a**) in the laboratory and (**b**) in the field. They discover basic information about the natural world, including the effects of technology on human health and the environment.

Polar Bears: How Long Will They Last?

Polar bears are keystone species, which means that their extinction will send a ripple effect throughout the entire Arctic community. Right now, the polar bear is listed as vulnerable to extinction. The chief reasons are these:

1. **Global warming.** Global warming decreases the amount of ice pack available for hunting and does not allow the bears to build the necessary layer of insulation to survive hibernation or nurse their young. Global warming is due to burning of fossil fuels, which releases carbon dioxide into the atmosphere (see page 501).
2. **Chemical pollution.** Polychlorinated biphenyls (PCBs) and organochlorine pesticides have entered the Arctic ecosystem, and they have been found in polar bears, beluga whales, seals, and seabirds. Such pollutants might negatively affect the reproductive abilities of polar bears and other animals.
3. **Oil and gas exploration.** Drilling in the Arctic National Wildlife refuge in Alaska will impact about half of the onshore denning sites that are used by pregnant females. Therefore, drilling will significantly lower the survival rate of polar bear females, as well as their young.

Polar bears prey on ringed seals; therefore, as polar bears decrease in number, the ringed seal population is expected to increase in size. This might sound good except that the prey of ringed seals—namely, cod, shrimp, and other crustacean populations—will decrease, and the Inuit tribes living in the Arctic are dependent on these same species for survival.

Our industrialized society is dependent on the consumption of oil and gas for its way of life. Therefore, even though people know that the burning of fossil fuels is contributing to global warming, they would most likely find it very difficult to reduce their consumption of these fuels. Also, the drilling for oil and gas in the refuge will help decrease our dependency on foreign sources of oil, as well as provide hundreds of jobs for the citizens of Alaska.

The possibility of the loss of a keystone species can serve as a warning light for concern. When the light turns red, it is an indication there are serious problems within an ecosystem. The human species depends on all the other species of the biosphere for survival. As we lose species, we only increase the problems that we will face in the near future. Therefore, it might behoove us to take all possible steps to conserve species and to reduce pollutants, including those that lead to global warming.

Figure 1B **Polar bear mother and cub.**

Decide Your Opinion

1. What selfish reasons can you give for wanting the Arctic community to continue functioning as before?
2. Is drilling for oil worth the risk of contributing to polar bear extinction?
3. What is more valuable, maintaining our lifestyle or the Arctic community?

Everyone Is Responsible

Science, technology, and society have interacted throughout human history, and scientific investigation and technology have always been affected by human values. Studying science, such as human biology, can give citizens the background they need to participate fully in ethical debates. All citizens should assume this responsibility because everyone, not just scientists, need to be involved in making value judgments about the proper use of technology.

Should windmills be placed in Nantucket Sound, or other bodies of water, to reduce our use of oil, which causes global warming? In order to participate in this debate, you have to know what windmills might do to the ecology of Nantucket Sound, what global warming is, and what the evidence for global warming is. In other words, you need to review the data of scientists and only then participate in the debate about how technology should be used in this particular instance.

Should the rise in human population be curtailed in order to help preserve biodiversity? You might hear on the news that we are in a biodiversity crisis—the number of extinctions expected to occur in the near future is unparalleled in the history of the Earth. In order to help answer this question, you will want to know how large the human population is, what the benefits of biodiversity are, how it is threatened, and what the threat is to humankind if biodiversity is not preserved.

Scientists can inform and educate us, but they need not bear the burden of making these decisions alone because science does not make value judgments. This is the job of all of us.

☑ Check Your Progress 1.5

1. What are some examples to show that technology has both benefits and risks?
2. Which members of society should be responsible for deciding how technology should be used?
3. Is this a good reason for everyone to become scientifically competent?

Summarizing the Concepts

1.1   The Characteristics of Life

Living things, often called organisms, share common characteristics. Organisms

- have levels of organization—atoms, molecules, cells, tissues, organs, organ systems, organisms, populations, community, ecosystem, and biosphere;
- take materials and energy from the environment;
- reproduce;
- grow and develop;
- are homeostatic;
- respond to stimuli; and
- have an evolutionary history and are adapted to a way of life.

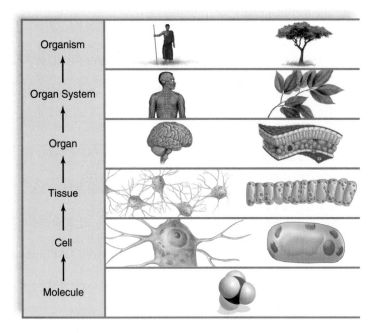

1.2   Humans Are Related to Other Animals

The classification of living things mirrors their evolutionary relationships. Humans are mammals, a type of vertebrate in domain Eukarya. Humans differ from other mammals, including apes, by their

- highly developed brains;
- completely upright stance;
- creative language; and
- ability to use a wide variety of tools.

Humans Have a Culture Heritage

Language, tool use, values, and information are passed on from one generation to the next.

Humans Are Members of the Biosphere

Humans depend on the biosphere for its many services, such as absorption of pollutants, sources of water and food, prevention of soil erosion, and natural beauty.

Humans Threaten the Biosphere

Unfortunately, humans threaten the biodiversity of the biosphere, most likely to their own detriment.

1.3   Science as a Process

The scientific method consists of

- making an observation;
- formulating a hypothesis;
- carrying out experiments and observations;
- coming to a conclusion; and
- developing a scientific theory.

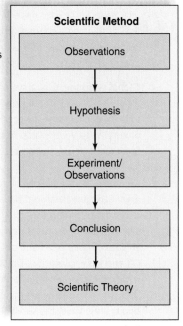

How the Cause of Ulcers Was Discovered

Dr. Marshall followed Koch's postulates to show that *Helicobacter pylori* causes ulcers. When he was unable to infect an animal with the bacterium, he and another volunteer infected themselves. His persistence led to clinical studies that showed antibiotics can cure ulcers.

How to Do a Controlled Study

- a large number of subjects are divided randomly into groups;
- the test group(s) is exposed to an experimental variable;
- a control group is not exposed to an experimental variable and is given a placebo;
- all groups are otherwise treated the same, and it is best if the subjects and the technicians do not know what group they are in; and
- the results and conclusion are published in a scientific journal.

Scientific Journals Versus Other Sources of Information

Primary sources of information are best, and if second-hand sources are used, the reader needs to carefully evaluate the source. The Internet is not regulated, although URLs that end in .edu, .gov, and .org most likely provide reliable scientific information.

1.4   Making Sense of a Scientific Study

- Beware of anecdotal and correlation data because they need further study to be substantiated.
- Both bar and line graphs clearly show the relationship between quantities.

Statistical data. The standard error tells us how uncertain a particular value is. The statistical significance tells us how trustworthy the results are. The lower the statistical significance, the less likely the results are due to chance and the more likely they are due to the experimental variable.

1.5   Science and Social Responsibility

- Scientific information is based on observation and experimentation. Therefore, scientists need not make value judgments for us.
- The use of modern technology has it risks, and all citizens need to be able to make informed decisions regarding how and when technology should be used.

Understanding Key Terms

adaptation 5
atom 2
biodiversity 7
biology 2
biosphere 2
cell 2
community 2
conclusion 8
control group 10
culture 7
data 8
development 4
domain 6
ecosystem 2
evolution 5
experiment 10
experimental variable 10
extinction 7
gene 4
homeostasis 4
hypothesis 8

kingdom 6
metabolism 4
molecule 2
multicellular 2
organ 2
organism 2
organ system 2
placebo 10
population 2
principle 8
reproduce 4
science 8
scientific method 8
scientific theory 8
species 2
standard error 12
technology 14
test group 10
tissue 2
vertebrate 6

Match the key terms to these definitions.

a. —————— Zone of air, land, and water at the surface of the Earth in which living organisms are found.

b. —————— Smallest unit of a human being and all living things.

c. —————— Concept supported by a broad range of observations, experiments, and conclusions.

d. —————— An internal environment that normally varies within only certain limits.

e. —————— An artificial situation devised to test a hypothesis.

Testing Your Knowledge of the Concepts

1. Name the 11 levels of biological organization from simple to complex, and briefly describe each. (pages 2–3)

2. What is homeostasis, and how is it maintained? Give some examples that show how systems work together to maintain homeostasis. (pages 4–5)

3. How do human activities threaten the biosphere? (page 7)

4. Discuss the importance of a scientific theory, and name several theories that are basic to understanding biological principles. (page 8)

5. With reference to the steps of the scientific method, explain how scientists arrive at a theory. (page 8)

6. What are Koch's postulates, and what are they used for? (page 9)

7. What is a control group, and what is the importance of a control group in a controlled study? (page 10)

8. How is technology different from scientific knowledge? (page 14)

In questions 9–12, match each description with the correct characteristic of life from the key.

Key:
   a. Life is organized.
   b. Living things reproduce and grow.
   c. Living things respond to stimuli.
   d. Living things have an evolutionary history.
   e. Living things acquire materials and energy.

9. Human heart rate increases when scared.

10. Humans produce only humans.

11. Humans need to eat for building blocks and energy.

12. Similar cells form tissues in the human body.

13. The level of organization that includes two or more tissues that work together is a(n)
   a. organ.   c. organ system.
   b. tissue.   d. organism.

14. The level of organization most responsible for the maintenance of homeostasis is the _____ level.
   a. cellular.
   b. organ.
   c. organ system.
   d. tissue.

15. The level of organization that includes all the populations in a given area along with the physical environment would be a(n)
   a. community.   c. biosphere.
   b. ecosystem.   d. tribe.

16. Which best describes the evolutionary relationship between humans and apes?
   a. Humans evolved from apes.
   b. Humans and chimpanzees evolved from apes.
   c. Humans and apes evolved from a common apelike ancestor.
   d. Chimpanzees evolved from humans.

17. The kingdom that contains humans is the kingdom
   a. Animalia.   c. Plantae.
   b. Fungi.   d. Protista.

In questions 18–20, match the explanation with a theory in the key.

Key:
   a. homeostasis   c. evolution
   b. cell   d. gene

18. All living things share a common ancestor.

19. The internal environment remains relatively stable.

20. Organisms contain coded information that dictates form and function.

In questions 21–25, match each description with a step in the scientific method from the key.

Key:

a. conclusion   d. observation
b. experiment   e. theory
c. hypothesis

21. Rejection or acceptance of a hypothesis based on collected data

22. Statement to be tested

23. Procedure designed to test a hypothesis

24. Hypothesis supported by many studies

25. Fact that may lead to the development of a hypothesis

Thinking Critically About the Concepts

The opening story describes the role of the sympathetic nervous system (NS) when you are in danger or believe that you are. When you are relaxing and/or digesting your food, another division of the nervous system, the parasympathetic NS, is active and in control. While the overall goal of the sympathetic NS is to save you from a threat by encouraging you to run away or put up a good fight, the overall goal of the parasympathetic NS is to carry on body activities in a normal manner.

1. Ashley described the increased heart rate she experienced during a scary movie scene. What other effects does the sympathetic NS have on your body? Consider what you read about homeostasis and blood pressure (page 5) and also your responses to scary events in your life.

2. Prey animals also experience these sympathetic NS responses when they're being stalked by a predator. Why would you expect their responses to be similar to those of humans?

3. During what perceived threat do many students experience responses stimulated by the sympathetic NS (particularly when they've not been the most studious of students)?

4. Name a couple of responses associated with the activity of the parasympathetic NS.

CHAPTER 2

Chemistry of Life

In spite of getting the recommended 8 hours of sleep every night for a couple of weeks, Rita felt tired and run down. Her friends commented on how pale she looked, but she brushed off these comments with the thought that everyone looks paler in the winter, after their summer color has faded.

When a close friend was diagnosed with mononucleosis, Rita decided it was time to visit her doctor to rule out a similar reason for her symptoms. Her doctor ordered a complete blood count and other tests for thyroid disease and infectious diseases. The results of Rita's blood work indicated she had a mild case of iron-deficiency anemia.* Iron is an element and a critical component of hemoglobin, the protein inside red blood cells. When oxygen is transported from the lungs to the body cells, it binds to iron for transport. Rita's doctor advised her to increase her daily intake of iron by eating iron-rich foods.

Good health and homeostasis, which we learned about in Chapter 1, depend heavily on the intake of certain chemicals, preferably in our food. An imbalance or deficiency of chemicals, such as iron, hemoglobin, or oxygen, can lead to serious health problems.

CHAPTER CONCEPTS

2.1 From Atoms to Molecules
All matter is composed of atoms, which react with one another to form molecules.

2.2 Water and Living Things
The properties of water make life, as we know it, possible. Living things are affected adversely by water that is too acidic or too basic.

2.3 Molecules of Life
Carbohydrates, lipids, proteins, and nucleic acids are macromolecules with specific functions in cells.

2.4 Carbohydrates
Glucose is blood sugar, and humans store glucose as glycogen. Cellulose is plant material that is a source of fiber in the diet.

2.5 Lipids
Fats and oils, which provide long-term energy, differ by consistency and can have a profound effect on our health. Other lipids function differently in the body.

2.6 Proteins
Proteins have numerous and varied functions in cells. Their shape suits their function.

2.7 Nucleic Acids
DNA that makes up our genes and RNA that serves as a helper to DNA are nucleic acids. ATP releases energy that is used by the cell to do metabolic work.

---

* anemia: a lower than normal hemoglobin amount or number of red blood cells.

2.1 From Atoms to Molecules

Matter refers to anything that takes up space and has mass. It is helpful to remember that matter can exist as a solid, a liquid, or a gas. Then we can realize that not only are we humans matter, but so are the water we drink and the air we breathe.

Elements

An **element** is one of the basic building blocks of matter; an element cannot be broken down by chemical means. Considering the variety of living and nonliving things in the world, it's quite remarkable that there are only 92 naturally occurring elements. It is even more surprising that over 90% of the human body is composed of just four elements: carbon, nitrogen, oxygen, and hydrogen. Even so, other elements, such as iron, are very important to our health. As the opening story describes, Rita had iron-deficiency anemia because her diet didn't contain enough iron for the making of hemoglobin. Hemoglobin serves an important function in the body because it transports oxygen to our cells.

Every element has a name and a symbol; for example, carbon has been assigned the atomic symbol C, and iron has been assigned the symbol Fe. Some of the symbols we use for elements are derived from Latin. For example, the symbol for sodium is Na because *natrium*, in Latin, means sodium, and *ferrum*, in Latin, means iron. Chemists arrange the elements in a periodic table which has this name because all the elements in a column react similarly. For example, all the elements in column VII (7) undergo the same type of chemical reactions, for reasons we will soon explore. Figure 2.1 shows only the first 36 elements in the periodic table, but a complete table is available in an appendix at the end of the book.

Atoms

An **atom** is the smallest unit of an element that still retains the chemical and physical properties of the element. The same name is given to the element and its atoms. While it is possible to split an atom, an atom is the smallest unit to enter into chemical reactions. For our purposes, it is satisfactory to think of each atom as having a central nucleus and pathways about the nucleus called shells. Even though an atom is extremely small, it contains even smaller subatomic particles. The subatomic particles, called **protons** and **neutrons,** are located in the nucleus, and **electrons** orbit about the nucleus in the shells (Fig. 2.2). Most of an atom is empty space. If we could draw an atom the size of a football stadium, the nucleus would be like a gumball in the center of the field, and the electrons would be tiny specks whirling about in the upper stands.

Protons carry a positive (+) charge, and electrons have a negative (−) charge. The **atomic numbers** given in Figure 2.1 tell you how many protons, and therefore how many electrons, an atom has when it is electrically neutral. For example, the atomic number of carbon is six; therefore, when carbon is neutral, it has six protons and six electrons.

How many electrons are there in each shell of an atom? The inner shell has the lowest energy level and can hold only two electrons; after that, each shell for the atoms noted in Figure 2.1 can hold up to eight electrons. Using this information, we can determine how many electrons are in the outer shells of the atoms shown in Figure 2.2. Notice that hydrogen (H) has only one shell that contains one electron. Carbon (atomic number of six) has two shells and the outer shell has 4 electrons. Nitrogen (atomic number of seven) has two shells and the outer shell has 5 electrons. Oxygen (atomic number of eight) has two shells and the outer shell has 6 electrons.

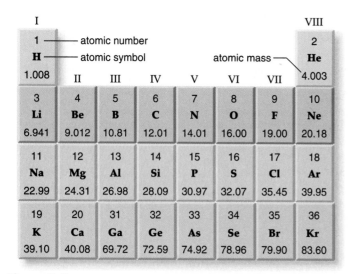

Figure 2.1   A portion of the periodic table of the elements.
The atomic symbol, atomic number, and atomic mass are given in this table, which shows only the elements through number 36.

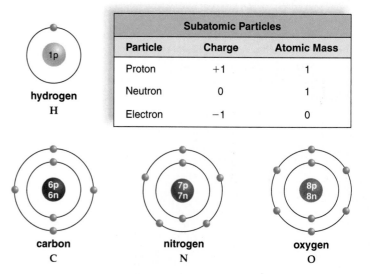

Figure 2.2   Atomic models.
The number of protons, neutrons, and electrons differ in hydrogen, carbon, nitrogen, and oxygen.

The **mass** of an atom represents its quantity of matter. The subatomic particles are so light that their mass is indicated by special designations called atomic mass units. Protons and neutrons are assigned one atomic mass unit, and electrons have almost no mass. Therefore, the **atomic mass** for each element in Figure 2.2 allows you to determine the number of neutrons these atoms have. For example, carbon, with an atomic mass of 12 and six protons, must have six neutrons.

Isotopes

Isotopes of the same type of atom differ in the number of neutrons and, therefore, mass. For example, the element carbon 12 has six neutrons, carbon 13 has seven neutrons, and carbon 14 has eight neutrons. Unlike the other two isotopes of carbon, carbon 14 is unstable and breaks down over time. As carbon 14 decays, it releases various types of energy in the form of rays and subatomic particles, and therefore it is a **radioisotope.** The radiation given off by radioisotopes can be detected in various ways. You may be familiar with the use of a Geiger counter to detect radiation.

Low Levels of Radiation

The importance of chemistry to biology and medicine is nowhere more evident than in the many uses of radioisotopes.

A radioisotope behaves the same as the stable isotopes of an element. This means that you can put a small amount of radioisotope in a sample and it becomes a **tracer,** by which to detect molecular changes.

Specific tracers are used in imaging the body's organs and tissues. For example, after a patient drinks a solution containing a minute amount of iodine 131, it becomes concentrated in the thyroid—the only organ to take it up to make the hormone thyroxine. A subsequent image of the thyroid indicates whether it is healthy in structure and function (Fig. 2.3*a*). Positron-emission tomography (PET) is a way to determine the comparative activity of tissues. Radioactively labeled glucose, which emits a subatomic particle known as a positron, can be injected into the body. The radiation given off is detected by sensors and analyzed by a computer. The result is a color image that shows which tissues took up glucose and are metabolically active (Fig. 2.3*b*). A PET scan of the brain can help diagnose a brain tumor, Alzheimer disease, epilepsy, or whether a stroke has occurred.

High Levels of Radiation

Radioactive substances in the environment can harm cells, damage DNA, and cause cancer. The release of radioactive particles following a nuclear power plant accident can have far-reaching and long-lasting effects on human health. The harmful effects of radiation can also be put to good use, however. Radiation from radioisotopes has been used for many years to sterilize medical and dental products. The possibility exists that it can be used to sterilize the U.S. mail, in order to free it of possible pathogens, such as anthrax spores.

The ability of radiation to kill cells is often applied to cancer cells. Radioisotopes can be introduced into the body in a way that allows radiation to destroy only cancer cells, with little risk to the rest of the body (Fig. 2.4).

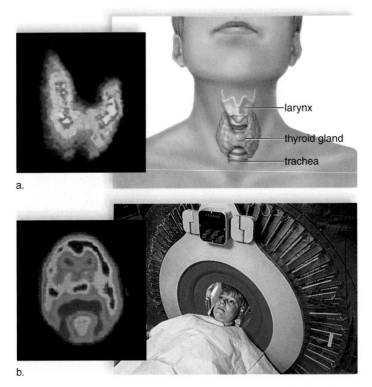

larynx

thyroid gland

trachea

a.

b.

Figure 2.3 Low levels of radiation.
a. The missing area (upper right) in this thyroid scan indicates the presence of a tumor that does not take up radioactive iodine. **b.** A PET (positron-emission tomography) scan reveals which portions of the brain are most active (red surrounded by light green).

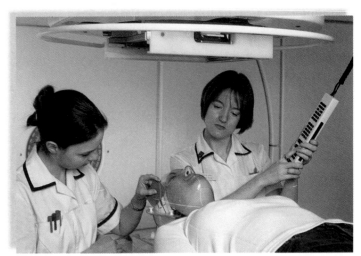

Figure 2.4 High levels of radiation.
Radiation causes cancer, but it can also be used to cure cancer.

Molecules and Compounds

Atoms often bond with one another to form a chemical unit called a **molecule.** A molecule can contain atoms of the same kind, as when an oxygen atom joins with another oxygen atom to form oxygen gas. Or, the atoms can be different, as when an oxygen atom joins with two hydrogen atoms to form water. When the atoms are different, a **compound** is present.

Two types of bonds join atoms: the ionic bond and the covalent bond.

Ionic Bonding

Atoms with more than one shell are most stable when the outer shell contains eight electrons. During an ionic reaction, atoms give up or take on an electron(s) in order to achieve a stable outer shell.

Figure 2.5 depicts a reaction between a sodium (Na) atom and a chlorine (Cl) atom. Sodium, with one electron in the outer shell, reacts with a single chlorine atom. Why? Because once the reaction is finished and sodium loses one electron to chlorine, its outer shell will have eight electrons. Similarly, a chlorine atom, which has seven electrons already, needs to acquire only one more electron to have a stable outer shell.

Ions are particles that carry either a positive (+) or negative (−) charge. When the reaction between sodium and chlorine is finished, the sodium ion carries a positive charge because it now has one less electron than protons, and the chloride ion carries a negative charge because it now has one more electron than protons:

Sodium Ion (Na⁺)	Chloride Ion (Cl⁻)
11 protons (+)	17 protons (+)
10 electrons (−)	18 electrons (−)
One (+) charge	One (−) charge

The attraction between oppositely charged sodium ions and chloride ions forms an **ionic bond.** The resulting compound, sodium chloride, is table salt, which we use to enliven the taste of foods.

In contrast to sodium, why would calcium, with two electrons in the outer shell, react with two chlorine atoms? Because, whereas calcium needs to lose two electrons, each chlorine, with seven electrons already, requires only one more electron to have a stable outer shell. The resulting salt ($CaCl_2$) is called calcium chloride.

The balance of various ions in the body is important to our health. Too much sodium in the blood can contribute to high blood pressure; not enough calcium leads to rickets (a bowing of the legs) in children; too much or too little potassium results in heartbeat irregularities. Bicarbonate, hydrogen, and hydroxide ions are all involved in maintaining the acid-base balance of the body (see pages 26–27).

Figure 2.5   Ionic bonding.

a. During the formation of sodium chloride, an electron is transferred from the sodium atom to the chlorine atom. At the completion of the reaction, each atom has eight electrons in the outer shell, but each also carries a charge as shown. **b.** In a sodium chloride crystal, ionic bonding between Na⁺ and Cl⁻ causes the ions to form a three-dimensional lattice configuration in which each sodium ion is surrounded by six chloride ions, and each chloride ion is surrounded by six sodium ions.

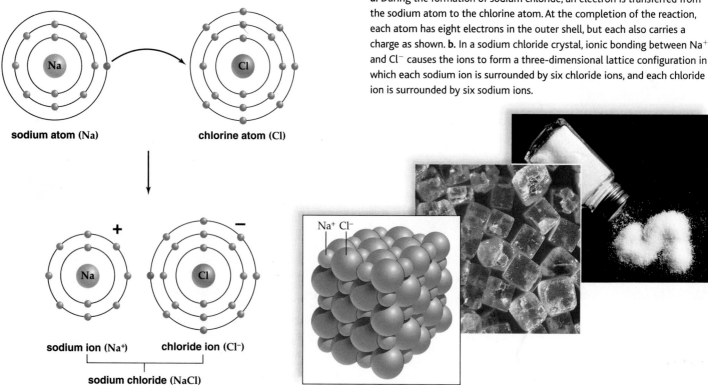

sodium atom (Na)   chlorine atom (Cl)

sodium ion (Na⁺)   chloride ion (Cl⁻)

sodium chloride (NaCl)

Na⁺ Cl⁻

a.   b.

Covalent Bonding

Atoms share electrons in **covalent bonds.** The overlapping outermost shells in Figure 2.6 indicate that the atoms are sharing electrons. Just as two hands participate in a handshake, each atom contributes one electron to the pair that is shared. These electrons spend part of their time in the outer shell of each atom; therefore, they are counted as belonging to both bonded atoms.

Double and Triple Bonds   Besides a single bond, in which atoms share only a pair of electrons, a double or a triple bond can form. In a double bond, atoms share two pairs of electrons, and in a triple bond, atoms share three pairs of electrons between them. For example, in Figure 2.6*b*, each oxygen atom (O) requires two more electrons to achieve a total of eight electrons in the outer shell. Notice that four electrons are placed in the outer overlapping shells in the diagram. Blood transports oxygen because the iron (Fe) in hemoglobin binds loosely to oxygen in the lungs and then gives it up in the tissues.

Structural and Molecular Formulas   Covalent bonds can be represented in a number of ways. In contrast to the diagrams in Figure 2.6, structural formulas use straight lines to show the covalent bonds between the atoms. Each line rep-

resents a pair of shared electrons. Molecular formulas indicate only the number of each type of atom making up a molecule. A comparison follows:

Structural formula: H—O—H, O=O

Molecular formula: H_2O, O_2

What would be the structural and molecular formulas for carbon dioxide? Carbon, with four electrons in the outer shell, requires four more electrons to complete its outer shell. Each oxygen, with six electrons in the outer shell, needs only two electrons to complete its outer shell. Therefore, carbon shares two pairs of electrons with each oxygen atom, and the formulas are as follows:

Structural formula: O=C=O

Molecular formula: CO_2

☑ Check Your Progress 2.1

1. How is an atom organized?
2. a. What are radioisotopes, and (b) how can they be used for the biological benefit of humans?
3. What are the two basic types of bondings between atoms?

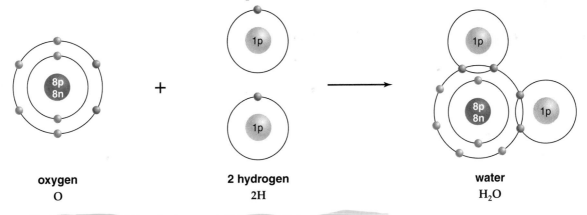

oxygen
O

2 hydrogen
2H

water
H_2O

a. When an oxygen and two hydrogen atoms covalently bond, water results.

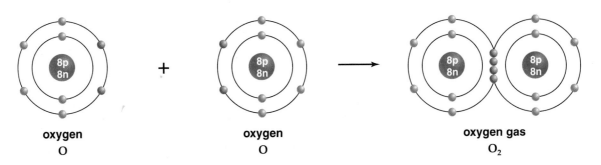

oxygen
O

oxygen
O

oxygen gas
O_2

b. When two oxygen atoms covalently bond, oxygen gas results.

Figure 2.6   Covalent bonding.
An atom can fill its outer shell by sharing electrons. To determine this, it is necessary to count the shared electrons as belonging to both bonded atoms. Hydrogen is most stable with two electrons in the outer shell; oxygen is most stable with eight electrons in the outer shell. Therefore, the molecular formula for water is **(a)** H_2O, and for oxygen gas it is **(b)** O_2.

2.2 Water and Living Things

Water is the most abundant molecule in living organisms, usually making up about 60–70% of the total body weight. Further, the physical and chemical properties of water make life as we know it possible.

In water, the electrons spend more time circling the oxygen (O) atom than the hydrogens because oxygen, the larger atom, has a greater ability to attract electrons than do the smaller hydrogen (H) atoms. Because the negatively charged electrons are closer to the oxygen atom, the oxygen atom becomes slightly negative, and the hydrogens, in turn, are slightly positive. Therefore, water is a **polar** molecule; the oxygen end of the molecule has a slight negative charge (δ^-), and the hydrogen end has a slight positive charge (δ^+):

The diagram on the left shows a structural formula of water, and the one on the right is called a space-filling model.

Hydrogen Bonds

A **hydrogen bond** occurs whenever a covalently bonded hydrogen is slightly positive and attracted to a negatively charged atom some distance away. A hydrogen bond is represented by a dotted line because it is relatively weak and can be broken rather easily.

In Figure 2.7, you can see that each hydrogen atom, being slightly positive, bonds to the slightly negative oxygen atom of another water molecule.

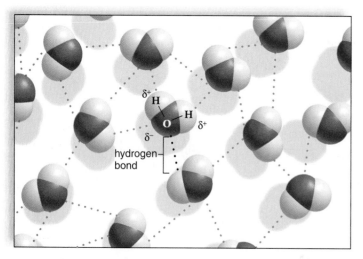

Figure 2.7   Hydrogen bonding between water molecules.
The polarity of the water molecules allows hydrogen bonds (dotted lines) to form between the molecules.

Properties of Water

Because of their polarity and hydrogen bonding, water molecules are cohesive, meaning that they cling together. Polarity and hydrogen bonding cause water to have many characteristics beneficial to life.

1. Water is a liquid at room temperature. Therefore, we are able to drink it, cook with it, and bathe in it.

Compounds with a low molecular weight are usually gases at room temperature. For example, oxygen (O_2), with a molecular weight of 32, is a gas; but water, with a molecular weight of 18, is a liquid. The hydrogen bonding between water molecules keeps water a liquid and not a gas at room temperature. Water does not boil and become a gas until 100°C, one of the reference points for the Celsius temperature scale (see inside back cover). Without hydrogen bonding between water molecules, our body fluids—and indeed our bodies—would be gaseous!

2. The temperature of liquid water rises and falls slowly, preventing sudden or drastic changes.

The many hydrogen bonds that link water molecules cause water to absorb a great deal of heat before it boils (Fig. 2.8a). A **calorie** of heat energy raises the temperature of 1 g of water 1°C. This is about twice the amount of heat required for other covalently bonded liquids. On the other hand, water holds heat, and its temperature falls slowly. Therefore, water protects us and other organisms from rapid temperature changes and helps us maintain our normal internal temperature. This property also allows great bodies of water, such as oceans, to maintain a relatively constant temperature. Water is a good temperature buffer.

3. Water has a high heat of vaporization, keeping the body from overheating.

It takes a large amount of heat to change water to steam (Fig. 2.8a). (Converting 1 g of the hottest water to steam requires an input of 540 calories of heat energy.) This property of water helps moderate the Earth's temperature so that life can continue to exist. Also, in a hot environment, most mammals sweat, and the body cools as body heat is used to evaporate sweat, which is mostly liquid water (Fig. 2.8b).

4. Frozen water is less dense than liquid water, so ice floats on water.

Remarkably, water is more dense at 4°C than at 0°C. Most substances contract when they solidify, but water expands when it freezes because in ice, water molecules form a lattice in which the hydrogen bonds are farther apart than in liquid water (Fig. 2.8c). This is why cans of soda burst when placed in a freezer or why frost heaves make northern roads bumpy in the winter. Also, because ice is lighter than cold water, bodies of water freeze from the top down. The ice acts as an

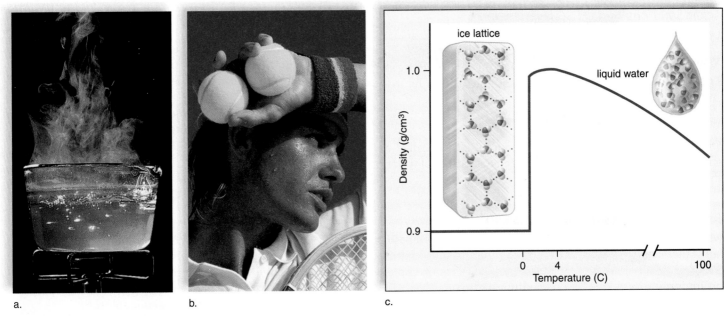

Figure 2.8 **Characteristics of water.**
a. Water becomes gaseous at 100°C. If it was a gas at a lower temperature, life could not exist. **b.** Body heat vaporizes sweat and, in this way, bodies cool when the temperature rises. **c.** Because ice is less dense than water, it forms a protective layer on top of ponds during the winter.

insulator to prevent the water below it from freezing. Thus, aquatic organisms are protected and have a better chance of surviving the winter.

5. Water molecules are cohesive, and, therefore, liquids fill vessels, such as blood vessels.

Water molecules cling together because of hydrogen bonding, and yet, water flows freely. This property allows dissolved and suspended molecules to be evenly distributed throughout a system. Therefore, water is an excellent transport medium. Within our bodies, blood fills our arteries and veins because it is 92% water. After blood transports oxygen and nutrients to cells, these molecules are used to produce cellular energy. Because Rita lacked sufficient iron-containing hemoglobin, her blood could not transport enough oxygen to her cells. Therefore, she felt tired and run down. Blood also removes wastes, such as carbon dioxide, from cells.

6. Water is a solvent for polar (charged) molecules, and thereby facilitates chemical reactions both outside and within our bodies.

When ions and molecules disperse in water, they move about and collide, allowing reactions to occur. Therefore, water is a solvent that facilitates chemical reactions. For example, when a salt such as sodium chloride (NaCl) is put into water, the negative ends of the water molecules are attracted to the sodium ions, and the positive ends of the water molecules are attracted to the chloride ions. This causes the sodium ions and the chloride ions to separate and to dissolve in water:

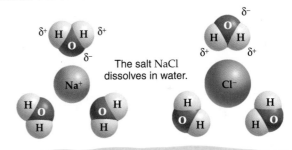

The salt NaCl dissolves in water.

Ions and molecules that interact with water are said to be **hydrophilic.** Nonionized and nonpolar molecules that do not interact with water are said to be **hydrophobic.**

Acids and Bases

When water molecules dissociate (break up), they release an equal number of hydrogen ions (H^+) and hydroxide ions (OH^-):

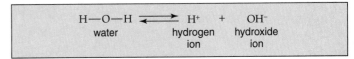

$$H—O—H \rightleftharpoons H^+ + OH^-$$
water   hydrogen   hydroxide
                ion   ion

Only a few water molecules at a time dissociate, and the actual number of H^+ and OH^- is 10^{-7} moles/liter. A mole is a unit of scientific measurement for atoms, ions, and molecules.[1]

[1]In chemistry, a mole is defined as the amount of matter that contains as many objects (atoms, molecules, ions) as the number of atoms in exactly 12 grams of ^{12}C.

Acidic Solutions (High H⁺ Concentrations)

Lemon juice, vinegar, tomatoes, and coffee are all acidic solutions. What do they have in common? **Acids** are substances that dissociate in water, releasing hydrogen ions (H^+). For example, an important inorganic acid is hydrochloric acid (HCl), which dissociates in this manner:

$$HCl \rightarrow H^+ + Cl^-$$

Dissociation is almost complete; therefore, HCl is called a strong acid. If hydrochloric acid is added to a beaker of water, the number of hydrogen ions (H^+) increases greatly.

Basic Solutions (Low H⁺ Concentrations)

Milk of magnesia and ammonia are commonly known basic substances. **Bases** are substances that either take up hydrogen ions (H^+) or release hydroxide ions (OH^-). For example, an important inorganic base is sodium hydroxide (NaOH), which dissociates in this manner:

$$NaOH \rightarrow Na^+ + OH^-$$

Dissociation is almost complete; therefore, sodium hydroxide is called a strong base. If sodium hydroxide is added to a beaker of water, the number of hydroxide ions increases.

You should not taste a strong acid or base because they are quite destructive to cells. Any container of household cleanser, such as ammonia, has a poison symbol and carries a strong warning not to ingest the product.

pH Scale

The **pH scale**[2] is used to indicate the acidity and basicity (alkalinity) of a solution. Pure water with an equal number of hydrogen ions (H^+) and hydroxide ions (OH^-) has a pH of exactly 7.

The pH scale was devised to simplify discussion of the hydrogen ion concentration [H^+] and consequently of the hydroxide ion concentration [OH^-]; it eliminates the use of cumbersome numbers. In order to understand the relationship between hydrogen ion concentration and pH, consider the following:

	[H^+] (moles per liter)		pH
0.000001	=	1×10^{-6}	6
0.0000001	=	1×10^{-7}	7
0.00000001	=	1×10^{-8}	8

Of the two values above and below pH 7, which one indicates a higher hydrogen ion concentration than pH 7 and, therefore, refers to an acidic solution? A number with a smaller negative exponent indicates a greater quantity of hydrogen ions (H^+) than one with a larger negative exponent. Therefore, the pH 6 solution is an acidic solution.

Basic solutions have fewer hydrogen ions (H^+) compared with hydroxide ions. Of the three values, pH 8 solution is a basic solution because it indicates a lower hydrogen ion concentration [H^+] (greater hydroxide ion concentration) than the pH 7 solution.

The pH scale (Fig. 2.9) ranges from 0 to 14. As we move toward a higher pH, each unit has 10 times the basicity of the previous unit, and as we move toward a lower pH, each unit has 10 times the acidity of the previous unit.

Buffers

In living things, the pH of body fluids needs to be maintained within a narrow range, or else health suffers. Normally, pH stability is possible because the body and the environment have **buffers** to prevent pH changes. A problem arises when precipitation in the form of rain or snow becomes so acidic,

Figure 2.9   The pH scale.
The dial of this pH meter indicates that pH ranges from 0 to 14, with 0 being the most acidic and 14 being the most basic. pH 7 (neutral pH) has equal amounts of hydrogen ions (H^+) and hydroxide ions (OH^-). An acidic pH has more H^+ than OH^- and a basic pH has more OH^- than H^+.

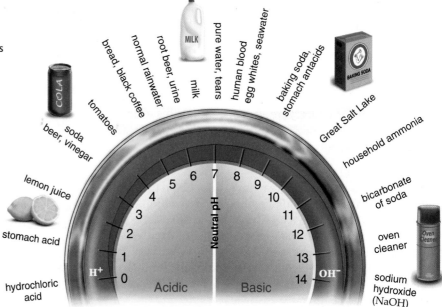

[2]pH is defined as the negative log of the hydrogen ion concentration [H^+]. A log is the power to which 10 must be raised to produce a given number.

a. b.

Figure 2.10 Effects of acid deposition.
The burning of gasoline derived from oil, a fossil fuel, leads to acid rain, which causes (**a**) statues to deteriorate and (**b**) trees to die.

the environment runs out of natural buffers in the soil or water. Rain normally has a pH of about 5.7, but rain with a pH of 2.1 is recorded every year in the northeastern Appalachian Mountains. Rain becomes acidic because the burning of gasoline emits sulfur dioxides (SO_X) and nitrogen oxides (NO_X) into the atmosphere. They combine with water to produce sulfuric acid (H_2SO_4) and nitric acid (HNO_3). Acid deposition destroys statues and kills forests (Fig. 2.10). It also leads to fish kills in lakes and streams.

Buffers help keep the pH within normal limits because they are chemicals or combinations of chemicals that take up excess hydrogen ions (H^+) or hydroxide ions (OH^-). For example, carbonic acid (H_2CO_3) is a weak acid that minimally dissociates and then re-forms in the following manner:

H_2CO_3 carbonic acid	dissociates ⇌ re-forms	H^+ hydrogen ion	+	HCO_3^- bicarbonate ion

The pH of our blood when we are healthy is always about 7.4—that is, just slightly basic (alkaline). Blood always contains a combination of some carbonic acid and some bicarbonate ions. When hydrogen ions (H^+) are added to blood, the following reaction occurs:

$$H^+ + HCO_3^- \longrightarrow H_2CO_3$$

When hydroxide ions (OH^-) are added to blood, this reaction occurs:

$$OH^- + H_2CO_3 \longrightarrow HCO_3^- + H_2O$$

These reactions prevent any significant change in blood pH.

✓ Check Your Progress 2.2

1. What characteristics of water help support life?
2. How does the hydrogen ion concentration and the pH change among water, acids, and bases?
3. Why does the pH of the environment (soil and water) and the body rarely change?

2.3 Molecules of Life

Four categories of **organic molecules,** called carbohydrates, lipids, proteins, and nucleic acids, are unique to cells. In biology, **organic** doesn't refer to how food is grown; it refers to a molecule that contains carbon (C) and hydrogen (H) and is usually associated with living things.

Each type of organic molecule in cells is composed of subunits. When a cell constructs a **macromolecule,** a molecule that contains many subunits, it uses a **dehydration reaction,** a type of synthesis reaction. During a dehydration reaction, an —OH (hydroxyl group) and an —H (hydrogen atom), the equivalent of a water molecule, are removed as the molecule forms (Fig. 2.11*a*). The reaction is reminiscent of a train whose length is determined by how many boxcars it has hitched together. To breakdown macromolecules, the cell uses a **hydrolysis reaction** in which the components of water are added (Fig. 2.11*b*).

✓ Check Your Progress 2.3

1. What are the four classes of molecules unique to cells?
2. What type of reaction occurs during synthesis of macromolecules?

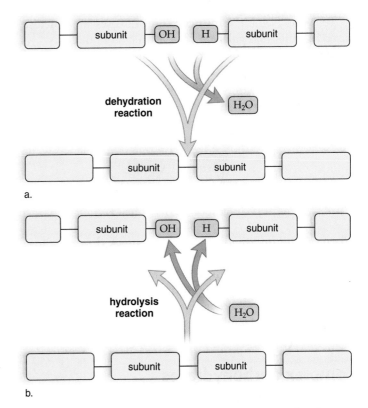

a.

b.

Figure 2.11 Synthesis and degradation of macromolecules.
a. Synthesis occurs when subunits bond and H_2O is removed.
b. Subunits of a macromolecule separate after the addition of H_2O.

2.4 Carbohydrates

Carbohydrate molecules are characterized by the presence of the atomic grouping H—C—OH, in which the ratio of hydrogen atoms (H) to oxygen atoms (O) is approximately 2:1. Since this ratio is the same as the ratio in water, the name "hydrates of carbon" seems appropriate. **Carbohydrates,** first and foremost, function for quick and short-term energy storage in all organisms, including humans.

Simple Carbohydrates

If the number of carbon atoms in a carbohydrate is low (from three to seven), it is called a simple sugar, or **monosaccharide.** The designation **pentose** means a 5-carbon sugar, and the designation **hexose** means a 6-carbon sugar. **Glucose,** the hexose our bodies use as an immediate source of energy, can be written in any one of these ways:

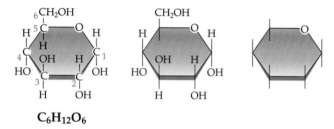

$$C_6H_{12}O_6$$

After Rita's cells receive sufficient oxygen, her cells will be able to break down glucose to give her the energy she needs to feel peppy again. Other common hexoses are fructose, found in fruits, and galactose, a constituent of milk.

A **disaccharide** (*di*, two; *saccharide*, sugar) is made by joining only two monosaccharides together by a dehydration reaction (see Fig. 2.11*a*). Maltose is a disaccharide that contains two glucose molecules:

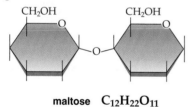

maltose $C_{12}H_{22}O_{11}$

When our hydrolytic digestive juices break down maltose, the result is two glucose molecules. When glucose and fructose join, the disaccharide sucrose forms. Sucrose, which is ordinarily derived from sugarcane and sugar beets, is commonly known as table sugar. As discussed in the Health Focus on page 29, an intake of refined carbohydrates high in sugar content is harmful to our good health.

Complex Carbohydrates (Polysaccharides)

Macromolecules such as starch, glycogen, and cellulose are **polysaccharides** that contain many glucose units.

Starch and **glycogen** are readily stored forms of glucose in plants and animals, respectively. Some of the macromolecules in starch are long chains of up to 4,000 glucose units. Starch has fewer side branches, or chains, of glucose that branch off from the main chain, than does glycogen, as shown in Figures 2.12 and 2.13. Flour, which we usually acquire by grinding wheat and use for baking, is high in starch, and so are potatoes.

After we eat starchy foods, such as potatoes, bread, and cake, glucose enters the bloodstream, and normally the liver stores glucose as glycogen. The release of the hormone insulin from the pancreas promotes the storage of glucose as glycogen. In between eating, the liver releases glucose so that, normally, the blood glucose concentration is always about 0.1%.

The polysaccharide **cellulose** is found in plant cell walls. In cellulose, the glucose units are joined by a slightly different type of linkage than that in starch or glycogen. While this might seem to be a technicality, actually it is important because we are unable to digest foods containing this type of linkage; therefore, cellulose largely passes through our digestive tract as fiber, or roughage.

Low-Carb Diet

If you, or someone you know, has lost weight by following the Atkins or South Beach diet, you may think "carbs" are

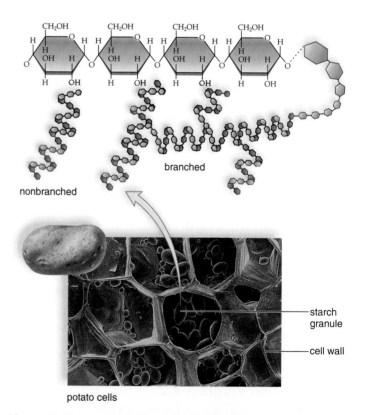

Figure 2.12   Starch structure and function.
Starch has straight chains of glucose molecules. Some chains are also branched, as indicated. The electron micrograph shows starch granules in potato cells. Starch is the storage form of glucose in plants.

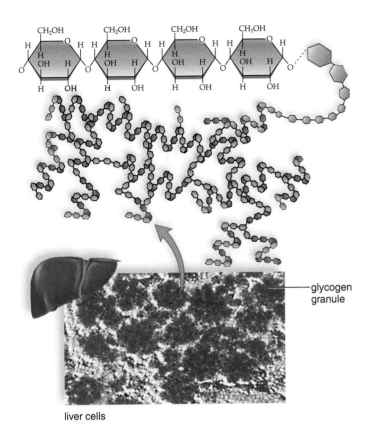

glycogen granule

liver cells

Figure 2.13 Glycogen structure and function.
Glycogen is more branched than starch. The electron micrograph shows glycogen granules in liver cells. Glycogen is the storage form of glucose in humans.

unhealthy and should be avoided. But nutritionists can point to countries in which the diet is 60–70% high-fiber carbohydrates, and these people have a low incidence of heart disease and cancer. Fiber includes various nondigestible carbohydrates derived from plants. Food sources rich in fiber include beans, peas, nuts, fruits, and vegetables. Whole-grain products should be preferred over foods made from refined grains. During refinement, fiber and also vitamins and minerals are removed from grains, so that primarily starch remains. So-called insoluble fiber adds bulk to fecal material, which stimulates movements of the large intestine, preventing constipation. Soluble fiber combines with cholesterol in the small intestine and prevents it from being absorbed. This may be protective against heart disease. In general, Americans should just about double the amount of fiber in their diet.

✓ Check Your Progress 2.4

1. What is the usual function of various carbohydrates in humans?
2. What is the difference in structure between a simple carbohydrate and the various complex carbohydrates?
3. Of what benefit is fiber in the diet?

Health Focus

Sugar: Sweet Poison

The explosion of presweetened cereals, sugary sodas, and tempting candies has driven the consumption of sugar to an all-time high. While it is not surprising that a high intake of sugar contributes to our current obesity epidemic, many are dismayed to learn that our society's insatiable sweet tooth may lead to many serious public health problems. Here are some recent findings:

- Researchers at the Harvard School of Public Health released the results of a 2004 study that followed the dietary habits and health of more than 91,000 women over the course of eight years. They found that women who consumed sugar-sweetened sodas more than once per day exhibited an alarming 80% increased risk of diabetes. These women tended to gain much more weight than other participants—17 pounds, on average—over the course of the study, further cementing the link between sugar consumption and obesity.
- The American Heart Association stated in 1998 that "short-term studies show consistent adverse effects of sugar consumption on cardiovascular health." Components of blood associated with cardiovascular health decrease and components that lead to cardiovascular disease increase.
- Immune system dysfunction is another possible health problem linked to ingestion of refined sugar. One study had test subjects drink two cans of sugary soda. Then, blood was drawn at regular intervals, and the ability of the subjects' white blood cells to neutralize pathogens was tested in the laboratory. A dramatic 50% decrease in this function over a period of about five hours was observed.
- In the journal *Nutritional Neuroscience*, in 2005, researchers at John Carroll University reported that laboratory animals who were fed large amounts of a glucose solution exhibited withdrawal symptoms after glucose was withheld. When the glucose was reintroduced, the animals guzzled more than usual, mimicking the relapse behavior of human drug addicts. Perhaps we are compelled to seek sweetened foods and beverages if our diet routinely contains a lot of sugar, much as an alcoholic craves liquor.

We have no nutritional need for the added sugar found in everyday foods, such as breakfast cereals, bread, or yogurt. Even foods considered healthy, such as juices, granola, trail mix, and canned fruit, may contain added sugar. Most health professionals now recommend that their patients avoid foods that contain added sugar and read food labels very carefully when shopping.

2.5 Lipids

Lipids contain more energy per gram than other biological molecules, and therefore, fats in animals and oils in plants function well as energy storage molecules. Others (phospholipids) form a membrane so that the cell is separated from its environment and has inner compartments as well. Steroids are a large class of lipids that includes, among other molecules, the sex hormones.

Lipids are diverse in structure and function, but they have a common characteristic: They do not dissolve in water. Their low solubility in water is due to an absence of polar groups. They contain little oxygen and consist mostly of carbon and hydrogen atoms.

Fats and Oils

The most familiar lipids are those found in fats and oils. **Fats,** which are usually of animal origin (e.g., lard and butter), are solid at room temperature. **Oils,** which are usually of plant origin (e.g., corn oil and soybean oil), are liquid at room temperature. Fat has several functions in the body: It is used for long-term energy storage, it insulates against heat loss, and it forms a protective cushion around major organs.

Emulsifiers can cause fats to mix with water. They contain molecules with a nonpolar end and a polar end. The molecules position themselves about an oil droplet so that their polar ends project outward. Now the droplet disperses in water, which means that **emulsification** has occurred. Emulsification takes place when dirty clothes are washed with soaps or detergents. Also, prior to the digestion of fatty foods, fats are emulsified by bile. The gallbladder stores bile for emulsifying fats prior to the digestive process.

Fats and oils form when one glycerol molecule reacts with three fatty acid molecules (Fig. 2.14). A fat is sometimes called a **triglyceride** because of its three-part structure, or the term neutral fat can be used because the molecule is nonpolar and carries no charges.

Saturated, Unsaturated, and Trans-Fatty Acids

A **fatty acid** is a carbon–hydrogen chain that ends with the acidic group —COOH (Fig. 2.14, *left*). Most of the fatty acids in cells contain 16 or 18 carbon atoms per molecule, although smaller ones with fewer carbons are also known. Fatty acids are either saturated or unsaturated. **Saturated fatty acids** have no double bonds between the carbon atoms. The chain is saturated, so to speak, with all the hydrogens it can hold. **Unsaturated fatty acids** have double bonds in the carbon chain wherever the number of hydrogens is less than two per carbon.

In general, oils, present in cooking oils and bottle margarines, are liquids at room temperature because the presence of a double bond creates a bend in the fatty acid chain. Such kinks prevent close packing between the hydrocarbon chains and account for the fluidity of oils. On the other hand, butter, which contains saturated fatty acids and no double bonds, is a solid at room temperature.

Saturated fats, in particular, are associated with cardiovascular disease because they cause lipid material, called plaque, to accumulate inside blood vessels. Even more harmful than naturally occurring saturated fats are the so-called **trans fats,** which are in vegetable oils that have been partially hydrogenated to make them semisolid. Complete hydrogenation of oils causes all double bonds to become saturated. Partial hydrogenation does not saturate all bonds. It reconfigures some double bonds,

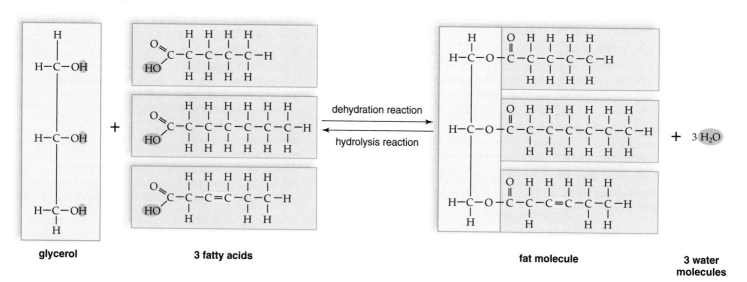

Figure 2.14 **Synthesis and degradation of a fat molecule.**
Fatty acids can be saturated (no double bonds between carbon atoms) or unsaturated (have double bonds, colored yellow, between carbon atoms). When a fat molecule forms, three fatty acids combine with glycerol, and three water molecules are produced.

and the hydrogen atoms end up on different sides of the chain. (*Trans* in Latin means across):

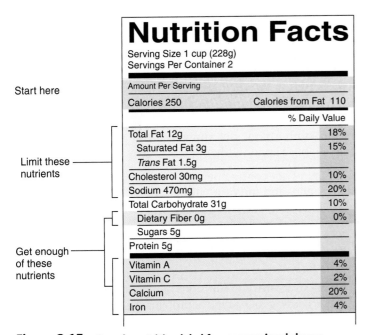

Unsaturated
(oils)

Saturated
(butter)

Trans fats
(hydrogenated oils)

Trans fats are found in shortenings and solid margarines. They also occur in processed foods (snack foods, baked goods, and fried foods).

Unsaturated oils, particularly monounsaturated (one double bond) oils, but also polyunsaturated (many double bonds), have been found to be protective against cardiovascular disease.

Dietary Fat

For good health, the diet should include some fat but, for the reasons stated, the first thing to do when looking at a nutrition label is to check the total amount of fat per serving. The total recommended amount of fat in a 2,000 calorie diet is 65 g. That information results in the % daily value (DV) given in the sample nutrition label for macaroni and cheese in Figure 2.15. As of January 2006, food manufacturers are required to list the amount of trans fats greater than 0.5 g in the nutrition label for a food.

In the meantime, the manufacturers of Crisco, a favorite shortening, have eliminated trans fats by using fully hydrogenated cottonseed oil blended with liquid soy or sunflower oils to improve the texture.

Nutrition Facts

Serving Size 1 cup (228g)
Servings Per Container 2

Start here

Amount Per Serving

Calories 250	Calories from Fat 110

	% Daily Value
Total Fat 12g	18%
Saturated Fat 3g	15%
Trans Fat 1.5g	
Cholesterol 30mg	10%
Sodium 470mg	20%
Total Carbohydrate 31g	10%
Dietary Fiber 0g	0%
Sugars 5g	
Protein 5g	

Vitamin A	4%
Vitamin C	2%
Calcium	20%
Iron	4%

Limit these nutrients

Get enough of these nutrients

Figure 2.15   Sample nutrition label for macaroni and cheese.
As of January 2006, foods must include the quantity of trans fats. There is no % daily value for trans fats, and they should be avoided.

Phospholipids

Phospholipids have a phosphate group (Fig. 2.16). Essentially, they are constructed like fats, except that in place of the third fatty acid, there is a phosphate group or a grouping that contains both phosphate and nitrogen. These molecules are not electrically neutral, as are fats, because the phosphate and nitrogen-containing groups are ionized. They form the so-called polar head of the molecule, while the rest of the molecule becomes the hydrophobic tails. Phospholipids are the primary components of cellular membranes; they spontaneously form a bilayer in which the hydrophilic heads face outward toward watery solutions and the tails form the hydrophobic interior.

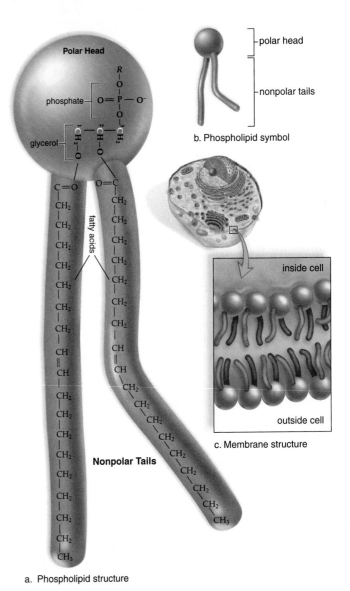

b. Phospholipid symbol

inside cell

outside cell

c. Membrane structure

a. Phospholipid structure

Figure 2.16   Phospholipid structure and function.
a. Phospholipids are structured like fats, but one fatty acid is replaced by a polar phosphate group. **b.** Therefore, the head is polar, while the tails are nonpolar. **c.** This causes the molecules to arrange themselves as shown when exposed to water inside and outside a cell.

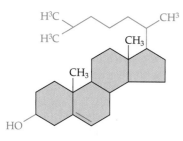

a. Cholesterol

Figure 2.17   Steroids.
a. All steroids are derived from cholesterol and have four rings. The effects of (**b**) testosterone in males and (**c**) estrogen in females largely depend on the difference in the attached groups, shown in blue.

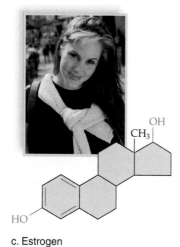

b. Testosterone

c. Estrogen

Steroids

Steroids are lipids that have an entirely different structure from those of fats. Steroid molecules have a backbone of four fused carbon rings. Each one differs primarily by the functional groups attached to the rings. Cholesterol is a component of an animal cell's plasma membrane and is the precursor of several other steroids, such as the sex hormones estrogen and testosterone. The body usually makes the cholesterol it needs; therefore, a dietary source should be restricted because, along with saturated fats and trans fats, cholesterol also causes fatty material to accumulate inside the lining of blood vessels, thereby reducing blood flow.

The male sex hormone, testosterone, is formed primarily in the testes, and the female sex hormone, estrogen, is formed primarily in the ovaries. Testosterone and estrogen differ only by the functional groups attached to the same carbon backbone, yet they have a profound effect on the body and the sexuality of humans and other animals (Fig. 2.17). The taking of anabolic steroids, usually to build muscle strength, is illegal because the side effects are quite harmful to the body (see page 311).

☑ Check Your Progress 2.5

1. **a.** What is the main function of fats and oils, and (b) of what are they composed?
2. What are the uses of phospholipids and steroids in the body?

2.6 Proteins

Proteins are of primary importance in the structure and function of cells. Here are some of their many functions in humans.

Support   Some proteins—such as keratin, which makes up hair and nails, and collagen, which lends support to ligaments, tendons, and skin—are structural proteins.

Enzymes   Enzymes bring reactants together and thereby speed chemical reactions in cells. They are specific for one particular type of reaction and can function at body temperature.

Hair is a protein.

Transport   Channel and carrier proteins in the plasma membrane allow substances to enter and exit cells. Some other proteins transport molecules in the blood of animals; **hemoglobin** in red blood cells is a complex protein that transports oxygen.

Defense   Antibodies are proteins. They combine with foreign substances, called antigens. In this way, they prevent antigens from destroying cells and upsetting homeostasis.

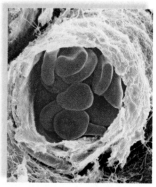

Hemoglobin is a protein.

Hormones   Hormones are regulatory proteins. They serve as intercellular messengers that influence the metabolism of cells. The hormone insulin regulates the content of glucose in the blood and in cells; the presence of growth hormone determines the height of an individual.

Motion   The contractile proteins actin and myosin allow parts of cells to move and cause muscles to contract. Muscle contraction accounts for the movement of animals from place to place.

Muscle contains protein.

The structures and functions of vertebrate cells and tissues differ according to the type of proteins they contain. For example, muscle cells contain actin and myosin; red blood cells contain hemoglobin; and support tissues contain collagen.

Amino Acids: Subunits of Proteins

Proteins are macromolecules with **amino acid** subunits. The central carbon atom in an amino acid bonds to a hydrogen atom and also to three other groups of atoms. The name amino acid is appropriate because one of these groups is an —NH_2 (amino group) and another is a —COOH (an acid group). The third group is the R group for an amino acid:

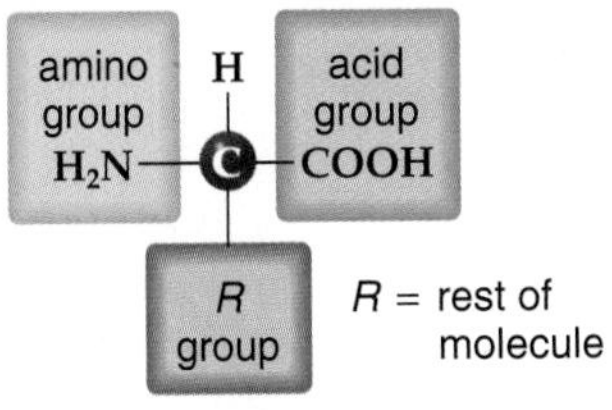

Amino acid

Amino acids differ according to their particular R group. The R groups range in complexity from a single hydrogen atom to a complicated ring compound. Some R groups are polar and some are not. Also, the amino acid cysteine has a group that ends with an —SH group, which often serves to connect one chain of amino acids to another by a disulfide bond, —S—S—. Several amino acids commonly found in cells are shown in Figure 2.18.

Peptides

Figure 2.19 shows how two amino acids join by a dehydration reaction between the carboxyl group of one and the amino group of another. The resulting covalent bond between two amino acids is called a **peptide bond.** The atoms associated with the peptide bond share the electrons unevenly because oxygen attracts electrons more than nitrogen. Therefore, the hydrogen attached to the nitrogen has a slightly positive charge (δ^+), while the oxygen has a slightly negative charge (δ^-):

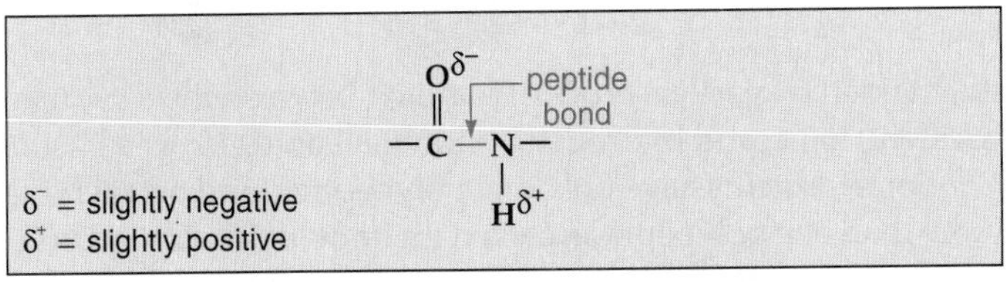

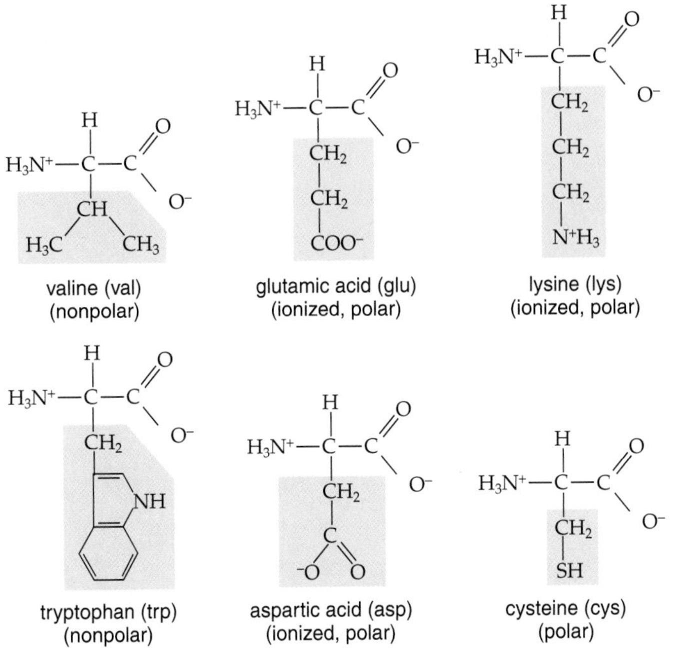

Figure 2.18 Amino acids.
Proteins contain 20 different kinds of amino acids, which differ by the particular R group, screened in blue, attached to the central carbon. Some R groups are nonpolar and hydrophobic; some are polar and hydrophilic; and some are ionized and hydrophilic.

Shape of Proteins

Proteins cannot function unless they have their usual shape. When proteins are exposed to extremes in heat and pH, they undergo an irreversible change in shape called **denaturation.** For example, we are all aware that the addition of acid to milk causes curdling and that heating causes egg whites, which contain a protein called albumin, to coagulate. Denaturation occurs because the normal bonding between the R groups has been disturbed. Once a protein loses its normal shape, it is no longer able to perform its usual function. Researchers hypothesize that an alteration in protein organization has occurred when Alzheimer disease and Creutzfeldt-Jakob disease (the human form of mad cow disease) develop.

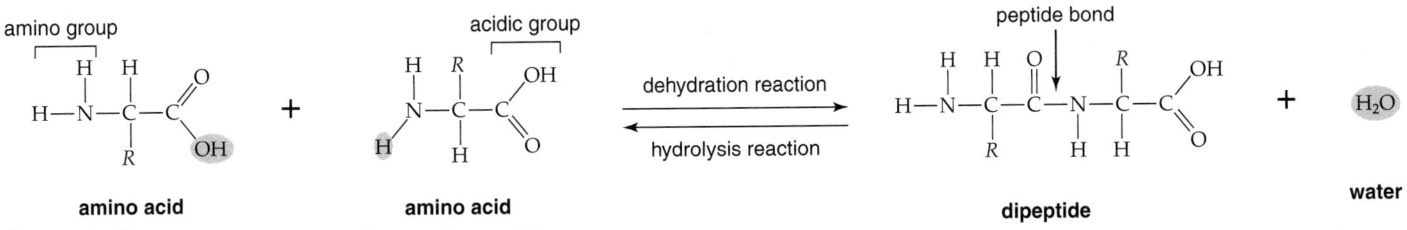

Figure 2.19 Synthesis and degradation of a peptide.
Following a dehydration reaction, a peptide bond joins two amino acids, and water is given off. Following a hydrolysis reaction, the bond is broken due to the addition of water.

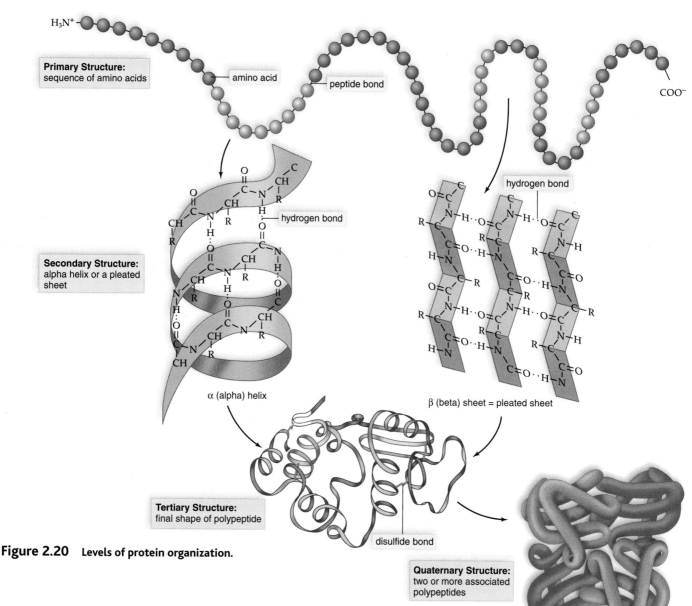

Primary Structure:
sequence of amino acids

amino acid

peptide bond

H_3N^+

COO^-

Secondary Structure:
alpha helix or a pleated sheet

hydrogen bond

hydrogen bond

α (alpha) helix

β (beta) sheet = pleated sheet

Tertiary Structure:
final shape of polypeptide

disulfide bond

Quaternary Structure:
two or more associated polypeptides

Figure 2.20 Levels of protein organization.

Levels of Protein Organization

The structure of a protein has at least three levels of organization and can have four levels (Fig. 2.20). The first level, called the *primary structure,* is the linear sequence of the amino acids joined by peptide bonds. Each particular **polypeptide** has its own sequence of amino acids.

The *secondary structure* of a protein comes about when the polypeptide takes on a certain orientation in space. Because the peptide bond is polar, and hydrogen bonding is possible between the C=O of one amino acid and the N—H of another amino acid in a polypeptide, a coiling of the chain results in an α (alpha) helix, or a right-handed spiral, and a folding of the chain results in a pleated sheet. Hydrogen bonding between peptide bonds holds the shape in place.

The *tertiary structure* of a protein is its final three-dimensional shape. In muscles, myosin molecules have a rod shape ending in globular (globe-shaped) heads. In enzymes, the polypeptide bends and twists in different ways. Invariably, the hydrophobic portions are packed mostly on the inside, and the hydrophilic portions are on the outside, where they can make contact with water. The tertiary shape of a polypeptide is maintained by various types of bonding between the *R* groups; covalent, ionic, and hydrogen bonding all occur.

Some proteins have only one polypeptide, and others have more than one polypeptide, each with its own primary, secondary, and tertiary structures. These separate polypeptides are arranged to give these proteins a fourth level of structure, termed the *quaternary structure.* Hemoglobin is a complex protein having a quaternary structure; many enzymes also have a quaternary structure. Each of four polypeptides in hemoglobin are tightly associated with a nonprotein heme group. A heme group contains an iron (Fe) atom that binds to oxygen, and in that way, hemoglobin transports O_2 to the tissues. Rita was unable to make sufficient hemoglobin and transport enough oxygen to her cells because her diet was lacking in iron.

✓ Check Your Progress 2.6

1. What are the major functions of proteins in organisms?

2. How does an amino acid get its name?

3. How does the shape of a protein relate to its function?

2.7 Nucleic Acids

The two types of nucleic acids are **DNA (deoxyribonucleic acid)** and **RNA (ribonucleic acid)** (Fig. 2.21). Early investigators called them nucleic acids because they were first detected in the nucleus of cells. The discovery of the structure of DNA has had an enormous influence on biology and on society in general. DNA stores genetic information in the cell and in the organism. DNA replicates and transmits this information when a cell reproduces and when an organism reproduces. We now know not only how genes work, but we can manipulate them. The science of biotechnology is largely devoted to altering the genes in living organisms.

Function of DNA and RNA

Each DNA molecule contains many genes, and genes specify the sequence of the amino acids in proteins. RNA is an intermediary that conveys DNA's instructions regarding the amino acid sequence in a protein. If DNA's information is faulty, illness can result. The relationship between a gene, a protein, and an illness is illustrated by sickle-cell disease. In sickle-cell disease, the individual's red blood cells are sickle-shaped because in one particular spot, an amino acid called valine appears where the amino acid called glutamate should be in hemoglobin. This substitution makes red blood cells lose their normally round, flexible shape and become hard and rigid. When these abnormal red blood cells go through small blood vessels, they clog the flow of blood and break apart. Sickle-cell disease is another cause of anemia, and it also results in pain and organ damage.

How the Structure of DNA and RNA Differs

Both DNA and RNA are polymers of nucleotides.

Nucleotide Structure

Every **nucleotide** is a molecular complex of three types of subunit molecules—phosphate (phosphoric acid), a pentose sugar, and a nitrogen-containing base:

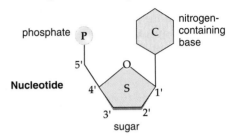

The nucleotides in DNA contain the sugar deoxyribose, and the nucleotides in RNA contain the sugar ribose; this difference accounts for their respective names (Table 2.1). There are four different types of bases in DNA: **adenine (A), thymine (T), guanine (G),** and **cytosine (C).** The base can have two rings (adenine or guanine) or one ring (thymine or cytosine). In RNA, the base **uracil (U)** replaces the base thymine. These structures are called bases because their presence raises the pH of a solution.

Polynucleotide Structure

The nucleotides form a polynucleotide called a strand, which has a backbone made up of phosphate-sugar-phosphate-sugar,

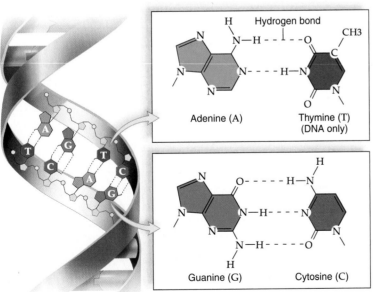

a. DNA structure with base pairs: A with T and G with C

b. RNA structure with bases G, U, A, C

Figure 2.21 DNA and RNA structure.
a. Structure of DNA and its bases. **b.** Structure of RNA, in which uracil replaces thymine.

with the bases projecting to one side of the backbone. Since the nucleotides of a gene occur in a definite order, so do the bases. After many years of work, researchers now know the sequence of the bases in human DNA—the human genome. This breakthrough is expected to lead to improved genetic counseling, gene therapy, and medicines to treat the causes of many human illnesses.

DNA is double-stranded, with the two strands twisted about each other in the form of a *double helix* (see Fig. 2.21*a*). In DNA, the two strands are held together by hydrogen bonds between the bases. When unwound, DNA resembles a stepladder. The uprights (sides) of the ladder are made entirely of phosphate and sugar molecules, and the rungs of the ladder are made only of **complementary paired bases.** Thymine (T) always pairs with adenine (A), and guanine (G) always pairs with cytosine (C). Complementary bases have shapes that fit together.

Complementary base pairing allows DNA to replicate in a way that ensures the sequence of bases will remain the same. This is important because it is the sequence of bases that determine the sequence of amino acids in a protein. RNA is single-stranded, and when RNA forms, complementary base pairing with one DNA strand, passes the correct sequence of bases to RNA (Fig. 2.21*b*). RNA is the nucleic acid that is directly involved in protein synthesis.

ATP: An Energy Carrier

In addition to being the subunits of nucleic acids, nucleotides have metabolic functions. When adenosine (adenine plus ribose) is modified, by the addition of three phosphate groups instead of one, it becomes **ATP (adenosine triphosphate),** an energy carrier in cells.

Structure of ATP Suits Its Function

ATP is a high-energy molecule because the last two phosphate bonds are unstable and easily broken. Usually in cells, the terminal phosphate bond is hydrolyzed, leaving the molecule **ADP (adenosine diphosphate)** and a molecule of inorganic phosphate ℗ (Fig. 2.22). The energy released by ATP breakdown is used by the cell to synthesize macromolecules,

Table 2.1	DNA Structure Compared to RNA Structure	
	DNA	**RNA**
Sugar	Deoxyribose	Ribose
Bases	Adenine, guanine, thymine, cytosine	Adenine, guanine, uracil, cytosine
Strands	Double-stranded with base pairing	Single-stranded
Helix	Yes	No

such as carbohydrates and proteins. In muscle cells, the energy is used for muscle contraction; and in nerve cells, it is used for the conduction of nerve impulses. After ATP breaks down, it is rebuilt by the addition of ℗ to ADP. Notice in Figure 2.22 that an input of energy is required to re-form ATP.

Glucose Breakdown Leads to ATP Buildup

A glucose molecule contains too much energy to be used as a direct energy source in cellular reactions. Instead, the energy of glucose is converted to that of ATP molecules. ATP contains an amount of energy that makes it usable to supply energy for chemical reactions in cells. Muscles use ATP energy and produce heat when they contract. This is the heat that warms the body.

As we shall see in Chapter 3, oxygen is involved in the breakdown of glucose. Insufficient oxygen limits glucose breakdown and limits ATP buildup. Rita's diet did not contain enough iron; therefore, her body could not make sufficient hemoglobin, so she had a reduced capacity to carry oxygen. Without sufficient oxygen, her body's cells could not produce enough ATP. Therefore, she felt tired and run down.

☑ Check Your Progress 2.7

1. How does the (a) function and (b) structure of DNA and RNA differ?
2. What type of bonding joins the bases within a DNA double helix?
3. What is the structure and function of ATP?

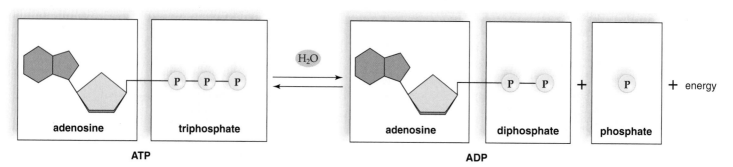

Figure 2.22   ATP reaction.
ATP, the universal energy currency of cells, is composed of adenosine and three phosphate groups (called a triphosphate). When cells require energy, ATP undergoes hydrolysis, producing ADP + ℗, with a release of energy.

Pesticide: An Asset and a Liability

A synthetic pesticide is any one of 55,000 chemical products —modified organic molecules—used to kill insects, plants, fungi, or rodents that interfere with human activities. The use of synthetic pesticides in the United States began in the 1930s and became widespread after World War II. The pesticide DDT was widely used in urban aerial sprays to control mosquitoes, gypsy moths, Japanese beetles, and other insects in the 1940s. By 1972, DDT was banned from the United States due to the development of resistance and evidence that DDT has a harmful affect on the environment and humans.

Figure 2A **Pesticide warning signs indicate that pesticides are harmful to our health.**

Harmful Effects of Pesticides

It is now apparent that pesticides are harmful to both the environment and to humans (Fig. 2A). Contact or respiratory allergic responses are among the most immediately obvious toxic effects of pesticides. But pesticides can also cause suppression of the immune system, leading to increased susceptibility to infection or tumor development. Lymphocyte impairment was found following the worst industrial accident in the world, which took place in India in 1984. Nerve gas used to produce an insecticide was released into the air; 3,700 people were killed, and 30,000 were injured. Vietnam veterans claim a variety of health effects due to contact with a herbicide called Agent Orange that was used as a defoliant during the Vietnam War. The Environmental Protection Agency (EPA) says that pesticide residues on food possibly cause suppression of the immune system, disorders of the nervous system, birth defects, and cancer. The immune, nervous, and endocrine systems communicate with one another by way of hormones, and what affects one system can affect the other. In men, testicular atrophy, low sperm counts, and abnormal sperm may be due to the ability of pesticides to mimic the effects of estrogen, a female sex hormone.

Alternatives to the Use of Pesticides

There are alternatives to the use of pesticides. Integrated pest management uses a diversified environment, mechanical and physical means, natural enemies, disruption of reproduction, and resistant plants to control, rather than eradicate, pest populations. Chemicals could be used only as a last resort. Maintaining hedgerows, weedy patches, and certain trees can provide diverse habitats and food for predators and parasites that help control pests. The use of strip farming and crop rotation by farmers denies pests a continuous food source.

Natural enemies abound in the environment. When lacewings were released in cotton fields, they reduced the boll weevil population by 96% and increased cotton yield threefold. A predatory moth was used to reclaim 60 million acres in Australia overrun by the prickly pear cactus, and another type of moth is now used to control alligator weed in the southern United States. Also in the United States, the Chrysolina beetle controls Klamath weed, except in shady places where a root-boring beetle is more effective.

Sex attractants and sterile insects are used in other types of biological control. In Sweden and Norway, scientists synthesized the sex pheromone of the Ips bark beetle, which attacks spruce trees. Almost 100,000 baited traps collected females normally attracted to males emitting this pheromone. Sterile males have also been used to reduce pest populations. The screwworm fly parasitizes cattle in the United States. Flies raised in a laboratory were made sterile by exposure to radiation. The entire Southeast was freed from this parasite when female flies mated with sterile males and then laid eggs that did not hatch.

In general, biological control is a more sophisticated method of controlling pests than the use of pesticides. It requires an in-depth knowledge of pests and/or their life cycles. Because it does not have an immediate effect on the pest population, the effects of biological control may not be apparent to the farmer or gardener who eventually benefits from it.

Malaria and Pesticides

Malaria is a devastating disease in Africa. By the early 1960s, malaria, which is transmitted from person to person by a mosquito, had been brought under control by using DDT to kill off mosquito populations. However, in 1996, South Africa bowed to international pressure and began using less effective and more expensive sprays instead of DDT. Malaria has now made a comeback. In 1999, South Africa's malaria rates suddenly skyrocketed to 50,000 cases a year from just a few thousand. Now South Africa and other African countries have gone back to using DDT, and DDT is again bringing malaria under control. So in May 2001, government officials from around the world met and gave South Africa and certain other countries an exemption from the global band against the use of DDT.

Decide Your Opinion

1. Should synthetic pesticide use be banned in the United States? Would you be willing to use only biological controls to kill pests in your lawn and garden?
2. Do you approve of letting Africa use DDT to control malaria? How far should rich nations go in imposing their own values and risk standards on poor nations?

Summarizing the Concepts

2.1 From Atoms to Molecules
- Matter is composed of elements; each element is made up of just one type of atom.
- An atom's weight is based on the number of protons and neutrons in the nucleus.
- An atom's chemical properties depend on the number of electrons in the outer shell.
- Atoms react by forming ionic bonds or covalent bonds.

2.2 Water and Living Things
- Water is a liquid, instead of a gas, at room temperature.
- Water heats and freezes slowly, moderating temperatures and allowing bodies to cool by vaporizing water.
- Frozen water is less dense than liquid water, so ice floats on water.
- Water is cohesive and fills tubular vessels, such as blood vessels.
- Water is the universal solvent because of its polarity.
- Water has a neutral pH.

  Acids increase H^+ but decrease the pH of water.

  Bases decrease H^+ but increase the pH of water.

2.3 Molecules of Life
Carbohydrates, lipids, proteins, and nucleic acids are macromolecules with specific functions in cells.

2.4 Carbohydrates
Simple carbohydrates are monosaccharides or disaccharides.
- Glucose is a 6-carbon sugar utilized by cells for quick energy.

  Complex carbohydrates are polysaccharides. Starch, glycogen, and cellulose are polysaccharides containing many glucose units.
- Plants store glucose as starch.
- Animals store glucose as glycogen.
- Cellulose forms plant cell walls. Cellulose is fiber.

2.5 Lipids
- Fats and oils, which function in long-term energy storage, contain glycerol and three fatty acids.
- Fatty acids can be saturated or unsaturated.
- Plasma membranes contain phospholipids.
- Testosterone and estrogen are steroids.

2.6 Proteins
- Some proteins are structural (keratin, collagen), hormones, or enzymes that speed chemical reactions.
- Some proteins account for cell movement (actin, myosin), enable muscle contraction (actin, myosin), or transport molecules in blood (hemoglobin).

  Proteins are macromolecules with amino acid subunits.
- A peptide is composed of two amino acids.
- A polypeptide contains many amino acids.

  A protein has levels of structure:
- A primary structure is determined by the sequence of amino acids that form a polypeptide.
- A secondary structure is an α (alpha) helix or pleated sheet.
- A tertiary structure occurs when the secondary structure forms a three-dimensional, globular shape.
- A quaternary structure occurs when two or more polypeptides join to form a single protein.

2.7 Nucleic Acids
Nucleic acids are macromolecules composed of nucleotides. Nucleotides are composed of a sugar, a base, and a phosphate. DNA and RNA are polymers of nucleotides.
- DNA contains the sugar deoxyribose; contains the bases adenine, guanine, thymine, and cytosine; is double-stranded; and forms a helix.
- RNA contains the sugar ribose; contains the bases adenine, guanine, uracil, and cytosine; and does not form a helix.
- ATP is a high-energy molecule because its bonds are unstable.
- ATP undergoes hydrolysis to ADP + ⓟ, which releases energy that is used by cells to do metabolic work.

Organic molecules	Examples	Monomers	Functions
Carbohydrates	Monosaccharides, disaccharides, polysaccharides	**Glucose**	Immediate energy and stored energy; structural molecules
Lipids	Fats, oils, phospholipids, steroids	**Glycerol** **Fatty acid**	Long-term energy storage; membrane components
Proteins	Structural, enzymatic, carrier, hormonal, contractile	amino group H_2N — **C** — acid group **COOH** R group **Amino acid**	Support, metabolic, transport, regulation, motion
Nucleic acids	DNA, RNA	phosphate **P** base **C** **S** **Nucleotide**	Storage of genetic information

Understanding Key Terms

acid 26
adenine (A) 35
ADP (adenosine diphosphate) 36
amino acid 33
atom 20
atomic mass 21
atomic number 20
ATP (adenosine triphosphate) 36
base 26
buffer 26
calorie 24
carbohydrate 28
cellulose 28
complementary paired bases 36
compound 22
covalent bond 23
cytosine (C) 35
dehydration reaction 27
denaturation 33
disaccharide 28
DNA (deoxyribonucleic acid) 35
electron 20
element 20
emulsification 30
fat 30
fatty acid 30
glucose 28
glycogen 28
guanine (G) 35
hemoglobin 32
hexose 28
hydrogen bond 24
hydrolysis reaction 27

hydrophilic 25
hydrophobic 25
ion 22
ionic bond 22
isotope 21
lipid 30
macromolecule 27
mass 21
matter 20
molecule 22
monosaccharide 28
neutron 20
nucleotide 35
oil 30
organic 27
organic molecule 27
pentose 28
peptide bond 33
phospholipid 31
pH scale 26
polar 24
polypeptide 34
polysaccharide 28
protein 33
proton 20
radioisotope 21
RNA (ribonucleic acid) 35
saturated fatty acid 30
starch 28
steroid 32
thymine (T) 35
tracer 21
trans fat 30
triglyceride 30
unsaturated fatty acid 30
uracil (U) 35

Match the key terms to these definitions.

a. —————— Breaking up of fat globules into smaller droplets by the action of bile salts or any other emulsifier.

b. —————— Charged particle that carries a negative or positive charge.

c. —————— Chemical bond in which atoms share one pair of electrons.

d. —————— Type of molecule that interacts with water by dissolving in water and/or forming hydrogen bonds with water molecules.

e. —————— Weak bond that arises between a slightly positive hydrogen atom of one molecule and a slightly negative atom of another molecule or between parts of the same molecule.

Testing Your Knowledge of the Concepts

1. Name the subatomic particles of the atom. Describe their charge, atomic mass, and location in the atom. (pages 20–21)

2. Why can a radioisotope be used as a tracer in the human body? Give an example. (page 21)

3. Explain the difference between an ionic bond and a covalent bond. (pages 22–23)

4. Relate the properties of water to its polarity and hydrogen bonding between water molecules. (pages 24–25)

5. On the pH scale, which numbers indicate a basic solution? An acidic solution? A neutral solution? What makes a solution basic, acidic, or neutral? (page 26)

6. What are buffers, and why are they important to life? (pages 26–27)

7. Nitrogen would be found in which categories of molcules unique to organisms? (pages 28–36)

8. Name some monosaccharides, disaccharides, and polysaccharides, and state some general functions of each. What is the most common subunit for polysaccharides? (page 28)

9. What are the subunits of a triglyceride? What is the difference between a saturated and an unsaturated fatty acid? What are the functions of fats in the body? (pages 30–31)

10. How does the structure of a phospholipid differ from a triglyceride? Describe the arrangement of phospholipids in the plasma membrane. (page 31)

11. What is the subunit of a protein, and how do two subunits join to form a peptide bond? (page 33)

12. Discuss the primary, secondary, and tertiary structures of proteins. Why are these structures so important? (page 34)

13. Describe the double-helix structure of DNA and the single-stranded structure of RNA. (pages 35–36)

14. What type of reaction releases the energy of an ATP molecule? Explain. (page 36)

15. The atomic number gives the

a. number of neutrons in the nucleus.
b. number of protons in the nucleus.
c. weight of the atom.
d. number of protons in the outer shell.

16. Isotopes differ in their
a. number of protons.
b. atomic number.
c. number of neutrons.
d. number of electrons.

17. Which type of bond results from the complete transfer of electrons from one atom to another?
    a. covalent
    b. ionic
    c. hydrogen
    d. neutral

18. Which of the following properties of water is not due to hydrogen bonding between water molecules?
    a. Water prevents large temperature changes.
    b. Ice floats on water.
    c. Water is a solvent for polar molecules and ionic compounds.
    d. Water flows and fills spaces because it is cohesive.

19. If a chemical accepted H^+ from the surrounding solution, the chemical would be a(n)
    a. base.
    b. acid.
    c. buffer.
    d. Both a and c are correct.

20. What is true of a solution that goes from pH 5 to pH 8?
    a. The H^+ concentration decreases as the solution becomes more basic.
    b. The H^+ concentration increases as the solution becomes more acidic.
    c. The H^+ concentration decreases as the solution becomes more acidic.

21. An example of a polysaccharide used for energy storage in humans is
    a. cellulose.
    b. glycogen.
    c. cholesterol.
    d. starch.

22. Saturated and unsaturated fatty acids differ in
    a. the number of carbon-to-carbon double bonds.
    b. the consistency at room temperature.
    c. the number of hydrogen atoms present.
    d. All of these are correct.

23. The difference between one amino acid and another is found in the
    a. amino group.
    b. carboxyl group.
    c. *R* group.
    d. All of these are correct.

24. An example of a hydrolysis reaction is
    a. amino acid + amino acid $\rightarrow$ dipeptide + H_2O.
    b. dipeptide + H_2O $\rightarrow$ amino acid + amino acid.
    c. denaturation of a polypeptide.
    d. Both b and c are correct.

25. The helix and pleated sheet form of a protein is its _____ structure.
    a. secondary.
    b. tertiary.
    c. primary.
    d. quaternary.

26. An RNA nucleotide differs from a DNA molecule in that RNA has
    a. ribose sugar.
    b. phosphate molecule.
    c. uracil base.
    d. Both a and c are correct.

In questions 27–30, match each subunit with the molecule in the key.

Key:
    a. fat
    b. polysaccharide
    c. polypeptide
    d. DNA, RNA

27. Glucose

28. Nucleotide

29. Glycerol and fatty acids

30. Amino acid

In questions 31–34, match the molecular category with a molecule in the preceding key.

31. Lipids

32. Proteins

33. Nucleic acids

34. Carbohydrates

35. Label this diagram using these terms: dehydration reaction, hydrolysis reaction, subunits, macromolecule.

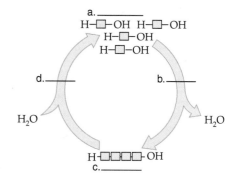

Thinking Critically About the Concepts

In the chapter opening story, Rita learned that she had iron-deficiency anemia associated with low blood iron. When oxygen is transported from your lungs to your cells, it binds to the iron portion of hemoglobin for transport. Once oxygen gets to your cells, it is used to make ATP (the cells' form of energy). Rita's low iron levels meant her cells weren't getting sufficient oxygen needed to make lots of energy, hence her tiredness.

1. Explain, in terms of ATP production, why a symptom of Rita's anemia was feeling tired.

2. Explain why people suffering from anemia often complain about feeling cold.

3. What are some foods that Rita can consume to prevent iron deficiency?

4. There are many other causes of anemia. What is a genetic cause discussed in this chapter?

5. What happens to the red blood cells when someone has hemolytic anemia (a word hint is the "lytic" of hemolytic or a similar word "lyse")?

6. Blood pH is slightly higher in the lungs and slightly lower in the tissues. Do you think basic or acidic conditions would cause hemoglobin to release oxygen?

CHAPTER 3

Cell Structure and Function

Calvin arrived home after a long day at school. The smell of freshly baked chocolate chip cookies greeted him as he walked through the front door. He dropped his backpack by the door and ran toward the kitchen. There, on the countertop, were rows of soft, warm cookies just waiting to be eaten.

As he reached for a cookie, he heard his mom say, "you can have only one cookie because I made them for tomorrow's school bake sale." Without hesitation, Calvin wolfed down one cookie and then reached for a second and third one. "I said ONE cookie," his mom said loudly when she saw Calvin reaching for more. Calvin grabbed the second and third cookie and dashed toward the back door before his mom could stop him. "Thanks Mom," he yelled loudly. "You make the best chocolate chip cookies!"

Calvin made use of all sorts of specialized cells in order to achieve his goal: chocolate chip cookies. Nerve cells made him aware of the cookies and allowed him to plan and carry out his course of action. Muscle cells moved his bones, allowing him to race to the kitchen, grab the cookies, and dash out the back door. Intestinal cells absorbed the glucose, and his lungs absorbed the oxygen he needed to convert glucose to ATP molecules. Nerves and muscles require ATP in order to function.

Cells are specialized to perform particular functions, and it's said that "structure suits function." Not surprising, skeletal muscle cells have a great number of mitochondria, where ATP is produced.

CHAPTER CONCEPTS

3.1 What Is a Cell?
Cells are the basic units of life, and new cells only come from preexisting cells. Cells are quite small (it usually takes a microscope to see them) because it gives them a favorable surface-to-volume ratio.

3.2 How Cells Are Organized
Human cells have a plasma membrane, cytoplasm, and a nucleus. The cytoplasm contains several types of organelles that carry on specific functions.

3.3 The Plasma Membrane and How Substances Cross It
The plasma membrane regulates its permeability. Diffusion, transport by carriers, and formation of vesicles at the plasma membrane allow substances to get into and out of cells.

3.4 The Nucleus and the Production of Proteins
DNA within the chromatin of a nucleus specifies the sequence of amino acids in a protein. Proteins are manufactured at ribosomes, many of which are attached to rough endoplasmic reticulum. The Golgi apparatus modifies proteins before secretion occurs.

3.5 The Cytoskeleton and Cell Movement
The cytoskeleton is composed of fibers that maintain the shape of the cell and assist the movement of organelles. Cilia and flagella contain microtubules and can move, utilizing ATP energy.

3.6 Mitochondria and Cellular Metabolism
Mitochondria are the site of cellular respiration, a process that requires oxygen for the complete enzymatic breakdown of glucose (and fatty acids and amino acids) with the buildup of ATP molecules. Fermentation does not require oxygen, but it results in only two ATP molecules per glucose molecule.

3.1 What Is a Cell?

Organisms, including humans, are composed of cells, a fact that isn't apparent until you compare unicellular organisms with the tissues of multicellular ones under the microscope. The cell theory, one of the fundamental principles of modern biology, wasn't formulated until the invention of the microscope in the 17th century.

Most cells are quite small and can be seen only under a microscope. The small size of cells means that they are measured using the smaller units of the metric system. Cells are about 100 micrometers (μm) in diameter, about the width of a human hair, and their contents are smaller yet. The units of the metric system, which are explained on the inside back cover, are common to people who use microscopes professionally.

The Cell Theory

As stated by the **cell theory,** *a cell is the basic unit of life.* Nothing smaller than a cell is alive. A unicellular organism exhibits the seven characteristics of life we discussed in Chapter 1. There is no smaller unit of life that is able to reproduce, respond to stimuli, remain homeostatic, grow and develop, take in and use materials from the environment, and become adapted to the environment. In short, life has a cellular nature.

All living things are made up of cells. While it may be apparent that a unicellular organism is necessarily a cell, what about multicellular ones? Humans are multicellular. Is there any tissue in the human body that is not composed of cells? At first, you might be inclined to say that bone is not composed of cells. But, if you were to examine bone tissue under the microscope, you would be able to see that it, too, is composed of cells surrounded by material they have deposited. Cells look quite different—a blood cell looks quite different from a nerve cell, and they both look quite different from a cartilage cell. (Fig. 3.1a). Despite having certain parts in common, cells in a multicellular organism are specialized in structure and function. Think about how many different types of cells Calvin utilized when he stole cookies from the kitchen, as mentioned in the opening story. Aside from those mentioned, Calvin needed the cells of his eyes and ears, for example. Any others?

New cells arise only from preexisting cells. This statement also requires some understanding, since it certainly wasn't readily apparent to early investigators who believed that organisms could arise from dirty rags, for example. Today, we know you can't produce a new mouse, without there being preexisting mice. When mice or humans reproduce, a sperm cell joins with an egg cell to form a zygote, which is the first cell of a new multicellular organism. Parents pass a copy of their genes onto their offspring, and the genes contain the instructions that allow the zygote to grow and develop into the full-blown organism.

Figure 3.1 Cells.
a. Cells vary in structure and function, but they all exchange substances with their environment. **b.** As the volume of a cube increases from 1 mm³ to 2 mm³, the surface-area-to-volume ratio decreases.

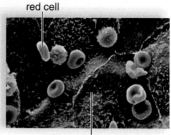
red cell

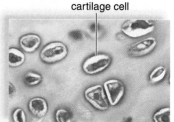
blood vessel cell

nerve cell
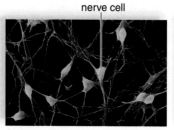

cartilage cell

a.

	1 mm	2 mm
Surface area (square mm)	$6 \times 1\ mm^2 = 6\ mm^2$	$6 \times 4\ mm^2 = 24\ mm^2$
Volume (cubic mm)	$1\ mm^3 = 1\ mm^3$	$2\ m^3 = 8\ mm^3$
Surface area / Volume	$\dfrac{6}{1}$	$\dfrac{24}{8} = \dfrac{3}{1}$

b.

Cell Size

A few cells, such as a hen's egg or a frog's egg, are large enough to be seen by the naked eye, but most are not. The small size of cells is explained by considering the surface-area-to-volume ratio of cells. Nutrients enter a cell, and wastes exit a cell at its surface; therefore, the greater the amount of surface, the greater the ability to get material in and out of the cell. A large cell requires more nutrients and produces more wastes than a small cell. In other words, the volume represents the needs of the cell. Yet, as cells get larger in volume, the proportionate amount of surface area actually decreases, as you can see by comparing the *two cubes* in Figure 3.1b.

We would expect, then, that there would be a limit to how large an actively metabolizing cell can become. Once a hen's egg is fertilized and starts metabolizing, it divides repeatedly without growth. Cell division restores the amount of surface area needed for adequate exchange of materials. Also important is the decrease in the distance a substance must travel to reach a destination, when a cell is small compared with one that is large.

Microscopy and Cell Structure

Micrographs are photographs of objects most often obtained by using the compound light microscope, the transmission electron microscope, or the scanning electron microscope (Fig. 3.2).

A compound light microscope uses a set of glass lenses and light rays passing through the object to magnify objects, and the image can be viewed directly by the human eye. The transmission electron microscope makes use of a stream of electrons to produce magnified images. The human eye can't see the image, and therefore, it is projected onto a fluorescent screen or photographic film to produce an image that can be viewed.

The magnification produced by a transmission electron microscope is much higher than that of a light microscope. Also, the ability of this microscope to make out detail in enlarged images is much greater. In other words, the transmission electron microscope has a higher resolving power—that is, the ability to distinguish between two adjacent points. Following is a comparison of the resolving power of the eye, the light microscope, and the transmission electron microscope:

Eye:	0.2 mm	= 200 μm	= 200,000 nm
Light microscope: (1,000×)	0.0002 mm	= 0.200 μm	= 200 nm
Transmission electron microscope: (50,000×)	0.00001 mm	= 0.01 μm	= 10 nm

A scanning electron microscope provides a three-dimensional view of the surface of an object. A narrow beam of electrons is scanned over the surface of the specimen, which has been coated with a thin layer of metal. The metal gives off secondary electrons, which are collected to produce a television-type picture of the specimen's surface on a screen.

As you no doubt will discover in the laboratory, the light microscope has the ability to view living specimens—this is not true of the electron microscope. Because electrons cannot travel very far in air, a strong vacuum must be maintained along the entire path of the electron beam. Usually for light microscopy and always for electron microscopy, cells are fixed so that they do not decompose, and are embedded into a matrix that allows the specimen to be thinly sliced. Cells are ordinarily transparent; therefore, sections are often stained with colored dyes before they are viewed by a light microscope. Certain components take up the dye more than other components; therefore, contrast is enhanced.

The application of electron-dense metals provides contrast with electron microscopy. The latter provides no color, so electron micrographs can be colored after the micrograph is obtained. The expression "falsely colored" means that the original micrograph has been colored after it was produced.

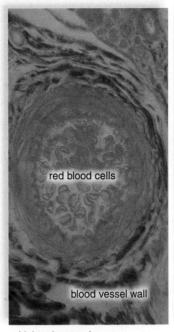

a. Light micrograph

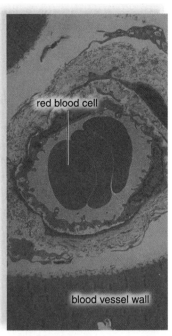

b. Transmission electron micrograph

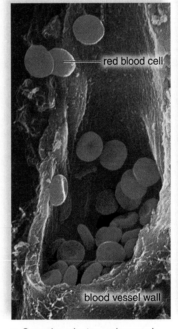

c. Scanning electron micrograph

d. Microscopist

Figure 3.2   **Micrographs of blood vessels and red blood cells.**
a. Light micrograph (LM) of many cells in a large vessel. **b.** Transmission electron micrograph (TEM) of just three cells in a small vessel. **c.** Scanning electron micrograph (SEM) gives a three-dimensional view of cells and vessels. **d.** Scientist using an electron microscope.

☑ Check Your Progress 3.1

1. What does the cell theory state?

2. Why are cells so tiny?

3. How do the light microscope and electron microscopes differ from one another?

3.2 How Cells Are Organized

Today's animal cell, such as the human cell, has an evolutionary history.

Evolutionary History of the Animal Cell

Figure 3.3 shows that the first cells to arise were **prokaryotic cells**. Prokaryotic cells lack a **nucleus**, a membrane-enclosed structure where DNA is found. Prokaryotic cells today are represented by the bacteria and archaea, which differ mainly by their chemistry. Bacteria are well known for causing diseases in humans, but they also have great environmental and commercial importance. We use them to help us manufacture all sorts of products, including biotechnology products. The **eukaryotic cell,** which does have a nucleus, is believed to have evolved from the archaea, which are best known for living in extreme environments that may mirror the first environments on Earth, being too hot, too salty, and/or too acidic for the survival of most cells.

Eukaryotic and prokaryotic cells have a **plasma membrane,** an outer membrane that regulates what enters and exits a cell. The plasma membrane is a phospholipid bilayer that is said to be selectively permeable because it allows certain molecules, but not others, to enter the cell. Proteins present in the plasma membrane play important roles in allowing substances to enter the cell.

All types of cells also contain **cytoplasm.** The matrix of the cytoplasm is a semifluid medium that contains water and various types of molecules suspended or dissolved in the medium. The presence of proteins accounts for the semifluid nature of the cytoplasm. The cytoplasm contains **organelles,** which in this context we will define as small structures that perform specific functions. Eukaryotic cells have many different types of organelles, but prokaryotic cells have only one or two types.

Internal Structure of Eukaryotic Cells

The internal structure of eukaryotic cells is believed to have evolved, as shown in Figure 3.3. The nucleus could have formed by an invagination of the plasma membrane. While the DNA of prokaryotes is centrally placed, it is not surrounded by membrane as is eukaryotic DNA. In the process of invaginating, the plasma membrane may have given rise to an internal system of membrane, called the **endomembrane system.** Surprisingly, some of the organelles in eukaryotic cells may have arisen by engulfing prokaryotic cells. One of these events would have given the eukaryotic cell a mitochondrion, an organelle that carries on cellular respiration; and another such event may have produced the chloroplast, which is an organelle, found only in plant cells, that carries on photosynthesis. Each type of eukaryotic organelle has a specific function. Since many organelles are composed of membrane, we can observe that membrane compartmentalizes the cell, keeping the various cellular activities separated from one another (Fig. 3.4).

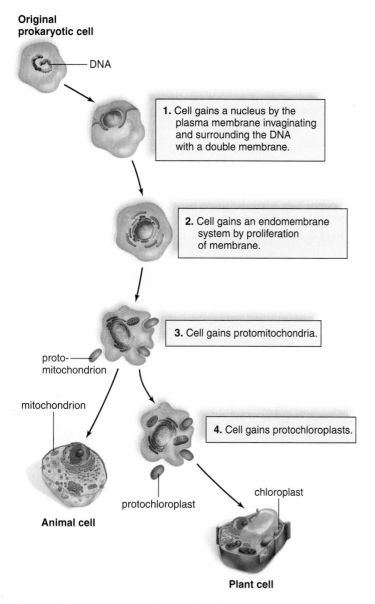

Original prokaryotic cell

—— DNA

1. Cell gains a nucleus by the plasma membrane invaginating and surrounding the DNA with a double membrane.

2. Cell gains an endomembrane system by proliferation of membrane.

3. Cell gains protomitochondria.

proto-mitochondrion

mitochondrion

4. Cell gains protochloroplasts.

Animal cell

protochloroplast

chloroplast

Plant cell

Figure 3.3   Origin of the eukaryotic cell.
Invagination of the plasma membrane could have created the nucleus and an endomembrane system. Later, the cell gained organelles, some of which may have been independent prokaryotes.

✓ Check Your Progress 3.2

1. a. What are the three main parts of a eukaryotic cell? b. Which two of these also occur in prokaryotes? c. Which one does not occur?

2. How might the nucleus, the endomembrane system, and two of the organelles in eukaryotic cells have arisen?

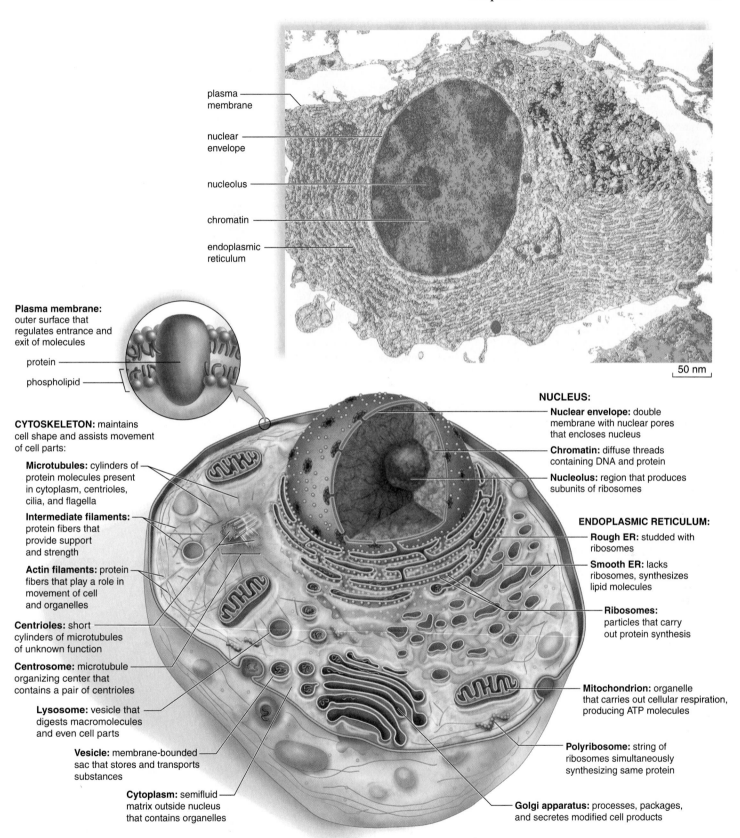

plasma membrane

nuclear envelope

nucleolus

chromatin

endoplasmic reticulum

50 nm

Plasma membrane: outer surface that regulates entrance and exit of molecules

protein

phospholipid

CYTOSKELETON: maintains cell shape and assists movement of cell parts:

Microtubules: cylinders of protein molecules present in cytoplasm, centrioles, cilia, and flagella

Intermediate filaments: protein fibers that provide support and strength

Actin filaments: protein fibers that play a role in movement of cell and organelles

Centrioles: short cylinders of microtubules of unknown function

Centrosome: microtubule organizing center that contains a pair of centrioles

Lysosome: vesicle that digests macromolecules and even cell parts

Vesicle: membrane-bounded sac that stores and transports substances

Cytoplasm: semifluid matrix outside nucleus that contains organelles

NUCLEUS:

Nuclear envelope: double membrane with nuclear pores that encloses nucleus

Chromatin: diffuse threads containing DNA and protein

Nucleolus: region that produces subunits of ribosomes

ENDOPLASMIC RETICULUM:

Rough ER: studded with ribosomes

Smooth ER: lacks ribosomes, synthesizes lipid molecules

Ribosomes: particles that carry out protein synthesis

Mitochondrion: organelle that carries out cellular respiration, producing ATP molecules

Polyribosome: string of ribosomes simultaneously synthesizing same protein

Golgi apparatus: processes, packages, and secretes modified cell products

Figure 3.4   Animal cell.

Generalized drawing (*below*) and transmission electron micrograph (*above*).

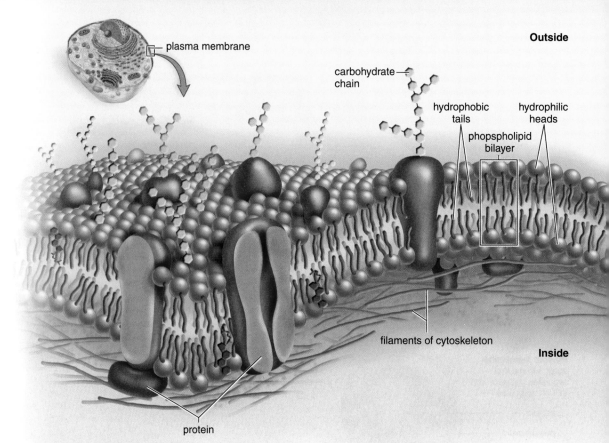

Outside

carbohydrate chain

hydrophobic tails

hydrophilic heads

phopspholipid bilayer

plasma membrane

filaments of cytoskeleton

Inside

protein

Figure 3.5
Fluid-mosaic model of plasma membrane structure.
The membrane is composed of a phospholipid bilayer in which proteins are embedded. The hydrophilic heads of phospholipids are a part of the outside surface and the inside surface of the membrane. The hydrophobic tails make up the interior of the membrane. Note the plasma membrane's asymmetry—carbohydrate chains are attached to the outside surface, and cytoskeleton filaments are attached to the inside surface.

3.3 The Plasma Membrane and How Substances Cross It

A human cell, like all cells, is surrounded by an outer plasma membrane (Fig. 3.5). The plasma membrane marks the boundary between the outside and the inside of the cell. Plasma membrane integrity and function are necessary to the life of the cell.

The plasma membrane is a phospholipid bilayer with attached or embedded proteins. A phospholipid molecule has a polar head and nonpolar tails (see Fig. 2.16). When phospholipids are placed in water, they naturally form a spherical bilayer. The polar heads, being charged, are hydrophilic (attracted to water) and they position themselves to face toward the watery environment outside and inside the cell. The nonpolar tails are hydrophobic (not attracted to water) and they turn inward toward one another, where there is no water.

At body temperature, the phospholipid bilayer is a liquid; it has the consistency of olive oil; and the proteins are able to change their position by moving laterally. The **fluid-mosaic model,** a working description of membrane structure, says that the protein molecules form a shifting pattern within the fluid phospholipid bilayer. Cholesterol lends support to the membrane.

Short chains of sugars are attached to the outer surface of some protein and lipid molecules (called glycoproteins and glycolipids, respectively). These carbohydrate chains, specific to each cell, help mark it as belonging to a particular individual. They account for why people have different blood types, for example. Other glycoproteins have a special configuration that allows them to act as a receptor for a chemical messenger, such as a hormone. Some plasma membrane proteins form channels through which certain substances can enter cells; others are enzymes that catalyze reactions or carriers involved in the passage of molecules through the membrane.

Plasma Membrane Functions

The plasma membrane keeps a cell intact. It allows only certain molecules and ions to enter and exit the cytoplasm freely; therefore, the plasma membrane is said to be **selectively permeable** (Fig. 3.6). Small molecules that are lipid-soluble, such as oxygen and carbon dioxide, can pass through the membrane easily. The small size of water molecules

Figure 3.6   Selective permeability.
The curved arrows indicate that these substances cannot diffuse across the plasma membrane, and the long back-and-forth arrows indicate that these substances can diffuse across the plasma membrane.

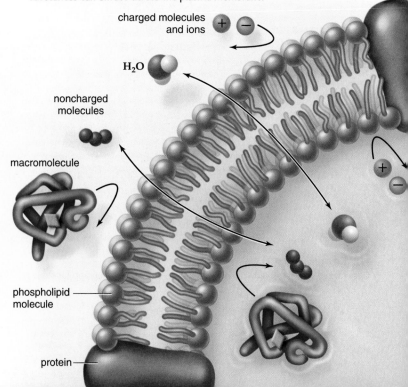

charged molecules and ions

H_2O

noncharged molecules

macromolecule

phospholipid molecule

protein

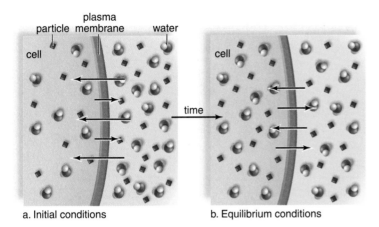

a. Initial conditions

b. Equilibrium conditions

Figure 3.7 **Diffusion of molecules across the membrane.**
a. When a substance can diffuse across the plasma membrane, it will move back and forth across the membrane, but the net movement will be toward the region of lower concentration. **b.** At equilibrium, an equal number of particles and water have crossed in both directions.

allows them to freely cross the membrane by utilizing protein channels called *aquaporins.* Ions and large molecules cannot cross the membrane without more direct assistance to be discussed later.

Diffusion **Diffusion** is the random movement of molecules from the area of higher concentration to the area of lower concentration, until they are equally distributed. Diffusion is a passive way for molecules to enter or exit a cell, meaning that no cellular energy is needed to bring it about.

To illustrate diffusion, remember that the aroma of chocolate chip cookies permeated the house and that wonderful smell greeted Calvin when he opened the front door. The smell filled the house because the cookies gave off gaseous molecules that moved from where they were concentrated (the kitchen) to all the rooms of the house. When they are equally distributed, the molecules will still be moving randomly in all directions. When smells reach the nose, molecules bind to plasma membrane receptors in olfactory cells of the nose, and this starts nerve impulses that go to the brain. Only then would Calvin become aware of the cookies.

Certain molecules can freely cross the plasma membrane by diffusion. When molecules can cross a plasma membrane, which way will they go? The molecules will move in both directions, but the *net movement* will be from the region of higher concentration to the region of lower concentration, until equilibrium is achieved. At equilibrium, as many molecules of the substance will be entering as leaving the cell (Fig. 3.7). Oxygen diffuses across the plasma membrane, and the net movement is toward the inside of the cell because a cell uses oxygen when it produces ATP molecules for energy purposes.

Osmosis **Osmosis** is the diffusion of water across a plasma membrane. Osmosis involves water and a solute (dissolved substance) that cannot readily cross the plasma membrane. **Tonicity** is simply the concentration of the sol-

ute in a solution, which can be stated as a percentage, such as 2% solute. As the amount of salt, for example, increases, the amount of water in a solution decreases. Normally, body fluids are isotonic to cells (Fig. 3.8a)—that is, there is the same concentration of nondiffusible solutes and water on both sides of the plasma membrane, and cells maintain their usual size and shape. Intravenous solutions medically administered are usually isotonic.

Solutions that cause cells to swell or even to burst due to an intake of water are said to be hypotonic. If red blood cells are placed in a hypotonic solution, which has a lower concentration of solute and a higher concentration of water than do the cells, water enters the cells and they swell to bursting (Fig. 3.8b). The term *lysis* is used to refer to disrupted cells; hemolysis, then, means disrupted red blood cells.

Solutions that cause cells to shrink or shrivel due to a loss of water are said to be hypertonic. If red blood cells are placed in a hypertonic solution, which has a higher concentration of solute and a lower concentration of water than do the cells, water leaves the cells and they shrink (Fig. 3.8c). The term *crenation* refers to red blood cells in this condition. These changes have occurred due to osmotic pressure. **Osmotic pressure** is the force exerted on a selectively permeable membrane because water has moved from the area of higher to lower concentration of water (higher concentration of solute).

Facilitated Transport Many solutes do not simply diffuse across a plasma membrane; rather, they are transported by means of protein carriers within the membrane. During **facilitated transport,** a molecule is transported at a rate

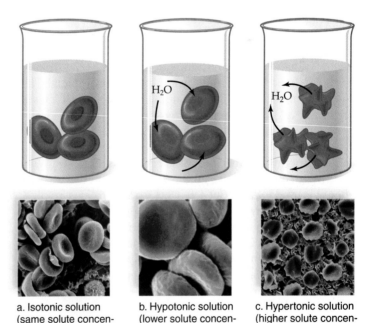

a. Isotonic solution (same solute concentration as in cell)

b. Hypotonic solution (lower solute concentration than in cell)

c. Hypertonic solution (higher solute concentration than in cell)

Figure 3.8 **Effect of tonicity on red blood cells.**
a. In an isotonic solution, cells remain the same. **b.** In a hypotonic solution, cells gain water and may burst (lysis). **c.** In a hypertonic solution, cells lose water and shrink (crenation).

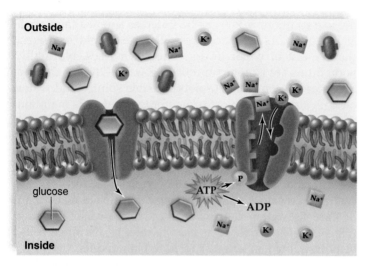

Figure 3.9   Transport by carriers.
A substance (red) does not enter the cell because it has no carrier. Glucose (green) enters by facilitated transport, and the result will be an equal distribution on both sides of the membrane. Na⁺ exits and K⁺ enters the cell by active transport, so Na⁺ will be concentrated outside and K⁺ will be concentrated inside the cell.

higher than usual across the plasma membrane from the side of higher concentration to the side of lower concentration (Fig. 3.9). This is a passive means of transport because the cell does not need to expend energy to move a substance down its concentration gradient. Each protein carrier, sometimes called a transporter, binds only to a particular molecule, such as glucose (green). Diabetes type 2 results when cells lack a sufficient number of glucose transporters.

Active Transport   During **active transport,** a molecule is moving contrary to the normal direction—that is, from lower to higher concentration. For example, iodine collects in the cells of the thyroid gland; sugar is completely absorbed from the gut by cells that line the digestive tract; and sodium (Na⁺) is sometimes almost completely withdrawn from urine by cells lining kidney tubules.

Active transport requires a protein carrier and the use of cellular energy obtained from the breakdown of ATP. When ATP is broken down, energy is released, and in this case, the energy is used to carry out active transport. Proteins involved in active transport often are called pumps, because just as a water pump uses energy to move water against the force of gravity, energy is used to move substances against their concentration gradients. One type of pump that is active in all cells, but is especially associated with nerve and muscle cells, moves sodium ions (Na⁺) to the outside and potassium ions (K⁺) to inside the cell (Fig. 3.9).

The passage of salt (NaCl) across a plasma membrane is of primary importance in cells. First, sodium ions are pumped across a membrane; then, chloride ions simply diffuse through channels that allow their passage. Chloride ion channels malfunction in persons with cystic fibrosis, and this leads to the symptoms of this inherited (genetic) disorder.

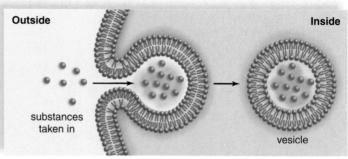

a. Endocytosis

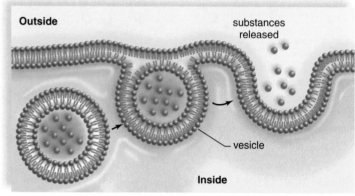

b. Exocytosis

Figure 3.10   Endocytosis and exocytosis.
a. Large substances enter a cell by endocytosis, and (**b**) large substances exit a cell by exocytosis.

Endocytosis and Exocytosis   During endocytosis, a portion of the plasma membrane invaginates, or forms a pouch, to envelop a substance and fluid. Then, the membrane pinches off to form an endocytic vesicle inside the cell (Fig. 3.10a). Some white blood cells are able to take up pathogens (disease-causing agents) by endocytosis, and the process is given a special name: **phagocytosis.** Usually, cells take up molecules and fluid, and then the process is called *pinocytosis.* An inherited form of cardiovascular disease occurs when cells fail to take up a combined lipoprotein and cholesterol molecule from the blood by pinocytosis.

During exocytosis, a vesicle fuses with the plasma membrane as secretion occurs (Fig. 3.10b). Later in the chapter, we will see that a steady stream of vesicles move between certain organelles, before finally fusing with the plasma membrane. This is the way that signaling molecules, called neurotransmitters, leave one nerve cell to excite the next nerve cell or a muscle cell. When Calvin dashed out the door, his nerves were exciting his muscle cells to contract due to exocytosis.

✅ Check Your Progress 3.3

1. What is the structure and overall function of the plasma membrane?

2. a. What are the various ways substances can enter and exit cells?
   b. Which are passive, and which are active ways of crossing the plasma membrane?

3. a. How do isotonic, hypotonic, and hypertonic solutions differ, and b. how do they affect cells?

3.4 The Nucleus and the Production of Proteins

The nucleus and several organelles are involved in the production and processing of proteins. In order for Calvin to digest his cookies, cells in nearby glands and lining his digestive tract will produce digestive enzymes, which are proteins.

The Nucleus

The **nucleus,** a prominent structure in cells, stores genetic information (Fig. 3.11). Every cell in the body contains the same genes, but each type of cell has certain genes, which are segments of DNA, turned on, while others are turned off. DNA, with RNA acting as an intermediary, specifies the proteins in a cell. Proteins have many functions in cells, and they help determine a cell's specificity.

When you look at the nucleus, even in an electron micrograph, you cannot see DNA molecules, but you can see **chromatin.** Chromatin undergoes coiling into rodlike structures called chromosomes just before the cell divides. Each **chromosome** contains a DNA molecule and associated proteins. Chromatin is immersed in a semifluid medium called the **nucleoplasm.** A difference in pH suggests that nucleoplasm has a different composition from cytoplasm.

Micrographs of a nucleus do show one or more dark regions of the chromatin. These are nucleoli (sing., **nucleolus**), where another type of RNA, called ribosomal RNA (rRNA), is produced, and where rRNA joins with proteins to form the subunits of ribosomes.

The nucleus is separated from the cytoplasm by a double membrane known as the **nuclear envelope,** which is continuous with the **endoplasmic reticulum (ER),** a membranous system of saccules and channels discussed in the next section. The nuclear envelope has **nuclear pores** of sufficient size to permit the passage of ribosomal subunits out of the nucleus and proteins into the nucleus.

Ribosomes

Ribosomes are organelles composed of proteins and rRNA. Protein synthesis occurs at the ribosomes. Ribosomes are often attached to the endoplasmic reticulum; they also occur free within the cytoplasm, either singly or in groups called **polyribosomes.**

Proteins synthesized by cytoplasmic ribosomes are used inside the cell for various purposes. Those produced by ribosomes attached to endoplasmic reticulum may eventually be secreted from the cell or become part of the plasma membrane.

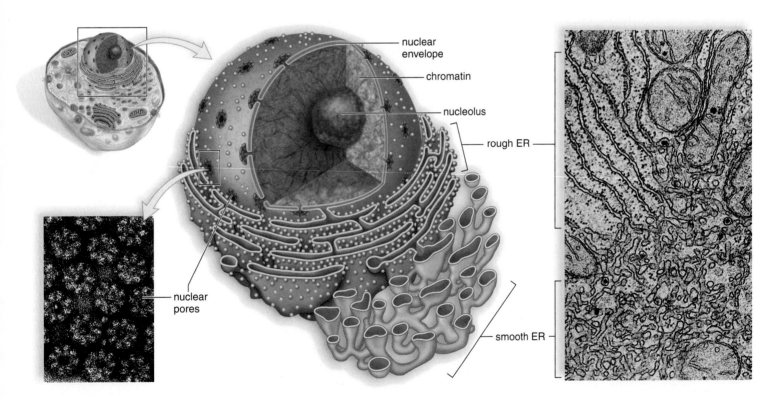

Figure 3.11 The nucleus and the ribosomes.
The nucleus contains chromatin. Chromatin has a special region called the nucleolus, which is where rRNA is produced and ribosome subunits are assembled. The nuclear envelope contains pores (TEM, *left*) that allow substances to enter and exit the nucleus to and from the cytoplasm. The nuclear envelope is attached to the endoplasmic reticulum (TEM, *right*), which often has attached ribosomes, where protein synthesis occurs.

The Endomembrane System

The **endomembrane system** consists of the nuclear envelope, the endoplasmic reticulum, the Golgi apparatus, lysosomes, and **vesicles** (tiny membranous sacs) (Fig. 3.12).

The Endoplasmic Reticulum

The endoplasmic reticulum has two portions. Rough ER is studded with ribosomes on the side of the membrane that faces the cytoplasm. Here, proteins are synthesized and enter the ER interior, where processing and modification begin. Some of these proteins are incorporated into membrane, and some are for export. Smooth ER, which is continuous with rough ER, does not have attached ribosomes. Smooth ER synthesizes the phospholipids that occur in membranes and has various other functions, depending on the particular cell. In the testes, it produces testosterone, and in the liver, it helps detoxify drugs.

The ER forms transport vesicles in which large molecules are transported to other parts of the cell. Often, these vesicles are on their way to the plasma membrane or the Golgi apparatus.

The Golgi Apparatus

The **Golgi apparatus** is named for Camillo Golgi, who discovered its presence in cells in 1898. The Golgi apparatus consists of a stack of slightly curved saccules, whose appearance can be compared to a stack of pancakes. Here, proteins and lipids received from the ER are modified. For example, a chain of sugars may be added to them, thereby making them glycoproteins and glycolipids, which are molecules found in the plasma membrane.

The vesicles that leave the Golgi apparatus move to other parts of the cell. Some vesicles proceed to the plasma membrane, where they discharge their contents. Altogether, the Golgi apparatus is involved in processing, packaging, and secretion. Other vesicles that leave the Golgi apparatus are lysosomes.

Lysosomes

Lysosomes, membranous sacs produced by the Golgi apparatus, contain *hydrolytic enzymes.* Lysosomes are found in all cells of the body but are particularly numerous in white blood cells that engulf disease-causing microbes. When a lysosome fuses with such an endocytic vesicle, its contents are digested by lysosomal enzymes into simpler subunits that then enter the cytoplasm. Even parts of a cell are digested by its own lysosomes (called autodigestion). Some human diseases are caused by the lack of a particular lysosome enzyme. Tay-Sachs disease occurs when an undigested substance collects in nerve cells, leading to retarded development of a child and early death.

☑ Check Your Progress 3.4

1. The nucleus, ribosomes, and rough endoplasmic reticulum make what contribution to protein synthesis?
2. The endoplasmic reticulum and the Golgi apparatus make what contribution to the processing of proteins?
3. What is the function of smooth endoplasmic reticulum and lysosomes?
4. What role do transport vesicles play in the endomembrane system?

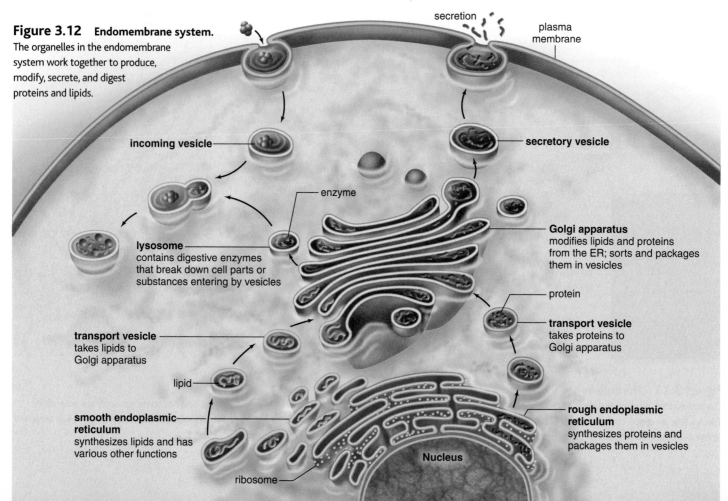

Figure 3.12   Endomembrane system. The organelles in the endomembrane system work together to produce, modify, secrete, and digest proteins and lipids.

3.5 The Cytoskeleton and Cell Movement

It took a high-powered electron microscope to discover that the cytoplasm of the cell is crisscrossed by several types of protein fibers collectively called the **cytoskeleton** (see Fig. 3.4). The cytoskeleton helps maintain a cell's shape and either anchors the organelles or assists their movement, as appropriate.

In the cytoskeleton, **microtubules** are much larger than actin filaments. Each is a cylinder that contains 13 longitudinal rows of a protein called tubulin. The regulation of microtubule assembly is under the control of a microtubule organizing center called the **centrosome** (see Fig. 3.4). Microtubules help maintain the shape of the cell and act as tracks along which organelles move. During cell division, microtubules form spindle fibers, which assist the movement of chromosomes. **Actin filaments,** made of a protein called actin, are long, extremely thin fibers that usually occur in bundles or other groupings. Actin filaments are involved in movement; microvilli, which project from certain cells and can shorten and extend, contain actin filaments. **Intermediate filaments,** as their name implies, are intermediate in size between microtubules and actin filaments. Their structure and function differ according to the type of cell.

Cilia and Flagella

Cilia (sing., **cilium**) and **flagella** (sing., **flagellum**) are both involved in movement. The ciliated cells that line our respiratory tract sweep debris trapped within mucus back up the throat, which helps keep the lungs clean. Similarly, ciliated cells move an egg along the oviduct, where it will be fertilized by a flagellated sperm cell (Fig. 3.13).

A cilium is about 20 times shorter than a flagellum but both have the same organization of microtubules within a plasma membrane covering. Motor molecules, powered by ATP, allow the microtubules in cilia and flagella to interact and bend, and thereby move. Cilia and flagella grow from **basal bodies,** structures that have the same organization as **centrioles,** which are structures located in centrosomes (see Fig. 3.4). Another possible relationship between centrioles and microtubules is mentioned again in Chapter 18, which concerns cell division.

The importance of normal cilia and flagella is illustrated by the occurrence of a genetic disorder. Some individuals have an inherited genetic defect that leads to malformed microtubules in cilia and flagella. Not surprisingly, they suffer from recurrent and severe respiratory infections, because the ciliated cells lining respiratory passages fail to keep their lungs clean. They are also unable to reproduce naturally due to the lack of ciliary action to move the egg in a female, or the lack of flagella action by sperm in a male.

✓ Check Your Progress 3.5

1. The cytoskeleton contains what three types of fibers?
2. Which of these three is found in cilia and flagella?
3. Compare the function of the cytoskeleton to that of cilia.

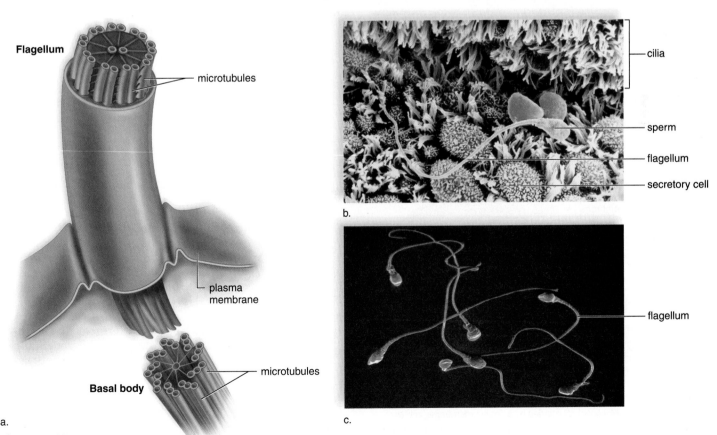

Figure 3.13   Cilia and flagella.
Human reproduction is dependent on the normal activity of cilia and flagella. **a.** Both cilia and flagella have an inner core of microtubules within a covering of plasma membrane. **b.** Cilia within the oviduct move the egg to where it is fertilized by a flagellated sperm. **c.** Sperm have very long flagella.

3.6 Mitochondria and Cellular Metabolism

Calvin was being very energetic when he quickly grabbed his cookies and dashed for the back door. Where did he get the energy from? We already know his blood delivers glucose and oxygen to his cells, where glucose breakdown provides the energy for ATP buildup. This section will explain how mitochondria make use of glucose and oxygen to produce ATP.

Mitochondria (sing., **mitochondrion**) are often called the powerhouses of the cell—just as a powerhouse burns fuel to produce electricity, the mitochondria convert the chemical energy of glucose products into the chemical energy of ATP molecules. In the process, mitochondria use up oxygen and give off carbon dioxide. Therefore, the process of producing ATP is called **cellular respiration**. The structure of mitochondria is appropriate to the task. The inner membrane is folded to form little shelves called cristae, which project into the matrix, an inner space filled with a gel-like fluid (Fig. 3.14). The matrix of a mitochondrion contains enzymes for breaking down glucose products. ATP production then occurs at the cristae. Protein complexes that aid in the conversion of energy are located in an assembly-line fashion on these membranous shelves.

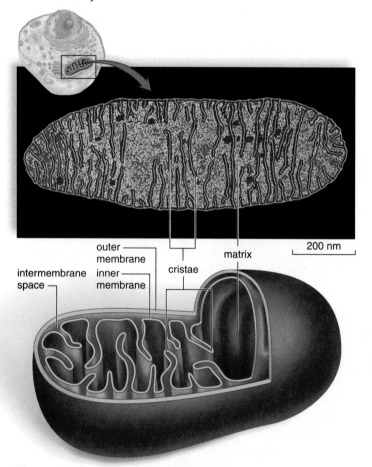

intermembrane space
outer membrane
inner membrane
cristae
matrix
200 nm

Figure 3.14 **Structure of mitochondria.**
A mitochondrion (TEM, *above*) is bounded by a double membrane, and the inner membrane folds into projections called cristae. The cristae project into a semifluid matrix that contains many enzymes.

Mitochondria and Aging

Do mitochondria cause us to age? Evidence suggests they do.

Free Radical Theory of Aging

The free radical theory of aging says that aging is caused by damage done to cells by reactive oxygen species (ROS), such as free radical O_2^- that runs around the cell destroying any molecule it touches. And it's known that the majority of ROS are produced by the mitochondria! It's been observed that ROS-induced cell damage does increase with age, and this lends support to the free radical theory of aging. Therefore, reducing the amount of ROS produced or boosting defenses against ROS-induced damage should correlate with increased life span. Studies have indeed shown that yeast, worms, flies, or mice with enhanced antioxidant defenses live longer. While many support the free radical theory of aging, more research is needed before it can be determined whether ROS are the cause of, or just correlate with, aging.

Mitochondrial Theory of Aging

The mitochondrial theory of aging suggests that it is the buildup of defects in mitochondrial DNA (mtDNA) that causes aging. Although mtDNA mutations accumulate over time, again it is also possible that mutations simply correlate with, and are not a cause of, aging. Mice engineered to have a high number of mtDNA mutations live about half as long as their normal counterparts. They also show a number of ill effects, similar to those in other aging animals (i.e., weight and muscle loss, balding, and bone changes). It's possible, though, that the mitochondrial gene changes in these animals simply result in poor health. Interestingly, these mutant mice were not found to have elevated levels of ROS.

Conclusion

In conclusion, we can say that the numbers of dysfunctional mitochondria in cells does increase with increasing age. And mitochondrial degeneration does result in a shortened life span for the model organisms already mentioned. Also, mtDNA defects can cause a number of degenerative human diseases, and mutations in mtDNA accumulate with age in humans. Certainly, any decline in ATP production by damaged mitochondria is likely to affect cell function. Nevertheless, our understanding of the complex concept of aging remains rudimentary, and there is much left to learn about its molecular mechanisms. In the coming years, research from genetically modified organisms may provide exciting clues about aging. So, stay tuned.

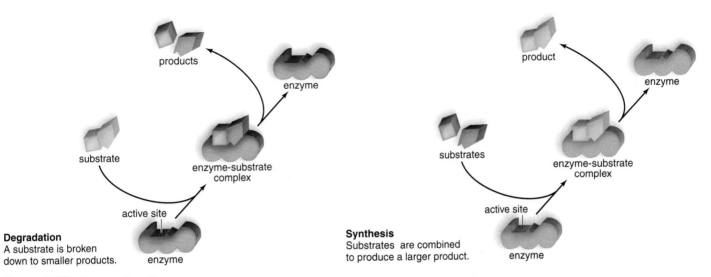

Figure 3.15 Enzymatic action.
An enzyme has an active site, where the substrates and enzyme fit together in such a way that the substrates are oriented to react. Following the reaction, the products are released, and the enzyme is free to act again. Some enzymes carry out degradation; the substrate is broken down to smaller products. Other enzymes carry out synthesis; the substrates are combined to produce a larger product.

The structure of a mitochondrion supports the hypothesis that they were originally prokaryotes engulfed by a cell. Mitochondria are bounded by a double membrane as a prokaryote would be if taken into a cell by endocytosis. Even more interesting is the observation that mitochondria have their own genes, and they reproduce themselves!

Cellular Respiration and Metabolism

Cellular respiration is a very important component of **metabolism**, which includes all the chemical reactions that occur in a cell. Quite often, metabolism requires metabolic pathways, which are carried out by enzymes sequentially arranged in cells:

$$A \xrightarrow{1} B \xrightarrow{2} C \xrightarrow{3} D \xrightarrow{4} E \xrightarrow{5} F \xrightarrow{6} G$$

The letters, except A and G, are **products** of the previous reaction and the **reactants** for the next reaction. A represents the beginning reactant(s), and G represents the end product(s). The numbers in the pathway refer to different enzymes. *Each reaction in a metabolic pathway requires a specific enzyme.* In effect, no reaction in the pathway occurs unless its enzyme is present. For example, if enzyme 2 in the diagram is missing, the pathway cannot function; it will stop at B. Since enzymes are so necessary in cells, their mechanism of action has been studied extensively.

Enzymes and Coenzymes

When an enzyme speeds up a reaction, the reactant(s) that participates in the reaction is (are) called the enzyme's **substrate(s).** Enzymes are often named for their substrates. For example, lipids are broken down by lipase, maltose by maltase, and lactose by lactase.

Enzymes have a specific region, called an **active site,** where the substrates are brought together so they can react. An enzyme's specificity is caused by the shape of the active site, where the enzyme and its substrate(s) fit together in a specific way, much as the pieces of a jigsaw puzzle fit together (Fig. 3.15). After one reaction is complete, the product or products are released, and the enzyme is ready to be used again. Therefore, a cell requires only a small amount of a particular enzyme to carry out a reaction, which can be summarized in the following manner:

$$E + S \; \rightarrow \; ES \; \rightarrow \; E + P$$

(where E = enzyme, S = substrate, ES = enzyme-substrate complex, and P = product). Notice that an enzyme can be used over and over again.

Coenzymes   **Coenzymes** are nonprotein molecules that assist the activity of an enzyme and may even accept or contribute atoms to the reaction. It is interesting that vitamins are often components of coenzymes. The vitamin niacin is a part of the coenzyme **NAD (nicotinamide adenine dinucleotide),** which carries hydrogen (H) and electrons after an enzyme called a **dehydrogenase** removes them from a substrate. When hydrogens, H^+, and electrons, e^-, are passed to NAD^+, it becomes NADH. A hydrogen atom consists of a hydrogen ion (H^+) and an electron (e^-). NAD^+ accepts two electrons, but only one hydrogen ion from a substrate:

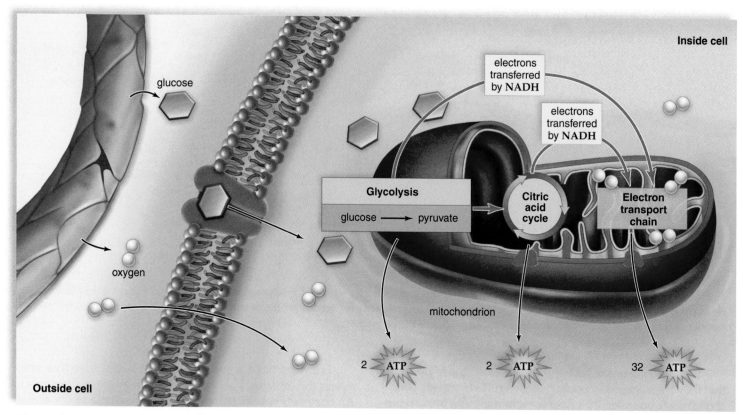

Figure 3.16   Production of ATP during cellular respiration.
Glucose enters a cell from the bloodstream by facilitated transport. The three main pathways of cellular respiration (glycolysis, citric acid cycle, and electron transport chain) all produce ATP, but most is produced by the electron transport chain. NADH carries electrons to the electron transport chain from glycolysis and the citric acid cycle. ATP exits a mitochondrion by facilitated transport.

Cellular Respiration

After blood transports glucose and oxygen to cells, cellular respiration, which breaks down glucose to carbon dioxide and water, begins. Three pathways are involved in the breakdown of glucose, and they are called glycolysis, the citric acid cycle, and the electron transport chain (Fig. 3.16). These metabolic pathways allow the energy within a glucose molecule to be released slowly, so that ATP can be produced gradually. Cells would lose a tremendous amount of energy if glucose breakdown occurred all at once—much energy would be lost as heat. When humans burn wood or coal, the energy escapes all at once as heat; but a cell "burns" glucose gradually and energy is captured as ATP.

Glycolysis   The term **glycolysis** means sugar splitting, and during glycolysis, glucose, a 6-carbon (C_6) molecule, is split so that the result is two 3-carbon (C_3) molecules of *pyruvate*. Glycolysis, which occurs in the cytoplasm, is found in most every type of cell; therefore, this pathway is believed to have evolved early in the history of life. The genes coding for the enzymes of the glycolytic pathway could have been present in the very first cell(s) and then passed on from that time forward.

Glycolysis is termed **anaerobic**, because it requires no oxygen. This pathway can occur in microbes that live in bogs or swamps or our intestinal tract, where there is no oxygen. During glycolysis, hydrogens and electrons are removed from glucose, and NADH results. The breaking of bonds releases enough energy for a net yield of two ATP molecules.

Pyruvate is a pivotal molecule in cellular respiration. When oxygen is available, the molecule enters mitochondria and is broken down completely. When oxygen is not available, fermentation occurs (see page 56); and, in this way, the cell can acquire some ATP, even though oxygen is not present.

Citric acid cycle   The **citric acid cycle** completes the breakdown of glucose. As this cyclical series of enzymatic reactions occurs in the matrix of mitochondria, carbon dioxide is released. Hydrogen and electrons are carried away by NADH. In addition, the citric acid cycle also produces two ATP per glucose molecule.

So far, we have considered only carbohydrates as a possible fuel for cellular respiration, but what about fats and proteins? Fats are digested to glycerol and fatty acids. When our cells run out of glucose, they primarily substitute fatty acids for glucose as an energy source. Fatty acids yield C_2 molecules that can enter the citric acid cycle.

Proteins are digested to amino acids whose carbon chain can easily enter the citric acid cycle. However, first the amino group has to be removed from the carbon chain. This step is primarily carried out in the liver because the liver has the appropriate enzymes to process amino groups. The liver converts the amino groups to urea, which is excreted by the kidneys. When you eat a lot of protein, your body begins to use amino acids as an energy source, and your kidneys work overtime excreting all that nitrogen.

Electron transport chain NADH molecules from glycolysis and the citric acid cycle deliver electrons to the electron transport chain. The members of the electron transport chain are carrier proteins that are grouped into complexes, which are embedded in the cristae of a mitochondrion. Each carrier of the **electron transport chain** accepts two electrons and passes them on to the next carrier. The hydrogens carried by NADH molecules will be used later.

High-energy electrons enter the chain and, as they are passed from carrier to carrier, the electrons lose energy so that low-energy electrons emerge from the chain. Oxygen serves as the final acceptor of the electrons at the end of the chain. After oxygen receives the electrons, it combines with hydrogens and becomes water.

The presence of oxygen makes the citric acid cycle and the electron transport chain **aerobic,** but notice that oxygen does not combine with any substrates during cellular respiration. Think about it—breathing is necessary to our existence, and the sole purpose of oxygen is to receive electrons at the end of the electron transport chain.

The energy, released as electrons pass from carrier to carrier, is used for ATP production. It took many years for investigators to determine exactly how this occurs, but the details are beyond the scope of this text. Suffice it to say that the inner mitochondrial membrane contains an ATP synthase complex that combines ADP + Ⓟ to produce ATP. The ATP synthase complex produces about 32 ATP per glucose molecule.

Each cell produces ATP within its mitochondria, and therefore, each cell uses ATP for its own specific purposes. Figure 3.17 shows the ATP cycle. Glucose breakdown leads to ATP buildup, and then ATP is used for the metabolic work of the cell. Muscle cells use ATP for contraction, and nerve cells use it for conduction of nerve impulses. When Calvin reached for cookies, he had sufficient ATP on hand to dash for the back door and keep going. ATP breakdown releases heat, and it could be that Calvin broke out into a sweat as he kept on running.

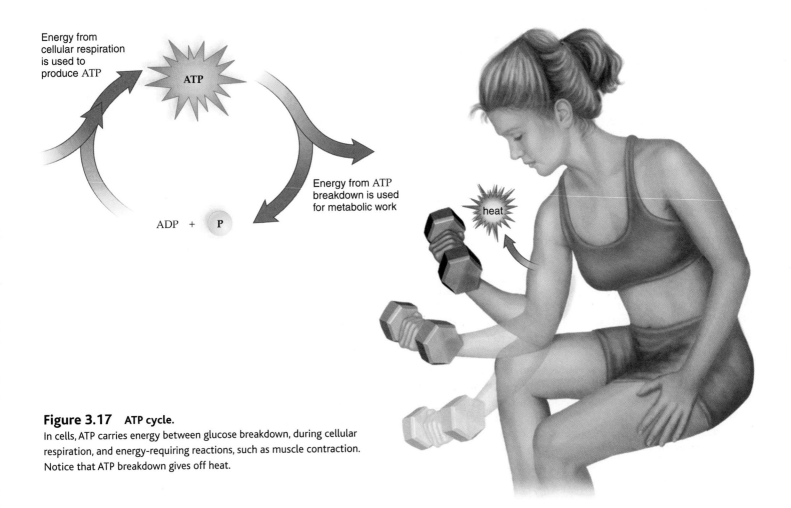

Figure 3.17 ATP cycle.
In cells, ATP carries energy between glucose breakdown, during cellular respiration, and energy-requiring reactions, such as muscle contraction. Notice that ATP breakdown gives off heat.

Fermentation

Fermentation is an anaerobic process, meaning that it does not require oxygen. When oxygen is not available to cells, the electron transport chain soon becomes inoperative because oxygen is not present to accept electrons. In this case, most cells have a safety valve so that some ATP can still be produced. Glycolysis operates as long as it is supplied with "free" NAD—that is, NAD that can pick up hydrogens and electrons. Normally, NADH takes electrons to the electron transport chain and, thereby, becomes "free" of them. However, if the system is not working due to a lack of oxygen, NADH passes its hydrogens and electrons to pyruvate molecules, as shown in the following reaction:

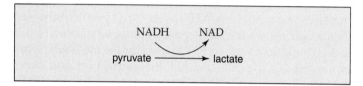

This means that the citric acid cycle and the electron transport chain do not function as part of fermentation. When oxygen is available again, lactate can be converted back to pyruvate, and metabolism can proceed as usual.

Fermentation can give us a burst of energy for a short time, but it produces only two ATP per glucose molecule. Also, fermentation results in the buildup of lactate. Lactate is toxic to cells and causes muscles to cramp and fatigue. If fermentation continues for any length of time, death follows.

It is of interest to know that fermentation takes its name from yeast fermentation. Yeast fermentation produces alcohol and carbon dioxide (instead of lactate). When yeast is used to leaven bread, carbon dioxide production makes the bread rise. When yeast is used to produce alcoholic beverages, it is the alcohol that humans make use of.

Lactate and the Athlete

Exercise is a dramatic test of homeostatic mechanisms. During exercise, the mitochondria of our muscle cells require much oxygen, and they produce an increased amount of carbon dioxide. No doubt, if you run as fast as you can, even for a short time, you notice that you get out of breath. You are in oxygen deficit—your muscles have run out of oxygen and have started fermenting instead. Aerobic exercise occurs when you can manage to get a steady supply of oxygen to your muscle cells so that oxygen deficit does not occur. Athletes are better at this than nonathletes. Why?

First of all, the number of mitochondria is higher in the muscles of persons who train. Therefore, an athlete is more likely to rely on the citric acid cycle and the electron transport chain to generate ATP. The citric acid cycle can be powered by fatty acids, instead of glucose; therefore, the level of glucose in the blood remains at a normal level, even though exercise is occurring.

Muscle cells with few mitochondria don't start consuming O_2 until they are out of ATP and ADP concentration is high. After endurance training, the large number of mitochondria start consuming O_2 as soon as the ADP concentration starts rising, due to muscle contraction and breakdown of ATP. This faster rise in O_2 uptake at the onset of exercise (Fig. 3.18) means that the O_2 deficit is less, and the formation of lactate due to fermentation is less.

As mentioned, the body is able to process lactate and change it back to pyruvate. This is the oxygen deficit—the amount of oxygen it takes to rid the body of lactate. Athletes incur less of an oxygen deficit than nonathletes do.

✓ Check Your Progress 3.6

1. **a.** Which of the three pathways of cellular respiration occurs in the matrix of a mitochondrion? **b.** Which occurs at the cristae?

2. Why does cellular respiration have so many small enzymatic reactions instead of one big step?

3. **a.** What's the difference between an enzyme and a coenzyme? **b.** What does NADH do?

4. **a.** What's the role of oxygen in cellular respiration? **b.** Where does the carbon dioxide we breathe out come from?

5. **a.** What is the benefit of fermentation? **b.** What's the drawback?

Figure 3.18 **Training reduces fermentation.**
In athletes, there is an increase in fat metabolism that keeps blood glucose at a normal level; a smaller O_2 deficit due to a more rapid increase in O_2 uptake at the onset of work; a reduction in lactate formation; and an increase in lactate removal.

Stem Cell Research

Some human illnesses, such as diabetes type 1, Alzheimer disease, and Parkinson disease, are clearly due to a loss of specialized cells. In diabetes type 1, there is a loss of insulin-secreting cells in the pancreas, and in Alzheimer disease and Parkinson disease there is a loss of brain cells. Specific types of cells are needed to cure these conditions.

Adult Stem Cells Versus Embryonic Stem Cells

Stem cells are cells that continuously divide to produce new cells that go on to become specialized cells. The bone marrow of adults and the umbilical cord of infants (Fig. 3A) contain stem cells for each type of blood cell in the body. It is relatively easy to retrieve blood stem cells from either of these sources. Researchers report that they have injected blood stem cells into the heart and liver only to find that they became cardiac cells and liver cells, respectively! The skin, gastrointestinal lining, and the brain also have stem cells, but the technology to retrieve them has not been perfected. Also, it has not been possible to change adult stem cells into a fully developed specific type of cell outside the body. If the technique is perfected, it might be possible to, say, change a blood stem cell into the dopamine-secreting cell needed by a Parkinson patient. But the fact the brain is not now doing it—or else the patient wouldn't have Parkinson disease—makes us less hopeful.

Today, young, relatively infertile couples seek assistance in achieving pregnancy and having children. During in vitro fertilization, several eggs and sperm are placed in laboratory glass, where fertilization occurs and development begins. A physician places two or three embryos in the woman's uterus for further development, but may hold back some in case these fail to take hold. Embryos that are never used remain frozen indefinitely unless they are made available to researchers. Each cell of an embryo is called an embryonic stem cell because it can become any kind of specialized cell in the body. Researchers have already used nonhuman embryonic cells to create supplies of nonhuman specialized cells. Therefore, they think the same will hold true with human embryonic stem cells. If so, medicine would undergo an advancement of enormous proportions.

The Ethical Dilemma

Even so, there is a bioethical dilemma. What about the embryos that have been forced to give up a chance of becoming an adult in order to extend the health span of those already living? Would this be ethical? In Great Britain, researchers can work with embryos that are 14 days or younger because embryos usually implant in the uterus around day 14. Robert George of Princeton University doesn't agree with this solution to the problem. He says, "I believe that all human beings are equal, and ought not to be harmed or considered to be less than human on the basis of age or size or stage of development or condition of dependency." He believes that embryos should not be used as a means to an end, even good ends, such as a cure for

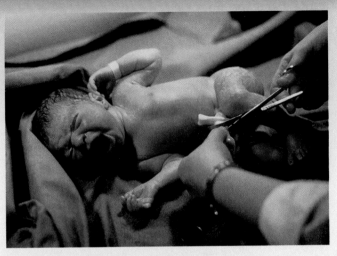

Figure 3A **Umbilical cords are valuable.**
The blood from a baby's umbilical cord can be banked, and then used as a source of blood stem cells. Investigators are hopeful they will be able to convert blood stem cells to various types of mature cells some day.

diseases or to save another human life. President George W. Bush agrees and signed an executive order that forbids the use of federal funds for the purpose of creating new cell lines derived from embryos in the United States. The order does not affect any embryonic stem cell lines previously established nor any work with adult stem cells. Nevertheless, some researchers have left the United States to work in countries where stem cell research is freely allowed without governmental restrictions.

Michael Sandel of Harvard University offers a way out of the bioethical dilemma. He says that to think in dualistic terms is not helpful—it isn't that an embryo is a human being or is not a human being—it's that a fully developed human being comes about gradually. He offers this situation to illustrate his point. What would you do if there was a fire in a fertility clinic and you were faced with the choice of saving a five year-old girl or a tray of ten embryos? Which would you choose? He still believes that "life is a gift that commands our reverence and restricts our use." While he believes that stem cell research is ethical, he does not believe that humans should be cloned.

Decide Your Opinion

1. Should researchers have access to embryonic stem cells, or only adult stem cells? What is your reasoning?
2. Do you believe, as Sandel does, that while it is ethical to do research with embryonic stem cells to cure human ills, it is not ethical to clone humans? What is your reasoning?
3. Some researchers are mixing nonhuman with human embryonic stem cells in order to study developmental differences. Is this ethical? Why or why not?

Summarizing the Concepts

3.1 What Is a Cell?
- Cells, the basic units of life, come from preexisting cells.
- Microscopes are used to view cells, which must remain small in order to have a more favorable surface-area-to-volume ratio.

3.2 How Cells Are Organized
The human cell is surrounded by a plasma membrane and has a central nucleus. Between the plasma membrane and the nucleus is the cytoplasm, which contains various organelles. Organelles in the cytoplasm have specific functions.

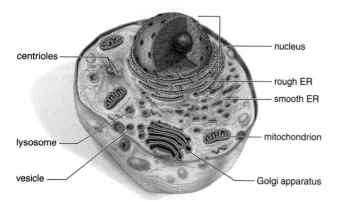

3.3 The Plasma Membrane and How Substances Cross It
The plasma membrane is a phospholipid bilayer that
- selectively regulates the passage of molecules and ions into and out of the cell.
- contains embedded proteins, which allow certain substances to cross the plasma membrane.

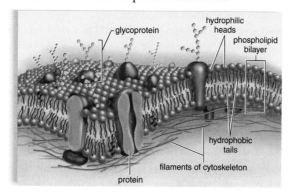

Passage of molecules into or out of cells can be passive or active.

Passive mechanisms (no energy required) are
- diffusion (osmosis) and facilitated transport.

Active mechanisms (energy required) are
- active transport and endocytosis and exocytosis.

3.4 The Nucleus and the Production of Proteins
- The nucleus houses DNA, which specifies the order of amino acids in proteins.

- In the nucleus, chromatin condenses to become chromosomes during cell division.
- The nucleolus produces ribosomal RNA (rRNA).
- Protein synthesis occurs in ribosomes, which are small organelles composed of proteins and rRNA.

The Endomembrane System
The endomembrane system consists of the nuclear envelope, endoplasmic reticulum (ER), Golgi apparatus, lysosomes, and vesicles.
- Rough ER has ribosomes, where protein synthesis occurs.
- Smooth ER has no ribosomes and has various functions, including lipid synthesis.
- The Golgi apparatus processes and packages proteins and lipids into vesicles for secretion or movement into other parts of the cell.
- Lysosomes are specialized vesicles produced by the Golgi apparatus. They fuse with incoming vesicles to digest enclosed material, and they autodigest old cell parts.

3.5 The Cytoskeleton and Cell Movement
The cytoskeleton consists of microtubules, actin filaments, and intermediate filaments that give cells their shape and allows organelles to move about the cell. Cilia and flagella, which contain microtubules, allow a cell to move.

3.6 Mitochondria and Cellular Metabolism
- Mitochondria have an inner membrane that forms cristae, which project into the matrix.
- Mitochondria are involved in cellular respiration, which uses oxygen and releases carbon dioxide.
- During cellular respiration, mitochondria convert the energy of glucose into the energy of ATP molecules.

Cellular Respiration and Metabolism
- A metabolic pathway is a series of reactions, each of which has its own enzyme.
- Enzymes speed reactions by forming an enzyme-substrate complex.
- Sometimes enzymes require coenzymes (such as NAD^+), nonprotein molecules that participate in the reaction.
- Cellular respiration is the enzymatic breakdown of glucose to carbon dioxide and water.
- Cellular respiration includes three pathways: glycolysis (occurs in the cytoplasm and is anaerobic), the citric acid cycle (releases carbon dioxide), and the electron transport chain (passes electrons to oxygen).

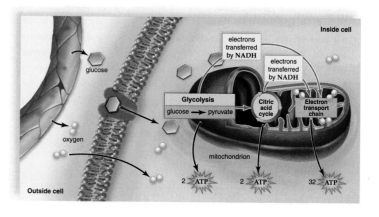

Fermentation
- If oxygen is not available in cells, the electron transport chain is inoperative, and fermentation (which does not require oxygen) occurs.
- Fermentation produces very little ATP.
- Lactate buildup puts the individual in oxygen deficit.

Understanding Key Terms

actin filament 51
active site 53
active transport 48
aerobic 55
anaerobic 54
basal body 51
cell theory 42
cellular respiration 52
centriole 51
centrosome 51
chromatin 49
chromosome 49
cilium 51
citric acid cycle 54
coenzyme 53
cytoplasm 44
cytoskeleton 51
dehydrogenase 53
diffusion 47
electron transport chain 55
endomembrane system 44, 50
endoplasmic reticulum (ER) 49
eukaryotic cell 44
facilitated transport 47
fermentation 56
flagellum 51
fluid-mosaic model 46

glycolysis 54
Golgi apparatus 50
intermediate filament 51
lysosome 50
metabolism 53
microtubule 51
mitochondrion 52
NAD (nicotinamide adenine dinucleotide) 53
nuclear envelope 49
nuclear pore 49
nucleolus 49
nucleoplasm 49
nucleus 44, 49
organelle 44
osmosis 47
osmotic pressure 47
phagocytosis 48
plasma membrane 44
polyribosome 49
product 53
prokaryotic cell 44
reactant 53
ribosome 49
selectively permeable 46
substrate 53
tonicity 47
vesicle 50

Match the key terms to these definitions.

a. _____ Protein molecules form a shifting pattern within the fluid phospholipid bilayer.

b. _____ Diffusion of water through a selectively permeable membrane.

c. _____ The cell will allow some substances to pass through while not permitting others.

d. _____ Anaerobic breakdown of glucose that results in a gain of two ATP and end products, such as alcohol and lactate.

e. _____ Metabolic pathways that use energy from carbohydrate, fatty acid, and protein break down to produce ATP molecules.

Testing Your Knowledge of the Concepts

Choose the best answer for each question.

1. Explain the three tenets of the cell theory. (page 42)

2. Which type of microscope would you use to observe the swimming behavior of a flagellated protozoan? Explain. (page 43)

3. Describe how the eukaryotic cell gained mitochondria and chloroplasts. (page 44)

4. Invagination of plasma membrane produced what structures in eukaryotic cells that are not present in prokaryotic cells? (page 44)

5. What are glycoproteins, and what functions do proteins, including glycoproteins, have in the plasma membrane? (page 46)

6. A plant wilts if its roots are placed in which type solution (i.e., a hypertonic, a hypotonic, or an isotonic solution)? Explain. (page 47)

7. What is endocytosis and exocytosis, and how do they occur? (page 48)

8. For the following cell organelles, describe the structure and function of each: nucleus, nucleolus, ribosomes, endoplasmic reticulum (rough and smooth), Golgi apparatus, lysosomes, centrioles, and mitochondria. (pages 49–52)

9. Describe the structure and function of the cytoskeleton. (page 51)

10. Describe an enzyme and coenzyme. Explain the mechanism of enzyme function, particularly the relationship of shape to its activity. (page 53)

11. Which stage of cellular respiration produces the most ATP? Explain. (pages 54–55)

12. Running for the bus may produce an oxygen deficit. Explain. (page 56)

13. The cell theory states
   a. cells form as organelles, and molecules become grouped together in an organized manner.
   b. the normal functioning of an organism does not depend on its individual cells.
   c. the cell is the basic unit of life for all living things.
   d. only animals are made of cells.

14. The small size of cells is best correlated with
   a. the fact that they are self-reproducing.
   b. an adequate surface area for exchange of materials.
   c. their vast versatility.
   d. All of these are correct.

15. A phospholipid has a head and two tails. The tails are found
   a. at the surfaces of the membrane.
   b. in the interior of the membrane.
   c. spanning the membrane.
   d. where the environment is hydrophobic.
   e. Both b and d are correct.

16. Facilitated diffusion differs from diffusion in that facilitated diffusion
   a. involves the passive use of a carrier protein.
   b. involves the active use of a carrier protein.
   c. moves a molecule from a low to high concentration.
   d. involves the use of ATP molecules.

17. When a cell is placed in a hypotonic solution,
    a. solute exits the cell to equalize the concentration on both sides of the membrane.
    b. water exits the cell toward the area of lower solute concentration.
    c. water enters the cell toward the area of higher solute concentration.
    d. solute exits and water enters the cell.

In questions 18–21, match each function to the proper organelle in the key.

Key:
    a. mitochondrion   c. Golgi apparatus
    b. nucleus   d. rough ER

18. Packaging and secretion

19. Powerhouse of cell

20. Protein synthesis

21. Control center for the cell

22. Vesicles carrying proteins for secretion move from the ER to the
    a. smooth ER   c. Golgi apparatus.
    b. lysosomes.   d. nucleolus.

23. Lysosomes function in
    a. protein synthesis.
    b. processing and packaging.
    c. intracellular digestion.
    d. lipid synthesis.

24. Mitochondria
    a. are involved in cellular respiration.
    b. break down ATP to release energy for cells.
    c. contain hemoglobin and cristae.
    d. have a convoluted outer membrane.
    e. All of these are correct.

25. The active site of an enzyme
    a. is identical to that of any other enzyme.
    b. is the part of the enzyme where the substrate can fit.
    c. can be used over and over again.
    d. is not affected by environmental factors such as pH and temperature.
    e. Both b and c are correct.

26. The metabolic process that produces the most ATP molecules is
    a. glycolysis.
    b. the citric acid cycle.
    c. the electron transport chain.
    d. fermentation.

27. The oxygen required by cellular respiration becomes part of which molecule?
    a. ATP.
    b. H_2O
    c. pyruvate
    d. CO_2

28. Use these terms to label the following diagram of the plasma membrane: carbohydrate chain, filaments of the cytoskeleton, hydrophilic heads, hydrophobic tails, protein (used twice), phospholipid bilayer.

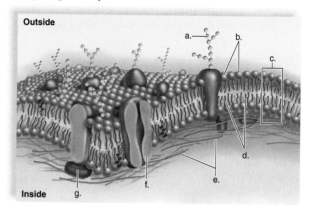

29. Use these terms to label the following diagram: substrates, enzyme (used twice), active site, product, and enzyme-substrate complex.

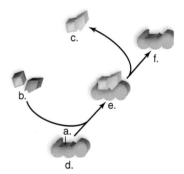

Thinking Critically About the Concepts

Remember Ashley's response to the scary movie in the opening story for Chapter 1? She was scared by the movie, so her sympathetic nervous system responded by initiating responses to save her life. This same series of responses would occur in Calvin if his mom was really angered by his cookie-snatching caper and she ran after him. The increase in heart rate experienced by Ashley would also happen to Calvin.

The sympathetic nervous system can also cause an increase in the breathing rate. With an increased heart and breathing rate, more oxygen would be delivered to Calvin's skeletal muscle cells. Then the mitochondria in his skeletal muscles could make more ATP, enabling him to escape the wrath of his mom.

1a. When you breathe, oxygen enters the lungs. The higher concentration of oxygen is in the lungs. By what process will oxygen cross over from the lungs into blood vessels?

 b. When glucose is delivered to active muscle cells by blood vessels, by what process will glucose enter cells so they can continue to synthesize ATP?

2a. Based on your understanding of how structure suits function, why do the testosterone-producing cells of the testes have an abundance of smooth endoplasmic reticulum?

 b. What organelles are characteristic of a pituitary gland cell that is making growth hormone? (Review Chapter 15 to learn more about the chemical nature of growth hormone.)

Organization and Regulation of Body Systems

Bladder cancer. The words sounded so ominous when Barbara heard her doctor say them. Given her long-term history of smoking, she would not have been surprised by a diagnosis of lung cancer, but bladder cancer? It's highly correlated with smoking, her doctor said.

The next thing Barbara knew, she was scheduled for surgery to have her bladder removed. Fortunately, her team of surgeons planned to create a substitute "bladder" from a section of her large intestine. Having a new internal "bladder" meant that wearing an external bag for collecting urine would be unnecessary.

In Chapter 3, we learned that cells are specialized to carry out specific functions; now it is time to consider that tissues and organs are specialized, too. The specific tissues of an organ, such as the bladder, enables it to perform its intended function.

Organs are a part of organ systems, all of which contribute to the maintenance of homeostasis. The bladder is part of the urinary system that works to maintain the water-salt balance and the acid-base balance of the blood, among other functions. When the urinary system doesn't function properly, imbalances will occur, and the results could be devastating or life threatening.

In this chapter, you will learn about the characteristics of different tissues and where they are found in the body. You'll also read about the various organ systems and what they do to maintain homeostasis.

CHAPTER CONCEPTS

4.1 Types of Tissues
The body contains four types of tissues: connective tissue, muscular tissue, nervous tissue, and epithelial tissue.

4.2 Connective Tissue Connects and Supports
Connective tissues, which join together other tissues, can be classified into three types: fibrous connective tissue, supportive connective tissue, and fluid connective tissue.

4.3 Muscular Tissue Moves the Body
Skeletal muscle is attached to the skeleton; smooth muscle is in the walls of internal organs; and cardiac muscle is in the wall of the heart.

4.4 Nervous Tissue Communicates
Nervous tissue, found in the brain, spinal cord, and nerves, contains neurons and support cells (neuroglia).

4.5 Epithelial Tissue Protects
Epithelial tissues, which line cavities and cover surfaces, are named according to the shape of the cell and can occur in a single layer or multiple layers.

4.6 Cell Junctions
Cell junctions, consisting of tight junctions, adhesion junctions, and gap junctions, help tissues perform their various functions.

4.7 Integumentary System
The skin and its accessory organs have many functions important to homeostasis.

4.8 Organ Systems
Each of the organ systems have functions that contribute to homeostasis, the relative constancy of the internal environment.

4.9 Homeostasis
Homeostasis is maintained by various physiological mechanisms.

4.1 Types of Tissues

Recall the biological levels of organization (Fig. 1.2). Cells are composed of molecules; a tissue has like cells; an organ contains several types of tissues; and several organs are found in an organ system. In this chapter, we consider the tissue, organ, and organ system levels of organization.

A **tissue** is composed of specialized cells of the same type that perform a common function in the body. The tissues of the human body can be categorized into four major types:

Connective tissue binds and supports body parts.
Muscular tissue moves the body and its parts.
Nervous tissue receives stimuli and conducts nerve
    impulses.
Epithelial tissue covers body surfaces and lines body
    cavities.

Cancers are classified according to the type of tissue from which they arise. Sarcomas are cancers arising in muscle or connective tissue (especially bone or cartilage); leukemias are cancers of the blood; lymphomas are cancers of lymphoid tissue, and carcinomas, the most common type, are cancers of epithelial tissue. The chance of developing cancer in a particular tissue shows a positive correlation to the rate of cell division; epithelial cells reproduce at a high rate, and 2,500,000 new blood cells appear each second. Thus, carcinomas and leukemias are common types of cancers.

☑ Check Your Progress 4.1

1. What are the four major tissue types found in the human body?

4.2 Connective Tissue Connects and Supports

Connective tissue is quite diverse in structure and function but, even so, all types have three components: spe-cialized cells, ground substance, and protein fibers. These components are shown in Figure 4.1, a diagrammatic representation of loose fibrous connective tissue. The ground substance is a noncellular material that separates the cells and varies in consistency from solid to semifluid to fluid.

The fibers are of three possible types. White **collagen fibers** contain collagen, a protein that gives them flexibility and strength. **Reticular fibers** are very thin collagen fibers that are highly branched and form delicate supporting networks. Yellow **elastic fibers** contain elastin, a protein that is not as strong as collagen but is more elastic. Inherited connective tissue disorders arise when people inherit genes that lead to malformed fibers. The result can be skin that is too loose, blood vessels that are too fragile, bones that are too brittle, and joints that are too loose, for example. Overly long or short bones can affect height.

Fibrous Connective Tissue

Both loose fibrous and dense fibrous connective tissues have cells called **fibroblasts** that are located some distance from one another and are separated by a jellylike ground substance containing white collagen fibers and yellow elastic fibers (Fig. 4.2). **Matrix** is a term that includes ground substance and fibers.

Loose fibrous connective tissue, also called areolar tissue, supports epithelium and many internal organs. Its presence in lungs, arteries, and the urinary bladder allows these organs to expand. It forms a protective covering enclosing many internal organs, such as muscles, blood vessels, and nerves.

Adipose tissue is a special type of loose connective tissue in which the cells enlarge and store fat. The body uses this stored fat for energy, insulation, and organ protection. Adipose tissue is found beneath the skin, around the kidneys, and on the surface of the heart.

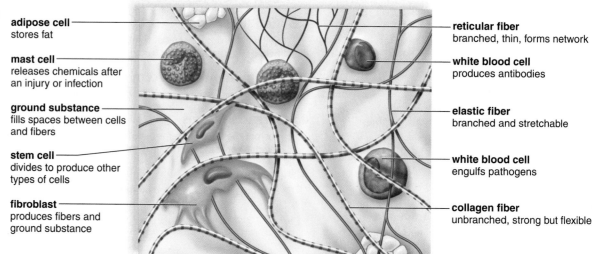

Figure 4.1 Diagram of loose fibrous connective tissue.

adipose cell
stores fat

mast cell
releases chemicals after
an injury or infection

ground substance
fills spaces between cells
and fibers

stem cell
divides to produce other
types of cells

fibroblast
produces fibers and
ground substance

reticular fiber
branched, thin, forms network

white blood cell
produces antibodies

elastic fiber
branched and stretchable

white blood cell
engulfs pathogens

collagen fiber
unbranched, strong but flexible

Dense fibrous connective tissue contains many collagen fibers that are packed together. This type of tissue has more specific functions than does loose connective tissue. For example, dense fibrous connective tissue is found in **tendons,** which connect muscles to bones, and in **ligaments,** which connect bones to other bones at joints.

Supportive Connective Tissue

In **cartilage,** the cells lie in small chambers called lacunae (sing., **lacuna**), separated by a matrix that is solid yet flexible. Unfortunately, because this tissue lacks a direct blood supply, it heals very slowly. There are three types of cartilage, distinguished by the type of fiber.

Hyaline cartilage (Fig. 4.2), the most common type of cartilage, contains only very fine collagen fibers. The matrix has a glassy, translucent appearance. Hyaline cartilage is found in the nose and at the ends of the long bones and the ribs, and it forms rings in the walls of respiratory passages. The fetal skeleton also is made of this type of cartilage. Later, the cartilaginous fetal skeleton is replaced by bone.

Elastic cartilage has more elastic fibers than hyaline cartilage. For this reason, it is more flexible and is found, for example, in the framework of the outer ear.

Fibrocartilage has a matrix containing strong collagen fibers. Fibrocartilage is found in structures that withstand tension and pressure, such as the disks between the vertebrae in the backbone and the wedges in the knee joint.

Bone

Bone is the most rigid connective tissue. It consists of an extremely hard matrix of inorganic salts, notably calcium salts, deposited around protein fibers, especially collagen fibers. The inorganic salts give bone rigidity, and the protein fibers provide elasticity and strength, much as steel rods do in reinforced concrete.

Compact bone makes up the shaft of a long bone (Fig. 4.2). It consists of cylindrical structural units called osteons (Haversian systems). The central canal of each osteon is surrounded by rings of hard matrix. Bone cells are located in spaces called lacunae between the rings of matrix. In the central canal, nerve fibers carry nerve impulses and blood vessels carry nutrients that allow bone to renew itself. Thin extensions of bone cells within canaliculi (minute canals) connect the cells to each other and to the central canal.

The ends of a long bone contain spongy bone, which has an entirely different structure. **Spongy bone** appears as an open, bony latticework with numerous bony bars and plates, separated by irregular spaces. Although lighter than compact bone, spongy bone is still designed for strength. Just as braces are used for support in buildings, the solid portions of spongy bone follow lines of stress.

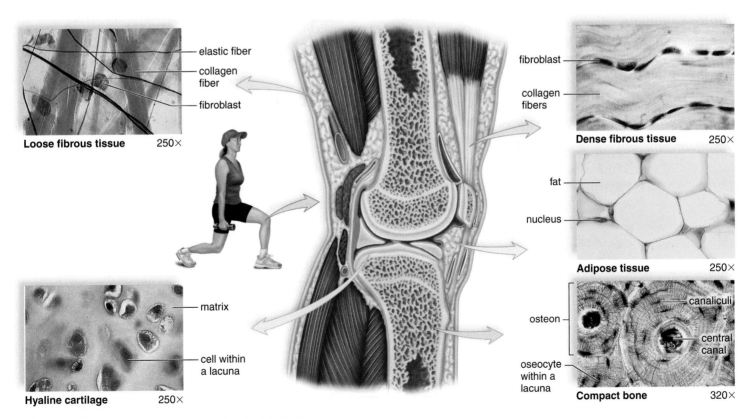

Figure 4.2   Connective tissues associated with the knee.
The human knee provides examples of most types of connective tissue.

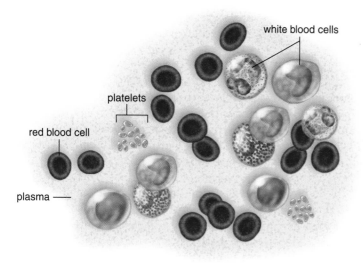

Figure 4.3   Formed elements in blood.
Red blood cells, which lack a nucleus, transport oxygen. Each type of white blood cell has a particular way to fight infections. Platelets, which are fragments of a particular cell, function in helping to seal injured blood vessels.

Fluid Connective Tissues

The body has two fluid connective tissues: blood and lymph.

Blood

Some people do not classify blood as connective tissue; instead, they suggest a separate tissue category called vascular tissue. **Blood,** which consists of formed elements (Fig. 4.3) and plasma, is located in blood vessels. Blood transports nutrients and oxygen to **tissue fluid,** a fluid that bathes the body's cells, and removes carbon dioxide and other wastes. Blood helps distribute heat and also plays a role in fluid, ion, and pH balance. The systems of the body help keep blood composition and chemistry within normal limits.

The formed elements each have specific functions. The **red blood cells (erythrocytes)** are small, biconcave, disk-shaped cells without nuclei. The presence of the red pigment hemoglobin makes the cells red and, in turn, makes the blood red. Hemoglobin is composed of four units; each unit is composed of the protein globin and a complex iron-containing structure called heme. The iron forms a loose association with oxygen and, in this way, red blood cells transport oxygen.

White blood cells (leukocytes) may be distinguished from red blood cells by the fact that they are usually larger, have a nucleus, and without staining would appear translucent. White blood cells characteristically vary from mostly bluish to pink because they have been stained. White blood cells fight infection, primarily in two ways. Some white blood cells are phagocytic and engulf infectious **pathogens.** Other white blood cells either produce antibodies, molecules that combine with foreign substances to inactivate them, or they kill cells outright.

Platelets (thrombocytes) are not complete cells; rather, they are fragments of giant cells present only in bone marrow. When a blood vessel is damaged, platelets form a plug that seals the vessel, and injured tissues release molecules that help the clotting process.

Lymph

Lymph is also a fluid connective tissue. Lymph is a clear, watery, sometimes faintly yellowish fluid derived from tissue fluid that contains white blood cells. Lymphatic vessels absorb excess tissue fluid and various dissolved solutes in the tissues and transport lymph to particular vessels of the cardiovascular system. Lymphatic vessels absorb fat molecules from the small intestine. Lymph nodes, composed of fibrous connective tissue, occur along the length of lymphatic vessels. Lymph is cleansed as it passes through lymph nodes, in particular, because white blood cells congregate there. Lymph nodes enlarge when you have an infection.

☑ Check Your Progress 4.2

1. What are the three types of connective tissue, and what are some examples of each type (Fig. 4.4)?

2. Contrast loose fibrous connective tissue with dense fibrous connective tissue.

3. Contrast spongy bone with compact bone.

4. Contrast blood with lymph.

Types of Connective Tissue

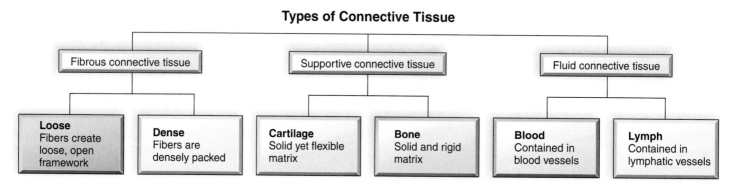

Figure 4.4   Types of connective tissue.

4.3 Muscular Tissue Moves the Body

Muscular (contractile) tissue is composed of cells called muscle fibers. Muscle fibers contain protein filaments, called actin and myosin filaments, whose interaction accounts for movement. The three types of vertebrate muscular tissue are skeletal, smooth, and cardiac.

Skeletal muscle, also called voluntary muscle (Fig. 4.5a), is attached by tendons to the bones of the skeleton, and when it contracts, body parts move. Contraction of skeletal muscle is under voluntary control and occurs faster than in the other muscle types. Skeletal muscle fibers are cylindrical and quite long—sometimes they run the length of the muscle. They arise during development when several cells fuse, resulting in one fiber with multiple nuclei. The nuclei are located at the periphery of the cell, just inside the plasma membrane. The fibers have alternating light and dark bands that give them a **striated,** or striped, appearance. These bands are due to the placement of actin filaments and myosin filaments in the cell.

Smooth (visceral) muscle is so named because the cells lack striations. The spindle-shaped cells, each with a single nucleus, form layers in which the thick middle portion of one cell is opposite the thin ends of adjacent cells. Consequently, the nuclei form an irregular pattern in the tissue (Fig. 4.5b). Smooth muscle is not under voluntary control, and therefore is said to be involuntary. Smooth muscle, found in the walls of viscera (intestine, bladder, and other internal organs) and blood vessels, contracts more slowly than skeletal muscle but can remain contracted for a longer time. When the smooth muscle of the bladder contracts, urine is sent into a tube called the urethra, which takes it to the outside. When the smooth muscle of the blood vessels contracts, blood vessels constrict, helping to raise blood pressure.

Cardiac muscle (Fig. 4.5c) is found only in the walls of the heart. Its contraction pumps blood and accounts for the heartbeat. Cardiac muscle combines features of both smooth and skeletal muscle. Like skeletal muscle, it has striations, but the contraction of the heart is involuntary for the most part. Cardiac muscle cells also differ from skeletal muscle cells in that they usually have a single, centrally placed nucleus. The cells are branched and seemingly fused one with another, and the heart appears to be composed of one large interconnecting mass of muscle cells. Actually, cardiac muscle cells are separate and individual, but they are bound end to end at **intercalated disks,** areas where folded plasma membranes between two cells contain adhesion junctions and gap junctions (see page 70).

☑ Check Your Progress 4.3

1. How does the structure and function of skeletal, smooth, and cardiac muscle differ?

2. Where do you find these muscles in the body?

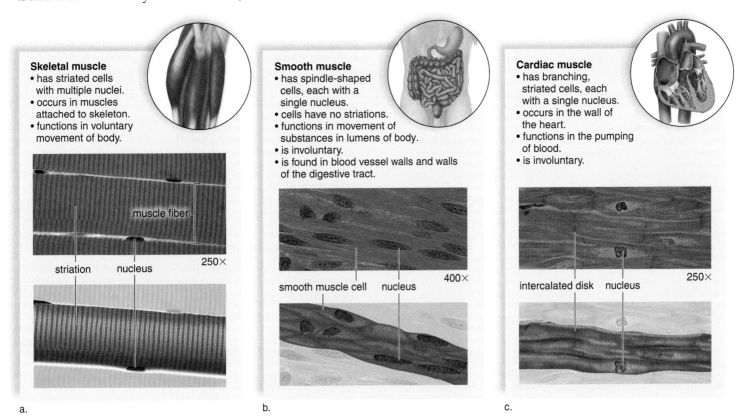

Skeletal muscle
- has striated cells with multiple nuclei.
- occurs in muscles attached to skeleton.
- functions in voluntary movement of body.

muscle fiber

striation   nucleus   250×

Smooth muscle
- has spindle-shaped cells, each with a single nucleus.
- cells have no striations.
- functions in movement of substances in lumens of body.
- is involuntary.
- is found in blood vessel walls and walls of the digestive tract.

smooth muscle cell   nucleus   400×

Cardiac muscle
- has branching, striated cells, each with a single nucleus.
- occurs in the wall of the heart.
- functions in the pumping of blood.
- is involuntary.

intercalated disk   nucleus   250×

a.   b.   c.

Figure 4.5 Muscular tissue.
a. Skeletal muscle is voluntary and striated. **b.** Smooth muscle is involuntary and nonstriated. **c.** Cardiac muscle is involuntary and striated. Cardiac muscle cells branch and fit together at intercalated disks.

66

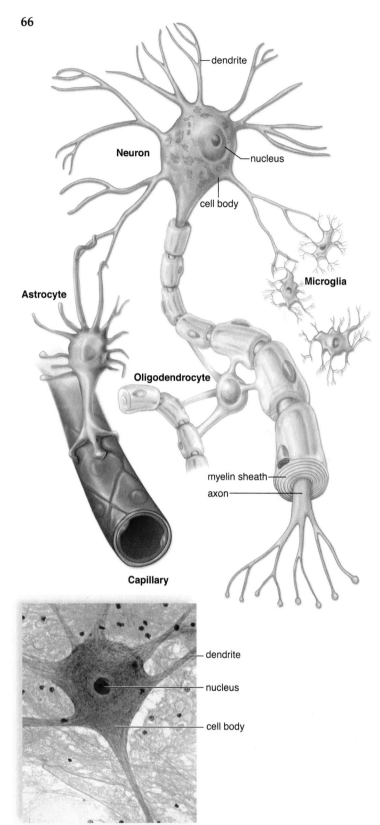

dendrite

Neuron

nucleus

cell body

Microglia

Astrocyte

Oligodendrocyte

myelin sheath

axon

Capillary

dendrite

nucleus

cell body

Micrograph of neuron

Figure 4.6 **Neurons and neuroglia in the brain.**
Neurons conduct nerve impulses. Neuroglia consists of cells that support and
service neurons and have various functions: Microglia are a type of neuroglia that
become mobile in response to inflammation and phagocytize debris. Astrocytes
lie between neurons and a capillary; therefore, substances entering neurons from
the blood must first pass through astrocytes. Oligodendrocytes form the myelin
sheaths around fibers in the brain and spinal cord.

4.4 Nervous Tissue Communicates

Nervous tissue consists of nerve cells, called neurons, and
neuroglia, the cells that support and nourish the neurons.

Neurons

A **neuron** is a specialized cell that has three parts: dendrites, a
cell body, and an axon (Fig. 4.6). A dendrite is an extension
that receives signals from sensory receptors or other neurons.
The cell body contains most of the cell's cytoplasm and the
nucleus. An axon is an extension that conducts nerve impulses.
Long axons are covered by myelin, a white fatty substance.
The term *fiber*[1] is used here to refer to an axon along with its
myelin sheath, if it has one. Outside the brain and spinal cord,
fibers bound by connective tissue form **nerves.**

The nervous system has just three functions: sensory
input, integration of data, and motor output. Nerves conduct
impulses from sensory receptors to the spinal cord and the
brain, where integration occurs. The phenomenon called sensation occurs only in the brain, however. Nerves also conduct
nerve impulses away from the spinal cord and brain to the
muscles and glands, causing them to contract and secrete,
respectively. In this way, a coordinated response to the stimulus is achieved.

Neuroglia

In addition to neurons, nervous tissue contains neuroglia.
Neuroglia are cells that outnumber neurons nine to one and
take up more than half the volume of the brain. Although the
primary function of neuroglia is to support and nourish neurons, research is currently being conducted to determine how
much they directly contribute to brain function. Types of neuroglia found in the brain are, for example, microglia, astrocytes, and oligodendrocytes (Fig. 4.6, *top*). Microglia, in
addition to supporting neurons, engulf bacterial and cellular
debris. Astrocytes provide nutrients to neurons and produce
a hormone known as glia-derived growth factor, which some-
day might be used as a cure for Parkinson disease and other
diseases caused by neuron degeneration. Oligodendrocytes
form the myelin sheaths around fibers in the brain and spinal
cord. Outside the brain, Schwann cells are the type of neuroglia that encircle long nerve fibers and form a myelin sheath.
Neuroglia do not have a long extension, but even so, researchers are now beginning to gather evidence that they do communicate among themselves and with neurons.

☑ Check Your Progress 4.4

1. a. What are the three parts of a neuron, and (b) what does each
part do?

2. What are some specific functions of neuroglia?

[1]In connective tissue, a fiber is a component of the matrix; in muscle tissue, a fiber is a
muscle cell; in nervous tissue, a fiber is an axon.

Science Focus

Nerve Regeneration

In humans, axons outside the brain and spinal cord can regenerate, but not those inside these organs (Fig. 4A). After injury, axons in the human central nervous system (CNS) degenerate, resulting in permanent loss of nervous function. Not so in cold-water fishes and amphibians, where axon regeneration in the CNS does occur. So far, investigators have identified several proteins that seem to be necessary to axon regeneration in the CNS of these animals, but even so, it may be a long time before biochemistry can offer a way to bring about axon regeneration in the human CNS. Still, it's possible that these proteins might one day be used as drugs or, following gene therapy, humans might produce these same proteins when CNS injuries occur.

In the meantime, some accident victims are trying other ways to bring about a cure. In 1995, Christopher Reeve, best known for his acting role as "Superman," was thrown head-first from his horse, crushing the spinal cord just below the neck's top two vertebrae (Fig. 4B*a*). Immediately, his brain lost almost all communication with the portion of his body below the site of damage, and he could not move his arms and legs. Many years later, Reeve could move his left index finger slightly and could take tiny steps while being held upright in a pool. He had sensation throughout his body and could feel his wife's touch.

Reeve's improvement was not the result of cutting-edge drugs or gene therapy—it was due to exercise (Fig. 4B*b*)! Reeve exercised as much as five hours a day, especially using a recumbent bike outfitted with electrodes that made his leg muscles contract and relax. The bike cost him $16,000. It could cost less if commonly used by spinal cord injury patients in their own homes. Reeve, who was an activist for the disabled, was pleased that insurance would pay for the bike about 50% of the time.

It's possible that Reeve's advances were the result of improved strength and bone density, which lead to stronger nerve signals. Normally, nerve cells are constantly signaling one another, but after a spinal cord injury, the signals cease. Perhaps Reeve's intensive exercise brought back some of the normal communication between nerves. Reeve's physician,

Figure 4B   Treatment today for spinal cord injuries.
a. Christopher Reeve suffered a spinal cord injury when horseback riding in 1995.
b. He exercised many hours a day, including aqua therapy. Reeve died in 2004.

a.

b.

John McDonald, a neurologist at Washington University in St. Louis, is convinced that his axons were regenerating. Fred Gage, a neuroscientist at the Salk Institute in La Jolla, California, has shown that exercise does enhance the growth of new cells in adult brains.

For himself, Reeve was convinced that stem cell therapy would one day allow him to be off his ventilator and functioning normally; however, Reeve died in 2004. The Bioethical Focus on page 57 discusses the promise of stem cell research.

Figure 4A   Nerve regeneration. Outside the CNS, nerves regenerate because new neuroglia called Schwann cells form a pathway for axons to reach a muscle. In the CNS, comparable neuroglia called oligodencrocytes do not do this.

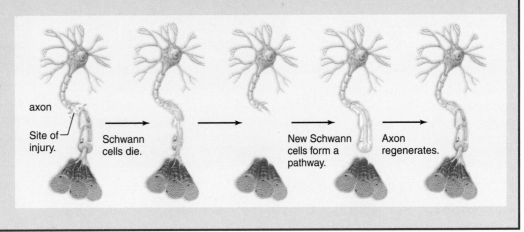

axon

Site of injury.

Schwann cells die.

New Schwann cells form a pathway.

Axon regenerates.

4.5 Epithelial Tissue Protects

Epithelial tissue, also called epithelium (pl., epithelia), consists of tightly packed cells that form a continuous layer. Epithelial tissue covers surfaces and lines body cavities. Usually, it has a protective function, but it can also be modified to carry out secretion, absorption, excretion, and filtration.

Epithelial cells are exposed to the environment on one side, and on the other side, they are bounded by a **basement membrane.** The basement membrane should not be confused with the plasma membrane, or the body membranes we will be discussing. It is simply a thin layer of various types of carbohydrates and proteins that anchors the epithelium to underlying connective tissue.

Simple Epithelia

Epithelial tissue is either simple or stratified. Simple epithelia have only a single layer of cells (Fig. 4.7) and are classified according to cell type. **Squamous epithelium,** which is composed of flattened cells, is found lining the air sacs of lungs and walls of blood vessels. Its shape and arrangement permit exchanges of substances in these locations. Oxygen and carbon dioxide exchange occurs in the lungs, and nutrient-for-waste exchange occurs across blood vessels in the tissues.

Cuboidal epithelium consists of a single layer of cube-shaped cells. This type of epithelium is frequently found in glands, such as the salivary glands, the thyroid gland, and the pancreas. Simple cuboidal epithelium also covers the ovaries and lines kidney tubules, the portion of the kidney in which urine is formed. When cuboidal cells are involved in absorption, they have microvilli (minute cellular extensions of the plasma membrane), which increase the surface area of the cells. And when cuboidal cells function in active transport, they contain many mitochondria.

Columnar epithelium has cells resembling rectangular pillars or columns, with nuclei usually located near the bottom of each cell. This epithelium is found lining the digestive tract, where microvilli expand the surface area and aid in absorbing the products of digestion. Ciliated columnar epithelium is found lining the oviducts, where it propels the egg toward the uterus, or womb.

Pseudostratified Columnar Epithelium

Pseudostratified columnar epithelium is so named because it appears to be layered (*pseudo,* false; *stratified,* layers). However, true layers do not exist because each cell touches the basement membrane. In particular, the irregular placement of the nuclei creates the appearance of several layers, where only one exists. The lining of the windpipe, or trachea, is pseudostratified ciliated columnar epithelium. A secreted covering of mucus traps foreign particles, and the upward motion of the cilia carries the mucus to the back of the throat, where it may either be swallowed or expectorated. Smoking can cause a change in mucous secretion and inhibit ciliary action, resulting in a chronic inflammatory condition called bronchitis.

Figure 4.7   Types of epithelia.
Basic epithelial tissues found in humans are shown, along with locations of the tissue and the primary function of the tissue at these locations.

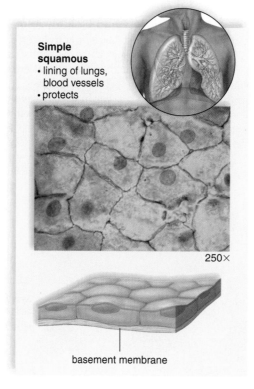

Simple squamous
• lining of lungs, blood vessels
• protects

250×

basement membrane

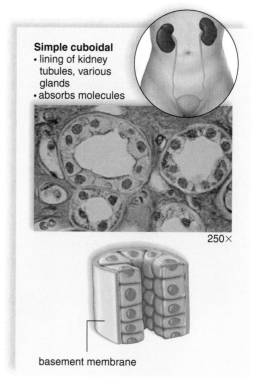

Simple cuboidal
• lining of kidney tubules, various glands
• absorbs molecules

250×

basement membrane

Transitional Epithelium

The term transitional epithelium implies changeability, and this tissue changes in response to tension. It forms the lining of the urinary bladder, the ureters (tubes that carry urine from the kidneys to the bladder), and part of the urethra (the single tube that carries urine to the outside). All are organs that may need to stretch. When the bladder is distended, this epithelium stretches, and the outer cells take on a squamous appearance.

Transitional epithelium is enough like columnar epithelium, and vice versa, that surgeons were able to reconstruct a bladder for Barbara, of the opening story, from a piece of intestine sewn to the urethra. This allowed Barbara to urinate normally.

Stratified Epithelia

Stratified epithelia have layers of cells piled one on top of the other. Only the bottom layer touches the basement membrane. The nose, mouth, esophagus, anal canal, the outer portion of the cervix (adjacent to the vagina), and vagina are all lined with stratified squamous epithelium. Cancer of the cervix is detectable by doing a *pap smear*. Cells lining the cervix are smeared onto a slide that is later examined to detect any abnormalities.

As we shall see, the outer layer of skin is also stratified squamous epithelium, but the cells have been reinforced by keratin, a protein that provides strength. Stratified cuboidal and stratified columnar epithelia also are found in the body.

Glandular Epithelia

When an epithelium secretes a product, it is said to be glandular. A **gland** can be a single epithelial cell, as in the case of mucus secreting goblet cells, or a gland can contain many cells. Glands with ducts that secrete their product onto the outer surface (e.g., sweat glands and mammary glands) or into a cavity (e.g., pancreas) are called **exocrine glands.** Ducts can be simple or compound:

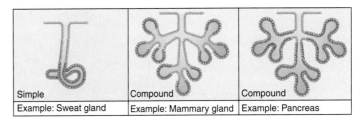

| Simple | Compound | Compound |
| Example: Sweat gland | Example: Mammary gland | Example: Pancreas |

Glands that have no duct are appropriately known as the ductless glands, or endocrine glands. **Endocrine glands** (e.g., pituitary gland and thyroid) secrete hormones internally, so they are transported by the bloodstream.

☑ Check Your Progress 4.5

1. Distinguish between the three types of simple epithelium and give a location for each.
2. a. How would you recognized pseudostratified epithelium, and (b) what function does this tissue perform in the trachea?
3. a. What is stratified epithelium, and (b) where would it be found?

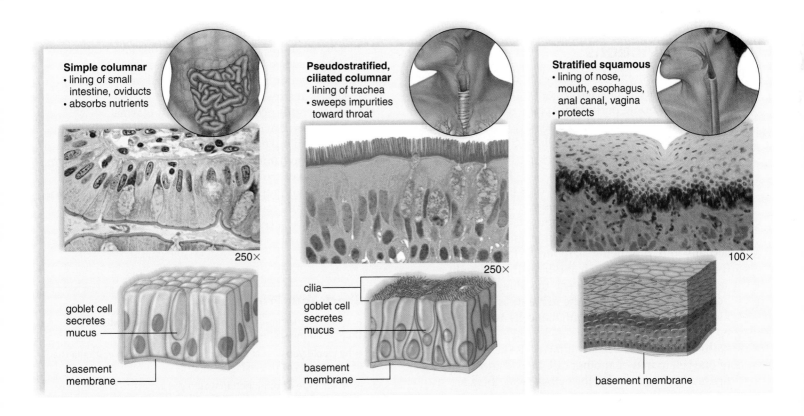

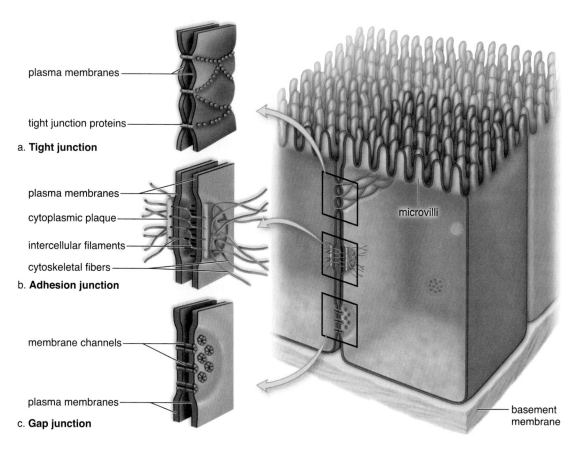

Figure 4.8   **Cell junction types.**
a. Tight junctions allow epithelial cells to form a layer that prevents leakage from one side of the sheet to the other. **b.** Adhesion junctions keep cells anchored to one another, but the layer of cells can still bend and stretch. **c.** Gap junctions allow the passage of small molecules and ions from one cell to the other. Not all epithelial cells are joined by all three types of junctions illustrated.

4.6 Cell Junctions

The epithelial cells, and sometimes the muscle and nerve cells, of a tissue are connected by cell junctions, and these junctions help a tissue perform its particular function. Just as cement holds bricks together, so do cell junctions join cells together into a cohesive hold. Cell junctions arise when plasma membranes are joined in these particular ways:

- **Tight junctions** allow epithelial cells to form a layer that covers the surface of organs and lines body cavities. The layer of cells becomes an impermeable barrier because adjacent plasma membrane proteins actually join, producing a zipperlike fastening (Fig. 4.8*a*). In the stomach and intestines, digestive secretions do not leak between the cells into the body, and in the kidneys, the urine stays within kidney tubules because epithelial cells are joined by tight junctions.
- **Adhesion junctions** firmly attach cytoskeletal fibers of one cell to that of another cell. In a desmosome, featured in Figure 4. 8*b*, the cytoskeletal fibers are anchored to a cytoplasmic plaque adjacent to the

plasma membrane. Adhesion junctions are common in tissues subject to mechanical stress. They allow the skin, for example, to stretch and bend in response to mechanical stress.
- **Gap junctions** occur when adjacent plasma membranes converge and leave a tiny channel between them. Small molecules and ions can diffuse through a gap junction from the cytoplasm of one cell to another. During development, signaling molecules pass between embryonic cells at gap junctions and influence the differentiation of cells. Gap junctions between cardiac muscle cells at intercalated disks allow the heart to beat as a coordinated whole.

☑ Check Your Progress 4.6

1. **a.** What are the three types of junctions between cells, and (b) how do they differ in structure and function?
2. Why would you expect cell junctions more often in epithelia than in other types of tissues?
3. **a.** Which type of cell junction would you expect to find between muscle cells in the heart wall? **b.** Why?

4.7 Integumentary System

Specific tissues are associated with particular organs. For example, nervous tissue is associated with the brain. But actually, an **organ** is composed of two or more types of tissues working together to perform particular functions. The skin is comprised of all four tissue types: epithelial, connective, muscle, and nervous tissue. An **organ system** contains many different organs that cooperate to carry out a process, such as the digestion of food. The skin has several accessory organs (hair, nails, sweat glands, and sebaceous glands), and, therefore, it is sometimes referred to as the **integumentary system.**

Skin is the most conspicuous system in the body because it covers the body. In an adult, the skin has a surface area of about 1.8 square meters (19.51 square feet) and accounts for nearly 15% of the weight of an average human. The skin has numerous functions. It protects underlying tissues from physical trauma, pathogen invasion, and water loss; it also helps regulate body temperature. Therefore, skin plays a significant role in homeostasis. The skin even synthesizes certain chemicals that affect the rest of the body. Because skin contains sensory receptors, it helps us be aware of our surroundings and communicate with others through touch.

Regions of the Skin

The skin has two regions: the epidermis and the dermis (Fig. 4.9). A **subcutaneous layer** is found between the skin and any underlying structures, such as muscle or bone.

The Epidermis

The **epidermis** is made up of stratified squamous epithelium. New epidermal cells for the renewal of skin are derived from stem (basal) cells. The importance of these stem cells is observed when an injury to the skin is deep enough to destroy stem cells. As soon as possible, the damaged tissue

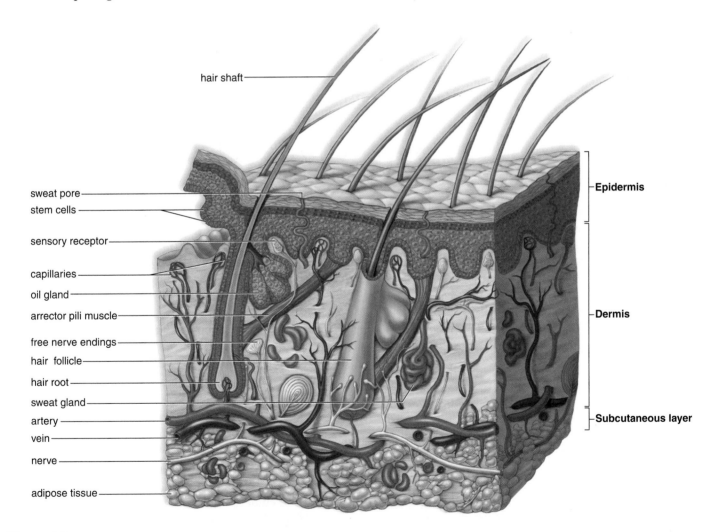

Figure 4.9   Human skin anatomy.
Skin consists of two regions: the epidermis and the dermis. A subcutaneous layer lies below the dermis. Skin has numerous functions. For example, it forms a protective covering over the entire body, safeguarding underlying parts from trauma, pathogen invasion, and water loss. The skin contains sensory receptors that communicate with the central nervous system and make us aware of external conditions. The skin also produces vitamin D, which has important metabolic functions, and the skin helps regulate body temperature.

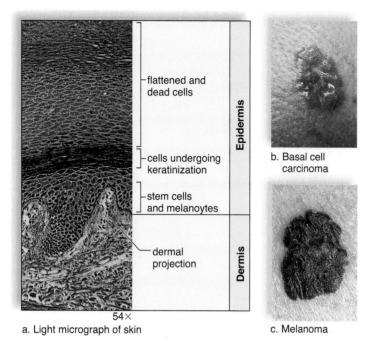

flattened and dead cells

Epidermis

cells undergoing keratinization

stem cells and melanoytes

dermal projection

Dermis

54×

a. Light micrograph of skin

b. Basal cell carcinoma

c. Melanoma

Figure 4.10   The epidermis.
a. Epidermal ridges following dermal projections are clearly visible. Stem cells and melanocytes are in this region. **b.** Basal cell carcinoma derived from stem cells and (**c**) melanoma derived from melanocytes are types of skin cancer.

is removed, and skin grafting is begun. The skin needed for grafting is usually taken from other parts of the patient's body. This is called autografting, as opposed to heterografting, in which the graft is received from another person. Autografting is preferred because rejection rates are very low. If the damaged area is quite extensive, it may be difficult to acquire enough skin for autografting. In that case, small amounts of epidermis are removed and cultured in the laboratory to produce thin sheets of skin that can be transplanted back to the patient.

Newly generated skin cells become flattened and hardened as they push to the surface (Fig. 4.10a). Hardening takes place because the cells produce keratin, a waterproof protein. Dandruff occurs when the rate of keratinization in the skin of the scalp is two or three times the normal rate. A thick layer of dead keratinized cells, arranged in spiral and concentric patterns, forms fingerprints and footprints that are genetically unique. Since outer skin cells are dead and keratinized, the skin is waterproof, thereby preventing water loss. The skin's waterproofing also prevents water from entering the body when the skin is immersed.

Two types of specialized cells are located deep in the epidermis. **Langerhans cells** are macrophages, a type of white blood cell that phagocytize pathogens and then travel to lymphatic organs, where they stimulate the immune system to react to the pathogen. **Melanocytes,** lying deep in the epidermis, produce melanin, the main pigment responsible

for skin color. Since the number of melanocytes is about the same in all individuals, variation in skin color is due to the amount of melanin produced and its distribution. When skin is exposed to the sun, melanocytes produce more melanin to protect the skin from the damaging effects of the ultraviolet (UV) radiation in sunlight. The melanin is passed to other epidermal cells, and the result is tanning, or in some people, the formation of patches of melanin called freckles. Another pigment, called carotene, is present in epidermal cells and in the dermis and gives the skin of certain Asians its yellowish hue. The pinkish color of fair-skinned people is due to the pigment hemoglobin in the red blood cells in the blood vessels of the dermis.

Some ultraviolet radiation does serve a purpose, however. Certain cells in the epidermis convert a steroid related to cholesterol into **vitamin D** with the aid of ultraviolet radiation. Only a small amount of UV radiation is needed. Vitamin D leaves the skin and helps regulate both calcium and phosphorus metabolism in the body. Calcium and phosphorus are very important to the proper development and mineralization of the bones.

Skin Cancer   While we tend to associate a tan with health, actually it signifies that the body is trying to protect itself from the dangerous rays of the sun. Too much ultraviolet radiation is dangerous and can lead to skin cancer. Basal cell carcinoma (Fig. 4.10b), derived from stem cells gone awry, is the more common type of skin cancer and the most curable. Melanoma (Fig. 4.10c), the type of skin cancer derived from melanocytes, is extremely serious. To prevent skin cancer, you should stay out of the sun altogether between the hours of 10 A.M. and 3 P.M. When you are in the sun

- use a broad-spectrum sunscreen that protects from both UV-A and UV-B radiation and has a sun protection factor (SPF) of at least 15. (This means that if you usually burn, for example, after a 20-minute exposure, it will take 15 times longer, or 5 hours, before you will burn.)
- wear protective clothing. Choose fabrics with a tight weave and wear a wide-brimmed hat.
- wear sunglasses that have been treated to absorb both UV-A and UV-B radiation.
- avoid tanning machines because, even if they use only high levels of UV-A radiation, the deep layers of the skin will become more vulnerable to UV-B radiation.

The Dermis

The **dermis** is a region of dense fibrous connective tissue beneath the epidermis. (Dermatology is a branch of medicine that specializes in diagnosing and treating skin disorders.) The dermis contains collagen and elastic fibers. The collagen fibers are flexible but offer great resistance to overstretching; they prevent the skin from being torn. The elastic

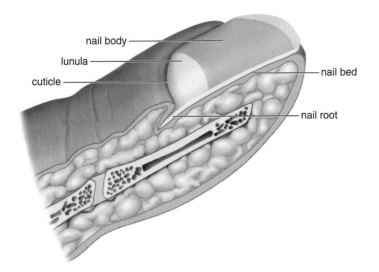

Figure 4.11   Nail anatomy.
Cells produced by the nail root become keratinized, forming the nail body.

fibers maintain normal skin tension but also stretch to allow movement of underlying muscles and joints. (The number of collagen and elastic fibers decreases with age and with exposure to the sun, causing the skin to become less supple and more prone to wrinkling.) The dermis also contains blood vessels that nourish the skin. When blood rushes into these vessels, a person blushes, and when blood is minimal in them, a person turns "blue." Blood vessels in the dermis play a role in temperature regulation. If body temperature starts to rise, the blood vessels in the skin will dilate. As a result, more blood is brought to the surface of the skin for cooling. If the outer temperature cools, the blood vessels constrict, so less blood is brought to the skin's surface.

The sensory receptors primarily in the dermis are specialized for touch, pressure, pain, hot, and cold. These receptors supply the central nervous system with information about the external environment. The sensory receptors also account for the use of the skin as a means of communication between people. For example, the touch receptors play a major role in sexual arousal.

The Subcutaneous Layer
Technically speaking, the subcutaneous layer beneath the dermis is not a part of skin. A common site for injections, this layer is composed of loose connective tissue and adipose tissue, which stores fat. Fat is a stored source of energy in the body. Adipose tissue helps to thermally insulate the body from either gaining heat from the outside or losing heat from the inside. A well-developed subcutaneous layer gives the body a rounded appearance and provides protective padding against external assaults. Excessive development of the subcutaneous layer accompanies obesity.

Accessory Organs of the Skin

Nails, hair, and glands are structures of epidermal origin, even though some parts of hair and glands are largely found in the dermis.

Nails are a protective covering of the distal part of fingers and toes, collectively called digits (Fig. 4.11). Nails grow from special epithelial cells at the base of the nail in the portion called the nail root. The cuticle is a fold of skin that hides the nail root. The whitish color of the half-moon-shaped base, or *lunula,* results from the thick layer of cells in this area. The cells of a nail become keratinized as they grow out over the nail bed.

Hair follicles begin at a bulb in the dermis and continue through the epidermis where the hair shaft extends beyond the skin (see Fig. 4.9). A dark hair color is largely due the production of true melanin by melanocytes present in the bulb. If the melanin contains iron and sulfur, hair is blond or red. Graying occurs when melanin cannot be produced, but white hair is due to bubbles in the hair shaft.

Contraction of the arrector pili muscles attached to hair follicles causes the hairs to "stand on end" and goosebumps to develop. Epidermal cells form the root of a hair, and their division causes a hair to grow. The cells become keratinized and die as they are pushed farther from the root.

Each hair follicle has one or more **oil glands** (see Fig. 4.9), also called sebaceous glands, which secrete sebum, an oily substance that lubricates the hair within the follicle and the skin itself. The oil secretions from sebaceous glands are acidic and retard the growth of bacteria. If the sebaceous glands fail to discharge, the secretions collect and form "whiteheads" or "blackheads." The color of blackheads is due to oxidized sebum. **Acne** is an inflammation of the sebaceous glands that most often occurs during adolescence due to hormonal changes.

Sweat glands (see Fig. 4.9), also called sudoriferous glands, are quite numerous and are present in all regions of skin. A sweat gland is a tubule that begins in the dermis and either opens into a hair follicle or, more often, opens onto the surface of the skin. Sweat glands play a role in modifying body temperature. When body temperature starts to rise, sweat glands become active. Sweat absorbs body heat as it evaporates. Once body temperature lowers, sweat glands are no longer active.

✔ Check Your Progress 4.7

1. Compare the structure and function of the epidermis and dermis.

2. a. What factors are involved in skin color, and (b) how can you protect skin from the ultraviolet rays of the sun?

3. a. Contrast the location and function of sweat glands and oil glands. b. How do oil glands protect us from bacteria?

4. a. Explain why the subcutaneous layer is not considered a part of the skin. b. What is its function?

4.8 Organ Systems

This text has several illustrations such as the one on page 79 that show how the various systems cooperate to maintain homeostasis, the relative constancy of the internal environment. In one sense, it is arbitrary to assign a particular organ to one system when it also assists the functioning of many other systems. The functions of the various systems of the body are listed in Figure 4.12.

Integumentary System

The integumentary system contains skin and also includes nails, hairs, muscles that move hairs, the oil and sweat glands, blood vessels, and nerves leading to sensory receptors. As discussed in Section 4.7 this system has many homeostatic functions.

Cardiovascular System

In the **cardiovascular system,** the heart pumps blood and sends it out under pressure into the blood vessels. In humans, blood is always contained in blood vessels, never running free unless the body suffers an injury.

While blood is moving throughout the body, it distributes heat produced by the muscles. Blood transports nutrients and oxygen to the cells and removes their waste molecules, including carbon dioxide. Despite the movement of molecules into and out of the blood, it has a fairly constant volume and pH, particularly due to exchanges in the lungs, the digestive tract, and the kidneys. The red blood cells in blood transport oxygen, while the white blood cells fight infections. Platelets are involved in blood clotting.

Lymphatic and Immune Systems

The **lymphatic system** consists of lymphatic vessels, lymph nodes, the spleen, and other lymphatic organs. This system collects excess tissue fluid and plays a role in absorbing fats and transporting lymph to cardiovascular veins. It also purifies lymph and stores lymphocytes, the white blood cells that produce antibodies.

The **immune system** consists of all the cells in the body that protect us from disease. The lymphocytes, in particular, belong to this system.

Digestive System

The **digestive system** consists of the mouth, esophagus, stomach, small intestine, and large intestine (colon), along with these associated organs: teeth, tongue, salivary glands, liver, gallbladder, and pancreas. This system receives food and digests it into nutrient molecules, which can enter the cells of the body. The nondigested remains are eventually eliminated.

Respiratory System

The **respiratory system** consists of the lungs and the tubes that take air to and from them. The respiratory system brings oxygen into the body and removes carbon dioxide from the

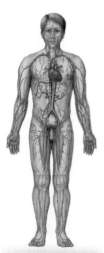
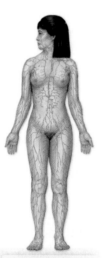
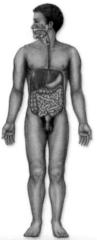
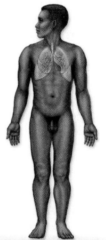
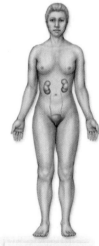

Integumentary system	**Cardiovascular system**	**Lymphatic and immune systems**	**Digestive system**	**Respiratory system**	**Urinary system**
• protects body. • receives sensory input. • helps control temperature. • synthesizes vitamin D.	• transports blood, nutrients, gases, and wastes. • defends against disease. • helps control temperature, fluid, and pH balance.	• helps control fluid balance. • absorbs fats. • defends against infectious disease.	• ingests food. • digests food. • absorbs nutrients. • eliminates waste.	• maintains breathing. • exchanges gases at lungs and tissues. • helps control pH balance.	• excretes metabolic wastes. • helps control fluid balance. • helps control pH balance.

Figure 4.12   Organ systems of the body.

body at the lungs. The removal of carbon dioxide helps adjust the acid-base balance of the blood.

Urinary System

The **urinary system** contains the kidneys, the urinary bladder, and the tubes that carry urine. The kidneys rid the body of metabolic wastes, particularly nitrogenous wastes, and help regulate the salt-water balance and acid-base balance of the blood. Fortunately for Barbara, a substitute bladder made from a piece of the intestine, will store urine and not affect the kidney's ability to carry out its homeostatic functions.

Skeletal System

The bones of the **skeletal system** protect body parts. For example, the skull forms a protective encasement for the brain, as does the rib cage for the heart and lungs. The skeleton helps move the body because it serves as a place of attachment for the skeletal muscles.

The skeletal system also stores minerals, notably calcium, and produces blood cells within red bone marrow.

Muscular System

In the **muscular system,** skeletal muscle contraction maintains posture and accounts for the movement of the body and its parts. Cardiac muscle contraction results in the heartbeat. The walls of internal organs, such as the bladder, contract due to the presence of smooth muscle. Muscle contraction releases heat, which helps warm the body.

Nervous System

The **nervous system** consists of the brain, spinal cord, and associated nerves. The nerves conduct nerve impulses from sensory receptors to the brain and spinal cord, where integration occurs. Nerves also conduct nerve impulses from the brain and spinal cord to the muscles and glands, allowing us to respond to both external and internal stimuli.

Endocrine System

The **endocrine system** consists of the hormonal glands, which secrete chemical messengers called hormones into the bloodstream. Hormones have a wide range of effects, including regulation of cellular metabolism, regulation of fluid and pH balance, and helping us respond to stress. Both the nervous and endocrine systems coordinate and regulate the functioning of the body's other systems. The endocrine system also helps maintain the functioning of the male and female reproductive organs.

Reproductive System

The **reproductive system** has different organs in the male and female. The male reproductive system consists of the testes, other glands, and various ducts that conduct semen to and through the penis. The testes produce sex cells called sperm. The female reproductive system consists of the ovaries, oviducts, uterus, vagina, and external genitals. The ovaries produce sex cells called eggs. When a sperm fertilizes an egg, an offspring begins development.

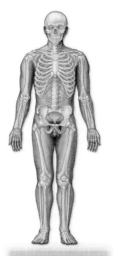

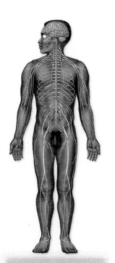
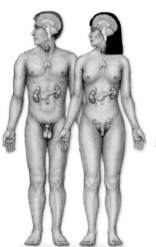

Skeletal system
- supports the body.
- protects body parts.
- helps move the body.
- stores minerals.
- produces blood cells.

Muscular system
- maintains posture.
- moves body and internal organs.
- produces heat.

Nervous system
- receives sensory input.
- integrates and stores input.
- initiates motor output.
- helps coordinate organ systems.

Endocrine system
- produces hormones.
- helps coordinate organ systems.
- responds to stress.
- helps regulate fluid and pH balance.
- helps regulate metabolism.

Reproductive system
- produces gametes.
- transports gametes.
- produces sex hormones.
- nurtures and gives birth to offspring in females.

Figure 4.12   Organ systems of the body—continued.

Health Focus

To Have or Not Have Botox

More and more studies indicate that a youthful, virile appearance is important for success in all aspects of life, especially in the workplace. In a never-ending quest to stop or even turn back the hands of time, millions of people have turned to Botox.

Botox Is a Drug

Botox is a drug used to reduce the appearance of facial wrinkles and lines. Botox is the trade name for a derivative of botulinum toxin A, a protein toxin produced by the bacterium *Clostridium botulinum.* This bacterium was once only associated with the scourge of botulism, a type of food poisoning, which is no longer a serious problem with modern food-processing techniques.

Botox interferes with the ability of the nervous system to properly communicate with the muscles of the body, causing muscle paralysis. Scientists isolated the toxin, and using modern biotechnology, mass produced it and marketed it as the drug Botox. Although originally developed for the treatment of muscular spasms in patients with disorders such as cerebral palsy and muscular dystrophy, the entrepreneurial spirit took hold when it was realized that it could alleviate facial lines and wrinkles. After numerous studies, Botox was approved by the U.S. Food and Drug Administration (FDA) for use as a cosmetic treatment for facial and neck wrinkles in 2002.

Botox treatments are performed by direct injection of the toxin under the skin with a syringe, where it causes facial muscle paralysis. The injections reduce the appearance of wrinkles and lines that appear as a result of normal facial muscle movement. This muscle movement is normally used to create the vast array of facial expressions that we humans are capable of producing. Botox treatment weakens these muscles to provide a cosmetic benefit. The effects are noticeable within days, reach a maximum benefit after one to two weeks, and can last for nearly three to six months in most cases. Since the cosmetic use of Botox has been studied for well over 20 years, dosage and injection techniques are well developed. Botox has also been studied for the treatment of other conditions, including eczema, excessive sweating, and other skin conditions.

The Benefits and Risks of Botox

Botox has become so popular that over 2.5 million women and 300,000 men underwent treatment in 2003, and numbers are still increasing even today. It has become so popular in some areas that "Botox parties" are conducted in doctors' offices, workplaces, and individual homes, where friends and co-workers are treated in a relaxed group setting under the supervision of a doctor at a discount price. Because of its low price, rapid results, and relatively low risk, many people have turned to Botox as a reasonable alternative to more costly, invasive procedures such as cosmetic surgery.

Despite its benefits when administered by a qualified doctor, Botox treatment is not without side effects. Fortunately, most are mild, such as excessive drooling or a slight rash around the injection site that may reduce the cosmetic benefits of the treatment. Spreading of Botox from the injection site may also paralyze facial muscles unintended for treatment, and in a few cases, muscle pain and weakness have resulted.

While quite rare, more serious side effects may also occur. Because it is capable of stimulating a response from the immune system, there is a risk of severe allergic reactions. Botox treatment is also risky for women who are pregnant, individuals who are taking certain medications, or patients with certain musculoskeletal disorders.

Even though Botox has been approved by the FDA, not all in the medical profession are convinced that it should be available for such widespread use. Since overuse of Botox can lessen the body's response to it requiring more frequent injections of larger amounts of the drug, some health-care professionals are concerned about the possibility that some users may become psychologically addicted or may be harmed by excessive doses. Others are concerned that the procedure is sometimes performed outside of a health-care facility, where an unmonitored patient might have a severe reaction or emergency that could be fatal or cause permanent harm. Still others protest the possibility of unqualified personnel administering the treatment. Finally, cases have been documented in which diluted or fake Botox was used, raising the possibility of fraud.

When performed in a medical facility by a licensed physician, Botox treatment is generally considered safe and effective, despite certain risks. Ultimately, it is up to the individual to weigh the benefits against the risks when considering such treatments. But in many segments of society where appearance is of the utmost importance, many people are willing to take these risks to obtain the more youthful facial appearance that this newfound "fountain of youth" promises.

Body Cavities

The human body is divided into two main cavities: the ventral cavity and the dorsal cavity (Fig. 4.13a). Called the coelom in early development, the ventral cavity later becomes the thoracic, abdominal, and pelvic cavities. The thoracic cavity contains the lungs and the heart. The thoracic cavity is separated from the abdominal cavity by a horizontal muscle called the **diaphragm.** The stomach, liver, spleen, pancreas, gallbladder, and most of the small and large intestines are in the abdominal cavity. The pelvic cavity contains the rectum, the urinary bladder, the internal reproductive organs, and the rest of the small and large intestine. Males have an external extension of the abdominal wall, called the scrotum, containing the testes.

The dorsal cavity has two parts: The cranial cavity within the skull contains the brain, while the vertebral canal, formed by the vertebrae, contains the spinal cord.

Body Membranes

Body membranes line cavities and the internal spaces of organs and tubes that open to the outside. The body membranes are of four types: mucous, serous, synovial, and meninges.

Mucous membranes line the tubes of the digestive, respiratory, urinary, and reproductive systems. They are composed of an epithelium overlying a loose fibrous connective tissue layer. The epithelium contains goblet cells that secrete mucus. This mucus ordinarily protects the body from invasion by bacteria and viruses; hence, more mucus is secreted and expelled when a person has a cold and has to blow her/his nose. In addition, mucus usually protects the walls of the stomach and small intestine from digestive juices, but this protection breaks down when a person develops an ulcer.

Serous membranes line and support the lungs, the heart, and the abdominal cavity and its internal organs (Fig. 4.13b). They secrete a watery fluid that keeps the membranes lubricated. Serous membranes support the internal organs and compartmentalize the large thoracic and abdominal cavities.

Serous membranes have specific names according to their location. The pleurae (sing., **pleura**) line the thoracic cavity and cover the lungs; the pericardium forms the pericardial sac and covers the heart; the peritoneum lines the abdominal cavity and covers its organs. A double layer of peritoneum, called mesentery, supports the abdominal organs and attaches them to the abdominal wall. **Peritonitis** is a life-threatening infection of the peritoneum.

Synovial membranes composed only of loose connective tissue line the cavities of freely movable joints. They secrete synovial fluid into the joint cavity; this fluid lubricates the ends of the bones so that they can move freely. In rheumatoid arthritis, the synovial membrane becomes inflamed and grows thicker, restricting movement.

The **meninges** are membranes found within the dorsal cavity. They are composed only of connective tissue and serve as a protective covering for the brain and spinal cord. **Meningitis** is a life-threatening infection of the meninges.

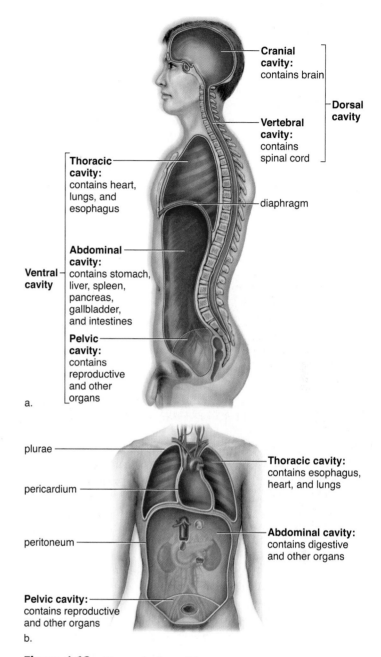

Figure 4.13 Human body cavities.
a. Side view. The dorsal (toward the back) cavity contains the cranial cavity and the vertebral canal. The brain is in the cranial cavity, and the spinal cord is in the vertebral canal. In the ventral (toward the front) cavity, the diaphragm separates the thoracic cavity from the abdominal cavity. The heart and lungs are in the thoracic cavity; the other internal organs are either in the abdominal cavity or in the pelvic cavity. **b.** Frontal view of the thoracic cavity, showing serous membranes.

☑ Check Your Progress 4.8

1. What is the overall function of each of the body systems?
2. **a.** What are the two major body cavities? **b.** What two cavities are in each of these?
3. What are four types of body membranes?

4.9 Homeostasis

Homeostasis is the body's ability to maintain a relative constancy of its internal environment by adjusting its physiological processes. Even though external conditions may change dramatically, we have physiologic mechanisms that respond to disturbances and limit the amount of internal change so that conditions usually stay within a narrow range of normality (Fig. 4.14). For example, blood glucose, pH levels, and body temperature typically fluctuate during the day, but not greatly. If internal conditions should change to any great degree, illness results.

Figure 4.14   Liv Arnesen and Ann Bancroft.
Liv Arnesen and Ann Bancroft needed all their body systems to function properly when they decided to hike to the South Pole. Antarctica is very inhospitable to people. It's the coldest place on Earth, the windiest, and, strange to say, the driest. There are no plants and no animals to eat. It's desolate, dangerous, and nothing but ice. Ice bridges just large enough for a sled stretch across deep crevasses, and if you fall into the water below, you freeze in minutes. You can get frostbitten if your skin is exposed to the subfreezing air. The mechanisms for maintaining homeostasis can be overwhelmed, so it's best to take protective measures to assist the body.

You can be sure that Liv and Ann wore the latest in protective clothing, drank plenty of water, and ate regularly to keep their muscles supplied with glucose. Sleep was necessary to refresh the brain, because while all the body's systems contribute, the brain coordinates their functioning in order to maintain homeostasis, the relative constancy of the internal environment. Swim the English Channel, climb Mt. Washington, cross the Sahara desert by camel, or hike the South Pole; if you are healthy, your body temperature will stay at just about 37°C (98.6°F) because the brain has mechanisms that maintain homeostasis.

The Internal Environment

The internal environment has two parts: blood and tissue fluid. Blood delivers oxygen and nutrients to the tissues and carries carbon dioxide and wastes away. Tissue fluid, not blood, actually bathes all of the body's cells. Therefore, tissue fluid is the medium through which substances are exchanged between cells and blood. Oxygen and nutrients pass through tissue fluid on their way to tissue cells from the blood, and then carbon dioxide and wastes are carried away from the tissue cells by the tissue fluid, where they are brought back into the blood. The cooperation of body systems is required to keep these substances within the range of normality in blood and tissue fluid.

The Body Systems and Homeostasis

The nervous and endocrine systems are particularly important in coordinating the activities of all the other organ systems as they function to maintain homeostasis (Fig. 4.15). The nervous system is able to bring about rapid responses to any changes in the internal environment. The nervous system issues commands by electrochemical signals that are rapidly transmitted to effector organs, which can be muscles, such as skeletal muscles, or glands, such as sweat and salivary glands. The endocrine system brings about responses that are slower to occur but generally have more lasting effects. Glands of the endocrine system, such as the pancreas or the thyroid, release hormones. Hormones, such as insulin from the pancreas, are chemical messengers that must travel through the blood and tissue fluid in order to reach their targets.

The nervous and endocrine systems together direct numerous activities that maintain homeostasis, but all the organ systems must do their part in order to keep us alive and healthy. Picture what would happen if, say, the cardiovascular, respiratory, digestive, or urinary system failed (Fig. 4.15). If someone is having a heart attack, the heart is unable to pump the blood to supply cells with oxygen. Or think of a person who is choking. Since the trachea (or windpipe) is blocked, no air can reach the lungs for uptake by the blood. Unless the obstruction is removed quickly, cells will begin to die as the blood's supply of oxygen is depleted. When the lining of the digestive tract is damaged, as in a severe bacterial infection, nutrient absorption is impaired and cells face an energy crisis. It is important not only to maintain adequate nutrient levels in the blood, but also to eliminate wastes and toxins. The liver makes urea, a nitrogenous end product of protein metabolism, but urea and other metabolic wastes are excreted by the kidneys, the urine-producing organs of the body. The kidneys rid the body of nitrogenous wastes and also help to adjust the blood's water-salt and acid-base balances.

A closer examination of how the blood glucose level is maintained helps us understand homeostatic mechanisms. When a healthy person consumes a meal and glucose enters the blood, the pancreas secretes the hormone insulin. Now glucose is removed from the blood as cells take it up. In the liver, glucose is stored in the form of glycogen. This storage is beneficial because later, if blood glucose levels drop, glycogen can be broken down to ensure that the blood level remains constant. Homeostatic mechanisms can fail. In diabetes mellitus, the pancreas cannot produce enough insulin, or the body cells cannot respond appropriately to it. Therefore, glucose does not enter the cells and they must turn to other molecules, such as fats and proteins, in order to survive. This, along with too much glucose in the blood, leads to the numerous complications of diabetes mellitus.

Another example of homeostasis is the ability of the body to regulate the acid-base balance of the body. When carbon dioxide enters the blood, it combines with water to form carbonic acid. However, the blood is buffered, and pH stays within normal range as long as the lungs are busy excreting carbon dioxide. These two mechanisms are backed up by the kidneys, which can rid the body of a wide range of acidic and basic substances and, in that way, adjust the pH.

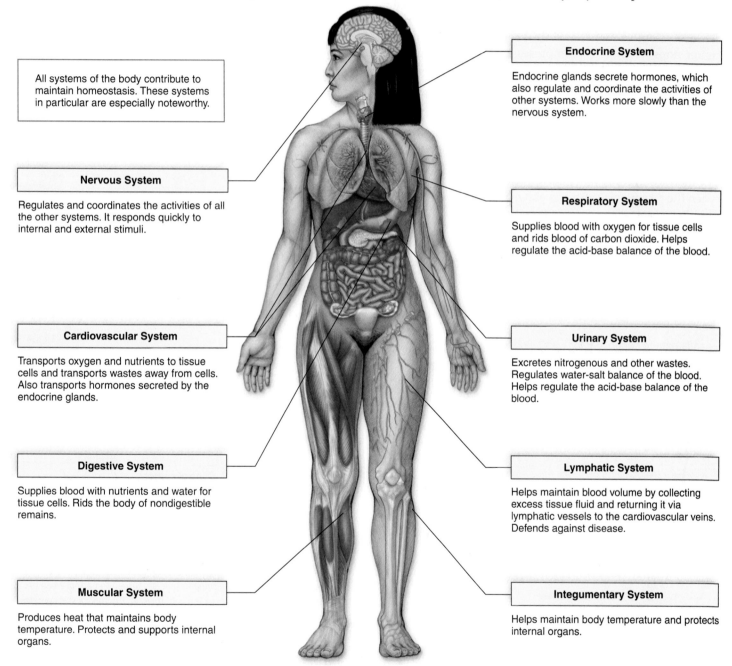

All systems of the body contribute to maintain homeostasis. These systems in particular are especially noteworthy.

Endocrine System

Endocrine glands secrete hormones, which also regulate and coordinate the activities of other systems. Works more slowly than the nervous system.

Nervous System

Regulates and coordinates the activities of all the other systems. It responds quickly to internal and external stimuli.

Respiratory System

Supplies blood with oxygen for tissue cells and rids blood of carbon dioxide. Helps regulate the acid-base balance of the blood.

Cardiovascular System

Transports oxygen and nutrients to tissue cells and transports wastes away from cells. Also transports hormones secreted by the endocrine glands.

Urinary System

Excretes nitrogenous and other wastes. Regulates water-salt balance of the blood. Helps regulate the acid-base balance of the blood.

Digestive System

Supplies blood with nutrients and water for tissue cells. Rids the body of nondigestible remains.

Lymphatic System

Helps maintain blood volume by collecting excess tissue fluid and returning it via lymphatic vessels to the cardiovascular veins. Defends against disease.

Muscular System

Produces heat that maintains body temperature. Protects and supports internal organs.

Integumentary System

Helps maintain body temperature and protects internal organs.

Figure 4.15   How the organ systems of the body contribute to homeostasis.
All the organ systems contribute to homeostasis in many ways. Some of the main contributions of each system are given in this illustration.

Negative Feedback

Negative feedback is the primary homeostatic mechanism that keeps a variable, such as the blood glucose level, close to a particular value, or set point. A homeostatic mechanism has at least two components: a sensor and a control center (Fig. 4.16). The sensor detects a change in the internal environment; the control center then brings about an effect to bring conditions back to normal again. Now, the sensor is no longer activated. In other words, a **negative feedback** mechanism is present when the output of the system dampens the original stimulus. For example, when blood pressure rises, sensory receptors signal a control center in the brain. The center stops sending nerve impulses to the arterial walls and they relax. Once the blood pressure drops, signals no longer go to the control center.

Mechanical Example

A home heating system is often used to illustrate how a more complicated negative feedback mechanism works (Fig. 4.17). You set the thermostat at, say, 68°F. This is the *set point*. The thermostat contains a thermometer, a sensor that detects when the room temperature is above or below the set point. The thermostat also contains a control center; it turns the

furnace off when the room is warm and turns it on when the room is cool. When the furnace is off, the room cools a bit, and when the furnace is on, the room warms a bit. In other words, typical of negative feedback mechanisms, there is a fluctuation above and below normal.

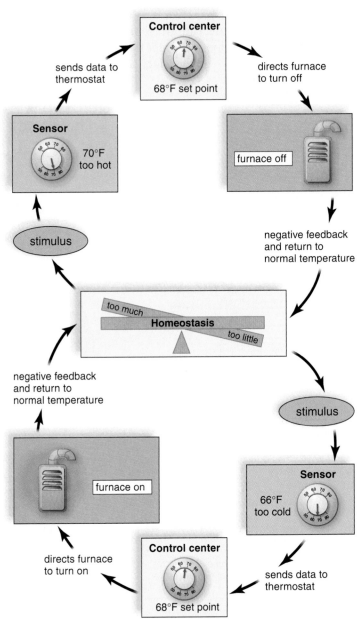

Figure 4.17 Complex negative feedback mechanism.
This diagram shows how room temperature is returned to normal when the room becomes too hot (*above*) or too cold (*below*). The thermostat contains both the sensor and the control center. *Above:* The sensor detects that the room is too hot, and the control center turns the furnace off. The stimulus is no longer present when the temperature returns to normal. *Below:* The sensor detects that the room is too cold, and the control center turns the furnace on. The stimulus is no longer present when the temperature returns to normal.

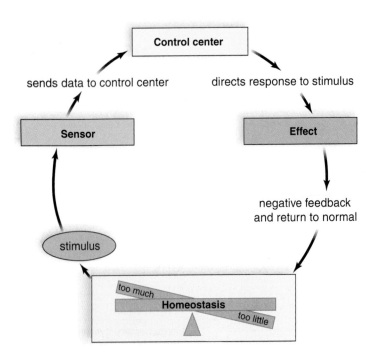

Figure 4.16 Negative feedback mechanism.
This diagram shows how the basic elements of a feedback mechanism work. A sensor detects the stimulus, and a control center brings about an effect that dampens the stimulus.

Human Example: Regulation of Body Temperature

The sensor and control center for body temperature is located in a part of the brain called the hypothalamus. Notice that a negative feedback mechanism prevents change in the same direction; body temperature does not get warmer and warmer because warmth brings about a change toward a lower body temperature. Also, body temperature does not get colder and colder because a body temperature below normal brings about a change toward a warmer body temperature.

Above Normal Temperature   When the body temperature is above normal, the control center directs the blood vessels of the skin to dilate (Fig. 4.18, *above*). This allows more blood to flow near the surface of the body, where heat can be lost to the environment. In addition, the nervous system activates the sweat glands, and the evaporation of sweat helps lower body temperature. Gradually, body temperature decreases to 98.6°F.

Below Normal Temperature   When the body temperature falls below normal, the control center directs (via nerve impulses) the blood vessels of the skin to constrict (Fig. 4.18, *below*). This conserves heat. If body temperature falls even lower, the control center sends nerve impulses to the skeletal muscles, and shivering occurs. Shivering generates heat, and gradually body temperature rises to 98.6°F. When the temperature rises to normal, the control center is inactivated.

Positive Feedback

Positive feedback is a mechanism that brings about an ever greater change in the same direction. When a woman is giving birth, the head of the baby begins to press against the cervix, stimulating sensory receptors there. When nerve impulses reach the brain, the brain causes the pituitary gland to secrete the hormone oxytocin. Oxytocin travels in the blood and causes the uterus to contract. As labor continues, the cervix is ever more stimulated, and uterine contractions become ever stronger until birth occurs.

A positive feedback mechanism can be harmful, as when a fever causes metabolic changes that push the fever still higher. Death occurs at a body temperature of 113°F because cellular proteins denature at this temperature and metabolism stops. Still, positive feedback loops such as those involved in childbirth, blood clotting, and the stomach's digestion of protein assist the body in completing a process that has a definite cut-off point.

✓ Check Your Progress 4.9

1. a. What is homeostasis, and (b) why is it important to body function?

2. How do body systems contribute to homeostasis?

3. How do negative feedback and positive feedback contribute to homeostasis?

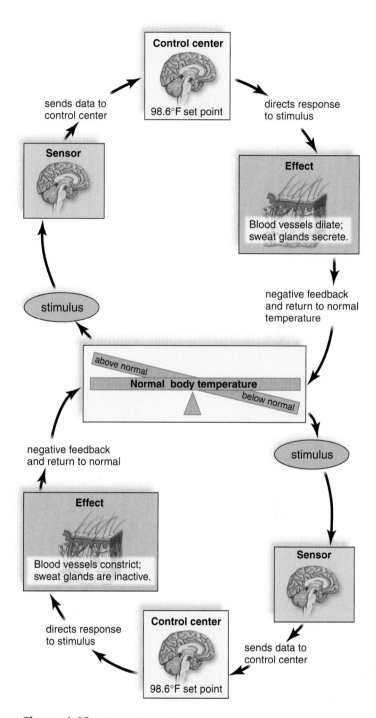

Figure 4.18   Regulation of body temperature.
Above: When body temperature rises above normal, the hypothalamus senses the change and causes blood vessels to dilate and sweat glands to secrete so that temperature returns to normal. *Below:* When body temperature falls below normal, the hypothalamus senses the change and causes blood vessels to constrict. In addition, shivering may occur to bring temperature back to normal. In this way, the original stimulus was removed.

Summarizing the Concepts

4.1 Types of Tissues

Human tissues are categorized into four groups:

| Connective | Muscular | Nervous | Epithelial |

4.2 Connective Tissue Connects and Supports

Connective tissues have cells separated by a matrix that contains ground substance and fibers (e.g., collagen fibers). The four types are

- loose fibrous connective tissue, including adipose tissue;
- dense fibrous connective tissue (tendons and ligaments);
- cartilage and bone (matrix for cartilage is solid, yet flexible; matrix for bone is solid and rigid); and
- blood (matrix is a liquid called plasma) and lymph.

4.3 Muscular Tissue Moves the Body

Muscular tissue is of three types: skeletal, smooth, and cardiac.

- Skeletal and cardiac muscle are striated.
- Cardiac and smooth muscle are involuntary.
- Skeletal muscle is found in muscles attached to bones.
- Smooth muscle is found in internal organs.
- Cardiac muscle makes up the heart.

4.4 Nervous Tissue Communicates

- Nervous tissue is composed of neurons and several types of neuroglia.
- Each neuron has dendrites, a cell body, and an axon. Axons conduct nerve impulses.

4.5 Epithelial Tissue Protects

Epithelial tissue covers the body and lines its cavities.

- Types of simple epithelia are squamous, cuboidal, and columnar.
- Certain of these tissues may have cilia or microvilli.
- Stratified epithelia have many layers of cells, with only the bottom layer touching the basement membrane.
- Glandular epithelia secretes a product either into ducts or into the blood.

4.6 Cell Junctions

Three types of junctions are common between epithelial cells:

- Tight junctions are zipperlike fastenings between cells.
- Adhesion junctions permit cells to stretch and bend.

- Gap junctions allow small molecules and signals to pass between cells.

4.7 Integumentary System

Skin and its accessory organs comprise the integumentary system. Skin has two regions:

- The epidermis contains stem cells, which produce new epithelial cells.
- The dermis contains epidermally derived glands and hair follicles, nerve endings, blood vessels, and sensory receptors.

A subcutaneous layer lies beneath the skin.

4.8 Organ Systems

Organs make up organ systems, which are summarized in the table below. Some organs are found in particular body cavities.

4.9 Homeostasis

Homeostasis is the relative constancy of the internal environment, which is tissue fluid and blood. All organ systems contribute to homeostasis.

- The cardiovascular, respiratory, digestive, and urinary systems directly regulate the amount of gases, nutrients, and wastes in the blood, keeping tissue fluid constant.
- The lymphatic system absorbs excess tissue fluid and functions in immunity.
- The nervous system and endocrine system regulate the other systems.

Negative Feedback

Negative feedback mechanisms keep the environment relatively stable. When a sensor detects a change above or below a set point, a control center brings about an effect that reverses the change and brings conditions back to normal again. Examples include

- regulation of blood glucose level by insulin,
- regulation of room temperature by a thermostat and furnace, and
- regulation of body temperature by the brain and sweat glands.

Positive Feedback

In contrast to negative feedback, a positive feedback mechanism brings about rapid change in the same direction as the stimulus and does not achieve relative stability. These mechanisms are useful under certain conditions, such as during birth.

Organ Systems	
Transport Cardiovascular (heart and blood vessels) Lymphatic and Immune (lymphatic vessels)	*Integumentary* Skin and accessory organs
Maintenance Digestive (e.g., stomach, intestines) Respiratory (tubes and lungs) Urinary (tubes and kidneys)	*Motor* Skeletal (bones and cartilage) Muscular (muscles)
Control Nervous (brain, spinal cord, and nerves) Endocrine (glands)	*Reproduction* Reproductive (tubes and testes in males; tubes and ovaries in females)

Understanding Key Terms

acne 73
adhesion junction 70
adipose tissue 62
basement membrane 68
blood 64
bone 63
cardiac muscle 65
cardiovascular system 74
cartilage 63
collagen fiber 62
columnar epithelium 68
compact bone 63
connective tissue 62
cuboidal epithelium 68
dense fibrous connective
    tissue 63
dermis 72
diaphragm 77
digestive system 74
elastic cartilage 63
elastic fiber 62
endocrine gland 69
endocrine system 75
epidermis 71
epithelial tissue 68
exocrine gland 69
fibroblast 62
fibrocartilage 63
gap junction 70
gland 69
hair follicle 73
homeostasis 78
hyaline cartilage 63
immune system 74
integumentary system 71
intercalated disks 65
lacuna 63
Langerhans cell 72
ligament 63
loose fibrous connective
    tissue 62
lymph 64
lymphatic system 74
matrix 62
melanocyte 72

meninges 77
meningitis 77
mucous membrane 77
muscular system 75
muscular (contractile) tissue 65
nail 73
negative feedback 80
nerve 66
nervous system 75
nervous tissue 66
neuroglia 66
neuron 66
oil gland 73
organ 71
organ system 71
pathogen 64
peritonitis 77
platelet (thrombocyte) 64
pleura 77
positive feedback 81
pseudostratified columnar
    epithelium 68
red blood cell (erythrocyte) 64
reproductive system 75
respiratory system 74
reticular fiber 62
serous membrane 77
skeletal muscle 65
skeletal system 75
skin 71
smooth (visceral) muscle 65
spongy bone 63
squamous epithelium 68
striated 65
subcutaneous layer 71
sweat gland 73
synovial membrane 77
tendon 63
tight junction 70
tissue 62
tissue fluid 64
urinary system 75
vitamin D 72
white blood cell (leukocyte) 64

Match the key terms to these definitions.

a. _____ Dense fibrous connective tissue that joins bone to
    bone at a joint.

b. _____ Outer region of the skin composed of keratinized
    stratified squamous epithelium.

c. _____ Cancer of epithelial tissue.

d. _____ Relative constancy of the body's internal
    environment.

e. _____ Porous bone found at the ends of long bones where
    blood cells are formed.

Testing Your Knowledge of the Concepts

1. What are the functions of the four major tissue types?
   (page 62)

2. What features do all connective tissues have in common?
   (page 62)

3. Explain why you would have expected skeletal muscle
   to be voluntary, but smooth muscle and cardiac muscle to
   be involuntary. (page 65)

4. What are the two types of cells found in nervous tissue?
   Briefly describe each type. (page 66)

5. How are epithelial tissues classified? Describe each
   major type, and give at least one location for each type.
   (pages 68–69)

6. Which of the three types of cell junctions permits
   communication between cells? Explain. (page 70)

7. Explain why the skin is sometimes referred to as the
   integumentary system. (page 71)

8. Referring to Figure 4.12, list each organ system, the major
   organs, and major functions of each. (pages 74–75)

9. What organs of the body are found in the thoracic cavity?
   The abdominal cavity? (page 77)

10. List the types of membranes found in the body, their
    functions, and their locations. (page 77)

11. Why is homeostasis defined as the "*relative* constancy of
    the internal environment?" Does negative feedback or
    positive feedback tend to promote homeostasis?
    Explain. (pages 78–81)

12. Which of the following is not a type of fibrous
    connective tissue?
    a. hyaline cartilage   c. tendons and ligaments
    b. areolar tissue   d. adipose tissue

13. What type of cartilage is found in the rib cage and walls
    of the respiratory passages?
    a. fibrocartilage   c. ligamentous cartilage
    b. elastic cartilage   d. hyaline cartilage

14. Blood is a(n) _____ tissue because it has a _____.
    a. connective; gap junction
    b. muscular; ground substance
    c. epithelial; gap junction
    d. connective; ground substance

15. Skeletal muscle is
    a. striated.   c. multinucleated.
    b. under voluntary control.   d. All of these are correct.

16. Which is true of both cardiac and smooth muscle?
    a. striated   c. involuntary control
    b. multinucleated   d. spindle-shaped cells

17. Which of the following forms the myelin sheath around
    nerve fibers outside the brain and spinal cord?
    a. microglia   c. Schwann cells
    b. neurons   d. astrocytes

18. Which of these is not a type of epithelial tissue?
    a. simple cuboidal and stratified columnar
    b. bone and cartilage
    c. stratified squamous and simple squamous
    d. pseudostratified and transitional
    e. All of these are epithelial tissues.

19. What type of epithelial tissue is found in the walls of the urinary bladder to provide it with the ability to distend?
    a. simple cuboidal epithelium
    b. transitional epithelium
    c. pseudostratified columnar epithelium
    d. stratified squamous epithelium

20. Tight junctions are associated with
    a. connective tissue.   c. cardiac muscle.
    b. adipose tissue.   d. epithelial tissue.

21. Without melanocytes, skin would
    a. be too thin.   c. not tan.
    b. lack nerves.   d. not be waterproof.

22. Which of the following is a function of skin?
    a. temperature regulation
    b. manufacture of vitamin D
    c. protection from invading pathogens
    d. All of these are correct.

23. Fluid balance is a primary goal of which system?
    a. endocrine   c. digestive
    b. lymphatic   d. integumentary

24. The skeletal system functions in
    a. blood cell production.   c. movement.
    b. mineral storage.   d. All of these are correct.

25. Which system helps control pH balance?
    a. digestive   c. respiratory
    b. urinary   d. Both b and c are correct.

26. Which type of membrane is found lining systems that are open to the outside environment, such as the respiratory system?
    a. serous   c. synovial
    b. mucous   d. meningeal

27. The correct order for homeostatic processing is
    a. sensory detection, control center, effect brings about change in body.
    b. control center, sensory detection, effect brings about change in environment.
    c. sensory detection, control center, effect causes no change in environment.
    d. None of these is correct.

28. Which allows rapid change in one direction and does not achieve stability?
    a. homeostasis   c. negative feedback
    b. positive feedback   d. All of these are correct.

29. Which of the following is an example of negative feedback?
    a. Uterine contractions increase as labor progresses.
    b. Insulin decreases blood sugar levels after eating a meal.
    c. Sweating increases as body temperature drops.
    d. Platelets continue to plug an opening in a blood vessel until blood flow stops.

30. Label the body cavities plus the diaphragm in this diagram.

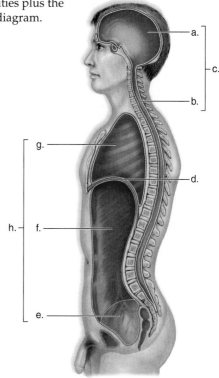

Thinking Critically About the Concepts

As Barbara from the opening story learned, the need for transplant organs is very high and far outnumbers the organs available for donation. Recently, natural bladder cells were grown in the laboratory, creating an actual bladder. When the lab-grown bladder was attached to a damaged bladder, it performed as well as a bladder repaired by more conventional surgical methods. Hopefully, lab-grown organs will one day be widely available for transplant. A lab-grown organ that duplicates the structure and function of a natural organ should increase the success of the transplant, integration of the transplanted organ into its organ system, and the ability of the organ system to do its part in homeostasis.

1. Do you think Barbara's new "bladder" (formerly a section of her large intestine) will function just like the actual cancerous bladder that was removed? Why or why not?

2a. What two systems are especially important in coordinating the efforts of the various organ systems as they work to maintain homeostasis?

  b. How do these two systems issue their "commands" to the other systems?

3a. What different tissues enable the trachea to perform its function of acting as an open tube for the passage of inhaled air?

  b. The bladder has a layer of smooth muscle like other hollow organs, but it also contains transitional epithelium, found nowhere else in our body. Transitional epithelial cells are able to elongate themselves. Why would you expect to find this kind of tissue in the bladder based on its function?

4a. What specific proteins that make our body's reactions possible are affected by an increased body temperature?

  b. What other homeostasis condition affects the function of these proteins?

  c. What two systems are involved in maintaining the second condition that affects these specific proteins?

5. When you are cold, you may shiver.

  a. How will shivering increase your body temperature?

  b. Once body temperature is normal again, will shivering continue? Why or why not?

CHAPTER 5

Cardiovascular System: Heart and Blood Vessels

Sharon was studying for an anatomy and physiology exam. "There is SO much to remember," she groaned. "How am I ever going to remember all of this for the test next week?" she wondered. Allen, her study partner, said the concept of structure suits function might help. You were introduced to this concept in Chapter 3 when you learned that cells have organelles enabling them to carry out their specific function. If you know the function of a cell or organ, you can often describe the structure of it, and vice versa. Allen suggested that Sharon use this technique when studying the organs of the cardiovascular system. For example, if Sharon learns that the aorta, the major artery in the body, receives blood under high pressure from the heart, she should be able to remember that the aorta is thick walled and stretchy.

Allen also suggested that Sharon think of analogies to help her remember the functions of various organs. He told her that arteries remind him of interstates because they carry a large volume of blood, just as interstates accommodate a large number of cars. As you study the organs of this system, you too might think of everyday examples to help you remember the functions of cardiovascular organs for the exam, when your knowledge will be assessed.

CHAPTER CONCEPTS

5.1 Overview of the Cardiovascular System
In the cardiovascular system, the heart pumps the blood, and the blood vessels transport blood about the body. Exchanges at capillaries refresh the blood and then tissue fluid.

5.2 The Types of Blood Vessels
Arteries, which branch into arterioles, take blood away from the heart to capillaries, which empty into venules. Venules join to form veins that take blood back to the heart.

5.3 The Heart Is a Double Pump
The right side of the heart pumps blood to the lungs, and the left side of the heart pumps blood to the rest of the body.

5.4 Features of the Cardiovascular System
Blood pressure, which moves blood in the arteries, drops off in the capillaries where exchange takes place. Skeletal muscle contraction largely accounts for the movement of blood in the veins.

5.5 Two Cardiovascular Pathways
One cardiovascular pathway takes blood from the heart to the lungs and back to the heart. The other pathway takes blood from the heart to all the other organs in the body and back to the heart.

5.6 Exchange at the Capillaries
At tissue capillaries, nutrients and oxygen are exchanged for carbon dioxide and other wastes. In this way, tissue fluid and the body's cells remain alive and healthy.

5.7 Cardiovascular Disorders
Hypertension and atherosclerosis are two cardiovascular disorders that can lead to stroke, heart attack, and aneurysm. Ways to prevent and treat cardiovascular disease are constantly being improved.

5.1 Overview of the Cardiovascular System

The cardiovascular system consists of (1) the heart, which pumps blood, and (2) the blood vessels, through which the blood flows. The beating of the heart sends blood into the blood vessels and, in humans, blood is always contained within blood vessels.

Circulation Performs Exchanges

Even though circulation of blood depends on the beating of the heart, the actual purpose of circulation is to service the cells. Recall that cells are always bathed by **tissue fluid**—in other words, blood exchanges substances with tissue fluid and not directly with cells. On the one hand, blood removes waste products from tissue fluid and, on the other, it brings tissue fluid the oxygen and nutrients cells need to continue

their existence. Blood would not be able to continue to perform this function if it did not become refreshed in the lungs, at the intestines, and at the kidneys (Fig. 5.1).

Following up on Allen's suggestion in the opening story, we can liken the blood vessels to streets that allow cars to move about. Just imagine that you are driving about town to return birthday gifts that are not useful for items you do find useful. Just like blood, at some stops you give up the useless item (wastes) and at these or other stops you acquire something of use (nutrients and oxygen). Similarly, at the lungs, carbon dioxide leaves blood and oxygen enters it, as indicated by the two arrows in and out of the lungs in Figure 5.1. Blood is purified of its wastes at the kidneys, while water and salts are retained as needed. Finally, nutrients enter blood at the intestines. The liver, the largest organ of the body, is very important because it takes up amino acids from the blood and returns proteins to it, which helps transport substances such as fats in the blood. The liver removes any poisons that have entered the blood at the intestines.

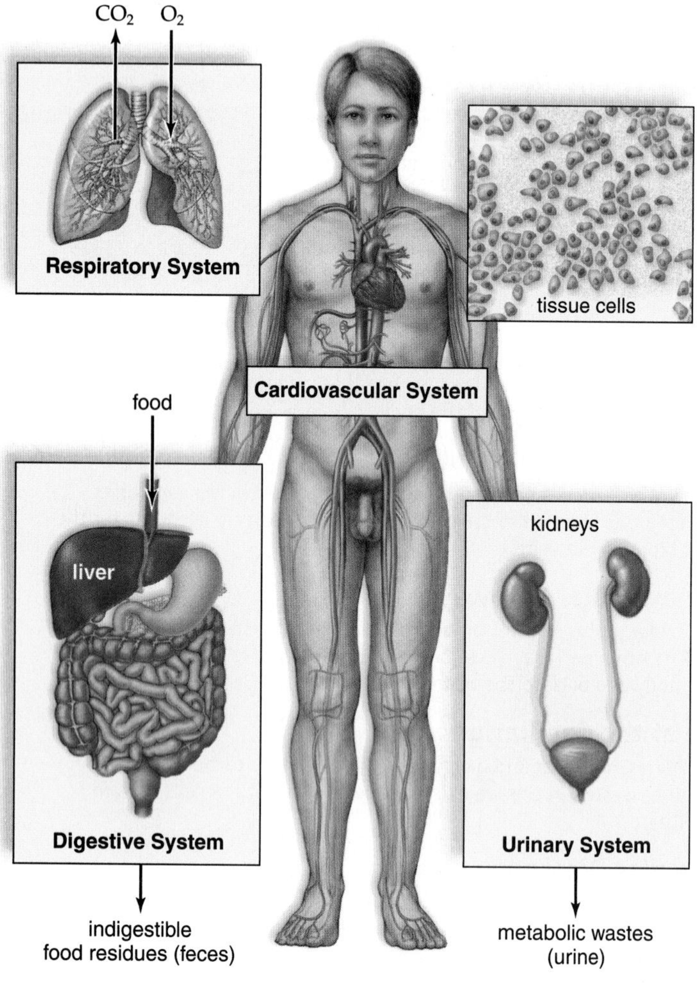

Figure 5.1   The cardiovascular system.
The cardiovascular system transports blood about the body and, with the help of other systems in the body, it maintains favorable conditions for the cells of the body.

Functions of the Cardiovascular System

In this chapter, we will see that

1. contractions of the heart generates blood pressure, which moves blood through blood vessels.
2. blood vessels transport the blood, which moves from the heart into arteries, capillaries, and veins, before returning to the heart.
3. exchanges at the capillaries (the smallest of the blood vessels) refreshes blood and then tissue fluid, sometimes called interstitial fluid.
4. the heart and blood vessels regulate blood flow, according to the needs of the body.

Lymphatic System

The **lymphatic system** is of assistance to the cardiovascular system because lymphatic vessels collect excess tissue fluid and return it to the cardiovascular system. When exchanges occur in the tissues between blood and tissue fluid, water collects in the tissues. This water enters lymphatic vessels, which start in the tissues and end at cardiovascular veins in the shoulders. As soon as fluid enters lymphatic vessels, it is called lymph. Lymph, you will recall, is a fluid tissue, as is blood (see p. 64).

✓ Check Your Progress 5.1

1. What are the two parts of the cardiovascular system?
2. What are the functions of the cardiovascular system?
3. How does the lymphatic system help the cardiovascular system?

5.2 The Types of Blood Vessels

As Allen advised Sharon in the opening story, the structure of the three types of blood vessels (arteries, capillaries, and veins) is appropriate to their function (Fig. 5.2).

The Arteries: From the Heart

The arterial wall has three layers. The innermost layer is a thin layer of cells called endothelium, the middle layer is a relatively thick layer of smooth muscle and elastic tissue, while the outer layer is connective tissue. The strong walls of an artery give it support when blood enters under pressure; the elastic tissue allows an artery to expand to absorb the pressure. **Arterioles** are small arteries just visible to the naked eye. The middle layer of arterioles has some elastic tissue but is composed mostly of smooth muscle, whose fibers encircle the arteriole. When these muscle fibers contract, the vessel constricts; when these muscle fibers relax, the vessel dilates. Whether arterioles are constricted or dilated regulates blood pressure. The greater the number of vessels dilated, the lower the blood pressure.

The Capillaries: Exchange

Arterioles branch into capillaries. Each capillary is an extremely narrow, microscopic tube with a wall composed only of endothelium (a single layer of epithelial cells) with a basement membrane. Although each capillary is small, their total surface area in humans is about 6,300 square meters. Capillary beds (networks of many capillaries) are present in all regions of the body, so no cell is far from a capillary. In the tissues, only certain capillaries are open at any given time. For example, after eating, the capillaries that serve the digestive system are open, and those that serve the muscles are closed. When a capillary bed is closed, the precapillary sphincters contract, and the blood moves from arteriole to venule by way of an arteriovenous shunt.

The Veins: To the Heart

Venules are small veins that drain blood from the capillaries and then join to form a vein. The walls of venules (and veins) have the same three layers as arteries, but there is less smooth muscle in the middle layer and less connective tissue in the outer layer. Therefore, the wall of a vein is thinner than that of an artery.

Veins often have **valves,** which allow blood to flow only toward the heart when open and prevent the backward flow of blood when closed. Valves are found in the veins that carry blood against the force of gravity, especially the veins of the lower extremities.

Since the walls of veins are thinner, they can expand to a greater extent. At any one time, about 70% of the blood is in the veins. In this way, the veins act as a blood reservoir. If blood is lost due to hemorrhaging, nervous stimulation causes the veins to constrict, providing more blood to the rest of the body.

☑ Check Your Progress 5.2

1. What are the types of blood vessels?
2. What is the structure and function of these vessels?

Figure 5.2   Anatomy of a capillary bed.
A capillary bed forms a maze of capillary vessels that lies between an arteriole and a venule. When sphincter muscles are relaxed, the capillary bed is open, and blood flows through the capillaries. Otherwise, blood flows through a shunt that carries blood directly from an arteriole to a venule. As blood passes through a capillary in the tissues, it gives up its oxygen (O_2). Therefore, blood goes from being O_2-rich in the arteriole (red color) to being O_2-poor in the vein (blue color).

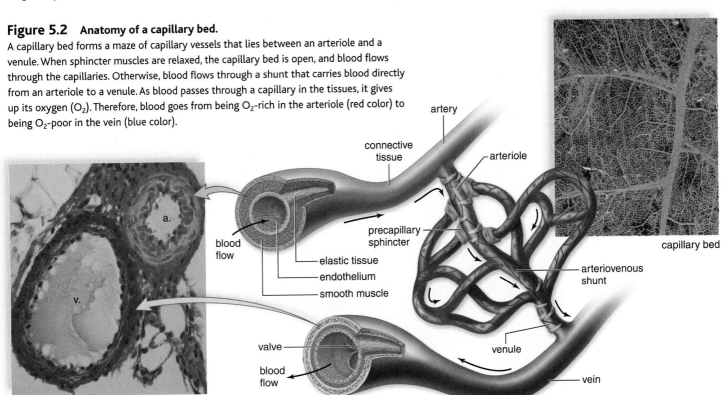

artery

connective tissue

arteriole

precapillary sphincter

capillary bed

blood flow

elastic tissue

endothelium

smooth muscle

arteriovenous shunt

a.

v.

valve

blood flow

venule

vein

v = vein; a = artery

5.3 The Heart Is a Double Pump

The **heart** is a cone-shaped, muscular organ located between the lungs, directly behind the sternum (breastbone). The heart is tilted so that the apex (the pointed end) is oriented to the left (Fig. 5.3). To approximate the size of your heart, make a fist, then clasp the fist with your opposite hand. The major portion of the heart, called the **myocardium,** consists largely of cardiac muscle tissue. Myocardium is serviced by the coronary artery and cardiac vein and not by the blood it pumps. The muscle fibers of myocardium are branched and tightly joined to one another at intercalated disks. The heart is surrounded by the **pericardium,** a thick, membranous sac that supports and protects the heart. The inside of the pericardium secretes a lubrication fluid, and the pericardium slides smoothly over the surface as the heart pumps the blood.

Internally, a wall called the **septum** separates the heart into a right side and a left side (Fig. 5.4*a*). The heart has four chambers. The two upper, thin-walled atria (sing., **atrium**) are called the right atrium and the left atrium. Each atrium has a wrinkled, protruding appendage called an auricle. The two lower chambers are the thick-walled **ventricles,** called the right ventricle and the left ventricle (Fig. 5.4*a*).

Heart valves keep blood flowing in the right direction and prevent its backward movement. The valves that lie between the atria and the ventricles are called the **atrioventricular valves (AV valves).** These valves are supported by strong fibrous strings called **chordae tendineae.** The chordae, which are attached to muscular projections of the ventricular walls, called papillary muscles, support the valves and prevent them from inverting when the heart contracts. The AV valve on the right side is called the tricuspid valve because it has three flaps, or cusps. The AV valve on the left side is called the bicuspid valve because it has two flaps. The bicuspid valve is commonly referred to as the mitral valve, because it has a shape like a bishop's hat, or mitre. The remaining two valves are the **semilunar valves,** with flaps shaped like half-moons, that lie between the ventricles and their attached vessels. The semilunar valves are named for their attached vessels: The pulmonary semilunar valve lies between the right ventricle and the pulmonary trunk. The aortic semilunar valve lies between the left ventricle and the aorta.

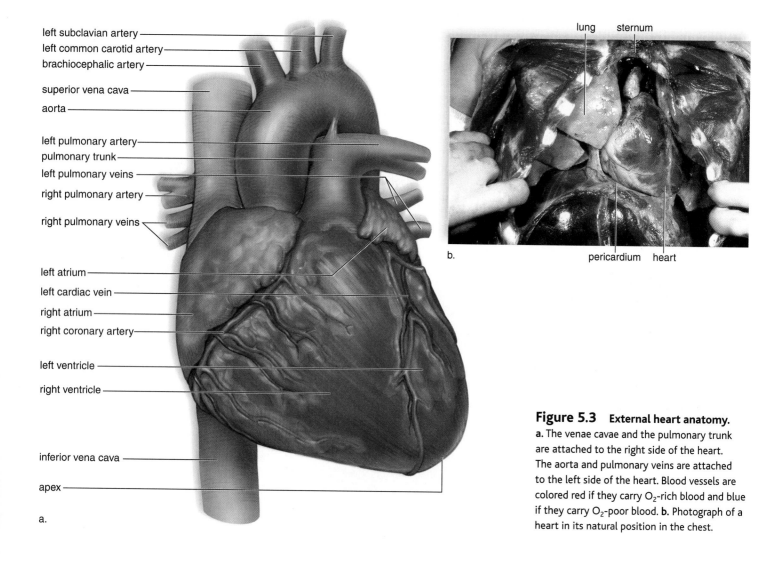

left subclavian artery
left common carotid artery
brachiocephalic artery
superior vena cava
aorta
left pulmonary artery
pulmonary trunk
left pulmonary veins
right pulmonary artery
right pulmonary veins
left atrium
left cardiac vein
right atrium
right coronary artery
left ventricle
right ventricle
inferior vena cava
apex

a.

lung   sternum

b.

pericardium   heart

Figure 5.3   External heart anatomy.
a. The venae cavae and the pulmonary trunk are attached to the right side of the heart. The aorta and pulmonary veins are attached to the left side of the heart. Blood vessels are colored red if they carry O_2-rich blood and blue if they carry O_2-poor blood. **b.** Photograph of a heart in its natural position in the chest.

Passage of Blood Through the Heart

Even though the presence of intercalated disks (Fig. 5.4b) between cardiac muscle cells allows both atria and then both ventricles to contract simultaneously, we can trace the path of blood through the heart in the following manner:

- The superior vena cava and the inferior vena cava, which carry O₂-poor blood, enter the right atrium.
- The right atrium sends blood through an atrioventricular valve (the tricuspid valve) to the right ventricle.
- The right ventricle sends blood through the pulmonary semilunar valve into the pulmonary trunk. The pulmonary trunk, which carries O₂-poor blood, divides into two **pulmonary arteries,** which go to the lungs.
- Four **pulmonary veins,** which carry O₂-rich blood, enter the left atrium.
- The left atrium sends blood through an atrioventricular valve (the bicuspid [mitral] valve) to the left ventricle.
- The left ventricle sends blood through the aortic semilunar valve into the aorta to the body proper.

From this description, you can see that O₂-poor blood never mixes with O₂-rich blood and that blood must go through the lungs in order to pass from the right side to the left side of the heart. The heart is a double pump because the right ventricle of the heart sends blood through the lungs, and the left ventricle sends blood throughout the body. Sharon had no trouble deciding that the left ventricle must have the harder job of pumping blood. She showed Allen that the atria have thin walls, and each pumps blood into the ventricle right below it. The ventricles are thicker, and they pump blood into blood vessels that travel to other parts of the body. The thinner myocardium of the right ventricle pumps blood to the lungs, which are nearby in the thoracic cavity. The left ventricle has a thicker wall than the right ventricle, and this enables it to pump blood to all the other parts of the body.

The pumping of the heart sends blood out under pressure into the arteries. Because the left side of the heart is the stronger pump, blood pressure is greatest in the aorta. Blood pressure then decreases as the total cross-sectional area of arteries and then arterioles increases (see Fig. 5.8). A different mechanism, aside from blood pressure, is used to move blood in the veins.

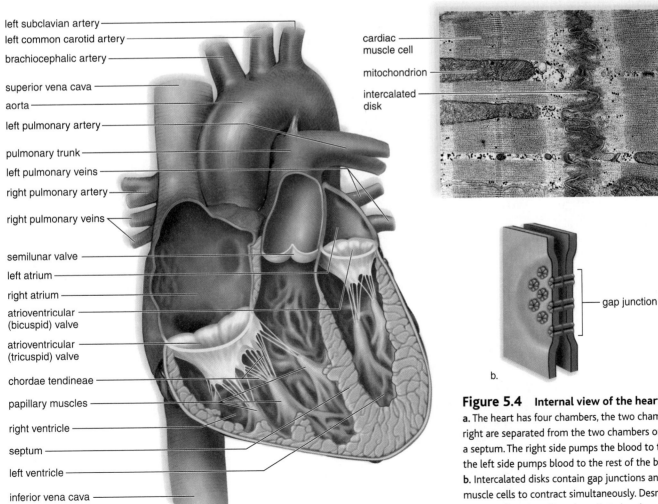

left subclavian artery
left common carotid artery
brachiocephalic artery
superior vena cava
aorta
left pulmonary artery
pulmonary trunk
left pulmonary veins
right pulmonary artery
right pulmonary veins
semilunar valve
left atrium
right atrium
atrioventricular (bicuspid) valve
atrioventricular (tricuspid) valve
chordae tendineae
papillary muscles
right ventricle
septum
left ventricle
inferior vena cava

a.

cardiac muscle cell
mitochondrion
intercalated disk

gap junction

b.

Figure 5.4 **Internal view of the heart.**
a. The heart has four chambers, the two chambers on the right are separated from the two chambers on the left by a septum. The right side pumps the blood to the lungs and the left side pumps blood to the rest of the body.
b. Intercalated disks contain gap junctions and these allow muscle cells to contract simultaneously. Desmosomes at the same location allow the cells to bend and stretch.

The Heartbeat Is Controlled

Each heartbeat is called a **cardiac cycle** (Fig. 5.5). When the heart beats, first the two atria contract at the same time; next, the two ventricles contract at the same time. Then, all chambers relax. **Systole,** the working phase, refers to contraction of the chambers, and the word **diastole,** the resting phase, refers to relaxation of the chambers. The heart contracts, or beats, about 70 times a minute, and each heartbeat lasts about 0.85 second. A normal adult rate at rest can vary from 60 to 80 beats per minute.

Allen told Sharon, in order to understand the "lubdup" sounds of a heartbeat, concentrate on the ventricles. The first sound, "lub," occurs when increasing pressure of blood inside a ventricle forces the cusps of the AV valve to slam shut. The force of the blood is just like a strong wind that blows a door (the valve cusps) shut, Allen thought. In contrast, the pressure of blood inside a ventricle causes the semilunar valves (pulmonary and aortic) to open. The "dup" occurs when the ventricles relax, and blood in the arteries pushes back, causing the semilunar valves to close.

A heart murmur, or a slight swishing sound after the "lub," is often due to ineffective valves, which allow blood to pass back into the atria after the AV valves have closed. Rheumatic fever resulting from a bacterial infection is one possible cause of a faulty valve, particularly the bicuspid valve. Faulty valves can be surgically corrected.

Internal Control of Heartbeat

The rhythmical contraction of the atria and ventricles is due to the internal (intrinsic) conduction system of the heart. Nodal tissue, which has both muscular and nervous characteristics, is a unique type of cardiac muscle located in two regions of the heart. The **SA (sinoatrial) node** is located in the upper dorsal wall of the right atrium; the **AV (atrioventricular) node** is located in the base of the right atrium very near the septum (Fig. 5.6a). The SA node initiates the heartbeat and automatically sends out an excitation impulse every 0.85 second; this causes the atria to contract. When impulses reach the AV node, there is a slight delay that allows the atria to finish their contraction before the ventricles begin their contraction. The signal for the ventricles to contract travels from the AV node through the two branches of the **atrioventricular (AV) bundle** before reaching the numerous and smaller **Purkinje fibers.** The AV bundle, its branches, and the Purkinje fibers work efficiently because gap junctions at intercalated disks allow electrical current to flow from cell to cell (see Fig. 5.4b).

The SA node is called the **pacemaker** because it usually keeps the heartbeat regular. If the SA node fails to work properly, the heart still beats due to impulses generated by the AV node. But the beat is slower (40 to 60 beats per minute). To correct this condition, it is possible to implant an artificial pacemaker, which automatically gives an electrical stimulus to the heart every 0.85 second.

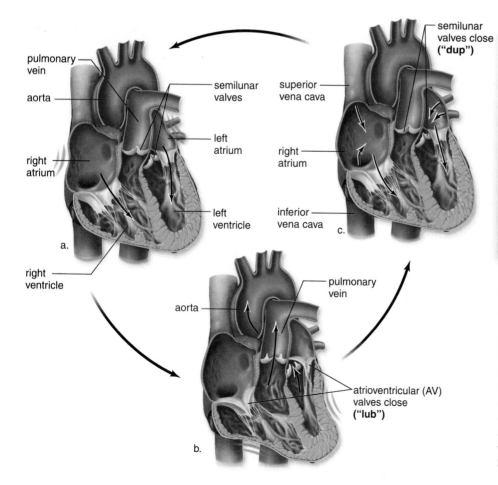

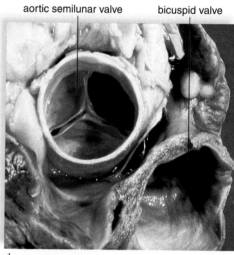

Figure 5.5   **Stages in the cardiac cycle.**
a. When the atria contract, the ventricles are relaxed and filling with blood. The atrioventricular valves are open, and the semilunar valves are closed.
b. When the ventricles contract, the atrioventricular valves are closed, the semilunar valves are open, and the blood is pumped into the pulmonary trunk and aorta. **c.** When the heart is relaxed, both the atria and the ventricles are filling with blood. The atrioventricular valves are open, and the semilunar valves are closed. **d.** Aortic semilunar valve and bicuspid valve.

External Control of Heartbeat

The body has an external (extrinsic) way to regulate the heartbeat. A cardiac control center in the medulla oblongata, a portion of the brain that controls internal organs, can alter the beat of the heart by way of the parasympathetic and sympathetic portions of the nervous system. Allen reminded Sharon that, as studied in Chapter 1, the parasympathetic division promotes those functions associated with a resting state, and the sympathetic division brings about those responses associated with fight or flight. It makes sense, then, said Allen, that the parasympathetic division decreases SA and AV nodal activity when we are inactive, and the sympathetic division increases SA and AV nodal activity when we are active or excited.

The hormones epinephrine and norepinephrine, which are released by the adrenal medulla, also stimulate the heart. During exercise, for example, the heart pumps faster and stronger due to sympathetic stimulation and due to the release of epinephrine and norepinephrine.

The Electrocardiogram Is a Record of the Heartbeat

An **electrocardiogram (ECG)** is a recording of the electrical changes that occur in myocardium during a cardiac cycle. Body fluids contain ions that conduct electric currents, and therefore, the electrical changes in myocardium can be detected on the skin's surface. When an ECG is being taken, electrodes placed on the skin are connected by wires to an instrument that detects the myocardium's electrical changes. Thereafter, a pen rises or falls on a moving strip of paper. Figure 5.6*b* depicts the pen's movements during a normal cardiac cycle.

When the SA node triggers an impulse, the atrial fibers produce an electrical change called the P wave. The P wave indicates the atria are about to contract. After that, the QRS complex signals that the ventricles are about to contract. The electrical changes that occur as the ventricular muscle fibers recover produce the T wave.

Various types of abnormalities can be detected by an ECG. One of these, called ventricular fibrillation, is an uncoordinated contraction of the ventricles (Fig. 5.6*c*). Ventricular fibrillation is of special interest because it can be caused by an injury or drug overdose. It is the most common cause of sudden cardiac death in a seemingly healthy person over age 35. Once the ventricles are fibrillating, they have to be defibrillated by applying a strong electric current for a short period of time. Then the SA node may be able to reestablish a coordinated beat.

✔ Check Your Progress 5.3

1. Why is the heart a double pump?

2. What causes the "lub" and the "dup" of the heart sounds?

3. What keeps the heartbeat regular?

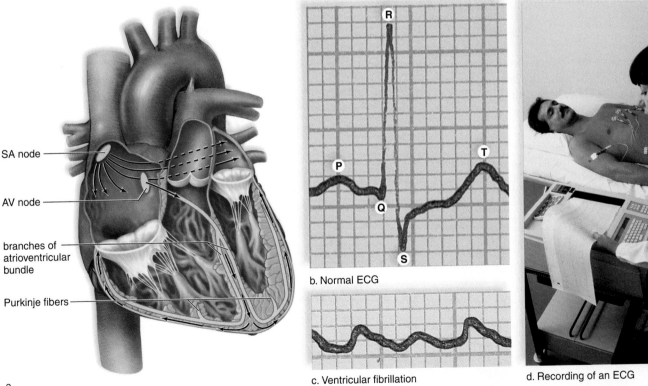

b. Normal ECG

c. Ventricular fibrillation

d. Recording of an ECG

Figure 5.6 Conduction system of the heart.
a. The SA node sends out a stimulus (black arrows), which causes the atria to contract. When this stimulus reaches the AV node, it signals the ventricles to contract. Impulses pass down the two branches of the atrioventricular bundle to the Purkinje fibers, and thereafter, the ventricles contract. **b.** A normal ECG indicates that the heart is functioning properly. The P wave occurs just prior to atrial contraction; the QRS complex occurs just prior to ventricular contraction; and the T wave occurs when the ventricles are recovering from contraction. **c.** Ventricular fibrillation produces an irregular ECG due to irregular stimulation of the ventricles. **d.** The recording of an ECG.

SA node

AV node

branches of atrioventricular bundle

Purkinje fibers

a.

5.4 Features of the Cardiovascular System

When the left ventricle contracts, blood is sent out into the aorta under pressure. A progressive decrease in pressure occurs as blood moves through the arteries, arterioles, capillaries, venules, and finally the veins. Blood pressure is highest in the aorta and lowest in the venae cavae, which enter the right atrium.

Pulse Rate Equals Heart Rate

The surge of blood entering the arteries causes their elastic walls to stretch, but then they almost immediately recoil. This rhythmic expansion and recoil of an arterial wall can be felt as a **pulse** in any artery that runs close to the body's surface. It is customary to feel the pulse by placing several fingers on the radial artery, which lies near the outer border of the palm side of a wrist. A carotid artery, on either side of the trachea in the neck, is another accessible location for feeling the pulse.

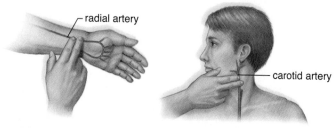

radial artery

carotid artery

Normally, the pulse rate indicates the heart rate because the arterial walls pulse whenever the left ventricle contracts. The pulse rate is usually 70 beats per minute but can vary between 60 and 80 beats per minute.

Blood Flow Is Regulated

The beating of the heart is necessary to homeostasis because it creates the pressure that propels blood in the arteries and the arterioles. Arterioles lead to the capillaries where exchange with tissue fluid takes place.

Blood Pressure Moves Blood in Arteries

Blood pressure is the pressure of blood against the wall of a blood vessel. A sphygmomanometer (blood pressure instrument) can be used to measure blood pressure, usually in the brachial artery of the arm (Fig. 5.7). The highest arterial pressure, called the **systolic pressure,** is reached during ejection of blood from the heart. The lowest arterial pressure, called the **diastolic pressure,** occurs while the heart ventricles are relaxing. Normal resting blood pressure for a young adult is said to be 120 mm mercury (Hg) over 80 mm Hg, or simply 120/80, but these values can vary somewhat and still be within the range of normal blood pressure (Table 5.1). The higher number is the systolic pressure, and the lower number is the diastolic pressure. Actually, blood pressure is higher than 120/80 in the aorta and lower than 120/80 in the venae cavae. High blood pressure is called hypertension, and low blood pressure is called hypotension.

Table 5.1	Normal Values for Adult Blood Pressure (mm)	
	Systolic Pressure	**Diastolic Pressure**
Normal	95–135	50–90
Hypertension	Greater than 135	Greater than 90
Hypotension	Less than 95	Less than 50

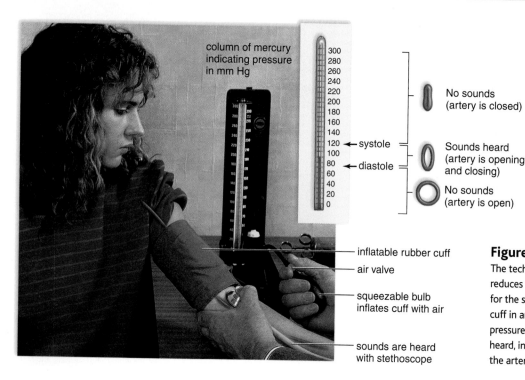

column of mercury indicating pressure in mm Hg

300
280
260
240
220
200
180
160
140
120 ← systole
100
80 ← diastole
60
40
20
0

No sounds (artery is closed)

Sounds heard (artery is opening and closing)

No sounds (artery is open)

inflatable rubber cuff

air valve

squeezable bulb inflates cuff with air

sounds are heard with stethoscope

Figure 5.7   Use of a sphygmomanometer.
The technician inflates the cuff with air, gradually reduces the pressure, and listens with a stethoscope for the sounds that indicate blood is moving past the cuff in an artery. This is systolic blood pressure. The pressure in the cuff is further reduced until no sound is heard, indicating that blood is flowing freely through the artery. This is diastolic pressure.

Both systolic and diastolic blood pressure decrease with distance from the left ventricle because the total cross-sectional area of the blood vessels increases—there are more arterioles than arteries. The decrease in blood pressure causes the blood velocity to gradually decrease as it flows toward the capillaries.

Blood Flow Is Slow in the Capillaries

There are many more capillaries than arterioles, and blood moves slowly through the capillaries (Fig. 5.8). This is important because the slow progress allows time for the exchange of substances between the blood in the capillaries and the surrounding tissues.

Blood Flow in Veins Returns Blood to Heart

While they were studying, Sharon pointed to Figure 5.8 and showed Allen that blood pressure is minimal in venules and veins. Even so, velocity of flow increases in veins. "There must be some influence, rather than blood pressure, that causes blood to flow in the veins," said Sharon. Indeed, venous return is dependent upon three factors:

1. the **skeletal muscle pump,** which is dependent on skeletal muscle contraction;

2. the **respiratory pump,** which is dependent on breathing;
3. valves in veins.

The skeletal muscle pump works like this: When skeletal muscles contract, they compress the weak walls of the veins. This causes the blood to move past a valve (Fig. 5.9). Once past the valve, blood cannot flow backward. The importance of the skeletal muscle pump in moving blood in the veins can be demonstrated by forcing a person to stand rigidly still for an hour or so. Fainting may occur because blood collects in the limbs, depriving the brain of needed oxygen. In this case, fainting is beneficial because the resulting horizontal position aids in getting blood to the head.

The respiratory pump works like this: When we inhale, the chest expands, and this reduces pressure in the thoracic cavity. Blood will flow from higher pressure (in the abdominal cavity) to lower pressure (in the thoracic cavity). When we exhale, the pressure reverses, but again the valves in the veins prevent backward flow. "Another example of structure suits function," said Allen, "because the function of the veins is to return blood to the heart."

✓ Check Your Progress 5.4

1. What does the pulse rate of a person indicate?
2. What accounts for the flow of blood in the arteries?
3. What accounts for the flow of blood in the veins?

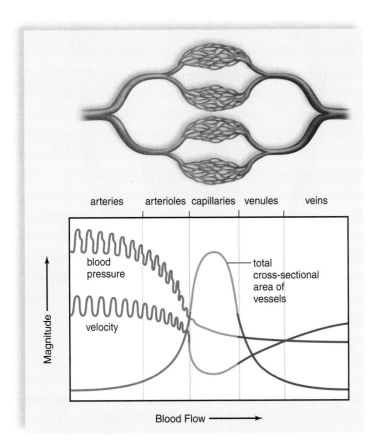

Figure 5.8   Velocity and blood pressure related to vascular cross-sectional area.
In capillaries, blood is under minimal pressure and has the least velocity. Blood pressure and blood velocity drop off because capillaries have a greater total cross-sectional area than arterioles.

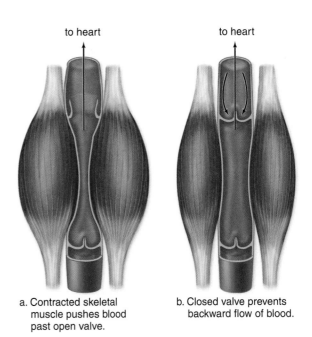

a. Contracted skeletal muscle pushes blood past open valve.

b. Closed valve prevents backward flow of blood.

Figure 5.9   Cross section of a valve in a vein.
a. Pressure on the walls of a vein, exerted by skeletal muscles, increases blood pressure within the vein and forces a valve open. **b.** When external pressure is no longer applied to the vein, blood pressure decreases, and back pressure forces the valve closed. Closure of the valves prevents the blood from flowing backward, and therefore, veins return blood to the heart.

5.5 Two Cardiovascular Pathways

The blood flows in two circuits: the **pulmonary circuit,** which circulates blood through the lungs, and the **systemic circuit,** which serves the needs of body tissues (Fig. 5.10). Both circuits, as we shall see, are necessary to homeostasis.

The Pulmonary Circuit: Exchange of Gases

The path of blood through the lungs can be traced as follows: Blood from all regions of the body first collects in the right atrium and then passes into the right ventricle, which pumps it into the pulmonary trunk. The pulmonary trunk divides into the right and left pulmonary arteries, which branch as they approach the lungs. The arterioles take blood to the pulmonary capillaries, where carbon dioxide is given off and oxygen is picked up. Blood then passes through the pulmonary venules, which lead to the four pulmonary veins that enter the left atrium. Since blood in the pulmonary arteries is O_2-poor but blood in the pulmonary veins is O_2-rich, it is not correct to say that all arteries carry blood that is high in oxygen and all veins carry blood that is low in oxygen (as people tend to believe). Just the reverse is true in the pulmonary circuit.

The Systemic Circuit: Exchanges with Tissue Fluid

The systemic circuit includes all of the arteries and veins shown in Figure 5.10. The heart pumps blood through 60,000 miles of blood vessels in order to deliver nutrients and oxygen and remove wastes from all body cells.

   The largest artery in the systemic circuit, the **aorta,** receives blood from the heart, and the largest veins, the **superior** and **inferior venae cavae,** return blood to the heart. The superior vena cava collects blood from the head, the chest, and the arms, and the inferior vena cava collects blood from the lower body regions. Both enter the right atrium.

Tracing the Path of Blood

Allen told Sharon that it's easy to trace the path of blood in the systemic circuit because you always begin with the left ventricle, which pumps blood into the aorta. Branches from the aorta go to the organs and major body regions. For example, this is the path of blood to and from the lower legs:

> left ventricle—aorta—common iliac artery—
> femoral artery—lower leg capillaries—
> femoral vein—common iliac vein—
> inferior vena cava—right atrium

   "So," said Allen, "when tracing blood, you need only mention the aorta, the proper branch of the aorta, the region, and the vein returning blood to the vena cava." In most instances, the artery and the vein that serve the same region are given the same name (Fig. 5.11). What happens in be-

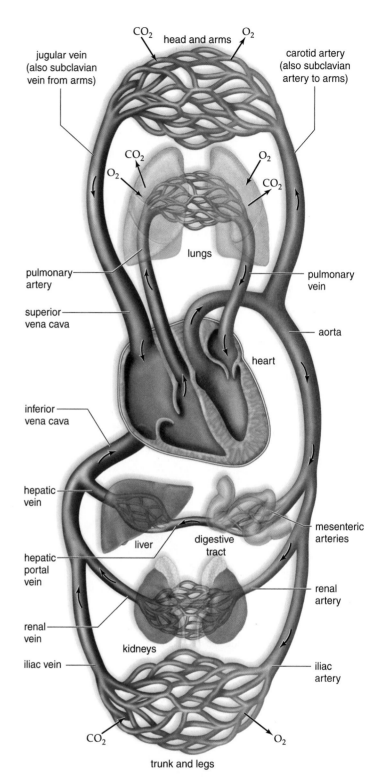

Figure 5.10   Cardiovascular system diagram.
The blue-colored vessels carry blood high in carbon dioxide, and the red-colored vessels carry blood high in oxygen; the arrows indicate the flow of blood. Compare this diagram, useful for learning to trace the path of blood, with Figure 5.11 to realize that arteries and veins go to all parts of the body. Also, there are capillaries in all parts of the body. No cell is located far from a capillary.

tween the artery and the vein? Arterioles from the artery branch into capillaries, where exchange takes place, and then venules join into the vein that enters a vena cava.

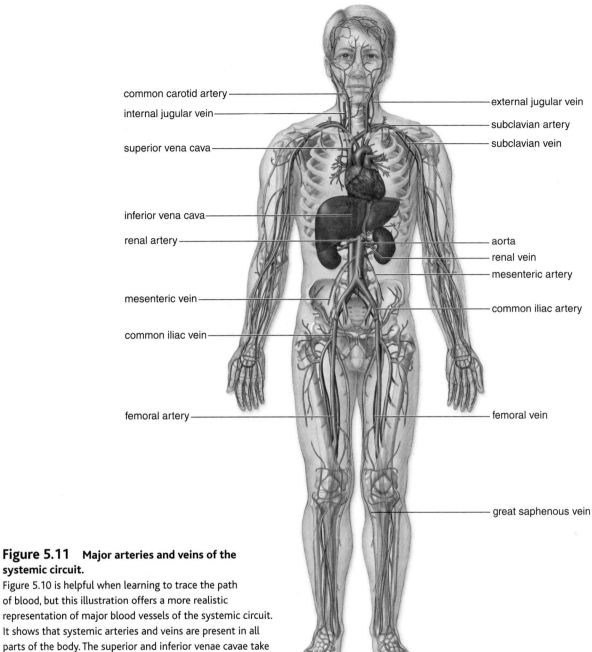

common carotid artery

internal jugular vein

superior vena cava

inferior vena cava

renal artery

mesenteric vein

common iliac vein

femoral artery

external jugular vein

subclavian artery

subclavian vein

aorta

renal vein

mesenteric artery

common iliac artery

femoral vein

great saphenous vein

Figure 5.11 Major arteries and veins of the systemic circuit.
Figure 5.10 is helpful when learning to trace the path of blood, but this illustration offers a more realistic representation of major blood vessels of the systemic circuit. It shows that systemic arteries and veins are present in all parts of the body. The superior and inferior venae cavae take their names from their relationship to which organ?

Coronary Circulation

The **coronary arteries** (see Fig. 5.3) serve the heart muscle itself. (The heart is not nourished by the blood in its chambers.) The coronary arteries are the first branches off the aorta. They originate just above the aortic semilunar valve, and they lie on the exterior surface of the heart, where they divide into diverse arterioles. Because they have a very small diameter, the coronary arteries may become clogged, as discussed in the Health Focus on page 98. The coronary capillary beds join to form venules. The venules converge to form the cardiac veins, which empty into the right atrium.

Hepatic Portal System

The **hepatic portal vein** takes blood from the capillary bed of the digestive tract to a capillary bed in the liver. A so-called

portal system always lies between capillary beds. The blood in the hepatic portal vein is O_2 poor, but rich in glucose and amino acids. The liver stores glucose as glycogen, and it synthesizes blood proteins from the amino acids or else stores them. The liver also purifies the blood of pathogens that have entered the body by way of the intestinal capillaries. The **hepatic vein** leaves the liver and enters the inferior vena cava.

✔ Check Your Progress 5.5

1. What is the relative oxygen content of the blood flowing in the pulmonary artery compared with that in the pulmonary vein? Explain.

2. What is the pathway of blood in the pulmonary circuit?

3. Trace the path of blood from the heart to the digestive tract and back to the heart by way of the hepatic portal vein.

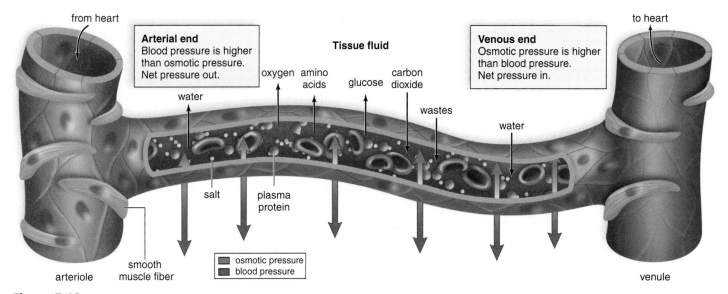

Figure 5.12   Capillary exchange.
A capillary, illustrating the exchanges that take place and the forces that aid the process. At the arterial end of a capillary, the blood pressure is higher than the osmotic pressure; therefore, water (H_2O) tends to leave the bloodstream. In the midsection, molecules, including oxygen (O_2) and carbon dioxide (CO_2), follow their concentration gradients. At the venous end of a capillary, the osmotic pressure is higher than the blood pressure; therefore, water tends to enter the bloodstream. Notice that the red blood cells and the plasma proteins are too large to exit a capillary.

5.6 Exchange at the Capillaries

Two forces primarily control movement of fluid through the capillary wall: blood pressure, which tends to cause water to move from capillary to tissue fluid, and osmotic pressure, which tends to cause water to move in the opposite direction. At the arterial end of a capillary, blood pressure (30 mm Hg) is higher than the osmotic pressure of blood (21 mm Hg) (Fig. 5.12). Osmotic pressure is created by the presence of salts and the plasma proteins. Because blood pressure is higher than osmotic pressure at the arterial end of a capillary, water exits a capillary at this end.

Midway along the capillary, where blood pressure is lower, the two forces essentially cancel each other, and there is no net movement of water. Solutes now diffuse according to their concentration gradient: Oxygen and nutrients (glucose and amino acids) diffuse out of the capillary; carbon dioxide and wastes diffuse into the capillary. Red blood cells and almost all plasma proteins remain in the capillaries. The substances that leave a capillary contribute to tissue fluid, the fluid between the body's cells. Since plasma proteins are too large to readily pass out of the capillary, tissue fluid tends to contain all components of plasma, except much lesser amounts of protein.

At the venule end of a capillary, where blood pressure has fallen even more, osmotic pressure is greater than blood pressure, and water tends to move into the capillary. Almost the same amount of fluid that left the capillary returns to it, although some excess tissue fluid is always collected by the lymphatic capillaries (Fig. 5.13). Tissue fluid contained

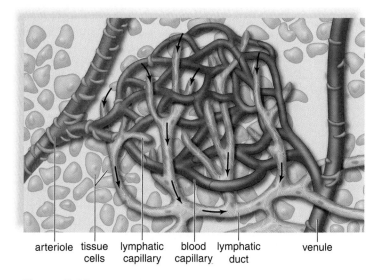

Figure 5.13   Capillary bed.
A lymphatic capillary bed, near a blood capillary bed. When lymphatic capillaries take up excess tissue fluid, it becomes lymph. Precapillary sphincters can shut down a blood capillary, and blood then flows through the shunt.

within lymphatic vessels is called **lymph.** Lymph is returned to the systemic venous blood when the major lymphatic vessels enter the subclavian veins in the shoulder region.

✓ Check Your Progress 5.6

1. How does the exchange of materials take place across a capillary wall in the tissues?

2. What happens to excess tissue fluid created by capillary exchange?

5.7 Cardiovascular Disorders

Cardiovascular disease (CVD) is the leading cause of untimely death in the Western countries. Modern research efforts have resulted in improved diagnosis, treatment, and prevention. This section discusses the range of advances that have been made, first in correcting vascular disorders, and then in correcting heart disorders. The Health Focus on page 98 emphasizes how to prevent CVD from developing in the first place.

Disorders of the Blood Vessels

Hypertension and atherosclerosis lead, most often, to stroke due to an artery blocked by a blood clot, or a heart attack due to a coronary artery clogged by plaque. Treatment involves doing away with the blood clot or prying open the coronary artery. Another possible outcome is an **aneurysm,** a burst blood vessel, which can be prevented by replacing a blood vessel about to burst with an artificial one.

High Blood Pressure

Hypertension occurs when blood moves through the arteries at a higher pressure than normal. Also called high blood pressure, hypertension is sometimes called a silent killer because it may not be detected until it has caused a heart attack, stroke, or even kidney failure. Hypertension is present when the systolic blood pressure is 140 or greater or the diastolic blood pressure is 90 or greater. While both systolic and diastolic pressures are considered important, it is the diastolic pressure that is emphasized when medical treatment is being considered.

The best safeguard against developing hypertension is regular blood pressure checks and a lifestyle that lowers the risk of CVD. If already present, a physician can prescribe various drugs that help lower blood pressure. Diuretics cause the kidneys to excrete water; beta blockers and angiotensin-converting enzyme (ACE) inhibitors counteract hormones that tend to raise the blood pressure.

Hypertension is often seen in individuals who have **atherosclerosis,** an accumulation of soft masses of fatty materials, including cholesterol, beneath the inner linings of arteries. Such deposits are called **plaque.** As it develops, plaque tends to protrude into the lumen of the vessel and interfere with the flow of blood (Fig. 5.14). In most instances, atherosclerosis begins in early adulthood and develops progressively through middle age, but symptoms may not appear until an individual is 50 or older. To prevent the onset and development of plaque, the American Heart Association and other organizations recommend a diet low in saturated fat and cholesterol but rich in omega-3 polyunsaturated fatty acids, as discussed in the Health Focus on page 98.

Plaque can cause a clot to form on the irregular arterial wall. As long as the clot remains stationary, it is called a **thrombus,** but when and if it dislodges and moves along

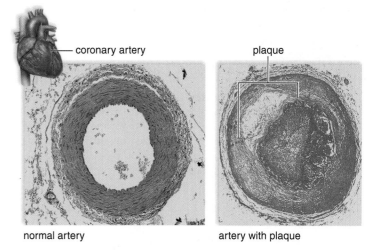

Figure 5.14   Coronary arteries and plaque.
When plaque is present in a coronary artery, a heart attack is more apt to occur because of restricted blood flow.

with the blood, it is called an **embolus.** If **thromboembolism** (a clot that has been carried in the bloodstream but is now stationary) is not treated, complications can arise, as mentioned in the following section.

Research has suggested several possible causes for atherosclerosis aside from hypertension. Chief among these, as discussed in the Health Focus reading on page 98, are smoking and a diet rich in lipids and cholesterol. Research also indicates that a low-level bacterial or viral infection that spreads to the blood may cause an injury that starts the process of atherosclerosis. This infection may surprisingly originate with gum diseases or be due to *Helicobacter pylori* (the bacterium that causes ulcers.) People who have high levels of C-reactive protein, a protein that occurs in the blood following a cold or injury, are more likely to have a heart attack.

Stroke, Heart Attack, and Aneurysm

Stroke, heart attack, and aneurysm are associated with hypertension and atherosclerosis. A cerebrovascular accident (CVA), also called a **stroke,** often results when a small cranial arteriole bursts or is blocked by an embolus. A lack of oxygen causes a portion of the brain to die, and paralysis or death can result. A person is sometimes forewarned of a stroke by a feeling of numbness in the hands or the face, difficulty in speaking, or temporary blindness in one eye.

A myocardial infarction (MI), also called a **heart attack,** occurs when a portion of the heart muscle dies due to a lack of oxygen. If a coronary artery becomes partially blocked, the individual may then suffer from **angina pectoris,** characterized by a radiating pain in the left arm. Nitroglycerin or related drugs dilate blood vessels and help relieve the pain. When a coronary artery is completely blocked, perhaps because of thromboembolism, a heart attack occurs.

Health Focus

Prevention of Cardiovascular Disease

Certain genetic factors predispose an individual to cardiovascular disease, such as family history of heart attack under age 55, male gender, and ethnicity (African Americans are at greater risk). Those with one or more of these risk factors need not despair, however. It means only that they should pay particular attention to the following guidelines for a heart-healthy lifestyle.

The Don'ts

Smoking

When a person smokes, the drug nicotine, present in cigarette smoke, enters the bloodstream. Nicotine causes the arterioles to constrict and the blood pressure to rise. Restricted blood flow and cold hands are associated with smoking in most people. More serious is the need for the heart to pump harder to propel the blood through the lungs at a time when the blood's oxygen-carrying capacity is reduced.

Drug Abuse

Stimulants, such as cocaine and amphetamines, can cause an irregular heartbeat and lead to heart attacks in people who are using drugs even for the first time. Intravenous drug use may also result in a cerebral blood clot and stroke.

Too much alcohol can destroy just about every organ in the body, the heart included. But investigators have discovered that people who take an occasional drink have a 20% lower risk of heart disease than do teetotalers. Two to four drinks a week is the recommended limit for men, one to three drinks for women.

Weight Gain

Hypertension (high blood pressure) is prevalent in persons who are more than 20% above the recommended weight for their height. More tissues require servicing, and the heart sends the blood out under greater pressure in those who are overweight. It may be harder to lose weight once it is gained, and therefore, it is recommended that weight control be a lifelong endeavor. Even a slight decrease in weight can bring with it a reduction in hypertension.

Being overweight increases the risk of diabetes type 2, in which glucose damages blood vessels and makes them prone to the development of plaque. Diabetics especially need to follow the diet discussed next.

The Do's

Healthy Diet

A diet low in saturated fats and cholesterol is protective against cardiovascular disease. Cholesterol is ferried in the blood by two types of plasma proteins, called LDL (low-density lipoprotein) and HDL (high-density lipoprotein). LDL (called "bad" lipoprotein) takes cholesterol from the liver to the tissues, and HDL (called "good" lipoprotein) transports cholesterol out of the tissues to the liver. When the LDL level in blood is abnormally high or the HDL level

Figure 5A   Exercising and good health.
Regular exercise helps prevent and control cardiovascular disease.

is abnormally low, cholesterol accumulates in the cells. When cholesterol-laden cells line the arteries, plaque develops, which interferes with circulation.

It is recommended that everyone know his or her blood cholesterol level. Individuals with a high blood cholesterol level (200 mg/100 mL) should be further tested to determine their LDL-cholesterol level. The LDL-cholesterol level, together with other risk factors such as age, family history, presence of diabetes, and whether the patient smokes, will determine who needs dietary therapy to lower their LDL. Eating foods high in saturated fat (red meat, cream, and butter) and foods containing so-called trans fats (margarine, commercially baked goods, and deep-fried foods) raises the LDL-cholesterol level. Unsaturated fatty acids in olive and canola oils, most nuts, and cold-water fish tend to lower LDL-cholesterol levels. Cold-water fish (e.g., halibut, sardines, tuna, and salmon) contain polyunsaturated fatty acids and especially omega-3 polyunsaturated fatty acids that can reduce plaque. If dietary changes are not sufficient, drugs called statins can be used to lower the LDL level.

Exercise

People who exercise are less apt to have cardiovascular disease (Fig. 5A). One study found that moderately active men who spent an average of 48 minutes a day on a leisure-time activity, such as gardening, bowling, or dancing, had one-third fewer heart attacks than peers who spent an average of only 16 minutes each day on such activities. Exercise, which helps keep weight under control, may also help minimize stress and reduce hypertension.

An aneurysm is a ballooning of a blood vessel, most often the abdominal artery or the arteries leading to the brain. Atherosclerosis and hypertension can weaken the wall of an artery to the point that an aneurysm develops. If a major vessel such as the aorta should burst, death is likely. It is possible to replace a damaged or diseased portion of a vessel, such as an artery, with a plastic tube. Cardiovascular function is preserved because exchange with tissue cells can still take place at the capillaries. In the future, it may be possible to use vessels made in the laboratory by injecting a patient's cells inside an inert mold.

Dissolving Blood Clots

Medical treatment for thromboembolism includes the use of t-PA, a biotechnology drug. This drug converts plasminogen, a molecule found in blood, into plasmin, an enzyme that dissolves blood clots. In fact, t-PA, which stands for tissue plasminogen activator, is the body's own way of converting plasminogen to plasmin. t-PA is also being used for thrombolytic stroke patients, but with limited success because some patients experience life-threatening bleeding in the brain. A better treatment might be new biotechnology drugs that act on the plasma membrane to prevent brain cells from releasing and/or receiving toxic chemicals caused by the stroke.

If a person has symptoms of angina or a stroke, aspirin may be prescribed. Aspirin reduces the stickiness of platelets and, thereby, lowers the probability that a clot will form.

There is evidence that aspirin protects against first heart attacks, but there is no clear support for taking aspirin every day to prevent strokes in symptom-free people. Physicians warn that long-term use of aspirin might have harmful effects, including bleeding in the brain.

Treating Clogged Arteries

Cardiovascular disease used to require open-heart surgery and, therefore, a long recuperation time and a long unsightly scar that could occasionally ache. Now, bypass surgery can be accomplished by using robotic technology (Fig. 5.15a). A video camera and instruments are inserted through small cuts, while the surgeon sits at a console and manipulates interchangeable grippers, cutters, and other tools attached to movable arms above the operating table. Looking through two eyepieces, the surgeon gets a three-dimensional view of the operating field. Robotic surgery is also used in valve repairs and other heart procedures.

A **coronary bypass operation** is one way to treat an artery clogged with plaque. A surgeon takes a blood vessel—usually a vein from the leg—and stitches one end to the aorta and the other end to a coronary artery past the point of obstruction. Figure 5.15b shows a triple bypass in which three blood vessel segments have been used to allow blood to flow freely from the aorta to cardiac muscle by way of the coronary artery. Instead of a coronary bypass, gene therapy has been used since 1997 to grow new blood vessels that will carry blood to

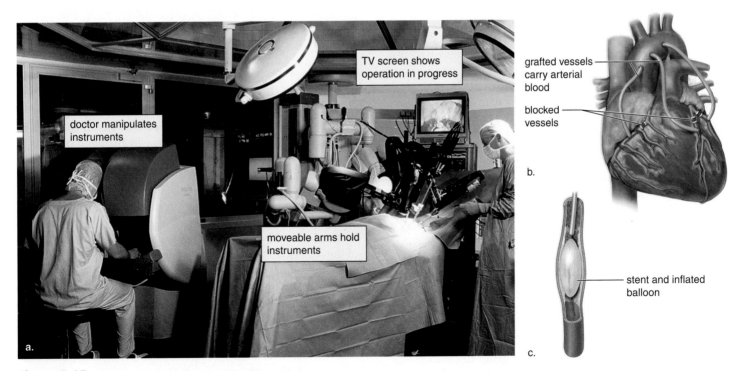

Figure 5.15   Treatment for clogged coronary arteries.
a. Many heart procedures, such as coronary bypass and stent insertion, can be performed using robotic surgery techniques. **b.** During a coronary bypass operation, blood vessels (usually veins from the leg) are stitched to the heart, taking blood past the region of obstruction. **c.** During stenting, a cylinder of expandable metal mesh is positioned inside the coronary artery by using a catheter. Then, a balloon is inflated so that the stent expands and opens the artery.

cardiac muscle. The surgeon need only make a small incision and inject many copies of the gene that codes for VEGF (vascular endothelial growth factor) between the ribs directly into the area of the heart that most needs improved blood flow. VEGF encourages new blood vessels to sprout out of an artery. If collateral blood vessels do form, they transport blood past clogged arteries, making bypass surgery unnecessary. About 60% of all patients who undergo the procedure do show signs of vessel growth within two to four weeks.

Another alternative to bypass surgery is available; namely, the stent (Fig. 5.15c). A stent is a small metal mesh cylinder that holds a coronary artery open after a blockage has been cleared. Until a couple of years ago, stenting was a second step following angioplasty. During **angioplasty,** a plastic tube is inserted into an artery of an arm or a leg and then guided through a major blood vessel toward the heart. When the tube reached the region of plaque in an artery, a balloon attached to the end of the tube was inflated, forcing the vessel open. Now, instead, a stent plus an inner balloon is pushed into the blocked area. When the balloon inside the stent is inflated, it expands, locking itself in place. Some patients go home the same day; at most, after an overnight stay. The stent is more successful when it is coated with a drug that seeps into the artery lining and discourages cell growth. Uncoated stents can close back up in a few months but not ones coated with a drug that discourages closure. Because blood clotting might occur, recipients have to take anticlotting medications.

Disorders of the Heart

When a person has **heart failure,** the heart no longer pumps as it should. Heart failure is a growing problem because people who used to die from heart attacks now survive, but are left with damaged hearts. Often the heart is oversized, not because the cardiac wall is stronger but, because it is sagging and swollen. One idea is to wrap the heart in a fabric sheath to prevent it from getting too big and to allow better pumping, similar to the way a weight lifter's belt restricts and reinforces stomach muscles. But a failing heart can have other problems, such as an abnormal heart rhythm. To counter that condition, it's possible to implant a cardioverter-defibrillator (ICD) just beneath the skin of the chest. These devices can sense both an abnormally slow and an abnormally fast heartbeat. If the former, they generate the missing beat like a pacemaker does. If the latter, they send the heart a sharp jolt of electricity to slow it down. If the heart rhythm should become erratic, it sends an even stronger shock like a defibrillator does.

Heart Transplants

Although heart transplants are now generally successful, many more people are waiting for new hearts than there are organs available. Today, only about 2,200 heart transplants are done annually, and many more thousands could

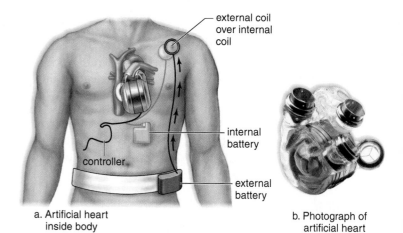

a. Artificial heart inside body

b. Photograph of artificial heart

Figure 5.16   Total artificial heart.
The AbioCor is a quiet, pulsatile device that moves the blood in the same manner as a natural heart. It is powered by an external battery pack that is small and portable so the recipient can be fairly active.

use one. Because of the shortage of human hearts, genetically altered pigs may one day be used as a source of hearts. Also, bone marrow stem cells have been injected into the heart, and researchers report that they apparently become cardiac muscle!

Today, a left ventricular assist device (LVAD), implanted in the abdomen, is an alternative to a heart transplant. A tube passes blood from the left ventricle to the device, which pumps it on to the aorta. A cable passes from the device through the skin to an external battery the patient must tote around. Soon, as many as 20,000 people a year may be outfitted with an LVAD. Instead of an LVAD, a few pioneers have volunteered to receive a Jarvik 2000, a pump that is inserted inside the left ventricle. The Jarvik is powered by an external battery no larger than a C-size battery.

Fewer patients still have received a so-called **total artificial heart (TAH).** The only model now available to patients is the AbioCor model (Fig. 5.16), which has no wires or tubes protruding from the chest. An internal battery and controller regulate the pumping speed, and an external battery powers the device by passing electricity through the skin via external and internal coils. A rotating centrifugal pump moves silicon hydraulic fluid between left and right sacs to force blood out of the heart into the pulmonary trunk and the aorta. All recipients, thus far, have been near death, and most have lived for only a short time. It's possible that once healthier patients receive an AbioCor, survival rate will improve. Different types of TAHs are being investigated in animals.

☑ Check Your Progress 5.7

1. a. What cardiovascular disorders are common in humans and (b) what treatments are available?

Paying for an Unhealthy Lifestyle

Cardiovascular disease is not only the number one killer in the United States today, it is also one of the most expensive. This disorder now accounts for more than $360 billion annually in spending on treatment and research, according to the American Heart Association and the National Heart, Lung, and Blood Institute (NHLBI). However, most cases of cardiovascular disease are preventable. Many well-known risk factors put an individual at high risk for developing cardiovascular disease. Among these are a sedentary lifestyle, obesity, smoking, and poor dietary habits. With few exceptions, these risk factors are largely a matter of choice.

Who Pays for Treatment?

As treatment costs and insurance premiums continue to climb at an unprecedented rate, public debate has raged: Should individuals suffering from cardiovascular disease receive the benefits of treatment at public expense? As the cost of health care soars, patients now rely much more on third-party payers to cover the cost of treatment than in the past. With much of the population covered by some type of health insurance, the risk is spread out over most of the population. This has created a system under which everyone, healthy or not, pays in part for treatment of the unhealthy. Under this system, everyone who pays into this system has a vested interest in ensuring that others avoid risky behaviors that contribute to rising health-care costs (Fig. 5B).

To an extent, people who practice risky behaviors already pay more for their habits. For example, unhealthy individuals often pay more for life or health insurance than healthy individuals. Taxes on risky behaviors, such as smoking and drinking alcohol, used to help defray health-care costs, are commonplace. Despite these policies, most studies show that, ultimately, we all pay for the treatment of cardiovascular disease. For example, insurance companies may refuse to cover preexisting conditions, including cardiovascular disease. Furthermore, more than 40 million people in the United States, many of them elderly and at risk for developing cardiovascular disease, do not have health insurance. In both cases, treatment costs are usually borne by the government. This creates an incentive, both within government and within the general public, to either ensure that such individuals adopt healthy lifestyle practices or bear more of the cost for their own treatment if they refuse to do so.

Is Legislation Needed?

Since obesity is a major cardiovascular disease risk factor, several organizations, including the Center for Science in the Public Interest (CSPI) and the World Health Organization (WHO), have pushed for the adoption of a "fat tax." This tax would be levied on foods with high fat content and/or poor nutritional value to pay, in part, for treatment of obesity-related diseases, including heart disease. Although taxing unhealthy behaviors to help defray treatment costs are generally popular with the public, the "fat tax" has proved surprisingly unpopular with voters. Thus, this measure seems unlikely to ever become public policy in the United States.

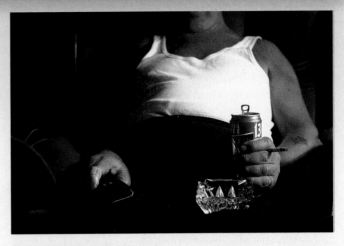

Figure 5B Unhealthy lifestyle.
An unhealthy lifestyle contributes to the development of cardiovascular disease.

Senator Tom Harkin recently introduced the Help America Act, which emphasizes healthy lifestyles and prevention. Smoking, one of the largest risk factors for cardiovascular disease, would be taxed to help pay for the program. Many studies show that every dollar invested in cardiovascular disease prevention results in a three-dollar savings in long-term health-care costs. With this in mind, many medical associations have endorsed this bill or other similar approaches to health care.

Keep the Status Quo?

However, not everyone agrees that cardiovascular disease is as rampant as reports indicate or that the current public policy initiatives are needed. One such critic points out that throughout the 1990s, despite large increases in obesity and marginal declines in risky behaviors, such as smoking, the incidence of cardiovascular disease has declined nearly 25%. Other critics insist that such intervention programs, while well intentioned, infringe on individual freedom and that rewarding healthy behaviors and penalizing risky behaviors are beyond the scope of government and public policy.

Regardless of whether or not you agree with shifting more of the cost of treatment to the unhealthy, it is up to individuals to adopt healthy practices that minimize the risk of cardiovascular disease. The ultimate human cost greatly outweighs the financial costs.

Decide Your Opinion

1. Would you support charging higher insurance premiums or taxes on people who do not practice a healthy lifestyle?
2. Do you support the use of public money for prevention programs targeted toward unhealthy people if there is the possibility of saving money in the future?
3. Since the public does subsidize health care that disproportionately benefits the less healthy, do you believe that financial interests trump personal freedom in this matter? Why or why not?

Summarizing the Concepts

5.1 Overview of the Cardiovascular System

The cardiovascular system consists of the heart and blood vessels.

The heart pumps blood and blood vessels take blood to and from the capillaries, where exchanges of nutrients for wastes occur with tissue cells. Blood is refreshed at the lungs, where gas exchange occurs; at the digestive tract, where nutrients enter the blood; and the kidney, where wastes are removed from the blood.

5.2 The Types of Blood Vessels

The Arteries   Arteries (and arterioles) take blood away from the heart. Arteries have the thickest walls, which allows them to withstand blood pressure.

The Capillaries   Capillaries are where exchange of substances occurs.

The Veins   Veins (and venules) take blood to the heart. Veins have relatively weak walls with valves that keep the blood flowing in one direction.

capillary bed

5.3 The Heart Is a Double Pump

The heart has a right and left side. Each side has an atrium and a ventricle. Valves keep the blood moving in the correct direction.

Passage of Blood Through the Heart

- The atrium receives O$_2$-poor blood from the body, and the ventricle pumps it into the pulmonary circuit (to the lungs).
- The atrium receives O$_2$-rich blood from the lungs, and a ventricle pumps it into the systemic circuit.

The Heart beat is Controlled

During the cardiac cycle, the SA node (pacemaker) initiates the heartbeat by causing the atria to contract. The AV node conveys the stimulus to the ventricles, causing them to contract. The heart sounds, "lub-dup," are due to the closing of the atrioventricular valves, followed by the closing of the semilunar valves.

5.4 Features of the Cardiovascular System

Pulse   The pulse rate indicates the heartbeat rate.

Blood Pressure Moves Blood in Arteries   Blood pressure caused by the beating of the heart accounts for the flow of blood in the arteries.

Blood Flow Is Slow in the Capillaries   The reduced velocity of blood flow in capillaries facilitates exchange of nutrients and wastes in the tissues.

Blood Flow in Veins Returns Blood to the Heart   Blood flow in veins is caused by skeletal muscle contraction, the presence of valves, and respiratory movements.

5.5 Two Cardiovascular Pathways

The cardiovascular system is divided into the pulmonary circuit and the systemic circuit.

The Pulmonary Circuit: Exchange of Gases

In the pulmonary circuit, blood travels to and from the lungs.

The Systemic Circuit: Exchanges with Tissue Fluid

In the systemic circuit, the aorta divides into blood vessels that serve the body's organs and cells. Venae cavae return O$_2$-poor blood to the heart.

5.6 Exchange at the Capillaries

This diagram pertains to capillary exchange in tissues of body parts—not including the gas-exchanging surfaces of the lungs.

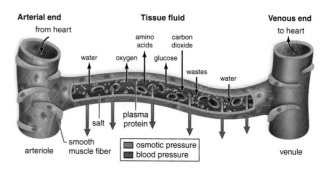

- At the arterial end of a cardiovascular capillary, blood pressure is greater than osmotic pressure; therefore, water leaves the capillary.
- In the midsection, oxygen and nutrients diffuse out of the capillary, while carbon dioxide and other wastes diffuse into the capillary.
- At the venous end, osmotic pressure created by the presence of proteins exceeds blood pressure, causing water to enter the capillary.

The fluid that is not picked up at the venous end of the cardiovascular capillary is excess tissue fluid. It enters the lymphatic capillaries.

- Lymph is tissue fluid contained within lymphatic vessels.
- The lymphatic system is a one-way system, and fluid is returned to blood by way of a cardiovascular vein.

5.7 Cardiovascular Disorders

Cardiovascular disease is the leading cause of death in the Western countries.

- Hypertension and atherosclerosis can lead to stroke, heart attack, or an aneurysm.
- Following a heart-healthy diet, getting regular exercise, maintaining a proper weight, and not smoking are protective against cardiovascular disease.

Understanding Key Terms

aneurysm 97
angina pectoris 97
angioplasty 100
aorta 94
arteriole 87
atherosclerosis 97
atrioventricular (AV) bundle 90
atrioventricular (AV) valve 88
atrium 88

AV (atrioventricular) node 90
blood pressure 92
cardiac cycle 90
chordae tendineae 88
coronary artery 95
coronary bypass operation 99
diastole 90
diastolic pressure 92
electrocardiogram (ECG) 91

embolus 97
heart 88
heart attack 97
heart failure 100
hepatic portal vein 95
hepatic vein 95
hypertension 97
inferior vena cava 94
lymph 96
lymphatic system 86
myocardium 88
pacemaker 90
pericardium 88
plaque 97
pulmonary artery 89
pulmonary circuit 94
pulmonary vein 89
pulse 92

Purkinje fibers 90
respiratory pump 93
SA (sinoatrial) node 90
semilunar valve 88
septum 88
skeletal muscle pump 93
stroke 97
superior vena cava 94
systemic circuit 94
systole 90
systolic pressure 92
thromboembolism 97
thrombus 97
tissue fluid 86
total artificial heart (TAH) 100
valve 87
ventricle 88
venule 87

Match the key terms to these definitions.

a. _____ Relaxation of a heart chamber.

b. _____ Large systemic vein that returns blood from body areas below the diaphragm.

c. _____ Rhythmic expansion and recoil of arteries resulting from heart contraction; can be felt from outside the body.

d. _____ Vessel that takes blood from capillaries to a vein.

e. _____ That part of the cardiovascular system that serves body parts and does not include the gas-exchanging surfaces in the lungs.

Testing Your Knowledge of the Concepts

1. What are the two parts of the cardiovascular system, and what are the functions of each part? (page 86)

2. Explain where exchanges occur in the body and the importance of those exchanges. (page 86)

3. Which of the three types of blood vessels are most numerous? Explain. (page 87)

4. Describe the structure of the heart, including the chambers and valves. (page 88)

5. What is the function of the septum in the heart? What would happen if the heart had no septum? (pages 88–89)

6. Trace the path of blood through the heart, including chambers, valves, and vessels the blood travels through. (page 89)

7. Describe the cardiac cycle, using the terms systole and diastole. What is the role of the SA node and the AV node in the cardiac cycle? (page 90)

8. Distinguish between the internal and external controls of the heartbeat. Explain how an ECG relates to the cardiac cycle. (pages 90–91)

9. In what vessel is the blood pressure the highest? The lowest? In what vessel is blood flow rate the fastest? The slowest?

Why is the pressure and rate in the capillaries important to capillary function? (pages 92–93)

10. Explain why skeletal muscle contraction has an affect on venous flow but not arterial flow. (page 93)

11. Distinguish between the two cardiovascular pathways. (page 94)

12. Trace the pathway of blood to and from the brain in the systemic circuit. (pages 94–95)

13. Describe the process by which nutrients are exchanged for wastes across a capillary, using glucose and carbon dioxide as examples. (page 96)

14. What is the most probable association between high blood pressure and a heart attack? With this association in mind, what type of diet might help prevent a heart attack? (page 97)

In questions 15–18, match the descriptions to the circuit in the key. Answers may be used more than once.

Key:
a. pulmonary circuit
b. systemic circuit
c. both pulmonary and systemic

15. Arteries carry O_2-rich blood.

16. Carbon dioxide leaves the capillaries, and oxygen enters the capillaries.

17. Arteries carry blood away from the heart, and veins carry blood toward the heart.

18. This contains the hepatic portal system.

In questions 19–24, match the descriptions to the blood vessel in the key. Answers may be used more than once.

Key:
a. venules   d. arteries
b. veins   e. arterioles
c. capillaries

19. Drain blood from capillaries

20. Empty into capillaries

21. May contain valves

22. Take blood away from the heart

23. Sites for exchange of substances between blood and tissue fluid

24. Rate of blood flow is the slowest

25. During ventricular diastole,
   a. blood flows into the aorta.
   b. the ventricles contract.
   c. the semilunar valves are closed.
   d. Both a and b are correct.

26. When the atria contract, the blood flows
   a. into the attached blood vessels.
   b. into the ventricles.
   c. through the atrioventricular valves.
   d. to the lungs.
   e. Both b and c are correct.

27. Heart valves located at the bases of the pulmonary trunk and aorta are called
    a. atrioventricular valves.
    b. semilunar valves.
    c. mitral valves.
    d. chordae tendineae.

28. Which of these associations is mismatched?
    a. left ventricle—aorta
    b. right ventricle—pulmonary trunk
    c. right atrium—vena cava
    d. left atrium—pulmonary artery
    e. Both b and c are incorrectly matched.

29. Which statement is not correct concerning the heartbeat?
    a. The atria contract at the same time.
    b. The ventricles relax at the same time.
    c. The AV valves open at the same time
    d. The semilunar valves open at the same time.
    e. First the right side contracts; then the left side contracts.

30. If a person's blood pressure is 115 mm Hg over 75 mm Hg, the 75 represents
    a. systolic pressure.
    b. diastolic pressure.
    c. pressure during ventricular relaxation.
    d. Both b and c are correct.

31. Accumulation of plaque in an artery wall is
    a. an aneurysm.
    b. angina pectoris.
    c. atherosclerosis.
    d. hypertension.
    e. a thromboembolism.

32. Label the following diagram of the cardiovascular system using this alphabetized list:

aorta	jugular vein
carotid artery	mesenteric arteries
hepatic portal vein	pulmonary artery
hepatic vein	pulmonary vein
iliac artery	renal artery
iliac vein	renal vein
inferior vena cava	superior vena cava

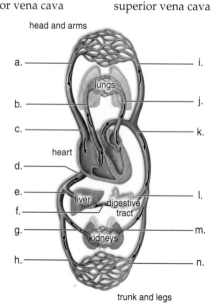

33. Label the following diagram showing the forces involved with capillary exchange. Use either blood pressure or osmotic pressure to label arrows a–d.

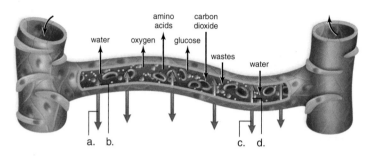

Thinking Critically About the Concepts

You should have noticed, as Sharon of the opening story did, several examples illustrating the underlying concept of structure supports function in this chapter. There were different kinds of blood vessels that each have a specific job to do and physical characteristics that enable them to do that job. The muscle walls of the right and left ventricle vary in thickness depending on where they pump the blood. When the structure of an organ is damaged or changed as arteries are in atherosclerosis, the ability of the organ to perform its function may be compromised as well. When organs' structures are damaged or changed, homeostatic conditions such as blood pressure may be affected. Dietary and lifestyle choices can either prevent damage or harm the cardiovascular system's organs.

1. The author and composer of the musical Rent died from an aortic aneurysm.
   a. Why don't aneurysms of arteries occur more frequently?
   b. Why do aortic aneurysms typically result in death?

2a. What happens to the ventricular walls of someone who has hypertrophic cardiomyopathy (consider what you know about the word hyperactive)?
   b. How would the ability of the heart to efficiently pump blood be affected?

3. Lymphatic vessels transport tissue fluid back to the cardiovascular system. The pressure in the lymphatic vessels is low. Do you expect lymphatic vessels to have valves? Why?

4. What dietary and lifestyle choices adversely impact the structure of arteries and their ability to transport blood?

5. Think of an analogy for capillaries. Remember they are small blood vessels with very thin walls and this facilitates their ability to perform exchanges with body cells. What are the similarities between capillaries and the object you have chosen to compare them to?

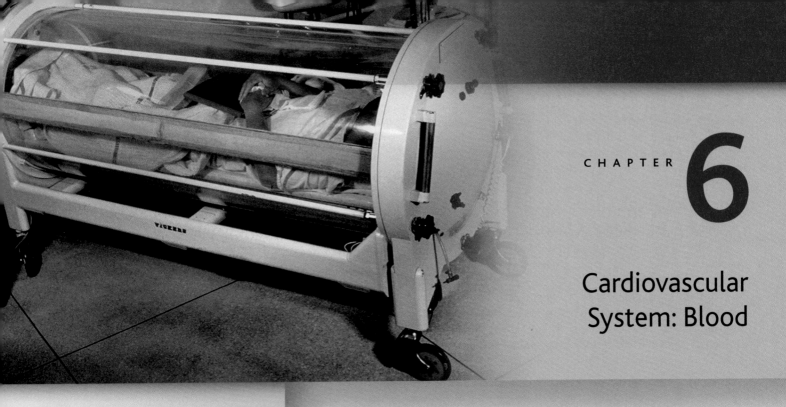

CHAPTER 6

Cardiovascular System: Blood

The fire had been a bad one, but the fire-fighting crew finally got it under control. Joe, a rookie firefighter, stumbled from the house carrying a young woman. "She's unconscious; get the paramedics and some oxygen," yelled Joe. Paramedics rushed to give Linda oxygen and take her to the hospital. Once there, a physician put her in a hyperbaric chamber because she had been exposed to a considerable amount of carbon monoxide (CO). CO binds tightly to hemoglobin and prevents it from binding to oxygen. The compressed air in a hyperbaric chamber allows a patient to take in six times the amount of oxygen with each breath. Even so, it would take a while to dislodge all the CO from Linda's blood.

In this chapter, you will learn about how bones make red blood cells and how a hormone made by the kidneys stimulates the production of more red blood cells. Any failure of various organs involved in the synthesis of red blood cells can adversely affect blood oxygen. The blood's white blood cells defend us against the infectious agents we encounter. Clotting, initiated by the platelets, prevents the loss of blood that would interrupt the distribution of vital molecules. Many homeostatic conditions depend on the ability of blood to transport substances, including glucose, oxygen, calcium, and heat (to regulate body temperature). Blood also contains buffers that maintain pH. As you read the homeostasis section of this chapter, you should begin to appreciate the numerous interactions between components of the cardiovascular systems and other organ systems.

CHAPTER CONCEPTS

6.1 Blood: An Overview
Blood, a liquid tissue, is a transport medium with various other functions: It fights infections and helps regulate body temperature and the pH of body fluids.

6.2 Red Blood Cells and Transport of Oxygen
Red blood cells contain hemoglobin, which transports oxygen and helps transport carbon dioxide. Red blood cell production is regulated by a hormone that is sometimes abused by athletes.

6.3 White Blood Cells and Defense Against Disease
White blood cells collectively fight infection. Each of five types has specific functions.

6.4 Platelets and Blood Clotting
Platelets are cell fragments that clump and seal a break in a blood vessel. Blood clotting, which follows, prevents loss of blood.

6.5 Blood Typing and Transfusions
Before a blood transfusion can be given, it must be typed because only certain type(s) of blood can be received by each person without ill effects.

6.6 Homeostasis
Various systems assist the cardiovascular system, whose proper functioning is critical to homeostasis.

6.1 Blood: An Overview

In Chapter 5, we learned that the cardiovascular system consists of the heart, which pumps the blood, and the blood vessels that conduct blood around the body. In this chapter, we learn about the functions and composition of blood.

Functions of Blood

The human body contains about 5 liters of blood, and the heart pumps this amount of blood with every beat! The functions of blood fall into three categories: transport, defense, and regulation.

Blood is the primary transport medium. Blood delivers oxygen from the lungs and nutrients from the digestive tract to the tissues, where an exchange takes place. It picks up and transports carbon dioxide and wastes away from the tissues to exchange surfaces in the lungs and kidneys, respectively.

In this way, capillary exchanges keep the composition of tissue fluid within normal limits (see Fig. 5.12).

Various organs and tissues secrete hormones into the blood, and blood transports these to other organs and tissues, where they serve as signals that influence cellular metabolism.

When Linda in the opening story was exposed to CO from the fire, her blood was unable to perform its usual function of transporting oxygen. Much of the hemoglobin in her red blood cells was bound to CO, and therefore unable to combine with oxygen. Hundreds of people die accidentally every year from CO poisoning caused by malfunctioning or improperly used fuel-burning appliances. Therefore, the government recommends that for safety's sake, everyone should make sure all fire-burning appliances are working properly.

Blood defends the body against invasion by pathogens in several ways. Certain blood cells are capable of phagocytizing and destroying pathogens, and others produce and

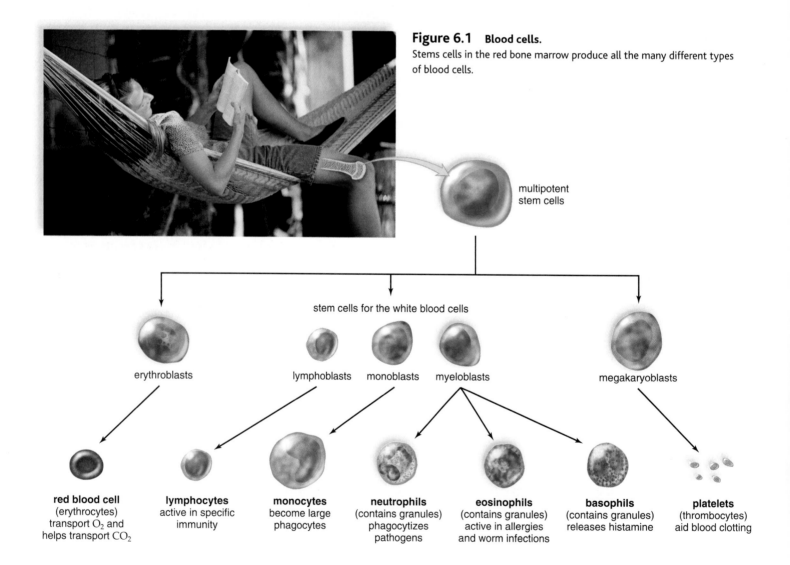

Figure 6.1   Blood cells.
Stems cells in the red bone marrow produce all the many different types of blood cells.

multipotent stem cells

stem cells for the white blood cells

erythroblasts   lymphoblasts   monoblasts   myeloblasts   megakaryoblasts

red blood cell
(erythrocytes)
transport O_2 and
helps transport CO_2

lymphocytes
active in specific
immunity

monocytes
become large
phagocytes

neutrophils
(contains granules)
phagocytizes
pathogens

eosinophils
(contains granules)
active in allergies
and worm infections

basophils
(contains granules)
releases histamine

platelets
(thrombocytes)
aid blood clotting

secrete antibodies into the blood. Antibodies incapacitate pathogens, making them subject to destruction, sometimes by white blood cells.

When an injury occurs, blood clots, and so prevents blood loss. Blood clotting involves platelets and a plasma protein, fibrinogen. Without blood clotting, we could bleed to death even from a small cut.

Blood has regulatory functions. Blood helps regulate body temperature by picking up heat, mostly from active muscles, and transporting it about the body. If the blood is too warm, the heat dissipates from dilated blood vessels in the skin.

The salts and plasma proteins in blood act to keep the liquid content of blood high. In this way, blood plays a role in helping to maintain its own water-salt balance.

Because blood contains buffers, which are chemicals that stabilize blood pH, it helps regulate body pH and keep it relatively constant.

Composition of Blood

Blood is a tissue, and, like any tissue, it contains cells but also cell fragments (Fig. 6.1). Collectively, the cells and cell fragments are called the **formed elements.** The cell and cell fragments are suspended in a liquid called **plasma,** and therefore, blood is classified as a liquid tissue.

The Formed Elements

The formed elements are red blood cells, white blood cells, and platelets. The formed elements are produced in red bone marrow, which occurs in most bones of a child but only in certain bones of an adult. Red bone marrow contains stem cells, which divide and give rise to the various types of blood cells (Fig. 6.1). The stem cells are of interest because it's thought that maybe under the right conditions in the laboratory they could be "coaxed" into becoming a greater variety of cells—cells that could cure humans of various ills, such as diabetes and Alzheimer disease, among many more.

Red blood cells are two to three times smaller than white blood cells, but there are many more of them, there being tens of millions of red blood cells and only tens of thousands of white blood cells in a mm³, which is about this size: ▪

Plasma

Plasma is the liquid medium for carrying various substances in the blood, and it also distributes the heat generated as a by-product of metabolism, particularly muscle contraction. About 91% of plasma is water (Fig. 6.2). The remaining 9% of plasma consists of various salts (ions) and organic molecules. The salts, which are simply dissolved in plasma, are a part of the buffers that help maintain the pH of the blood. Small organic molecules such as glucose and amino acids are nutrients for cells; urea is a nitrogenous waste product on its way to the kidneys for excretion.

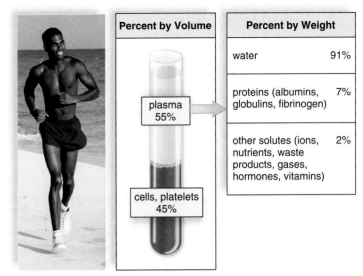

blood 8% of body weight

Figure 6.2 Plasma.
Plasma, the liquid portion of blood, is mainly water and proteins. However, many solutes such as nutrients, vitamins, and hormones are transported in plasma.

The most abundant organic molecules in blood are called the **plasma proteins.** The liver produces the plasma proteins, with one exception to be mentioned. The plasma proteins have many functions that help maintain homeostasis. They are able to take up and release hydrogen ions; therefore, they help keep blood pH around 7.4. Plasma proteins are too large to pass through capillary walls; therefore, they remain in the blood and establish an osmotic gradient between blood and tissue fluid. This **osmotic pressure** is a force that prevents excessive loss of plasma from the capillaries into tissue fluid.

Three major types of plasma proteins are the **albumins, globulins,** and **fibrinogen.** Albumins are the most abundant of the plasma proteins and contribute most to plasma's osmotic pressure. They also combine with and help transport other organic molecules. The globulins are of three types called alpha, beta, and gamma globulins. Alpha and beta globulins also combine with and help transport substances in the blood such as hormones, cholesterol, and iron. The gamma globulins are produced by the white blood cells called lymphocytes and therefore are not produced by the liver. The gamma globulins are important in fighting disease. Fibrinogen is a plasma protein that is active in formation of blood clots.

✔ Check Your Progress 6.1

1. a. What are the functions of blood, and (b) what are its two main portions?
2. a. What is the composition of plasma, and (b) what are the functions of the plasma proteins?

6.2 Red Blood Cells and Transport of Oxygen

Red blood cells (erythrocytes) are small, biconcave disks that lack a nucleus when mature. They occur in great quantity; there are 4–6 million red blood cells per mm^3 of whole blood.

How Red Blood Cells Carry Oxygen

Red blood cells (RBCs) are highly specialized for oxygen (O_2) transport. They lack a nucleus and, instead, contain many copies of hemoglobin. **Hemoglobin (Hb)** is a pigment that makes red blood cells and blood a red color. The globin portion of hemoglobin is a protein that contains four highly folded polypeptide chains. The heme part of hemoglobin is an iron-containing group in the center of each polypeptide chain (Fig. 6.3). The iron combines reversible with oxygen—this means that heme accepts O_2 in the lungs and then lets go of it in the tissues. Recall that, in contrast, Linda had to remain for hours in the hyperbaric chamber because CO combines with the iron of heme and then will not let go easily.

How many molecules of O_2 can each red blood cell transport? Each hemoglobin molecule can transport four molecules of O_2, and each RBC contains about 280 million hemoglobin molecules. This means that each red blood cell can carry over a billion copies of oxygen.

Red blood cells are an excellent example of structure suits function. They have no nucleus; in fact, their biconcave shape comes about because they lose a nucleus during maturation (see Fig. 6.1). The biconcave shape of RBCs gives them a greater surface area for the diffusion of gases into and out of the cell. All of the internal space of RBCs is used for transport of oxygen; and,

aside from having no nucleus, they also lack most organelles, including mitochondria. RBCs produce ATP anaerobically, and they do not consume any of the oxygen they transport.

When oxygen binds to heme in the lungs, hemoglobin assumes a slightly different shape and is called **oxyhemoglobin.** In the tissues, heme gives up this oxygen, and hemoglobin resumes its former shape, which is called **deoxyhemoglobin.** This oxygen diffuses out of the blood into tissue fluid and then into cells.

How Red Blood Cells Help Transport Carbon Dioxide

After blood picks up carbon dioxide (CO_2) in the tissues, about 7% is dissolved in plasma—any more and plasma would be carbonated and bubble just as soft drinks do. Instead, hemoglobin directly transports about 25% of CO_2. Carbon dioxide transport doesn't involve heme; instead, CO_2 combines with the terminal amino groups of globin molecules.

All the rest of the CO_2 (about 68%) is transported as the bicarbonate ion (HCO_3^-) in the plasma. Consider this equation:

$$CO_2 + H_2O \rightleftharpoons H_2CO_3 \rightleftharpoons H^+ + HCO_3^-$$

| carbon dioxide | water | carbonic acid | hydrogen ion | bicarbonate ion |

Carbon dioxide moves into RBCs where an enzyme, called carbonic anhydrase, is present that speeds this reaction. The bicarbonate ion diffuses out of RBCs to be carried in the plasma. The H^+ from this equation binds to amino acids of the globin portion of hemoglobin; and, in this way, hemoglobin

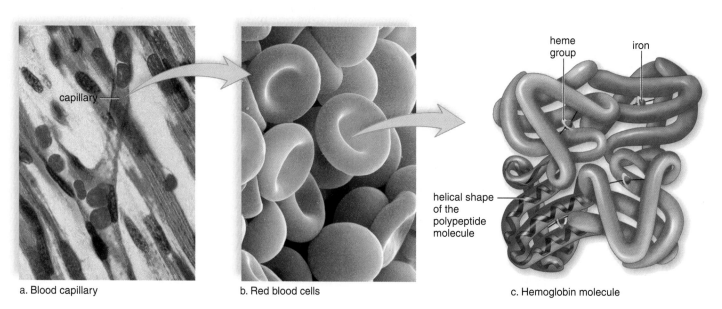

a. Blood capillary   b. Red blood cells   c. Hemoglobin molecule

Figure 6.3   Red blood cells.
a. Red blood cells move single file through the capillaries. **b.** Red blood cells are biconcave disks containing many molecules of hemoglobin. **c.** Hemoglobin contains four polypeptide chains (blue), which are the globin portion of the molecule. There is an iron-containing heme group in the center of each chain. Oxygen combines loosely with iron when hemoglobin is oxygenated.

also assists in keeping the pH of the blood constant. Once blood reaches the lungs, carbonic anhydrase in RBCs reverses this reaction, and CO_2 diffuses out of the blood.

Red Blood Cells Are Produced in Bone Marrow

The RBC stem cell in the bone marrow divides and produces new cells that differentiate into mature RBCs (see Fig. 6.1). As red blood cells mature, they lose their nucleus and acquire hemoglobin. Possibly because they lack a nucleus, red blood cells live only about 120 days. When they age, red blood cells are destroyed in the liver and spleen by macrophages, white blood cells derived from monocytes.

It is estimated that about 2 million RBCs are destroyed per second, and therefore, an equal number must be produced to keep the red blood cell count in balance. When red blood cells are broken down, hemoglobin is released. The globin portion of hemoglobin is broken down into its component amino acids, which are recycled by the body. The iron is recovered and returned to the bone marrow for reuse. (Even so, iron is a needed nutrient in the diet.) The rest of the heme portion of the molecule undergoes chemical degradation and is excreted by the liver. Chemical breakdown of heme in the skin causes a bruise to change color from red/purple to blue to green to yellow.

Blood Doping

The body has a way to boost the number of RBCs when insufficient oxygen is being delivered to the cells. The kidneys release a hormone called **erythropoietin (EPO)**, which stimulates the stem cells in the bone marrow to produce more red blood cells (Fig. 6.4). The liver and other tissues also produce EPO for the same purpose.

Blood doping is any method of increasing the normal supply of RBCs for the purpose of delivering oxygen more efficiently, reducing fatigue, and giving athletes a competitive edge. To accomplish blood doping, athletes can inject themselves some months before the competition with EPO to increase the number of RBCs in their blood. Several weeks later, four units of their blood are removed and centrifuged to concentrate the RBCs. The concentrated RBCs are reinfused shortly before the race. Several cyclists died in the 1990s from heart failure, probably due to blood that was too thick with cells for the heart to pump.

Disorders Involving Red Blood Cells

When there is an insufficient number of red blood cells or the cells do not have enough hemoglobin, the individual suffers from **anemia** and has a tired, run-down feeling. Iron, vitamin B_{12}, and the B vitamin folic acid are necessary for the production of red blood cells. Rita, from the Chapter 2 opening story, had iron-deficiency anemia and insufficient hemoglobin synthesis because of an inadequate dietary intake of iron. A lack of vitamin B_{12} causes pernicious anemia, in which stem cell activity to produce red blood cells is reduced due to inadequate DNA production. Folic acid deficiency anemia also

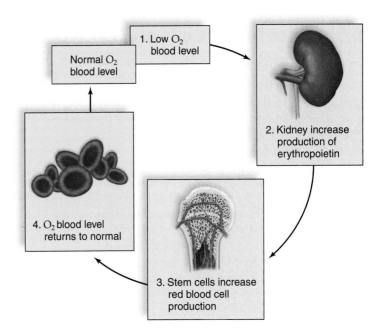

Figure 6.4 Action of erythropoietin.
The kidneys release increased amounts of erythropoietin whenever the oxygen capacity of the blood is reduced. Erythropoietin stimulates the red bone marrow to speed up its production of red blood cells, which carry oxygen.

leads to a reduced number of RBCs, particularly during pregnancy. Pregnant women should consult with their health-care provider about the need to increase their intake of folic acid, because a deficiency can lead to birth defects in the newborn.

Hemolysis is the rupturing of red blood cells. In hemolytic anemia, the rate of red blood cell destruction increases. **Sickle-cell disease** is a hereditary condition in which the individual has sickle-shaped red blood cells that tend to rupture as they pass through the narrow capillaries. The RBCs look like these to the right: The problem arises because the amino acid chain

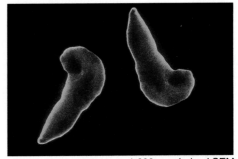

1,600×, colorized SEM

in two of the four chains making up hemoglobin is abnormal. The life expectancy of sickle red blood cells is about 90 days instead of 120 days.

Hemolytic disease of the newborn, which is discussed on page 115, is also a type of hemolytic anemia.

☑ Check Your Progress 6.2

1. What substance allows red blood cells (RBCs) to transport oxygen?

2. Why do RBCs have a biconcave shape?

3. Name three disorders associated with RBCs.

6.3 White Blood Cells and Defense Against Disease

White blood cells (leukocytes) differ from red blood cells in that they are usually larger, have a nucleus, lack hemoglobin, and are translucent unless stained. White blood cells are not as numerous as red blood cells; there are only 5,000–11,000 per mm³ of blood. White blood cells are derived from stem cells in the red bone marrow, where most types mature. There are several types of white blood cells (Fig. 6.5), and the production of each type is regulated by a protein called a **colony-stimulating factor (CSF).** In a person with normally functioning bone marrow, the numbers of white blood cells can double within hours, if needed. White blood cells are able to squeeze through pores in the capillary wall, and therefore they are also found in tissue fluid and lymph (Fig. 6.6).

White blood cells fight infection, and are an important part of the immune system. The **immune system**, discussed in Chapter 7, consists of a variety of cells, tissues, and organs that defend the body against pathogens, cancer cells, and foreign proteins. Many white blood cells live only a few days—they probably die while engaging pathogens. Others live for months or even years.

White blood cells have various ways to fight infection. Certain ones are very good at phagocytosis (see Fig. 3.10a). During phagocytosis, a projection from the cell surrounds a pathogen and literally engulfs it. The resulting vesicle moves toward and fuses with a lysosome where enzymes digest the pathogen to debris that leaves the cell. Other white blood cells produce **antibodies**, proteins that combine with **antigens,** proteins foreign to the individual, and

mark them for destruction. We will be describing antigens and antibodies that are involved in blood typing and coagulation in Section 6.5.

Types of White Blood Cells

White blood cells are classified into the **granular leukocytes** and the **agranular leukocytes** because some have noticeable granules and some do not have such noticeable granules. The granules, like lysosomes, contain various enzymes and protein. They help white blood cells defend against diseases.

Granular Leukocytes

The granular leukocytes include neutrophils, eosinophils, and basophils.

Among granular leukocytes, **neutrophils** account for 50–70% of all white blood cells; therefore, they are the most abundant of the white blood cells. Because they have a multilobed nucleus, they are called polymorphonuclear leukocytes, or "polys." The granules of neutrophils do not significantly take up the stain eosin, a pink to red acidic stain, or a basic stain that is blue to purple. (This accounts for their name, neutrophil.)

Neutrophils are usually first responders to bacterial infection, and their intense phagocytic activity is essential to overcoming an invasion by a pathogen. It could be said that they are like "vacuum cleaners" because they suck up unwanted substances, for example, pathogens. At times, their death in large numbers results in **pus**.

Eosinophils have a bilobed nucleus, and their large, abundant granules take up eosin and become a red color. (This accounts for their name, eosinophil.) Not much is known specifically about the function of eosinophils, but they increase in number in the event of a parasitic worm infection or an allergic reaction.

Basophils have a U-shaped or lobed nucleus. Their granules take up the basic stain and become a dark blue color. (This accounts for their name, basophil.) In the connective

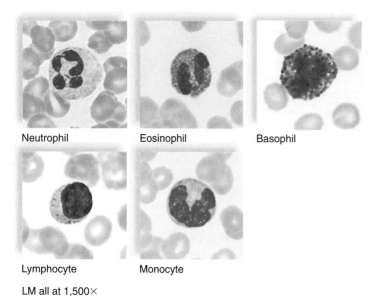

Neutrophil   Eosinophil   Basophil

Lymphocyte   Monocyte

LM all at 1,500×

Figure 6.5   White blood cells.
Eosinophils, basophils, and neutrophils have granules. Lymphocytes and monocytes do not have granules.

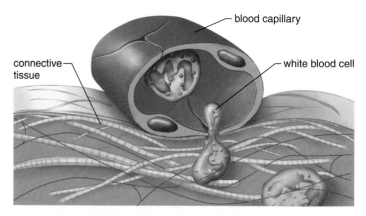

blood capillary

connective tissue

white blood cell

Figure 6.6   Mobility of white blood cells.
White blood cells can squeeze between the cells of a capillary wall and enter the tissues of the body.

tissues, basophils, and also similar type cells called **mast cells**, release histamine associated with allergic reactions. Histamine dilates blood vessels, but constricts the air tubes that lead to the lungs. This is what happens during an asthma attack when someone has difficulty breathing.

Agranular Leukocytes

The agranular leukocytes include the lymphocytes and the monocytes. Lymphocytes and monocytes do not have granules and have nonlobular nuclei. They are sometimes called the mononuclear leukocytes.

Lymphocytes account for 25–35% of all white blood cells, and therefore, they are the second most abundant type of white blood cells. Lymphocytes are responsible for specific immunity to particular pathogens and their toxins (poisonous substances). The lymphocytes are of two types: B cells and T cells. B-cell descendants (plasma cells) are the cells that protect us by producing antibodies, the proteins that combine with target pathogens and mark them for destruction. Some T cells (cytotoxic T cells) directly destroy pathogens. The AIDS virus attacks one of several types of T cells; and, in this way, the virus causes immune deficiency, an inability to defend the body against pathogens. B lymphocytes and T lymphocytes are discussed more fully in Chapter 7.

Monocytes are the largest of the white blood cells, and after taking up residence in the tissues, they differentiate into even larger macrophages except in the skin, where they become dendritic cells. The "vacuum cleaner" function of neutrophils is shared by macrophages and dendritic cells, which phagocytize pathogens, old cells, and cellular debris. Macrophages and dendritic cells also stimulate other white blood cells, including lymphocytes, to defend the body.

Disorders Involving White Blood Cells

Immune deficiencies are sometimes inherited. Children have **severe combined immunodeficiency disease (SCID)** when the stem cells of white blood cells lack an enzyme called adenosine deaminase. Without this enzyme, the body cannot fight infections of any sort. About 100 children are born with the disease each year, but the most famous SCID patient was a boy named David, who lived for 12 years in a germ-free plastic bubble (Fig. 6.7) in order to keep him from getting an infection. He died in 1984 when doctors injected his bone marrow with donated blood cells and the treatment failed. Today, there are two ways to treat SCID; one way is to give repeated injections of the missing enzyme, and the other way is to use gene therapy (see page 456). Gene therapy results in a cure if it is successful. Stem cells are removed from red bone marrow, given the gene, and then reinjected into the body. The hope is that these stem cells will settle permanently in the bone marrow and produce normal lymphocytes.

Cancer is due to uncontrolled cell growth, and **leukemia**, which means "white blood," refers to a group of cancerous conditions that involve uncontrolled white blood cell proliferation. Most of these white blood cells are abnormal or immature, therefore, they are incapable of performing their normal defense functions. Each type of leukemia is named for the type of cell that is dividing out of control. For example, lymphocytic leukemia involves abnormal lymphocyte proliferation.

An Epstein-Barr virus (EBV) infection of lymphocytes is the cause of **infectious mononucleosis.** It's called infectious mononucleosis because lymphocytes are mononuclear. EBV, a member of the herpes virus family, is one of the most common human viruses. Symptoms of infectious mononucleosis are fever, sore throat, and swollen lymph glands. Although the symptoms of infectious mononucleosis usually disappear in one or two months without medication, EBV remains dormant and hidden in a few cells in the throat and blood for the rest of a person's life. Stress can reactivate the virus, as when Linda almost died due to CO inhalation. Reactivation means that a person's saliva can pass on the infection to someone else as with intimate kissing. This is why mononucleosis is called the "kissing disease."

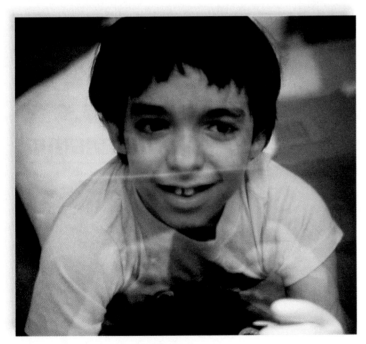

Figure 6.7 David, the bubble boy.
Children, like David, who are unable to produce functioning white blood cells are in danger of dying from an infection. To prevent this, David lived for 12 years in a germ-free bubble. He died when a red bone marrow transplant failed.

☑ Check Your Progress 6.3

1. What are the different types of white blood cells?

2. What is the structure and function of each type of white blood cell?

3. Name and describe three disorders of white blood cells.

Health Focus

What to Know When Giving Blood

The Procedure

After you register to give blood, an attendant will ask private and confidential questions about your health history and your lifestyle, and answer any questions you may have.

He or she will also check your temperature, blood pressure, and pulse, and test a drop of your blood to ensure that you are not anemic (Fig. 6A).

You will have several opportunities prior to giving blood and even afterwards to let Red Cross officials know whether your blood is safe to give to another person.

All of the supplies, including the needle, are sterile and are used only once—for *you*. You cannot get infected with HIV (the virus that causes AIDS) or any other disease by donating blood.

When the actual donation is started, you may feel a brief "sting." The procedure takes about 10 minutes, and you will have given about a pint of blood. Your body replaces the liquid part (plasma) in hours and the cells in a few weeks.

After you donate, you are given a card with a number to call if you decide after you leave that your blood may not be safe to give to another person.

An area is provided in which to relax after donating blood. Most people feel fine while they give blood and afterward. A few may have an upset stomach; a faint or dizzy feeling; or bruising, redness, and pain where the needle was inserted. Very rarely, a person may faint, have muscle spasms, and/or suffer nerve damage.

Your blood is tested for syphilis, AIDS antibodies, hepatitis, and other viruses. You are notified if tests give a positive result. Your blood won't be used if it could make someone ill.

The Cautions

DO NOT GIVE BLOOD

If you have

> ever had hepatitis;
>
> had malaria, taken drugs to prevent malaria in the last three years, or traveled to a country where malaria is common;
>
> been treated for syphilis or gonorrhea in the last 12 months.

If you have AIDS or one of its symptoms:

> unexplained weight loss (4.5 kg or more in less than two months);

> night sweats;
>
> blue or purple spots on or under skin;
>
> long-lasting white spots or unusual sores in mouth;
>
> lumps in neck, armpits, or groin for over a month;
>
> diarrhea lasting over a month;
>
> persistent cough and shortness of breath;
>
> fever higher than 37°C for more than ten days.

If you are at risk for AIDS—that is, if you have

> taken illegal drugs by needle, even once;
>
> taken clotting factor concentrates for a bleeding disorder such as hemophilia;
>
> tested positive for any AIDS virus or antibody;
>
> been given money or drugs for sex since 1977;
>
> had a sexual partner within the last 12 months who did any of the above things;
>
> (for men) had sex *even once* with another man since 1977, *or* had sex with a female prostitute within the last 12 months;
>
> (for women) had sex with a male or female prostitute within the last 12 months, *or* had a male sexual partner who had sex with another man *even once* since 1977.

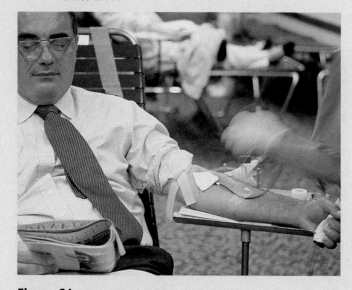

Figure 6A   Donating blood.
Donating blood can help save a life—perhaps even that of a family member or friend.

6.4 Platelets and Blood Clotting

Platelets (thrombocytes) result from fragmentation of certain large cells, called **megakaryocytes,** in the red bone marrow. Platelets are produced at a rate of 200 billion a day, and the blood contains 150,000–300,000 per mm³. These formed elements are involved in the process of blood **clotting,** or coagulation. Also involved are the plasma proteins prothrombin and fibrinogen, which are manufactured in the liver and deposited in the blood. Vitamin K is necessary to the production of prothrombin.

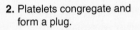

1. Blood vessel is punctured.

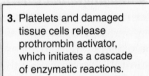

2. Platelets congregate and form a plug.

3. Platelets and damaged tissue cells release prothrombin activator, which initiates a cascade of enzymatic reactions.

Prothrombin activator

$$\text{Prothrombin} \xrightarrow{\text{Ca}^{2+}} \text{Thrombin}$$

$$\text{Fibrinogen} \xrightarrow{\text{Ca}^{2+}} \text{Fibrin threads}$$

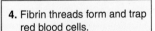

4. Fibrin threads form and trap red blood cells.

a. Blood-clotting process

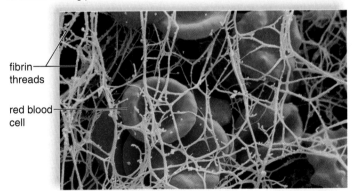

fibrin threads

red blood cell

b. Blood clot   4,400×

Figure 6.8   Blood clotting.
a. Platelets and damaged tissue cells release prothrombin activator, which acts on prothrombin in the presence of Ca²⁺ (calcium ions) to produce thrombin. Thrombin acts on fibrinogen in the presence of Ca²⁺ to form fibrin threads. **b.** A scanning electron micrograph of a blood clot shows red blood cells caught in the fibrin threads.

Blood Clotting

When a blood vessel in the body is damaged, platelets clump at the site of the puncture and seal the break, if it is not too extensive (Fig. 6.8). A large break may also require a blood clot to stop the bleeding. At least 12 clotting factors participate in the formation of a blood clot, but we will mention only a few of these. Also needed are calcium ions (Ca²⁺).

Clot formation is initiated when platelets and damaged tissue release **prothrombin activator,** which converts the plasma protein **prothrombin** to thrombin. **Thrombin,** in turn, acts as an enzyme that severs two short amino acid chains from each fibrinogen molecule. These activated fragments then join end to end, forming long threads of **fibrin.** Fibrin threads wind around the platelet plug in the damaged area of the blood vessel and provide the framework for the clot. Red blood cells also are trapped within the fibrin threads; these cells make a clot appear red. A fibrin clot is present only temporarily. As soon as blood vessel repair is initiated, an enzyme called plasmin destroys the fibrin network and restores the fluidity of plasma.

After blood clots, a yellowish fluid escapes from the clot. This fluid is called **serum,** and it contains all the components of plasma except fibrinogen and prothrombin.

Disorders Related to Blood Clotting

An insufficient number of platelets is called **thrombocytopenia.** Thrombocytopenia is either due to low production of platelets in the bone marrow or increased breakdown of platelets outside the marrow. A number of conditions can lead to thrombocytopenia, but it can also be drug-induced. Symptoms include bruising, nosebleeds or bleeding in the mouth, and a rash. Gastrointestinal bleeding or bleeding in the brain are possible complications.

Particularly if a blood vessel is rough due to the presence of plaque (see page 97), a clot can form in an unbroken blood vessel. Such a clot is called a thrombus if it remains stationary. Should the clot dislodge and travel in the blood, it is called an embolus. If **thromboembolism** is not treated, a heart attack can occur, as discussed in Chapter 5.

Hemophilia is an inherited clotting disorder due to a deficiency in a clotting factor. Hemophilia A, due to the lack of clotting factor VIII, is more apt to be inherited by boys than girls (see page 437). In hemophilia, the slightest bump can cause bleeding into the joints. Cartilage degeneration in the joints and resorption of underlying bone can follow. The most frequent cause of death is bleeding into the brain with accompanying neurological damage.

☑ Check Your Progress 6.4

1. Name the major participants in blood clotting and their functions.

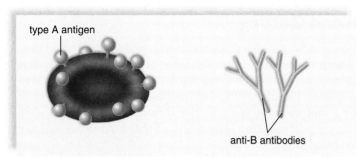

Type A blood. Red blood cells have type A surface antigens. Plasma has anti-B antibodies.

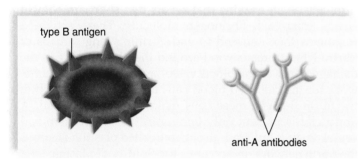

Type B blood. Red blood cells have type B surface antigens. Plasma has anti-A antibodies.

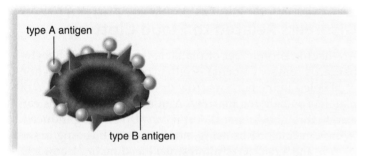

Type AB blood. Red blood cells have type A and type B surface antigens. Plasma has neither anti-A nor anti-B antibodies.

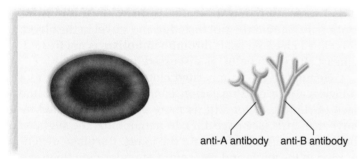

Type O blood. Red blood cells have neither type A nor type B surface antigens. Plasma has both anti-A and anti-B antibodies.

Figure 6.9   Types of blood.
In the ABO system, blood type depends on the presence or absence of antigens A and B on the surface of red blood cells. In these drawings, A and B antigens are represented by different shapes on the red blood cells. The possible anti-A and anti-B antibodies in the plasma are shown for each blood type. Notice that an anti-B antibody cannot bind to an A antigen, and vice versa.

6.5 Blood Typing and Transfusions

A **blood transfusion** is the transfer of blood from one individual into the blood of another. In order for transfusions to be safely done, it is necessary for blood to be typed so that **agglutination** (clumping of red blood cells) does not occur. Blood typing usually involves determining the ABO blood group and whether the individual is Rh⁻ (negative) or Rh⁺ (positive).

ABO Blood Groups

Only certain types of blood transfusions are safe because the plasma membranes of red blood cells carry glycoproteins that can be antigens to others. ABO blood typing is based on the presence or absence of two possible antigens, called type A antigen and type B antigen. Whether these antigens are present or not depends on the particular inheritance of the individual.

In Figure 6.9, type A antigen and type B antigen are given different shapes. Study the figure to see what type blood has what type antigen on the red blood cells. Notice that type O blood has no antigens on the red blood cells.

It so happens that an individual with type A blood has anti-B antibodies in the plasma; a person with type B blood has anti-A antibodies in the plasma; and a person with type O blood has both antibodies in the plasma (Fig. 6.9). These antibodies are not present at birth, but they appear over the course of several months after birth. The presence of these antibodies can cause agglutination.

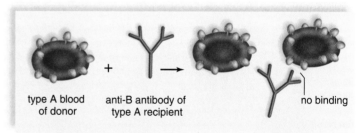

a. No agglutination

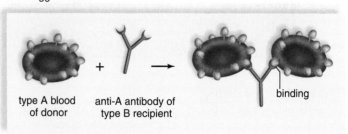

b. Agglutination

Figure 6.10   Blood transfusions.
No agglutination (**a**) versus agglutination (**b**) is determined by whether antibodies are present that can combine with antigens.

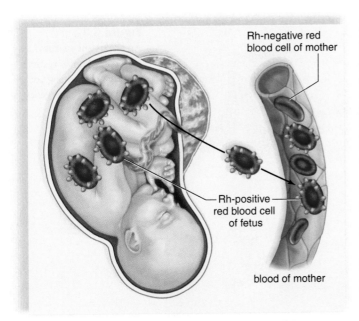

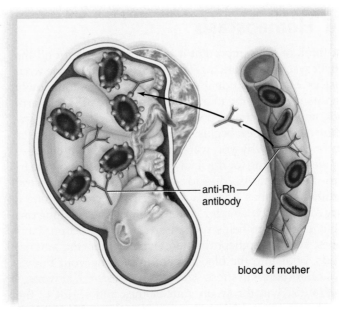

a. Fetal Rh-positive red blood cells leak across placenta into mother's bloodstream.

b. Mother forms anti-Rh antibodies that cross the placenta and attack fetal Rh-positive red blood cells.

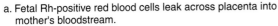

Figure 6.11   Hemolytic disease of the newborn.
a. Due to a pregnancy in which the child is Rh-positive, an Rh-negative mother can begin to produce antibodies against Rh-positive red blood cells. **b.** Usually in a subsequent pregnancy, these antibodies can cross the placenta and cause hemolysis of an Rh-positive child's red blood cells.

Blood Compatibility

Blood compatibility is very important when transfusions are done. The antibodies in the plasma must not combine with the antigens on the surface of the red blood cells, or else agglutination occurs. With agglutination, anti-A antibodies have combined with type A antigens, and anti-B antibodies have combined with type B antigens. Therefore, agglutination is expected if the donor has type A blood and the recipient has type B blood (Fig. 6.10). What about other combinations of blood types? Try out all other possible donors and recipients to see if agglutination will occur.

Type O blood is sometimes called the universal donor because agglutination will not occur with any other type blood. In practice, however, there are other possible blood groups, aside from ABO blood groups, so it is necessary to physically put the donor's blood on a slide with the recipient's blood and observe whether agglutination occurs before blood can be safely given from one person to another. This procedure, called blood-type matching, is done before a blood transfusion is performed.

Rh Blood Groups

The designation of blood type usually also includes whether the person has or does not have the Rh factor on the red blood cell. Rh⁻ individuals normally do not have antibodies to the Rh factor, but they make them when exposed to the Rh factor.

If a mother is Rh⁻ and the father is Rh⁺, a child can be Rh⁺. During a pregnancy, Rh⁺ can leak across the placenta into the mother's bloodstream. The presence of these Rh⁺ antigens causes the mother to produce anti-Rh antibodies (Fig. 6.11). Usually in a subsequent pregnancy with another Rh⁺ baby, the anti-Rh antibodies may cross the placenta and destroy the child's red blood cells. This is called hemolytic disease of the newborn (HDN) because hemolysis continues after the baby is born. Due to red blood cell destruction, excess hemoglobin breakdown products in the blood can lead to brain damage and mental retardation, or even death.

The Rh problem is prevented by giving Rh⁻ women an Rh immunoglobulin injection no later than 72 hours after giving birth to an Rh⁺ child. This injection contains anti-Rh antibodies that attack any of the baby's red blood cells in the mother's blood before these cells can stimulate her immune system to produce her own antibodies. This injection is not beneficial if the woman has already begun to produce antibodies; therefore, the timing of the injection is most important.

✔ Check Your Progress 6.5

1. **a.** What are the different blood types, and **(b)** what determines blood type?
2. Among the ABO types of blood, who can give blood to whom? Why?
3. When does hemolytic disease of the newborn occur?

6.6 Homeostasis

Just think of how important it is to homeostasis for all the human systems to work together. For example, imagine what must be one of a student's worst nightmares. You slowly open your eyes in the morning, blearily gaze at your alarm clock, and realize that you've overslept. In fact, it's just a few minutes before class begins. Then, you snap wide awake as you remember why you were up so late the night before—you were studying for the final exam! Most likely, your heart starts beating faster, due to symphathetic *nervous system* stimulation, just as Ashley's did when she was watching a scary movie in the Chapter 1 opening story. The resulting rise in blood pressure helps prepare the body for action. Hormones, such as epinephrine (adrenaline) from the adrenal glands, pour into the bloodstream to prolong your body's physical response to stress. The *endocrine system* is at work.

Frantic, you throw on your clothes and sprint to the bus. Your *muscular* and *respiratory systems* meet the demand. As you run, your muscles require more oxygen to support their activity. Your breathing and heart rate increase, in order to deal with the demand. The hard-working muscles generate carbon dioxide and heat. Carbon dioxide is carried via the blood to the lungs so it can be expelled. To keep you from overheating, you not only sweat, but blood flow to your skin increases so that heat radiates away.

The nightmare continues. In your mad dash up the steps to get to your exam, you fall, skinning your knee (Fig. 6.12). At first, blood flows from the wound, but a clot will soon form to seal off the injured area and allow a scab to form so that healing can begin. Any injury that breaks the skin can allow harmful bacteria into the body, but the immune system usually deals with them in short order. Later, you might notice that your knee is not only skinned, but also black and blue. The bruise is where blood leaked out of damaged vessels underneath the skin, and then clotted.

You make it to class just in time to get your exam from the instructor. As you settle down to take the exam, your heartbeat and breathing gradually slow, since the need of your muscles is not as great. Now that you are in thinking mode, blood flow increases in certain highly active parts of your brain. The brain is a very demanding organ in terms of both oxygen and glucose.

You start to feel hungry. Since there is little glucose in your bloodstream to provide fuel for your body's cells, glycogen stored in the liver is broken down to make glucose available to the blood. Finally, you hand in the exam and go to the dining hall for breakfast. When most of the nutrients from your meal are absorbed by the *digestive system*, they enter the bloodstream. However, the fats enter lacteals, which are vessels of the *lymphatic system*, for transport to the blood.

Notice in this scenario how many organ systems had to interact with the cardiovascular system. The nervous, endocrine, muscular, respiratory, digestive, and lymphatic systems helped to serve the cells' needs. Throughout all of the challenges to homeostasis described above, the urinary sys-

Figure 6.12
An emergency situation.
How many human systems participate in helping you get through an emergency situation?

tem was busy making urine for excretion by the kidneys. No organ system works alone where homeostasis is concerned!

How Body Systems Work Together

Figure 6.13 provides more information about how these and other systems cooperate with the cardiovascular system to maintain homeostasis. You learned previously that the body's internal environment contains blood and tissue fluid. Tissue fluid actually originates from blood. In fact, it is very similar to the plasma component of blood, except that it contains no plasma proteins. Tissue fluid is taken up from the space between the tissue cells by lymphatic capillaries, after which it is referred to as lymph. The lymph courses through lymphatic vessels and is eventually returned to the venous system. Thus, the cardiovascular and lymphatic systems are intimately linked. Fluid cycles continuously through blood plasma, tissue fluid, and lymph, and back to blood plasma again.

Homeostasis is possible only if (1) the cardiovascular system delivers oxygen from the lungs and nutrients from the digestive system to, and takes away metabolic wastes from, the tissue fluid that surrounds cells; and (2) the lymphatic system returns tissue fluid to the bloodstream.

The muscular system plays an essential contribution to the movement of blood in the cardiovascular system because the heart is cardiac muscle and skeletal muscle contraction moves blood in the veins back to the heart. Also, skeletal muscle contraction is essential to the movement of lymph in the lymphatic veins. Without this contribution, tissue fluid would collect in the tissues (a condition called **edema,** which results in swelling), and blood pressure would fall. Red bone marrow in the bones produces the blood cells, and this produc-

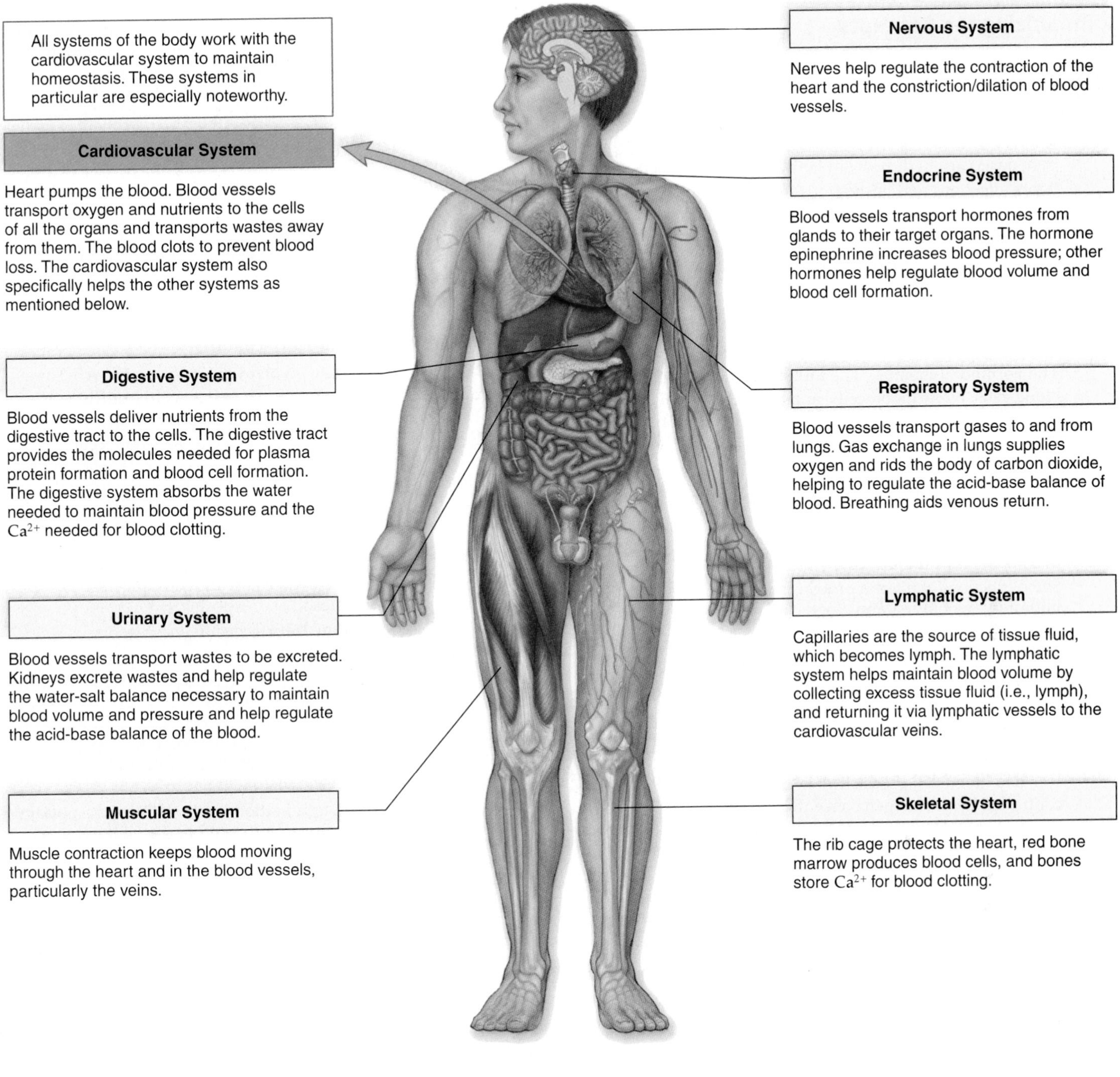

All systems of the body work with the cardiovascular system to maintain homeostasis. These systems in particular are especially noteworthy.

Cardiovascular System

Heart pumps the blood. Blood vessels transport oxygen and nutrients to the cells of all the organs and transports wastes away from them. The blood clots to prevent blood loss. The cardiovascular system also specifically helps the other systems as mentioned below.

Digestive System

Blood vessels deliver nutrients from the digestive tract to the cells. The digestive tract provides the molecules needed for plasma protein formation and blood cell formation. The digestive system absorbs the water needed to maintain blood pressure and the Ca^{2+} needed for blood clotting.

Urinary System

Blood vessels transport wastes to be excreted. Kidneys excrete wastes and help regulate the water-salt balance necessary to maintain blood volume and pressure and help regulate the acid-base balance of the blood.

Muscular System

Muscle contraction keeps blood moving through the heart and in the blood vessels, particularly the veins.

Nervous System

Nerves help regulate the contraction of the heart and the constriction/dilation of blood vessels.

Endocrine System

Blood vessels transport hormones from glands to their target organs. The hormone epinephrine increases blood pressure; other hormones help regulate blood volume and blood cell formation.

Respiratory System

Blood vessels transport gases to and from lungs. Gas exchange in lungs supplies oxygen and rids the body of carbon dioxide, helping to regulate the acid-base balance of blood. Breathing aids venous return.

Lymphatic System

Capillaries are the source of tissue fluid, which becomes lymph. The lymphatic system helps maintain blood volume by collecting excess tissue fluid (i.e., lymph), and returning it via lymphatic vessels to the cardiovascular veins.

Skeletal System

The rib cage protects the heart, red bone marrow produces blood cells, and bones store Ca^{2+} for blood clotting.

Figure 6.13   Human systems work together.
Each of these systems makes critical contributions to the functioning of the cardiovascular system and, therefore, to homeostasis. See if you can suggest contributions by each system before looking at the information given.

tion is regulated by hormones. Bones contribute calcium ions (Ca^{2+}) to the process of blood clotting. Without blood clotting, the loss of blood due to an accident, such as skinning a knee, would lead to death.

We must not forget the urinary system has other functions besides producing and excreting urine. The kidneys help regulate the acid-base and the salt-water balance of the internal environment—blood and tissue fluid.

✔ Check Your Progress 6.6

1. Tell how the functions of the cardiovascular system contribute to homeostasis.

2. Give one significant contribution of each system listed in Figure 6.13 to the cardiovascular system.

3. In what way does the liver contribute to the functioning of the cardiovascular system?

Summarizing the Concepts

6.1 Blood: An Overview

The functions of blood are

- transports hormones, oxygen, and nutrients to cells;
- transports carbon dioxide and other wastes from cells;
- fights infections and has various regulatory functions;
- maintains blood pressure;
- regulates body temperature; and
- keeps the pH of body fluids within normal limits.

All of these functions help maintain homeostasis. Blood has two main components called formed elements and plasma.

The Formed Elements The formed elements are red blood cells, white blood cells, and platelets.

Plasma

- 91% of blood plasma is water.
- Plasma proteins (albumins, globulins, and fibrinogen) are mostly produced by the liver.
- Plasma proteins maintain osmotic pressure, help regulate pH, and transport molecules.
- Some functions of plasma proteins are transport (albumins), immunity (globulins), and blood clotting (fibrinogen).

6.2 Red Blood Cells and Transport of Oxygen

Red blood cells lack a nucleus and contain hemoglobin, which combines with oxygen and transports it to the tissues.

red blood cells

Red blood cell production is controlled by the oxygen concentration of the blood. When the oxygen concentration decreases, the kidneys increase their production of erythropoietin, and more red blood cells are produced.

6.3 White Blood Cells and Defense Against Disease

White blood cells are larger than red blood cells. They have a nucleus and are translucent unless stained. White blood cells are either granular leukocytes or agranular leukocytes.

- The granular leukocytes are eosinophils, basophils, and neutrophils. Neutrophils are abundant and respond first to infections. They phagocytize pathogens.

eosinophil   basophil   neutrophil

- The agranular leukocytes include monocytes and lymphocytes. Monocytes are the largest of the white blood cells. They can become macrophages that phagocytize pathogens and cellular debris. Lymphocytes (B cells and T cells) are responsible for specific immunity.

monocyte   lymphocyte

All blood cells are produced within red bone marrow from stem cells. They live about 120 days and are eventually destroyed in the liver and spleen.

6.4 Platelets and Blood Clotting

Platelets result from fragmentation of megakaryocytes in the red bone marrow. Platelets function in blood clotting.

platelets

Blood Clotting

Platelets and two plasma proteins, prothrombin and fibrinogen, function in blood clotting, an enzymatic process that results in fibrin threads, which trap red blood cells.

6.5 Blood Typing and Transfusions

Blood typing usually involves determining the ABO blood group and whether the person is Rh⁻ or Rh⁺. Determining blood type is necessary for transfusions so that agglutination of red blood cells does not occur.

ABO Blood Groups

ABO blood typing determines the presence or absence of type A antigen and type B antigen on the surface of red blood cells.

- **Type A Blood** Type A surface antigens; plasma has anti-B antibodies.
- **Type B Blood** Type B surface antigens; plasma has anti-A antibodies.
- **Type AB Blood** Both type A and type B surface antigens; plasma has neither anti-A nor anti-B antibodies (universal recipient).
- **Type O Blood** Neither type A nor type B surface antigens. Plasma has both anti-A and anti-B antigens (universal donor).
- **Agglutination** Agglutination occurs if the corresponding antigen and antibody are put together (i.e., if the donor has type A blood and the recipient has type B blood).

Rh Blood Groups

The Rh antigen must also be considered when transfusing blood, and it is very important during pregnancy because an Rh⁻ mother may form antibodies to the Rh antigen while carrying or after the birth of an Rh⁺ child. These antibodies can cross the placenta to destroy the red blood cells of any subsequent Rh⁺ child.

6.6 Homeostasis

Homeostasis depends upon the cardiovascular system because it serves the needs of the cells. Several other body systems are also critical to cardiovascular system function:

- The digestive system supplies nutrients.
- The respiratory system supplies oxygen and removes carbon dioxide from the blood.
- The nervous and endocrine systems are involved in maintaining blood pressure.
- The lymphatic system returns tissue fluid to the veins.
- Skeletal muscle contraction (skeletal system) and breathing movements (respiratory system) propel blood in the veins.

Understanding Key Terms

agglutination 114
agranular leukocyte 110
albumin 107
anemia 109
antibody 110
antigen 110
basophil 110
blood doping 109
blood transfusion 114
clotting 113
colony-stimulating factor
    (CSF) 110
deoxyhemoglobin 108
edema 116
eosinophil 110
erythropoietin (EPO) 109
fibrin 113
fibrinogen 107
formed element 107
globulin 107
granular leukocyte 110
hemoglobin (Hb) 108
hemolysis 109
hemolytic disease of the
    newborn 109
hemophilia 113
immune system 110

infectious mononucleosis 111
leukemia 111
lymphocyte 111
mast cell 111
megakaryocyte 113
monocyte 111
neutrophil 110
osmotic pressure 107
oxyhemoglobin 108
plasma 107
plasma protein 107
platelet (thrombocyte) 113
prothrombin 113
prothrombin activator 113
pus 110
red blood cell
    (erythrocyte) 108
serum 113
severe combined
    immunodeficiency disease
    (SCID) 111
sickle-cell disease 109
thrombin 113
thrombocytopenia 113
thromboembolism 113
white blood cell (leukocyte) 110

Match the key terms to these definitions.

a. _____ Iron-containing protein in red blood cells that combines with and transports oxygen.

b. _____ Component of blood that is either cellular or derived from a cell.

c. _____ Liquid portion of blood.

d. _____ A group of cancerous conditions that involve uncontrolled white blood cell proliferation.

e. _____ Plasma protein that is converted to thrombin during the steps of blood clotting.

Testing Your Knowledge of the Concepts

1. The transport function of blood is dependent on what components? The defense function of blood is dependent on what components? The regulatory functions of blood are dependent on what components? (pages 106–7)

2. The osmotic pressure of blood is dependent on what components of blood? (page 107)

3. Describe the structure of a red blood cell, including the molecule hemoglobin. What is the role of red blood cells in the blood? How is the production of red blood cells regulated? (pages 108–9)

4. What are the two main categories of white blood cells? Which of the white blood cells are phagocytes? Explain. (pages 110–11)

5. What formed element is crucial to blood clotting? What other substances are necessary for clotting? Explain the steps that take place when blood clots. (page 113)

6. List and describe three disorders involving RBCs, three disorders involving WBCs, and two disorders involving platelets. (pages 109, 111, and 113)

7. For each type of ABO blood, give the antigen(s) and antibody(ies) present. List which blood type each can receive and to which type each can be given. (pages 114–15)

8. Explain what occurs in hemolytic disease of the newborn. (page 115)

9. Choose five body systems and briefly explain how they are critical to cardiovascular system function. (pages 116–17)

In questions 10–14, match each description with a component of blood in the key. Answers can be used more than once.

Key:
a. red blood cells
b. white blood cells
c. red and white blood cells and platelets
d. plasma

10. Antigens in plasma membrane determine blood type

11. Transport oxygen

12. Includes monocytes

13. Contains plasma proteins

14. Formed elements

In questions 15–19, match each description with a white blood cell in the key.

Key:
a. lymphocytes
b. neutrophils
c. basophils
d. monocytes
e. eosinophils

15. U-shaped nucleus, blue-stained granules that release histamine

16. Includes B cells and T cells that provide specific immunity

17. Bilobed nucleus, red-stained granules, allergic reactions, and parasitic worms

18. Largest, no granules, become macrophages

19. Most abundant, multilobed nucleus, first responders to invasion

20. Which of the following is not a formed element of blood?
a. leukocyte
b. eosinophil
c. fibrinogen
d. platelet

21. Which of the plasma proteins contributes most to osmotic pressure?
    a. albumin
    b. globulins
    c. erythrocytes
    d. fibrinogen

22. Stem cells are responsible for
    a. red blood cell production.
    b. white blood cell production.
    c. platelet production.
    d. the production of all formed elements.

23. Which hemoglobin component is recovered for reuse following red blood cell destruction?
    a. heme
    b. globin
    c. iron
    d. Both b and c are correct.

24. When the oxygen capacity of the blood is reduced,
    a. the liver produces more bile.
    b. the kidneys release erythropoietin.
    c. the bone marrow produces more red blood cells.
    d. sickle-cell disease occurs.
    e. Both b and c are correct.

25. Which of the following conditions can cause anemia?
    a. lack of iron, folic acid, or vitamin B_{12} in the diet
    b. red blood cell destruction
    c. lack of hemoglobin
    d. All of these are correct.
    e. All but a are correct.

26. Which of the following is not true of white blood cells?
    a. formed in red bone marrow
    b. carry oxygen and carbon dioxide
    c. can leave the bloodstream and enter tissues
    d. can fight disease and infection

27. Megakaryocytes give rise to
    a. basophils.
    b. lymphocytes.
    c. monocytes.
    d. platelets.

28. Which of the following is in the correct sequence for blood clotting?
    a. prothrombin activator, prothrombin, thrombin
    b. fibrin threads, prothrombin activator, thrombin
    c. thrombin, fibrinogen, fibrin threads
    d. prothrombin, clotting factors, fibrinogen
    e. Both a and c are correct.

29. Theoretically, a person with type AB blood should be able to receive
    a. type B and type AB blood.
    b. type O and type B blood.
    c. type A and type O blood.
    d. All of these are correct.

30. If a person has type B⁻ blood, it means there are
    a. anti-A antibodies in the plasma.
    b. no B antigens on the red blood cells.
    c. Rh antigens on the red blood cells.
    d. no Rh antigens on the red blood cells.
    e. Both a and d are correct.

Thinking Critically About the Concepts

As does Linda, from the opening story, you should now better appreciate the danger posed by exposure to carbon monoxide. The red blood cells' hemoglobin would much rather bind to CO than oxygen so hemoglobin's ability to transport oxygen is compromised in the presence of CO. Remember from Chapter 3 that oxygen is needed to transform the energy in glucose into ATP, the cells' energy. Without ATP, body cells won't be able to perform the functions they're intended to do to maintain homeostasis and sustain life.

After reading the homeostasis section of this chapter, you should better understand the complexity of maintaining internal conditions and the various interactions between the different organ systems involved in homeostasis. No system ever operates alone, so as you continue your study of the organ systems, keep in mind what you've learned in Chapters 5 and 6 about the cardiovascular system. You may be surprised by how often you hear about concepts from Chapters 5 and 6 in upcoming chapters.

1. A fire is just one way you could be exposed to carbon monoxide. What are some other situations in which you could be exposed to carbon monoxide?

2. Think back to Chapter 2 when you read about Rita and her diagnosis of iron-deficiency anemia.

a. What kind of organic molecule is hemoglobin?

b. How is someone likely to feel when they've been exposed to low (not lethal) levels of CO? Consider what you learned about cellular respiration in Chapter 3 and Rita's complaints prior to her diagnosis.

3. How does the appearance of a person with carbon monoxide poisoning differ from the appearance of someone suffering from hypoxia (low oxygen levels unrelated to CO exposure)?

4. Athlete's who abuse erythropoietin have many more red blood cells than usual.

a. After examining Figure 6.3 and imagining many more red blood cells than usual in this capillary, explain why an athlete might die from having too many red blood cells.

b. Why is death more likely at night when an athlete is sleeping than during the day when they are active? Base your explanation on heartbeat rate.

5. After reading the homeostasis section of this chapter, you should be able to think of at least two other organ systems that interact with the cardiovascular system (blood in particular) to regulate blood oxygen. List two organ systems and how they work with the cardiovascular system to make oxygen available to body cells.

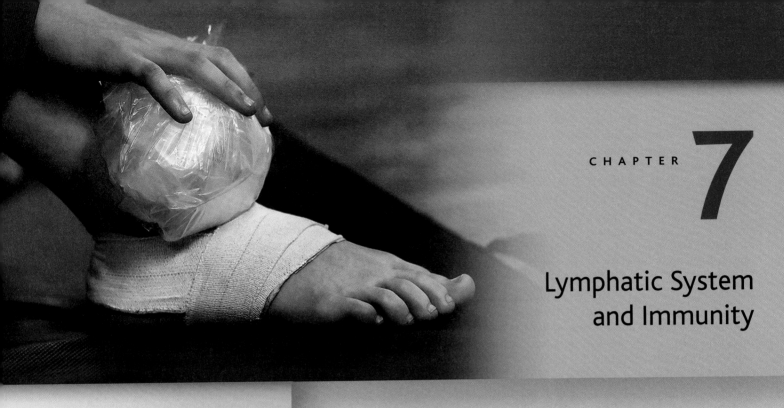

CHAPTER 7

Lymphatic System and Immunity

Lia puzzled over the order of ice, then heat, or was it heat, then ice, as she contemplated how to treat her injured ankle. She had been out running over rough terrain and had twisted her ankle near the end of her run. Now her ankle was rapidly swelling and beginning to hurt.

Any injury initiates an increased flow of blood to that location. An increased flow results in more tissue fluid than the lymphatic capillaries can immediately absorb. As a result, the injured area becomes swollen.

The blood brings the "troops" (the white blood cells that fight infections) to where they may be needed to combat various pathogens or mop up cellular debris. You should not be surprised that immunity is dependent on the cardiovascular system because you previously learned that white blood cells are a component of blood, and that blood vessels are the channels through which blood travels about the body.

In this chapter, you will see that there are several layers of defense against pathogens, just like a bank has a sequence of doors that can be shut to prevent you from reaching the vault. In the Chapter 5 opening story, Allen suggested to Sharon that she use analogies to remember the material in the chapter. Here, too, it will be easier to remember the function of the various levels of defense and kinds of white blood cells if you think of good analogies. The use of the word, "troops" to collectively stand for white blood cells in this opening story might get you started.

CHAPTER CONCEPTS

7.1 Microbes, Pathogens, and You

Bacteria and viruses are pathogens that cause human diseases. Emerging viruses are of special concern of late.

7.2 The Lymphatic System

The lymphatic vessels return excess tissue fluid to cardiovascular veins. The lymphatic organs (red bone marrow, thymus, lymph nodes, and spleen) are important to immunity.

7.3 Nonspecific Defenses

Nonspecific defenses are barriers that prevent pathogens from entering the body and mechanisms that are able to deal with minor invasions.

7.4 Specific Defenses

Specific defenses specifically counteract an invasion in two ways: by producing antibodies and by killing abnormal cells outright.

7.5 Acquired Immunity

The two main types of acquired immunity are immunization by vaccines and the administration of prepared antibodies.

7.6 Hypersensitivity Reactions

The immune system is associated with allergies, tissue reaction, and autoimmune disorders. Treatment is available for these, but research goes forward into finding new and better cures.

7.1 Microbes, Pathogens, and You

Microbes (microscopic organisms, such as bacteria) are widely distributed in the environment. They cover inanimate objects and the surfaces of plants and animals; they are plentiful on and within our bodies. Many of the activities of microbes are useful to humans. We actually eat foods produced by bacteria every day. They contribute to the production of yogurt, cheese, bread, beer, wine, and many pickled foods. Today also, drugs available through biotechnology are produced by bacteria. Microbes help us in still another way. Without the activity of decomposers, the biosphere, including ourselves, would cease to exist. When a tree falls to the forest floor, it eventually rots because decomposers, including bacteria and fungi, break down the remains of dead organisms to inorganic nutrients, which plants need to make the many molecules that become food for us.

Unfortunately, however, human infectious diseases are typically caused by bacteria and viruses, collectively called **pathogens.** The body has three lines of defense against invasion:

1. Barriers to entry, such as the skin and mucous membranes of body cavities, act to prevent pathogens from gaining entrance into the body.
2. First responders, such as the phagocytic white blood cells, act to prevent an infection after an invasion has occurred.
3. Specific defenses overcome an infection by killing the particular disease-causing agent that has entered the body. Specific defenses also protect us against cancer.

Bacteria

Bacteria are single-celled prokaryotes, and they don't have a nucleus. Figure 7.1 illustrates the main features of bacterial anatomy and shows the three common shapes: a **bacillus** has a rod shape; a **coccus** has a spherical shape; and a **spirillum** is curved. Bacteria have a cell wall that contains a unique amino-disaccharide. The "cillin" antibiotics interfere with the production of the cell wall. The cell wall of some bacteria is surrounded by a **capsule** that has a thick, gummy consistency. Capsules often allow bacteria to stick to surfaces such as teeth. They also prevent phagocytic white blood cells from taking them up and destroying them.

Motile bacteria usually have long, very thin appendages called flagella (sing., **flagellum**). The flagella rotate 360° and cause the bacterium to move backwards. Some bacteria have **fimbriae,** stiff fibers that allow bacteria to adhere to surfaces such as host cells. Fimbriae allow a bacterium to cling to and gain access to the body. In contrast, a **pilus** is an elongated hollow appendage used to transfer DNA from one cell to another. Genes that allow bacteria to be resistant to antibiotics can be passed from one to the other through a pilus.

Bacteria are independent cells that are metabolically competent. Their DNA is packaged in a chromosome that

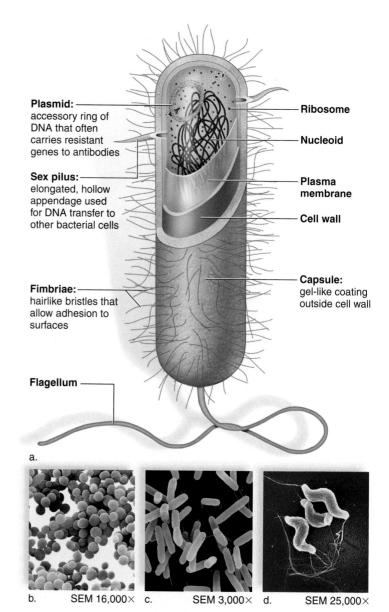

Plasmid: accessory ring of DNA that often carries resistant genes to antibodies

Sex pilus: elongated, hollow appendage used for DNA transfer to other bacterial cells

Fimbriae: hairlike bristles that allow adhesion to surfaces

Flagellum

Ribosome

Nucleoid

Plasma membrane

Cell wall

Capsule: gel-like coating outside cell wall

a.

b.   SEM 16,000×   c.   SEM 3,000×   d.   SEM 25,000×

Figure 7.1   Anatomy of bacteria.
a. Bacterial features, especially the four with definitions, contribute to the ability of bacteria to cause disease. Bacteria occur in three shapes. **b.** *Staphylococcus aureus,* a sphere-shaped bacterium that causes toxic shock syndrome. **c.** *Pseudomonas aeruginosa,* a rod-shaped bacterium that causes urinary tract infections. **d.** *Campylobacter jejuni,* a curve-shaped bacterium that causes food poisoning.

occupies the center of the cell. Many bacteria also have accessory rings of DNA called **plasmids.** Genes that allow bacteria to be resistant to antibiotics are often located in a plasmid. Abuse of antibiotic therapy also leads to resistant bacterial strains that are difficult to cure, even with antibiotics.

Bacteria reproduce by a process called binary fission, which can be diagrammed as in Figure 7.2. The single, circular chromosome attached to the plasma membrane is copied, and then the chromosomes are separated as the cell enlarges. The newly formed plasma membrane and cell wall separate the cell into two cells. Bacteria can reproduce rapidly under favorable conditions, doubling their numbers every 12 minutes.

Figure 7.2 **Binary fission.**
When bacteria reproduce, division (fission) produces two (binary) cells that are identical to the original cell.

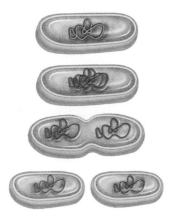

Strep throat, tuberculosis, botulism, food poisoning, gangrene, gonorrhea, and syphilis are well-known bacterial diseases. Growth of bacteria causes disease and also bacteria release molecules called **toxins** that inhibit cellular metabolism. For example, it is important to have a tetanus shot because the bacteria that causes this disease (*Clostridium tetani*) produces a toxin that prevents relaxation of muscles. In time, the body contorts because all the muscles have contracted. Eventually, suffocation occurs.

Viruses

Viruses bridge the gap between the living and the nonliving. Outside a host, viruses are essentially chemicals that can be stored on a shelf. But when the opportunity arises, viruses replicate inside cells, and during this period of time, they clearly appear to be alive.

Viruses are acellular—not composed of cells. They are obligate parasites and they do not live independently. Viruses cause disease, such as colds, flu, measles, chicken pox, polio, rabies, AIDS, genital warts, and genital herpes.

Virus particles are about four times smaller than a bacterium, which is about a hundred times smaller than a eukaryotic cell (Fig. 7.3). A virus always has two parts: an outer capsid composed of protein units and an inner core of nucleic acid (Fig. 7.4). A virus carries the genetic information needed to reproduce itself. In contrast to cellular organisms, the viral genetic

Figure 7.3 **Size comparison of a virus, prokaryotic cell, and eukaryotic cell.**
Viruses are tiny noncellular particles; bacteria are small independent cells; and eukaryotic cells are complex because they contain a nucleus and many organelles.

material need not be double-stranded DNA, nor even DNA. Indeed, some viruses, such as HIV, have an RNA genome. A virus may also contain various enzymes that help it reproduce.

Viruses are microscopic pirates, commandeering the metabolic machinery of a host cell. Viruses gain entry into and are specific to a particular host cell because portions of the virus adhere in a lock-and-key manner with a receptor on the host cell's outer surface. The viral nucleic acid then enters the cell. Once inside, the nucleic acid codes for the protein units in the capsid. In addition, the virus may have genes for special enzymes needed for the virus to reproduce and exit from the host cell. In large measure, however, a virus relies on the host's enzymes and ribosomes for its own reproduction.

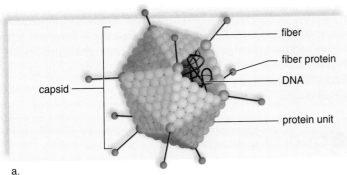

a.

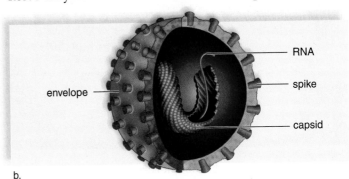

b.

Figure 7.4 **Anatomy of viruses.**
Despite their diversity, all viruses have an outer capsid, composed of protein subunits, and a nucleic acid core, composed of either DNA or RNA, but not both.
a. Adenoviruses cause colds and (**b**) influenza viruses cause the flu. Influenza viruses are surrounded by an envelope with spikes.

Emerging Viruses

Where do apparently new (emerging) viruses come from? Diseases with animal reservoirs are hard to control. In a best-selling novel, *The Hot Zone,* the Drs. Jaax had to contain an outbreak of new Ebola virus to a colony of monkeys in Reston, Virginia. Ebola is one of a group of terrible viruses that cause hemorrhagic fevers. The Ebola viruses quickly cause extensive tissue damage, leading to internal bleeding, multiple organ failure, and death. The Sudan and Zaire strains of Ebola are named for outbreaks in those places. Lassa virus, which can cause a hemorrhagic fever, was named after Lassa, Nigeria. Presently there is concern about the avian (bird) flu, which is passed to humans from birds, as discussed in the Science Focus on page 125. Smallpox could be eradicated only because it infects only humans. Otherwise it may have been impossible.

In some cases, a virus emerges by being transported from one location to another where it has not been before. Flu strains move from Southeast Asia to the United States each year. The West Nile virus is making headlines because it has changed its range, being transported into the United States and taking hold in bird and mosquito populations. Severe acute respiratory syndrome (SARS) was clearly transported from Southeast Asia to Toronto, Canada. A world where you can wake up in Tokyo and sleep in New York is a world where disease can spread at unprecedented rates (Fig. 7.5).

Many viral diseases are transmitted by vectors, usually insects, that carry disease from an infected individual to a healthy individual. Diseases spread by insects may require massive efforts to control the vector population. A disease can emerge when a mutation allows a virus to use a new and

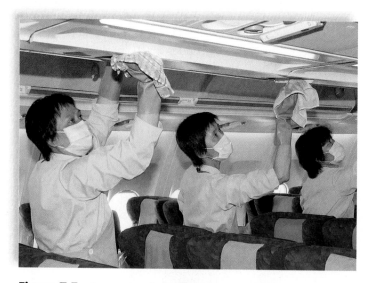

Figure 7.5 **Preventing the transmission of SARS.**
These cleaning women are protecting themselves from SARS (severe acute respiratory syndrome), an easily transmitted emerging virus that is highly contagious.

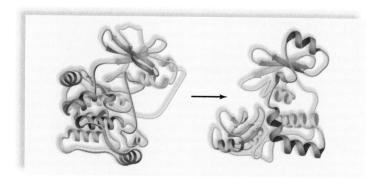

Figure 7.6 **Normal protein versus a prion.**
These models illustrate how a change in protein configuration can also change the shape of a protein so that it would no longer be able to function as it once did.

different insect vector. If a virus transmitted by a rare species of mosquito can now be transmitted by a more common species, a disease could spread more easily. This illustrates that the mode of transmission can be very important. What if the HIV virus was transmitted like the common cold instead of sexually? How many more people might be infected?

Viruses can emerge because of the inability of the immune system to recognize a change in the virus—its new "face," so to speak—that allows it to cause disease. Viruses that continually change their faces, such as HIV, are far more difficult to eradicate.

Prions

Prions, proteinaceous infectious particles, cause a group of degenerative diseases of the nervous system, also called wasting diseases. Originally thought to be viral diseases, prions cause Creutzfeldt-Jakob disease (CJD) in humans, scrapie in sheep, and bovine spongiform encephalopathy (BSE), commonly called **mad cow disease.** These infections are apparently transmitted by ingestion of brain and nerve tissues from infected animals. Prions are proteins of unknown function in the brains of healthy individuals. Disease occurs when certain prion proteins change their shape into a "rogue" form that can convert normal prion proteins into the rogue configuration that does not function because it has the wrong shape. Two molecular models of a protein are shown in Figure 7.6. We can imagine the one on the left is the "normal" protein, and the one on the right is the prion. Nervous tissue is lost and calcified plaques show up in the brain due to activity of prions. Thankfully, the incidence of prion diseases in humans is very low.

✔ Check Your Progress 7.1

1. Which anatomical structures in bacteria can be associated with virulence, the ability to cause disease?

2. Why are viruses considered obligate parasites and always cause disease?

Science Focus

Avian Flu, a Disease of Birds

Imagine waking up one morning with a sore throat, slight fever, and achy muscles. After going to the doctor, you are diagnosed with a case of the flu. As you sit at home resting, you notice the latest news about the avian bird flu and you begin to worry, thinking that it might be avian flu.

Jumping the Species Barrier

Birds can become infected by coming into contact with the saliva, feces, or mucus of another infected bird. Many forms of avian flu will cause no or mild symptoms in infected birds. However, some strains are producing a highly contagious and rapidly fatal disease. Since 1997, a deadly strain of avian flu (H5N1) has been showing up in Asia. Vietnam, Cambodia, China, and several other Asian countries have experienced serious outbreaks of avian flu among their domesticated poultry (chickens, geese, and ducks). The ability of a virus to "jump the species" barrier and infect humans is what concerns the scientific community. In Hong Kong, in 1997, the virus was discovered in several people who had been in close contact with infected birds. Six of the infected people died. In early February 2006, the confirmed death toll as a result of avian flu was 103 people. The lethal strain of bird flu has killed approximately one-third of the people that have been infected.

Possibility of a Pandemic

There is a growing concern among the scientific community that the bird flu could merge with a human flu virus. The concern is that a new hybrid virus could form from the merging of the other two. This new virus could be highly infectious, fatal, and easily transmitted from person to person. Rapid travel between countries would enable people to spread the flu virus across the globe at an unprecedented rate, potentially triggering a global pandemic (a disease that occurs worldwide).

There are two possibilities for the avian flu virus and human flu virus to merge and become a hybrid virus. The first possibility is when a human is infected with the human flu and then comes into contact with the avian flu. The two viruses could meet in the person's body and swap genes with each other. If the new hybrid virus has the avian flu genes for lethality and the human flu genes that allow it to be passed from person to person, a pandemic could be the result. The second possibility is if the two viruses infect a different host species and swap their genes in that host species. Pigs are seen as the potential alternate host species because they are

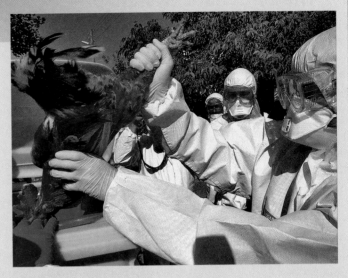

Figure 7A Infected birds.
In an attempt to contain avian flu, infected birds are gathered and killed.

susceptible to both the avian and human flu virus. If a new hybrid virus forms in pigs, it could easily be passed back to human farmers who come into contact with infected pigs.

In January 2005, the first case of human-to-human transmission of avian flu was recorded. A Thai woman contracted the virus from her sick daughter, and both women died from the avian flu. Every time the avian flu virus jumps from a bird to a person, the risk of the hybrid virus forming increases.

Action Taken

Currently, there are no known vaccinations that work against the bird flu. Several drugs are being trialed and have produced mixed results. Scientists are racing to find a potential vaccination but face the challenge of a virus that is constantly evolving. In countries that have been affected by the avian flu, governments have begun programs to remove the infected and potentially infected birds (Fig. 7A). In Asia, millions of domesticated birds have been killed in an attempt to control the spread of the virus. It is hoped that by eliminating the birds, it will contain the virus and decrease the potential of spreading. The containment efforts by the infected Asian countries have been somewhat successful at keeping the virus from spreading among domesticated livestock.

The problem is preventing the spread of the virus among wild waterfowl as they migrate across the globe. As of February 2006, confirmations of the deadly strain of avian flu in waterfowl have been reported in Asia, Africa, the Middle East, and Europe. It is only a matter of time before it reaches the United States and Central and South America.

7.2 The Lymphatic System

The lymphatic system consists of lymphatic vessels and the lymphatic organs. This system, which is closely associated with the cardiovascular system, has four main functions that contribute to homeostasis: (1) Lymphatic capillaries absorb excess tissue fluid and return it to the bloodstream; (2) in the small intestines, lymphatic capillaries called lacteals absorb fats in the form of lipoproteins and transport them to the bloodstream; (3) the lymphatic system is responsible for the production, maintenance, and distribution of lymphocytes; and (4) the lymphatic system helps defend the body against pathogens.

Lymphatic Vessels

Lymphatic vessels form a one-way system of, first, capillaries (side streets), then, vessels (streets), and, finally, ducts (highways) that take lymph to cardiovascular veins in the shoulders (Fig. 7.7). Lymphatic capillaries, as you know, take up excess tissue fluid. Tissue fluid is mostly water, but it also contains solutes (i.e., nutrients, electrolytes, and oxygen) derived from plasma and cellular products (i.e., hormones, enzymes, and wastes) secreted by cells. The fluid inside lymphatic vessels is called **lymph.** Lymph is usually a colorless liquid, but after a meal, it appears creamy because of its lipid content.

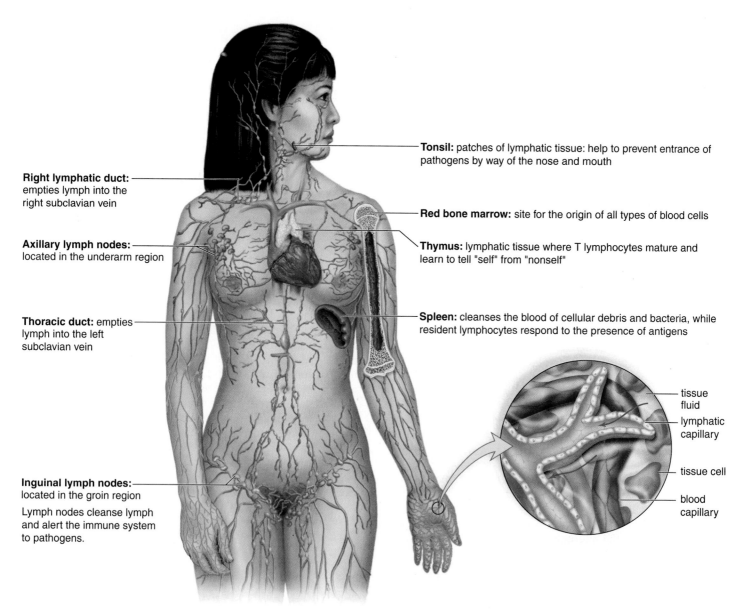

Right lymphatic duct: empties lymph into the right subclavian vein

Axillary lymph nodes: located in the underarm region

Thoracic duct: empties lymph into the left subclavian vein

Inguinal lymph nodes: located in the groin region

Lymph nodes cleanse lymph and alert the immune system to pathogens.

Tonsil: patches of lymphatic tissue: help to prevent entrance of pathogens by way of the nose and mouth

Red bone marrow: site for the origin of all types of blood cells

Thymus: lymphatic tissue where T lymphocytes mature and learn to tell "self" from "nonself"

Spleen: cleanses the blood of cellular debris and bacteria, while resident lymphocytes respond to the presence of antigens

tissue fluid

lymphatic capillary

tissue cell

blood capillary

Figure 7.7   Lymphatic system.
Lymphatic vessels drain excess fluid from the tissues and return it to the cardiovascular system. The enlargement shows that lymphatic vessels, like cardiovascular veins, have valves to prevent backward flow. The lymph nodes, spleen, thymus gland, and red bone marrow are the main lymphatic organs that assist immunity.

The lymphatic system has two ducts (highways): the thoracic duct and the right lymphatic duct. The larger thoracic duct returns lymph collected from the body below the thorax, the left arm, and left side of the head and neck into the left subclavian vein. The right lymphatic duct returns lymph from the right arm and right side of the head and neck into the right subclavian vein.

The construction of the larger lymphatic vessels is similar to that of cardiovascular veins, including the presence of valves. The movement of lymph within lymphatic capillaries is largely dependent upon skeletal muscle contraction. Lymph forced through lymphatic vessels, as a result of muscular compression, is prevented from flowing backward by one-way valves.

Lymphatic Organs

Lymphatic organs are divided into those that are primary: red bone marrow and the thymus gland; and those that are secondary: lymph nodes and spleen (Fig. 7.8). To continue the suggestion made on page 121 about analogies, the red bone marrow is like a central fort that puts the troops (white blood cells) through their basic training (maturation), with one exception. Lymphocytes are either B lymphocytes (B cells) or T lymphocytes (T cells). The B cells mature in the bone marrow, but the T cells need special training, and lymphocytes that will become T cells have to pass through the thymus, another fort.

The secondary lymphatic organs are like staging areas for the special forces, the lymphocytes and the macrophages. These two types of cells (soldiers) are always found together for reasons that will be explored later in this chapter. The secondary lymphatic organs purify and stand guard over the lymph and, also, the blood.

The Primary Lymphatic Organs

Red bone marrow produces all types of blood cells. In a child, most bones have red bone marrow, but in an adult, it is limited to the sternum, vertebrae, ribs, part of the pelvic girdle, and the upper ends of the humerus and femur. In addition to the red blood cells, bone marrow produces the various types of white blood cells: neutrophils, eosinophils, basophils, lymphocytes, and monocytes. B cells mature in the bone marrow, but the T cells mature in the thymus.

The soft, bilobed **thymus gland** is located in the thoracic cavity between the trachea and the sternum, superior to the heart. The thymus varies in size, but it is largest in children and shrinks as we get older.

The thymus has two functions: (1) The thymus gland produces thymic hormones, such as thymosin, that are thought to aid in the maturation of T lymphocytes. Thymosin may also have other functions in immunity. (2) Immature T lymphocytes migrate from the bone marrow through the bloodstream to the thymus, where they mature. Only about 5% of these cells ever leave the thymus. These T lymphocytes have survived a critical test: If any show the ability to react with the individual's cells, they die. If they have potential to attack a pathogen, they leave the thymus. The thymus is absolutely critical to immunity because without mature T cells, the body's response to pathogens is poor or absent.

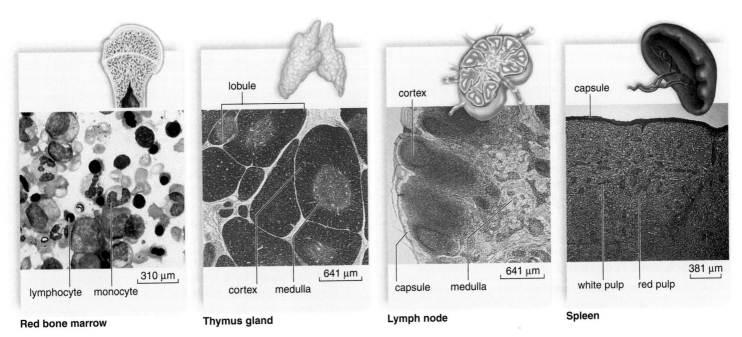

lobule	cortex	capsule

310 μm   641 μm   641 μm   381 μm

lymphocyte   monocyte   cortex   medulla   capsule   medulla   white pulp   red pulp

Red bone marrow   **Thymus gland**   **Lymph node**   **Spleen**

Figure 7.8 **The lymphatic organs.**
Left: Red bone marrow and the thymus gland are the primary lymphatic organs. Blood cells, including lymphocytes, are produced in red bone marrow. B cells mature in the bone marrow, but T cells mature in the thymus. *Right:* Lymph nodes and the spleen are secondary lymphatic organs. Lymph is cleansed in lymph nodes, while blood is cleansed in the spleen.

Secondary Lymphatic Organs

The secondary lymphatic organs are the spleen, the lymph nodes, and other organs, such as the tonsils, Peyer's patches, and the appendix.

The **spleen** filters blood. The spleen, the largest lymphatic organ, is located in the upper left region of the abdominal cavity posterior to the stomach. Connective tissue divides the spleen into regions known as white pulp and red pulp (Fig. 7.8). The red pulp, which surrounds venous sinuses (cavities), is involved in filtering the blood. Blood entering the spleen must pass through the sinuses before exiting. Here, macrophages are like powerful vacuum cleaners that engulf pathogens and debris, such as worn-out red blood cells.

The spleen's outer capsule is relatively thin, and an infection or a blow can cause the spleen to burst. Although the spleen's functions are replaced by other organs, a person without a spleen is often slightly more susceptible to infections and may have to receive antibiotic therapy indefinitely.

Lymph nodes, which occur along lymphatic vessels, filter lymph. Connective tissue forms a capsule and also divides a lymph node into compartments. Each compartment contains a sinus that increases in size toward the center of the node. As lymph courses through the sinuses, it is filtered by macrophages, which engulf pathogens and debris. Lymphocytes, also present in sinuses, fight infections and attack cancer cells.

Lymph nodes are named for their location. For example, inguinal nodes are in the groin, and axillary nodes are in the armpits. Physicians often feel for the presence of swollen, tender lymph nodes in the neck as evidence that the body is fighting an infection. This is a noninvasive, preliminary way to help make such a diagnosis.

Lymphatic nodules are concentrations of lymphatic tissue not surrounded by a capsule. The **tonsils** are patches of lymphatic tissue located in a ring about the pharynx (see Fig. 8.4). The tonsils perform the same functions as lymph nodes, but because of their location, they are the first to encounter pathogens and antigens that enter the body, by way of the nose and mouth.

Peyer's patches located in the intestinal wall and the appendix attached to the cecum, a blind pouch of the large intestine, encounter pathogens that enter the body by way of the intestinal tract.

✔ Check Your Progress 7.2

1. **a.** Which organs are primary lymphatic organs? **b.** Secondary lymphatic organs?

2. How do the primary and secondary lymphatic organs function in immunity?

3. Which organ produces blood cells, which filters blood, which filters lymph, and which causes T cells to mature?

7.3 Nonspecific Defenses

Immunity, the ability to combat diseases and cancer, includes lines of defense. These two lines of defense are nonspecific because they act indiscriminately against all pathogens:

- Barriers to entry. The body puts up barriers that can be likened to doors that are usually shut tight, and many of which have special features that bar the possible entry of pathogens.

- Phagocytic white blood cells, the neutrophils and macrophages. We will consider these two phagocytic white blood cells in the context of the **inflammatory response,** a special reaction the body has when first invaded. Protective proteins are also part of this line of defense.

Barriers to Entry

The body has built-in barriers, both physical and chemical, that serve as the first line of defense against an infection by pathogens.

Skin and Mucous Membranes   The intact skin is generally a very effective physical barrier that prevents infection. Mucous membranes lining the respiratory, digestive, reproductive, and urinary tracts are also physical barriers to entry by pathogens. The ciliated cells that line the upper respiratory tract sweep mucus and trapped particles up into the throat, where they can be swallowed or expectorated (spit out).

Chemical Barriers   The chemical barriers to infection include the secretions of sebaceous (oil) glands of the skin. These secretions contain chemicals that weaken or kill certain bacteria on the skin.

Perspiration, saliva, and tears contain an antibacterial enzyme called **lysozyme.** Saliva also helps to wash microbes off the teeth and tongue, and tears wash the eyes. Similarly, as urine is voided from the body, it flushes bacteria from the urinary tract.

The acid pH of the stomach inhibits growth or kills many types of bacteria. At one time, it was thought that no bacterium could survive the acidity of the stomach. But now we know that ulcers are caused by the bacterium *Helicobacter pylori* (see page 149). Similarly, the acidity of the vagina and its thick walls discourage the presence of pathogens. For this reason, vaginal sexual intercourse provides protection against the AIDS virus.

Resident Bacteria   Finally, a significant chemical barrier to infection is created by the normal flora, microbes that usually reside in the mouth, intestine, and other areas. By using up available nutrients and releasing their own waste, these resident bacteria prevent potential pathogens from taking up residence. For this reason, abusing antibiotics can make a person susceptible to pathogenic infection by killing off the normal flora.

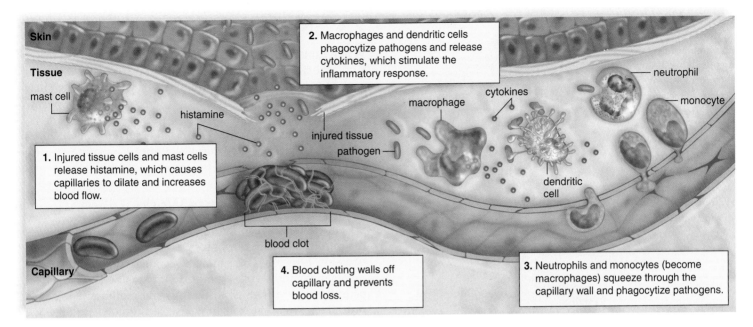

Skin

Tissue

mast cell

histamine

1. Injured tissue cells and mast cells release histamine, which causes capillaries to dilate and increases blood flow.

2. Macrophages and dendritic cells phagocytize pathogens and release cytokines, which stimulate the inflammatory response.

neutrophil

cytokines

macrophage

monocyte

injured tissue

pathogen

dendritic cell

blood clot

Capillary

4. Blood clotting walls off capillary and prevents blood loss.

3. Neutrophils and monocytes (become macrophages) squeeze through the capillary wall and phagocytize pathogens.

Figure 7.9 Inflammatory response.
① Due to capillary changes in a damaged area and the release of chemical mediators, such as histamine by mast cells, an inflamed area exhibits redness, heat, swelling, and pain. The inflammatory response can be accompanied by other reactions to the injury. ② Macrophages and dendritic cells, present in the tissues, release cytokines, which stimulate the inflammatory and other immune responses. ③ Macrophages and dendritic cells phagocytize pathogens, as do neutrophils, which squeeze through capillary walls from the blood.④ A blood clot can form a seal in a break in a blood vessel.

Inflammatory Response

The inflammatory response exemplifies the second line of defense against invasion by a pathogen. Like an army defending its nation against an invading force, inflammation employs its foot soldiers, mainly the neutrophils and macrophages, to surround and kill (engulf by phagocytosis) pathogens that are trying to get a foothold inside the body. Inflammation is usually recognized by its four hallmark symptoms: redness, heat, swelling, and pain (Fig. 7.9).

When Lia in the opening story twisted her ankle, the inflammatory response began to occur. The four signs of the inflammatory response are due to capillary changes in the damaged area, and all serve to protect the body. Chemical mediators, such as **histamine,** released by damaged tissue cells and **mast cells,** cause the capillaries to dilate and become more permeable. Excess blood flow due to enlarged capillaries causes the skin to redden and become warm. Increased temperature in an inflamed area tends to inhibit growth of some pathogens, and increased blood flow brings white blood cells to the area. Increased permeability of capillaries allow fluids and proteins, including blood clotting factors, to escape into the tissues. Clot formation in the injured area prevents blood loss. The excess fluid in the area presses on nerve endings, causing the familiar pain that is associated with swelling. Together, these events send out a call to arms and summon white blood cells to the area.

As soon as the white blood cells arrive, they move out of the bloodstream into the surrounding tissue. The neutro-phils are first and act as scouts—they assess the situation and actively devour debris, dead cells, and bacteria they encounter. The many neutrophils that are attracted to the area can usually localize any infection and keep it from spreading. If neutrophils die off in great quantity, they become a yellow-white substance we call **pus.**

When an injury is not serious, the inflammatory response is short-lived, and the healing process will quickly return the affected area to a normal state. Nearby cells secrete growth factors to ensure the growth and repair of blood vessels and growth of new cells to fill in the damaged area.

If, on the other hand, the neutrophils are overwhelmed, they call for reinforcements by secreting chemical mediators called **cytokines.** Cytokines attract more white blood cells to the area, including monocytes. Monocytes are longer-lived cells that become **macrophages,** which are even more powerful phagocytes than neutrophils. Not only that, macrophages can enlist the help of lymphocytes to carry out specific defense mechanisms.

Despite the obvious misery and pain that Lia endured as the inflammatory response occurred, it is the body's natural response to an irritation or injury and serves an important role. Once the tide of the battle has turned, inflammation rapidly subsides. However, in some cases, chronic inflammation may last for weeks, months, or even years if an irritation or infection cannot be overcome. Inflammatory chemicals are often like bombs that may cause collateral damage to the body, in addition to killing the invaders. Should an inflammation persist, anti-inflammatory medications, such as aspirin, ibuprofen, or cortisone, can minimize the effects of various chemical mediators.

**Figure 7.10
Action of the
complement system
against a bacterium.**
When complement
proteins in the blood
plasma are activated by
an immune response, they
form a membrane attack
complex that makes holes
in bacterial cell walls
and plasma membranes,
allowing fluids and salts
to enter until the cell
eventually bursts.

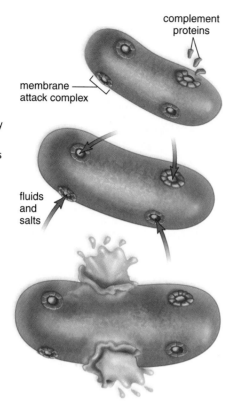

Protective Proteins

The **complement system,** often simply called complement,
is composed of a number of blood plasma proteins designated by the letter C and a subscript. The complement proteins "complement" certain immune responses, which
accounts for their name. For example, they are involved in
and amplify the inflammatory response because certain
complement proteins can bind to mast cells and trigger histamine release, and others can attract phagocytes to the
scene. Some complement proteins bind to the surface of
pathogens already coated with antibodies, which ensures
that the pathogens will be phagocytized by a neutrophil,
dendritic cell, or macrophage.

Certain other complement proteins join to form a **membrane attack complex** that produces holes in the surface of
bacteria and some viruses. Fluids and salts then enter the bacterial cell or virus to the point that they burst (Fig. 7.10).

Interferons are proteins produced by virus-infected cells
as a warning to noninfected cells in the area. Interferon binds
to receptors of noninfected cells, causing them to prepare for
possible attack by producing substances that interfere with
viral replication. Interferons are used as treatment in certain
viral infections, such as hepatitis C.

☑ Check Your Progress 7.3

1. What are some examples of the body's nonspecific defenses?

2. Which blood cells, in particular, should you associate with
   nonspecific defenses, and how do they function?

3. Why are the complement proteins so named?

7.4 Specific Defenses

When nonspecific defenses have failed to prevent an infection, specific defenses comes into play. Specific defenses
overcome an infection by doing away with the particular
disease-causing agent that has entered the body. Specific
defenses also protect us against cancer.

The specific defenses can be likened to special forces
that can attack selected targets without harming nearby
residents (cells).

How Specific Defense Works

Specific defenses respond to **antigens,** which are molecules
the immune system recognizes as foreign to the body. Antigens are typically large molecules, such as proteins. Fragments of bacteria, viruses, molds, or parasitic worms can all
be antigenic; further, abnormal plasma membrane proteins
produced by cancer cells may also be antigens. Because we
do not ordinarily become immune to our own normal cells,
it is said that the immune system is able to distinguish "self"
from "nonself."

Specific defenses primarily depend on the action of lymphocytes, which differentiate as either **B cells (B lymphocytes)** or **T cells (T lymphocytes).** B cells and T cells are
capable of recognizing antigens because they have specific
antigen receptors—plasma membrane receptor proteins,
whose shape allows them to combine with particular antigens. Each lymphocyte has only one type of receptor. It is
often said that the receptor and the antigen fit together like a
lock and a key. Because we encounter a million different
antigens during our lifetime, we need a diversity of B cells
and T cells to protect us against them. Remarkably, diversifi-
cation occurs to such an extent during the maturation process that there are specific B cells and/or T cells for any
possible antigen.

B cells differentiate into other types of cells, and so do
T cells. The several types of cells we will be discussing are
listed in Table 7.1 for easy reference.

Table 7.1	Immunocell Type and Function
Cell	**Function**
B cells	Produce plasma cells and memory cells
Plasma cells	Produce specific antibodies
Memory cells	Ready to produce antibodies in the future
T cells	Regulate immune response; produce cytotoxic T cells and helper T cells
Cytotoxic T cells	Kill virus-infected cells and cancer cells
Helper T cells	Regulate immunity
Memory T cells	Ready to kill in the future

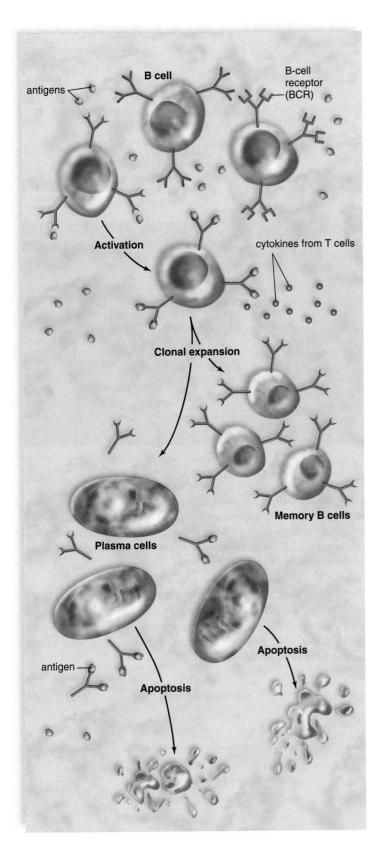

Figure 7.11 **Clonal selection model as it applies to B cells.**
Activation of a B cell occurs when its B-cell receptor (BCR) can combine with
an antigen (colored green). In the presence of cytokines, the B cell undergoes
clonal expansion, producing many plasma cells that secrete antibodies
specific to the antigen.

B Cells and Antibody-Mediated Immunity

The receptor on a B cell is called a **B-cell receptor (BCR).** The *clonal selection model* states that an antigen *selects,* then binds to the BCR of only one type B cell and then this B cell produces multiple copies of itself. The resulting group of identical cells is called a *clone.* (Similarly, an antigen can bind to a T-cell receptor (TCR) and this T cell will clone.)

B Cells Become Plasma Cells and Memory B Cells   Note in Figure 7.11 that each B cell has a specific BCR represented by shape. Only the B cell with a BCR that has a shape that fits the antigen (green circle) undergoes clonal expansion. During clonal expansion, cytokines secreted by helper T cells (see page 135) stimulate B cells to clone. Most of the cloned B cells become **plasma cells,** which circulate in the blood and lymph. Plasma cells are larger than regular B cells because they have extensive rough endoplasmic reticulum for the mass production and secretion of antibodies to a specific antigen. Antibodies are identical to the BCR of the B cell that was activated. Some cloned B cells become memory cells, which are the means by which long-term immunity is possible. If the same antigen enters the system again, memory B cells quickly divide and give rise to more plasma cells capable of quickly producing the correct type antibody.

Once the threat of an infection has passed, the development of new plasma cells ceases, and those present undergo apoptosis. **Apoptosis** is the process of programmed cell death (PCD), involving a cascade of specific cellular events leading to the death and destruction of the cell.

Defense by B cells is called **antibody-mediated immunity** because activated B cells become plasma cells that produce antibodies. Collectively, plasma cells probably produce as many as two million different antibodies. A human being doesn't have two million genes, so there cannot be a separate gene for each type antibody. As discussed in the Science Focus on page 133, it has been found that scattered DNA segments can be shuffled and combined in various ways to produce the DNA sequence coding for the BCR unique to each type of B cell.

Characteristics of B Cells
- Antibody-mediated immunity against pathogens
- Produced and mature in bone marrow
- Directly recognize antigen and then undergo clonal selection
- Clonal expansion produces antibody-secreting plasma cells as well as memory B cells

Structure of an Antibody The basic unit which composes antibody molecules is a Y-shaped protein molecule with two arms. Each arm has a "heavy" (long) polypeptide chain and a "light" (short) polypeptide chain (Fig. 7.12). These chains have constant regions, located at the trunk of the Y, where the sequence of amino acids is set. The class of antibody of each individual molecule is determined by the structure of the antibody's constant region. The variable regions form an antigen-binding site, and their shape is specific to a particular antigen. The antigen combines with the antibody at the antigen-binding site in a lock-and-key manner. Antibodies may consist of single Y-shaped molecules, called monomers, or be paired together in a molecule termed a dimer. Very large antibodies belonging to the M class are pentamers—that is, clusters of five Y-shaped molecules linked together.

The antigen-antibody reaction can take several forms. Antigens can be a part of a pathogen and therefore, antibodies sometimes react with viruses and toxins by coating them completely, a process called neutralization. Often, the reaction produces a clump of antigens combined with antibodies, termed an immune complex. The antibodies in an immune complex are like a beacon that attracts white blood cells that move in for the kill.

Classes of Antibodies There are five different classes of circulating antibodies (Table 7.2). IgG antibodies are the major type in blood, and lesser amounts are found in lymph and tissue fluid. IgG antibodies bind to pathogens and their toxins. IgG antibodies can cross the placenta from a mother to her fetus, so the newborn has a temporary, partial immune response. As mentioned, IgM antibodies are pentamers. IgM antibodies are the first antibodies produced by a newborn's body. IgM antibodies are the first to appear in blood soon after an infection begins and the first to disappear before the infection is over. They are good activators of the complement system. IgA antibodies are monomers or dimers containing two Y-shaped structures. They are the main type of antibody found in body secretions: saliva, tears, mucus, and breast milk. IgA molecules bind to pathogens and prevent

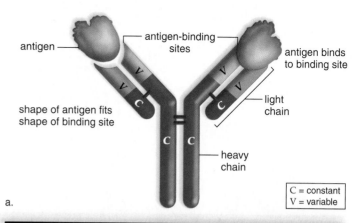

a.

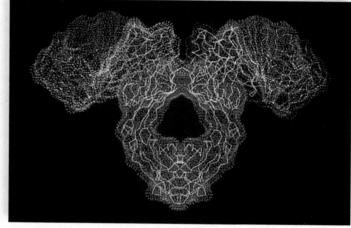

b.

Figure 7.12 **Structure of an antibody.**
a. An antibody contains two heavy (long) polypeptide chains and two light (short) chains arranged so there are two variable regions, where a particular antigen is capable of binding with an antibody (*V* = variable region, *C* = constant region). **b.** Computer model of an antibody molecule. The antigen combines with the two side branches.

them from reaching the bloodstream. The main function of IgD molecules seems to be to serve as antigen receptors on immature B cells. IgE antibodies are responsible for prevention of parasitic worm infections, but they can also cause immediate allergic responses.

Table 7.2	Antibodies	
Class	**Presence**	**Function**
IgG	Main antibody type in circulation; crosses the placenta from mother to fetus	Binds to pathogens, activates complement, and enhances phagocytosis by white blood cells
IgM	Antibody type found in circulation; largest antibody; first antibody formed by a newborn; first antibody formed with any new infection	Activates complement; clumps cells
IgA	Main antibody type in secretions such as saliva and milk	Prevents pathogens from attaching to epithelial cells in digestive and respiratory tract
IgD	Antibody type found on surface of immature B cells	Presence signifies readiness of B cell
IgE	Antibody type found as antigen receptors on basophils in blood and on mast cells in tissues	Responsible for immediate allergic response and protection against certain parasitic worms

Science Focus

Antibody Diversity

In 1987, Susumu Tonegawa (Fig. 7B*a*) became the first Japanese scientist to win the Nobel Prize in Physiology or Medicine after dedicating himself to finding the solution to an engrossing puzzle. Immunologists knew that each B cell makes an antibody especially equipped to recognize the specific shape of a particular antigen. But they did not know how the human DNA contained enough genetic information to permit the production of up to two million different antibody types needed to combat all of the pathogens we are likely to encounter during our lives.

The Puzzle

An antibody is composed of two light and two heavy polypeptide chains, which are divided into constant and variable regions. The constant region determines the antibody class, and the variable region determines the specificity of the antibody, because this is where an antigen binds to a specific antibody (see Fig. 7.12). Each B cell must have a genetic way to code for the variable regions of both the light and heavy chains.

The Experiment

Tonegawa's colleagues say that he is a creative genius who intuitively knows how to design experiments to answer specific questions. In this instance, he examined the DNA sequences of lymphoblasts (immature lymphocytes) and compared them to mature B cells. He found that the DNA segments coding for the variable and constant regions were scattered throughout the DNA in lymphoblasts, and that only certain of these segments were present in each mature antibody-secreting B cell, where they randomly came together and coded for a specific variable region. Later, the variable and constant regions are joined to give a specific antibody (Fig. 7B*b*). As an analogy, consider that each person entering a supermarket chooses various items for purchase, and that the possible combination of items in any particular grocery bag is astronomical. Tonegawa also found that mutations occur as the variable segments are undergoing rearrangements. Such mutations are another source of antibody diversity.

A Prize Winner

Tonegawa received his BS in chemistry in 1963 at Kyoto University and earned his PhD in biology from the University of California at San Diego (UCSD) in 1969. After that, he worked as a research fellow at UCSD and the Salk Institute. In 1971, he moved to the Basel Institute for Immunology and began the experiments that eventually led to his Nobel Prize-winning discovery. Tonegawa also contributed to the effort to decipher the receptors of T cells. This was an even more challenging area of research than the diversity of antibodies produced by B cells. Since 1981, he has been a full professor at Massachusetts Institute of Technology (MIT), where he has a reputation for being an "aggressive, determined researcher" who often works late into the night.

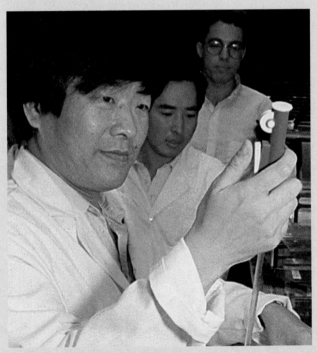

a. Susumu Tonegawa in the laboratory

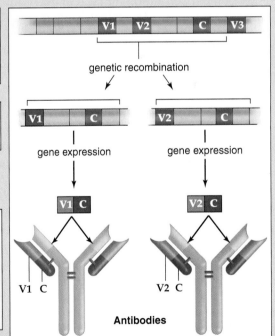

Genetic recombination brings certain DNA segments together.

Functional DNA in mature B cells is different.

Light chains of antibody molecules are different. The same mechanism to introduce variability also applies to heavy chains.

genetic recombination

gene expression   gene expression

Antibodies

b. Antibody variable regions

Figure 7B   Antibody diversity.
a. Susumu Tonegawa received a Nobel Prize for his findings regarding antibody diversity. **b.** Different genes for the variable regions of heavy and light chains are brought together during the production of B lymphocytes so that the antigen receptors of each one can combine with only a particular antigen.

T Cells and Cell-Mediated Immunity

Cell-mediated immunity is named for the action of T cells that directly attack diseased cells and cancer cells. Other T cells, however, release cytokines that stimulate both nonspecific and specific defenses.

How T Cells Recognize an Antigen When a T cell leaves the thymus, it has a unique **T-cell receptor (TCR),** just as B cells have. Unlike B cells, however, T cells are unable to recognize an antigen without help. The antigen must be displayed to them by an **antigen-presenting cell (APC),** such as a macrophage. After phagocytizing a pathogen, such as a bacterium, APCs travel to a lymph node or spleen, where T cells also congregate. In the meantime, the APC has broken the pathogen apart in a lysosome. A piece of the pathogen is then displayed in the groove of an MHC (major histocompatibility complex) protein on the cell's surface. Human MHC proteins are called **HLA (human leukocyte antigens).** These proteins are found on all of our body cells. More than 50 different proteins have been identified, and each individual human being has a unique combination—no two sets are exactly alike (with the exception of the set belonging to identical twins. Because identical twins arise from division of a single zygote, their HLA proteins are identical.). Because they mark the cell as belonging to a particular individual, HLA antigens are **self** proteins. The importance of self proteins in plasma membranes was first recognized when it was discovered that they contribute to the specificity of tissues and make it difficult to transplant tissue from one human to another. Comparison studies of HLA antigens must always be carried out before a transplant is attempted. The more of the 50+ proteins that the donor and recipient share in common, the better the tissue match. In other words, when the donor and the recipient are histo (tissue)-compatible, a transplant is more likely to be successful.

When an antigen-presenting cell links a foreign antigen to the self protein on its plasma membrane, it carries out an important safeguard for the rest of the body. Now, the T cell to be activated can compare the antigen and self protein side by side. The activated T cell and all of the daughter cells, which it will form, can recognize "foreign" from "self," and go on to destroy cells carrying foreign antigens, while leaving normal body cells unharmed.

Clonal Expansion In Figure 7.13, the T cells have specific TCRs, represented by their different shapes. A macrophage is presenting an antigen to a T cell that has the specific TCR that will combine with this particular antigen, represented by a green circle. Now, the T cell is activated and undergoes clonal expansion. Many copies of the activated T cell are produced during clonal expansion. There are two classes of HLA proteins (called HLA I and HLA II). If an APC displays an antigen within the groove of an HLA I protein, the activated T cell will form cytotoxic T cells. If an APC displays an antigen within the groove of an HLA II protein, the activated T cell will form helper T cells.

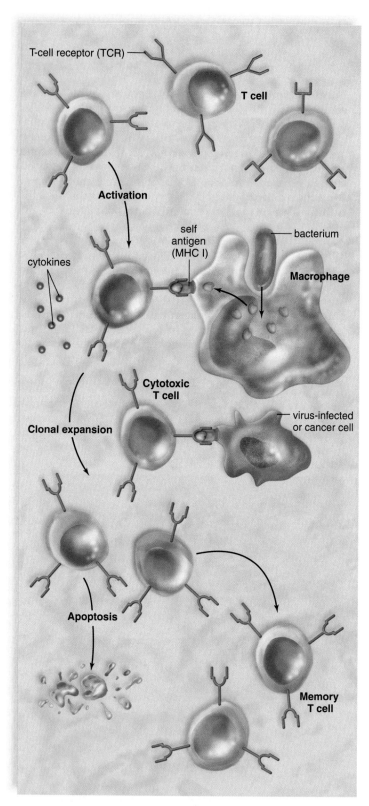

Figure 7.13 Clonal selection model as it applies to T cells.
Activation of a T cell occurs when its T-cell receptor (TCR) can combine with an antigen presented by a macrophage. In this example, cytotoxic T cells are produced, and when the immune response is finished, they undergo apoptosis. A small number of memory T cells remain.

As the illness disappears, the immune reaction wanes, and activated T cells become susceptible to apoptosis. As mentioned previously, apoptosis contributes to homeostasis by regulating the number of cells present in an organ, or in this case, in the immune system. When apoptosis does not occur as it should, the potential exists for an autoimmune response (see page 139) or for T-cell cancers (i.e., lymphomas and leukemias).

Cytotoxic T cells are like paratroopers that carry rifles with bayonets. They can go behind the enemy lines and seek out a specific enemy for destruction. Cytotoxic T cells have storage vacuoles containing perforins and storage vacuoles containing enzymes called granzymes. After a cytotoxic T cell binds to a virus-infected cell or tumor cell, it releases perforin molecules, which punch holes into the plasma membrane, forming a pore. Cytotoxic T cells then deliver granzymes into the pore, and these cause the cell to undergo apoptosis and die. Once cytotoxic T cells have released the perforins and granzymes, they move on to the next target cell. Cytotoxic T cells are responsible for so-called **cell-mediated immunity** (Fig. 7.14).

Helper T cells, despite their name, are like generals because they do not fight directly. Instead, they regulate immunity by secreting **cytokines,** the chemicals (orders) that enhance the response of all types of immune cells. B cells cannot be activated without T cell help. Because the human immunodeficiency virus (HIV), the virus that causes AIDS, infects helper T cells and other cells of the immune system, it inactivates the immune response and makes HIV-infected individuals susceptible to the opportunistic infections that eventually will kill them.

Notice in Figure 7.13 that a few of the clonally expanded T cells are **memory T cells.** They remain in the body and can jump-start an immune reaction to an antigen previously present in the body.

Characteristics of T Cells

- Cell-mediated immunity against virus-infected cells and cancer cells
- Produced in bone marrow, mature in thymus
- Antigen must be presented in groove of an HLA molecule
- Cytotoxic T cells destroy nonself antigen-bearing cells
- Helper T cells secrete cytokines that control the immune response

---

✓ Check Your Progress 7.4

1. How does nonspecific defense differ from specific defense?
2. What are some examples of the body's specific defenses?
3. Which blood cells are mainly responsible for specific defense, and how do they function?

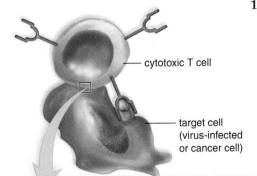
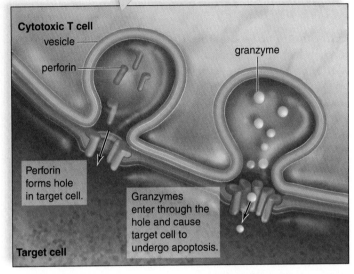

cytotoxic T cell

target cell (virus-infected or cancer cell)

Cytotoxic T cell
vesicle
perforin
granzyme

Perforin forms hole in target cell.

Granzymes enter through the hole and cause target cell to undergo apoptosis.

Target cell

a.

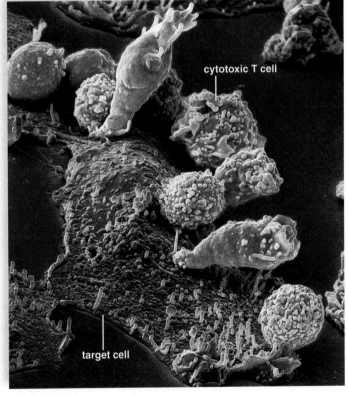

cytotoxic T cell

target cell

b.

Figure 7.14 Cell-mediated immunity.
The art (**a**) and photo (**b**) show how a cytotoxic T cell destroys a target cell.

7.5 Acquired Immunity

Immunity occurs naturally through infection or is brought about artificially by medical intervention. The two types of acquired immunity are active and passive. In **active immunity,** the individual alone produces antibodies against an antigen; in **passive immunity,** the individual is given prepared antibodies via an injection.

Active Immunity

Active immunity sometimes develops naturally after a person is infected with a pathogen. However, active immunity is often induced when a person is well so that future infection will not take place. To prevent infections, people can be artificially immunized against them. The United States is committed to immunizing all children against the common types of childhood disease (Fig. 7.15).

Immunization involves the use of **vaccines,** substances that contain an antigen to which the immune system responds. Traditionally, vaccines are the pathogens themselves, or their products, that have been treated so they are no longer virulent (able to cause disease). Today, it is possible to genetically engineer bacteria to mass-produce a protein from pathogens, and this protein can be used as a vaccine. This method has now produced a vaccine against hepatitis B, a viral-induced disease, and is being used to prepare a vaccine against malaria, a protozoan-induced disease.

After a vaccine is given, it is possible to follow an immune response by determining the amount of antibody present in a sample of plasma—this is called the **antibody titer.** After the first exposure to a vaccine, a primary response occurs. For a period of several days, no antibodies are present; then the titer rises slowly, levels off, and gradually declines as the antibodies bind to the antigen or simply break down (Fig. 7.15). After a second exposure to the vaccine, a secondary response is expected. The titer rises rapidly to a level much greater than before; then it slowly declines. The second exposure is called a "booster" because it boosts the antibody titer to a high level. The high antibody titer now is expected to help prevent disease symptoms, even if the individual is exposed to the disease-causing antigen.

Active immunity depends upon the presence of memory B cells and memory T cells that are capable of responding to lower doses of antigen. Active immunity is usually long-lasting, although a booster may be required every so many years.

Passive Immunity

Passive immunity occurs when an individual is given prepared antibodies or immune cells to combat a disease. Since these antibodies are not produced by the individual's

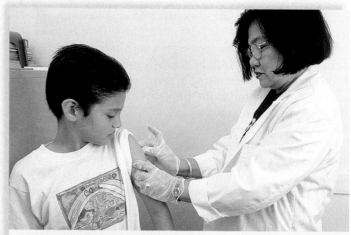

Suggested Immunization Schedule

Vaccine	Age (months)	Age (years)
Hepatitis B	Birth–18	11–12
Diphtheria, tetanus, pertussis (DTP)	2, 4, 6, 15–18	4–6
Tetanus and pertussis only		11–12
Haemophilus influenzae, type b	2, 4, 6, 12–15	
Polio	2, 4, 6–18	4–6, 11–12
Pneumococcal	2, 4, 6, 12–15	
Measles, mumps, rubella (MMR)	12–15	4–6, 11–12
Varicella (chicken pox)	12–18	11–12
Meningococcal		11–12
Hepatitis A (in selected areas)	12–23	11–12 2–18

a.

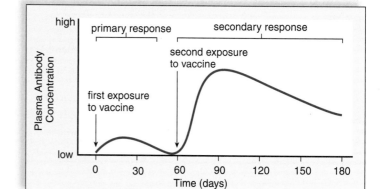

b.

Figure 7.15   Active immunity due to immunizations.
a. Suggested immunization schedule for infants and children. **b.** During immunization, the primary response, after the first exposure to a vaccine, is minimal, but the secondary response, which may occur after the second exposure, shows a dramatic rise in the amount of antibody present in plasma.

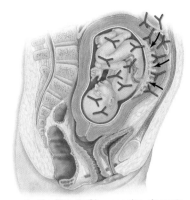

a. Antibodies (IgG) cross the placenta.

b. Antibodies (IgG, IgA) are secreted into breast milk.

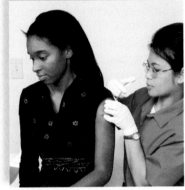

c. Antibodies can be injected by a physician.

Figure 7.16 Passive immunity.
During passive immunity, antibodies are received (**a**) by crossing the placenta, (**b**) in breast milk, or (**c**) by injection. Because the body is not producing the antibodies, passive immunity is short-lived.

plasma cells, passive immunity is temporary. For example, newborn infants are passively immune to some diseases because IgG antibodies have crossed the placenta from the mother's blood (Fig. 7.16*a*). These antibodies soon disappear, however, so within a few months, infants become more susceptible to infections. Breast-feeding prolongs the natural passive immunity an infant receives from the mother because IgG and IgA antibodies are present in the mother's milk (Fig. 7.16*b*).

Even though passive immunity does not last, it is sometimes used to prevent illness in a patient who has been unexpectedly exposed to an infectious disease. Usually, the patient receives a **gamma globulin** injection (serum that contains antibodies), perhaps taken from individuals who have recovered from the illness (Fig. 7.16*c*). For example, a health-care worker who suffers an accidental needle stick may come into contact with the blood from a patient infected with hepatitis virus. Immediate treatment with a gamma-globulin injection (along with simultaneous vaccination against the virus) can typically prevent the viruses from causing infection in the health-care worker.

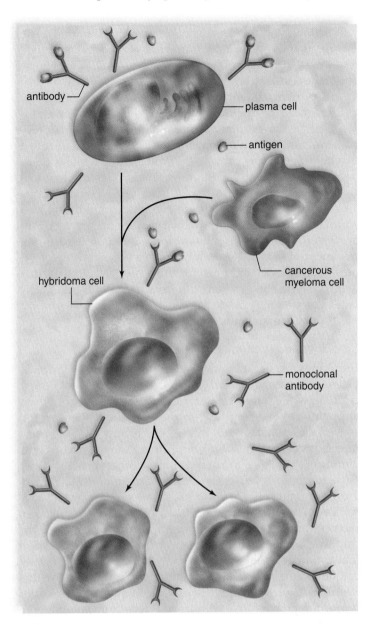

Figure 7.17 Production of monoclonal antibodies.
Plasma cells of the same type (derived from immunized mice) are fused with myeloma (cancerous) cells, producing hybridoma cells that are "immortal." Hybridoma cells divide and continue to produce the same type of antibody, called monoclonal antibodies.

Monoclonal Antibodies

Every plasma cell derived from the same B cell secretes antibodies against a specific antigen. These are **monoclonal antibodies** because all of them are the same type and because they are produced by plasma cells derived from the same B cell. One method of producing monoclonal antibodies in vitro (outside the body in glassware) is depicted in Figure 7.17. B lymphocytes are removed from an animal (today, usually mice are used) and are exposed to a particular antigen. The resulting plasma cells are fused with myeloma cells (malignant plasma cells that live and divide indefinitely; they are immortal cells). The fused cells are called hybridomas—*hybrid-* because they result from the fusion of two different cells, and *-oma* because one of the cells is a cancer cell.

At present, monoclonal antibodies are being used for quick and certain diagnosis of various conditions. For example, a particular hormone is present in the urine of a pregnant woman. A monoclonal antibody can be used to detect this hormone; if it is present, the woman knows she is pregnant. Monoclonal antibodies are also used to identify infections. And because they can distinguish between cancerous and normal tissue cells, they are used to carry radioisotopes or toxic drugs to tumors, which can then be selectively destroyed. Herceptin is a monoclonal antibody used in the treatment of breast cancer. It binds to a protein receptor on breast cancer cells and prevents the cancer cells from dividing so fast. Antibodies that bind to cancer cells also can activate complement and can increase phagocytosis by macrophages and neutrophils.

Cytokines and Immunity

Cytokines are signaling molecules produced by T lymphocytes, macrophages, and other cells. Because cytokines regulate white blood cell formation and/or function, they are being investigated as a possible adjunct therapy for cancer and AIDS. Both interferon, produced by virus-infected cells, and **interleukins,** produced by various white blood cells, have been used as immunotherapeutic drugs, particularly to enhance the ability of the individual's own T cells to fight cancer.

Because most cancer cells carry an altered protein on their cell surface, they should be attacked and destroyed by cytotoxic T cells. Whenever cancer develops, it is possible that cytotoxic T cells have not been activated. In that case, cytokines might awaken the immune system and lead to the destruction of the cancer. In one technique being investigated, researchers first withdraw T cells from the patient, present cancer cell antigens to them, and then activate the cells by culturing them in the presence of an interleukin. The T cells are reinjected into the patient, who is given doses of interleukin to maintain the killer activity of the T cells.

Scientists who are actively engaged in interleukin research believe that interleukins soon will be used as adjuncts for vaccines, for the treatment of chronic infectious disease, and perhaps for the treatment of cancer. Interleukin antagonists also may prove helpful in preventing skin and organ rejection, autoimmune diseases, and allergies.

✅ Check Your Progress 7.5

1. a. What is acquired immunity? b. What are the two different types of acquired immunity? c. What are some examples of each?

2. What are two types of immune therapies that can assist passive immunity?

7.6 Hypersensitivity Reactions

Sometimes, the immune system responds in a manner that harms the body, as when individuals develop allergies, receive an incompatible blood type, suffer tissue rejection, or have an autoimmune disease.

Allergies

Allergies are hypersensitivities to substances, such as pollen, food, or animal hair, that ordinarily would do no harm to the body. The response to these antigens, called **allergens,** usually includes some degree of tissue damage (Fig. 7.18).

An **immediate allergic response** can occur within seconds of contact with the antigen. The response is caused by antibodies known as IgE (see Table 7.2). IgE antibodies are attached to receptors on the plasma membrane of mast cells in the tissues and also to basophils in the blood. When an allergen attaches to the IgE antibodies on these cells, they release histamine and other substances that bring about the allergic symptoms. When pollen is an allergen, histamine stimulates the mucous membranes of the nose and eyes to release fluid, causing the runny nose and watery eyes typical of hay fever. If a person has asthma, the airways leading to the lungs constrict, resulting in difficult breathing accompanied by wheezing. When food contains an allergen, nausea, vomiting, and diarrhea often result.

Anaphylactic shock is an immediate allergic response that occurs because the allergen has entered the bloodstream. Bee stings and penicillin shots are known to cause this reaction because both inject the allergen into the blood. Anaphy-

SEM of pollen

Figure 7.18   Allergies.
When people are allergic to pollen, they develop symptoms that include watery eyes, sinus headaches, increased mucous production, labored breathing, and sneezing.

lactic shock is characterized by a sudden and life-threatening drop in blood pressure due to increased permeability of the capillaries by histamine. Taking epinephrine can counteract this reaction until medical help is available.

People with allergies produce ten times more IgE than those people without allergies. A new treatment using injections of monoclonal IgG antibodies for IgEs is currently being tested in individuals with severe food allergies. More routinely, injections of the allergen are given so that the body will build up high quantities of IgG antibodies. The hope is that these will combine with allergens received from the environment before they have a chance to reach the IgE antibodies located in the membrane of mast cells and basophils.

A **delayed allergic response** is initiated by memory T cells at the site of allergen contact in the body. The allergic response is regulated by the cytokines secreted by both T cells and macrophages. A classic example of a delayed allergic response is the skin test for tuberculosis (TB). When the test result is positive, the tissue where the antigen was injected becomes red and hardened. This shows that there was prior exposure to tubercle bacilli, the cause of TB. Contact dermatitis, which occurs when a person is allergic to poison ivy, jewelry, cosmetics, and many other substances that touch the skin, is also an example of a delayed allergic response.

Tissue Rejection

Certain organs, such as the skin, the heart, and the kidneys, could be transplanted easily from one person to another if the body did not attempt to reject them. Rejection of transplanted tissue results because the recipient's immune system recognizes that the transplanted tissue is not "self." Cytotoxic T cells respond by attacking the cells of the transplanted tissue.

Organ rejection can be controlled by carefully selecting the organ to be transplanted and administering **immunosuppressive** drugs. It is best if the transplanted organ has the same type of MHC antigens as those of the recipient, because cytotoxic T cells recognize foreign MHC antigens. Two well-known immunosuppressive drugs, cyclosporine and tacrolimus, both act by inhibiting the production of certain T-cell cytokines.

Xenotransplantation is the use of animal organs instead of human organs in transplant patients. Scientists have chosen to use the pig because animal husbandry has long included the raising of pigs as a meat source, and pigs are prolific. Genetic engineering can make pig organs less antigenic. The ultimate goal is to make pig organs as widely accepted as type O blood.

An alternative to xenotransplantation exists because tissue engineering is making organs in the laboratory. Scientists are ready to test a lab-grown urinary bladder in human patients. They hope that production of organs lacking HLA antigens will one day do away with the problem of rejection.

Figure 7.19
Rheumatoid arthritis.
Rheumatoid arthritis is due to recurring inflammation in skeletal joints. Complement proteins, T cells, and B cells all participate in deterioration of the joints, which eventually become immobile.

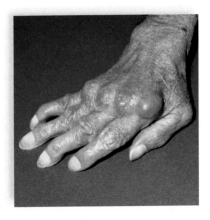

Disorders of the Immune System

When a person has an **autoimmune disease,** cytotoxic T cells or antibodies mistakenly attack the body's own cells as if they bear foreign antigens. Exactly what causes autoimmune diseases is not known. However, sometimes they occur after an individual has recovered from an infection.

In the autoimmune disease **myasthenia gravis,** antibodies attach to and interfere with the functioning of neuromuscular junctions, and muscular weakness results. In **multiple sclerosis (MS),** T cells attack the myelin sheath of nerve fibers, and this causes various neuromuscular symptoms. A person with **systemic lupus erythematosus (SLE)** has various symptoms prior to death due to kidney damage from the deposition of excessive antigen-antibody complexes. In **rheumatoid arthritis,** the joints are affected (Fig. 7.19). Researchers suggest that heart damage following rheumatic fever and diabetes type 1 are also autoimmune illnesses. As yet, there are no cures for autoimmune disorders, but they can sometimes be controlled with immunosuppressive drugs.

When a person has an immune deficiency, the immune system is unable to protect the body against disease. Acquired immunodeficiency syndrome (AIDS) is an example of an acquired immune deficiency. As a result of a weakened immune system, AIDS patients show a greater susceptibility to infections and also have a higher risk of cancer. Immune deficiency may also be congenital (i.e., inherited). Infrequently, a child may be born with an impaired immune system, caused by a defect in lymphocyte development. In **severe combined immunodeficiency disease (SCID),** both antibody- and cell-mediated immunity are lacking or inadequate. Without treatment, even common infections can be fatal. Gene therapy has been successful in SCID patients.

☑ Check Your Progress 7.6

1. What types of complications and disorders are associated with the functioning of the immune system? Explain each of these.

Summarizing the Concepts

7.1 Microbes, Pathogens, and You

Microbes perform valuable services, but they also cause disease.

Bacteria

- are prokaryotic cells that live and reproduce independently of host cells; and
- cause disease by multiplying in hosts and also by producing toxins.

Viruses

- are noncellular particles, consisting of a protein coat and a nucleic acid core;
- take over the machinery of the host in order to reproduce; and
- can emerge and cause new diseases the human body has difficulty combating.

7.2 The Lymphatic System

The lymphatic system consists of lymphatic vessels that return lymph to cardiovascular veins.

The primary lymphatic organs are

- the red bone marrow, where all blood cells are made and the B lymphocytes mature; and
- the thymus gland, where T lymphocytes mature.

The secondary lymphatic organs are

- the spleen, lymph nodes, and other organs such as the tonsils, Peyer's patches, and the appendix. Blood is cleansed of pathogens and debris in the spleen. Lymph is cleansed of pathogens and debris in the nodes.

7.3 Nonspecific Defenses

Immunity involves nonspecific and specific defenses. The nonspecific defenses include

- barriers to entry;
- the inflammatory reaction, which involves the phagocytic neutrophils and macrophages; and
- protective proteins.

7.4 Specific Defenses

Specific defenses require B cells and T cells, also called B lymphocytes and T lymphocytes.

B Cells and Antibody-Mediated Immunity

- Activated B cells undergo clonal selection with production of plasma cells and memory B cells, after their B-cell receptor combines with a specific antigen.
- Plasma cells secrete antibodies and eventually undergo apoptosis. Plasma cells are responsible for antibody-mediated immunity.
- An antibody is usually a Y-shaped molecule that has two binding sites for a specific antigen.
- Memory B cells remain in the body and produce antibodies if the same antigen enters the body at a later date.

T Cells and Cell-Mediated Immunity

- For a T cell to recognize an antigen, the antigen must be presented by an antigen-presenting macrophage, along with an HLA (human leukocyte antigen).

- Activated T cells undergo clonal expansion until the illness has been stemmed. Then, most of the activated T cells undergo apoptosis. A few cells remain, however, as memory T cells.
- The two main types of T cells are cytotoxic T cells and helper T cells.
- Cytotoxic T cells kill virus-infected cells or cancer cells on contact because they bear a nonself protein.
- Helper T cells produce cytokines and stimulate other immune cells.

7.5 Acquired Immunity

- Active (long-lived) immunity can be induced by vaccines when a person is well and in no immediate danger of contracting an infectious disease. Active immunity depends upon the presence of memory cells in the body.
- Passive immunity is needed when an individual is in immediate danger of succumbing to an infectious disease. Passive immunity is short-lived because the antibodies are administered to, and not made by, the individual.
- Monoclonal antibodies, which are produced by the same plasma cell, have various functions, from detecting infections to treating cancer.
- Cytokines, including interferon, are a form of passive immunity used to treat AIDS and to promote the body's ability to recover from cancer.

7.6 Hypersensitivity Reactions

Allergic responses occur when the immune system reacts vigorously to substances not normally recognized as foreign.

- Immediate allergic responses, usually consisting of coldlike symptoms, are due to the activity of antibodies.
- Delayed allergic responses, such as contact dermatitis, are due to the activity of T cells.
- Tissue rejection occurs when the immune system recognizes a tissue as foreign.
- Autoimmune disorders occur when the immune system reacts to tissues/organs of the individual as if they were foreign.

Understanding Key Terms

active immunity 136
allergen 138
allergy 138
anaphylactic shock 138
antibody-mediated
  immunity 131
antibody titer 136
antigen 130
antigen-presenting cell
  (APC) 134
apoptosis 131
autoimmune disease 139
bacillus 122
bacteria 122
B cell (B lymphocyte) 130
B-cell receptor (BCR) 131
capsule 122
cell-mediated immunity 135
coccus 122
complement system 130
cytokine 129, 135, 138

cytotoxic T cell 135
delayed allergic response 139
fimbriae 122
flagellum 122
gamma globulin 137
helper T cell 135
histamine 129
HLA (human leukocyte
  antigen) 134
immediate allergic
  response 138
immunity 128
immunization 136
immunosuppressive 139
inflammatory response 128
interferon 130
interleukin 138
lymph 126
lymphatic nodule 128
lymphatic organ 127
lymph node 128

lysozyme 128
macrophage 129
mad cow disease 124
mast cell 129
membrane attack complex 130
memory T cell 135
monoclonal antibody 137
multiple sclerosis (MS) 139
myasthenia gravis 139
passive immunity 136
pathogen 122
Peyer's patches 128
pilus 122
plasma cell 131
plasmid 122
prion 124
pus 129
red bone marrow 127

rheumatoid arthritis 139
self 134
severe combined
    immunodeficiency disease
    (SCID) 139
spirillum 122
spleen 128
systemic lupus erythematosus
    (SLE) 139
T cell (T lymphocyte) 130
T cell receptor (TCR) 134
thymus gland 127
tonsils 128
toxin 123
vaccine 136
virus 123
xenotransplantation 139

Match the key terms to these definitions.

a. —————— Antigens prepared in such a way that they can promote active immunity without causing disease.

b. —————— Series of proteins in plasma that form a nonspecific defense mechanism against pathogen invasion.

c. —————— Foreign substance, usually a protein, that stimulates the immune system to react, such as by producing antibodies.

d. —————— Process of programmed cell death involving a cascade of specific cellular events leading to the death and destruction of the cell.

e. —————— Type of lymphocyte that matures in the thymus gland and exists in three varieties, one of which kills antigen-bearing cells outright.

Testing Your Knowledge of the Concepts

1. Describe the basic characteristics of bacteria. Explain how five particular features contribute to the ability of bacteria to cause disease. (pages 122–23)

2. What is the structure of a virus? Is a virus living? Explain how a virus is able to reproduce. (page 123)

3. What are some ways that can explain the emergence of new viral diseases? (page 124)

4. What are prions, and how do they cause disease? (page 124)

5. Why are the red bone marrow and the thymus gland termed primary lymphatic organs? (page 127)

6. In what ways are the spleen and lymph nodes similar, and in what ways are they different? (pages 127–28)

7. How do nonspecific defenses differ from specific defenses? (pages 128 and 130)

8. What is the first line of defense? Describe several methods of protection. (page 128)

9. What is the second line of defense? Describe the various cells and chemicals and their roles in protecting the body.

What are the four signs of inflammation, and what causes them? (page 129)

10. What two types of cells are involved in providing specific defenses against pathogens? What type of immunity does each provide? (pages 130–35)

11. What is the clonal selection model as it applies to B cells? What becomes of the clones that are produced? (page 131)

12. Describe the structure of an antibody. What are the five main classes, where are they found, and what are their functions? (page 132)

13. Explain how T cells recognize an antigen. What are the types of T cells, and how do they function in immunity? (pages 134–35)

14. How is active immunity achieved? How is passive immunity achieved? (pages 136–38)

15. Discuss the production of monoclonal antibodies and their applications. (pages 137–38)

16. Discuss allergies, tissue rejection, and autoimmune diseases as they relate to the immune system. (pages 138–39)

17. Which of the following is a function of the spleen?
    a. produces T cells
    b. removes worn-out red blood cells
    c. produces immunoglobulins
    d. produces macrophages
    e. regulates the immune system

18. Which of the following is a function of the thymus gland?
    a. production of red blood cells
    b. secretion of antibodies
    c. production and maintenance of stem cells
    d. site for the maturation of T lymphocytes

For questions 19–23, match the lymphatic organs in the key to the description of its location.

Key:
    a. lymph nodes
    b. Peyer's patches
    c. spleen
    d. thymus gland
    e. tonsils

19. Arranged in a ring around the pharynx

20. In the intestinal wall

21. Occur along lymphatic vessels

22. Upper left abdominal cavity, posterior to the stomach

23. Superior to the heart, posterior to the sternum

24. Which of the following is a function of the secondary lymphatic organs?
    a. transport of lymph
    b. clonal selection of B cells
    c. located where lymphocytes encounter antigens
    d. All of these are correct.

25. Defense mechanisms that function to protect the body against many infectious agents are called
    a. specific.
    b. nonspecific.
    c. barriers to entry.
    d. immunity.

26. Which of the following is most directly responsible for the increase in capillary permeability during the inflammatory reaction?
    a. pain
    b. white blood cells
    c. histamine
    d. tissue damage

27. Which of the following is not a goal of the inflammatory reaction?
    a. bring more oxygen to damaged tissues
    b. decrease blood loss from a wound
    c. decrease the number of white blood cells in the damaged tissues
    d. prevent entry of pathogens into damaged tissues

28. Which of the following is not correct concerning interferon?
    a. Interferon is a protective protein.
    b. Virus-infected cells produce interferon.
    c. Interferon has no effect on viruses.
    d. Interferon can be used to treat certain viral infections.

29. Which one of these does not pertain to B cells?
    a. Have passed through the thymus.
    b. Have specific receptors.
    c. Are responsible for antibody-mediated immunity.
    d. Synthesize and liberate antibodies.

30. Which of these pertain(s) to T cells?
    a. Have specific receptors.
    b. Are of more than one type.
    c. Are responsible for cell-mediated immunity.
    d. Stimulate antibody production by B cells.
    e. All of these are correct.

31. During a secondary immune response,
    a. antibodies are made quickly and in great amounts.
    b. antibody production lasts longer than in a primary response.
    c. B cells become plasma cells.
    d. All of these are correct.

32. Active immunity may be produced by
    a. having a disease.
    b. receiving a vaccine.
    c. receiving gamma globulin injections.
    d. Both a and b are correct.
    e. Both b and c are correct.

33. Which of the following is not a nonspecific body defense?
    a. complement
    b. the skin and mucous membranes
    c. antibodies
    d. lysozyme

Thinking Critically About the Concepts

Remember from Chapters 5 and 6 that tissue fluid is created when capillary exchange occurs. Lymphatic capillaries are responsible for taking up that fluid, carrying it to larger lymphatic vessels that then return it to general circulation with the blood. If tissue fluid accumulates because larger quantities are created from increased blood flow to an injured area, as in Lia's case from the opening story, or because the lymphatic vessels are blocked, as with elephantiasis, swelling of the extremities results.

The inflammatory response initiated by Lia's ankle injury is one component of the nonspecific defenses of your immune system. Lymphocyte responses that result in antibody production, direct cellular attack, or coordination of other white blood cell responses are part of the immune system's specific responses. Our knowledge of the function of the immune system enables us to prepare it for future encounters with pathogens with vaccinations (active immunity), save someone from a poisonous snake bite with antivenom (passive immunity), and prevent tissue rejections or treat autoimmune diseases with immunosuppressive drugs.

1. Slowing down the flow of blood to Lia's ankle will minimize the swelling and pain. Will ice or heat minimize the swelling of Lia's ankle?

2. How would applying heat to an injured area benefit it?

3. Why do pregnant women (especially those who are in their third trimester of pregnancy) experience and complain about their swollen feet and ankles?

4. Lymphatic vessels transport tissue fluid back to the cardiovascular system. The pressure in the lymphatic vessels is low. Which blood vessels are lymphatic vessels most similar to with regard to structure?

5. Think of an analogy (something you're already familiar with) for the barrier defenses like your skin and mucous membranes. Remember the opening story indicated the different layers of defense are similar to a bank's efforts to keep people out of its vault.

6. Someone bitten by a poisonous snake should be given some antivenom (antibodies) to prevent them from dying. If they are bitten by the same kind of snake three years after the initial bite, will they have immunity to the venom, or should they get another shot of antivenom? Justify your response with an explanation of the kind of immunity someone gains from a shot of antibodies.

CHAPTER 8

Digestive System and Nutrition

Now weighing more than 350 pounds, 17-year-old Monica sat on the edge of her bed, staring at her reflection in the mirror on the wall. Disgusted with what she saw, she mulled her options silently in her head. She had tried dieting—Jenny Craig, Weight Watchers, the Atkins diet, and many others—and had lost a few pounds here and there, but the weight loss never seemed to last. She had even consulted a nutritionist, to no avail. Unable to control her weight by any other means, she decided to ask her doctor about surgical remedies.

Monica is considering bariatric surgery, a term used to describe several procedures that are performed on the digestive system to promote weight loss. One common procedure is known as "stomach stapling." In this procedure, the stomach is reduced to the size of a golf ball, and the digestive tract is rerouted so that food bypasses the first two feet of the small intestine. Bariatric surgery compels patients to eat less because they become very ill if they overeat. After bariatric surgery is performed, patients remain at a lifelong risk of nutritional deficiencies.

With nearly two-thirds of Americans overweight and one-third obese, we are mired in an obesity epidemic. Bariatric surgery is becoming more widely used, and an increasing number of obese teens, like Monica, are asking for these procedures. As you read this chapter, you will learn how the digestive system is organized, how it ordinarily works, and how its function is regulated by other organs of the body.

CHAPTER CONCEPTS

8.1 Overview of Digestion
The gastrointestinal tract (GI tract) contains a number of organs that carry out several processes as they digest food. The wall of the tract typically has four layers and is modified in each of the organs.

8.2 First Part of the Digestive Tract
The mouth receives food where it is chewed as chemical digestion begins before food enters the pharynx and the esophagus, which takes it to the stomach.

8.3 The Stomach and Small Intestine
The stomach stores food and continues chemical digestion, which is completed in the small intestine. The products of digestion are absorbed by the small intestine into the blood or into the lymph.

8.4 Three Accessory Organs and Regulation of Secretions
The pancreas produces pancreatic juice, which is sent to the small intestine for the chemical digestion of food, and the hormone insulin, which ordinarily causes cells, including liver cells, to take up glucose from the blood. The liver is a major metabolic organ, which stores glucose as glycogen and produces bile. Bile is stored in the gallbladder before it is sent to the small intestine for emulsification of fats.

8.5 The Large Intestine and Defecation
The large intestine absorbs water and vitamins produced by bacteria that normally reside in the large intestine, and it carries out defecation.

8.6 Nutrition and Weight Control
Planning nutritious meals and snacks to control one's weight involves making healthy and informed food choices.

8.1 Overview of Digestion

The organs of the digestive system are located within a tube called the gastrointestinal tract (GI tract), which is depicted in Figure 8.1. Food, whether it is a Big Mac or a shrimp salad, consists of the organic macromolecules you studied in Chapter 2, mainly carbohydrates, fats, and proteins. These molecules are too big to cross plasma membranes, and the purpose of digestion is to hydrolyze these macromolecules to their unit molecules. The unit molecules, namely sugars, amino acids, fatty acids, and glycerol, can cross plasma membranes. These are the nutrients that are carried by the blood to our cells. Our food also contains water, salts, vitamins, and minerals that help the body function normally. Specifically, these processes are necessary to the digestive process.

- **Ingestion** occurs when the mouth takes in food. Ingestion can be associated with our diet. The expression "we dig our graves with our teeth" recognizes that our diet is very important to our health. By learning about good nutrition, you increase your likelihood of enjoying a

longer, more active and productive life. Unfortunately, poor diet and lack of physical activity now rivals smoking as a major cause of preventable death in the United States.

- **Digestion** can be mechanical or chemical. Mechanical digestion occurs when food is divided into pieces that can be acted on by the digestive enzymes. Actually, we aid mechanical digestion when we cut up our food prior to ingestion. Mechanical digestion occurs primarily in the mouth and stomach.

  All parts of the tract, except the large intestine, contain digestive enzymes that hydrolyze particular foods to molecular nutrients. Chemical digestion begins in the mouth and is not completed until food reaches the small intestine. The thick semifluid mass of partly digested food that is passed from the stomach to the small intestine is termed **chyme.**

- **Movement** of GI tract contents along the digestive tract is very important in order for the tract to fulfill its other functions. For example, food must be passed along from one organ to the next, and indigestible remains must be expelled.

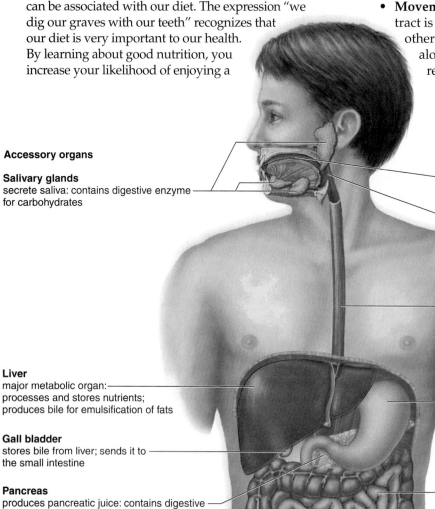

Accessory organs

Salivary glands
secrete saliva: contains digestive enzyme for carbohydrates

Liver
major metabolic organ: processes and stores nutrients; produces bile for emulsification of fats

Gall bladder
stores bile from liver; sends it to the small intestine

Pancreas
produces pancreatic juice: contains digestive enzymes, and sends it to the small intestine; produces insulin and secretes it into the blood after eating

Digestive tract organs

Mouth
teeth chew food; tongue tastes and pushes food for chewing and swallowing

Pharynx
passageway where food is swallowed

Esophagus
passageway where peristalsis pushes food to stomach

Stomach
secretes acid and digestive enzyme for protein; churns, mixing food with secretions, and sends chyme to small intestine

Small intestine
mixes chyme with digestive enzymes for final breakdown; absorbs nutrient molecules into body; secretes digestive hormones into blood

Large intestine
absorbs water and salt to form feces

Rectum
stores and regulates elimination of feces

Anus

Figure 8.1
The human gastrointestinal (GI) tract.
Food is digested within the organs of the GI tract; the accessory organs assist this process in the ways noted.

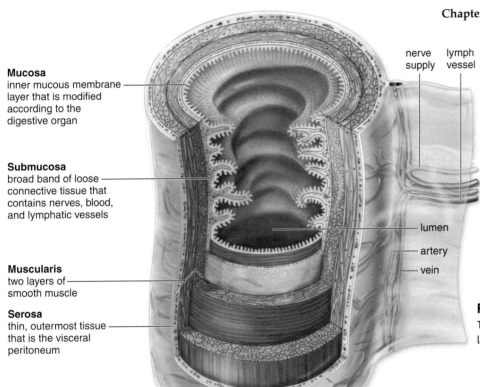

Mucosa
inner mucous membrane layer that is modified according to the digestive organ

Submucosa
broad band of loose connective tissue that contains nerves, blood, and lymphatic vessels

Muscularis
two layers of smooth muscle

Serosa
thin, outermost tissue that is the visceral peritoneum

nerve supply

lymph vessel

lumen

artery

vein

Figure 8.2   Wall of the gastrointestinal tract.
The wall of the gastrointestinal tract contains the four layers noted.

- **Absorption** occurs as unit molecules produced by digestion (i.e., nutrients) cross the wall of the GI tract and enter the cells lining the tract. From there, the nutrients enter the blood for delivery to the cells.
- **Elimination:** Molecules that cannot be digested need to be eliminated from the body. The removal of indigestible wastes through the anus, in the form of feces, is defecation.

Wall of the Digestive Tract

We can compare the GI tract to a garden hose that has a beginning (mouth) and an end (anus). The so-called **lumen** is the central space that contains water (food being digested). The wall of the GI tract has four layers (Fig. 8.2), and we will associate each layer with a particular disorder.

The first layer of the wall next to the lumen is called the **mucosa.** The mucosa is more familiarly called a mucous membrane and, of course, it produces mucus, which protects the wall from the digestive enzymes inside the lumen. In the mouth, stomach, and small intestine, the mucosa either contains glands that secrete and/or receive digestive enzymes from glands that secrete digestive enzymes.

Diverticulosis is a condition in which portions of the mucosa literally have pushed through the other layers and formed pouches, where food can collect. The pouches can be likened to an inner tube that pokes through weak places in a tire. When the pouches become infected or inflamed, the condition is called *diverticulitis.* This happens in 10–25% of people with diverticulosis.

The second layer in the GI wall is called the **submucosa.** The submucosal layer is a broad band of loose connective tissue that contains blood vessels, lymphatic vessels, and nerves. These are the vessels that will carry the nutrients absorbed by the mucosa. Lymph nodules, called Peyer's patches, are also in the submucosa. Like the tonsils, they help protect us from disease. Because the submucosa contains blood vessels, it can be the site of an inflammatory response (see page 129) that leads to *inflammatory bowel disease (IBD),* characterized by chronic diarrhea, abdominal pain, fever, and weight loss.

The third layer is termed the **muscularis,** and it contains two layers of smooth muscle. The inner, circular layer encircles the tract; the outer, longitudinal layer lies in the same direction as the tract. The contraction of these muscles, which are under nervous control, accounts for movement of digested food from the esophagus to the anus. The muscularis can be associated with *irritable bowel syndrome (IBS),* in which contractions of the wall cause abdominal pain, constipation, and/or diarrhea. The underlying cause of IBS is not known, although some suggest stress as an underlying cause.

The fourth layer of the tract is the **serosa** (serous membrane layer), which secrete a serous fluid. The serosa is a part of the peritoneum, the internal lining of the abdominal cavity. The **appendix** is a worm-shaped blind tube projecting from the first part of the large intestine on the right side of the abdomen. An inflamed appendix *(appendicitis)* has to be removed because, should the appendix burst, the result can be **peritonitis,** a life-threatening infection of the peritoneum.

✓ Check Your Progress 8.1

1. Name and describe the processes that occur during the digestive process.

2. What are the four layers of the GI tract? Associate an illness with each of the layers.

8.2 First Part of the Digestive Tract

The mouth, the pharynx, and the esophagus are in the first part of the GI tract.

The Mouth

The mouth receives food and begins the process of mechanical and chemical digestion. The mouth is bounded externally by the lips and cheeks. The lips extend from the base of the nose to the start of the chin. The red portion of the lips is poorly keratinized, and this allows blood to show through.

The roof of the mouth separates the nasal cavities from the oral cavity. The roof has two parts: an anterior (toward the front) **hard palate** and a posterior (toward the back) **soft palate** (Fig. 8.3*a*). The hard palate contains several bones, but the soft palate is composed entirely of muscle. The soft palate ends in a finger-shaped projection called the uvula. The tonsils, which are lymphatic tissue and help protect us from disease, are also in the back of the mouth, on either side of the tongue, and in the nasopharynx (called adenoids).

Three pairs of **salivary glands** send juices (saliva) by way of ducts to the mouth. One pair of salivary glands lies at the sides of the face immediately below and in front of the ears. These glands swell when a person has the mumps, a disease caused by a viral infection. Salivary glands have ducts that open on the inner surface of the cheek at the location of the second upper molar. Another pair of salivary glands lies beneath the tongue, and still another pair lies beneath the floor of the oral cavity. The ducts from these salivary glands open under the tongue. You can locate the openings if you use your tongue to feel for small flaps on the inside of your cheek and under your tongue. Saliva is a solution of mucus and water, which also contains bicarbonate and an enzyme called **salivary amylase** that begins the process of digesting starch.

The Teeth and Tongue

Mechanical digestion occurs when our teeth chew food into pieces convenient for swallowing. Monica, in the opening story, will have to chew her food thoroughly after her operation because her stomach will no longer be able to continue the process of mechanical digestion. Monica might even have to put her food in a blender before eating it.

During the first two years of life, the 20 smaller deciduous, or baby, teeth appear. These are eventually replaced by 32 adult teeth (Fig. 8.3*a*). The third pair of molars, called the wisdom teeth, sometimes fail to erupt. If they push on the other teeth and/or cause pain, they can be removed by a dentist or oral surgeon. Each tooth has two main divisions: a crown and a root (Fig. 8.3*b*). The crown has a layer of enamel, an extremely hard outer covering of calcium compounds; dentin, a thick layer of bonelike material; and an inner pulp, which contains the nerves and the blood vessels. Dentin and pulp are also found in the root.

Tooth decay, called **dental caries,** or cavities, occurs when bacteria within the mouth metabolize sugar and give off acids, which erode teeth. Tooth decay can be painful when it is severe enough to reach the nerves of the inner pulp. Two measures can prevent tooth decay: eating a limited amount of sweets, and daily brushing and flossing of teeth. Fluoride treatments, particularly in children, can make the enamel stronger and more resistant to decay. Gum disease, which is now known to be linked to cardiovascular disease, is more apt to occur with aging. Inflammation of the gums (gingivitis) can spread to the periodontal membrane, which lines the tooth socket. A person then has **periodontitis,** characterized by a loss of bone and loosening of the teeth so that extensive dental work may be required. Stimulation of the gums in a manner advised by your dentist is helpful in controlling this condition. Medications are also available.

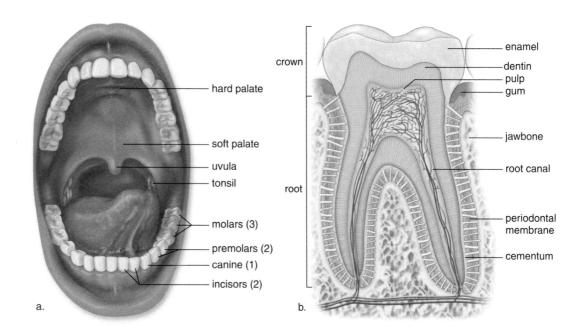

Figure 8.3   Adult mouth and teeth.

a. The chisel-shaped incisors bite; the pointed canines tear; the fairly flat premolars grind; and the flattened molars crush food. **b.** Longitudinal section of a tooth. The crown is the portion that projects above the gum line and can be replaced by a dentist if damaged. When a "root canal" is done, the nerves are removed. When the periodontal membrane is inflamed, the teeth can loosen.

hard palate

soft palate

uvula

tonsil

molars (3)

premolars (2)

canine (1)

incisors (2)

a.

crown

root

enamel

dentin

pulp

gum

jawbone

root canal

periodontal membrane

cementum

b.

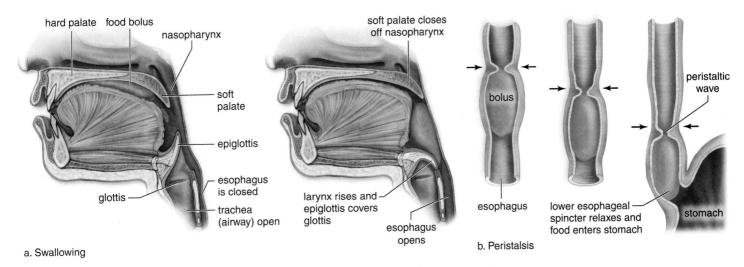

Figure 8.4 **Swallowing of food and peristalsis.**
a. When food is swallowed, the tongue pushes a bolus of food up against the soft palate (*left*). Then, the soft palate closes off the nasal cavities, and the epiglottis closes off the larynx so the bolus of food enters the esophagus (*right*). **b.** Peristalsis moves food through a sphincter into the stomach.

The tongue is covered by mucous membrane, which contains the sensory receptors called taste buds (see page 249). When taste buds are activated by the presence of food, nerve impulses travel by way of nerves to the brain. The tongue is composed of skeletal muscle, and it assists the teeth in carrying out mechanical digestion by moving food around in the mouth. In preparation for swallowing, the tongue forms chewed food into a mass called a **bolus** that it pushes toward the pharynx.

The Pharynx and Esophagus

Both the mouth and the nasal passages lead to a cavity called the **pharynx** (Fig. 8.4). The food passage and air passage cross in the pharynx because the trachea (windpipe) is anterior to (in front of) the **esophagus,** a long, narrow tube that takes food to the stomach.

Swallowing

Swallowing has a voluntary phase—from day-to-day living, you know that you can swallow voluntarily. (Try it!) However, once food or drink is pushed back into the pharynx, swallowing becomes a reflex action performed automatically (without you willing it). During swallowing, food normally enters the esophagus, a muscular tube that takes food to the stomach, because other possible avenues are blocked. The soft palate moves back to close off the nasal passages, and the trachea moves up under the **epiglottis** to cover the glottis. The **glottis** is the opening to the larynx (voice box), and therefore, the air passage. We do not breathe when we swallow. The up-and-down movement of the Adam's apple, the front part of the larynx, is easy to observe when a person swallows.

Unfortunately, we have all had the unpleasant experience of having food "go the wrong way." The wrong way may be either into the nasal cavities or into the trachea. If it is the latter, coughing will most likely force the food out of the trachea and into the pharynx again.

A rhythmic contraction called **peristalsis** pushes the food along the esophagus and continues in all the organs of the digestive tract. The esophagus plays no role in the chemical digestion of food. Its sole purpose is to move the food bolus from the mouth to the stomach. **Sphincters** are muscles that encircle tubes and act as valves; tubes close when sphincters contract, and they open when sphincters relax. The entrance of the esophagus to the stomach is marked by a constriction, often called a *lower gastroesophageal sphincter,* although the muscle is not as developed as in a true sphincter. Relaxation of the sphincter allows food to pass into the stomach, while contraction prevents the acidic contents of the stomach from backing up into the esophagus. When food or saliva is swallowed, the sphincter relaxes for a few seconds to allow the food or saliva to pass from the esophagus into the stomach, and then it closes again.

Heartburn, discussed in the Health Focus on page 148, occurs due to acid reflux, when some of the stomach's contents escape into the esophagus. When vomiting occurs, the abdominal muscles and the **diaphragm,** a muscle that separates the thoracic and abdominal cavities, contract.

✔ Check Your Progress 8.2

1. Describe the mechanical digestion and the chemical digestion that occurs in the mouth.

2. What ordinarily prevents food from entering the nose or entering the trachea when you swallow?

Heartburn (GERD)

In the absence of other symptoms, that burning sensation in your chest may have nothing to do with your heart. Instead, it is likely due to acid reflux. The stomach contents are more acidic than those of the esophagus. When the stomach contents pass upward into the esophagus, the acidity begins to erode the lining of the esophagus, producing the burning sensation associated with heartburn.

Almost everyone has had acid reflux and heartburn at some time. The burning sensation occurs in the area of the esophagus that lies behind the heart, and that is why it is termed heartburn. We might think it means we have a heart problem.

When heartburn or acid reflux becomes a chronic condition, the patient is diagnosed with gastroesophageal reflux disease (GERD). The term signifies that the stomach (gastric refers to the stomach) and the esophagus are involved in the disease. In GERD, patients' reflux is more frequent, remains in the esophagus longer, and will often contain higher levels of acid than in a patient with typical heartburn or acid reflux. People diagnosed with GERD may also experience pain in their chest, feel like they are choking, and have trouble swallowing. An anatomical disorder may be present. For example, normally during swallowing, wavelike contractions (peristalsis) of the esophagus will move food down the esophagus and into the stomach. Patients with weak or abnormal esophageal contractions will have difficulty pushing food into the stomach. These weak contractions can also prevent reflux from being pushed back into the stomach after it has entered the esophagus, thus producing GERD. When patients are lying down, the effects of abnormal esophageal contractions become more severe. This is because gravity is not helping to return reflux to the stomach.

Many people take over-the-counter medications for acid reflux. These drugs, being basic, neutralize stomach acid. Some progress to taking prescribed medications, which shut down and reduce acid production. However, persons with acid reflux could try modifying their eating habits first.

Diet and Exercise

It has been found that diet can help control acid reflux. Here are some tips:

- Avoid high-fat meals, such as those served by fast-food chains. Fatty foods stay in the stomach

Figure 8A   Acid reflux.
Which person is more likely to suffer from acid reflux? A healthy diet and light exercise are helpful to someone who has GERD (gastroesophageal reflux disease).

longer and, therefore, cause the stomach to secrete more acid.

- Don't overeat. In the movie *Supersize Me*, Morgan Spurlock doesn't just have acid reflux after eating a meal consisting of a supersize hamburger, coke, and fries. He throws up! Vomiting involves the same processes as does acid reflux.

- Eat several small meals instead of three large meals a day. Avoid foods that lead to stomach acidity, including alcohol, any form of fat, cakes and candy, tomato sauces, caffeinated beverages, and citrus fruits. Eat complex carbohydrates such as multigrain bread, brown rice, and pasta, instead of foods high in refined sugar. Besides providing you with vitamins and minerals, complex carbohydrates can help you control acid reflux!

- Light exercise such as riding a bike at a slow pace, walking, yoga, and light weight lifting are usually extremely helpful in heartburn patients. Both diet and exercise should help control weight. The pressure applied to the abdominal wall from obesity places pressure on the stomach and causes the normal acid contents of the stomach to reflux upward into the esophagus.

8.3 The Stomach and Small Intestine

The stomach and small intestine complete the digestion of food, which began in the mouth.

The Stomach

The **stomach** (Fig. 8.5) is a thick-walled, J-shaped organ that lies on the left side of the body beneath the diaphragm. The stomach is continuous with the esophagus above and the duodenum of the small intestine below. The stomach stores food, initiates the digestion of protein, and controls the movement of chyme into the small intestine. Notice that the stomach does not absorb nutrients; however, it does absorb alcohol because alcohol is fat soluble and can pass through the membrane for that reason.

The stomach wall has the usual four layers, but two of them are modified for particular functions. The muscularis contains three layers of smooth muscle (Fig. 8.5a). In addition to the circular and longitudinal layers, the stomach also contains a layer of smooth muscle that runs obliquely to the other two. The oblique layer also allows the stomach to stretch and to mechanically break down food into smaller fragments that are mixed with gastric juice. Since Monica's stomach will be the size of a golf ball after surgery, her stomach will not be able to carry out mechanical digestion, it will not be able to store food, and she might be subject to GERD, as described in the Health Focus on page 148.

The mucosa of the stomach has deep folds, the **rugae,** which disappear as the stomach fills to an approximate capacity of 1 liter. The mucosa of the stomach has millions of gastric pits, which lead into **gastric glands** (Fig. 8.5*b* and *c*). The gastric glands produce gastric juice. Gastric juice contains an enzyme called **pepsin,** which digests protein, plus hydrochloric acid (HCl) and mucus. HCl causes the stomach to have a high acidity with a pH of about 2, and this is beneficial because it kills most bacteria present in food. Although HCl does not digest food, it does break down the connective tissue of meat and activates pepsin.

Normally, the stomach empties in about 2–6 hours. When food leaves the stomach, it is a thick, soupy liquid called chyme. Chyme enters the small intestine in squirts.

The peristaltic waves move the chyme toward the pyloric sphincter, which closes and squeezes most of the chyme back, letting only a small amount to enter the small intestine at one time (Fig. 8.5*d*).

Historical Aspects

Not much was known, in general, about digestion in 1822 when an opportunity arose for a surgeon, named William Beaumont, to study how the stomach functions. After a patient, Alexis St. Martin, was shot, leaving a hole in his stomach, Beaumont began performing experiments on digestion. Most of the experiments consisted of tying a piece of food to a string and inserting it through the hole in St. Martin's stomach. Every few hours, Beaumont would remove the food and observe how well it had been digested. Through careful observations, he discovered most of what we know today about the stomach.

Also, about a 150 years later, in 1982, Dr. Marshall and Dr. Warren began their work on the cause of ulcers. As related in Chapter 1, they discovered that the bacterium *Helicobacter pyloris* is able to overcome the protective function of mucus in the stomach and the result is a gastric ulcer. (**Gastric** always refers to the stomach.)

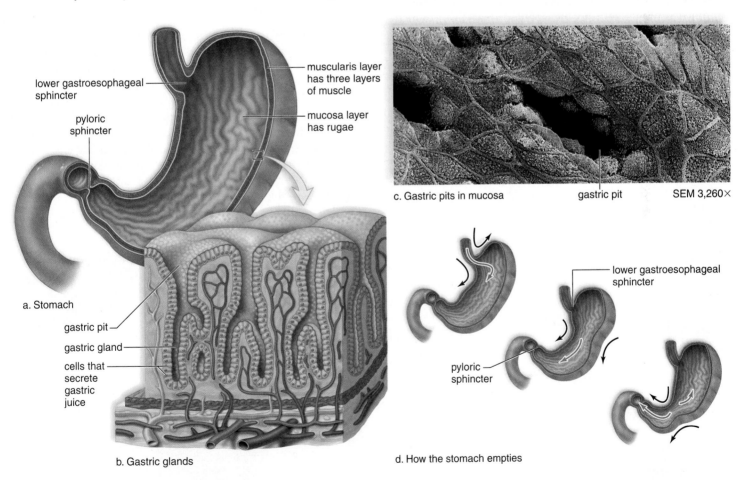

lower gastroesophageal sphincter

pyloric sphincter

muscularis layer has three layers of muscle

mucosa layer has rugae

a. Stomach

gastric pit

gastric gland

cells that secrete gastric juice

b. Gastric glands

c. Gastric pits in mucosa   gastric pit   SEM 3,260×

lower gastroesophageal sphincter

pyloric sphincter

d. How the stomach empties

Figure 8.5 Structure and function of the stomach.
a. Structure of the stomach showing that the muscularis has three muscle layers and the mucosa has folds called rugae. **b.** Gastric glands present in the mucosa secrete mucus, HCl, and pepsin, an enzyme that digests protein. **c.** Micrograph of gastric pits. **d.** Peristalsis in the stomach controls the secretion of chyme into the small intestine at the pyloric sphincter.

The Small Intestine

The **small intestine** is named for its small diameter (compared with that of the large intestine), but perhaps it should be called the long intestine. The small intestine averages about 6 m (18 ft) in length, compared with the large intestine, which is about 1.5 m (4½ ft) in length.

Digestion Is Completed in the Small Intestine

Notice in Table 8.1 that the small intestine contains enzymes to digest all types of foods, primarily carbohydrates, proteins, and fats. These enzymes are secreted by the pancreas and enter via a duct at the **duodenum,** the name for the first 25 cm of the small intestine. Also, a duct brings bile from the liver and gallbladder into the duodenum (see Fig. 8.8). **Bile** emulsifies fat—emulsification causes fat droplets to disperse in water. After fat is mechanically broken down to fat droplets by bile, it is hydrolyzed to glycerol and fatty acids by **lipase** present in pancreatic juice. Pancreatic amylase begins and an intestinal enzyme finishes the digestion of carbohydrates to glucose. Similarly, pancreatic trypsin begins and an intestinal enzyme finishes the digestion of proteins to amino acids. The intestine has a slightly basic pH because pancreatic juice contains sodium bicarbonate ($NaHCO_3$), which neutralizes chyme. Monica's surgery is associated with nutritional deficiencies because her GI tract now bypasses the duodenum.

Nutrients Are Absorbed in the Small Intestine

The wall of the small intestine absorbs the molecules, namely sugars, amino acids, fatty acids, and glycerol, which are the products of the digestive process. The mucosa of the small intestine is modified for absorption. It has been suggested that the surface area of the small intestine is approximately that of a tennis court. What factors contribute to increasing its surface area? The mucosa of the small intestine contains fingerlike projections called villi (sing., **villus**), which give the intestinal wall a soft, velvety appearance (Fig. 8.6). A villus has an outer layer of columnar epithelial cells, and each of these cells has thousands of microscopic extensions called microvilli. Collectively, in electron micrographs, microvilli give the villi a fuzzy border known as a "brush border." Since the microvilli bear the intestinal enzymes, these enzymes are called brush-border enzymes. The microvilli greatly increase the surface area of the villus for the absorption of nutrients.

Nutrients are absorbed into the vessels of a villus (Fig. 8.7). A villus contains blood capillaries and a small lymphatic capillary, called a **lacteal.** As you know, the lymphatic system is an adjunct to the cardiovascular system; its vessels carry a fluid called lymph to the cardiovascular veins. Sugars (digested from carbohydrates) and amino acids (digested from proteins) enter the blood capillaries of a villus. Glycerol and fatty acids (digested from fats) enter the epithelial cells of the villi, and within these cells are joined and packaged as lipoprotein droplets, called chylomicrons, which enter a lacteal. After nutrients are absorbed, they are eventually carried to all the cells of the body by the bloodstream.

Lactose Intolerance

Lactose is the primary sugar in milk. People who do not have the brush border enzyme called lactase cannot digest lactose. The result is a condition called **lactose intolerance,** characterized by diarrhea, gas, bloating, and abdominal cramps after drinking milk and other dairy products. Diarrhea occurs because the indigestible lactose causes fluid retention in the small intestine. Gas, bloating, and cramps occur when bacteria break down the lactose anaerobically, producing gas.

Persons with lactose intolerance can eat dairy products such as cheese and yogurt, in which the lactose has already been broken down; lactose-free milk; and/or take a dietary supplement that aids in the digestion of lactose.

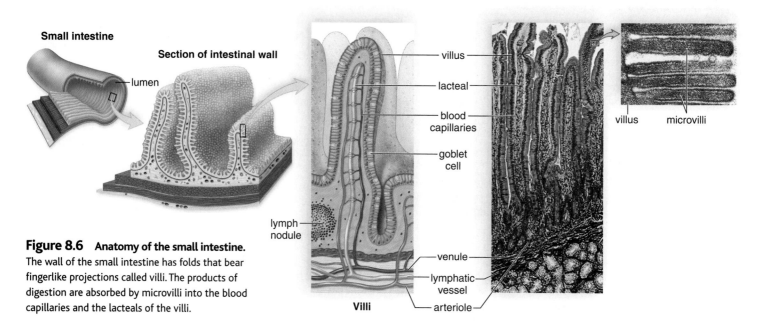

Figure 8.6 Anatomy of the small intestine. The wall of the small intestine has folds that bear fingerlike projections called villi. The products of digestion are absorbed by microvilli into the blood capillaries and the lacteals of the villi.

Table 8.1	Major Digestive Enzymes			
Enzyme	**Produced By**	**Site of Action**	**Optimum pH**	**Digestion**
CARBOHYDRATE DIGESTION:				
Salivary amylase	Salivary glands	Mouth	Neutral	Starch + H_2O → maltose
Pancreatic amylase	Pancreas	Small intestine	Basic	Starch + H_2O → maltose
Maltase	Small intestine	Small intestine	Basic	Maltose + H_2O → glucose + glucose
PROTEIN DIGESTION:				
Pepsin	Gastric glands	Stomach	Acidic	Protein + H_2O → peptides
Trypsin	Pancreas	Small intestine	Basic	Protein + H_2O → peptides
Peptidases	Small intestine	Small intestine	Basic	Peptide + H_2O → amino acids
NUCLEIC ACID DIGESTION:				
Nuclease	Pancreas	Small intestine	Basic	RNA and DNA + H_2O → nucleotides
Nucleosidases	Small intestine	Small intestine	Basic	Nucleotide + H_2O → base + sugar + phosphate
FAT DIGESTION:				
Lipase	Pancreas	Small intestine	Basic	Fat droplet + H_2O → monoglycerides + fatty acids

Figure 8.7   Digestion and absorption of nutrients.
a. Starch is digested to glucose, which is actively transported into the cells of intestinal villi. From there, glucose moves into the bloodstream. **b.** Proteins are digested to amino acids, which are actively transported into the cells of intestinal villi. From there, amino acids move into the bloodstream. **c.** Fats are emulsified by bile and digested to monoglycerides and fatty acids. These diffuse into cells, where they recombine and join with proteins to form lipoproteins, called chylomicrons. Chylomicrons enter a lacteal.

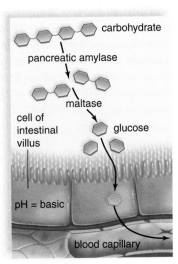

a. Carbohydrate digestion

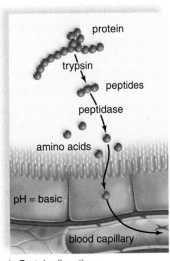

b. Protein digestion

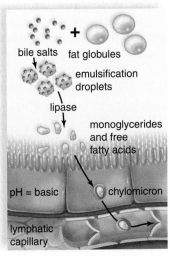

c. Fat digestion

Obesity: Diabetes Type 2 and Cardiovascular Disease

The nutrients absorbed at the small intestine have a profound effect on the body because these molecules can affect our health. For example, the intake of too much sugar and fat can result in obesity, which is associated with diabetes type 2 and cardiovascular disease. Nutritionists point out that consuming too much food from any source contributes to body fat, which increases a person's risk of obesity and associated illness. Still, foods such as doughnuts, cakes, pies, cookies, and white bread, which are high in refined carbohydrates (starches and sugars), and fried foods, which are high in fat, may very well be responsible for the current epidemic of obesity among Americans.

Switching to a healthy diet, increasing physical activity, and losing weight improves the ability of the hormone insulin to function properly in type 2 diabetics. The high fatty acid levels that accompany diabetes type 2 can lead to an increased risk of cardiovascular disease. It is well worthwhile to prevent diabetes type 2 by switching to a healthy diet and engaging in daily exercise because all diabetics are at risk for blindness, kidney disease, and cardiovascular disease.

✓ Check Your Progress 8.3

1. **a.** What are the functions of the stomach, and **(b)** how is the wall of the stomach modified to perform these functions?

2. **a.** What are the functions of the small intestine, and **(b)** how is the wall of the small intestine modified to perform these functions?

8.4 Three Accessory Organs and Regulation of Secretions

First, we will take a look at three accessory organs of digestion before considering how the secretions of these organs and those of the GI tract are regulated.

Three Accessory Organs

The **pancreas** is a fish-shaped, spongy, grayish pink organ that stretches across the back of the abdomen behind the stomach. Most pancreatic cells produce pancreatic juice, which enters the duodenum via the pancreatic duct (Fig. 8.8). Pancreatic juice contains sodium bicarbonate ($NaHCO_3$) and digestive enzymes for all types of food. Sodium bicarbonate neutralizes acid chyme from the stomach. **Pancreatic amylase** digests starch, **trypsin** digests protein, and **lipase** digests fat.

The pancreas is also an endocrine gland that secretes the hormone insulin into the blood. A **hormone** is a substance produced by one set of cells that affects a different set of cells, the so-called target cells. When the blood glucose level rises rapidly, the pancreas produces an overload of insulin to bring the level under control. Over the years, the body's cells can become insulin resistant with the result that diabetes type 2 can occur.

The **liver,** which is the largest and a major metabolic gland in the body, lies mainly in the upper-right section of the abdominal cavity, under the diaphragm (see Fig. 8.1). The liver contains approximately 100,000 lobules that serve as its structural and functional units (Fig. 8.8b). The liver receives blood via the hepatic portal vein from the capillary bed of the GI tract and filters blood in the capillaries of the lobules. In a sense, the liver acts like a sewage treatment plant when it removes poisonous substances from the blood and detoxifies them, while blood courses through its lobules (Table 8.2).

The liver is also a storage organ. It removes iron and the vitamins A, D, E, K, and B_{12} from blood and stores them. After we eat and in the presence of insulin, the liver stores glucose as glycogen, and between eating, it breaks down glycogen to keep the blood glucose level relatively constant. If need be, the liver converts glycerol (from fats) and amino acids to glucose molecules. As amino acids are converted to glucose, the liver combines their amino groups with carbon dioxide to form urea, the usual nitrogenous waste product in humans.

On the other hand, the liver makes the plasma proteins and helps regulate the quantity of cholesterol in the blood. It does this by producing bile salts, which are derived from cholesterol. **Bile** has a yellowish green color because it also contains bilirubin, derived from the breakdown of hemoglobin, another function of the liver. Bile is stored in the **gallbladder,** a pear-shaped organ just below the liver, until it is sent via the bile ducts to the duodenum. **Gallstones** form when liquid stored in the gallbladder hardens into pieces of stonelike material. Monica may have a tendency toward gallstones after her surgery because her gallbladder is no longer ridding itself of bile. In the small intestine, bile salts emulsify fat. When fat is emulsified, it breaks up into droplets, providing a much larger surface area, which can be acted upon by a digestive enzyme from the pancreas.

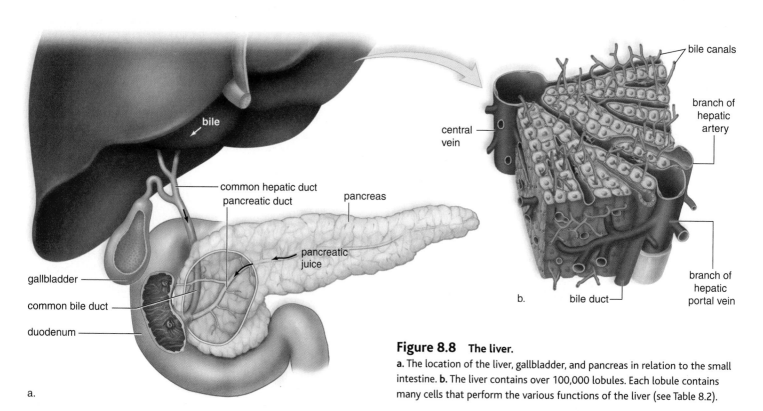

Figure 8.8   The liver.
a. The location of the liver, gallbladder, and pancreas in relation to the small intestine. **b.** The liver contains over 100,000 lobules. Each lobule contains many cells that perform the various functions of the liver (see Table 8.2).

Table 8.2	Functions of the Liver

1. Destroys old red blood cells; excretes bilirubin, a breakdown product of hemoglobin in bile, a liver product

2. Detoxifies blood by removing and metabolizing poisonous substances

3. Stores iron (Fe^{2+}) and the fat-soluble vitamins A, D, E, and K

4. Makes plasma proteins, such as albumins and fibrinogen, from amino acids

5. Stores glucose as glycogen after a meal, and breaks down glycogen to glucose to maintain the glucose concentration of blood between eating periods

6. Produces urea after breaking down amino acids

7. Helps regulate blood cholesterol level, converting some to bile salts

Liver Disorders

Hepatitis and cirrhosis are two serious diseases that affect the entire liver and hinder its ability to repair itself. Therefore, they are life-threatening diseases. When a person has a liver ailment, **jaundice** may occur in which bile pigments leak into the blood. Jaundice is a yellowish tint to the whites of the eyes and also to the skin of light-pigmented persons. Jaundice can result from **hepatitis,** inflammation of the liver. Viral hepatitis occurs in several forms. Hepatitis A is usually acquired from sewage-contaminated drinking water. Hepatitis B, which is usually spread by sexual contact, can also be spread by blood transfusions or contaminated needles. The hepatitis B virus is more contagious than the AIDS virus, which is spread in the same way. Vaccines are now available for hepatitis A and hepatitis B. Hepatitis C, for which there is no vaccine, is usually acquired by contact with infected blood and can lead to chronic hepatitis, liver cancer, and death.

Cirrhosis is another chronic disease of the liver. First, the organ becomes fatty, and then liver tissue is replaced by inactive fibrous scar tissue. Cirrhosis of the liver is often seen in alcoholics, due to malnutrition and to the excessive amounts of alcohol (a toxin) the liver is forced to break down. Physicians are now beginning to see cirrhosis of the liver in obese people who are overweight due to a diet high in fatty foods.

The liver has amazing regenerative powers and can recover if the rate of regeneration exceeds the rate of damage. During liver failure, however, there may not be enough time to let the liver heal itself. Liver transplantation is usually the preferred treatment for liver failure, but artificial livers have been developed and tried in a few cases. The liver is a vital organ, and its failure leads to death.

Regulation of Digestive Secretions

The secretions of digestive juices are controlled by the nervous system and by digestive hormones. When you look at

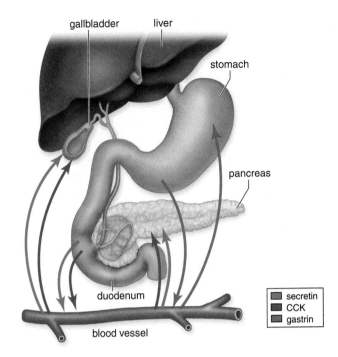

Figure 8.9 **Hormonal control of digestive gland secretions.** Gastrin (blue), from the lower stomach, feeds back to stimulate the upper part of the stomach to produce digestive juice. Secretin (green) and CCK (purple) from the duodenal wall stimulate the pancreas to secrete digestive juice and the gallbladder to release bile.

or smell food, the parasympathetic nervous system automatically stimulates gastric secretion. Also, when a person has eaten a meal particularly rich in protein, the stomach produces the hormone gastrin. Gastrin enters the bloodstream, and soon the secretory activity of gastric glands is increasing.

Cells of the duodenal wall produce two other hormones that are of particular interest—secretin and CCK (cholecystokinin). Acid, especially hydrochloric acid (HCl) present in chyme, stimulates the release of secretin, while partially digested protein and fat stimulate the release of CCK. Soon after these hormones enter the bloodstream, the pancreas increases its output of pancreatic juice. Pancreatic juice buffers the acidic chyme entering the intestine from the stomach and helps digest food. CCK also causes the liver to increase its production of bile, and the gallbladder to contract and release stored bile. Figure 8.9 summarizes the actions of gastrin, secretin, and CCK.

☑ Check Your Progress 8.4

1. What are the three main accessory organs that assist with the digestive process?

2. How does each accessory organ contribute to the digestion of food?

3. How are digestive secretions regulated in the body?

8.5 The Large Intestine and Defecation

The **large intestine** includes the cecum, the colon, the rectum, and the anal canal (Fig. 8.10). The large intestine is larger in diameter than the small intestine (6.5 cm compared with 2.5 cm), but it is shorter in length (see Fig. 8.1).

The **cecum,** which lies below the junction with the small intestine, is the blind end of the large intestine. The cecum usually has a small projection called the **vermiform appendix** (*vermiform* means wormlike) (Fig. 8.10). In humans, the appendix also may play a role in fighting infections. As mentioned previously, the appendix can become infected, a condition called appendicitis.

The **colon** includes the ascending colon, which goes up the right side of the body to the level of the liver; the transverse colon, which crosses the abdominal cavity just below the liver and the stomach; the descending colon, which passes down the left side of the body; and the sigmoid colon, which enters the **rectum,** the last 20 cm of the large intestine. The rectum opens at the **anus,** where **defecation,** the expulsion of feces, occurs.

Functions of the Large Intestine

The large intestine absorbs water, which is an important function that prevents dehydration of the body. Note that the large intestine does not produce any digestive enzymes and it does not absorb any nutrients.

The large intestine absorbs vitamins produced by bacteria called the intestinal flora. For many years, it was believed

Escherichia coli were the major inhabitants of the colon, but new culture methods show that over 99% of the colon bacteria are other types of bacteria. The bacteria in the large intestine break down indigestible material, and also produce B complex vitamins and most of the vitamin K needed by our bodies. In this way, they perform a service for us.

The large intestine forms feces. Feces are three-quarters water and one-quarter solids. Bacteria, dietary **fiber** (indigestible remains), and other indigestible materials are in the solid portion. Bacterial action on indigestible materials causes the odor of feces and also accounts for the presence of gas. A breakdown product of bilirubin (see page 152) and the presence of oxidized iron cause the brown color of feces.

Defecation, which is ridding the body of feces, is also a function of the large intestine. Peristalsis occurs infrequently in the large intestine, but, when it does, feces are forced into the rectum. Feces collect in the rectum until it is appropriate to defecate. At that time, stretching of the rectal wall initiates nerve impulses to the spinal cord, and shortly thereafter, the rectal muscles contract and the anal sphincters relax, allowing the feces to exit the body through the anus (Fig. 8.10). A person can inhibit defecation by contracting the external anal sphincter. Ridding the body of indigestible remains is another way the digestive system helps maintain homeostasis.

Water is considered unsafe for swimming when the coliform (nonpathogenic intestinal) bacterial count reaches a certain number. A high count indicates that a significant amount of feces has entered the water. The more feces present, the greater the possibility that disease-causing bacteria are also present.

Disorders of the Colon and Rectum

The large intestine is subject to a number of disorders. Many of these can be prevented or minimized by a good diet and good bowel habits.

Diarrhea The major causes of **diarrhea** are infection of the lower intestinal tract and nervous stimulation. In the case of infection, such as food poisoning caused by eating contaminated food, the intestinal wall becomes irritated, and peristalsis increases. Water is not absorbed, and the diarrhea that results rids the body of the infectious organisms. In nervous diarrhea, the nervous system stimulates the intestinal wall, and diarrhea results. Prolonged diarrhea can lead to dehydration because of water loss and to disturbances in the heart's contraction due to an imbalance of salts in the blood.

Constipation When a person is constipated, the feces are dry and hard. One reason for this condition is that socialized persons have learned to ignore the urge to defecate. An intake of water and fiber can help regularity of defecation. The frequent use of laxatives is discouraged. If, however, it is necessary to take a laxative, a bulk laxative is the most natural because, like fiber, it produces a soft mass of cellulose in the colon. Lubricants, such as mineral oil, make the colon slippery; saline laxatives, such as milk of magnesia, act osmoti-

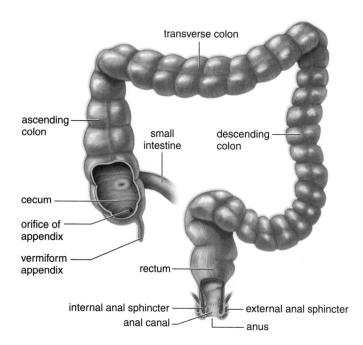

Figure 8.10 Large intestine.
The regions of the large intestine are the cecum, colon, rectum, and anal canal.

Health Focus

Swallowing a Camera

PillCam has become a viable alternative to traditional endoscopy procedure, during which a doctor uses a retractable tubelike instrument with an embedded camera, called an endoscope, to visualize the patient's GI tract. With a gulp of water, PillCam is swallowed, and its travels through the digestive system. Instead of spending an uncomfortable half day or more at the doctor's office, a patient simply visits the doctor in the morning, swallows the camera, dons the recording device, and goes about his or her daily routine.

Propelled by the normal muscular movement of the digestive system, PillCam embarks on a four- to eight-hour journey through the digestive system. As it travels through the stomach, the twists and turns of the small intestine, and the large intestine, PillCam continuously captures high-quality, wide-angle film footage of its journey and continuously beams this information to the recording device worn by the patient.

At the end of the day, PillCam reaches the end of its journey, and it is simply defecated normally. Later, the recording device is returned to the doctor's office so that the data may be retrieved. The doctor can view PillCam's journey as a 90-minute movie.

Like the food that we eat, PillCam traverses the numerous twists and turns of the intestine with ease. This provides a more accurate diagnosis, since a larger portion of the GI tract can be examined. Since PillCam does not create discomfort in the patient, the considerable risks involved with using anesthetics and painkillers are eliminated. And finally, the doctor does not need to be present for the entire procedure, saving valuable time and money for both doctor and patient.

Colonoscopy is a routine procedure used to examine and diagnose colon cancer. It employs a colonoscope, an instrument similar to an endoscope, that is inserted through the anus into the large intestine and is capable of removing precancerous tissue before it becomes invasive. Since PillCam cannot be used to remove tissue samples for analysis, its use as a replacement for colonoscopy is somewhat limited.

Figure 8B Patient swallowing PillCam.
PillCam traverses through the entire GI tract, taking pictures after it is swallowed.

cally—they prevent water from being absorbed. Some laxatives are irritants that increase peristalsis.

Chronic constipation is associated with the development of **hemorrhoids,** enlarged and inflamed blood vessels at the anus. Other contributing factors include pregnancy, aging, and anal intercourse.

Diverticulosis   As mentioned previously (see page 145), diverticulosis is the occurrence of little pouches of mucosa that have pushed out through weak spots in the muscularis. A frequent site is the last part of the descending colon.

Irritable Bowel Syndrome (IBS) (spastic colon)   Also mentioned previously (see page 145), IBS is a condition in which the muscularis contracts powerfully but without its normal coordination. The symptoms are abdominal cramps, gas, constipation, and urgent, explosive stools (feces discharge).

Inflammatory Bowel Disease (IBD) (colitis)   IBD is a collective term for a number of inflammatory disorders. Ulcerative colitis and Crohn's disease are the most common of these. Ulcerative colitis affects the large intestine and rectum, but in Crohn's disease, the inflammation can be more widespread and ulcers can penetrate more deeply. Ulcers are painful and cause bleeding because they erode the submucosal layer, where nerves and blood vessels are. Colitis causes diarrhea, rectal bleeding, abdominal cramps, and urgency.

Polyps and Cancer   The colon is subject to the development of **polyps,** small growths arising from the epithelial lining. Polyps, whether benign or cancerous, can be removed surgically. If colon cancer is detected while still confined to a polyp, the expected outcome is a complete cure. Some investigators believe that dietary fat increases the likelihood of colon cancer because dietary fat causes an increase in bile secretion. It could be that intestinal bacteria convert bile salts to substances that promote the development of cancer. On the other hand, fiber in the diet seems to inhibit the development of colon cancer. Regular elimination reduces the time that the colon wall is exposed to any cancer-promoting agents in feces.

As discussed in the Health Focus above, endoscopy by means of a flexible tube (and camera) inserted into the GI tract, usually from the anus, is gradually being replaced by the PillCam, a camera you swallow by mouth!

Check Your Progress 8.5

1. What are the different parts of the large intestine?
2. What are the functions of the large intestine?

8.6 Nutrition and Weight Control

Obesity—that is, being grossly overweight, has doubled in the United States in only 20 years. Nearly one-third (33%) of adults are now obese. Furthermore, being overweight is now prevalent among children and adolescents. These statistics are of great concern because excess body fat is associated with a higher risk for premature death, diabetes type 2, hypertension, cardiovascular disease, stroke, gallbladder disease, respiratory disfunction, osteoarthritis, and certain kinds of cancers.

Obesity is also on the rise throughout the world. The term globesity has been coined in recognition of the worldwide problem. In all regions, obesity appears to escalate as income increases. In Brazil and Colombia, the figure of overweight people is about 40%—comparable with a number of European countries.

The Health Focus on page 157 tells us about the various unhealthy ways people have tried to keep their weight under control. The conclusion is that, while dieting, *"Eat a variety of foods, watch your weight, and exercise."* To reverse a trend toward obesity, eat fewer calories, be more active, and make wiser food choices.

How Obesity Is Defined

Today, obesity is often defined as a **body mass index (BMI)** of 32 or greater. The table in Figure 8.11 shows your BMI, as long as you already know your weight and height. The BMI can also be calculated by taking the weight of the individual in kilograms and then dividing the weight by the square of the height in meters. Most people find that using the table is a lot easier. As a general rule,

healthy BMIs = 19.1 to 26.4;
overweight BMIs = 26.5 to 31.1;
obese BMIs = 32.3 to 39.9; and
morbidly obese BMIs = 40 or more.

Your BMI is suppose to give you an idea of how much of your weight is due to adipose tissue—that is, fat. In general, the taller you are, the more you could weigh without it being due to fat. Using BMI in this way works for most people, especially if they tend to be sedentary. But your BMI number should be used as a general guide only because it does not take into account fitness, bone structure, or gender. For example, a weight lifter might have an obese BMI, not because he has body fat, but because he is very muscular.

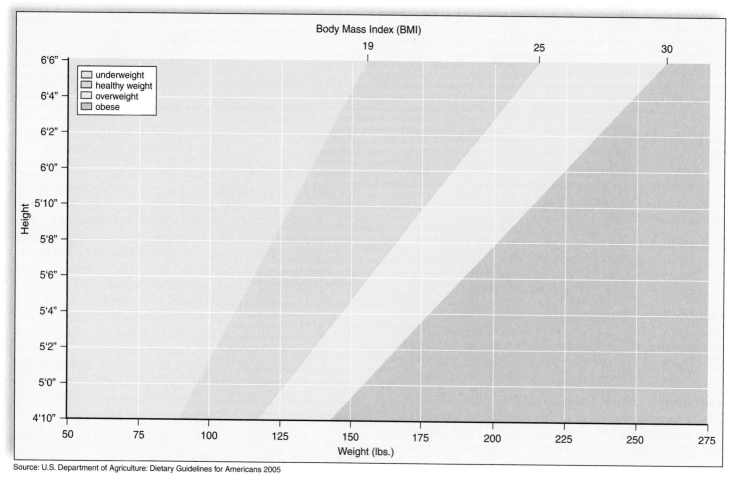

Source: U.S. Department of Agriculture: Dietary Guidelines for Americans 2005

Figure 8.11 **Determining your BMI.**
Match your weight with your height, then determine your body mass index (BMI). Healthy BMIs = 19.1 to 26.4; overweight BMIs = 26.5 to 31.1; obese BMIs = 32.3 to 39.9; morbidly obese BMIs = 40 or more.

Health Focus

Searching for the Magic Weight-Loss Bullet

While dieting, *"Eat a variety of foods, watch your weight, and exercise"* doesn't sound very glamorous, and not only that, you can't sell it to the public and make a lot of money. No wonder the public, which is always looking for the magic weight-loss bullet, is offered so many solutions to being overweight, most of which are not healthy. The solutions involve trendy diet programs, new prescription medications, and even surgery. The latter two options are for people who, like Monica from the opening story, have tried a low-calorie diet and, hopefully, regular physical activity but have been unsuccessful in losing weight. Prescription medications should only be taken under a physician's supervision, and, of course, surgeries are only done by physicians.

TRENDY DIET PROGRAMS: Various diets for the overweight have been around for many years, and here are some recently touted:

> The Pritikin Diet   This diet encourages the consumption of large amounts of carbohydrates and fiber in the form of whole grains and vegetables. The diet is so low-fat that the dieter may not be able to consume a sufficient amount of "healthy" fats.
>
> The Atkin's Diet   This diet is just the opposite of the Pritikin Diet because it is a low-carbohydrate (carb) diet. It is based on the assumption that if we eat more protein and fat, our bodies lose weight by burning stored body fat. The Atkin's diet is thought by many to be a serious threat to homeostasis because it puts a strain on the body to maintain the blood glucose level, the breakdown of fat lowers blood pH, and the excretion of nitrogen from protein breakdown stresses the kidneys.
>
> The Zone Diet and the South Beach Diet   As a reaction to the Atkin's diet, these diets recommend only "healthy" fats and permit low-sugar carbs. In other words, these diets are bringing us back, once again, to "Eat a variety of foods, maintain your weight, and exercise."

You may also have heard of the caveman diet. A professor at Colorado State University goes back in time, claiming to have found the lifestyle change necessary for weight management. If you would rather just flush your system of toxins and keep to your own lifestyle, some nutritionists claim that by consuming certain foods, such as cayenne pepper, mustard, cinnamon, green vegetables, and omega-3-rich fish, you can boost your metabolic rate and flush fat away.

Then again, according to a professor at Brigham Young University, the cure for endless dieting and the key to reaching a healthy weight is to listen to your body. Using a hunger scale of 1 to 10, with 1 being starving and 10 being very overfull, keep around 3 to 5 and you will eat less. Unfortunately, such sensible advice doesn't seem to have caught on yet.

PRESCRIPTION DRUGS: Ever since Health and Human Services Secretary Tommy Thompson said that Medicare and Medicaid will consider obesity a disease, drugmakers have revved up the search for a magic weight-loss pill. Despite law-suits against the manufacturers of the prescription drug fen-phen for causing heart problems and the belief that a drug called Meridia elevates blood pressure, new drugs continue to enter clinical trials. A drug called L-Marc is now being tested around the nation, and there is no lack of volunteers. Another one called Acomplia, which is suppose to block pleasure receptors in the brain, is in late-stage clinical trials. Another idea is a weight-loss nasal spray. The availability of so many possible prescription drugs make it seem likely that the magic bullet may not be synonymous with the magic pill.

SURGICAL PROCEDURES: For people who are morbidly obese, surgery may be recommended. There are three common types of weight-loss surgery; gastroplasty, gastric bypass, and gastric banding. Gastroplasty involves partitioning or stapling a small section of the stomach to reduce total food intake. Gastric bypass surgery involves attaching the *lower* part of the small intestine to the stomach, so that most of the food bypasses both the stomach and small intestine, decreasing the surface area for absorption. Gastric banding is a new procedure in which stomach size is reduced using a constricting band. About one-third to one-half of people who received obesity surgery lose significant amounts of weight and keep this weight off for at least five years. The downside of these operations is that people face the possibility of a lifetime of post-surgery problems, including chronic diarrhea, vomiting, intolerance to dairy products, dehydration, and nutritional deficiencies, due to alterations in nutrient digestion and absorption. Careful consideration of the potential benefits and risks must be weighed.

Many people also try liposuction to remove fat cells from localized areas in the body, but this is not recommended to treat obesity or morbid obesity. This procedure is clearly not the solution to long-term weight loss, as the millions of fat cells that remain in the body after liposuction enlarge if the person continues to overeat.

We are facing an obesity epidemic and must do something about it. Are these weight-loss options the answer? There are signs that the public is no longer searching for the magic bullet and, instead, is focusing on "To achieve or maintain a healthy weight, eat a variety of healthy foods and exercise."

Classes of Nutrients

A **nutrient** can be defined as a component of food that performs a physiological function in the body. Nutrients provide us with energy, promote growth and development, and regulate cellular metabolism.

Carbohydrates

Carbohydrates are either simple or complex. Glucose is a simple sugar preferred by the body as an energy source. Complex carbohydrates with several sugar units are digested to glucose. While body cells can utilize fatty acids as an energy source, brain cells require glucose. For this reason alone, it is necessary to include carbohydrates in the diet because the body is unable to convert fatty acids to glucose.

Any product made from refined grains, such as white bread, cake, and cookies, should be minimized in the diet. During refinement, fiber and also vitamins and minerals are removed from grains, so primarily starch remains. In contrast, sources of complex carbohydrates, such as beans, peas, nuts, fruits, and whole-grain products, are recommended as a good source of vitamins, minerals, and also fiber (Fig. 8.12). Insoluble fiber adds bulk to fecal material and stimulates movements of the large intestine, preventing constipation. Soluble fiber combines with bile salts and cholesterol in the small intestine and prevents them from being absorbed.

Can Carbohydrates Be Harmful?   Some nutritionists hypothesize that the high intake of refined carbohydrates and fructose sweeteners processed from cornstarch may be responsible for obesity in the United States. In addition, these foods are said to have a high **glycemic index (GI),** because the blood glucose response to these foods is high. When the blood glucose level rises rapidly, the pancreas produces an overload of insulin to bring the level under control. Investi-

Table 8.3	Reducing High-Glycemic Index Carbohydrates

To reduce dietary sugar:

1. Eat fewer sweets, such as candy, soft drinks, ice cream, and pastry.
2. Eat fresh fruits or fruits canned without heavy syrup.
3. Use less sugar—white, brown, or raw—and less honey and syrups.
4. Avoid sweetened breakfast cereals.
5. Eat less jelly, jam, and preserves.
6. Eat fresh fruit; especially avoid artificial fruit juices.
7. When cooking, use spices, such as cinnamon, instead of sugar to flavor foods.
8. Do not put sugar in tea or coffee.
9. Avoid potatoes and processed foods made from refined carbohydrates, such as white bread, rice, and pasta.

gators tell us that a chronically high insulin level may lead to insulin resistance, diabetes type 2, and increased fat deposition. Deposition of fat is associated with coronary heart disease, liver ailments, and several types of cancer.

Table 8.3 gives suggestions on how to reduce your intake of dietary sugars.

> Carbohydrates are the preferred energy source for the body, but select complex carbohydrates in whole grains, beans, nuts, and fruits, contain fiber, vitamins, and minerals in addition to carbohydrate.

Proteins

Dietary proteins are digested to amino acids, which cells use to synthesize hundreds of cellular proteins. Of the 20 different amino acids, eight are **essential amino acids** that must be present in the diet. Children will not grow if their diets lack the essential amino acids. Eggs, milk products, meat, poultry, and most other foods derived from animals contain all nine essential amino acids and are "complete" or "high-quality" protein sources.

Legumes (beans and peas) (Fig. 8.13), other types of vegetables, seeds and nuts, and also grains supply us with amino acids, but each of these alone is an incomplete protein source, because of a deficiency in at least one of the essential amino acids. Absence of one essential amino acid prevents utilization of the other 19 amino acids. Therefore, vegetarians are counseled to combine two or more incomplete types of plant products to acquire all the essential amino acids. Also, tofu, soymilk, and other foods made from processed soybeans are complete protein sources. A balanced vegetarian diet is quite possible with a little planning.

Amino acids are not stored in the body, and a daily supply is needed. However, it does not take very much

Figure 8.12   Fiber-rich foods.
Plants provide a good source of carbohydrates. They also provide a good source of vitamins, minerals , and fiber when they are not processed (refined).

protein to meet the daily requirement. Two servings of meat a day (one serving is equal in size to a deck of cards) is usually plenty.

Can Proteins Be Harmful? While the body is harmed if the amount of protein in the diet is severely limited, it is also likely harmed if the diet contains an overabundance of protein. The average American eats about twice as much protein as needed, and some people may be on a diet that encourages the intake of proteins instead of carbohydrates as an energy source. Also, bodybuilders, weight lifters, and other athletes may include amino acid or protein supplements in the diet because they think these supplements will increase muscle mass. However, excess amino acids are not always converted into muscle tissue. The liver removes the nitrogen portion during deamination and uses it to form urea, which is excreted in urine. The water needed for excretion of urea can cause dehydration when a person is exercising and losing water by sweating. High-protein diets, especially those rich in animal proteins, can also increase calcium loss in urine. Excretion of calcium can lead to kidney stones.

Certain types of meat, especially red meat, are known to be high in saturated fats, while other sources of protein, such as chicken, fish, and eggs, are more likely to be low in saturated fats. As discussed next, an intake of saturated fats leads to cardiovascular disease.

> Sufficient proteins are needed to supply the essential amino acids. Meat and dairy sources of protein may supply unwanted saturated fat, but vegetable sources do not.

Lipids

Fats, oils, and cholesterol are lipids. Saturated fats, which are solids at room temperature, usually have an animal origin. Two well-known exceptions are palm oil and coconut oil, which contain mostly saturated fats and come from the plants mentioned (Fig. 8.14). Butter and meats, such as marbled red meats and bacon, contain saturated fats.

Oils contain unsaturated fatty acids, which do not promote cardiovascular disease. Corn oil and safflower oil are high in polyunsaturated fatty acids. Polyunsaturated oils are nutritionally essential because they are the only type of fat that contains linoleic acid and linolenic acid, two fatty acids the body cannot make. Since these fatty acids must be supplied by diet, they are called **essential fatty acids.**

Olive oil and canola oil are well known to contain a larger percentage of monounsaturated fatty acids than other types of cooking oils. Omega-3 fatty acids—with a double bond in the third position—are believed to be especially protective against heart disease. Some cold-water fishes, such as salmon, sardines, and trout, are rich sources of omega-3, but they contain only about half that of flaxseed oil, the best source from plants.

Can Lipids Be Harmful? Cardiovascular disease is often due to arteries blocked by **plaque,** which contains saturated fats and cholesterol. Cholesterol is carried in the blood by two types of lipoproteins: low-density lipoprotein (LDL) and high-density lipoprotein (HDL). LDL is thought of as "bad" because it carries cholesterol from the liver to the cells, while HDL is thought of as "good" because it carries cholesterol to the liver, which takes it up and converts it to bile salts. Saturated fats tend to raise LDL cholesterol levels, while unsaturated fats lower LDL cholesterol levels.

Figure 8.13 Beans as food.
Beans are a good source of complex carbohydrates and protein. But beans don't supply all nine of the essential amino acids. To ensure a complete source of protein in the diet, beans should be eaten in combination with a grain, such as rice.

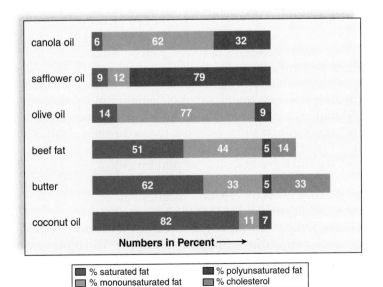

Figure 8.14 Fat and oils.
This illustration gives the percentages of saturated and unsaturated fatty acids in fats and oils. Which one do you judge to be the most healthy to use?

Trans-fatty acids (trans fats) arise when unsaturated fatty acids are hydrogenated to produce a solid fat. Found largely in commercially produced products, trans fats may reduce the function of the cell-membrane receptors that clear cholesterol from the bloodstream. Trans fats are found in commercially packaged goods, such as cookies and crackers; commercially fried foods, such as french fries from some fast-food chains; packaged snacks, such as microwaved popcorn; as well as in vegetable shortening and some margarines. Any packaged goods that contain partially hydrogenated vegetable oils or "shortening" most likely contain trans fats.

Table 8.4 gives suggestions on how to reduce dietary saturated fat and cholesterol. It is not a good idea to rely on commercially produced low-fat foods. In some products, sugars have replaced the fat, and in others, protein is used.

> Unsaturated fats such as those in oils do not lead to cardiovascular disease and are preferred. Fats and oils contain many more calories per gram than do carbohydrates and protein.

Minerals

Minerals are divided into major minerals and trace minerals. The body contains more than 5 grams of each major mineral and less than 5 grams of each trace mineral. Table 8.5 lists the selected minerals and gives their functions and food sources. The major minerals are constituents of cells and body fluids and are structural components of tissues.

The trace minerals are parts of larger molecules. For example, iron (Fe^{2+}) is present in hemoglobin, and iodine (I^-) is a part of thyroxine and triiodothyronine, hormones produced by the thyroid gland. Zinc (Zn^{2+}), copper (Cu^{2+}), and manganese (Mn^{2+}) are present in enzymes that catalyze a variety of reactions. As research continues, more and more elements are added to the list of trace minerals considered essential. During the past three decades, for example, very small amounts of selenium, molybdenum, chromium, nickel, vanadium, silicon, and even arsenic have been found to be essential to good health. Table 8.5 lists the functions of various minerals and gives their food sources and signs of deficiency and toxicity.

Occasionally, individuals do not receive enough iron (especially women), calcium, magnesium, or zinc in their diets. Adult females need more iron in their diet than males (18 mg compared with 10 mg) because they lose hemoglobin each month during menstruation. Stress can bring on a magnesium deficiency, and due to its high-fiber content, a vegetarian diet may make zinc less available to the body. However, a varied and complete diet usually supplies enough of each type of mineral.

Table 8.4	Reducing Certain Lipids

To reduce saturated fats and trans fats in the diet:

1. Choose poultry, fish, or dry beans and peas as a protein source.
2. Remove skin from poultry, and trim fat from red meats before cooking; place on a rack so that fat drains off.
3. Broil, boil, or bake rather than fry.
4. Limit your intake of butter, cream, trans fats, shortenings, and tropical oils (coconut and palm oils).*
5. Use herbs and spices to season vegetables instead of butter, margarine, or sauces. Use lemon juice instead of salad dressing.
6. Drink skim milk instead of whole milk, and use skim milk in cooking and baking.

To reduce dietary cholesterol:

1. Avoid cheese, egg yolks, liver, and certain shellfish (shrimp and lobster). Preferably, eat white fish and poultry.
2. Substitute egg whites for egg yolks in both cooking and eating.
3. Include soluble fiber in the diet. Oat bran, oatmeal, beans, corn, and fruits, such as apples, citrus fruits, and cranberries, are high in soluble fiber.

Calcium

Calcium (Ca^{2+}) is a major mineral needed for the construction of bones and teeth and for nerve conduction and muscle contraction. Many people take calcium supplements to counteract **osteoporosis,** a degenerative bone disease that afflicts an estimated one-fourth of older men and one-half of older women in the United States. Osteoporosis develops because bone-eating cells called osteoclasts are more active than bone-forming cells called osteoblasts. Therefore, the bones are porous, and they break easily because they lack sufficient calcium. Due to recent studies that show consuming more calcium does slow bone loss in elderly people, the guidelines have been revised. A calcium intake of 1,000 mg a day is recommended for men and for women who are premenopausal, and 1,300 mg a day is recommended for postmenopausal women. To achieve this amount, supplemental calcium is most likely necessary.

Vitamin D is an essential companion to calcium in preventing osteoporosis. Other vitamins may also be helpful; for example, magnesium has been found to suppress the cycle that leads to bone loss. In addition to adequate calcium and vitamin intake, exercise helps prevent osteoporosis. Risk factors for osteoporosis include drinking more than nine cups of caffeinated coffee per day and smoking. Medications are also available that slow bone loss while increasing skeletal mass. These are still being studied for their effectiveness and possible side effects.

Table 8.5	Minerals

Mineral	Functions	Food Sources	Conditions with	
			Too Little	*Too Much*
MAJOR (MORE THAN 100 MG/DAY NEEDED)				
Calcium (Ca^{2+})	Strong bones and teeth, nerve conduction, muscle contraction	Dairy products, leafy green vegetables	Stunted growth in children, low bone density in adults	Kidney stones; interferes with iron and zinc absorption
Phosphorus (PO_4^{3-})	Bone and soft tissue growth; part of phospholipids, ATP, and nucleic acids	Meat, dairy products, sunflower seeds, food additives	Weakness, confusion, pain in bones and joints	Low blood and bone calcium levels
Potassium (K^+)	Nerve conduction, muscle contraction	Many fruits and vegetables, bran	Paralysis, irregular heartbeat, eventual death	Vomiting, heart attack, death
Sulfur (S^{2-})	Stabilizes protein shape, neutralizes toxic substances	Meat, dairy products, legumes	Not likely	In animals, depresses growth
Sodium (Na^+)	Nerve conduction, pH and water balance	Table salt	Lethargy, muscle cramps, loss of appetite	Edema, high blood pressure
Chloride (Cl^-)	Water balance	Table salt	Not likely	Vomiting, dehydration
Magnesium (Mg^{2+})	Part of various enzymes for nerve and muscle contraction, protein synthesis	Whole grains, leafy green vegetables	Muscle spasm, irregular heartbeat, convulsions, confusion, personality changes	Diarrhea
TRACE (LESS THAN 20 MG/DAY NEEDED)				
Zinc (Zn^{2+})	Protein synthesis, wound healing, fetal development and growth, immune function	Meats, legumes, whole grains	Delayed wound healing, night blindness, diarrhea, mental lethargy	Anemia, diarrhea, vomiting, renal failure, abnormal cholesterol levels
Iron (Fe^{2+})	Hemoglobin synthesis	Whole grains, meats, prune juice	Anemia, physical and mental sluggishness	Iron toxicity disease, organ failure, eventual death
Copper (Cu^{2+})	Hemoglobin synthesis	Meat, nuts, legumes	Anemia, stunted growth in children	Damage to internal organs if not excreted
Iodine (I^-)	Thyroid hormone synthesis	Iodized table salt, seafood	Thyroid deficiency	Depressed thyroid function, anxiety
Selenium (SeO_4^{2-})	Part of antioxidant enzyme	Seafood, meats, eggs	Vascular collapse, possible cancer development	Hair and fingernail loss, discolored skin
Manganese (Mn^{2+})	Part of enzymes	Nuts, legumes, green vegetables	Weakness and confusion	Confusion, coma, death

Sodium

Sodium plays a major role in regulating the body's water balance, as does chloride (Cl^-). The recommended amount of sodium intake per day is 500 mg, although the average American takes in 4,000–4,700 mg every day. In recent years, this imbalance has caused concern because sodium in the form of salt intensifies hypertension (high blood pressure) if you already have it. About one-third of the sodium we consume occurs naturally in foods; another third is added during commercial processing; and we add the last third either during home cooking or at the table in the form of table salt.

Clearly, it is possible to cut down on the amount of sodium in the diet. Table 8.6 gives recommendations for doing so.

Table 8.6	Reducing Dietary Sodium

To reduce dietary sodium:

1. Use spices instead of salt to flavor foods.
2. Add little or no salt to foods at the table, and add only small amounts of salt when you cook.
3. Eat unsalted crackers, pretzels, potato chips, nuts, and popcorn.
4. Avoid hot dogs, ham, bacon, luncheon meats, smoked salmon, sardines, and anchovies.
5. Avoid processed cheese and canned or dehydrated soups.
6. Avoid brine-soaked foods, such as pickles or olives.
7. Read nutrition labels to avoid high-salt products.

Vitamins

Vitamins are organic compounds (other than carbohydrate, fat, and protein) that the body uses for metabolic purposes but is unable to produce in adequate quantity. Many vitamins are portions of coenzymes, which are enzyme helpers. For example, niacin is part of the coenzyme NAD, and riboflavin is part of another dehydrogenase, FAD, discussed in Chapter 3. Coenzymes are needed in only small amounts because each can be used over and over again. Not all vitamins are coenzymes; vitamin A, for example, is a precursor for the visual pigment that prevents night blindness. If vitamins are lacking in the diet, various symptoms develop (Fig. 8.15). Altogether, there are 13 vitamins, which are divided into those that are fat soluble (Table 8.7) and those that are water soluble (Table 8.8).

Antioxidants

Over the past 20 years, numerous statistical studies have been done to determine whether a diet rich in fruits and vegetables can protect against cancer. Cellular metabolism generates free radicals, unstable molecules that carry an extra electron. The most common free radicals in cells are superoxide (O_2^-) and hydroxide (OH^-). In order to stabilize themselves, free radicals donate an electron to DNA, to proteins (including enzymes), or to lipids, which are found in plasma membranes. Such donations most likely damage these cellular molecules and thereby may lead to disorders, perhaps even cancer.

Vitamins C, E, and A are believed to defend the body against free radicals, and therefore, they are termed antioxidants. These vitamins are especially abundant in fruits and vegetables. The food pyramid in Figure 8.16 suggests we eat 2½ cups of vegetables and 2 cups of fruit a day. To achieve this goal, think in terms of salad greens, raw or cooked vegetables, dried fruit, and fruit juice, in addition to apples and oranges and other fresh fruits.

Dietary supplements may provide a potential safeguard against cancer and cardiovascular disease, but nutritionists do not think people should take supplements instead of improving their intake of fruits and vegetables. There are many beneficial compounds in fruits that cannot be obtained from a vitamin pill. These compounds enhance one another's absorption or action and also perform independent biological functions.

Vitamin D

Skin cells contain a precursor cholesterol molecule that is converted to vitamin D after UV exposure. Vitamin D leaves the skin and is modified first in the kidneys and then in the liver until finally it becomes calcitriol. Calcitriol promotes the absorption of calcium by the intestines. The lack of vitamin D leads to rickets in children (Fig. 8.15a). Rickets, characterized by bowing of the legs, is caused by defective mineralization of the skeleton. Most milk today is fortified with vitamin D, which helps prevent the occurrence of rickets.

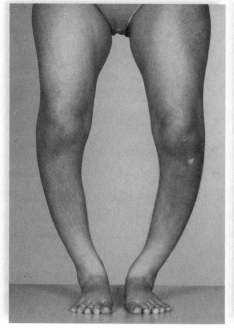

a.

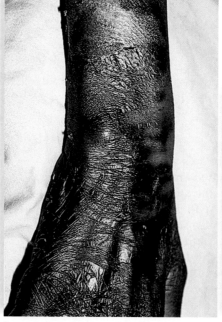

b.

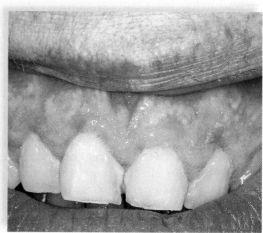

c.

Figure 8.15   Illnesses due to vitamin deficiency.
a. Bowing of bones (rickets) due to vitamin D deficiency. **b.** Dermatitis (pellagra) of areas exposed to light due to niacin (vitamin B₃) deficiency. **c.** Bleeding gums are another early sign of vitamin C deficiency.

| Table 8.7 | Fat-Soluble Vitamins |

Vitamin	Functions	Food Sources	Conditions with	
			Too Little	*Too Much*
Vitamin A	Antioxidant synthesized from beta-carotene; needed for healthy eyes, skin, hair, and mucous membranes, and for proper bone growth	Deep yellow/orange and leafy, dark green vegetables, fruits, cheese, whole milk, butter, eggs	Night blindness, impaired growth of bones and teeth	Headache, dizziness, nausea, hair loss, abnormal development of fetus
Vitamin D	A group of steroids needed for development and maintenance of bones and teeth	Milk fortified with vitamin D, fish liver oil; also made in the skin when exposed to sunlight	Rickets, decalcification and weakening of bones	Calcification of soft tissues, diarrhea, possible renal damage
Vitamin E	Antioxidant that prevents oxidation of vitamin A and polyunsaturated fatty acids	Leafy green vegetables, fruits, vegetable oils, nuts, whole-grain breads and cereals	Unknown	Diarrhea, nausea, headaches, fatigue, muscle weakness
Vitamin K	Needed for synthesis of substances active in clotting of blood	Leafy green vegetables, cabbage, cauliflower	Easy bruising and bleeding	Can interfere with anticoagulant medication

| Table 8.8 | Water-Soluble Vitamins |

Vitamin	Functions	Food Sources	Conditions with	
			Too Little	*Too Much*
Vitamin C	Antioxidant; needed for forming collagen; helps maintain capillaries, bones, and teeth	Citrus fruits, leafy green vegetables, tomatoes, potatoes, cabbage	Scurvy, delayed wound healing, infections	Gout, kidney stones, diarrhea, decreased copper
Thiamine (vitamin B_1)	Part of coenzyme needed for cellular respiration; also promotes activity of the nervous system	Whole-grain cereals, dried beans and peas, sunflower seeds, nuts	Beriberi, muscular weakness, enlarged heart	Can interfere with absorption of other vitamins
Riboflavin (vitamin B_2)	Part of coenzymes, such as FAD;* aids cellular respiration, including oxidation of protein and fat	Nuts, dairy products, whole-grain cereals, poultry, leafy green vegetables	Dermatitis, blurred vision, growth failure	Unknown
Niacin (nicotinic acid)	Part of coenzymes NAD;+ needed for cellular respiration, including oxidation of protein and fat	Peanuts, poultry, whole-grain cereals, leafy green vegetables, beans	Pellagra, diarrhea, mental disorders	High blood sugar and uric acid, vasodilation, etc.
Folacin (folic acid)	Coenzyme needed for production of hemoglobin and formation of DNA	Dark leafy green vegetables, nuts, beans, whole-grain cereals	Megaloblastic anemia, spina bifida	May mask B_{12} deficiency
Vitamin B_6	Coenzyme needed for synthesis of hormones and hemoglobin; CNS control	Whole-grain cereals, bananas, beans, poultry, nuts, leafy green vegetables	Rarely, convulsions, vomiting, seborrhea, muscular weakness	Insomnia, neuropathy
Pantothenic acid	Part of coenzyme A needed for oxidation of carbohydrates and fats; aids in the formation of hormones and certain neurotransmitters	Nuts, beans, dark green vegetables, poultry, fruits, milk	Rarely, loss of appetite, mental depression, numbness	Unknown
Vitamin B_{12}	Complex, cobalt-containing compound; part of the coenzyme needed for synthesis of nucleic acids and myelin	Dairy products, fish, poultry, eggs, fortified cereals	Pernicious anemia	Unknown
Biotin	Coenzyme needed for metabolism of amino acids and fatty acids	Generally in foods, especially eggs	Skin rash, nausea, fatigue	Unknown

*FAD = flavin adenine dinucleotide
+NAD = nicotinamide adenine dinucleotide

How to Plan Nutritious Meals

Many serious disorders in Americans are linked to a diet that results in excess body fat. While genetics is a factor in being overweight, a person cannot become fat without taking in more food energy (Calories, Kcal) than is needed. A person needs calories (energy) for their basal metabolism, which is the number calories your body burns at rest to maintain normal body functions, and a person needs calories for exercise. The less exercise, the less calories that are needed beyond the basal metabolic rate. So, the first step in planning a diet is to limit the number of calories to an amount that you will use up each day. Let's say you do all the necessary calculations (which are beyond the scope of this book) and discover that, as a woman, the maximum number of calories you can take in each day is 2,000, and as a man, you can afford 2,500 calories without the anxiety of gaining weight.

Now, the new food pyramid developed by the U.S. Department of Agriculture (USDA) can be used to help you decide how those calories should be distributed among the foods to eat (Fig. 8.16). The new pyramid emphasizes food that should be eaten often and omits foods that should not be eaten on a regular basis. Additionally, the USDA provides recommendations concerning the minimum quantity of foods in each group that should be eaten daily. In general, we should:

- eat a variety of foods. Foods from all food groups should be included in the diet.
- eat more of these foods: fruits, vegetables, whole grains, and fat-free or low-fat milk products. Choose dark-green vegetables, orange vegetables, and dry beans and peas over potatoes, corn, or green peas more often. When eating grains, choose whole grains such as brown rice, oatmeal, whole wheat bread. Choose fruit as a snack or a topping for foods, instead of sugar.
- eat less of foods high in saturated or trans fats, added sugars, cholesterol, salt, and alcohol. Remember that

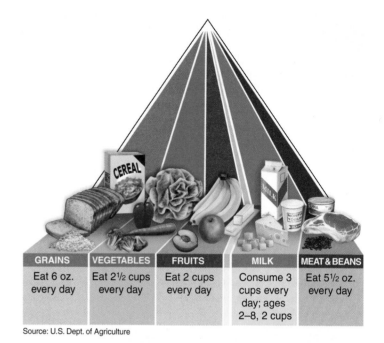

GRAINS	VEGETABLES	FRUITS	MILK	MEAT & BEANS
Eat 6 oz. every day	Eat 2½ cups every day	Eat 2 cups every day	Consume 3 cups every day; ages 2–8, 2 cups	Eat 5½ oz. every day

Source: U.S. Dept. of Agriculture

Figure 8.16   **The 2005 USDA food guide pyramid.**
This food guide pyramid can be helpful when planning nutritious meals and snacks.

dry beans and peas can be used to meet protein needs. Choose lean meats such as poultry, fish high in omega-3 fatty acids, such as salmon, trout, and herring, in moderate sized portions. Include oils that are rich in monounsaturated and polyunsaturated fatty acids in the diet. Choose and prepare foods and beverages with little added sugars and salt.

- be physically active every day. If weight loss is needed, decrease calorie intake slowly, while maintaining adequate nutrient intake and increasing physical activity.

Table 8.9	Preparing Meals and Snacks Using the Food Guide Pyramid				

Breakfast	Snack	Lunch	Snack	Dinner	Dessert
½ cup oat flakes cereal	Whole-wheat muffin	Toasted American cheese sandwich (whole-grain bread)	Apple	3 oz. hamburger patty	½ cup frozen vanilla yogurt topped with sliced strawberries
8 oz. nonfat milk	2 Tbsp peanut butter mixed with 1 tsp sunflower seeds	1 cup vegetable salad	4 graham crackers	Whole-wheat bun	
½ cup blueberries	8 oz. nonfat milk	1 Tbsp olive oil and vinegar salad dressing	½ cup baked beans		
1 slice whole-grain toast		4 oz. tomato juice		1 cup steamed broccoli spears	
1 tsp soft maragarine				1 tsp soft maragarine	
4 oz. orange juice					

a. Anorexia nervosa

b. Bulimia nervosa

c. Muscle dysmorphia

Figure 8.17   Eating disorders.
a. People with anorexia nervosa have a mistaken body image and think they are fat, even though they are thin. **b.** Those with bulimia nervosa overeat and then purge their bodies of the food they have eaten. **c.** People with muscle dysmorphia think their muscles are underdeveloped. They spend hours at the gym and are preoccupied with diet as a way to gain muscle mass.

Table 8.9 gives you a sample menu, developed by a nutritionist, so you can see what a good diet for one day would look like. Table 8.9 also gives you an idea of what a nutritionist means by a "serving size."

Eating Disorders

People with eating disorders are dissatisfied with their body image. Social, cultural, emotional, and biological factors all contribute to the development of an eating disorder. Serious conditions such as obesity, anorexia nervosa, and bulimia nervosa can lead to malnutrition, disability, and death. Regardless of the eating disorder, early recognition and treatment are crucial.

Anorexia nervosa is a severe psychological disorder characterized by an irrational fear of getting fat that results in the refusal to eat enough food to maintain a healthy body weight (Fig. 8.17*a*). A self-imposed starvation diet is often accompanied by occasional binge eating that is followed by purging and extreme physical activity to avoid weight gain. Binges usually include large amounts of high-calorie foods, and purging episodes involve self-induced vomiting and laxative abuse. About 90% of people suffering from anorexia nervosa are young women; an estimated 1 in 200 teenage girls is affected.

A person with **bulimia nervosa** binge-eats, and then purges to avoid gaining weight (Fig. 8.17*b*). The binge-purge cycle behavior can occur several times a day. People with bulimia nervosa can be difficult to identify because their body weights are often normal and they tend to conceal their binging and purging practices. Women are more likely than men to develop bulimia; an estimated 4% of young women suffer from this condition.

Other abnormal eating practices include binge-eating disorder and muscle dysmorphia. Many obese people suffer from **binge-eating disorder,** a condition characterized by episodes of overeating that are not followed by purging. Stress, anxiety, anger, and depression can trigger food binges. A person suffering from **muscle dysmorphia** (Fig. 8.17*c*) thinks his or her body is underdeveloped. Body-building activities and a preoccupation with diet and body form accompany this condition. Each day, the person may spend hours in the gym working out on muscle-strengthening equipment. Unlike anorexia nervosa and bulimia, muscle dysmorphia affects more men than women.

☑ Check Your Progress 8.6

1. **a.** What is the rationale behind using BMI to judge obesity?
   b. What disorders are associated with obesity?

2. Why might carbohydrates and why might fats be the cause of the obesity epidemic today?

3. **a.** What type fats are dangerous to our health, and **(b)** what type is protective for cardiovascular disease?

4. Proteins provide the essential amino acids; why shouldn't they be eaten as an energy source?

5. In general, what does the new USDA food pyramid tell you?

6. Name and describe four eating disorders.

Summarizing the Concepts

8.1 Overview of Digestion

The organs of the digestive system are located within the GI tract. The processes of digestion require ingestion, digestion, movement, absorption, and elimination. All parts of the tract have four layers, called the mucosa, submucosa, muscularis, and serosa.

8.2 First Part of the Digestive Tract

- In the mouth, teeth chew the food, saliva contains salivary amylase for digesting starch, and the tongue forms a bolus for swallowing.
- The air passage and food passages cross in the pharynx. During swallowing, the air passage is blocked off by the soft palate and epiglottis; food enters the esophagus; and peristalsis begins. The esophagus moves food to the stomach.

8.3 The Stomach and Small Intestine

- The stomach expands and stores food and also churns, mixing food with the acidic gastric juices. This juice contains pepsin, an enzyme that digests protein.
- The duodenum of the small intestine receives bile from the liver and pancreatic juice from the pancreas. Bile emulsifies fat and readies it for digestion by lipase.
- The pancreas produces enzymes that digest starch (pancreatic amylase), protein (trypsin), and fat (lipase). The intestinal enzymes finish the process of chemical digestion.
- Small nutrient molecules are absorbed at the villi in the walls of the small intestine.

8.4 Three Accessory Organs and Regulation of Secretions

Three accessory organs of digestion send secretions to the duodenum via ducts. These organs are the pancreas, liver, and gallbladder.

- The pancreas produces pancreatic juice, which contains digestive enzymes for carbohydrate, protein, and fat.
- The liver produces bile, destroys old blood cells, detoxifies blood, stores iron, makes plasma proteins, stores glucose as glycogen, breaks down glycogen to glucose, produces urea, and helps regulate blood cholesterol levels.
- The gallbladder stores bile, which is produced by the liver.

The secretions of digestive juices are controlled by the nervous system and by hormones.

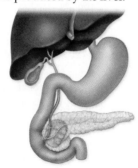

- Gastrin produced by the lower part of the stomach stimulates via the bloodstream the upper part of the stomach to secrete pepsin.
- Secretin and CCK produced by the duodenal wall stimulate the pancreas to secrete its juices and the gallbladder to release bile.

8.5 The Large Intestine and Defecation

- The large intestine consists of the cecum, the colon (including the ascending, transverse, and descending colon), and the rectum, which ends at the anus.
- The large intestine absorbs water, salts, and some vitamins; forms the feces; and carries out defecation.
- Disorders of the large intestine include diverticulosis, irritable bowel syndrome, inflammatory bowel disease, polyps, and cancer.

8.6 Nutrition and Weight Control

The nutrients released by the digestive process should provide us with adequate energy, essential amino acids and fatty acids, and all necessary vitamins and minerals. Today, obesity is on the increase, possibly because people eat too much food and make improper choices of food. Obesity is associated with many illnesses, including diabetes type 2 and cardiovascular disease. The food guide pyramid shows foods to emphasize and foods to minimize for good health.

- Carbohydrates are necessary in the diet, but simple sugars and refined starches cause a rapid release of insulin that can lead to diabetes type 2.
- Proteins supply essential amino acids.
- Unsaturated fatty acids, particularly the omega-3 fatty acids, are protective against cardiovascular disease.
- Saturated fatty acids lead to plaque, which occludes blood vessels.
- Vitamins and minerals are also required by the body in certain amounts.

Understanding Key Terms

absorption 145
anorexia nervosa 165
anus 154
appendix 145
bile 150, 152
binge-eating disorder 165
body mass index (BMI) 156
bolus 147
bulimia nervosa 165
cecum 154
chyme 144
cirrhosis 153
colon 154
constipation 154
defecation 154
dental caries 146
diaphragm 147
diarrhea 154
digestion 144
duodenum 150
elimination 145

epiglottis 147
esophagus 147
essential amino acids 158
essential fatty acids 159
fiber 154
gallbladder 152
gallstone 152
gastric 149
gastric gland 149
glottis 147
glycemic index (GI) 158
hard palate 146
heartburn 147
hemorrhoid 155
hepatitis 153
hormone 152
ingestion 144
jaundice 153
lacteal 150
lactose intolerance 150
large intestine 154

lipase 152
liver 152
lumen 145
mineral 160
movement 144
mucosa 145
muscle dysmorphia 165
muscularis 145
nutrient 158
obesity 156
osteoporosis 160
pancreas 152
pancreatic amylase 152
pepsin 149
periodontitis 146
peristalsis 147
peritonitis 145

pharynx 147
plaque 159
polyp 155
rectum 154
rugae 149
salivary amylase 146
salivary gland 146
serosa 145
small intestine 150
soft palate 146
sphincter 147
stomach 148
submuscosa 145
trypsin 152
vermiform appendix 154
villus 150
vitamin 162

Match the key terms to these definitions.

a. _____ Essential requirement in the diet, needed in small amounts. Often a part of a coenzyme.

b. _____ Fat-digesting enzyme secreted by the pancreas.

c. _____ Lymphatic vessel in an intestinal villus; it aids in the absorption of fats.

d. _____ Muscular tube for moving swallowed food from the pharynx to the stomach.

e. _____ Organ attached to the liver that serves to store and concentrate bile.

Testing Your Knowledge of the Concepts

1. Argue that absorption is the most important of the five processes of digestion over the other four processes. (pages 144–45)

2. List the main organs of the digestive tract, and state the contribution of each to the digestive process. (pages 146–54)

3. Discuss the absorption of the products of digestion into the lymphatic and cardiovascular systems. (page 150)

4. Name the enzymes involved in the digestion of starch, protein, and fat, and tell where these enzymes are active and what they do. (page 151)

5. Why are the pancreas, liver, and gallbladder considered accessory organs of digestion and not an organ of digestion? (page 152)

6. Name and state the functions of the hormones that assist the nervous system in regulating digestive secretions. (page 153)

7. What is the chief contribution of each of these in the body—carbohydrates, proteins, fats, fruits, and vegetables? (pages 158–59, 162)

8. Which three eating disorders involve binge eating? How are these three disorders different from one another? (pages 164–65)

9. Tracing the path of food in the following list (a–f), which step is out of order first?
   a. mouth
   b. pharynx
   c. esophagus
   d. small intestine
   e. stomach
   f. large intestine

10. The appendix connects to the
    a. cecum.
    b. small intestine.
    c. esophagus.
    d. liver.
    e. All of these are correct.

11. Which association is incorrect?
    a. mouth—starch digestion
    b. esophagus—protein digestion
    c. small intestine—starch, lipid, protein digestion
    d. stomach—food storage
    e. liver—production of bile

12. Why can a person not swallow food and talk at the same time?
    a. In order to swallow, the epiglottis must close off the trachea.
    b. The brain cannot control two activities at once.
    c. In order to speak, air must come through the larynx to form sounds.
    d. A swallowing reflex is only initiated when the mouth is closed.
    e. Both a and c are correct.

13. Which association is incorrect?
    a. pancreas—produces alkaline secretions and enzymes
    b. salivary glands—produce saliva and amylase
    c. gallbladder—produces digestive enzymes
    d. liver—produces bile

14. Which of the following could be absorbed directly without need of digestion?
    a. glucose
    b. fat
    c. polysaccharides
    d. protein
    e. nucleic acid

15. Peristalsis occurs
    a. from the mouth to the small intestine.
    b. from the beginning of the esophagus to the anus.
    c. only in the stomach.
    d. only in the small and large intestine.
    e. only in the esophagus and stomach.

16. An organ is a structure made of two or more tissues performing a common function. Which of the four tissue types are present in the wall of the digestive tract?
    a. epithelium
    b. connective tissue
    c. nervous tissue
    d. muscle tissue
    e. All of these are correct.

17. Which association is incorrect?
    a. protein—trypsin
    b. fat—bile
    c. fat—lipase
    d. maltose—pepsin
    e. starch—amylase

18. Most of the products of digestion are absorbed across the
    a. squamous epithelium of the esophagus.
    b. striated walls of the trachea.
    c. convoluted walls of the stomach.
    d. fingerlike villi of the small intestine.
    e. smooth wall of the large intestine.

19. Bile
    a. is an important enzyme for the digestion of fats.
    b. cannot be stored.
    c. is made by the gallbladder.
    d. emulsifies fat.
    e. All of these are correct.

20. Which of the following is not a function of the liver in adults?
    a. produces bile   d. produces urea
    b. detoxifies alcohol   e. makes red blood cells
    c. stores glucose

21. The large intestine
    a. digests all types of food.
    b. is the longest part of the intestinal tract.
    c. absorbs water.
    d. is connected to the stomach.
    e. is subject to hepatitis.

In questions 22–26, match each function to an organ in the key.

Key:
    a. mouth   d. small intestine
    b. esophagus   e. large intestine
    c. stomach

22. Removes nondigestible remains

23. Serves as a passageway

24. Stores food

25. Absorbs nutrients

26. Receives food

27. The amino acids that must be consumed in the diet are called essential. Nonessential amino acids
    a. can be produced by the body.
    b. are only needed occasionally.
    c. are stored in the body until needed.
    d. can be taken in by supplements.

28. Which of the following are often organic portions of important coenzymes?
    a. minerals   c. protein
    b. vitamins   d. carbohydrates

29. The products of digestion are
    a. large macromolecules needed by the body.
    b. enzymes needed to digest food.
    c. small nutrient molecules that can be absorbed.
    d. regulatory hormones of various kinds.
    e. the food we eat.

In questions 30–34, match each statement to a layer of the wall of the esophagus in the key. Answers are used more than once.

Key:
    a. mucosa   c. muscularis
    b. submucosa   d. serosa

30. Loose connective tissue that contains lymph nodules

31. Contains a layer of epithelium that lines the lumen

32. Very thin layer of squamous epithelium that secretes a fluid, keeping the organ moist

33. Contains digestive glands and mucus-secreting goblet cells

34. Two layers of smooth muscle

In questions 35–40, match each statement to an answer in the key. Answers are used more than once. Some may have more than one answer.

Key:
    a. gastrin   d. All of these are correct.
    b. secretin   e. None of these is correct.
    c. CCK

35. Stimulates gallbladder to release bile

36. Hormone carried in bloodstream

37. Stimulates the stomach to digest protein

38. Enzyme that digests food

39. Secreted by duodenum

40. Secreted by the stomach

In questions 41–45, match each statement to a vitamin or mineral in the key.

Key:
    a. calcium   d. iodine
    b. vitamin K   e. vitamin A
    c. sodium

41. Needed to make thyroid hormone

42. Needed for night vision

43. Needed for bones, teeth, and muscle contraction

44. Needed for nerve conduction, pH, and water balance

45. Needed for making clotting proteins

Thinking Critically About the Concepts

Monica, from the opening story, is considering bariatric surgery, which reduces the size of the stomach and enables food to bypass a section of the small intestine. The surgery is generally done when obese individuals have unsuccessfully tried numerous ways to lose weight and their health is compromised by their weight. There are many risks associated with the surgery, but it helps a number of people lose a considerable amount of weight. After people undergo the surgery, there are several lifestyle changes they must make to avoid nutritional deficiencies and to compensate for the small size of their stomach.

1a. Why do some people who had bariatric surgery process their food in a blender (or have to chew thoroughly) before swallowing their food?

  b. Why should people who had bariatric surgery drink liquids between meals rather than with meals?

2. What risk is there to the esophagus after bariatric surgery?

3. The text on page 149 describes how William Beaumont was able to observe the workings of the stomach through a fistula in St. Martin's stomach.

a. What do you think Beaumont observed after he inserted a piece of meat into St. Martin's stomach?

b. What do you think Beaumont observed after he inserted a piece of bread into St. Martin's stomach?

CHAPTER 9

Respiratory System

Heather had been babysitting for a number of years, but this was the first time she'd taken care of Olivia. She was use to older children and was a little nervous about babysitting a four-year-old.

Right away, Olivia was clamoring for a snack, so Heather gave her some dry cereal and a glass of juice. Olivia grabbed a handful of the cereal and shoved it in her mouth. A few seconds later, Olivia began to cough, and within a few seconds, pieces of cereal flew out of her mouth. Heather wanted to yell at Olivia to be more careful, but she forced herself to ask calmly, "Are you okay?" Olivia nodded and began to cry. "I want my mamma," sobbed Olivia. "She'll be back soon," said Heather, as she tried to console the distraught child. "NO," screamed Olivia, "I want her NOW!" Before Heather could say or do anything else, Olivia began holding her breath in an effort to get her way. Heather did her best to cajole Olivia into exhaling to no avail. Just as Heather picked up the phone to call 911, she heard Olivia exhale with a gasp. Before Olivia could start crying again, Heather swept her off to the playroom and the welcome distraction of Olivia's toys. Heather was relieved the situation with Olivia ended well, but vowed to stick to babysitting older children in the future.

After studing the material in this chapter, you will understand why Olivia choked on the dry cereal. You'll also learn how changes in blood pH prevent someone from holding their breath indefinitely.

CHAPTER CONCEPTS

9.1 The Respiratory System

The organs of the respiratory system consist of those that form a pathway to the lungs and the lungs, where gas exchange occurs.

9.2 The Upper Respiratory Tract

Air is warmed and filtered in the nose, then moves across the pharynx to pass through the glottis into the larynx.

9.3 The Lower Respiratory Tract

The wall of the trachea sweeps impurities toward the pharynx, where they can be swallowed. The trachea leads to the bronchial tree, which ends with little tubes that enter the lungs. Gas exchange occurs in the air sacs called alveoli.

9.4 Mechanism of Breathing

During inspiration, air comes into the lungs because the chest has expanded, and during expiration, air leaves the lungs because the chest has returned to its normal size. It is possible to measure the volumes of air exchanged during inspiration and expiration.

9.5 Control of Ventilation

The respiratory center in the brain regulates breathing, which occurs automatically. When the pH of the blood lowers due to the presence of carbon dioxide, the center speeds up breathing.

9.6 Gas Exchanges in the Body

In the lungs, carbon dioxide leaves blood, and oxygen enters the blood to be carried by hemoglobin. In the tissues, oxygen leaves the blood, and carbon dioxide, to be carried largely as the bicarbonate ion, enters the blood.

9.7 Respiration and Health

A number of illnesses are associated with the respiratory tract. Those that affect the upper respiratory tract are not as serious as those that affect the lower respiratory tract.

9.1 The Respiratory System

The organs of the respiratory system ensure that oxygen enters the body and carbon dioxide leaves the body (Fig. 9.1). During **inspiration,** or inhalation (breathing in), air is conducted from the atmosphere to the lungs by a series of cavities, tubes, and openings, illustrated in Figure 9.1. During **expiration,** or exhalation (breathing out), air is conducted from the lungs to the atmosphere by way of the same structures.

Ventilation is another term for breathing that includes both inspiration and expiration. Once ventilation has occurred, the respiratory system works with the cardiovascular system to transport oxygen (O_2) from the lungs to the tissues and carbon dioxide (CO_2) from the tissues to the lungs.

Gas exchange is necessary because the cells of the body carry out cellular respiration for the purpose of producing ATP. During cellular respiration, cells use up O_2 and produce CO_2.

☑ Check Your Progress 9.1

1. Trace the path of air from the nasal cavities to the lungs.
2. What is the function of the respiratory system?

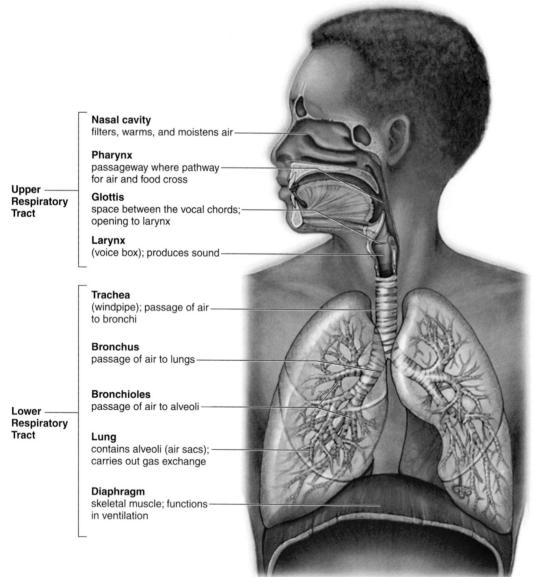

Nasal cavity
filters, warms, and moistens air

Pharynx
passageway where pathway
for air and food cross

Glottis
space between the vocal chords;
opening to larynx

Larynx
(voice box); produces sound

Upper Respiratory Tract

Trachea
(windpipe); passage of air
to bronchi

Bronchus
passage of air to lungs

Bronchioles
passage of air to alveoli

Lung
contains alveoli (air sacs);
carries out gas exchange

Diaphragm
skeletal muscle; functions
in ventilation

Lower Respiratory Tract

Figure 9.1   The respiratory tract.
The respiratory tract extends from the nose to the lungs. Note the organs that are in the upper respiratory tract and the ones that are in the lower respiratory tract.

9.2 The Upper Respiratory Tract

The nasal cavities, pharynx, and larynx are the organs of the upper respiratory tract (Fig. 9.2).

The Nose

The nose opens at the nares (nostrils) that lead to the **nasal cavities.** The nasal cavities are narrow canals separated from each other by a septum composed of bone and cartilage (Fig. 9.2).

Air entering the nasal cavities is met by large stiff hairs that act as a screening device, filtering the air of dust so it doesn't enter air passages. The rest of the nasal cavities are lined by mucous membrane with a submucosa that has plentiful capillaries. These capillaries help warm and moisten the air, but they also make us more susceptible to nose bleeds if the nose suffers an injury. The mucus secreted by the mucous membrane also helps trap dust and move it to the pharynx, where it can be swallowed.

Moistened air leaves the nose, and when we breathe out on a cold day, it condenses so that we can see our breath.

Special ciliated cells in the narrow upper recesses of the nasal cavities act as odor receptors. Nerves lead from these cells to the brain, where the impulses generated by the odor receptors are interpreted as smell.

The tear (lacrimal) glands drain into the nasal cavities by way of tear ducts. When Olivia in the opening story began to cry, she got a runny nose because the water from her eyes ended up in her nose.

The nasal cavities also communicate with the cranial sinuses, air-filled mucosa-lined spaces in the skull. If inflammation due to a cold or an allergic reaction blocks the ducts leading from the sinuses, fluid may accumulate, causing a sinus headache.

Air in the nasal cavities passes into the nasopharynx, the upper portion of the pharynx. The middle ears have tubes, called **auditory (Eustachian) tubes,** that empty into the nasopharynx. Some people chew gum as a plane takes off and lands in an effort to move air from the ears, otherwise, the auditory tube openings may "pop" when air pressure inside the middle ears equalizes with that in the nasopharynx.

The Pharynx

The **pharynx** is a funnel-shaped passageway that connects the nasal and oral cavities to the larynx. Therefore, the pharynx, which is commonly referred to as the "throat," has three parts: the nasopharynx, where the nasal cavities open above the soft palate; the oropharynx, where the oral cavity opens; and the laryngopharynx, which opens into the larynx.

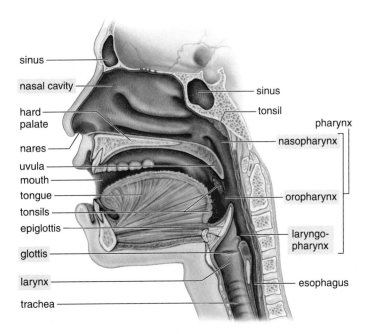

Figure 9.2   The upper respiratory tract and associated organs. This drawing shows the path of air from the nasal cavities to the trachea, which is in the lower respiratory tract. The designated structures are in the upper respiratory tract.

The tonsils form a protective ring at the junction of the oral cavity and the pharynx. Inhaled air passes directly over the tonsils, making them the primary lymphatic defense during breathing. Being lymphatic tissue, the tonsils contain lymphocytes that protect against invasion of foreign antigens that are inhaled. In the tonsils, B cells and T cells are prepared to respond to antigens that may subsequently invade internal tissues and fluids. Therefore, the respiratory tract assists the immune system in maintaining homeostasis.

In the pharynx, the air passage and the food passage cross because the larynx, which receives air, is in front of the esophagus, which receives food. The larynx is normally open, allowing air to pass, but the esophagus is normally closed and opens only when a person swallows. When Olivia swallowed, some of the dry cereal went into the larynx, causing her to cough until the cereal was dislodged and flew out of her mouth. A **cough** is a sudden, noisy explosion of air from the lungs. If Olivia hadn't coughed up the cereal on her own, Heather should have performed the Heimlich maneuver, which is shown in Figure 9.3.

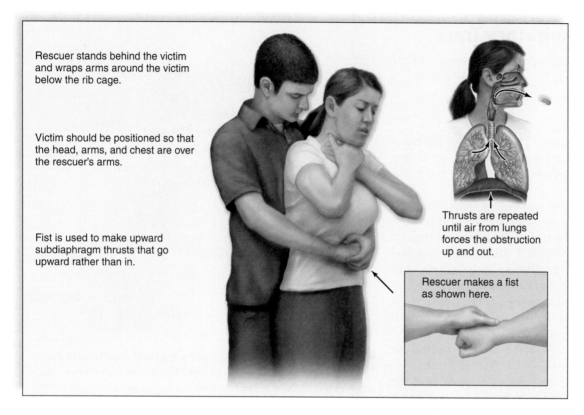

Rescuer stands behind the victim and wraps arms around the victim below the rib cage.

Victim should be positioned so that the head, arms, and chest are over the rescuer's arms.

Fist is used to make upward subdiaphragm thrusts that go upward rather than in.

Thrusts are repeated until air from lungs forces the obstruction up and out.

Rescuer makes a fist as shown here.

Figure 9.3
The Heimlich maneuver.

Larynx

The **larynx** is a cartilaginous structure that serves as a passageway for air between the pharynx and the trachea. The larynx can be pictured as a triangular box whose apex, the Adam's apple, is located at the front of the neck. The larynx is called the *voice box* because it houses the vocal cords. The **vocal cords** are mucosal folds supported by elastic ligaments, and the slit between the vocal cords is an opening called the **glottis** (Fig. 9.4). When air is expelled through the glottis, the vocal cords vibrate, producing sound. At the time

of puberty, the growth of the larynx and the vocal cords is much more rapid and accentuated in the male than in the female, causing the male to have a more prominent Adam's apple and a deeper voice. The voice "breaks" in the young male due to his inability to control the longer vocal cords. These changes cause the lower pitch of the voice in males.

The high or low pitch of the voice is regulated when speaking and singing by changing the tension on the vocal cords. The greater the tension, as when the glottis becomes narrower, the higher the pitch. When the glottis is wider, the pitch is lower (Fig. 9.4, *right*). The loudness, or intensity, of the voice depends upon the amplitude of the vibrations—that is, the degree to which the vocal cords vibrate.

Ordinarily when food is swallowed, the larynx moves upward against the **epiglottis,** a flap of tissue that prevents food from passing into the larynx. You can detect the movement of the larynx by placing your hand gently on your larynx and swallowing.

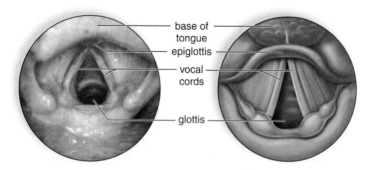

base of tongue
epiglottis
vocal cords
glottis

Figure 9.4   **Placement of the vocal cords.**
Viewed from above, the vocal cords can be seen to stretch across the glottis, the opening to the trachea. When air is expelled through the glottis, the vocal cords vibrate, producing sound. The glottis is narrow when we produce a high-pitched sound, and it widens as the pitch deepens.

✓ Check Your Progress 9.2

1. What organs are a part of the upper respiratory tract? What are the functions of each organ?

2. Several different organs have connections with the pharynx. What are they?

3. a. What two pathways cross in the pharynx? b. Why is it correct to say that they "cross"?

9.3 The Lower Respiratory Tract

Whereas the nasal cavities, pharynx, and larynx are a part of the upper respiratory tract, the trachea and the rest of the respiratory system are in the lower respiratory tract.

The Trachea

The **trachea,** commonly called the windpipe, is a tube connecting the larynx to the primary bronchi. Its walls consist of connective tissue and smooth muscle reinforced by C-shaped cartilaginous rings.

The trachea lies ventral to the esophagus. The open part of the C-shaped rings faces the esophagus, and this allows the esophagus to expand when swallowing. The mucous membrane that lines the trachea has an outer layer of pseudostratified ciliated columnar epithelium. (*Pseudostratified* means that while the epithelium appears to be layered, actually each cell touches the basement membrane.) The cilia that project from the epithelium keep the lungs clean by sweeping mucus, produced by goblet cells, and debris toward the pharynx (Fig. 9.5). When one coughs, the tracheal wall contracts, narrowing its diameter. Therefore, coughing causes air to move more rapidly through the trachea, helping to expel

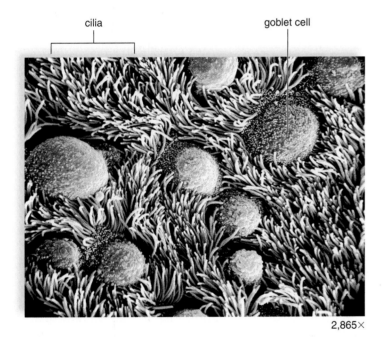

cilia   goblet cell

2,865×

Figure 9.5 Surface of the trachea.
Scanning electron micrograph of the surface of the mucous membrane lining the trachea consisting of goblet cells and ciliated cells. The cilia sweep mucus and debris embedded in it toward the pharynx, where it is swallowed or expectorated. Smoking causes the cilia to disappear; consequently, debris now enters the bronchi and lungs.
© Dr. Kessel & Dr. Kardon/Tissues & Organs/Visuals Unlimited

mucus and foreign objects. Smoking is known to destroy the cilia, and consequently, the soot in cigarette smoke collects in the lungs. Smoking cigarettes is discussed more fully in the Health Focus on page 183.

If the trachea is blocked because of illness or the accidental swallowing of a foreign object, it is possible to insert a breathing tube by way of an incision made in the trachea. This tube acts as an artificial air intake and exhaust duct. The operation is called a **tracheostomy.**

The Bronchial Tree

The trachea divides into right and left primary bronchi (sing., **bronchus**), which lead into the right and left lungs (see Fig. 9.1). The bronchi branch into a few secondary bronchi that also branch, until the branches become about 1 mm in diameter and are called **bronchioles.** The bronchi resemble the trachea in structure, but as the bronchial tubes divide and subdivide, their walls become thinner, and the small rings of cartilage are no longer present. During an asthma attack, the smooth muscle of the bronchioles contracts, causing bronchiolar constriction and characteristic wheezing. Each bronchiole leads to an elongated space enclosed by a multitude of air pockets, or sacs, called alveoli (sing., **alveolus**). The components of the bronchiole tree beyond the primary bronchi compose the lungs.

The Lungs

The **lungs** are paired, cone-shaped organs that occupy the thoracic cavity, except for the central area that contains the trachea, the heart, and esophagus. The right lung has three lobes, and the left lung has two lobes, allowing room for the heart, which points left. A lobe is further divided into lobules, and each lobule has a bronchiole serving many alveoli.

The lungs follow the contours of the thoracic cavity, including the diaphragm, the muscle that separates the thoracic cavity from the abdominal cavity. Each lung is enclosed by pleura, a double layer of serous membrane that produces serous fluid. The parietal pleura adheres to the thoracic cavity wall, and the visceral pleura adheres to the surface of the lung. Surface tension is the tendency for water molecules to cling to one another due to hydrogen bonding between molecules. Surface tension holds the two pleural layers together, and therefore, the lungs must follow the movement of the thorax when breathing occurs.

The Alveoli

The lungs have about 300 million alveoli, with a total cross-sectional area of 50–70 m². Each alveolar sac is surrounded by blood capillaries. The wall of the sac and the wall of the capillary are largely simple squamous epithelium—thin flattened

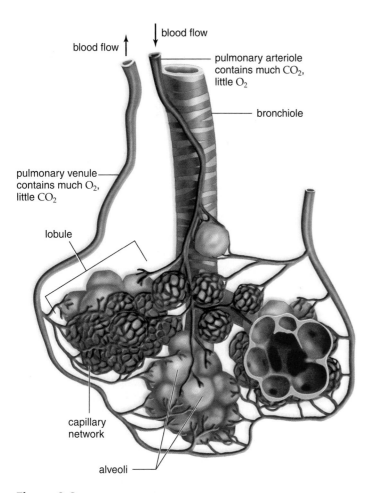

blood flow

blood flow

blood flow

pulmonary arteriole
contains much CO_2,
little O_2

bronchiole

pulmonary venule
contains much O_2,
little CO_2

lobule

capillary
network

alveoli

Figure 9.6   **Gas exchange in the lungs.**
The lungs consist of alveoli surrounded by an extensive capillary network.
Notice that the pulmonary artery carries O_2-poor (CO_2-rich) blood (colored
blue), and the pulmonary vein carries O_2-rich blood (colored red).

cells—and this facilitates gas exchange. Gas exchange occurs
between air in the alveoli and blood in the capillaries. Oxygen
diffuses across the alveolar wall and enters the bloodstream,
while carbon dioxide diffuses from the blood across the alveolar wall to enter the alveoli (Fig. 9.6) .

The alveoli of human lungs are lined with a **surfactant,** a
film of lipoprotein that lowers the surface tension of water
and prevents them from closing. The lungs collapse in some
newborn babies, especially premature infants, who lack this
film. The condition, called **infant respiratory distress syndrome,** is now treatable by surfactant replacement therapy.

☑ Check Your Progress 9.3

1. What organs are a part of the lower respiratory tract? What are
   the functions of each organ?
2. What part of the respiratory tract functions in gas exchange?

9.4 Mechanism of Breathing

Ventilation, or breathing, has two phases. The process of **inspiration,** also called inhalation, moves air into the lungs,
and the process of **expiration,** also called exhalation, moves
air out of the lungs. To understand ventilation, the manner
in which air enters and exits the lungs, it is necessary to remember the following facts:

1. Normally, there is a continuous column of air from the
   pharynx to the alveoli of the lungs.
2. The lungs lie within the sealed off thoracic cavity. The
   rib cage, consisting of the ribs joined to the vertebral
   column posteriorly and to the sternum anteriorly, forms
   the top and sides of the thoracic cavity. The intercostal
   muscles lie between the ribs. The diaphragm and
   connective tissue form the floor of the thoracic cavity.
3. The lungs adhere to the thoracic wall by way of the pleura.
   Any space between the two pleurae is minimal due to the
   surface tension of the fluid between them.

Inspiration

Inspiration is the active phase of ventilation because this is
the phase in which the diaphragm and the external intercostal
muscles contract (Fig. 9.7*a*). In its relaxed state, the diaphragm
is dome-shaped; during inspiration, it contracts and becomes
a flattened sheet of muscle. Also, the external intercostal muscles contract, and the rib cage moves upward and outward.

Following contraction of the diaphragm and the external intercostal muscles, the volume of the thoracic cavity
will be larger than it was before. As the thoracic volume
increases, the lungs increase in volume as well because the
lung adheres to the wall of the thoracic cavity. As the lung
volume increases, the air pressure within the alveoli
decreases, creating a partial vacuum. In other words, alveolar pressure is now less than atmospheric pressure (air pressure outside the lungs). Because a continuous column of air
reaches into the lungs, air will naturally flow from outside
the body into the respiratory passages and into the alveoli.

It is important to realize that air comes into the lungs
because they have already opened up; air does not force
the lungs open. This is why it is sometimes said that *humans
inhale by negative pressure.* The creation of a partial vacuum
in the alveoli causes air to enter the lungs. While inspiration is the active phase of breathing, the actual flow of air
into the alveoli is passive. Just as with bellows that are
used to fan a fire, when the handles of the bellows are
pulled apart, air automatically flows into the bellows.

Expiration

Usually, expiration is the passive phase of breathing, and no
effort is required to bring it about. During expiration, the elastic
properties of the thoracic wall and lungs cause them to recoil.

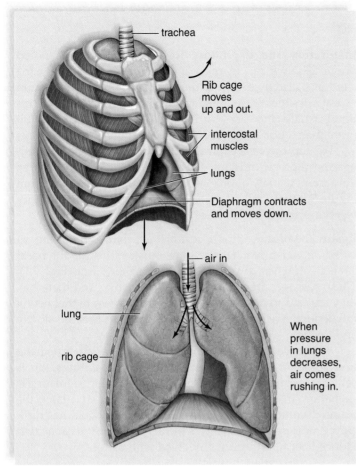

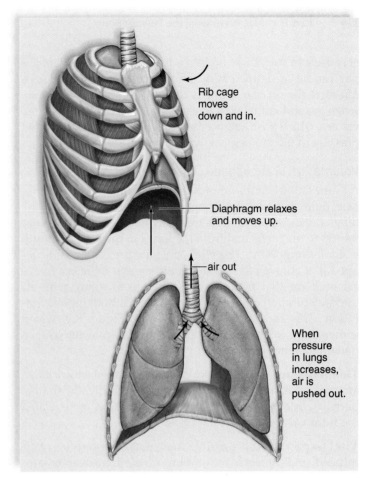

a. Inspiration

b. Expiration

Figure 9.7 **Inspiration versus expiration.**
a. During inspiration, the thoracic cavity and lungs expand so that air is drawn in. **b.** During expiration, the thoracic cavity and lungs resume their original positions and pressures. Now air is forced out.

In addition, the lungs recoil because the surface tension of the fluid lining the alveoli tends to draw them closed. During expiration, the abdominal organs press up against the diaphragm, and the rib cage moves down and inward (Fig. 9.7*b*). What keeps the alveoli from collapsing as a part of expiration? Recall that the presence of surfactant lowers the surface tension within the alveoli. Also, as the lungs recoil, the pressure between the

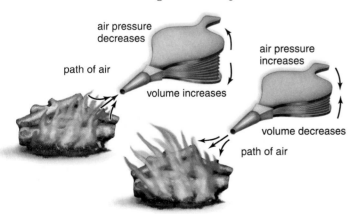

Figure 9.8 **How bellows work.**

pleura decreases, and this tends to make the alveoli stay open. The importance of the reduced intrapleural pressure is demonstrated when, by design or accident, air enters the intrapleural space. Now the lung collapses. We can liken this phase of breathing to pushing the handles of bellows together, increasing the pressure inside and causing air to flow out (Fig. 9.8).

Maximum Inspiratory Effort and Forced Expiration

If you recall the last time you exercised vigorously—perhaps running in a race, or even just climbing all those stairs to your classroom—you probably remember that you were breathing a lot harder than normal, during and immediately after that heavy exercise. Maximum inspiratory effort involves muscles of the back, chest, and neck, which increases the size of the thoracic cavity larger than normal, thus allowing maximum expansion of the lungs.

While inspiration is always the active phase of breathing, expiration is usually passive—that is, the diaphragm and external intercostal muscles are simply allowed to relax, the lungs recoil, and expiration occurs. However, expiration can also be forced. Forced expiration accompanies the maximum inspiratory efforts of heavy exercise. Forced expiration is also necessary to sing, blow air into a trumpet, or blow out birth-day candles. Contraction of the internal intercostal muscles can force the rib cage to move downward and inward. Also, when the abdominal wall muscles contract, they push on the viscera, which push against the diaphragm, and the increased pressure in the thoracic cavity helps expel air.

Volumes of Air Exchanged During Ventilation

As ventilation occurs, air moves into the lungs from the nose or mouth during inspiration; and then moves out of the lungs during expiration. A free flow of air from the nose or mouth to the lungs and from the lungs to the nose or mouth is vitally important. Therefore, a technique has been developed that allows physicians to determine if there is a medical problem that prevents the lungs from filling with air upon inspiration and releasing it from the body upon expiration. This technique is illustrated in Figure 9.9, which shows the measurements recorded by a spirometer when a person breathes as directed by a technician.

Tidal Volume   Normally when we are relaxed, only a small amount of air moves in and out with each breath, similar, perhaps, to the tide at the beach. This amount of air, called the **tidal volume,** is only about 500 mL.

Vital Capacity   It is possible to increase the amount of air inhaled, and therefore the amount exhaled, by deep breathing. The maximum volume of air that can be moved in plus the maximum amount that can be moved out during a single breath is called the **vital capacity.** It is called vital capacity because your life depends on breathing, and the more air you can move, the better off you are. A number of different illnesses, discussed at the end of this chapter, can decrease vital capacity.

Inspiratory and Expiratory Reserve Volume   As noted previously, we can increase inspiration by expanding the chest and also by lowering the diaphragm to the maximum extent possible. Forced inspiration (**inspiratory reserve volume**) usually increases by 2,900 mL, and that's quite a bit more than a tidal volume of only 500 mL!

We can increase expiration by contracting the abdominal and thoracic muscles. This so-called **expiratory reserve volume** is usually about 1,400 mL of air. You can see from Figure 9.9 that vital capacity is the sum of tidal, inspiratory reserve, and expiratory reserve volumes.

Residual Volume   It is a curious fact that some of the inhaled air never reaches the lungs; instead, it fills the nasal cavities, trachea, bronchi, and bronchioles (see Fig. 9.1). These passages are not used for gas exchange, and therefore, they are said to contain **dead air space.** To ensure that newly inhaled air reaches the lungs, it is better to breathe slowly and deeply.

Also, note in Figure 9.9 that even after a very deep exhalation, some air (about 1,000 mL) remains in the lungs; this is called the **residual volume.** This air is no longer useful for gas exchange. In some lung diseases to be discussed later, the residual volume builds up because the individual has difficulty emptying the lungs. This means that the vital capacity is reduced because the lungs are filled with useless air.

☑ Check Your Progress 9.4

1. **a.** How is breathing achieved? **b.** Is pressure within the lungs greatest during inspiration or expiration?

2. **a.** Why do physicians measure our ability to move air into and out of the lungs? **b.** What measurements do they take?

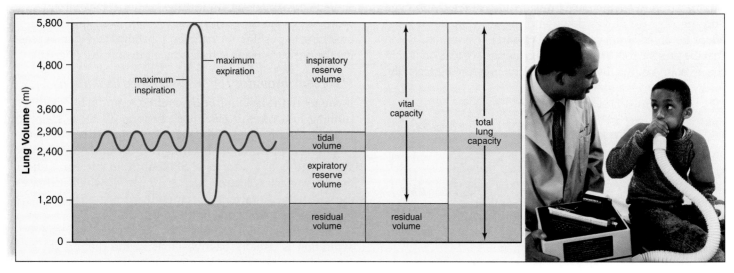

Figure 9.9   Vital capacity.
A spirometer is an instrument that measures the amount of air inhaled and exhaled with each breath. During inspiration, there is an upswing, and during expiration, there is a downswing. Vital capacity (red) is measured by taking the deepest breath and then exhaling as much as possible.

9.5 Control of Ventilation

Breathing is controlled in two ways. It is under nervous and, also, chemical control.

Nervous Control of Breathing

Normally, adults have a breathing rate of 12 to 20 ventilations per minute. The rhythm of ventilation is controlled by a **respiratory control center** located in the medulla oblongata of the brain. The respiratory control center automatically sends out impulses by way of nerves to the diaphragm and the external intercostal muscles of the rib cage, causing inspiration to occur (Fig. 9.10). When the respiratory center stops sending neuronal signals to the diaphragm and the rib cage, the diaphragm relaxes and resumes its dome shape. Now expiration occurs.

Sudden infant death syndrome (SIDS), or crib death, claims the life of about 2,000 infants a year in the United States. An infant under one year of age is put to bed seemingly healthy, and sometime while sleeping, the child stops breathing. The cause of SIDS is not known, and it apparently is due to a respiratory center that just stops sending out the impulses that cause us to breath.

Although the respiratory center automatically controls the rate and depth of breathing, its activity can be influenced by nervous input. We can voluntarily change our breathing pattern to accommodate activities such as speaking, singing, eating, swimming under water, and so forth. Following forced inspiration, stretch receptors in the alveolar walls initiate inhibitory nerve impulses that travel from the inflated lungs to the respiratory center. This temporarily stops the respiratory center from sending out nerve impulses.

Chemical Control of Breathing

As you know, working cells produce carbon dioxide, which enters the blood. There, carbon dioxide combines with water, forming an acid, which breaks down and gives off hydrogen ions (H^+). These hydrogen ions can change the pH of the blood. **Chemoreceptors** are sensory receptors in the body that are sensitive to chemical composition of body fluids. Two sets of chemoreceptors that are sensitive to pH can cause breathing to speed up. A centrally placed set is located in the medulla oblongata of the brain stem, and a peripherally placed set is in the circulatory system. Carotid bodies, located in the carotid arteries, and aortic bodies, located in the aorta, are sensitive to blood pH. These chemoreceptors are not affected directly by low oxygen (O_2) levels; instead, they are stimulated when the carbon dioxide entering the blood is sufficient to change blood pH.

When the pH of the blood decreases, the respiratory center increases the rate and depth of breathing. With an increased breathing rate, more carbon dioxide is removed from the blood, and the hydrogen ion concentration returns to normal, and the breathing rate returns to normal.

Recall that Olivia, in a fit of rage, tried to hold her breath, but soon gave up. Most people are unable to hold their breath for more than one minute, because when you hold your breath, metabolically produced carbon dioxide begins accumulating in the blood. The respiratory center, stimulated by the chemoreceptors, is able to override a person's voluntary inhibition of respiration, and breathing resumes, despite attempts to prevent it. Heather didn't have to call 911 to get help.

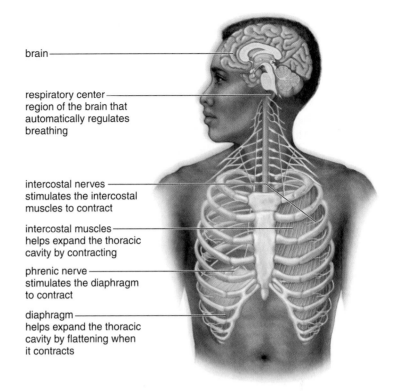

brain

respiratory center
region of the brain that
automatically regulates
breathing

intercostal nerves
stimulates the intercostal
muscles to contract

intercostal muscles
helps expand the thoracic
cavity by contracting

phrenic nerve
stimulates the diaphragm
to contract

diaphragm
helps expand the thoracic
cavity by flattening when
it contracts

Figure 9.10   Nervous control of breathing.
During inspiration, the respiratory center, located in the medulla oblongata, stimulates the external intercostal (rib) muscles to contract via the intercostal nerves and stimulates the diaphragm to contract via the phrenic nerve. The thoracic cavity, and then the lungs, expand and air comes rushing in. Expiration occurs due to a lack of stimulation from the respiratory center to the diaphragm and intercostal muscles. Now, as the thoracic cavity, and then the lungs, resume their original size, air is pushed out.

✅ Check Your Progress 9.5

1. Explain why we automatically breath 12–20 times a minute.
2. Explain why Olivia was unable to hold her breath more than about a minute, so Heather did not have to call 911.

9.6 Gas Exchanges in the Body

Gas exchange is critical to homeostasis. The act of breathing brings oxygen in air to the lungs and carbon dioxide from the lungs to outside the body. As mentioned previously, respiration includes not only the exchange of gases in the lungs, but also the exchange of gases in the tissues (Fig. 9.11).

The principles of diffusion, alone, govern whether O_2 or CO_2 enters or leaves the blood in the lungs and in the tissues. Gases exert pressure, and the amount of pressure each gas exerts is called its partial pressure, symbolized as P_{O_2} and P_{CO_2}. If the partial pressure of oxygen differs across a membrane, oxygen will diffuse from the higher to lower partial pressure.

External Respiration

External respiration refers to the exchange of gases between air in the alveoli and blood in the pulmonary capillaries (see Fig. 9.6). Blood in the pulmonary capillaries has a higher P_{CO_2} than atmospheric air. Therefore, *CO_2 diffuses out of the plasma into the lungs*. Most of the CO_2 is carried in plasma as **bicarbonate ions** (HCO_3^-). In the low P_{CO_2} environment of the lungs, the reaction proceeds to the right:

$$\underset{\substack{\text{hydrogen}\\\text{ion}}}{H^+} + \underset{\substack{\text{bicarbonate}\\\text{ion}}}{HCO_3^-} \longrightarrow \underset{\substack{\text{carbonic}\\\text{acid}}}{H_2CO_3} \overset{\substack{\text{carbonic}\\\text{anhydrase}}}{\longrightarrow} \underset{\text{water}}{H_2O} + \underset{\substack{\text{carbon}\\\text{dioxide}}}{CO_2}$$

The enzyme **carbonic anhydrase** speeds the breakdown of carbonic acid (H_2CO_3) in red blood cells.

What happens if you hyperventilate (breathe at a high rate), and therefore push this reaction far to the right? The blood will have fewer hydrogen ions, and alkalosis, a high blood pH, results. In that case, breathing will be inhibited, and you may suffer various symptoms from dizziness to tetanic contractions of the muscles. What happens if you hypoventilate (breathe at a low rate) and this reaction does not occur? Hydrogen ions build up in the blood, and acidosis will occur. Buffers may compensate for the low pH, and breathing will most likely increase. Otherwise, you may become comatose and die.

The pressure pattern for O_2 during external respiration is the reverse of that for CO_2. Blood in the pulmonary capillaries is low in oxygen, and alveolar air contains a higher partial pressure of oxygen. Therefore, *O_2 diffuses into plasma and then into red blood cells in the lungs*. Hemoglobin takes up this oxygen and becomes **oxyhemoglobin** (HbO_2):

$$\underset{\text{deoxyhemoglobin}}{Hb} + \underset{\text{oxygen}}{O_2} \longrightarrow \underset{\text{oxyhemoglobin}}{HbO_2}$$

Internal Respiration

Internal respiration refers to the exchange of gases between the blood in systemic capillaries and the tissue cells. In Figure 9.11, internal respiration is shown at the bottom. Blood entering systemic capillaries is a bright red color because red blood cells contain oxyhemoglobin. Because the temperature in the tissues is higher and the pH is slightly lower, oxyhemoglobin naturally gives up oxygen. After oxyhemoglobin gives up O_2, it diffuses out of the blood into the tissues:

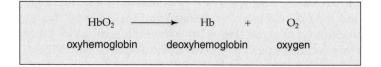

$$\underset{\text{oxyhemoglobin}}{HbO_2} \longrightarrow \underset{\text{deoxyhemoglobin}}{Hb} + \underset{\text{oxygen}}{O_2}$$

Oxygen diffuses out of the blood into the tissues because the P_{O_2} of tissue fluid is lower than that of blood. The lower P_{O_2} is due to cells continuously using up oxygen in cellular respiration. *Carbon dioxide diffuses into the blood from the tissues* because the P_{CO_2} of tissue fluid is higher than that of blood. Carbon dioxide, produced continuously by cells, collects in tissue fluid.

After CO_2 diffuses into the blood, most enters the red blood cells, where a small amount is taken up by hemoglobin, forming **carbaminohemoglobin** ($HbCO_2$). In plasma, CO_2 combines with water, forming carbonic acid (H_2CO_3), which dissociates to hydrogen ions (H^+) and bicarbonate ions (HCO_3^-):

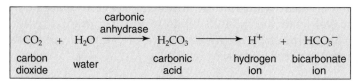

$$\underset{\substack{\text{carbon}\\\text{dioxide}}}{CO_2} + \underset{\text{water}}{H_2O} \overset{\substack{\text{carbonic}\\\text{anhydrase}}}{\longrightarrow} \underset{\substack{\text{carbonic}\\\text{acid}}}{H_2CO_3} \longrightarrow \underset{\substack{\text{hydrogen}\\\text{ion}}}{H^+} + \underset{\substack{\text{bicarbonate}\\\text{ion}}}{HCO_3^-}$$

The enzyme carbonic anhydrase, mentioned previously, speeds the reaction in red blood cells. Bicarbonate ions (HCO_3^-) diffuse out of red blood cells and are carried in the plasma. The globin portion of hemoglobin combines with excess hydrogen ions produced by the overall reaction, and Hb becomes HHb, called **reduced hemoglobin.** In this way, the pH of blood remains fairly constant. Blood that leaves the systemic capillaries is a dark maroon color because red blood cells contain reduced hemoglobin.

☑ Check Your Progress 9.6

1. What are the differences between external respiration and internal respiration?

2. a. What respiratory pigment is found in human red blood cells?
   b. How does it function in the transport of oxygen; carbon dioxide?

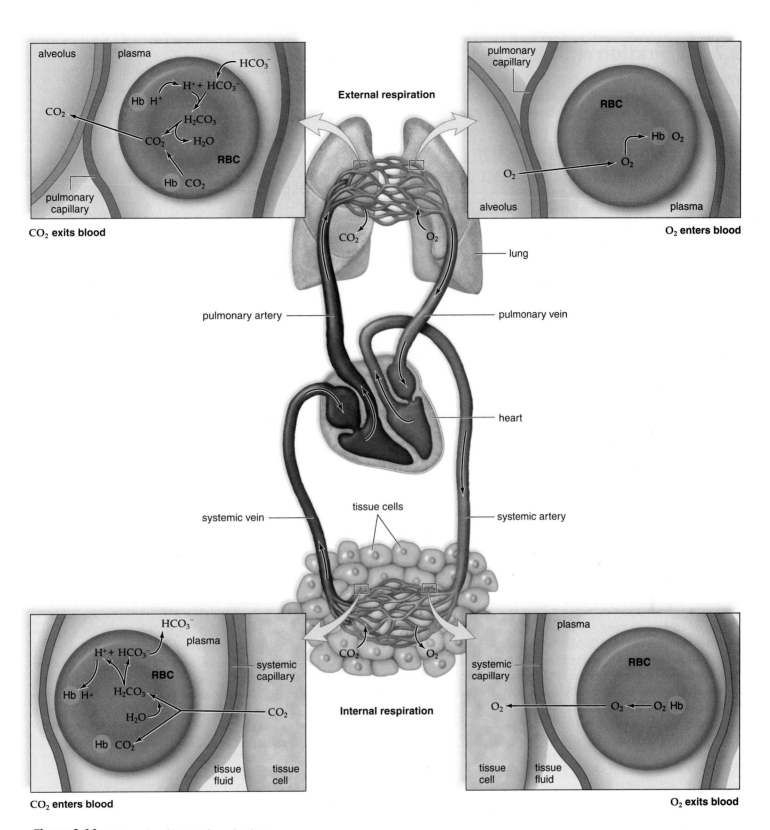

Figure 9.11 **External and internal respiration.**

During external respiration in the lungs, HCO_3^- is converted to CO_2, which exits the blood. O_2 enters the blood and hemoglobin (Hb) carries O_2 to the tissues. During internal respiration in the tissues, O_2 exits the blood, and CO_2 enters the blood. Most of the CO_2 enters red blood cells, where it becomes the bicarbonate ion, which is carried in the plasma. Some hemoglobin combines with CO_2 and some combines with H^+.

9.7 Respiration and Health

The respiratory tract is constantly exposed to environmental air. The quality of this air, and whether it contains infectious pathogens such as bacteria and viruses, can affect our health.

Upper Respiratory Tract Infections

Upper respiratory infections (URI) can spread from the nasal cavities to the sinuses, middle ears, and larynx. What we call "strep throat" is a primary bacterial infection caused by *Streptococcus pyogenes* that can lead to a generalized URI and even a systemic (affecting the body as a whole) infection. Although antibiotics have no effect on viral infections, they are successfully used to treat most bacterial infections, including strep throat. The symptoms of strep throat are severe sore throat, high fever, and white patches on a dark red throat.

Sinusitis

Sinusitis develops when nasal congestion blocks the tiny openings leading to the sinuses. Symptoms include postnasal discharge, as well as facial pain that worsens when the patient bends forward. Pain and tenderness usually occur over the lower forehead or over the cheeks. If the latter, toothache is also a complaint. Successful treatment depends on restoring proper drainage of the sinuses. Even a hot shower and sleeping upright can be helpful. Otherwise, spray decongestants are preferred over oral antihistamines, which thicken rather than liquefy the material trapped in the sinuses.

Otitis Media

Otitis media is an infection of the middle ear. This infection is considered here because it is a complication often seen in children who have a nasal infection. Infection can spread by way of the auditory tube from the nasopharynx to the middle ear. Pain is the primary symptom of a middle ear infection. A sense of fullness, hearing loss, vertigo (dizziness), and fever may also be present. Antibiotics are prescribed if necessary, but physicians are aware today that overuse of antibiotics can lead to resistance of bacteria to antibiotics. Tubes (called tympanostomy tubes) are sometimes placed in the eardrums of children with multiple recurrences to help prevent the buildup of pressure in the middle ear and the possibility of hearing loss. Normally, the tubes fall out with time.

Tonsillitis

Tonsillitis occurs when the **tonsils** become inflamed and enlarged. The tonsils in the posterior wall of the nasopharynx are often called adenoids. If tonsillitis occurs frequently and enlargement makes breathing difficult, the tonsils can be removed surgically in a **tonsillectomy.** Fewer tonsillectomies are performed today than in the past because we now know that the tonsils trap many of the pathogens that enter the pharynx; therefore, they are a first line of defense against invasion of the body.

Laryngitis

Laryngitis is an infection of the larynx with accompanying hoarseness, leading to the inability to talk in an audible voice. Usually, laryngitis disappears with treatment of the URI. Persistent hoarseness without the presence of a URI is one of the warning signs of cancer and therefore should be looked into by a physician.

Lower Respiratory Tract Disorders

Lower respiratory tract disorders include infections, restrictive pulmonary disorders, obstructive pulmonary disorders, and lung cancer.

Lower Respiratory Infections

Acute bronchitis is an infection of the primary and secondary bronchi. Usually, it is preceded by a viral URI that has led to a secondary bacterial infection. Most likely, a nonproductive cough has become a deep cough that expectorates mucus and perhaps pus.

Pneumonia is a viral or bacterial infection of the lungs in which the bronchi and alveoli fill with thick fluid (Fig. 9.12). Most often, it is preceded by influenza. High fever and chills, with headache and chest pain, are symptoms of pneumonia. Rather than being a generalized lung infection, pneumonia may be localized in specific lobules of the lungs; obviously, the more lobules involved, the more serious is the infection. Pneumonia can be caused by a bacterium that is usually held in check but has gained the upper hand due to stress and/or reduced immunity. AIDS patients are subject to a particularly rare form of pneumonia caused by the protozoan *Pneumocystis jiroveci* (formerly *Pneumocystis carinii*). Pneumonia of this type is almost never seen in individuals with a healthy immune system.

Pulmonary tuberculosis is caused by a bacterium. When the bacteria *(Mycobacterium tuberculosis)* invade the lung tissue, the cells build a protective capsule around the foreigners, isolating them from the rest of the body. This tiny capsule is called a tubercle. If the resistance of the body is high, the imprisoned organisms die, but if the resistance is low, the organisms eventually can be liberated. If a chest X-ray detects active tubercles, the individual is put on appropriate drug therapy to ensure the localization of the disease and the eventual destruction of any live bacteria. It is possible to tell if a person has ever been exposed to tuberculosis with a test in which a highly diluted extract of the bacillus is injected into the skin of the patient. A person who has never been in contact with *M. tuberculosis* shows no reaction, but one who has had or is fighting an infection shows an area of inflammation that peaks in about 48 hours.

Restrictive Pulmonary Disorders

In restrictive pulmonary disorders, vital capacity is reduced because the lungs have lost their elasticity. Inhaling particles such as silica (sand), coal dust, asbestos, and, now it

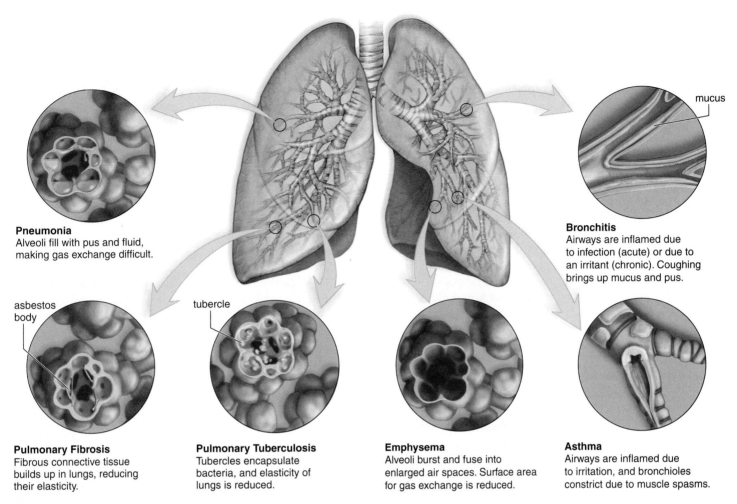

Pneumonia
Alveoli fill with pus and fluid,
making gas exchange difficult.

mucus

Bronchitis
Airways are inflamed due
to infection (acute) or due to
an irritant (chronic). Coughing
brings up mucus and pus.

asbestos
body

tubercle

Pulmonary Fibrosis
Fibrous connective tissue
builds up in lungs, reducing
their elasticity.

Pulmonary Tuberculosis
Tubercles encapsulate
bacteria, and elasticity of
lungs is reduced.

Emphysema
Alveoli burst and fuse into
enlarged air spaces. Surface area
for gas exchange is reduced.

Asthma
Airways are inflamed due
to irritation, and bronchioles
constrict due to muscle spasms.

Figure 9.12   **Common bronchial and pulmonary diseases.**
Exposure to infectious pathogens and/or polluted air, including tobacco smoke, causes the diseases and disorders shown here.

seems, fiberglass can lead to **pulmonary fibrosis,** a condition in which fibrous connective tissue builds up in the lungs. The lungs cannot inflate properly and are always tending toward deflation. Breathing asbestos is also associated with the development of cancer. Because asbestos was formerly used widely as a fireproofing and insulating agent, unwarranted exposure has occurred. It has been projected that two million deaths caused by asbestos exposure—mostly in the workplace—will occur in the United States between 1990 and 2020.

Obstructive Pulmonary Disorders

In obstructive pulmonary disorders, air does not flow freely in the airways, and the time it takes to inhale or exhale maximally is greatly increased. Several disorders, including chronic bronchitis, emphysema, and asthma, are collectively referred to as chronic obstructive pulmonary disease (COPD) because they tend to recur.

In **chronic bronchitis,** the airways are inflamed and filled with mucus. A cough that brings up mucus is common. The bronchi have undergone degenerative changes, including the loss of cilia and their normal cleansing action. Under these conditions, an infection is more likely to occur. Smoking is the most frequent cause of chronic bronchitis. Exposure to other pollutants can also cause chronic bronchitis.

Emphysema is a chronic and incurable disorder in which the alveoli are distended and their walls damaged, so that the surface area available for gas exchange is reduced. Emphysema, which is most often caused by smoking, is often preceded by chronic bronchitis. Air trapped in the lungs leads to alveolar damage and a noticeable ballooning of the chest. The elastic recoil of the lungs is reduced, so not only are the airways narrowed, but the driving force behind expiration is also reduced. The victim is breathless and may have a cough. Because the surface area for gas exchange is reduced, less oxygen reaches the heart and the brain. Even so, the heart works furiously to force more blood through the lungs, and an increased workload on the heart can result. Lack of oxygen to the brain can make the person feel depressed, sluggish, and irritable. Exercise, drug therapy, supplemental oxygen, and giving up smoking may relieve the symptoms and possibly slow the progression of emphysema.

Asthma is a disease of the bronchi and bronchioles that is marked by wheezing, breathlessness, and sometimes a cough and expectoration of mucus. The airways are unusually sensitive to specific irritants, which can include a wide range of allergens such as pollen, animal dander, dust, tobacco smoke, and industrial fumes. Even cold air can be an irritant. When exposed to the irritant, the smooth muscle in the bronchioles undergoes spasms. It now appears that chemical mediators given off by immune cells in the bronchioles cause the spasms. Most asthma patients have some degree of bronchial inflammation that further reduces the diameter of the airways and contributes to the seriousness of an attack. Asthma is not curable, but it is treatable. Special inhalers can control the inflammation and hopefully prevent an attack, while other types of inhalers can stop the muscle spasms should an attack occur.

Lung Cancer

Lung cancer is more prevalent in men than in women, but recently lung cancer has surpassed breast cancer as a cause of death in women. The recent increase in the incidence of lung cancer in women is directly correlated to increased numbers of women who smoke. Autopsies on smokers have revealed the progressive steps by which the most common form of lung cancer develops. The first event appears to be thickening and callusing of the cells lining the bronchi. (Callusing occurs whenever cells are exposed to irritants.) Then cilia are lost, making it impossible to prevent dust and dirt from settling in the lungs. Following this, cells with atypical nuclei appear in the callused lining. A tumor consisting of disordered cells with atypical nuclei is considered cancer in situ (at one location). A normal lung versus a lung with cancerous tumors is shown in Figure 9.13. A final step occurs when some of these cells break loose and penetrate other tissues, a process called metastasis. Now the cancer has spread. The original tumor may grow until a bronchus is blocked, cutting off the supply of air to that lung. The entire lung then collapses, the secretions trapped in the lung spaces become infected, and pneumonia or a lung abscess (localized area of pus) results. The only treatment that offers a possibility of cure is to remove a lobe or the whole lung before metastasis has had time to occur. This operation is called **pneumonectomy.** If the cancer has spread, chemotherapy and radiation are also required.

The Health Focus on page 183 lists the various illnesses, including cancer, that are apt to occur when a person smokes. Current research indicates that passive smoking (second-hand smoke)—exposure to smoke related by others who are smoking—can also cause lung cancer and other illnesses associated with smoking. If a person stops voluntary smoking and avoids passive smoking, and if the body tissues are not already cancerous, the lungs may return to normal over time.

☑ Check Your Progress 9.7

1. **a.** What are some common respiratory infections and disorders of the upper respiratory tract? **b.** Of the lower respiratory tract?

2. What are three respiratory disorders commonly associated with smoking tobacco?

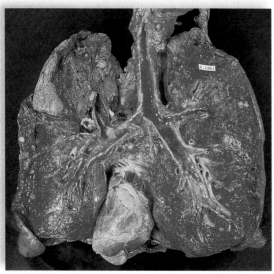

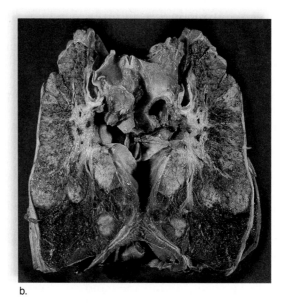

Figure 9.13
Normal lung versus cancerous lung.
a. Normal lung with heart in place. Note the healthy red color. **b.** Lungs of a heavy smoker. Notice how black the lungs are except where cancerous tumors have formed.

a.   b.

Health Focus

Questions About Smoking, Tobacco, and Health

Is there a safe way to smoke?

No. All cigarettes can damage the human body. Any amount of smoke is dangerous. Cigarettes are perhaps the only legal product whose advertised and intended use—smoking—is harmful to the body and causes cancer.

Is cigarette smoking really addictive?

Yes. The nicotine in cigarette smoke causes addiction to smoking. Nicotine is an addictive drug (just like heroin and cocaine) for three main reasons: small amounts make the smoker want to smoke more; smokers usually suffer withdrawal symptoms when they stop; and nicotine can affect the mood and nature of the smoker. The younger a person is when he or she begins to smoke, the more likely he or she is to develop an addiction to nicotine.

Does smoking cause cancer?

Yes. Tobacco use accounts for about one-third of all cancer deaths in the United States. Smoking causes almost 90% of lung cancers. Smoking also causes cancers of the larynx (voice box), oral cavity, pharynx (throat), and esophagus; and contributes to the development of cancers of the bladder, pancreas, cervix, kidney, and stomach. Smoking is also linked to the development of some leukemias.

How does cigarette smoke affect the lungs?

All cigarette smokers have a lower level of lung function than nonsmokers. Cigarette smoking causes several lung diseases that can be just as dangerous as lung cancer. For example, smoking is a major cause of chronic bronchitis in which the airways produce excess mucus, forcing the smoker to cough more; emphysema, a disease that slowly destroys a person's ability to breathe; and chronic obstructive pulmonary disease, the name used to describe both chronic bronchitis and emphysema.

Why do smokers have "smoker's cough?"

Cigarette smoke contains chemicals that irritate the air passages and lungs. When a smoker inhales these substances, the body tries to protect itself by producing mucus and coughing. Normally, cilia (tiny hair-like formations that line the airways) beat outwards and "sweep" harmful material out of the lungs. Smoke, however, decreases this sweeping action, so some of the poisons in the smoke remain in the lungs.

If you smoke but do not inhale, is there any danger?

Yes. Wherever smoke touches living cells, it does harm. Even if smokers don't inhale, they are breathing the smoke as secondhand smoke and are still at risk for lung cancer. Pipe and cigar smokers, who often do not inhale, are at an increased risk for lip, mouth, tongue, and several other cancers.

Does cigarette smoking affect the heart?

Yes. Smoking increases the risk of heart disease, which is the number one cause of death in the United States. Smoking, high blood pressure, high cholesterol, physical inactivity, obesity, and diabetes are all risk factors for heart disease, but cigarette smoking is the biggest risk factor for sudden heart death. Smokers who have a heart attack are more likely to die within an hour of the heart attack than nonsmokers. Cigarette smoke can cause harm to the heart at very low levels, much lower than what causes lung disease.

How does smoking affect pregnant women and their babies?

Smoking during pregnancy is linked with a greater chance of miscarriage, premature delivery, stillbirth, infant death, low birth weight, and sudden infant death syndrome (SIDS). Up to 10% of infant deaths would be prevented if pregnant women did not smoke. When a pregnant woman smokes, she really is smoking for two because the nicotine, carbon monoxide, and other dangerous chemicals in smoke enter her bloodstream and then pass into the baby's body, preventing the baby from getting essential nutrients and oxygen for growth.

What are some of the short-term and long-term effects of smoking cigarettes?

Short-term effects include shortness of breath and nagging coughs, diminished ability to smell and taste, premature aging of the skin, and increased risk of sexual impotence in men. Smokers tend to tire easily during physical activity. Long-term effects include many types of cancer, heart disease, aneurysms, bronchitis, emphysema, and stroke. Smoking contributes to the severity of pneumonia and asthma.

What are the dangers of environmental tobacco smoke (ETS)?

ETS causes about 3,000 lung cancer deaths and about 35,000 to 40,000 deaths from heart disease each year in healthy nonsmokers. Children whose parents smoke are more likely to suffer from asthma, pneumonia or bronchitis, ear infections, coughing, wheezing, and increased mucus production in the first two years of life than children who come from smoke-free households.

Are chewing tobacco and snuff safe alternatives to cigarette smoking?

No. The juice from smokeless tobacco is absorbed directly through the lining of the mouth. This creates sores and white patches that often lead to cancer of the mouth. Smokeless tobacco users greatly increase their risk of other cancers, including those of the pharynx (throat). Other effects of smokeless tobacco include harm to teeth and gums.

Bans on Smoking

In 1964, the surgeon general of the United States made it known to the general public that smoking was hazardous to our health, and thereafter, a health warning was placed on packs of cigarettes. At that time, 40.4% of adults smoked, but by 1990, only about 26% of adults smoked. In the meantime, however, the public became aware that passive smoking—that is, just being in the vicinity of someone who is smoking—can also lead to cancer and other health problems. By now, many state and local governments have passed legislation that bans smoking in public places such as restaurants, elevators, public meeting rooms, and in the workplace.

Is legislation that restricts the freedom to smoke ethical? Or is such legislation akin to racism and creating a population of second-class citizens who are segregated from the majority on the basis of a habit? Are the desires of nonsmokers being allowed to infringe on the rights of smokers? Or is this legislation one way to help smokers become nonsmokers? One study showed that workplace bans on smoking reduce the daily consumption of cigarettes among smokers by 10%.

Is legislation that disallows smoking in family-style restaurants fair, especially if bars and restaurants associated with casinos are not included in the ban on smoking? The selling of tobacco and even the increased need for health care it generates helps the economy. One smoker writes, "Smoking causes people to drink more, eat more, and leave larger tips. Smoking also powers the economy of Wall Street." Is this a reason to allow smoking to continue? Or should we simply require all places of business to put in improved air filtration systems? Would that do away with the dangers of passive smoking?

Does legislation that bans smoking in certain areas represent government invasion of our privacy? If yes, is reducing the chance of cancer a good enough reason to allow the government to invade our privacy? Some people are prone to cancer more than others. Should we all be regulated by the same legislation? Are we our brother's keeper, meaning that we have to look out for one another?

Decide Your Opinion

1. Is legislation that bans smoking in public places creating a group of second-class citizens whose rights are being denied?
2. Should we be concerned about passing and following legislation that possibly puts a damper on the economy, even if it does improve the health of people?
3. Are bans on smoking an invasion of our privacy? If so, is prevention of cancer in certain persons a good enough reason to risk a possible invasion of our privacy?

Summarizing the Concepts

9.1 The Respiratory System

The respiratory tract consists of the nose, the pharynx, the larynx, the trachea, the bronchi, the bronchioles, and the lungs.

9.2 The Upper Respiratory Tract

Air from the nose enters the pharynx and passes through the glottis into the larynx:

- nose: filters and warms the air
- pharynx: air and food passageways cross
- larynx: the voice box that houses the vocal cords

9.3 The Lower Respiratory Tract

- The trachea (windpipe) consists of goblet cells and ciliated cells.
- The bronchi, along with the pulmonary arteries and veins, enter the lungs.
- The lungs consist of the alveoli, air sacs surrounded by a capillary network.

9.4 Mechanism of Breathing

Breathing involves inspiration and expiration of air.

Inspiration

The diaphragm lowers, and the rib cage moves upward and outward; the lungs expand, and air rushes in.

Expiration

The diaphragm relaxes and moves up. The rib cage moves down and in; pressure in the lungs increases; air is pushed out of the lungs.

Respiratory volumes can be measured:

- Tidal volume: the amount of air that normally enters and exits with each breath.
- Vital capacity: the amount of air that moves in plus the amount that moves out with maximum effort.
- Inspiratory reserve volume and the expiratory reserve volume: the difference between normal amounts and the maximum effort amounts of air moved.
- Residual volume: the amount of air that stays in the lungs when we breathe.

9.5 Control of Ventilation

The respiratory center in the brain automatically causes us to breath 12–20 times a minute. Extra carbon dioxide in the blood can decrease the pH, and if so, chemoreceptors alert the respiratory center, which increases the rate of breathing.

9.6 Gas Exchanges in the Body

Both external and internal respiration depend on diffusion. Hemoglobin activity is essential to the transport of gases and, therefore, to external and internal respiration.

External Respiration

- CO_2 diffuses out of plasma into lungs.
- O_2 diffuses into the plasma and then into red blood cells in the capillaries. O_2 is carried by hemoglobin.

Internal Respiration
- O_2 diffuses out of the blood into the tissues.
- CO_2 diffuses into the blood from the tissues. CO_2 is carried in the plasma as the bicarbonate ion (HCO_3^-).

9.7 Respiration and Health

A number of illnesses are associated with the respiratory tract.

Upper Respiratory Tract Infections
- Infections of the nasal cavities, sinuses, throat, tonsils, and larynx are all upper respiratory tract infections.
- These include sinusitis, otitis media, tonsillitis, and laryngitis.

Lower Respiratory Tract Disorders
- Lower respiratory tract infections include acute bronchitis, pneumonia, and pulmonary tuberculosis.
- Restrictive pulmonary disorders are exemplified by pulmonary fibrosis.
- Obstructive pulmonary disorders are exemplified by chronic bronchitis, emphysema, and asthma.
- Smoking can eventually lead to lung cancer.

Understanding Key Terms

acute bronchitis 180	lung cancer 182
alveolus 173	lungs 173
asthma 182	nasal cavity 171
auditory (Eustachian) tube 171	otitis media 180
bicarbonate ion 178	oxyhemoglobin 178
bronchiole 173	pharynx 171
bronchus 173	pneumonectomy 182
carbaminohemoglobin 178	pneumonia 180
carbonic anhydrase 178	pulmonary fibrosis 181
chemoreceptor 177	pulmonary tuberculosis 180
chronic bronchitis 181	reduced hemoglobin 178
cough 171	residual volume 176
dead air space 176	respiratory control center 177
emphysema 181	sinusitis 180
epiglottis 172	sudden infant death syndrome
expiration 170, 174	(SIDS) 177
expiratory reserve	surfactant 174
volume 176	tidal volume 176
external respiration 178	tonsil 180
glottis 172	tonsillectomy 180
infant respiratory distress	tonsillitis 180
syndrome 174	trachea 173
inspiration 170, 174	tracheostomy 173
inspiratory reserve volume 176	ventilation 170
internal respiration 178	vital capacity 176
laryngitis 180	vocal cord 172
larynx 172	

Match the key terms to these definitions.

a. _____ Common passageway for both food intake and air movement, located between the mouth and the esophagus.

b. _____ Chemical in the lungs that reduces the surface tension of water to keep the alveoli from collapsing.

c. _____ Fold of tissue across the glottis within the larynx; creates vocal sounds when it vibrates.

d. _____ Form in which most of the carbon dioxide is transported in the bloodstream.

e. _____ Stage during breathing when air is pushed out of the lungs.

Testing Your Knowledge of the Concepts

1. Name as many structures as you can that connect to the pharynx. (page 171)

2. How is the structure of the trachea important for respiration, as well as digestion? (page 173)

3. Describe the structure of an alveolus, and explain how it is suited for gas exchange. (pages 173–74)

4. What are the steps of inspiration and expiration? How is breathing controlled? (pages 174–75, 177)

5. Using Figure 9.9, list and define the volumes and capacities of air movement. (page 176)

6. Describe what occurs during external respiration, utilizing two important equations. What is the driving force for the gas exchange? (pages 178–79)

7. Describe what occurs during internal respiration, utilizing two important equations. What is the driving force for the gas exchange? (pages 178–79)

8. Describe several upper and lower respiratory tract disorders (other than cancer). If appropriate, explain why breathing is difficult with these conditions. (pages 180–82)

9. List the steps by which lung cancer develops. (page 182)

10. Which of these is anatomically incorrect?
    a. The nose has two nasal cavities.
    b. The pharynx connects the nasal and oral cavities to the larynx.
    c. The larynx contains the vocal cords.
    d. The trachea enters the lungs.
    e. The lungs contain many alveoli.

11. How is inhaled air modified before it reaches the lungs?
    a. It must be humidified.   c. It must be filtered.
    b. It must be warmed.   d. All of these are correct.

12. What is the name of the structure that prevents food from entering the trachea?
    a. glottis   c. epiglottis
    b. septum   d. Adam's apple

In questions 13–17, match each description with a structure in the key.

Key:
   a. pharynx   d. trachea
   b. glottis   e. bronchi
   c. larynx   f. bronchioles

13. Branched tubes that lead from bronchi to the alveoli

14. Reinforced tube that connects larynx with bronchi

15. Chamber behind oral cavity and between nasal cavity and larynx

16. Opening into larynx

17. Divisions of the trachea that enter lungs

18. Which of these is incorrect concerning inspiration?
    a. Rib cage moves up and out.
    b. Diaphragm contracts and moves down.
    c. Pressure in lungs decreases, and air comes rushing in.
    d. The lungs expand because air comes rushing in.

19. Air enters the human lungs because
    a. atmospheric pressure is lower than the pressure inside the lungs.
    b. atmospheric pressure is greater than the pressure inside the lungs.
    c. although the pressures are the same inside and outside, the partial pressure of oxygen is lower within the lungs.
    d. the residual air in the lungs causes the partial pressure of oxygen to be lower than it is outside.

20. The maximum volume of air that can be moved in and out during a single breath is called the
    a. expiratory and inspiratory reserve volume.
    b. residual volume.
    c. tidal volume.
    d. vital capacity.
    e. functional residual capacity.

21. If air enters the intrapleural space (the space between the pleura),
    a. a lobe of the lung can collapse.
    b. the lungs could swell and burst.
    c. the diaphragm will contract.
    d. nothing will happen because air is needed in the intrapleural space.

22. Internal respiration refers to
    a. the exchange of gases between alveolar air and the blood in the lungs.
    b. the movement of air into the lungs.
    c. the exchange of gases between the blood and tissue fluid.
    d. cellular respiration, resulting in the production of ATP.

23. The chemical reaction that converts carbon dioxide to a bicarbonate ion takes place in
    a. the blood plasma.
    b. red blood cells.
    c. the alveolus.
    d. the hemoglobin molecule.

24. The enzyme carbonic anhydrase
    a. causes the blood to be more basic in the tissues.
    b. speeds up the conversion of carbonic acid to carbon dioxide and water, and the reverse.
    c. actively transports carbon dioxide out of capillaries.
    d. is active only at high altitudes.
    e. All of these are correct.

25. Hemoglobin assists transport of gases by
    a. combining with oxygen.
    b. combining with CO_2.
    c. combining with H^+.
    d. being present in red blood cells.
    e. All of these are correct.

26. In humans, the respiratory center
    a. is stimulated by carbon dioxide.
    b. is located in the medulla oblongata.
    c. controls the rate of breathing.
    d. All of these are correct.

27. Which of the following is not true of obstructive pulmonary disorders?
    a. Air does not flow freely in the airways.
    b. Vital capacity is reduced due to loss of lung elasticity.
    c. Disorders may include chronic bronchitis, emphysema, and asthma.
    d. Ventilation takes longer to occur.

28. Label this diagram of the human respiratory tract.

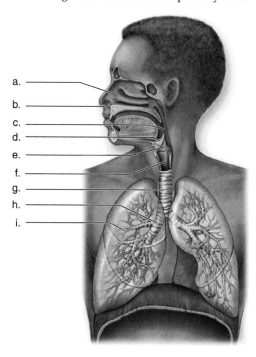

a. _____
b. _____
c. _____
d. _____
e. _____
f. _____
g. _____
h. _____
i. _____

Thinking Critically About the Concepts

Heather was fortunate that, although Olivia, in the opening story, choked on her cereal, she was able to cough it up. Sometimes when people swallow a (large) piece of food, it enters the larynx and blocks the trachea, preventing them from breathing. If they are to be saved, the Heimlich maneuver must be done to dislodge the food.

1. Explain the expression, "the food went the wrong way," by referring to structures along the path of air?

2. Why couldn't Olivia hold her breath for an extended period of time?

3. Long-term smokers often develop a chronic cough.
   a. What purpose does the smokers' cough serve?
   b. Why are nonsmokers less likely to develop a chronic cough?

4. Professional singers who need to hold a long note while singing must exert a great deal of control over their breathing. What muscle(s) needs/need conditioning to acquire better control over their breathing?

5. Someone who has been breathing rapidly (called hyperventilation) may stop breathing (called apnea). Why would you have this person breathe into a paper bag (or some type of closed container)?

CHAPTER 10

Urinary System and Excretion

Kathy sighed as she contemplated the assignment her biology professor had just given the class. "Where do you think she comes up with stuff?" she asked Brad who was sitting next to her. "Who knows," he groaned.

Their assignment was to watch TV and identify at least three commercials that pertain to the urinary system. They also needed at least one commercial with references to an organ system that works with the urinary system to achieve homeostasis.

Later that day, Kathy made a point of watching TV, just to study the commercials she so often ignored. After just a few minutes of commercial watching, Kathy heard a catchy phrase, "gotta go, gotta go, gotta go right now…" in a commercial about a woman with an overactive bladder. "Bingo," she thought, "one down, two to go." About an hour later, a commercial referred to a man with a "growing problem," when actually he was having trouble urinating.

When she called Brad later to find out what he'd seen, he said he had seen only one commercial that counted. It was about a medication for patients undergoing chemotherapy. Their fatigue could be combated by the medication because it stimulates red blood cell production. "Good thing we just finished the endocrine system," he said. "Otherwise, I don't think I'd have connected that commercial with the urinary system."

As you study the material in this chapter, keep an eye out for the manner in which the three commercials mentioned are connected to the urinary system.

CHAPTER CONCEPTS

10.1 Urinary System
In the urinary system, kidneys produce urine, which is stored in the bladder before being discharged from the body. The kidneys are major organs of homeostasis.

10.2 Kidney Structure
Microscopically, the kidneys are composed of kidney tubules (nephrons), which have a blood supply that interacts with parts of the tubule as they produce urine.

10.3 Urine Formation
Urine is composed primarily of nitrogenous waste products and salts in water. Urine formation is a step-wise process.

10.4 Regulatory Functions of the Kidneys
The kidneys are involved in the salt-water balance and the acid-base balance of the blood, in addition to excreting nitrogenous wastes.

10.5 Disorders with Kidney Function
Various types of illnesses, including diabetes, kidney stones, and infections, can lead to renal failure, which necessitates undergoing hemodialysis.

10.6 Homeostasis
The kidneys work with other systems in the body to maintain the composition of the blood within normal limits.

10.1 Urinary System

The kidneys are the primary organs of excretion. **Excretion** is the removal of metabolic wastes from the body. People sometimes confuse the terms excretion and defecation, but they do not refer to the same process. Defecation, the elimination of feces from the body, is a function of the digestive system. Excretion, on the other hand, is the elimination of metabolic wastes, which are the products of metabolism. For example, the undigested food and bacteria that make up feces have never been a part of the functioning of the body, while the substances excreted in urine were once metabolites in the body.

Organs of the Urinary System

The urinary system consists of the kidneys, ureters, urinary bladder, and urethra (Fig. 10.1).

Kidneys

The **kidneys** are paired organs located near the small of the back, on either side of the vertebral column. They lie in depressions beneath the peritoneum, where they receive some protection from the lower rib cage. A sharp blow to the back can dislodge a kidney, which is then called a **floating kidney.**

The kidneys are bean-shaped and reddish brown in color. The fist-sized organs are covered by a tough capsule of fibrous connective tissue, called a renal capsule. Masses of adipose tissue adhere to each kidney. The concave side of a kidney has a depression, where a **renal artery** enters and a **renal vein** and a ureter exit the kidney.

Ureters

The **ureters** conduct urine from the kidneys to the bladder. They are small, muscular tubes about 25 cm long and 5 mm in diameter. The wall of a ureter has three layers: an inner mucosa (mucous membrane), a smooth muscle layer, and an outer fibrous coat of connective tissue. Peristaltic contractions cause urine to enter the bladder even if a person is lying down. Urine enters the bladder in spurts that occur at the rate of one to five per minute.

Urinary Bladder

The **urinary bladder** stores urine until it is expelled from the body. The bladder has three openings: two for the ureters and one for the urethra, which drains the bladder (Fig. 10.2).

The bladder wall is expandable because it contains a middle layer of circular fibers and two layers of longitudinal muscle. The epithelium of the mucosa becomes thinner, and folds in the mucosa called *rugae* disappear as the bladder enlarges.

The bladder has other features that allow it to retain urine. After urine enters the bladder from a ureter, small folds of bladder mucosa act like a valve to prevent backward flow. Two sphincters in close proximity are found where the urethra exits the bladder. The internal sphincter occurs around the opening to the urethra. It is composed of smooth muscle and is involuntarily controlled. An external sphincter is composed of skeletal muscle that can be voluntarily controlled.

When the urinary bladder fills to about 250 mL with urine, stretch receptors send sensory nerve impulses to the spinal cord. Subsequently, motor nerve impulses from the spinal cord cause the urinary bladder to contract and the sphincters to relax so that urination, also called **micturi-**

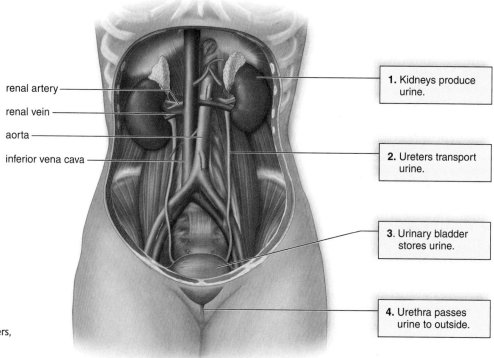

renal artery
renal vein
aorta
inferior vena cava

1. Kidneys produce urine.

2. Ureters transport urine.

3. Urinary bladder stores urine.

4. Urethra passes urine to outside.

Figure 10.1   The urinary system.
Urine is found only within the kidneys, the ureters, the urinary bladder, and the urethra.

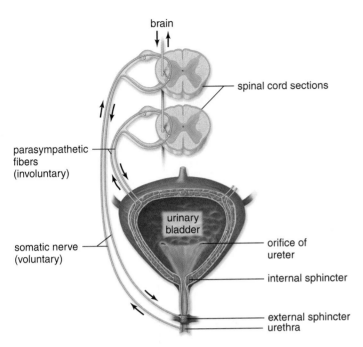

brain

spinal cord sections

parasympathetic
fibers
(involuntary)

urinary
bladder

orifice of
ureter

internal sphincter

somatic nerve
(voluntary)

external sphincter
urethra

Figure 10.2 Urination.
As the bladder fills with urine, sensory impulses go to the spinal cord and then
to the brain. The brain can override the urge to urinate. When urination occurs,
motor nerve impulses cause the bladder to contract and the sphincters to relax.

tion, is possible (Fig. 10.2). In the opening story, Kathy saw
a commercial for a drug that inhibits the ability of the nervous system to bring about contraction of the bladder.

Urethra

The **urethra** is a small tube that extends from the urinary
bladder to an external opening. Therefore, its function is to
remove urine from the body. The urethra has a different length
in females than in males. In females, the urethra is only about
4 cm long. The short length of the female urethra makes bacterial invasion easier. In males, the urethra averages 20 cm when
the penis is flaccid (limp, nonerect). As the urethra leaves the
male urinary bladder, it is encircled by the prostate gland. As
Kathy learned when doing her assignment, the prostate sometimes enlarges, restricting the flow of urine in the urethra. The
Health Focus on page 190 discusses this problem in men.

In females, the reproductive and urinary systems are not
connected. In males, the urethra carries urine during urination and sperm during ejaculation.

Functions of the Urinary System

As the kidneys produce urine, they carry out the following
four functions that contribute to homeostasis:

1. *Excretion of Metabolic Wastes*   The kidneys excrete
   metabolic wastes, notably nitrogenous wastes. Urea is
   the primary nitrogenous end product of metabolism
   in human beings, but humans also excrete some
   ammonium, creatinine, and uric acid.

Urea is a by-product of amino acid metabolism.
The breakdown of amino acids in the liver releases
ammonia, which the liver rapidly combines with
carbon dioxide to produce urea. Ammonia is very
toxic to cells, but urea is much less toxic.

Creatine phosphate is a high-energy phosphate
reserve molecule in muscles. The metabolic breakdown
of creatine phosphate results in **creatinine.**

The breakdown of nucleotides, such as those
containing adenine and thymine, produces **uric acid.**
Uric acid is rather insoluble. If too much uric acid is
present in blood, crystals form and precipitate out.
Crystals of uric acid sometimes collect in the joints,
producing a painful ailment called **gout.**

2. *Maintenance of Water-Salt Balance*   A principal function
   of the kidneys is to maintain the appropriate water-
   salt balance of the blood. As we shall see, blood
   volume is intimately associated with the salt balance
   of the body. As you know, salts, such as NaCl, have
   the ability to cause osmosis, the diffusion of water—in
   this case, into the blood. The more salts there are
   in the blood, the greater the blood volume and the
   greater the blood pressure. In this way, the kidneys
   are involved in regulating blood pressure.

   The kidneys also maintain the appropriate level of
   other ions, such as potassium ions (K^+), bicarbonate
   ions (HCO_3^-), and calcium ions (Ca^{2+}), in the blood.

3. *Maintenance of Acid-Base Balance*   The kidneys regulate
   the acid-base balance of the blood. In order for a
   person to remain healthy, the blood pH should be just
   about 7.4. The kidneys monitor and help control blood
   pH, mainly by excreting hydrogen ions (H^+) and
   reabsorbing the bicarbonate ions (HCO_3^-) as needed
   to keep blood pH at 7.4. Urine usually has a pH of 6
   or lower because our diet often contains acidic foods.

4. *Secretion of Hormones*   The kidneys assist the endocrine
   system in hormone secretion. The kidneys release
   renin, a substance that leads to the secretion of the
   hormone aldosterone from the adrenal glands, which
   lie atop the kidneys. As described in Section 10.4,
   aldosterone is involved in regulating the water-salt
   balance of the blood.

   One of the commercials Brad saw on TV triggered
   his memory that whenever the oxygen-carrying
   capacity of the blood is reduced (resulting in fatigue),
   the kidneys secrete the hormone **erythropoietin,**
   which stimulates red blood cell production.

   The kidneys also help activate vitamin D from the
   skin. Vitamin D is a molecule that promotes calcium
   (Ca^{2+}) absorption from the digestive tract.

✔ Check Your Progress 10.1

1. What are the organs of the urinary system, and what are their
   functions ?

2. In what four ways do the kidneys help maintain homeostasis?

Health Focus

Urinary Difficulties Due to an Enlarged Prostate

The prostate gland, which is part of the male reproductive system, surrounds the urethra at the point where the urethra leaves the urinary bladder (Fig. 10A). The prostate gland produces and adds a fluid to semen as semen passes through the urethra within the penis. At about age 50, the prostate gland often begins to enlarge, growing from the size of a walnut to that of a lime or even a lemon. This condition is called benign prostatic hyperplasia (BPH). As it enlarges, the prostate squeezes the urethra, causing urine to back up—first into the bladder, then into the ureters, and finally, perhaps, into the kidneys.

Treatment Emphasis Is on Early Detection

The treatment for BPH can involve (1) a more invasive procedure to reduce the size of the prostate or (2) taking a drug that is expected to shrink the prostate and/or improve urine flow. Prostate tissue can be destroyed by applying microwaves to a specific portion of the prostate. In many cases, however, a physician may decide that prostate tissue should be removed surgically. In some cases, rather than performing abdominal surgery, which requires an incision of the abdomen, the physician gains access to the prostate via the urethra. This operation, called transurethral resection of the prostate (TURP), requires careful consideration because one study found that the death rate during the five years following TURP is much higher than that following abdominal surgery.

Some drug treatments recognize that prostate enlargement is due to a prostate enzyme (5a-reductase) that acts on the male sex hormone testosterone, converting it into a substance that promotes prostate growth. Growth is fine during puberty, but continued growth in an adult is undesirable. Three substances, one a nutrient supplement and two prescription drugs, interfere with the action of the enzyme that promotes growth. Saw palmetto, which is sold in tablet form as an over-the-counter nutrient supplement, is derived from a plant of the same name. This drug should not be taken unless the need for it is confirmed by a physician, but it is particularly effective during the early stages of prostate enlargement. Finasteride and dutasteride, prescription drugs, are more powerful inhibitors of the enzyme, but patients complain of erectile dysfunction and loss of libido while on the drugs.

Two other medications have a different mode of action. Nafarelin prevents the release of LH, which leads to testosterone production. When it is administered, approximately half of the patients report relief of urinary symptoms even after drug treatment is halted. However, again the patients experience erectile dysfunction and other side effects, such as hot flashes. The drug terazosin, which is

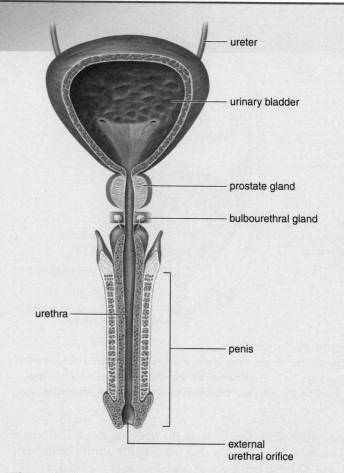

Figure 10A **Longitudinal section of a male urethra leaving the bladder.**
Note the position of the prostate gland, which can enlarge to obstruct urine flow.

on the market for hypertension because it relaxes arterial walls, also relaxes muscle tissue in the prostate. Improved urine flow was experienced by 70% of the patients taking this drug. However, the drug has no effect on the prostate's overall size.

Many men are concerned that BPH may be associated with prostate cancer, but the two conditions are not necessarily related. BPH occurs in the inner zone of the prostate, whereas cancer tends to develop in the outer area. If prostate cancer is suspected, blood tests and a biopsy, in which a tiny sample of prostate tissue is surgically removed, will confirm the diagnosis.

Enlarged Prostate and Cancer

Although prostate cancer is the second most common cancer in men, it is not a major killer. Typically, prostate cancer is so slow growing that the survival rate is about 98% if the condition is detected early.

10.2 Kidney Structure

When a kidney is sliced lengthwise, it is possible to see that many branches of the renal artery and vein reach inside the kidney (Fig. 10.3a). If the blood vessels are removed, it is easier to identify the three regions of a kidney. (1) The **renal cortex** is an outer, granulated layer that dips down in between a radially striated inner layer called the renal medulla. (2) The **renal medulla** consists of cone-shaped tissue masses called renal pyramids. (3) The **renal pelvis** is a central space, or cavity, that is continuous with the ureter (Fig. 10.3c).

Microscopically, the kidney is composed of over one million **nephrons,** sometimes called renal, or kidney, tubules (Fig. 10.3d). The nephrons produce urine and are positioned so that the urine flows into a collecting duct. Several nephrons enter the same collecting duct; the collecting ducts eventually enter the renal pelvis.

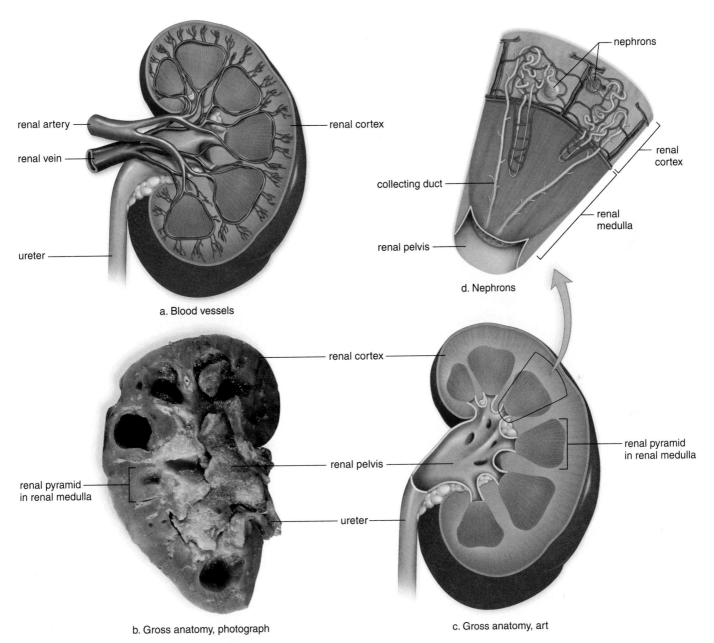

a. Blood vessels

d. Nephrons

b. Gross anatomy, photograph

c. Gross anatomy, art

Figure 10.3 **Gross anatomy of the kidney.**
a. A longitudinal section of the kidney showing the blood supply. Note that the renal artery divides into smaller arteries, and these divide into arterioles. Venules join to form small veins, which join to form the renal vein. **b.** and **c.** The same section without the blood supply. Now it is easier to distinguish the renal cortex, the renal medulla, and the renal pelvis, which connects with the ureter. The renal medulla consists of the renal pyramids. **d.** An enlargement showing the placement of nephrons.

Anatomy of a Nephron

Each nephron has its own blood supply, including two capillary regions (Fig. 10.4). From the renal artery, an afferent arteriole leads to the **glomerulus,** a knot of capillaries inside the glomerular capsule. Blood leaving the glomerulus enters the efferent arteriole. Blood pressure is higher in the glomerulus because the efferent arteriole is narrower than the afferent arteriole. The efferent arteriole takes blood

to the **peritubular capillary network,** which surrounds the rest of the nephron. From there, the blood goes into a venule that carries blood into the renal vein.

Parts of a Nephron

Each nephron is made up of several parts (Fig. 10.4). Some functions are shared by all parts of the nephron; however, the specific structure of each part is especially suited to a particular function.

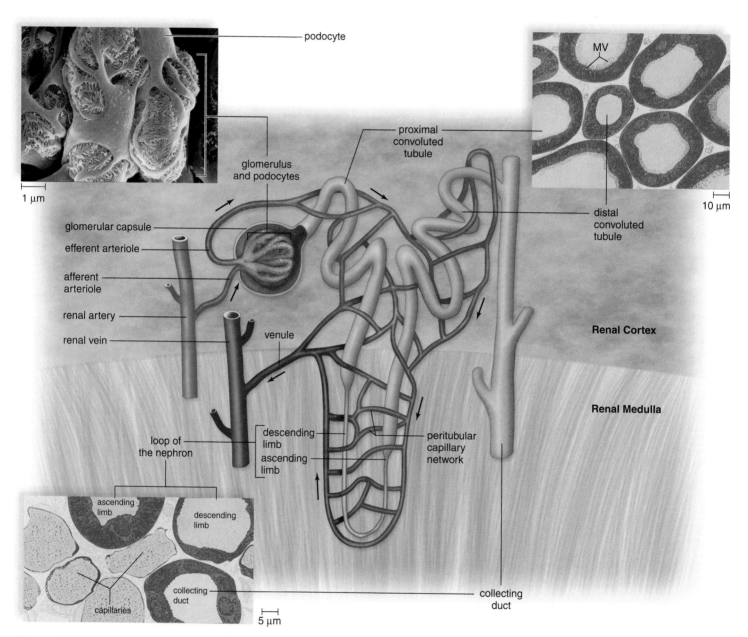

Figure 10.4   Nephron anatomy.

A nephron is made up of a glomerular capsule, the proximal convoluted tubule, the loop of the nephron, the distal convoluted tubule, and the collecting duct. The photomicrographs show the microscopic anatomy of these structures. You can trace the path of blood about the nephron by following the arrows. (MV = microvilli)

First, the closed end of the nephron is pushed in on itself to form a cuplike structure called the **glomerular capsule** (Bowman's capsule). The outer layer of the glomerular capsule is composed of squamous epithelial cells; the inner layer is made up of podocytes that have long cytoplasmic extensions. The podocytes cling to the capillary walls of the glomerulus and leave pores that allow easy passage of small molecules from the glomerulus to the inside of the glomerular capsule. This process, called glomerular filtration, produces a filtrate of the blood.

Next, there is a **proximal convoluted tubule.** The cuboidal epithelial cells lining this part of the nephron have numerous microvilli, about 1 μm in length, that are tightly packed and form a brush border (Fig. 10.5). A brush border greatly increases the surface area for the tubular reabsorption of filtrate components. Each cell also has many mitochondria, which can supply energy for active transport of molecules from the lumen to the peritubular capillary network.

Simple squamous epithelium appears as the tube narrows and makes a U-turn called the **loop of the nephron** (loop of Henle). Each loop consists of a descending limb that allows water to leave and an ascending limb that extrudes salt (NaCl). Indeed, as we shall see, this activity facilitates the reabsorption of water by the nephron and collecting duct.

The cuboidal epithelial cells of the **distal convoluted tubule** have numerous mitochondria, but they lack microvilli. This means that the distal convoluted tubule is not specialized for reabsorption but can participate in moving molecules from the blood into the tubule, a process called tubular secretion. The distal convoluted tubules of several nephrons enter one collecting duct. Many **collecting ducts** carry urine to the renal pelvis.

As shown in Figure 10.4, the glomerular capsule and the convoluted tubules always lie within the renal cortex. The loop of the nephron dips down into the renal medulla; a few nephrons have a very long loop of the nephron, which penetrates deep into the renal medulla. Collecting ducts are also located in the renal medulla, and together they give the renal pyramids their appearance.

☑ Check Your Progress 10.2

1. What are the three major areas of a kidney?
2. What microscopic structure is responsible for the production of urine?
3. How do the parts of a nephron differ?

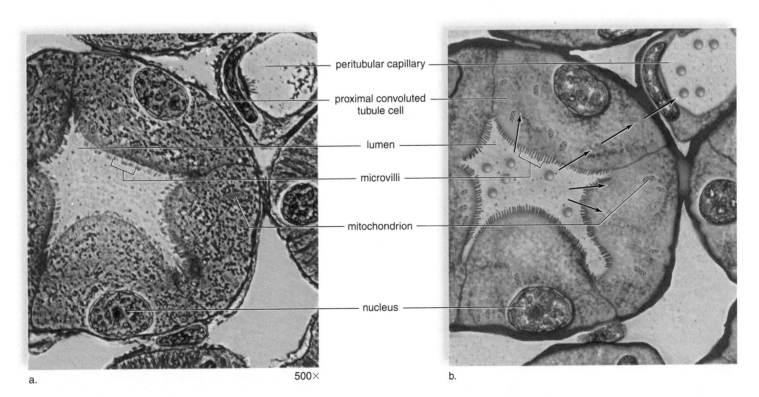

a.   500×   b.

Figure 10.5   Proximal convoluted tubule.
a. This photomicrograph shows that the cells lining the proximal convoluted tubule have a brushlike border composed of microvilli, which greatly increase the surface area exposed to the lumen. The peritubular capillary network surrounds the cells. **b.** Diagrammatic representation of (**a**) shows that each cell has many mitochondria, which supply the energy needed for active transport, the process that moves molecules (green) from the lumen of the tubule to the capillary, as indicated by the arrows.

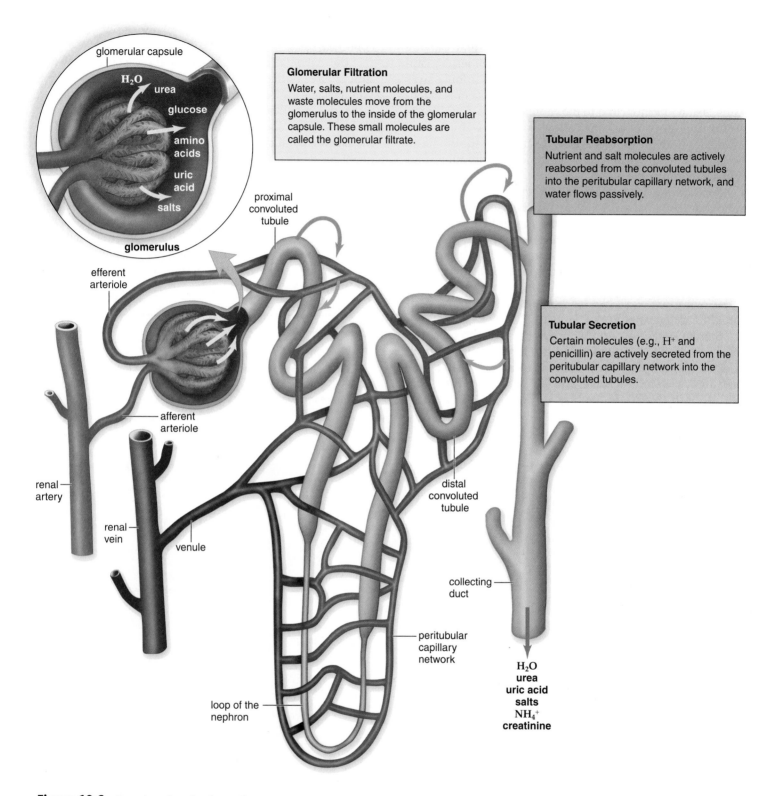

Glomerular Filtration

Water, salts, nutrient molecules, and waste molecules move from the glomerulus to the inside of the glomerular capsule. These small molecules are called the glomerular filtrate.

Tubular Reabsorption

Nutrient and salt molecules are actively reabsorbed from the convoluted tubules into the peritubular capillary network, and water flows passively.

Tubular Secretion

Certain molecules (e.g., H^+ and penicillin) are actively secreted from the peritubular capillary network into the convoluted tubules.

glomerular capsule

H_2O

urea

glucose

amino acids

uric acid

salts

glomerulus

efferent arteriole

afferent arteriole

renal artery

renal vein

venule

proximal convoluted tubule

distal convoluted tubule

collecting duct

peritubular capillary network

loop of the nephron

H_2O
urea
uric acid
salts
NH_4^+
creatinine

Figure 10.6   Processes in urine formation.
The three main processes in urine formation are described in boxes and color coded to arrows that show the movement of molecules into or out of the nephron at specific locations. In the end, urine is composed of the substances within the collecting duct (see brown arrow).

10.3 Urine Formation

Figure 10.6 gives an overview of urine formation, which is divided into three processes:

Glomerular Filtration

Glomerular filtration occurs when whole blood enters the glomerulus by way of the afferent arteriole. Due to glomerular blood pressure, water and small molecules move from the glomerulus to the inside of the glomerular capsule. This is a filtration process because large molecules and formed elements are unable to pass through the capillary wall. In effect, then, blood in the glomerulus has two portions, the filterable components and the nonfilterable components:

Filterable Blood Components	Nonfilterable Blood Components
Water	Formed elements (blood cells and platelets)
Nitrogenous wastes	Plasma proteins
Nutrients	
Salts (ions)	

The nonfilterable components leave the glomerulus by way of the efferent arteriole. The **glomerular filtrate** inside the glomerular capsule now contains the filterable blood components in approximately the same concentration as plasma.

As indicated in Table 10.1, nephrons in the kidneys filter 180 liters of water per day, along with a considerable amount of small molecules (such as glucose) and ions (such as sodium). If the composition of urine were the same as that of the glomerular filtrate, the body would continually lose water, salts, and nutrients. Therefore, we can conclude that the composition of the filtrate must be altered as this fluid passes through the remainder of the tubule.

Tubular Reabsorption

Tubular reabsorption occurs as molecules and ions are both passively and actively reabsorbed from the nephron into the blood of the peritubular capillary network. The osmolarity of the blood is maintained by the presence of both plasma proteins and salt. When sodium ions (Na^+) are actively reabsorbed, chloride ions (Cl^-) follow passively. The reabsorption of salt (Na^+Cl^-) increases the osmolarity of the blood compared with the filtrate, and therefore, water moves passively from the tubule into the blood. About 65% of Na^+ is reabsorbed at the proximal convoluted tubule.

Nutrients such as glucose and amino acids return to the peritubular capillaries almost exclusively at the proximal convoluted tubule. This is a selective process because only molecules recognized by carrier proteins are actively reabsorbed. Glucose is an example of a molecule that ordinarily is completely reabsorbed because there is a plentiful supply of carrier proteins for it. However, every substance has a maximum rate

Table 10.1	Reabsorption from Nephrons		
Substance	Amount Filtered (per day)	Amount Excreted (per day)	Reabsorption (%)
Water, L	180	1.8	99.0
Sodium, g	630	3.2	99.5
Glucose, g	180	0.0	100.0
Urea, g	54	30.0	44.0

L = liters, g = grams

of transport, and after all its carriers are in use, any excess in the filtrate will appear in the urine. In **diabetes mellitus,** because the liver and muscles fail to store glucose as glycogen, the blood glucose level is above normal, and glucose appears in the urine. The presence of excess glucose in the filtrate raises its osmolarity, and therefore, less water is reabsorbed into the peritubular capillary network. The frequent urination and increased thirst experienced by untreated diabetics are due to the fact that less water is being reabsorbed from the filtrate into the blood.

We have seen that the filtrate that enters the proximal convoluted tubule is divided into two portions, components that are reabsorbed from the tubule into blood, and components that are not reabsorbed and continue to pass through the nephron to be further processed into urine:

Reabsorbed Filtrate Components	Nonreabsorbed Filtrate Components
Most water	Some water
Nutrients	Much nitrogenous waste
Required salts (ions)	Excess salts (ions)

The substances that are not reabsorbed become the tubular fluid, which enters the loop of the nephron.

Tubular Secretion

Tubular secretion is a second way by which substances are removed from blood and added to the tubular fluid. Hydrogen ions (H^+), creatinine, and drugs such as penicillin are some of the substances that are moved by active transport from blood into the kidney tubule. In the end, urine contains substances that have undergone glomerular filtration but have not been reabsorbed, and substances that have undergone tubular secretion. Tubular secretion is now known to occur along the length of the kidney tubule.

✓ Check Your Progress 10.3

1. What are the major processes in urine formation?
2. How does the nephron carry out these processes?

Health Focus

Urine Analysis

A routine urine analysis can detect abnormalities in urine color, concentration, and content, while providing significant information about the general state of one's health, particularly in the diagnosis of renal and metabolic diseases. A complete urine analysis consists of three phases of examination: physical, chemical, and microscopic.

Physical Examination

During a physical examination, a clinician may examine urine color, clarity, and odor. The color of normal, fresh urine is usually pale yellow; however, the color may vary from almost colorless (dilute) to dark yellow (concentrated). Pink, red, or smoky brown urine is usually a sign of bleeding due to a kidney, bladder, or urinary tract infection. Liver disorders can also produce dark brown urine. The clarity of normal urine may be clear or cloudy. However, a cloudy urine sample is also characteristic of abnormal levels of bacteria. Finally, urine odor is usually "nutty" or aromatic; but a foul-smelling odor is characteristic of a urinary tract infection. A sweet, fruity odor is characteristic of glucose in the urine or diabetes.

Chemical Examination

The chemical examination is often done with a dipstick, which is a thin strip of plastic, impregnated with chemicals that change color upon reaction with certain substances present in urine (Fig. 10B). The color change on each segment of the dipstick is compared with a color chart. Dipsticks can be used to determine urine's specific gravity; pH; and content of glucose, bilirubin, urobilinogen, ketone, protein, nitrite, blood, and white blood cells (WBCs) in the urine.

1. *Specific gravity* is an indicator of how well the kidneys are able to adjust tonicity in urine. Normal values for urine specific gravity range from 1.002 to 1.035. A high value (concentrated urine) may be a result of dehydration or diabetes mellitus.
2. Normal *urine pH* can be as low as 4.5 and as high as 8.0. In patients with kidney stone disease, urine pH has a direct effect on the type of stones formed.
3. *Glucose* is normally not present in urine, and if present, diabetes mellitus is suspected.
4. *Bilirubin* (a by-product of hemoglobin degradation) is not normally present in the urine. *Urobilinogen* (a by-product of bilirubin degradation) is normally present in very small amounts. High levels of bilirubin or urobilinogen may indicate liver disease.
5. *Ketones* are not normally found in urine. Urine ketones are a by-product of fat metabolism and may develop when a person's body metabolizes fat, instead of the preferred carbohydrates, as an energy source.

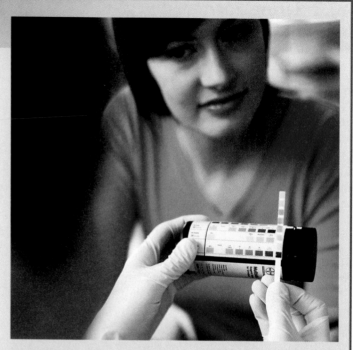

Figure 10B   **Urine analysis.**
This patient's urine has been tested and the multiple test stick is being compared to a reference chart. The dark brown pad shows that the patient has glucose in her urine and this indicates she has diabetes mellitus.

6. Plasma *proteins* should not be present in urine. A significant amount of urine protein (proteinuria) is usually a sign of kidney damage.
7. Urine typically does not contain *nitrates*. The presence of nitrites is a sign of urinary tract infection.
8. The chemical test for *WBCs* is normally negative. A high urine WBC count usually indicates a bacterial infection somewhere in the urinary tract.

Microscopic Examination

Finally, in a microscopic examination, the urine is centrifuged, and the sediment (solid material) is examined under a microscope. When renal disease is present, the urine will often contain an abnormal amount of cellular material. The presence of WBCs in urine may indicate some type of infection or injury to the kidneys, ureters, bladder, or urethra. Urinary casts are sediments formed by the coagulation of protein material in the distal convoluted tubule or the collecting duct (distal nephron) and may be a sign of many different disorders, depending on the type of cast. Lastly, the presence of crystals in the urine is characteristic of kidney stones, kidney damage, or problems with metabolism.

There are more than 100 different tests that can be done on urine. Many different factors such as diet, kidney function, and other health disorders can affect what ends up in one's urine. Consequently, a urine analysis can provide a clinician with important information about a patient's overall health.

10.4 Regulatory Functions of the Kidneys

The kidneys maintain the water-salt balance of the blood within normal limits. In this way, they also maintain the blood volume and blood pressure. The kidneys are also involved in regulating the pH of the blood, so that it stays within normal limits.

Water-Salt Balance

Most of the water found in the filtrate is reabsorbed into the blood before urine leaves the body. All parts of a nephron and the collecting duct participate in the reabsorption of water. The reabsorption of salt always precedes the reabsorption of water. In other words, water is returned to the blood by the process of osmosis. During the process of reabsorption, water passes through recently discovered water channels, called **aquaporins**, within a plasma membrane protein.

Sodium (Na^+) is an important ion in plasma, but the kidneys also excrete or reabsorb other ions, such as potassium ions (K^+), bicarbonate ions (HCO_3^-), and magnesium ions (Mg_2^+), as needed. Usually, more than 99% of sodium (Na^+) filtered at the glomerulus is returned to the blood.

Reabsorption of Salt and Water from Cortical Portions of the Nephron

The proximal convoluted tubule, the distal convoluted tubule, and the cortical portion of the collecting ducts are present in the renal cortex. Most (65%) of the water that enters the glomerular capsule is reabsorbed from the nephron into the blood at the proximal convoluted tubule. Na^+ is actively reabsorbed, and Cl^- follows passively. Aquaporins are always open and water is reabsorbed osmotically into the blood.

Hormones regulate the reabsorption of sodium and water in the distal convoluted tubule. **Aldosterone** is a hormone secreted by the adrenal glands that promotes the excretion of K^+ and the reabsorption of Na^+ from this portion of the nephron into the blood. The release of aldosterone is set in motion by the kidneys themselves. The **juxtaglomerular apparatus** is a region of contact between the afferent arteriole and the distal convoluted tubule (Fig. 10.7). When blood volume, and therefore blood pressure, seems (perhaps mistakenly in people with high blood pressure) insufficient for filtration, the juxtaglomerular apparatus secretes renin. **Renin** is a substance that leads to the secretion of aldosterone by the adrenal glands, which sit atop the kidneys. Now Na^+ is reabsorbed.

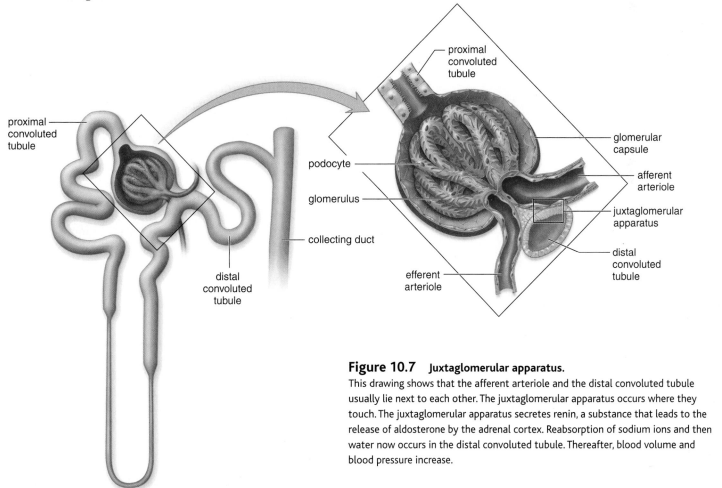

Figure 10.7 Juxtaglomerular apparatus.
This drawing shows that the afferent arteriole and the distal convoluted tubule usually lie next to each other. The juxtaglomerular apparatus occurs where they touch. The juxtaglomerular apparatus secretes renin, a substance that leads to the release of aldosterone by the adrenal cortex. Reabsorption of sodium ions and then water now occurs in the distal convoluted tubule. Thereafter, blood volume and blood pressure increase.

Aquaporins are not always open in the distal convoluted tubule. Another hormone called **antidiuretic hormone (ADH)** must be present. ADH is secreted by the posterior pituitary according to the osmolarity of the blood. If our intake of water has been low, ADH is secreted by the posterior pituitary and water moves from the distal convoluted tubule and, also the collecting duct, into the blood.

Atrial natriuretic hormone (ANH) is a hormone secreted by the atria of the heart when cardiac cells are stretched due to increased blood volume. ANH inhibits the secretion of renin by the juxtaglomerular apparatus and the secretion of aldosterone by the adrenal glands. Its effect, therefore, is to promote the excretion of Na^+, called natriuresis. If ANH is present, less water will be reabsorbed, even if ADH is present, because the reabsorption of water is dependent on a salt concentration gradient.

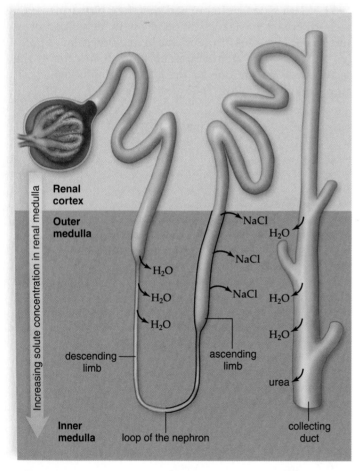

Figure 10.8   Countercurrent mechanism.
Salt (NaCl) diffuses and is actively transported out of the ascending limb of the loop of the nephron into the renal medulla; also, urea is believed to leak from the collecting duct and to enter the tissues of the renal medulla. This creates a hypertonic environment, which draws water out of the descending limb and the collecting duct. This water is returned to the cardiovascular system. (The thick black outline of the ascending limb means that it is impermeable to water.)

Reabsorption of Salt and Water From the Medulla Portions of the Nephron

The ability of humans to regulate the tonicity of their urine is dependent on the work of the medullary portions of the nephron (loop of the nephron) and the collecting duct.

The Loop of the Nephron   A long loop of the nephron, which typically penetrates deep into the renal medulla, is made up of a descending limb and an ascending limb. Salt (NaCl) passively diffuses out of the lower portion of the *ascending limb,* but the upper, thick portion of the limb actively extrudes any remaining salt into the tissue of the outer medulla (Fig. 10.8). In the end, the concentration of salt is greater in the direction of the inner medulla. Surprisingly, however, the inner medulla has an even higher concentration of solutes than expected. It is believed that urea leaks from the lower portion of the collecting duct, and this molecule contributes to the high solute concentration of the inner medulla.

Because of the osmotic gradient within the medulla, *water leaves the descending limb* along its entire length. This is a countercurrent mechanism: although water is reabsorbed as soon as fluid enters the descending limb, the remaining fluid within the limb always encounters an ever greater osmotic concentration of solute; therefore, water continues to reabsorbed, even to the bottom of the descending limb. (The ascending limb does not reabsorb water—it has no aquaporins as indicated by the dark line in Figure 10.8—it's job is to help establish the solute concentration gradient.)

The Collecting Duct   Fluid within the collecting duct encounters the same osmotic gradient established by the ascending limb of the nephron. Therefore, water will diffuse from the entire length of the collecting duct into the blood if aquaporins are open, as they will be if ADH is present.

To understand the action of ADH, consider its name, antidiuretic hormone. Diuresis means increased amount of urine, and antidiuresis means decreased amount of urine. When ADH is present, more water is reabsorbed (blood volume and pressure rise), and a decreased amount of urine results. ADH is the ultimate fine tuner of the tonicity of urine according to the needs of the body. For example, ADH is secreted at night when we are not drinking water, and this explains why the first urine of the day is more concentrated.

Diuretics

Diuretics are chemicals that increase the flow of urine. Drinking alcohol causes diuresis because it inhibits the secretion of ADH. The dehydration that follows is believed to contribute to the symptoms of a hangover. Caffeine is a diuretic because it increases the glomerular filtration rate and decreases the tubular reabsorption of Na^+. Diuretic drugs developed to counteract high blood pressure also decrease the tubular reabsorption of Na^+. A decrease in water reabsorption and a decrease in blood volume follow.

Acid-Base Balance of Body Fluids

The pH scale, as discussed in Chapter 2, page 26, can be used to indicate the basicity (alkalinity) or the acidity of body fluids. A basic solution has a lesser hydrogen ion concentration [H^+] than the neutral pH of 7.0, and an acidic solution has a greater [H^+] than neutral pH. The normal pH for body fluids is about 7.4. This is the pH at which our proteins, such as cellular enzymes, function properly. If the blood pH rises above 7.4, a person is said to have **alkalosis,** and if the blood pH decreases below 7.4, a person is said to have **acidosis.** Alkalosis and acidosis are abnormal conditions that may need medical attention.

The foods we eat add basic or acidic substances to the blood, and so does metabolism. For example, cellular respiration adds carbon dioxide that combines with water to form carbonic acid, and fermentation adds lactic acid. The pH of body fluids stays at just about 7.4 via several mechanisms, primarily acid-base buffer systems, the respiratory center, and the kidneys.

Acid-Base Buffer Systems

The pH of the blood stays near 7.4 because the blood is buffered. A **buffer** is a chemical or a combination of chemicals that can take up excess hydrogen ions (H^+) or excess hydroxide ions (OH^-). One of the most important buffers in the blood is a combination of carbonic acid (H_2CO_3) and bicarbonate ions (HCO_3^-). When hydrogen ions (H^+) are added to blood, the following reaction occurs:

$$H^+ + HCO_3^- \rightarrow H_2CO_3$$

When hydroxide ions (OH^-) are added to blood, this reaction occurs:

$$OH^- + H_2CO_3 \rightarrow HCO_3^- + H_2O$$

These reactions temporarily prevent any significant change in blood pH. A blood buffer, however, can be overwhelmed unless some more permanent adjustment is made. The next adjustment to keep the pH of the blood constant occurs at pulmonary capillaries.

Respiratory Center

As discussed in Chapter 9, the respiratory center in the medulla oblongata increases the breathing rate if the hydrogen ion concentration of the blood rises. Increasing the breathing rate rids the body of hydrogen ions because the following reaction takes place in pulmonary capillaries:

$$H^+ + HCO_3^- \rightleftharpoons H_2CO_3 \rightleftharpoons H_2O + CO_2$$

In other words, when carbon dioxide is exhaled, this reaction shifts to the right, and the amount of hydrogen ions is reduced.

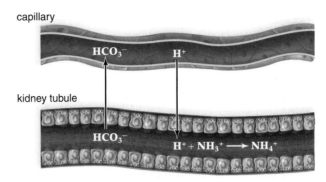

capillary

kidney tubule

Figure 10.9   Acid-base balance.
In the kidneys, bicarbonate ions (HCO_3^-) are reabsorbed, and hydrogen ions (H^+) are excreted as needed to maintain the pH of the blood. Excess hydrogen ions are buffered, for example, by ammonia (NH_3), which becomes ammonium (NH_4^+). Ammonia is produced in tubule cells by the deamination of amino acids.

It is important to have the correct proportion of carbonic acid and bicarbonate ions in the blood. Breathing readjusts this proportion so that this particular acid-base buffer system can continue to absorb both H^+ and OH^- as needed.

The Kidneys

As powerful as the acid-base buffer and the respiratory center mechanisms are, only the kidneys can rid the body of a wide range of acidic and basic substances and, otherwise, adjust the pH. The kidneys are slower acting than the other two mechanisms, but they have a more powerful effect on pH. For the sake of simplicity, we can think of the kidneys as reabsorbing bicarbonate ions and excreting hydrogen ions as needed to maintain the normal pH of the blood (Fig. 10.9). If the blood is acidic, hydrogen ions are excreted, and bicarbonate ions are reabsorbed. If the blood is basic, hydrogen ions are not excreted, and bicarbonate ions are not reabsorbed. Because the urine is usually acidic, it follows that an excess of hydrogen ions is usually excreted. Ammonia (NH_3) provides another means of buffering and removing the hydrogen ions in urine: ($H^+ + NH_3 \rightarrow NH_4^+$). Ammonia (whose presence is quite obvious in the diaper pail or kitty litter box) is produced in tubule cells by the deamination of amino acids. Phosphate provides another means of buffering hydrogen ions in urine.

The importance of the kidneys' ultimate control over the pH of the blood cannot be overemphasized. As mentioned, the enzymes of cells cannot continue to function if the internal environment does not have near-normal pH.

✓ Check Your Progress 10.4

1. **a.** Where is water reabsorbed, and **(b)** how is water reabsorbed so that urine is concentrated?

2. **a.** What three hormones influence urine production and kidney function? **b.** How do these hormones work together?

3. How do the kidneys regulate the pH of body fluids?

10.5 Disorders with Kidney Function

Many types of illnesses, especially diabetes, hypertension, and inherited conditions, cause progressive renal disease and renal failure. Infections are also contributory. If the infection is localized in the urethra, it is called **urethritis.** If the infection invades the urinary bladder, it is called **cystitis.** Finally, if the kidneys are affected, the infection is called **pyelonephritis.**

Urinary tract infections, an enlarged prostate gland, pH imbalances, or simply an intake of too much calcium can lead to kidney stones. Kidney stones are hard granules made of calcium, phosphate, uric acid, and protein. Kidney stones form in the renal pelvis and usually pass unnoticed in the urine flow. If they grow to several centimeters and block the renal pelvis or ureter, a reverse pressure builds up and destroys nephrons. When a large kidney stone passes, strong contractions within a ureter can be excruciatingly painful.

One of the first signs of nephron damage is albumin, white blood cells, or even red blood cells in the urine and can be detected when a urinalysis is done. If damage is so extensive that more than two-thirds of the nephrons are inoperative, urea and other waste substances accumulate in the blood. This condition is called **uremia.** Although nitrogenous wastes can cause serious damage, the retention of water and salts is of even greater concern. The latter causes edema, fluid accumulation in the body tissues. Imbalance in the ionic composition of body fluids can lead to loss of consciousness and to heart failure.

Hemodialysis

Patients with renal failure can undergo **hemodialysis,** utilizing either an artificial kidney machine or continuous ambulatory peritoneal dialysis (CAPD). *Dialysis* is defined as the diffusion of dissolved molecules through a semipermeable natural or synthetic membrane that has pore sizes that allow only small molecules to pass through. In an artificial kidney machine (Fig. 10.10), the patient's blood is passed through a membranous tube, which is in contact with a dialysis solution, or **dialysate.** Substances more concentrated in the blood diffuse into the dialysate, and substances more concentrated in the dialysate diffuse into the blood. The dialysate is continuously replaced to maintain favorable concentration gradients. In this way, the artificial kidney can be utilized either to extract substances from blood, including waste products or toxic chemicals and drugs, or to add substances to blood—for example, bicarbonate ions (HCO_3^-) if the blood is acidic. In the course of a 3–6 hour hemodialysis, from 50 to 250 g of urea can be removed from a patient, which greatly exceeds the amount excreted by normal kidneys. Therefore, a patient needs to undergo treatment only about twice a week.

CAPD is so named because the peritoneum is the dialysis membrane. A fresh amount of dialysate is introduced directly into the abdominal cavity from a bag that is temporarily attached to a permanently implanted plastic tube. The dialysate flows into the peritoneal cavity by gravity. Waste and salt molecules pass from the blood vessels in the abdominal wall into the dialysate before the fluid is collected 4–8 hours later. The solution is drained into a bag from the abdominal cavity by gravity, and then it is discarded. One advantage of CAPD over an artificial kidney machine is that the individual can go about his or her normal activities during CAPD.

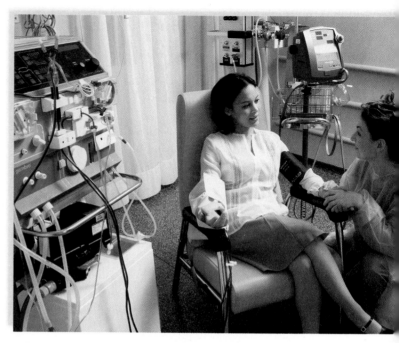

Figure 10.10   An artificial kidney machine.
As the patient's blood is pumped through dialysis tubing, it is exposed to a dialysate (dialysis solution). Wastes exit from blood into the solution because of a preestablished concentration gradient. In this way, blood is not only cleansed, but its water-salt and acid-base balances can also be adjusted.

Replacing a Kidney

Patients with renal failure sometimes undergo a kidney transplant operation, during which a functioning kidney from a donor is received. As with all organ transplants, there is the possibility of organ rejection. Receiving a kidney from a close relative has the highest chance of success. The current one-year survival rate is 97% if the kidney is received from a relative and 90% if it is received from a nonrelative. In the future, it may be possible to use kidneys from pigs, especially bred so that they are not antigenic to humans, or kidneys created in the laboratory.

☑ Check Your Progress 10.5

1. What are the most common causes of renal disease, and how can it be treated?

10.6 Homeostasis

Figure 10.11 tells us how the kidneys assist the work of our systems and/or how these systems help the urinary system carry out its functions. Our final discussion of the urinary system centers around these functions of the kidney.

The Kidneys Excrete Waste Molecules

In Chapter 8, we compared the liver to a sewage treatment plant because it removes poisonous substances from the blood and prepares them for excretion. Similarly, the liver produces urea, the primary nitrogenous end product of humans, which is excreted by the kidneys. If the liver is a sewage treatment plant, the tubules of the kidney are like the trucks that take the sludge, prepared waste, away from the town (the body).

Metabolic waste removal is absolutely necessary for maintaining homeostasis. The blood must constantly be cleansed of the nitrogenous wastes, which are end products of metabolism. The liver produces urea, and muscles make creatinine. These wastes, and also uric acid from the cells, are carried by the cardiovascular system to the kidneys (Fig. 10.12). The urine producing kidneys are responsible for the excretion of nitrogenous wastes. They are assisted to a limited degree by the sweat glands in the skin, which excrete perspiration, a mixture of water, salt, and some urea. In times of kidney failure, urea is excreted by the sweat glands and forms a so-called urea frost on the skin.

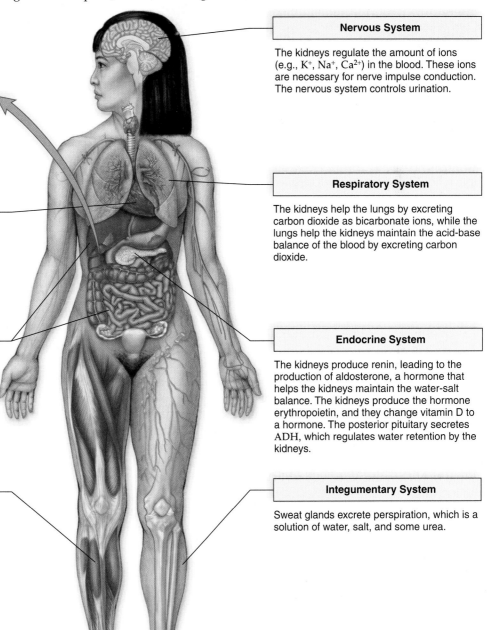

All systems of the body work with the urinary system to maintain homeostasis. These systems are especially noteworthy.

Urinary System

As an aid to all the systems, the kidneys excrete nitrogenous wastes and maintain the water-salt balance and the acid-base balance of the blood. The urinary system also specifically helps the other systems, as mentioned below.

Cardiovascular System

Production of renin by the kidneys helps maintain blood pressure. Blood vessels transport nitrogenous wastes to the kidneys and carbon dioxide to the lungs. The buffering system of the blood helps the kidneys maintain the acid-base balance.

Digestive System

The liver produces urea excreted by the kidneys. The yellow pigment found in urine, called urochrome (breakdown product of hemoglobin), is produced by the liver. The digestive system absorbs nutrients, ions, and water. These help the kidneys maintain the proper level of ions and water in the blood.

Muscular System

The kidneys regulate the amount of ions in the blood. These ions are necessary to the contraction of muscles, including those that propel fluids in the ureters and urethra.

Nervous System

The kidneys regulate the amount of ions (e.g., K^+, Na^+, Ca^{2+}) in the blood. These ions are necessary for nerve impulse conduction. The nervous system controls urination.

Respiratory System

The kidneys help the lungs by excreting carbon dioxide as bicarbonate ions, while the lungs help the kidneys maintain the acid-base balance of the blood by excreting carbon dioxide.

Endocrine System

The kidneys produce renin, leading to the production of aldosterone, a hormone that helps the kidneys maintain the water-salt balance. The kidneys produce the hormone erythropoietin, and they change vitamin D to a hormone. The posterior pituitary secretes ADH, which regulates water retention by the kidneys.

Integumentary System

Sweat glands excrete perspiration, which is a solution of water, salt, and some urea.

Figure 10.11   Human systems work together.
The urinary system particularly works with these systems, in order to bring about homeostasis.

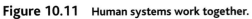

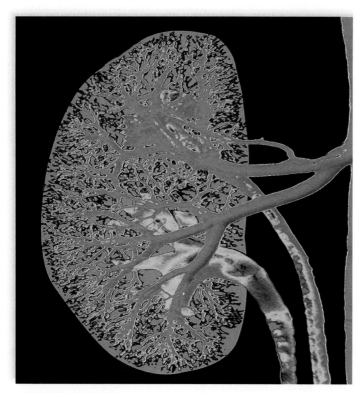

Figure 10.12 Kidneys are well supplied with blood, as shown in this angiogram (X-ray).
Angiography involves injecting a contrast medium opaque to X-rays into the blood vessels, which highlights them when an X-ray of the area is taken.

Water-Salt Balance

If blood does not have the usual water-salt balance, blood volume and blood pressure are affected. Without adequate blood pressure, exchange across capillary walls cannot take place, nor is glomerular filtration possible in the kidneys themselves.

What happens if you have insufficient Na^+ in your blood and tissue fluid? This can occur due to prolonged heavy sweating, as in athletes running a marathon. And the problem can be worsened by consuming too much water, which dilutes the remaining Na^+ in the body. This is why sports beverages contain sodium. Too low a concentration of Na^+ in the blood causes blood pressure to lower and activates the renin-aldosterone sequence, and then the kidneys increase Na^+ reabsorption, in order to conserve as much as possible. Subsequently, the osmolarity of the blood and the blood pressure return to normal. One cause of hypertension could be the release of renin by the juxtaglomerula apparatus when it is not actually needed.

Think what happens if you eat a big tub of salty popcorn at the movies. When salt (NaCl) is absorbed from the digestive tract, the Na^+ content of the blood increases above normal, and special cells in the hypothalamus detect that the osmolarity of the blood has increased. The hypothalamus

directs the posterior pituitary to release ADH. ADH causes aquaporins to open in the distal convoluted tubule and the collecting duct. More water is reabsorbed; osmolarity is restored, but blood pressure rises. Restricting salt in the diet is one way to prevent this cause of hypertension.

Acid-Base Balance of Blood

How is the acid-base balance of the blood maintained? Two closely associated mechanisms work to offset short-term challenges to the acid-base balance. These are the blood bicarbonate (HCO_3^-) buffering system and the process of breathing. Usually, the excretion of carbon dioxide (CO_2) by the lungs helps to keep blood pH within normal limits. Remember what happened in the Chapter 9 opening story, when Olivia held her breath? Since she wasn't expelling CO_2 from her lungs, blood pH decreased, and her blood became more acidic. Then, the chemoreceptors in the carotid bodies (located in the carotid arteries) and in the aortic bodies (located in the aorta) stimulated her respiratory center, and she had to succumb to the overwhelming urge to take a breath.

As powerful as the combined bicarbonate buffer and breathing systems are, only the kidneys can rid the body of a wide range of acidic and basic substances—thus, the importance of the kidneys' ultimate control over blood pH cannot be overemphasized. The kidneys are slower acting than the buffer/breathing mechanism, but their effect on blood pH is stronger. They do this by adjusting the secretion of hydrogen ions and bicarbonate ions as needed.

The Kidneys Assist Other Systems

Aside from producing renin, the kidneys assist the endocrine system and also the cardiovascular system by producing erythropoietin (see Fig. 10.11). By doing his biology assignment, Brad learned that erythropoietin is now a commercial product. It is used to stimulate red bone marrow production in patients who are recovering from chemotherapy, for example. The kidneys assist the skeletal, nervous, and muscular systems by helping to regulate the amount of calcium ions (Ca^{2+}) in the blood. The kidneys convert vitamin D to its active form needed for Ca^{2+} absorption by the digestive tract, and they regulate the excretion of electrolytes, including Ca^{2+}. The kidneys also regulate the sodium (Na^+) and potassium (K^+) content of the blood. These ions, needed for nerve conduction, are necessary to the contraction of the heart and other muscles in the body.

✔ Check Your Progress 10.6

1. What are three main functions of the kidneys, and how do they carry them out?

2. a. Why do we have to have at least one kidney in order to live without kidney dialysis? b. Why would one be enough?

The Right To Die

Writer and humorist Art Buchwald passed away on January 17, 2007. He elected to not spend many hours undergoing kidney dialysis that would have extended his life, and instead to live out the remainder of his days always surrounded by friends and family. Because Buchwald is a popular columnist, his choice led to much media coverage and discussion about end-of-life decisions. Should individuals make their own decisions about medical treatments and can they choose to forego treatment entirely? Do we have the right to die?

Making a Decision to Prolong Life

Today, most Americans die in a hospital where advance medical technologies are available, instead of at home. These technologies include life-support machines to prolong life, as well as long-term therapies, such as hemodialysis. With the use of new technologies, however, comes new questions. By prolonging life, might we also be prolonging suffering? Just because the life of a seriously ill person can be extended, should it be—especially if the patient may be in discomfort? Furthermore, who should make such decisions, the patient or the physician?

Death is often regarded as a taboo subject, perhaps because we fear death. However, unlike other social issues, such as reproductive rights, death will come to everyone. Many have been through the difficult or painful death of a loved one and do not wish to have such an experience themselves. Such considerations influence our feelings about our own end-of-life care. Religious views also color our opinions. Some people who believe that we have the right to refuse life-extending medical treatment often base their viewpoint on personal autonomy—the idea that one should be allowed to make their own choices, even about death. The courts tend to generally agree with this viewpoint—most states honor living wills as legally binding documents, and end-of-life care is usually guided by patient wishes.

Making Decisions About Medical Care

Mr. Buchwald reported receiving letters of support, and also letters disagreeing with his description of dialysis as a negative treatment option. Of course, a lot of patients would not make the same decision because many people who are dependent on hemodialysis live full lives. Decisions about health care are highly personal and often influenced by other factors. Buchwald's choice was preceded and influenced by the amputation of his leg.

Decision making isn't a simple process for anyone. Many want to avoid situations that would cause pain to family members and friends. Care options can be openly discussed with a physician, and often family members can participate in these discussions. A full understanding of the patient's prognosis of the illness is needed and all possible treatments need to be assessed. Will a treatment cure the condition, make it tolerable and/or slow its progress? What are the downsides to treatment, and are they worth the potential benefits? In the case of hemodialysis, patients might have to spend several hours a week undergoing treatment. In other cases, treatment may be painful. For example, it is not uncommon for cancer patients to end or rule out chemotherapy, unless there is a strong chance of it leading to a positive outcome, because chemotherapy can cause severe, unpleasant side affects.

It is not always possible for a physician to predict the outcome of treatment. For example, a physician cannot always say that a disease is hopeless or give an accurate prognoses. If a patient requests the right to die, a physician would like to know what is prompting this decision. Perhaps it is due to depression, or inadequately treated pain. If so, if the physician could improve pain management, would this change the patient's mind?

Unfortunately, not all doctors and nurses are adequately trained to help patients make end-of-life decisions. Discussing death does not necessarily come any easier to health care workers than to the rest of the population. Mr. Buchwald had positive things to say about hospice care. Unfortunately, however, hospice care may not be available to everyone when they need it. No matter how one feels about the right to die, the public discussion about it has led to positive outcomes. It has turned a spotlight on the need for good end-of-life care and what changes that can be implemented to increase physical and emotional comfort for the dying.

Decide Your Opinion

1. Does a doctor have a responsibility to administer treatment to a patient that may extend his or her life, even if it is against the patient's wishes?
2. How might a physician feel sure that they understand their patient's wishes?
3. If a patient is incapacitated such that they cannot direct their medical treatment, who do you think should be responsible for making decisions about their care?

Figure 10C Art Buchwald presiding over a charity auction. Buchwald refused hemodialysis treatment and lived much longer than expected by his physicians.

Summarizing the Concepts

10.1 Urinary System

Organs of the Urinary System

Only the urinary system contains urine.

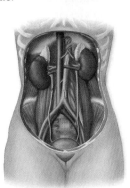

- Kidneys produce urine.
- Ureters take urine to the bladder.
- Urinary bladder stores urine.
- Urethra releases urine to the outside.

Functions of the Urinary System

- Excrete nitrogenous wastes, including urea, uric acid, and creatinine.
- Maintain the normal water-salt balance of the blood.
- Maintain the the acid-base balance of the blood.
- Assist endocrine system in hormone secretion by secreting erythropoietin (stimulates red blood cell production) and releasing renin (leads to the secretion of aldosterone).

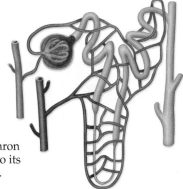

10.2 Kidney Structure

- Macroscopic structures are the renal cortex, renal medulla, and renal pelvis.
- Microscopic structures are the nephrons.

Anatomy of a Nephron

- Each nephron has its own blood supply: the afferent arteriole divides to become the glomerulus. The efferent arteriole branches into the peritubular capillary network.
- Each region of the nephron is anatomically suited to its task in urine formation.

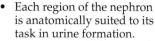

10.3 Urine Formation

Urine is composed primarily of nitrogenous waste products and salts in water. The steps in urine formation include the following:

Glomerular Filtration

Water, salts, nutrients, and wastes move from the glomerulus to the inside of the glomerular capsule.

Tubular Reabsorption

Nutrients and salt molecules are reabsorbed from the proximal convoluted tubules into the peritubular capillary network; water follows.

Tubular Secretion

Certain molecules are actively secreted from the peritubular capillary network into the convoluted tubules.

10.4 Regulatory Functions of the Kidneys

The kidneys maintain the water-salt balance of the blood and, therefore, participate in blood volume and pressure control.

Water-Salt Balance

- In the renal cortex, most of the reabsorption of salts and water occurs in the proximal convoluted tubule. Aldosterone controls the reabsorption of sodium and antidiuretic hormone (ADH) controls the reabsorption of water in the distal convoluted tubule. Atrial natriuretic hormone (ANH) acts contrary to aldosterone.
- In the renal medulla, the ascending limb of the loop of the nephron establishes a solute gradient that increases toward the inner medulla. The solute gradient draws water from the descending limb of the loop of the nephron and from the collecting duct. When ADH is present, more water is reabsorbed from the collecting duct, and a decreased amount of urine results.

Acid-Base Balance of Body Fluids

- The kidneys maintain the acid-base balance of blood (blood pH).
- They reabsorb HCO_3^- and excrete H^+ as needed to maintain the pH at about 7.4. Ammonia buffers H^+ in the urine.

10.5 Disorders with Kidney Function

Various types of problems, including diabetes, kidney stones, and infections, can lead to renal failure, which necessitates undergoing hemodialysis by utilizing a kidney machine or CAPD, or by receiving a kidney transplant.

10.6 Homeostasis

- The kidneys maintain the water-salt balance of blood.
- The kidneys also keep blood pH within normal limits.
- The urinary system works with the other systems of the body to maintain homeostasis.

Understanding Key Terms

Match the key terms to these definitions.

a. —————— Drug used to counteract hypertension by causing the excretion of water.

b. —————— Removal of metabolic wastes from the body.

c. —————— Tubular structure that receives urine from the bladder and carries it to the outside of the body.

d. —————— Hollow chamber in the kidney that lies inside the renal medulla and receives freshly prepared urine from the collecting ducts.

e. —————— Filtered portion of blood contained within the glomerular capsule.

Testing Your Knowledge of the Concepts

1. Describe the path of urine and the structure of each organ mentioned. (pages 188–89)

2. In what ways do the four functions of the urinary system contribute to homeostasis? (page 189)

3. Describe the macroscopic structure of the kidney. (page 191)

4. Trace the path of blood into and out of the kidney. (pages 191–92)

5. Name the parts of a nephron, and describe the structure of each part. (pages 192–93)

6. Why would it be proper to associate glucose with the first two processes involved in urine formation but not the third process? (pages 194–95)

7. Explain how a hypertonic urine can be formed, detailing where and how salt and water move, and the influence of hormones on the process. (pages 197–98)

8. Describe three ways that the body maintains the acid–base balance of body fluids. (page 199)

9. Explain how an artificial kidney machine and continuous ambulatory peritoneal dialysis work to cleanse the blood. (page 200)

10. How do the kidneys assist other body systems? (pages 201–2)

11. Which of the following is a structural difference between the urinary systems of males and females?
    a. Males have a longer urethra than females.
    b. In males, the urethra passes through the prostate.
    c. In males, the urethra serves both the urinary and reproductive systems.
    d. All of these are correct.

12. The renal medulla has a striped appearance due to the presence of which structure(s)?
    a. loops of the nephron
    b. collecting ducts
    c. peritubular capillaries
    d. Both a and b are correct.

13. Which of these functions of the kidneys are mismatched?
    a. excretes metabolic wastes—rids the body of urea
    b. maintains the water-salt balance—helps regulate blood pressure
    c. maintains the acid-base balance—rids the body of uric acid
    d. secretes hormones—secretes erythropoietin
    e. All of these are correct.

14. Which of the following is not correct?
    a. Uric acid is produced from the breakdown of amino acids.
    b. Creatinine is produced from breakdown reactions in the muscles.
    c. Urea is the primary nitrogenous waste of humans.
    d. Ammonia results from the deamination of amino acids.

15. Which part of a nephron is out of sequence first?
    a. glomerular capsule
    b. proximal convoluted tubule
    c. distal convoluted tubule
    d. loop of the nephron
    e. collecting duct

16. Which portion of the nephron has cells with a brush border and many mitochondria?
    a. glomerular capsule
    b. proximal convoluted tubule
    c. loop of the nephron
    d. distal convoluted tubule
    e. collecting duct

17. When tracing the path of filtrate, the loop of the nephron follows which structure?
    a. collecting duct
    b. distal convoluted tubule
    c. proximal convoluted tubule
    d. glomerulus
    e. renal pelvis

18. When tracing the path of blood, the blood vessel that follows the renal artery is the
    a. peritubular capillary.
    b. efferent arteriole.
    c. afferent arteriole.
    d. renal vein.
    e. glomerulus.

19. The function of the descending limb of the loop of the nephron in the process of urine formation is
    a. reabsorption of water.
    b. production of filtrate.
    c. reabsorption of solutes.
    d. secretion of solutes.

20. Which of the following materials would not normally be filtered from the blood at the glomerulus?
    a. water
    b. urea
    c. protein
    d. glucose
    e. sodium ions

21. Which of the following materials would not be maximally reabsorbed from the filtrate?
    a. water
    b. glucose
    c. sodium ions
    d. urea
    e. amino acids

22. By what process are most molecules secreted from the blood into the tubule?
    a. osmosis
    b. diffusion
    c. active transport
    d. facilitated diffusion

23. Reabsorption of the glomerular filtrate occurs primarily at the
    a. proximal convoluted tubule.
    b. distal convoluted tubule.
    c. loop of the nephron.
    d. collecting duct.

24. A countercurrent mechanism draws water from the
    a. proximal convoluted tubule.
    b. descending limb of the loop of the nephron.
    c. distal convoluted tubule.
    d. collecting duct.
    e. Both b and d are correct.

25. Sodium is actively extruded from which part of the nephron?
    a. descending portion of the proximal convoluted tubule
    b. ascending portion of the loop of the nephron
    c. ascending portion of the distal convoluted tubule
    d. descending portion of the collecting duct

26. Excretion of a hypertonic urine in humans is best associated with the
    a. glomerular capsule and the tubules.
    b. proximal convoluted tubule only.
    c. loop of the nephron and collecting duct.
    d. distal convoluted tubule and peritubular capillary.

27. Which of these hormones is most likely to directly cause a drop in blood pressure?
    a. aldosterone
    b. antidiuretic hormone (ADH)
    c. erythropoietin
    d. atrial natriuretic hormone (ANH)

28. The presence of ADH (antidiuretic hormone) causes an individual to excrete
    a. sugars.   c. more water.
    b. less water.   d. Both a and c are correct.

29. To lower blood acidity,
    a. hydrogen ions are excreted, and bicarbonate ions are reabsorbed.
    b. hydrogen ions are reabsorbed, and bicarbonate ions are excreted.
    c. hydrogen ions and bicarbonate ions are reabsorbed.
    d. hydrogen ions and bicarbonate ions are excreted.
    e. urea, uric acid, and ammonia are excreted.

30. The function of erythropoietin is
    a. reabsorption of sodium ions.
    b. excretion of potassium ions.
    c. reabsorption of water.
    d. to stimulate red blood cell production.
    e. to increase blood pressure.

In questions 31–34, match the function of the urinary system to the human organ system in the key.

Key:

    a. muscular system   e. respiratory system
    b. nervous system   f. digestive system
    c. endocrine system   g. reproductive system
    d. cardiovascular system

31. Liver synthesizes urea.

32. Smooth muscular contraction assists voiding of urine.

33. ADH, aldosterone, and atrial natriuretic hormone regulate reabsorption of Na^+ by kidneys.

34. Blood vessels deliver waste to be excreted.

35. Label this diagram of a nephron.

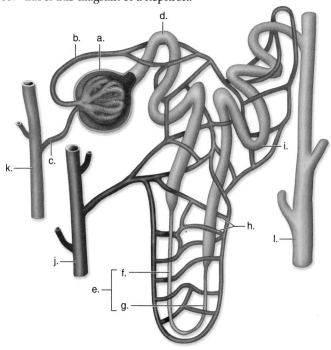

Thinking Critically About the Concepts

Kathy, from the opening story, learned that synthetic erythropoietin (EPO) is prescribed for people in kidney failure. Failing kidneys produce less erythropoietin, resulting in fewer red blood cells and symptoms of fatigue. Supplemental EPO will increase red blood cell synthesis and energy levels.

Athletes have been known to use EPO in an attempt to enhance their competitive performance. Athletes with a higher number of red blood cells than normal should be able to transport more oxygen, make more ATP, and have an edge over their competitors. This "new" form of blood doping is illegal (in sporting events), and several athletes have lost their Olympic medals or comparable titles after they were found to have used EPO.

1. Can you identify at least one other system connected to the three commercial storylines covered in the opening story?

2. Why would increasing the production of red blood cells in people undergoing chemotherapy alleviate their symptom of fatigue?

3. A side effect of using Detrol LA to calm an overactive bladder is constipation. Look back to Chapter 8 and identify the type of muscle found throughout the digestive system. Speculate as to why constipation might occur after the use of Detrol LA.

4. Men who have prostate cancer often have some or all of the prostate removed. The surgeon tries very hard not to damage the sphincter muscles associated with the bladder. What problem might a man experience if the bladder's sphincter muscles were damaged during the surgery?

5a. How would they identify athletes who have used EPO to enhance their performance?

 b. What risks are there to using EPO when there is no health-related reason to do so?

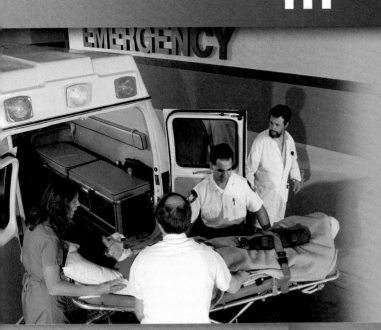

C H A P T E R **11**

Skeletal System

Katie was a bit dazed after falling off her horse. She finished landing in a sitting position, but remembered hitting her back, which now hurt quite a bit. Her friends quickly decided an ambulance was in order and called for one to come to the farm. During the wait for the ambulance, Katie was relieved to find that her fingers and toes all wiggled when she tried to move them. When the medics arrived, they quickly stabilized her neck and back on a backboard and carted her off to an area hospital.

After numerous X-rays and scans at the hospital, Katie was told she had a burst fracture of the T12 (12th thoracic) vertebra, but there wasn't any apparent damage to Katie's spinal cord. Treatment options, as presented by the orthopedic surgeon, were to stay in bed for up to six weeks or have surgery to fuse the burst vertebra to the ones above and below it. Katie opted for surgery because she could be out of bed in days. During the surgery, titanium rods and pins were inserted along Katie's vertebral column to stabilize the burst vertebra and facilitate the fusion of the T11, T12, and L1 (1st lumbar) vertebrae. As promised, Katie was up and walking within a day of the surgery.

As you study the material covered in this chapter, you'll learn about the cells and tissues associated with the skeletal system and what occurs to repair a broken bone. You'll discover the importance of the skeletal system to maintaining blood calcium (a contribution to homeostasis!) and the connections between the skeletal and endocrine systems that keep calcium levels within an acceptable range.

C H A P T E R   C O N C E P T S

11.1 Overview of Skeletal System
Bones are the organs of the skeletal system. The tissues of the system are compact and spongy bone, various types of cartilage, and fibrous connective tissue in the ligaments that hold bones together.

11.2 Bone Growth, Remodeling, and Repair
Bone is a living tissue that grows, remodels, and repairs itself. In all of these processes, some bone cells break down bone, and some repair bone.

11.3 Bones of the Axial Skeleton
The axial skeleton lies in the midline of the body and consists of the skull, the hyoid bone, the vertebral column, and the rib cage.

11.4 Bones of the Appendicular Skeleton
The appendicular skeleton consists of the bones of the pectoral girdle, upper limbs, pelvic girdle, and lower limbs.

11.5 Articulations
Joints are classified according to their degree of movement. Synovial joints are freely moveable.

11.1 Overview of Skeletal System

The skeletal system consists of the bones (206 in adults), along with the cartilage and fibrous connective tissue, which occurs in the ligaments at the joints.

Functions of the Skeleton

The skeleton supports the body. The bones of the legs support the entire body when we are standing, and bones of the pelvic girdle supports the abdominal cavity.

The skeleton protects soft body parts. The bones of the skull protect the brain; the rib cage protects the heart and lungs. The vertebrae protect the spinal cord, which makes nervous connections to all the muscles of the limbs. Katie, in the opening story, wiggled her fingers and toes to make sure her spinal cord was working properly. Her vertebrae, even though one was now compressed, had protected her spinal cord.

The skeleton produces blood cells. All bones in the fetus have red bone marrow that produces blood cells. In the adult, only certain bones produce blood cells.

The skeleton stores minerals and fat. All bones have a matrix that contains calcium phosphate, a source of calcium ions and phosphate ions in the blood. Fat is stored in yellow bone marrow.

The skeleton, along with the muscles, permits flexible body movement. While articulations (joints) occur between all the bones, we associate body movement in particular with the bones of limbs.

Anatomy of a Long Bone

Figure 11.1 shows the anatomy of a long bone. The shaft, or main portion of the bone, is called the diaphysis. The diaphysis has a large **medullary cavity,** whose walls are composed of compact bone. The medullary cavity is lined with a thin, vascular membrane (the endosteum) and is filled with yellow bone marrow that stores fat.

The expanded region at the end of a long bone is called an epiphysis (pl., epiphyses). The epiphyses are composed largely of spongy bone that contains red bone marrow, where blood cells are made. The epiphyses are coated with a thin layer of hyaline cartilage, which is called **articular cartilage** because it occurs at a joint.

Except for the articular cartilage on its ends, a long bone is completely covered by a layer of fibrous connective tissue called the **periosteum.** This covering contains blood vessels, lymphatic vessels, and nerves. Note in Figure 11.1 how a blood vessel penetrates the periosteum and enters the bone, where it gives off branches within the central canals. The periosteum is continuous with ligaments and tendons that are connected to a bone.

Bone

Compact bone is highly organized and composed of tubular units called osteons. In a cross section of an osteon, bone cells called **osteocytes** lie in lacunae, which are tiny chambers arranged in concentric circles around a central canal (Fig. 11.1).

Matrix fills the space between the rows of lacunae. Tiny canals called canaliculi (sing., canaliculus) run through the matrix, connecting the lacunae with one another and with the central canal. The cells stay in contact by strands of cytoplasm that extend into the canaliculi. Osteocytes nearest the center of an osteon exchange nutrients and wastes with the blood vessels in the central canal. These cells then pass on nutrients and collect wastes from the other cells via the gap junctions.

Compared with compact bone, **spongy bone** has an unorganized appearance (Fig. 11.1). It contains numerous thin plates (called trabeculae) separated by unequal spaces. Although this makes spongy bone lighter than compact bone, spongy bone is still designed for strength. Just as braces are used for support in buildings, the trabeculae follow lines of stress. The spaces of spongy bone are often filled with **red bone marrow,** a specialized tissue that produces all types of blood cells. The osteocytes of spongy bone are irregularly placed within the trabeculae, and canaliculi bring them nutrients from the red bone marrow.

Cartilage

Cartilage is not as strong as bone, but it is more flexible because the matrix is gel-like and contains many collagenous and elastic fibers. The cells, called **chondrocytes,** lie within lacunae that are irregularly grouped. Cartilage has no nerves, making it well suited for padding joints where the stresses of movement are intense. Cartilage also has no blood vessels, making it slow to heal.

The three types of cartilage differ according to the type and arrangement of fibers in the matrix. *Hyaline cartilage* is firm and somewhat flexible. The matrix appears uniform and glassy, but actually it contains a generous supply of collagen fibers. Hyaline cartilage is found at the ends of long bones, in the nose, at the ends of the ribs, and in the larynx and trachea.

Fibrocartilage is stronger than hyaline cartilage because the matrix contains wide rows of thick, collagen fibers. Fibrocartilage is able to withstand both tension and pressure, and is found where support is of prime importance—in the disks located between the vertebrae and also in the cartilage of the knee.

Elastic cartilage is more flexible than hyaline cartilage because the matrix contains mostly elastin fibers. This type of cartilage is found in the ear flaps and the epiglottis.

Fibrous Connective Tissue

Fibrous connective tissue contains rows of cells called fibroblasts separated by bundles of collagenous fibers. This tissue makes up the **ligaments** that connect bone to bone and the **tendons** that connect muscles to a bone at **joints,** which are also called articulations.

☑ Check Your Progress 11.1

1. a. What is the anatomy of compact bone? b. Of spongy bone?
2. What are the three types of cartilage, and where are they found in the body?
3. What type of connective tissue makes up ligaments and tendons?
4. What is the structure of a typical bone, such as a long bone?

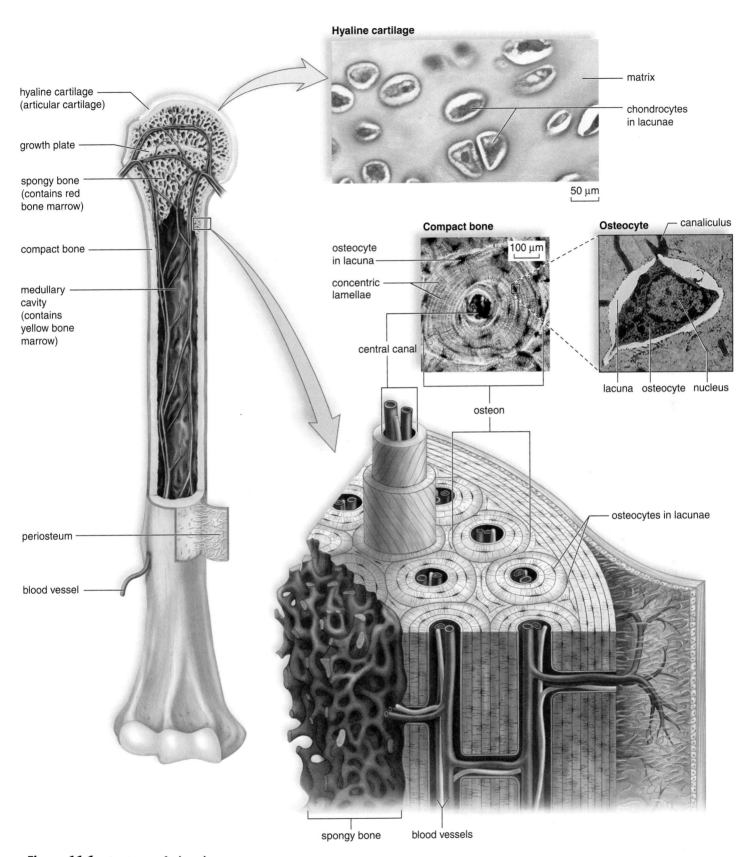

Hyaline cartilage

matrix

chondrocytes in lacunae

50 μm

hyaline cartilage (articular cartilage)

growth plate

spongy bone (contains red bone marrow)

compact bone

medullary cavity (contains yellow bone marrow)

periosteum

blood vessel

Compact bone

osteocyte in lacuna

concentric lamellae

central canal

100 μm

Osteocyte

canaliculus

lacuna   osteocyte   nucleus

osteon

osteocytes in lacunae

spongy bone   blood vessels

Figure 11.1 Anatomy of a long bone.

Left: A long bone is encased by fibrous membrane (periosteum) except where it is covered at the ends by hyaline cartilage (see micrograph). Spongy bone located beneath the cartilage may contain red bone marrow. The central shaft contains yellow bone marrow and is bordered by compact bone, which is shown in the enlargement and micrograph (*right*).

11.2 Bone Growth, Remodeling, and Repair

The importance of the skeleton to the human form is evident by its early appearance during development. The skeleton starts forming at about six weeks, when the embryo is only about 12 mm (0.5 in.) long. Most bones grow in length and width through adolescence, but some continue enlarging until about age 25. In a sense, bones can grow throughout a lifetime because they are able to respond to stress by changing size, shape, and strength. This process is called remodeling. If a bone fractures, it can heal by what is called bone repair.

Bones are composed of living tissues, as exemplified by their ability to grow, remodel, and undergo repair. Several different types of cells are involved in bone growth, remodeling, and repair:

Osteoblasts are bone-forming cells. They secrete the organic matrix of bone and promote the deposition of calcium salts into the matrix.

Osteocytes are mature bone cells derived from osteoblasts. They maintain the structure of bone.

Osteoclasts are bone-absorbing cells. They break down bone and assist in depositing calcium and phosphate in the blood.

Throughout life, osteoclasts are removing the matrix of bone, and osteoblasts are building it up. When osteoblasts are surrounded by calcified matrix, they become the osteocytes within lacunae.

Bone Development and Growth

The term **ossification** refers to the formation of bone. The bones of the skeleton form during embryonic development in two distinctive ways: intramembranous ossification and endochondral ossification (Fig. 11.2).

Intramembranous Ossification

Flat bones, such as the bones of the skull, are examples of intramembranous bones. In **intramembranous ossification,** bones develop between sheets of fibrous connective tissue. Here, cells derived from connective tissue cells become osteoblasts located in ossification centers. The osteoblasts secrete the organic matrix of bone consisting of mucopolysaccharides and collagen fibrils. Calcification occurs when calcium salts are added to the organic matrix. The osteoblasts promote calcification. Ossification results in the trabeculae of spongy bone, and spongy bone remains on the inside. The spongy bone of flat bones, such as those of the skull and clavicles (collarbones), contains red bone marrow.

A periosteum forms outside the spongy bone and osteoblasts, derived from the periosteum, carry out further ossification. Trabeculae form and fuse to become compact bone. The compact bone forms a bone collar that surrounds the spongy bone on the inside.

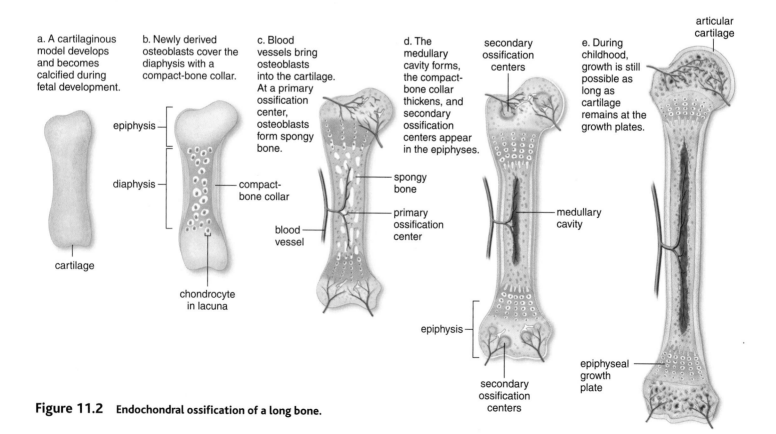

a. A cartilaginous model develops and becomes calcified during fetal development.

b. Newly derived osteoblasts cover the diaphysis with a compact-bone collar.

c. Blood vessels bring osteoblasts into the cartilage. At a primary ossification center, osteoblasts form spongy bone.

d. The medullary cavity forms, the compact-bone collar thickens, and secondary ossification centers appear in the epiphyses.

e. During childhood, growth is still possible as long as cartilage remains at the growth plates.

epiphysis

diaphysis

cartilage

compact-bone collar

chondrocyte in lacuna

blood vessel

spongy bone

primary ossification center

secondary ossification centers

epiphysis

secondary ossification centers

medullary cavity

articular cartilage

epiphyseal growth plate

Figure 11.2   Endochondral ossification of a long bone.

Endochondral Ossification

Most of the bones of the human skeleton are formed by **endochondral ossification.** During endochondral ossification, bone replaces the cartilaginous models of the bones. Gradually, the cartilage is replaced by the calcified bone matrix that makes these bones capable of bearing weight.

Inside, bone formation spreads from the center to the ends, and this accounts for the term used for this type of ossification. The long bones, such as the tibia, provide examples of endochondral ossification (Fig. 11.2).

1. *The cartilage model.* Chondrocytes lay down hyaline cartilage, which is shaped like the future bones. Therefore, they are called cartilage models of the future bones. As the cartilage models calcify, the chondrocytes die off.
2. *The bone collar.* Osteoblasts, derived from the newly formed periosteum, secrete the organic bone matrix, and the matrix undergoes calcification. The result is a bone collar, which covers the diaphysis (Fig. 11.2). The bone collar is composed of compact bone. In time, the bone collar thickens.
3. *The primary ossification center.* Blood vessels bring osteoblasts to the interior, and they begin to lay down spongy bone. This region is called a primary ossification center because it is the first center for bone formation.
4. *The medullary cavity and secondary ossification sites.* The spongy bone of the diaphysis is absorbed by osteoclasts, and the cavity created becomes the medullary cavity. Shortly after birth, secondary ossification centers form in the epiphyses. Spongy

bone persists in the epiphyses, and they contain red bone marrow for quite some time. Cartilage is present at two locations: the epiphyseal (growth) plate and articular cartilage.

5. *The epiphyseal (growth) plate.* A band of cartilage called a **growth plate** remains between the primary ossification center and each secondary center. The limbs keep increasing in length as long as growth plates are still present.

Figure 11.3 shows that the epiphyseal plate contains four layers. The layer nearest the epiphysis is the resting zone, where cartilage remains. The next layer is the proliferating zone, in which chondrocytes are producing new cartilage cells. In the third layer, the degenerating zone, the cartilage cells are dying off; and in the fourth layer, the ossification zone, bone is forming. Bone formation here causes the length of the bone to increase. The inside layer of articular cartilage also undergoes ossification in the manner described.

The diameter of a bone enlarges as a bone lengthens. Osteoblasts derived from the periosteum are active in new bone deposition as osteoclasts enlarge the medullary cavity from inside.

Final Size of the Bones   When the epiphyseal plates close, bone length can no longer occur. The epiphyseal plates in the arms and legs of women close at about age 18, while they do not close in men until about age 20. Portions of other types of bones may continue to grow until age 25. Hormones are chemical messengers secreted by the endocrine glands and distributed about the body by the bloodstream. Hormones control the activity of the epiphyseal plate, as is discussed next.

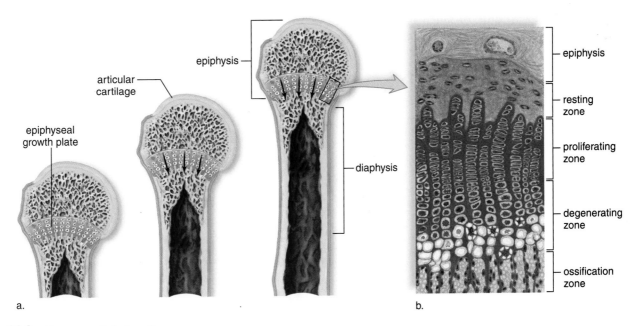

Figure 11.3   Bone growth in length.
a. Length of a bone increases when cartilage is replaced by bone at the growth plate. **b.** Chondrocytes produce new cartilage in the proliferating zone, and cartilage becomes bone in the ossification zone closest to the diaphysis. Arrows show the direction of ossification.

Hormones Affect Bone Growth

The importance of bone growth is signified by the involvement of several hormones in bone growth. A **hormone** is a chemical messenger, produced by one part of the body, which acts on a different part of the body.

Vitamin D is formed in the skin when it is exposed to sunlight, but it can also be consumed in the diet. Milk, in particular, is often fortified with vitamin D today. In the kidneys, vitamin D is converted to a hormone that acts on the intestinal tract. The chief function of vitamin D is intestinal absorption of calcium. In the absence of vitamin D, children can develop rickets, a condition marked by bone deformities including bowed long bones (see Fig. 8.15*a*).

Growth hormone (GH) directly stimulates growth of the epiphyseal plate, as well as bone growth in general. However, growth hormone will be somewhat ineffective if the metabolic activity of cells is not promoted. Thyroid hormone, in particular, promotes the metabolic activity of cells. Too much growth hormone during childhood can produce excessive growth and even gigantism, while too little can cause dwarfism. Too much GH produces gigantism in youths prior to epiphyseal fusion and acromegaly in adults following epiphyseal fusion (see Fig. 15.7 and Fig. 15.8).

Adolescents usually experience a dramatic increase in height, called the growth spurt, due to an increased level of sex hormones, which apparently stimulate osteoblast activity to the point that the epiphyseal plates become "paved over" by the faster growing bone tissue, within one or two years of the onset of puberty.

Bone Remodeling and Its Role in Homeostasis

Bone is constantly being broken down by osteoclasts and reformed by osteoblasts in the adult. As much as 18% of bone is recycled each year. This process of bone renewal, often called **bone remodeling,** normally keeps bones strong (Fig. 11.4). In Paget's disease, new bone is generated at a faster-than-normal rate. This rapid remodeling produces bone that's softer and weaker than normal bone and can cause bone pain, deformities, and fractures.

Bone recycling allows the body to regulate the amount of calcium in the blood. To illustrate that the blood calcium level is critical, recall that calcium is required for blood to clot. Also, if the blood calcium concentration is too high, neurons and muscle cells no longer function, and if it falls too low, they become so excited, convulsions occur. The bones are the storage sites for calcium—if the blood calcium rises above normal, at least some of the excess is deposited in the bones, and if the blood calcium dips too low, calcium is removed from the bones to bring it back up to the normal level.

Two hormones, in particular, are involved in regulating the blood calcium level. Parathyroid hormone (PTH) accelerates bone recycling and, in that way, increases the blood calcium level. Calcitonin is a hormone that acts opposite to PTH. The female sex hormone estrogen can actually increase the number of osteoblasts, and the reduction of estrogen in older women is often given as reason for the development of weak bones, called osteoporosis. Osteoporosis is discussed in the Health Focus on page 213. In the young adult, the activity of osteoclasts is matched by the activity of osteoblasts, and bone mass remains stable until about age 45 years in women. After that age, bone mass starts to decrease. Men also experience osteoporosis later in life.

Bone remodeling also accounts for why bones can respond to stress (Fig. 11.4). If you engage in an activity that calls upon the use of a particular bone, it will enlarge in diameter at the region most affected by the activity. During this process, osteoblasts in the periosteum form compact bone around the external bone surface, and osteoclasts break down bone on the internal bone surface, around the medullary cavity. This prevents the bones from getting too heavy and thick. Today, exercises, such as walking, jogging, and weight lifting, are recommended ways of keeping the bones strong because they stimulate the work of osteoblasts instead of osteoclasts.

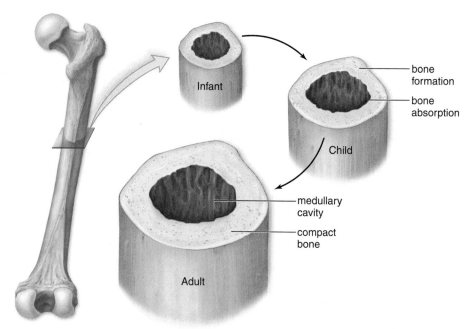

Figure 11.4   Bone growth and remodeling.
Diameter of a bone increases as bone absorption occurs inside the shaft and is matched by bone formation outside the shaft.

Health Focus

You Can Avoid Osteoporosis

Osteoporosis is a condition in which the bones are weakened due to a decrease in the bone mass that makes up the skeleton. The skeletal mass continues to increase until ages 20 to 30. After that, there is an equal rate of formation and breakdown of bone mass until ages 40 to 50. Then, reabsorption begins to exceed formation, and the total bone mass slowly decreases.

Over time, men are apt to lose 25% and women to lose 35% of their bone mass. But we have to consider that men, unless they have taken asthma medications that decrease bone formation, tend to have denser bones than women anyway, and their testosterone (male sex hormone) level generally does not begin to decline significantly until after age 65. In contrast, the estrogen (female sex hormone) level in women begins to decline at about age 45. Since sex hormones play an important role in maintaining bone strength, this difference means that women are more likely than men to suffer a higher incidence of fractures, involving especially the hip, vertebrae, long bones, and pelvis. Although osteoporosis may at times be the result of various disease processes, it is essentially a disease that occurs as we age.

How to Avoid Osteoporosis

Everyone can take measures to avoid having osteoporosis when they get older. Adequate dietary calcium throughout life is an important protection against osteoporosis. The U.S. National Institutes of Health recommend a calcium intake of 1,200–1,500 mg per day during puberty. Males and females require 1,000 mg per day until the age of 65 and 1,500 mg per day after age 65. In older women, 1,500 mg per day is especially desirable.

A small daily amount of vitamin D is also necessary for the body to use calcium correctly. Exposure to sunlight is required to allow skin to synthesize a precursor to vitamin D. If you reside on or north of a "line" drawn from Boston to Milwaukee, to Minneapolis, to Boise, chances are you're not getting enough vitamin D during the winter months. Therefore, you should avail yourself of vitamin D present in fortified foods such as low-fat milk and cereal.

Very inactive people, such as those confined to bed, lose bone mass 25 times faster than people who are moderately active. On the other hand, a regular, moderate, weight-bearing exercise such as walking or jogging is another good way to maintain bone strength (Fig. 11A).

Diagnosis and Treatment

Postmenopausal women with any of the following risk factors should have an evaluation of their bone density:

- white or Asian race
- thin body type
- family history of osteoporosis
- early menopause (before age 45)
- smoking
- a diet low in calcium, or excessive alcohol consumption and caffeine intake
- sedentary lifestyle

Presently, bone density is measured by a method called dual energy X-ray absorptiometry (DEXA). This test measures bone density based on the absorption of photons generated by an X-ray tube. Soon there may be blood and urine tests to detect the biochemical markers of bone loss. Then it will be made easier for physicians to screen older women and at-risk men for osteoporosis.

If the bones are thin, it is worthwhile to take all possible measures to gain bone density because even a slight increase can significantly reduce fracture risk. Although estrogen therapy does reduce the incidence of hip fractures, long term estrogen therapy is rarely recommended today for osteoporosis because estrogen is known to increase the risk of breast cancer, heart disease, stroke, and blood clots. Other medications are available, however. Calcitonin, a thyroid hormone, has been shown to increase bone density and strength, while decreasing the rate of bone fractures. Also, the bisphosphonates are a family of non-hormonal drugs used to prevent and treat osteoporosis. To achieve optimal results with calcitonin or one of the bisphosphonates, patients should also receive adequate amounts of calcium and vitamin D.

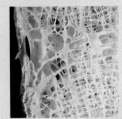

a. normal bone

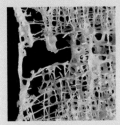

b. osteoporosis

c.

Figure 11A **Preventing osteoporosis.**
Weight-lifting exercise, when we are young, can help prevent osteoporosis when we are older. **a.** Normal bone growth compared with bone from a person with (**b**) osteoporosis. **c.** An elderly person with osteoporosis.

Bone Repair

Repair of a bone is required after it breaks or fractures. Fracture repair takes place over a span of several months in a series of four steps, shown in Figure 11.5 and listed here:

1. *Hematoma.* After a fracture, blood escapes from ruptured blood vessels and forms a hematoma (mass of clotted blood) in the space between the broken bones within 6–8 hours.
2. *Fibrocartilaginous callus.* Tissue repair begins, and a fibrocartilaginous callus fills the space between the ends of the broken bone for about three weeks.
3. *Bony callus.* Osteoblasts produce trabeculae of spongy bone and convert the fibrocartilage callus to a bony callus that joins the broken bones together and lasts about three to four months.
4. *Remodeling.* Osteoblasts build new compact bone at the periphery, and osteoclasts absorb the spongy bone, creating a new medullary cavity.

In some ways, bone repair parallels the development of a bone except that the first step, hematoma, indicates that injury has occurred, and then a fibrocartilaginous callus precedes the production of compact bone.

The naming of fractures tells you what kind of break occurred. A fracture is complete if the bone is broken clear through and incomplete if the bone is not separated into two parts. A fracture is simple if it does not pierce the skin and compound if it does pierce the skin. Impacted means that the broken ends are wedged into each other, and a spiral fracture occurs when the break is ragged due to twisting of a bone.

Vertebral Fusion

Katie had not actually broken a bone; her vertebra had been compressed when she fell from the horse. A compressed vertebra is called a *burst* vertebra because it spreads out in all directions. During fusion surgery, bone grafts are placed around the spine. Then, like healing a fracture, the body joins the grafts to the vertebrae, "welding," or fusing, the vertebrae together. After fusion, the rods and pins that held the vertebrae together are no longer needed and are removed.

☑ Check Your Progress 11.2

1. What are the types of cells involved in bone growth, remodeling, and repair?
2. How does bone growth occur during development?
3. What hormones are involved in bone growth?
4. How does remodeling affect the blood calcium concentration?
5. How does bone repair itself?

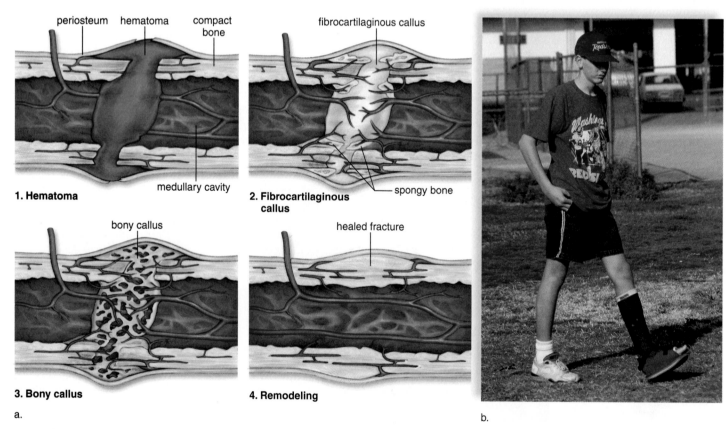

1. **Hematoma**   2. **Fibrocartilaginous callus**

3. **Bony callus**   4. **Remodeling**

a.   b.

Figure 11.5 **Bone fracture and repair.**
a. Steps in the repair of a fracture. **b.** A cast helps stabilize the bones while repair takes place. Fiberglass casts are now replacing plaster of Paris as the usual material for a cast.

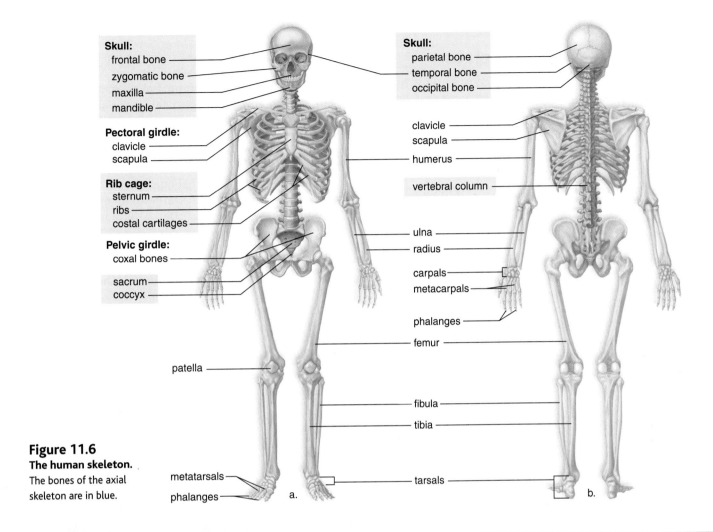

Figure 11.6
The human skeleton.
The bones of the axial skeleton are in blue.

Science **Focus**

Identifying Skeletal Remains

Regardless of how, when, and where human bones are unexpectedly found, many questions must be answered. How old was this person at the time of death? Are these the bones of a male or female? What was the ethnicity? Are there any signs that this person was murdered?

Clues about the identity and history of the deceased person are available throughout the skeleton. Age is approximated by *dentition,* or the structure of the teeth in the upper jaw (maxilla) and lower jaw (mandible). For example, infants aged 0–4 months will have no teeth present; children aged approximately 6–10 years will have missing deciduous, or "baby teeth;" young adults acquire their last molars, or "wisdom teeth," around age 20. The age of older adults may be approximated by the number and location of missing or broken teeth. Studying areas of bone ossification also gives clues to the age of the deceased at the time of death. In older adults, signs of joint breakdown provide additional information about age. Hyaline cartilage becomes worn, yellowed, and brittle with age, and the hyaline cartilages covering bone ends wear down over time. The amount of yellowed, brittle, or missing cartilage helps scientists guess the person's age.

If skeletal remains include the individual's pelvic bones, these provide the best method for determining an adult's gender. The pelvis is more shallow and wider in the female than in the male. The long bones, particularly the humerus and femurs, give information about gender, as well. Long bones will be thicker and denser in males, and points of muscle attachment will be bigger and more prominent. The skull of a male will have a square chin and more prominent ridges above the eye sockets or orbits.

Determining ethnic origin of skeletal remains can be difficult, since so many people have a mixed racial heritage. Forensic anatomists rely on observed racial characteristics of the skull. In general, individuals of African or African American descent will have a greater distance between the eyes, eye sockets that are roughly rectangular, and a jaw that is large and prominent. Skulls of Native Americans will typically have round eye sockets, prominent cheek (zygomatic) bones, and a rounded palate. Caucasian skulls will usually have a U-shaped palate, and a suture line between the frontal bones will often be visible. Additionally, the external ear canals in Caucasians are long and straight, so that the auditory ossicles can be seen.

Once the identity of the individual has been determined, the skeletal remains can be returned to the family for proper burial.

11.3 Bones of the Axial Skeleton

The 206 bones of the skeleton are classified according to whether they occur in the axial skeleton or the appendicular skeleton (see Fig. 11.6). The **axial skeleton** lies in the midline of the body and consists of the skull, hyoid bone, vertebral column, and the rib cage.

The Skull

The **skull** is formed by the cranium (braincase) and the facial bones. It should be noted, however, that some cranial bones contribute to the face.

The Cranium   The cranium protects the brain. In adults, it is composed of eight bones fitted tightly together. In newborns, certain cranial bones are not completely formed and instead are joined by membranous regions called **fontanels.** The fontanels usually close by the age of 16 months by the process of intramembranous ossification.

Some of the bones of the cranium contain the **sinuses,** air spaces lined by mucous membrane. The sinuses reduce the weight of the skull and give a resonant sound to the voice. Two sinuses, called the mastoid sinuses, drain into the middle ear. **Mastoiditis,** a condition that can lead to deafness, is an inflammation of these sinuses.

The major bones of the cranium have the same names as the lobes of the brain: frontal, parietal, occipital, and temporal. On the top of the cranium (Fig. 11.7a), the **frontal bone** forms the forehead, the **parietal bones** extend to the sides, and the **occipital bone** curves to form the base of the skull. Here there is a large opening, the **foramen magnum** (Fig. 11.7b), through which the spinal cord passes and becomes the brain stem. Below the much larger parietal bones, each **temporal bone** has an opening (external auditory canal) that leads to the middle ear.

The **sphenoid bone,** which is shaped like a bat with outstretched wings, extends across the floor of the cranium from one side to the other. The sphenoid is the keystone of the cranial bones because all the other bones articulate with it. The sphenoid completes the sides of the skull and also contributes to forming the orbits (eye sockets). The **ethmoid bone,** which lies in front of the sphenoid, also helps form the orbits and the nasal septum. The orbits are completed by various facial bones. The eye sockets are called orbits because we can rotate our eyes.

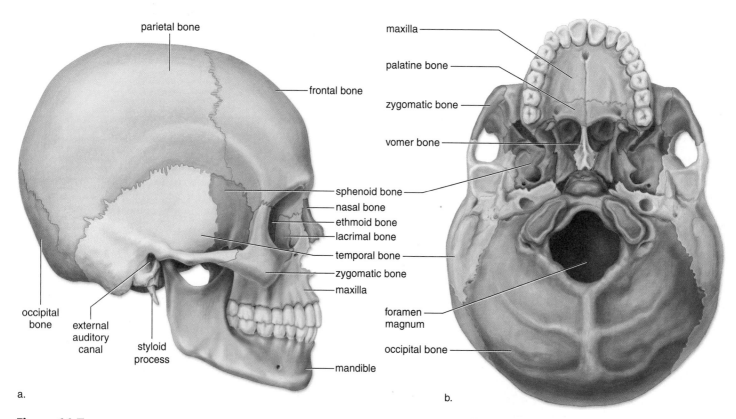

Figure 11.7   Bones of the skull.
a. Lateral view. **b.** Inferior view.

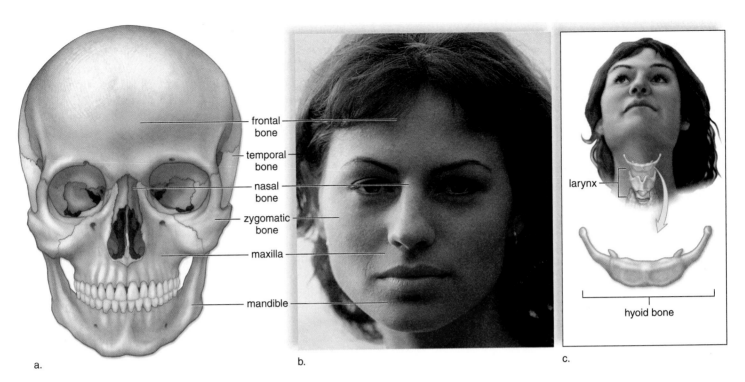

Figure 11.8 **Bones of the face and the hyoid.**
a. The frontal bone forms the forehead and eyebrow ridges; the zygomatic bones form the cheekbones; and the maxillae have numerous functions. They assist in the formation of the eye sockets and the nasal cavity. They form the upper jaw and contain sockets for the upper teeth. The mandible is the lower jaw with sockets for the lower teeth. The mandible has a projection we call the chin. **b.** The maxillae, frontal, and nasal bones help form the external nose. **c.** The hyoid bone is located as shown.

The Facial Bones The most prominent of the facial bones are the mandible, the maxillae (sing., maxilla), the zygomatic bones, and the nasal bones.

The **mandible,** or lower jaw, is the only movable portion of the skull (Fig. 11.8a,b), and its action permits us to chew our food. It also forms the chin. Tooth sockets are located on the mandible and on the **maxillae,** the bones that form the upper jaw and the anterior portion of the hard palate. The palatine bones make up the posterior portion of the hard palate and the floor of the nose (see Fig. 11.7b).

The lips and cheeks have a core of skeletal muscle. The **zygomatic bones** are the cheekbone prominences, and the **nasal bones** form the bridge of the nose. Other bones (e.g., ethmoid and vomer) are a part of the nasal septum, which divides the interior of the nose into two nasal cavities. The lacrimal bone (see Fig. 11.7a) contains the opening for the nasolacrimal canal, which drains tears from the eyes to the nose.

Certain cranial bones contribute to the face. The sphenoid wings account for the flattened areas we call the temples. The frontal bone forms the forehead and has supraorbital ridges, where the eyebrows are located. Glasses sit where the frontal bone joins the nasal bones.

The exterior portion of ears are formed only by cartilage and not by bone, while the nose is a mixture of bones, cartilages, and connective tissues. The cartilages complete the tip of the nose, and fibrous connective tissue forms the flared sides of the nose.

The Hyoid Bone

Although the **hyoid bone** is not part of the skull, it will be mentioned here because it is a part of the axial skeleton. It is the only bone in the body that does not articulate with another bone (Fig. 11.8c). It is attached to the temporal bones by muscles and ligaments and to the larynx by a membrane. The larynx is the voice box at the top of the trachea in the neck region. The hyoid bone anchors the tongue and serves as the site for the attachment of muscles associated with swallowing. Due to its position, the hyoid bone is not usually easy to fracture in most situations. In cases of suspicious death, however, a fractured hyoid is a strong sign of *manual strangulation.*

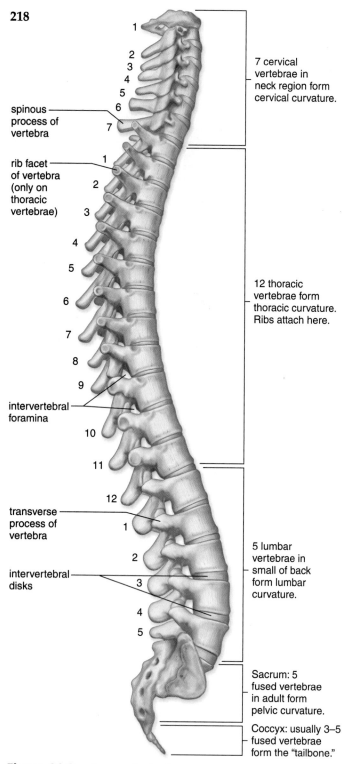

spinous process of vertebra

rib facet of vertebra (only on thoracic vertebrae)

intervertebral foramina

transverse process of vertebra

intervertebral disks

1
2
3
4
5
6
7

1
2
3
4
5
6
7
8
9
10
11
12

1
2
3
4
5

7 cervical vertebrae in neck region form cervical curvature.

12 thoracic vertebrae form thoracic curvature. Ribs attach here.

5 lumbar vertebrae in small of back form lumbar curvature.

Sacrum: 5 fused vertebrae in adult form pelvic curvature.

Coccyx: usually 3–5 fused vertebrae form the "tailbone."

Figure 11.9 **The vertebral column.**
The vertebral column is made up of 33 vertebrae separated by intervertebral disks. The intervertebral disks make the column flexible. The vertebrae are named for their location in the vertebral column. For example, the thoracic vertebrate are located in the thorax. Note that humans have a coccyx, which is also called a tailbone.

The Vertebral Column

The **vertebral column** consists of 33 vertebrae (Fig. 11.9). Normally, the vertebral column has four curvatures that provide more resilience and strength for an upright posture than a straight column could provide. **Scoliosis** is an abnormal lateral (sideways) curvature of the spine. Two other well-known abnormal curvatures are kyphosis, an abnormal posterior curvature that often results in a hunchback, and lordosis, an abnormal anterior curvature resulting in a swayback.

The vertebral column forms when the vertebrae join. The spinal cord, which passes through the vertebral canal, gives off the spinal nerves at the intervertebral foramina. Spinal nerves function to control skeletal muscle contraction among other functions. You'll recall that Katie wiggled her fingers and toes to make sure her spinal cord was still functioning after she fell from the horse and suffered a back injury. Katie was very lucky because when a vertebra is compressed (called a burst vertebra), it spreads out in all directions, and the spinal cord is liable to be injured; the result can be complete paralysis.

The spinous processes of the vertebrae can be felt as bony pojections along the midline of the back. The spinous processes and also the transverse processes, which extend laterally, serve as attachment sites for the muscles that move the vertebral column.

Types of Vertebrae The various vertebrae are named according to their location in the vertebral column. The cervical vertebrae are located in the neck. The first cervical vertebra, called the **atlas,** holds up the head. It is so named because Atlas, of Greek mythology, held up the world. Movement of the atlas permits the "yes" motion of the head. It also allows the head to tilt from side to side. The second cervical vertebra is called the **axis** because it allows a degree of rotation, as when we shake the head "no." The thoracic vertebrae have long, thin, spinous processes and articular facets for the attachment of the ribs (Fig. 11.10*a*). Katie damaged her twelfth thoracic vertebra (T12), which is the last one before the lumbar vertebrae. Lumbar vertebrae have a large body and thick processes. The five sacral vertebrae are fused together in the sacrum. The coccyx, or tailbone, is usually composed of four fused vertebrae.

Intervertebral Disks Between the vertebrae are **intervertebral disks** composed of fibrocartilage that acts as a kind of padding. They prevent the vertebrae from grinding against one another and absorb shock caused by movements such as running, jumping, and even walking. The presence of the disks allows the vertebrae to move as we bend forward,

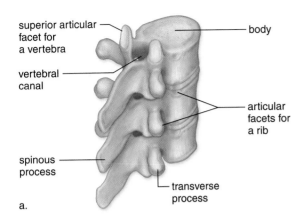

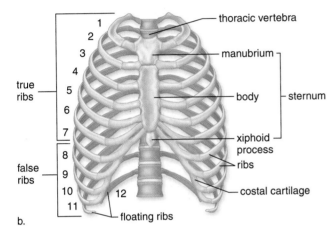

Figure 11.10 Thoracic vertebrae and the rib cage.
a. The thoracic vertebrae articulate with one another and also with the ribs at articular facets. A thoracic vertebra has two facets for articulation with a rib; one is on the body, and the other is on the transverse process. **b.** The rib cage consists of the 12 thoracic vertebrae, the 12 pairs of ribs, the costal cartilages, and the sternum. The rib cage protects the lungs and the heart.

backward, and from side to side. Unfortunately, these disks become weakened with age and can herniate and rupture. Pain results if a disk presses against the spinal cord and/or spinal nerves. If that occurs, surgical removal of the disk may relieve the pain.

The Rib Cage

The rib cage, also called the thoracic cage, is composed of the thoracic vertebrae, the ribs and their associated cartilages, and the sternum (Fig. 11.10b). The rib cage is part of the axial skeleton.

The rib cage demonstrates how the skeleton is protective but also flexible. The rib cage protects the heart and lungs; yet it swings outward and upward upon inspiration and then downward and inward upon expiration.

The Ribs

A rib is a flattened bone that originates at the thoracic vertebrae and proceeds toward the anterior thoracic wall. There are 12 pairs of ribs. All 12 pairs connect directly to the thoracic vertebrae in the back. A rib articulates with the body and transverse process of its corresponding thoracic vertebra. Each rib curves outward and then forward and downward.

The upper seven pairs of ribs connect directly to the sternum by means of costal cartilages. These are called the "true ribs." The "false ribs" include the next three pairs of ribs, which connect to the sternum by means of a common cartilage and also the last two pairs called "floating ribs" because they do not attach to the sternum at all. A floating rib is attached to T12, the thoracic vertebra that Katie damaged when she fell from the horse.

The Sternum

The **sternum** lies in the midline of the body. It, along with the ribs, helps protect the heart and lungs. The sternum, or breastbone, is a flat bone that has the shape of a knife.

The sternum is composed of three bones. These bones are the manubrium (the handle), the body (the blade), and the xiphoid process (the point of blade). The manubrium articulates with the clavicles of the appendicular skeleton and the first pair of ribs. The manubrium joins with the body of the sternum at an angle. This is an important anatomical landmark because it occurs at the level of the second rib and therefore allows the ribs to be counted. Counting the ribs is sometimes done to determine where the apex of the heart is located—usually between the fifth and sixth ribs.

The xiphoid process is the third part of the sternum. The variably shaped xiphoid process serves as an attachment site for the diaphragm, which separates the thoracic cavity from the abdominal cavity.

☑ Check Your Progress 11.3

1. What are the bones of the axial skeleton?
2. What are the bones of the cranium, and how are they situated?
3. What are the bones of the face, and how do they contribute to facial features?
4. **a.** Describe the various types of vertebrae and state their function. **b.** What is the structure and function of the rib cage?

11.4 Bones of the Appendicular Skeleton

The **appendicular skeleton** consists of the bones within the pectoral and pelvic girdles and their attached limbs. A pectoral (shoulder) girdle and upper limb are specialized for flexibility; the pelvic (hip) girdle and lower limbs are specialized for strength.

The Pectoral Girdle and Upper Limb

The body has left and right **pectoral girdles;** each consists of a scapula (shoulder blade) and a clavicle (collarbone) (Fig. 11.11). The **clavicle** extends across the top of the thorax; it articulates with (joins with) the sternum and the acromion process of the **scapula,** a visible bone in the back. The muscles of the arm and chest attach to the coracoid process of the scapula. The **glenoid cavity** of the scapula articulates with, and is much smaller than, the head of the humerus. This allows the arm to move in almost any direction, but reduces stability. This is the joint that is most apt to dislocate. Ligaments, and also tendons, stabilize this joint. Tendons that extend to the humerus from four small muscles originating on the scapula form the **rotator cuff.** Vigorous circular movements of the arm can lead to rotator cuff injuries.

The components of a pectoral girdle follow freely the movements of the upper limb, which consists of the humerus within the arm and the radius and ulna within the forearm. The **humerus,** the single long bone in the arm, has a smoothly rounded head that fits into the glenoid cavity of the scapula as mentioned. The shaft of the humerus has a tuberosity (protuberance) where the deltoid, a shoulder muscle, attaches. You can determine, even after death, that the person did a lot of heavy lifting during their lifetime by the size of their deltoid tuberosity.

The far end of the humerus has two protuberances, called the capitulum and the trochlea, which articulate respectively with the **radius** and the **ulna** at the elbow. The bump at the back of the elbow is the olecranon process of the ulna.

When the upper limb is held so that the palm is turned forward, the radius and ulna are about parallel to each other. When the upper limb is turned so that the palm is turned backward, the radius crosses in front of the ulna, a feature that contributes to the easy twisting motion of the forearm.

The hand has many bones, and this increases its flexibility. The wrist has eight **carpal** bones, which look like small pebbles. From these, five **metacarpal** bones fan out to form a framework for the palm. The metacarpal bone that leads to the thumb is opposable to the other digits. (**Digits** is a term that refers to either fingers or toes.) The knuckles are the enlarged distal ends of the metacarpals. Beyond the metacarpals are the **phalanges,** the bones of the fingers and the thumb. The phalanges of the hand are long, slender, and lightweight.

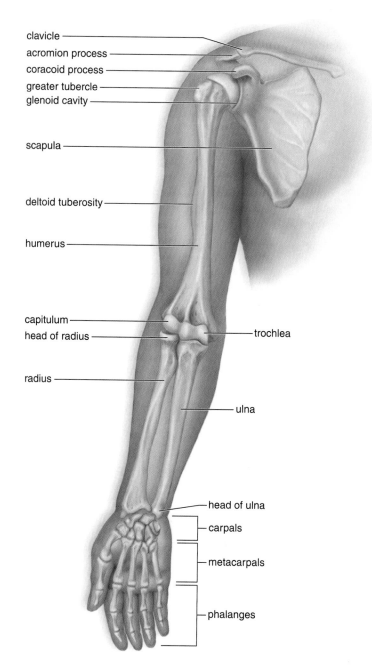

Figure 11.11   **Bones of the pectoral (shoulder) girdle and upper limb.**

Pectoral girdle
- scapula and clavicle

Upper limb
- arm contains humerus
- forearm contains radius and ulna
- hand contains carpals, metacarpals, and phalanges

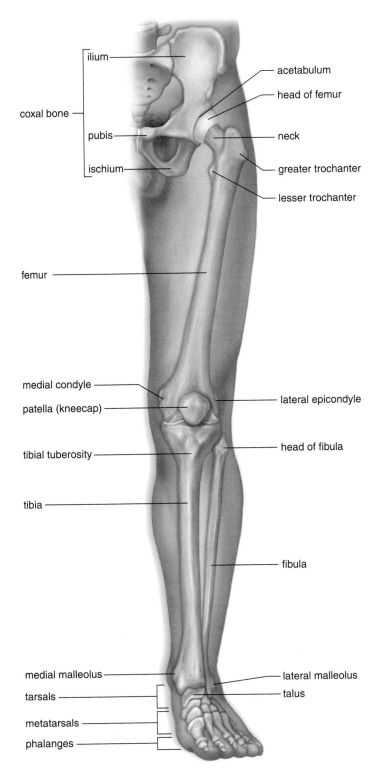

ilium

acetabulum

head of femur

coxal bone

pubis

neck

ischium

greater trochanter

lesser trochanter

femur

medial condyle

patella (kneecap)

lateral epicondyle

tibial tuberosity

head of fibula

tibia

fibula

medial malleolus

lateral malleolus

tarsals

talus

metatarsals

phalanges

Figure 11.12 **Bones of the pelvic girdle and lower limb.**

Pelvic girdle
- coxal bones

Lower limb
- thigh contains femur
- leg contains tibia and fibula
- foot contains tarsals, metatarsals, and phalanges

The Pelvic Girdle and Lower Limb

Figure 11.12 shows how the lower limb is attached to the pelvic girdle. The **pelvic girdle** (hip girdle) consists of two heavy, large coxal bones (hip bones). The **pelvis** is a basin composed of the pelvic girdle, sacrum, and coccyx. The pelvis bears the weight of the body, protects the organs within the pelvic cavity, and serves as the place of attachment for the legs.

Each **coxal bone** has three parts: the ilium, the ischium, and the pubis, which are fused in the adult (Fig. 11.12). The hip socket, called the acetabulum, occurs where these three bones meet. The ilium is the largest part of the coxal bones, and our hips occur where it flares out. We sit on the ischium, which has a posterior spine called the ischial spine for muscle attachment. The pubis, from which the term pubic hair is derived, is the anterior part of a coxal bone. The two pubic bones are joined together by a fibrocartilaginous joint, the **pubic symphysis.**

The male and female pelves differ from each other. In the female, the iliac bones are more flared; the pelvic cavity is more shallow, but the outlet is wider. These adaptations facilitate giving birth.

The **femur** (thighbone) is the longest and strongest bone in the body. The head of the femur articulates with the coxal bones at the acetabulum, and the short neck better positions the legs for walking. The femur has two large processes, the greater and lesser trochanters, which are places of attachment for thigh muscles, buttock muscles, and hip flexors. At its distal end, the femur has medial and lateral condyles that articulate with the **tibia** of the leg. This is the region of the knee and the **patella,** or kneecap. The patella is held in place by the quadriceps tendon, which continues as a ligament that attaches to the tibial tuberosity. At the distal end, the medial malleolus of the tibia causes the inner bulge of the ankle. The **fibula** is the more slender bone in the leg. The fibula has a head that articulates with the tibia and a distal lateral malleolus that forms the outer bulge of the ankle.

Each foot has an ankle, an instep, and five toes. The many bones of the foot give it considerable flexibility, especially on rough surfaces. The ankle contains seven **tarsal** bones, one of which (the talus) can move freely where it joins the tibia and fibula. Strange to say, the calcaneus, or heel bone, is also considered part of the ankle. The talus and calcaneus support the weight of the body.

The instep has five elongated **metatarsal** bones. The distal end of the metatarsals forms the ball of the foot. If the ligaments that bind the metatarsals together become weakened, flat feet are apt to result. The bones of the toes are called **phalanges,** just like those of the fingers, but in the foot, the phalanges are stout and extremely sturdy.

☑ Check Your Progress 11.4

1. a. What are the bones of the pectoral girdle and (b) of the upper limb?

2. a. What are the bones of the pelvic girdle and (b) of the lower limb?

11.5 Articulations

Bones are joined at the joints, which are classified as fibrous, cartilaginous, or synovial. Many fibrous joints, such as the **sutures** between the cranial bones, are immovable. Cartilaginous joints are connected by hyaline cartilage, as in the costal cartilages that join the ribs to the sternum, or by fibrocartilage, as in the intervertebral disks. Cartilaginous joints tend to be slightly movable, but after the three vertebrae were fused in Katie's back, no movement between these vertebrae was possible. Synovial joints are freely movable.

Figure 11.13 illustrates the anatomy of a freely movable **synovial joint**—that is, a joint having a cavity filled with synovial fluid, a lubricant for the joint. Ligaments connect bone to bone and support or strengthen a joint. Fluid-filled sacs called bursae (sing., bursa) ease friction between bare areas of bone and overlapping muscles, or between skin and tendons. The full joint contains menisci (sing., meniscus),

C-shaped pieces of hyaline cartilage between the bones. These give added stability and act as shock absorbers.

The ball-and-socket joints at the hips and shoulders allow movement in all planes, even rotational movement. The elbow and knee joints are synovial joints called hinge joints, because like a hinged door, they largely permit movement in one direction only.

Movements Permitted by Synovial Joints

Intact skeletal muscles are attached to bones by tendons that span joints. When a muscle contracts, one bone moves in relation to another bone. The more common types of movements are described in Figure 11.14.

✅ Check Your Progress 11.5

1. **a.** What are the different types of articulations, and **(b)** what types of movements are permitted by synovial joints?

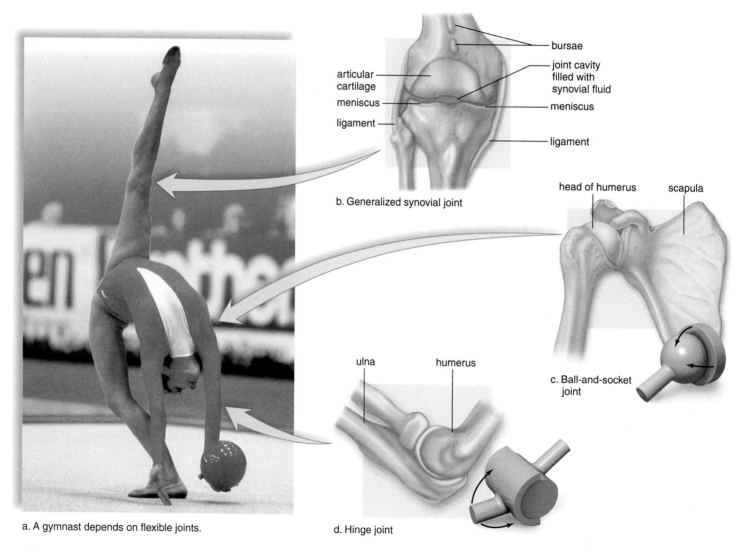

a. A gymnast depends on flexible joints.

b. Generalized synovial joint

    bursae
    joint cavity filled with synovial fluid
    articular cartilage
    meniscus
    meniscus
    ligament
    ligament

head of humerus   scapula

c. Ball-and-socket joint

ulna   humerus

d. Hinge joint

Figure 11.13   **Synovial joints.**

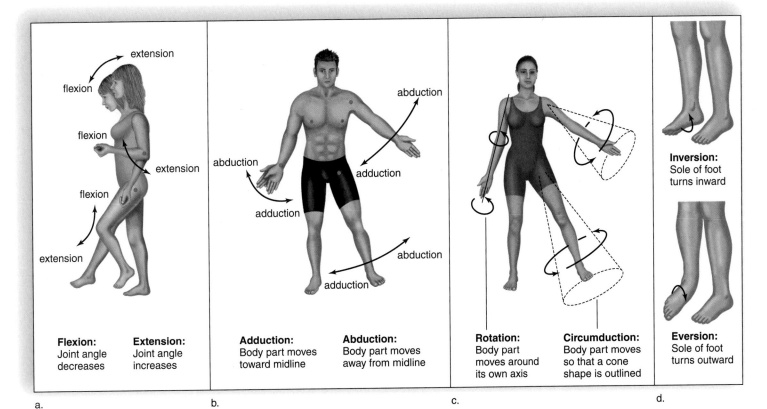

Figure 11.14 Synovial joint movements.
a. Flexion and extension. **b.** Adduction and abduction. **c.** Rotation and circumduction. **d.** Inversion and eversion. Red dots indicate pivot points.

Health Focus

Joint Replacement Surgery

Osteoarthritis is a type of arthritis that occurs when the cartilage within the joint breaks down. Rheumatoid arthritis is a type of inflammatory arthritis that occurs when the body causes inflammation within the joint. The end result of arthritis is the loss of the natural smoothness to the joint-bearing surfaces. This is what causes the pain and stiffness experienced by millions of arthritis suffers in America. Fortunately, joint replacement surgery has become a safe and effective method of dealing with painful arthritic joints.

During surgery, the doctor will replace the damaged parts of the joint with synthetic components. In knee replacement surgery, the damaged ends of bones are removed and replaced with artificial components that resemble the original bone ends. In hip replacement surgery, a plastic socket is inserted into the pelvis to replace the worn out bone socket. Polyethylene is a durable and wear-resistant plastic that is often used for this plastic component of the replacement joint. The damaged ball of the femur is replaced with a metal ball attached to a stem inserted into the shaft of the femur. Stainless steel, cobalt, and chrome alloys, as well as titanium, are commonly used for this metal component of the prosthesis. Hip and knee replacements are, by far, the most common type of joint replacement surgery, but ankles, feet, shoulders, elbows, and fingers can also be replaced.

After surgery, the patient is encouraged to begin using their new joint as soon as possible. The day after a hip or knee replacement, patients will often begin standing and walking, in order to start strengthening the replaced joint. Most patients can expect to experience some degree of pain in the replaced joint. This is due to the weakness of the muscles surrounding the joint, as well as the healing of the damaged tissues. Complete recovery from pain varies from patient to patient, but most can expect recovery to take several months.

Exercise and activity are a critical part of the recovery process. Your orthopedic surgeon will often discuss an exercise program, depending on the joint that was replaced and the specific needs of the patients. Most patients will be able to have a moderate level of activity after the surgery; however, strenuous sports may be discouraged. The extent of improvement and range of motion of the joint depends on how stiff the joint was before the surgery, as well as the amount of patient effort during therapy following surgery.

Many older patients can expect their replacements to last about ten years and give them many years of pain-free living with an improved quality of life that would not have been possible otherwise. Younger patients may need a second replacement later in life, if they happen to wear out their first prosthesis. Still, individuals who have joint replacement surgery have a bright future with an improved quality of life through greater independence and healthier pain-free activity.

Summarizing the Concepts

11.1 Overview of Skeletal System

Functions of the skeletal system:

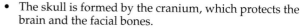

- Supports and protects the body.
- Produces blood cells.
- Stores mineral salts, particularly calcium phosphate. It also stores fat.
- Along with the muscles, permits flexible body movement.

The bones of the skeleton are composed of bone tissues and cartilage. Ligaments composed of fibrous connective tissue connect bones at joints.

In a long bone,

- hyaline cartilage covers the ends of a long bone.
- periosteum (fibrous connective tissue) covers the rest of the bone.
- spongy bone (containing red bone marrow) is in the epiphyses.
- yellow bone marrow is in the medullary cavity of the diaphysis.
- compact bone makes up the wall of the diaphysis.

11.2 Bone Growth, Remodeling, and Repair

Cells involved in growth, remodeling, and repair of bone are:

- osteoblasts, which are bone-forming cells;
- osteocytes, which are mature bone cells derived from osteoblasts; and
- osteoclasts, which break down and absorb bone.

Bone Development and Growth

- Intramembranous ossification: Bones develop between sheets of fibrous connective tissue. Examples are flat bones such as bones of the skull.
- Endochondral ossification: Cartilaginous models of the bones are replaced by calcified bone matrix.
- Bone growth is affected by vitamin D, growth hormone, and sex hormones.

Bone Remodeling and Its Role in Homeostasis

- Bone remodeling is the renewal of bone. Osteoclasts break down bone and osteoblasts re-form bone. Some bone is recycled each year.
- Bone recycling allows the body to regulate blood calcium.

Bone Repair

Repair of a fracture requires four steps:

- hematoma formation,
- fibrocartilaginous callus,
- bony callus, and
- remodeling.

11.3 Bones of the Axial Skeleton

The axial skeleton consists of the skull, the hyoid bone, the vertebral column, and the rib cage.

- The skull is formed by the cranium, which protects the brain and the facial bones.
- The hyoid bone anchors the tongue and is the site of attachment of muscles involved with swallowing.
- The vertebral column is composed of vertebrae separated by shock-absorbing disks, which make the column flexible. It supports the head and truck, protects the spinal cord, and is a site for muscle attachment.
- The rib cage is composed of the thoracic vertebrae, ribs, costal cartilages, and sternum. It protects the heart and lungs.

11.4 Bones of the Appendicular Skeleton

The appendicular skeleton consists of the bones of the pectoral girdles, upper limbs, pelvic girdle, and lower limbs.

- The pectoral girdles and upper limbs are adapted for flexibility.
- The pelvic girdle and the lower limbs are adapted for supporting weight; the femur is the longest and strongest bone in the body.

11.5 Articulations

Bones are joined at joints, of which there are three types:

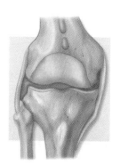

- Fibrous joints (such as the sutures of the cranium) are immovable.
- Cartilaginous joints (such as those between the ribs and sternum and the pubic symphysis) are slightly movable.
- Synovial joints (which have a synovial membrane) are freely movable.

Understanding Key Terms

appendicular skeleton 220
articular cartilage 208
axial skeleton 216
bone remodeling 212
cartilage 208
chondrocyte 208
compact bone 208
endochondral ossification 211
fibrous connective tissue 208
fontanel 216
foramen magnum 216
growth plate 211
hormone 212
intervertebral disk 218
intramembranous
    ossification 210
joint 208
ligament 208
mastoiditis 216

medullary cavity 208
ossification 210
osteoblast 210
osteoclast 210
osteocyte 208, 210
pectoral girdle 220
pelvic girdle 221
pelvis 221
periosteum 208
red bone marrow 208
rotator cuff 220
scoliosis 218
sinus 216
skull 216
spongy bone 208
suture 222
synovial joint 222
tendon 208
vertebral column 218

Match the key terms to these definitions.

a. ———— Fibrous connective tissue that connects bone to bone.

b. —————— Type of bone that contains osteons consisting of concentric layers of matrix and osteocytes in lacunae.

c. —————— Cavity or hollow space in cranial and facial bones.

d. —————— Membranous regions located between certain cranial bones in the skull of an infant or fetus.

e. —————— Mature bone cell located within the lacunae of bone.

Testing Your Knowledge of the Concepts

1. What are the five functions of the skeletal system? (page 208)

2. Where would you expect to find spongy bone in a long bone? What are its functions? (page 208)

3. Differentiate between compact and spongy bone. (page 208)

4. Describe the structure of hyaline cartilage. (page 208)

5. Why are osteoclasts needed, even though they break down bone? (page 210)

6. Explain how bone is formed by intramembranous ossification. (page 210)

7. Describe the process of endochondral ossification. (page 211)

8. Describe the actions of the hormones involved in bone growth and bone remodeling. (page 212)

9. Provide the rationale for the four steps of bone repair. (page 214)

10. Explain the terms axial skeleton and appendicular skeleton. (pages 214, 216, 220)

11. List the bones that make up the cranium and the face. (pages 216–17)

12. What are the types of vertebrae, and how many of each are there? (page 218)

13. What bones make up the rib cage? What are the functions of the rib cage? (page 219)

14. Name the bones of the pectoral girdle and upper limb. (page 220)

15. Name the bones of the pelvic girdle and lower limb. (page 221)

16. What are the two types of cartilaginous joints, and what type of movement do cartilaginous joints have? (page 222)

17. Describe the different types of movements made by synovial joints. (pages 222–23)

In questions 18–23, match each bone to a location in the key.

Key:
- a. forehead
- b. chin
- c. cheekbone
- d. collarbone
- e. shoulder blade
- f. hip
- g. arm

18. Zygomatic bone

19. Clavicle

20. Frontal bone

21. Humerus

22. Coxal bone

23. Scapula

In questions 24–29, match each bone to a feature in the key.

Key:
- a. glenoid cavity
- b. trochlea
- c. acetabulum
- d. spinous process
- e. greater and lesser trochanters
- f. xiphoid process

24. Femur

25. Scapula

26. Ulna

27. Coxal bone

28. Sternum

29. Vertebra

30. Spongy bone
   a. contains osteons.
   b. contains red bone marrow, where blood cells are formed.
   c. lends no strength to bones.
   d. takes up most of a leg bone.
   e. All of these are correct.

31. Which of these associations is mismatched?
   a. slightly movable joint—vertebrae
   b. hinge joint—hip
   c. synovial joint—elbow
   d. immovable joint—sutures in cranium

32. The bone cell that is responsible for breaking down bone tissue is the _____ , while the bone cell that produces new bone tissue is the _____ .
   a. osteoclast, osteoblast
   b. osteocyte, osteoclast
   c. osteoblast, osteocyte
   d. osteocyte, osteoblast
   e. osteoclast, osteocyte

33. All blood cells—red, white, and platelets—are produced by which of the following?
   a. yellow bone marrow
   b. red bone marrow
   c. periosteum
   d. medullary cavity

34. This bone is the only movable bone of the skull.
   a. sphenoid
   b. frontal
   c. mandible
   d. maxilla
   e. temporal

35. Which of the following is not a function of the skeletal system?
   a. production of blood cells
   b. storage of minerals
   c. involved in movement
   d. storage of fat
   e. production of body heat

36. When you bend your arm, the bump seen as your elbow is part of the
   a. humerus.
   b. radius.
   c. ulna.
   d. carpal.

37. The bump seen on the outside of the ankle is part of the
    a. femur.　　　　　　c. fibula.
    b. tibia.　　　　　　d. tarsal bones.

38. Which of the following is not a bone of the appendicular skeleton?
    a. the scapula　　　　c. a metatarsal bone
    b. a rib　　　　　　d. the patella

39. Which of the following statements is incorrect?
    a. A growth plate occurs between the primary ossification center and a secondary center.
    b. Each temporal bone has an opening that leads to the middle ear.
    c. Intervertebral disks are composed of fibrocartilage.
    d. Bone cells are rigid and hard because they are dead.
    e. The sternum is composed of three bones that fuse during fetal development.

40. The clavicle articulates with
    a. the scapula and humerus.
    b. the humerus and manubrium.
    c. the sternum and scapula.
    d. the manubrium, scapula, and humerus.
    e. the femur and the tibia.

41. The vertebrae that articulate with the ribs are the
    a. lumbar vertebrae.
    b. sacral vertebrae.
    c. thoracic vertebrae.
    d. cervical vertebrae.
    e. coccyx.

In questions 42–46, indicate whether the statement is true (T) or false (F).

42. The pectoral girdle is specialized for weight-bearing, while the pelvic girdle is specialized for flexibility of movement. _____

43. The term phalanges refers to the bones in both the fingers and the toes. _____

44. Bones synthesize vitamin D for the body. _____

45. Bones store minerals and fat. _____

46. Most bones develop through endochondral ossification. _____

In questions 47–51, match each term with the correct characteristic given in the key.

Key:
    a. closing the angle at a joint
    b. movement in all planes
    c. made of cartilage
    d. movement toward the body
    e. fluid-filled sac

47. Ball-and-socket joints

48. Flexion

49. Bursa

50. Meniscus

51. Adduction

52. Label this diagram of a skeleton.

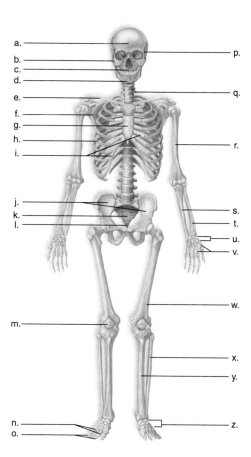

a. _____
b. _____
c. _____
d. _____
e. _____
f. _____
g. _____
h. _____
i. _____
j. _____
k. _____
l. _____
m. _____
n. _____
o. _____
p. _____
q. _____
r. _____
s. _____
t. _____
u. _____
v. _____
w. _____
x. _____
y. _____
z. _____

Thinking Critically About the Concepts

Nine months after the surgery to initiate the vertebral fusion and insert the stabilizing rods and pins, Katie, from the opening story, was scheduled for surgery again. This time it was to remove the rods and pins because fusion of the three vertebrae had been successful. The four-inch rods, one-inch screws, and miscellaneous pieces of metal were requested by Katie as a souvenir of her ordeal.

According to the physician's assistant, Katie was lucky to have the hardware removed after nine months. Often people have stabilizing hardware in for a year or more, and in some folks, it is a permanent addition. It is likely that Katie's age, good health prior to her accident, nutritional habits, and lack of smoking contributed to a quicker-than-normal rate of fusion.

1. What nutritional habits would have contributed to Katie's quicker-than-normal rate of vertebral fusion?

2. Why do the broken bones of older people take much longer to mend than broken bones of children and young adults?

3a. What stimulates the releases of parathyroid hormone (PTH)?

b. What disease(s) is/are likely to result from hyperparathyroidism?

4. Why is the spinal cord almost always damaged when someone breaks one or more cervical vertebrae?

5. Athletes sometimes abuse a synthetic form of erythropoietin (EPO), which stimulates increased synthesis of red blood cells by the spongy bone marrow.

a. How would blood oxygen be affected by this abuse of EPO?

b. Why are athletes interested in affecting blood oxygen in this manner?

CHAPTER 12

Muscular System

Jason groaned when the call came in to the crime lab. It was 8:30 A.M., his first day on the job, and already it was necessary to leave the office because a body had been found in a dumpster. The team gathered up their equipment and headed out to collect evidence before transporting the body to the morgue.

When Jason and his colleagues arrived at the crime scene, they began taking pictures of the dumpster and surrounding area. Being the "newbie," Jason was elected to climb into the dumpster and take pictures of the body before it was removed from the container. Jason made a note of the smell of the garbage and the absence of the smell of decay. "I don't think he's been here long," Jason shouted to his coworkers. "OK, take the pictures. We have to get the body out of there," said the senior CSI. "The medical examiner has arrived."

Once the body was removed, the medical examiner began her examination. She flexed the victim's arms and legs to determine the extent of rigor mortis, inserted a thermometer to determine body temperature, and examined the visible skin for lividity. "I don't think he's been dead for more than a couple of hours, but I'll know more once we have him back at the morgue," she said. The team quickly gathered their collections, wrapped them, and prepared the body for transport to the lab where they would continue their investigation.

In this chapter, you will learn about rigor mortis and how muscles function while we are alive. The contributions of both the skeletal and muscular systems are reviewed.

CHAPTER CONCEPTS

12.1 Overview of Muscular System

The muscular system includes three types of muscles; this chapter concentrates on skeletal muscle, which are attached to the bones of the skeleton. The skeletal muscles are named for their size, shape, location, and other attributes.

12.2 Skeletal Muscle Fiber Contraction

A skeletal muscle contains muscle fibers that are highly organized to carry out contraction. Muscle contraction occurs when the protein myosin interacts with the protein actin.

12.3 Whole Muscle Contraction

Degree of whole muscle contraction depends on how many motor units are maximally contracted. The energy for whole muscle contraction comes from carbohydrates and fats. Submaximal contraction for an extended period of time offers the best opportunity to burn fat.

12.4 Muscular Disorders

Muscular disorders include spasms and injuries, as well as such diseases as lockjaw, muscular dystrophy, and myasthenia gravis.

12.5 Homeostasis

The actions of the muscular and skeletal systems working together help maintain homeostasis. Together these systems move and protect body parts. The skeletal system is also involved in blood cell formation and in maintaining blood calcium levels. The muscular system also helps maintain body temperature.

12.1 Overview of Muscular System

All muscles, regardless of their particular type, can contract—that is, shorten—and when muscles contract, some part of the body or the entire body moves.

Types of Muscles

Humans have three types of muscle tissue: smooth, cardiac, and skeletal (Fig. 12.1). The cells of these tissues are called **muscle fibers.**

Smooth muscle fibers are spindle-shaped cells, each with a single nucleus (uninucleated). The cells are usually arranged in parallel lines, forming sheets. Striations (bands of light and dark) are seen in cardiac and skeletal muscle but not in smooth muscle. Smooth muscle is located in the walls of hollow internal organs, and it causes these walls to contract. Contraction of smooth muscle is involuntary, occurring without conscious control. Although smooth muscle is slower to contract than skeletal muscle, it can sustain prolonged contractions and does not fatigue easily.

Cardiac muscle forms the heart wall. Its fibers are generally uninucleated, striated, tubular, and branched, which allows the fibers to interlock at **intercalated disks.** The plasma membranes at intercalated disks contain gap junctions that permit contractions to spread quickly throughout the heart wall. Cardiac fibers relax completely between contractions, which prevents fatigue. Contraction of cardiac muscle is rhythmical; it occurs without outside nervous stimulation and without conscious control. Thus, cardiac muscle contraction is involuntary.

Skeletal muscle fibers are tubular, multinucleated, and striated. They make up the skeletal muscles attached to the skeleton. They run the length of the muscle and can be quite long. Skeletal muscle is voluntary because we can decide to move a particular part of the body, such as the arms and legs.

Functions of Skeletal Muscles

Skeletal muscles have numerous functions.

Skeletal muscles support the body. Skeletal muscle contraction opposes the force of gravity and allows us to remain upright.

Skeletal muscles make bones move. Muscle contraction accounts not only for the movement of arms and legs but also for movements of the eyes, facial expressions, and breathing. The contraction of muscles at death, observed by Jason in the opening story, is called **rigor mortis.**

Skeletal muscles help maintain a constant body temperature. Skeletal muscle contraction causes ATP to break down, releasing heat that is distributed about the body.

Skeletal muscle contraction assists movement in cardiovascular and lymphatic vessels. The pressure of skeletal muscle contraction keeps blood moving in cardiovascular veins and lymph moving in lymphatic vessels.

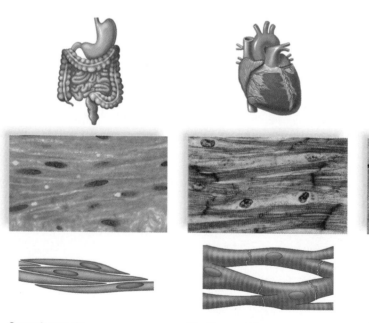

Figure 12.1 Types of muscles.
Human muscles are of three types: smooth, cardiac, and skeletal. These muscles have different characteristics, as noted.

Smooth muscle
- has spindle-shaped, nonstriated, uninucleated fibers.
- occurs in walls of internal organs.
- is involuntary.

Cardiac muscle
- has striated, branched, generally uninucleated fibers.
- occurs in walls of heart.
- is involuntary.

Skeletal muscle
- has striated, tubular, multinucleated fibers.
- is usually attached to skeleton.
- is voluntary.

Figure 12.2 **Skeletal muscles of the arm are being used when you throw a ball.**
a. A fascia covers the surface of a muscle and contributes to the tendon, which attaches the muscle to a bone. Connective tissue also separates bundles of muscle fibers that make up a skeletal muscle. **b.** When the biceps brachii contracts, the forearm flexes, and (**c**) when the triceps brachii contracts, the forearm extends. Therefore, these two muscles are antagonistic. The origin of a skeletal muscle is on a bone that remains stationary, and the insertion of a muscle is on a bone that moves when the muscle contracts.

Skeletal muscles help protect internal organs and stabilize joints. Muscles pad the bones, and the muscular wall in the abdominal region protects the internal organs. Muscle tendons help hold bones together at joints.

Skeletal Muscles of the Body

In the animal kingdom, humans are vertebrates, animals whose skeletal muscles lie outside an internal skeleton that has jointed appendages. Our skeletal muscles are attached to the skeleton, and their contraction causes the movement of bones at a joint.

Basic Structure of Skeletal Muscles

Skeletal muscles are well organized. A whole muscle contains bundles of skeletal muscle fibers called fascicles (Fig. 12.2*a*). These are the strands of muscle that we see when we cut red meat and poultry. Within a fascicle, each fiber is surrounded by connective tissue, and the fascicle itself is also surrounded by connective tissue. Muscles are covered with fascia, a type of connective tissue that extends beyond the muscle and becomes its **tendon.** Tendons quite often extend past a joint before anchoring a muscle to a bone.

Skeletal Muscles Work in Pairs

In general, each muscle is concerned with the movement of only one bone. For the sake of discussion, we can imagine the movement of that bone and no others. The **origin** of a muscle is on a stationary bone, and the **insertion** of a muscle is on a bone that moves. When a muscle contracts, it pulls on the tendons at its insertion, and the bone moves. For example, when the biceps brachii contracts, it raises the forearm. Anatomists say that the upper limb is made up of the arm above the elbow and the forearm below the elbow. (They also say that the lower limb is made up of the thigh above the knee and the leg below the knee.)

Skeletal muscles usually function in groups. Consequently, to make a particular movement, your nervous system does not stimulate a single muscle, it stimulates an appropriate group of muscles. Even so, for any particular movement, one muscle does most of the work and is called the prime mover. While a prime mover is working, however, certain muscles, called the synergists, are assisting the prime mover and making its action more effective.

When muscles contract, they shorten. Therefore, muscles can only pull; they cannot push. This means that muscles work in opposite pairs. The muscle that acts opposite to a prime mover is called an antagonist. For example, the biceps brachii and the triceps brachii are antagonists. The biceps flex the forearm (Fig. 12.2*b*), and the triceps extend the forearm (Fig. 12.2*c*). If both of these muscles contracted at once, the forearm would remain rigid. Smooth body movements depend on an antagonist relaxing when a prime mover is acting.

⚛ Science **Focus**

Rigor Mortis

When a person dies, the physiological events that accompany death occur in an orderly progression. Respiration ceases, the heart ultimately stops beating, and tissue cells begin to die. The first tissues to die are those with the highest oxygen requirement. Brain and nervous tissue have an extremely high requirement for oxygen. Deprived of oxygen, these cells typically die after only 6 minutes because of a lack of ATP. However, tissues that can produce ATP by fermentation (which does not require oxygen) can "live" for an hour or more before ATP is completely depleted. Muscle is capable of generating ATP by fermentation; therefore, muscle cells can survive for a time after clinical death occurs. Muscle death is signaled by a process termed rigor mortis, the "stiffness of death."

Stiffness occurs because, for biochemical reasons we will discuss later, muscles cannot relax unless they have a supply of ATP. Without ATP, the muscles remain fixed in their last state of contraction. If, for example, a murder victim dies while sitting at a desk, the body in rigor mortis will be frozen in the sitting position. Rigor mortis resolves approximately 24–36 hours after death. Muscles lose their stiffness because lysosomes inside the cell eventually rupture, releasing enzymes that break the bonds between the muscle proteins, actin and myosin.

Body temperature and the presence or absence of rigor mortis allows the time of death to be estimated. For example, the body of someone dead for 3 hours or less will still be warm (close to body temperature, 98.6°F, 37°C), and rigor mortis will be absent. After approximately 3 hours, the body will be significantly cooler than normal, and rigor mortis will begin to develop. The corpse of an individual who has been dead at least 8 hours will be in full rigor mortis, and the temperature of the body will be the same as the surroundings. Forensic pathologists know that a person has been dead for more than 24 hours if the body temperature is the same as the environment, and there is no longer a trace of rigor mortis.

Names and Actions of Skeletal Muscles

Figure 12.3 shows that contraction of the facial muscles produce the facial expressions that tell us about the emotions and mood of a person. Figure 12.4*a* and *b* illustrates the location of some of the major skeletal muscles and gives their actions. When learning the names of muscles, considering what the name means will help you remember it. The names of the various skeletal muscles are often combinations of the following terms used to characterize muscles. Not all the muscles mentioned are featured in Figure 12.4, but most are.

1. **Size.** For example, the gluteus maximus is the largest muscle that makes up the buttocks. The gluteus minimus is the smallest of the gluteal muscles. Other terms used to indicate size are vastus (huge), longus (long), and brevis (short).
2. **Shape.** The deltoid is shaped like a triangle. (The Greek letter delta has this appearance: Δ.) The trapezius is shaped like a trapezoid. Other terms used to indicate shape are latissimus (wide) and terres (round).
3. **Location.** For example, the external obliques are located outside the internal obliques. The frontalis overlies the frontal bone. Other terms used to indicate location are pectoralis (chest), gluteus (buttock), brachii (arm), and sub (beneath).
4. **Direction of muscle fibers.** For example, the rectus abdominis is a longitudinal muscle of the abdomen (rectus means straight). The orbicularis oculi is a circular muscle around the eye. Other terms used to indicate direction are transverse (across) and oblique (diagonal).
5. **Attachment.** For example, the sternocleidomastoid is attached to the sternum, clavicle, and mastoid process. The brachioradialis is attached to the brachium (arm) and the radius.

Figure 12.3   Facial expressions.
Our facial expressions and hand movements are due to muscle contractions.

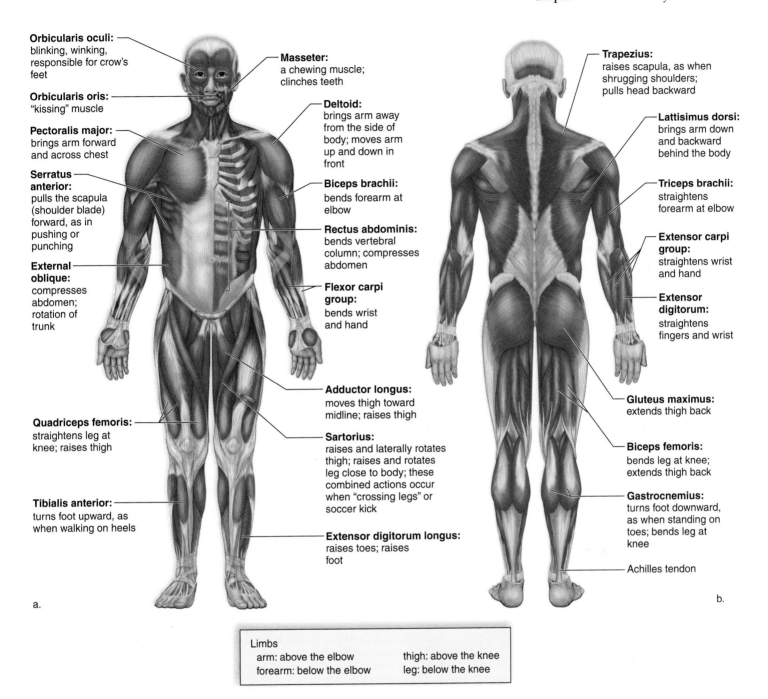

Orbicularis oculi:
blinking, winking, responsible for crow's feet

Orbicularis oris:
"kissing" muscle

Pectoralis major:
brings arm forward and across chest

Serratus anterior:
pulls the scapula (shoulder blade) forward, as in pushing or punching

External oblique:
compresses abdomen; rotation of trunk

Quadriceps femoris:
straightens leg at knee; raises thigh

Tibialis anterior:
turns foot upward, as when walking on heels

a.

Masseter:
a chewing muscle; clinches teeth

Deltoid:
brings arm away from the side of body; moves arm up and down in front

Biceps brachii:
bends forearm at elbow

Rectus abdominis:
bends vertebral column; compresses abdomen

Flexor carpi group:
bends wrist and hand

Adductor longus:
moves thigh toward midline; raises thigh

Sartorius:
raises and laterally rotates thigh; raises and rotates leg close to body; these combined actions occur when "crossing legs" or soccer kick

Extensor digitorum longus:
raises toes; raises foot

Trapezius:
raises scapula, as when shrugging shoulders; pulls head backward

Lattisimus dorsi:
brings arm down and backward behind the body

Triceps brachii:
straightens forearm at elbow

Extensor carpi group:
straightens wrist and hand

Extensor digitorum:
straightens fingers and wrist

Gluteus maximus:
extends thigh back

Biceps femoris:
bends leg at knee; extends thigh back

Gastrocnemius:
turns foot downward, as when standing on toes; bends leg at knee

Achilles tendon

b.

Limbs
arm: above the elbow
forearm: below the elbow
thigh: above the knee
leg: below the knee

Figure 12.4 Human musculature.
Superficial skeletal muscles in (**a**) anterior and (**b**) posterior view.

6. **Number of attachments**. For example, the biceps brachii has two attachments, or origins, and is located on the arm. The quadriceps femoris has four origins and is located on the femur.

7. **Action.** For example, the extensor digitorum extends the fingers or digits. The adductor longus is a large muscle that adducts the thigh. Adduction is the movement of a body part toward the midline. Other terms used to indicate action are flexor (to flex or bend), masseter (to chew), and levator (to lift).

With every muscle you learn, try to understand its name.

☑ **Check Your Progress 12.1**

1. What are the three types of muscles in the human body?
2. What are the functions of skeletal muscles?
3. How do skeletal muscles work together to cause bones to move?
4. What are the major skeletal muscles in the body?

12.2 Skeletal Muscle Fiber Contraction

We have already examined the structure of skeletal muscle, as seen with the light microscope. As you know, skeletal muscle tissue has alternating light and dark bands, giving it a striated appearance. Now we will see that these bands are due to the arrangement of myofilaments in a muscle fiber.

Muscle Fibers and How They Slide

A muscle fiber is a cell containing the usual cellular components, but special names have been assigned to some of these components (Table 12.1 and Fig. 12.5). The plasma membrane is called the **sarcolemma;** the cytoplasm is the sarcoplasm; and the endoplasmic reticulum is the **sarcoplasmic reticulum.** A muscle fiber also has some unique anatomical characteristics. One feature is its T (for transverse) system; the sarcolemma forms **T (transverse) tubules** that penetrate, or dip down, into the cell so that they come into contact—but do not fuse—with expanded portions of the sarcoplasmic reticulum. The expanded portions of the sarcoplasmic reticulum are calcium storage sites. Calcium ions (Ca^{2+}), as we shall see, are essential for muscle contraction.

The sarcoplasmic reticulum encases hundreds and sometimes even thousands of **myofibrils,** each about 1 μm in diameter, which are the contractile portions of the muscle fibers. Any other organelles, such as mitochondria, are located in the sarcoplasm between the myofibrils. The sarcoplasm also

| Table 12.1 | Microscopic Anatomy of a Muscle Fiber |

Name	Function
Sarcolemma	Plasma membrane of a muscle fiber that forms T tubules
Sarcoplasm	Cytoplasm of a muscle fiber that contains the organelles, including myofibrils
Glycogen	A polysaccharide that stores energy for muscle contraction
Myoglobin	A red pigment that stores oxygen for muscle contraction
T tubule	Extension of the sarcolemma that extends into the muscle fiber and conveys impulses that cause Ca^{2+} to be released from the sarcoplasmic reticulum
Sarcoplasmic reticulum	The smooth ER of a muscle fiber that stores Ca^{2+}
Myofibril	A bundle of myofilaments that contracts
Myofilament	Actin filaments or myosin filaments, whose structure and functions account for muscle striations and contractions

contains glycogen, which provides stored energy for muscle contraction, and the red pigment myoglobin, which binds oxygen until it is needed for muscle contraction.

Myofibrils and Sarcomeres

Myofibrils are cylindrical in shape and run the length of the muscle fiber. The light microscope shows that skeletal muscle fibers have light and dark bands called striations. The electron microscope shows that the striations of skeletal muscle fibers are formed by the placement of myofilaments within units of myofibrils called **sarcomeres.** A sarcomere extends between two dark lines called the Z lines. A sarcomere contains two types of protein myofilaments. The thick filaments are made up of a protein called **myosin,** and the thin filaments are made up of a protein called **actin.** The I band is light colored because it contains only actin filaments attached to a Z line. The dark regions of the A band contain overlapping actin and myosin filaments, and its H zone has only myosin filaments.

Myofilaments

The thick and thin filaments differ in the following ways:

Thick Filaments A thick filament is composed of several hundred molecules of the protein myosin. Each myosin molecule is shaped like a golf club, with the straight portion of the molecule ending in a globular head, or *cross-bridge.* The cross-bridges occur on each side of a sarcomere but not in the middle.

Thin Filaments Primarily, a thin filament consists of two intertwining strands of the protein actin. Two other proteins, called tropomyosin and troponin, also play a role, as we will discuss later in this section.

Sliding Filaments We will also see that when muscles are stimulated, impulses travel down a T tubule, and calcium is released from the sarcoplasmic reticulum. Now the muscle fiber contracts as the sarcomeres, within the myofibrils, shorten. When a sarcomere shortens, the actin (thin) filaments slide past the myosin (thick) filaments and approach one another. This causes the I band to shorten, the Z line to move inward, and the H zone to almost or completely disappear. The movement of actin filaments in relation to myosin filaments is called the **sliding filament model** of muscle contraction. During the sliding process, the sarcomere shortens, even though the filaments themselves remain the same length. ATP supplies the energy for muscle contraction. Although the actin filaments slide past the myosin filaments, it is the myosin filaments that do the work. Myosin filaments break down ATP, and their cross-bridges pull the actin filaments toward the center of the sarcomere.

As an analogy, think of yourself and a group of friends as myosin. Collectively, your hands are the cross-bridges, and you are pulling on a rope (actin) in order to get an object tied to the end of the rope (the Z line). As you pull the rope, you grab, pull, release, and then grab further along on the rope.

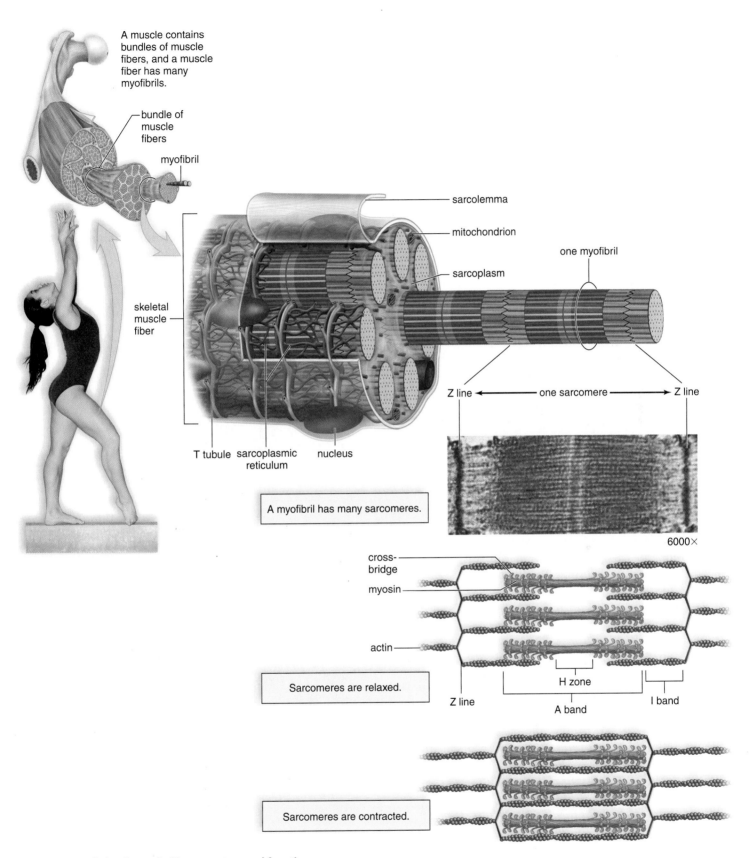

A muscle contains bundles of muscle fibers, and a muscle fiber has many myofibrils.

bundle of muscle fibers

myofibril

skeletal muscle fiber

sarcolemma

mitochondrion

sarcoplasm

one myofibril

T tubule   sarcoplasmic reticulum   nucleus

Z line ← one sarcomere → Z line

A myofibril has many sarcomeres.

6000×

cross-bridge

myosin

actin

Sarcomeres are relaxed.

Z line

H zone

A band

I band

Sarcomeres are contracted.

Figure 12.5   Skeletal muscle fiber structure and functions.
A muscle fiber contains many myofibrils divided into sarcomeres, which are contractile. When the myofibrils of a muscle fiber contract, the sarcomeres shorten: The actin (thin) filaments slide past the myosin (thick) filaments toward the center. Notice that the Z lines have moved and the H zone has gotten smaller, to the point of disappearing.

Control of Muscle Fiber Contraction

Muscle fibers are stimulated to contract by motor neurons whose axons are in nerves. The axon of one motor neuron can stimulate from a few to several muscle fibers of a muscle because each axon has several branches (Fig. 12.6*a*). Each branch of an axon ends in an axon terminal that lies in close proximity to the sarcolemma of a muscle fiber. A small gap, called a synaptic cleft, separates the axon terminal from the sarcolemma (Fig. 12.6*b*). This entire region is called a **neuromuscular junction.**

Axon terminals contain synaptic vesicles that are filled with the neurotransmitter acetylcholine (ACh). When nerve impulses, traveling down a motor neuron, arrive at an axon terminal, the synaptic vesicles release ACh into the synaptic

cleft (Fig. 12.6*c*). Botox is the trade name for botulinum toxin A, a neurotoxin produced by a bacterium. Botox prevents wrinkling of the brow and skin about the eyes because it blocks the release of ACh into the synaptic cleft, and therefore muscle contraction never occurs. Another point of interest concerns rigor mortis. We can well imagine that upon death the synaptic vesicles lose their integrity and ACh pours into the synaptic cleft. This might be the reason why rigor mortis is seen in muscles some time after a person is brain dead.

When ACh is released, it quickly diffuses across the cleft and binds to receptors in the sarcolemma. Now, the sarcolemma generates impulses that spread over the sarcolemma and down T tubules to the sarcoplasmic reticulum. The release of Ca^{2+} from the sarcoplasmic reticulum leads to sarcomere contraction, as explained in Figure 12.7.

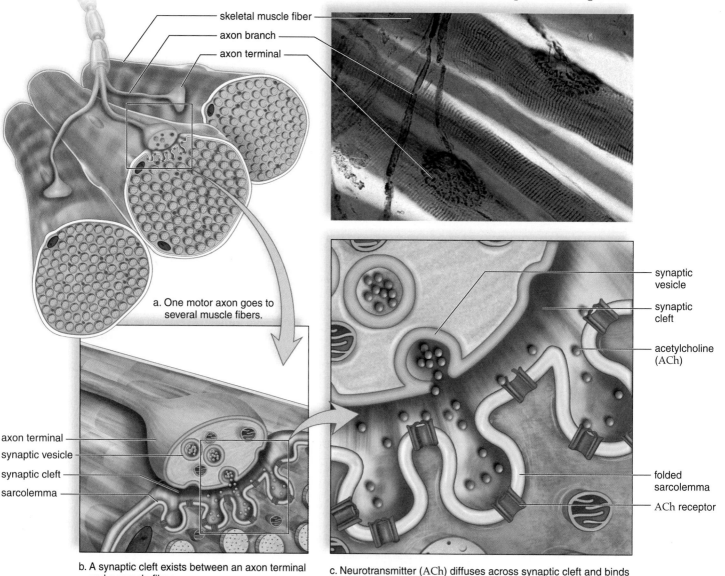

a. One motor axon goes to several muscle fibers.

b. A synaptic cleft exists between an axon terminal and a muscle fiber.

c. Neurotransmitter (ACh) diffuses across synaptic cleft and binds to receptors in sarcolemma.

Figure 12.6 **Neuromuscular junction.**

a. The branch of a motor nerve fiber terminates in an axon terminal that meets. **b.** A synaptic cleft separates the axon terminal from the sarcolemma of the muscle fiber. **c.** Nerve impulses traveling down a motor fiber cause synaptic vesicles to discharge acetylcholine, which diffuses across the synaptic cleft and binds to ACh receptors. Impulses travel down the T system of a muscle fiber and the muscle fiber contracts.

Two other proteins are associated with an actin filament. Threads of **tropomyosin** wind about an actin filament, and **troponin** occurs at intervals along the threads. When Ca^{2+} ions are released from the sarcoplasmic reticulum, they combine with troponin, and this causes the tropomyosin threads to shift their position, exposing myosin binding sites. In other words, myosin can now bind to actin (Fig. 12.7*a*).

In order for Jason to fully understand rigor mortis, he needs to study Figure 12.7*b*. 1. The heads of a myosin filament have ATP binding sites. At this site, ATP is split to ADP and Ⓟ. 2. The ADP and Ⓟ remain on the myosin heads, while the heads attach to an actin filament, forming cross-bridges. 3. Now, ADP and Ⓟ are released, and the cross-bridges bend sharply. This is the *power stroke* that pulls the actin filament toward the center of the sarcomere. 4. When ATP molecules again bind to the myosin heads,

the cross-bridges are broken, and heads detach from the actin filament. This is the step that does not happen during rigor mortis. Relaxing the muscle is impossible, because ATP is needed to break the bond between an actin binding site and the myosin cross-bridge.

In living muscle the cycle begins again, and myosin reattaches further along the actin filament. The cycle recurs until calcium ions are actively returned to the calcium storage sites. This step also requires ATP.

☑ Check Your Progress 12.2

1. What are the microscopic levels of structure in a skeletal muscle?
2. How does contraction of a skeletal muscle come about?
3. What is the role of ATP in muscle contraction?

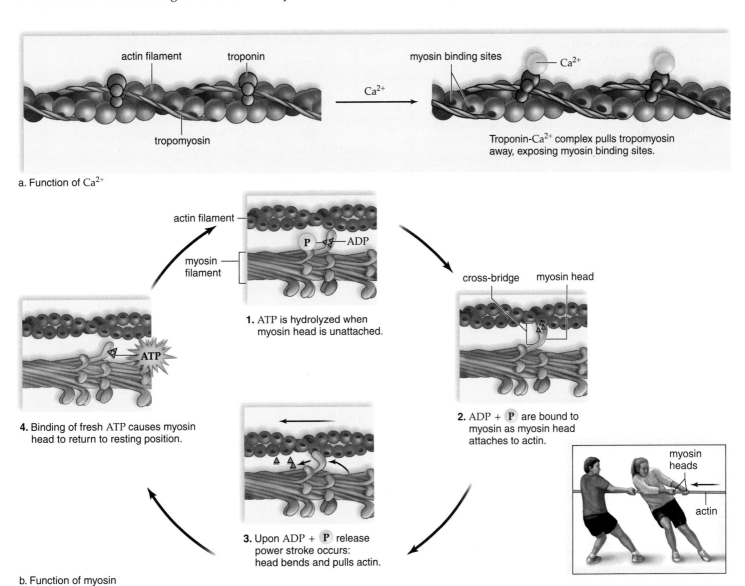

a. Function of Ca^{2+}

actin filament

troponin

myosin binding sites — Ca^{2+}

Ca^{2+}

tropomyosin

Troponin-Ca^{2+} complex pulls tropomyosin away, exposing myosin binding sites.

actin filament

myosin filament

P ◁—ADP

1. ATP is hydrolyzed when myosin head is unattached.

cross-bridge   myosin head

2. ADP + **P** are bound to myosin as myosin head attaches to actin.

ATP

4. Binding of fresh ATP causes myosin head to return to resting position.

3. Upon ADP + **P** release power stroke occurs: head bends and pulls actin.

myosin heads

actin

b. Function of myosin

Figure 12.7 **Function of calcium (Ca²⁺) and myosin in muscle contraction.**
a. Calcium (Ca²⁺) binds to troponin, exposing myosin binding sites. **b.** Follow 1–3 to see how myosin utilizes ATP and does the work of pulling actin toward the center of the sarcomere, much as a group of people pulling a rope (*right*).

12.3 Whole Muscle Contraction

Whole muscle contraction is dependent on muscle fiber contraction.

Muscles Have Motor Units

As mentioned, each axon within a nerve stimulates a number of muscle fibers. A nerve fiber, together with all of the muscle fibers it innervates, is called a **motor unit.** A motor unit obeys the **all-or-none law**. Why? Because all the muscle fibers in a motor unit are stimulated at once, and they all either contract or do not contract. A variable of interest is the number of muscle fibers within a motor unit. For example, in the ocular muscles that move the eyes, the innervation ratio is one motor axon per 23 muscle fibers, while in the gastrocnemius muscle of the leg, the ratio is about one motor axon per 1,000 muscle fibers. Thus, moving the eyes requires finer control than moving the legs.

When a motor unit is stimulated by infrequent electrical impulses, a single contraction occurs that lasts only a fraction of a second. This response is called a **muscle twitch.** A muscle twitch is customarily divided into three stages: the latent period, or the period of time between stimulation and initiation of contraction; the contraction period, when the muscle shortens; and the relaxation period, when the muscle returns to its former length (Fig 12.8a). We can use our knowledge of muscle fiber contraction to understand these events. For example, we know that a muscle fiber contracts when calcium leaves sarcoplasmic reticulum and relaxes when calcium returns to sarcoplasmic reticulum.

If a motor unit is given a rapid series of stimuli, it can respond to the next stimulus without relaxing completely. Summation is increased muscle contraction until maximal sustained contraction, called **tetanus,** is achieved (Fig. 12.8b). Tetanus continues until the muscle fatigues due to depletion of energy reserves. Fatigue is apparent when a muscle relaxes, even though stimulation continues. The tetanus of muscle cells is not the same as the infection called tetanus. The infection called *tetanus* is caused by the bacterium *Clostridium tetan,* and death occurs because the muscles, including the respiratory muscles, become fully contracted and do not relax.

A whole muscle typically contains many motor units. As the intensity of nervous stimulation increases, more and more motor units in a muscle are activated. This phenomenon is known as recruitment. Maximum contraction of a muscle would require that all motor units be undergoing tetanic contraction. This rarely happens because they could all fatigue at the same time. Instead, some motor units are contracting maximally, while others are resting, allowing sustained contractions to occur.

Muscle Tone

One desirable effect of exercise is to have good "muscle tone," which is dependent on muscle contraction. When some motor units are always contracted but not enough to cause movement, the muscle is firm and solid—has good tone—as opposed to being soft and flabby.

Energy for Muscle Contraction

Muscles can use various fuel sources for energy, and they have various ways of producing ATP during muscle contraction.

Fuel Sources for Exercise

A muscle has four possible energy sources (Fig. 12.9). Two of these are stored in muscle, and two are acquired from

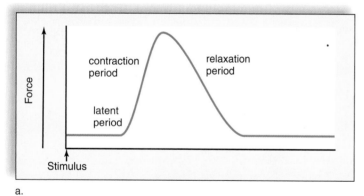

a.

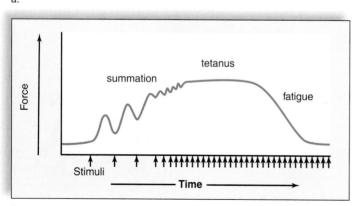

b.

Figure 12.8 **Physiology of skeletal muscle contraction.**
a. Stimulation of a muscle by an electric shock results in a simple muscle twitch. **b.** Repeated stimulation results in summation and tetanus.

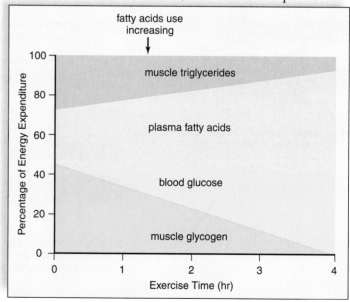

Figure 12.9 **Fuel sources for muscle contraction.**
Percentage of energy derived from the four major fuel sources during submaximal exercise (65–75% of effort). Notice that the amount from plasma fatty acids increases during the time span shown.

Figure 12.10 **Muscle cells produce ATP in three ways.**

a.When contraction begins, muscle cells break down creatine phosphate to produce ATP. When resting, muscle cells rebuild their supply of creatine phosphate (red arrow). **b.** Muscle cells also use fermentation to produce ATP quickly. When resting, muscle cells metabolize lactate, reforming as much glucose and then glycogen as possible (red arrow). **c.** For the long term, muscle cells switch to cellular respiration to produce ATP aerobically.

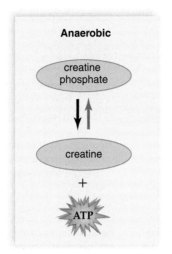
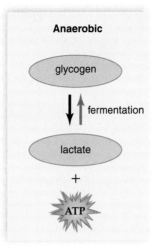
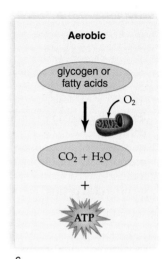

a.   b.   c.

blood. Both glycogen and fat (triglycerides) are stored in muscles. Which of these are used depends on exercise intensity and duration. Figure 12.9 shows the percentage of energy derived from these sources due to submaximal exercise (65–75% of effort) over time.

Muscles also make use of blood glucose and plasma fatty acids as an energy source. Both of these are delivered to muscles by circulating blood. Some of us specifically exercise in order to lose weight, and so we are particularly interested in the use of plasma fatty acids by exercising muscles. Adipose tissue, which tends to make us look fat, is the source of plasma fatty acids that muscle burns as an energy source. If we are on a diet that restricts the amount of fat eaten, exercise will decrease body fat. Figure 12.9 suggests that there is a positive relationship between the amount of fat burned and the length of the exercise. Submaximal exercise burns fat better than maximal exercise for reasons we will now explore.

Sources of ATP for Muscle Contraction

Muscle cells store limited amounts of ATP, but they have three ways to acquire more ATP needed for contraction once stored ATP has been used up (Fig. 12.10). The three ways include (1) formation of ATP by the creatine phosphate (CP) pathway; (2) formation of ATP by fermentation (see page 56); and (3) formation of ATP by cellular respiration, which, of course, involves the use of oxygen by mitochondria. Aerobic exercising depends on cellular respiration to supply ATP. Neither formation of ATP by the CP pathway or by fermentation involves the need for oxygen.

The CP Pathway The simplest and most rapid way for muscle to produce ATP is to use the CP pathway because it only consists of one reaction (Fig. 12.10a):

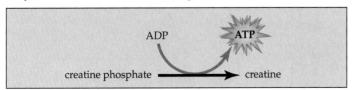

This reaction occurs in the midst of sliding filaments, and therefore, this method of supplying ATP is the speediest energy source available to muscles. Creatine phosphate is

formed only when a muscle cell is resting, and only a limited amount is stored. The CP pathway is used at the beginning of submaximal exercise and during short-term, high-intensity exercise that lasts less than 5 seconds. The energy to complete a single play in a football game comes principally from the CP system. Intense activities lasting longer than 5 seconds will also make use of fermentation.

Fermentation Fermentation, as you know, produces two ATP from the breakdown of glucose to lactate anaerobically. This pathway is the one most likely to begin with glycogen. Hormones provide the signal to muscle cells to break down glycogen, making glucose available as an energy source.

Fermentation, like the CP pathway, is fast-acting, but it results in the buildup of lactate (Fig. 12.10b). Formation of lactate is noticeable because it produces short-term muscle aches and fatigue upon exercising. We have all had the experience of needing to continue breathing following strenuous exercise. This continued intake of oxygen, called **oxygen debt,** is required, in part, to complete the metabolism of lactate and restore cells to their original energy state. The lactate is transported to the liver, where 20% of it is completely broken down to carbon dioxide and water. The ATP gained by this respiration is then used to reconvert 80% of the lactate to glucose and the glycogen. In persons who train, the number of mitochondria increases, and there is a greater reliance on them rather than on fermentation to produce ATP.

Cellular Respiration Although muscle cells contain **myoglobin,** a molecule that combines with and stores oxygen, cellular respiration, as we have seen, does not immediately supply ATP for muscle contraction. Also, cellular respiration is more likely to supply ATP when exercise is submaximal in intensity. Cellular respiration can make use of glucose from the breakdown of glycogen stored in muscle, glucose taken up from blood, and also *fatty acids* (Fig. 12.10c). According to Figure 12.9, if you are interested in exercising to lose weight, you should do so at submaximal intensity and for a generous amount of time. This means that your aerobic exercise class at the local gym is designed to have your body utilize cellular respiration to produce ATP for muscle contraction and, therefore, burn fat.

☤ Health Focus

Exercise, Exercise, Exercise

Exercise programs improve muscular strength, muscular endurance, and flexibility. Muscular strength is the force a muscle group (or muscle) can exert against a resistance in one maximal effort. Muscular endurance is judged by the ability of a muscle to contract repeatedly or to sustain a contraction for an extended period. Flexibility is tested by observing the range of motion about a joint.

Exercise also improves cardiorespiratory endurance. The heart rate and capacity increase, and the air passages dilate so that the heart and lungs are able to support prolonged muscular activity. The blood level of high-density lipoprotein (HDL), the molecule that prevents the development of plaque in blood vessels, increases. Also, body composition—that is, the proportion of protein to fat—changes favorably when you exercise.

Exercise also seems to help prevent certain kinds of cancer. Cancer prevention involves eating properly, not smoking, avoiding cancer-causing chemicals and radiation, undergoing appropriate medical screening tests, and knowing the early warning signs of cancer. However, studies show that people who exercise are less likely to develop colon, breast, cervical, uterine, and ovarian cancers.

Physical training with weights can improve the density and strength of bones and the strength and endurance of muscles in all adults, regardless of age. Even men and women in their eighties and nineties can make substantial gains in bone and muscle strength that help them lead more independent lives. Exercise helps prevent osteoporosis, a condition in which the bones are weak and tend to break. Exercise promotes the activity of osteoblasts in young as well as older people. The stronger the bones when a person is young, the less chance of osteoporosis as that person ages. Exercise helps prevent weight gain, not only because the level of activity increases but also because muscles metabolize faster than other tissues. As a person becomes more muscular, the body is less likely to accumulate fat.

Exercise relieves depression and enhances the mood. Some people report that exercise actually makes them feel more energetic, and that after exercising, particularly in the late afternoon, they sleep better that night. Self-esteem rises because of improved appearance, as well as other factors that are not well understood. For example, vigorous exercise releases endorphins, hormonelike chemicals that are known to alleviate pain and provide a feeling of tranquility.

A sensible exercise program is one that provides all of these benefits without the detriments of a too-strenuous program. Overexertion can actually be harmful to the body and might result in sports injuries, such as lower back strains or torn ligaments of the knees. The beneficial programs suggested in Table 12A are tailored according to age.

Dr. Arthur Leon at the University of Minnesota performed a study involving 12,000 men, and the results showed that only moderate exercise is needed to lower the risk of a heart attack by one-third. In another study conducted by the Institute for Aerobics Research in Dallas, Texas, which included 10,000 men and more than 3,000 women, even a little exercise was found to lower the risk of death from cardiovascular diseases and cancer. Increasing daily activity by walking to the corner store instead of driving and by taking the stairs instead of the elevator can improve your health.

Table 12A   Staying Fit

Exercise	Children, 7–12	Teenagers, 13–18	Adults, 19–55
Amount	Vigorous activity 1–2 hrs daily	Vigorous activity 1 hr 3–5 days a week; otherwise, ½ hr daily moderate activity	Vigorous activity 1 hr 3 days a week; otherwise, ½ hr daily moderate activity
Purpose	Free play	Build muscle with calisthenics	Exercise to prevent lower back pain: aerobics, stretching, or yoga
Organized	Build motor skills through team sports, dancing, or swimming	Continue team sports: dancing, hiking, or swimming	Do aerobic exercise to control buildup of fat cells
Group	Enjoy more exercise outside of physical education classes	Pursue tennis, swimming, horseback riding: sports that can be enjoyed for a lifetime	Find exercise partners: join a running club, bicycle club, or outing group
Family	Participate in family outings: bowling, boating, camping, or hiking	Take active vacations: hike, bicycle, or cross-country ski	Initiate family outings: bowling, boating, camping, or hiking

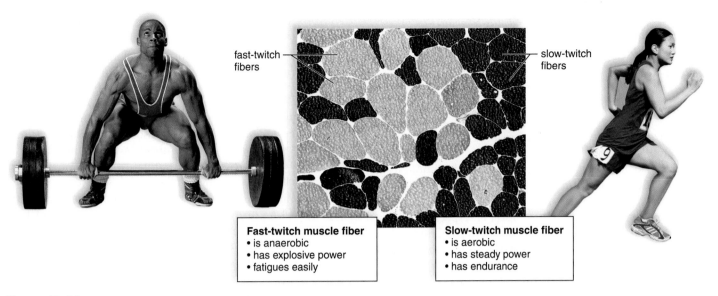

Figure 12.11   **Fast-and slow twitch muscle fibers.**
If your muscles contain many fast-twitch fibers (light color), you would probably do better at a sport like weight lifting. If your muscles contain many slow-twitch fibers (dark color), you would probably do better at a sport like cross-country running.

Fast-Twitch and Slow-Twitch Muscle Fibers

We have seen that all muscle fibers metabolize both aerobically and anaerobically. Some muscle fibers, however, utilize one method more than the other to provide myofibrils with ATP. Fast-twitch fibers tend to rely on the creatine phosphate pathway and fermentation, which are anaerobic means of supplying ATP to muscles. Slow-twitch fibers tend to prefer cellular respiration, which is aerobic (Fig. 12.11).

Fast-Twitch Fibers

Fast-twitch fibers are usually anaerobic and seem to be designed for strength because their motor units contain many fibers. They provide explosions of energy and are most helpful in sports activities such as sprinting, weight lifting, swinging a golf club, or throwing a shot. Fast-twitch fibers are light in color because they have fewer mitochondria, little or no myoglobin, and fewer blood vessels than slow-twitch fibers do. Fast-twitch fibers can develop maximum tension more rapidly than slow-twitch fibers can, and their maximum tension is greater. However, their dependence on anaerobic energy leaves them vulnerable to an accumulation of lactate, which causes them to fatigue quickly.

Slow-Twitch Fibers

Slow-twitch fibers have a steadier tug and have more endurance despite more units with smaller number of fibers. These muscle fibers are most helpful in sports such as long-distance running, biking, jogging, and swimming. Because they produce most of their energy aerobically, they tire only when their fuel supply is gone. Slow-twitch fibers have many mitochondria and are dark in color because they contain myoglobin, the respiratory pigment found in muscles. They are also surrounded by dense capillary beds and draw more blood and oxygen than fast-twitch fibers. Slow-twitch fibers have a low maximum tension, which develops slowly, but the muscle fibers are highly resistant to fatigue. Because slow-twitch fibers have a substantial reserve of glycogen and fat, their abundant mitochondria can maintain a steady, prolonged production of ATP when oxygen is available.

Delayed Onset Muscle Soreness

Many of us have experienced delayed onset muscle soreness (DOMS), which generally appears some 24–48 hours after strenuous exercise. It is thought that DOMS is due to tissue injury that takes several days to heal. Any movement you aren't use to can lead to DOMS, but it is especially associated with any activity that causes muscles to contract while they are lengthening. Examples include walking down stairs, running downhill, lowering weights, and the downward motion of squats and push-ups. To prevent DOMS, try warming up thoroughly and cooling down completely. Stretch after exercising. When beginning a new activity, start gradually and build up your endurance gradually. Avoid making sudden major changes in your exercise routine.

✔ Check Your Progress 12.3

1. **a.** What is a motor unit, and **(b)** why does a motor unit obey the all-or-none law?

2. How does a whole muscle avoid fatigue caused by tetanus?

3. **a.** What are the sources of ATP for muscle contraction, and **(b)** which type of muscle fibers rely primarily on anaerobic ATP production?

12.4 Muscular Disorders

Muscular disorders are divided into those that are commonly seen and those that are more serious in nature.

Common Muscular Conditions

Spasms are sudden and involuntary muscular contractions most often accompanied by pain. Spasms can occur in smooth and skeletal muscles. A spasm of the intestinal tract is a type of colic sometimes called a bellyache. Multiple spasms of skeletal muscles are called a seizure, or **convulsion. Cramps** are strong, painful spasms, especially of the leg and foot, usually due to strenuous activity. Cramps can even occur when sleeping after a strenuous workout. **Facial tics,** such as periodic eye blinking, head turning, or grimacing, are spasms that can be controlled voluntarily, but only with great effort.

A **strain** is caused by stretching or tearing of a muscle. A **sprain** is a twisting of a joint leading to swelling and injury, not only of muscles but also of ligaments, tendons, blood vessels, and nerves. The ankle and knee are often subject to sprains.

Tendinitis and Bursitis

In **tendinitis,** the normal, smooth gliding motion of a tendon is impaired, the tendon is inflamed, and movement of a joint becomes painful. The most common cause of tendinitis is overuse. A new exercise program or an increased level of exercise can bring on the symptoms of tendinitis. As we age, tendinitis is usually associated with the shoulder, elbow, hip, and knee.

Every person has hundreds of bursae located in the joints throughout the body. A **bursa** has been likened to a plastic bag filled with a small amount of oil, and it provides a smooth, slippery surface where muscles and tendons glide over bones. **Bursitis** is an inflammation of a bursa. Bursitis usually results from a repetitive movement or from prolonged and excessive pressure. Patients who rest on their elbows for long periods or those who bend their elbows frequently and repetitively (for example, using a vacuum for hours at a time) can develop elbow bursitis, also called olecranon bursitis. Similarly in other parts of the body, repetitive use or frequent pressure can irritate a bursa and lead to inflammation.

Over-the-counter and prescribed nonsteroidal anti-inflammatory drugs, commonly referred to as NSAIDs (pronounced en-sayds) are the most commonly used medications for treating arthritis, bursitis, and tendinitis. Medicines such as ibuprofen (Motrin) and naproxen (Aleve) are available over the counter, whereas other medications must be prescribed. Cortisone injections are also a common, nonsurgical way to treat tendinitis and bursitis.

Muscular Diseases

These conditions are more serious and always require close medical care.

Myalgia and Fibromyalgia

Myalgia refers to achy muscles. The most common cause for myalgia is either overuse or overstretching of a muscle or group of muscles. Myalgia without a traumatic history is often due to viral infections. Myalgia may accompany myositis, which is inflammation of the muscles, either in response to viral infection or as an immune system disorder. **Fibromyalgia** is a chronic condition whose symptoms include achy pain, tenderness, and stiffness of muscles. Its precise cause is not known and may also be due to a underlying infection that is not obvious at first.

Muscular Dystrophy

Muscular dystrophy is a broad term applied to a group of disorders that are characterized by a progressive degeneration and weakening of muscles. As muscle fibers die, fat and connective tissue take their place. **Duchenne muscular dystrophy,** the most common type, is inherited through a flawed gene carried by the mother. It is now known that the lack of a protein called dystrophin causes the condition. When dystrophin is absent, calcium leaks into the cell and activates an enzyme that dissolves muscle fibers. In an attempt to treat the condition, muscles have been injected with immature muscle cells that do produce dystrophin.

Myasthenia Gravis

Myasthenia gravis is an autoimmune disease characterized by weakness that especially affects the muscles of the eyelids, face, neck, and extremities. Muscle contraction is impaired because the immune system mistakenly produces antibodies that destroy acetylcholine (ACh) receptors. In many cases, the first sign of the disease is a drooping of the eyelids and double vision. Treatment includes drugs that inhibit the enzyme that digests acetylcholine so that ACh accumulates in neuromuscular junctions.

Amyotrophic Lateral Sclerosis

Amyotrophic lateral sclerosis (ALS) is better known as Lou Gehrig's disease because Lou Gehrig is a famous 1930s-era baseball player who died of the disease. ALS sufferers experience gradual loss of the ability to walk, talk, chew, and swallow. Mental abilities and sensations are not affected, however. Drugs are available that slow the progression of the disease, but it cannot be cured.

✔ Check Your Progress 12.4

1. What are tendinitis and bursitis?
2. What are four more serious muscular conditions?

Anabolic Steroid Use

In 1998, baseball sluggers Mark McGwire and Sammy Sosa were embroiled in a season-long home run slugfest, which resulted in the smashing of Roger Maris's long-standing single season home run record. Although largely credited for reviving interest in baseball, the great home run competition drew unwanted attention to the darker side of professional sports when it was alleged that both McGwire and Sosa were using anabolic steroids. The controversy still continues today—despite having smashed Babe Ruth's home run record with his 715th home run, baseball legend Barry Bonds has also been accused of steroid abuse.

Although use of anabolic steroids in professional sports is denied by most athletes and officials, many people from both inside and outside the industry allege that this abuse has been going on for many years, and that it continues despite the negative publicity. As finger-pointing and accusations escalated over the next few years, Congress joined the fray by holding hearings during the summer of 2005.

What Are Anabolic Steroids?

Steroids encompass a large category of substances, both beneficial and harmful. Anabolic steroids are a class of steroids that generally cause tissue growth by promoting protein production. They are naturally occurring hormones that are created by the body and are commonly used to regulate many physiological processes, from growth to sexual function. Most anabolic steroids are closely related to male sex hormones, such as testosterone.

Because of their many side effects and vast potential for abuse, these metabolically potent drugs are controlled substances that are available only by prescription under the close supervision of a physician. However, a few anabolic steroids are still legal due to loopholes in drug laws, despite being banned by most professional sports organizations.

Robust muscle growth is not possible from prescription doses of anabolic steroids, so large doses must be used to obtain that effect. Athletes may take dangerously large amounts of anabolic steroids to enhance athletic performance or to increase strength, often with serious consequences. Steroid abusers may vary the type and quantity of drug taken (called "stacking") or may take the drugs and then stop for a period of time, only to resume later (called "cycling"). Stacking and cycling are done in order to minimize serious side effects while maximizing the desired effects of the drugs. Even so, dangerous health consequences can occur.

Dangerous Health Consequences

The most common health consequences include high blood pressure, jaundice (yellowing of the skin), acne, and a greatly increased risk of cancer. In women, anabolic steroid abuse may cause masculinization, including a deepened voice, excessive facial and body hair, coarsening of the hair, menstrual cycle irregularities, and enlargement of the clitoris. Anabolic steroid abuse is even more dangerous during adolescence. When taken prior to or during the teenage growth spurt, steroids may result in permanently shortened height or early onset of puberty. Ironically, while proper use of anabolic steroids has been helpful in treating many cases of impotence in males, abuse of these drugs may cause impotence and even shrinking of the testicles.

Perhaps the most frightening aspect of anabolic steroid abuse are the reports of increased aggressive behavior and violent mood swings. Furthermore, many users have reported extremely severe withdrawal symptoms upon quitting. Also, many anabolic steroids have also been identified as a "gateway drug," leading abusers to escalate their drug habit to more dangerous drugs such as heroin and cocaine.

Widespread Use

Unfortunately, anabolic steroid use is not confined to professional sports. A 1999 study by the National Institute on Drug Abuse (NIDA) found that use of anabolic steroids has increased at an alarming rate among adolescents in middle and high school. The study also found that hundreds of thousands of adults abuse steroids on occasion. Nearly half of the individuals surveyed were unaware of the danger posed by these drugs, indicating the need for better education and awareness. In imitating their heroes, many young athletes are turning to anabolic steroids for the competitive edge that they offer, despite the many serious risks posed. Unfortunately, it may cause these young athletes to strike out.

Decide Your Opinion

1. Do you believe the techniques athletes use to train and enhance their performance should be regulated in any way? Why or why not? Does this apply to high-school athletes as well?
2. Is it acceptable for athletes to endanger their health by practicing excessively? By gaining or losing weight? By taking drugs? Why or why not?
3. Who, if anyone, should be in charge of regulating the behavior of athletes so that they do not harm themselves?

Figure 12A   Mark McGwire hits a home run.

12.5 Homeostasis

In this section, our discussion centers on the contribution of both the muscular and skeletal systems to homeostasis (Fig. 12.12).

Both Systems Produce Movement

Movement is essential to maintaining homeostasis. The skeletal and muscular systems work together to enable body movement (Fig. 12.13). This is most evidently illustrated by what happens when skeletal muscles contract and pull on the bones to which they are attached, causing movement at joints. Body movement of this sort allows us to respond to certain types of changes in the environment. For instance, if you are sitting in the sun and start to feel hot, you can get up and move to a shady spot.

The muscular and skeletal systems work for other types of movements that are just as important for maintaining homeostasis. Contraction of skeletal muscles associated with the jaw and tongue allow you to grind food

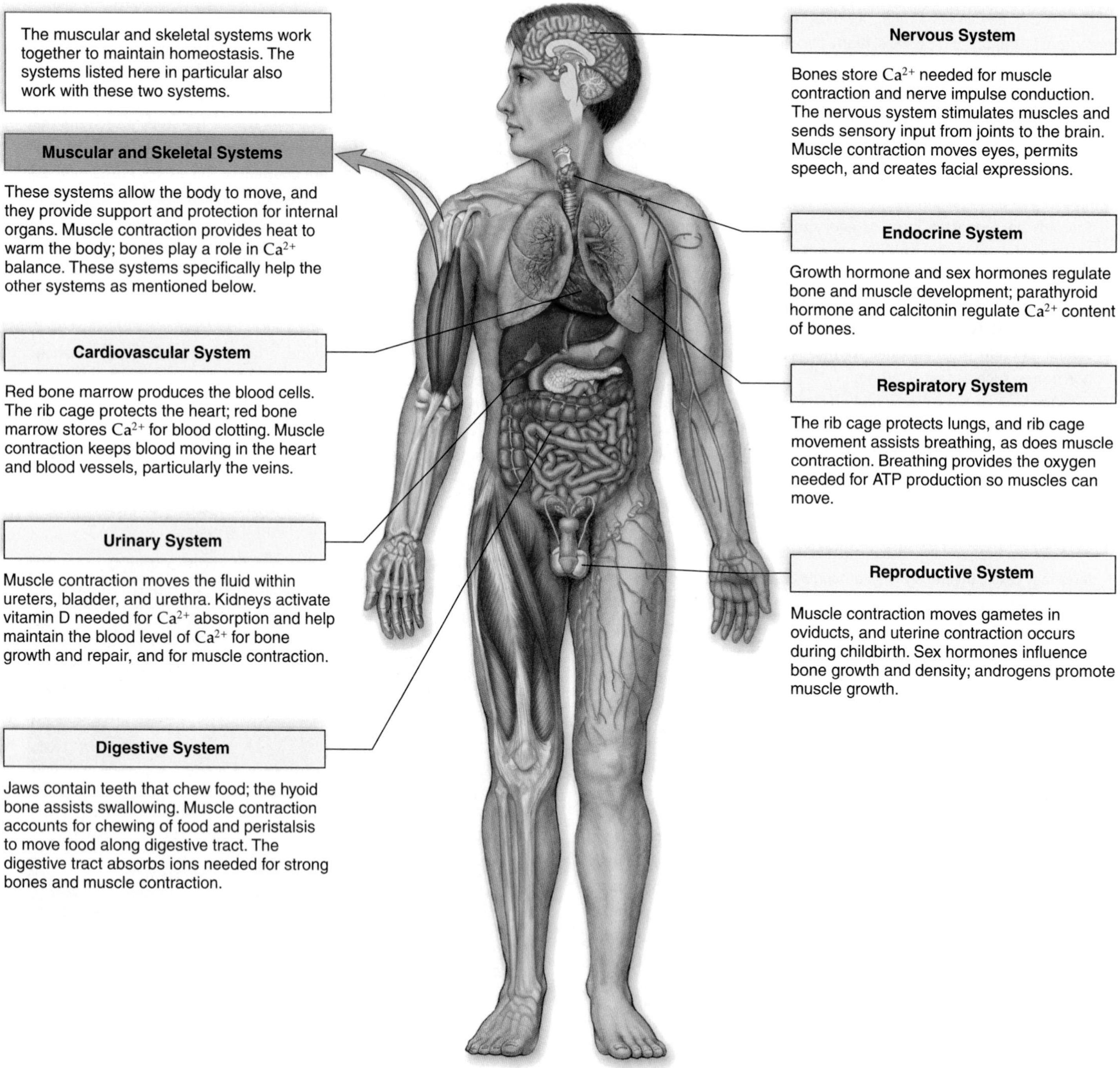

> The muscular and skeletal systems work together to maintain homeostasis. The systems listed here in particular also work with these two systems.

Muscular and Skeletal Systems

These systems allow the body to move, and they provide support and protection for internal organs. Muscle contraction provides heat to warm the body; bones play a role in Ca^{2+} balance. These systems specifically help the other systems as mentioned below.

Cardiovascular System

Red bone marrow produces the blood cells. The rib cage protects the heart; red bone marrow stores Ca^{2+} for blood clotting. Muscle contraction keeps blood moving in the heart and blood vessels, particularly the veins.

Urinary System

Muscle contraction moves the fluid within ureters, bladder, and urethra. Kidneys activate vitamin D needed for Ca^{2+} absorption and help maintain the blood level of Ca^{2+} for bone growth and repair, and for muscle contraction.

Digestive System

Jaws contain teeth that chew food; the hyoid bone assists swallowing. Muscle contraction accounts for chewing of food and peristalsis to move food along digestive tract. The digestive tract absorbs ions needed for strong bones and muscle contraction.

Nervous System

Bones store Ca^{2+} needed for muscle contraction and nerve impulse conduction. The nervous system stimulates muscles and sends sensory input from joints to the brain. Muscle contraction moves eyes, permits speech, and creates facial expressions.

Endocrine System

Growth hormone and sex hormones regulate bone and muscle development; parathyroid hormone and calcitonin regulate Ca^{2+} content of bones.

Respiratory System

The rib cage protects lungs, and rib cage movement assists breathing, as does muscle contraction. Breathing provides the oxygen needed for ATP production so muscles can move.

Reproductive System

Muscle contraction moves gametes in oviducts, and uterine contraction occurs during childbirth. Sex hormones influence bone growth and density; androgens promote muscle growth.

Figure 12.12   Human systems work together.

Figure 12.13   Muscles and bones.
Muscles and bones work closely together to create movements.

with the teeth. The rhythmic smooth-muscle contractions of peristalsis move ingested materials through the digestive tract. These processes are necessary for supplying the body's cells with nutrients. The ceaseless beating of your heart, which propels blood into the arterial system, is the contraction of cardiac muscle. Contractions of skeletal muscles in the body, especially those associated with breathing and leg movements, aid in the process of venous return by pushing blood back toward the heart. This is why soldiers and marching bands are cautioned not to lock their knees when standing at attention—the reduction in venous return causes a drop in blood pressure that can result in fainting. The pressure exerted by skeletal muscle contraction also helps to squeeze tissue fluid into the lymphatic capillaries, where it is referred to as lymph.

Both Systems Protect Body Parts

The skeletal system plays an important role just by protecting the soft internal organs of your body, such as the brain, heart, lungs, spinal cord, kidneys, liver, and most of the endocrine glands. All of these, particularly the nervous and endocrine organs, must be protected so that they can carry out activities necessary for homeostasis.

The muscular system pads bones and this alone offers protection for the abdominal organs.

Bones Store and Release Calcium

Under the direction of the endocrine system, the skeletal system performs tasks that are vital for calcium homeostasis. Calcium ions (Ca^{2+}) are needed for a variety of processes in your body, such as muscle contraction and nerve conduction. They are also necessary for the regulation of cellular metabolism by acting in cellular messenger systems. Thus, it is important to always maintain an adequate level of Ca^{2+} in the blood. When you have plenty of Ca^{2+} in your blood, the hormone calcitonin from the thyroid gland ensures that calcium salts are deposited in bone tissue.

Thus, the skeleton acts as a reservoir for storage of this important mineral. If your blood Ca^{2+} level starts to fall, parathyroid hormone secretion stimulates osteoclasts to break down bone tissue and thereby make Ca^{2+} available to the blood. Vitamin D is needed for the absorption of Ca^{2+} from the digestive tract, which is why vitamin D deficiency can result in weak bones. It is easy to get enough of this vitamin, since your skin produces it when exposed to sunlight, and the milk you buy at the grocery store is fortified with vitamin D.

Blood Cells Are Produced in Bones

Red bone marrow is the site of blood cell production. The red blood cells are the carriers of oxygen in the blood. Oxygen is necessary for the production of ATP by aerobic cellular respiration. White blood cells also originate in the red bone marrow. The white cells are involved in defending your body against pathogens and cancerous cells; without them, you would soon succumb to disease and die.

The bones of your skeleton contain two types of marrow: yellow and red. Fat is stored in yellow bone marrow, thus making it part of the body's energy reserves.

Muscles Help Maintain Body Temperature

The muscular system contributes to body temperature. When you are very cold, smooth muscle in the blood vessels that supply your skin constrict, reducing the amount of blood that is close to the surface of the body. This helps to conserve heat in the body's core, where vital organs lie. If you are cold enough, you may start to experience involuntary skeletal muscle contractions, known as shivering. This is initiated by temperature-sensitive neurons in the hypothalamus of the brain. Skeletal muscle contraction requires ATP, and using ATP generates heat. You may also notice that you get goose bumps when you are cold. This is because arrector pili muscles, tiny bundles of smooth muscle attached to the hair follicles, contract and cause the hairs to stand up. This is not very helpful in keeping humans warm, but it is quite effective in our furrier fellow mammals. Think of a cat or dog outside on a cold winter day. Its fur is a better insulator when standing up than lying flat. Goose bumps can also be a sign of fear. Although a human with goose bumps may not look very impressive, a frightened or aggressive animal whose fur is standing on end looks bigger and (hopefully) more intimidating to a predator or rival.

☑ Check Your Progress 12.5

1. What functions important to homeostasis are performed by both the skeletal and muscular systems working together?

2. a. What functions important to homeostasis are especially carried out by the skeletal system? b. The muscular system?

Summarizing the Concepts

12.1 Overview of Muscular System

Humans have three types of muscle tissue:

- Smooth muscle is involuntary and occurs in walls of internal organs.
- Cardiac muscle is involuntary and occurs in walls of the heart.
- Skeletal muscle is voluntary, contains bundles of muscle fibers called fascicles, and is usually attached by tendons to the skeleton.

Skeletal muscle functions:

- Helps maintain posture.
- Provide movement and heat.
- Protect underlying organs.

Skeletal Muscles of the Body

When achieving movement, some muscles are prime movers, some are synergists, and others are antagonists.

Names and Actions of Skeletal Muscles

Muscles are named for their size, shape, location, direction of fibers, number of attachments, and action.

12.2 Skeletal Muscle Fiber Contraction

Muscle fibers contain myofibrils, and myofibrils contain actin and myosin filaments. Muscle contraction occurs when sarcomeres shorten and actin filaments slide past myosin filaments.

- Nerve impulses travel down motor neurons and stimulate muscle fibers at neuromuscular junctions.
- The sarcolemma of a muscle fiber forms T tubules that almost touch the sarcoplasmic reticulum, which stores calcium ions.
- When calcium ions are released into muscle fibers, actin filaments slide past myosin filaments.
- At a neuromuscular junction, synaptic vesicles release acetylcholine (neurotransmitter), which diffuses across the synaptic cleft.
- When acetylcholine (Ach) is received by the sarcolemma, impulses begin and lead to the release of calcium.
- Calcium ions bind to troponin, exposing myosin binding sites.
- Myosin filaments break down ATP and attach to actin filaments, forming cross-bridges.
- When ADP and ℗ are released, cross-bridges change their positions.
- This pulls actin filaments to the center of a sarcomere.

12.3 Whole Muscle Contraction

Muscles Have Motor Units

- A muscle contains motor units: several fibers under the control of a single motor axon.
- Motor unit contraction is described in terms of a muscle twitch, summation, and tetanus.
- The strength of muscle contraction varies according to recruitment of motor units.
- In the body, a continuous slight tension (called muscle tone) is maintained by muscle motor units that take turns contracting.

Energy for Muscle Contraction

A muscle fiber has three ways to acquire ATP for muscle contraction.

- Creatine phosphate transfers a phosphate to ADP, and ATP results. This CP pathway is the most rapid.

- Fermentation also produces ATP quickly. Fermentation is associated with an oxygen debt because oxygen is needed to metabolize the lactate that accumulates.

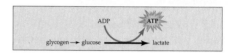

- Cellular respiration provides most of the muscle's ATP, but takes longer because much of the glucose and oxygen must be transported in blood to mitochondria. Cellular respiration occurs during aerobic exercise and burns fatty acids in addition to glucose.

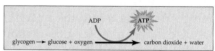

Fast-Twitch and Slow-Twitch Muscle Fibers

- Fast-twitch fibers, for sports like weight lifting, rely on an anaerobic means of acquiring ATP; have few mitochondria and myoglobin, but motor units contain more muscle fibers; and are known for explosive power, but fatigue quickly.
- Slow-twitch fibers, for sports like running and swimming, rely on aerobic respiration to acquire ATP; and have a plentiful supply of mitochondria and myoglobin, which gives them a dark color.

12.4 Muscular Disorders

Muscular disorders include spasms and injuries, as well as diseases such as muscular dystrophy and myasthenia gravis.

12.5 Homeostasis

- The muscles and bones produce movement and protect body parts.
- The bones produce red blood cells and are involved in the regulation of blood calcium levels.
- The muscles produce the heat that gives us a constant body temperature.

Understanding Key Terms

actin 232	intercalated disk 228
all-or-none law 236	motor unit 236
amylotrophic lateral sclerosis (ALS) 240	muscle fiber 228
	muscle twitch 236
bursa 240	muscular dystrophy 240
bursitis 240	myalgia 240
cardiac muscle 228	myasthenia gravis 240
convulsion 240	myofibril 232
cramp 240	myoglobin 237
Duchenne muscular dystrophy 240	myosin 232
	neuromuscular junction 234
facial tic 240	origin 229
fibromyalgia 240	oxygen debt 237
insertion 229	rigor mortis 228

sarcolemma 232

sarcomere 232

sarcoplasmic reticulum 232

skeletal muscle 228

sliding filament model 232

smooth muscle 228

spasm 240

sprain 240

strain 240

T (transverse) tubule 232

tendinitis 240

tendon 229

tetanus 236

tropomyosin 235

troponin 235

Match the key terms to these definitions.

a. _____ Structural and functional unit of a myofibril; contains actin and myosin filaments.

b. _____ End of a muscle that is attached to a movable bone.

c. _____ Sustained maximal muscle contraction.

d. _____ Contraction of muscles at death due to lack of ATP.

e. _____ Stretching or tearing of a muscle.

Testing Your Knowledge of the Concepts

1. What are the characteristics of the three types of muscles in the human body? Where is each type found? (page 228)

2. Give an example to show that skeletal muscles work in antagonistic pairs. Explain. (page 229)

3. What criteria are used to name muscles? Give an example of each one. (pages 230–31)

4. What are the functions of a muscle fiber's components? (page 232)

5. Describe the sliding filament model of muscle contraction within a sarcomere. Begin with the nerve impulse and end with the relaxation of the muscle. (pages 232–35)

6. What are the four possible energy sources for a muscle and the three sources of ATP for muscle contraction? (pages 236–37)

7. Compare fast- and slow-twitch muscle fibers. (page 239)

8. What are some common muscular disorders and some more serious muscular diseases? (page 240)

9. How does the muscular system help maintain homeostasis? (pages 242–43)

10. Impulses that move down the T system of a muscle fiber most directly cause
    a. movement of tropomyosin.
    b. attachment of the cross-bridges to myosin.
    c. release of Ca^{2+} from the sarcoplasmic reticulum.
    d. splitting of ATP.

11. Which of the following statements about cross-bridges is false?
    a. They are composed of myosin.
    b. They bind to ATP after they detach from actin.
    c. They contain an ATPase.
    d. They split ATP before they attach to actin.

12. Which statement about sarcomere contraction is incorrect?
    a. The A bands shorten.
    b. The H zones shorten.
    c. The I bands shorten.
    d. The sarcomeres shorten.

13. Which of the following muscles would have motor units with the lowest innervation ratio?
    a. leg muscles
    b. arm muscles
    c. muscles that move the fingers
    d. muscles of the trunk

14. The thick filaments of a muscle fiber are made up of
    a. actin.
    b. troponin.
    c. fascia.
    d. myosin.

15. As ADP and Ⓟ are released from a myosin head,
    a. actin filaments move toward the H zone.
    b. myosin cross-bridges pull the thin filaments.
    c. a sarcomere shortens.
    d. Only a and c are correct.
    e. All of these are correct.

16. Which of these is a direct source of energy for muscle contraction?
    a. ATP
    b. creatine phosphate
    c. lactic acid
    d. glycogen
    e. Both a and b are correct.

17. When muscles contract,
    a. sarcomeres increase in length.
    b. actin breaks down ATP.
    c. myosin slides past actin.
    d. the H zone disappears.
    e. calcium is taken up by the sarcoplasmic reticulum.

18. Nervous stimulation of muscles
    a. occurs at a neuromuscular junction.
    b. results in an impulse that travels down the T system.
    c. causes calcium to be released from expanded regions of the sarcoplasmic reticulum.
    d. All of these are correct.

19. In a muscle fiber,
    a. the sarcolemma is connective tissue holding the myofibrils together.
    b. the T system causes release of Ca^{2+} from the sarcoplasmic reticulum.
    c. both actin and myosin filaments have cross-bridges.
    d. there is no endoplasmic reticulum.
    e. All of these are correct.

20. To increase the force of muscle contraction,
    a. individual muscle cells have to contract with greater force.
    b. motor units have to contract with greater force.
    c. motor units need to be recruited.
    d. All of these are correct.
    e. None of these is correct.

21. Lack of calcium in muscles will
    a. result in no contraction.
    b. cause weak contraction.
    c. cause strong contraction.
    d. will have no effect.
    e. None of these is correct.

22. Which of these energy relationships are mismatched?
    a. creatine phosphate—anaerobic
    b. cellular respiration—aerobic
    c. fermentation—anaerobic
    d. oxygen debt—anaerobic
    e. All of these are properly matched.

23. During muscle contraction,
    a. ATP is hydrolyzed when the myosin head is unattached.
    b. ADP and ⓅP are released as the myosin head attaches to actin.
    c. ADP and ⓅP release causes the head to change position and actin filaments to move.
    d. release of ATP causes the myosin head to return to resting position.

24. Proper functioning of a neuromuscular junction requires the
    a. presence of acetylcholine.
    b. presence of a synaptic cleft.
    c. presence of a motor terminal.
    d. sarcolemma of a muscle cell.
    e. All of these are correct.

In questions 25–28, match each muscle to a region of the body in the key.

Key:
    a. neck and back   c. arm
    b. abdomen   d. thigh

25. Biceps femoris

26. Trapezius

27. Rectus abdominis

28. Triceps brachii

In questions 29–32, match each function to a muscle of the buttocks and legs in the key.

Key:
    a. tibialis anterior   c. biceps femoris
    b. gluteus maximus   d. gastrocnemius

29. Bends leg at knee and extends thigh back

30. Bends ankle so that foot is upward

31. Bends leg at knee and bends sole of foot

32. Extends thigh back

In questions 33–35, match each function to a muscle of the head and neck in the key.

Key:
    a. trapezius   c. masseter
    b. orbicularis oculi

33. A chewing muscle

34. Moves head and scapula

35. Closes eye

In questions 36–38, match the functions to a muscle of the upper limb and trunk in the key.

Key:
    a. external oblique   c. pectoralis major
    b. triceps brachii

36. Compresses abdomen

37. Brings arm across chest

38. Straightens forearm at elbow

In questions 39–41, match each function to a muscle of the lower limb in the key.

Key:
    a. adductor longus   c. sartorius
    b. quadriceps femoris

39. Straightens leg at knee

40. Moves thigh toward the body

41. Lateral rotates thigh

42. Label this diagram of a muscle fiber, using these terms: myofibril, T tubule, sarcomere, sarcolemma, sarcoplasmic reticulum, Z line.

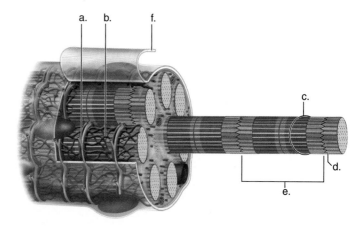

Thinking Critically About the Concepts

You may be a fan of the popular CSI series of TV shows and have heard the characters discuss the extent of rigor mortis in a recently discovered body. Or, perhaps you've read *The Body Farm* by Patricia Cornwell and are familiar with the processes involved with determining the time of death.

The extent of rigor mortis or absence of it helps officials determine time of death. However, the onset of rigor mortis is influenced by a number of environmental variables. Crime scene investigators, such as Jason from the opening story, must collect environmental data as well as consider other indicators of time of death and personal variables such as the size of the victim, the victim's clothing, and the victim's activity prior to death.

1a. If a body was rapidly cooled after death, how would the onset of rigor mortis change?

  b. Explain how rapid cooling affects the onset of rigor mortis.

2. Rigor mortis is usually complete within 1–2 days after death (depending on environmental variables). Why would rigor mortis wear off after a couple of days?

3. The Anthropological Research Facility (Body Farm) was created in Knoxville, Tennessee, by Dr. William Bass to investigate the variables associated with the decomposition of a body.

  a. What kinds of environmental variables would you expect them to be investigating at the Body Farm?

  b. What kinds of limitations would the research findings at the Body Farm have?

4a. Why would rigor mortis be of concern to the meat industry?

  b. What happens during the "aging" of meat (beef), and why is it done?

  c. Why are lamb and pork not commonly aged?

CHAPTER **13**

Nervous System

D r. Smith reminded his class that the term *neurons* means nerve cells. "Today, I want to discuss how important the myelin sheath is that occurs around many axons," he said. "The myelin sheath," he related, "is interrupted by gaps called nodes. When myelin is present, the impulse travels down an axon by 'jumping' from node to node. This is called saltatory conduction. *Saltar* is a Spanish verb that means 'to jump.' Now when you go to your Spanish classes, be sure to impress your professors and friends with your new vocabulary."

Myelination, the laying down of the myelin sheath, begins during fetal development and continues during early childhood. That's one reason children's motor skills can improve with age and practice. Alternately, certain diseases, such as multiple sclerosis and leukodystrophies, destroy the myelin sheath. The movie, *Lorenzo's Oil*, was about a child with a leukodystrophy and his parents' search for a cure. When the myelin sheath is lost, nerve impulses are transmitted more slowly, and their control of muscle contractions lessens. Currently, there is no cure for multiple sclerosis and leukodystrophies, but there are medications and nutritional regimens that slow down the progression of the diseases.

CHAPTER CONCEPTS

13.1 Overview of the Nervous System
Of three main functions of the nervous system, reception of stimuli can be associated with sensory neurons; integration can be associated with interneurons; and motor output can be associated with motor neurons. All neurons use the same methods to transmit nerve impulses along neurons and across synapses.

13.2 The Central Nervous System
The central nervous system consists of the brain and the spinal cord. The brain is divided into portions, each with specific functions, and the spinal cord communicates with the brain by sending it input from, and receiving output for, most of the body.

13.3 The Limbic System and Higher Mental Functions
The limbic system involves many parts of the brain, and it gives emotional overtones to the activities of the brain. It is also important to the processes of learning and memory.

13.4 The Peripheral Nervous System
The peripheral nervous system consists of nerves that project from the CNS. Cranial nerves project from the brain, and spinal nerves project from the spinal cord.

13.5 Drug Abuse
Although neurological drugs are quite varied, each type has been found to either promote or prevent the action of a particular neurotransmitter at a synapse.

13.1 Overview of the Nervous System

The nervous system has two major divisions (Fig. 13.1*a*). The **central nervous system (CNS)** consists of the brain and spinal cord, which are located in the midline of the body. The **peripheral nervous system (PNS)** consists of nerves. Nerves lie outside the CNS. The division between the CNS and the PNS is arbitrary; the two systems work together and are connected to each other (Fig. 13.1*b*).

The nervous system has three specific functions:

1. The nervous system receives sensory input—sensory receptors in skin and other organs respond to external and internal stimuli by generating nerve impulses that travel by way of the PNS to the CNS. When Calvin opened the front door in the opening story for Chapter 3, the smell of the cookies his mother had just baked caused his olfactory receptors to send nerve impulses to the CNS.

2. The CNS performs integration—the CNS sums up the input it receives from all over the body. Calvin's brain quickly realized that his mother was upstairs and conceived the idea of swiping a few cookies.

3. The CNS generates motor output—nerve impulses from the CNS go by way of the PNS to the muscles and glands. Calvin ran into the kitchen, grabbed the cookies, and scampered out the back door, avoiding a confrontation with his mother, who was calling to him.

Nervous Tissue

Nervous tissue contains two types of cells: neurons and neuroglia (neuroglial cells). **Neurons** are the cells that transmit nerve impulses between parts of the nervous system; **neuroglia** support and nourish neurons. Neuroglia were discussed in Chapter 4, and this section discusses the structure and function of neurons.

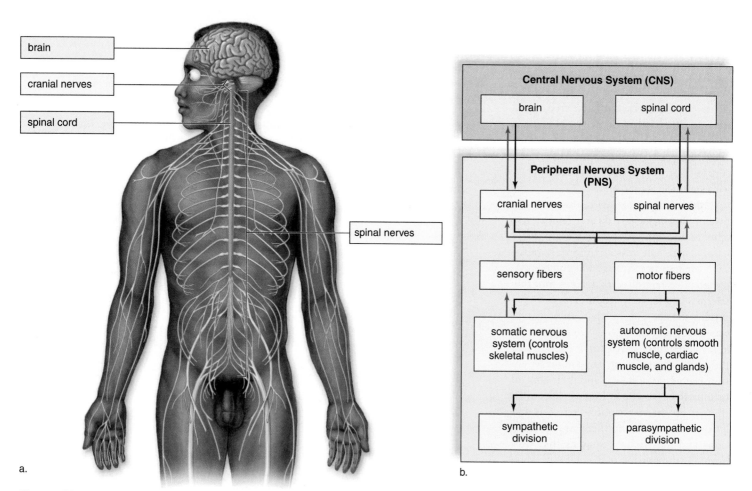

a.

b.

Figure 13.1   Organization of the nervous system.

a. The central nervous system (CNS) consists of the brain and spinal cord. The peripheral nervous system (PNS) consists of the nerves, which lie outside the CNS.
b. The red arrows are the pathway by which the CNS receives sensory information. The black arrows are the pathway by which the CNS communicates with the somatic nervous system and the autonomic coervous system, two divisions of the PNS.

Neuron Structure

Classified according to function, the three types of neurons are sensory neurons, interneurons, and motor neurons (Fig. 13.2). Their functions are best described in relation to the CNS. A **sensory neuron** takes nerve impulses (messages) from a sensory receptor to the CNS. **Sensory receptors** are special structures that detect changes in the environment. An **interneuron** lies entirely within the CNS. Interneurons can receive input from sensory neurons and also from other interneurons in the CNS. Thereafter, they sum up all the nerve impulses received from these neurons before they communicate with motor neurons. A **motor neuron** takes nerve impulses away from the CNS to an effector (muscle fiber or gland). **Effectors** carry out our responses to environmental changes, whether they are external or internal.

Neurons vary in appearance, but all of them have just three parts: a cell body, dendrites, and an axon. The **cell body** contains the nucleus, as well as other organelles. **Dendrites** are the many short extensions that receive signals from sensory receptors or other neurons. These signals can result in nerve impulses that are then conducted by an axon. The **axon** is the portion of a neuron that conducts nerve impulses. An axon can be quite long, and when present in nerves, an axon is called a nerve fiber.

Notice that in sensory neurons, a very long axon carries nerve impulses from the dendrites associated with a sensory receptor to the CNS and that this axon is interrupted by the cell body. In interneurons and motor neurons, on the other hand, multiple dendrites take signals to the cell body, and then an axon conducts nerve impulses away from the cell body.

Myelin Sheath

Many axons are covered by a protective **myelin sheath.** In the PNS, this covering is formed by a type of neuroglia called **Schwann cells,** which contain myelin (a lipid substance) in their plasma membranes. In the CNS, oligodendrocytes perform this function. The myelin sheath develops when these cells wrap themselves around an axon many times. Because each neuroglia cell covers only a portion of an axon, the myelin sheath is interrupted. The gaps where there is no myelin sheath are called **nodes of Ranvier** (Fig. 13.2).

Long axons tend to have a myelin sheath, but short axons do not. The gray matter of the CNS is gray because it contains no myelinated axons; the white matter of the CNS is white because it does. In the PNS, myelin gives nerve fibers their white, glistening appearance and serves as an excellent insulator. The myelin sheath also plays an important role in nerve regeneration within the PNS. If an axon is accidentally severed, the myelin sheath remains and serves as a passageway for new fiber growth.

As mentioned in the opening story, multiple sclerosis (MS) and leukodystrophies are caused by loss of myelin from

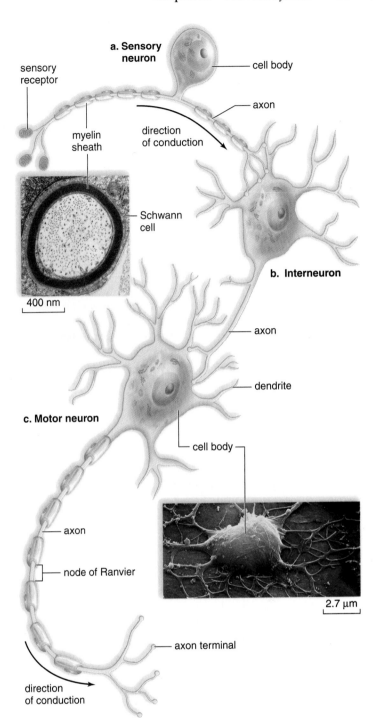

400 nm

Figure 13.2   Types of neurons.

a. A sensory neuron has a long axon covered by a myelin sheath that takes nerve impulses all the way from dendrites to the CNS. **b.** In the CNS, some interneurons, such as this one, have a short axon that is not covered by a myelin sheath. **c.** A motor neuron has a long axon covered by a myelin sheath that takes nerve impulses from the CNS to an effector.

the axons. Leukodystrophies are generally caused by a defect in one of the genes involved with the growth and maintenance of myelin. MS is believed to be caused by an attack on the myelin by the body's immune system.

The Nerve Impulse

Nerve impulses convey information within the nervous system. The nerve impulse is studied by using excised axons and a voltmeter to measure voltage. The voltmeter allows us to measure the potential difference between two sides of the axonal membrane (plasma membrane of the axon), expressed in terms of a voltage.

Resting Potential

When the axon is not conducting an impulse, the voltmeter records a membrane potential of about −65 mV (millivolts), indicating that the inside of the neuron is more negative than the outside (Fig. 13.3a). This is called the **resting potential** because the axon is not conducting an impulse.

The resting potential correlates with a difference in ion distribution on either side of the axonal membrane. As Figure 13.3a shows, the concentration of sodium ions (Na⁺) is greater outside the axon than inside, and the concentration of potassium ions (K⁺) is greater inside the axon than outside. The unequal distribution of these ions is due to the action of the **sodium-potassium pump** that actively transports Na⁺ out of and K⁺ into the axon. The work of the pump maintains the unequal distribution of Na⁺ and K⁺ across the membrane.

The membrane is permeable to K⁺ but not Na⁺; therefore, there are always more positive ions outside the membrane than inside. Large, negatively charged organic ions in the axoplasm also contribute to the polarity across a resting axonal membrane.

Action Potential

An **action potential** is a rapid change in polarity across an axonal membrane as the nerve impulse occurs. If a stimulus causes the axonal membrane to depolarize to a certain level, called **threshold,** an action potential occurs in an all-or-none manner. The strength of an action potential does not change; an intense stimulus can cause an axon to fire (start an axon potential) more often in a given time interval.

The action potential requires two types of gated channel proteins in the membrane. One gated channel protein opens to allow Na⁺ to pass through the membrane, and another opens to allow K⁺ to pass through the membrane.

Sodium Gates Open   When an action potential occurs, the gates of sodium channels open first, and Na⁺ flows into the axon. As Na⁺ moves to inside the axon, the membrane potential changes from −65 mV to +40 mV. This is a *depolarization* because the charge inside the axon changes from negative to positive (Fig. 13.3b).

Potassium Gates Open   Second, the gates of potassium channels open, and K⁺ flows to outside the axon. As K⁺ moves to outside the axon, the action potential changes from +40 mV back to −65 mV. This is a *repolarization* because the inside of the axon resumes a negative charge as K⁺ exits the axon (Fig. 13.3c).

Figure 13.3   **Resting and action potential of the axonal membrane.** **a.** Resting potential occurs when a neuron is not conducting a nerve impulse. During an action potential, **(b)** depolarization is followed by **(c)** repolarization. **d.** Visualizing an action potential.

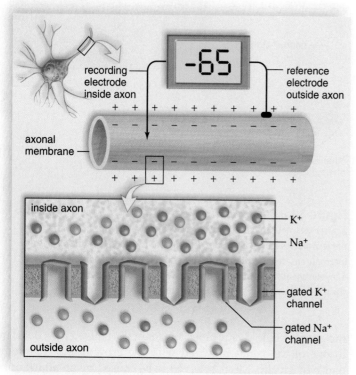

a. Resting potential: more Na⁺ outside the axon and more K⁺ inside the axon causes polarization.

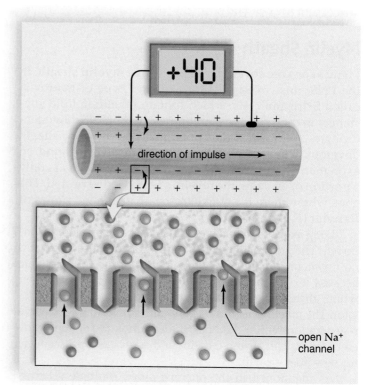

b. Action potential begins: depolarization occurs when Na⁺ gates open and Na⁺ moves to inside the axon.

Visualizing an Action Potential

In order to visualize such rapid fluctuations in voltage across the axonal membrane, researchers generally find it useful to plot the voltage changes over time (Fig. 13.3*d*).

After an action potential has passed by, the sodium-potassium pump restores the resting potential by moving the potassium back to the inside and sodium back to the outside.

Propagation of an Action Potential

When an axon is unmyelinated, the action potential at one locale stimulates an adjacent part of the axon's membrane to produce an action potential. In myelinated axons, an action potential at one Ranvier node causes an action potential at the next node:

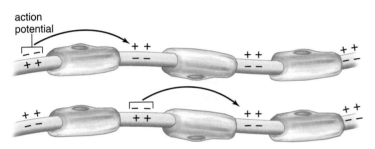

As mentioned in the opening story, this type of conduction is called **saltatory conduction,** because the nerve impulse jumps from node to node. Saltatory conduction is much faster than otherwise. In thin, unmyelinated axons, the nerve impulse travels about 1.0 m/second, and in thick, myelinated axons, the rate is more than 100 m/second. In any case,

action potentials are self-propagating; each action potential generates another along the length of an axon.

Demyelination occurs in persons with multiple sclerosis and leukodystrophies, and then maximum speed decreases sometimes by tenfold or more. The loss of myelin leads to significant disruption in the proper functioning of the nervous system. The movie *Lorenzo's Oil* tells the real life story of Lorenzo Odone who was initially a normal, healthy boy until the condition adrenoleukodystrophy (ALD) began to take its toll. Eventually, Lorenzo became unresponsive as his condition worsened. There are at least 30 different types of leukodystrophies, each of them due to the inheritance of a faulty gene for the production and maintenance of phospholipids in myelin. The altered phospholipids contain very long chain, saturated fatty acids instead of fatty acids of moderate length that are unsaturated. Lorenzo's parents developed a diet they felt helped their son, and they have shared this knowledge with other parents.

The conduction of a nerve impulse (action potential) is an all-or-none event—that is, either an axon conducts a nerve impulse or it does not. The intensity of a message is determined by how many nerve impulses are generated within a given time span. An axon can conduct a volley of nerve impulses because only a small number of ions are exchanged with each impulse. As soon as an impulse has passed by each successive portion of an axon, it undergoes a **refractory period,** during which the sodium gates are unable to open. This ensures that the action potential cannot move backward and instead always moves down an axon toward its branches.

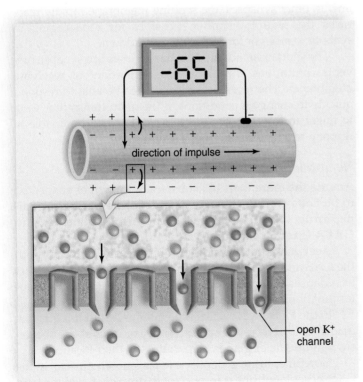

c. Action potential ends: repolarization occurs when K⁺ gates open and K⁺ moves to outside the axon.

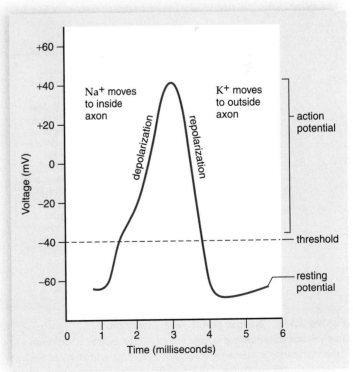

d. An action potential can be visualized if voltage changes are graphed over time.

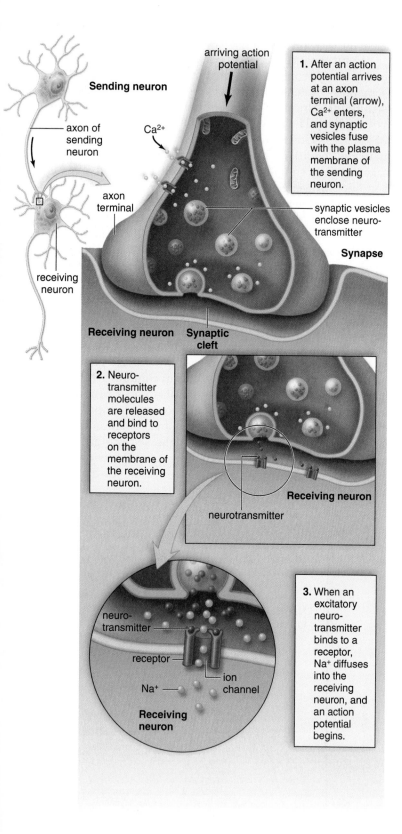

Sending neuron

arriving action potential

axon of sending neuron

Ca^{2+}

axon terminal

receiving neuron

1. After an action potential arrives at an axon terminal (arrow), Ca^{2+} enters, and synaptic vesicles fuse with the plasma membrane of the sending neuron.

synaptic vesicles enclose neuro-transmitter

Synapse

Receiving neuron **Synaptic cleft**

2. Neuro-transmitter molecules are released and bind to receptors on the membrane of the receiving neuron.

Receiving neuron

neurotransmitter

neuro-transmitter

receptor

Na^+

ion channel

Receiving neuron

3. When an excitatory neuro-transmitter binds to a receptor, Na^+ diffuses into the receiving neuron, and an action potential begins.

Figure 13.4 **Synapse structure and function.**
Transmission across a synapse from one neuron to another occurs when a neurotransmitter is released and diffuses across a synaptic cleft and binds to a receptor in the membrane of the receiving neuron.

The Synapse

Every axon branches into many fine endings, each tipped by a small swelling called an **axon terminal.** Each terminal lies very close to either the dendrite or the cell body of another neuron. This region of close proximity is called a **synapse** (Fig. 13.4). At a synapse, a small gap called the **synaptic cleft** separates the sending neuron from the receiving neuron. The nerve impulse is unable to jump the cleft, and therefore, another means is needed to pass the nerve impulse from the sending neuron to the receiving neuron.

Transmission across a synapse is carried out by molecules called **neurotransmitters,** which are stored in synaptic vesicles in the axon terminals. These events occur: (1) Nerve impulses traveling along an axon reach an axon terminal. (2) Calcium ions enter the terminal, and they stimulate synaptic vesicles to merge with the sending membrane. (3) Neurotransmitter molecules are released into the synaptic cleft, and they diffuse across the cleft to the receiving membrane, where they bind with specific receptor proteins. Depending on the type of neurotransmitter, the response of the receiving neuron can be toward excitation or toward inhibition. In Figure 13.5, excitation occurs because the neurotransmitter, such as acetycholine, has caused the sodium gate to open, and Na^+ diffuses into the receiving neuron. Inhibition would occur if a neurotransmitter caused K^+ to enter the receiving neuron.

Once a neurotransmitter has been released into a synaptic cleft and has initiated a response, it is removed from the cleft. In some synapses, the receiving membrane contains enzymes that rapidly inactivate the neurotransmitter. For example, the enzyme **acetylcholinesterase (AChE)** breaks down acetylcholine. In other synapses, the sending membrane rapidly reabsorbs the neurotransmitter, possibly for repackaging in synaptic vesicles or for molecular breakdown.

The short existence of neurotransmitters at a synapse prevents continuous stimulation (or inhibition) of receiving membranes. The receiving cell needs to be able to respond quickly to changing conditions; if the neurotransmitter were to linger in the cleft, the receiving cell would be unable to respond to a new signal from a sending cell.

Neurotransmitter Molecules

Among the more than 100 substances known or suspected to be neurotransmitters are **acetylcholine (ACh), norepinephrine (NE), dopamine, serotonin, glutamate,** and **GABA (gamma aminobutyric acid).**

Acetylcholine and norepinephrine are active in both the CNS and PNS. In the PNS, these neurotransmitters act at synapses called neuromuscular junctions. Neuromuscular junctions are discussed on page 234.

In the PNS, ACh excites skeletal muscle but inhibits cardiac muscle. It has either an excitatory or inhibitory effect on smooth muscle or glands, depending on their location.

Norepinephrine generally excites smooth muscle. In the CNS, norepinephrine is important to dreaming, waking, and mood. Serotonin is involved in thermoregulation, sleeping, emotions, and perception. Reduced levels of norepinephrine and serotonin seem to be linked to depression, and the antidepressant drug Prozac blocks the removal of serotonin from a synapse.

Many of the drugs that affect the nervous system act by interfering with or potentiating the action of neurotransmitters. As described on pages 267–69, drugs can enhance or block the release of a neurotransmitter, mimic the action of a neurotransmitter or block the receptor, or interfere with the removal of a neurotransmitter from a synaptic cleft. GABA is an abundant inhibitory neurotransmitter in the CNS. Valium binds to the receptors of GABA, thereby increasing its effects.

Neuromodulators are molecules that block the release of a neurotransmitter or modify a neuron's response to a neurotransmitter. The caffeine in coffee, chocolate, and tea keeps us awake by interfering with the effects of inhibitory neurotransmitters in the brain. Two well-known neuromodulators are substance P and endorphins. Substance P is released by sensory neurons when pain is present. Endorphins block the release of substance P and, therefore, serve as natural painkillers. They are associated with the "runner's high" of joggers and are produced by the brain, not only when physical stress, but also emotional stress, is present. The opiates—namely, codeine, heroin, and morphine—bind to endorphin receptors, and in this way reduce pain and produce a feeling of well-being.

Synaptic Integration

A single neuron has many dendrites plus the cell body, and all of these can have synapses with many other neurons. Therefore, a neuron is on the receiving end of many signals (potential change), which can be either excitatory or inhibitory. An excitatory neurotransmitter produces a potential change called a signal that drives the neuron closer to an action potential; an inhibitory neurotransmitter produces a signal that drives the neuron farther from an action potential.

Neurons integrate these incoming signals. **Integration** is the summing up of excitatory and inhibitory signals (Fig. 13.5). If a neuron receives enough excitatory signals (either from different synapses or at a rapid rate from one synapse) to outweigh the inhibitory ones, chances are the axon will transmit a nerve impulse. On the other hand, if a neuron receives more inhibitory than excitatory signals, the summing up of these signals may prohibit the axon from firing.

☑ Check Your Progress 13.1

1. What are the two major divisions of the nervous system?
2. a. What are the three types of neurons, and (b) what are the three parts of a neuron?
3. a. What is a nerve impulse, and (b) how is it propagated?
4. Since neurons don't physically touch, how is an impulse transmitted from one neuron to the next?

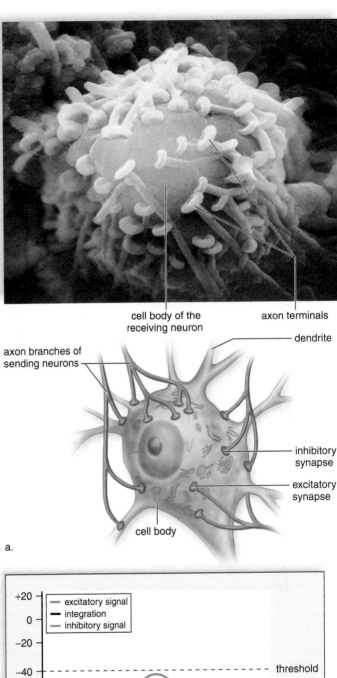

a.

b.

Figure 13.5 Synaptic integration.
a. Inhibitory signals and excitatory signals are summed up in the dendrite and cell body of the postsynaptic neuron. Only if the combined signals cause the membrane potential to rise above threshold does an action potential occur. b. In this example, threshold was not reached.

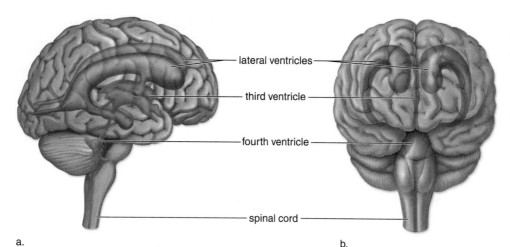

a.   b.

Figure 13.6 **Ventricles of the brain.** The brain has four ventricles, two are called the lateral ventricles, and then these are followed by the third and fourth ventricles. **a.** Lateral view of ventricles seen through a transparent brain. **b.** Anterior view of ventricles seen through a transparent brain.

13.2 The Central Nervous System

The spinal cord and the brain make up the CNS, where sensory information is received and motor control is initiated. Both the spinal cord and the brain are protected by bone; the spinal cord is surrounded by vertebrae, and the brain is enclosed by the skull. Also, both the spinal cord and the brain are wrapped in protective membranes known as **meninges** (sing., meninx). The spaces between the meninges are filled with **cerebrospinal fluid,** which cushions and protects the CNS. A small amount of this fluid is sometimes withdrawn from around the spinal cord for laboratory testing when a spinal tap (lumbar puncture) is performed. Meningitis is an infection of the meninges.

Cerebrospinal fluid is also contained within the ventricles of the brain and in the central canal of the spinal cord. The brain has four **ventricles,** interconnecting chambers that produce and serve as a reservoir for cerebrospinal fluid (Fig. 13.6). Normally, any excess cerebrospinal fluid drains away into the cardiovascular system. However, blockages can occur. In an infant, the brain can enlarge due to cerebrospinal fluid accumulation, resulting in a condition called hydrocephalus ("water on the brain"). If cerebrospinal fluid collects in an adult, the brain cannot enlarge, and instead is pushed against the skull, possibly becoming injured.

The CNS is composed of two types of nervous tissue—gray matter and white matter. **Gray matter** contains cell bodies and short, nonmyelinated fibers. **White matter** contains myelinated axons that run together in bundles called **tracts.**

The Spinal Cord

The **spinal cord** extends from the base of the brain through a large opening in the skull called the foramen magnum and into the vertebral canal formed by openings in the vertebrae.

Structure of the Spinal Cord

A cross section of the spinal cord shows a central canal, gray matter, and white matter (Fig. 13.7a). Figure 13.7b shows how an individual vertebra protects the spinal cord.

The spinal nerves project from the cord between the vertebrae that make up the vertebral column. Intervertebral disks separate the vertebrae, and if a disk slips a bit and presses on the spinal cord, pain results.

The central canal contains cerebrospinal fluid, as do the meninges that protect the spinal cord. The gray matter is centrally located and shaped like the letter H (Fig. 13.7a, b, and c). Portions of sensory neurons and motor neurons are found there, as are interneurons that communicate with these two types of neurons. The dorsal root of a spinal nerve contains sensory fibers entering the gray matter, and the ventral root of a spinal nerve contains motor fibers exiting the gray matter. The dorsal and ventral roots join before the spinal nerve leaves the vertebral canal as a mixed nerve (Fig. 13.7c, d). Spinal nerves are a part of the PNS.

The white matter of the spinal cord occurs in areas around the gray matter. The white matter contains ascending tracts taking information to the brain (primarily located dorsally) and descending tracts taking information from the brain (primarily located ventrally). Because many tracts cross just after they enter and exit the brain, the left side of the brain controls the right side of the body, and the right side of the brain controls the left side of the body.

Functions of the Spinal Cord

The spinal cord provides a means of communication between the brain and the peripheral nerves that leave the cord. When someone touches your hand, sensory receptors generate nerve impulses that pass through sensory fibers to the spinal cord and up ascending tracts to the brain (see Fig. 13.1b, red arrows).

The gate control theory of pain proposes that the tracts in the spinal cord have "gates," and that these "gates" control the flow of pain messages from the peripheral nerves to the brain. Depending on how the gates process a pain signal, the pain message can be allowed to pass directly to the brain or can be prevented from reaching the brain. As mentioned earlier, endorphins can temporarily block pain messages and so can other messages, such as those received from touch receptors.

When the brain initiates voluntarily control over our limbs, motor impulses originating in the brain pass down descending tracts to the spinal cord and out to our muscles by way of motor fibers (see Fig. 13.1*b*, black arrows). Therefore, if the spinal cord is severed, we suffer a loss of sensation and a loss of voluntary control—that is, paralysis. If the cut occurs in the thoracic region, the lower body and legs are paralyzed, a condition known as paraplegia. If the injury is in the neck region, all four limbs are usually affected, a condition called quadriplegia.

Reflex Actions The spinal cord is the center for thousands of reflex arcs (see Fig. 13.16). A stimulus causes sensory receptors to generate nerve impulses that travel in sensory axons to the spinal cord. Interneurons integrate the incoming data and relay signals to motor neurons. A response to the stimulus occurs when motor axons cause skeletal muscles to contract. Each interneuron in the spinal cord has synapses with many other neurons, and therefore, they send signals to several other interneurons and motor neurons.

The spinal cord plays a similar role for the internal organs. For example, when blood pressure falls, internal receptors in the carotid arteries and aorta generate nerve impulses that pass through sensory fibers to the cord and then up an ascending tract to a cardiovascular center in the brain. Thereafter, nerve impulses pass down a descending tract to the spinal cord. Motor impulses then cause blood vessels to constrict so that the blood pressure rises.

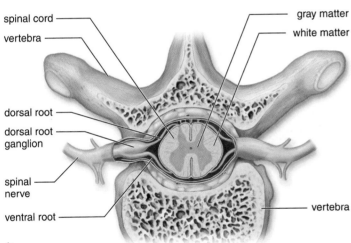

a.

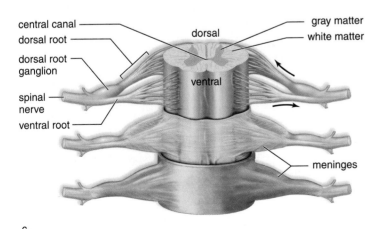

b.

c.

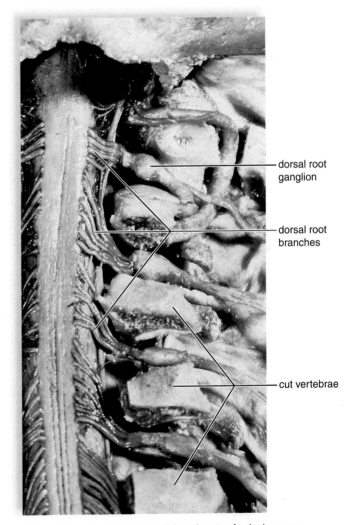

d. Dorsal view of spinal cord and dorsal roots of spinal nerves.

Figure 13.7 Spinal cord.
a. Cross section of spinal cord. **b.** The spinal cord is protected by vertebrae. **c.** The spinal cord and spinal nerves. **d.** Spinal nerves emerging from cord.

The Brain

The human **brain** has been called the last great frontier of biology. The goal of modern neuroscience is to understand the structure and function of the brain's various parts so well that it will be possible to prevent or correct the thousands of mental disorders that rob human beings of a normal life. This section gives only a glimpse of what is known about the brain and the modern avenues of research.

We will discuss the parts of the brain with reference to the cerebrum, the diencephalon, the cerebellum, and the brain stem. The brain's four ventricles (described on page 254) are called, in turn, the two lateral ventricles, the third ventricle, and the fourth ventricle. It may be helpful for you to associate the cerebrum with the two lateral ventricles, the diencephalon with the third ventricle, and the brain stem and the cerebellum with the fourth ventricle (Fig. 13.8a).

The Cerebrum

The **cerebrum,** also called the telencephalon, is the largest portion of the brain in humans. The cerebrum is the last center to receive sensory input and carry out integration before commanding voluntary motor responses. It communicates with and coordinates the activities of the other parts of the brain.

Cerebral Hemispheres   Just as the human body has two halves, so does the cerebrum. These halves are called the left and right **cerebral hemispheres** (Fig. 13.8b). A deep groove called the longitudinal fissure divides the left and right cerebral hemispheres. Still, the two cerebral hemispheres are connected by a bridge of tracts within the corpus callosum.

Shallow grooves called sulci (sing., sulcus) divide each hemisphere into lobes (Fig. 13.9). The *frontal lobe* is the most ventral of the lobes (directly behind the forehead). The *parietal lobe* is dorsal to the frontal lobe. The *occipital lobe* is dorsal to the parietal lobe (at the rear of the head). The *temporal lobe* lies inferior to the frontal and parietal lobes (at the temple and the ear).

Each lobe is associated with particular functions, as indicated in Figure 13.9.

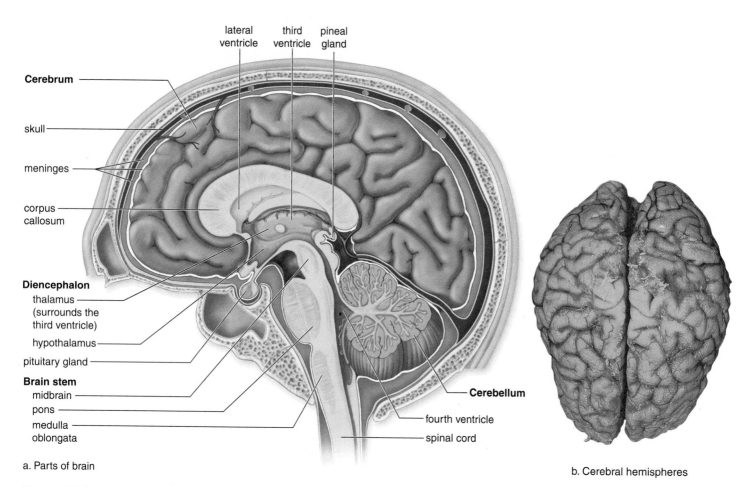

a. Parts of brain

b. Cerebral hemispheres

Figure 13.8   The human brain in longitudinal section.
a. The cerebrum is the largest part of the brain in humans. **b.** Viewed from above, the cerebrum has a left and right cerebral hemisphere. The hemispheres are connected by the corpus callosum.

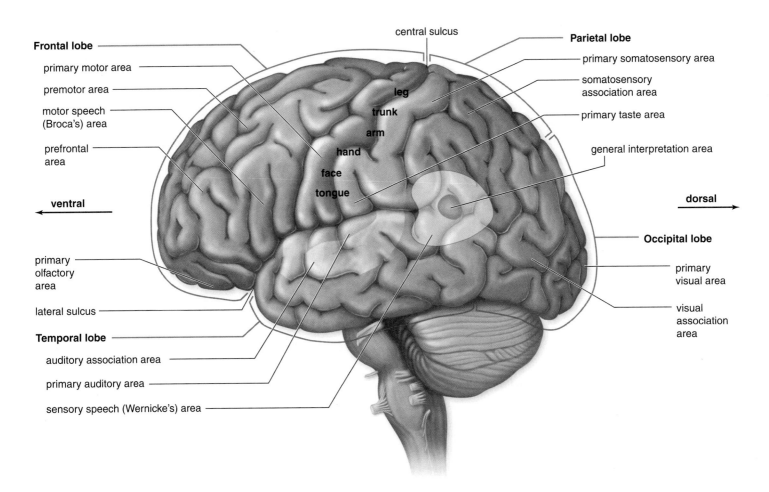

Figure 13.9 **The lobes of a cerebral hemisphere.**
Each cerebral hemisphere is divided into four lobes: frontal, parietal, temporal, and occipital. These lobes contain centers for reasoning and movement, somatic sensing, hearing, and vision, respectively.

The Cerebral Cortex The **cerebral cortex** is a thin but highly convoluted outer layer of gray matter that covers the cerebral hemispheres. The cerebral cortex contains over one billion cell bodies and is the region of the brain that accounts for sensation, voluntary movement, and all the thought processes we associate with consciousness.

Primary Motor and Sensory Areas of the Cortex The cerebral cortex contains motor areas and sensory areas, as well as association areas. The **primary motor area** is in the frontal lobe just ventral to (before) the central sulcus. Voluntary commands to skeletal muscles begin in the primary motor area, and each part of the body is controlled by a certain section. For example, the versatile human hand takes up an especially large portion of the primary motor area (Fig. 13.10*a*).

The **primary somatosensory area** is just dorsal to the central sulcus in the parietal lobe. Sensory information from the skin and skeletal muscles arrives here, where each part of the body is sequentially represented (Fig. 13.10*b*). A *primary taste area* (pink) accounts for taste sensations. A *primary visual area* in the occipital lobe receives information

from our eyes, and a *primary auditory area* in the temporal lobe receives information from our ears. A *primary olfactory area* for smell is located in the temporal lobe (see Fig. 13.9).

Association Areas **Association areas** are places where integration occurs. Ventral to the primary motor area is a premotor area. The premotor area organizes motor functions for skilled motor activities, such as walking and talking at the same time, and then the primary motor area sends signals to the cerebellum, which integrates them. A momentary lack of oxygen during birth can damage the motor areas of the cerebral cortex so that cerebral palsy, a condition characterized by a spastic weakness of the arms and legs, develops. The *somatosensory association area*, located just dorsal to the primary somatosensory area, processes and analyzes sensory information from the skin and muscles. The *visual association area* in the occipital lobe associates new visual information with previously received visual information. It might "decide," for example, if we have seen this face, scene, or symbol before. The *auditory association area* in the temporal lobe performs the same functions with regard to sounds.

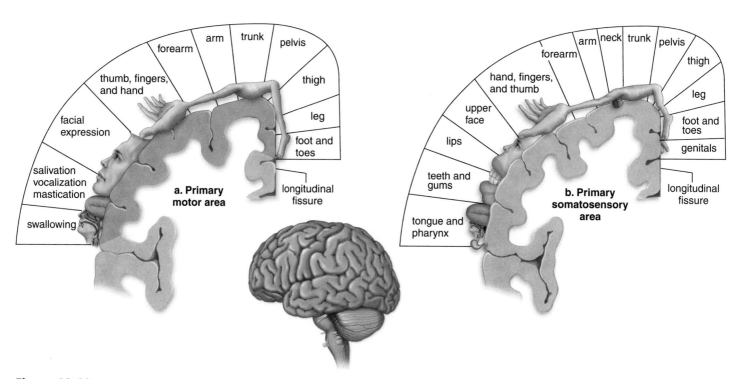

Figure 13.10 **The primary motor area and the primary somatosensory area.**
a. The primary motor area and (**b**) the primary somatosensory area. The primary taste area is colored pink. The size of each body region shown indicates the relative amount of cortex devoted to that body region.

Processing Centers Processing centers of the cortex receive information from the other association areas and perform higher-level analytical functions. The **prefrontal area,** an association area in the frontal lobe, receives information from the other association areas and uses this information to reason and plan our actions. Integration in this area accounts for our most cherished human abilities; to think critically and to formulate appropriate behaviors.

The unique ability of humans to speak is partially dependent upon two processing centers found only in the left cerebral cortex. **Wernicke's area** is located in the dorsal part of the left temporal lobe, and **Broca's area** is located in the left frontal lobe. Broca's area is located just ventral to the portion of the primary motor area for speech musculature (lips, tongue, larynx, and so forth) (see Fig. 13.9). Wernicke's area helps us understand both the written and spoken word and sends the information to Broca's area. Broca's area adds grammatical refinements and directs the primary motor area to stimulate the appropriate muscles for speaking and writing.

Central White Matter Much of the rest of the cerebrum is composed of white matter. As mentioned in the opening story, myelination occurs and, therefore, white matter develops as a child grows. As neurons become myelinated within these tracts, children become more capable, for example,

of speech. Descending tracts from the primary motor area communicate with lower brain centers, and ascending tracts from lower brain centers send sensory information up to the primary somatosensory area. Because the tracts cross over in the medulla, the left side of the cerebrum controls the right side of the body, and vice versa. Tracts within the cerebrum also take information between the different sensory, motor, and association areas pictured in Figure 13.9. As previously mentioned, the corpus callosum contains tracts that join the two cerebral hemispheres.

The Diencephalon

The hypothalamus and the thalamus are in the **diencephalon,** a region that encircles the third ventricle. The **hypothalamus** forms the floor of the third ventricle. The hypothalamus is an integrating center that helps maintain homeostasis by regulating hunger, sleep, thirst, body temperature, and water balance. The hypothalamus controls the pituitary gland and thereby, serves as a link between the nervous and endocrine systems.

The **thalamus** consists of two masses of gray matter located in the sides and roof of the third ventricle. The thalamus is on the receiving end for all sensory input except smell. Visual, auditory, and somatosensory information arrives at the thalamus via the cranial nerves and tracts from the spinal cord. The thalamus integrates this informa-

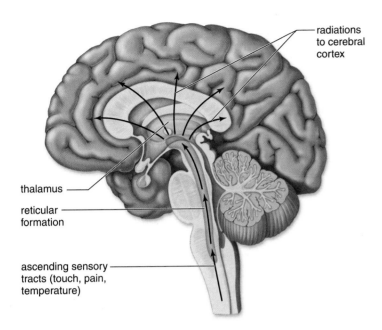

radiations
to cerebral
cortex

thalamus

reticular
formation

ascending sensory
tracts (touch, pain,
temperature)

Figure 13.11   The reticular activating system.
The reticular formation receives and sends on motor and sensory
information to various parts of the CNS. One portion, the reticular
activating system (RAS; see arrows), arouses the cerebrum and, in this way,
controls alertness versus sleep.

tion and sends it on to the appropriate portions of the cerebrum. The thalamus is involved in arousal of the cerebrum, and it also participates in higher mental functions such as memory and emotions.

The pineal gland, which secretes the hormone melatonin, is located in the diencephalon. Presently there is much popular interest in the role of melatonin in our daily rhythms; some researchers believe it can help ameliorate jet lag or insomnia. Scientists are also interested in the possibility that the hormone may regulate the onset of puberty.

The Cerebellum

The **cerebellum** lies under the occipital lobe of the cerebrum and is separated from the brain stem by the fourth ventricle. The cerebellum has two portions that are joined by a narrow median portion. Each portion is primarily composed of white matter, which in longitudinal section has a treelike pattern called arbor vitae. Overlying the white matter is a thin layer of gray matter that forms a series of complex folds.

The cerebellum receives sensory input from the eyes, ears, joints, and muscles about the present position of body parts, and it also receives motor output from the cerebral cortex about where these parts should be located. After integrating this information, the cerebellum sends motor impulses by way of the brain stem to the skeletal muscles.

In this way, the cerebellum maintains posture and balance. It also ensures that all of the muscles work together to produce smooth, coordinated, voluntary movements. The cerebellum assists the learning of new motor skills such as playing the piano or hitting a baseball.

The Brain Stem

The tracts cross in the **brain stem,** which contains the midbrain, the pons, and the medulla oblongata (see Fig. 13.8*a*). The **midbrain** acts as a relay station for tracts passing between the cerebrum and the spinal cord or cerebellum. It also has reflex centers for visual, auditory, and tactile responses. The word **pons** means "bridge" in Latin, and true to its name, the pons contains bundles of axons traveling between the cerebellum and the rest of the CNS. In addition, the pons functions with the medulla oblongata to regulate breathing rate and has reflex centers concerned with head movements in response to visual and auditory stimuli.

The **medulla oblongata** contains a number of reflex centers for regulating heartbeat, breathing, and vasoconstriction (blood pressure). It also contains the reflex centers for vomiting, coughing, sneezing, hiccuping, and swallowing. The medulla oblongata lies just superior to the spinal cord, and it contains tracts that ascend or descend between the spinal cord and higher brain centers.

The Reticular Formation   The **reticular formation** is a complex network of **nuclei** (masses of gray matter) and fibers that extend the length of the brain stem (Fig. 13.11). The reticular formation is a major component of the reticular activating system (RAS), which receives sensory signals and sends them up to higher centers, and motor signals, which it sends to the spinal cord.

The RAS arouses the cerebrum via the thalamus and causes a person to be alert. If you want to awaken the RAS, surprise it with sudden stimuli, such as an alarm clock ringing, bright lights, smelling salts, or splashing cold water on your face. Apparently, the RAS can filter out unnecessary sensory stimuli, explaining why you can study with the TV on. To inactivate the RAS, remove visual or auditory stimuli, allowing yourself to become drowsy and drop off to sleep. General anesthetics function by artificially suppressing the RAS. A severe injury to the RAS can cause a person to be comatose, from which recovery may be impossible.

✅ Check Your Progress 13.2

1. What structures compose the central nervous system?
2. What are the two main functions of the spinal cord?
3. What are the four major parts of the brain and the general function of each?
4. What brain structure is responsible for a person being awake and alert, asleep, or comatose?

13.3 The Limbic System and Higher Mental Functions

The limbic system is intimately involved in our emotions and higher mental functions. After a short description, we will discuss the functions of the limbic system.

Limbic System

The **limbic system** is an evolutionary ancient group of linked structures deep within the cerebrum that is a functional grouping rather than an anatomical one (Fig. 13.12). The limbic system blends primitive emotions and higher mental functions into a united whole. It accounts for why activities such as sexual behavior and eating seem pleasurable and also why, say, mental stress can cause high blood pressure.

Two significant structures within the limbic system are the amygdala and the hippocampus. The **amygdala,** in particular, can cause experiences to have emotional overtones, and it creates the sensation of fear. This center can use past knowledge fed to it by association areas to assess a current situation and, if necessary, trigger the fight-or-flight reaction. So if you are out late at night, and you turn to see someone in a ski mask following you, the amygdala may immediately cause you to start running. The frontal cortex can override the limbic system and cause us to rethink the situation and prevent us from acting out strong reactions.

The **hippocampus** is believed to play a crucial role in learning and memory. The hippocampal region acts as an information gateway during the learning process, determining what information about the world is to be sent to memory, and how this information is to be encoded and stored by other regions in the brain. Most likely, the hippocampus can communicate with the frontal cortex because we know that memories are an important part of our decision-making processes.

It's hoped that research into the functioning of the hippocampus will help in understanding Alzheimer disease, a brain disorder characterized by gradual loss of memory (see page 266).

Higher Mental Functions

As in other areas of biological research, brain research has progressed due to technological breakthroughs. Neuroscientists now have a wide range of techniques at their disposal for studying the human brain, including modern technologies that allow us to record its functioning.

Memory and Learning

Just as the connecting tracts of the corpus callosum are evidence that the two cerebral hemispheres work together, so the limbic system indicates that cortical areas may work with lower centers to produce learning and memory. **Memory** is the ability to hold a thought in mind or to recall events from the past, ranging from a word we learned only yesterday to an early emotional experience that has shaped our lives. **Learning** takes place when we retain and utilize past memories.

Types of Memory   We have all tried to remember a seven-digit telephone number for a short period of time. If we say we are trying to keep it in the forefront of our brain, we are exactly correct. The prefrontal area, which is active during **short-term memory,** lies just dorsal to our forehead! There are some telephone numbers that we have memorized; in other words, they have gone into **long-term memory.** Think of a telephone number you know by heart, and try to bring it to mind without also thinking about the place or person associated with that number. Most likely you cannot, because, typically, long-term memory is a mixture of what is called **semantic memory** (numbers, words, etc.) and **episodic memory** (persons, events, etc.). Due to brain damage, some people lose one type of memory but not the other. For example, without a working episodic memory, they can carry on a conversation but have no recollection of recent events. If you are talking to them and then leave the room, they don't remember you when you come back!

Skill memory is another type of memory that can exist independent of episodic memory. Skill memory is involved in performing motor activities such as riding a bike or playing ice hockey. When a person first learns a skill, more areas of the cerebral cortex are involved than after the skill is perfected. In other words, you have to think about what you are doing when you learn a skill, but later the actions become automatic.

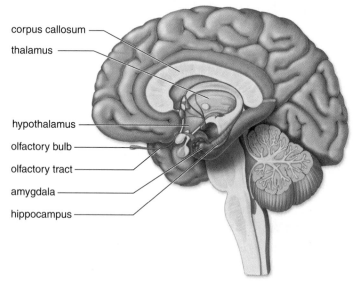

corpus callosum
thalamus
hypothalamus
olfactory bulb
olfactory tract
amygdala
hippocampus

Figure 13.12   The limbic system.
In the limbic system (purple), structures deep within each cerebral hemisphere and surrounding the diencephalon join higher mental functions, such as reasoning, with more primitive feelings, such as fear and pleasure. Therefore, primitive feelings can influence our behavior, but reason can also keep them in check.

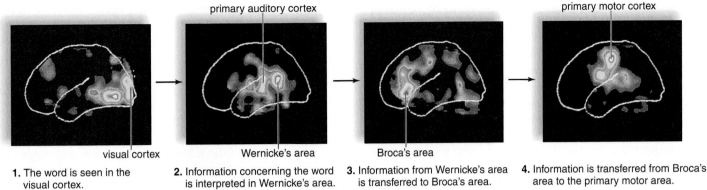

1. The word is seen in the visual cortex.
2. Information concerning the word is interpreted in Wernicke's area.
3. Information from Wernicke's area is transferred to Broca's area.
4. Information is transferred from Broca's area to the primary motor area.

Figure 13.13   Language and speech.
These functional images were captured by a high-speed computer during PET (positron emission tomography) scanning of the brain. A radioactively labeled solution is injected into the subject, and then the subject is asked to perform certain activities. Cross-sectional images of the the brain generated by the computer reveal where activity is occurring because the solution is preferentially taken up by active brain tissue and not by inactive brain tissue. These PET images show the cortical pathway for reading words and then speaking them. Red indicates the most active areas of the brain, and blue indicates the least active areas.

Skill memory involves all the motor areas of the cerebrum below the level of consciousness.

Long-Term Memory Storage and Retrieval   Our long-term memories are apparently stored in bits and pieces throughout the sensory association areas of the cerebral cortex. Visions are stored in the vision association area, sounds are stored in the auditory association area, and so forth. As previously mentioned, the hippocampus serves as a bridge between the sensory association areas, where memories are stored, and the prefrontal area, where memories are utilized. The prefrontal area communicates with the hippocampus, when memories are stored and when these memories are brought to mind. Why are some memories so emotionally charged? As you know, the amygdala seems to be responsible for fear conditioning and associating danger with sensory stimuli received from various parts of the brain.

Long-Term Potentiation   While it is helpful to know the memory functions of various portions of the brain, an important step toward curing mental disorders is understanding memory on the cellular level. After synapses have been used intensively for a short period of time, they release more neurotransmitters than before. This phenomenon, called **long-term potentiation (LTP),** may very well be involved in memory storage.

Language and Speech

Language depends on semantic memory. Therefore, we would expect some of the same areas in the brain to be involved in both memory and language. Any disruption of these pathways could very well contribute to an inability to comprehend our environment and use speech correctly.

Seeing and hearing words depends on sensory centers in the occipital and temporal lobes, respectively. Damage to Wernicke's area, discussed earlier, results in the inability to comprehend speech. Damage to Broca's area, on the other hand, results in the inability to speak and write. The functions of the visual cortex, Wernicke's area, and Broca's area are shown in Figure 13.13.

One interesting aside pertaining to language and speech is the recognition that the left brain and the right brain may have different functions. Consistent with the recognition that the left hemisphere, not the right, contains Broca's area and Wernicke's area, it appears that the left hemisphere plays a role of great importance in language functions. In an attempt to cure epilepsy in the early 1940s, the corpus callosum was surgically severed in some patients. If viewed only by the right hemisphere, these so-called split-brain patients could choose the proper object for a particular use but were unable to name it. Later, the left brain was contrasted with the right brain along these lines:

Left Hemisphere	Right Hemisphere
Verbal	Nonverbal, visuospatial
Logical, analytical	Intuitive
Rational	Creative

Researchers now believe that the hemispheres process the same information differently. The left hemisphere is more global, whereas the right hemisphere is more specific in its approach.

✔ Check Your Progress 13.3

1. What is the function of the limbic system?

2. What limbic system structures are involved in the fight-or-flight reaction, learning, and long-term memory?

3. a. What areas of the cerebrum are involved in language and speech? b. Where are they located?

13.4 The Peripheral Nervous System

The peripheral nervous system (PNS), which lies outside the central nervous system, contains the nerves. Nerves are designated as cranial nerves when they arise from the brain and spinal nerves when they arise from the spinal cord. In any case, all nerves take impulses to and from the CNS. So right now, your eyes are sending messages by way of a cranial nerve to the brain, allowing you to read a page, and your brain is directing the muscles in your fingers to turn a page by way of the spinal cord and a spinal nerve.

Figure 13.14 (*left*) gives the anatomy of a nerve. The cell body and the dendrites of neurons are in the CNS or ganglia. The axons of neurons project from the CNS and form the spinal cord. In other words, *nerves,* whether cranial or spinal, are composed of axons, the long part of neurons.

Humans have 12 pairs of **cranial nerves** attached to the brain. By convention, the pairs of cranial nerves are referred to by roman numerals (Fig. 13.14, *right*). Some cranial nerves are sensory nerves—that is, they contain only sensory fibers; some are motor nerves that contain only motor fibers; and others are mixed nerves that contain both sensory and motor fibers. Cranial nerves are largely concerned with the head, neck, and facial regions of the body. However, the vagus nerve (X) has branches not only to the pharynx and larynx, but also to most of the internal organs. From which part of the brain do you think the vagus arises? It arises from the brain stem, specifically the medulla oblongata that communicates so well with the hypothalamus. These two parts of the brain control the internal organs.

As you know, the **spinal nerves** of humans emerge in 31 pairs from either side of the spinal cord (see Fig. 13.7). The roots of a spinal nerve physically separate the axons of sensory neurons from the axons of motor neurons. The dorsal root of a spinal nerve contains sensory fibers that conduct impulses inward (toward the spinal cord) from sensory receptors. The cell body of a sensory neuron is in a **dorsal-root ganglion.** A **ganglion** is a collection of cell bodies outside the CNS. The ventral root of a spinal nerve contains motor fibers that conduct impulses outward (away from the cord) to effectors. All spinal nerves are called mixed nerves because they contain both sensory and motor fibers. Each

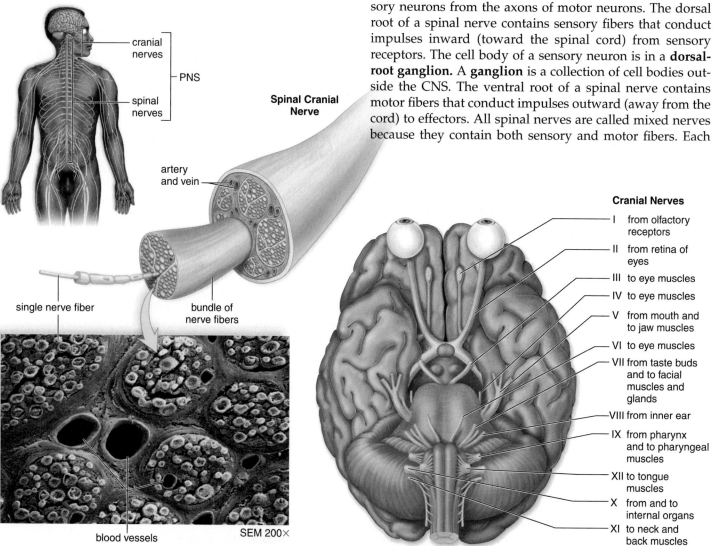

Cranial Nerves

I   from olfactory receptors
II from retina of eyes
III to eye muscles
IV to eye muscles
V   from mouth and to jaw muscles
VI to eye muscles
VII from taste buds and to facial muscles and glands
VIII from inner ear
IX from pharynx and to pharyngeal muscles
XII to tongue muscles
X   from and to internal organs
XI to neck and back muscles

Figure 13.14   Peripheral nervous system.
The peripheral nervous system consists of the cranial nerves and the spinal nerves. A nerve (*left*) is composed of bundles of axons separated from one another by connective tissue. In large measure, the cranial nerves (*right*) receive input from and control the head region, and the spinal nerves receive input from and control the rest of the body. An important exception is cranial nerve X, the vagus nerve, which does communicate with internal organs.

spinal nerve serves the particular region of the body in which it is located. For example, the intercostal muscles of the rib cage are innervated by thoracic nerves.

Somatic System

The PNS has divisions, and right now we are going to consider the somatic system. The nerves in the **somatic system** serve the skin, skeletal muscles, and tendons (see Fig. 13.1). The somatic system includes nerves that take sensory information from external sensory receptors to the CNS and motor commands away from the CNS to the skeletal muscles. Calvin, from the Chapter 3 opening story, was making use of his somatic system when his olfactory receptors sent a message to his brain that chocolate chip cookies were in the vicinity, and then after thinking it over, his brain ordered his muscles to snatch a few and make a beeline for the back door.

Not all actions are voluntary as was Calvin's. Some actions are automatic. Automatic responses to a stimulus in the somatic system are called **reflexes.** A reflex occurs quickly, without our even having to think about it. So as Calvin hurried out, the swinging of the the back door may have made his eyes blink without him willing it. We will study the path of a reflex because it allows us to study in detail the path of nerve impulses to and from the CNS.

The Reflex Arc

Figure 13.15 illustrates the path of a reflex that involves only the spinal cord. If your hand touches a sharp pin, sensory receptors in the skin generate nerve impulses that move along sensory fibers through the dorsal-root ganglia toward the spinal cord. Sensory neurons that enter the cord dorsally pass signals on to many interneurons. Some of these interneurons synapse with motor neurons whose short dendrites and cell bodies are in the spinal cord. Nerve impulses travel along these motor fibers to an effector, which brings about a response to the stimulus. In this case, the effector is a muscle, which contracts so that you withdraw your hand from the pin. Various other reactions are also possible—you will most likely look at the pin, wince, and cry out in pain. This whole series of responses occurs because some of the interneurons involved carry nerve impulses to the brain. The brain makes you aware of the stimulus and directs these other reactions to it. You don't feel pain until the brain receives the information and interprets it!

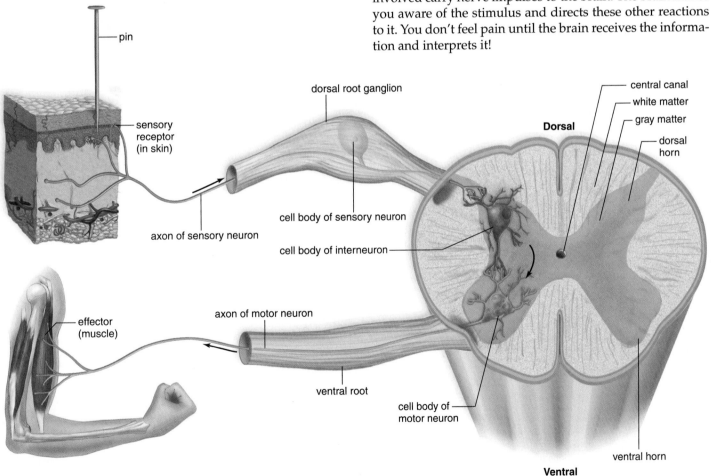

Figure 13.15   **A reflex arc showing the path of a spinal reflex.**
A stimulus (e.g., sharp pin) causes sensory receptors in the skin to generate nerve impulses that travel in sensory axons to the spinal cord. Interneurons integrate data from sensory neurons and then relay signals to motor neurons. Motor axons convey nerve impulses from the spinal cord to a skeletal muscle, which contracts. Movement of the hand away from the pin is the response to the stimulus.

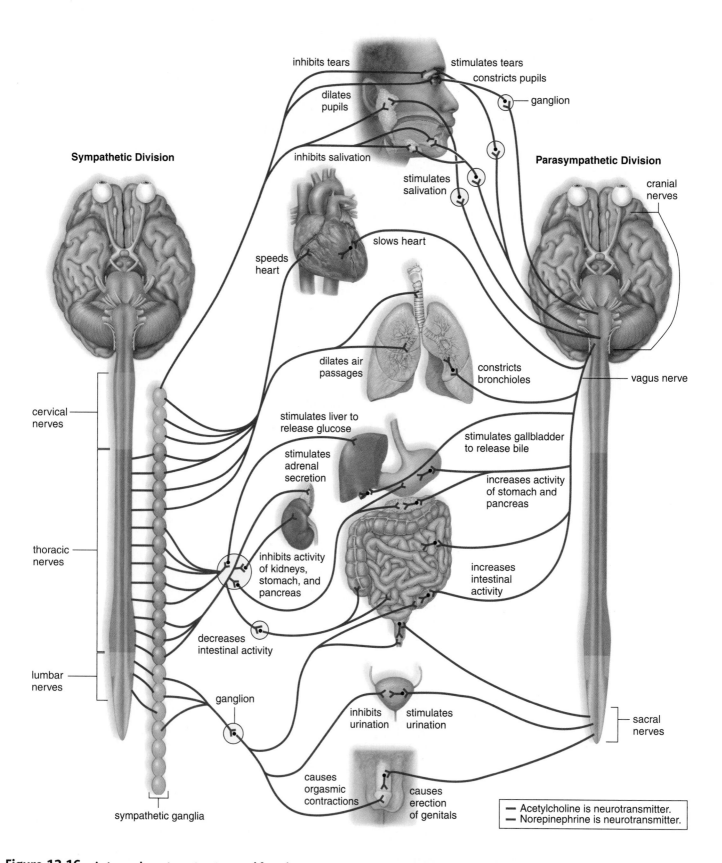

Sympathetic Division

inhibits tears

dilates pupils

inhibits salivation

speeds heart

cervical nerves

thoracic nerves

lumbar nerves

stimulates liver to release glucose

stimulates adrenal secretion

inhibits activity of kidneys, stomach, and pancreas

decreases intestinal activity

ganglion

sympathetic ganglia

causes orgasmic contractions

stimulates tears

constricts pupils

ganglion

stimulates salivation

Parasympathetic Division

cranial nerves

slows heart

dilates air passages

constricts bronchioles

vagus nerve

stimulates gallbladder to release bile

increases activity of stomach and pancreas

increases intestinal activity

sacral nerves

inhibits urination

stimulates urination

causes erection of genitals

— Acetylcholine is neurotransmitter.
— Norepinephrine is neurotransmitter.

Figure 13.16   Autonomic system structure and function.
Sympathetic preganglionic fibers (*left*) arise from the thoracic and lumbar portions of the spinal cord; parasympathetic preganglionic fibers (*right*) arise from the cranial and sacral portions of the spinal cord. Each system innervates the same organs but has contrary effects.

Autonomic System

The **autonomic system** is also in the PNS (see Fig. 13.1). The autonomic system regulates the activity of cardiac and smooth muscles and glands. The system is divided into the sympathetic and parasympathetic divisions (Fig. 13.16). Activation of these two systems generally causes opposite responses.

Although their functions are different, the two divisions share some features: (1) They function automatically and usually in an involuntary manner; (2) they innervate all internal organs; and (3) they utilize two neurons and one ganglion for each impulse. The first neuron has a cell body within the CNS and a preganglionic fiber. The second neuron has a cell body within a ganglion and a postganglionic fiber.

Reflex actions, such as those that regulate blood pressure and breathing rate, are especially important to the maintenance of homeostasis. These reflexes begin when the sensory neurons in contact with internal organs send messages to the CNS. They are completed by motor neurons within the autonomic system.

Sympathetic Division

Most preganglionic fibers of the **sympathetic division** arise from the middle, or thoracolumbar, portion of the spinal cord and almost immediately terminate in ganglia that lie near the cord. Therefore, in this division, the preganglionic fiber is short, but the postganglionic fiber that makes contact with an organ is long.

The sympathetic division is especially important during emergency situations when you might be required to *fight or take flight*. It accelerates the heartbeat and dilates the bronchi—active muscles, after all, require a ready supply of glucose and oxygen. Remember Ashley in the opening story for Chapter 1? A scary scene in the movie made her heart pound, and she became so agitated that she stood right up. Ashley wasn't really in danger, but her CNS interpreted the situation as a dangerous one. Calvin may have experienced some of the same reactions when he heard his mother's voice calling after him as he scampered away.

The sympathetic division inhibits the digestive tract—digestion is not an immediate necessity if you are under attack. The neurotransmitter released by the postganglionic axon is primarily norepinephrine (NE). The structure of NE is like that of epinephrine (adrenaline), an adrenal medulla hormone that usually increases heart rate and contractility.

Parasympathetic Division

The **parasympathetic division** includes a few cranial nerves (e.g., the vagus nerve) as well as fibers that arise from the sacral (bottom) portion of the spinal cord. Therefore, this division is often referred to as the craniosacral portion of the autonomic system. In the parasympathetic division, the preganglionic fiber is long, and the postganglionic fiber is short because the ganglia lie near or within the organ.

The parasympathetic division, sometimes called the housekeeper division, promotes all the internal responses we associate with a relaxed state; for example, it causes the pupil of the eye to contract, promotes digestion of food, and retards the heartbeat. It's been suggested that the parasympathetic system could be called the *rest and digest* system. The neurotransmitter utilized by the parasympathetic division is acetylcholine (ACh).

The Somatic Versus the Autonomic Systems

Recall that the PNS includes the somatic system and the autonomic system. Table 13.1 summarizes the features and functions of the somatic motor pathway with the motor pathways of the autonomic system.

☑ Check Your Progress 13.4

1. a. How many cranial nerves are there? b. How many spinal nerves are there?
2. What is the fastest way for you to react to a stimulus?
3. a. What is the autonomic system, and (b) how does it function?

Table 13.1	Comparison of Somatic Motor and Autonomic Motor Pathway		
	Somatic Motor Pathway	**Autonomic Motor Pathways**	
		Sympathetic	**Parasympathetic**
Type of control	Voluntary/involuntary	Involuntary	Involuntary
Number of neurons per message	One	Two (preganglionic shorter than postganglionic)	Two (preganglionic longer than postganglionic)
Location of motor fiber	Most cranial nerves and all spinal nerves	Thoracolumbar spinal nerves	Cranial (e.g., vagus) and sacral spinal nerves
Neurotransmitter	Acetylcholine	Norepinephrine	Acetylcholine
Effectors	Skeletal muscles	Smooth and cardiac muscle, glands	Smooth and cardiac muscle, glands

Health Focus

Degenerative Brain Disorder

Many researchers are now engaged in studying Alzheimer and Parkinson diseases, which are degenerative brain disorders.

Alzheimer Disease

Signs of **Alzheimer disease (AD)** may appear before the age of 50, but it is usually seen in individuals past the age of 65. One of the early symptoms is loss of memory, particularly for recent events. Characteristically, the person asks the same question repeatedly and becomes disoriented even in familiar places. Signs of mental disturbance are frequently seen. Gradually, the person loses the ability to perform any type of daily activity and becomes bedridden. Death is usually due to pneumonia or some other such illness.

In AD patients, abnormal neurons are present throughout the brain but especially in the hippocampus and amygdala. Several genes predispose a person to AD. One of them, called $APOE_4$, is found in 65% of persons with AD. But it is unknown why inheritance of this gene leads to AD neurons. Plasma membrane enzymes, called secretases, snip beta amyloid from a larger molecule, even in healthy cells. People with AD produce too much beta amyloid, particularly one very sticky type. Beta amyloid plaques grow so dense that they trigger an inflammatory reaction that ends in neuron death. Some researchers are trying to block a type of secretase enzyme, thinking that this will stop the formation of beta amyloid plaques.

Aside from beta amyloid plaques, AD neurons have neurofibrillary tangles in axons that extend upward to surround the nucleus (Fig. 13A). The neurofibrillary tangles arise because a protein called tau no longer has the correct shape. Tau holds microtubules in place so that they support the structure of neurons. When tau changes shape, it grabs onto other tau molecules, and the tangles result.

The presence of plaques precedes the formation of tangles. Researchers injected mutant mice that developed both plaques and misshapen tau with antibodies against beta amyloid. In some mice, this treatment removed, first, the plaques and, then, the tau. This bolsters the hypothesis that beta amyloid blocks the work of proteosomes, complexes that ordinarily clear the cell of any excess protein, including tau. If the tau had already formed tangles, injection of antibodies was too late, and only the plaques disappeared. Also, the treatment didn't last long, and the plaques reappeared within two weeks.

Frank M. LaFerla, of the University of California, Irvine, who led a team doing this work, suggests that amyloid works with tau to kill neurons and trigger the confusion and memory loss so characteristic of AD. Based on this research,

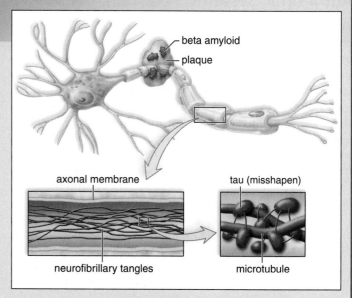

Figure 13A   Alzheimer disease.
Some of the neurons of Alzheimer disease (AD) patients have beta amyloid plaques and neurofibrillary tangles. AD neurons are present throughout the brain but concentrated in the hippocampus and amygdala.

it should appear that treatment for AD in humans should begin as early as possible, and that it should be continued indefinitely.

Parkinson Disease

Parkinson disease is characterized by a gradual loss of motor control beginning between the ages of 50 and 60. Eventually, the person has a wide-eyed, unblinking expression, involuntary tremors of the fingers and thumbs, muscular rigidity, and a shuffling gait. Speaking and performing ordinary daily tasks become laborious.

In Parkinson patients, the basal nuclei function improperly because of a degeneration of the dopamine-releasing neurons in the brain. Without dopamine, which is an inhibitory neurotransmitter, the excessive excitatory signals from the motor cortex and other brain areas result in the symptoms of Parkinson disease.

Unfortunately, it is not possible to give Parkinson patients dopamine directly because of the impermeability of the capillaries serving the brain. However, symptoms can be alleviated by giving patients l-dopa, a chemical that can be changed to dopamine in the body, until too few cells are left to do the job. Then, patients must turn to a number of controversial surgical procedures. Implantation of dopamine-secreting tissue from various sources has been tried with mixed results.

13.5 Drug Abuse

A wide variety of drugs, either natural molecules or synthetically derived compounds, affect the nervous system and can alter the mood and/or emotional state. These drugs have two general effects: (1) They affect the limbic system, and (2) they either promote or decrease the action of a particular neurotransmitter. Stimulants are drugs that increase the likelihood of neuron excitation, and depressants decrease the likelihood of excitation.

The neurotransmitter *dopamine* plays a central role in the workings of the brain's built-in *reward circuit* and regulation of mood. The reward circuit is a collection of neurons that, under normal circumstances, promotes healthy, pleasurable activities, such as consuming food. It's possible to abuse behaviors, such as eating, because the behavior stimulates the reward circuit and make us feel good. Drug abusers take drugs that artificially affect the reward circuit to the point that they neglect their basic physical needs in favor of continued drug use.

Drug abuse is apparent when a person takes a drug at a dose level and under circumstances that increase the potential for a harmful effect. Drug abusers are apt to display a psychological and/or physical dependence on the drug. Psychological dependence is apparent when a person craves the drug, spends time seeking the drug, and takes it regularly. With physical dependence, formerly called "addiction," the person has become tolerant to the drug—that is, more of the drug is needed to get the same effect, and withdrawal symptoms occur when he or she stops taking the drug. This not only is true for teenagers and adults, but it is also true for newborn babies of mothers who abuse and are addicted to drugs (Fig. 13.17). Alcohol, drugs, and tobacco can all adversely affect the developing embryo, fetus, or newborn.

Alcohol

Alcohol consumption is the most socially accepted form of drug use worldwide. Its widespread use probably correlates with its ancient origin and its derivation from grains or fruits. The approximate number of adults that consume alcohol in the United States on a regular basis is 65%. Of those who are drinkers, 5% are heavy drinkers. Notably, 80% of college-age young adults drink.

Alcohol (ethanol) has known harmful effects on the body and brain. Alcohol readily crosses cell membranes, including the blood-brain barrier and denatures protein structures, causing damage to several tissues, including vital organs of the body such as the liver and brain. The liver is the major detoxification organ of the body and prolonged alcohol consumption scars the liver and impairs its function. Depending on the amount consumed, the effects of alcohol on the brain can lead to a feeling of relaxation, a lowering of inhibitions, impaired concentration and coordination, slurred speech, and vomiting. If the alcohol level of blood becomes too high, coma or death can occur.

In the central nervous system (CNS), alcohol acts as a *depressant* (Table 13.2) and influences many brain regions

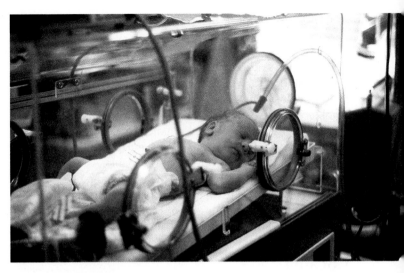

Figure 13.17   Addicted newborn.
This newborn is addicted to cocaine.

and neurotransmitter systems. For example, alcohol increases the action of GABA and increases the release of beta-endorphins in the hypothalamus. Chronic alcohol consumption can damage the frontal lobes, decrease overall brain size, and increase the size of the ventricles. Brain damage is manifested by permanent memory loss, amnesia, confusion, apathy, disorientation, or lack of motor coordination.

There is no effective treatment to cure alcoholism. However, some drugs have been approved by the FDA to help reduce cravings in those who seriously wish to quit drinking.

Nicotine

Nicotine is a small molecule, not normally produced by the body, that acts as a *stimulant*. During the smoking of tobacco, nicotine is rapidly delivered to the CNS, especially the midbrain. When nicotine binds to neurons in the CNS, dopamine is released. In the peripheral nervous system, nicotine mimics acetylcholine and increases skeletal muscle activity, heart rate, and blood pressure, as well as digestive tract motility, which may account for weight loss in smokers.

Table 13.2	Drug Influence on CNS and Route	
Substance	**Effect**	**Mode of Transmission**
Alcohol	Depressant	Drinking
Nicotine	Stimulant	Smoking or smokeless tobacco
Cocaine	Stimulant	Sniffing/snorting, injected, or smoked
Methamphetamine	Stimulant	Smoked or pill form
Heroin	Depressant	Sniffed/snorted, injected, or smoked
Marijuana	Psychoactive	Smoked or consumed

The physiological and psychological addictive nature of nicotine is well known. Approximately 70 million Americans smoke (Fig. 13.18). The addiction rate of smokers is about 70%. The failure rate in those who try to quit smoking is about 80–90% of smokers. Withdrawal symptoms include irritability, headache, insomnia, poor cognitive performance, the urge to smoke, and weight gain. The attempt to quit smoking can include the use of nicotine skin patches, nicotine gum, or the taking of drugs by mouth that block the actions of acetylcholine. The effectiveness of these therapies is variable. An experimental therapy involves "immunizing" the brain of smokers against nicotine. Injections cause the production of antibodies that bind to nicotine and prevent it from passing the blood-brain barrier. The effectiveness of this new therapy is not yet known.

Cocaine

Cocaine is an alkaloid derived from the shrub *Erythroxylon coca*. Approximately 35 million Americans have used cocaine by sniffing/snorting, injecting, or smoking. Cocaine is a powerful *stimulant* in the CNS that interferes with the re-uptake of dopamine at synapses. The result is a rush of well-being that lasts from 5–30 minutes. Although it seemed at first that cocaine could relieve depression, it soon became apparent that cocaine is extremely addictive and its use is very harmful. During cocaine sprees (or binges), the drug is taken repeatedly and at ever-higher doses. The result is sleeplessness, lack of appetite, increased sex drive, tremors, and "cocaine psychosis," a condition that resembles paranoid schizophrenia. During the crash period, fatigue, depression, and irritability are common, along with memory loss and a confused state of cognition.

"Crack" is the street name given to cocaine processed to a free base for smoking. Approximately eight million Americans use crack. The term *crack* refers to the crackling sound heard when smoking, allowing extremely high doses of the drug to reach the brain rapidly, so that there is an intense and immediate high or rush.

With continued use, the body makes less dopamine to compensate for the apparent excess at synapses. Drug tolerance leads to withdrawal symptoms, during which cravings for cocaine increase. Cocaine-related deaths are usually due to cardiac and/or respiratory arrest. The combination of cocaine and alcohol dramatically increase the risk of sudden death. Currently, there are no effective treatments for cocaine addiction, although cognitive-behavioral therapies have met with some success.

Methamphetamine

Methamphetamine is a synthetic drug made from amphetamine by the addition of a methyl group. Over nine million U.S. residents have used methamphetamine at least once in their lifetime; teenagers and young adults represent approximately one-fourth of these. Because the addition of the methyl group is fairly simple, methamphetamine is often produced from amphetamine in makeshift home laborato-

Figure 13.18   Why do people use drugs?
Social motivations may be at play if a person smokes and drinks only in the presence of others. Other factors may be at work if the activity occurs to excess when the person is alone.

ries. It is available as a powder (speed) or as crystals (crystal meth or ice). The crystals are smoked, and the effects are almost instantaneous, and nearly as quick as when it is snorted. The effects last 4 to 8 hours when smoked.

Methamphetamine has a structure similar to that of dopamine, and its *stimulatory* effect mimics cocaine. It reverses the effects of fatigue, maintains wakefulness, and temporarily elevates the mood of the user. After the initial rush, there is typically a state of high agitation that, in some individuals, leads to violent behavior. Chronic use can lead to what is called an amphetamine psychosis resulting in paranoia, auditory and visual hallucinations, self-absorption, irritability, and aggressive, erratic behavior. Drug tolerance, dependence, and addiction are common, and hyperthermia, convulsions, and death can occur.

Ecstasy is the street name for MDMA (methylenedioxymethamphetamine), a drug that has the same effects as methamphetamine, but without hallucinations. Ecstasy comes in a pill form, and it often contains other drugs besides MDMA, some of which are more dangerous.

There are no pharmacologic treatments for methamphetamine addiction. The most common therapy is cognitive-behavioral interventions.

Heroin

Heroin is derived from the resin or sap of the opium poppy plant, which is grown from Turkey to Southeast Asia and in parts of Latin America. Heroin is a highly addictive drug that acts as a *depressant* in the nervous system. The opiates, which also include morphine and codeine, have pain-killing effects.

Heroin is the most abused opiate because it is rapidly delivered to the brain, where it is converted to morphine. Morphine binds promptly to opioid receptors, and the result is a

Medical Marijuana Use

The plant *Cannabis sativa* has been used in many cultures throughout history as a general-purpose medicine, pain reliever, and salve. In recent years, there has been a renewed interest in therapeutic applications of marijuana in the United States.

Marijuana Is Illegal

In 2005, the Supreme Court ruled that the constitutional authority of Congress to regulate interstate market in drugs extends to small quantities of doctor-recommended marijuana. So, patients using marijuana prescribed by a physician can be criminally prosecuted by federal law enforcement agencies.

Those opposed to medical marijuana (1) express reservations about its toxicity, possible interactions with other drugs, and potential for dependence; (2) believe that smoking marijuana might cause damage to the respiratory system; (3) feel that medical use of marijuana will make it more accessible to drug abusers and lead to use of other, more harmful drugs; (4) cite animal studies that show marijuana or its constituents may deteriorate motor coordination, negatively affect a fetus, and lower sperm count or motility—potentially negative side effects.

Benefits of Marijuana Use

The goal of medical marijuana use is to relieve suffering in seriously ill patients, for which other treatments have not worked. To this end, marijuana or its ingredients have been reported to reduce anxiety, relax muscles, increase appetite, and modulate the immune system and cardiovascular systems, all which could be exploited in a therapeutic setting. Currently, some applications for medical marijuana include treating pain, preventing nausea and vomiting in patients undergoing chemotherapy, and lowering intraocular pressure in glaucoma patients.

While the Food and Drug administration (FDA) dismisses claims of marijuana's medical benefits, it has paradoxically approved certain THC-based drug therapies, meant for use by cancer patients and AIDS patients experiencing extreme weight loss.

How to Make Medical Use of Marijuana Legal

Many researchers agree that more studies using doses of fixed purities and strength are needed to probe marijuana's safety and effectiveness. However, the legal status of marijuana poses significant hurdles to this research. Following the Supreme Court's 2005 ruling, advocates for medical marijuana use and continued research may need to bring their fight to Congress.

Decide Your Opinion

1. Should marijuana be available for medical use by all patients? Why or why not?
2. Do you agree with the Supreme Court's ruling that lets federal authorities prosecute medical marijuana users, even if such use is legal in the state they live in?
3. How should the use of medical marijuana be regulated?

---

rush sensation and euphoric experience. Opiates depress breathing, activate the reward circuit, block pain pathways, cloud mental function, and sometimes cause nausea and vomiting. Long-term effects of heroin use are addiction, hepatitis, HIV/AIDS, and various bacterial infections because of shared needles. As with other drugs of abuse, tolerance and dependence is common, and heavy users may experience convulsions and death by respiratory arrest.

Heroin can be injected, snorted, or smoked. Abusers typically inject heroin up to four times a day. It is estimated that four million Americans have used heroin some time in their lives, and over 300,000 people use heroin annually. The many available treatments for heroin addiction include synthetic opiate compounds, such as methadone or suboxone. They decrease withdrawal symptoms and block heroin's effects.

Marijuana

The dried flowering tops, leaves, and stems of the Indian hemp plant *Cannabis sativa* contain and are covered by a resin that is rich in THC (tetrahydrocannabinol). The names *cannabis* and *marijuana* apply to either the plant or THC. Usually, marijuana is smoked in a cigarette called a "joint," or it can be consumed. An estimated 22 million Americans use marijuana. Although the drug was banned in the United States in 1937, several states have legalized its use for medical purposes.

A neurotransmitter called anandamide was recently discovered. Both THC and anandamide belong to a class of chemicals called cannabinoids. It would seem, then, that THC mimics the actions of anadamide. Receptors that bind cannabinoids are located in the hippocampus, cerebellum, basal ganglia, and cerebral cortex, brain areas that are important for memory, orientation, balance, motor coordination, and perception.

When THC reaches the CNS by smoking a "joint," mild euphoria, along with alterations in vision and judgment, occur. Distortions of space and time can also occur in occasional users. In heavy users, hallucinations, anxiety, depression, rapid flow of ideas, body image distortions, paranoia, and psychotic symptoms can result. The terms cannabis psychosis and cannabis delirium describe such reactions of marijuana's influence on the brain. Regular usage can cause cravings that make it difficult to stop marijuana use.

It is estimated that 100,000 people each year seek treatment for marijuana abuse. There are no effective therapies, but, similar to cocaine abuse, cognitive-behavioral therapies are making inroads to address dependency problems.

✓ Check Your Progress 13.5

1. How can the abuse of drugs, including alcohol and nicotine, affect the nervous system?

Summarizing the Concepts

13.1 Overview of the Nervous System

The nervous system

- is divided into the central nervous system (CNS) and the peripheral nervous system (PNS).
- has three functions: (1) reception of input; (2) integration of data; and (3) generates motor output.

Nervous Tissue

Nervous tissue contains two types of cells: neurons and neuroglia.

- Neurons transmit nerve impulses.
- Neuroglia nourish and support neurons.

Neuron Structure

A neuron is composed of dendrites, a cell body, and an axon. There are three types of neurons:

- Sensory neurons take nerve impulses from sensory receptors to the CNS.
- Interneurons occur within the CNS.
- Motor neurons take nerve impulses from the CNS to effectors (muscles or glands).

Myelin Sheath

- Long axons are covered by a myelin sheath.

The Nerve Impulse

Resting Potential   More Na^+ outside the axon and more K^+ inside the axon. The axon does not conduct an impulse.

Action Potential   A change in polarity across the axonal membrane as a nerve impulse occurs: When Na^+ gates open, Na^+ moves to the inside of the axon, and a depolarization occurs. When K^+ gates open and K^+ moves to outside the axon, a repolarization occurs.

The Synapse

- When a neurotransmitter is released into a synaptic cleft, transmission of a nerve impulse occurs.
- Binding of the neurotransmitter to receptors in the receiving membrane causes excitation or inhibition.
- Integration is the summing of excitatory and inhibitory signals.

13.2 The Central Nervous System

The CNS receives and integrates sensory input and formulates motor output. The CNS consists of the spinal cord and brain.

The Spinal Cord

- Gray matter of the spinal cord contains neuron cell bodies.
- White matter consists of myelinated axons that occur in tracts.
- Conduction to and from brain; carries out reflex actions.

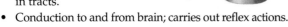

The Brain

Cerebrum   The cerebrum has two cerebral hemispheres connected by the corpus callosum.

- Sensation, reasoning, learning and memory, and language and speech take place in the cerebrum.

- The cerebral cortex covers the cerebrum. The cerebral cortex of each cerebral hemisphere has four lobes: frontal, parietal, occipital, and temporal.
- The primary motor area in the frontal lobe sends out motor commands to lower brain centers, which pass them on to motor neurons.
- The primary somatosensory area in the parietal lobe receives sensory information from lower brain centers in communication with sensory neurons.
- Association areas are located in all the lobes.

The Diencephalon   The hypothalamus controls homeostasis. The thalamus sends sensory input on to the cerebrum.

The Cerebellum   The cerebellum coordinates skeletal muscle contractions.

The Brain Stem   The medulla oblongata and the pons have centers for breathing and the heartbeat.

13.3 The Limbic System and Higher Mental Functions

- The limbic system lying deep in the brain is involved in determining emotions.
- The amygdala determines when a situation deserves the emotion we call "fear."
- The hippocampus is particularly involved in storing and retrieving memories.

13.4 The Peripheral Nervous System

- The PNS contains only nerves and ganglia.
- Cranial nerves take impulses to and from the brain.
- Spinal nerves take impulses to and from the spinal cord.
- The PNS is divided into the somatic system and the autonomic system.

Somatic System

The somatic system serves the skin, skeletal muscles, and tendons.

- Some actions are due to reflexes, which are automatic and involuntary.
- Other actions are voluntary; originate in cerebral cortex.

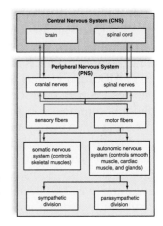

Autonomic System

Two divisions in this system are the sympathetic division and the parasympathetic division.

Sympathetic Division   Responses that occur during times of stress.

Parasympathetic Division   Responses that occur during times of relaxation.

- Actions in these divisions are involuntary and automatic.
- These divisions innervate internal organs.
- Two neurons and one ganglion are utilized for each impulse.

13.5 Drug Abuse

- Neurological drugs either promote or prevent the action of a particular neurotransmitter.
- Dependency occurs when the body compensates for the presence of neurological drugs.

Understanding Key Terms

acetylcholine (ACh) 252
acetylcholinesterase
   (AChE) 252
action potential 250
Alzheimer disease (AD) 266
amygdala 260
association area 257
autonomic system 265
axon 249
axon terminal 252
brain 256
brain stem 259
Broca's area 258
cell body 249
central nervous system
   (CNS) 248
cerebellum 259
cerebral cortex 257
cerebral hemisphere 256
cerebrospinal fluid 254
cerebrum 256
cranial nerve 262
dendrite 249
diencephalon 258
dopamine 252
dorsal-root ganglion 262
drug abuse 267
effector 249
episodic memory 260
GABA (gamma aminobutyric
   acid) 252
ganglion 262
glutamate 252
gray matter 254
hippocampus 260
hypothalamus 258
integration 253
interneuron 249
learning 260
limbic system 260
long-term memory 260
long-term potentiation
   (LTP) 261
medulla oblongata 259
memory 260

meninges (sing., meninx) 254
midbrain 259
motor neuron 249
myelin sheath 249
nerve impulse 250
neuroglia 248
neuron 248
neurotransmitter 252
node of Ranvier 249
norepinephrine (NE) 252
nuclei 259
parasympathetic division 265
Parkinson disease 266
peripheral nervous system
   (PNS) 248
pons 259
prefrontal area 258
primary motor area 257
primary somatosensory
   area 257
reflex 263
refractory period 251
resting potential 250
reticular formation 259
saltatory conduction 251
Schwann cell 249
semantic memory 260
sensory neuron 249
sensory receptor 249
serotonin 252
short-term memory 260
skill memory 260
sodium-potassium pump 250
somatic system 263
spinal cord 254
spinal nerve 262
sympathetic division 265
synapse 252
synaptic cleft 252
thalamus 258
threshold 250
tract 254
ventricle 254
Wernicke's area 258
white matter 254

Match the key terms to these definitions.

a. _____ Automatic, involuntary response of an organism to a stimulus.

b. _____ Chemical stored at the ends of axons that is responsible for transmission across a synapse.

c. _____ Part of the peripheral nervous system that regulates internal organs.

d. _____ Collection of neuron cell bodies, usually outside the central nervous system.

e. _____ Neurotransmitter active in the somatic system of the peripheral nervous system.

Testing Your Knowledge of the Concepts

1. What are the three functions of the nervous system? (page 248)

2. What are the functions performed by the three types of neurons? Describe the structure and functions of the three parts of a neuron. (page 249)

3. What is the resting potential, and how is it created and maintained? (page 250)

4. What is an action potential? Describe the two parts of the process. (page 250)

5. Explain transmission of a nerve impulse across a synapse. (page 252)

6. List and describe the functions of several neurotransmitters. (pages 252–53)

7. Describe the structure and functions of the spinal cord. (pages 254–55)

8. Name the major parts of the brain and give a function for each part. (pages 256–59)

9. What are the lobes of the cerebrum? Give the location and function of the following areas: primary motor area, primary somatosensory area, primary visual area, primary auditory area, primary taste area, and primary olfactory area. (pages 256–58)

10. What are association areas, and what roles do they play? (page 257)

11. What is the reticular formation, and what is the role of the RAS? (page 259)

12. What is the limbic system? What are the two main parts and their functions? (page 260)

13. Describe the various types of memory. (pages 260–61)

14. What types of nerves make up the PNS? How many of each type are there, and what areas of the body do each serve? (page 262)

15. Distinguish between the somatic and the autonomic nervous systems as to structure, effectors, neurotransmitters, and functions. (pages 263–65)

16. Trace the path of a reflex arc. (page 263)

17. Describe the physiological effects and mode of action of alcohol, nicotine, cocaine, methamphetamine, heroin, and marijuana. (pages 267–69)

18. What type of neuron lies completely in the CNS?
   a. motor neuron
   b. interneuron
   c. sensory neuron

19. Which of the following neuron parts receive(s) signals from sensory receptors of other neurons?
   a. cell body   c. dendrites
   b. axon   d. Both a and c are correct.

20. The neuroglial cells that form myelin sheaths in the PNS are
   a. oligodendrocytes.   d. astrocytes.
   b. ganglionic cells.   e. microglia.
   c. Schwann cells.

21. Which of these correctly describes the distribution of ions on either side of an axon when it is not conducting a nerve impulse?
    a. more sodium ions (Na^+) outside and more potassium ions (K^+) inside
    b. more K^+ outside and less Na^+ inside
    c. charged protein outside; Na^+ and K^+ inside
    d. Na^+ and K^+ outside and water only inside
    e. chlorine ions (Cl^-) on outside and K^+ and Na^+ on inside

22. When the action potential begins, sodium gates open, allowing Na^+ to cross the membrane. Now the polarity changes to
    a. negative outside and positive inside.
    b. positive outside and negative inside.
    c. neutral outside and positive inside.
    d. All of these are correct.

23. Repolarization of an axon during an action potential is produced by
    a. inward diffusion of Na^+.
    b. outward diffusion of K^+.
    c. inward active transport of Na^+.
    d. active extrusion of K^+.

24. Transmission of the nerve impulse across a synapse is accomplished by the
    a. movement of Na^+ and K^+.
    b. release of a neurotransmitter by a dendrite.
    c. release of a neurotransmitter by an axon.
    d. release of a neurotransmitter by a cell body.
    e. All of these are correct.

25. Synaptic vesicles are
    a. at the ends of dendrites and axons.
    b. at the ends of axons only.
    c. along the length of all long fibers.
    d. All of these are correct.

26. Which of the following cerebral areas is not correctly matched with its function?
    a. occipital lobe—vision
    b. parietal lobe—somatosensory area
    c. temporal lobe—primary motor area
    d. frontal lobe—Broca's motor speech area

27. The hypothalamus does not
    a. control skeletal muscles.
    b. regulate thirst.
    c. control the pituitary gland.
    d. regulate body temperature.

28. Which of the following brain regions is not correctly matched to its function?
    a. medulla oblongata—regulated heartbeat, breathing, and blood pressure
    b. cerebellum—coordinated voluntary muscle movements
    c. thalamus—secretes melatonin that regulates daily body rhythms
    d. midbrain—reflex centers for visual, auditory, and tactile responses

29. A spinal nerve takes nerve impulses
    a. to the CNS.
    b. away from the CNS.
    c. both to and away from the CNS.
    d. from the CNS to the spinal cord.

30. Which of these are the first and last elements in a spinal reflex?
    a. axon and dendrite
    b. sensory receptor and muscle effector
    c. ventral horn and dorsal horn

d. brain and skeletal muscle
e. motor neuron and sensory neuron

31. The autonomic system has two divisions, called the
    a. CNS and PNS.
    b. somatic and skeletal divisions.
    c. efferent and afferent divisions.
    d. sympathetic and parasympathetic divisions.

32. The sympathetic division of the autonomic system does not cause
    a. the liver to release glycogen.
    b. dilation of bronchioles.
    c. the gastrointestinal tract to digest food.
    d. an increase in the heart rate.

33. Label this diagram.

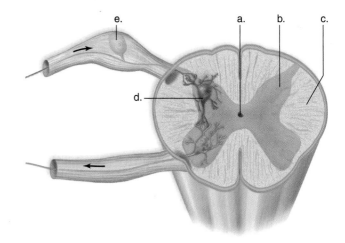

Thinking Critically About the Concepts

As Dr. Smith, from the opening story points out to his class, myelin and the cells that create it (Schwann in the PNS; oligodendrocytes in the CNS) are the subject of numerous research projects. The goal of some projects is to determine how to restore lost myelin, which might help (or possibly cure) folks living with MS or a form of leukodystrophy. Other projects are investigating the effects of myelin on nerve regeneration. A better understanding of myelin's effects on nerve repair could offer hope to victims of spinal cord injury.

1. Why are impulses transmitted more quickly down a myelinated axon than an unmyelinated axon?

2. A buildup of very long chain saturated fatty acids are believed to be the cause of myelin loss in someone with adrenoleukodystrophy (ALD).

a. From your study of chemistry in Chapter 2, a fatty acid is a part of what kind of molecule?

b. What distinguishes a saturated fatty acid from an unsaturated fatty acid?

c. From your study of nutrition in Chapter 8, what kinds of foods would contain saturated fatty acids?

3. Why would you expect the motor skills of a child to improve as myelination continues during early childhood development?

4. If you do some research on the Internet, you should be able to find more information about ALD. How does a child get ALD?

CHAPTER 14

Senses

an and his friends planned to have a big day celebrating the end of school at the local amusement park. There was a new roller coaster called the "Twisted Tornado" and they were anxious to try it out. As they approached the park, they could see the shiny, new tracks gleaming in the sunlight. Tickets in hand, Ian and his friends headed straight for the new "Tornado" on the far side of the park.

On the way to the coaster, they passed the games of chance offered at the park. There were opportunities to throw darts at balloons, shoot water from squirt guns to make cars race up a track, and more. "Come, try your luck, and win a prize," yelled the people working at the booths.

The food court was in the center of the park. As the boys ran by it, the smell of hot dogs, popcorn, and other tasty treats made their mouths water. "I'm starving," shouted Ian, "but if we get to the coaster now, we shouldn't have to wait in line to ride it."

The line to ride the new coaster was short, and the ride worker quickly secured the boys to the car with padded harnesses. The car took off like it was shot out of a cannon and was soon zooming, whirling, and spinning around the track. Minutes later, the car screeched to a stop, and the boys stumbled towards the closest bench to sit down. "Geez," moaned Ian, "I think I'm going to be sick." The boys sat down, closed their eyes, and tried to make their heads stop spinning. After several minutes, Ian jumped up and shouted, "Let's go again!"

CHAPTER CONCEPTS

14.1 Sensory Receptors and Sensations

Each type of sensory receptor detects a particular kind of stimulus. When stimulation occurs, sensory receptors initiate nerve impulses that are transmitted to the CNS. Sensation occurs when nerve impulses reach the cerebral cortex.

14.2 Proprioceptors and Cutaneous Receptors

Proprioception, an awareness of the part of the body, is dependent on sensory receptors in muscles and joints. Touch, pressure, temperature, and pain are dependent on sensory receptors in the skin.

14.3 Senses of Taste and Smell

Taste and smell involve the activity of sensory receptors in the mouth and nose, respectively.

14.4 Sense of Vision

Vision depends on the sensory receptors in the eyes, the optic nerves, and the visual cortex.

14.5 Sense of Hearing

Hearing depends on sensory receptors in the ears, the cochlear nerve, and the auditory cortex.

14.6 Sense of Equilibrium

The inner ear contains sensory receptors for our sense of equilibrium. Normally they help us keep our balance, but the brain is challenged when it receives conflicting messages from the other senses.

14.1 Sensory Receptors and Sensations

Sensory receptors are dendrites specialized to detect certain types of stimuli (sing., **stimulus**). **Exteroceptors** are sensory receptors that detect stimuli from outside the body, such as those that result in taste, smell, vision, hearing, and equilibrium (Table 14.1). **Interoceptors** receive stimuli from inside the body. For example, among interoceptors, pressoreceptors respond to changes in blood pressure, osmoreceptors detect changes in the water-salt balance, and chemoreceptors monitor the pH of the blood.

Interoceptors are directly involved in homeostasis and are regulated by a negative feedback mechanism. For example, when blood pressure rises, pressoreceptors signal a regulatory center in the brain, which sends out nerve impulses to the arterial walls, causing them to relax; the blood pressure then falls. Therefore, the pressoreceptors are no longer stimulated, and the system shuts down.

Exteroceptors such as those in the eyes and ears are not directly involved in homeostasis and continually send messages to the central nervous system regarding environmental conditions.

Types of Sensory Receptors

Sensory receptors in humans can be classified into just four categories: chemoreceptors, photoreceptors, mechanoreceptors, and thermoreceptors. When Ian and his friends, mentioned in the opening story, went to a local amusement park, they utilized all of these receptors. The various smells of hot dogs, popcorn, and sweaty people, and the sweet taste of cotton candy or salty pretzels would have stimulated their chemoreceptors.

Chemoreceptors respond to chemical substances in the immediate vicinity. As Table 14.1 indicates, taste and smell, which detect external stimuli, utilize chemoreceptors; but so do various other organs that are sensitive to internal stimuli. Chemoreceptors that monitor blood pH are located in the carotid arteries and aorta. If the pH lowers, the breathing rate increases. As more carbon dioxide is expired, the blood pH rises.

Pain receptors (nociceptors) are a type of chemoreceptor. They are naked dendrites that respond to chemicals released by damaged tissues. Pain receptors are protective because they alert us to possible danger. For example, without the pain of appendicitis, we might never seek the medical help needed to avoid a ruptured appendix.

Photoreceptors respond to light energy. Our eyes contain photoreceptors that are sensitive to light rays and thereby provide us with a sense of vision. Stimulation of the photoreceptors known as rod cells results in black-and-white vision, while stimulation of the photoreceptors known as cone cells results in color vision. The sights at the amusement park, including colorful balloons, spinning rides, and rings being tossed, would have stimulated the boys' photoreceptors.

Mechanoreceptors are stimulated by mechanical forces, which most often result in pressure of some sort. When we hear, airborne sound waves are converted to fluid-borne pressure waves that can be detected by mechanoreceptors in the inner ear. The boys at the park would have heard people yelling and screaming, music playing, bells ringing, and popcorn popping, for example.

Mechanoreceptors are responding to fluid-borne pressure waves when we detect changes in gravity and motion, helping us keep our balance. These receptors are in the vestibule and semicircular canals of the inner ear, respectively. The "Twisted Tornado" tested the ability of the boys to maintain their balance, despite the ordeals of their ride.

The sense of touch depends on pressure receptors that are sensitive to either strong or slight pressures. Pressoreceptors located in certain arteries detect changes in blood pressure, and stretch receptors in the lungs detect the degree of lung inflation. Proprioceptors, which respond to the stretching of muscle fibers, tendons, joints, and ligaments, make us aware of the position of our limbs.

Thermoreceptors located in the hypothalamus and skin are stimulated by changes in temperature. Those that respond

Table 14.1	Exteroceptors			
Sensory Receptor	**Stimulus**	**Category**	**Sense**	**Sensory Organ**
Taste cells	Chemicals	Chemoreceptor	Taste	Taste buds
Olfactory cells	Chemicals	Chemoreceptor	Smell	Olfactory epithelium
Rod cells and cone cells in retina	Light rays	Photoreceptor	Vision	Eye
Hair cells in spiral organ	Sound waves	Mechanoreceptor	Hearing	Ear
Hair cells in semicircular canals	Motion	Mechanoreceptor	Rotational equilibrium	Ear
Hair cells in vestibule	Gravity	Mechanoreceptor	Gravitational equilibrium	Ear

when temperatures rise are called warmth receptors, and those that respond when temperatures lower are called cold receptors. The heat of the sun would have activated the thermoreceptors of the boys, and the rise in their body temperature would most likely have had them seeking some shade.

How Sensation Occurs

Sensory receptors respond to environmental stimuli by generating nerve impulses. When the nerve impulses arrive at the cerebral cortex of the brain, **sensation,** which is the conscious perception of stimuli, occurs.

As we discussed in Chapter 13, sensory receptors are the first element in a reflex arc. We are aware of a reflex action only when sensory information reaches the brain. At that time, the brain integrates this information with other information received from other sensory receptors. After all, if you burn yourself and quickly remove your hand from a hot stove, the brain receives information not only from your skin, but also from your eyes, nose, and all sorts of sensory receptors.

Some sensory receptors are free nerve endings or encapsulated nerve endings, while others are specialized cells closely associated with neurons. Often the plasma membrane of a sensory receptor contains receptor proteins that react to the stimulus. For example, the receptor proteins in the plasma membrane of chemoreceptors bind to certain molecules. When this happens, ion channels open, and ions flow across the plasma membrane. If the stimulus is sufficient, nerve impulses begin and are carried by a sensory nerve fiber within the PNS to the CNS (Fig. 14.1). The stronger the stimulus, the greater the frequency of nerve impulses. Nerve impulses that reach the spinal cord first are conveyed to the brain by ascending tracts. If nerve impulses finally reach the cerebral cortex, sensation and perception occur.

All sensory receptors initiate nerve impulses; the sensation that results depends on the part of the brain receiving the nerve impulses. Nerve impulses that begin in the optic nerve eventually reach the visual areas of the cerebral cortex, and, thereafter, we see objects. Nerve impulses that begin in the auditory nerve eventually reach the auditory areas of the cerebral cortex, and, thereafter, we hear sounds. If it were possible to switch these nerves, stimulation of the eyes would result in hearing! On the other hand, when a blow to the eye stimulates photoreceptors, we "see stars" because nerve impulses from the eyes can only result in sight.

Before sensory receptors initiate nerve impulses, they also carry out **integration,** the summing up of signals. One type of integration is called **sensory adaptation,** a decrease in response to a stimulus. We have all had the experience of smelling an odor when we first enter a room and then later not being aware of it at all. Some authorities believe that when sensory adaptation occurs, sensory receptors have stopped sending impulses to the brain. Others believe that the reticular activating system (RAS) has filtered out the

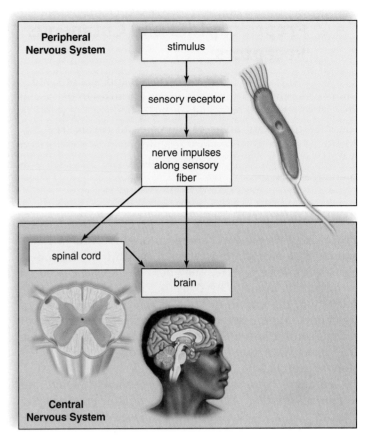

Figure 14.1 Sensation and perception.
After detecting a stimulus, sensory receptors initiate nerve impulses within the PNS. These impulses give the central nervous system (CNS) information about the external and internal environment. The CNS integrates all incoming information and then initiates a motor response to the stimulus.

ongoing stimuli. You will recall that sensory information is conveyed from the brain stem through the thalamus to the cerebral cortex by the RAS. The thalamus acts as a gatekeeper and passes on information only of immediate importance. Just as we gradually become unaware of particular environmental stimuli, we can suddenly become aware of stimuli that may have been present for some time. This can be attributed to the workings of the RAS, which has synapses with all the great ascending sensory tracts.

The functioning of our sensory receptors makes a significant contribution to homeostasis. Without sensory input, we would not receive information about our internal and external environment. This information leads to appropriate reflex and voluntary actions to keep the internal environment constant.

✅ Check Your Progress 14.1

1. What is the function of a sensory receptor?

2. What are the different types of sensory receptors?

3. Explain sensation.

14.2 Proprioceptors and Cutaneous Receptors

Sensory receptors in the muscles, joints and tendons, other internal organs, and skin send nerve impulses to the spinal cord. From there, they travel up the spinal cord in tracts to the somatosensory areas of the cerebral cortex (see Fig. 13.10). These general sensory receptors can be categorized into three types: proprioceptors, cutaneous receptors, and pain receptors.

Proprioceptors

Proprioceptors are mechanoreceptors involved in reflex actions that maintain muscle tone, and thereby the body's equilibrium and posture. They help us know the position of our limbs in space by detecting the degree of muscle relaxation, the stretch of tendons, and the movement of ligaments. Muscle spindles act to increase the degree of muscle contraction, and Golgi tendon organs act to decrease it. The result is a muscle that has the proper length and tension, or muscle tone.

Figure 14.2 illustrates the activity of a muscle spindle. In a muscle spindle, sensory nerve endings are wrapped around thin muscle cells within a connective tissue sheath.

When the muscle relaxes and undue stretching of the muscle spindle occurs, nerve impulses are generated. The rapidity of the nerve impulses generated by the muscle spindle is proportional to the stretching of a muscle. A reflex action then occurs, which results in contraction of muscle fibers adjoining the muscle spindle. The knee-jerk reflex, which involves muscle spindles, offers an opportunity for physicians to test a reflex action. The information sent by muscle spindles to the CNS is used to maintain the body's equilibrium and posture, despite the force of gravity always acting upon the skeleton and muscles.

Cutaneous Receptors

The skin is composed of two layers: the epidermis and the dermis. In Figure 14.3, the artist has dramatically indicated these two layers by separating the epidermis from the dermis in one location. The epidermis is stratified squamous epithelium in which cells become keratinized as they rise to the surface, where they are sloughed off. The dermis is a thick connective tissue layer. The dermis contains **cutaneous receptors,** which make the skin sensitive to touch, pressure, pain, and temperature (warmth and cold). The dermis is a mosaic of these tiny receptors, as you can determine by slowly passing a metal probe

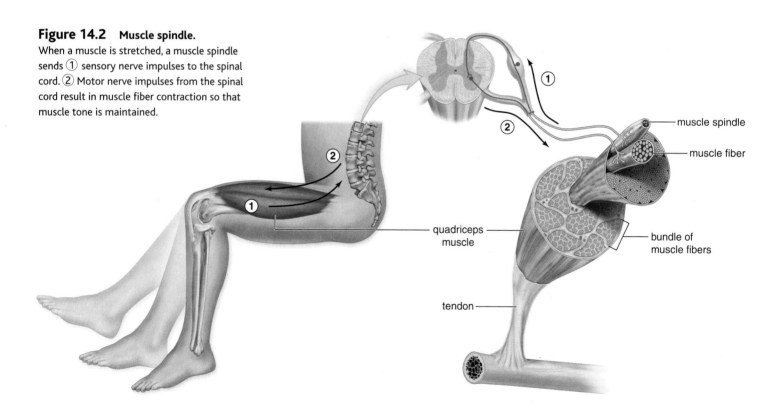

Figure 14.2 Muscle spindle.
When a muscle is stretched, a muscle spindle sends ① sensory nerve impulses to the spinal cord. ② Motor nerve impulses from the spinal cord result in muscle fiber contraction so that muscle tone is maintained.

quadriceps muscle

tendon

muscle spindle

muscle fiber

bundle of muscle fibers

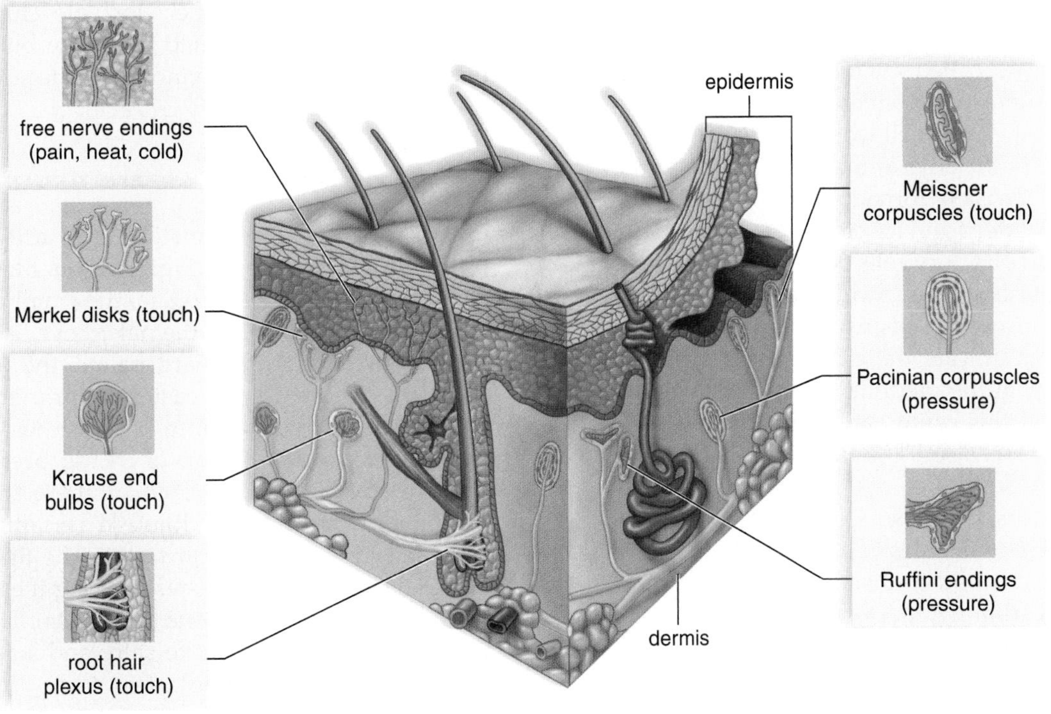

Figure 14.3 **Sensory receptors in human skin.**
The classical view is that each sensory receptor has the main function shown here. However, investigators report that matters are not so clear-cut. For example, microscopic examination of the skin of the ear shows only free nerve endings (pain receptors), and yet the skin of the ear is sensitive to all sensations. Therefore, it appears that the receptors of the skin are somewhat, but not completely, specialized.

over your skin. At certain points, you will feel touch or pressure, and at others, you will feel heat or cold (depending on the probe's temperature).

Three types of cutaneous receptors are sensitive to fine touch, which gives a person specific information such as the location of the touch plus its shape, size, and texture. *Meissner corpuscles* and Krause end bulbs are concentrated in the fingertips, the palms, the lips, the tongue, the nipples, the penis, and the clitoris. *Merkel disks* are found where the epidermis meets the dermis. A free nerve ending called a *root hair plexus* winds around the base of a hair follicle and fires if the hair is touched.

Two types of cutaneous receptors that are sensitive to pressure are Pacinian corpuscles and Ruffini endings. *Pacinian corpuscles* are onion-shaped sensory receptors that lie deep inside the dermis. *Ruffini endings* are encapsulated by sheaths of connective tissue and contain lacy networks of nerve fibers.

Temperature receptors are simply free nerve endings in the epidermis. Some free nerve endings are responsive to cold; others are responsive to warmth. Cold receptors are far more numerous than warmth receptors, but the two types have no known structural differences.

Pain Receptors

Like the skin, many internal organs have pain receptors, also called *nociceptors*, which are sensitive to chemicals released by damaged tissues. When inflammation occurs, due to mechanical, thermal, or electrical stimuli or toxic substances, cells release chemicals that stimulate pain receptors. Aspirin and ibuprofen reduce pain by inhibiting the synthesis of one class of these chemicals.

Sometimes, stimulation of internal pain receptors is felt as pain from the skin, as well as the internal organs. This is called **referred pain.** Some internal organs have a referred pain relationship with areas located in the skin of the back, groin, and abdomen; for example, pain from the heart is felt in the left shoulder and arm. This most likely happens when nerve impulses from the pain receptors of internal organs travel to the spinal cord and synapse with neurons also receiving impulses from the skin.

☑ Check Your Progress 14.2

1. What is the function of proprioceptors?
2. What are the functions of cutaneous receptors in the skin?

14.3 Senses of Taste and Smell

Taste and smell are called chemical senses because their receptors are sensitive to molecules in the food we eat and the air we breathe. Taste and smell would have played a big role in the experience of the boys at the park as soon as they finished riding the "Twisted Tornado" and headed toward the food stands for hot dogs and ice cream (Fig. 14.4*a*).

Taste cells and olfactory cells bear chemoreceptors. Chemoreceptors are also in the carotid arteries and in the aorta, where they are primarily sensitive to the pH of the blood. These bodies communicate via sensory nerve fibers with the respiratory center in the medulla oblongata. When the pH drops, they signal this center, and immediately thereafter, the breathing rate increases. The expiration of CO_2 raises the pH of the blood.

Chemoreceptors are plasma membrane receptors that bind to particular molecules. They are divided into two types: those that respond to distant stimuli and those that respond to direct stimuli. Olfactory cells act from a distance and taste cells act directly. Calvin, from the Chapter 3 open-ing story, could smell the chocolate chip cookies in the kitchen as soon as he entered the house, but he couldn't taste them until they were in his mouth. pH receptors also respond to direct stimuli.

Sense of Taste

In adult humans, approximately 3,000 **taste buds** are located primarily on the tongue (Fig. 14.4*b–e*). Many taste buds lie along the walls of the papillae, the small elevations on the tongue that are visible to the naked eye. Isolated taste buds are also present on the hard palate, the pharynx, and the epiglottis. There are at least four primary types of taste (sweet, sour, salty, and bitter). A fifth taste, called umami, may exist for certain flavors of cheese, beef broth, and some seafood. Taste buds for each of these tastes are located throughout the tongue, although certain regions may be most sensitive to particular tastes: The tip of the tongue is most sensitive to sweet tastes, making it especially pleasurable to lick an ice cream cone (Fig. 14.4*a*). The margins of the tongue are most sensitive to salty and sour tastes; and the rear of the tongue to bitter tastes.

How the Brain Receives Taste Information

Taste buds open at a taste pore. They have supporting cells and a number of elongated taste cells that end in microvilli. When molecules bind to receptor proteins of the microvilli, nerve impulses are generated in sensory nerve fibers that go to the brain. When they reach the gustatory (taste) cortex, they are interpreted as particular tastes.

Since we can respond to a range of sweet, sour, salty, and bitter tastes, the brain appears to survey the overall pattern of incoming sensory impulses and to take a "weighted

a.

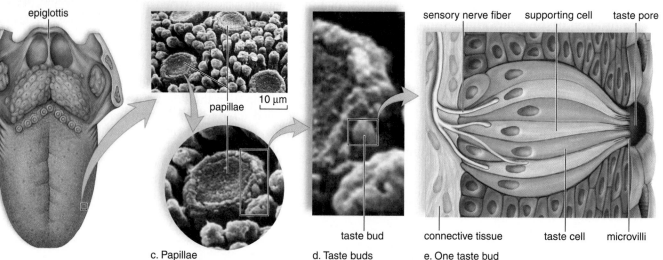
b. Tongue   c. Papillae   d. Taste buds   e. One taste bud

Figure 14.4   Taste buds in humans.
a. Without a tongue, it would be impossible to lick an ice cream cone and its taste would be minimal. **b.** Papillae on the tongue contain taste buds that are sensitive to sweet, sour, salty, bitter, and perhaps umami. **c.** Photomicrograph and enlargement of papilla. **d.** Taste buds occur along the walls of the papillae. **e.** Taste cells end in microvilli that bear receptor proteins for certain molecules. When molecules bind to the receptor proteins, nerve impulses are generated and go to the brain, where the sensation of taste occurs.

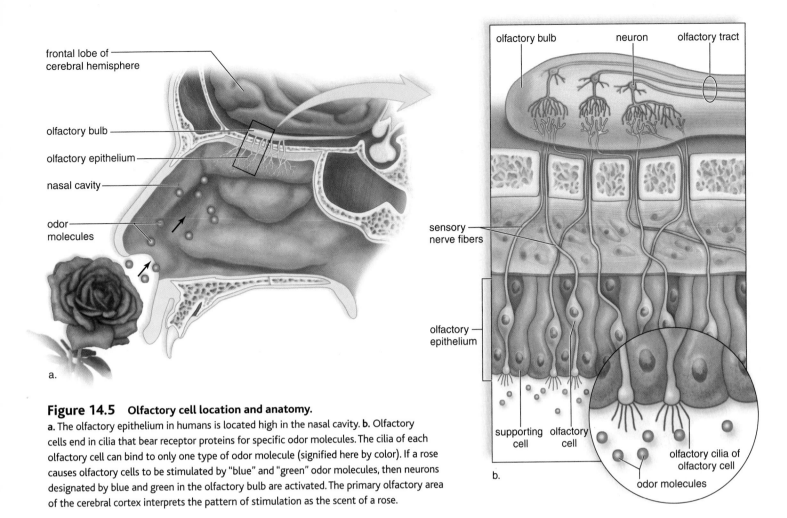

Figure 14.5 **Olfactory cell location and anatomy.**
a. The olfactory epithelium in humans is located high in the nasal cavity. **b.** Olfactory cells end in cilia that bear receptor proteins for specific odor molecules. The cilia of each olfactory cell can bind to only one type of odor molecule (signified here by color). If a rose causes olfactory cells to be stimulated by "blue" and "green" odor molecules, then neurons designated by blue and green in the olfactory bulb are activated. The primary olfactory area of the cerebral cortex interprets the pattern of stimulation as the scent of a rose.

average" of their taste messages as the perceived taste. Again, we can note that even though our senses depend on sensory receptors, the cortex integrates the incoming information and gives us our sensations.

Sense of Smell

Approximately 80–90% of what we perceive as "taste" actually is due to the sense of smell, which accounts for how dull food tastes when we have a head cold or a stuffed-up nose. Our sense of smell depends on between 10 and 20 million **olfactory cells** located within olfactory epithelium high in the roof of the nasal cavity (Fig. 14.5*a* and *b*). Olfactory cells are modified neurons. Each cell ends in a tuft of about five olfactory cilia, which bear receptor proteins for odor molecules.

How the Brain Receives Odor Information

Each olfactory cell has only one out of several hundred different types of receptor proteins. Nerve fibers from like olfactory cells lead to the same neuron in the olfactory bulb, an extension of the brain. An odor contains many odor molecules, which activate a characteristic combination of receptor proteins. For example, a rose might stimulate olfactory cells, designated by blue and green in Figure 14.5, while a gardenia might stimulate a different combination. An odor's signature

in the olfactory bulb is determined by which neurons are stimulated. When the neurons communicate this information via the olfactory tract to the olfactory areas of the cerebral cortex, we know we have smelled a rose or a gardenia.

Have you ever noticed that a certain aroma vividly brings to mind a certain person or place? A whiff of a perfume may remind you of a specific person, or the smell of boxwood may remind you of your grandfather's farm. The olfactory bulbs have direct connections with the limbic system and its centers for emotions and memory. One investigator showed that when subjects smelled an orange while viewing a painting, they not only remembered the painting when asked about it later, but they also had many deep feelings about it.

The number of olfactory cells declines with age, and the remaining ones become less sensitive. Therefore, older people tend to apply excessive amounts of perfume or aftershave before they can detect its smell. This can be dangerous if these individuals cannot smell smoke or a gas leak.

✅ Check Your Progress 14.3

1. **a.** Taste cells are what type of sensory receptor? **b.** What are the types of taste that result from stimulating these receptors?

2. **a.** What are the receptors responsible for smell called, and **(b)** where are they located?

3. Explain why a rose smells different than a gardenia.

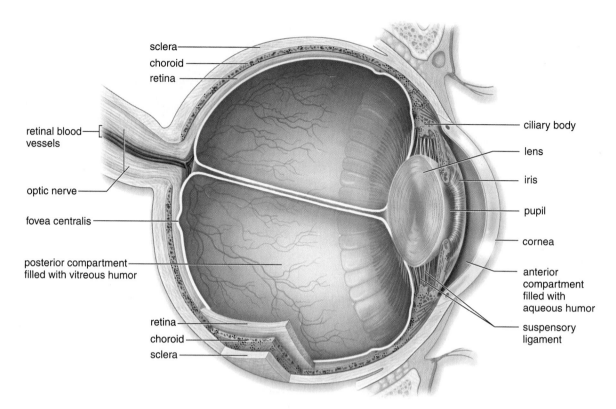

sclera
choroid
retina
retinal blood vessels
optic nerve
fovea centralis
posterior compartment filled with vitreous humor
retina
choroid
sclera

ciliary body
lens
iris
pupil
cornea
anterior compartment filled with aqueous humor
suspensory ligament

Figure 14.6
Anatomy of the human eye.
Notice that the sclera (the outer layer of the eye) becomes the cornea and that the choroid (the middle layer) is continuous with the ciliary body and the iris. The retina (the inner layer) contains the photoreceptors for vision. The fovea centralis is the region where vision is most acute.

14.4 Sense of Vision

Vision requires the work of the eyes and the brain. As we shall see, much processing of stimuli occurs in the eyes before nerve impulses are sent to the brain. Still, researchers estimate that at least a third of the cerebral cortex takes part in processing visual information.

Anatomy and Physiology of the Eye

The eyeball, which is an elongated sphere about 2.5 cm in diameter, has three layers, or coats: the sclera, the choroid, and the retina (Fig. 14.6 and Table 14.2). The outer layer, the **sclera,** is white and fibrous except for the **cornea,** which is made of transparent collagen fibers. The cornea is the window of the eye.

The middle, thin, darkly pigmented layer, the **choroid,** is vascular and absorbs stray light rays that photoreceptors have not absorbed. Toward the front, the choroid becomes the donut-shaped **iris.** The iris regulates the size of the **pupil,** a hole in the center of the iris through which light enters the eyeball. The color of the iris (and therefore the color of your eyes) correlates with its pigmentation. Heavily pigmented eyes are brown, while lightly pigmented eyes are green or blue. Behind the iris, the choroid thickens and forms the circular ciliary body. The **ciliary body** contains the ciliary muscle, which controls the shape of the lens for near and far vision.

The **lens,** attached to the ciliary body by suspensory ligaments, divides the eye into two compartments; the one in front of the lens is the anterior compartment, and the one behind the lens is the posterior compartment. The anterior compartment is filled with a clear, watery fluid called the **aqueous humor.** A small amount of aqueous humor is continually produced each day. Normally, it leaves the anterior compartment by way of tiny ducts. When a person has **glaucoma,** these drainage ducts are blocked, and aqueous humor

Table 14.2	Reabsorption from Nephrons
Part	**Function**
SCLERA	Protects and supports eyeball
Cornea	Refracts light rays
Pupil	Admits light
CHOROID	Absorbs stray light
Ciliary body	Holds lens in place, accommodation
Iris	Regulates light entrance
RETINA	Contains sensory receptors for sight
Rod cells	Make black-and-white vision possible
Cone cells	Make color vision possible
Fovea centralis	Makes acute vision possible
OTHER	
Lens	Refracts and focuses light rays
Humors	Transmit light rays and support eyeball
Optic nerve	Transmits impulse to brain

Figure 14.7   Winning the prize.
Ian made use of his good eye-hand coordination to win a prize at the amusement park. He was able to throw a ball and hit the target on the first try. His eyes began a chain of events that resulted in a good throw. From his eyes, nerve impulses traveled to the visual cortex and then on to the motor cortex, which directed his cerebellum to fine-tune his movements initiated after the spinal cord sent messages to the muscles in his throwing arm.

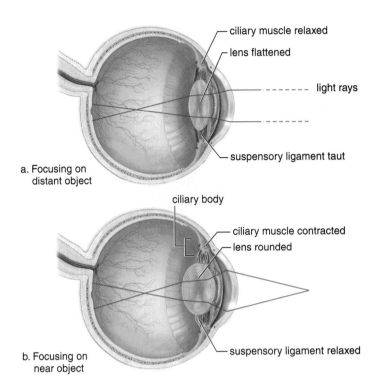

a. Focusing on distant object

ciliary muscle relaxed
lens flattened
light rays
suspensory ligament taut

ciliary body

ciliary muscle contracted
lens rounded

b. Focusing on near object

suspensory ligament relaxed

Figure 14.8   Focusing of the human eye.
Light rays from each point on an object are bent by the cornea and the lens in such a way that an inverted and reversed image of the object forms on the retina. **a.** When focusing on a distant object, the lens is flat because the ciliary muscle is relaxed and the suspensory ligament is taut. **b.** When focusing on a near object, the lens accommodates—that is, it becomes rounded because the ciliary muscle contracts, causing the suspensory ligament to relax.

builds up. If glaucoma is not treated, the resulting pressure compresses the arteries that serve the nerve fibers of the retina, where photoreceptors are located. The nerve fibers begin to die due to lack of nutrients, and the person becomes partially blind. Eventually, total blindness can result.

The third layer of the eye, the **retina,** is located in the posterior compartment, which is filled with a clear, gelatinous material called the **vitreous humor.** The retina contains photoreceptors called rod cells and cone cells. The rods are very sensitive to light, but they do not see color; therefore, at night or in a darkened room, we see only shades of gray. The cones, which require bright light, are sensitive to different wavelengths of light, and therefore, we have the ability to distinguish colors. The retina has a very special region called the **fovea centralis** where cone cells are densely packed. Light is normally focused on the fovea when we look directly at an object. This is helpful because vision is most acute in the fovea centralis. Sensory fibers from the retina form the **optic nerve,** which takes nerve impulses to the visual cortex (Fig. 14.7).

Function of the Lens

The cornea, assisted by the lens and the humors, focuses images on the retina. Focusing starts with the cornea and continues as the rays pass through the lens and the humors. The image produced is much smaller than the object because light rays are bent (refracted) when they are brought into **focus.** If the eyeball is too long or too short, the person may need corrective lenses to bring the image into focus. The image on the retina is inverted (upside down) and reversed from left to right.

Visual accommodation occurs for close vision. During visual accommodation, the lens rounds up, in order to bring the image to focus on the retina. The shape of the lens is controlled by the ciliary muscle, within the ciliary body. When we view a distant object, the ciliary muscle is relaxed, causing the suspensory ligaments attached to the ciliary body to be taut; therefore, the lens remains relatively flat (Fig. 14.8a). When we view a near object, the ciliary muscle contracts, releasing the tension on the suspensory ligaments, and the lens rounds up due to its natural elasticity (Fig. 14.8b). Now the image is focused on the retina. Because close work requires contraction of the ciliary muscle, it very often causes muscle fatigue, known as eyestrain. Usually after the age of 40, the lens loses some of its elasticity and is unable to accommodate. Bifocal lenses may then be necessary for those who already have corrective lenses.

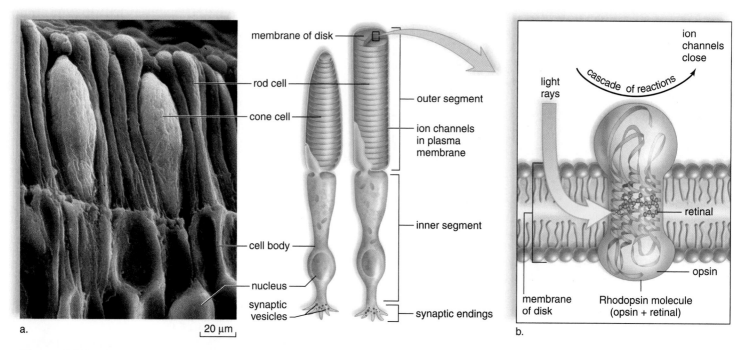

Figure 14.9   **Photoreceptors in the eye.**
a. The outer segment of rods and cones contains stacks of membranous disks, which contain visual pigments. **b.** In rods, the membrane of each disk contains rhodopsin, a complex molecule containing the protein opsin and the pigment retinal. When rhodopsin absorbs light energy, it splits, releasing retinal, which sets in motion a cascade of reactions that cause ion channels in the plasma membrane to close. Thereafter, nerve impulses go to the brain.

Visual Pathway to the Brain

The pathway for vision begins once light has been focused on the photoreceptors in the retina. Some integration occurs in the retina, where nerve impulses begin before the optic nerve transmits them to the brain.

Function of Photoreceptors   Figure 14.9*a* illustrates the structure of the photoreceptors called **rod cells** and **cone cells.** Both rods and cones have an outer segment joined to an inner segment by a stalk. Pigment molecules are embedded in the membrane of the many disks present in the outer segment. Synaptic vesicles are located at the synaptic endings of the inner segment.

The visual pigment in rods is a deep purple pigment called rhodopsin (Fig. 14.9*b*). **Rhodopsin** is a complex molecule made up of the protein opsin and a light-absorbing molecule called **retinal,** which is a derivative of vitamin A. When a rod absorbs light, rhodopsin splits into opsin and retinal, leading to a cascade of reactions and the closure of ion channels in the rod cell's plasma membrane. The release of inhibitory transmitter molecules from the rod's synaptic vesicles ceases. Thereafter, signals go to other neurons in the retina. Rods are very sensitive to light and, therefore, are suited to night vision. (Because carrots are rich in vitamin A, it is true that eating carrots can improve your night vision.) Rod cells are plentiful throughout the entire retina, except the fovea. Therefore, rods also provide us with peripheral vision and perception of motion.

The cones, on the other hand, are located primarily in the fovea and are activated by bright light. They allow us to detect the fine detail and the color of an object. **Color vision** depends on three different kinds of cones, which contain pigments called the B (blue), G (green), and R (red) pigments. Each pigment is made up of retinal and opsin, but there is a slight difference in the opsin structure of each, which accounts for their individual absorption patterns. Various combinations of cones are believed to be stimulated by in-between shades of color.

Function of the Retina   The retina has three layers of neurons (Fig. 14.10). The layer closest to the choroid contains the rod cells and cone cells; the middle layer contains bipolar cells; and the innermost layer contains ganglion cells, whose sensory fibers become the optic nerve. Only the rod cells and the cone cells are sensitive to light, and therefore, light must penetrate to the back of the retina before they are stimulated.

The rod cells and the cone cells synapse with the bipolar cells, which, in turn, synapse with ganglion cells whose axons become the optic nerve. Notice in Figure 14.10 that there are many more rod cells and cone cells than ganglion cells. In fact, the retina has as many as 150 million rod cells and 6 million cone cells but only 1 million ganglion cells. The sensitivity of cones versus rods is mirrored by how directly they connect to ganglion cells. As many as 150 rods may activate the same ganglion cell. No wonder stimulation of rods results in vision that is blurred and indistinct. In contrast, some cone cells in

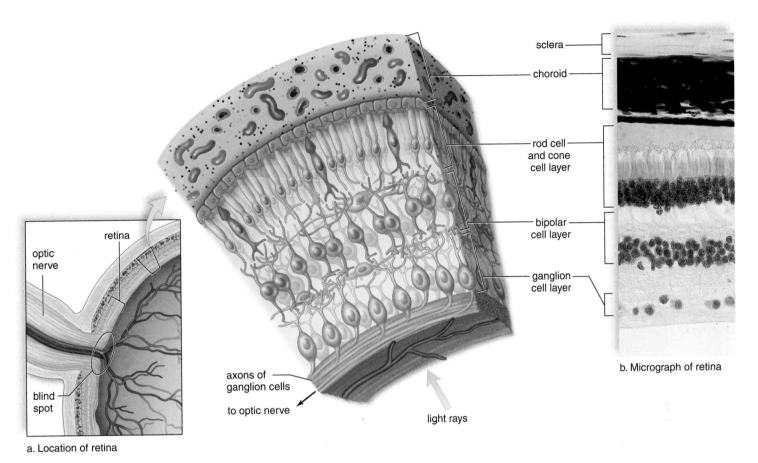

sclera

choroid

rod cell and cone cell layer

bipolar cell layer

ganglion cell layer

optic nerve

retina

blind spot

a. Location of retina

axons of ganglion cells

to optic nerve

light rays

b. Micrograph of retina

Figure 14.10 **Structure and function of the retina.**
a. The retina is the inner layer of the eyeball. Rod and cone cells, located at the back of the retina nearest the choroid, synapse with bipolar cells, which synapse with ganglion cells. Integration of signals occurs at these synapses; therefore, much processing occurs in bipolar and ganglion cells. Further, notice that many rod cells share one bipolar cell, but cone cells do not. Certain cone cells synapse with only one ganglion cell. Cone cells, in general, distinguish more detail than do rod cells. **b.** Micrograph shows that the sclera and choroid are relatively thin compared to the retina, which has several layers of cells.

the fovea centralis activate only one ganglion cell. This explains why cones, especially in the fovea centralis, provide us with a sharper, more detailed image of an object.

As signals pass to bipolar cells and ganglion cells, integration occurs. Therefore, considerable processing occurs in the retina before ganglion cells generate nerve impulses, which are carried in the optic nerve to the visual cortex. Additional integration occurs in the visual cortex.

Blind Spot   Figure 14.10 also provides an opportunity to point out that there are no rods and cones where the optic nerve exits the retina. Therefore, no vision is possible in this area. You can prove this to yourself by putting a dot to the right of center on a piece of paper. Use your right hand to move the paper slowly toward your right eye, while you look straight ahead. The dot will disappear at one point—this is your right eye's **blind spot.** The two eyes together provide complete vision because the blind spot for the right eye is not the same as the blind spot for the left eye. The blind spot for the right eye is right of center and the blind spot for the left eye is left of center.

From the Retina to the Visual Cortex   To reach the visual cortex, the optic nerves carry nerve impulses from the eyes to the optic chiasma (Fig. 14.11). The **optic chiasma** has an X shape, formed by a crossing-over of optic nerve fibers. Fibers from the right half of each retina converge and continue on together in the *right optic tract*, and fibers from the left half of each retina converge and continue on together in the *left optic tract.*

The optic tracts sweep around the hypothalamus, and most fibers synapse with neurons in nuclei (masses of neuron cell bodies) within the thalamus. Axons from the thalamic nuclei form optic radiations that take nerve impulses to the *visual cortex* within the occipital lobe. The image has been split because the right visual cortex receives information originally carried by the right optic tract, and the left visual cortex receives information originally carried by the left optic tract. For good depth perception, the right and left visual cortices communicate with each other. Also, because the image is inverted and reversed, it must be righted in the brain for us to correctly perceive the visual field.

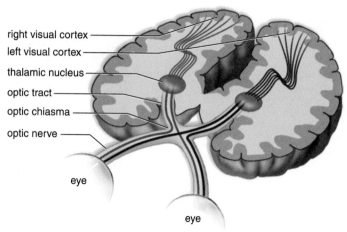

Figure 14.11   Optic chiasma.
Because of the optic chiasma, data from the right half of each retina go to the right visual cortex, and data from the left half of the retina go to the left visual cortex. These data are then combined to allow us to see the entire visual field.

Abnormalities of the Eye

Color blindness and misshapen eyeballs are two common abnormalities of the eyes. Complete color blindness is extremely rare. In most instances, only one type of cone is defective or deficient in number. The most common mutation is the inability to see the colors red and green. This abnormality affects 5–8% of the male population. If the eye lacks cones that respond to red wavelengths, green colors are accentuated, and vice versa.

Distance Vision

If you can see from 20 feet what a person with normal vision can see from 20 feet, you are said to have 20/20 vision. Persons who can see close objects but cannot see the letters from this distance are said to be nearsighted. **Nearsighted** people can see close objects better than they can see objects at a distance. These individuals have an elongated eyeball, and when they attempt to look at a distant object, the image is brought to focus in front of the retina (Fig. 14.12*a*). They can see close objects because their lens can compensate for the long eyeball. To see distant objects, these people can wear concave lenses, which diverge the light rays so that the image focuses on the retina.

Persons who can easily see the optometrist's chart but cannot see close objects well are **farsighted;** these individuals can see distant objects better than they can see close objects. They have a shortened eyeball, and when they try to see close objects, the image is focused behind the retina (Fig. 14.12*b*). When the object is distant, the lens can compensate for the short eyeball. When the object is close, these persons can wear convex lenses to increase the bending of light rays so that the image can be focused on the retina.

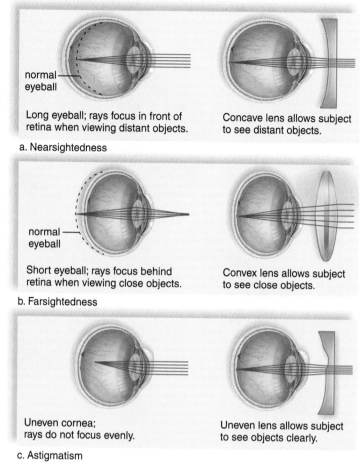

a. Nearsightedness

Long eyeball; rays focus in front of retina when viewing distant objects.

Concave lens allows subject to see distant objects.

b. Farsightedness

Short eyeball; rays focus behind retina when viewing close objects.

Convex lens allows subject to see close objects.

c. Astigmatism

Uneven cornea; rays do not focus evenly.

Uneven lens allows subject to see objects clearly.

Figure 14.12   Common abnormalities of the eye with possible corrective lenses.
a. A concave lens in nearsighted persons focuses light rays on the retina. **b.** A convex lens in farsighted persons focuses light rays on the retina. **c.** An uneven lens in persons with astigmatism focuses light rays on the retina.

When the cornea or lens is uneven, the image is fuzzy. The light rays cannot be evenly focused on the retina. This condition, called **astigmatism,** can be corrected by an unevenly ground lens to compensate for the uneven cornea (Fig. 14.12*c*).

Many people today opt to have LASIK surgery instead of wearing lenses. LASIK surgery is discussed in the Health Focus on page 285.

✔ Check Your Progress 14.4

1. What parts of the eye assist in focusing an image on the retina?

2. What are the two types of photoreceptors?

3. Once photoreceptors initiate a visual signal, what other cells in the eye integrate the signal and pass it on to the visual cortex?

Health Focus

Correcting Vision Problems

Poor vision can be due to a number of problems, some more serious than others. For example, cataracts and glaucoma are two conditions that need medical attention.

Cataracts and Glaucoma

Cataracts develop when the lens of the eye becomes cloudy. Normally the lens is clear and allows light to pass through easily. A cloudy lens will decrease light levels that reach the retina and slowly cause vision loss. Fortunately, a doctor can surgically remove the cloudy lens and replace it with a clear plastic lens. This will often restore the light level passing through the lens and improve the patients' vision.

When fluid pressure builds up inside the eye, a patient is diagnosed with glaucoma. Glaucoma can lead to a decrease in vision and eventually blindness. Special eye drops and oral medications are often prescribed to help reduce the intraocular pressure. If eye drops and medications are not capable of controlling the pressure, surgery may be the only option. During glaucoma surgery, the doctor will use a laser to create tiny holes in the eye where the cornea and iris meet. This will hopefully increase fluid drainage from the eye and decrease the intraocular pressure.

The Benefits of LASIK Surgery

For many people, a sign of aging is the slow and steady decrease of their ability to see close-up. A difficulty in focusing on small print is usually the first sign of presbyopia, a condition that usually begins in the late 30s. By age 55, nearly 100% of the population is affected. Reading in low light situations becomes more difficult, and letters begin to look fuzzy when reading close-up. Many people suffering from presbyopia will often experience headaches while reading. In order to accommodate for their deteriorating vision, presbyopia sufferers will hold reading materials at arm's length. One solution to presbyopia is to wear bifocal lenses. Bifocals are designed to correct vision at a distance of 12" to 18". These lenses work well for most people while reading, but pose a problem for people who use a computer. Computer monitors are usually 19" to 24" away. This forces bifocal lens wearers to constantly move their head up and down in an attempt to switch between the close and distant viewing sections of the bifocals. Another solution is to wear contact lenses with one eye corrected to see close objects and the other eye corrected to see distant objects. The same type of correction can be done with LASIK surgery. LASIK, which stands for laser in-situ keratomileusis is generally a safe and effective treatment option for a wide array of vision problems.

LASIK is a quick and painless procedure that involves the use of a laser to permanently change the shape of the cornea. For the majority of patients, LASIK will improve their vision and reduce their dependency on corrective lenses. The

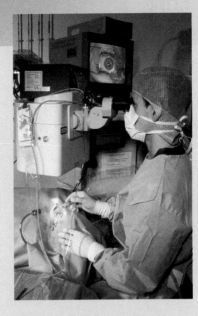

Figure 14A
LASIK surgery.

ideal LASIK candidate is over 18 years of age and has had a stable contact or glasses prescription for at least two years. Patients need to have a cornea that is thick enough to allow the surgeon to safely create a clean corneal flap of appropriate depth. Typically, patients affected by common vision problems (nearsightedness, astigmatism, or farsightedness) will respond well to LASIK. Anyone who suffers from any disease that will decrease their ability to heal properly after surgery is not a viable candidate for LASIK. Candidates should thoroughly discuss the procedure with their eye-care professional before electing to have LASIK. People need to realize that the goal of LASIK is to reduce their dependency on glasses or contact lenses, not completely eliminate them.

Individuals who suffer from cataracts, advanced glaucoma, corneal or other eye diseases are not considered viable candidates for LASIK. Patients who expect LASIK to completely correct their visual problems and make them completely independent of their corrective lenses are not a viable candidate either.

The LASIK Procedure

During the LASIK procedure, a small flap of tissue (the conjunctiva) is cut away from the front of the eye. The flap is folded back exposing your cornea, allowing the surgeon to remove a defined amount of tissue from your cornea. Each pulse of the laser will remove a small amount of corneal tissue, allowing the surgeon to flatten or increase the steepness of the curve of your cornea. After the procedure, the flap of tissue is put back into place and allowed to heal on its own. LASIK patients receive eye drops or medications to help relieve the pain of the procedure. Improvements to your vision will begin as early as the day after the surgery, but typically take two to three months. Most patients will have vision that is close to 20/20, but your chances for improved vision are based in part on how good your eyes were before the surgery.

14.5 Sense of Hearing

The ear has two sensory functions: hearing and balance (equilibrium). The sensory receptors for both of these are located in the inner ear, and each consists of **hair cells** with stereocilia (actually long microvilli) that are sensitive to mechanical stimulation. They are mechanoreceptors.

Anatomy and Physiology of the Ear

Figure 14.13 shows that the ear has three divisions: outer, middle, and inner. The **outer ear** consists of the **pinna** (external flap) and the **auditory canal.** The opening of the auditory canal is lined with fine hairs and sweat glands. Modified sweat glands are located in the upper wall of the canal; they secrete earwax, a substance that helps guard the ear against the entrance of foreign materials, such as air pollutants.

The **middle ear** begins at the **tympanic membrane** (eardrum) and ends at a bony wall containing two small openings covered by membranes. These openings are called the **oval window** and the **round window.** Three small bones are found between the tympanic membrane and the oval window. Collectively called the **ossicles,** individually they are the **malleus** (hammer), the **incus** (anvil), and the **stapes** (stirrup) because their shapes resemble these objects. The malleus adheres to the tympanic membrane, and the stapes touches the oval window. An **auditory tube** (eustachian tube), which extends from the middle ear to the nasopharynx, permits equalization of air pressure. Chewing gum, yawning, and swallowing in elevators and airplanes help move air through the auditory tubes upon ascent and descent. As this occurs, we often hear the ears "pop."

Whereas the outer ear and the middle ear contain air, the inner ear is filled with fluid. Anatomically speaking, the **inner ear** has three areas: The **semicircular canals** and the **vestibule** are both concerned with equilibrium; the **cochlea** is concerned with hearing. The cochlea resembles the shell of a snail because it spirals.

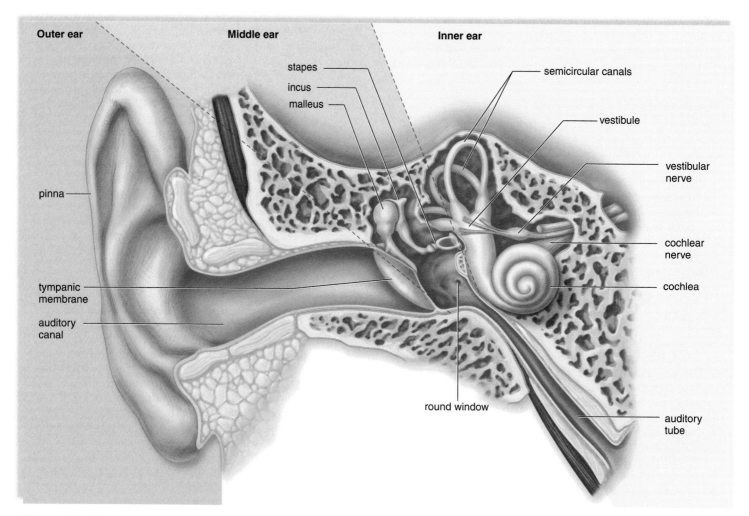

Figure 14.13   Anatomy of the human ear.
In the middle ear, the malleus (hammer), the incus (anvil), and the stapes (stirrup) amplify sound waves. In the inner ear, the mechanoreceptors for equilibrium are in the semicircular canals and the vestibule, and the mechanoreceptors for hearing are in the cochlea.

Auditory Pathway to the Brain

The sound pathway begins with the auditory canal. Thereafter, hearing requires the other parts of the ear, the cochlear nerve, and the brain.

Through the Auditory Canal and Middle Ear   The process of hearing begins when sound waves enter the auditory canal. Just as ripples travel across the surface of a pond, sound waves travel by the successive vibrations of molecules. Ordinarily, sound waves do not carry much energy, but when a large number of waves strike the tympanic membrane, it moves back and forth (vibrates) ever so slightly. The malleus then takes the pressure from the inner surface of the tympanic membrane and passes it, by means of the incus, to the stapes in such a way that the pressure is multiplied about 20 times as it moves. The stapes strikes the membrane of the oval window, causing it to vibrate, and in this way, the pressure is passed to the fluid within the cochlea.

From the Cochlea to the Auditory Cortex   If the cochlea is examined in cross section (Fig. 14.14), you can see that it has three canals. The sense organ for hearing, called the **spiral organ** (organ of Corti), is located in the cochlear canal. The spiral organ consists of little hair cells and a gelatinous material called the **tectorial membrane.** The hair cells sit on the basilar membrane, and their stereocilia are embedded in the tectorial membrane.

When the stapes strikes the membrane of the oval window, pressure waves move from the vestibular canal to the tympanic canal across the basilar membrane. The basilar membrane moves up and down, and the stereocilia of the hair cells embedded in the tectorial membrane bend. Then, nerve impulses begin in the **cochlear nerve** and travel to the brain. When they reach the auditory cortex in the temporal lobe, they are interpreted as a sound.

Each part of the spiral organ is sensitive to different wave frequencies, or pitch. Near the tip, the spiral organ responds to low pitches, such as a tuba, and near the base (beginning), it responds to higher pitches, such as a bell or a whistle. The nerve fibers from each region along the length of the spiral organ lead to slightly different areas in the auditory cortex. The pitch sensation we experience depends upon which region of the basilar membrane vibrates and which area of the auditory cortex is stimulated.

Volume is a function of the amplitude of sound waves. Loud noises cause the fluid within the vestibular canal to exert more pressure and the basilar membrane to vibrate to a greater extent. The resulting increased stimulation is interpreted by the brain as volume. It is believed that the brain interprets the tone of a sound based on the distribution of the hair cells stimulated.

✓ Check Your Progress 14.5

1. What parts in the ear assist in amplifying sound waves?

2. **a.** What receptors allow us to hear? **b.** Where are these receptors located? **c.** How do they function?

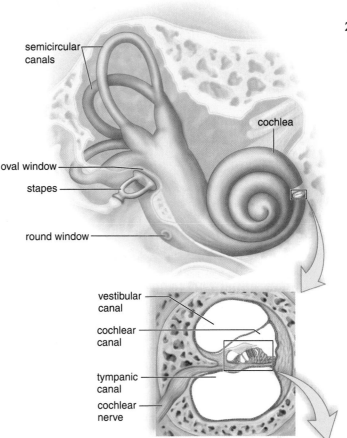

Cochlea cross section

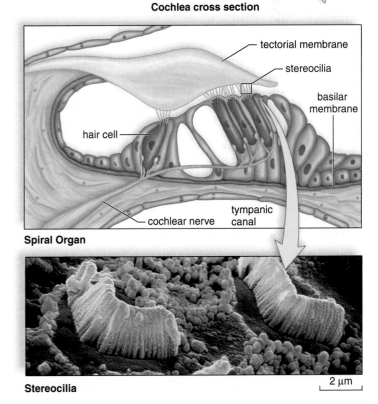

Spiral Organ

Stereocilia

2 μm

Figure 14.14   Mechanoreceptors for hearing.
The spiral organ (organ of Corti) is located within the cochlea. Note that the spiral organ consists of hair cells resting on the basilar membrane, with the tectorial membrane above. Pressure waves moving through the canals cause the basilar membrane to vibrate. This causes the stereocilia embedded in the tectorial membrane to bend. Nerve impulses traveling in the cochlear nerve result in hearing.

Noise Pollution

Though we can sometimes tune its presence out, noise—unwanted sounds—is all around us. Noise pollution is noise from the environment that is annoying, distracting, and potentially harmful. It comes from airplanes, cars, lawn mowers, machinery, our own and our neighbors' stereos. It is present at our workplaces, in public spaces like amusement parks, and at home, and its prevalence allows it to have a potentially high impact on our welfare.

How Does Noise Affect Us?

How does noise affect human health? Perhaps the greatest worry about noise pollution is that exposure to loud (>85 decibels) or chronic noises can damage cells of the inner ear and cause hearing loss (Fig. 14B). When we are young, we often do not consider the damage that noise may be doing to our spiral organ. For example, Ian and his friends were having too much fun at the amusement park to consider that the noise of the bell ringing when someone had won a prize could possibly damage their hearing. The stimulation of loud music is often sought by young people at rock concerts without regard to the possibility that their hearing may be diminished as a result. Over the years, loud noises can bring deafness and accompanying depression when we are seniors.

Noise can affect well-being by other means, too. Data from studies of environmental noise can be difficult to interpret because of the presence of other confounding factors, including physical or chemical pollution. The tolerance level for noise also varies from person to person. Nonetheless, laboratory and field studies show that noise may be detrimental in nonauditory ways. Its effects on mental health include annoyance, inability to concentrate, and increased irritability. Long-term noise exposure from air or car traffic may impair cognitive ability, language learning, and memory in children. Noise is often the grounds for losses in sleep and productivity and can induce stress. Additionally, several studies have suggested that it negatively affects cardiovascular health, though more research needs to be done to firm up this link.

Federal and Local Control

Noise pollution has been a concern for several decades. In 1972, the Noise Control Act was passed as a means for coordinating federal noise control and research and to develop noise emission standards. The aim was to protect Americans from "noise that jeopardizes their health or welfare." The Environmental Protection Agency (EPA) had federal authority to regulate noise pollution, and their Office of Noise Abatement and Control (ONAC) worked on establishing noise guidelines. However, funding for ONAC was cut off in 1981. Today, there is no national noise policy, though several federal governmental departments have some oversight.

Workplace noise exposure is controlled by the Occupational Safety and Health Administration (OSHA). They have set guidelines for workplace noise and require that protective gear be provided if sound levels exceed certain values. This may include noise-reducing ear muffs and other equipment for people who work around big equipment. However, OSHA guidelines don't cover things like telephone ringing and computer or typewriter noise that may be present in a nonindustrial environment such as an open-plan office. Aviation noise and traffic noise reduction plans are overseen by the Department of Transportation, the Federal Aviation Administration (FAA), and the Federal Highway Administration (FHA), respectively.

Some cities and local governments have taken steps to decrease unwanted or annoying noise. Several have noise ordinances and control noise levels during particular times of day. Persons throwing loud house parties might get a knock on their door from the police, and owners of barking dogs, a visit from animal control. Business enterprises may also be regulated, and concerts at outdoor venues may be required to finish by a certain time in the evening, for example. There is often opposition to this type of control, though, on the basis that it impedes commerce.

Many people believe that much more should be done to curb noise pollution. There are difficulties in regulating noise pollution generated by private persons or business owners, however. Individuals often feel that they should be free to do as they wish on private property, including playing music at full volume or running machinery. For example, club and café owners may believe that their livelihood depends on having reasonably loud sound systems; the same may hold true of sports fields and stadiums. There are also challenges to regulating airport and traffic noise. The main responsibilities of the regulating agencies (FAA and FHA)—promoting aviation industry growth and maintaining highway infrastructure—can be at odds with controlling noise for the public's sake and, therefore, noise abatement isn't necessarily their top priority.

Finding neighbor- and commerce-friendly solutions to noise problems will likely require individual citizens, communities, and governments to work together. Stopping the source of noise may not be the only answer to this issue. The implementation of new technology could also make things quieter by preventing sounds from traveling too far away from their sources.

Decide Your Opinion

1. Do you think that more should be done to curb noise pollution in your neighborhood? If so, why?
2. Do you believe that the government has the right to regulate what someone does on their own property? Why or why not?
3. Should regulation of airport and traffic noise fall to the Department of Transportation? If not, who may better control airport or highway noise?
4. If technology is available to reduce noise on highways, who do you think should pay for implementing it—tax payers, car owners, or others?

Table 14A	Noises That Affect Hearing	
Type of Noise	**Sound Level (decibels)**	**Effect**
"Boom car," jet engine, shotgun, rock concert	Over 125	Beyond threshold of pain; potential for hearing loss high
Nightclub, "boom box," thunderclap	Over 120	Hearing loss likely
Chain saw, pneumatic drill, jackhammer, symphony orchestra, snowmobile, garbage truck, cement mixer	100–200	Regular exposure of more than 1 min risks permanent hearing loss
Farm tractor, newspaper press, subway, motorcycle	90–100	Fifteen minutes of unprotected exposure potentially harmful
Lawn mower, food blender	85–90	Continuous daily exposure for more than 8 hrs can cause hearing damage
Diesel truck, average city traffic noise	80–85	Annoying; constant exposure may cause hearing damage

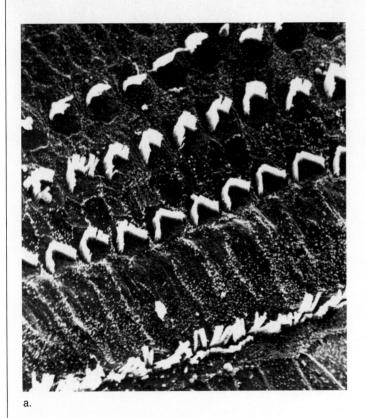

a.

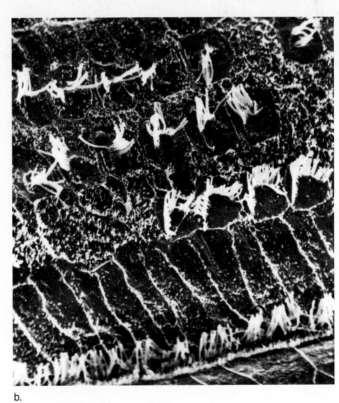

b.

Figure 14B   Hair cell damage.
a. Normal hair cells in the spiral organ of a guinea pig. **b.** Damaged cells. This damage occurred after 24-hour exposure to a noise level equivalent to that at a heavy-metal rock concert (see Table 14A). Hearing is permanently impaired because lost cells will not be replaced, and damaged cells may also die.

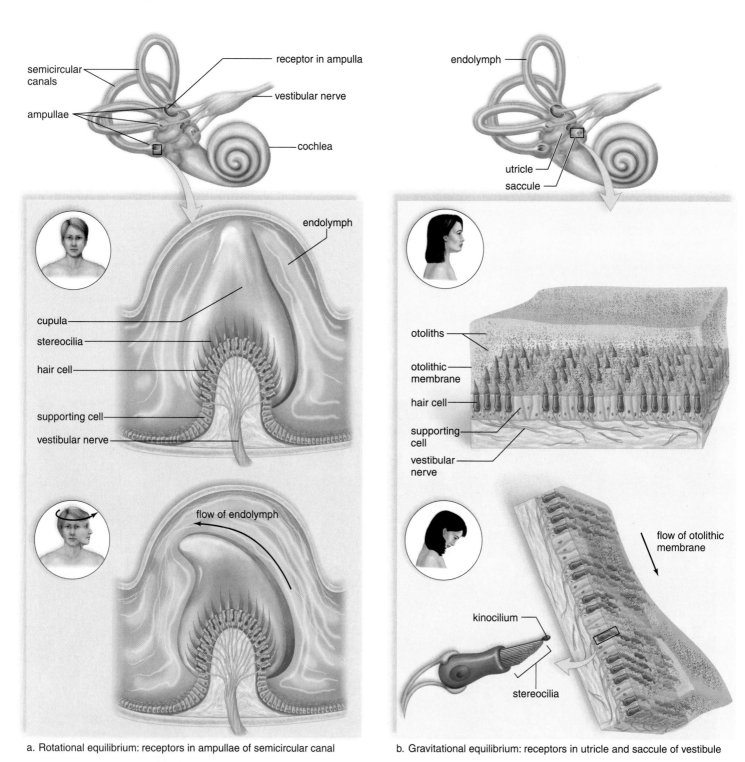

a. Rotational equilibrium: receptors in ampullae of semicircular canal

b. Gravitational equilibrium: receptors in utricle and saccule of vestibule

Figure 14.15   The mechanoreceptors for equilibrium.

a. Rotational equilibrium. The ampullae of the semicircular canals contain hair cells with stereocilia embedded in a cupula. When the head rotates, the cupula is displaced, bending the stereocilia. Thereafter, nerve impulses travel in the vestibular nerve to the brain. **b.** Gravitational equilibrium. The utricle and the saccule contain hair cells with stereocilia embedded in an otolithic membrane. When the head bends, otoliths are displaced, causing the membrane to sag and the stereocilia to bend. If the stereocilia bend toward the kinocilium, the longest of the stereocilia, nerve impulses increase in the vestibular nerve. If the stereocilia bend away from the kinocilium, nerve impulses decrease in the vestibular nerve. This difference tells the brain, by way of the vestibular nerve, in which direction the head moved.

14.6 Sense of Equilibrium

The vestibular nerve, which arises in the semicircular canals, saccule, and utricle, takes nerve impulses to the brain stem and cerebellum (Fig. 14.15). Through its communication with the brain, the vestibular nerve helps us achieve equilibrium, but other structures in the body are also involved. For example, we already mentioned that proprioceptors are necessary for maintaining our equilibrium. Vision, if available, usually provides extremely helpful input the brain can act upon. Not so for Ian and his friends after experiencing the "Twisted Tornado" ride at the amusement park. To explain, let's take a look at the two sets of mechanoreceptors for equilibrium.

Rotational Equilibrium Pathway

Mechanoreceptors in the **semicircular canals** detect rotational and/or angular movement of the head **(rotational equilibrium)** (Fig. 14.15*a*). Notice that the three semicircular canals are arranged so that there is one in each dimension of space. The base of each of the three canals, called the **ampulla,** is slightly enlarged. Little hair cells, whose stereocilia are embedded within a gelatinous material called a cupula, are found within the ampullae. Because of the way the semicircular canals are arranged, each ampulla responds to head rotation in a different plane of space. As fluid within a semicircular canal flows over and displaces a cupula, the stereocilia of the hair cells bend, and the pattern of impulses carried by the vestibular nerve to the brain changes. The brain uses information from the hair cells within ampulla of the semicircular canals to maintain equilibrium through appropriate motor output to various skeletal muscles that can right our present position in space as need be.

What made Ian and his friends dizzy? When we spin, the cupula slowly begins to move in the same direction we are spinning, and bending of the stereocilia causes hair cells to send messages to the brain. As time goes by, the cupula catches up to the rate we are spinning, and the hair cells no longer send messages to the brain. When we stop spinning, the slow-moving cupula continues to move in the direction of the spin and the stereocilia bend again, indicating that we are moving. Yet, the eyes know we have stopped. The mixed messages sent to the brain cause us to feel dizzy (Fig. 14.6).

Gravitational Equilibrium Pathway

The mechanoreceptors in the utricle and saccule detect movement of the head in the vertical or horizontal planes **(gravitational equilibrium).** The **utricle** and **saccule** are two membranous sacs located in the inner ear near the semicircular canals. Both of these sacs contain little hair cells, whose stereocilia are embedded within a gelatinous material called an otolithic membrane (Fig. 14.15*b*). Calcium carbonate ($CaCO_3$) granules, or **otoliths,** rest on this membrane. The utricle is especially sensitive to horizontal (back-forth) move-

Figure 14.16   Testing balance.
Using playground equipment, children love to test their rotational equilibrium. Adults are not too interested.

ments and the bending of the head, while the saccule responds best to vertical (up-down) movements.

When the body is still, the otoliths in the utricle and saccule rest on the otolithic membrane above the hair cells. When the head bends or the body moves in the horizontal and vertical planes, the otoliths are displaced and the otolithic membrane sags, bending the stereocilia of the hair cells beneath. If the stereocilia move toward the largest stereocilium, called the kinocilium, nerve impulses increase in the vestibular nerve. If the stereocilia move away from the kinocilium, nerve impulses decrease in the vestibular nerve. The frequency of nerve impulses in the vestibular nerve indicates whether you are moving up or down.. These data reach the brain, which uses them to determine the direction of the movement of the head at the moment. The brain uses this information to maintain gravitational equilibrium through appropriate motor output to various skeletal muscles that can right our present position in space as need be. Continuous stimulation of the stereocilia, as when the boys went on the "Twisted Tornado" ride, can contribute to motion sickness, especially when messages reaching the brain conflict with visual information from the eyes. The tendency toward motion sickness can be tested in the laboratory.

✅ Check Your Progress 14.6

1. a. Where are the receptors that detect rotation located?
   b. How do they function?

2. a. Where are the receptors that detect gravity located?
   b. How do they function?

Summarizing the Concepts

14.1 Sensory Receptors and Sensations

There are four types of sensory receptors: chemoreceptors; photoreceptors; mechanoreceptors; and thermoreceptors.

- Sensory receptors initiate nerve impulses that are transmitted to the spinal cord and/or brain.
- Sensation occurs when nerve impulses reach the cerebral cortex.
- Perception is an interpretation of sensations.

14.2 Proprioceptors and Cutaneous Receptors

Proprioceptors
- are mechanoreceptors involved in reflex actions.
- help maintain equilibrium and posture.

Cutaneous Receptors
- are found in the skin.
- are for touch, pressure, temperature, and pain.

14.3 Senses of Taste and Smell

Taste and smell are due to chemoreceptors that are stimulated by molecules in the environment.

Sense of Taste   Microvilli of taste cells have receptor proteins for molecules that cause the brain to distinguish sweet, sour, salty, and bitter tastes.

Sense of Smell   The cilia of olfactory cells have receptor proteins for molecules that cause the brain to distinguish odors.

14.4 Sense of Vision

Vision depends on the eye, the optic nerves, and the visual areas of the cerebral cortex.

Anatomy and Physiology of the Eye
The eye has three layers:

- The sclera (outer layer) protects and supports the eyeball.
- The choroid (middle, pigmented layer) absorbs stray light rays.
- The retina (inner layer) contains the rod cells (sensory receptors for dim light) and cone cells (sensory receptors for bright light and color).

Function of the Lens   The lens (assisted by the cornea and the humors) brings the light rays to focus on the retina. To see a close object, visual accommodation occurs as the lens rounds up.

Visual Pathway to the Brain   The visual pathway begins when light strikes photoreceptors (rod cells and cone cells) in the retina. The optic nerves carry nerve impulses from the eyes to the optic chiasma, then pass through the thalamus before reaching the primary vision area in the occipital lobe of the brain.

Abnormalities of the Eye
- color blindness
- misshapen eyeballs (cause of nearsightedness, farsightedness, or astigmatism)

14.5 Sense of Hearing

Hearing depends on the ear, the cochlear nerve, and the auditory areas of the cerebral cortex.

Anatomy and Physiology of the Ear
The ear has three parts:

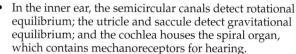

- In the outer ear, the pinna and the auditory canal direct sound waves to the middle ear.
- In the middle ear, the tympanic membrane and the ossicles (malleus, incus, and stapes) amplify sound waves.
- In the inner ear, the semicircular canals detect rotational equilibrium; the utricle and saccule detect gravitational equilibrium; and the cochlea houses the spiral organ, which contains mechanoreceptors for hearing.

Auditory Pathway to the Brain   The auditory pathway begins when the outer ear receives and the middle ear amplifies sound waves that then strike the oval window membrane.

- The mechanoreceptors for hearing are hair cells on the basilar membrane of the spiral organ.
- Nerve impulses begin in the cochlear nerve and are carried to the primary auditory area in the temporal lobe of the cerebral cortex.

14.6 Sense of Equilibrium

The ear also contains mechanoreceptors for equilibrium.

Rotational Equilibrium Pathway
- Mechanoreceptors (hair cells) in the semicircular canals detect rotational and/or angular movement of the head.

Gravitational Equilibrium Pathway
- Mechanoreceptors (hair cells) in the utricle and saccule detect head movement in the vertical or horizontal planes.

Understanding Key Terms

ampulla 291	hair cell 286
aqueous humor 280	incus 286
astigmatism 284	inner ear 286
auditory canal 286	integration 275
auditory tube 286	interoceptor 274
blind spot 283	iris 280
chemoreceptor 274	lens 280
choroid 280	malleus 286
ciliary body 280	mechanoreceptor 274
cochlea 286	middle ear 286
cochlear nerve 287	nearsighted 284
color vision 282	olfactory cell 279
cone cell 282	optic chiasma 283
cornea 280	optic nerve 281
cutaneous receptor 276	ossicle 286
exteroceptor 274	otolith 291
farsighted 284	outer ear 286
focus 281	oval window 286
fovea centralis 281	pain receptor 274
glaucoma 280	photoreceptor 274
gravitational equilibrium 291	pinna 286

proprioceptor 276
pupil 280
referred pain 277
retina 281
retinal 282
rhodopsin 282
rod cell 282
rotational equilibrium 291
round window 286
saccule 291
sclera 280
semicircular canal 286, 291
sensation 275

sensory adaptation 275
sensory receptor 274
spiral organ 287
stapes 286
stimulus 274
taste bud 278
tectorial membrane 287
thermoreceptor 274
tympanic membrane 286
utricle 291
vestibule 286
visual accommodation 281
vitreous humor 281

Match the key terms to these definitions.

a. _____ Structure that receives sensory stimuli and is a part of a sensory neuron or transmits signals to a sensory neuron.

b. _____ Inner layer of the eyeball containing the photoreceptors—rod cells and cone cells.

c. _____ Outer, white, fibrous layer of the eye that surrounds the eye except for the transparent cornea.

d. _____ Receptor that is sensitive to chemical stimulation—for example, receptors for taste and smell.

e. _____ Specialized region of the cochlea containing the hair cells for sound detection and discrimination.

Testing Your Knowledge of the Concepts

1. Contrast exteroceptors and interoceptors. (page 274)

2. What is sensory adaptation? (page 275)

3. What is proprioception, and what is the role of muscle spindles? (page 276)

4. List the cutaneous receptors and the type of stimulus each responds to. (pages 276–77)

5. Describe the structure of a taste bud and explain how a taste cell functions. (page 278)

6. Describe the structure and function of the olfactory epithelium. How does the sense of smell come about? (page 279)

7. Describe the anatomy of the eye and the function of each part. (pages 280–81)

8. How does the eye respond when viewing an object far away? When viewing a close object? (page 281)

9. Describe the sequence of events after rhodopsin absorbs light. (page 282)

10. What are three abnormalities caused by a misshapened eyeball, and what type lens can be helpful? (page 284)

11. Describe the anatomy of the ear and the function of each part. (page 286)

12. Explain the pathway of sound and how sound is produced. (pages 286–87)

13. What structures are involved in equilibrium, their structures, and functions? (pages 290–91)

14. A sensory receptor
  a. is the first portion of a reflex arc.
  b. initiates nerve impulses.
  c. can be internal or external.
  d. All of these are correct.

15. Receptors that are sensitive to changes in blood pressure are
  a. interoceptors.   c. proprioceptors.
  b. exteroceptors.   d. nociceptors.

16. Conscious interpretation of changes in the internal and external environment is called
  a. responsiveness.   c. sensation.
  b. perception.   d. accommodation.

17. Which of these is an incorrect difference between proprioceptors and cutaneous receptors?

Proprioceptors	**Cutaneous Receptors**
a. located in muscles and tendons	located in the skin
b. chemoreceptors	mechanoreceptors
c. respond to tension	respond to pain, hot, cold, touch, pressure
d. interoceptors	exteroceptors

  e. All of these contrasts are correct.

18. Pain perceived as coming from another location is known as
  a. intercepted pain.   c. referred pain.
  b. phantom pain.   d. parietal pain.

19. Tasting something "sweet" versus "salty" is a result of
  a. activating different sensory receptors.
  b. activating many versus few sensory receptors.
  c. activating no sensory receptors.
  d. None of these are correct.

20. Which structure of the eye is incorrectly matched with its function?
  a. lens—focusing
  b. cones—color vision
  c. iris—regulation of amount of light
  d. choroid—location of cones
  e. sclera—protection

21. Which of the following gives the correct path for light rays entering the human eye?
  a. sclera, retina, choroid, lens, cornea
  b. fovea centralis, pupil, aqueous humor, lens
  c. cornea, pupil, lens, vitreous humor, retina
  d. cornea, fovea centralis, lens, choroid, rods
  e. optic nerve, sclera, choroid, retina, humors

22. The thin, darkly pigmented layer that underlies most of the sclera is
  a. the conjunctiva.   c. the retina.
  b. the cornea.   d. the choroid.

23. Adjustment of the lens to focus on objects close to the viewer is called
  a. convergence.   c. focusing.
  b. visual accommodation.   d. constriction.

24. Retinal is
  a. a derivative of vitamin A.
  b. sensitive to light energy.
  c. a part of rhodopsin.
  d. found in both rods and cones.
  e. All of these are correct.

25. In order to focus on objects that are close to the viewer,
    a. the suspensory ligaments must be pulled tight.
    b. the lens needs to become more rounded.
    c. the ciliary muscle will be relaxed.
    d. the image must focus on the area of the optic nerve.

26. Which abnormality of the eye is incorrectly matched with its cause?
    a. astigmatism—either the lens or cornea is not even
    b. farsightedness—eyeball is shorter than usual
    c. nearsightedness—image focuses behind the retina
    d. color blindness—genetic disorder in which certain types of cones may be missing

27. Which of these associations is incorrectly matched?
    a. semicircular canals—inner ear
    b. utricle and saccule—outer ear
    c. auditory canal—outer ear
    d. cochlea—inner ear
    e. ossicles—middle ear

28. Which of the following wouldn't you mention if you were tracing the path of sound vibrations?
    a. auditory canal   d. semicircular canals
    b. tympanic membrane   e. cochlea
    c. ossicles

29. The middle ear is separated from the inner ear by
    a. the oval window.   c. the round window.
    b. the tympanic membrane. d. Both a and c are correct.

30. Which one of these correctly describes the location of the spiral organ?
    a. between the tympanic membrane and the oval window in the inner ear
    b. in the utricle and saccule within the vestibule
    c. between the tectorial membrane and the basilar membrane in the cochlear canal
    d. between the nasal cavities and the throat
    e. between the outer and inner ear within the semicircular canals

31. Which of the following could result in hearing loss?
    a. symphony orchestra
    b. earphone use
    c. consistent use of loud equipment such as a jackhammer
    d. use of firearms
    e. All of these are correct.

32. Which of the following structures would allow you to know that you were upside down, even if you were in total darkness?
    a. utricle and saccule   c. semicircular canals
    b. cochlea   d. tectorial membrane

33. Which of these is an incorrect difference between olfactory receptors and equilibrium receptors?

Olfactory Receptors	Equilibrium Receptors
a. located in nasal cavities	located in the inner ear
b. chemoreceptors	mechanoreceptors
c. respond to molecules in the air	respond to movements of the body
d. communicate with brain via a tract	communicate with brain via vestibular nerve
e. All of these contrasts are correct.	

34. Both olfactory receptors and sound receptors
    a. are chemoreceptors.   d. initiate nerve impulses.
    b. are a part of the brain. e. All of these are correct.
    c. are mechanoreceptors.

35. Label this diagram of a human eye.

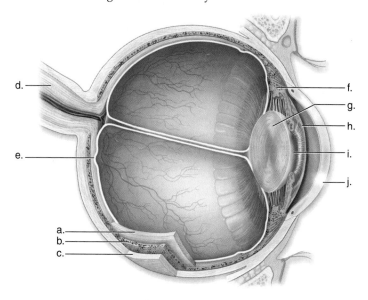

Thinking Critically About the Concepts

Ian, from the opening story, and his friends' trip to the amusement park focused primarily on the role of sensory receptors in gathering information about external surroundings. Keep in mind the hugely important role of the interoreceptors in homeostasis. When Olivia (from the opening story in Chapter 9) held her breath during her temper tantrum, her blood pH dropped (became more acidic) when blood CO_2 increased. Chemoreceptors in her blood vessels then signaled her brain about the change in pH. Olivia eventually had to exhale when her blood pH fell out of the acceptable range for pH. Any unacceptable changes in homeostatic conditions is reported to the brain by some kind of interoreceptor.

1. What kind of receptors would have recognized the changes in blood pressure stimulated by Ashley's (opening story from Chapter 1) sympathetic nervous system response to the scary movie?

2. Besides the blood pH mentioned above, what other homeostasis conditions will be monitored by chemoreceptors?

3. Ian and his friends were exposed to a great deal of environmental stimuli during their day at the amusement park.

a. What different kinds of receptors were responsible for their ability to sense the wide variety of stimuli?

b. What different parts of the brain would have interpreted the information about the variety of environmental stimuli at the amusement park?

4. Amusement-park workers are likely to be exposed to continuous, loud noises. What would you predict the long-term effect on their hearing to be? Why?

CHAPTER 5

Endocrine System

A t 4' 11", Adam Casey was the shortest boy in eighth grade at his school. The growth spurt most boys his age experience never happened to him. Classmates, who towered over him, called him "shrimp" and teased him about being a midget or dwarf. His younger brother was already 2 in. taller than Adam. Adam's parents worried about how their son would be treated if he failed to reach what is considered to be a respectable height for a man.

The Caseys decided to investigate how growth hormone (GH) treatment might help Adam gain some additional inches. They made an appointment to see Dr. Drake, an endocrinologist, who could advise them of their options. Dr. Drake told them that Adam might be a candidate for GH therapy if the epiphyseal plates of his long bones had not yet fused. Blood work to assess Adam's GH levels would help determine the course of action as well.

X-rays of Adam's long bones showed unfused epiphyseal plates, and his blood work results indicated lower than normal levels of GH. Dr. Drake seemed to think GH injections could help Adam gain a significant number of inches. Some of Dr. Drake's other patients had grown 5–7 in. following two years of injections. Dr. Drake cautioned them that the injections might cause headaches and nausea and that long-term risks were unknown. The injections would be very expensive ($30,000 per year), and health insurance may or may not cover the cost. The Caseys decided to proceed with the injections in the hopes that any increase in height would improve Adam's self-esteem and potential success as an adult.

C H A P T E R   C O N C E P T S

15.1 Endocrine Glands
Endocrine glands produce hormones that are secreted into the bloodstream and distributed to target cells where they alter cellular metabolism.

15.2 Hypothalamus and Pituitary Gland
The hypothalamus controls the secretions of the pituitary gland, which, in turn, controls the secretion of certain other glands. Growth hormone produced by the anterior pituitary determines our height.

15.3 Thyroid and Parathyroid Glands
Some hormones secreted by the thyroid gland stimulate cellular metabolism, and another helps control Ca^{2+} blood levels, as does parathyroid hormone.

15.4 Adrenal Glands
The adrenal glands respond to both long-term and short-term stress.

15.5 Pancreas
The pancreas secretes insulin and glucagon, which together keep the blood glucose level fairly constant.

15.6 Other Endocrine Glands
Other endocrine glands include the testes and ovaries, which are discussed in Chapter 16, the thymus gland, which was discussed in Chapter 7, and the pineal gland.

15.7 Homeostasis
The nervous and endocrine systems work together to control and regulate the other systems.

15.1 Endocrine Glands

The nervous system and the endocrine system work together to regulate the activities of the other systems. Both systems use chemical signals when they respond to changes that might threaten homeostasis, but they have different means of delivering these signals (Fig. 15.1). As discussed, the nervous system is composed of neurons. In this system, sensory receptors (specialized dendrites) detect changes in the internal and external environment. The CNS integrates the information and responds by stimulating muscles and glands. Communication depends on nerve impulses, conducted in axons, and neurotransmitters, which cross synapses. Axon conduction occurs rapidly and so does diffusion of a neurotransmitter across the short distance of a synapse. In other words, the nervous system is organized to respond rapidly to stimuli. This is particularly useful if the stimulus is an external event that endangers our safety—we can move quickly to avoid being hurt.

The endocrine system functions differently. The endocrine system is largely composed of glands (Fig. 15.2). These glands secrete **hormones,** which are carried by the bloodstream to target cells throughout the body. In the opening story, Adam was short for a boy because his blood didn't contain enough growth hormone produced by the anterior pituitary. Growth hormone stimulates the growth of long bones,

and in this way affects the height of an individual. It takes time to deliver hormones, and it takes time for cells to respond, but the effect is longer lasting. In other words, the endocrine system is organized for a slow but prolonged response.

Endocrine glands can be contrasted with exocrine glands. Exocrine glands have ducts and secrete their products into these ducts, which take them to the lumens of other organs or outside the body. For example, the salivary glands send saliva into the mouth by way of the salivary ducts. **Endocrine glands,** as stated, secrete their products into the bloodstream, which delivers them throughout the body. It must be stressed that only certain cells, called target cells, can respond to certain hormones. If a cell can respond to a hormone, the hormone and receptor proteins bind together as a key fits a lock.

It is of interest to note that both the nervous system and the endocrine system make use of negative feedback mechanisms. If the blood pressure falls, sensory receptors signal a control center in the brain. This center sends out nerve impulses to the arterial walls so that they constrict, and blood pressure rises. Now the sensory receptors are no longer stimulated, and the feedback mechanism is inactivated. Similarly, a rise in blood glucose level causes the pancreas to release insulin, which, in turn, promotes glucose uptake by the liver, muscles, and other cells of the body (Fig. 15.1). When the blood glucose level falls, the pancreas no longer secretes insulin.

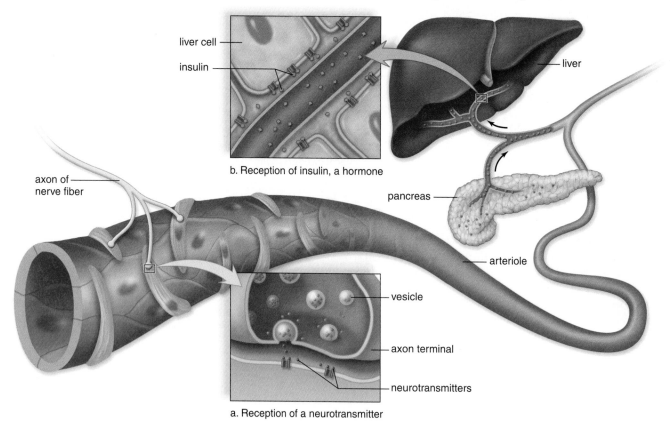

liver cell

insulin

liver

b. Reception of insulin, a hormone

axon of nerve fiber

pancreas

arteriole

vesicle

axon terminal

neurotransmitters

a. Reception of a neurotransmitter

Figure 15.1   Modes of action of the nervous and endocrine systems.
a. Nerve impulses passing along an axon cause the release of a neurotransmitter. The neurotransmitter, a chemical signal, causes the wall of an arteriole to constrict.
b. The hormone insulin, a chemical signal, travels in the cardiovascular system from the pancreas to the liver, where it causes liver cells to store glucose as glycogen.

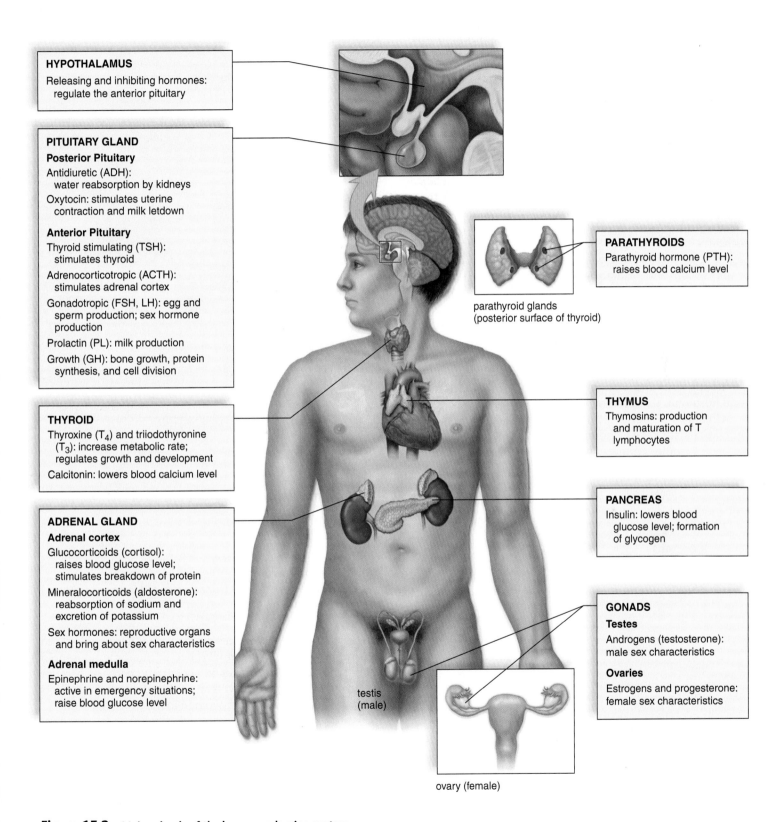

HYPOTHALAMUS

Releasing and inhibiting hormones: regulate the anterior pituitary

PITUITARY GLAND

Posterior Pituitary

Antidiuretic (ADH): water reabsorption by kidneys

Oxytocin: stimulates uterine contraction and milk letdown

Anterior Pituitary

Thyroid stimulating (TSH): stimulates thyroid

Adrenocorticotropic (ACTH): stimulates adrenal cortex

Gonadotropic (FSH, LH): egg and sperm production; sex hormone production

Prolactin (PL): milk production

Growth (GH): bone growth, protein synthesis, and cell division

THYROID

Thyroxine (T_4) and triiodothyronine (T_3): increase metabolic rate; regulates growth and development

Calcitonin: lowers blood calcium level

ADRENAL GLAND

Adrenal cortex

Glucocorticoids (cortisol): raises blood glucose level; stimulates breakdown of protein

Mineralocorticoids (aldosterone): reabsorption of sodium and excretion of potassium

Sex hormones: reproductive organs and bring about sex characteristics

Adrenal medulla

Epinephrine and norepinephrine: active in emergency situations; raise blood glucose level

PARATHYROIDS

Parathyroid hormone (PTH): raises blood calcium level

parathyroid glands
(posterior surface of thyroid)

THYMUS

Thymosins: production and maturation of T lymphocytes

PANCREAS

Insulin: lowers blood glucose level; formation of glycogen

GONADS

Testes

Androgens (testosterone): male sex characteristics

Ovaries

Estrogens and progesterone: female sex characteristics

testis
(male)

ovary (female)

Figure 15.2   Major glands of the human endocrine system.

Major glands and the hormones they produce are depicted. Also, the endocrine system includes other organs such as the kidneys, gastrointestinal tract, and the heart, which also produce hormones but not as a primary function of these organs.

Hormones Are Chemical Signals

Like other **chemical signals,** hormones are a means of communication between cells, between body parts, and even between individuals. They typically affect the metabolism of cells that have receptors to receive them (Fig. 15.3). In a condition called androgen insensitivity, an individual has X and Y sex chromosomes, and the testes, which remain in the abdominal cavity, produce the sex hormone testosterone. However, the body cells lack receptors that are able to combine with testosterone, and the individual appears to be a normal female.

Like testosterone, most hormones act at a distance between body parts. They travel in the bloodstream from the gland that produced them to their target cells. Also counted as hormones are the secretions produced by neurosecretory cells in the hypothalamus, a part of the brain. They travel in the capillary network that runs between the hypothalamus and the pituitary gland. Some of these secretions stimulate the pituitary to secrete its hormones, and others prevent it from doing so.

Not all hormones act between body parts. As we shall see, prostaglandins are a good example of a *local hormone.* After prostaglandins are produced, they are not carried in the bloodstream; instead, they affect neighboring cells, sometimes promoting pain and inflammation. Also, growth factors are local hormones that promote cell division and mitosis.

Chemical signals that influence the behavior of other individuals are called **pheromones.** Pheromones have been released by other animals and a researcher has isolated one

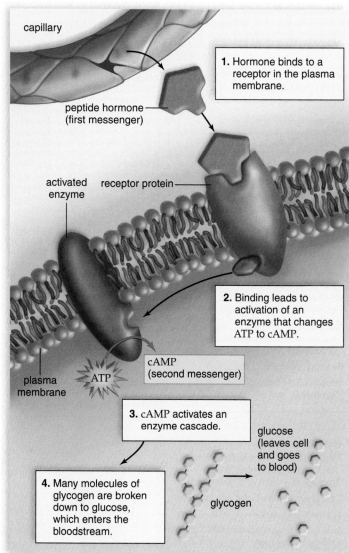

Figure 15.4   Peptide hormone.
A peptide hormone (first messenger) binds to a receptor in the plasma membrane. Thereafter, cyclic AMP (second messenger) forms and activates an enzyme cascade.

released by men that reduces premenstrual nervousness and tension in women. Women who live in the same household often have menstrual cycles in synchrony, perhaps because the armpit secretions of a woman who is menstruating affects the menstrual cycle of other women in the household.

The Action of Hormones

Hormones have a wide range of effects on cells. Some of these effects induce a target cell to increase its uptake of particular substances, such as glucose, or ions, such as calcium. Some bring about an alteration of the target cell's structure in some way. Some simply influence cell metabolism. Growth hormone is a peptide hormone that influences cell metabolism

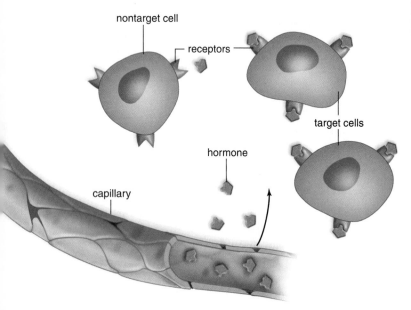

Figure 15.3   Target cell concept.
Most hormones are distributed by the bloodstream to target cells. Target cells have receptors for the hormone, and the hormone combines with the receptor as a key fits a lock.

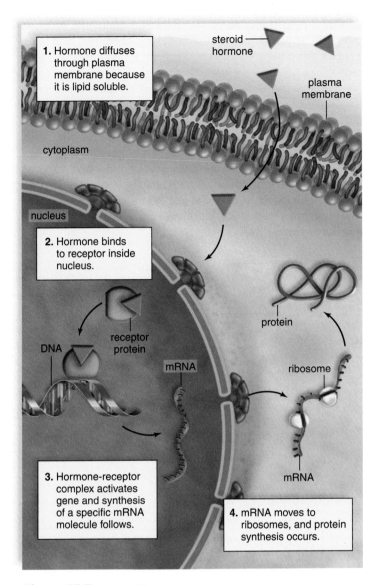

1. Hormone diffuses through plasma membrane because it is lipid soluble.

steroid hormone

plasma membrane

cytoplasm

nucleus

2. Hormone binds to receptor inside nucleus.

DNA

receptor protein

protein

mRNA

ribosome

3. Hormone-receptor complex activates gene and synthesis of a specific mRNA molecule follows.

mRNA

4. mRNA moves to ribosomes, and protein synthesis occurs.

Figure 15.5 Steroid hormone.
A steroid hormone passes directly through the target cell's plasma membrane before binding to a receptor in the nucleus or cytoplasm. The hormone-receptor complex binds to DNA, and gene expression follows.

leading to a change in the structure of bone. This hormone, mentioned in the opening story, is a **peptide hormone.** The term peptide hormone is used to include hormones that are peptides, proteins, glycoproteins, and modified amino acids. Growth hormone is a protein produced and secreted by the anterior pituitary. **Steroid hormones** have the same complex of four carbon rings because they are all derived from cholesterol (see Fig. 2.17).

The Action of Peptide Hormones Most hormonal glands secrete peptide hormones. The actions of peptide hormones can vary, and we will concentrate on what happens in muscle cells after the hormone epinephrine binds to a receptor in the plasma membrane (Fig. 15.4). In muscle cells, the re-

ception of epinephrine leads to the breakdown of glycogen to glucose, which provides energy for ATP production. The immediate result of binding is the formation of **cyclic adenosine monophosphate (cAMP).** Cyclic AMP contains one phosphate group attached to adenosine at two locations. Therefore, the molecule is cyclic. Cyclic AMP activates a protein kinase enzyme in the cell, and this enzyme, in turn, activates another enzyme, and so forth. The series of enzymatic reactions that follows cAMP formation is called an enzyme cascade. Because each enzyme can be used over and over again at every step of the cascade, more enzymes are involved. Finally, many molecules of glycogen are broken down to glucose, which enters the bloodstream.

Typical of a peptide hormone, epinephrine never enters the cell. Therefore, the hormone is called the **first messenger,** while cAMP, which sets the metabolic machinery in motion, is called the **second messenger.** To explain this terminology, let's imagine that the adrenal medulla, which produces epinephrine, is like the home office that sends out a courier (i.e., the hormone epinephrine is the first messenger) to a factory (the cell). The courier doesn't have a pass to enter the factory, so when he arrives at the factory, he tells a foreman through the screen door that the home office wants the factory to produce a particular product. The foreman (i.e., cAMP, the second messenger) walks over and flips a switch that starts the machinery (the enzymatic pathway), and a product is made.

The Action of Steroid Hormones Only the adrenal cortex, the ovaries, and the testes produce steroid hormones. Thyroid hormones act similarly to steroid hormones, even though they have a different structure. Steroid hormones do not bind to plasma membrane receptors, and instead they are able to enter the cell because they are lipids (Fig. 15.5). Once inside, a steroid hormone binds to a receptor, usually in the nucleus but sometimes in the cytoplasm. Inside the nucleus, the hormone-receptor complex binds with DNA and activates certain genes. Messenger RNA (mRNA) moves to the ribosomes in the cytoplasm and protein (e.g. enzyme) synthesis follows. To continue our analogy, a steroid hormone is like a courier that has a pass to enter the factory (the cell). Once inside, it makes contact with the plant manager (DNA) who sees to it that the factory (cell) is ready to produce a product.

Steroids act more slowly than peptides because it takes more time to synthesize new proteins than to activate enzymes already present in cells. Their action lasts longer, however.

✓ Check Your Progress 15.1

1. How can the nervous system be contrasted with the endocrine system?

2. What is a hormone?

3. How do hormones affect the metabolism of cells?

15.2 Hypothalamus and Pituitary Gland

The **hypothalamus** regulates the internal environment through the autonomic system. For example, it helps control heartbeat, body temperature, and water balance. The hypothalamus also controls the glandular secretions of the **pituitary gland.** The pituitary, a small gland about 1 cm in diameter, is connected to the hypothalamus by a stalklike structure. The pituitary has two portions: the posterior and the anterior pituitary.

Posterior Pituitary

Neurons in the hypothalamus called neurosecretory cells produce the hormones **antidiuretic hormone (ADH)** and oxytocin (Fig. 15.6, *left*). These hormones pass through axons into the **posterior pituitary** where they are stored in axon endings. Certain neurons in the hypothalamus are sensitive to the water-salt balance of the blood. When these cells determine that the blood is too concentrated, ADH is released from the posterior pituitary. Upon reaching the kidneys, ADH causes more water to be reabsorbed into kidney capillaries. As the blood becomes dilute, ADH is no longer released. This is an example of control by negative feedback because the effect of the hormone (to dilute blood) acts to shut down the release of the hormone. Negative feedback maintains stable conditions and homeostasis.

Inability to produce ADH causes diabetes insipidus (watery urine), in which a person produces copious amounts of urine with a resultant loss of ions from the blood. The condition can be corrected by the administration of ADH.

Oxytocin, the other hormone made in the hypothalamus, causes uterine contraction during childbirth and milk letdown when a baby is nursing. The more the uterus contracts during labor, the more nerve impulses reach the hypothalamus, causing oxytocin to be released. The sound of a baby crying may also stimulate the release of oxytocin. Similarly, the more a baby suckles, the more oxytocin is released. In both instances, the release of oxytocin from the posterior pituitary is controlled by **positive feedback**—that is, the stimulus continues to bring about an effect that ever increases in intensity. Positive feedback terminates due to some external event as when a baby is full and stops suckling. Positive feedback is not a way to maintain stable conditions and homeostasis.

Anterior Pituitary

A portal system, consisting of two capillary systems connected by a vein, lies between the hypothalamus and the **anterior pituitary** (Fig. 15.6, *right*). The hypothalamus controls the anterior pituitary by producing **hypothalamic-releasing** and **hypothalamic-inhibiting hormones** which pass from the hypothalamus to the anterior pituitary by way of the portal system. To take an example, there is a thyroid-releasing hormone (TRH) and a thy-roid-inhibiting hormone (TIH). TRH stimulates the anterior pituitary to secrete thyroid-stimulating hormone, and TIH inhibits the pituitary from secreting thyroid-stimulating hormone.

Three of the six hormones produced by the anterior pituitary have an effect on other glands: **Thyroid-stimulating hormone (TSH)** stimulates the thyroid to produce the thyroid hormones; **adrenocorticotropic hormone (ACTH)** stimulates the adrenal cortex to produce cortisol; and **gonadotropic hormones** stimulate the gonads—the testes in males and the ovaries in females—to produce gametes and sex hormones. In each instance, the blood level of the last hormone in the sequence exerts negative feedback control over the secretion of the first two hormones:

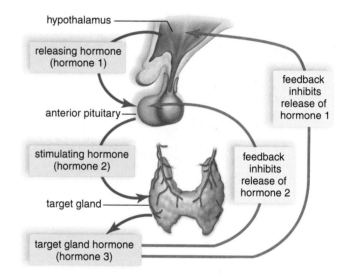

The other three hormones produced by the anterior pituitary do not affect other endocrine glands. **Prolactin (PRL)** is produced in quantity only after childbirth. It causes the mammary glands in the breasts to develop and produce milk. It also plays a role in carbohydrate and fat metabolism.

Melanocyte-stimulating hormone (MSH) causes skin-color changes in many fishes, amphibians, and reptiles having melanophores, special skin cells that produce color variations. The concentration of this hormone in humans is very low.

Growth hormone (GH), or somatotropic hormone, promotes skeletal and muscular growth. It stimulates the rate at which amino acids enter cells and protein synthesis occurs. It also promotes fat metabolism as opposed to glucose metabolism.

In the 1980s, growth hormone became a biotechnology product, and it was possible to treat short children and those diagnosed as pituitary dwarfs. A growth hormone blood test can be done in order to tell if a child is able to produce the normal amount of growth hormone. If not, like Adam in the opening story, growth hormone can be injected as a medication.

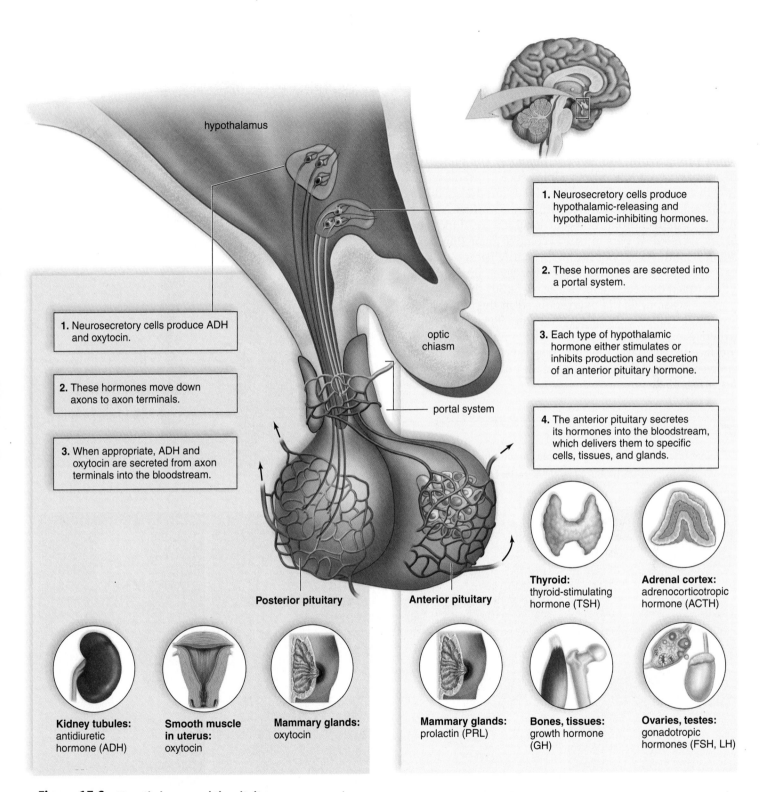

1. Neurosecretory cells produce hypothalamic-releasing and hypothalamic-inhibiting hormones.

2. These hormones are secreted into a portal system.

3. Each type of hypothalamic hormone either stimulates or inhibits production and secretion of an anterior pituitary hormone.

4. The anterior pituitary secretes its hormones into the bloodstream, which delivers them to specific cells, tissues, and glands.

hypothalamus

optic chiasm

portal system

1. Neurosecretory cells produce ADH and oxytocin.

2. These hormones move down axons to axon terminals.

3. When appropriate, ADH and oxytocin are secreted from axon terminals into the bloodstream.

Posterior pituitary

Anterior pituitary

Thyroid:
thyroid-stimulating hormone (TSH)

Adrenal cortex:
adrenocorticotropic hormone (ACTH)

Kidney tubules:
antidiuretic hormone (ADH)

Smooth muscle in uterus:
oxytocin

Mammary glands:
oxytocin

Mammary glands:
prolactin (PRL)

Bones, tissues:
growth hormone (GH)

Ovaries, testes:
gonadotropic hormones (FSH, LH)

Figure 15.6   Hypothalamus and the pituitary.
Left: The hypothalamus produces two hormones, ADH and oxytocin, which are stored and secreted by the posterior pituitary. *Right:* The hypothalamus controls the secretions of the anterior pituitary, and the anterior pituitary controls the secretions of the thyroid, adrenal cortex, and gonads, which are also endocrine glands.

Figure 15.7
Effect of growth hormone.
a. The amount of growth hormone produced by the anterior pituitary during childhood affects the height of an individual. Plentiful growth hormone produces very tall basketball players. **b.** Too much growth hormone can lead to gigantism, while an insufficient amount results in limited stature and even pituitary dwarfism.

Effects of Growth Hormone

GH is produced by the anterior pituitary. The quantity is greatest during childhood and adolescence, when most body growth is occurring. If too little GH is produced during childhood, the individual has **pituitary dwarfism,** characterized by perfect proportions but small stature. If too much GH is secreted, a person can become a giant (Fig. 15.7). Giants usually have poor health, primarily because GH has a secondary effect on the blood sugar level, promoting an illness called diabetes mellitus (see pages 309–10).

On occasion, GH is overproduced in the adult, and a condition called **acromegaly** results. Since long bone growth is no longer possible in adults, only the feet, hands, and face (particularly the chin, nose, and eyebrow ridges) can respond, and these portions of the body become overly large (Fig. 15.8).

✅ Check Your Progress 15.2

1. What role does the hypothalamus play in the endocrine system?
2. What hormones are produced by the anterior pituitary?

a.

b.

Age 9   Age 16   Age 33   Age 52

Figure 15.8 Acromegaly.
Acromegaly is caused by overproduction of GH in the adult. It is characterized by enlargement of the bones in the face, the fingers, and the toes as a person ages.

15.3 Thyroid and Parathyroid Glands

The **thyroid gland** is a large gland located in the neck, where it is attached to the trachea just below the larynx (see Fig. 15.2). The parathyroid glands are embedded in the posterior surface of the thyroid gland.

Thyroid Gland

The thyroid gland is composed of a large number of follicles, each a small spherical structure made of thyroid cells filled with triiodothyronine (T_3), which contains three iodine atoms, and **thyroxine (T_4),** which contains four.

Effects of Thyroid Hormones

To produce triiodothyronine and thyroxine, the thyroid gland actively acquires iodine. The concentration of iodine in the thyroid gland can increase to as much as 25 times that of the blood. If iodine is lacking in the diet, the thyroid gland is unable to produce the thyroid hormones. In response to constant stimulation by the anterior pituitary, the thyroid enlarges, resulting in a **simple goiter** (Fig. 15.9a). Some years ago, it was discovered that the use of iodized salt allows the thyroid to produce the thyroid hormones and, therefore, helps prevent simple goiter.

Thyroid hormones increase the metabolic rate. They do not have a target organ; instead, they stimulate all cells of the body to metabolize at a faster rate. More glucose is broken down, and more energy is utilized.

If the thyroid fails to develop properly, a condition called **congenital hypothyroidism** results (Fig. 15.9b). Individuals with this condition are short and stocky and have had extreme hypothyroidism (undersecretion of thyroid hormone) since infancy or childhood. Thyroid hormone therapy can initiate growth, but unless treatment is begun within the first two months of life, mental retardation results. The occurrence of hypothyroidism in adults produces the condition known as **myxedema,** which is characterized by lethargy, weight gain, loss of hair, slower pulse rate, lowered body temperature, and thickness and puffiness of the skin. The administration of adequate doses of thyroid hormones restores normal function and appearance.

In the case of hyperthyroidism (oversecretion of thyroid hormone), the thyroid gland is overactive, and a goiter forms. This type of goiter is called **exophthalmic goiter** (Fig. 15.9c). The eyes protrude because of edema in eye socket tissues and swelling of the muscles that move the eyes. The patient usually becomes hyperactive, nervous and irritable, and suffers from insomnia. Removal or destruction of a portion of the thyroid by means of radioactive iodine is sometimes effective in curing the condition. Hyperthyroidism can also be caused by a thyroid tumor, which is usually detected as a lump during physical examination. Again, the treatment is surgery in combination with administration of radioactive iodine. The prognosis for most patients is excellent.

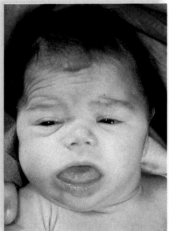
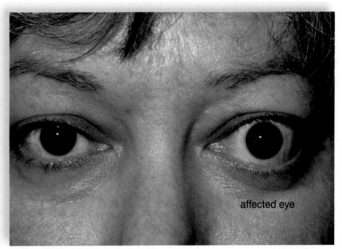

a. Simple goiter   b. Congenital hypothyroidism   c. Exophthalmic goiter

Figure 15.9 Abnormalities of the thyroid.
a. An enlarged thyroid gland is often caused by a lack of iodine in the diet. Without iodine, the thyroid is unable to produce its hormones, and continued anterior pituitary stimulation causes the gland to enlarge. **b.** Individuals who develop hypothyroidism during infancy or childhood do not grow and develop as others do. Unless medical treatment is begun, the body is short and stocky; mental retardation is also likely. **c.** In exophthalmic goiter, a goiter is due to an overactive thyroid, and the eyes protrude because of edema in eye socket tissue.

Calcitonin

Calcium (Ca^{2+}) plays a significant role in both nervous conduction and muscle contraction. It is also necessary for blood clotting. The blood Ca^{2+} level is regulated in part by **calcitonin,** a hormone secreted by the thyroid gland when the blood Ca^{2+} level rises (Fig. 15.10). The primary effect of calcitonin is to bring about the deposit of Ca^{2+} in the bones. It also temporarily reduces the activity and number of osteoclasts. When the blood Ca^{2+} level lowers to normal, the release of calcitonin by the thyroid is inhibited, but a low level stimulates the release of **parathyroid hormone (PTH)** by the parathyroid glands.

Parathyroid Glands

Many years ago, the four parathyroid glands were sometimes mistakenly removed during thyroid surgery because of their size and location. PTH, the hormone produced by the **parathyroid glands,** causes the blood Ca^{2+} level to increase.

A low blood Ca^{2+} level stimulates the release of PTH. PTH promotes the activity of osteoclasts and the release of calcium from the bones. PTH also promotes the reabsorption of calcium by the kidneys, where it activates vitamin D. Activated vitamin D is a hormone sometimes called calcitriol, which stimulates the absorption of Ca^{2+} from the intestine. These effects bring the blood Ca^{2+} level back to the normal range so that the parathyroid glands no longer secrete PTH.

When insufficient PTH production leads to a dramatic drop in the blood calcium level, tetany results. In **tetany,** the body shakes from continuous muscle contraction. This effect is brought about by increased excitability of the nerves, which initiate nerve impulses spontaneously and without rest.

Healing a Fracture

Remember Katie, who suffered a burst vertebra in the opening story for Chapter 11? In order for her injury to heal, osteoclasts will have to destroy old bone, and osteoblasts will have to lay down new bone. Many factors influence the growth of forming new bone, including parathyroid hormone, calcitonin, and vitamin D. Working together, the calcium needed to fuse her vertebrae would be made readily available as new blood capillaries penetrated the area.

---

⑥ Check Your Progress 15.3

1. **a.** What hormones are produced by the thyroid gland? **b.** How do thyroid hormones affect the metabolic rate?

2. **a.** What hormone is produced by the parathyroid gland? **b.** What is the function of this hormone? **c.** What hormone is secreted by the thyroid gland when the blood Ca^{2+} levels are high?

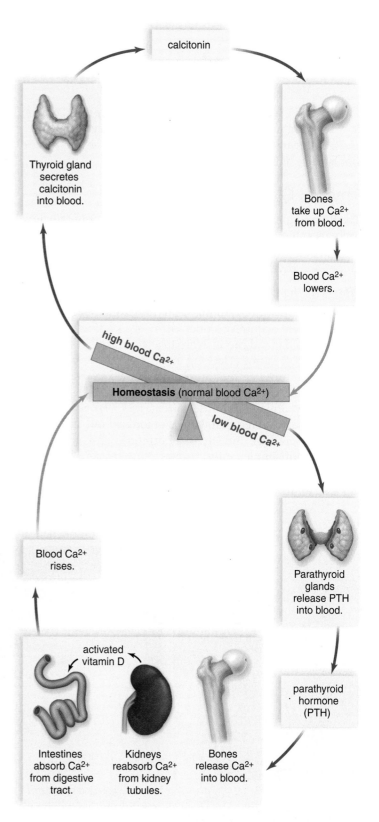

Figure 15.10 **Regulation of blood calcium level.**
Top: When the blood calcium (Ca^{2+}) level is high, the thyroid gland secretes calcitonin. Calcitonin promotes the uptake of Ca^{2+} by the bones, and therefore, the blood Ca^{2+} level returns to normal. *Bottom:* When the blood Ca^{2+} level is low, the parathyroid glands release parathyroid hormone (PTH). PTH causes the bones to release Ca^{2+}. It also causes the kidneys to reabsorb Ca^{2+} and activate vitamin D; thereafter, the intestines absorb Ca^{2+}. Therefore, the blood Ca^{2+} level returns to normal.

15.4 Adrenal Glands

The **adrenal glands** sit atop the kidneys (see Fig. 15.2). Each adrenal gland consists of an inner portion called the **adrenal medulla** and an outer portion called the **adrenal cortex.** These portions, like the anterior and the posterior pituitary, are two functionally distinct endocrine glands. The adrenal medulla is under nervous control, and portions of the adrenal cortex are under the control of ACTH, an anterior pituitary hormone. Stress of all types, including emotional and physical trauma, prompts the hypothalamus to stimulate a portion of the adrenal glands (Fig. 15.11).

Adrenal Medulla

The hypothalamus initiates nerve impulses that travel by way of the brain stem, spinal cord, and preganglionic sympathetic nerve fibers to the adrenal medulla, which then secretes its hormones. The cells of the adrenal medulla are thought to be modified postganglionic neurons.

Epinephrine (adrenaline) and **norepinephrine** (noradrenaline) produced by the adrenal medulla rapidly bring about all the body changes that occur when an individual reacts to an emergency situation in a fight-or-flight manner. The effect of these hormones provide a short-term response to stress.

Adrenal Cortex

In contrast, the hormones produced by the adrenal cortex provide a long-term response to stress (Fig. 15.11). The two major types of hormones produced by the adrenal cortex are the mineralocorticoids and the glucocorticoids. The **mineralocorticoids** regulate salt and water balance, leading to increases in blood volume and blood pressure. The **glucocorticoids,** whose secretion is controlled by ACTH, regulate carbohydrate, protein, and fat metabolism, leading to an increase in blood glucose level. Glucocorticoids also suppress the body's inflammatory response. Cortisone, the medication often administered for inflammation of joints, is a glucocorticoid.

The adrenal cortex also secretes a small amount of male sex hormones and a small amount of female sex hormones in both sexes. That is, in both males and females, both male and female sex hormones are produced by the adrenal cortex.

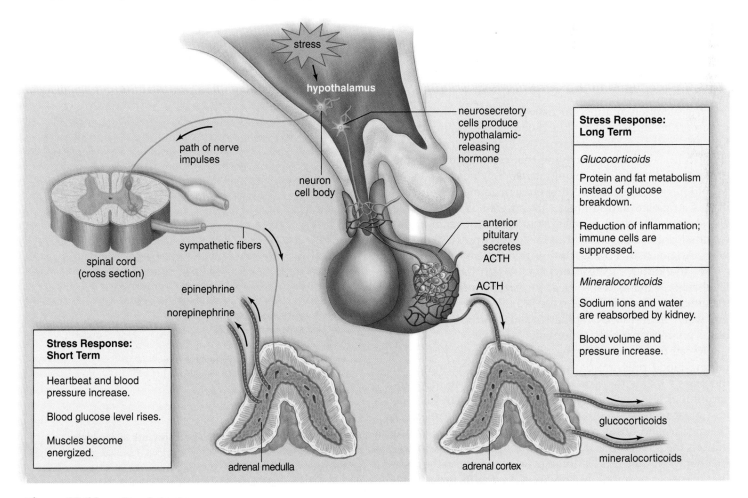

Figure 15.11 **Adrenal glands.**
Both the adrenal cortex and the adrenal medulla are under the control of the hypothalamus when they help us respond to stress. *Left:* Nervous stimulation causes the adrenal medulla to provide a rapid, but short-term, stress response. *Right:* The adrenal cortex provides a slower, but long-term, stress response. ACTH causes the adrenal cortex to release glucocorticoids. Independently, the adrenal cortex releases mineralocorticoids.

Glucocorticoids

ACTH stimulates those portions of the adrenal cortex which secrete the glucocorticoids. **Cortisol** and also cortisone are biologically significant glucocorticoids. Glucocorticoids raise the blood glucose level in at least two ways: (1) They promote the breakdown of muscle proteins to amino acids, which are taken up by the liver from the bloodstream. The liver then breaks down these excess amino acids to glucose, which enters the blood. (2) They promote the metabolism of fatty acids rather than carbohydrates, and this spares glucose.

The glucocorticoids also counteract the inflammatory response that leads to the pain and swelling of joints in arthritis and bursitis. The administration of cortisone aids these conditions because it reduces inflammation. Very high levels of glucocorticoids in the blood can suppress the body's defense system, including the inflammatory response that occurs at infection sites. Cortisone and other glucocorticoids can relieve swelling and pain from inflammation, but by suppressing pain and immunity, they can also make a person highly susceptible to injury and infection.

Mineralocorticoids

Aldosterone is the most important of the mineralocorticoids. The aldosterone primarily targets the kidney, where it promotes renal absorption of sodium (Na^+) and renal excretion of potassium (K^+).

The secretion of mineralocorticoids is not controlled by the anterior pituitary. When the blood Na^+ level and, therefore, the blood pressure are low, the kidneys secrete **renin** (Fig. 15.12). Renin is an enzyme that converts the plasma protein angiotensinogen to angiotensin I, which is changed to angiotensin II by a converting enzyme found in lung capillaries. Angiotensin II stimulates the adrenal cortex to release aldosterone. The effect of this system, called the renin-angiotensin-aldosterone system, is to raise blood pressure in two ways: Angiotensin II constricts the arterioles, and aldosterone causes the kidneys to reabsorb Na^+. When the blood Na^+ level rises, water is reabsorbed, in part, because the hypothalamus secretes ADH (see page 300). Reabsorption means that water enters kidney capillaries and thus the blood. Then blood pressure increases to normal.

Recall that we studied the role of the kidneys in maintaining blood pressure on pages 197–98. At that time, we mentioned that if the blood pressure rises due to the reabsorption of Na^+, the atria of the heart are apt to stretch. Due to a great increase in blood volume, cardiac cells release a chemical called **atrial natriuretic hormone (ANH)**, which inhibits the secretion of aldosterone from the adrenal cortex. In other words, the heart is among various organs in the body that releases a hormone, but obviously not as its major function. (Therefore, the heart is not included as an endocrine gland in Figure 15.2.) The effect of this ANH is to cause the excretion of Na^+—that is, *natriuresis*. When Na^+ is excreted, so is water, and therefore, blood pressure lowers to normal.

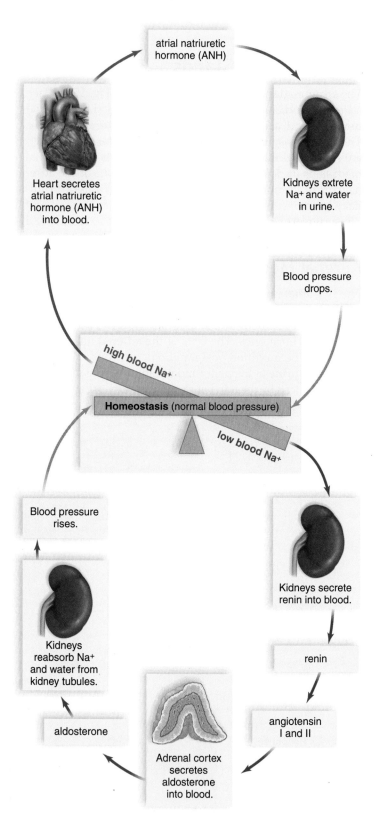

Figure 15.12   Regulation of blood pressure and volume.
Bottom: When the blood sodium (Na^+) level is low, a low blood pressure causes the kidneys to secrete renin. Renin leads to the secretion of aldosterone from the adrenal cortex. Aldosterone causes the kidneys to reabsorb Na^+, and water follows, so that blood volume and pressure return to normal. *Top:* When a high blood Na^+ level accompanies a high blood volume, the heart secretes atrial natriuretic hormone (ANH). ANH causes the kidneys to excrete Na^+, and water follows. The blood volume and pressure return to normal.

Malfunction of the Adrenal Cortex

When the blood level of glucocorticoids is low due to hyposecretion, a person develops **Addison disease.** The presence of excessive but ineffective ACTH causes a bronzing of the skin because ACTH, like MSH, can lead to a buildup of melanin (Fig. 15.13). Without the glucocorticoids, glucose cannot be replenished when a stressful situation arises. Even a mild infection can lead to death. In some cases, hyposecretion of aldosterone results in a loss of sodium and water, the development of low blood pressure, and possibly severe dehydration. Left untreated, Addison disease can be fatal.

When the level of glucocorticoids is high due to hypersecretion, a person develops **Cushing syndrome.** The excess glucocorticoids result in a tendency toward diabetes mellitus, as muscle protein is metabolized and subcutaneous fat is deposited in the midsection. The result is a swollen "moon" face and an obese trunk, with arms and legs of normal size. Children will show obesity and poor growth in height (Fig. 15.14). Depending on the cause and duration of the Cushing syndrome, some people may have more dramatic changes, including masculinization with increased blood pressure and weight gain.

☑ Check Your Progress 15.4

1. What are the two major hormones produced by the adrenal cortex, and what do they regulate?

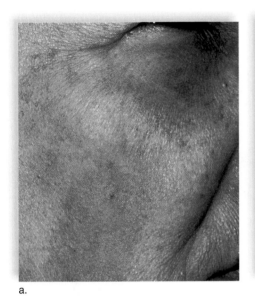

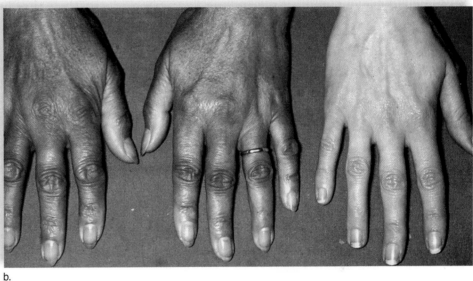

a.

b.

Figure 15.13   Addison disease.
Addison disease is characterized by a peculiar bronzing of the skin, particularly noticeable in these light-skinned individuals. Note the color of **(a)** the face and **(b)** the hands compared with the hand of an individual without the disease.

Figure 15.14   Cushing syndrome.
Cushing syndrome results from hypersecretion of adrenal cortex hormones. *Left:* Patient first diagnosed with Cushing syndrome. *Right:* Four months later, after therapy.

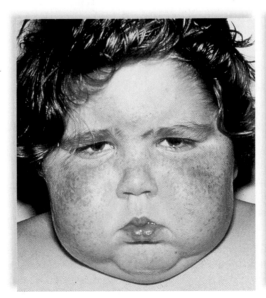

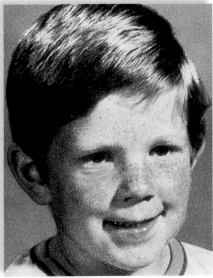

15.5 Pancreas

The **pancreas** is a fish-shaped organ that stretches across the abdomen behind the stomach and near the duodenum of the small intestine. It is composed of two types of tissue. Exocrine tissue produces and secretes digestive juices that go by way of ducts to the small intestine. Endocrine tissue, called the **pancreatic islets** (islets of Langerhans), produces and secretes the hormones **insulin** and **glucagon** directly into the blood (Fig. 15.15).

The pancreas is not under pituitary control. Insulin is secreted when the blood glucose level is high, which usually occurs just after eating. Insulin stimulates the uptake of glucose by cells, especially liver cells, muscle cells, and adipose tissue cells. In liver and muscle cells, glucose is then stored as glycogen. In muscle cells, the glucose supplies energy for muscle contraction, and in fat cells, glucose enters the metabolic pool and thereby supplies glycerol for the formation of fat. In these various ways, insulin lowers the blood glucose level (Fig. 15.16, *top*)

Glucagon is secreted from the pancreas, usually between eating, when the blood glucose level is low. The major target tissues of glucagon are the liver and adipose tissue. Glucagon stimulates the liver to break down glycogen to glucose and to use fat and protein in preference to glucose as energy sources. Adipose tissue cells break down fat to glycerol and fatty acids. The liver takes these up and uses them as substrates for glucose formation. In these ways, glucagon raises the blood glucose level (Fig. 15.16, *bottom*).

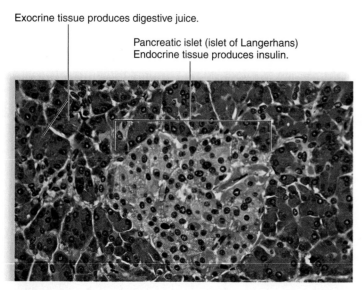

Figure 15.15 **Pancreas.**

This light micrograph shows that the pancreas has two types of cells. The exocrine tissue produces a digestive juice and the endocrine tissue produces the hormone insulin.

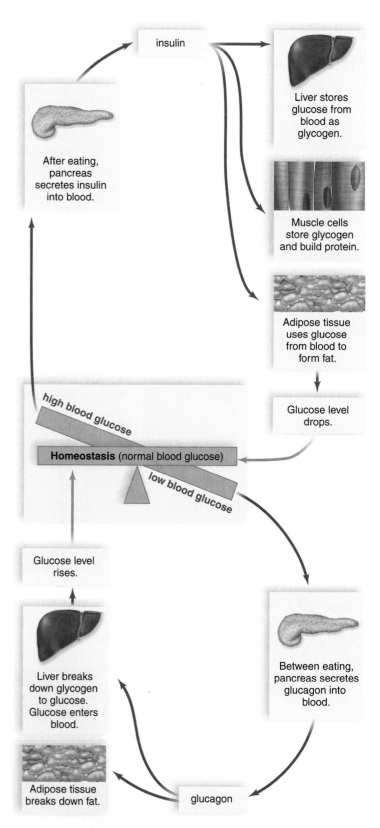

Figure 15.16 **Regulation of blood glucose level.**

Top: When the blood glucose level is high, the pancreas secretes insulin. Insulin promotes the storage of glucose as glycogen and the synthesis of proteins and fats. Therefore, insulin lowers the blood glucose level. *Bottom:* When the blood glucose level is low, the pancreas secretes glucagon. Glucagon acts opposite to insulin; therefore, glucagon raises the blood glucose level to normal.

Health Focus

Do You Have Diabetes?

The world is experiencing a diabetes epidemic—more and more people are being diagnosed with the condition. In the United States, 18 million people now have either diabetes mellitus type 1 or diabetes type 2. The number of those with diabetes type 2, in particular, is expected to rise at an alarming rate. By 2025, the incidence of diabetes type 2 is expected to double over what it is now.

Diagnosing Diabetes

People with diabetes have to check their blood sugar level, usually before meals and at bedtime. Today, automatic lancets prick the finger and computerized devices can record readings automatically (Fig. 15A). In people without diabetes, the device records a blood sugar level between 70 and 110 mg/dL (milligrams per deciliter which is .10 liters). In diabetics, the blood sugar level is more than 140 mg/dL. What's wrong? In the nondiabetic, the hormone insulin secreted by the pancreas causes tissue cells (particularly fat, muscle, and liver cells) to take up glucose so that the blood sugar level remains within the normal range. In diabetes type 1, the pancreas fails to secrete insulin—the cells that make insulin have died off! In diabetes type 2, the pancreas usually fails to secrete enough insulin, but the real problem is that the cells are resistant to insulin. Receptors in the plasma membrane don't bind insulin properly, and then the plasma membrane doesn't have enough carrier proteins to transport the glucose into the cell.

Most people don't have a means to check their blood glucose level, so how do they become aware that they have diabetes? These symptoms can help you decide if a physician should check you out:

- Frequent urination, especially at night
- Unusual hunger and/or thirst
- Unexplained change of weight
- Blurred vision
- Sores that do not heal
- Excessive fatigue

A physician has a number of other ways to diagnose diabetes. One of the most common is to test for sugar in the urine. When the blood sugar level is high, the kidneys excrete sugar, and a lab test can detect its presence in the urine. Excessive thirst accompanies untreated diabetes because the kidneys use large amounts of water to flush the excess glucose out of the body. Hunger and exhaustion occur because the cells are starving for glucose in the midst of plenty. Without glucose, some cells cannot produce ATP, the energy currency of cells.

What Causes Diabetes

Diabetes type 1 runs in families, and researchers have identified about 20 genes that can increase your risk of diabetes. Diabetes type 1 usually occurs after a viral infection. The immune system gears up to fight the infection by killing off cells that are

Figure 15A   Testing the blood sugar level.
Diabetics have to test their blood sugar level during the day and just before retiring to make sure their treatment is keeping it within the normal range.

harboring the virus. When the infection is over, the immune system keeps on killing cells—this time the pancreatic cells that produce insulin. Diabetes type 1 is clearly a self-inflicted disease, but in another way, so is diabetes type 2. Diabetes type 2 used to be thought of as an adult-onset disorder. Now more and more children have it. The risk factors for getting diabetes type 2 are excessive food intake leading to obesity, especially fat located in the abdominal region, and physical inactivity.

Treatment for Diabetes

Those with diabetes type 1 must have insulin injections, but those with diabetes type 2 usually receive metformin, a medication that makes cells more likely to respond to the presence of insulin.

A healthy diet and exercise are emphasized as an important part of their regimen. Even limited weight loss can often lead to better blood glucose regulation. Foods rich in complex carbohydrates, including dietary fiber and fruit, are preferred over easily digested foods such as many junk foods (for example, potato chips). Building up the muscles and using them regularly by, say, walking, riding a bike, or dancing uses up glucose and improves insulin efficiency. All diabetics must work closely with a physician to tailor the correct regimen of medications, diet, and exercise for them.

Diabetes is associated with the possibility of blindness, cardiovascular diseases, and kidney disease. Nerve deterioration can lead to an inability to feel pain, particularly in the hands and feet. If so, treatment of an infection may be delayed to the point that limb amputation is required. Diabetes is definitely a disorder to avoid if at all possible. The very best course of action is to adopt a healthy lifestyle—the younger the better—in order to keep this condition at bay. This is the way you can help your body maintain homeostasis, the relative constancy of the internal environment; in this case, a normal blood glucose level.

Diabetes Mellitus

Diabetes mellitus is a fairly common hormonal disease in which liver cells, and indeed most body cells, are unable to take up glucose as they should. Therefore, cellular famine exists in the midst of plenty, and the person becomes extremely hungry. As the blood glucose level rises, glucose, along with water, is excreted in the urine. Urination is frequent, and the loss of water causes the diabetic to be extremely thirsty.

The glucose tolerance test assists in the diagnosis of diabetes mellitus. After the patient is given 100 grams of glucose, the blood glucose concentration is measured at intervals. In a diabetic, the blood glucose level rises greatly and remains elevated for several hours (Fig. 15.17). In the meantime, glucose appears in the urine. In a nondiabetic, the blood glucose level rises somewhat and then returns to normal after about two hours.

Types of Diabetes

There are two types of diabetes mellitus. In *diabetes type 1,* the pancreas is not producing insulin. This condition is believed to be brought on by exposure to an environmental agent, most likely a virus, whose presence causes cytotoxic T cells to destroy the pancreatic islets. The body turns to the metabolism of fat, which leads to the buildup of ketones in the blood and, in turn, to acidosis (acid blood), which can lead to coma and death. As a result, the individual must have daily insulin injections. These injections control the diabetic symptoms but can still cause inconveniences because the blood sugar level may swing between hypoglycemia (low blood glucose level) and hyperglycemia (high blood glucose level). Without testing the blood glucose level, it is difficult to be certain which of these is present because the symptoms are similar. The symptoms include perspiration, pale skin, shallow breathing, and anxiety. Whenever these symptoms appear, immediate attention is required to bring the blood glucose level to its proper level. If the problem is hypoglycemia, the cure is a cube of sugar, and if hyperglycemia, the cure is insulin.

Some diabetics have learned to use an insulin pump to better regulate their blood sugar level. The pump is worn outside the body, usually attached to a belt or waistband. Insulin is pumped from a reservoir through a tube inserted under the skin of the abdominal wall. It is also possible to transplant a working pancreas into patients with diabetes type 1. To do away with the necessity of taking immunosuppressive drugs after the transplant, fetal pancreatic islet cells have been injected into patients. Another experimental procedure is to place pancreatic islet cells in a capsule that allows insulin to get out but prevents antibodies and T lymphocytes from getting in. This artificial organ is implanted in the abdominal cavity.

Of the 16 million people who now have diabetes in the United States, most have *diabetes type 2.* Often, the patient is obese. Usually, after insulin binds to a plasma membrane receptor, the number of protein carriers for glucose increases, and more glucose than usual enters the cell. In the case of diabetes type 2, glucose binds to the receptor, but the number of carriers does not increase. Therefore, the cell is said to be insulin resistant.

It is possible to prevent or at least control diabetes type 2 by adhering to a low-fat, low-sugar diet and exercising regularly. If this fails, oral drugs that stimulate the pancreas to secrete more insulin and enhance the metabolism of glucose in the liver and muscle cells are available. It is projected as many as seven million Americans may have diabetes type 2 without being aware of it. Yet, the effects of untreated diabetes type 2 are as serious as those of diabetes type 1.

Long-term complications of both types of diabetes are blindness, kidney disease, and cardiovascular disorders, including atherosclerosis, heart disease, stroke, and reduced circulation. The latter can lead to gangrene in the arms and legs. Pregnancy carries an increased risk of diabetic coma, and the child of a diabetic is somewhat more likely to be stillborn or to die shortly after birth. These complications of diabetes are not expected to appear if the mother's blood glucose level is carefully regulated and kept within normal limits.

✔ Check Your Progress 15.5

1. **What hormones are produced by the pancreas and how does each affect blood glucose?**

2. **What is diabetes mellitus?**

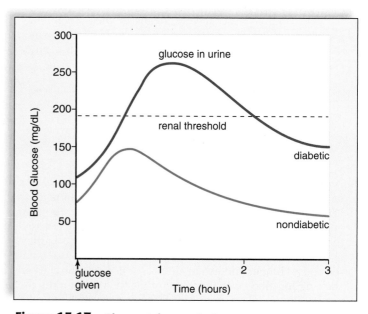

Figure 15.17   Glucose tolerance test.
Following the administration of 100 grams of glucose, the blood glucose level rises dramatically in the diabetic, and glucose appears in the urine. Also, the blood glucose level at 2 hours is equal to more than 200 mg/dL.

15.6 Other Endocrine Glands

The **gonads** are the testes in males and the ovaries in females. The gonads are endocrine glands. Other lesser known glands and some tissues also produce hormones.

Testes and Ovaries

The activity of the testes and ovaries are controlled by the hypothalamus and pituitary. The **testes** are located in the scrotum, and the **ovaries** are located in the pelvic cavity. The testes produce **androgens** (e.g., **testosterone**), which are the male sex hormones, and the ovaries produce **estrogens** and **progesterone,** the female sex hormones. These hormones feedback to control the hypothalamus secretion of GnRH (gonadotropic releasing hormone) and the pituitary gland secretion of FSH and LH, the gonadotropic hormones (Fig. 15.18). The activities of FSH and LH are discussed in Chapter 16.

Under the influence of the gonadotropic hormones, the testes begin to release increased amounts of testosterone at the time of puberty, and testosterone stimulates the growth of the penis and the testes. Testosterone also brings about and maintains the male secondary sex characteristics that develop during puberty, including the growth of a beard, axillary (underarm) hair, and pubic hair. It prompts the larynx and the vocal cords to enlarge, causing the voice to lower. Testosterone also stimulates oil and sweat glands in the skin; therefore, it is largely responsible for acne and body odor. Another side effect of testosterone is baldness. Although females, like males, do inherit genes for baldness, baldness is seen more often in males because of the presence of testosterone. Testosterone is partially responsible for the muscular strength of males, and this is why some athletes take supplemental amounts of **anabolic steroids,** which are either testosterone or related chemicals. The Bioethical Focus on page 241 discusses the detrimental effect anabolic steroids can have on the body.

The female sex hormones, estrogens (often referred to in the singular) and progesterone, have many effects on the body. In particular, estrogen secreted at the time of puberty stimulates the growth of the uterus and the vagina. Estrogen is necessary for egg maturation and is largely responsible for the secondary sex characteristics in females, including female body hair and fat distribution. In general, females have a more rounded appearance than males because of a greater accumulation of fat beneath the skin. Also, the pelvic girdle is wider in females than in males, resulting in a larger pelvic cavity. Both estrogen and progesterone are required for breast development and for regulation of the uterine cycle, which includes monthly menstruation (discharge of blood and mucosal tissues from the uterus).

Thymus Gland

The lobular **thymus gland** lies just beneath the sternum (see Fig. 15.2). This organ reaches its largest size and is most active during childhood. With aging, the organ gets smaller and becomes fatty. Lymphocytes that originate in the bone marrow and then pass through the thymus are transformed into T lymphocytes. The lobules of the thymus are lined by epithelial cells that secrete hormones called **thymosins.** These hormones aid in the differentiation of lymphocytes packed inside the lobules. Although the hormones secreted by the thymus ordinarily work in the thymus, there is hope that these hormones could be injected into AIDS or cancer patients, where they would enhance T lymphocyte function.

Pineal Gland

The **pineal gland,** which is located in the brain (see Fig. 15.2), produces the hormone **melatonin,** primarily at night. Melatonin is involved in our daily sleep-wake cycle; nor-

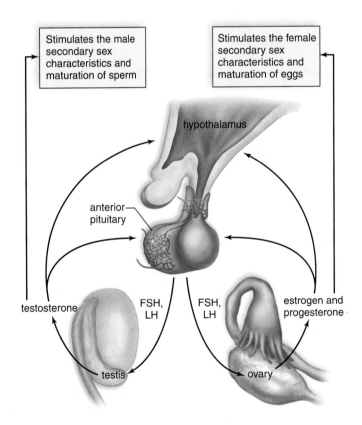

Figure 15.18 **Regulation of the gonadotropic and sex hormones.** The testes and ovaries secrete the sex hormones. The testes secrete testosterone, and the ovaries secrete estrogens and progesterone. In each sex, secretion of GnRH from the hypothalamus and secretion of FSH and LH from the pituitary are controlled by their respective hormones.

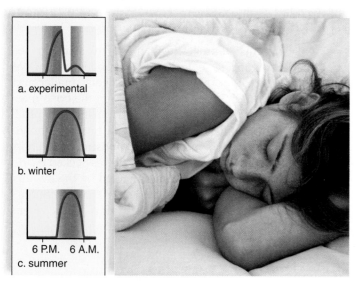

a. experimental

b. winter

6 P.M.   6 A.M.
c. summer

Figure 15.19 Melatonin production.
Melatonin production is greatest at night when we are sleeping. Light suppresses melatonin production (**a**), so it is secreted for a longer time in the winter (**b**) than in the summer (**c**).

mally we grow sleepy at night when melatonin levels increase and awaken once daylight returns and melatonin levels are low (Fig. 15.19). Daily 24-hour cycles such as this are called **circadian rhythms,** and these rhythms are controlled by a biological clock located in the hypothalamus.

Animal research suggests that melatonin also regulates sexual development. In keeping with these findings, it has been noted that children whose pineal gland has been destroyed due to a brain tumor experience early puberty.

Hormones from Other Organs/Tissues

Some organs that are not usually considered endocrine glands do indeed secrete hormones (Table 15.1). We have already

Table 15.1	Hormones from Other Organs/Tissues	
	Hormone	**Function**
Kidney	Renin	Leads to aldosterone and higher blood pressure
	Erythropoietin	Stimulates red bone marrow to produce red blood cells
Heart	Atrial natriuretic hormone	Lowers blood pressure
GI tract	Secretin, gastrin, cholecystokinin	Control secretion of some GI enzymes and bile
Fat cells	Leptin	Helps regulate appetite
Tissue cells	Prostaglandins	Various functions

mentioned that the kidneys secrete renin and that the heart produces atrial natriuretic hormone (see page 306). And you will recall that the stomach and the small intestine produce peptide hormones that regulate digestive secretions. A number of other types of tissues produce hormones.

Erythropoietin

In response to a low oxygen blood level, the kidneys secrete erythropoietin, which stimulates red blood cell formation in the red bone marrow. A number of different types of organs and cells also produce peptide growth factors, which stimulate cell division and mitosis. Growth factors can be considered hormones because they act on cell types with specific receptors to receive them. Some are released into the blood; others diffuse to nearby cells.

Leptin

Leptin is a protein hormone produced by adipose tissue. Leptin acts on the hypothalamus, where it signals satiety—that is, the individual has had enough to eat. Strange to say, the blood of obese individuals may be rich in leptin. It is possible that the leptin they produce is ineffective because of a genetic mutation, or else their hypothalamic cells lack a suitable number of receptors for leptin.

Prostaglandins

Prostaglandins are potent chemical signals produced within cells from arachidonate, a fatty acid. Prostaglandins are not distributed in the blood; instead, they act locally, quite close to where they were produced. In the uterus, prostaglandins cause muscles to contract; therefore, they are implicated in the pain and discomfort of menstruation in some women. Also, prostaglandins mediate the effects of pyrogens, chemicals that are believed to reset the temperature regulatory center in the brain. Aspirin reduces body temperature and controls pain because of its effect on prostaglandins.

Certain prostaglandins reduce gastric secretion and have been used to treat gastric reflux; others lower blood pressure and have been used to treat hypertension; and yet others inhibit platelet aggregation and have been used to prevent thrombosis. However, different prostaglandins have contrary effects, and it has been very difficult to successfully standardize their use. Therefore, prostaglandin therapy is still considered experimental.

✓ Check Your Progress 15.6

1. a. What other organs are considered major endocrine glands?
   b. Which of these are under the control of the anterior pituitary?

2. A number of other organs/tissues secrete hormones. Give some specific examples.

3. a. What is a local hormone? b. Give an example.

Hormone Replacement Therapy

Menopause is the time in a woman's life when menstruation comes to an end. During menopause, which may begin as early as 35 years of age, the ovaries gradually produce lower levels of the sex hormones. Doctors may recommend using hormone replacement therapy (HRT) to counter some of the problems often associated with menopause (hot flashes, night sweats, sleeplessness, mood swings, and vaginal dryness) or to prevent some long-term conditions that are more common in postmenopausal women, such as osteoporosis. HRT used to be quite common. Data from a 1997 national survey showed that 45% of U.S. women born between 1897 and 1950 used menopausal hormones for at least one month, and 20% continued use for five or more years.

In recent years, a number of long-term studies have attempted to assess the risks and benefits of HRT. The best data comes from the Women's Health Initiative (WHI), a large, randomized clinical trial of over 16,000 healthy women, ages 50 through 79, in which half of the participants took hormones and the other half took a placebo pill (which does not contain any drug). The trial, sponsored by the National Institutes of Health (NIH), was halted early when, in July 2002, investigators reported that the overall risks of HRT outweighed the benefits. What were the results of this study with regard to quality of life, cancer and cardiovascular disease risk, and osteoporosis?

Quality of Life On the basis of the WHI study and others, HRT has no significant effects on the general health, vitality, mental health, depressive symptoms, or sexual satisfaction of women. Although hormone use was associated with a small benefit in terms of sleep disturbance, physical functioning, and bodily pain after one year of use, the effect was too small to be considered clinically significant. However, the risk of developing dementia (including Alzheimer disease) was double in women age 65 and older.

Cancer HRT significantly increased the risk of developing breast cancer in women ages 50 to 64. Current hormone users were also more likely to die from breast cancer than women who did not use them. Within about five years of stopping use, the increased risk largely disappeared. Although not as definite as breast cancer, the possibility exists that HRT also increases a woman's risk of endometrial cancer (cancer of the lining of the uterus) and ovarian cancer.

As a benefit, women on HRT have fewer cases of colorectal cancer compared with women taking a placebo.

Cardiovascular Disease HRT may increase the risk of heart disease among generally healthy postmenopausal women. The greatest increased risk occurred in the first year. Women on HRT have double the combined rate of blood clots in the lungs and legs. Studies have consistently reported increased risks of blood clots in the lungs (pulmonary embolisms) and deep veins in the legs with hormone use. The WHI study also indicated that the risk of a stroke increases when women use HRT.

Osteoporosis Osteoporosis is the loss of bone mass and density, which causes bones to become fragile and increases the chance of bone fractures. As a benefit, estrogen alone and estrogen combined with progestin have been shown to protect against osteoporosis. Results from the WHI showed that estrogen plus progestin can prevent fractures of the hip, vertebrae, and other bones.

Conclusion The WHI found that use of this estrogen plus progestin pill increases the risk of breast cancer, heart disease, stroke, and blood clots. The study also found that there were fewer cases of hip fractures and colon cancer among women using estrogen plus progestin than in those taking a placebo. Therefore, women are now advised to discuss the pros and cons of HRT with their physician and decide for themselves if they wish to use it (Fig. 15B). In other words, the patient has to be aware of the risks and accept the responsibility for taking replacement hormones.

Decide Your Opinion

1. Should physicians wait to recommend medications until they are certain about their benefits? Or, are they duty-bound to make recommendations based on incomplete information if there is a possibility of improving the health of millions of people?
2. Should patients accept the responsibility of deciding for themselves if they should take a medicine, or should physicians assume this responsibility?
3. HRT must be prescribed by a physician. Should women who suffered health consequences after undergoing HRT be encouraged to sue the physicians who wrote prescriptions that allowed them to use the therapy?

Figure 15B Thinking it through.
Menopausal women are now advised to consult their physicians and then to decide for themselves if they wish to be on hormone replacement therapy.

15.7 Homeostasis

The nervous and endocrine systems exert control over the other systems and thereby maintain homeostasis (Fig. 15.20).

Responding to External Changes

The nervous system is particularly able to respond to changes in the external environment. Some responses are automatic as you can testify by trying this: Take a piece of clear plastic and hold it just in front of your face. Get someone to gently toss a soft object, such as a wadded-up piece of paper, at the plastic. Can you prevent yourself from blinking? This reflex protects your eyes.

The eyes and other organs that have sensory receptors provide us with valuable information about the external environment. The central nervous system, which is on the receiving end of millions of bits of information, integrates

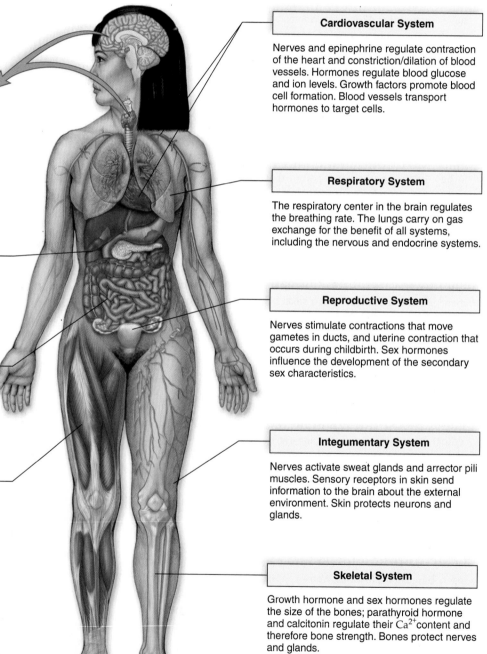

The nervous and endocrine systems work together to maintain homeostasis. The systems listed here in particular also work with these two systems.

Nervous and Endocrine Systems

The nervous and endocrine systems coordinate the activities of the other systems. The brain receives sensory input and controls the activity of muscles and various glands. The endocrine system secretes hormones that influence the metabolism of cells, the growth and development of body parts, and homeostasis.

Urinary System

Nerves stimulate muscles that permit urination. Hormones (ADH and aldosterone) help kidneys regulate the water-salt balance and the acid-base balance of the blood.

Digestive System

Nerves stimulate smooth muscle and permit digestive tract movements. Hormones help regulate digestive juices that break down food to nutrients for neurons and glands.

Muscular System

Nerves stimulate muscles, whose contractions allow us to move out of danger. Androgens promote growth of skeletal muscles. Sensory receptors in muscles and joints send information to the brain. Muscles protect neurons and glands.

Cardiovascular System

Nerves and epinephrine regulate contraction of the heart and constriction/dilation of blood vessels. Hormones regulate blood glucose and ion levels. Growth factors promote blood cell formation. Blood vessels transport hormones to target cells.

Respiratory System

The respiratory center in the brain regulates the breathing rate. The lungs carry on gas exchange for the benefit of all systems, including the nervous and endocrine systems.

Reproductive System

Nerves stimulate contractions that move gametes in ducts, and uterine contraction that occurs during childbirth. Sex hormones influence the development of the secondary sex characteristics.

Integumentary System

Nerves activate sweat glands and arrector pili muscles. Sensory receptors in skin send information to the brain about the external environment. Skin protects neurons and glands.

Skeletal System

Growth hormone and sex hormones regulate the size of the bones; parathyroid hormone and calcitonin regulate their Ca^{2+} content and therefore bone strength. Bones protect nerves and glands.

Figure 15.20   Human systems work together.
The nervous and endocrine systems work together to regulate and control the other systems.

information, compares it with previously stored memories, and "decides" on the proper course of action. Suppose you have spent the morning skiing; your muscles ache, your stomach churns, and your toes are cold. The brain will send out the motor impulses that will soon have you recovering from your exertions. The nervous system often responds to changes in the external environment through body movement. It gives us the ability to stay in as moderate an environment as possible—one that is not too hot or cold, for example. Otherwise, we test the ability of the nervous system to maintain homeostasis despite extreme conditions.

Responding to Internal Changes

The governance of internal organs usually requires that the nervous and endocrine systems work together, usually below the level of consciousness. Subconscious control often depends on reflex actions that involve the hypothalamus and the medulla oblongata. Let's take blood pressure as an example. You've just run three miles to raise money for hunger relief and decide to sit down under a tree to rest a bit. When you stand up to push off again, you feel faint, but it quickly passes because the medulla oblongata responds to input from the baroreceptors in the aortic arch and carotid arteries and immediately acts through the sympathetic system to increase heart rate and constrict the blood vessels so that your blood pressure rises. Sweating may have upset the water-salt balance of your blood. If so, the hormone aldosterone from the adrenal cortex will act on the kidney tubules to conserve Na^+, and water reabsorption will follow. The hypothalamus can also help by sending antidiuretic hormone (ADH) to the posterior pituitary gland, which releases it into the blood. ADH actively promotes water reabsorption by the kidney tubules.

Recall from Chapter 13 that certain drugs, such as alcohol, can affect ADH secretion. When you consume alcohol, it is quickly absorbed across the stomach lining into the bloodstream, where it travels to the hypothalamus and inhibits ADH secretion. When ADH levels fall, the kidney tubules absorb less water. The result is increased production of dilute urine. Excessive water loss, or dehydration, is a disturbance of homeostasis. This is why drinking alcohol when you are exercising, or simply perspiring heavily on a hot day, is not a good idea. Instead of keeping you hydrated, an alcoholic beverage, such as beer, will have just the opposite effect.

Controlling the Reproductive System

Few systems intrigue us more than the reproductive system, which couldn't function without nervous and endocrine control (Fig. 15.21). The hypothalamus controls the anterior pituitary, which, in turn, controls the release of hormones from the testes and the ovaries and the produc-

Figure 15.21   Control of reproduction.
Successful reproduction requires the participation of both the nervous and endocrine systems.

tion of their gametes. The nervous system directly controls the muscular contractions of the ducts, which propel the sperm, and the oviducts, which move a developing embryo to the uterus, where development continues. Without the positive feedback cycle involving oxytocin, produced by the hypothalamus and released by the posterior pituitary, birth might not occur.

The Neuroendocrine System

The nervous and endocrine systems work so closely together, they form what is sometimes called the neuroendocrine system. As we have seen, the hypothalamus certainly bridges the regulatory activities of both the nervous and endocrine systems. In addition to producing the hormones released by the posterior pituitary, the hypothalamus produces hormones that control the anterior pituitary. And the hypothalamus acts directly through the nerves of the autonomic system to control other organs. The hypothalamus truly belongs to both the nervous and endocrine systems. Indeed, it is often and appropriately referred to as a neuroendocrine organ.

✔ Check Your Progress 15.7

1. If sweating has caused you to lose Na^+ and water, how will your body restore its salt-water balance?

2. Give examples to show that the hypothalamus belongs to both the nervous system and the endocrine system.

Summarizing the Concepts

15.1 Endocrine Glands

Endocrine glands secrete hormones into the bloodstream, and from there they are distributed to target organs or tissues.

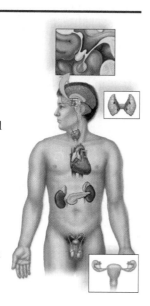

- Hormones are a type of chemical signal that usually act at a distance between body parts.
- Hormones are either peptides or steroids.
- Reception of a peptide hormone at the plasma membrane activates an enzyme cascade inside the cell.
- Steroid hormones combine with a receptor, and the complex attaches to and activates DNA. Protein synthesis follows.

15.2 Hypothalamus and Pituitary Gland

Neurosecretory cells in the hypothalamus produce antidiuretic hormone and oxytocin, which are stored in axon endings in the posterior pituitary until they are released.

- The hypothalamus produces hypothalamic-releasing and hypothalamic-inhibiting hormones, which pass to the anterior pituitary by way of a portal system.
- The anterior pituitary produces at least six types of hormones, and some of these stimulate other hormonal glands to secrete hormones.

15.3 Thyroid and Parathyroid Glands

The thyroid gland requires iodine to produce triiodothyronine and thyroxine, which increase the metabolic rate.

- If iodine is available in limited quantities, a simple goiter develops.
- If the thyroid is overactive, an exophthalmic goiter develops.
- The thyroid gland produces calcitonin, which helps lower the blood calcium level.
- The parathyroid glands secrete parathyroid hormone, which raises the blood calcium level.

15.4 Adrenal Glands

The adrenal glands respond to stress:

- **Adrenal Medulla**   The adrenal medulla immediately secretes epinephrine and norepinephrine. Heartbeat and blood pressure increase; blood glucose level rises; muscles become energized.
- **Adrenal Cortex**   The adrenal cortex produces the glucocorticoids (e.g., cortisol) and the mineralocorticoids (e.g., aldosterone). The glucocorticoids regulate carbohydrate, protein, and fat metabolism and also suppress the inflammatory response. Mineralocorticoids regulate salt and water balance, leading to increases in blood volume and blood pressure.

15.5 Pancreas

The pancreatic islets secrete the hormones insulin and glucagon.

- Insulin lowers the blood glucose level.
- Glucagon raises the blood glucose level.
- Diabetes mellitus is due to the failure of the pancreas to produce insulin or the failure of the cells to take it up.

15.6 Other Endocrine Glands

Other endocrine glands produce hormones.

- The testes and ovaries produce the sex hormones. Male sex hormones are the androgens (e.g., testosterone); female sex hormones are the estrogens and progesterone.
- The thymus gland secretes thymosins, which stimulate T-lymphocyte production and maturation.
- The pineal gland produces melatonin, which may be involved in circadian rhythms and the development of the reproductive organs.

Tissues also produce hormones.

- Kidneys produce erythropoietin.
- Adipose tissue produces leptin, which acts on the hypothalamus.
- Prostaglandins are produced within cells and act locally.

15.7 Homeostasis

The nervous and endocrine systems exert control over the other systems and thereby maintain homeostasis.

- The nervous system is able to respond to the external environment after receiving data from the sensory receptors. Sensory receptors are present in such organs as the eyes and ears.
- The nervous and endocrine systems work together to govern the subconscious control of internal organs. This control often depends on reflex actions involving the hypothalamus and medulla oblongata.
- The nervous and endocrine systems work so closely together that they form what is sometimes called the neuroendocrine system.

Understanding Key Terms

acromegaly 302	Cushing syndrome 307
Addison disease 307	cyclic adenosine monophosphate (cAMP) 299
adrenal cortex 305	
adrenal gland 305	diabetes mellitus 310
adrenal medulla 305	endocrine gland 296
adrenocorticotropic hormone (ACTH) 300	epinephrine 305
	estrogen 311
aldosterone 306	exophthalmic goiter 303
anabolic steroid 311	first messenger 299
androgen 311	glucagon 308
anterior pituitary 300	glucocorticoid 305
antidiuretic hormone (ADH) 300	gonad 311
	gonadotropic hormone 300
atrial natriuretic hormone (ANH) 306	growth hormone (GH) 300
	hormone 296
calcitonin 304	hypothalamic-inhibiting hormone 300
chemical signal 298	
circadian rhythm 312	hypothalamic-releasing hormone 300
congenital hypothyroidism 303	
cortisol 306	hypothalamus 300

insulin 308
leptin 312
melanocyte-stimulating
  hormone (MSH) 300
melatonin 311
mineralocorticoid 305
myxedema 303
norepinephrine 305
ovary 311
oxytocin 300
pancreas 308
pancreatic islets 308
parathyroid gland 304
parathyroid hormone
  (PTH) 304
peptide hormone 299
pheromone 298
pineal gland 311
pituitary dwarfism 302

pituitary gland 300
positive feedback 300
posterior pituitary 300
progesterone 311
prolactin (PRL) 300
prostaglandin 312
renin 306
second messenger 299
simple goiter 303
steroid hormone 299
testes 311
testosterone 311
tetany 304
thymosin 311
thymus gland 311
thyroid gland 303
thyroid-stimulating
  hormone (TSH) 300
thyroxine (T$_4$) 303

Match the key terms to these definitions.

a. _____ Organ that is in the neck and secretes several important hormones, including thyroxine and calcitonin.

b. _____ Condition characterized by high blood glucose level and the appearance of glucose in the urine.

c. _____ Hormone secreted by the anterior pituitary that stimulates portions of the adrenal cortex.

d. _____ Type of hormone that causes the activation of an enzyme cascade in cells.

e. _____ Hormone released by the posterior pituitary that causes contraction of the uterus and milk letdown.

Testing Your Knowledge of the Concepts

1. Compare and contrast the nervous and endocrine systems. (page 296)

2. How does the action of a peptide hormone differ from that of a steroid hormone? (pages 298–99)

3. Explain the relationship between the hypothalamus and the posterior pituitary gland and to the anterior pituitary gland. List the hormones secreted by the anterior pituitary and the posterior pituitary glands and their actions. (pages 300–301)

4. Give an example of the negative feedback relationship among the hypothalamus, the anterior pituitary, and other endocrine glands. (page 300)

5. Discuss the action of growth hormone on the body. What occurs if there is too much or too little GH during the growing years? What occurs if there if too much GH in an adult? (pages 300, 302–3)

6. What types of goiters and other conditions are associated with a malfunctioning thyroid gland? Explain each type. (page 303)

7. Explain how the thyroid and parathyroid glands work together to maintain blood calcium homeostasis. (pages 303–4)

8. What type of tissue is the adrenal medulla made of? What are its hormones and their actions? (page 305)

9. What hormones are secreted by the adrenal cortex, and what are their actions? (pages 305–6)

10. What are the causes and symptoms of Addison disease and Cushing syndrome? (page 307)

11. Explain how insulin and glucagon maintain blood glucose homeostasis. What are the two types of diabetes mellitus, and what are the major symptoms? (pages 308–10)

12. Name the other endocrine glands and tissues mentioned in the chapter, and discuss the actions of the hormones they secrete. (pages 311–12)

13. How does the neuroendocrine system work with other systems to maintain homeostasis? (pages 314–15)

14. Hormones are never
  a. steroids.
  b. amino acids.
  c. glycoproteins.
  d. fats (triglycerides).

15. Which type of glands are ductless?
  a. exocrine
  b. endocrine
  c. Both a and b are correct.
  d. Neither a nor b is correct.

16. Which hormones can cross cell membranes?
  a. peptide hormones
  b. steroid hormones
  c. Both a and b are correct.
  d. Neither a nor b is correct.

17. The anterior pituitary controls the secretion(s) of
  a. both the adrenal medulla and the adrenal cortex.
  b. both thyroid and adrenal cortex.
  c. both ovaries and testes.
  d. Both b and c are correct.

18. Growth hormone is produced by the
  a. posterior adrenal gland.
  b. posterior pituitary.
  c. anterior pituitary.
  d. kidneys.
  e. None of these is correct.

19. _____ is released through positive feedback and causes _____.
  a. Insulin, stomach contractions
  b. Oxytocin, stomach contractions
  c. Oxytocin, uterine contractions
  d. None of these is correct.

20. PTH causes the blood levels of calcium to _____, and calcitonin causes it to _____.
  a. increase, increase
  b. increase, decrease
  c. decrease, increase
  d. decrease, decrease

21. Bodily response to stress includes
  a. water reabsorption by the kidneys.
  b. blood pressure increase.
  c. increase in blood glucose levels.
  d. heart rate increase.
  e. All of these are correct.

22. Anabolic steroid use can cause
  a. liver damage.
  b. severe acne.
  c. balding.
  d. reduced testicular size.
  e. All of these are correct.

23. Lack of aldosterone will cause a blood imbalance of
    a. sodium.
    b. potassium.
    c. water.
    d. All of these are correct.
    e. None of these is correct.

24. Glucagon causes
    a. use of fat for energy.
    b. glycogen to be converted to glucose.
    c. use of amino acids to form fats.
    d. Both a and b are correct.
    e. None of these is correct.

25. Long-term complications of diabetes include
    a. blindness.
    b. kidney disease.
    c. circulatory disorders.
    d. All of these are correct.
    e. None of these is correct.

26. Diabetes mellitus is associated with
    a. too much insulin in the blood.
    b. too high a blood glucose level.
    c. blood that is too dilute.
    d. All of these are correct.

27. Which of these is not a pair of antagonistic hormones?
    a. insulin—glucagon
    b. calcitonin—parathyroid hormone
    c. cortisol—epinephrine
    d. aldosterone—atrial natriuretic hormone (ANH)

28. Which hormone and condition is mismatched?
    a. growth hormone—acromegaly
    b. thyroxine—goiter
    c. parathyroid hormone—tetany
    d. cortisol—myxedema
    e. insulin—diabetes

In questions 29–33, match the hormones to the correct gland in the key.

Key:

    a. glucagon
    b. prostaglandin
    c. melatonin
    d. insulin
    e. leptin

29. Raises blood glucose levels

30. Conversion of glucose to glycogen

31. Hunger control

32. Controls circadian rhythms

33. Causes uterine contractions

In questions 34–38, match the hormones to the correct gland in the key.

Key:

    a. pancreas
    b. anterior pituitary
    c. posterior pituitary
    d. thyroid
    e. adrenal medulla
    f. adrenal cortex

34. Cortisol

35. Growth hormone (GH)

36. Oxytocin storage

37. Insulin

38. Epinephrine

39. Complete the diagram below.

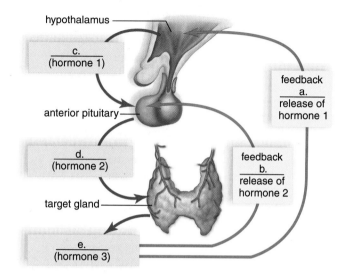

Thinking Critically About the Concepts

Adam's treatments were considered a success because after a year of GH treatments, he was found to be 5'4". Adam, from the opening story, was thrilled by his new stature because he was now taller than most of the girls in his class!

1. How is follicle-stimulating hormone similar to growth hormone?

2. Adults continue to produce very small quantities of growth hormone (much lower than in someone diagnosed with acromegaly). That growth hormone doesn't make people taller since their epiphyseal plates have ossified. How does this continued production of growth hormone affect people's appearance after a long time?

3. Growth hormone has been touted by some as the "fountain of youth."
   a. What do you think elderly people would gain from supplemental growth hormone?
   b. What potentially harmful side effects might result from the use of supplemental GH?

4a. Now that GH (for treatment of dwarfism or low GH levels) is easier to obtain, what abuses of its use would you predict?
   b. Do you think insurance companies should be expected to pay for GH treatment if a child is not expected to be a pituitary dwarf?
   c. Do you think humans (our culture/society) will lose anything if pituitary dwarfs become "extinct," due to GH treatment/therapy?

5. Gigantism is thought to be caused by a pituitary tumor. What other medical/developmental problems would you expect to result from a pituitary tumor?

C H A P T E R

16

Reproductive System

Jessica grimly contemplated her calendar. She methodically counted the days from now until her wedding day and the start of her honeymoon. Her fiance had planned a fabulous trip to Tahiti, and they anticipated relaxing on the beach for a considerable amount of time. "It's just my luck," she moaned to her roommate, Abbie. "It looks like I'll start my period on my wedding day and have it while I'm on my honeymoon."

"Don't pout," replied Abbie. "Read this article about preventing a period." Abbie handed the latest issue of *Modern Woman* to Jessica. She located the article and started reading. The article included an interview with a gynecologist who said that it's not harmful if a period is missed. The gynecologist said to continue the use of birth control pills (the active ones and omit the inactive—placebo—ones), and a woman would skip her period when it was likely to occur at an inconvenient time, like a wedding day or during a beach vacation.

Jessica also learned another helpful bit of information from the gynecologist's interview. Antibiotics can compromise the effectiveness of the pill in preventing pregnancy. In case she needed to take antibiotics during their honeymoon, Jessica decided to pack some condoms for her husband to use. She intended to thoroughly enjoy being a newlywed before starting the family she and her fiance intended to have in the future.

In this chapter, you will learn about the male and female reproductive systems and how the birth control pill and other contraceptives work.

C H A P T E R   C O N C E P T S

16.1 Human Life Cycle
The male reproductive system produces sperm, and the female reproductive system produces eggs. Whereas body cells have 46 chromosomes, sperm and egg have only 23 chromosomes each due to reduction division that takes place as they develop.

16.2 Male Reproductive System
The male reproductive system consists of the testes and a series of ducts that deliver sperm by way of the penis to a female sexual partner. The testes produce sperm and also the male sex hormones that maintain the sex organs and the secondary sex characteristics of males.

16.3 Female Reproductive System
In the female, the ovaries produce eggs and the sex hormones. The sex hormones maintain the sex organs and the secondary sex characteristics of females.

16.4 Female Hormone Levels
The sex hormones fluctuate in monthly cycles, resulting in ovulation once a month followed by menstruation if pregnancy does not occur. Pregnancy and the birth control pill prevent these events from occurring.

16.5 Control of Reproduction
Numerous birth control methods are available for those who wish to prevent pregnancy. Some couples are infertile, and if so, they may use assisted reproductive technologies in order to have a child.

16.6 Sexually Transmitted Diseases
Medications have been developed to control AIDS and genital herpes, but these STDs are not curable. A vaccine is now available for the most common form of genital warts and hepatitis A and B.

STDs caused by bacteria are curable with antibiotic therapy, but resistance is making this more and more difficult.

16.1 Human Life Cycle

Unlike the other systems of the body, the reproductive system is quite different in males and females. *Puberty* is the sequence of events by which a child becomes a sexually competent young adult. The reproductive system does not begin to fully function until puberty is complete. Sexual maturity occurs between the age of 11 and 13 in girls, and 14 and 16 in boys. At the completion of puberty, the individual is capable of producing children.

The reproductive organs (genitals) have the following functions:

1. Males produce sperm within testes, and females produce eggs within ovaries.

2. Males nurture and transport the sperm in ducts until they exit the penis, and females transport the eggs in uterine tubes to the uterus.

3. The male penis functions to deliver sperm to the female vagina, which functions to receive the sperm. The vagina also transports menstrual fluid to the exterior and is the birth canal.

4. The uterus of the female allows the fertilized egg to develop within her body. After birth, the female breast provides nourishment in the form of milk.

5. The testes and ovaries produce the sex hormones that maintain the testes and ovaries and have a profound effect on the body because they bring about masculinization and feminization of various features. In females, the sex hormones also allow a pregnancy to continue.

Mitosis and Meiosis

Our DNA is distributed among 46 chromosomes within the nucleus. Each and every cell in the body has 46 chromosomes, and ordinarily when a cell divides by a process called mitosis, the new cells also have 46 chromosomes. Mitosis is *duplication division*. (As an analogy, imagine the cell producing exact copies of itself during mitosis, much like a duplicating machine does with a page of notes.) In the life cycle of a human being, mitosis is the type of cell division that takes place during growth and repair of tissues (Fig. 16.1).

In addition to mitosis, the human includes a type of cell division called meiosis, which is *reduction division*. Meiosis takes place only in the testes of males during the production of sperm and in the ovaries of females during the production of eggs. During meiosis, the chromosome number is reduced from the normal 46 chromosomes, called the diploid, or 2n, number, down to 23 chromosomes, called the haploid, or n, number of chromosomes. Meiosis requires two successive divisions, called meiosis I and meiosis II.

The flagellated sperm is quite small compared to the egg. It is specialized to carry only chromosomes as it swims to the egg. The egg is specialized to await the arrival of a sperm and to provide the new individual with cytoplasm in addition to chromosomes. The first cell of a new human being is called the zygote. Because a sperm has 23 chromosomes and the egg has 23 chromosomes, the zygote has 46 chromosomes altogether. Without meiosis, the chromosome number in each generation of human beings would double, and the cells would no longer be able to function.

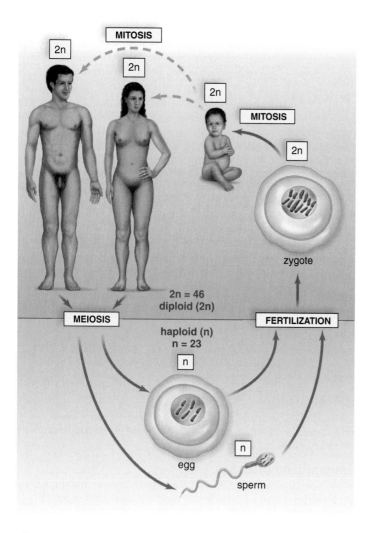

Figure 16.1   The human life cycle.
The human life cycle has two types of cell divisions: mitosis, in which the chromosome number stays constant, and meiosis, in which the chromosome number is reduced. During growth or cell repair, mitosis ensures that each new cell has 46 chromosomes. During production of sex cells, the chromosome number is reduced from 46 to 23. Therefore, an egg and a sperm each have 23 chromosomes so that when the sperm fertilizes the egg, the new cell called a zygote has 46 chromosomes.

☑ Check Your Progress 16.1

1. Contrast the two types of cell division in the human life cycle.

2. a. How many chromosomes does a mother contribute to the new individual? b. How many does the father contribute?

3. a. Where does meiosis occur in males? b. In females?

16.2 Male Reproductive System

The male reproductive system includes the organs depicted in Figure 16.2 and listed in Table 16.1. The male gonads, or primary sex organs, are paired **testes** (sing., testis), which are suspended within the sacs of the **scrotum.**

Sperm produced by the testes mature within the **epididymis** (pl., epididymides), which is a tightly coiled duct lying just outside each testis. Maturation seems to be required in order for sperm to swim to the egg. When sperm leave an epididymis, they enter a **vas deferens** (pl., vasa deferentia), also called the ductus deferens, where they may also be stored for a time. Each vas deferens passes into the abdominal cavity, where it curves around the bladder and empties into an ejaculatory duct. The ejaculatory ducts enter the **urethra.**

At the time of ejaculation, sperm leave the penis in a fluid called **semen.** The seminal vesicles, the prostate gland, and the bulbourethral glands (Cowper glands) add secretions to seminal fluid. The pair of **seminal vesicles** lie at the base of the bladder, and each has a duct that joins with a vas deferens. The **prostate gland** is a single, donut-shaped gland that surrounds the upper portion of the urethra just below the bladder. In older men, the prostate can enlarge and squeeze off the urethra, making urination painful and difficult. The condition can be treated medically. **Bulbourethral glands** are pea-sized organs that lie posterior to the prostate on either side of the urethra. Their secretion makes the seminal fluid gelatinous.

Table 16.1	Male Reproductive Organs
Organ	**Function**
Testes	Produce sperm and sex hormones
Epididymides	Ducts where sperm mature and some sperm are stored
Vasa deferentia	Conduct and store sperm
Seminal vesicles	Contribute nutrients and fluid to semen
Prostate gland	Contributes fluid to semen
Urethra	Conducts sperm
Bulbourethral glands	Contribute mucus-containing fluid to semen
Penis	Organ of sexual intercourse

Each component of seminal fluid seems to have a particular function. Sperm are more viable in a basic solution, and seminal fluid, which is milky in appearance, has a slightly basic pH (about 7.5). Swimming sperm require energy, and seminal fluid contains the sugar fructose, which presumably serves as an energy source. Semen also contains prostaglandins, chemicals that cause the uterus to contract. Some investigators believe that uterine contractions help propel the sperm toward the egg.

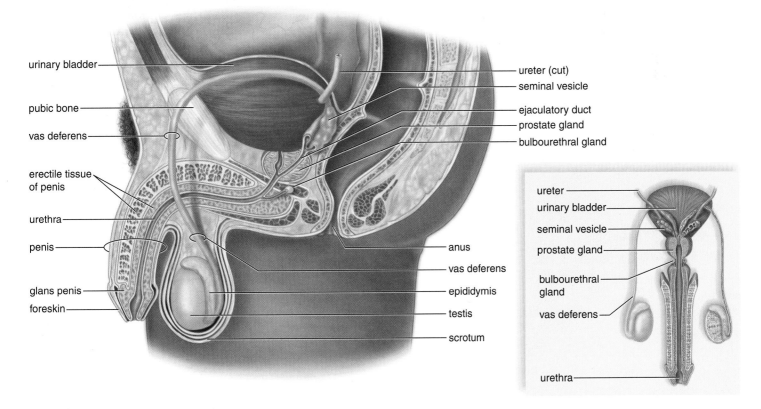

Figure 16.2   The male reproductive system.
The testes produce sperm. The seminal vesicles, the prostate gland, and the bulbourethral glands provide a fluid medium for the sperm, which move from the vas deferens through the ejaculatory duct to the urethra in the penis. The foreskin (prepuce) is removed when a penis is circumcised.

Orgasm in Males

The **penis** (Fig. 16.3) is the male organ of sexual intercourse. The penis has a long shaft and an enlarged tip called the glans penis. The glans penis is normally covered by a layer of skin called the foreskin. **Circumcision,** the surgical removal of the foreskin, is usually done soon after birth.

Spongy, erectile tissue containing distensible blood spaces extends through the shaft of the penis. During sexual arousal, autonomic nerves release nitric oxide, NO. This stimulus leads to the production of cGMP (cyclic guanosine monophosphate), which causes the smooth muscle of incoming arterial walls to relax and the erectile tissue to fill with blood. The veins that take blood away from the penis are compressed, and the penis becomes erect. **Erectile dysfunction** (formerly called impotency) exists when the erectile tissue doesn't expand enough to compress the veins. Medications for the treatment of erectile dysfunction inhibit the enzyme that breaks down cGMP, ensuring that a full erection will take place. Certain of these medications can cause vision problems because the same enzyme occurs in the retina. During an erection, a sphincter closes off the bladder so that no urine enters the urethra. (The urethra carries either urine or semen at different times.)

In the opening story, Jessica decided to pack condoms for her husband to use just in case they were needed as a backup method of birth control on her honeymoon. A condom is a sheath that fits over the shaft of the erect penis. If a condom is used correctly each time, it is estimated to be about 97% effective in preventing pregnancy.

As sexual stimulation intensifies, sperm enter the urethra from each vas deferens, and the glands contribute secretions to the seminal fluid. Once seminal fluid is in the urethra, rhythmic muscle contractions cause it to be expelled from the penis in spurts (ejaculation). A condom catches seminal fluid so that it does not enter the vagina.

The contractions that expel seminal fluid from the penis are a part of male orgasm, the physiological and psychological sensations that occur at the climax of sexual stimulation. The psychological sensation of pleasure is centered in the brain, but the physiological reactions involve the genital (reproductive) organs and associated muscles, as well as the entire body. Marked muscular tension is followed by contraction and relaxation. Following ejaculation and/or loss of sexual arousal, the penis returns to its normal flaccid state. Usually a period of time, called the refractory period, follows during which stimulation does not bring about an erection. The length of the refractory period increases with age.

There may be in excess of 400 million sperm in the 3.5 ml of semen expelled during ejaculation. The sperm count can be much lower than this, however, and fertilization of the egg by a sperm can still take place.

Male Gonads, the Testes

The testes, which produce sperm and also the male sex hormones, lie outside the abdominal cavity of the male, within the scrotum. The testes begin their development inside the abdominal cavity but descend into the scrotal sacs during the last two months of fetal development. If, by chance, the testes do not descend and the male is not treated or operated on to place the testes in the scrotum, sterility—the inability to produce offspring—usually follows. This is because the internal temperature of the body is too high to produce viable sperm. The scrotum helps regulate the temperature of the testes by holding them closer or farther away from the body.

Seminiferous Tubules and Interstitial Cells

A longitudinal section of a testis shows that it is composed of compartments called lobules, each of which contains one to

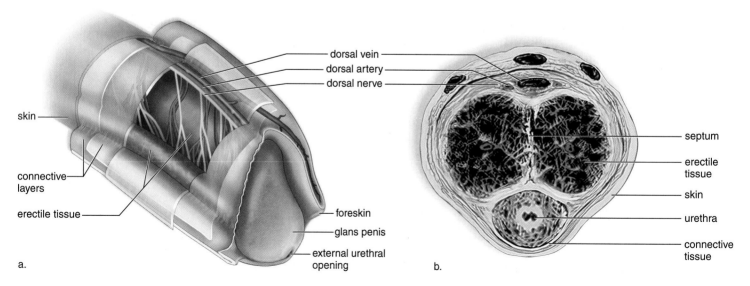

Figure 16.3 **Penis anatomy.**
a. Penis shaft. The shaft of the penis ends in an enlarged tip called the glans penis, which, in uncircumsized males, is partially covered by a foreskin (prepuce). The penis contains columns of erectile tissue. **b.** Micrograph of shaft in cross section showing location of erectile tissue. One column surrounds the urethra.

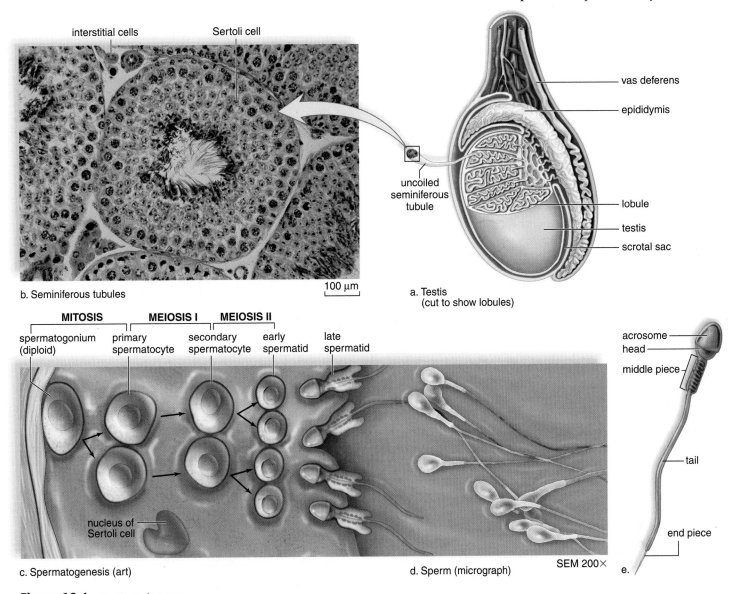

Figure 16.4   **Testis and sperm.**

a. The lobules of a testis contain seminiferous tubules. **b.** Electron micrograph of a cross section of the seminiferous tubules, where spermatogenesis occurs. Note the location of interstitial cells in clumps among the seminiferous tubules. **c.** Diagrammatic representation of spermatogenesis, which occurs in wall of tubules. **d.** Micrograph of sperm. **e.** A sperm has a head, a middle piece, and a tail. The nucleus is in the head, which is capped by the enzyme-containing acrosome.

three tightly coiled **seminiferous tubules** (Fig. 16.4*a*). A microscopic cross section of a seminiferous tubule reveals that it is packed with cells undergoing **spermatogenesis** (Fig. 16.4*b*), the production of sperm.

During the production of sperm, spermatogonia divide to produce primary spermatocytes (2n) that move away from the outer wall, increase in size, and undergo meiosis I to produce secondary spermatocytes, each with only 23 chromosomes (Fig. 16.4*c*). Secondary spermatocytes (n) undergo meiosis II to produce four spermatids, each of which also has 23 chromosomes. Spermatids then differentiate into sperm. Note the presence of **Sertoli cells** (purple), which support, nourish, and regulate the process of spermatogenesis. It takes approximately 74 days for sperm to undergo development from spermatogonia to sperm.

Mature **sperm,** or spermatozoa, have three distinct parts: a head, a middle piece, and a tail (Fig. 16.4*d*). Mitochondria in the middle piece provide energy for the movement of the tail, which is a flagellum. The head contains a nucleus covered by a cap called the **acrosome,** which stores enzymes needed to penetrate the egg. The ejaculated semen of a normal human male contains several hundred million sperm, but only one sperm normally enters an egg. Sperm usually do not live more than 48 hours in the female genital tract.

Interstitial Cells

The male sex hormones, the androgens, are secreted by cells that lie between the seminiferous tubules. Therefore, they are called **interstitial cells.** The most important of the androgens is testosterone, whose functions are discussed next.

Hormonal Regulation in Males

The hypothalamus has ultimate control of the testes' sexual function because it secretes a hormone called **gonadotropin-releasing hormone,** or **GnRH,** that stimulates the anterior pituitary to secrete the gonadotropic hormones. There are two gonadotropic hormones, **follicle-stimulating hormone (FSH)** and **luteinizing hormone (LH),** which are present in both males and females. In males, FSH promotes the production of sperm in the seminiferous tubules. LH in males controls the production of testosterone by the interstitial cells.

All these hormones are involved in a negative feedback relationship that maintains the fairly constant production of sperm and testosterone (Fig. 16.5). When the amount of testosterone in the blood rises to a certain level, it causes the hypothalamus and anterior pituitary to decrease their respective secretion of GnRH and LH. As the level of testosterone begins to fall, the hypothalamus increases its secretion of GnRH, and the anterior pituitary increases its secretion of LH, which stimulates the interstitial cells to produce testosterone. A similar feedback mechanism maintains the continuous production of sperm. The Sertoli cells in the wall of the seminiferous tubules produce a hormone called *inhibin* that blocks GnRH and FSH secretion when appropriate (Fig. 16.4).

Testosterone, the main sex hormone in males, is essential for the normal development and functioning of the organs listed in Table 16.1. Testosterone also brings about and maintains the male secondary sex characteristics that develop at the time of puberty. Males are generally taller than females and have broader shoulders and longer legs relative to trunk length. The deeper voices of males compared with those of females are due to a larger larynx with longer vocal cords. Since the so-called Adam's apple is a part of the larynx, it is usually more prominent in males than in females. Testosterone causes males to develop noticeable hair on the face, chest, and occasionally other regions of the body, such as the back. A related chemical also leads to the receding hairline and pattern baldness that occur in males.

Testosterone is responsible for the greater muscular development in males. Knowing this, both males and females sometimes take anabolic steroids, which are either testosterone or related steroid hormones resembling testosterone. Health problems involving the kidneys, the cardiovascular system, and hormonal imbalances can arise from such use. The testes shrink in size, and feminization of other male traits occurs (see page 241).

☑ Check Your Progress 16.2

1 a. What male structure functions to produce sperm? b. What structures are involved in the production of seminal fluid? c. Which structure is involved in transport of semen out of the body?

2. a. Which endocrine glands are involved in promoting and maintaining the sex characteristics of males? b. Which hormones are involved?

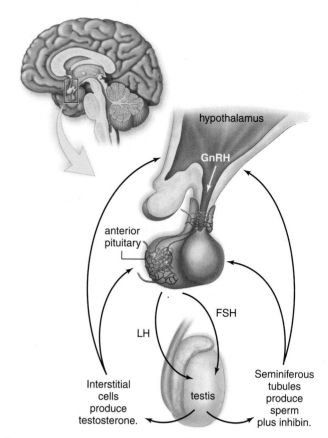

Figure 16.5 Hormonal control of testes.
GnRH (gonadotropin-releasing hormone) stimulates the anterior pituitary to secrete the gonadotropic hormones: Follicle-stimulating hormone (FSH) stimulates the production of sperm, and luteinizing hormone (LH) stimulates the production of testosterone. Testosterone and inhibin exert negative feedback control over the hypothalamus and the anterior pituitary, and this regulates the level of testosterone in the blood and the production of sperm by the testes.

16.3 Female Reproductive System

The female reproductive system includes the organs depicted in Figure 16.6 and listed in Table 16.2. The female gonads are paired **ovaries** that lie in shallow depressions, one on each side of the upper pelvic cavity. The ovaries produce **eggs** and the female sex hormones, estrogen and progesterone.

The Genital Tract

The **oviducts,** also called the uterine or fallopian tubes, extend from the uterus to the ovaries; however, the oviducts are not attached to the ovaries. Instead, they have finger-like projections called fimbriae (sing., **fimbria**) that sweep over the ovaries. When an egg bursts from an ovary during ovulation, it usually is swept into an oviduct by the combined action of the fimbriae and the beating of cilia that line the oviducts.

Once in the oviduct, the egg is propelled slowly by ciliary movement and tubular muscle contraction toward the uterus. An egg lives approximately only 6–24 hours, unless fertilization occurs. Fertilization, and therefore **zygote** formation, usually takes place in the oviduct. A developing embryo normally arrives at the uterus after several days, and then

implantation occurs—the embryo embeds in the uterine lining, which has been prepared to receive it.

The **uterus** is a thick-walled, muscular organ about the size and shape of an inverted pear. Normally, it lies above and is tipped over the urinary bladder. The oviducts join the uterus at its upper end, while at its lower end, the **cervix** enters the vagina nearly at a right angle.

Cancer of the cervix is a common form of cancer in women. Early detection is possible by means of a **Pap test,** which requires the removal of a few cells from the region of the cervix for microscopic examination. If the cells are cancerous, a physician may recommend a hysterectomy. A hysterectomy is the removal of the uterus, including the cervix. Removal of the ovaries in addition to the uterus is technically termed an ovariohysterectomy (radical hysterectomy). Because the vagina remains, the woman can still engage in sexual intercourse.

Development of the embryo and fetus normally takes place in the uterus. This organ, sometimes called the womb, is approximately 5 cm wide in its usual state but is capable of stretching to over 30 cm wide to accommodate a growing fetus. The lining of the uterus, called the **endometrium,** participates in the formation of the placenta (see page 330), which supplies nutrients needed for embryonic and fetal development. The endometrium has two layers: a basal layer and an inner, functional layer. In the nonpregnant female, the functional layer of the endometrium varies in thickness according to a monthly reproductive cycle called the uterine cycle.

A small opening in the cervix leads to the vaginal canal. The **vagina** is a tube that lies at a 45° angle to the small of the back. The mucosal lining of the vagina lies in folds and can extend. This is especially important when the vagina serves as the birth canal, and it facilitates sexual intercourse, when the vagina receives the penis. The vagina also acts as an exit for menstrual flow. Several different types of bacteria normally reside in the vagina and create an acidic environment. This environment is protective against the possible growth of pathogenic bacteria, but sperm prefer the basic environment provided by seminal fluid.

Table 16.2	Female Reproductive Organs
Organ	**Function**
Ovaries	Produce eggs and sex hormones
Oviducts	Conduct eggs; location of fertilization (uterine or fallopian tubes)
Uterus (womb)	Houses developing fetus
Cervix	Contains opening to uterus
Vagina	Receives penis during sexual intercourse; serves as birth canal and as an exit for menstrual flow

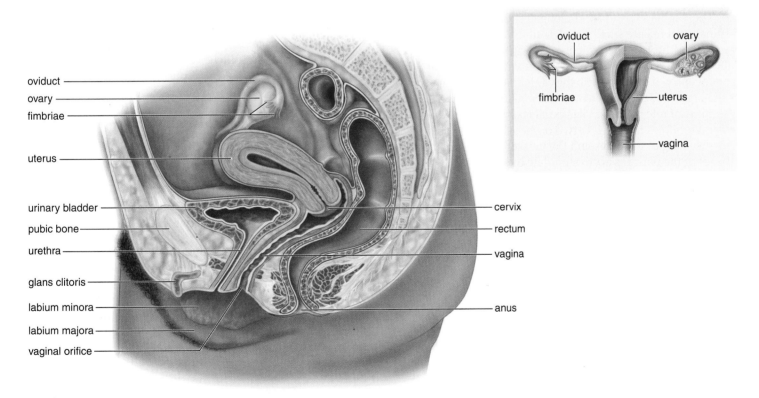

Figure 16.6   The female reproductive system.
The ovaries usually release one egg a month; fertilization occurs in the oviduct, and development occurs in the uterus. The vagina is the birth canal, as well as the organ of sexual intercourse and outlet for menstrual flow.

External Genitals

The external genital organs of the female are known collectively as the **vulva** (Fig. 16.7). The vulva includes two large, hair-covered folds of skin called the labia majora. The labia majora extend backward from the mons pubis, a fatty prominence underlying the pubic hair. The labia minora are two small folds lying just inside the labia majora. They extend forward from the vaginal opening to encircle and form a foreskin for the glans clitoris. The glans clitoris is the organ of sexual arousal in females and, like the penis, contains a shaft of erectile tissue that becomes engorged with blood during sexual stimulation.

The cleft between the labia minora contains the openings of the urethra and the vagina. The vagina may be partially closed by a ring of tissue called the hymen. The hymen is ordinarily ruptured by sexual intercourse or by other types of physical activities. If remnants of the hymen persist after sexual intercourse, they can be surgically removed.

Notice that the urinary and reproductive systems in the female are entirely separate. For example, the urethra carries only urine, and the vagina serves only as the birth canal and the organ for sexual intercourse.

Orgasm in Females

Upon sexual stimulation, the labia minora, the vaginal wall, and the clitoris become engorged with blood. The breasts also swell, and the nipples become erect. The labia majora enlarge, redden, and spread away from the vaginal opening.

The vagina expands and elongates. Blood vessels in the vaginal wall release small droplets of fluid that seep into the vagina and lubricate it. Mucus-secreting glands beneath the labia minora on either side of the vagina also provide lubrication for entry of the penis into the vagina. Although the vagina is the organ of sexual intercourse in females, the clitoris plays a significant role in the female sexual response. The extremely sensitive clitoris can swell to two or three times its usual size. The thrusting of the penis and the pressure of the pubic symphyses of the partners act to stimulate the clitoris.

Orgasm occurs at the height of the sexual response. Blood pressure and pulse rate rise, breathing quickens, and the walls of the uterus and oviducts contract rhythmically. A sensation of intense pleasure is followed by relaxation when organs return to their normal size. Females have no refractory period, and multiple orgasms can occur during a single sexual experience.

✔ Check Your Progress 16.3

1. Among the female structures, which (a) produce the egg, (b) transport the egg, (c) house a developing embryo, and (d) serve as the birth canal?

2. Among the external genitalia, which plays a significant role in female sexual response?

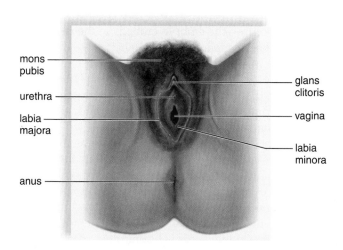

Figure 16.7   External genitals of the female.
At birth, the opening of the vagina is partially blocked by a membrane called the hymen. Physical activities and sexual intercourse rupture the hymen.

16.4 Female Hormone Levels

Hormone levels cycle in the female on a monthly basis, and the ovarian cycle drives the uterine cycle, as discussed in this section.

Ovarian Cycle: Nonpregnant

An ovary contains many **follicles,** and each one contains an immature egg, called an oocyte. A female is born with as many as two million follicles, but the number is reduced to 300,000–400,000 by the time of puberty. Only a small number of follicles (about 400) ever mature because a female usually produces only one egg per month during her reproductive years. As the follicle matures during the **ovarian cycle,** it changes from a primary to a secondary to a vesicular (Graafian) follicle (Fig. 16.8). Epithelial cells of a primary follicle surround a primary oocyte. Pools of follicular fluid bathe the oocyte in a secondary follicle. In a vesicular follicle, the fluid-filled cavity increases to the point that the follicle wall balloons out on the surface of the ovary.

Figure 16.8*b* traces the steps of **oogenesis.** A primary oocyte undergoes meiosis I, and the resulting cells are haploid with 23 chromosomes each. One of these cells is called a polar body. A polar body is a sort of cellular "trash can" because its function is simply to hold discarded chromosomes. The secondary oocyte undergoes meiosis II, but only if it is first fertilized by a sperm cell. If the secondary oocyte remains unfertilized, it never completes meiosis and will die shortly after being released from the ovary.

When appropriate, the vesicular follicle bursts, releasing the oocyte (often called an egg) surrounded by a clear membrane. This process is referred to as **ovulation.** Once a

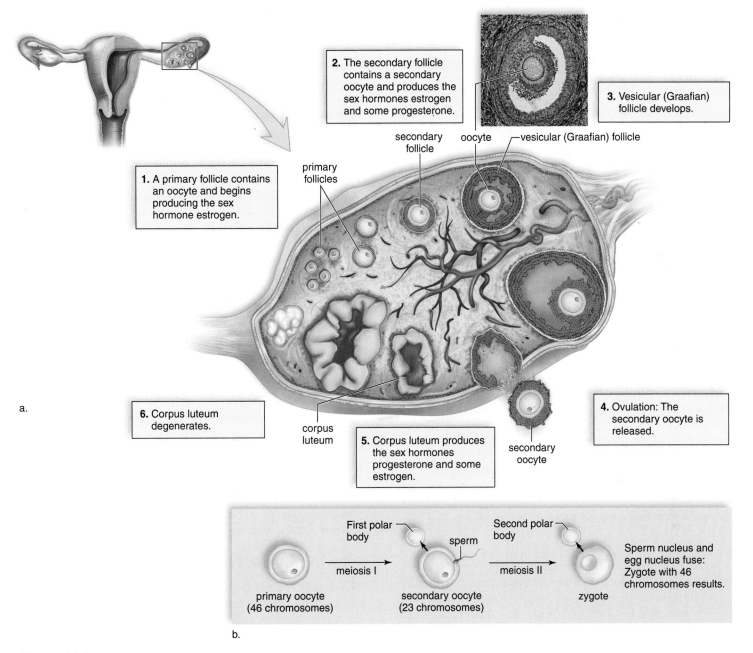

2. The secondary follicle contains a secondary oocyte and produces the sex hormones estrogen and some progesterone.

3. Vesicular (Graafian) follicle develops.

secondary follicle

oocyte

vesicular (Graafian) follicle

1. A primary follicle contains an oocyte and begins producing the sex hormone estrogen.

primary follicles

a.

6. Corpus luteum degenerates.

corpus luteum

5. Corpus luteum produces the sex hormones progesterone and some estrogen.

secondary oocyte

4. Ovulation: The secondary oocyte is released.

First polar body

sperm

Second polar body

primary oocyte (46 chromosomes)

meiosis I

secondary oocyte (23 chromosomes)

meiosis II

zygote

Sperm nucleus and egg nucleus fuse: Zygote with 46 chromosomes results.

b.

Figure 16.8   **Ovarian cycle.**

a. A single follicle actually goes through all stages (1–6) in one place within the ovary. As the follicle matures, layers of follicle cells surround a secondary oocyte. Eventually, the mature follicle ruptures, and the secondary oocyte is released. The follicle then becomes the corpus luteum, which eventually disintegrates. **b.** During oogenesis, the chromosome number is reduced from 46 to 23. Fertilization restores the full number of chromosomes.

vesicular follicle has lost the oocyte, it develops into a **corpus luteum,** a glandlike structure. If the egg is not fertilized, the corpus luteum disintegrates.

As mentioned previously, the ovaries produce eggs and also the female sex hormones estrogen and progesterone. A primary follicle produces estrogen, and a secondary follicle produces estrogen and some progesterone. The corpus luteum produces progesterone.

Phases of the Ovarian Cycle

Similar to the testes, the hypothalamus has ultimate control of the ovaries' sexual function because it secretes gonadotropin-releasing hormone, or GnRH. GnRH stimulates the anterior pituitary to produce FSH and LH, and these hormones control the ovarian cycle. The gonadotropic hormones are not present in constant amounts, and instead are secreted at different rates during the cycle. For

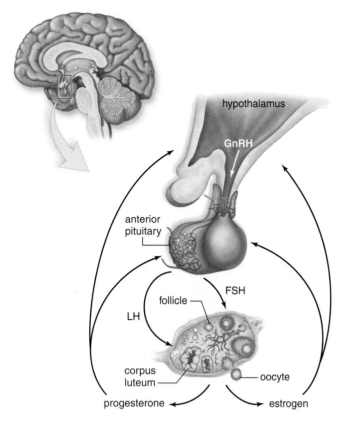

Figure 16.9 Hormonal control of ovaries.
The hypothalamus produces GnRH (gonadotropin-releasing hormone). GnRH stimulates the anterior pituitary to produce FSH (follicle-stimulating hormone) and LH (luteinizing hormone). FSH stimulates the follicle to produce primarily estrogen, and LH stimulates the corpus luteum to produce primarily progesterone. Estrogen and progesterone maintain the sexual organs (e.g., uterus) and the secondary sex characteristics, and they exert feedback control over the hypothalamus and the anterior pituitary. Feedback control regulates the relative amounts of estrogen and progesterone in the blood.

simplicity's sake, it is convenient to emphasize that during the first half, or *follicular phase,* FSH promotes the development of follicles that primarily secrete estrogen (Fig. 16.9). As the estrogen level in the blood rises, it exerts feedback control over the anterior pituitary secretion of FSH so that the follicular phase comes to an end.

Then, due to a positive feedback effect, an estrogen spike causes a sudden secretion of a large amount of GnRH from the hypothalamus. This leads to a surge of LH production by the anterior pituitary and to ovulation at about the 14th day of a 28-day cycle.

Now, the *luteal phase* begins. During the luteal phase of the ovarian cycle, LH promotes the development of the corpus luteum, which secretes primarily progesterone. When pregnancy does not occur, the corpus luteum regresses, and a new cycle begins with menstruation (Fig. 16.10).

Estrogen and Progesterone

Estrogen and progesterone affect not only the uterus but other parts of the body as well. Estrogen is largely responsible for the secondary sex characteristics in females, including body hair and fat distribution. In general, females have a more rounded appearance than males because of a greater accumulation of fat beneath the skin. Like males, females develop axillary and pubic hair during puberty. In females, the upper border of pubic hair is horizontal, but in males, it tapers toward the navel. Both estrogen and progesterone are also required for breast development. Other hormones are involved in milk production following pregnancy and milk letdown when a baby begins to nurse.

The pelvic girdle is wider and deeper in females, so the pelvic cavity usually has a larger relative size compared with that of males. This means that females have wider hips than males and their thighs converge at a greater angle toward the knees. Because the female pelvis tilts forward, females tend to have more of a lower back curve than males, an abdominal bulge, and protruding buttocks.

Menopause, the period in a woman's life during which the ovarian cycle ceases, is likely to occur between ages 45 and 55. The ovaries are no longer responsive to the gonadotropic hormones produced by the anterior pituitary, and the ovaries no longer secrete estrogen or progesterone. At the onset of menopause, menstruation becomes irregular, but as long as it occurs, it is still possible for a woman to conceive. Therefore, a woman is usually not considered to have completed menopause until menstruation is absent for a year.

Until recently, many women took combined estrogen-progestin drugs to ease menopausal symptoms. However, a new study conducted by the Women's Health Initiative (WHI) found that the long-term use of the combined drugs by most menopausal women caused increases in breast cancer, heart attacks, strokes, and blood clots. Those risks outweigh the drugs' actual benefits—a small decrease in hip fractures and a decrease in cases of colorectal cancer.

Uterine Cycle: Nonpregnant

The female sex hormones, **estrogen** and **progesterone,** have numerous functions. One function of these hormones affects the endometrium, causing the uterus to undergo a cyclical series of events known as the **uterine cycle** (Fig. 16.9). Twenty-eight-day cycles are divided as follows:

During *days 1–5,* a low level of estrogen and progesterone in the body causes the endometrium to disintegrate and its blood vessels to rupture. On day 1 of the cycle, a flow of blood and tissues, known as the menses, passes out of the vagina during **menstruation,** also called the menstrual period.

During *days 6–13,* increased production of estrogen by a new ovarian follicle in the ovary causes the endometrium to thicken and become vascular and glandular. This is called the proliferative phase of the uterine cycle.

On *day 14* of a 28-day cycle, ovulation usually occurs.

During *days 15–28,* increased production of progesterone by the corpus luteum in the ovary causes the endometrium of the uterus to double or triple in thickness (from 1 mm to 2–3 mm) and the uterine

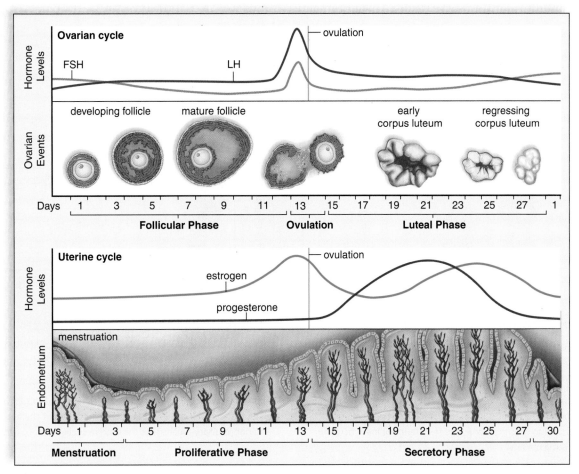

Figure 16.10   Female hormone levels.
During the follicular phase, FSH released by the anterior pituitary promotes the maturation of a follicle in the ovary. The ovarian follicle produces increasing levels of estrogen, which causes the endometrium to thicken during the proliferative phase of the uterine cycle. After ovulation and during the luteal phase of the ovarian cycle, LH promotes the development of the corpus luteum. Progesterone in particular causes the endometrial lining to become secretory. Menses, due to the breakdown of the endometrium, begins when progesterone production declines to a low level.

glands to mature, producing a thick mucoid secretion. This is called the secretory phase of the uterine cycle. The endometrium is now prepared to receive the developing embryo. If this does not occur, the corpus luteum in the ovary regresses, and the low level of sex hormones in the female body results in the endometrium breaking down during menstruation.

Table 16.3 compares the stages of the uterine cycle with those of the ovarian cycle when pregnancy does not occur.

Table 16.3	Ovarian and Uterine Cycles: Nonpregnant		
Ovarian Cycle	**Events**	**Uterine Cycle**	**Events**
Follicular phase—Days 1–13	FSH secretion begins.	Menstruation—Days 1–5	Endometrium breaks down.
	Follicle maturation occurs.	Proliferative phase—Days 6–13	Endometrium rebuilds.
	Estrogen secretion is prominent.		
Ovulation—Day 14*	LH spike occurs.		
Luteal phase—Days 15–28	LH secretion continues.	Secretory phase—Days 15–28	Endometrium thickens, and glands are secretory.
	Corpus luteum forms.		
	Progesterone secretion is prominent.		

*Assuming a 28-day cycle.

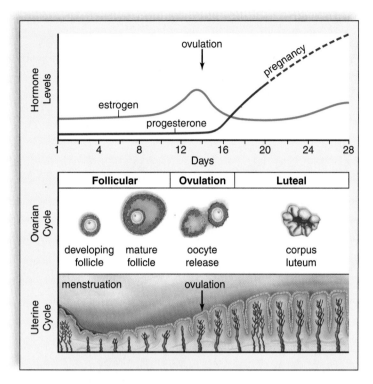

Figure 16.11   **Effect of pregnancy.**
If pregnancy occurs, the corpus luteum does not regress. Instead, the corpus luteum is maintained and secretes increasing amounts of progesterone. Therefore, menstruation does not occur, and the uterine lining, where the embryo resides, is maintained.

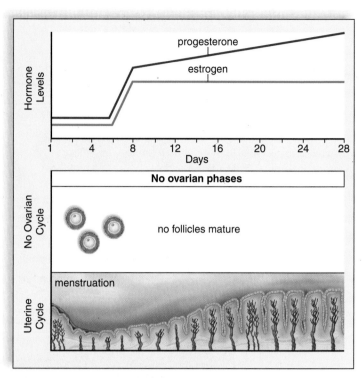

Figure 16.12   **Effect of birth control pills.**
Active pills cause the uterine lining to build up, and this lining is shed when inactive pills are taken. Feedback inhibition of the hypothalamus and anterior pituitary means that the ovarian cycle does not occur.

Fertilization and Pregnancy

Following unprotected sexual intercourse, many sperm will most likely make their way into the oviduct, where the egg is located following ovulation. Only one sperm fertilizes the egg, and it becomes a zygote, which begins development even as it travels down the oviduct to the uterus. The endometrium is now prepared to receive the developing embryo, which becomes implanted in the lining several days following fertilization. Pregnancy has now begun; an abortion is the removal of an implanted embryo.

The **placenta,** which sustains the developing embryo and later the fetus, originates from both maternal and fetal tissues. It is the region of exchange of molecules between fetal and maternal blood, although the two rarely mix. At first, the placenta produces **human chorionic gonadotropin (HCG),** which maintains the corpus luteum in the ovary. (A pregnancy test detects the presence of HCG in the blood or urine.) Rising amounts of HCG stimulate the corpus luteum to produce increasing amounts of progesterone, and this progesterone shuts down the hypothalamus and anterior pituitary so that no new follicles begin in the ovary. The progesterone maintains the uterine lining where the embryo now resides. The absence of menstruation is a signal to the woman that she may be pregnant (Fig. 16.11).

Eventually, the placenta itself produces progesterone and some estrogen. The corpus luteum is no longer needed and it regresses.

Birth Control Pill   Jessica was on the birth control pill to prevent the occurrence of pregnancy. For 21 days she took active pills that contain a synthetic estrogen and progesterone, and these were followed by seven pills that were inactive because they contained no estrogen and progesterone. The active pills supplied her body with both sex hormones for a large part of the cycle. These hormones feed back to inhibit the hypothalmus and the anterior pituitary; therefore, no new follicles began in the ovary, and ovulation did not occur. In effect, the birth control pills fool the body into acting as if pregnancy had occurred (Fig. 16.12).

Birth control pills also thicken the cervical mucus, preventing sperm entry into the uterus. The uterine lining does build up to a degree, and when Jessica takes the inactive pills, a mini-menstruation occurs. Because Jessica skipped taking the inactive pills and continued taking the active pills, menstruation (her period) did not occur on her wedding day and honeymoon. Skipping a period is all right once in a while, but for continued health of the uterine lining, menstruation should not be skipped routinely.

☑ Check Your Progress 16.4

1. **a.** Which hormones control the ovarian cycle? **b.** Which hormones control the uterine cycle?

2. **a.** What changes occur in these cycles when a woman becomes pregnant? **b.** When she takes the birth control pill?

16.5 Control of Reproduction

Several means are available to dampen or enhance our reproductive potential. **Birth control methods** are used to regulate the number of children an individual or couple will have.

Birth Control Methods

The most reliable method of birth control is abstinence—that is, not engaging in sexual intercourse. This form of birth control has the added advantage of preventing transmission of a sexually transmitted disease. Table 16.4 lists other means of birth control used in the United States, and rates their effectiveness. For example, with the birth control pill, we expect 100% effectiveness and no sexually active women will get pregnant within the year. On the other hand, with natural family planning, one of the least effective methods given in the table, we expect that 70% will not get pregnant and 30% will get pregnant within the year.

Figure 16.13 features some of the most effective and commonly used means of birth control. **Contraceptives** are medications and devices that reduce the chance of pregnancy. As discussed, the oral contraception **(birth control pills)** taken by Jessica was typical in that it contains a combination of estrogen and progesterone for the first 21 days, followed by seven inactive pills (Fig. 16.13*h*). The estrogen and progesterone in the birth control pill or a patch (Fig. 16.13*b*) applied to the skin effectively shuts down the pituitary production of both FSH and LH so that no follicle in the ovary begins to develop in the ovary;

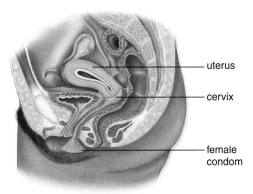

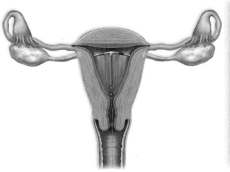

a. Intrauterine device

b. Hormone skin patch

c. Depo-Provera

d. Diaphragm and spermicidal jelly

e. Female condom

uterus

cervix

female condom

f. Male condom

g. Implant

h. Oral contraception

Figure 16.13   Various birth control devices.
a. Intrauterine device. **b.** Hormone skin patch. **c.** Depo-Provera (a progesterone injection). **d.** Diaphragm and spermicidal jelly. **e.** Female condom.
f. Male condom. **g.** Implant. **h.** Oral contraception (birth control pills).

since ovulation does not occur, pregnancy cannot take place. Because of possible side effects, women taking birth control pills should see a physician regularly.

An **intrauterine device (IUD)** is a small piece of molded plastic that is inserted into the uterus by a physician (Fig. 16.13a). IUDs are believed to alter the environment of the uterus and oviducts so that fertilization probably will not occur—but if fertilization should occur, implantation cannot take place.

The **diaphragm** is a soft latex cup with a flexible rim that lodges behind the pubic bone and fits over the cervix (Fig. 16.13d). Each woman must be properly fitted by a physician, and the diaphragm can be inserted into the vagina no more than 2 hours before sexual relations. Also, it must be used with spermicidal jelly or cream and should be left in place at least 6 hours after sexual relations. The cervical cap is a minidiaphragm.

There has been a renewal of interest in barrier methods of birth control, because these methods offer some protection against sexually transmitted diseases. A **female condom,** now available, consists of a large polyurethane tube with a flexible ring that fits onto the cervix (Fig. 16.13e). The open end of the tube has a ring that covers the external genitals. A **male condom** is most often a latex sheath that fits over the erect penis (Fig. 16.13f). The ejaculate is trapped inside the sheath, and thus does not enter the vagina. When used in conjunction with a spermicide, the protection is better than with the condom alone.

Contraceptive implants utilize a synthetic progesterone to prevent ovulation by disrupting the ovarian cycle. The older version of the implant consists of six match-sized, time-release capsules that are surgically implanted under the skin of a woman's upper arm. The newest version consists of a single capsule that remains effective for about three years (Fig. 16.13g).

Contraceptive injections are available as progesterone only (Fig. 16.13c) or a combination of estrogen and progesterone. The length of time between injections can vary from three months to a few weeks.

Contraceptive vaccines are now being developed. For example, a vaccine intended to immunize women against HCG, the hormone so necessary to maintaining the implantation of the embryo, was successful in a limited clinical trial. Since HCG is not normally present in the body, no autoimmune reaction is

expected, but the immunization does wear off with time. Others believe that it would also be possible to develop a safe anti-sperm vaccine that could be used in women.

Vasectomy and Tubal Ligation

Vasectomy and tubal ligation are two methods to bring about sterilization, the inability to reproduce (Fig. 16.14). **Vasectomy** consists of cutting and sealing the vas deferens on each side so that the sperm are unable to reach the seminal fluid that is ejected at the time of orgasm. The sperm are then largely reabsorbed. Following this operation, which can be done in a doctor's office, the amount of ejaculate remains normal because sperm account for only about 1% of the volume of semen. Also, there is no effect on the secondary sex characteristics since testosterone continues to be produced by the testes.

Tubal ligation consists of cutting and sealing the oviducts. Pregnancy rarely occurs because the passage of the egg through the oviducts has been blocked. Using a method called laparoscopy, which requires only two small incisions, the surgeon inserts a small, lighted telescope to view the oviducts and a small surgical blade to sever them.

It is best to view a vasectomy or tubal ligation as permanent. Even following successful reconnection, fertility is usually reduced by about 50%.

Morning-After Pills

A morning-after pill, or emergency contraception, refers to a medication that will prevent pregnancy after unprotected intercourse. The expression "morning-after pill" is a misnomer in that the medication can begin one to several days after unprotected intercourse.

One type, a kit called Preven, contains four synthetic progesterone pills; two are taken up to 72 hours after unprotected intercourse, and two more are taken 12 hours later. The medication upsets the normal uterine cycle, making it difficult for an embryo to implant itself in the endometrium. In a recent study, it was estimated that the medication was 85% effective in preventing unintended pregnancies.

Mifepristone, better known as RU-486, is a pill that is presently used to cause the loss of an implanted embryo by

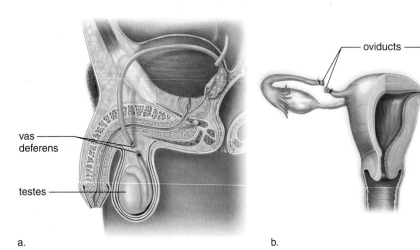

a.   b.

vas deferens

testes

oviducts

Figure 16.14   Sterilization.
a. Vasectomy involves making two small cuts in the skin of the scrotum. Each vas deferens is lifted out and cut. The cut ends are tied or sealed with an electric current. The openings in the scrotum are closed with stitches. **b.** During tubal ligation, one or two small incisions are made in the abdomen. Using instruments that are inserted through the incisions, the oviducts are coagulated (burned), sealed shut with cautery, or cut and tied. The skin incision is then stitched closed.

Table 16.4	Common Methods of Contraception			
Name	**Procedure**	**How Does It Work?**	**Effectiveness?**	**Health Risk**
Abstinence	Refrain from sexual intercourse	No sperm in vagina	100%	None; also protects against STDs
Natural family planning	Determine day of ovulation by keeping records	Intercourse is avoided during the time while ovum is viable	80%	None
Withdrawal method	Penis withdrawn from vagina just before ejaculation	Ejaculation outside the woman's body; no sperm in vagina	75%	None
Douching	Vagina cleansed after intercourse	Washes sperm out of vagina	≥ 70%	May cause inflammation
Male condom	Sheath of latex, polyurethane, or natural material fitted over erect penis	Prevents entry of sperm into vagina; latex and polyurethane forms protect against STDs	89%	Latex allergy with latex forms; no protection against STDs with natural material condoms
Female condom	Polyurethane liner fitted inside vagina	Prevents entry of sperm into vagina; some protection against STDs	79%	Possible allergy or irritation, urinary tract infection
Spermicide: jellies, foams, creams	Spermicidal products inserted into vagina before intercourse	Spermicide nonoxynol-9 kills large numbers of sperm cells	50-80%	Irritation, allergic reaction, urinary tract infection
Contraceptive sponges	Sponge containing spermicide inserted into vagina and placed against cervix	Spermicide nonoxynol-9 kills large numbers of sperm cells	72-86%	Irritation, allergic reaction, urinary tract infection, toxic shock syndrome
Combined hormone: vaginal ring	Flexible plastic ring inserted into vagina; releases hormones that are absorbed into the bloodstream	Combined hormonal methods suppress ovulation by the combined actions of the hormones estrogen and progestin.	98%	Combined hormonal method's can cause dizziness; nausea; changes in menstruation, mood, and weight; rarely, cardiovascular disease, including high blood pressure, blood clots, heart attack, and strokes.
Combined hormone pill	Pills are swallowed daily; chewable form also available		98%	
Combined hormone 91-day regimen	Pills are swallowed daily; user has 3-4 menstrual periods a year		98%	
Combined hormones injection (Lunelle)	Injection of long-acting hormone given once a month		99%	
Combined hormone patch	Patch is applied to skin and left in place for 1 week; new patch applied		98%	
Progestin-only mini-pill	Pills are swallowed daily	Thickens cervical mucus, preventing sperm from contacting egg	98%	Irregular bleeding, weight gain, breast tenderness
Progesterone-only injection (Depo-Provera)	Injection of progestin once every three months	Inhibits ovulation; prevents sperm from reaching the egg; prevents implantation.	99%	Irregular bleeding, weight gain, breast tenderness, osteoporosis possible
Emergency contraception	Must be taken within 72 hours after unprotected intercourse	Suppresses ovulation by the combined actions of the hormones estrogen and progestin; prevents implantation	80%	Nausea, vomiting, abdominal pain, fatigue, headache
Diaphragm	Latex cup, placed into vagina to cover cervix before intercourse	Blocks entrance of sperm into uterus, spermicide kills sperm	90% with spermicide	Irritation, allergic reaction, urinary tract inflection, toxic shock syndrome
Cervical cap	Latex cap held over cervix	Blocks entrance of sperm into uterus, spermicide kills sperm	90% with spermicide	Irritation, allergic reaction, toxic shock syndrome, abnormal Pap smear
Cervical shield	Latex cap in upper vagina, held in place by suction	Blocks entrance of sperm into uterus, spermicide kills sperm	90% with spermicide	Irritation, allergic reaction, urinary tract infection, toxic shock syndrome
Intrauterine device Copper T	Placed in uterus	Causes cervical mucus to thicken; fertilized embryo cannot implant	99%	Cramps, bleeding, infertility, perforation of uterus
Intrauterine device progesterone-releasing type	Placed in uterus	Prevents ovulation; causes cervical mucus to thicken; fertilized embryo cannot implant	99%	Cramps, bleeding, infertility, perforation of uterus

blocking the progesterone receptor proteins of endometrial cells. Without functioning receptors for progesterone, the endometrium sloughs off, carrying the embryo with it. When taken in conjunction with a prostaglandin to induce uterine contractions, RU-486 is 95% effective. It is possible that some day this medication will also be a "morning-after pill," taken when menstruation is late without evidence that pregnancy has occurred.

Infertility

Infertility is the failure of a couple to achieve pregnancy after one year of regular, unprotected intercourse. The American Medical Association (AMA) estimates that 15% of all couples are infertile. The cause of infertility can be attributed to the male (40%), the female (40%), or both (20%).

Causes of Infertility

The most frequent cause of infertility in males is low sperm count and/or a large proportion of abnormal sperm, which can be due to environmental influences. It appears that a sedentary lifestyle coupled with smoking and alcohol consumption is most often the cause of male infertility. When males spend most of the day sitting in front of a computer or the TV or driving, the testes temperature remains too high for adequate sperm production.

Body weight appears to be the most significant factor in causing female infertility. In women of normal weight, fat cells produce a hormone called leptin that stimulates the hypothalamus to release GnRH. In overweight women, the ovaries often contain many small follicles, and the woman fails to ovulate. Other causes of infertility in females are blocked oviducts due to pelvic inflammatory disease (see page 337) and endometriosis. Endometriosis is the presence of uterine tissue outside the uterus, particularly in the oviducts and on the abdominal organs. Backward flow of menstrual fluid allows living uterine cells to establish themselves in the abdominal cavity, where they go through the usual uterine cycle, causing pain and structural abnormalities that make it more difficult for a woman to conceive.

Sometimes the causes of infertility can be corrected by medical intervention so that couples can have children. If no obstruction is apparent and body weight is normal, it is possible to give females fertility drugs, which are gonadotropic hormones that stimulate the ovaries and bring about ovulation. Such hormone treatments may cause multiple ovulations and multiple births.

When reproduction does not occur in the usual manner, many couples adopt a child. Others sometimes try one of the assisted reproductive technologies discussed in the following paragraphs.

Assisted Reproductive Technologies

Assisted reproductive technologies (ART) consist of techniques used to increase the chances of pregnancy. Often, sperm and/or eggs are retrieved from the testes and ovaries, and fertilization takes place in a clinical or laboratory setting.

Artificial Insemination by Donor (AID) During artificial insemination, sperm are placed in the vagina by a physician. Sometimes a woman is artificially inseminated by her partner's sperm. This is especially helpful if the partner has a low sperm count, because the sperm can be collected over a period of time and concentrated so that the sperm count is sufficient to result in fertilization. Often, however, a woman is inseminated by sperm acquired from a donor who is a complete stranger to her. At times, a combination of partner and donor sperm is used.

Figure 16.15
In vitro fertilization.
A microscope connected to a television screen is used to carry out in vitro fertilization. A pipette holds the egg steady while a needle (not visible) introduces the sperm into the egg.

A variation of AID is *intrauterine insemination (IUI).* In IUI, fertility drugs are given to stimulate the ovaries, and then the donor's sperm are placed in the uterus, rather than in the vagina.

If the prospective parents wish, sperm can be sorted into those that are believed to be X-bearing or Y-bearing to increase the chances of having a child of the desired sex.

In Vitro Fertilization (IVF) During IVF, conception occurs in laboratory glassware. Ultrasound machines can now spot follicles in the ovaries that hold immature eggs; therefore, the latest method is to forgo the administration of fertility drugs and retrieve immature eggs by using a needle. The immature eggs are then brought to maturity in glassware, and then concentrated sperm are added (Fig. 16.15). After about two to four days, the embryos are ready to be transferred to the uterus of the woman, who is now in the secretory phase of her uterine cycle. If desired, the embryos can be tested for a genetic disease, and only those found to be free of disease will be used. If implantation is successful, development is normal and continues to term.

Gamete Intrafallopian Transfer (GIFT) The term **gamete** refers to a sex cell, either a sperm or an egg. GIFT was devised to overcome the low success rate (15–20%) of in vitro fertilization. The method is exactly the same as in vitro fertilization, except the eggs and the sperm are placed in the oviducts immediately after they have been brought together. GIFT has the advantage of being a one-step procedure for the woman—the eggs are removed and reintroduced all in the same time period. A variation on this procedure is to fertilize the eggs in the laboratory and then place the zygotes in the oviducts.

Surrogate Mothers In some instances, women are contracted and paid to have babies. These women are called surrogate mothers. The sperm and even the egg can be contributed by the contracting parents.

Intracytoplasmic Sperm Injection (ICSI) In this highly sophisticated procedure, a single sperm is injected into an egg. It is used effectively when a man has severe infertility problems.

✓ Check Your Progress 16.5

1. What are some common birth control methods available to men and women today?

2. If a couple has trouble conceiving a child, what options can they explore?

Should Infertility Be Treated

Every day, couples make plans to start or expand their families. Yet for many, their dreams might not be realized because conception is difficult or impossible. Before seeking medical treatment for infertility, a couple might want to decide how far they are willing to go in order to have a child. Here are some of the possible risks.

Some of the Procedures Used

If a man has low sperm count or motility, artificial or intrauterine insemination of his partner with a large number of specially selected sperm may be done to stimulate pregnancy. Although there are dangers in all medical procedures, artificial insemination is generally safe.

If a woman is infertile because of physical abnormalities in her reproductive system, she may be treated surgically. While surgeries are now very sophisticated, they nonetheless have risks, including bleeding, infection, organ damage, and adverse reactions to anesthesia. Similar risks are associated with collecting eggs for in vitro fertilization (IVF). In order to ensure the collection of several eggs, a woman may be placed on hormone-based medications that stimulate egg production. Such medications may cause ovarian hyperstimulation syndrome—enlarged ovaries and abdominal fluid accumulation. In mild cases, the only symptom is discomfort, but in severe cases (though rare), a woman's life may be endangered, and in any case, the fluid has to be drained.

Usually, IVF involves the creation of many embryos; the healthiest-looking ones are transferred into the woman's body. Others may be frozen for future attempts at establishing pregnancy, given to other infertile couples, donated for research, or destroyed. Of those that are transferred, none, one, or all might develop into fetuses. The significant increase of multifetal pregnancies in the United States in the last 15 years has been largely attributed to fertility treatment (Fig. 16A).

Figure 16A   Couple and children.
Are assisted reproductive technologies a boon to society? To individuals who wish to have children? Or a detriment to both?

While the number of triplet and higher number multiple pregnancies started to level off in 1999, twin pregnancies continue to climb.

While they might seem like a dream come true, multifetal pregnancies are difficult. The mother is more likely to develop complications such as gestational diabetes and high blood pressure than are women carrying single babies. Positioning of the babies in the uterus may make vaginal delivery less likely, and there is likely a chance of preterm labor; babies born prematurely face numerous hardships. Infant death and long-term disabilities are also more common with multiple births; this is true even of twins. Even if all babies are healthy, parenting multiples poses unique challenges.

What Happens to Frozen Embryos?

In spite of potential trials, thousands of people undergo fertility treatment every year. Its popularity has brought a number of ethical issues to light. For example, the estimated high numbers of stored frozen embryos (a few hundred thousand in the United States) has generated debate about their fate, complicated by the fact that the long-term viability of frozen embryos is not well understood. Scientists may worry that embryos donated to other couples are ones that were already screened out from one implantation, and not be likely to survive. Some religious groups strongly oppose destruction of these embryos or their use in research. Patients for whom the embryos were created generally feel that they should have sole rights to make decisions about their fate. However, a fertility clinic may no longer be receiving monetary compensation for the storage of frozen embryos and unable to contact the couples for whom they were produced. The question then becomes whether the clinic now has the right to determine their fate.

Who Should Be Treated?

Additionally, since fertility treatment is voluntary, is it ever acceptable to turn some people away? What if the prospect of a satisfactory outcome is very slim or almost nonexistent? This may happen when one of the partners is ill or the woman is at an advanced age. Should a physician go ahead with treatment even if it might endanger a woman's (or baby's) health? Those in favor of limiting treatment argue that a physician has a responsibility to prevent potential harm to a patient. On the other hand, there is concern that if certain people are denied fertility for medical reasons, might they be denied for other reasons also, such as race, religion, sexuality, or income?

Decide Your Opinion

1. Should couples go to all lengths to have children even if it could endanger the life of one or both spouses?
2. Should couples with multiples due to infertility treatment receive assistant from private and public services?
3. To what lengths should society go in order to protect frozen embryos?
4. Do you think that anyone should be denied fertility treatment? If so, what factors do you think a doctor should take into consideration when deciding whether to provide someone fertility treatment?

16.6 Sexually Transmitted Diseases

Sexually transmitted diseases (STDs) are caused by viruses, bacteria, protists, fungi, and animals.

STDs Caused by Viruses

Among those STDs caused by viruses, effective treatment is available for AIDS and genital herpes. However, it is important to note that treatment for HIV/AIDS and genital herpes cannot presently eliminate the virus from the person's body. Drugs used for treatment can merely slow replication of the viruses. Thus, neither viral disease is presently curable. Further, antiviral drugs have serious, debilitating side effects on the body.

HIV Infections

The AIDS (acquired immunodeficiency syndrome) supplement, which begins on page 343, discusses HIV infections at greater length than this brief summary. Since there is at present no vaccine to prevent an HIV infection nor a cure for AIDS, the best course of action is to follow the guidelines for preventing transmission outlined in the Health Focus on page 339.

AIDS is the last stage of an HIV infection. It's estimated that 38.6 people worldwide are now infected with the HIV virus and that nearly 25 million persons have died from AIDS. In the United States, AIDS was first seen among the homosexual community, and they still are the most exposed group. However, new HIV infections are more likely to occur in minority women. From this, you can conclude that HIV is now being transmitted between heterosexuals and more frequently between African Americans and Hispanics than Caucasians.

The primary host for HIV is a helper T lymphocyte (Fig. 16.16). Since these are the very cells that stimulate an immune response, the immune system becomes severely impaired in persons with AIDS. During the first stage of an HIV infec-

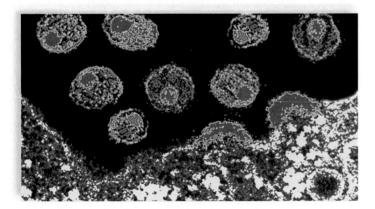

Figure 16.16   HIV, the AIDS virus.
False-colored micrographs showing HIV particles budding from an infected helper T cell. These viruses can infect helper T cells and also macrophages, which work with helper T cells to stem the infection.

tion, symptoms are few, but the individual is highly contagious. Several months to several years after infection, the helper T lymphocyte count falls, and infections, such as other sexually transmitted diseases, begin to appear. In the last stage of infection, called AIDS, the helper T cell count falls way below normal, and at least one opportunistic infection is present. Such diseases have the opportunity to occur only because the immune system is severely weakened. Persons with AIDS typically die from an opportunistic disease, such as *Pneumocystis* pneumonia.

There is no cure for AIDS, but a treatment called highly active antiretroviral therapy (HAART) is usually able to stop HIV reproduction to the extent that the virus becomes undetectable in the blood. The medications must be continued indefinitely because as soon as HAART is discontinued, the virus rebounds.

Genital Warts

Genital warts are caused by the human papillomaviruses (HPVs). Many times, carriers either do not have any sign of warts or merely have flat lesions. When present, the warts commonly are seen on the penis and foreskin of men and near the vaginal opening in women. A newborn can become infected while passing through the birth canal.

Individuals who are currently infected with visible growth may have those growths removed by surgery, freezing, or burning with lasers or acids. However, visible warts that are removed may recur. A new vaccine has been released for the human papillomaviruses that most commonly cause genital warts. This development is an extremely important step in the prevention of cancer, as well as in the prevention of warts themselves. Genital warts are associated with cancer of the cervix, as well as tumors of the vulva, vagina, anus, and penis. Researchers believe that these viruses may be involved in up to 90% of all cases of cancer of the cervix. Vaccination might make such cancers a thing of the past.

Genital Herpes

Genital herpes is caused by herpes simplex virus. Type 1 usually causes cold sores and fever blisters, while type 2 more often causes genital herpes (Fig. 16.17).

Persons usually get infected with herpes simplex virus type 2 when they are adults. Some people exhibit no symptoms; other may experience a tingling or itching sensation before blisters appear on the genitals. Once the blisters rupture, they leave painful ulcers that may take as long as three weeks or as little as five days to heal. The blisters may be accompanied by fever, pain on urination, swollen lymph nodes in the groin, and in women, a copious discharge. At this time, the individual has an increased risk of acquiring an HIV infection.

After the ulcers heal, the disease is only latent, and blisters can recur, although usually at less frequent intervals and with milder symptoms. Fever, stress, sunlight, and menstruation are associated with recurrence of symptoms. Exposure to

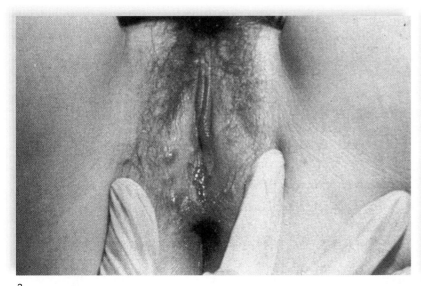

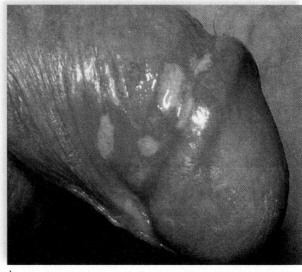

a.

b.

Figure 16.17   Genital herpes.
There are several types of herpes viruses, and usually the ones called herpes simplex-2 cause genital herpes. About one million persons become infected each year, most of them teens and young adults. Symptoms of genital herpes are due to an outbreak of blisters, which can be present on the labia of females **(a)** or on the penis of males **(b)**.

herpes in the birth canal can cause an infection in the new-born, which leads to neurological disorders and even death. Birth by cesarean section prevents this possibility.

Hepatitis

Hepatitis infects the liver and can lead to liver failure, liver cancer, and death. There are six known viruses that cause hepatitis, designated A-B-C-D-E-G. Hepatitis A is usually acquired from sewage-contaminated drinking water, but this infection can also be sexually transmitted through oral/anal contact. Hepatitis B is spread through sexual contact and by blood-borne transmission (accidental needle stick on the job; receiving a contaminated blood transfusion; a drug abuser sharing infected needles while injecting drugs; from mother to fetus, etc.). Simultaneous infection with hepatitis B and HIV is common, since both share the same routes of transmission. Fortunately, a combined vaccine is available for hepatitis A and B; it is currently recommended that all children receive the vaccine to prevent infection (see page 136). Hepatitis C (also called non-A, non-B hepatitis) causes most cases of posttransfusion hepatitis. Hepatitis D and G are sexually transmitted, while hepatitis E is acquired from contaminated water. Screening of blood and blood products can prevent transmission of hepatitis viruses during a transfusion. Proper water-treatment techniques can prevent contamination of drinking water.

STDs Caused by Bacteria

Only STDs caused by bacteria are curable with antibiotics. Antibiotic resistance acquired by these bacteria may require treatment with extremely strong drugs for an extended period, in order to achieve a cure.

Chlamydia

Chlamydia is named for the tiny bacterium that causes it *(Chlamydia trachomatis)*. The incidence of new chlamydia infections has steadily increased since 1984.

Chlamydia infections of the lower reproductive tract are usually mild or asymptomatic, especially in women. About 18 to 21 days after infection, men may experience a mild burning sensation on urination and a mucoid discharge. Women may have a vaginal discharge along with the symptoms of a urinary tract infection. Chlamydia also causes cervical ulcerations, which increase the risk of acquiring HIV.

If the infection is misdiagnosed or if a woman does not seek medical help, there is a particular risk of the infection spreading from the cervix to the uterine tubes so that pelvic inflammatory disease (PID) results. This very painful condition can result in blockage of the uterine tubes with the possibility of sterility and infertility. If a baby comes in contact with chlamydia during birth, inflammation of the eyes or pneumonia can result.

Gonorrhea

Gonorrhea is caused by the bacterium *Neisseria gonorrhoeae*. Diagnosis in the male is not difficult, since typical symptoms are pain upon urination and a thick, greenish yellow urethral discharge. In males and females, a latent infection leads to pelvic inflammatory disease (PID), which can also cause sterility in males. If a baby is exposed during birth, an eye infection leading to blindness can result. All newborns are given eyedrops to prevent this possibility.

Gonorrhea proctitis, an infection of the anus characterized by anal pain and blood or pus in the feces, also occurs in patients. Oral/genital contact can cause infection of the

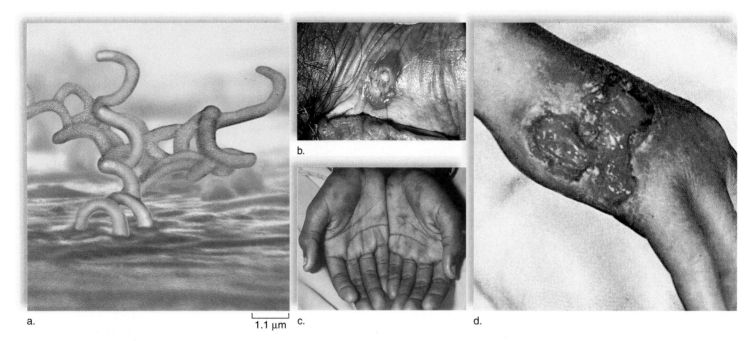

a.   1.1 μm   c.   d.

Figure 16.18   Syphilis.

a. Scanning electron micrograph of *Treponema pallidum,* the cause of syphilis. **b.** The three stages of syphilis. The primary stage of syphilis is a chancre at the site where the bacterium enters the body. **c.** The secondary stage is a body rash that occurs even on the palms of the hands and soles of the feet. **d.** In the tertiary stage, gummas may appear on the skin or internal organs.

mouth, throat, and tonsils. Gonorrhea can spread to internal parts of the body, causing heart damage or arthritis. If, by chance, the person touches infected genitals and then touches his or her eyes, a severe eye infection can result. Up to now, gonorrhea was curable by antibiotic therapy, but resistance to antibiotics is becoming more and more common, and 40% of all strains are now known to be resistant to therapy.

Syphilis

Syphilis is caused by a bacterium called *Treponema pallidum* (Fig. 16.18). As with many other bacterial diseases, penicillin is an effective antibiotic. Syphilis has three stages, often separated by latent periods, during which the bacteria are resting before multiplying again. During the primary stage, a hard **chancre** (ulcerated sore with hard edges) indicates the site of infection. The chancre usually heals spontaneously, leaving little scarring. During the secondary stage, the victim breaks out in a rash that does not itch and is seen even on the palms of the hands and the soles of the feet. Hair loss and infectious gray patches on the mucous membranes may also occur. These symptoms disappear of their own accord.

During the tertiary stage, which lasts until the patient dies, syphilis may affect the cardiovascular system by causing aneurysms, particularly in the aorta. In other instances, the disease may affect the nervous system, resulting in psychological disturbances. Also, gummas, large destructive ulcers, may develop on the skin or within the internal organs.

Congenital syphilis is caused by syphilitic bacteria crossing the placenta. The child is born blind and/or with numerous anatomical malformations. Control of syphilis depends on prompt and adequate treatment of all new cases; therefore, it is crucial for all sexual contacts to be traced so they can be treated. Diagnosis of syphilis can be made by blood tests or by microscopic examination of fluids from lesions.

Two Other Infections

Bacterial vaginosis is believed to account for 50% of vaginitis cases in American women. The overgrowth of the bacterium *Gardnerella vaginalis* and consequent symptoms can be due to nonsexual reasons, but symptomless males may pass on the bacteria to women, who do exhibit symptoms. Also, females often have an infection of the vagina, caused by either the flagellated protozoan *Trichomonas vaginalis* or the yeast *Candida albicans.* The protozoan infection causes a frothy white or yellow, foul-smelling discharge accompanied by itching, and the yeast infection causes a thick, white, curdy discharge, also accompanied by itching. Trichomoniasis is most often acquired though sexual intercourse, and the asymptomatic male is usually the reservoir of infection. *Candida albicans,* however, is a normal organism found in the vagina; its growth simply increases beyond normal under certain circumstances. For example, women taking birth control pills are sometimes prone to yeast infections. Also, the indiscriminate use of antibiotics can alter the normal balance of organisms in the vagina so that a yeast infection flares up.

⍾ Check Your Progress 16.6

1.   What condition can occur due to both a chlamydial and gonorrheal infection?

2.   Genital warts are associated with what other medical condition in women?

Health Focus

Preventing Transmission of STDs

Sexual Activities Transmit STDs

Abstain from sexual intercourse or develop a long-term monogamous (always the same partner) sexual relationship with a partner who is free of STDs (Fig. 16B).

Refrain from multiple sex partners or having relations with someone who has multiple sex partners. If you have sex with two other people and each of these has sex with two people, and so forth, the number of people who are relating is quite large.

Remember that the prevalence of AIDS is presently higher among homosexuals and bisexuals than among those who are heterosexual.

Be aware that having relations with an intravenous drug user is risky because the behavior of this group risks AIDS and hepatitis B. Be aware that anyone who already has another sexually transmitted disease is more susceptible to an HIV infection.

Avoid anal-rectal intercourse (in which the penis is inserted into the rectum) because this behavior increases the risk of an HIV infection. The lining of the rectum is thin, and infected CD4 T cells can easily enter the body there. Also, the rectum is supplied with many blood vessels, and insertion of the penis into the rectum is likely to cause tearing and bleeding that facilitate the entrance of HIV. The vaginal lining is thick and difficult to penetrate, but the lining of the uterus is only one cell thick at certain times of the month, and does allow CD4 T cells to enter.

Uncircumcised males are more likely to become infected than circumcised males because vaginal secretions can remain under the foreskin for a long time.

Practice Safer Sex

Always use a latex condom during sexual intercourse if you are not in a monogamous relationship. Be sure to follow the directions supplied by the manufacturer for the use of a condom. At one time, condom users were advised to use nonoxynol-9 in conjunction with a condom, but recent testing shows that this spermicide has no effect on viruses, including HIV.

Avoid fellatio (kissing and insertion of the penis into a partner's mouth) **and cunnilingus** (kissing and insertion of the tongue into the vagina) because they may be a means of transmission. The mouth and gums often have cuts and sores that facilitate catching an STD.

Practice penile, vaginal, oral, and hand cleanliness. Be aware that hormonal contraceptives make the female genital tract receptive to the transmission of sexually transmitted diseases, including HIV.

Be cautious about using alcohol or any drug that may prevent you from being able to control your behavior.

Drug Use Transmits HIV

Stop, if necessary, or do not start the habit of injecting drugs into your veins. Be aware that HIV and hepatitis B can be spread by blood-to-blood contact.

Always use a new sterile needle for injection or one that has been cleaned in bleach if you are a drug user and cannot stop your behavior (Fig. 16C).

Figure 16B   Sexual activities transmit STDs.

Figure 16C   Sharing needles transmits STDs.

Summarizing the Concepts

16.1 Human Life Cycle

The life cycle of higher organisms requires two types of cell division: mitosis and meiosis.

- **Mitosis:** growth and repair of tissues.
- **Meiosis:** gamete production.

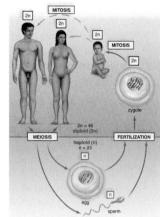

16.2 Male Reproductive System

The external genitals of males are

- the penis (organ of sexual intercourse).
- the scrotum (contains the testes).

Spermatogenesis, occurring in seminiferous tubules of the testes, produces sperm.

- Mature sperm are stored in the epididymides.
- Sperm pass from the vasa deferentia to the urethra.
- The seminal vesicles, prostate gland, and bulbourethral glands add fluids (by secretion) to sperm.
- Sperm and secretions are called semen or seminal fluid.

Orgasm in males results in ejaculation of semen from the penis.

Hormonal Regulation in Males

- Hormonal regulation, involving secretions from the hypothalamus, the anterior pituitary, and the testes, maintains a fairly constant level of testosterone.
- FSH from the anterior pituitary promotes spermatogenesis.
- LH from the anterior pituitary promotes testosterone production by interstitial cells.

16.3 Female Reproductive System

Oogenesis occurring within the ovaries typically produces one mature follicle each month.

- This follicle balloons out of the ovary and bursts, releasing an egg that enters an oviduct.
- The oviducts lead to the uterus, where implantation and development occur.

The female external genital area includes the vaginal opening, the clitoris, the labia minora, and the labia majora.

- The vagina is the organ of sexual intercourse and the birth canal in females.

Orgasm in females culminates in uterine and oviduct contractions.

16.4 Female Hormone Levels

Ovarian Cycle: Nonpregnant

- The ovarian cycle is under the hormonal control of the hypothalamus and the anterior pituitary.
- During the cycle's first half, FSH from the anterior pituitary causes maturation of a follicle that secretes estrogen and some progesterone.
- After ovulation and during the cycle's second half, LH from the anterior pituitary converts the follicle into the corpus luteum.
- The corpus luteum secretes progesterone and some estrogen.

Uterine Cycle: Nonpregnant

Estrogen and progesterone regulate the uterine cycle.

- Estrogen causes the endometrium to rebuild.
- Ovulation usually occurs on day 14 of a 28-day cycle.
- Progesterone produced by the corpus luteum causes the endometrium to thicken and become secretory.
- A low level of hormones causes the endometrium to break down as menstruation occurs.

Fertilization and Pregnancy

If fertilization takes place, the embryo implants itself in the thickened endometrium.

- The corpus luteum is maintained because of HCG production by the placenta, and therefore, progesterone production does not cease.
- Menstruation usually does not occur during pregnancy.

16.5 Control of Reproduction

Numerous birth control methods and devices are available.

- A few of these are the birth control pill, diaphragm, and condom.
- Effectiveness varies.

Assisted reproductive technologies may help infertile couples to have children. Some of these technologies are

- artificial insemination by donor (AID).
- in vitro fertilization (IVF).
- gamete intrafallopian transfer (GIFT).
- intracytoplasmic sperm injection (ICSI).

16.6 Sexually Transmitted Diseases

STDs are caused by viruses, bacteria, protists, fungi, and animals.

STDs Caused by Viruses

- AIDS is caused by HIV (human immunodeficiency virus).
- Genital warts are caused by human papillomaviruses; these viruses cause warts or lesions on genitals and are associated with certain cancers.
- Genital herpes is caused by herpes simplex virus type 2; causes blisters on genitals.
- Hepatitis is caused by hepatitis viruses A, B, C, D, E, and G. A and E are usually acquired from contaminated water; hepatitis B and C from bloodborne transmission; and B, D, and G are sexually transmitted.

STDs Caused by Bacteria

- Chlamydia is caused by *Chlamydia trachomatis;* PID can result.
- Gonorrhea is caused by *Neisseria gonorrhoeae;* PID can result.
- Syphilis is caused by *Treponema pallidum.* It has three stages, with the third stage resulting in death.

Two Other Infections
- Bacterial vaginosis is caused by *Gardnerella vaginalis*.
- Trichomoniasis is an overgrowth of *Candida albicans*.

Understanding Key Terms

acrosome 323
birth control method 331
birth control pill 331
bulbourethral gland 321
cervix 325
chancre 338
circumcision 322
contraceptive 331
contraceptive implant 332
contraceptive injection 332
contraceptive vaccine 332
corpus luteum 327
diaphragm 332
egg 324
endometrium 325
epididymis 321
erectile dysfunction 322
estrogen 328
female condom 332
fimbria 324
follicle 326
follicle-stimulating hormone
   (FSH) 324
gamete 334
gonadotropin-releasing
   hormone (GnRH) 324
human chorionic gonadotropin
   (HCG) 330
implantation 325
infertility 334
interstitial cell 323
intrauterine device (IUD) 332

luteinizing hormone (LH) 324
male condom 332
menopause 328
menstruation 328
oogenesis 326
ovarian cycle 326
ovary 324
oviduct 324
ovulation 326
Pap test 325
penis 322
placenta 330
progesterone 328
prostate gland 321
scrotum 321
semen 321
seminal vesicle 321
seminiferous tubule 323
Sertoli cell 323
sperm 323
spermatogenesis 323
testes 321
testosterone 324
tubal ligation 332
urethra 321
uterine cycle 328
uterus 325
vagina 325
vas deferens 321
vasectomy 332
vulva 326
zygote 324

Match the key terms to these definitions.

a. —————— Release of an oocyte from the ovary.

b. —————— Female sex hormone that causes the endometrium of the uterus to become secretory during the uterine cycle; along with estrogen, it maintains secondary sex characteristics in females.

c. —————— Thick, whitish fluid consisting of sperm and secretions from several glands of the male reproductive tract.

d. —————— Narrow end of the uterus, which projects into the vagina.

e. —————— Cap at the anterior end of a sperm that partially covers the nucleus and contains enzymes that help the sperm penetrate the egg.

Testing Your Knowledge of the Concepts

1. What type of cell division produces the gametes? Why is this type of cell division necessary? (page 320)

2. Trace the path of sperm. What glands contribute fluids to semen? (page 321)

3. Where are sperm produced in the testes? What is the process called? Where is testosterone produced in the testes? (pages 323–24)

4. Name the hormones involved in maintaining the sex characteristics of the male, and tell what each does. (page 324)

5. What are the organs of the female reproductive tract and their functions? (pages 324–25)

6. Describe the ovarian cycle in a nonpregnant female and the hormones involved. (pages 326–28)

7. Describe the uterine cycle in a nonpregnant female, and relate it to the ovarian cycle. (pages 328–29)

8. Describe the hormonal role of the placenta. (page 330)

9. Briefly describe various birth control methods, along with their effectiveness. (pages 331–33)

10. Briefly describe various types of assisted reproductive technologies. (page 334)

11. What STDs are caused by viruses? List the causative agent, symptoms, and treatments. What STDs are caused by bacteria? List the causative agent, symptoms, and treatments. (pages 336–38)

12. Label this diagram of the male reproductive system, and trace the path of sperm.

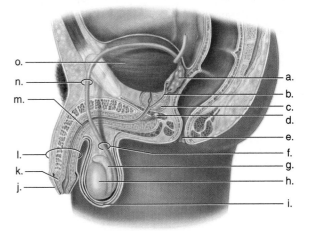

13. Which of these associations is mismatched?
    a. interstitial cells—testosterone
    b. seminiferous tubules—sperm production
    c. vasa deferentia—seminal fluid production
    d. urethra—conducts sperm

14. Follicle-stimulating hormone (FSH)
    a. is secreted by females but not by males.
    b. stimulates the seminiferous tubules to produce sperm.
    c. secretion is controlled by gonadotropin-releasing hormone (GnRH).
    d. Both b and c are correct.

15. In tracing the path of sperm, you would mention the vasa deferentia before the
    a. testes.   c. urethra.
    b. epididymides.   d. uterus.

16. Semen does not contain
    a. prostate fluid.   d. prostaglandins.
    b. urine.   e. Both b and d are correct.
    c. fructose.

17. The scrotum is
    a. part of the primary sex organ.
    b. important in regulating the temperature of the testes.
    c. poorly innervated.
    d. an extension of the spermatic cord.

18. Testosterone is produced and secreted by
    a. spermatogonia.   c. seminiferous tubules.
    b. sustentacular cells.   d. interstitial cells.

19. Luteinizing hormone in males
    a. stimulates sperm development.
    b. triggers ovulation.
    c. is responsible for secondary sex characteristics.
    d. controls testosterone production by interstitial cells.

20. Secondary sex characteristics are
    a. the only body features that regress as we age.
    b. the same in males and females.
    c. those characteristics that develop only after puberty.
    d. the only reproductive structures that are affected by hormones.

21. The release of the oocyte from the follicle is caused by
    a. a decreasing level of estrogen.
    b. a surge in the level of follicle-stimulating hormone.
    c. a surge in the level of luteinizing hormone.
    d. progesterone released from the corpus luteum.

22. Which of the following is not an event of the ovarian cycle?
    a. FSH promotes the development of a follicle.
    b. The endometrium thickens.
    c. The corpus luteum secretes progesterone.
    d. Ovulation of an egg occurs.

23. An oocyte is fertilized in the
    a. vagina.   c. oviduct.
    b. uterus.   d. ovary.

24. Following implantation, the corpus luteum is maintained by
    a. estrogen.
    b. progesterone.
    c. follicle-stimulating hormone.
    d. human chorionic gonadotropin.

25. During pregnancy,
    a. the ovarian and uterine cycles occur more quickly than before.
    b. GnRH is produced at a higher level than before.
    c. the ovarian and uterine cycles do not occur.
    d. the female secondary sex characteristics are not maintained.

26. Female oral contraceptives prevent pregnancy because
    a. the pill inhibits the release of luteinizing hormone.
    b. oral contraceptives prevent the release of an egg.
    c. follicle-stimulating hormone is not released.
    d. All of these are correct.

27. Which contraceptive method is most effective?
    a. abstinence   c. IUD
    b. diaphragm   d. Depo-Provera injection

In questions 28–30, match each method of protection with a means of birth control in the key.

Key:
    a. vasectomy   d. diaphragm
    b. oral contraception   e. male condom
    c. intrauterine device (IUD)

28. Blocks entrance of sperm to uterus

29. Traps sperm and also prevents STDs

30. Prevents implantation of an embryo

31. Which of the following may cause an eye infection in newborns delivered to infected mothers?
    a. chlamydia   c. AIDS
    b. gonorrhea   d. Both a and b are correct.

32. For which of the following infections is there currently no cure?
    a. chlamydia   c. syphilis
    b. gonorrhea   d. herpes

33. Which of the following is indicative of the third stage of a syphilis infection?
    a. chancre   c. gumma
    b. rash   d. generalized edema

Thinking Critically About the Concepts

Jessica, from the opening story, continually uses birth control pills to keep progesterone levels high. Maintaining progesterone will prevent the breakdown of the endometrium so she won't have a period on her wedding day or during her honeymoon. The progesterone in the pills is also what prevents follicle development, ovulation, and pregnancy because the progesterone is introduced early in the menstrual cycle with use of the pill.

If a female becomes pregnant, HCG maintains the corpus luteum, which continues to produce progesterone. This prevents the endometrium from breaking down, giving the developing embryo a place to implant. If a woman has a history of miscarriages, she may be given supplemental progesterone to maintain a new pregnancy. So progesterone can be used to prevent a pregnancy or maintain one.

1. Refer to Figure 16.10.
   a. Redraw the illustration and add progesterone early in the cycle (to represent use of birth control pills) and draw new lines to show how FSH/LH will be affected by progesterone's presence early in the cycle.
   b. Redraw the illustration and show the effects of HCG (produced when the female gets pregnant) on the corpus luteum, progesterone levels, and endometrium.

2. Women who use birth control pills appear to have a lower risk of developing ovarian cancer. Women who use fertility enhancing drugs (which increase the number of follicles that develop so many "eggs" can be harvested) may increase their risk of developing ovarian cancer. What do birth control pills and fertility-enhancing drugs have in common (with regard to a big picture function) that might explain how they affect a woman's risk of developing ovarian cancer?

3. It is fairly common for very serious female athletes to not have their periods. These athletes might have a 10–15% body fat composition. Explain the connection between low body fat composition and the absence of a menstrual cycle in these athletes.

4. Giving men additional testosterone has recently been shown to decrease sperm production enough to be considered a successful form of birth control. How would additional testosterone inhibit sperm production?

AIDS Supplement

Acquired immunodeficiency syndrome (AIDS) is caused by a virus known as a **human immunodeficiency virus (HIV)**. There are two main types of HIV: HIV-1 and HIV-2. HIV-1 is the more widespread, virulent form of HIV.

HIV infects and destroys cells of the immune system, particularly helper T cells and macrophages. As the number of helper T cells declines, the body's ability to fight an infection also declines, and the person becomes ill with various diseases. AIDS is the advanced stage of HIV infection, when a person develops one or more of a number of opportunistic infections. An **opportunistic infection** is one that has the *opportunity* to occur only because the immune system is severely weakened.

S.1 Origin of and Prevalence of HIV

It is generally accepted that HIV originated in Africa and then spread to the United States and Europe, by way of the Caribbean. HIV has been found in a preserved 1959 blood sample taken from a man who lived in an African country now called the Democratic Republic of the Congo. Even before this discovery, scientists speculated that an immunodeficiency virus may have evolved into HIV during the late 1950s.

Of the two types of HIV, HIV-2 corresponds to a type of immunodeficiency virus found in the green monkey, which lives in western Africa. Recently, it was announced that researchers have found a virus identical to HIV-1 in a subgroup of chimpanzees once common in west-central Africa. Perhaps HIV viruses were originally found only in nonhuman primates. They could have mutated to HIV after humans ate nonhuman primates for meat.

British scientists have been able to show that AIDS came to their country perhaps as early as 1959. They examined the preserved tissues of a Manchester seaman who died that year and concluded that he most likely died of AIDS. Similarly, it is thought that HIV entered the United States on numerous occasions as early as the 1950s. But the first documented case is a 15-year-old male who died in Missouri in 1969 with skin lesions now known to be characteristic of an AIDS-related cancer. Doctors froze some of his tissues because they could not identify the cause of death. Researchers also want to test the preserved tissue samples of a 49-year-old Haitian who died in New York in 1959 of the type of pneumonia now known to be AIDS-related.

Throughout the 1960s, it was customary in the United States to list leukemia as the cause of death in immunodeficient patients. Most likely, some of these people actually died of AIDS. Since HIV is not extremely infectious, it took several decades for the number of AIDS cases to increase to the point that AIDS became recognizable as a specific and separate disease. The name AIDS was coined in 1982, and HIV was found to be the cause of AIDS in 1983–1984.

Prevalence of HIV

AIDS is a **pandemic** because the disease is prevalent in the entire human population around the globe (Fig. S.1). Worldwide, as shown at the bottom of Table S.1, an estimated 38.6 million people (36.3 million adults and 2.3 million children) are now living with HIV infection. Among the 4.1 million new HIV infections, nearly 50% are among young adults ages 15–24—the persons who are the backbone of a nation's social, political, and economical structure and stability. Table S.1 also tells us that 2.8 million people died during 2005, but since the beginning of the epidemic, nearly 25 million people have died of AIDS. Today, at least 1% of the adults in the world have an HIV infection.

As we can deduce from studying Figure S.1 and Table S.1, most people infected with HIV live in the developing (poor, low- to middle-income) countries. The few of the hardest hit regions are as follows:

- **Sub-Saharan Africa and Asia:** The 24 million people living with HIV infection in sub-Saharan Africa represent 64% of all people living with HIV today. In many African nations, over 20% of the adult population are HIV-positive.

   In contrast, Asia has 8.3 million people living with HIV, 930,000 new HIV infections, and 600,000 AIDS deaths. Per country, India now has the highest number of people living with HIV infection (5.7 million) in the world, followed by South Africa (5.5 million).

- **Latin America and the Caribbean:** In Latin America and the Caribbean, the AIDS epidemic is being fueled by the sex industry, particularly female sex workers and their clients. The HIV prevalence among sex workers in some regions of Latin America is as high as 16%. AIDS is the leading cause of death among people ages 15–44 in the Caribbean. The Bahamas, Barbados, Haiti, Cuba, Trinidad, and Jamaica are a few of the hardest-hit countries in the Caribbean. The Caribbean has an HIV prevalence of 1.6%, making it the second most affected region of the world—Africa is number one with 6.1%. The HIV prevalence in Latin America is 0.5%.

- **North America, Europe, and Central Asia:** In 2005, the regions of North America, Europe, and Central Asia each had over 1 million people living with HIV. Russia's epidemic is fueled by injection drug users. Russia had the largest number of people living with HIV (950,000) in Eastern Europe and Central Asia. The number of people living with HIV in the United States is higher, being 1.3 million in 2005. African Americans and Hispanics made up almost 70% of all new HIV/AIDS cases diagnosed in the United States.

- **North Africa and the Middle East:** About 90% of the people living with HIV in North Africa and the Middle East live in Sudan, a country of North Africa. In 2005, the Middle East and North Africa had 64,000 new HIV infections and 37,000 AIDS deaths.

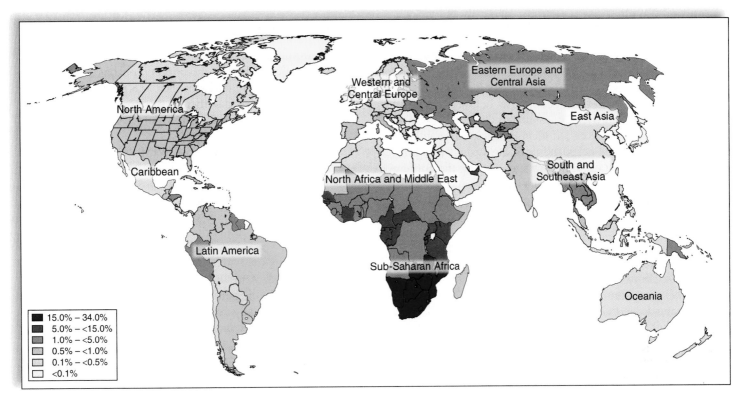

Figure S.1   **Adult HIV prevalence in 2005.**
HIV/AIDS occurs in all continents of the globe, and presently 36.3 million adults and 2.3 million children are living with HIV. Sub-Saharan Africa has the highest HIV prevalence in the world. The epidemic is spreading rapidly in Asia (see Table S.1).

- **Oceania:** Over 90% of all HIV infections in Oceania were in Papua New Guinea, where the rate of new HIV diagnoses has been increasing by 30% each year. Australia has an estimated 16,000 people living with HIV.

☑ Check Your Progress S.1

1. Which part of the world is hardest hit with HIV infections?

2. Explain why Russia and the Caribbean have a high incidence of HIV infections.

Table S.1	HIV Global Statistics, 2005			
	People Living with HIV	**New Infection**	**AIDS Deaths**	**Adult Prevalence (Percentage)**
Sub-Saharan Africa	24.5 million	2.7 million	2 million	6.1%
Asia	8.3 million	930,000	600,000	1.4%
Latin America	1.6 million	140,000	59,000	.5%
Caribbean	330,000	37,000	27,000	1.6%
North American, Western and Central Europe	2 million	65,000	30,000	.5%
Eastern Europe and Central Asia	1.5 million	220,000	53,000	.8%
North Africa and the Middle East	440,000	64,000	37,000	.2%
Oceania	78,000	7,200	3,400	0.3%
Total	**38.6 million**	**4.1 million**	**2.8 million**	**1%**

S.2 Phases of an HIV Infection

HIV occurs as several subtypes. HIV-1C is prominent in Africa, while HIV-1B causes most infections in the United States. The following description of the phases of HIV infection pertains to an HIV-1B infection. The helper T cells and macrophages infected by HIV are called CD4 cells because they display a molecule called CD4 on their surface. With the destruction of CD4 cells, the immune system is significantly impaired. After all, macrophages present the antigen to helper T cells, which coordinate the immune response by stimulating C lymphocytes to produce antibodies and cytotoxic T cells to destroy cells infected with a virus. In the United States, one of the most common causes of AIDS deaths is from *Pneumocystis jiroveci* pneumonia (PCP), while tuberculosis kills more HIV-infected people in Africa than any other AIDS-related illness.

In 1993, the Centers for Disease Control (CDC) in the United States issued clinical guidelines for the classification of HIV to help clinicians track the status, progression, and phases of HIV infection. The classification of HIV infection will be discussed in three categories (or phases), and the system is based on two aspects of a person's health—the CD4 T cell count and the history of AIDS-defining illnesses.

Category A: Acute Phase

A person in category A typically has no apparent symptoms (asymptomatic), is highly infectious, and has a CD4 T cell count that has never fallen below 500 cells per mm³ (500 cells/mm³) of blood, which is sufficient for the immune system to function normally (Fig. S.2). A normal CD4 T cell count is at least 800 cells/mm³.

Today, investigators are able to track not only the blood level of CD4 T cells, but also the viral load. The viral load is the number of HIV particles in the blood. At the start of an HIV-1B infection, the virus is replicating ferociously, and the killing of CD4 T cells is evident because the blood level of these cells drops dramatically. During first few weeks of infections, some people (1–2%) develop flulike symptoms (fever, chills, aches, swollen lymph nodes) that may last an average of two weeks. After this, a person may remain "symptom free" for years. During this acute phase of infection, an HIV antibody test is usually negative because it generally takes an average of 25 days before there are detectable levels of HIV antibodies in body fluids.

After a period of time, the body responds to the infection by increased activity of immune cells, and the HIV blood test becomes positive. During this phase, the number of CD4

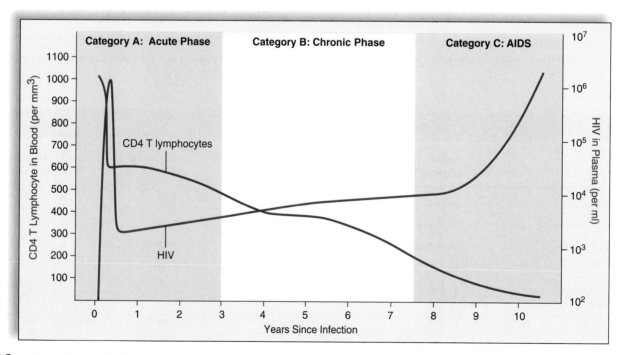

Figure S.2 **Stages of an HIV infection.**
In category A individuals, the number of HIV particles in plasma rises upon infection and then falls. The number of CD4 T lymphocytes falls, but stays above 400/mm³. In category B individuals, the number of HIV particles in plasma is slowly rising, and the number of T lymphocytes is decreasing. In category C individuals, the number of HIV particles in plasma rises dramatically as the number of T lymphocytes falls below 200/mm³.

T cells is greater than the viral load (Fig. S.2). But some investigators believe that a great unseen battle is going on. The body is staying ahead of the hordes of viruses entering the blood by producing as many as one to two billion new helper T lymphocytes each day. This is called the "kitchen sink model" for CD4 T cell loss. The sink's faucet (production of new CD4 T cells) and the sink's drain (destruction of CD4 T cells) are wide open. As long as the body can produce enough new CD4 T cells to keep pace with the destruction of these cells by HIV and by cytotoxic T cells, the person has a healthy immune system that can deal with the infection. In other words, a person in category A would have no history of conditions listed in categories B and C.

Category B: Chronic Phase

A person in category B would have a CD4 count between 499 and 200 cells/mm^3, and one or more of a variety of symptoms related to an impaired immune system, such as yeast infections of the mouth or vagina, cervical dysplasia, prolonged diarrhea, thick sores on the tongue (hairy leukoplakia), or shingles (to list a few). Others symptoms may appear such as swollen lymph nodes, unexplained persistent or recurrent fevers, fatigue, coughs, or diarrhea. During this chronic stage of infection, the number of HIV particles is on the rise (Fig. S.2). However, they do not as yet have any of the conditions listed for category C.

Category C: AIDS

A person in category C is diagnosed with AIDS. When a person has AIDS, the CD4 T cell count has fallen below 200 cells/mm^3, or the person has developed one or more of the 25 AIDS-defining illnesses (or opportunistic infections) described by the CDC's list of conditions in the 1993 AIDS surveillance case definition. Persons with AIDS die from one or more of the following opportunistic diseases rather than from the HIV infection itself:

- *Pneumocystis jiroveci* pneumonia—a fungal infection of the lungs.
- *Mycobacterium tuberculosis*—a bacterial infection usually of lymph nodes or lungs but may be spread to other organs.
- Toxoplasmic encephalitis—a protozoan parasitic infection, often seen in the brain of AIDS patients.
- Kaposi's sarcoma—an unusual cancer of the blood vessels, which gives rise to reddish purple, coin-sized spots and lesions on the skin.
- Invasive cervical cancer—a cancer of the cervix, which spreads to nearby tissues.

Once one or more of these opportunistic infections has occurred, the person will remain in category C. Newly developed drugs can treat opportunistic diseases, but most AIDS patients are repeatedly hospitalized due to weight loss, constant fatigue, and multiple infections (Fig. S.3). Death usually follows in two to four years. Although there is still no cure for AIDS, many people with HIV infection are living longer, healthier lives due to the expanding use of antiretroviral therapy.

☑ Check Your Progress S.2

1. Why might a person in the acute phase of AIDS have both a high viral load and a normal CD4 T cell count?

2. What CD4 T cell count should you associate with the three phases of an HIV infection?

3. Why does a person with AIDS have opportunistic infections? Explain.

a. AIDS patient Tom Moran, at diagnosis

b. AIDS patient Tom Moran, 6 months later

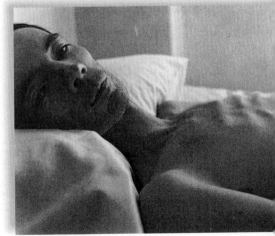

c. AIDS patient Tom Moran, in terminal stage of AIDS

Figure S.3 A patient with AIDS.
These photos show how the health of a patient with AIDS deteriorates.

S.3 HIV Structure and Life Cycle

HIV consists of two single strands of RNA (its nucleic acid genome), various proteins, and an envelope, which it acquires from its host cell in the manner to be described (Fig. S.4). The virus's genetic material is protected by a series of three protein coats, namely, the nucleocapsid, capsid, and the matrix. Within the matrix, there are three very important enzymes, reverse transcriptase, integrase, and protease.

- **Reverse transcriptase** catalyzes reverse transcription, which is the conversion of the viral RNA to viral DNA.
- **Integrase** catalyzes the integration of viral DNA into the DNA of the host cell.
- **Protease** catalyzes the breakdown of the newly synthesized viral polypeptides into functional viral proteins.

Embedded in HIV's envelope are protein spikes referred to as Gp120. These spikes must be present in order for HIV to gain entry into its target immune cells. Notice that the genome for HIV consists of RNA instead of DNA, which classifies HIV as a retrovirus. A **retrovirus** must use reverse transcription to convert its RNA into viral DNA, and then it can insert its genome into the host's genome (DNA).

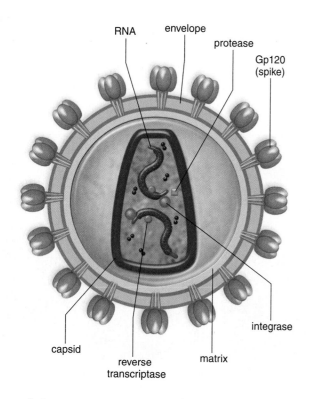

Figure S.4   The structure of HIV.
The components of HIV are necessary to its ability to infect CD4 T cells.

HIV Life Cycle

The events that occur in the reproductive cycle of an HIV virus (Fig. S.5) are these:

1. *Attachment.* During attachment, HIV binds to the plasma membrane of its target cell. Gp120, the spike located on the surface of HIV, binds to a CD4 receptor on the surface of a helper T cell or macrophage.
2. *Fusion.* After attachment occurs, HIV fuses with the plasma membrane, and the virus enters the cell.
3. *Entry.* During a process called uncoating, the capsid and protein coats are removed, releasing RNA and viral proteins into the cytoplasm of the host cell.
4. *Reverse transcription.* This event in the reproductive cycle is unique to retroviruses. During this phase, the enzyme called reverse transcriptase catalyzes the conversion of HIV's single-stranded RNA into double-stranded viral DNA. Usually in cells, DNA is transcribed into RNA. Retroviruses can do the opposite only because they have a unique enzyme from which they take their name. (*Retro* in Latin means reverse.)
5. *Integration.* The newly synthesized viral DNA, along with the viral enzyme integrase, migrates into the nucleus of the host cell. Then, with the help of integrase, the host cell's DNA is spliced, followed by the integration of the double-stranded viral DNA into the host cell's DNA (chromosome). Once viral DNA has integrated into the host cell's DNA, HIV is referred to as a **provirus,** meaning it is now a part of the cell's genetic material. HIV is usually transmitted to another person by means of cells that contain proviruses. Also, proviruses serve as a latent reservoir for HIV during drug treatment. Even if drug therapy results in an undetectable viral load, investigators know that there are still proviruses inside infected lymphocytes.
6. *Biosynthesis and cleavage.* When the provirus is activated, perhaps by a new and different infection, the normal cell machinery directs the production of more viral RNA. Some of this RNA becomes the genetic material for new viral particles. The rest of viral RNA brings about the synthesis of very long polypeptides. These polypeptides have to be cut up into smaller pieces. This cutting process, called cleavage, is catalyzed by a third HIV enzyme called protease.
7. *Assembly.* Capsid proteins, viral enzymes, and RNA can now be assembled to form new viral particles.
8. *Budding.* During budding, the virus gets its envelope and envelope marker coded for by the viral genetic material. Notice that the envelope is actually host plasma membrane.

The life cycle of an HIV virus includes transmission to a new host. Body secretions, such as semen from an infected male, contain proviruses inside CD4 T cells. When this semen is discharged into the vagina, rectum, or mouth, infected CD4 T cells migrate through the organ's lining and enter the body. The receptive partner in anal-rectal intercourse appears to be most at risk because the lining of the rectum is apparently prone to penetration. CD4 macrophages present in tissues are believed to be the first infected when proviruses enter the body. When these macrophages move to the lymph nodes, HIV begins to infect CD4 T cells. HIV can hide out in local lymph nodes for some time, but eventually the lymph nodes degenerate, and large numbers of HIV particles enter the general bloodstream. Now the viral load begins to increase; when it exceeds the CD4 T-cell count, the individual progresses to the final phase of an HIV infection.

Transmission and Prevention of HIV

HIV is transmitted by sexual contact with an infected person, including vaginal or rectal intercourse and oral/genital contact. Also, needle-sharing among intravenous drug users is high-risk behavior. A less common mode of transmission

(and now rare in countries where blood is screened for HIV) is through transfusions of infected blood or blood-clotting factors. Babies born to HIV-infected women may become infected before or during birth or through breast-feeding after birth. From a global perspective, heterosexual sex is the main mode of HIV transmission. In some nations, however, men who have sex with men, injection drug use, and sex industry workers (prostitution) are major modes of HIV transmission. Differences in cultures, sexual practices, and belief systems around the world influences the type of HIV prevention strategies needed to fight the spread of the disease.

Blood, semen, vaginal fluid, and breast milk are the body fluids known to have the highest concentrations of HIV. However, it is important to know that HIV is not transmitted through casual contact in the workplace, schools, or social settings. It is not transmitted through a casual kiss, hugging, or shaking hands or from toilet seats, door knobs, dishes, drinking glasses, food, or pets. The general message of HIV prevention across the globe is abstinence, sex with only one uninfected partner, and accurate and consistent condom use during each sexual encounter.

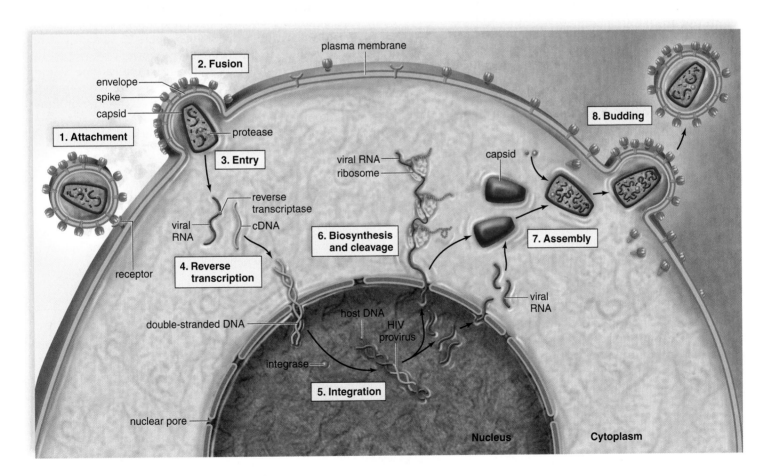

Figure S.5   Reproduction of HIV.
HIV is a retrovirus that utilizes reverse transcription to produce viral DNA. Viral DNA integrates into the cell's chromosomes before it reproduces and buds from the cell.

HIV Testing and Treatment for HIV

Generally, an HIV test does not test for the virus itself; instead, initial HIV tests are designed to detect the presence of HIV antibodies in the body. Most people will develop antibodies to HIV within two to eight weeks (average of 25 days), but it can take from three to six months. In addition to the conventional blood, oral, and urine HIV antibody tests where test results may not be available for weeks, rapid HIV antibody tests are now being used, with results available within 20 minutes.

Until a few years ago, an HIV infection almost invariably led to AIDS and an early death, because there were no drugs for controlling the progression of HIV disease in people infected with the virus. But since late 1995, scientists have gained a much better understanding of the structure of HIV and its life cycle. Now, therapy is available that successfully controls HIV replication and keeps patients in the chronic phase of infection for a variable number of years, so that the development of AIDS is postponed. However, this is not the case for those in poor nations where funds for HIV drugs are limited. In fact, only 20% of those world-wide who need antiretroviral therapy are receiving this treatment.

AIDS in the United States is presently caused by HIV-1B, and drug therapy has brought the condition under control. But the drug therapy has two dangers. People may become lax in their efforts to avoid infection because they know that drug therapy is available. Also, the use of drugs leads to drug-resistant viruses. Even now, some HIV-1B viruses have become drug-resistant when patients have failed to adhere to their drug regimens.

Another possible problem is that the present treatment is designed for HIV-1B. It's quite possible that in 5–20 years, the more-developed countries, including the United States, will experience a new epidemic of AIDS caused by HIV-1C. Therefore, it behooves the more-developed countries to do all they can to help African countries aggressively seek a solution to the HIV-1C epidemic.

Drug Therapy

There is no cure for AIDS, but a treatment called highly active antiretroviral therapy (HAART) is usually able to stop HIV replication to such an extent that the viral load becomes undetectable. HAART utilizes a combination of drugs that interfere with the life cycle of HIV. Entry inhibitors stop HIV from entering a cell by, for example, preventing the virus from binding to a receptor in the plasma membrane. Reverse transcriptase inhibitors, such as zidovudine (AZT), interfere with the operation of the reverse transcriptase enzyme. Integrase inhibitors prevent HIV from inserting its own genetic material into that of the host cells. Protease inhibitors prevent protease from cutting up newly created polypeptides. Assembly and budding inhibitors are in the experimental stage, and none are available as yet. The hope is that by giving a patient a combination of these drugs, the virus is less likely to replicate, successfully mutate in a particular way making drug therapy ineffective, or become resistant to drug therapy.

Investigators have found that when HAART is discontinued, the virus rebounds; therefore, therapy must be continued indefinitely. An HIV-positive pregnant woman who takes reverse transcriptase inhibitors during her pregnancy reduces the chances of HIV transmission to her newborn. If possible, drug therapy should be delayed until the tenth to twelfth week of pregnancy to minimize any adverse effects of AZT on fetal development. But, if treatment begins at this time, the chance of transmission is reduced by 66%. If treatment is delayed until the last few weeks of a pregnancy, transmission of HIV to an offspring is still cut by 50%.

Vaccines

The consensus is that control of the AIDS pandemic will not occur until a vaccine that prevents an HIV infection is developed. At the start of 2006, some 27 vaccines were undergoing human trials around the world. Unfortunately, none of these would be a cure; instead, the vaccine would be a preventive measure that helps people who are not yet infected escape infection (preventive vaccine) or slow down the progression of the disease upon future infection (therapeutic vaccines). Scientists have studied more than 50 different preventive vaccines and over 30 therapeutic vaccines. At the start of 2006, 27 preventive vaccines were undergoing human trials around the world. Scientist do not expect to have results until the year 2010.

The Science Focus on page 351 reviews the difficulties in developing an effective AIDS vaccine so that you can understand why we do not have an AIDS vaccine yet and, indeed, why we may never have an ideal one.

☑ Check Your Progress S.3

1. What are the three HIV enzymes, and what role does each play in the HIV life cycle?

2. Associate each of the three HIV enzymes with a means of drug therapy.

3. Why would an AIDS vaccine, like any vaccine, have to be given before a person is exposed to HIV?

Science Focus

Is an AIDS Vaccine Possible?

An ideal AIDS vaccine would be inexpensive and able to provide lifelong protection against all strains of HIV (Fig. S.A). Is this possible? Effective vaccines have been developed against diseases such as hepatitis B, smallpox, polio, tetanus, influenza, and the measles. But what about AIDS? The top 10 "setbacks" in AIDS vaccine development are these:

1. The ideal AIDS vaccine is one that would prevent HIV entry into human cells and, thereby, prevent the progression and transmission of the disease. However, no vaccine has ever proven to be 100% effective at blocking a virus from entry into cells. Instead, most vaccines simply prevent, modify, or weaken the disease caused by the infection.

2. Due to HIV's high rate of mutation, there are several genetically different types and subtypes of HIV. HIV strains may differ by 10% within one person and by 35% in people across the globe. Viruses that are genetically different may have different surface proteins, and HIV surface proteins are the focus of many AIDS vaccines. The question is, will scientists need a vaccine for each HIV subtype, or will one vaccine provide protection for all HIV variants?

3. The vaccine may produce only short-term protection, meaning people would need to continue to get booster shots similar to shots given yearly for the flu.

4. There are fears that an AIDS vaccine would make people more vulnerable to HIV infection. It is believed that in some diseases such as yellow fever and Rift Valley fever, the antibodies produced after receiving the vaccine actually helped the virus infect more cells.

5. HIV infects and destroys immune cells, particularly helper T cells. It would be hard to create an effective vaccine to stimulate the immune cells that HIV is targeting for destruction.

6. Because HIV can be transmitted as a free virus and in infected cells, perhaps a vaccine would need to stimulate both cellular and antibody-mediated responses. To date, most of the successful vaccines stimulate only antibody production.

7. Most vaccines in use today against other diseases are prepared from live-attenuated (weakened) forms of the infectious virus. There are concerns that an AIDS vaccine made using weakened forms of the live virus will cause HIV disease (AIDS).

8. No person has ever completely recovered from HIV infection.

9. HIV inserts its genetic material into human cells, where it can hide from the immune system.

10. Scientists do not have an ideal animal model for AIDS vaccine testing.

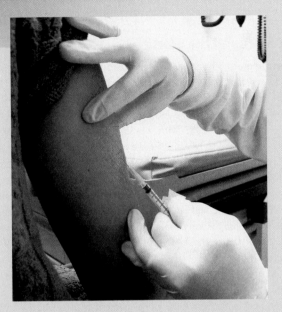

Figure S.A   HIV vaccine.
Will one vaccine be effective against HIV-1 and HIV-2? If an AIDS vaccine is developed, how often would it have to be altered to keep up with HIV's high rate of mutation?

Although there are many difficulties in vaccine development, AIDS vaccine trials are currently underway. The process can take many years. After a vaccine has been tested in animals, it must pass through three phases of clinical trials before it is marketed or administered to the public. In Phases I and II, the vaccine is tested from one to two years in a small number of HIV-uninfected volunteers. The most effective vaccines move into Phase III. In Phase III, the vaccine is tested three to four years in thousands of HIV-uninfected people. One possible AIDS vaccine called ALVAC (AIDSVAX) is in Phase III trials. The trials involve 16,000 people in Thailand, but the results will probably not be available until the year 2010.

Currently, there are no approved preventive or therapeutic vaccines for HIV infection. So, is an AIDS vaccine possible? Most scientists have many reasons to be optimistic that an AIDS vaccine can and will be developed. One reason is that scientists have had some success with vaccinating monkeys. But the most compelling reason for optimism is the human body's own ability to suppress the infection. The immune system is able to successfully and effectively decrease the HIV viral load in the body, helping to delay the onset of AIDS an average of ten years in 60% of people who are HIV-infected in the United States. Studies have shown a small number of people remain HIV-uninfected after repeated exposure to the virus, and a few HIV-infected individuals maintain a healthy immune system for over 15 years. It is these stories of the human body's ability to fight HIV infection that keep scientists hopeful that there is a way to help the body fight HIV infection. Most scientists agree that, despite the obstacles or the "setbacks," an AIDS vaccine is possible somewhere in our future.

AIDS—A Natural Disaster

AIDS Around the World

Approximately 95% of all HIV-infected persons live in developing (poor, low- to middle-income) countries. Millions are dying due to lack of antiviral drugs and other health and educational services. According to a recent report, the AIDS death toll in Africa could reach 100 million by the year 2025, with sub-Saharan Africa losing 10% of their entire population to HIV/AIDS. India now has the world's largest HIV/AIDS population, approximately 5.7 million HIV cases compared to 5.5 million in South Africa. Regions all over the world, Africa, Asia, Latin America, Europe, America, and the Middle East, are suffering from the social, economical, and political impact of AIDS. By the end of 2005, an estimated 65 million people had been HIV infected worldwide, and nine out of ten of these HIV-positive people were unaware of their infection.

Since 1981, AIDS has claimed more than 25 million lives. Each year, nearly 5 million people become infected with HIV, and over 3 million people die from AIDS. Eighty percent of these AIDS deaths are among people ages 20 to 50. In fact, young people ages 15–24 make up about 50% of all new HIV infections worldwide. Women, who account for 47% of all people living with HIV, have left over 12 million AIDS orphans in Africa.

AIDS in the United States

Three percent of the people living with HIV/AIDS live in the United States. Since 1981, AIDS has claimed more than a half million American lives, and each year there are approximately 40,000 new HIV infections. In 2004, 73% of all new HIV/AIDS diagnoses were among men and 27% among women. From an international perspective, the main mode of HIV transmission is through heterosexual contact. However, in the United States, male-to-male sexual contact is the main mode of HIV transmission. In fact, among men in the United States, HIV is transmitted primarily through male-to-male contact (65%), 16% through heterosexual contact, and 14% through injection drug use. For women, the statistics are quite different. Seventy-eight percent of HIV transmission among women is through heterosexual contact and 20% through injection drug use.

During the early years of the AIDS epidemic, from 1981 until about 1995, half of all U.S. AIDS cases were among whites, especially gay white males. Today, 50% of all new HIV/AIDS diagnoses are among African Americans and approximately 18% among Hispanics (Fig. S.B). During 2001–2004, the rate of new HIV diagnoses declined, but the rate of new infections among African Americans remained higher than any other race or ethnic group in the country. By 2002, AIDS had become one of the top three killers of African American men ages 25–54 and the number one killer of African American women ages 25–34. In 2004, the rate of AIDS diagnosis for African American women was 23 times the rate for white women, and African American men had a rate of diagnosis eight times that of white men. Reasons for such an increased rate of infections among African American women may be attributed, in part, to men who have sex with men without identifying themselves as gay or bisexual. By the end of 2004, there were over 400,000 people living with AIDS in the United States; 35% were white, 43% were African American, 20% were Hispanic, and 2% were of other race/ethnicity. Today, there are just over one million people living with HIV/AIDS in the United States, and some 25% of these people are unaware of their HIV infection.

As nations around the globe come together to fight the AIDS epidemic, there is hope that the "war on AIDS" will eventually be won as new drugs, treatments, and educational campaigns reach the hardest-hit regions of the world. With so much going on in the world, we often forget about those disasters that aren't making the front page of a newspaper, the cover of a magazine, or the headliner of CNN news. No war, natural disaster, or terrorist attack has ever killed more people than HIV/AIDS. AIDS is the worst and deadliest global epidemic that mankind has ever faced. Our ultimate hope is for an AIDS vaccine, for without it, AIDS will continue to claim millions of lives throughout the world.

Decide Your Opinion

1. What can nations do to help prevent the spread of HIV/AIDS?
2. What do you think can be done to decrease the number of HIV infections among minority groups in the United States, particularly African Americans?

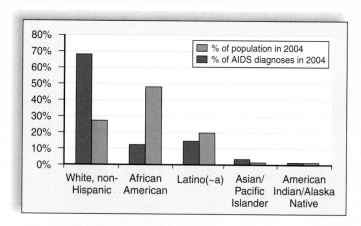

Figure S.B **AIDS diagnoses by ethnicity in 2004.**
African Americans and Hispanics make up 27% of the U.S. population, yet together account for nearly 70% of all AIDS diagnoses in 2004.

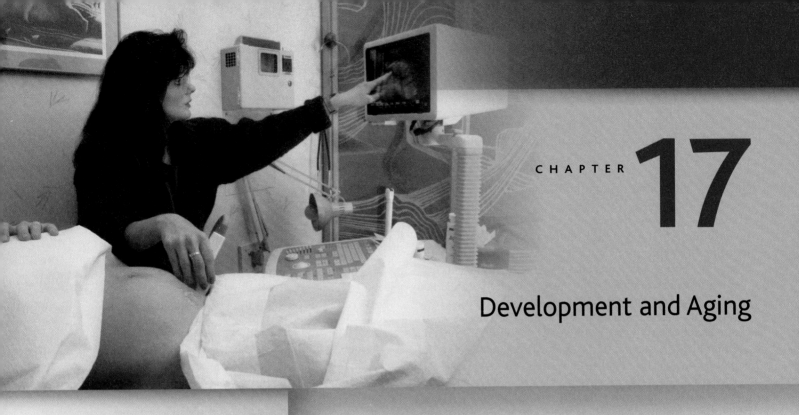

CHAPTER 17

Development and Aging

Amber contemplated the blue lines on the test stick in her hand and consulted the directions that came with the pregnancy test. As she compared the lines that appeared on the test stick with the directions' pictures, a smile began to appear on her face. "Woohoo," she yelled, "we're pregnant!"

A week later, Amber got confirmation of the good news from her physician, who told Amber she was probably about six or seven weeks pregnant. Dr. Davis cautioned Amber to be careful about her eating habits and to avoid exposure to cigarette smoke and alcohol because of critical development that occurs during the first trimester. Amber asked about her blood pressure medication because she'd read about a commonly prescribed type of medication that appeared to increase the risk of birth defects in the early stages of a pregnancy. Dr. Davis assured Amber that her medication was not an ACE inhibitor, and that she should continue to take it.

Amber was anxious to see the baby and asked when she would have her first ultrasound. "Let's schedule you for an ultrasound in a month when we may be able to determine whether its' a boy or girl. But in any case, we'll be able to hear the heartbeat. Later in the pregnancy if it is warranted, we may need to discuss using pitocin to induce labor or delivery via cesarean section, but for now, just enjoy your good news," he continued.

CHAPTER CONCEPTS

17.1 Fertilization
During fertilization, a sperm nucleus fuses with the egg nucleus. Once one sperm penetrates the plasma membrane, the egg undergoes changes that prevent any more sperm from entering.

17.2 Pre-Embryonic and Embryonic Development
Pre-embryonic development occurs between the time of fertilization and implantation in the uterine lining. By the end of embryonic development, all organ systems are established, and there is a mature and functioning placenta. The embryo is only about 38 mm (1½ in.) long.

17.3 Fetal Development
During fetal development, the gender becomes obvious, the skeleton continues to ossify, fetal movement begins, and the fetus gains weight.

17.4 Pregnancy and Birth
During pregnancy, the mother gains weight as the uterus comes to occupy most of the abdominal cavity. A positive feedback mechanism that involves uterine contractions and the oxytocin explains the onset and continuation of labor so that the child is born.

17.5 Development After Birth
Development after birth consists of infancy, childhood, adolescence, and adulthood. Aging is influenced by our genes, whole body changes, and extrinsic factors.

17.1 Fertilization

Fertilization is the union of a sperm and egg to form a **zygote,** the first cell of the new individual (Fig. 17.1). Fertilization is the first step for Amber in the opening story to get pregnant.

Steps of Fertilization

The tail of a sperm is a flagellum, which allows it to swim toward the egg. The middle piece contains energy-producing mitochondria, and the head contains a nucleus capped by a membrane-bound acrosome. Notice that only the nucleus from the sperm head fuses with the egg nucleus. Therefore, the zygote receives cytoplasm and organelles only from the mother.

The plasma membrane of the egg is surrounded by an extracellular matrix termed the zona pellucida. In turn, the zona pellucida is surrounded by a few layers of adhering follicular cells, collectively called the corona radiata. These cells nourished the egg when it was in a follicle of the ovary.

During fertilization, several sperm penetrate the corona radiata, several sperm attempt to penetrate the zona pellu-cida, but only one sperm enters the egg. The acrosome plays a role in allowing sperm to penetrate the zona pellucida. After a sperm head binds tightly to the zona pellucida, the acrosome releases digestive enzymes that forge a pathway for the sperm through the zona pellucida. When a sperm binds to the egg, their plasma membranes fuse, and this sperm (the head, the middle piece, and usually the tail) enters the egg. Fusion of the sperm nucleus and the egg nucleus follows.

To ensure proper development, only one sperm should enter an egg. Prevention of polyspermy (entrance of more than one sperm) depends on changes in the egg's plasma membrane and in the zona pellucida. As soon as a sperm touches an egg, the egg's plasma membrane depolarizes (from 265 mV to 10 mV), and this prevents the binding of any other sperm. Then, vesicles called cortical granules release enzymes that cause the zona pellucida to become an impenetrable fertilization mem-brane. Now sperm cannot bind to the zona pellucida either.

☑ Check Your Progress 17.1

1. How does fertilization occur?

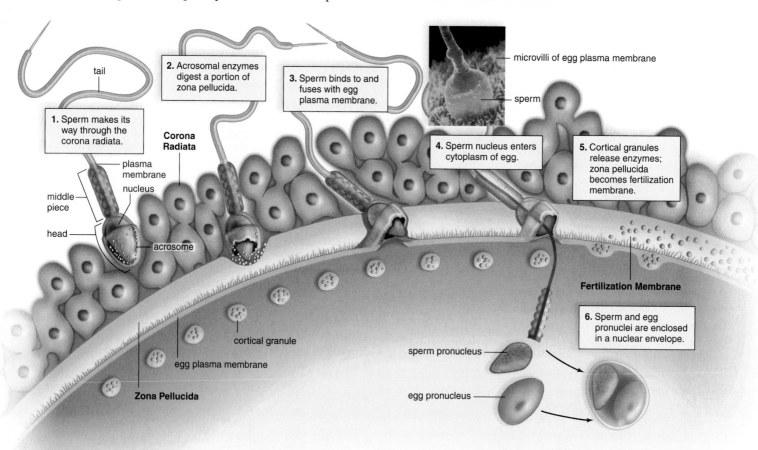

Figure 17.1   Fertilization.
During fertilization, a single sperm is drawn into the egg by microvilli of its plasma membrane (micrograph). With the help of enzymes from the acrosome, a sperm makes its way through the zona pellucida. After a sperm binds to the plasma membrane of the egg, changes occur that prevent other sperm from entering the egg. Fertilization is complete when the sperm pronucleus and the egg pronucleus contribute chromosomes to the zygote.

17.2 Pre-Embryonic and Embryonic Development

Before we discuss the stages of embryonic development, you will want to become familiar with the processes of development and the names and functions of the extraembryonic membranes.

Processes of Development

As a human being develops, these processes occur:

Cleavage   Immediately after fertilization, the zygote begins to divide so that there are first 2, then 4, 8, 16, and 32 cells, and so forth. Increase in size does not accompany these divisions (see Fig. 17.3). Cell division during cleavage is mitotic, and each cell receives a full complement of chromosomes and genes.

Growth   During embryonic development, cell division is accompanied by an increase in size of the daughter cells.

Morphogenesis   Morphogenesis refers to the shaping of the embryo and is first evident when certain cells are seen to move, or migrate, in relation to other cells. By these movements, the embryo begins to assume various shapes.

Differentiation   When cells take on a specific structure and function, differentiation occurs. The first system to become visibly differentiated is the nervous system.

Extraembryonic Membranes

The **extraembryonic membranes** are not part of the embryo and fetus; instead, as implied by their name, they are outside the embryo (Fig. 17.2). The names of the extraembryonic membranes in humans are strange to us because they are named for their function in shelled animals! In shelled animals, the chorion lies next to the shell and carries on gas exchange. The amnion contains the protective amniotic fluid, which bathes the developing embryo. The allantois collects nitrogenous wastes, and the yolk sac surrounds the yolk, which provides nourishment.

The functions of the extraembryonic membranes are different in humans because humans develop inside the uterus. The extraembryonic membranes have these functions in humans:

1. **Chorion.** The chorion develops into the fetal half of the **placenta,** the organ that provides the embryo/fetus with nourishment and oxygen and takes away its waste. Blood vessels within the chorionic villi are continuous with the umbilical blood vessels.

2. **Allantois.** The allantois, like the yolk sac, extends away from the embryo. It accumulates the small amount of urine produced by the fetal kidneys and later gives rise to the urinary bladder. For now, its blood vessels become the umbilical blood vessels, which take blood to and from the fetus. The umbilical arteries carry O_2-poor blood to the placenta, and the umbilical veins carry O_2-rich blood from the placenta.

3. **Yolk sac.** The yolk sac is the first embryonic membrane to appear. In shelled animals such as birds, the yolk sac contains yolk, which is food for the developing embryo. In mammals such as humans, this function is taken over by the placenta, and the yolk sac contains little yolk. But the yolk sac contains plentiful blood vessels. It is the first site of blood cell formation.

4. **Amnion.** The amnion enlarges as the embryo and then the fetus enlarges. It contains fluid to cushion and protect the embryo, which develops into a fetus.

Stages of Development

Development encompasses the events that occur from fertilization to birth. When Amber went to the doctor, he calculated the baby's due date by adding 280 days to the start of the last menstruation, a date that is usually known. (The 280 days is called the normal **gestation** period.) However, only about 5% of babies actually arrive on the predicted date.

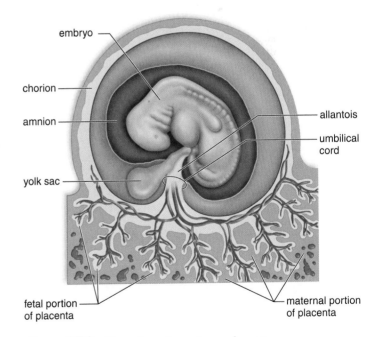

Figure 17.2   The extraembryonic membranes.
The extraembryonic membranes. The chorion and amnion surround the embryo. The two other extraembryonic membranes, the yolk sac and allantois, contribute to the umbilical cord.

Pre-Embryonic Development

We will subdivide development into pre-embryonic, embryonic, and fetal development. **Pre-embryonic development** encompasses the events of the first week, as shown in Figure 17.3.

Immediately after fertilization, the zygote divides repeatedly as it passes down the oviduct to the uterus. A **morula** is a compact ball of embryonic cells that becomes a **blastocyst.** The many cells of the blastocyst arrange themselves so that there is an **inner cell mass** surrounded by an outer layer of cells. The inner cell mass will become the embryo, and the layer of cells will become the **chorion.** The early appearance of the chorion emphasizes the complete dependence of the developing embryo on this extraembryonic membrane.

Each cell within the inner cell mass has the genetic capability of becoming any type of tissue. Sometimes during

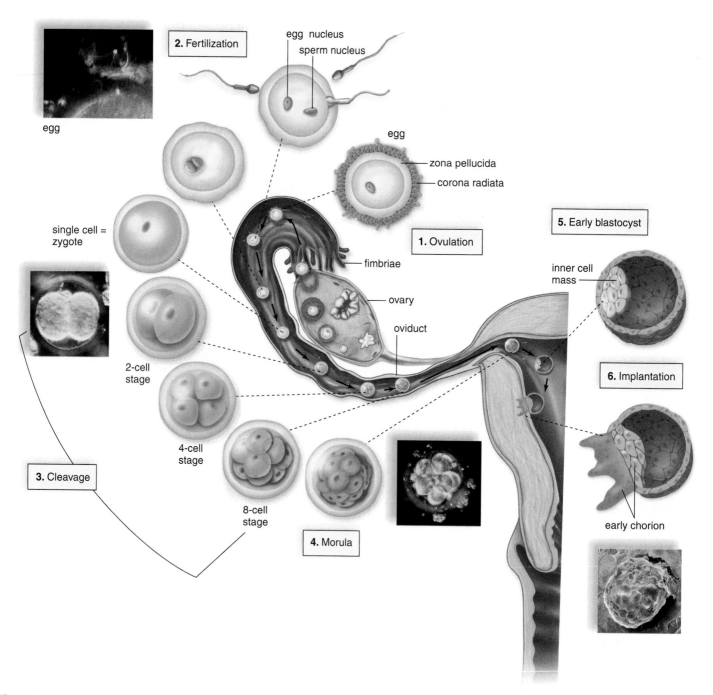

Figure 17.3 **Pre-embryonic development.**
Structures and events proceed clockwise. (1) At ovulation, the secondary oocyte leaves the ovary. A single sperm nucleus enters the egg, and (2) fertilization occurs in the oviduct. As the zygote moves along the oviduct, it undergoes (3) cleavage to produce (4) a morula. (5) The blastocyst forms and (6) implants itself in the uterine lining.

development, the cells of the morula separate, or the inner cell mass splits, and two pre-embryos are present rather than one. If all goes well, these two pre-embryos will be identical twins because they have inherited exactly the same chromosomes. Fraternal twins, which arise when two different eggs are fertilized by two different sperm, do not have identical chromosomes.

Embryonic Development

Embryonic development begins with the second week and lasts until the end of the second month of development.

Second Week At the end of the first week, the **embryo** usually begins the process of implanting itself in the wall of the uterus. When **implantation** was successful, Amber was clinically pregnant. On occasion, it happens that the embryo implants itself in a location other than the uterus—most likely, the oviduct. Such a so-called **ectopic pregnancy** cannot succeed because an oviduct is unable to support it.

During implantation, the chorion secretes enzymes to digest away some of the tissue and blood vessels of the endometrium of the uterus. The chorion also begins to secrete **human chorionic gonadotropin (HCG),** the hormone that is the basis for the pregnancy test. HCG acts like luteinizing hormone (LH) in that it serves to maintain the corpus luteum past the time it normally disintegrates. Because it is being stimulated, the corpus luteum secretes progesterone, the endometrium is maintained, and the expected menstruation does not occur.

The embryo is now about the size of the period at the end of this sentence. As the week progresses, the inner cell mass

becomes the **embryonic disk,** and two more extraembryonic membranes form (Fig. 17.4*a*). The yolk sac is the first site of blood cell formation. The amniotic cavity surrounds the embryo (and then the fetus) as it develops. In humans, amniotic fluid acts as an insulator against cold and heat and also absorbs shock, such as that caused by the mother exercising.

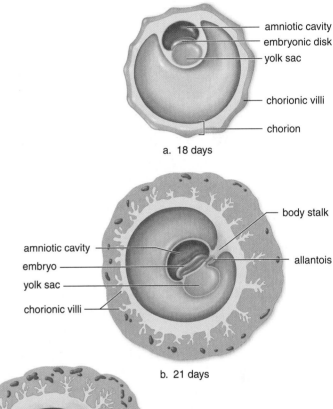

a. 18 days

b. 21 days

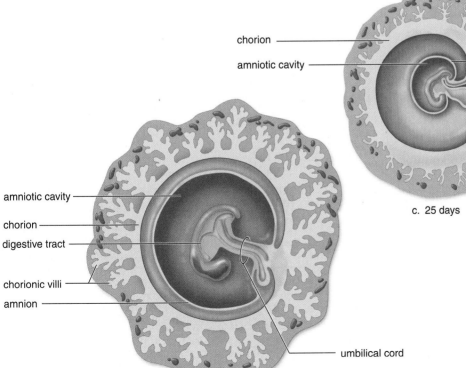

c. 25 days

d. 35+ days

Figure 17.4 Embryonic development.
a. At first, no organs are present in the embryo, only tissues. The amniotic cavity is above the embryonic disk, and the yolk sac is below. The chorionic villi are present. **b, c.** The allantois and yolk sac, two more extraembryonic membranes, are positioned inside the body stalk as it becomes the umbilical cord. **d.** At 35+days, the embryo has a head region and a tail region. The umbilical cord takes blood vessels between the embryo and the chorion (placenta).

The start of the major event called **gastrulation,** turns the inner cell mass into the embryonic disk. Gastrulation is an example of morphogenesis (see page 355) during which cells move or migrate, in this case to become tissue layers called the **primary germ layers.** By the time gastrulation is complete, the embryonic disk has become an embryo with three primary germ layers: ectoderm, mesoderm, and endoderm. Figure 17.5 shows the significance of the primary germ layers—all the organs of an individual can be traced back to one of the primary germ layers.

Third Week   Two important organ systems make their appearance during the third week. The nervous system

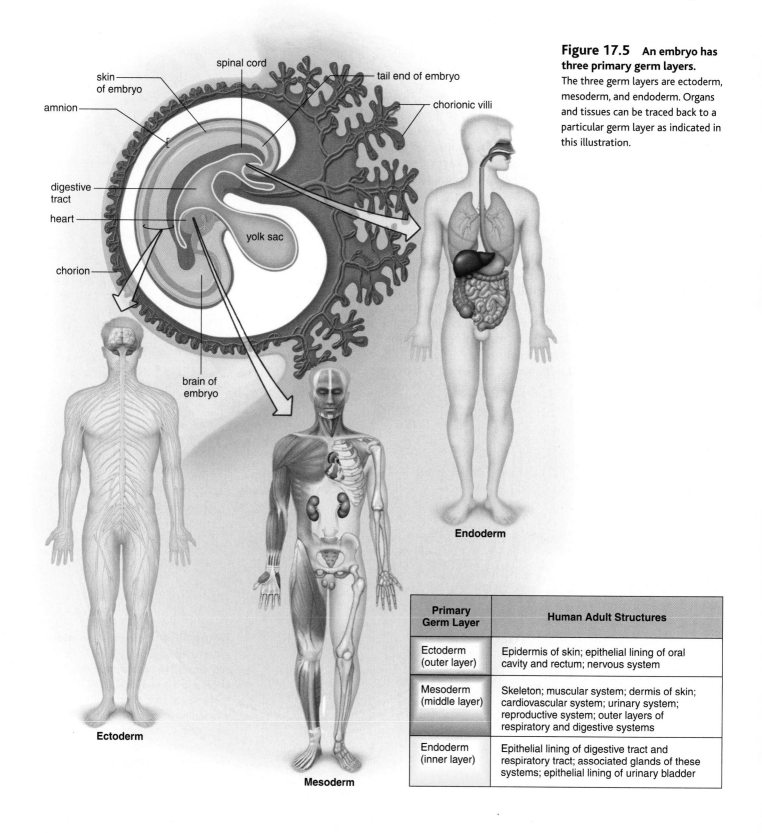

skin of embryo
spinal cord
amnion
digestive tract
heart
chorion
tail end of embryo
chorionic villi
yolk sac
brain of embryo
Ectoderm
Mesoderm
Endoderm

Figure 17.5   An embryo has three primary germ layers.
The three germ layers are ectoderm, mesoderm, and endoderm. Organs and tissues can be traced back to a particular germ layer as indicated in this illustration.

Primary Germ Layer	Human Adult Structures
Ectoderm (outer layer)	Epidermis of skin; epithelial lining of oral cavity and rectum; nervous system
Mesoderm (middle layer)	Skeleton; muscular system; dermis of skin; cardiovascular system; urinary system; reproductive system; outer layers of respiratory and digestive systems
Endoderm (inner layer)	Epithelial lining of digestive tract and respiratory tract; associated glands of these systems; epithelial lining of urinary bladder

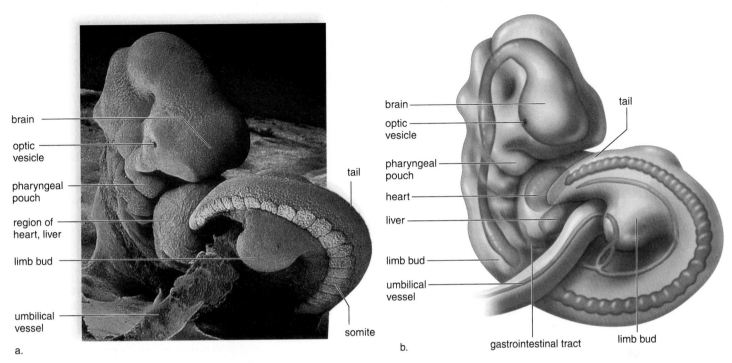

brain
optic vesicle
pharyngeal pouch
region of heart, liver
limb bud
umbilical vessel
a.
tail

brain
optic vesicle
pharyngeal pouch
heart
liver
limb bud
umbilical vessel
tail
b.
gastrointestinal tract
limb bud
somite

Figure 17.6   Human embryo at beginning of fifth week.
a. Scanning electron micrograph. **b.** The embryo is curled so that the head touches the heart and liver, the two organs whose development is farther along than the rest of the body. The organs of the gastrointestinal tract are forming, and the arms and the legs develop from the bulges called limb buds. The tail is an evolutionary remnant; its bones regress and become those of the coccyx (tailbone). The pharyngeal pouches become functioning gills in fishes and amphibian larvae; in humans, the first pair of pharyngeal pouches becomes the auditory tubes. The second pair becomes the tonsils, while the third and fourth become the thymus gland and the parathyroid glands.

is the first organ system to be visually evident. At first, a thickening appears along the entire posterior length of the embryo, and then invagination occurs as neural folds appear. When the neural folds meet at the midline, the neural tube, which later develops into the brain and the spinal cord, is formed.

Development of the heart begins in the third week and continues into the fourth week. At first, there are right and left heart tubes; when these fuse, the heart begins pumping blood, even though the chambers of the heart are not fully formed. The veins enter posteriorly, and the arteries exit anteriorly from this largely tubular heart, but later the heart twists so that all major blood vessels are located anteriorly.

Fourth and Fifth Weeks   At four weeks, the embryo is barely larger than the height of this print. A body stalk (future umbilical cord) connects the embryo to the chorion which has treelike projections called **chorionic villi** (see Fig. 17.4c,d). The fourth extraembryonic membrane, the allantois, lies within the body stalk, and its blood vessels become the umbilical blood vessels. The head and the tail then lift up, and the body stalk moves anteriorly by constriction. Once this process is complete, the **umbilical cord,** which connects the developing embryo to the placenta, is fully formed (see Fig. 17.4d).

Little flippers called limb buds appear (Fig. 17.6); later, the arms and the legs develop from the limb buds, and even the hands and the feet become apparent. At the same time—

during the fifth week—the head enlarges, and the sense organs become more prominent. It is possible to make out the developing eyes and ears, and even the nose.

Sixth Through Eighth Weeks   During the sixth through eighth weeks of development, the embryo changes to a form that is easily recognized as a human being. Concurrent with brain development, the head achieves its normal relationship with the body as a neck region develops. The nervous system is developed well enough to permit reflex actions, such as a startle response to touch. At the end of this period, the embryo is about 38 mm (1.5 in.) long and weighs no more than an aspirin tablet, even though all organ systems have been established. Amber was about six or seven weeks pregnant when she went to the doctor because she knew she was pregnant. All sexually active women who could become pregnant are advised to avoid smoking and alcohol because the embryo has formed by the time pregnancy is evident.

✔ Check Your Progress 17.2

1. Development involves what processes?
2. **a.** What are the extraembryonic membranes? **b.** What are their functions?
3. **a.** What happens during pre-embryonic development? **b.** During embryonic development?

17.3 Fetal Development

Table 17.1 outlines the major events during development for easy reference. The placenta is the source of progesterone and estrogen during pregnancy. These hormones have two functions: (1) because of negative feedback on the hypothalamus and anterior pituitary, they prevent any new follicles from maturing, and (2) they maintain the endometrium—menstruation does not usually occur during pregnancy.

The placenta has a fetal side contributed by the chorion and a maternal side consisting of uterine tissues (Fig. 17.7). The

Table 17.1	Human Development	
Time	**Events for Mother**	**Events for Baby**
Pre-Embryonic Development		
First week	Ovulation occurs.	Fertilization occurs. Cell division begins and continues. Chorion appears.
Embryonic Development		
Second week	Symptoms of early pregnancy (nausea, breast swelling and fatigue) are present. Blood pregnancy test is positive.	Implantation occurs. Amnion and yolk sac appear. Embryo has tissues. Placenta begins to form.
Third week	First menstruation is missed. Urine pregnancy test is positive. Symptoms of early pregnancy continue.	Nervous system begins to develop. Allantois and blood vessels are present. Placenta is well formed.
Fourth week		Limb buds form. Heart is noticeable and beating. Nervous system is prominent. Embryo has tail. Other systems form.
Fifth week	Uterus is the size of a hen's egg. Mother feels frequent need to urinate due to pressure of growing uterus on bladder.	Embryo is curved. Head is large. Limb buds show divisions. Nose, eyes, and ears are noticeable.
Sixth week	Uterus is the size of an orange.	Fingers and toes are present. Skeleton is cartilaginous.
Two months	Uterus can be felt above the pubic bone.	All systems are developing. Bone is replacing cartilage. Facial features are becoming refined. Embryo is about 38 mm (1½ in.) long.
Fetal Development		
Third month	Uterus is the size of a grapefruit.	Gender can be distinguished by ultrasound. Fingernails appear.
Fourth month	Fetal movement is felt by a mother who has previously been pregnant.	Skeleton is visible. Hair begins to appear. Fetus is about 150 mm (6 in.) long and weighs about 170 grams (6 oz).
Fifth month	Fetal movement is felt by a mother who has not previously been pregnant. Uterus reaches up to level of umbilicus, and pregnancy is obvious.	Protective cheesy coating, called vernix caseosa, begins to be deposited. Heartbeat can be heard.
Sixth month	Doctor can tell where baby's head, back, and limbs are. Breasts have enlarged, nipples and areolae are darkly pigmented, and colostrum is produced.	Body is covered with fine hair called lanugo. Skin is wrinkled and reddish.
Seventh month	Uterus reaches halfway between umbilicus and rib cage.	Testes descend into scrotum. Eyes are open. Fetus is about 300 mm (12 in.) long and weighs about 1,350 grams (3 lb).
Eighth month	Weight gain is averaging about a pound a week. Standing and walking are difficult for the mother because her center of gravity is thrown forward.	Body hair begins to disappear. Subcutaneous fat begins to be deposited.
Ninth month	Uterus is up to rib cage, causing shortness of breath and heartburn. Sleeping becomes difficult.	Fetus is ready for birth. It is about 530 mm (20½ in.) long and weighs about 3,400 grams (7½ lb).

blood of the mother and the fetus never mix since exchange always takes place across the villi. Carbon dioxide and other wastes move from the fetal side to the maternal side, and nutrients and oxygen move from the maternal side to the fetal side of the placenta by diffusion. Harmful chemicals can also cross the placenta, and this is of particular concern during the embryonic period, when various structures are first forming. Each organ or part seems to have a sensitive period during which a substance can alter its normal function. The Health Focus on page 366 concerns maternal behaviors that can prevent certain birth defects.

Path of Fetal Blood

The umbilical cord, which stretches between the placenta and the fetus, is the lifeline of the fetus because it contains the umbilical arteries and vein. Blood within the fetal aorta travels to its various branches, including the iliac arteries, which connect to the *umbilical arteries* carrying O_2-poor blood to the placenta. The *umbilical vein* carries blood rich in nutrients and O_2 away from the placenta to the fetus. The umbilical vein enters the liver and then joins the *venous duct*, which merges with the inferior vena cava, a vessel that returns blood to the right atrium. This mixed blood enters the heart and is shunted to the left atrium through the *oval opening*. The left ventricle pumps this blood into the aorta. O_2-poor blood that enters the right atrium is pumped into the pulmonary trunk, but it joins the aorta by way of the arterial duct. Therefore, all blood entering the right atrium by-passes the lungs.

Various circulatory changes occur at birth due to the tying of the cord and the expansion of the lungs. The lack of any one of these changes might cause the "blue baby" syndrome in which the skin is blue due to O_2-poor blood. (1) Return of this blood to the left side of the heart usually causes a flap to cover the oval opening. Even if this mechanism fails, passage of blood from the right atrium to the left atrium rarely occurs because either the opening is small or it closes when the atria contract. (2) The arterial duct closes at birth because endothelial cells divide and block off the duct. (3) Remains of the arterial duct and parts of the umbilical arteries and vein later are transformed into connective tissue.

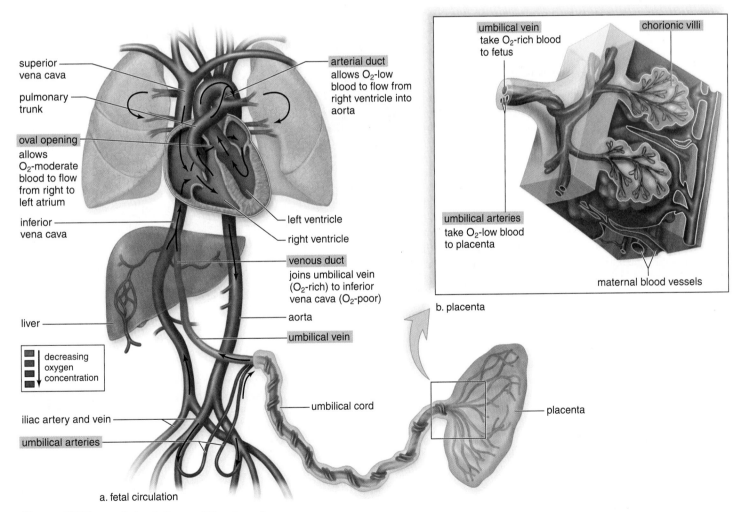

Figure 17.7 **Fetal circulation and the placenta.**

a. Trace the path of blood by following the arrows. The structures in red are unique to the fetus. **b.** At the placenta, an exchange of molecules between fetal and maternal blood takes place across the walls of the chorionic villi.

Science **Focus**

Cloning Humans: Can It Be Done?

In March 1997, Scottish investigators announced they had cloned a sheep called Dolly and their procedure is now routinely used (Fig. 17A). A donor 2n nucleus is substituted for the n nucleus of an egg; a stimulus is applied that triggers cell division, and the resulting embryo is implanted into a surrogate mother where it develops to term. Using the procedure developed by the Scottish researchers, it is now common practice to clone all sorts of farm animals (horses, cows, sheep, goats, pigs) and also cats and monkeys.

Success of Cloning

Even so, cloning of animals is still in its infancy, and many problems still exist. (1) The vast majority of pregnancies involving clones are not successful. To clone Dolly the sheep, it took 247 tries before one was successful. In many cases, the clone grows abnormally large, and the uterus enlarges with fluid to the point it can rip apart. Almost all clone pregnancies spontaneously abort. (2) Of the small number of animal clones born, most have severe abnormalities: malfunctioning livers, abnormal blood vessels and heart problems, underdeveloped lungs, diabetes, immune system deficiencies are all seen in newborns. Several cow clones had head deformities—none survived very long. (3) Even if the newborn clone appears healthy, it usually soon develops diseases seen in older animals. Dolly was euthanized in 2003 because she was suffering from lung cancer and crippling arthritis. She had lived only half the normal life span for a Dorset sheep.

Reproductive Cloning Versus Therapeutic Cloning

In most industrialized countries, including the United States, research into the cloning of humans is illegal. In under developed countries, there are no laws against it. Even if successful, would clones be exact copies? Only if 2n donor nucleus and the donor egg came from the same woman. Recall that mitochondria have genes, and these genes would be contributed by the egg even if the egg nucleus is removed. Even then, the clone would be subject to different environmental factors and a different upbringing to his/her genetic parent.

Reproductive cloning is quite different from therapeutic cloning. In **reproductive cloning,** the desired end is an individual, as shown on this page. In **therapeutic cloning,** the desired end is not an individual; rather, it is the embryonic cells that possibly can be "coaxed" into becoming various cell types. The purpose of therapeutic cloning is (1) to learn more about how specialization of cells occurs and (2) to provide cells and tissues that could be used to treat human illnesses, such as diabetes, spinal cord injuries, Parkinson disease, and so forth. In the United States, investigators can use only embryonic stem cell lines that already exist and cannot start new ones because of ethical concerns. However, researchers can freely make use of adult stem cells that reside in most tissues of the body. Investigators have been able to isolate adults stem cells and use them to regenerate nerve tissue for the treatment of Parkinson disease, for example.

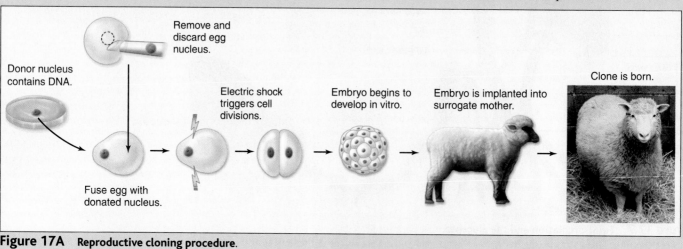

Remove and discard egg nucleus.

Donor nucleus contains DNA.

Electric shock triggers cell divisions.

Embryo begins to develop in vitro.

Embryo is implanted into surrogate mother.

Clone is born.

Fuse egg with donated nucleus.

Figure 17A   Reproductive cloning procedure.

Events of Fetal Development

Fetal development includes the third through the ninth months of development. At this time, the fetus is recognizably human (Fig. 17.8), but many refinements still need to be added. The fetus usually increases in size and gains the weight that will be needed to allow it to live as an independent individual.

Third and Fourth Months

At the beginning of the third month, the fetal head is still very large relative to the rest of the body, the nose is flat, the eyes are far apart, and the ears are well formed. Head growth now begins to slow down as the rest of the body increases in length. Epidermal refinements, such as fingernails, nipples, eyelashes, eyebrows, and hair on the head, appear.

Cartilage begins to be replaced by bone as ossification centers appear in most of the bones. Cartilage remains at the ends of the long bones, and ossification is not complete until age 18 or 20 years. The skull has six large membranous areas called **fontanels,** which permit a certain amount of flexibility as the head passes through the birth canal and allow rapid growth of the brain during infancy. Progressive fusion of the skull bones causes the fontanels to close, usually by 2 years of age.

Sometime during the third month, it is possible to distinguish males from females. As discussed on page 364, the presence of an *SRY* gene, usually on the Y chromosome, leads to the development of testes and male genitals. Otherwise, ovaries and females genitals develop. At this time, either testes or ovaries are located within the abdominal cavity, but later, in the last trimester of fetal development, the testes descend into the scrotal sacs (scrotum). Sometimes the testes fail to descend, and in that case, an operation may be done later to place them in their proper location.

During the fourth month, the fetal heartbeat is loud enough to be heard when a physician applies a stethoscope to the mother's abdomen. By the end of this month, the fetus is about 152 mm (6 in.) in length and weighs about 171 g (6 oz).

Fifth Through Seventh Months

During the fifth through seventh months (Fig. 17.8), the mother begins to feel movement. At first, there is only a fluttering sensation, but as the fetal legs grow and develop, kicks and jabs are felt. The fetus, though, is in the fetal position, with the head bent down and in contact with the flexed knees.

The wrinkled, translucent skin is covered by a fine down called **lanugo.** This, in turn, is coated with a white, greasy, cheeselike substance called **vernix caseosa,** which probably protects the delicate skin from the amniotic fluid. The eyelids are now fully open.

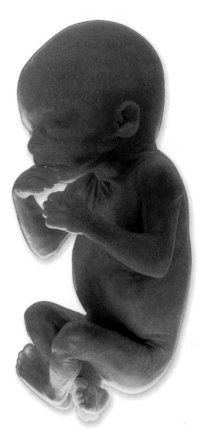

Figure 17.8   Five- to seven-month-old fetus.
Wrinkled skin is covered by fine hair.

At the end of this period, the fetus's length has increased to about 300 mm (12 in.), and it weighs about 1,380 g (3 lb). It is possible that, if born now, the baby will survive.

Eighth Through Ninth Months

At the end of nine months, the fetus is about 530 mm (20½ in.) long and weighs about 3,400 g (7½ lb). Weight gain is due largely to an accumulation of fat beneath the skin. Full-term babies have the best chance of survival, while premature babies are subject to various challenges, such as respiratory distress syndrome because their lungs are underdeveloped (see page 174), jaundice (see page 153), and infections.

As the end of development approaches, the fetus usually rotates so that the head is pointed toward the cervix. However, if the fetus does not turn, a **breech birth** (rump first) is likely. It is very difficult for the cervix to expand enough to accommodate this form of birth, and asphyxiation of the baby is more likely to occur. Thus, a **cesarean section** may be prescribed for delivery of the fetus (incision through the abdominal and uterine walls). Birth is discussed on page 368.

Development of Male and Female Genitals

The sex of an individual is determined at the moment of fertilization. Males have a pair of chromosomes designated as X and Y, and females have two X chromosomes.

Normal Development of the Genitals

Development of the internal and external genitals is shown in Figure 17.9.

Internal Genitals   During the first several weeks of development, it is impossible to tell by external inspection whether the unborn child is a boy or a girl. Gonads don't start developing until the seventh week of development.

The tissue that gives rise to the gonads is called *indifferent* because it can become testes or ovaries, depending on the action of hormones.

In Figure 17.9a, notice that ① at six weeks, both males and females have the same types of ducts. During this indifferent stage, an embryo has the potential to develop into a male or a female. If a gene called *SRY* is present, testes develop and **testosterone** produced by the testes stimulate the Wolffian ducts to become male genital ducts. ② The Wolffian ducts enter the urethra, which belongs to both the urinary and reproductive systems in males. An anti-Müllerian hormone causes the Müllerian ducts to regress. In the absence of an *SRY* gene, ovaries develop instead of testes from the same indifferent tissue. ③ Now the Wolffian ducts regress, and under the influence of estrogen from the ovaries, the

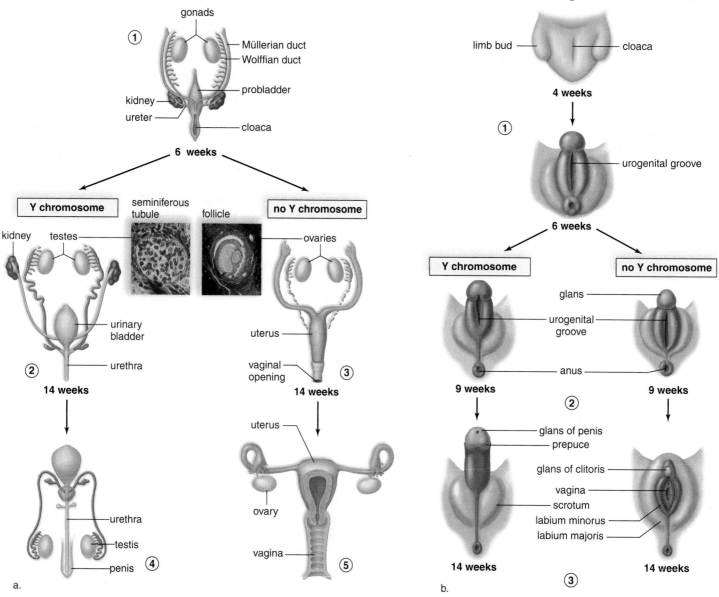

Figure 17.9   Male and female organs.
a. Development of gonads and ducts. **b.** Development of external genitals.

Müllerian ducts develop into the uterus and oviducts. Estrogen has no effect on the Wolffian duct, which degenerates in females. A developing vagina also extends from the uterus. There is no connection between the urinary and genital systems in females.

At 14 weeks, both the primitive testes and ovaries are located deep inside the abdominal cavity. An inspection of the interior of the testes would show that sperm are even now starting to develop (see micrograph), and similarly, the ovaries already contain large numbers of tiny follicles, each having an ovum (see micrograph). ④ Toward the end of development, the testes descend into the scrotal sac; ⑤ the ovaries remain in the abdominal cavity.

External Genitals   Figure 17.9*b* shows the development of the external genitals. These tissues are also indifferent at first—they can develop into either male or female genitals. ① At six weeks, a small bud appears between the legs; this can develop into the male penis or the female clitoris. ② At nine weeks, a urogenital groove bordered by two swellings appears. ③ By 14 weeks, this groove has disappeared in males, and the scrotum has formed from the original swellings. In females, the groove persists and becomes the vaginal opening. Labia majora and labia minora are present instead of a scrotum. These changes are due to the presence or absence of another hormone produced by the testes. It is called dihydrotestosterone (DHT), which is derived from testosterone.

Abnormal Development of Genitals

It's not correct to say that all XY individuals develop into males. Some XY individuals become females (XY female syndrome). Similarly, some XX individuals develop into males (XX male syndrome). In individuals with the XY female syndrome, a piece of the Y chromosome is missing. In individuals with the XX male syndrome, this same small piece is present on an X chromosome. The piece of a Y chromosome that causes male genitals to develop is called the *SRY* (sex determining region of the Y) gene. The *SRY* gene causes testes to form, and then the testes secrete these hormones: (1) Testosterone stimulates development of the epididymides, vasa deferentia, seminal vesicles, and ejaculatory duct. (2) Anti-Müllerian hormone prevents further development of female structures, and instead causes them to degenerate. (3) Dihydrotestosterone, which is derived from testosterone, directs the development of the urethra, prostate gland, penis, and scrotum.

Ambiguous Sex Determination   The absence of any one or more of these hormones results in ambiguous sex determination in which the individual has the external appearance of a female, although the gonads of a female are absent.

In *androgen insensitivity syndrome,* these three types of hormones are produced by testes during development, but the individual develops as a female because the plasma membrane receptors for testosterone are ineffective (Fig. 17.10). The external genitalia develop as female, and the

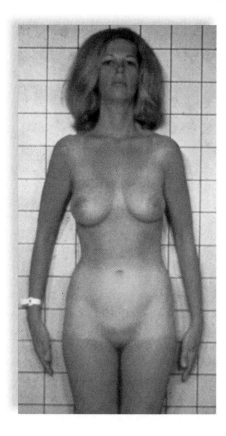

Figure 17.10
Androgen insensitivity. This individual has a female appearance but the XY chromosomes of a male. Testes instead of ovaries and a uterus are in the abdominal cavity. She developed as a female because her plasma membrane receptors for testosterone are ineffective.

Wolffian duct degenerates internally. Because the individual does not develop a scrotum, the testes fail to descend and instead remain deep within the body. The individual develops the secondary sex characteristics of a female, and no abnormality is suspected until the individual fails to menstruate.

True gonadal hermaphroditism in which a person has both ovarian and testicular tissue is rare. However, it is estimated that about 1% of the population may have *male pseudohermaphroditism.* The individual has testes but appears to be a normal female until puberty. The testes never produce testosterone or dihydrotestosterone. Anti-Müllerian hormone is produced and the female set of tubes degenerate. At puberty, the clitoris begins to enlarge as a response to testosterone produced by the adrenal cortex. The clitoris even looks like a penis; the voice deepens, and muscles enlarge. There is no breast development, and the individual does not menstruate. In the Dominican Republic, this syndrome is called guevedoces, which means "penis at age 12."

✓ Check Your Progress 17.3

1. a. Trace the path of blood in the fetus starting with the placenta; (b) name the structures unique to fetal circulation.

2. What are the major events during fetal development?

3. How does the development of the genitals differ in males and females?

Health Focus

Preventing Birth Defects

A woman can take certain steps to ensure the birth of a healthy baby. Most birth defects are not due to inheritance of an abnormal number of chromosomes or any other genetic abnormality. More women are having babies after the age of 35, and first births among women older than 40 have increased by 50% since 1980. The chance of an older woman bearing a child with a birth defect unrelated to genetic inheritance is no greater than that of a younger woman. Still, an older woman has a greater risk of having a child with a chromosomal abnormality leading to premature delivery, cesarean section, a low birth weight, or certain syndromes. Many of these disorders can be detected in utero so that therapy for these disorders can begin as soon as possible. Otherwise, all women should follow certain sensitive guidelines to help ensure the birth of a healthy baby.

Get a Physical Exam

For example, it is best if a woman has a physical exam even before she becomes pregnant. At that time, it can be determined if she has been immunized against rubella (German measles). Depending on exactly when a pregnant woman has the disease, rubella can cause blindness, deafness, mental retardation, heart malformations, and other serious problems in an unborn child. A vaccine to prevent the disease can be given to a woman before she gets pregnant, but cannot be given to a woman who is already pregnant because it contains live viruses. Also, the presence of human immunodeficiency virus (HIV) (the causative agent for acquired immunodeficiency syndrome, AIDS) should be tested for because preventative therapies are available to improve maternal and infant health.

Have Good Health Habits

Good health habits are a must during pregnancy, including proper nutrition, adequate rest, and exercise. Moderate exercise can usually continue throughout pregnancy and hopefully will contribute to ease of delivery. Basic nutrients are required in adequate amounts to meet the demands of both fetus and mother. A growing number of studies confirm that small, thin newborns are more likely to develop certain chronic diseases, such as diabetes and high blood pressure, when they become adults than are babies who are born heavier.

An increased amount of minerals, such as calcium for bone growth and iron for red blood cell formation, and certain vitamins, such as vitamin B6 for proper metabolism and folate (folic acid), are required. Pregnant women need more folate a day to meet an increased rate of cell division and DNA synthesis in their own bodies and that of the developing child. A maternal deficiency of folate has been linked to development of neural tube defects in the fetus. These defects include spina bifida (spinal cord or spinal fluid bulge through the back) and anencephaly (absence of a brain). Perhaps as many as 75% of these defects could be prevented by adequate folate intake even before pregnancy occurs. Consuming fortified breakfast cereals is a good way to meet folate needs, because they contain a more absorbable form of folate.

Avoid Smoking, Alcohol, and Drugs of Abuse   Good health habits include avoiding substances that can cross the placenta and harm the fetus (Table 17A). Cigarette smoke poses a serious threat to the health of a fetus because it contains not only carbon monoxide but also other fetotoxic chemicals. Children born to smoking mothers have a greater chance of a cleft lip or palate, increased incidence of respiratory diseases, and later on, more reading disorders than those born to mothers who did not smoke during their pregnancy.

Alcohol easily crosses the placenta, and even one drink a day appears to increase the chance of a miscarriage. The more alcohol consumed, the greater the chances of physical abnormalities if the pregnancy continues. Heavy consumption of alcohol puts a fetus at risk for a mental defect be-

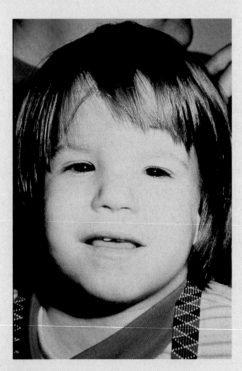

Figure 17B   A child with fetal alcohol syndrome.

Table 17A	Behaviors Harmful to the Unborn
Drinking alcohol	
Smoking cigarettes	
Taking illegal drugs	
Taking any medication not approved by a physician	
Exposure to environmental toxins and radiation	

cause alcohol enters the brain of the fetus. Babies born to heavy drinkers are apt to undergo delirium tremens after birth—shaking, vomiting, and extreme irritability—and to have fetal alcohol syndrome (FAS). Babies with FAS have decreased weight, height, and head size, with malformation of the head and face (Fig. 17B). Later, mental retardation is common, as are numerous other physical malformations.

Certainly, illegal drugs, such as marijuana, cocaine, and heroin, should be completely avoided during pregnancy. *Crack babies* now make up 60% of drug-affected babies. Cocaine use causes severe fluctuations in a mother's blood pressure that temporarily deprive the developing fetus's brain of oxygen. Cocaine babies have visual problems, lack coordination, and are mentally retarded.

Avoid Having X-rays

Children born to women who received X-ray treatment during pregnancy for, say, cancer are apt to have birth defects and/or to develop leukemia later. It takes a lower amount of X-rays to cause mutations in a developing embryo or fetus than in an adult. Dental and other diagnostic X-rays that result in only a small amount of radiation are probably safe. Still, a woman should be sure a physician knows that she is or may be pregnant. Similarly, toxic chemicals, such as pesticides, and many organic industrial chemicals, such as vinyl chloride, formaldehyde, asbestos, and benzenes, are mutagenic and can cross the placenta, resulting in abnormalities. Lead circulating in a pregnant woman's blood can

Figure 17C   Child care.
Proper care of a child begins before birth in order to prevent birth defects.

cause a child to be mentally retarded. Agents that produce abnormalities during development are called teratogens.

Avoid Certain Medications and Supplements

A woman has to be very careful about taking medications and supplements while pregnant. Excessive vitamin A, sometimes used to treat acne, may damage an embryo. In the 1950s and 1960s, DES (diethylstilbestrol), a synthetic hormone related to the natural female hormone estrogen, was given to pregnant women to prevent cramps, bleeding, and threatened miscarriage. But in the 1970s and 1980s, some adolescent girls and young women whose mothers had been treated with DES showed various abnormalities of the reproductive organs and an increased tendency toward cervical cancer. Other sex hormones, including birth control pills, can possibly cause abnormal fetal development, including abnormalities of the sex organs.

The drug thalidomide was a popular tranquilizer during the 1950s and 1960s in many European countries and to a degree in the United States. The drug, which was taken to prevent nausea in pregnant women, arrested the development of arms and legs in some children and also damaged heart and blood vessels, ears, and the digestive tract. Some mothers of affected children report that they took the drug for only a few days. Because of such experiences, physicians are generally very cautious about prescribing drugs during pregnancy, and no pregnant woman should take any drug—even ordinary cold remedies or aspirin—without checking first with her physician.

Help Prevent Conditions Associated with Birth

Unfortunately, immunization for sexually transmitted diseases is not possible. The HIV virus can cross the placenta and cause mental retardation. As mentioned, proper medication can greatly reduce the chance of this happening. When a mother has herpes, gonorrhea, or chlamydia, newborns can become infected as they pass through the birth canal. Blindness and other physical and mental defects may develop. Birth by cesarean section could prevent these occurrences.

An Rh-negative woman who has given birth to an Rh-positive child should receive an Rh immunoglobulin injection within 72 hours to prevent her body from producing Rh antibodies. She will start producing these antibodies when some of the child's Rh-positive red blood cells enter her bloodstream, possibly before but particularly at birth. Rh antibodies can cause nervous system and heart defects in a fetus. The first Rh-positive baby is not usually affected. But in subsequent pregnancies, antibodies created at the time of the first birth cross the placenta and begin to destroy the blood cells of the fetus, thereby causing anemia and other complications.

Now that physicians and laypeople are aware of the various ways birth defects can be prevented, it is hoped that the incidence of birth defects will decrease in the future (Fig. 17C).

17.4 Pregnancy and Birth

Major changes that take place in the mother's body during pregnancy are due to placental hormones.

The Energy Level Fluctuates

When first pregnant, the mother may experience nausea and vomiting, loss of appetite, and fatigue. These symptoms subside, and some mothers report increased energy levels and a general sense of well-being despite an increase in weight. During pregnancy, the mother gains weight due to breast and uterine enlargement, weight of the fetus, amount of amniotic fluid, size of the placenta, her own increase in total body fluid, and an increase in storage of proteins, fats, and minerals. The increased weight can lead to lordosis (swayback) and lower back pain.

The Uterus Relaxes

Aside from an increase in weight, many of the physiological changes in the mother are due to the presence of the placental hormones that support fetal development (Table 17.2). Progesterone decreases uterine motility by relaxing smooth muscle, including the smooth muscle in the walls of arteries. The arteries expand, and this leads to a low blood pressure that sets in motion the renin-angiotensin-aldosterone mechanism, which is promoted by estrogen. Aldosterone activity promotes sodium and water retention, and blood volume increases until it reaches its peak sometime during weeks 28–32 of pregnancy. Altogether, blood volume increases from 5 L to 7 L—a 40% rise. An increase in the number of red blood cells follows. With the rise in blood volume, cardiac output increases by 20–30%. Blood flow to the kidneys, placenta, skin, and breasts rises significantly. Smooth muscle relaxation also explains the common gastrointestinal effects of pregnancy. The heartburn experienced by many is due to relaxation of the esophageal sphincter and reflux of stomach contents into the esophagus. Constipation is caused by a decrease in intestinal tract motility.

The Pulmonary Values Increase

Of interest is the increase in pulmonary values in a pregnant woman. The bronchial tubes relax, but this alone cannot explain the typical 40% increase in vital capacity and tidal volume. The increasing size of the uterus from a nonpregnant weight of 60–80 g to 900–1,200 g contributes to an improvement in respiratory functions. The uterus comes to occupy most of the abdominal cavity, reaching nearly to the xiphoid process of the sternum. This increase in size not only pushes the intestines, liver, stomach, and diaphragm superiorly, but it also widens the thoracic cavity. Compared with nonpregnant values, the maternal oxygen level changes little, but blood carbon dioxide levels fall by 20%, creating a concentration gradient favorable to the flow of carbon dioxide from fetal blood to maternal blood at the placenta.

Still Other Effects

The enlargement of the uterus does result in some problems. In the pelvic cavity, compression of the ureters and urinary

Table 17.2	Effects of Placental Hormones on Mother
Hormone	**Chief Effects**
Progesterone	Relaxation of smooth muscle; reduced uterine motility; reduced maternal immune response to fetus
Estrogen	Increased uterine blood flow; increased renin-angiotensin-aldosterone activity; increased protein biosynthesis by the liver
Peptide hormones	Increased insulin resistance

Source: Moore, Thomas R., *Gestation Encyclopedia of Human Biology*, Vol. 7, 7th edition. Copyright © 1997 Academic Press.

bladder can result in stress incontinence. Compression of the inferior vena cava, especially when lying down, decreases venous return, and the result is edema and varicose veins.

Aside from the steroid hormones progesterone and estrogen, the placenta also produces some peptide hormones. One of these makes cells resistant to insulin, and the result can be pregnancy-induced diabetes. Some of the integumentary changes observed during pregnancy are also due to placental hormones. **Striae gravidarum,** commonly called "stretch marks," typically form over the abdomen and lower breasts in response to increased steroid hormone levels rather than stretching of the skin. Melanocyte activity also increases during pregnancy. Darkening of the areolae, skin in the line from the navel to the pubis, areas of the face and neck, and vulva is common.

Birth

The uterus has contractions throughout pregnancy. At first, these are light, lasting about 20–30 seconds and occurring every 15–20 minutes. Near the end of pregnancy, the contractions may become stronger and more frequent so that a woman thinks she is in labor. "False-labor" contractions are called **Braxton Hicks contractions.** However, the onset of true labor is marked by uterine contractions that occur regularly every 15–20 minutes and last for 40 seconds or longer.

A positive feedback mechanism can explain the onset and continuation of labor. Uterine contractions are induced by a stretching of the cervix, which also brings about the release of oxytocin from the posterior pituitary gland. Oxytocin stimulates the uterine muscles, both directly and through the action of prostaglandins. Uterine contractions push the fetus downward, and the cervix stretches even more. This cycle keeps repeating itself until birth occurs.

Prior to or at the first stage of **parturition,** which is the process of giving birth to an offspring, there can be a "bloody show" caused by expulsion of a mucous plug from the cervical canal. This plug prevents bacteria and sperm from entering the uterus during pregnancy.

Stage 1

During the first stage of labor, the uterine contractions of labor occur in such a way that the cervical canal slowly disappears as the lower part of the uterus is pulled upward toward the baby's head. This process is called effacement, or "taking up the cervix." With further contractions, the baby's head acts as a wedge to assist cervical dilation (Fig. 17.11*a*). If the amniotic membrane has not already ruptured, it is apt to do so during this stage, releasing the amniotic fluid, which leaks out of the vagina (an event sometimes referred to as "breaking water"). The first stage of parturition ends once the cervix is dilated completely.

Stage 2

During the second stage of parturition, the uterine contractions occur every 1–2 minutes and last about 1 minute each. They are accompanied by a desire to push, or bear down. As the baby's head gradually descends into the vagina, the desire to push becomes greater. When the baby's head reaches the exterior, it turns so that the back of the head is uppermost (Fig. 17.11*b*). To enlarge the vaginal orifice, an **episiotomy** is often performed. This incision, which enlarges the opening, is sewn together later.

As soon as the head is delivered, the physician may hold the head and guide it downward, while one shoulder and then the other emerges. The rest of the baby follows easily (Fig. 17.11*c*).

Once the baby is breathing normally, the umbilical cord is cut and tied, severing the child from the placenta. The stump of the cord shrivels and leaves a scar, which is the umbilicus.

Stage 3

The placenta, or **afterbirth,** is delivered during the third stage of parturition (Fig. 17.11*d*). About 15 minutes after delivery of the baby, uterine muscular contractions shrink the uterus and dislodge the placenta. The placenta then is expelled into the vagina. As soon as the placenta and its membranes are delivered, the third stage of parturition is complete.

✔ Check Your Progress 17.4

1. What chemical factors are responsible for the many physiological changes in a pregnant woman?

2. Maternal blood carbon dioxide levels fall by 20% during pregnancy. How does this benefit the fetus?

3. Describe the three stages of labor.

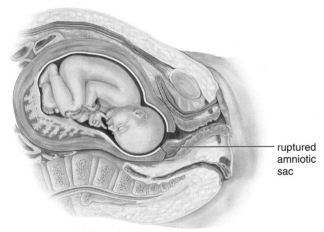

a. First stage of birth: cervix dilates

ruptured amniotic sac

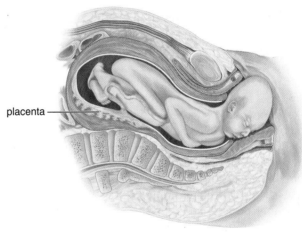

placenta

b. Second stage of birth: baby emerges

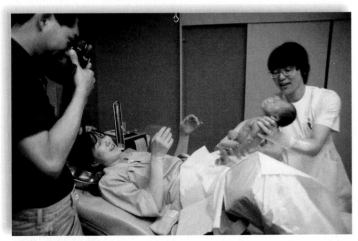

c. Baby has arrived

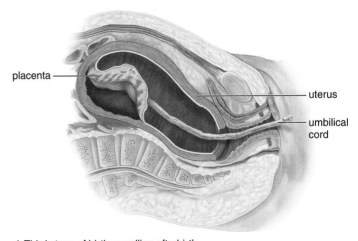

placenta

uterus

umbilical cord

d. Third stage of birth: expelling afterbirth

Figure 17.11   Process of birth.

17.5 Development After Birth

Development does not cease once birth has occurred but continues throughout the stages of life: infancy, childhood, adolescence, and adulthood. **Aging** encompasses these progressive changes that contribute to an increased risk of infirmity, disease, and death.

Today, **gerontology**, the study of aging, is of great interest because there are now more older individuals in our society than ever before, and the number is expected to rise dramatically. In the next half-century, the number of people over age 65 will increase 147%. The human life span is judged to be a maximum of 120–125 years. The present goal of gerontology is not necessarily to increase the life span, but to increase the health span, the number of years that an individual enjoys the full functions of all body parts and processes (Fig. 17.12).

Hypotheses of Aging

Of the many hypotheses about the cause of aging, three are considered here.

Genetic in Origin

Several lines of research indicate that aging has a genetic basis. Researchers working with simple organisms, such as yeast and roundworms, have identified a host of genes whose expression decreases the life span. If these genes are silenced through mutations or restricted food intake, the organism lives longer. What do these genes have in common? Apparently, when these genes are inactive, mitochondria do not produce energy—the cell uses alternative pathways. The current *mitochondrial hypothesis of aging* has been supported by engineering mice that have a defective DNA polymerase. (Recall that mitochondria have their own DNA.) These mice aged much faster than their peers. Why? Possibly because their defective mitochondria produced more free radicals than usual. Free radicals (see page 162) are unstable molecules that carry an extra electron. To become stable, free radicals donate an electron to another molecule, such as DNA or proteins (e.g., enzymes) or lipids, found in plasma membranes. Eventually, these molecules are unable to function, and the cell is destroyed. The well-known observation that a low-calorie diet can expand the life span is consistent with the mitochondrial hypothesis of aging. Caloric restriction also shuts down the genes that decrease the life span—the genes that turn on the activity of mitochondria!

Whole-Body Process

A decline in the hormonal system can affect many different organs of the body. For example, diabetes type 2 is common in older individuals. The pancreas makes insulin, but the cell's receptors are ineffective. Menopause in women and andropause in men occurs for similar reasons. The blood-

Figure 17.12 Aging.
Aging is a slow process during which the body undergoes changes that eventually bring about death, even if no marked disease or disorder is present. Medical science is trying to extend the human life span and the health span, the length of time the body functions normally.

stream contains adequate amounts of anterior pituitary hormones, but the ovaries and testes do not respond. Perhaps aging results from the loss of hormonal activities and a decline in the functions they control.

The immune system, too, no longer performs as it once did, and this can affect the body as a whole. The thymus gland gradually decreases in size, and eventually most of it is replaced by fat and connective tissue. The incidence of cancer increases among the elderly, which may signify that the immune system is no longer functioning as it should. This idea is also substantiated by the increased incidence of autoimmune diseases in older individuals.

It is possible, though, that aging is not due to the failure of a particular system that can affect the body as a whole, but to a specific type of tissue change that affects all organs and even the genes. It has been noticed for some time that proteins—such as the collagen fibers present in many support tissues—become increasingly cross-linked as people age. Undoubtedly, this cross-linking contributes to the stiffening and loss of elasticity characteristic of aging tendons and ligaments. It may also account for the inability of such organs as the blood vessels, the heart, and the lungs to function as they once did. Some researchers have now found that glucose has the tendency to attach to any type of protein, which is the first step in a cross-linking process. They are presently experimenting with drugs that can prevent cross-linking.

Extrinsic Factors

The current data about the effects of aging are often based on comparisons of the elderly to younger age groups. But perhaps today's elderly were not as aware when they were younger of the importance of, for example, diet and exercise to general health. It is possible, then, that much of what we attribute to aging is instead due to years of poor health habits.

Consider, for example, osteoporosis. This condition is associated with a progressive decline in bone density in both males and females, so fractures are more likely to occur after only minimal trauma. Osteoporosis is common in the elderly—by age 65, one-third of women will have vertebral fractures, and by age 81, one-third of women and one-sixth of men will have suffered a hip fracture. While there is no denying that a decline in bone mass occurs as a result of aging, certain extrinsic factors are also important. The occurrence of osteoporosis itself is associated with cigarette smoking, heavy alcohol intake, and inadequate calcium intake. Not only is it possible to eliminate these negative factors by personal choice, but it is also possible to add a positive factor. A moderate exercise program has been found to slow down the progressive loss of bone mass.

Even more important, a sensible exercise program and a proper diet that includes at least five servings of fruits and vegetables a day will most likely help eliminate cardiovascular disease. Experts no longer believe that the cardiovascular system necessarily suffers a large decrease in functioning ability with age. Persons 65 years of age and older can have well-functioning hearts and open coronary arteries if their health habits are good and they continue to exercise regularly.

Effect of Age on Body Systems

Data about how aging affects body systems are necessarily based on past events. It is possible that, in the future, age will not have these effects or at least not to the same degree as those described here.

Skin

As aging occurs, skin becomes thinner and less elastic because the number of elastic fibers decreases and the collagen fibers undergo cross-linking, as discussed previously. Also, there is less adipose tissue in the subcutaneous layer; therefore, older people are more likely to feel cold. The loss of thickness partially accounts for sagging and wrinkling of the skin.

Homeostatic adjustment to heat is also limited because there are fewer sweat glands for sweating to occur. Because of fewer hair follicles, the hair on the scalp and the extremities thins out. The number of oil (sebaceous) glands is reduced, and the skin tends to crack. Older people also experience a decrease in the number of melanocytes, making their hair gray and their skin pale. In contrast, some of the remaining pigment cells are larger, and pigmented blotches appear on the skin.

Processing and Transporting

Cardiovascular disorders are the leading cause of death today. The heart shrinks because of a reduction in cardiac muscle cell size. This leads to loss of cardiac muscle strength and reduced cardiac output. Still, the heart, in the absence of disease, is able to meet the demands of increased activity because it can double its rate or triple the amount of blood pumped each minute even though the maximum possible output declines.

Because the middle layer of arteries contains elastic fibers, which are most likely subject to cross-linking, the arteries become more rigid with time, and their size is further reduced by plaque, a buildup of fatty material. Therefore, blood pressure readings gradually rise. Such changes are common in individuals living in Western industrialized countries but not in agricultural societies. A diet low in cholesterol and saturated fatty acids has been suggested as a way to control degenerative changes in the cardiovascular system.

Blood flow to the liver is reduced, and this organ does not metabolize drugs as efficiently as before. This means that, as a person gets older, less medication is needed to maintain the same level of a drug in the bloodstream.

Cardiovascular problems are often accompanied by respiratory disorders, and vice versa. Growing inelasticity of lung tissue means that ventilation is reduced. Because we rarely use the entire vital capacity, these effects are not noticed unless the demand for oxygen increases.

Blood supply to the kidneys is also reduced. The kidneys become smaller and less efficient at filtering wastes. Salt and water balance is difficult to maintain, and the elderly dehydrate faster than young people do. Difficulties involving urination include incontinence (lack of bladder control) and the inability to urinate. In men, the prostate gland may enlarge and reduce the diameter of the urethra, making urination so difficult that surgery is often needed.

The loss of teeth, which is frequently seen in elderly people, is more apt to be the result of long-term neglect than aging. The digestive tract loses tone, and secretion of saliva and gastric juice is reduced, but there is no indication of reduced absorption. Therefore, an adequate diet, rather than vitamin and mineral supplements, is recommended. Elderly people commonly complain of constipation, increased gas, and heartburn; gastritis, ulcers, and cancer can also occur.

Integration and Coordination

While most tissues of the body regularly replace their cells, some at a faster rate than others, the brain and the muscles ordinarily do not. However, contrary to previous opinion, recent studies show that few neural cells of the cerebral cortex are lost during the normal aging process. This means that cognitive skills remain unchanged even though a loss in short-term memory characteristically occurs. Although the elderly learn more slowly than the young, they can acquire and remember new material. The results of tests indicate that, when more time is given for the subject to respond, age differences in learning decrease.

Neurons are extremely sensitive to oxygen deficiency, and if neuron death does occur, it may be due not to aging itself but to reduced blood flow in narrowed blood vessels. Specific disorders, such as depression, Parkinson disease, and Alzheimer disease, are sometimes seen in the elderly, but they are not common. Reaction time, however, does slow, and more stimulation is needed for hearing, taste, and smell receptors to function as before. After age 50, the ability to hear tones at higher frequencies decreases gradually, and this can make it difficult to identify individual voices and to understand conversation in a group. The lens of the eye does not accommodate as well and also may develop a cataract. Glaucoma, the buildup of pressure due to increased fluid, is more likely to develop because of a reduction in the size of the anterior cavity of the eye.

Loss of skeletal muscle mass is not uncommon, but it can be controlled by following a regular exercise program. The capacity to do heavy labor decreases, but routine physical work should be no problem. A decrease in the strength of the respiratory muscles and inflexibility of the rib cage contribute to the inability of the lungs to expand as before, and reduced muscularity of the urinary bladder contributes to an inability to empty the bladder completely, and therefore to the occurrence of urinary infections.

As noted before, aging is accompanied by a decline in bone density. Osteoporosis, characterized by a loss of calcium and other minerals from bone, is not uncommon, but evidence indicates that proper health habits can prevent its occurrence. Arthritis, which causes pain upon movement of a joint, is also seen.

Weight gain occurs because the basal metabolism decreases and inactivity increases. Muscle mass is replaced by stored fat and retained water.

The Reproductive System

As mentioned, females undergo menopause, and thereafter, the level of female sex hormones in the blood falls markedly. The uterus and the cervix decrease in size, and the walls of the oviducts and the vagina become thinner. The external genitals become less pronounced. Males undergo andropause, and the level of androgens falls gradually over the age span of 50–90, but some sperm production continues until death.

Figure 17.13   Remaining active.
The aim of gerontology is to allow seniors to enjoy living.

It is of interest that, as a group, females live longer than males. Males suffer a marked increase in heart disease in their forties, but an increase is not noted in females until after menopause, when women lead men in the incidence of stroke. Men are still more likely than women to have a heart attack, however. At one time it was thought that estrogen offers women some protection against cardiovascular disorders, but this hypothesis is called into question because post-menopausal administration of estrogen has been shown to increase the risk of cardiovascular disorders in women.

Conclusion

We have listed many adverse effects of aging, but it is important to emphasize that although such effects are seen, they are not inevitable (Fig. 17.13). We must discover any extrinsic factors that precipitate these adverse effects and guard against them. Just as it is wise to make the proper preparations to remain financially independent when older, it is also wise to realize that, biologically, successful old age begins with the health habits developed when we are younger.

✅ Check Your Progress 17.5

1. What are the different hypotheses of aging?
2. What is the effect of aging on the various body systems?
3. What is the best way to keep healthy, even though aging occurs?

Bioethical Focus

Maternal Health Habits

The fetus is subject to harm if the mother uses medicines and drugs of abuse, including nicotine and alcohol. Also, various sexually transmitted diseases, notably HIV infection, can be passed on to the fetus by way of the placenta. Women need to know how to protect their unborn children from harm, including eating right and exercising (Fig. 17D). Indeed, if they are sexually active, their behavior should be protective, even if they are using a recognized form of birth control. Harm can occur before a woman realizes she is pregnant!

Because maternal health habits can affect a child before it is born, there has been a growing acceptance of prosecuting women when a newborn has a condition, such as fetal alcohol syndrome, that could only have been caused by the drinking habits of the mother. Employers have also become aware that they might be subject to prosecution if the workplace exposes pregnant employees to toxins. To protect themselves, Johnson Controls, a U.S. battery manufacturer, developed a fetal protection policy. No woman who could bear a child was offered a job that might expose her to toxins that could negatively affect the development of her baby. To get such a job, a woman had to show that she had been sterilized or was otherwise incapable of having children. In 1991, the U.S. Supreme Court declared this policy unconstitutional on the basis of sexual discrimination. The decision was hailed as a victory for women, but was it? The decision was written in such a way that women alone, and not an employer, are responsible for any harm done to the fetus by workplace toxins.

Some have noted that prosecuting women for causing prenatal harm can itself have a detrimental effect. The women may tend to avoid prenatal treatment, thereby increasing the risk to their children. Or they may opt for an abortion in order to avoid the possibility of prosecution. Women feel they are in a no-win situation. If they have a

Figure 17D Health habits.
Pregnant women should practice good health habits.

child that has been harmed due to their behavior, they are bad mothers; if they abort, they are also bad.

Decide Your Opinion

1. Do you believe a woman should be prosecuted if her child is born with a preventable condition? Why or why not?
2. Is the woman or the physician responsible when a woman of childbearing age takes a prescribed medication that harms an unborn child? Is the employer or the woman responsible when a workplace toxin harms an unborn child?
3. Should sexually active women who can bear a child be expected to avoid substances or situations that could possibly harm an unborn child, even if they are using birth control? Why or why not?

Summarizing the Concepts

17.1 Fertilization

The acrosome of a sperm releases enzymes that digest a pathway for the sperm through the zona pellucida. The sperm nucleus enters the egg and fuses with the egg nucleus.

17.2 Pre-Embryonic and Embryonic Development

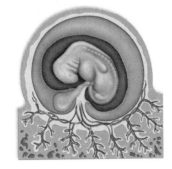

- Cleavage, growth, morphogenesis, and differentiation are the processes of development.
- The extraembryonic membranes (chorion, allantois, yolk sac, and amnion) function in internal development.

17.3 Fetal Development

- At the end of the embryonic period, all organ systems are established, and there is a mature and functioning placenta. The umbilical arteries and umbilical vein take blood to and from the placenta, where exchanges take place.
- Exchanges supply the fetus with oxygen and nutrients and rid the fetus of carbon dioxide and wastes.
- The venous duct joins the umbilical vein to the inferior vena cava.
- The oval duct and arterial duct allow the blood to pass through the heart without going to the lungs.

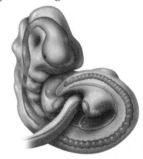

Fetal development extends from the third through the ninth months.

- During the third and fourth months, the skeleton is becoming ossified.
- The sex of the fetus becomes distinguishable. If an *SRY* gene is present, testes and male genitals develop. Otherwise, ovaries and female genitals develop.

- During the fifth through the ninth months, the fetus continues to grow and to gain weight.

17.4 Pregnancy and Birth

Major changes take place in the mother's body during pregnancy.

- Weight gain occurs as the uterus occupies most of the abdominal cavity.
- Many complaints, such as constipation, heartburn, darkening of certain skin areas, and pregnancy-induced diabetes, are due to the presence of placental hormones.

Birth

- A positive feedback mechanism that involves uterine contractions and oxytocin explains the onset and continuation of labor.
- During stage 1 of parturition (birth), the cervix dilates.
- During stage 2, the child is born.
- During stage 3, the afterbirth is expelled.

17.5 Development After Birth

Development after birth consists of infancy, childhood, adolescence, and adulthood.

- Aging encompasses progressive changes from about age 20 on that contribute to an increased risk of infirmity, disease, and death.

Hypotheses of Aging

- Aging may have a genetic basis.
- Aging may be due to changes that affect the whole body (e.g., decline of hormonal system).
- Aging may be due to extrinsic factors (e.g., diet and exercise).

Effect of Age on Body Systems

- Deterioration of organ systems can possibly be prevented or reduced in part by utilizing good health habits.

Understanding Key Terms

afterbirth 369
aging 370
allantois 355
amnion 355
blastocyst 356
Braxton Hicks contraction 368
breech birth 363
cesarean section 363
chorion 355, 356
chorionic villi 359
cleavage 355
differentiation 355
ectopic pregnancy 357
embryo 357
embryonic development 357

embryonic disk 357
episiotomy 369
extraembryonic membrane 355
fertilization 354
fetal development 363
fontanel 363
gastrulation 358
gerontology 370
gestation 355
growth 355
human chorionic gonadotropin (HCG) 357
implantation 357
inner cell mass 356
lanugo 363

morphogenesis 355
morula 356
parturition 368
placenta 355
pre-embryonic development 356
primary germ layer 358
reproductive cloning 362

striae gravidarum 368
testosterone 364
therapeutic cloning 362
umbilical cord 359
vernix caseosa 363
yolk sac 355
zygote 354

Match the key terms to these definitions.

a. _____ Short, fine hair that is present during the later portion of fetal development.

b. _____ The placenta that is delivered during the third stage of parturition.

c. _____ The study of aging.

d. _____ Union of a sperm nucleus and an egg nucleus, which creates a zygote with the diploid number of chromosomes.

e. _____ Mitotic cell division of the zygote with no increase in cell size.

Testing Your Knowledge of the Concepts

1. Describe how polyspermy is prevented during fertilization. (page 354)

2. Name the four embryonic membranes and give a human function for each one. (page 355)

3. Justify the division of development into pre-embryonic, embryonic, and fetal development. (pages 356–61)

4. What are the three primary germ layers, and what body structures come from each germ layer? (page 358)

5. Briefly summarize the weekly events of embryonic development. (pages 357–60)

6. Briefly summarize the monthly events of fetal development. (pages 360, 363)

7. Explain how blood circulates to and from the placenta and the fetus. How is blood shunted away from the lungs? (page 361)

8. List the hormones involved in the development of the male and female internal and external sex organs and state their functions. (pages 364–65)

9. Describe some of the changes that occur in the mother during pregnancy. (pages 360, 368)

10. What event marks the end of each stage of birth? (page 369)

11. Discuss three hypotheses concerning aging. How can you prevent the major changes that can occur in the body as we age? (pages 370–72)

12. Only one sperm enters an egg because
    a. sperm have an acrosome.
    b. the corona radiata gets larger.
    c. changes occur in the zona pellucida.
    d. the cytoplasm hardens.
    e. All of these are correct.

13. Which of these statements is correct?
    a. All major organs are formed during embryonic development.
    b. The hands and feet begin as paddlelike structures.
    c. The heart is at first tubular.
    d. The placenta functions until birth occurs.
    e. All of these are correct statements.

14. Which length can be best associated with the embryo at the end of two months when it becomes a fetus?
    a. less than 1 mm, which is microscopic
    b. 38 mm, which is about 1½ in.
    c. 1 ft, which is about ⅓ m
    d. same length as a fetus at birth, which is about 20 in.

15. When all three germ layers are present (ectoderm, endoderm, and mesoderm), what event has occurred?
    a. blastulation
    b. limb formation
    c. gastrulation
    d. morulation

16. Which of these is not a process of development?
    a. cleavage
    b. parturition
    c. growth
    d. morphogenesis
    e. differentiation

17. Which of these is mismatched?
    a. chorion—sense perception
    b. yolk sac—first site of blood cell formation
    c. allantois—umbilical blood vessels
    d. amnion—contains fluid that protects embryo

18. In human development, which part of the blastocyst will develop into a embryo?
    a. trophoblast
    b. inner cell mass
    c. chorion
    d. yolk sac

19. Which primary germ layer is not correctly matched to an organ system or organ that develops from it?
    a. ectoderm—the nervous system
    b. endoderm—lining of the digestive tract
    c. mesoderm—skeletal system
    d. endoderm—cardiovascular system

20. Human chorionic gonadotropin is a
    a. hormone.
    b. basis of pregnancy test.
    c. cause of ectopic pregnancy.
    d. Both a and b are correct.

21. Which is a correct sequence that ends with the stage that implants?
    a. morula, blastocyst, embryonic disk, gastrula
    b. ovulation, fertilization, cleavage, morula, early blastocyst
    c. embryonic disk, gastrula, primitive streak, neurula
    d. primitive streak, neurula, extraembryonic membranes, chorion

22. Differentiation is equivalent to which term?
    a. morphogenesis
    b. growth
    c. specialization
    d. gastrulation

23. Which process refers to the shaping of the embryo and involves cell migration?
    a. cleavage
    b. differentiation
    c. growth
    d. morphogenesis

24. Structures that develop and become part of the placenta are the
    a. amniotic villi.
    b. yolk sac villi.
    c. allantois villi.
    d. chorionic villi.
    e. Both a and c are correct.

25. Which association is not correct?
    a. third week—development of heart
    b. fourth and fifth weeks—limb buds appear
    c. sixth through eighth weeks—germ layers appear for the first time
    d. All of these are correct.

26. Which association is not correct?
    a. third and fourth months—fetal heart has formed, but it does not beat
    b. fifth through seventh months—mother feels movement
    c. eighth through ninth months—usually head is now pointed toward the cervix
    d. All of these are correct.

27. At 25 days, the embryo has
    a. become a fetus.
    b. all the extraembryonic membranes.
    c. a nervous system and a digestive system.
    d. already undergone gastrulation.
    e. Both b and d are correct.

28. At three months, the embryo has
    a. become a fetus.
    b. body systems already.
    c. a head, arms, and legs.
    d. ears and eyes which don't function.
    e. All but b are correct.

29. Which of these structures is not a circulatory feature unique to the fetus?
    a. arterial duct
    b. oval opening
    c. umbilical vein
    d. pulmonary trunk

30. Which of these statements is correct?
    a. Fetal circulation, like adult circulation, takes blood equally to a pulmonary circuit and a systemic circuit.
    b. Fetal circulation shunts blood away from the lungs but makes full use of the systemic circuit.
    c. Fetal circulation includes exchange of substances between fetal blood and maternal blood at the placenta.
    d. Unlike adult circulation, fetal blood always carries O_2-rich blood and therefore has no need for the pulmonary circuit.
    e. Both b and c are correct.

31. Which of these is a hormone involved in development of male and female sex organs?
    a. estrogen
    b. anti-Müllerian hormone
    c. dihydrotestosterone
    d. testosterone
    e. All of these hormones are involved.

32. The gender of the fetus can be observed when
    a. embryo is a zygote.
    b. germ layers have appeared.
    c. vertebrae have appeared.
    d. the fetus is about three months.

33. Which hormone can be administered to begin the process of childbirth?
    a. estrogen
    b. oxytocin
    c. prolactin
    d. testosterone
    e. Both b and d are correct.

34. Label this diagram illustrating the placement of the extraembryonic membranes, and give a function for each membrane in humans.

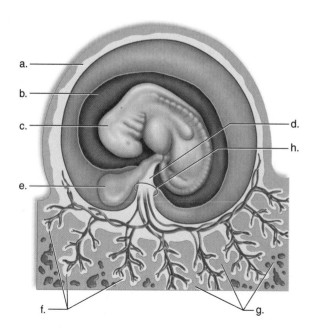

35. Label this diagram illustrating the fetal circulatory system.

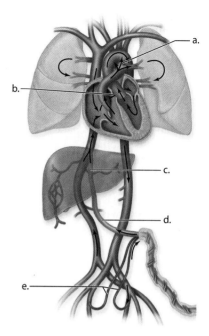

Thinking Critically About the Concepts

Years later, Amber, from the opening story, visited her physician to find out the cause of her sleeplessness and itchy skin. The results of blood tests done by Dr. Davis indicated Amber was beginning menopause. Dr. Davis recommended Amber eat low-fat foods and lots of fruits, vegetables, and whole grains. Regular exercise and avoiding cigarette smoke and limiting caffeine and alcohol were suggested by Dr. Davis as well. The irony of it all is that Amber's daughter recently heard similar recommendations when her pregnancy was confirmed.

1. At home, pregnancy tests check for the presence of HCG in a female's urine. Why is HCG found in a pregnant woman's urine?

2. A blood test at a doctor's office can also check for the presence of HCG in a female's blood.
   a. Why would you expect to find HCG circulating in a pregnant female's blood?
   b. HCG is a protein, so how does HCG affect its target cells?

3a. What recommendations would be made to a female (or male) who hopes to prevent osteoporosis?
   b. Why is it better/easier to prevent osteoporosis than treat it once someone has it?

4a. What pituitary hormone is checked with a blood test to diagnose menopause?
   b. Will levels of this hormone be increased or decreased if the female is in menopause?
   c. How does the changed (increased or decreased) level of this pituitary hormone cause the onset of menopause (cessation of menses)?

Patterns of Chromosome Inheritance

Eleven months separated brothers Nick and Josh. However, they were often mistaken for twins because they looked so much alike. As pranksters, the boys had great fun trading places with one another. Sometimes people never caught on that they were dealing with Nick instead of Josh, or vice versa. They even fooled their parents.

The only person the boys didn't try to fool was their younger sister, Amanda. Amanda was somewhat challenged, having been born with Down syndrome. Nick and Josh were very protective of their little sister and would never dream of trying to confuse her with their silly games. They were especially proud of her success in Special Olympics events, and they always helped when the games were held nearby.

It was at a Special Olympics event that Nick and Josh's ability to swap places with each other was compromised. Nick was hurt while helping with one of the events and got a pretty good gash below his ear. For a week his stitches were an easy way for folks to distinguish the two boys. Nick figured they'd be back to their old tricks once his stitches were gone and the scar faded. Much to his surprise, a keloid developed where the gash had been, leaving him with a permanent means by which people could distinguish him from Josh. As you study this chapter, you will learn how new cells are created when an injury occurs or when someone grows taller. How someone gets an extra chromosome that results in Down or Klinefelter syndrome will be explained too.

CHAPTER CONCEPTS

18.1 Chromosomes and the Cell Cycle
A karyotype is a picture of chromosomes about to divide. Cell division, called mitosis, is a part of the cell cycle, which consists of interphase and mitosis.

18.2 Mitosis
Mitosis is duplication division in which the daughter cells have the same number and kinds of chromosomes as the mother cell. Cell division, called mitosis, is a part of the cell cycle, which consists of interphase and mitosis.

18.3 Meiosis
Meiosis is reduction division in which the daughter cells have half the number of chromosomes as the parent cell. In the end, the daughter cells also have a different combination of chromosomes and genes than the parent cell.

18.4 Comparison of Meiosis and Mitosis
Meiosis I uniquely pairs and separates the paired chromosomes so that the daughter cells have half the number of chromosomes. Meiosis II is exactly like mitosis, except the cells have half the number of chromosomes.

18.5 Chromosome Inheritance
Abnormalities in chromosome inheritance occur due to changes in chromosome number and changes in chromosome structure. Changes in chromosome number lead to one less than normal and one more than normal. Changes in chromosome structure include deletions of some segments and duplications of other segments, for example.

18.1 Chromosomes and the Cell Cycle

A human nucleus is only about 5 to 8 μm, yet it holds all the chromatin that condenses to form the chromosomes when cells divide. Humans have 46 chromosomes that occur in 23 pairs. Twenty-two of these pairs are called autosomes. All of their genes control traits that have nothing to do with the gender of the individual. One pair of chromosomes is called the sex chromosomes because they do contain the genes that control gender. Males have the sex chromosomes X and Y, and females have two X chromosomes. (Recall that a Y chromosome contains the *SRY* gene that causes testes to develop.)

Suppose you wanted to see the chromosomes of an individual. What would you do? Such a happening is actually possible. Any cell in the body except red blood cells, which lack a nucleus, can be a source of chromosomes for examination. In adults, it is easiest to obtain and use white blood cells separated from a blood sample for the purpose of looking at the chromosomes. After a cell sample has been obtained, the cells are stimulated to divide in a culture medium. When a cell divides, chromatin becomes chromosomes, and the nuclear envelope fragments, liberating the chromosomes. Next, a chemical is used to stop the division process when the chromosomes are most compacted and visible microscopically. Stains are applied to the slides, and the cells can be photographed with a camera attached to a microscope. Staining causes the chromosomes to have dark

and light cross-bands of varying widths, and these can be used in addition to size and shape to help pair up the chromosomes. Today, a computer is used to arrange the chromosomes in pairs (Fig. 18.1). The display is called a karyotype. The karyotype in Figure 18.1 is that of a normal male. Nick and Josh, the brothers in the opening story, would each have a karyotype such as this one.

A Karyotype

A normal karyotype tells us a lot about a body cell. First, we could notice that a normal body cell is diploid—it has the full complement of 46 chromosomes. How did it happen that every body cell you ever had or will have has 46 chromosomes? **Mitosis**—duplication division—which began when the fertilized egg started dividing, ensures that every cell has 46 chromosomes.

Notice, too, that the enlargement of a pair of chromosomes shows that in dividing cells each chromosome is composed of two identical parts, called **sister chromatids.** These chromosomes are said to be replicated or duplicated chromosomes because the two sister chromatids contain the same genes, the units of heredity that control the cell. This is possible only because each chromatid contains a DNA double helix.

The chromatids are held together at a region called the centromere. A **centromere** has the function of holding the chromatids together until a certain phase of mitosis when the centromere splits. Once separated, each sister chromatid is a

Figure 18.1 Chromosomes. In body cells, the chromosomes occur in pairs. In a karyotype, the pairs have been numbered. These chromosomes are duplicated, and each one is composed of two sister chromatids.

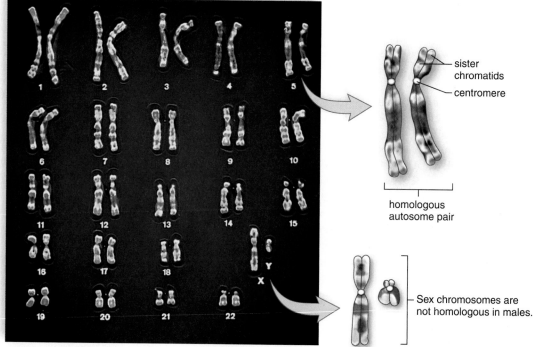

The 46 chromosomes of a male

chromosome. In this way, a duplicated chromosome gives rise to two individual daughter chromosomes. When daughter chromosomes separate, the new cell gets one of each kind and, therefore, a full complement of chromosomes.

The Cell Cycle

The **cell cycle** is an orderly process that has two parts: interphase and cell division. In order to understand the cell cycle, it is necessary to recall the structure of a cell (see Fig. 3.4). A human cell has a plasma membrane, which encloses the cytoplasm, the content of the cell outside the nucleus. In the cytoplasm are various organelles, which carry on various functions necessary to the life of the cell. When a cell is not undergoing division, the DNA (and associated proteins) within a nucleus is a tangled mass of thin threads called chromatin.

Interphase

As Figure 18.2 shows, most of the cell cycle is spent in **interphase.** This is the time when the organelles carry on their usual functions. Also, the cell gets ready to divide: It grows larger, the number of organelles doubles, and the amount of chromatin doubles as DNA synthesis occurs. For mammalian cells, interphase lasts for about 20 hours, which is 90% of the cell cycle.

The event of DNA synthesis permits interphase to be divided into three stages: the G_1 stage occurs before DNA synthesis, the S stage includes DNA synthesis, and the G_2 stage occurs after DNA synthesis. Originally, G stood for the "gaps"—that is, those times during interphase when DNA synthesis was not occurring. But now that we know growth happens during these stages, the G can be thought of as standing for growth. Let us see what specifically happens during these stages.

> G_1 *stage.* A cell doubles its organelles (e.g., mitochondria and ribosomes), and it accumulates the materials needed for DNA synthesis. Also, various proteins are needed to change chromatin into chromosomes that are visible under the microscope when stained. A chromosome contains both proteins called histones and DNA.
>
> *S stage.* DNA replication occurs; a copy is made of all the DNA in the cell. Because DNA replication occurs each chromosome consists of two identical DNA double helix molecules. These molecules occur in the strands called sister chromatids. Another way of expressing these events is to say that DNA replication has resulted in duplicated chromosomes.
>
> G_2 *stage.* The cell synthesizes the proteins needed for cell division, such as the protein found in microtubules. The role of microtubules in cell division is described in a later section.

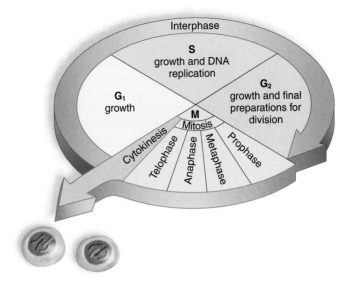

Figure 18.2   The cell cycle.
The cell cycle has four stages. During interphase, which consists of G_1, S, and G_2, the cell gets ready to divide, and during the mitotic stage, nuclear division and cytokinesis (cytoplasmic division) occur.

The amount of time the cell takes for interphase varies widely. Some cells, such as nerve and muscle cells, typically do not complete the cell cycle and are permanently arrested in G_1. These cells are said to have entered a G_0 stage. Embryonic cells spend very little time in G_1 and complete the cell cycle in a few hours.

Cell Division

Following interphase, the cell enters the cell division part of the cell cycle. Cell division has two stages: M (for mitotic) stage and cytokinesis. **Mitosis** is a type of nuclear division. Recall that mitosis is called *duplication division* because each new nucleus contains the same number and kind of chromosomes as the former cell. **Cytokinesis** is division of the cytoplasm.

During mitosis, the sister chromatids of each chromosome separate, becoming chromosomes that are distributed to two daughter nuclei. When cytokinesis is complete, two daughter cells are now present. Mammalian cells usually require only about 4 hours to complete the mitotic stage.

The cell cycle, including interphase and cell division occurs continuously in certain tissues. Right now your body is producing thousands of new red blood cells, skin cells, and cells that line your respiratory and digestive tracts. The process of **apoptosis,** programmed cell death, also occurs to do away with any cells that are dividing when they shouldn't. The control of the cell cycle is discussed on page 384.

☑ Check Your Progress 18.1

1. What are the three stages of interphase?
2. How does interphase prepare a cell for cell division?

Science **Focus**

Obtaining Fetal Chromosomes

Physicians and prospective parents sometimes want to view an unborn child's chromosomes to determine whether a chromosomal abnormality exists. For example, the opening story mentions that Amanda has Down syndrome. A **syndrome** is a group of symptoms that always occur together. People with Down syndrome have three number 21 chromosomes and several physical characteristics by which to recognize the condition (see page 394). Amanda's parents may have known ahead of time she would be born with Down syndrome if chorionic villi sampling or amniocentesis had been done in order to get a sample of Amanda's cells for the purpose of karyotyping them (Fig. 18A).

Chorionic Villi Sampling

Chorionic villi sampling (CVS) is usually performed from the eighth to the twelfth week of pregnancy. The doctor inserts a long, thin tube through the vagina into the uterus. With the help of ultrasound, which gives a picture of the uterine contents, the tube is placed between the uterine lining and the chorionic villi. Fetal cells are obtained from the villi by suction (Fig. 18A*a*). Enough cells are obtained to allow karyotyping to be done right away. CVS carries a greater risk of spontaneous abortion than amniocentesis—0.8% compared with 0.3%. The advantage of CVS is getting the results of karyotyping at an earlier date.

Amniocentesis

Amniocentesis is usually performed from the fifteenth to the seventeenth week of pregnancy. A long needle is passed through the abdominal wall to withdraw a small amount of amniotic fluid, along with a few fetal cells (Fig. 18A*b, c*). These few cells are allowed to undergo mitosis in the laboratory until there are enough for karyotyping to be done. This may take about four weeks time.

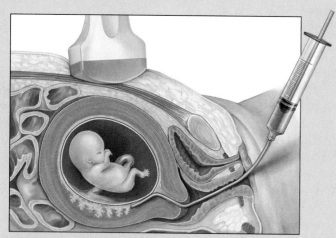

a. During chorionic villi sampling, a suction tube is used to remove cells from the chorion, where the placenta will develop.

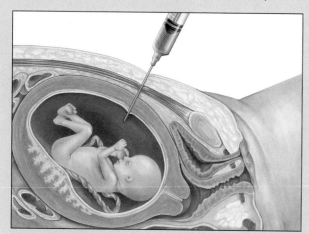

b. During amniocentesis, a long needle is used to withdraw amniotic fluid containing fetal cells.

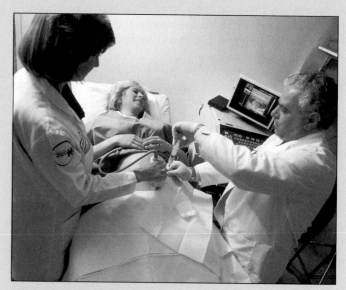

c. Amniocentesis procedure

Figure 18A Obtaining fetal chromosomes for karyotyping.

18.2 Mitosis

Mitosis *is duplication division. The nuclei of the two new cells have the same number and kinds of chromosomes as the cell that divides.* The cell that divides is called the **parent cell,** and the new cells are called the **daughter cells.** Because the parent cell and daughter cell have the same number and kinds of chromosomes, they are genetically identical.

Overview of Mitosis

As mentioned, when mitosis is going to occur, chromatin in the nucleus becomes highly condensed, and the chromosomes become visible. Because replication of DNA occurred, each chromosome is now duplicated and is composed of two identical parts, called sister chromatids, held together at a centromere. They are called sister chromatids because they contain the same genes:

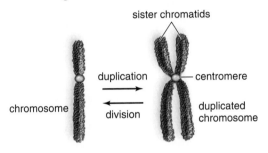

Figure 18.3 gives an overview of mitosis; for simplicity, only four chromosomes are depicted. (In determining the number of chromosomes, it is necessary to count only the number of independent centromeres.) As you know, the complete number of chromosomes is called the **diploid (2n)** number.

During mitosis, the centromeres divide and the sister chromatids separate. (Following separation, each chromatid is called a chromosome.) Each daughter cell gets a complete set of chromosomes and is 2n. Therefore, each daughter cell receives the same number and kinds of chromosomes as the parent cell, and each daughter cell is genetically identical to the other and to the parent cell.

The Spindle

Another event of importance during mitosis is the duplication of the **centrosome,** the microtubule organizing center of the cell. After centrosomes duplicate, they separate and form the poles of the **mitotic spindle,** where they assemble the microtubules that make up the spindle fibers. The chromosomes are attached to the spindle fibers at their centromeres (Fig. 18.4). An array of microtubules, called an **aster,** is also at the poles.

The **centrioles** are short cylinders of microtubules that are present in centrosomes. The centrioles lie at right angles to one another. Although it has not been show, they could possibly assist in the formation of the spindle that separates the chromatids during mitosis.

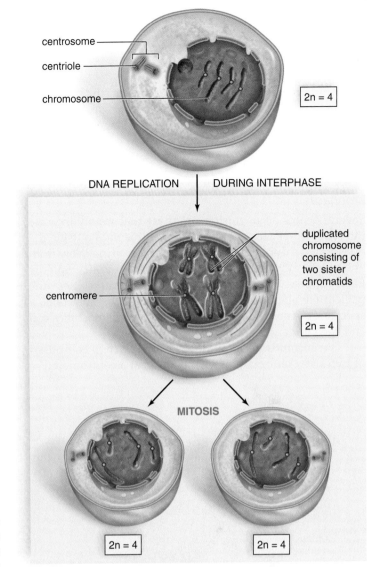

Figure 18.3 Mitosis overview.
Following DNA replication, each chromosome is duplicated. When the centromeres split, the sister chromatids, now called chromosomes, move into daughter nuclei. (The blue and red chromosomes were inherited from different parents.)

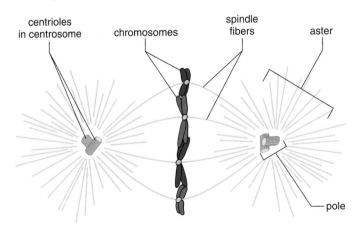

Figure 18.4 The mitotic spindle.
Before mitosis begins, the centrosome and the centrioles duplicate; during mitosis, they separate, and the mitotic spindle, composed of microtubules, forms between them.

Phases of Mitosis

As an aid in describing the events of mitosis, the process is divided into four phases: prophase, metaphase, anaphase, and telophase (Fig. 18.5). Although the stages of mitosis are depicted as if they were separate, they are actually continuous, and one stage flows from the other with no noticeable interruption.

Prophase

Several events occur during **prophase** that visibly indicate the cell is preparing to divide. The centrosomes outside the nucleus have duplicated, and they begin moving away from one another toward opposite ends of the nucleus. Spindle fibers appear between the separating centrosomes, the nuclear envelope begins to fragment, and the nucleolus, a special region of DNA, disappears as the chromosomes coil and become condensed.

The chromosomes are now visible. Each is composed of two sister chromatids held together at a centromere.

Spindle fibers attach to the centromeres as the chromosomes continue to shorten and to thicken. During prophase, chromosomes are randomly placed in the nucleus (Fig. 18.5).

Metaphase

During **metaphase,** the nuclear envelope is fragmented, and the spindle occupies the region formerly occupied by the nucleus. The chromosomes are now at the equator (center) of the spindle. Metaphase is characterized by a fully formed spindle, and the chromosomes, each with two sister chromatids, are aligned at the equator (Fig. 18.5).

Anaphase

At the start of **anaphase,** the centromeres uniting the sister chromatids divide. Then the sister chromatids separate, becoming chromosomes that move toward opposite poles of the spindle. Separation of the sister chromatids ensures that each cell receives a copy of each type of chromosome, and thereby has a full complement of genes. Anaphase is characterized by the 2n (diploid) number of chromosomes moving toward each pole.

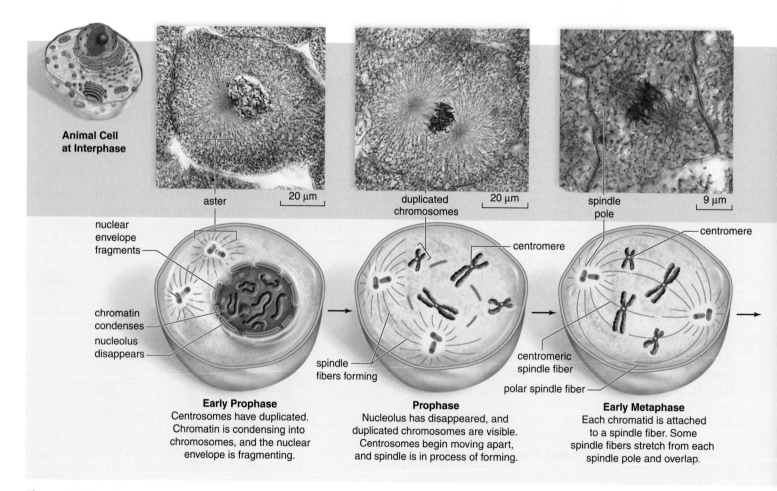

Animal Cell at Interphase

aster   20 μm

nuclear envelope fragments

chromatin condenses

nucleolus disappears

spindle fibers forming

Early Prophase
Centrosomes have duplicated. Chromatin is condensing into chromosomes, and the nuclear envelope is fragmenting.

duplicated chromosomes   20 μm

centromere

spindle

Prophase
Nucleolus has disappeared, and duplicated chromosomes are visible. Centrosomes begin moving apart, and spindle is in process of forming.

spindle pole   9 μm

centromere

centromeric spindle fiber

polar spindle fiber

Early Metaphase
Each chromatid is attached to a spindle fiber. Some spindle fibers stretch from each spindle pole and overlap.

Figure 18.5 **Phases of mitosis.**

Remember that counting the number of centromeres indicates the number of chromosomes. Therefore, in Figure 18.5, each pole receives four chromosomes: two are red and two are blue.

Function of the Spindle   The spindle brings about chromosomal movement. Two types of spindle fibers are involved in the movement of chromosomes during anaphase. One type extends from the poles to the equator of the spindle; there, they overlap. As mitosis proceeds, these fibers increase in length, and this helps push the chromosomes apart. The chromosomes themselves are attached to other spindle fibers that simply extend from their centromeres to the poles. These fibers (composed of microtubules that can disassemble) get shorter and shorter as the chromosomes move toward the poles. Therefore, they pull the chromosomes apart.

Spindle fibers, as stated earlier, are composed of microtubules. Microtubules can assemble and disassemble by the addition or subtraction of tubulin (protein) subunits. This is what enables spindle fibers to lengthen and shorten, and what ultimately causes the movement of the chromosomes.

Telophase

Telophase begins when the chromosomes arrive at the poles. During telophase, the chromosomes become indistinct chromatin again. The spindle disappears as the nuclear envelope components reassemble in each cell. Each nucleus has a nucleolus because each has a region of the DNA, where ribosomal subunits are produced. Telophase is characterized by the presence of two daughter nuclei.

Cytokinesis

Cytokinesis is the division of the cytoplasm and organelles. In human cells, a slight indentation, called a **cleavage furrow,** passes around the circumference of the cell. Actin filaments form a contractile ring, and as the ring gets smaller and smaller, the cleavage furrow pinches the cell in half. As a result, each cell becomes enclosed by its own plasma membrane.

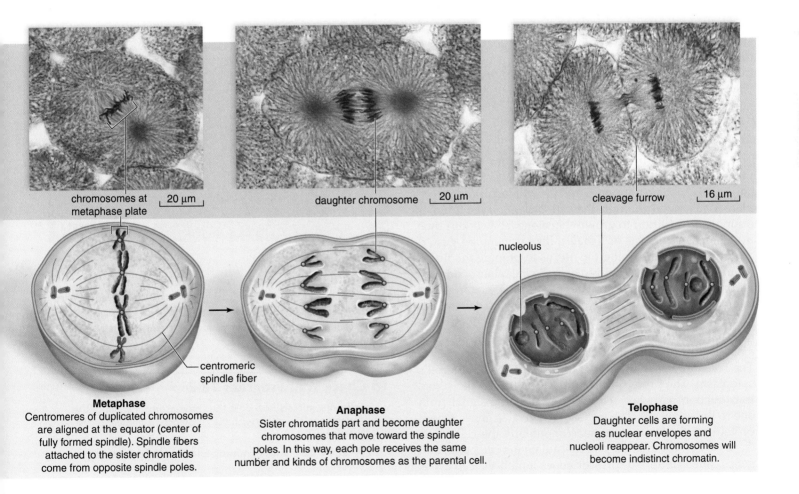

chromosomes at metaphase plate   20 μm

daughter chromosome   20 μm

cleavage furrow   16 μm

nucleolus

centromeric spindle fiber

Metaphase
Centromeres of duplicated chromosomes are aligned at the equator (center of fully formed spindle). Spindle fibers attached to the sister chromatids come from opposite spindle poles.

Anaphase
Sister chromatids part and become daughter chromosomes that move toward the spindle poles. In this way, each pole receives the same number and kinds of chromosomes as the parental cell.

Telophase
Daughter cells are forming as nuclear envelopes and nucleoli reappear. Chromosomes will become indistinct chromatin.

The Importance of the Cell Cycle and Mitosis

The cell cycle, which includes mitosis, is terribly important to the well-being of humans. Remember Adam in the opening story of Chapter 15? He never achieved the height of most grown men because his cells hadn't responded properly to growth hormone. In other words, although growth hormone was present in his blood, bone and tissue cells didn't undergo the cell cycle including mitosis. When we grow, the cell cycle, including mitosis occurs (Fig. 18.6a).

Repair of an injury is another time when the cell cycle, including mitosis, occurs (Fig. 18.6b). In this chapter's opening story, Nick was now distinguishable from his brother Josh because he developed a keloid when a gash below his ear healed. The word keloid is derived from Greek. In Greek, "kel" means tumor and and "eidois" means form. A keloid is a benign, noncancerous tumor that forms when a wound heals. When a keloid forms, the cell cycle keeps occurring and doesn't stop when it should.

Ordinarily, a cell cycle control system works perfectly to produce more cells only to the extent necessary (Fig. 18.6c). A benign tumor or a cancerous tumor develops when the cell cycle control system is not functioning properly. A benign tumor is confined to a particular location, such as Nick's keloid. A cancerous tumor is not confined. Cancer cells have a tendency to leave the original tumor by way of the lymphatic vessels and start new tumors in other parts of the body.

Cell Cycle Control System

The cell cycle control system extends from the plasma membrane to particular genes in the nucleus. Some external signals such as hormones and growth factors can stimulate a cell to go through the cell cycle. To take another example, in addition to growth hormone, at a certain time in the menstrual cycle of females, the hormone progesterone stimulates cells lining the uterus to prepare the lining for implantation of a fertilized egg. On the other hand, epidermal growth factor stimulates skin in the vicinity of an injury to finish the cell cycle, thereby repairing damage.

As shown in Figure 18.6c, ① during reception, an external signal delivers a message to a specific receptor embedded in the plasma membrane of a receiving cell. ② The receptor relays the signal to proteins inside the cell's cytoplasm. The proteins form a pathway called the signal transduction pathway because they pass the signal from one to the other. ③ The last signal activates genes whose protein product ④ stimulates or inhibits the cell cycle. Genes called proto-oncogenes stimulate the cell cycle, and genes called tumor-suppressor genes inhibit the cell cycle. These genes are discussed more in Chapter 19, which is about cancer.

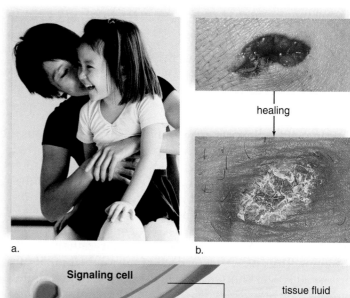

a.   b.

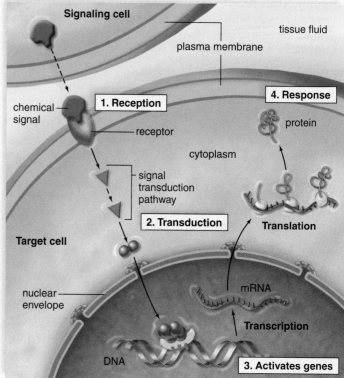

c.

Figure 18.6   The cell cycle and mitosis.
a. The cell cycle, including mitosis, occurs when humans grow and (**b**) when tissues undergo repair. **c.** Growth factors stimulate a cell-signaling pathway that stretches from the plasma membrane to the genes that regulate the occurrence of the cell cycle.

☑ Check Your Progress 18.2

1. Following mitosis, how does the chromosome number of the daughter cell compare with the chromosome number of the parent cell?

2. What are the phases of mitosis, and what happens during each phase?

3. How is the cytoplasm divided between the daughter cells following mitosis?

18.3 Meiosis

Meiosis is *reduction division. Because meiosis involves two divisions, there are four daughter cells. Each daughter cell has one of each kind of chromosome and, therefore, half as many chromosomes as the parent cell.* The parent cell has the diploid (2n) number of chromosomes, while the daughter cells have half this number, which is called the **haploid (n)** number of chromosomes. The daughter cells that result from meiosis go on to become the gametes.

In Figure 18.7, the diploid (2n) number of chromosomes is four chromosomes. The parent cell has the 2n number of chro-

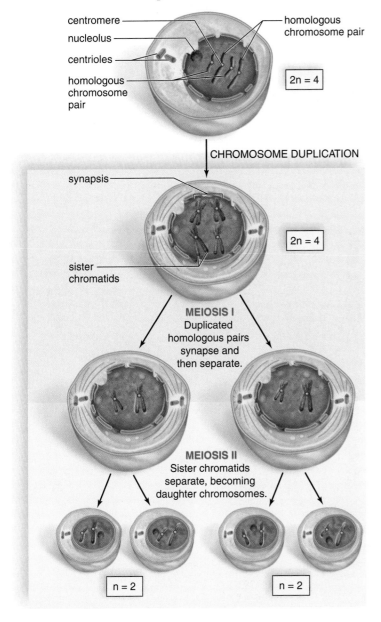

Figure 18.7   **Meiosis overview.**
DNA replication is followed by meiosis I when homologous chromosomes pair and then separate. During meiosis II, the sister chromatids become chromosomes that move into daughter nuclei.

mosomes, while the daughter cells have the haploid (n) number of chromosomes, which is equal to two chromosomes.

Overview of Meiosis

At the start of meiosis, the parent cell is 2n, or diploid, and the chromosomes occur in pairs. For simplicity's sake, there are only two pairs of chromosomes (Fig. 18.7). The short chromosomes are one pair, and the long chromosomes are another. The members of a pair are called **homologous chromosomes,** or **homologues,** because they look alike and carry genes for the same traits, such as type of hair or color of eyes.

Meiosis I

The two cell divisions of meiosis are called meiosis I and meiosis II. Prior to meiosis I, DNA replication has occurred, and the chromosomes are duplicated. Each chromosome consists of two chromatids held together at a centromere. During meiosis I, the homologous chromosomes come together and line up side by side. This so-called **synapsis** results in an association of four chromatids that stay in close proximity during the first two phases of meiosis I. Synapsis is quite significant because its occurrence leads to a reduction of the chromosome number.

Because of synapsis, there are pairs of homologous chromosomes at the equator during meiosis I. Notice that only during meiosis I is it possible to observe paired chromosomes at the equator. When the members of these pairs separate, each daughter nucleus receives one member of each pair. Therefore, each daughter cell now has the haploid (n) number of chromosomes, as you can verify by counting its centromeres. Each chromosome, however, is still duplicated, and no replication of DNA occurs between meiosis I and meiosis II. The period of time between meiosis I and meiosis II is called **interkinesis.**

Meiosis II and Fertilization

During meiosis II, the centromeres divide and the sister chromatids separate, becoming chromosomes that are distributed to daughter nuclei. In the end, each of four daughter cells has the n, or haploid, number of chromosomes, and each chromosome consists of one chromatid.

In humans, the daughter cells mature into **gametes** (sex cells—sperm and egg) that fuse during fertilization. **Fertilization** restores the diploid number of chromosomes in the **zygote,** the first cell of the new individual. If the gametes carried the diploid instead of the haploid number of chromosomes, the chromosome number would double with each fertilization. After several generations, the zygote would be nothing but chromosomes.

Figure 18.8 shows the phases of meiosis I and meiosis II. Meiosis I has four phases: prophase I, metaphase I, anaphase I, and telophase I. Some of the meiosis I stages are discussed more fully on page 388. The stages of meiosis II are named similarly to those of meiosis I, except the name is followed by a II.

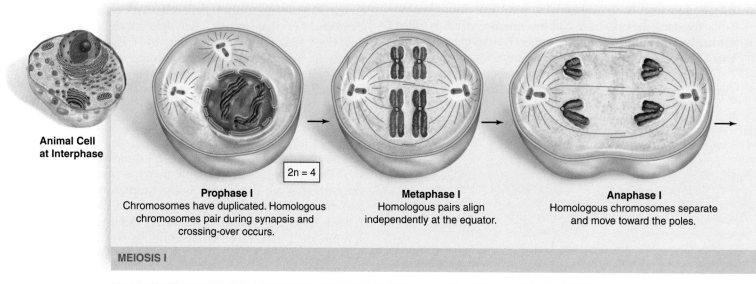

Animal Cell at Interphase

MEIOSIS I

2n = 4

Prophase I
Chromosomes have duplicated. Homologous chromosomes pair during synapsis and crossing-over occurs.

Metaphase I
Homologous pairs align independently at the equator.

Anaphase I
Homologous chromosomes separate and move toward the poles.

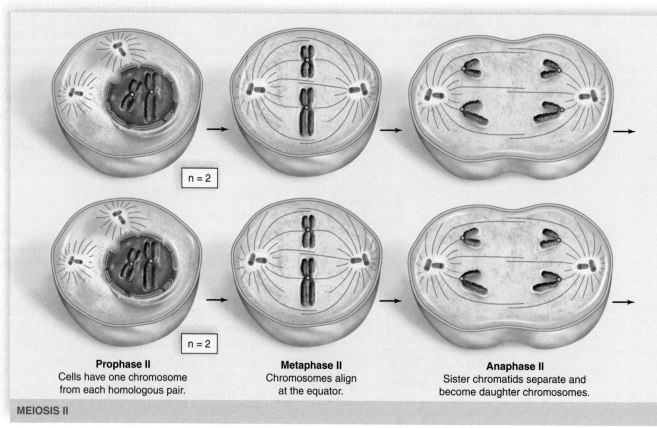

MEIOSIS II

n = 2

n = 2

Prophase II
Cells have one chromosome from each homologous pair.

Metaphase II
Chromosomes align at the equator.

Anaphase II
Sister chromatids separate and become daughter chromosomes.

Figure 18.8
Meiosis I and II in drawings of animal cells.
Homologous chromosomes pair and then separate during meiosis I. Crossing-over, which occurs during meiosis I, is discussed more fully on page 388. Chromatids separate, becoming daughter chromosomes during meiosis II. Following meiosis II, there are four haploid daughter cells.

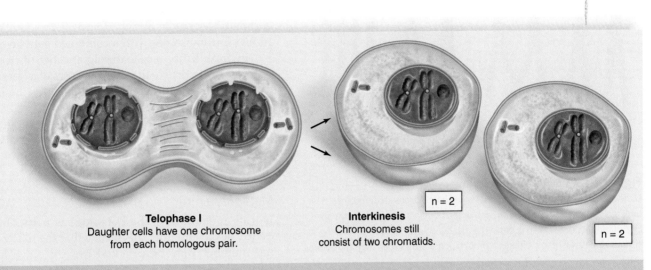

Telophase I
Daughter cells have one chromosome
from each homologous pair.

Interkinesis
Chromosomes still
consist of two chromatids.

n = 2

n = 2

MEIOSIS I cont'd

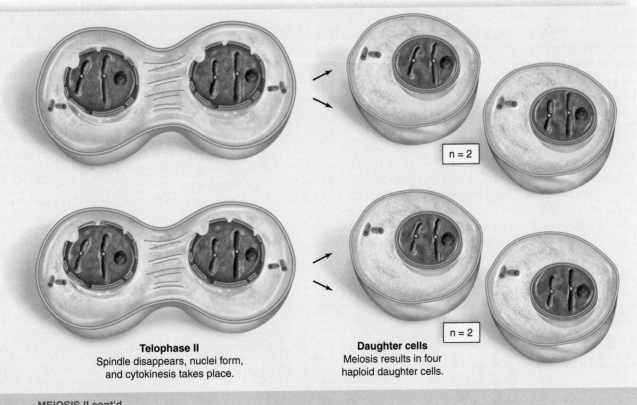

n = 2

Telophase II
Spindle disappears, nuclei form,
and cytokinesis takes place.

Daughter cells
Meiosis results in four
haploid daughter cells.

n = 2

MEIOSIS II cont'd

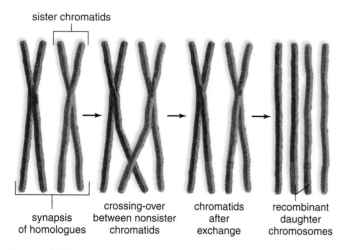

sister chromatids

synapsis of homologues

crossing-over between nonsister chromatids

chromatids after exchange

recombinant daughter chromosomes

Figure 18.9   Synapsis and crossing-over.

During meiosis I, *from left to right,* duplicated homologous chromosomes undergo synapsis and line up with each other. During crossing-over, nonsister chromatids break and then rejoin in the manner shown. Two of the resulting chromosomes will have a different combination of genes than they had before.

Stages of Meiosis

Meiosis is a part of sexual reproduction. The process of meiosis ensures that the next generation of individuals will have the diploid number of chromosomes and a combination of characteristics different from that of either parent. Both meiosis I and meiosis II have the same four stages of nuclear division as did mitosis—prophase, metaphase, anaphase, and telophase. Here we discuss only prophase I and metaphase I because special events occur during these phases.

Prophase I

In prophase I, synapsis occurs, and then the spindle appears, while the nuclear envelope fragments and the nucleolus disappears. During synapsis, the homologous chromosomes come together and line up side by side. Now, an exchange of genetic material may occur between the nonsister chromatids of the homologous pair (Fig. 18.9). This exchange, which can be as many as ten crossovers during one pairing, is called **crossing-over.** Crossing-over means that the chromatids held together by a centromere are no longer identical. When the chromatids separate during meiosis II, the daughter cells will receive chromosomes with recombined genetic material.

In order to appreciate the significance of crossing-over, it is necessary to realize that the members of a homologous pair can carry slightly different instructions for the same genetic trait. For example, one homologue may carry instructions for brown eyes, while the corresponding homologue may carry instructions for blue eyes. Therefore, crossing-over causes the offspring to receive a different combination of instructions than the mother or the father received.

Metaphase I

During metaphase I, the homologous pairs align independently at the equator. This means that the maternal or paternal member may be oriented toward either pole. Figure 18.10 shows the eight possible orientations for a cell that contains only three pairs of chromosomes. The first four orientations will result in gametes that have different combinations of maternal and paternal chromosomes. The next four will result in the same types of gametes as the first four. For example, the first cell and the last cell will both produce gametes with either three red or three blue chromosomes.

Once all possible orientations are considered, the result will be 2^3, or 8, possible combinations of maternal and paternal chromosomes in the resulting gametes from this cell. In humans, where there are 23 pairs of chromosomes, the number of possible chromosomal combinations in the gametes is a staggering 2^{23}, or 8,388,608. And this does not even consider the genetic variations that are introduced due to crossing-over.

The events of prophase I and metaphase I help ensure that gametes will not have the same combination of chromosomes and genes.

Figure 18.10   Independent alignment.

When a parent cell has three pairs of homologous chromosomes, there are eight possible chromosome alignments at the equator due to independent assortment. Among the 16 daughter nuclei resulting from these alignments, there are eight different combinations of chromosomes.

Significance of Meiosis

Meiosis is a part of gametogenesis, production of the sperm and egg. One function of meiosis is to keep the chromosome number constant from generation to generation. Because the gametes are haploid, the zygote has only the diploid number of chromosomes.

An easier way to keep the chromosome number constant is to reproduce asexually. Unicellular organisms such as bacteria, protozoans, and yeasts reproduce by binary fission:

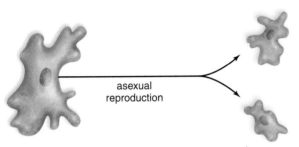

asexual
reproduction

This is a form of asexual reproduction because one parent has produced identical offspring. Binary fission is a quick and easy way to asexually reproduce many organisms within a short period of time. A bacterium can increase to over one million cells in about 7 hours, for example. Then why do organisms expend the energy to reproduce sexually?

It takes energy to find a mate, carry out a courtship and produce eggs or sperm that may never be used for reproductive purposes. A human male produces 300 million sperm per day, and very few of these will actually fertilize an egg.

Most likely, humans and other animals practice sexual reproduction that includes meiosis because it results in genetic recombination. Genetic recombination ensures that offspring will be genetically different compared to each other and to their parents. Genetic recombination occurs because of crossing-over and independent alignment of chromosomes, as discussed on page 388. Also, at the time of fertilization, parents contribute genetically different chromosomes to the offspring.

All environments are subject to a change in conditions. If the world is getting warmer as predicted, humans who sweat may be at an advantage compared to those who do not sweat as much, for example. Because environments are subject to change, sexual reproduction is advantageous because it generates the diversity needed so that at least a few will be suited to new and different environmental circumstances.

☑ Check Your Progress 18.3

1. Following meiosis, how does the chromosome number of the daughter cells compare to the chromosome number of the parent cell?

2. How does meiosis reduce the likelihood that gametes will have the same combination of chromosomes and genes?

3. What happens during the two cell divisions of meiosis?

18.4 Comparison of Meiosis and Mitosis

Mitosis and meiosis are both nuclear divisions, but there are several differences between them. You will want to be able to easily recognize how they differ. To this end, Figure 18.11 compares meiosis to mitosis. You will want to study each of the differences listed here until you are familiar with them.

- DNA replication takes place only once prior to both meiosis and mitosis. However, meiosis requires two nuclear divisions, but mitosis requires only one.
- Four daughter nuclei are produced by meiosis, and following cytokinesis, there are four daughter cells. Mitosis followed by cytokinesis results in two daughter cells.
- The four daughter cells following meiosis are haploid (n) and have half the chromosome number of the parent cell. The daughter cells following mitosis have the same chromosome number as the parent cell—that is, the 2n, or diploid, number.
- The daughter cells from meiosis are not genetically identical to each other or to the parent cell. The daughter cells from mitosis are genetically identical to each other and to the parent cell.

The specific differences between these nuclear divisions can be categorized according to occurrence and process.

Occurrence

Meiosis occurs only at certain times in the life cycle of sexually reproducing organisms. In humans, meiosis occurs only in the reproductive organs and produces the gametes. Mitosis is more common because it occurs in all tissues during growth and repair. Which one is out of control when cancer occurs? Mitosis can result in a proliferation of body cells.

Process

Comparison of Meiosis I with Mitosis

Notice that these events distinguish meiosis I from mitosis:

- Homologous chromosomes pair and undergo crossing-over during prophase I of meiosis, but not during mitosis.
- Paired homologous chromosomes align at the equator during metaphase I in meiosis. These paired chromosomes have four chromatids altogether. Individual chromosomes align at the equator during metaphase in mitosis. They each have two chromatids.

   This difference makes it easy to tell whether you are looking at mitosis, meiosis I, or meiosis II. For example, if a cell has 16 chromosomes, then 16

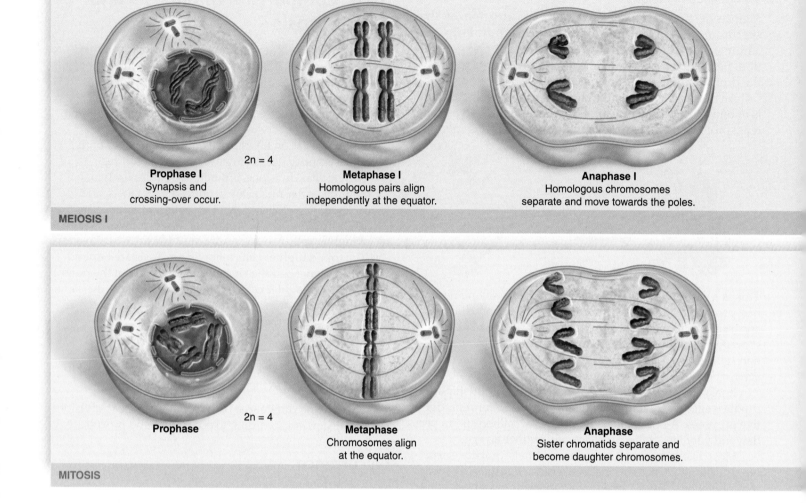

MEIOSIS I

Prophase I Synapsis and crossing-over occur.	**Metaphase I** Homologous pairs align independently at the equator.	**Anaphase I** Homologous chromosomes separate and move towards the poles.

2n = 4

MITOSIS

Prophase	**Metaphase** Chromosomes align at the equator.	**Anaphase** Sister chromatids separate and become daughter chromosomes.

2n = 4

chromosomes are at the equator during mitosis but only 8 chromosomes during meiosis II. Only meiosis I has paired duplicated chromosomes at the equator.

- Homologous chromosomes (with centromeres intact) separate and move to opposite poles during anaphase I of meiosis. Centromeres split, and sister chromatids, now called chromosomes, move to opposite poles during anaphase in mitosis.

Comparison of Meiosis II with Mitosis

The events of meiosis II are just like those of mitosis except in meiosis II, the nuclei contain the haploid number of chromosomes. If the parent cell has 16 chromosomes, then the cells undergoing meiosis II have 8 chromosomes, and the daughter cells have 8 chromosomes, for example.

Summary

To summarize the process, Tables 18.1 and 18.2 separately compare meiosis I and meiosis II with mitosis.

Table 18.1	Comparison of Meiosis I with Mitosis
Meiosis I	**Mitosis**
Prophase I Pairing of homologous chromosomes	*Prophase* No pairing of chromosomes
Metaphase I Homologous duplicated chromosomes at equator	*Metaphase* Duplicated chromosomes at equator
Anaphase I Homologous chromosomes separate.	*Anaphase* Sister chromatids separate, becoming daughter chromosomes that move to the poles.
Telophase I Two haploid daughter cells	*Telophase* Two daughter cells, identical to the parent cell

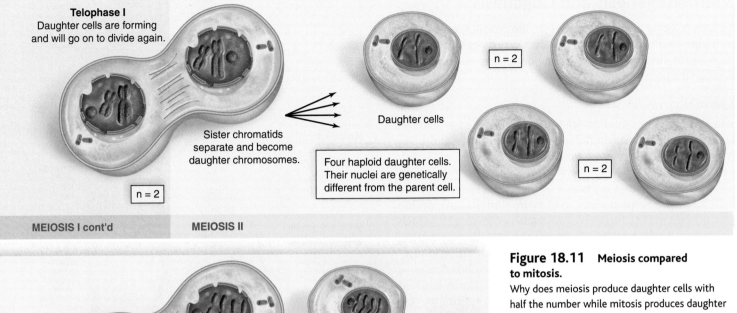

Telophase I
Daughter cells are forming and will go on to divide again.

n = 2

Sister chromatids separate and become daughter chromosomes.

Daughter cells

Four haploid daughter cells. Their nuclei are genetically different from the parent cell.

n = 2

n = 2

MEIOSIS I cont'd   **MEIOSIS II**

Daughter cells

Telophase
Daughter cells are forming.

Two diploid daughter cells. Their nuclei are genetically identical to the parent cell.

MITOSIS cont'd

Figure 18.11   Meiosis compared to mitosis.
Why does meiosis produce daughter cells with half the number while mitosis produces daughter cells with the same number of chromosomes as the parent cell? Compare metaphase I of meiosis to metaphase of mitosis. Only in metaphase I are the homologous chromosomes paired at the metaphase plate. Members of homologous chromosome pairs separate during anaphase I, and therefore, the daughter cells are haploid. The blue chromosomes were inherited from the paternal parent, and the red chromosomes were inherited from the maternal parent. The exchange of color between nonsister chromatids represents the crossing-over that occurs during meiosis I.

Table 18.2	Comparison of Meiosis II with Mitosis

Meiosis II	Mitosis
Prophase II	*Prophase*
No pairing of chromosomes	No pairing of chromosomes
Metaphase II	*Metaphase*
Haploid number of duplicated chromosomes at equator	Duplicated chromosomes at equator
Anaphase II	*Anaphase*
Sister chromatids separate, becoming daughter chromosomes that move to the poles.	Sister chromatids separate, becoming daughter chromosomes that move to the poles.
Telophase II	*Telophase*
Four haploid daughter cells	Two daughter cells, identical to the parent cell

☑ Check Your Progress 18.4

Key:
a. mitosis   c. neither
b. meiosis   d. both

1. Which one results in four daughter cells?

2. Which one has no pairing of chromosomes?

3. Which one is out of control when cancer occurs?

4. a. Which one is duplication division? b. Reduction division?

5. Which one results in cells that are exactly like the parent cell?

6. Which one results in cells that are genetically different from the parent cell?

7. Which one is a type of nuclear division?

8. Which one is involved in gametogenesis?

9. Which one is part of the cell cycle?

Spermatogenesis and Oogenesis

Meiosis is a part of **spermatogenesis,** the production of sperm in males, and **oogenesis,** the production of eggs in females (Fig. 18.12). Following meiosis, the daughter cells mature to become the gametes.

Spermatogenesis

After puberty, the time of life when the sex organs mature, spermatogenesis is continual in the testes of human males. As many as 300,000 sperm are produced per minute, or 400 million per day.

Spermatogenesis is shown in Figure 18.12 *top.* The so-called *primary spermatocytes,* which are diploid (2n), divide during meiosis I to form two *secondary spermatocytes,* which are haploid (n). Secondary spermatocytes divide during meiosis II to produce four *spermatids,* which are also haploid (n). What's the difference between the chromosomes in haploid secondary spermatocytes and those in haploid spermatids? The chromosomes in secondary spermatocytes are duplicated and consist of two chromatids, while those in spermatids consist of only one. Spermatids mature into sperm (spermatozoa). In human males, sperm have 23 chromosomes, which is the haploid number. The process of meiosis in males always results in four cells that become sperm. In other words, all four daughter cells—the spermatids—become sperm.

Oogenesis

As you know, the ovary of a female contains many immature follicles (see Fig. 16.8). Each of these follicles contains a primary oocyte arrested in prophase I. As shown in Figure 18.12 *bottom,* a primary oocyte, which is diploid (2n), divides during meiosis I into two cells, each of which is haploid. Note that the chromosomes are duplicated. One of these cells, termed the **secondary oocyte,** receives almost all the cytoplasm. The other is the first polar body. A **polar body** acts like a trash can to hold discarded chromosomes. The first polar body contains duplicated chromosomes and completes meiosis II occasionally. The secondary oocyte begins meiosis II but stops at metaphase II and doesn't complete it unless a sperm enters during the fertilization process.

The secondary oocyte (for convenience, called the egg) leaves the ovary during ovulation and enters an oviduct, where it may be fertilized by a sperm. If so, the oocyte is activated to complete the second meiotic division. Following meiosis II, there is one egg and two or possibly three polar bodies. The mature egg has 23 chromosomes. The polar bodies disintegrate. They are a way to discard unnecessary chromosomes, while retaining much of the cytoplasm in the egg.

One egg can be the source of identical twins if after one division of the fertilized egg during development, the cells separate and each one becomes a complete individual. On the other hand, the occurrence of fraternal twins requires that two eggs be fertilized separately.

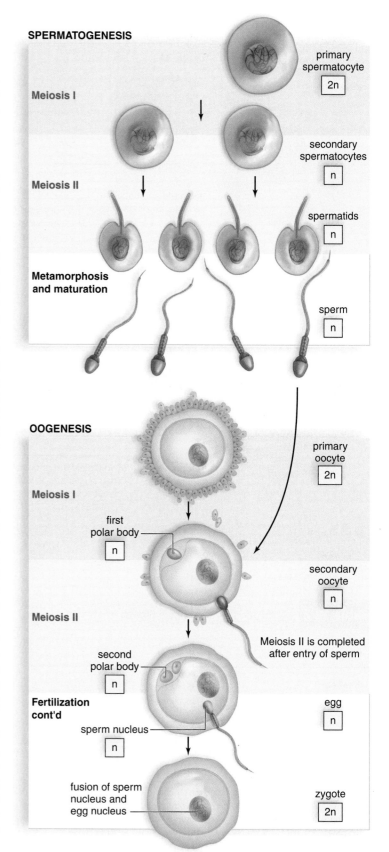

Figure 18.12   Spermatogenesis and oogenesis in mammals.
Spermatogenesis produces four viable sperm, whereas oogenesis produces one egg and at least two polar bodies. In humans, both sperm and egg have 23 chromosomes each; therefore, following fertilization, the zygote has 46 chromosomes.

18.5 Chromosome Inheritance

Normally, an individual receives 22 pairs of autosomes and two sex chromosomes. Each pair of autosomes carries alleles for particular traits. The alleles can be different as when one calls for freckles and one does not.

Changes in Chromosome Number

Sometimes individuals are born with either too many or too few autosomes or sex chromosomes, most likely due to nondisjunction during meiosis. **Nondisjunction** occurs during meiosis I, when both members of a homologous pair go into the same daughter cell, or during meiosis II, when the sister chromatids fail to separate and both daughter chromosomes go into the same gamete. Figure 18.13 assumes that nondisjunction has occurred during oogenesis; some abnormal eggs have 24 chromosomes, while others have only 22 chromosomes. If an egg with 24 chromosomes is fertilized with a normal sperm, the result is a **trisomy,** so called because one type of chromosome is present in three copies. If an egg with 22 chromosomes is fertilized with a normal sperm, the result is a **monosomy,** so called because one type of chromosome is present in a single copy.

Normal development depends on the presence of exactly two of each kind of chromosome. Too many chromosomes is tolerated better than a deficiency of chromosomes, and several trisomies are known to occur in humans. Among autosomal trisomies, only trisomy 21 (Down syndrome) has a reasonable chance of survival after birth. This is probably due to the fact that chromosome 21 is the smallest of the chromosomes.

The chances of survival are greater when trisomy or monosomy involves the sex chromosomes. In normal XX females, one of the X chromosomes becomes a darkly staining mass of chromatin called a **Barr body** (after the person who discovered it). A Barr body is an inactive X chromosome; therefore, we now know that the cells of females function with a single X chromosome just as those of males do. This is most likely the reason why a zygote with one X chromosome (Turner syndrome) can survive. Then, too, all extra X chromosomes beyond a single one become Barr bodies, and this explains why poly-X females and XXY males are seen fairly frequently. An extra Y chromosome, called Jacobs syndrome, is tolerated in humans, most likely because the Y chromosome carries few genes. Jacobs syndrome (XYY) is due to nondisjunction during meiosis II of spermatogenesis. We know this because two Ys are present only during meiosis II in males.

Down Syndrome, an Autosomal Trisomy

The most common autosomal trisomy seen among humans is Down syndrome, also called trisomy 21. Persons with Down syndrome usually have three copies of chromosome 21 because the egg had two copies instead of one. (In 23% of

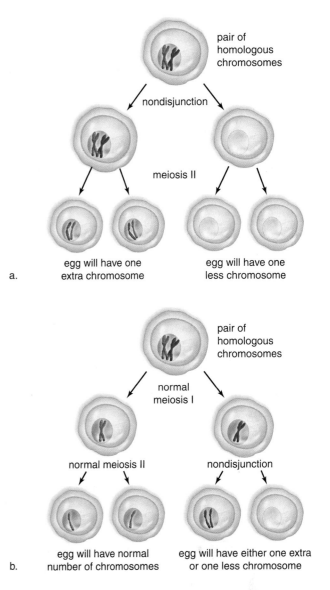

a. egg will have one extra chromosome / egg will have one less chromosome

b. egg will have normal number of chromosomes / egg will have either one extra or one less chromosome

Figure 18.13   Nondisjunction of chromosomes during oogenesis, followed by fertilization with normal sperm.
a. Nondisjunction can occur during meiosis II if the sister chromatids separate, but the resulting chromosomes go into the same daughter cell. Then the egg will have one more (24) or one less (22) than the usual number of chromosomes. Fertilization of these abnormal eggs with normal sperm produces an abnormal zygote with 47 or 45 chromosomes. **b.** Nondisjunction can also occur during meiosis I and result in abnormal eggs that also have one more or one less than the normal number of chromosomes. Fertilization of these abnormal eggs with normal sperm results in a zygote with an abnormal chromosome number.

the cases studied, however, the sperm had the extra chromosome 21.) The chances of a woman having a Down syndrome child increase rapidly with age, starting at about age 40, and the reasons for this are still being determined.

Although an older woman is more likely to have a Down syndrome child, most babies with Down syndrome are born to women younger than age 40 because this is the age group having the most babies. Karyotyping can detect a Down syndrome child. However, young women are not routinely encouraged to undergo the procedures necessary to get a sample of fetal cells (i.e., amniocentesis or chorionic villi sampling) because the risk of complications is greater than the risk of having a Down syndrome child. Fortunately, a test based on substances in maternal blood can help identify fetuses who may need to be karyotyped.

Down syndrome is easily recognized by these common characteristics: short stature; an eyelid fold; a flat face; stubby fingers; a wide gap between the first and second toes; a large, fissured tongue; a round head; a palm crease, the so-called simian line; and, unfortunately, mental retardation, which can vary in intensity. In the opening story, Amanda has Down syndrome, and every one in her family gives her loving attention. Similarly, in real life, Chris Burke (Fig. 18.14*a*) was born with Down syndrome, and his parents were advised to put him in an institution. But Chris's parents didn't do that. They gave him the same loving care and attention they gave their other children, and it paid off. Chris is remarkably talented. He is a playwright, actor, and musician. He starred in *Life Goes On*, a TV series written just for him, and he is sometimes asked to be a guest star in other TV shows. His love of music and collaboration with other musicians has led to the release of several albums—

like Chris, the songs are uplifting and inspirational. You can read more about this remarkable young man in his autobiography, *A Special Kind of Hero*.

The genes that cause Down syndrome are located on the bottom third of chromosome 21 (Fig. 18.14*b*), and extensive investigative work has been directed toward discovering the specific genes responsible for the characteristics of the syndrome. Thus far, investigators have discovered several genes that may account for various conditions seen in persons with Down syndrome. For example, they have located genes most likely responsible for the increased tendency toward leukemia, cataracts, accelerated rate of aging, and mental retardation. The gene for mental retardation, dubbed the *Gart* gene, causes an increased level of purines in the blood, a finding associated with mental retardation. One day, it may be possible to control the expression of the *Gart* gene even before birth so that at least this symptom of Down syndrome does not appear.

Changes in Sex Chromosome Number

An abnormal sex chromosome number is the result of inheriting too many or too few X or Y chromosomes. Figure 18.13 can be used to illustrate nondisjunction of the sex chromosomes during oogenesis, if you assume that the chromosomes

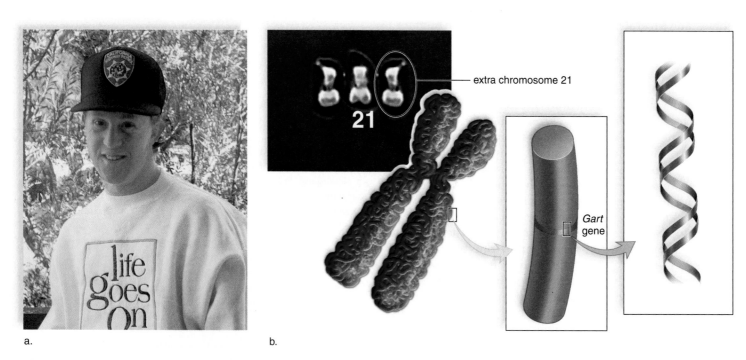

a.   b.

Figure 18.14   Abnormal autosomal chromosome number.
a. Chris Burke was born with Down syndrome. Common characteristics of the syndrome include a wide, rounded face and a fold on the upper eyelids. Mental retardation, along with an enlarged tongue, makes it difficult for a person with Down syndrome to speak distinctly. **b.** Karyotype of an individual with Down syndrome shows an extra chromosome 21. More sophisticated technologies allow investigators to pinpoint the location of specific genes associated with the syndrome. An extra copy of the Gart gene, which leads to a high level of purines in the blood, may account for the mental retardation seen in persons with Down syndrome.

shown represent X chromosomes. Nondisjunction during oogenesis or spermatogenesis can result in gametes that have too few or too many X or Y chromosomes.

A person with Turner syndrome (XO) is a female, and a person with Klinefelter syndrome (XXY) is a male. This shows that in humans, the presence of a Y chromosome, not the number of X chromosomes, determines maleness. The *SRY* (*sex-determining region of Y*) gene, on the short arm of the Y chromosome, produces a hormone called testis-determining factor, which plays a critical role in the development of male genitals.

Turner Syndrome   From birth, an individual with Turner syndrome has only one sex chromosome, an X (Fig. 18.15*a*). As adults, Turner females are short, with a broad chest and folds of skin on the back of the neck. The ovaries, oviducts, and uterus are very small and underdeveloped. Turner females do not undergo puberty or menstruate, and their breasts do not develop. However, some have given birth following in vitro fertilization using donor eggs. They usually are of normal intelligence

and can lead fairly normal lives if they receive hormone supplements.

Klinefelter Syndrome   About 1 in 650 live males are born with two X chromosomes and one Y chromosome (Fig. 26.4*b*). The symptoms of this condition (referred to as "47, XXY") are often so subtle that only 25% are ever diagnosed, and those are usually not diagnosed until after age 15. Earlier diagnosis opens the possibility for educational accommodations and other interventions that can help mitigate common symptoms, which include speech and language delays. Those 47, XXY males who develop more severe symptoms as adults are referred to as having "Klinefelter syndrome." All 47, XXY adults will require assisted reproduction in order to father children. Affected individuals commonly receive testosterone supplementation beginning at puberty.

The Health Focus on the next page tells of the experiences of a person with Klinefelter syndrome who suggests that it is best for parents to know right away that they have a child with this disorder because much can be done to help the child lead a normal life.

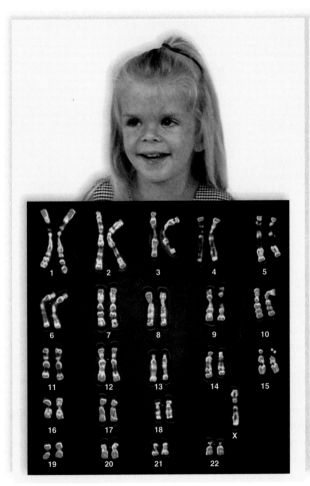

a. Turner syndrome   b. Klinefelter syndrome

Figure 18.15
Abnormal sex chromosome number.
People with
(a) Turner syndrome, who have only one sex chromosome, an X, as shown, and
(b) Klinefelter syndrome, who have more than one X chromosome plus a Y chromosome, as shown, can look relatively normal (especially as children) and can lead relatively normal lives.

Health Focus

Living with Klinefelter Syndrome

In 1996, at the age of 25, I was diagnosed with Klinefelter syndrome (KS). Being diagnosed has changed my life for the better.

I was a happy baby, but when I was still very young, my parents began to believe that there was something wrong with me. I knew something was different about me, too, as early on as five years old. I was very shy and had trouble making friends. One minute I'd be well behaved, and the next I'd be picking fights and flying into a rage. Many psychologists, therapists, and doctors tested me because of school and social problems and severe mood changes. Their only diagnosis was "learning disabilities" in such areas as reading comprehension, abstract thinking, word retrieval, and auditory processing. No one could figure out what the real problem was, and I hated the tutoring sessions I had. In the seventh grade, a psychologist told me that I was stupid and lazy, I would probably live at home for the rest of my life, and I would never amount to anything. For the next five years, he was basically right, and I barely graduated from high school.

I believe, though, that I have succeeded because I was told that I would fail. I quit the tutoring sessions when I enrolled at a community college; I decided I could figure things out on my own. I received an associate degree there, then transferred to a small liberal arts college. I never told anyone about my learning disabilities and never sought special help. However, I never had a semester below a 3.0, and I graduated with two B.S. degrees. I was accepted into a graduate program but decided instead to accept a job as a software engineer even though I did not have an educational background in this field. As I later learned, many KS'ers excel in computer skills. I had been using a computer for many years and had learned everything I needed to know on my own, through trial and error.

Around the time I started the computer job, I went to my physician for a physical. He sent me for blood tests because he noticed that my testes were smaller than usual. The results were conclusive: Klinefelter syndrome with sex chromosomes XXY. I initially felt denial, depression, and anger, even though I now had an explanation for many of the problems I had experienced all my life. But then I decided to learn as much as I could about the condition and treatments available. I now give myself a testosterone injection once every two weeks, and it has made me a different person, with improved learning abilities and stronger thought processes in addition to a more outgoing personality.

I found, though, that the best possible path I could take was to help others live with the condition. I attended my first support group meeting four months after I was diagnosed. By spring 1997, I had developed an interest in KS that was more than just a part-time hobby. I wanted to be able to work with this condition and help people forever. I have been very involved in KS conferences and have helped to start support groups in the U.S., Spain, and Australia.

Since my diagnosis, it has been my dream to have a son with KS, although when I was diagnosed, I found out it was unlikely that I could have biological children. Through my work with KS, I had the opportunity to meet my fiancee Chris. She has two wonderful children: a daughter, and a son who has the same condition that I do. There are a lot of similarities between my stepson and me, and I am happy I will be able to help him get the head start in coping with KS that I never had. I also look forward to many more years of helping other people seek diagnosis and live a good life with Klinefelter syndrome.

Stefan Schwarz
stefan13@cox.net
http://klinefeltersyndrome.org

Poly-X Females   A poly-X female has more than two X chromosomes and extra Barr bodies in the nucleus. Females with three X chromosomes have no distinctive phenotype, aside from a tendency to be tall and thin. Although some have delayed motor and language development, most poly-X females are not mentally retarded. Some may have menstrual difficulties, but many menstruate regularly and are fertile. Their children usually have a normal karyotype.

Females with more than three X chromosomes occur rarely. Unlike XXX females, XXXX females are more likely to be retarded. Various physical abnormalities are seen, but they may menstruate normally.

Jacobs Syndrome   XYY males with Jacobs syndrome can only result from nondisjunction during spermatogenesis. Affected males are usually taller than average, suffer from persistent acne, and tend to have speech and reading problems. At one time, it was suggested that these men were likely to be criminally aggressive, but it has since been shown that the incidence of such behavior among them may be no greater than among XY males.

Changes in Chromosome Structure

Changes in chromosome structure are another type of chromosomal mutation. Various agents in the environment, such as radiation, certain organic chemicals, or even viruses, can cause chromosomes to break. Ordinarily, when breaks occur in chromosomes, the two broken ends reunite to give the same sequence of genes. Sometimes, however, the broken ends of one or more chromosomes do not rejoin in the same pattern as before, and the result is various types of chromosomal mutation.

Changes in chromosome structure include deletions, translocations, duplications, and inversions of chromosome segments. A **deletion** occurs when an end of a chromosome breaks off or when two simultaneous breaks lead to the loss of an internal segment (Fig. 18.16a). Even when only one member of a pair of chromosomes is affected, a deletion often causes abnormalities.

A **duplication** is the presence of a chromosomal segment more than once in the same chromosome (Fig. 18.16b). An **inversion** has occurred when a segment of a chromosome is turned around 180° (Fig. 18.16c). This reversed sequence of genes can lead to altered gene activities and to deletions and duplications as described in Figure 18.17.

A **translocation** is the movement of a chromosome segment from one chromosome to another nonhomologous chromosome (Fig. 18.16d). In 5% of cases, a translocation that occurred in a previous generation between chromosomes 21 and 14 is the cause of Down syndrome. In other words, because a portion of chromosome 21 is now attached to a portion of chromosome 14, the individual has three copies of the alleles that bring about Down syndrome when they are present in triplet copy. In these cases, Down syndrome is not related to the age of the mother but instead tends to run in the family of either the father or the mother.

Human Syndromes

Changes in chromosome structure occur in humans and lead to various syndromes, many of which are just now being discovered.

Deletion Syndromes Williams syndrome occurs when chromosome 7 loses a tiny end piece (Fig. 18.18). Children who have this syndrome look like pixies, with turned-up noses, wide mouths, a small chin, and large ears. Although their academic skills are poor, they exhibit excellent verbal and musical abilities.

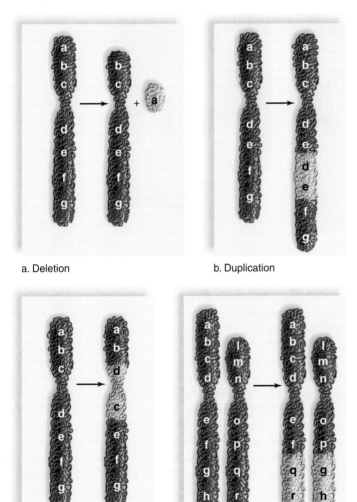

a. Deletion

b. Duplication

c. Inversion

d. Translocation

Figure 18.16 **Types of chromosomal mutations.**
a. Deletion is the loss of a chromosome piece. **b.** Duplication occurs when the same piece is repeated within the chromosome. **c.** Inversion occurs when a piece of chromosome breaks loose and then rejoins in the reversed direction. **d.** Translocation is the exchange of chromosome pieces between nonhomologous pairs.

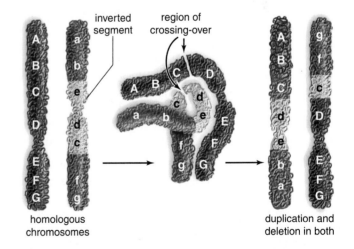

homologous chromosomes

duplication and deletion in both

Figure 18.17 **Inversion.**
Left: A segment of one homologue is inverted. Notice that in the shaded segment, *edc* occurs instead of *cde*. *Middle:* The two homologues can pair only when the inverted sequence forms an internal loop. After crossing-over, a duplication and a deletion can occur. *Right:* The homologue on the left has *AB* and *ab* sequences and neither *fg* nor *FG* genes. The homologue on the right has *gf* and *FG* sequences and neither *AB* nor *ab* genes.

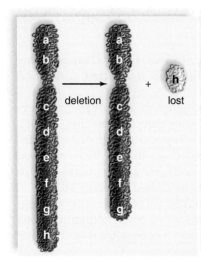

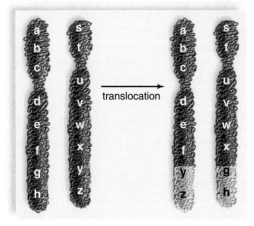

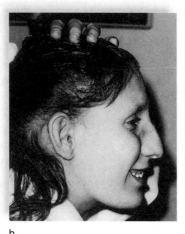

a.   b.

Figure 18.18
Deletion.
a. When chromosome 7 loses an end piece, the result is Williams syndrome.
b. These children, although unrelated, have the same appearance, health, and behavioral problems.

Figure 18.19
Translocation.
a. When chromosomes 2 and 20 exchange segments, (**b**) Alagille syndrome, with distinctive facial features, sometimes results because the translocation disrupts an allele on chromosome 20.

a.   b.

The gene that governs the production of the protein elastin is missing, and this affects the health of the cardiovascular system and causes their skin to age prematurely. Such individuals are very friendly but need an ordered life, perhaps because of the loss of a gene for a protein that is normally active in the brain.

Cri du chat (cat's cry) syndrome is seen when chromosome 5 is missing an end piece. The affected individual has a small head, is mentally retarded, and has facial abnormalities. Abnormal development of the glottis and larynx results in the most characteristic symptom—the infant's cry resembles that of a cat.

Translocation Syndromes A person who has both of the chromosomes involved in a translocation has the normal amount of genetic material and is healthy, unless the chromosome exchange broke an allele into two pieces. The person who inherits only one of the translocated chromosomes will no doubt have only one copy of certain alleles and three copies of certain other alleles. A genetic counselor begins to suspect a translocation has occurred when spontaneous abortions are commonplace and family members suffer from various syndromes.

Figure 18.19 shows a father and son who have a translocation between chromosomes 2 and 20. Although they have the normal amount of genetic material, they have the distinctive face, abnormalities of the eyes and internal organs, and severe itching characteristic of Alagille syndrome. People with this syndrome ordinarily have a deletion on chromosome 20;

therefore, it can be deduced that the translocation disrupted an allele on chromosome 20 in the father. The symptoms of Alagille syndrome range from mild to severe, so some people may not be aware they have the syndrome. This father did not realize it until he had a child with the syndrome.

Translocations can also be responsible for a variety of other disorders including certain types of cancer. In the 1970s, new staining techniques identified that a translocation from a portion of chromosome 22 to chromosome 9 was responsible for chronic myelogenous leukemia. This translocated chromosome was called Philadelphia chromosome. In Burkett lymphoma, a cancer common in children in equatorial Africa, a large tumor develops from lymph glands in the region of the jaw. This disorder involves a translocation from a portion of chromosome 8 to chromosome 14.

✅ Check Your Progress 18.5

1. What process usually causes an individual to have an abnormal number of chromosomes?

2. What is the specific chromosome abnormality of a person with Down syndrome?

3. What are some syndromes that result from inheritance of an abnormal sex chromosome number?

4. What are other types of chromosome mutations, aside from abnormal chromosome number?

Selecting Children

Human beings have always attempted to influence the characteristics of their children. For example, couples have attempted to determine the sex of their children for centuries through a variety of methods. Amniocentesis has allowed us to test fetuses for chromosomal abnormalities and debilitating developmental defects before birth. Modern genetic testing technology enables parents to directly select children bearing desired traits, even at the very earliest stages of development.

Fanconi's Anemia

Recently, preimplantation genetic diagnosis selected an embryo for a couple because the newborn could save the life of his sister (Fig. 18B). The couple, Jack and Lisa Nash, had a daughter with Fanconi's anemia, a rare inherited disorder in which affected persons cannot properly repair DNA damage that results from certain toxins. The disease primarily afflicts the bone marrow and, therefore, results in a reduction of all types of blood cells. Anemia occurs, due to a deficiency of red blood cells. Patients are also at high risk of infection, because of low white blood cell numbers, and of leukemia, because white blood cells cannot properly repair any damage to their DNA.

Fanconi's anemia may be treated by a traditional bone marrow transplant or by an adult stem cell transplant, preferably from a parent or sibling, because the risk of rejection is lower. Adult stem cells are almost always the preferred treatment option because stem cells are hardier and much less likely to be rejected than a bone marrow transplant. Recall that the umbilical cord of a newborn is a rich source of adult stem cells for all types of blood cells. (Called adult because they are not embryonic stem cells.)

Testing the Embryo

During preimplantation genetic diagnosis, an embryonic cell is removed and tested for a disorder that runs in the family (see Fig. 20A). In this case, doctors wanted to find an embryo that would become a healthy individual and also who would be able to benefit his sister. The parents underwent in vitro fertilization, and the 15 resulting embryos were screened to see which ones would not produce an individual with Fanconi's anemia and also had similar recognition proteins as their daughter. Two embryos met these requirements, but only one implanted in the uterus, and it developed into a healthy baby boy. Adult stem cells were harvested from the umbilical cord of the newborn and were successfully used to treat his sister's anemia. The physician who performed the genetic screening stated that he has received numerous inquiries about performing the procedure for other couples with diseased children.

Ethical Dilemma

This case, and other related cases, has raised a number of ethical issues surrounding prenatal selection of children based on genetic traits. While the AMA insists that selection based on traits not related to disease is unethical, the AMA's chair of the Council on Ethical and Judicial Affairs made an exception for this case because the child was selected for medical reasons. Dr. Jacques Montagut, who helped develop in vitro fertilization, believes that it is dangerous to bear children for the purpose of curing others, and compared it with "a new form of biological slavery." Still others think that, soon, children will be selected for less altruistic reasons, such as for their height, physical prowess, or intellectual abilities.

Decide Your Opinion

1. In general, do you think it is ethical to have children to cure medically related conditions, regardless of how fertilization occurred? If not, do you agree with the AMA that this case is an acceptable exception?
2. Because the brother was created ostensibly as a treatment for his sister's disease, do you believe that there is a moral obligation to provide him with compensation?
3. Would embryonic stem cells, derived from an aborted fetus and cultured in the laboratory, be an acceptable substitute?
4. Would you willingly donate sperm or eggs for in vitro fertilization to produce a healthy child for a couple who could not have one because of the risk of an inherited disease, such as Fanconi's anemia?

Figure 18B **The Nash family.**
Jack, Molly holding baby Adam, and Lisa Nash. Adam was genetically selected as an embryo because the stem cells of his umbilical cord would save the life of his sister, Molly.

Summarizing the Concepts

18.1 Chromosomes and the Cell Cycle

- Chromosomes occur in pairs in body cells.
- A karyotype is a visual display of a person's chromosomes.
  A normal human karyotype shows 22 homologous pairs of autosomes and one pair of sex chromosomes.
- Normal sex chromosomes in males: XY
- Normal sex chromosomes in females: XX

The Cell Cycle

- The cell cycle occurs continuously and has four stages: G_1, S, G_2, (the interphase stages), and M (the mitotic stage, which includes cytokinesis and the stages of mitosis).

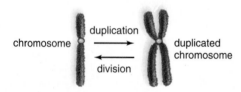

- In G_1, a cell doubles organelles and accumulates materials for DNA synthesis.
- In S, DNA replication occurs.
- In G_2, a cell synthesizes proteins needed for cell division.
- Apoptosis also occurs during the cell cycle.

18.2 Mitosis

Mitosis is duplication division that assures that all body cells have the diploid number and the same kinds of chromosomes as the cell that divides.

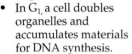

The phases of mitosis are prophase, metaphase, anaphase, and telophase.

- **Prophase**   Chromosomes attach to spindle fibers.
- **Metaphase**   Chromosomes align at the equator.
- **Anaphase**   Chromatids separate, becoming chromosomes that move toward the poles.
- **Telophase**   Nuclear envelopes form around chromosomes; cytokinesis begins.

  Cytokinesis is the division of cytoplasm and organelles following mitosis.

- The proper workings of the cell cycle and mitosis are critical to growth and tissue repair.

18.3 Meiosis

Meiosis involves two cell divisions: meiosis I and meiosis II.

- **Meiosis I**   Homologous chromosomes pair and then separate.
- **Meiosis II**   Sister chromatids separate, resulting in four cells with the haploid number of chromosomes that move into daughter nuclei.

  Meiosis results in genetic recombination due to crossing-over; gametes have all possible combinations of chromosomes. Upon fertilization, recombination of chromosomes occurs.

18.4 Comparison of Meiosis and Mitosis

- In prophase I, homologous chromosomes pair; there is no pairing in mitosis.
- In metaphase I, homologous duplicated chromosomes align at equator.
- In anaphase I, homologous chromosomes separate.

Spermatogenesis and Oogenesis

- **Spermatogenesis**   In males, produces four viable sperm.
- **Oogenesis**   In females, produces one egg and two or three polar bodies. Oogenesis goes to completion if the sperm fertilizes the developing egg.

18.5 Chromosome Inheritance

Meiosis is a part of gametogenesis (spermatogenesis in males and oogenesis in females) and contributes to genetic diversity.

Changes in Chromosome Number

- Nondisjunction changes the chromosome number in gametes, resulting in trisomy or monosomy.
- Autosomal syndromes include trisomy and Down syndrome.

Changes in Sex Chromosome Number

- Nondisjunction during oogenesis or spermatogenesis can result in gametes that have too few or too many X or Y chromosomes.
- Syndromes include Turner, Klinefelter, poly-X, and Jacobs.

Changes in Chromosome Structure

- Chromosomal mutations can produce chromosomes with deleted, duplicated, inverted, or translocated segments.
- These result in various syndromes such as Williams and cri du chat (deletion) and Alagille and certain cancers (translocation).

Understanding Key Terms

anaphase 382	interphase 379
apoptosis 379	inversion 397
aster 381	meiosis 385
Barr body 393	metaphase 382
cell cycle 379	mitosis 378, 379, 381
centriole 381	mitotic spindle 381
centromere 378	monosomy 393
centrosome 381	nondisjunction 393
cleavage furrow 383	oogenesis 392
crossing-over 388	parent cell 381
cytokinesis 379	polar body 392
daughter cell 381	prophase 382
deletion 397	secondary oocyte 392
diploid (2n) 381	sister chromatids 378
duplication 397	spermatogenesis 392
fertilization 385	synapsis 385
gamete 385	syndrome 380
haploid (n) 385	telophase 383
homologous chromosome 385	translocation 397
homologue 385	trisomy 393
interkinesis 385	zygote 385

Match the key terms to these definitions.

a. —————— Dark-staining nuclei of females that contains a condensed, inactive X chromosome.

b. —————— Member of a pair of chromosomes that are alike and come together in synapsis of prophase of meiosis I.

c. —————— Having one less chromosome than usual.

d. —————— A change in chromosome structure in which a segment is turned around 180°.

e. —————— Having the n number of chromosomes; for humans, n=23.

Testing Your Knowledge of the Concepts

1. Describe the two parts of the cell cycle. (page 379)

2. What is the function of the mitotic spindle? (pages 381, 383)

3. What is the importance of mitosis, and how is the process controlled? (page 384)

4. How do the terms diploid (2n) and haploid (n) relate to meiosis? (page 385)

5. List and describe the events in meiosis I and meiosis II. (pages 385–88)

6. What is the significance of meiosis? (page 389)

7. Contrast mitosis and meiosis I and meiosis II. (pages 389–91)

8. How does spermatogenesis differ from oogenesis? (page 392)

9. What is nondisjunction, when can it occur, and what are the results? (page 393)

10. What are some syndromes caused by changes in chromosome number, and what is the change? (pages 393–94)

11. Describe three changes that can occur in chromosome structure and some syndromes caused by such changes. (pages 397–98)

12. The point of attachment for two sister chromatids is the
    a. centriole.   c. chromosome.
    b. centromere.   d. karyotype.

13. Label the drawing of the cell cycle; then tell the main event of each stage.

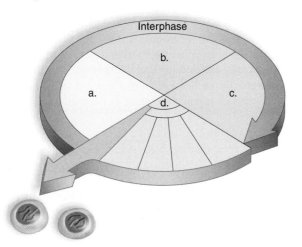

Interphase

b.

a.

d.

c.

In questions 14–20, match the statement to interphase or the phase of mitosis in the key.

Key:
 a. metaphase
 b. interphase
 c. telophase
 d. prophase
 e. anaphase

14. Spindle fibers begin to appear.

15. DNA synthesis occurs.

16. Chromosomes line up at the equator.

17. Duplicated chromosomes become visible.

18. Centromere splits and sister chromosomes move to opposite poles.

19. Cytokinesis occurs during this phase.

20. Chromosomes become chromatin.

21. If a parent cell has 18 chromosomes before mitosis, how many chromosomes will the daughter cells have?
    a. 18   c. 9
    b. 36   d. 27

22. In humans, mitosis is necessary to
    a. grow and repair damaged tissue.
    b. form the gametes.
    c. maintain the same chromosome number in body cells.
    d. Both a and c are correct.

23. If a parent cell has 18 chromosomes, how are they arranged at the equator during metaphase of mitosis?
    a. single file of 18 replicated chromosomes in random order
    b. double file of 9 replicated chromosomes in random order
    c. single file of 18 replicated chromosomes from number 1 to 18
    d. double file of 9 replicated chromosomes with 1–9 from the mother on one side and 1–9 from the father on the other side

24. Crossing-over occurs between
    a. sister chromatids of the same chromosome.
    b. chromatids of nonhomologous chromosomes.
    c. nonsister chromatids of a homologous pair.
    d. Both b and c are correct.

25. At the equator in metaphase of meiosis I, there are
    a. single chromosomes in random order.
    b. unpaired duplicated chromosomes in random order.
    c. homologous pairs in random order.
    d. homologous pairs with those from the mother on one side and from the father on the other side.

26. The end products of _____ are _____ cells.
    a. mitosis, diploid   c. meiosis, diploid
    b. meiosis, haploid   d. Both a and b are correct.

27. If a parent cell has 22 chromosomes, the daughter cells following meiosis II will have
    a. 22 chromosomes.
    b. 44 chromosomes.
    c. 11 chromosomes.
    d. Any of these could be correct.

28. Which of these drawings represents metaphase of meiosis I?

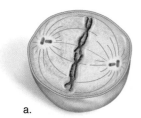 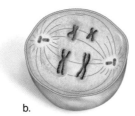

a.   b.

In questions 29–33, match the part of the diagram to the correct label.

29. Meiosis II

30. Primary spermatocyte

31. Sperm

32. Secondary spermatocyte

33. Spermatids

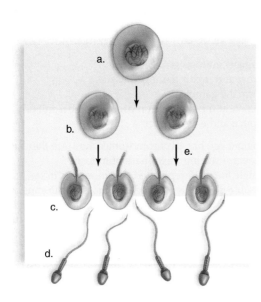

34. Which of these helps to ensure that genetic diversity will be maintained?
    a. independent alignment during metaphase I
    b. crossing-over during prophase I
    c. fusion of sperm and egg nuclei during fertilization
    d. All of these are correct.

35. How many viable cells are produced by oogenesis?
    a. four
    b. three
    c. one
    d. two

36. Monosomy or trisomy occurs because of
    a. crossing-over.
    b. inversion.
    c. translocation.
    d. nondisjunction.

37. Trisomy of chromosome 21 is called
    a. Down syndrome.
    b. Klinefelter syndrome.
    c. Turner syndrome.
    d. Jacobs syndrome.

38. A person with Klinefelter syndrome is _____ and has _____ sex chromosomes.
    a. male, XYY
    b. male, XXY
    c. female, XXY
    d. female, XO

Thinking Critically About the Concepts

Nick, from the opening story, was a little perturbed by the keloid that developed when the gash below his ear healed. His mom researched what treatment options were available and came to the conclusion that Nick's keloid might be better left alone than treated, since treatment could cause more scarring. They decided that an easy way to camouflage the scar was to simply let Nick wear his hair a little longer. The disappointment Nick had about the scarring and inability to swap places with Josh ended when he began a growth spurt typical of puberty. He was delighted that his extra inches meant he towered over his younger brother and sister.

1. Cell fragments created by apoptosis are phagocytized by white blood cells and surrounding tissue cells. From your study of the immune system in Chapter 7, what specific white blood cells will phagocytize the cell fragments?

2. Explain how the separation of homologous chromosomes during meiosis affects the appearance of siblings (that some resemble each other and others look very different from one another).

3a. What would you conclude about the ability of nervous and muscle tissue to repair themselves if nerve and muscle cells are typically arrested in G$_1$ of interphase?

b. What are the implications of this arrested state to someone who suffers a spinal cord injury or heart attack?

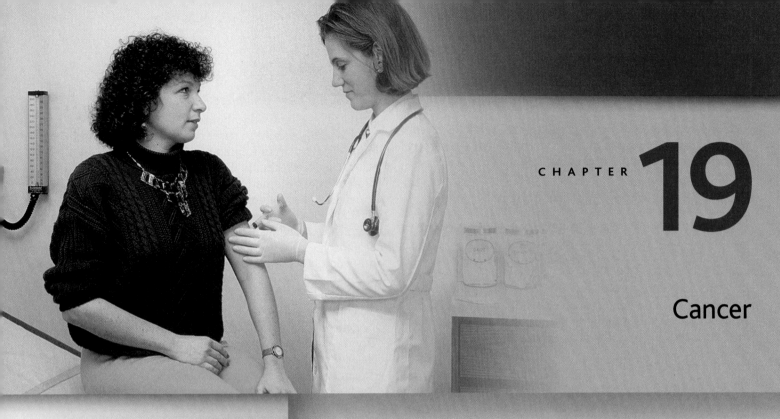

CHAPTER 19

Cancer

Caitlin contemplated the letter she'd just received from her ob-gyn's office. It was an informational letter about the recently approved vaccine to prevent most cases of cervical cancer. She was late for class, so she tossed the letter aside to read later that evening. When Caitlin returned home after several hours of classes and lab, she dumped her lab manual and notes on top of the letter and crashed on the sofa. The letter was forgotten.

She finally remembered the letter later while watching TV when she saw a commercial for a vaccine to prevent cervical cancer. "Geez," she thought, "I totally spaced reading that letter from my ob-gyn." Rummaging through a pile of papers and notes, she located the letter and sat down to read it. Its content was very much like the information she'd just heard during the commercial. The vaccine was for the human papillomavirus (HPV), implicated in the vast majority of cervical cancer cases. If the vaccine is given before sexual activity begins, it almost eliminates the risk of developing cervical cancer later in life. The vaccine is recommended for all females ages 11–26 because even sexually active women may not have been exposed to the HPV types in the vaccine yet. Caitlin was advised to call the office if she had any questions or to schedule an appointment to be vaccinated. In an ideal world, cancer would be prevented by vaccinations and/or healthy lifestyle choices. As you study the material in this chapter, you should consider what you do or don't do that could influence whether or not you experience cancer in the future.

CHAPTER CONCEPTS

19.1 Cancer Cells
Cancer cells have a number of abnormal characteristics that prevent them from functioning in the same manner as normal cells. They divide repeatedly and form tumors in the place of origin and in other parts of the body.

19.2 Causes and Prevention of Cancer
Whether cancer develops is partially due to inherited genes, but exposure to carcinogens such as UV radiation, tobacco smoke, pollutants, industrial chemicals, and certain viruses play a significant role also.

19.3 Diagnosis of Cancer
Cancer is usually diagnosed by certain screening procedures and by imaging the body and tissues, utilizing various techniques.

19.4 Treatment of Cancer
Surgery involving bone marrow transplants followed by radiation and/or chemotherapy has now become fairly routine. Other methods are under investigation.

19.1 Cancer Cells

Although **cancer** is actually over a hundred different diseases and each type of cancer can vary from another, these characteristics are common to cancer cells.

Characteristics of Cancer Cells

Cancer is a cellular disease, and cancer cells share characteristics that distinguish them from normal cells.

Cancer Cells Lack Differentiation

Cancer cells are nonspecialized and do not contribute to the functioning of a body part. A cancer cell does not look like a differentiated epithelial, muscle, nervous, or connective tissue cell; instead, it looks distinctly abnormal.

Cancer Cells Have Abnormal Nuclei

The nuclei of cancer cells are enlarged and may contain an abnormal number of chromosomes. The nuclei of the cervical cancer cells shown in Figure 19.1*b, c* have increased to the point that they take up most of the cell. As discussed in Chapter 1, cervical cancer is often caused by the human papillomavirus (HPV), a sexually transmitted virus. Therefore, young women, like Caitlin in the opening story, should get vaccinated with a new vaccine that can protect them against any new HPV infections.

Not only the nuclei but also chromosomes of cancerous cells are abnormal. Some portions of the chromosomes may be duplicated, and some may be deleted. In addition, gene amplification (extra copies of specific genes) is seen much more frequently than in normal cells. Ordinarily, cells with damaged DNA undergo **apoptosis,** or programmed cell death. Cancer cells fail to undergo apoptosis, even though they are abnormal cells.

Tissues that divide frequently, such as those that line the respiratory and digestive tracts, are more likely to become cancerous. Cell division gives them the opportunity to undergo genetic mutations, each one making the cell more abnormal and giving it the ability to produce more of its own kind.

Cancer Cells Have Unlimited Replicative Potential

Ordinarily, cells divide about 60–70 times and then just stop dividing and die. Cancer cells are immortal and keep on dividing for an unlimited number of times.

Just as shoelaces are capped by small pieces of plastic, chromosomes in human cells end with special repetitive DNA sequences called **telomeres.** Specific proteins bind to telomeres and protect the ends of chromosomes from DNA repair enzymes that always tend to bind together naked ends of chromosomes. The telomeres get shorter after each cell cycle, and eventually the chromosomes do bind together, causing the cell to undergo apopstosis and die. Telomerase is a special enzyme that can rebuild telomere sequences, and in that way prevent these cells from ever losing their potential to divide. The gene that codes for telomerase is turned on in cancer cells. Telomerase continuously rebuilds the telomeres in cancer cells so that the telomeres remain at a constant length, and the cell can keep dividing over and over again.

Cancer Cells Form Tumors

Normal cells anchor themselves to a substratum and/or adhere to their neighbors. They exhibit contact inhibition—when they come in contact with a neighbor, they stop dividing. Cancer cells have lost all restraint; they pile on top of one another and grow in multiple layers, forming a **tumor.** As cancer develops, the most aggressive cell becomes the dominant cell of the tumor (Fig. 19.2).

A benign tumor is usually encapsulated and, therefore, will never invade adjacent tissue. Cancer in situ is a tumor of epithelium in its place of origin not always surrounded by a capsule. As yet, the cancer has not penetrated beyond the basement membrane, a layer of nonliving material that anchors epithelial tissue to underlying connective tissue.

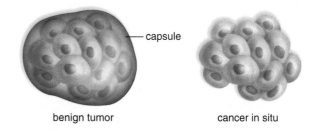

benign tumor   cancer in situ

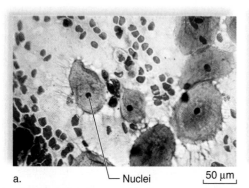
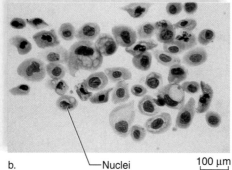
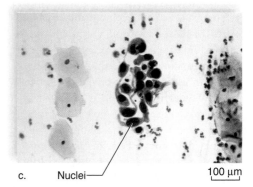

a. �262Nuclei   50 μm   b. �262Nuclei   100 μm   c.   Nuclei�   100 μm

Figure 19.1   Development of cervical cancer.
a. Normal cervical cells. **b.** Precancerous cervical cells. **c.** Cancerous cervical cells.

Cancer Cells Have No Need for Growth Factors

Chemical signals between cells tell them whether they should be dividing or not dividing. These chemical signals, called growth factors, are of two types: stimulatory growth factors and inhibitory growth factors. Cancer cells keep on dividing, even when stimulatory growth factors are absent, and they do not respond to inhibitory growth factors.

Cancer Cells Gradually Become Abnormal

Figure 19.2 illustrates that **carcinogenesis,** the development of cancer, is a multistage process that can be divided into these three phases:

Initiation: A single cell undergoes a mutation that causes it to begin to divide repeatedly (Fig. 19.2*a*).

Promotion: A tumor develops, and the tumor cells continue to divide. As they divide, they undergo mutations (Fig. 19.2*b,c*).

Progression: One cell undergoes a mutation that gives it a selective advantage over the other cells (Fig. 19.2*c*). This process is repeated several times, and, eventually, there is a cell that has the ability to invade surrounding tissues (Fig. 19.2*d-e*).

Cancer Cells Undergo Angiogenesis and Metastasis

To grow larger than about a million cells, about the size of a pea, a tumor must have a well-developed capillary network to bring it nutrients and oxygen. **Angiogenesis** is the formation of new blood vessels. The low oxygen content in the middle of a tumor may turn on genes coding for angiogenic growth factors that diffuse into the nearby tissues and cause new vessels to form.

Due to mutations, cancer cells tend to be motile because they have a disorganized internal cytoskeleton and lack intact actin filament bundles. To metastasize, cancer cells must make their way across the basement membrane and invade a blood vessel or lymphatic vessel. Invasive cancer cells are sperm-shaped (see Fig. 19.1*c*) and don't look at all like the normal cell seen nearby. Cancer cells produce proteinase enzymes that degrade the membrane and allow them to invade underlying tissues. Malignancy is present when cancer cells are found in nearby lymph nodes. When these cells begin new tumors far from the primary tumor, **metastasis** has occurred (Fig. 19.2*f*). Not many cancer cells achieve this feat (maybe 1 in 10,000), but those that successfully metastasize to various parts of the body make the prognosis (the predicted outcome of the disease) for recovery doubtful.

Cancer Is a Genetic Disease

Recall that the cell cycle, as discussed on page 379, consists of interphase, followed by mitosis. *Cyclin* is a molecule that has to be present for a cell to proceed from interphase to

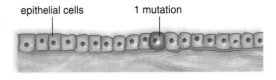

a. Cell (dark red) acquires a mutation for repeated cell division.

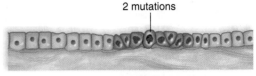

b. New mutations arise, and one cell (brown) has the ability to start a tumor.

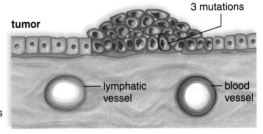

c. Cancer in situ. The tumor is at its place of origin. One cell (purple) mutates further.

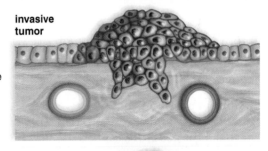

d. Cells have gained the ability to invade underlying tissues by producing a proteinase enzyme.

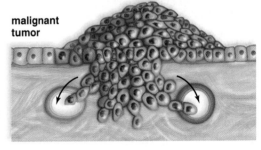

e. Cancer cells now have the ability to invade lymphatic and blood vessels.

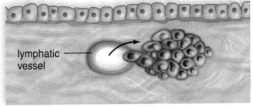

f. New metastatic tumors are found some distance from the original tumor.

Figure 19.2 Tumor progression.
a. One cell (dark pink) in a tissue mutates. **b.** This mutated cell divides repeatedly and a cell (brown) with two mutations appears. **c.** A tumor forms and a cell with three mutations (purple) appears. **d.** This cell (purple) which takes over the tumor, can invade underlying tissue. **e.** Tumor cells invade lymphatic and blood vessels. **f.** A new tumor forms at a distant location.

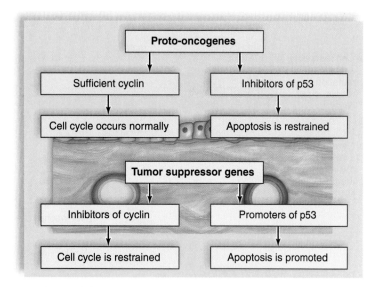

Figure 19.3 **Normal cells.**
In normal cells, proto-oncogenes code for sufficient cyclin to keep the cell cycle going normally, and they code for proteins that inhibit p53 and apoptosis. Tumor-suppressor genes code for proteins that inhibit cyclin, and promote p53 and apoptosis.

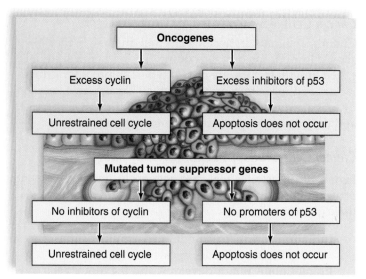

Figure 19.4 **Cancer cells.**
Notice that oncogenes and mutated tumor-suppressor genes have the same end effects: an unrestrained cell cycle, and apoptosis does not occur.

mitosis. When cancer develops, the cell cycle occurs repeatedly, in large part, due to mutations in two types of genes. Figure 19.3 shows that:

1. **Proto-oncogenes** code for proteins that promote the cell cycle and prevent apoptosis. They are often likened to the gas pedal of a car because they cause acceleration of the cell cycle.
2. **Tumor-suppressor genes** code for proteins that inhibit the cell cycle and promote apoptosis. They are often likened to the brakes of a car because they inhibit acceleration.

Proto-Oncogenes Become Oncogenes

When proto-oncogenes mutate, they become cancer-causing genes called **oncogenes.** These mutations can be called "gain-of-function" mutations because overexpression is the result (Fig. 19.4). Whatever a proto-oncogene does, an oncogene does it better.

A **growth factor** is a signal that activates a cell-signaling pathway, resulting in cell division. Some proto-oncogenes code for a growth factor or for a receptor protein that receives a growth factor. When these proto-oncogenes become oncogenes, receptor proteins are easy to activate and even may be stimulated by a growth factor produced by the receiving cell. Several proto-oncogenes code for Ras proteins that promote mitosis by activating **cyclin.** *Ras* oncogenes are typically found in many different types of cancers. Cyclin D is a proto-oncogene that codes for a cyclin directly. When this gene becomes an oncogene, cyclin is readily available all the time.

p53 is a protein that turns on genes that stop the cell cycle and activates repair enzymes. If repair is impossible, the p53 protein promotes *apoptosis,* programmed cell death. Apoptosis is an important way carcinogenesis is prevented. One proto-oncogene codes for a protein that functions to make p53 unavailable. When this proto-oncogene becomes an oncogene, no matter how much p53 is made, none will be available. Many tumors are lacking in p53 activity.

Tumor-Suppressor Genes Become Inactive

When tumor-suppressor genes mutate, their products no longer inhibit the cell cycle nor promote apoptosis. Therefore, these mutations can be called "loss of function" mutations (Fig. 19.4).

The retinoblastoma protein (RB) that turns on the gene for cyclin D and other genes whose products promote entry into the S phase of the cell cycle. When the tumor-suppressor gene *p16* mutates, the RB protein is always functional, and the result is, again, too much active cyclin D in the cell. The cell experiences repeated rounds of DNA synthesis without the occurrence of mitosis.

The protein Bax promotes apoptosis. When a tumor-suppressor gene *Bax* mutates, the protein Bax is not present, and apoptosis is less likely to occur. The gene contains a run of eight consecutive G bases, making it subject to mutations.

Types of Cancer

Statistics indicate that one in three Americans will deal with cancer in their lifetime. Therefore, this is a topic of considerable importance to the health and well-being of every individual. **Oncology** is the study of cancer, and a medical specialist in cancer is, therefore, known as an **oncologist.** The patient's prognosis (probable outcome) depends on (1) whether the tumor has invaded surrounding tissues and (2) whether there are metastatic tumors in distant parts of the body.

Tumors are classified according to their place of origin. **Carcinomas** are cancers of the epithelial tissues and adeno-carcinomas are cancers of glandular epithelial cells. Carcinomas include cancer of the skin, breast, liver, pancreas, intestines, lung, prostate, and thyroid. **Sarcomas** are cancers that arise in muscles and connective tissue, such as bone and fibrous connective tissue. **Leukemias** are cancers of the blood, and **lymphomas** are cancers of lymphatic tissue.

Common Cancers

Cancer occurs in all parts of the body, but some organs are more susceptible than others (Fig. 19.5). In the respiratory system, lung cancer is the most common type. Overall, this is one of the most common types of cancer, and smoking is known to increase a person's risk for this disease. In the digestive system, colorectal cancer (colon/rectum) is another common tumor. Other cancers of the digestive system include those of the pancreas, stomach, esophagus, and other organs. In the cardiovascular system, cancers include leukemia and plasma cell tumors. In the lymphatic system, cancers are classified as either Hodgkin or non-Hodgkin lymphoma. Thyroid cancer is the most common type of tumor in the endocrine system, whereas brain and spinal tumors are found in the central nervous system.

Breast cancer is one of the most common types of cancer. Although predominantly found in women, occasionally it is also found in men. Cancers of the cervix, ovaries, and other reproductive structures also occur in women. In males, prostate cancer is one of the most common cancers. Other cancers of the male reproductive system include cancer of the testis and the penis. Bladder and kidney cancers are associated with the urinary system. Skin cancers include melanoma and basal cell carcinoma. Oral cavity cancer is more frequent in men than women.

☑ Check Your Progress 19.1

1. What characteristics of cancer cells allow them to grow uncontrollably?

2. Mutations in what two types of genes lead to uncontrollable growth?

3. What are the most common types of cancer?

Male	Female
Prostate 234,460 (33%)	Breast 212,900 (31%)
Lung and bronchus 92,700 (13%)	Lung and bronchus 81,770 (12%)
Colon and rectum 72,800 (10%)	Colon and rectum 75,810 (11%)
Urinary bladder 44,690 (6%)	Uterine corpus 41,200 (6%)
Melanoma of the skin 34,260 (5%)	Non-Hodgkin lymphoma 28,190 (4%)
Non-Hodgkin lymphoma 30,680 (4%)	Melanoma of the skin 27,930 (4%)
Kidney and renal pelvis 24,650 (3%)	Thyroid 22,590 (3%)
Oral cavity and pharynx 20,180 (3%)	Ovary 20,180 (3%)
Leukemia 20,000 (3%)	Urinary bladder 16,730 (2%)
Pancreas 17,150 (2%)	Pancreas 16,580 (2%)
All sites 720,280 (100%)	All sites 679,510 (100%)

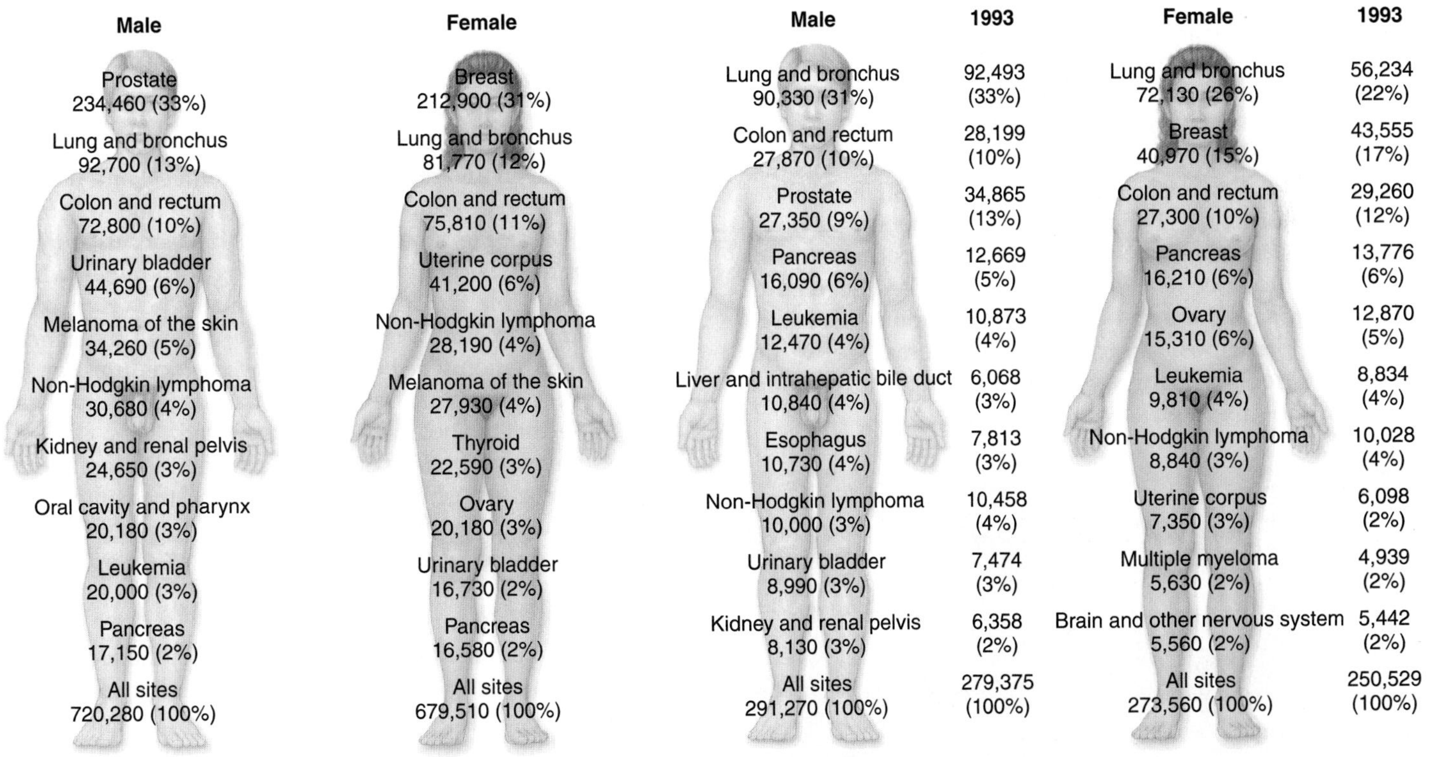

a. **Cancer cases by site and sex**

Male	1993	Female	1993
Lung and bronchus 90,330 (31%)	92,493 (33%)	Lung and bronchus 72,130 (26%)	56,234 (22%)
Colon and rectum 27,870 (10%)	28,199 (10%)	Breast 40,970 (15%)	43,555 (17%)
Prostate 27,350 (9%)	34,865 (13%)	Colon and rectum 27,300 (10%)	29,260 (12%)
Pancreas 16,090 (6%)	12,669 (5%)	Pancreas 16,210 (6%)	13,776 (6%)
Leukemia 12,470 (4%)	10,873 (4%)	Ovary 15,310 (6%)	12,870 (5%)
Liver and intrahepatic bile duct 10,840 (4%)	6,068 (3%)	Leukemia 9,810 (4%)	8,834 (4%)
Esophagus 10,730 (4%)	7,813 (3%)	Non-Hodgkin lymphoma 8,840 (3%)	10,028 (4%)
Non-Hodgkin lymphoma 10,000 (3%)	10,458 (4%)	Uterine corpus 7,350 (3%)	6,098 (2%)
Urinary bladder 8,990 (3%)	7,474 (3%)	Multiple myeloma 5,630 (2%)	4,939 (2%)
Kidney and renal pelvis 8,130 (3%)	6,358 (2%)	Brain and other nervous system 5,560 (2%)	5,442 (2%)
All sites 291,270 (100%)	279,375 (100%)	All sites 273,560 (100%)	250,529 (100%)

b. **Cancer deaths by site and sex**

Figure 19.5   Types of cancer in the United States.
a. Estimated new cancer cases; percentages for leading sites in 2006. **b.** Cancer deaths. Percentages of leading sites in 2006 compared to 1993. Changes have been minimal. Adapted with the permission of the American Cancer Society, Inc. All rights reserved.

19.2 Causes and Prevention of Cancer

Our current understanding of the causes of cancer are incomplete; however, by studying patterns of cancer development in populations, scientists have determined that, aside from heredity, there are environmental risk factors for developing cancer. Avoiding these risk factors and following dietary guidelines can help prevent cancer.

Heredity

The first gene associated with breast cancer was discovered in 1990. Scientists named this gene *breast cancer 1* or *BRCA1* (pronounced brak-uh). Later, they found that breast cancer could also be due to another breast cancer gene they called *BRCA2*. These genes are tumor-suppressor genes that follow a particular inheritance pattern. We inherit two copies of every gene; one from each parent. If a mutated copy of *BRCA1* or *2* is inherited from either parent, a mutation in the other copy is required before predisposition to cancer is increased. Each cell in the body has the mutated copy, but cancer develops wherever the second mutation occurs. If the second mutation occurs in the breast, breast cancer may develop. If the second mutation is in the ovary, ovarian cancer may develop if additional cancer-causing mutations occur.

The *RB* gene is also a tumor-suppressor gene. It takes its name from its association with an eye tumor called a *retinoblastoma tumor*, which first appears as a white mass in the retina. When a mutated copy of the *RB* gene is inherited, again it takes another mutation to increase the chance of cancer. A retinoblastoma tumor in one eye is more usual because it takes two mutations before cancer can develop (Fig. 19.6).

An abnormal *RET* gene, which predisposes an individual to thyroid cancer, can be passed from parent to child. *RET* is a proto-oncogene known to be inherited in an autosomal dominant manner—only one mutation is needed to increase a predisposition to cancer. The remainder of the mutations necessary for a thyroid cancer to develop are acquired (not inherited).

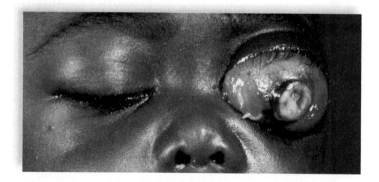

Figure 19.6   Inherited retinoblastoma.
A child is at risk for an eye tumor when a mutated copy of the *RB* gene is inherited, even though a second mutation in the normal copy is required before the tumor develops.

Environmental Carcinogens

A **mutagen** is an agent that causes mutations. A simple laboratory test, called an Ames test, is capable of testing if a substance is mutagenic. A **carcinogen** is a chemical that causes cancer, for example, by being mutagenic. Some carcinogens cause only initiation; others cause initiation and promotion.

Heredity can predispose a person to cancer, but whether it develops or not depends on environmental mutagens, such as the ones we will be discussing.

Radiation

Ionizing radiation, such as in ultraviolet light, radon gas, nuclear fuel, and X-rays, is capable of affecting DNA and causing mutations. Ultraviolet radiation in sunlight and tanning lamps are most likely responsible for the dramatic increases seen in skin cancer in the past several years. Today, at least six cases of skin cancer occur for every one case of lung cancer. Nonmelanoma skin cancers are usually curable through surgery, but melanoma skin cancer tends to metastasize and is responsible for 1–2% of total cancer deaths in the United States.

Another natural source of ionizing radiation is radon gas, which comes from the natural (radioactive) breakdown of uranium in soil, rock, and water. The Environmental Protection Agency recommends that every home be tested for radon because it is the second-leading cause of lung cancer in the United States. The combination of radon gas and smoking cigarettes can be particularly dangerous. A vent pipe system and fan, which pulls radon from beneath the house and vents it to the outside, is the most common way to rid a house of radon after the house is constructed.

We are well aware of the damaging effects of a nuclear bomb explosion or accidental emissions from nuclear power plants. For example, more cancer deaths have occurred in the vicinity of the Chernobyl Power Station (in Ukraine), which suffered a terrible accident in 1986. Usually, however, diagnostic X-rays account for most of our exposure to artificial sources of radiation. The benefits of these procedures can far outweigh the possible risk, but it is still wise to avoid X-ray procedures that are not medically warranted.

Despite much publicity, scientists have not been able to show a clear relationship between cancer and the nonionizing radiation that comes from electric power lines, household appliances, and cellular telephones.

Organic Chemicals

Certain organic chemicals, particularly synthetic organic chemicals, have been found to be risk factors for cancer. We will take two examples: the organic chemicals that are in tobacco and those that are pollutants in the environment.

Tobacco Smoke   Tobacco smoke contains a number of organic chemicals that are known mutagens, including nitrosonor-nicotine, vinyl chloride, and benzo[a]pyrenes (a known suppressor of p53).The greater the number of cigarettes smoked per day and the earlier the habit starts, the more

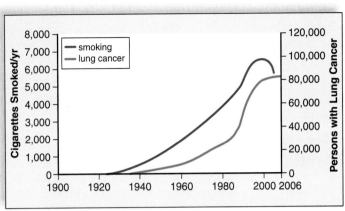

a. Men in the United States

b. Women in the United States

Figure 19.7   Smoking and cancer.
a. Note that as the incidence of smoking in men grew, so did the incidence of lung cancer. Lung cancer was rare in 1920, but as more men took up smoking, lung cancer became more common. **b.** As late as 1960, the incidence of lung cancer in women was low and so was the incidence of smoking. As women became smokers, the incidence of lung cancer in women also rose.

likely it is that cancer will develop. On the basis of data, such as shown in Figure 19.7, scientists estimate that about 80% of all cancers, including oral cancer and cancers of the larynx, esophagus, pancreas, bladder, kidney, and cervix, are related to the use of tobacco products. When smoking is combined with drinking alcohol, the risk of these cancers increases even more.

Passive smoking, or inhalation of someone else's tobacco smoke, is also dangerous and probably causes a few thousand deaths each year.

Pollutants   Being exposed to substances such as metals, dust, chemicals, or pesticides at work can increase the risk of cancer. Asbestos, nickel, cadmium, uranium, radon, vinyl chloride, benzidine, and benzene are well-known examples of carcinogens in the workplace. For example, inhaling asbestos fibers increases the risk of lung diseases, including cancer, and the cancer risk is especially high for asbestos workers who smoke.

Data show the incidence in soft tissue sarcomas (STS), malignant lymphomas, and non-Hodgkin lymphomas increases in farmers living in Nebraska and Kansas who used 2,4-D (a commonly used herbicidal agent) on crops and to clear weeds along railroad tracks.

Viruses

However, at least four types of DNA viruses—hepatitis B and C viruses, Epstein-Barr virus, and human papillomavirus—are directly believed to cause human cancers.

In China, almost all the people have been infected with the hepatitis B virus, and this correlates with the high incidence of liver cancer in that country. A combined vaccine is now available for hepatitis A and B. For a long time, circumstances suggested that cervical cancer was a sexually transmitted disease, and now human papillomaviruses (HPVs) are routinely isolated from cervical cancers. As Caitlin learned, a new vaccine is now available that will prevent new infections

of the HPV types that are more likely to cause cervical cancer. Burkitt lymphoma occurs frequently in Africa, where virtually all children are infected with the Epstein-Barr virus. In China, the Epstein-Barr virus is isolated in nearly all nasopharyngeal cancer specimens.

RNA-containing retroviruses, in particular, are known to cause cancers in animals (see AIDS supplement). In humans, the retrovirus HTLV-1 (human T-cell lymphotropic virus, type 1) has been shown to cause hairy cell leukemia. This disease occurs frequently in parts of Japan, the Caribbean, and Africa, particularly in regions where people are known to be infected with the virus. HIV, the virus that causes AIDS, and also Kaposi's sarcoma-associated herpesvirus (KSHV) are responsible for the development of Kaposi's sarcoma and certain lymphomas due to the suppression of proper immune system functions.

Dietary Choices

Nutrition is emerging as a way to help prevent cancer. The incidence of breast and prostate cancer parallels a high-fat diet, but so does obesity. The Health Focus on page 410 discusses how to protect yourself from cancer. The American Cancer Society recommends consumption of fruits and vegetables, whole grains instead of processed (refined) grains, and limited consumption of red meats (especially high-fat and processed meats). Moderate to vigorous activity for 30–45 minutes a day, five or more days a week, is also recommended.

☑ Check Your Progress 19.2

1. What evidence is there that you can inherit genes that lead to cancer?
2. What environmental carcinogens, in particular, are known to play a role in the development of cancer?
3. What protective steps can you take to reduce your risk of cancer?

Health Focus

Prevention of Cancer

Protective Behaviors

These behaviors help prevent cancer:

Don't smoke   Cigarette smoking accounts for about 30% of all cancer deaths. Smoking is responsible for 90% of lung cancer cases among men and 79% among women—about 87% altogether. People who smoke two or more packs of cigarettes a day have lung cancer mortality rates 15–25 times greater than those of nonsmokers. Smokeless tobacco (chewing tobacco or snuff) increases the risk of cancers of the mouth, larynx, throat, and esophagus.

Don't sunbathe   Almost all cases of basal-cell and squamous-cell skin cancers are considered sun-related. Further, sun exposure is a major factor in the development of melanoma, and the incidence of this cancer increases for people living near the equator.

Avoid alcohol   Cancers of the mouth, throat, esophagus, larynx, and liver occur more frequently among heavy drinkers, especially when accompanied by tobacco use (cigarettes or chewing tobacco).

Avoid radiation   Excessive exposure to ionizing radiation can increase cancer risk. Even though most medical and dental X-rays are adjusted to deliver the lowest dose possible, unnecessary X-rays should be avoided. Excessive radon exposure in homes increases the risk of lung cancer, especially in cigarette smokers. It is best to test your home and take the proper remedial actions.

Be tested for cancer   Do the shower check for breast cancer or testicular cancer. Have other exams done regularly by a physician. See Table 19.1.

Be aware of occupational hazards   Exposure to several different industrial agents (nickel, chromate, asbestos, vinyl chloride, etc.) and/or radiation increases the risk of various cancers.
Risk from asbestos is greatly increased when combined with cigarette smoking.

Be aware of postmenopausal hormone therapy   A new study conducted by the Women's Health Initiative found that estrogen-progestin combined therapy prescribed to ease the symptoms of menopause increased the incidence of breast cancer. And that the risk outweighed the possible decrease in the number of colorectal cancer cases.

Get vaccinated   Get vaccinated for HPV and hepatitis A and B.

The Right Diet

Statistical studies have suggested that people who follow certain dietary guidelines are less likely to have cancer. The following dietary guidelines greatly reduce your risk of developing cancer:

Avoid obesity   The risk of cancer (especially colon, breast, and uterine cancers) is 55% greater among obese women, and the risk of colon cancer is 33% greater among obese men, compared with people of normal weight.

Eat plenty of high-fiber foods   Studies have indicated that a high-fiber diet (whole-grain cereals, fruits, and vegetables) protects against colon cancer, a frequent cause of cancer deaths. It is worth noting that foods high in fiber also tend to be low in fat!

Increase consumption of foods that are rich in vitamins A and C   Beta-carotene, a precursor of vitamin A, is found in dark green, leafy vegetables; carrots; and various fruits. Vitamin C is present in citrus fruits. These vitamins are called antioxidants because in cells they prevent the formation of free radicals (organic ions having an unpaired electron) that can possibly damage DNA. Vitamin C also prevents the conversion of nitrates and nitrites into carcinogenic nitrosamines in the digestive tract.

Reduce consumption of salt-cured, smoked, or nitrite-cured foods   Salt-cured or pickled foods may increase the risk of stomach and esophageal cancers. Smoked foods, such as ham and sausage, contain chemical carcinogens similar to those in tobacco smoke. Nitrites are sometimes added to processed meats (e.g., hot dogs and cold cuts) and other foods to protect them from spoilage; as mentioned previously, nitrites are converted to nitrosamines in the digestive tract.

Include vegetables from the cabbage family in the diet   The cabbage family includes cabbage, broccoli, brussels sprouts, kohlrabi, and cauliflower. These vegetables may reduce the risk of gastrointestinal and respiratory tract cancers.

Be moderate in the consumption of alcohol   Risks of cancer development rise as the level of alcohol intake increases. The strongest cancer associations are with oral, pharyngeal, esophageal, and laryngeal cancer, but cancer of the breast and liver are also implicated. People who drink and smoke greatly enhance their risk for developing cancer.

19.3 Diagnosis of Cancer

The earlier a cancer is detected, the more likely it can be effectively treated. At present, physicians have ways to detect several types of cancer before they become malignant. But they are always looking for new and better detection methods. A growing number of researchers now believe the future of early detection lies in testing for the molecular fingerprints of a cancer. Several teams of scientists are working on blood, saliva, and urine tests to catch cancerous gene and protein patterns in these bodily fluids before a tumor develops. In the meantime, cancer is usually diagnosed by the methods discussed here.

Seven Warning Signs

At present, diagnosis of cancer before metastasis is difficult, although treatment at this stage is usually more successful. The American Cancer Society publicizes seven warning signals, which spell out the word CAUTION and which everyone should be aware of:

C hange in bowel or bladder habits
A sore that does not heal
U nusual bleeding or discharge
T hickening or lump in breast or elsewhere
I ndigestion or difficulty in swallowing
O bvious change in wart or mole
N agging cough or hoarseness

Keep in mind that these signs do not necessarily mean that you have cancer. However, they are an indication that something is wrong and a medical professional should be consulted. Unfortunately, some of these symptoms are not obvious until cancer has progressed to one of its later stages.

Routine Screening Tests

Self-examination, followed by examination by a physician, can help detect the presence of cancer. For example, the ABCDs of melanoma (the most serious form of skin cancer, Fig. 19.8) is helpful, as is the self-examination for breast and testicular cancers, which are discussed in the Health Focus on page 412.

An aim of medicine is to develop tests for cancer that are relatively easy to do, cost little, and are fairly accurate. So far, only the Pap test for cervical cancer fulfills these three requirements. A physician merely takes a sample of cells from the cervix, which are then examined microscopically for signs of abnormality (see Fig. 19.1). Even though Caitlin intends to get the sequence of three injections of the new HPV vaccine, she should still get regular Pap tests because (1) the vaccine will not protect against all types of HPV that cause cervical cancer, and (2) the vaccine does not protect

against any HPV infections she may already have if she is sexually active. Regular Pap tests are credited with preventing over 90% of deaths from cervical cancer. Screening for colon cancer also depends upon three types of testing. A digital rectal examination performed by a physician is actually of limited value because only a portion of the rectum can be reached by finger. With flexible sigmoidoscopy, the second procedure, a much larger portion of the colon can be examined by using a thin, pliable, lighted tube. Finally, a stool blood test (fecal occult blood test) consists of examining a stool sample to detect any hidden blood. The sample is smeared on a slide, and a chemical is added that changes color in the presence of hemoglobin. This procedure is based on the supposition that a cancerous polyp bleeds, although

A = Asymmetry, one half the mole does not look like the other half.

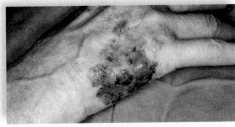

B = Border, irregular scalloped or poorly circumscribed border.

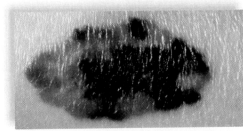

C = Color, varied from one area to another; shades of tan, brown, black, or sometimes white, red, or blue.

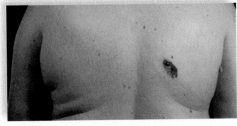

D = Diameter, larger than 6 mm (the diameter of a pencil eraser).

Figure 19.8 Detecting melanoma.
Suspicion of melanoma can begin by discovering a mole that has one or more of these characteristics.

✚ Health Focus

Shower Check for Cancer

The American Cancer Society urges women to do a breast self-exam and men to do a testicle self-exam every month. Breast cancer and testicular cancer are far more curable if found early, and we must all take on the responsibility of checking for one or the other.

Shower Self-Exam for Women

1. Check your breasts for any lumps, knots, or changes about one week after your period.
2. Place your right hand behind your head. Move your *left* hand over your *right* breast in a circle. Press firmly with the pads of your fingers (Fig. 19A). Also check the armpit.
3. Now place your left hand behind your head and check your *left* breast with your *right* hand in the same manner as before. Also check the armpit.
4. Check your breasts while standing in front of a mirror right after you do your shower check. First, put your hands on your hips and then raise your arms above your head (Fig. 19B). Look for any changes in the way your breasts look: dimpling of the skin, changes in the nipple, or redness or swelling.
5. If you find any changes during your shower or mirror check, see your doctor right away.

You should know that the best check for breast cancer is a mammogram. When your doctor checks your breasts, ask about getting a mammogram.

Shower Self-Exam for Men

1. Check your testicles once a month.
2. Roll each testicle between your thumb and finger as shown in Figure 19C. Feel for hard lumps or bumps.
3. If you notice a change or have aches or lumps, tell your doctor right away so he or she can recommend proper treatment.

Cancer of the testicles can be cured if you find it early. You should also know that prostate cancer is the most common cancer in men. Men over age 50 should have an annual health checkup that includes a prostate examination.

Information provided by the American Cancer Society. Used by permission.

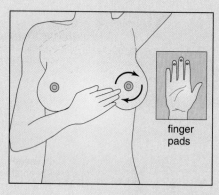

Figure 19A Shower check for breast cancer.
finger pads

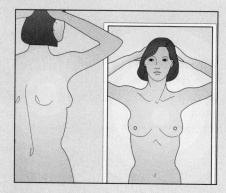

Figure 19B Mirror check for breast cancer.

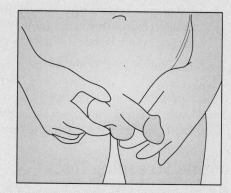

Figure 19C Shower check for testicular cancer.

some polyps do not bleed, and bleeding is not always due to a polyp. Therefore, the percentage of false negatives and false positives is high. All positive tests are followed up by a colonoscopy, an examination of the entire colon, or by X-ray after a barium enema. If the colonoscopy detects polyps, they can be destroyed by laser therapy. Blood tests can be used to detect leukemia and urinalysis for the diagnosis of bladder cancer.

Breast cancer is not as easily detected, but three procedures are recommended. First, every woman should do a monthly breast self-examination. But this is not a suffi-

cient screen for breast cancer. Therefore, all women should have an annual physical examination, especially women above age 40, when a physician will do the same procedure. However, even then, this type of examination may not detect lumps before metastasis has already taken place. That is the goal of the third recommended procedure, *mammography,* which is an X-ray study of the breast (Fig. 19.9). However, mammograms do not show all cancers, and new tumors may develop in the interval between mammograms. The objective is that a mammogram will reveal a lump that is too small to be felt and at a time

Figure 19.9
Mammogram.
An X-ray image of the breast can find tumors too small to be felt.

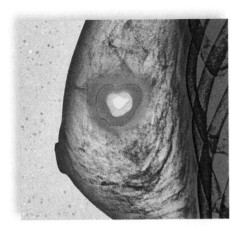

raphy (or CAT scan) uses computer analysis of scanning X-ray images to create cross-sectional pictures that portray a tumor's size and location (see Fig. 2.3). Magnetic resonance imaging (MRI) is another type of imaging technique that depends on computer analysis. MRI is particularly useful for analyzing tumors in tissues surrounded by bone, such as tumors of the brain or spinal cord. A radioactive scan obtained after a radioactive isotope is administered can reveal any abnormal isotope accumulation due to a tumor. During ultrasound, echoes of high-frequency sound waves directed at a part of the body are used to reveal the size, shape, and location of tissue masses. Ultrasound can confirm tumors of the stomach, prostate, pancreas, kidney, uterus, and ovary.

when the cancer is still highly curable. Table 19.1 outlines when routine screening tests should be done for various cancers, including breast cancer.

The diagnosis of cancer in other parts of the body may involve other types of imaging. Computerized axial tomog-

Aside from various imaging procedures, a diagnosis of cancer can be confirmed without major surgery by performing a *biopsy* or viewing body parts. Needle biopsies allow removal of a few cells for examination, and sophisticated techniques, such as laparoscopy, permit viewing of body parts.

Table 19.1	Recommendations for the Early Detection of Cancer in Average-Risk Asymptomatic People		
Cancer	**Population**	**Test or Procedure**	**Frequency**
Breast	Women, age ≥ 20 years	Breast self-examination Clinical breast examination and mammography	Monthly, starting at age 20 years; see Health Focus, p. 412. For women in 20s and 30s, every 3 years; aged 40 years and over, preferably annually.
Colorectal	Men and women, age ≥ 50 years	Fecal occult blood test (FOBT) or fecal immunochemical test (FIT)*, *or* FOBT (or FIT) and flexible sigmoidoscopy *or* double contrast barium enema *or* colonoscopy	Annually, starting at age 50 years. Annually, starting at age 50 years. Every 5 years, starting at age 50 years. Every 10 years, starting at age 50 years.
Prostate	Men, age ≥ 50 years	Digital rectal examination and prostate-specific antigen test (PSA)	Annually, starting at age 50 years, for men who have a life expectancy of at least 10 more years.
Cervical	Women, age ≥ 18 years	Pap test	3 years after vaginal intercourse begins but no later than 18 years; after age 30 years, every 2 to 3 years if three normal pap tests in a row. Women age 70 years may choose to stop cervical cancer screening if three normal pap tests in a row.
Endometrial	Women, after menopause	Report any unexpected bleeding or spotting to a physician.	
Testicular	Men, age 20 years	Testicle, self-examination	Monthly, starting at age 20 years, see Health Focus, p. 412..
Other cancers	Men and women, age 20 years	On the occasion of a periodic health examination, a cancer-related checkup should include examination for other cancers. See Health Focus, p. 410, for counseling about tobacco, sun exposure, diet and nutrition, risk factors, and environmental and occupational exposures.	

*FOBT as it is sometimes done in physicians' offices, with the single stool sample collected on a fingertip during a digital rectal examination, is not an adequate substitute for the recommended at-home procedure of collecting two samples from three consecutive specimens. Toilet-bowl FOBT tests also are not recommended. FIT is more patient-friendly, and are likely to be equal or better in sensitivity and specificity. *2006 ACS Guidelines for Early Cancer Detection*

Tumor Marker Tests

Tumor marker tests are blood tests for tumor antigens/antibodies. They are possible because tumors release substances that provoke an antibody response in the body. For example, if an individual has already had colon cancer, it is possible to use the presence of an antigen called *CEA* (for *carcinoembryonic antigen*) to detect any relapses. When the CEA level rises, additional tumor growth has occurred.

There are also tumor marker tests that can be used as an adjunct procedure to detect cancer in the first place. They are not reliable enough to count on solely, but in conjunction with physical examination (see page 413) and ultrasound, they are considered useful. For example, there is a *prostate-specific antigen (PSA) test* for prostate cancer, a *CA-125* test for ovarian cancer, and an *alpha-fetoprotein (AFP) test* for liver tumors.

Genetic Tests

Tests for genetic mutations in proto-oncogenes and tumor-suppressor genes are making it possible to detect the likelihood of cancer before the development of a tumor. Tests are available that signal the possibility of colon, bladder, breast, and thyroid cancers, as well as melanoma. Physicians now believe that a mutated *RET* gene means that thyroid cancer is present or may occur in the future, and a mutated *p16* gene appears to be associated with melanoma. Genetic testing can also be used to determine if cancer cells still remain after a tumor has been removed.

A genetic test is also available for the presence of *BRCA1* (breast cancer gene 1). A woman who has inherited this gene can choose to either have prophylactic surgery or to be frequently examined for signs of breast cancer. Physicians can use microsatellites to detect chromosomal deletions that accompany bladder cancer. Microsatellites are small regions of DNA that always have di-, tri-, or tetranucleotide repeats. They compare the number of nucleotide DNA repeats in a lymphocyte microsatellite with the number in a microsatellite of a cell found in urine. When the number of repeats is less in the cell from urine, a bladder tumor is suspected.

Telomerase, you will recall, is the enzyme that keeps telomeres a constant length in cells. The gene that codes for telomerase is turned off in normal cells but is active in cancer cells. Therefore, if the test for the presence of telomerase is positive, the cell is cancerous.

✓ Check Your Progress 19.3

1. What routine screening tests are available to detect and diagnose cancer?

2. What other types of tests are available to detect and diagnose cancer even before the routine tests indicate the presence of cancer?

19.4 Treatment of Cancer

Certain therapies for cancer have been available for some time. Other methods of therapy are in clinical trials, and if they prove successful, will become more generally available in the future.

Standard Therapies

Surgery, radiation, and chemotherapy are the standard methods of cancer therapy.

Surgery

Surgery alone is sufficient for cancer in situ. But because there is always the danger that some cancer cells were left behind, surgery is often preceded by and/or followed by radiation therapy.

Radiation Therapy

Ionizing radiation causes chromosomal breakage and cell cycle disruption. Therefore, dividing cells, such as cancer cells, are more susceptible to its effects than other cells. Powerful X-rays or gamma rays can be administered through an externally applied beam or, in some instances, by implanting tiny radioactive sources into the patient's body. Cancer of the cervix and larynx, early stages of prostate cancer, and Hodgkin disease are often treated with radiation therapy alone.

Although X-rays and gamma rays are the mainstays of radiation therapy, protons and neutrons also work well. Proton beams can be aimed at the tumor like a rifle bullet hitting the bull's-eye of a target.

Side effects of radiation therapy greatly depend upon which part of the body is being irradiated and how much radiation is used. Examples of short-term side effects, which in most cases are temporary, include diarrhea; dry, red, or irritated skin, including blistering burns, at the treatment site; dry mouth; fatigue or weakness; hair loss at the treatment site, which in some situations can be permanent; and nausea.

Chemotherapy

Radiation is localized therapy, while chemotherapy is a way to catch cancer cells that have spread throughout the body (Fig. 19.10 and Fig. 19.11). One of chemotherapy's main advantages is that—unlike radiation, which treats only the part of the body exposed to the radiation—chemotherapy treats the entire body. As a result, any cells that may have escaped from where the cancer originated are treated. Most chemotherapeutic drugs kill cells by damaging their DNA or interfering with DNA synthesis. The hope is that all cancer cells will be killed, while leaving untouched enough normal cells to allow the body to keep functioning. Combining drugs that have different actions at the cellular level may help destroy a greater number of cancer cells and might reduce the risk of the cancer developing resistance to one

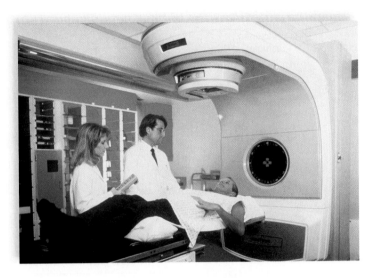

Figure 19.10 X-ray therapy for cancer.
Most people who receive radiation therapy for cancer have external radiation on an outpatient basis. This type of therapy is delivered by a machine to a specific part of the body.

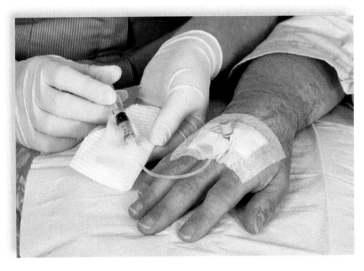

Figure 19.11 Chemotherapy.
The intravenous route is the most common, allowing chemotherapy drugs to spread quickly throughout the entire body by way of the bloodstream.

particular drug. What chemicals are used is generally based on the type of cancer and the patient's age, general health, and perceived ability to tolerate potential side effects. Some of the types of chemotherapy medications commonly used to treat cancer include

Alkylating agents. These medications interfere with the growth of cancer cells by blocking the replication of DNA.

Antimetabolites. These drugs block the enzymes needed by cancer cells to live and grow.

Antitumor antibiotics. These antibiotics—different from those used to treat bacterial infections—interfere with DNA, blocking certain enzymes and cell division and changing cell membranes.

Mitotic inhibitors. These drugs inhibit cell division or hinder certain enzymes necessary in the cell reproduction process.

Nitrosoureas. These medications impede the enzymes that help repair DNA.

Whenever possible, chemotherapy is specifically designed for the particular cancer. For example, in some cancers, a small portion of chromosome 9 is missing, and therefore, DNA metabolism differs in the cancerous cells compared with normal cells. Specific chemotherapy for the cancer can exploit this metabolic difference and destroy the cancerous cells.

One drug, *taxol*, extracted from the bark of the Pacific yew tree, was found to be particularly effective against advanced ovarian cancers, as well as breast, head, and neck tumors. Taxol interferes with microtubules needed for cell division. Now, chemists have synthesized a family of related drugs, called taxoids, which may be more powerful and have fewer side effects than taxol itself.

Certain types of cancer, such as leukemias, lymphomas, and testicular cancer, are now successfully treated by combination chemotherapy alone. The survival rate for children with childhood leukemia is 80%. Hodgkin disease, a lymphoma, once killed two out of three patients. Now, combination therapy, using four different drugs, can wipe out the disease in a matter of months in three out of four patients, even when the cancer is not diagnosed immediately. In other cancers—most notably, breast and colon cancer—chemotherapy can reduce the chance of recurrence after surgery has removed all detectable traces of the disease.

Chemotherapy sometimes fails because cancer cells become resistant to one or several chemotherapeutic drugs, a phenomenon called multidrug resistance. This occurs because all the drugs are capable of interacting with a plasma membrane carrier that pumps them out of the cell. Researchers are testing drugs known to poison the pump in an effort to restore the efficacy of the drugs. Another possibility is to use combinations of drugs with nonoverlapping patterns of toxicity, because cancer cells can't become resistant to many different types at once.

Bone marrow transplants are sometimes done in conjunction with chemotherapy. The red bone marrow contains large populations of dividing cells; therefore, red bone marrow is particularly prone to destruction by chemotherapeutic drugs. In bone marrow autotransplantation, a patient's stem cells are harvested and stored before chemotherapy begins. Quite high doses of radiation or chemotherapeutic drugs are then given within a relatively short time. This prevents multidrug resistance from occurring, and the treatment is more likely to catch each and every cancer cell. Then, the stored stem cells, which are needed to produce blood cells, are returned to the patient by injection. They automatically make their way to bony cavities, where they initiate blood cell formation.

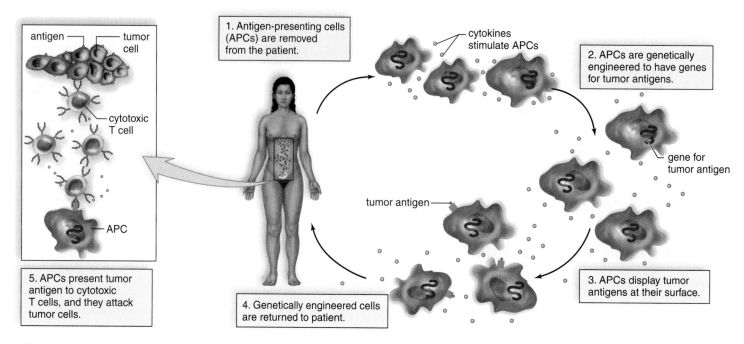

Figure 19.12   Immunotherapy.
1. Antigen presenting cells (APCs) are removed and 2. are genetically engineered to 3. display tumor antigens. 4. After these cells are returned to the patient, 5. they present the antigen to cytotoxic T cells, which then kill tumor cells.

Newer Therapies

Several therapies are now in clinical trials and are expected to be increasingly used to treat cancer.

Immunotherapy

When cancer develops, the immune system has failed to dispose of cancer cells, even though they bear antigens that make them different from the body's normal cells. A vaccine, called Melacine, which contains broken melanoma cells from two different sources is under investigation against melanoma.

Another idea is to use immune cells, genetically engineered to bear the tumor's antigens (Fig. 19.12). When these cells are returned to the body, they produce cytokines and present the antigen to cytotoxic T cells, which then go forth and destroy tumor cells in the body.

Passive immunotherapy is also possible. Monoclonal antibodies are antibodies of the same type because they are produced by the same plasma cell (see Fig. 7.17). Some monoclonal antibodies are designed to zero in on the receptor proteins of cancer cells. To increase the killing power of monoclonal antibodies, they are linked to radioactive isotopes or chemotherapeutic drugs. It is expected that soon they will be used as initial therapies, in addition to chemotherapy.

p53 Gene Therapy

Researchers believe that *p53* expression is needed for only 19 hours to trigger apoptosis, programmed cell death. And the *p53* gene seems to trigger cell death only in cancer cells—that is, elevating the *p53* level in a normal cell doesn't do any harm, possibly because apoptosis requires extensive DNA damage.

Ordinarily, when adenoviruses infect a cell, they first produce a protein that inactivates *p53*. In a cleverly designed procedure, investigators genetically engineered an adenovirus that lacks the gene for this protein. Now, the adenovirus can infect and kill only cells that lack a *p53* gene. Which cells are those? Tumor cells, of course. Another plus to this procedure is that the injected adenovirus spreads through the cancer, killing tumor cells as it goes. This genetically engineered virus is now in clinical trials.

Other Therapies

Many other therapies are now being investigated. Among them, drugs that inhibit angiogenesis are a proposed therapy under investigation. Antiangiogenic drugs confine and reduce tumors by breaking up the network of new capillaries in the vicinity of a tumor. A number of antiangiogenic compounds are currently being tested in clinical trials. Two highly effective drugs, called angiostatin and endostatin, have been shown to inhibit angiogenesis in laboratory animals and are expected to do the same in humans.

☑ Check Your Progress 19.4

1. What three types of therapy are presently the standard ways to treat cancer?

2. a. What is an active form of immunotherapy and (b) a passive form of immunotherapy for cancer?

3. What type of *p53* therapy is available?

4. What is the rationale for antiangiogenesis therapy?

Control of Tobacco

The 2004 Surgeon Generals' report on "The Health Consequences of Smoking" came to three major conclusions:

1. Smoking will diminish the health of the smoker and will damage nearly every major organ in the body (Fig. 19D).
2. Within minutes of quitting, a smoker's body will begin the healing process. Health improvements will continue for many years to come. These improvements include a decrease in the heart rate, improved circulation, and lowering the individual's risk of having a heart attack, lung cancer, or a stroke.
3. There is no difference to a smoker's health if they smoke low-tar or low-nicotine cigarettes.

The list of diseases caused by smoking continues to expand as more research is done. Smoking has been linked to an increased rate of wound infections following surgery, increased risk of cancer, cardiovascular disease, and respiratory disease, as well as reproductive problems. Over 4,000 different chemicals can be found in a cigarette, while cigarette smoke contains over 400 different toxic compounds. More than 440,000 Americans die every year due to smoking and smoking-related diseases. Smoking costs the United States approximately 97.2 billion dollars a year in medical-related expenses and lost productivity.

The Tobacco Industry Targets Young People

Statistics indicate that nine out of ten smokers start before they turn 18. Many teens start smoking in an attempt to look older, to feel more relaxed, or to try to fit in. When tempted to start smoking, many feel that they won't have a problem quitting. Once an individual starts, it often becomes very difficult to stop due to the addictiveness and eventual dependency a smoker will have upon the nicotine found in tobacco. Tobacco companies have been aware of the addictive nature of tobacco and health risks associated with tobacco for decades. Much of this information was hidden from the public in order to make a profit. A significant amount of the marketing done by the tobacco industry has been directed toward a younger audience. If they can get a child hooked, the company will most likely have a customer for life.

A majority of Americans are in favor of tobacco control plans that will help decrease the number of deaths and illnesses caused by tobacco. Most people also favor limiting a child's access to tobacco and believe that everyone should be able to breathe smoke-free air. In the latest attempt to hook kids on cigarettes, the tobacco industry has begun selling candy-flavored cigarettes that would appeal to a younger audience. Nine states have introduced legislation that banned the sale of candy-flavored cigarettes. The American Lung Association is calling on state and local policy makers to protect the public from the dangers of secondhand smoke by closing the existing loopholes in the smoke-free air laws. The demand for smoke-free public places is on the rise, and many local lawmakers have responded by passing smoke-free air laws at the state and local level.

Is Tax on Tobacco Helpful?

While billions of dollars are spent in medical costs associated with tobacco use, many states depend upon the tax revenue generated by tobacco sales. Every state imposes a cigarette tax, but the amount can range from 3 cents to as high as $2.46 per pack. The cigarette tax is often explained as a means of making up for the medical costs associated with tobacco use. Many lawmakers feel a high cigarette tax will also help discourage people from smoking. The cigarette tax not only supports the medical costs associated with tobacco but will fund other state-budget purposes. Many smokers in states that have a high cigarette tax will often make their purchases across state lines if the neighboring state has a lower tax, or over the Internet where there is no sales tax. It is estimated that over $1 billion is lost in state tax revenues to Internet sales and trips across state lines.

How to Quit

Smoking has been identified as the single most preventable cause of death and disease in the United States. There is a variety of smoking cessation programs available for anyone serious about quitting smoking. Plans range from nicotine patches to smoking cessation support groups. In order to be successful in quitting, smokers must be willing to recognize their addiction and have the willpower to overcome it.

Decide Your Opinion

1. Who should pay the medical bills associated with smoking—the individual, the tobacco industry, or the government?
2. With all of the information available about the dangers of tobacco, is it fair to sue the tobacco industry if someone becomes ill as the result of tobacco use?
3. Should the government prevent the sales of tobacco, or is it up to the individual not to use a cancer-causing substance?

Figure 19D Another life lost to lung cancer is remembered.

Summarizing the Concepts

19.1 Cancer Cells

Certain characteristics are common to cancer cells. Cancer cells

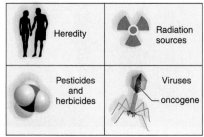

malignant tumor

- lack differentiation and do not contribute to function.
- do not undergo apoptosis—they enter the cell cycle an unlimited number of times.
- form tumors and do not need growth factors to signal them to divide.
- gradually become abnormal—carcinogenesis is comprised of initiation, promotion, and progression.
- undergo angiogenesis (the growth of blood vessels to support them) and can spread throughout the body (metastasis).

Cancer Is a Genetic Disease

Cells become increasingly abnormal due to mutations in proto-oncogenes and tumor-suppressor genes.

In normal cells: The cell cycle functions normally.

- **Proto-oncogenes** promote cell cycle activity and restrain apoptosis. Proto-oncogenes can mutate into oncogenes.
- **Tumor-suppressor genes** restrain the cell cycle and promote apoptosis.

In cancer cells: The cell cycle is accelerated and occurs repeatedly.

- **Oncogenes** cause an unrestrained cell cycle and prevent apoptosis.
- **Mutated tumor-suppressor genes** cause an unrestrained cell cycle and prevent apoptosis.

Types of Cancer

Cancers are classified according to their places of origin.

- Carcinomas originate in epithelial tissues.
- Sarcomas originate in muscle and connective tissues.
- Leukemias originate in blood.
- Lymphomas originate in lymphatic tissue.

Certain body organs are more susceptible to cancer than others.

19.2 Causes and Prevention of Cancer

Development of cancer is determined by a person's genetic profile, plus exposure to environmental carcinogens.

Heredity | Radiation sources

Pesticides and herbicides | Viruses — oncogene

Some sources contributing to development of cancer

- Cancers that run in families are most likely due to the inheritance of mutated genes (e.g., breast cancer, retinoblastoma tumor).

- Certain environmental factors are carcinogens (e.g., UV radiation, tobacco smoke, pollutants).
- Industrial chemicals (e.g., pesticides and herbicides) are carcinogenic.
- Certain viruses (e.g., hepatitis B and C, human papillomavirus, and Epstein-Barr virus) cause specific cancers.

19.3 Diagnosis of Cancer

The earlier a cancer is diagnosed, the more likely it can be effectively treated. Tests for cancer include

- Pap test for cervical cancer.
- mammogram for breast cancer.
- tumor marker tests—blood tests that detect tumor antigens/antibodies.
- tests for genetic mutations of oncogenes and tumor-suppressor genes.
- biopsy and imaging—used to confirm the diagnosis of cancer.

C hange in bowel or bladder habits
A sore that does not heal
U nusual bleeding or discharge
T hickening or lump in breast or elsewhere
I ndigestion or difficulty in swallowing
O bvious change in wart or mole
N agging cough or hoarseness

19.4 Treatment of Cancer

Surgery, radiation, and chemotherapy are traditional methods of treating cancer. Other methods include

- chemotherapy involving bone marrow transplants.
- immunotherapy.
- *p53* gene therapy (one type of *p53* gene therapy ensures that cancer cells undergo apoptosis).
- other therapies, such as inhibitory drugs for angiogenesis and metastasis, which are being investigated.

Understanding Key Terms

Match the key terms to these definitions.

a. _____ End of chromosome that prevents it from binding to another chromosome; normally get shorter with each cell division.

b. ——————— Environmental agent that contributes to the development of cancer.

c. ——————— Normal gene involved in cell growth and differentiation that becomes an oncogene through mutation.

d. ——————— Formation of new blood vessels, such as a capillary network.

e. ——————— Spread of cancer from the place of origin to throughout the body; caused by the ability of cancer cells to migrate and invade tissues.

Testing Your Knowledge of the Concepts

1. List and briefly discuss the seven characteristics of cancer cells that cause them to be abnormal. (pages 404–5)

2. What are the roles of the two genes that mutate causing cancer to develop? (page 406)

3. What are the four types of cancer based on place of origin? (page 407)

4. What are the top four types of new cancer cases in males? In females? (page 407)

5. What are the general causes of cancer? What are several types of environmental mutagens? (page 408–9)

6. What are the seven warning signals of cancer? (page 411)

7. List and describe three tests designed to detect cancer. (pages 411–14)

8. Describe three standard therapies for cancer treatment. (pages 414–15)

9. Describe some newer therapies that may be successful in cancer treatment. (page 416)

10. Whereas ————— stimulate the cell cycle, ————— inhibit the cell cycle.
    a. tumor-suppressor genes, oncogenes
    b. oncogenes, tumor-suppressor genes
    c. proto-oncogenes, oncogenes
    d. proto-oncogenes, tumor-suppressor genes

11. Growth factors lead to
    a. increased cell division.
    b. the functioning of cyclin proteins.
    c. progression through the cell cycle.
    d. All of these are correct.

12. Which of these is not true of the gene *p53*?
    a. Mutations of both proto-oncogenes and tumor-suppressor genes lead to inactivity of *p53*.
    b. Normally, *p53* functions to stop the cell cycle and initiate repair enzymes, when necessary.
    c. Normally, *p53* restores the length of telomeres.
    d. Cancer cells shut down the activity of *p53*, even though it may be present.

13. Which association is incorrect?
    a. proto-oncogenes—code for cyclin and proteins that inhibit the activity of *p53*
    b. oncogenes—"grain of function" genes
    c. mutated tumor-suppressor genes—code for cyclin and proteins that inhibit the activity of *p53*
    d. mutated tumor-suppressor genes—"loss of function" mutations
    e. Both a and c are incorrect.

14. To stimulate the immune system to fight cancer, which type cells is genetically engineered?
    a. Macrophages because they will devour cancer cells.
    b. Tumor cells because they can be engineered to display more antigens than normal.
    c. Antigen-presenting cells because they can present tumor antigens to cytotoxic T cells.
    d. Any type of cell can present tumor antigens to cytotoxic T cells.
    e. All of these are correct.

15. Following each cell cycle, telomeres
    a. get longer.   c. return to the same length.
    b. get shorter.   d. bind to cyclin proteins.

16. Bone marrow transplants are done in conjunction with chemotherapy because
    a. Smoking and pollutants can kill bone marrow stem cells.
    b. Blood cells will be engineered to have cancer antigens.
    c. This procedure is no longer done.
    d. Chemotherapy sometimes kills bone marrow stem cells.
    e. All of these are correct.

17. Angiogenic growth factors function to
    a. stimulate the development of new blood vessels.
    b. activate tumor-suppressor genes.
    c. change proto-oncogenes into oncogenes.
    d. promote metastasis.

18. A tumor with cells that spread to secondary locations is referred to as
    a. benign.
    b. cancer in situ.
    c. malignant.
    d. encapsulated.

19. Which of the following is not a type of carcinogen?
    a. tobacco smoke
    b. radiation
    c. pollutants
    d. viruses
    e. All of these are carcinogens.

20. What type of cancer is associated with human papillomaviruses?
    a. breast   c. kidney
    b. cervical   d. lymphatic

21. Why is cancer called a genetic disease?
    a. Cancer is always inherited.
    b. Carcinogenesis is accompanied by mutations.
    c. Cancer causes mutations that are passed on to offspring.
    d. All of these are correct.

22. Leukemia is a form of cancer that affects
    a. lymphatic tissue.
    b. bone tissue.
    c. blood-forming cells.
    d. nervous system structures.

23. Which is the name of the tumor-suppressor gene that causes retinoblastoma?
    a. *p21*
    b. *RB*
    c. *ras*
    d. *TGF-b*

24. What is the most common form of cancer in men?
    a. colon
    b. lung
    c. prostate
    d. testicular

25. Concerning the causes of cancer, which one is incorrect?
    a. Genetic mutations cause cancer.
    b. Genetic mutations can be caused by environmental influences, such as radiation, organic chemicals, and viruses.
    c. An active immune system, diet, and exercise can help prevent cancer.
    d. Heredity cannot be a cause of cancer.

26. Which of the following is not a warning signal for cancer?
    a. sore that does not heal
    b. change in bowel or bladder habits
    c. nagging cough or hoarseness
    d. shortness of breath or fatigue

27. Which of these tests for the particular cancer is mismatched?
    a. breast cancer—mammogram
    b. lung cancer—X-ray
    c. cervical cancer—Pap test
    d. prostate cancer—CA-125 test

28. Following a biopsy, what does a doctor look for to diagnose cancer?
    a. He looks for abnormal-appearing cells.
    b. He sees if the cells can divide.
    c. He sees if the cells will respond to growth factors.
    d. He sees if the chromosomes have telomeres.

29. Most chemotherapeutic drugs kill cells by
    a. producing pores in plasma membranes.
    b. interfering with protein synthesis.
    c. interfering with cellular respiration.
    d. interfering with DNA and/or enzymes.

30. Multidrug resistance to chemotherapeutic drugs occurs because
    a. cancer cells can use plasma membrane carriers to pump drugs out of the cell.
    b. one drug may interfere with the activity of another drug.
    c. using several drugs at once will overtax the patient's immune system.
    d. using several drugs at once decreases the effectiveness of radiation therapy.

31. *p53* gene therapy
    a. triggers cytotoxic T cells to destroy tumor cells.
    b. triggers apoptosis in cancer cells.
    c. produces monoclonal antibodies against the tumor cells.
    d. reduces tumors by breaking up their blood vessels.

Thinking Critically About the Concepts

Caitlin, from the opening story, is trying to live a cancer free life. One way that she can accomplish this is by having the vaccine for HPV.

1. Barbara, from the opening story in Chapter 4, had her cancerous bladder replaced by a "bladder" made from a section of her large intestine. Her doctor told her that the incidence of bladder cancer correlated highly with cigarette smoking.
   a. How would Barbara's smoking habit have caused her bladder cancer?
   b. What other lifestyle habits would have contributed to the occurrence of Barbara's cancer?

2. If women aren't vaccinated against HPV, how else could they protect themselves from infection by it?

3. The vaccine for HPV is recommended for females ages 11–26. There is some controversy over the vaccination of young girls because some people believe vaccinating them against a sexually transmitted disease will encourage them to engage in sexual activities.
   a. Why would the vaccine be recommended for young girls who may not be sexually active yet?
   b. What kind of immunity would women develop from getting vaccinated for HPV? You may need to revisit Chapter 7 on immunity to answer this question.

4. Why are lymph nodes surrounding a breast with an invasive cancer often removed when a mastectomy is performed?

5. What are some healthy lifestyle choices you could make that might prevent cancer later in your life?

Miss Lane typically used the characteristics of her mom, dad, and sister as examples while working genetics problems for her biology class. As a class, they'd just completed working on one-trait problems.

Hoping to challenge the class, Miss Lane told her students that she and her dad have blood type A. Both her mom and her sister have type O blood. "Can you figure out the genotypes of the people in my family?" Ryan, one of her star pupils, called out, "Your mom and sister are homozygous recessive because that's the only way to have type O blood." "Very good," replied Miss Lane. "So, if my mom is homozygous recessive and I have type A blood, what does my genotype have to be?" she continued. "You have to be heterozygous," shouted Michael. Miss Lane looked a little stunned for a moment because Michael was more apt to get in trouble than to get an answer correct. Then she replied with an "absolutely, good for you."

Miss Lane wrapped up the lesson on genetic inheritance with a problem about X-linked inheritance. "OK," she said to the class. "My dad is color-blind, my mom, sister, and I are not color-blind. What are the genotypes of my family members and me?" A few minutes later, students began raising their hands. Miss Lane paused for a moment before calling on Michael, who was vigorously waving his arm. "Does your fiancé know you're a carrier for color blindness?" was his answer. The class burst out laughing. Miss Lane replied quickly that it wasn't really necessary for her to tell her fiancé because color blindness is not life threatening.

CHAPTER CONCEPTS

20.1 Genotype and Phenotype
The genotype refers to the genes for a particular trait, and the phenotype refers to physical characteristics, such as hairline, blood type, color blindness, and even any cellular disorder.

20.2 One- and Two-Trait Inheritance
Each gene is represented by two alleles; some alleles are dominant and some are recessive. It is not possible to tell genotype by the phenotype when a person has one dominant and one recessive allele. Many disorders, including cellular disorders, are inherited in the same manner as normal dominant/recessive traits.

20.3 Beyond Simple Inheritance Patterns
There are other inheritance patterns beyond simple dominant/recessive ones. Some are especially affected by the environment, in addition to the genotype.

20.4 Sex-Linked Inheritance
Some traits are controlled by alleles on the X chromosome that have nothing to do with the gender of the person. The Y is blank for X-linked recessive alleles, and any on the X chromosome will be expressed. Therefore, males are more apt to have an X-linked disorder than are females.

20.1 Genotype and Phenotype

Genotype refers to the genes of the individual. Alternative forms of a gene having the same position (locus) on a pair of chromosomes and affecting the same trait are called **alleles.** It is customary to designate an allele by a letter, which represents the specific trait (characteristic) it controls; a **dominant allele** is assigned an uppercase (capital) letter, while a **recessive allele** is given the same letter but in lowercase. In humans, for example, unattached (free) earlobes are dominant over attached earlobes, so a suitable key would be *E* for unattached earlobes and *e* for attached earlobes.

Alleles occur in pairs; therefore, the individual normally has two alleles for a trait. Just as one of each pair of chromosomes is inherited from each parent, so too is one of each pair of alleles inherited from each parent. Indeed, the alleles occur in the same location, called their **locus** (plural, loci), on a pair of homologous chromosomes.

Figure 20.1 shows three possible fertilizations, the resulting genetic makeup of the zygote and, therefore, the individual. In the first instance, the chromosomes of both the sperm and the egg carry an *E*. Consequently, the zygote and subsequent individual have the alleles *EE*, which may be called a **homozygous dominant** genotype. A person with genotype *EE* obviously has unattached earlobes. The physical appearance of the individual—in this case, unattached earlobes—is called the **phenotype.** Notice in Figure 20.1 that the genotype (letters) and then the phenotype (description) are given after the drawing.

In the second fertilization, the zygote has received two recessive alleles (*ee*), and the genotype is called **homozygous recessive.** An individual with this genotype has the recessive phenotype, which is attached earlobes. In the third fertilization, the resulting individual has the alleles *Ee*, which is called a **heterozygous** genotype. A heterozygote shows the dominant phenotype; therefore, this individual has unattached earlobes.

How many dominant alleles must an individual inherit in order to have the dominant phenotype? These examples show that a dominant allele contributed from only one parent can bring about a particular dominant phenotype. How many recessive alleles must an individual inherit in order to have the recessive phenotype? A recessive allele must be received from both parents to bring about the recessive phenotype.

Phenotype

In our example, the phenotype was attached or unattached earlobes. From this you might get the impression that the phenotype has to be some trait that is easily observable. However, the phenotype can be any characteristic of the individual, including color blindness or a metabolic disorder such as the lack of an enzyme to metabolize the amino acid phenylalanine.

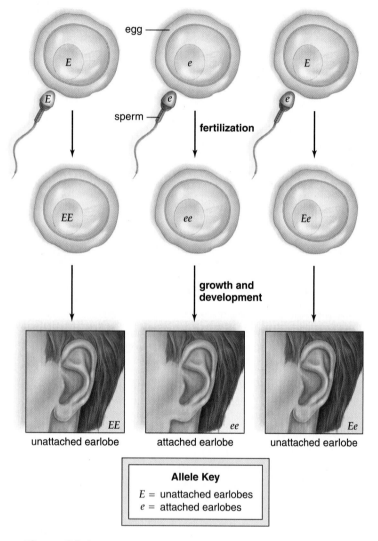

Allele Key

E = unattached earlobes
e = attached earlobes

Figure 20.1 **Genetic inheritance.**
Individuals inherit a minimum of two alleles for every characteristic of their anatomy and physiology. The inheritance of a single dominant allele (*E*) causes an individual to have unattached earlobes; two recessive alleles (*ee*) cause an individual to have attached earlobes. Notice that each individual receives one allele from the father (by way of a sperm) and one allele from the mother (by way of an egg).

✓ Check Your Progress 20.1

1. What is the difference between genotype and phenotype?

2. What are the three possible genotypes and the two possible phenotypes for a characteristic that is controlled by two alleles, one being dominant and the other recessive?

20.2 One- and Two-Trait Inheritance

In one-trait crosses, the inheritance of only one set of alleles is being considered, and in two-trait crosses, the inheritance of two sets of alleles is being considered. For both types of crosses, it will be necessary to determine the gametes of both individuals who are reproducing.

Forming the Gametes

During gametogenesis, the chromosome number is reduced. Whereas the individual has 46 chromosomes, a gamete has only 23 chromosomes. (If this did not happen, each new generation of individuals would have twice as many chromosomes as their parents.) Reduction of the chromosome number occurs when the homologous chromosomes separate as meiosis occurs. Since the alleles are on the chromosomes, they also separate during meiosis, and therefore, the gametes carry only one allele for each trait. *In the simplest of terms, no two letters in a gamete can be the same letter of the alphabet.*

If an individual carried the alleles *EE*, all the gametes would carry an *E* since that is the only choice. Similarly, if an individual carried the alleles *ee*, all the gametes would carry an *e*.

What if an individual were *Ee*? Half of the gametes would carry an *E* and half would carry an *e*. Figure 20.2 shows the genotypes and phenotypes for certain other traits in humans, and you can practice deciding what alleles the gametes would carry in order to produce these genotypes. If the genotype is *EeSs*, all combinations of any two different letters can be present. Therefore, *ES*, *eS*, *Es*, and *es* are all possible gametes.

☑ Check Your Progress 20.2a

1. For each of the following genotypes, give all possible gametes.

   a. *WW*

   b. *WWSs*

   c. *Tt*

   d. *Ttgg*

   e. *AaBb*

2. For each of the following, state whether a genotype or a gamete is represented.

   a. *D*

   b. *Ll*

   c. *Pw*

   d. *LlGg*

3. What is the genotype of the individual from the following crosses?

   a. *ES* × *es*

   b. *eS* × *eS*

a. Widow's peak: *WW* or *Ww*   b. Straight hairline: *ww*

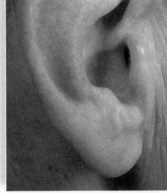
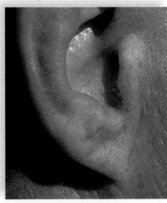

c. Unattached earlobes: *EE* or *Ee*   d. Attached earlobes: *ee*

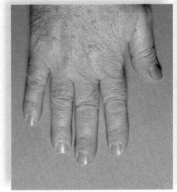
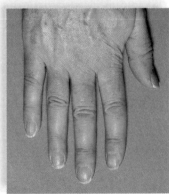

e. Short fingers: *SS* or *Ss*   f. Long fingers: *ss*

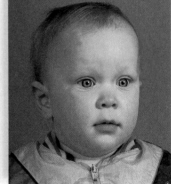

g. Freckles: *FF* or *Ff*   h. No freckles: *ff*

Figure 20.2   Common inherited traits in human beings.
The allele keys indicate which traits are dominant and which are recessive.

One-Trait Crosses

Many times, parents would like to know the chances of having a child with a certain genotype and, therefore, a certain phenotype. To illustrate, let us consider a cross involving freckles. In that way we will see why Miss Lane's students in the opening story could easily do one-trait problems. If both her parents were homozygous recessive, their genotype was *ff*. Obviously, in a cross of *ff* × *ff*, the children must also be *ff* and not have freckles.

To challenge you a bit more, let's consider the results when a homozygous dominant man with freckles reproduces with a woman with no freckles. Will the children of this couple have freckles? In solving the problem, we (1) will use the key provided in Figure 20.3 to indicate the genotype of each parent; (2) determine what the possible gametes are for each parent; (3) combine all possible gametes; and (4), finally, determine the genotypes and the phenotypes of all the offspring.

In the following diagram, the letters in the first row give the genotypes of the parents. Each parent has only one type of gamete with regard to freckles, and therefore, all the children have a similar genotype and phenotype. The children are heterozygous (*Ff*) and have freckles. When writing a heterozygous genotype, always put the capital letter first to avoid confusion.

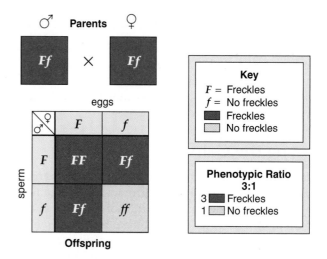

Figure 20.3   Monohybrid cross.
A Punnett square diagrams the results of a cross. When the parents are heterozygous, each child has a 75% chance of having the dominant phenotype and a 25% chance of having the recessive phenotype.

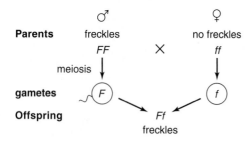

These children are **monohybrids;** that is, they are heterozygous with regard to one pair of alleles. If they reproduce with someone else of the same genotype, will their children have freckles? In this problem (*Ff* × *Ff*), each parent has two possible types of gametes (*F* or *f*), and we must ensure that all types of sperm have an equal chance to fertilize all possible types of eggs. One way to do this is to use a **Punnett square** (Fig. 20.3), in which all possible types of sperm are lined up vertically and all possible types of eggs are lined up horizontally (or vice versa), and every possible combination of gametes occurs within the squares.

After we determine the genotypes and the phenotypes of the offspring, we can first determine the genotypic and then the phenotypic ratio. The genotypic ratio is 1 *FF*: 2, *Ff*: 1 : *ff* or simply 1:2:1, but the phenotypic ratio is 3:1. Why?

Because three individuals will have freckles and one will not have freckles.

This 3:1 phenotypic ratio is always expected for a monohybrid cross when one allele is completely dominant over the other. The exact ratio is more likely to be observed if a large number of matings take place and if a large number of offspring result. Only then do all possible kinds of sperm have an equal chance of fertilizing all possible kinds of eggs. Naturally, we do not routinely observe hundreds of offspring from a single type of cross in humans. The best interpretation of Figure 20.3, in humans, is to say that each child has three chances out of four to have freckles, or one chance out of four to not have freckles.

It is important to realize that *chance has no memory;* for example, if two heterozygous parents already have three children with freckles and are expecting a fourth child, this child still has a 75% chance of having freckles and a 25% chance of not having freckles. Suppose a man and a women did not have freckles, what chance would children have for freckles? The answer would be 0% chance. How about their chance of not having freckles? The answer would be 100% chance. How about the cross *Ff* × *ff*? What are the chances of freckles? No freckles? In each instance, the child has a 50% chance for freckles and a 50% chance for no freckles.

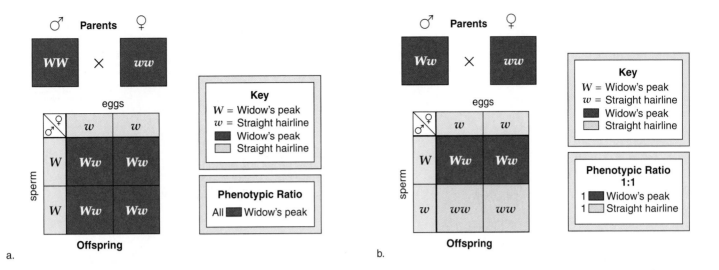

Figure 20.4 One-trait crosses.
This cross will determine if an individual with the dominant phenotype is homozygous or heterozygous. **a.** Because all offspring show the dominant characteristic, the individual is most likely homozygous, as shown. **b.** Because the offspring show a 1:1 phenotypic ratio, the individual is heterozygous, as shown.

Other One-Trait Crosses

It is not possible to tell by inspection if a person expressing a dominant allele is homozygous dominant or heterozygous. However, it is sometimes possible to tell by the results of a cross. For example, Figure 20.4 shows two possible results when a man with a widow's peak reproduces with a woman who has a straight hairline. If the man is homozygous dominant, all his children will have a widow's peak. If the man is heterozygous, each child has a 50% chance of having a straight hairline. The birth of just one child with a straight hairline indicates that the man is heterozygous.

Consider, also, that a person's parentage sometimes tells you that he is heterozygous. In Figure 20.4, each of the offspring with a widow's peak has to be heterozygous. Why? Because one of the parents was homozygous recessive, and therefore had to give each offspring a *w*. You will want to do all the practice problems in Check your Progress 20.2b in order to ensure that you can do one-trait genetics problems.

The Punnett Square and Probability

The Punnett square allows you to figure the chances or the probability that an offspring will have a particular genotype/phenotype. The two laws of probability are the product rule and the sum rule. The product rule says that the chance of two different events occurring together is the product (multiplication) of their chance occurring separately. The sum rule says the chance of any event that can occur in more than one way is the sum (addition) of the individual chances.

Notice that when you bring the allele(s) donated by the sperm and egg together into the same square, in all combinations, you are using the product rule because the father gives an offspring a certain chance of getting a particular allele and the mother gives a certain chance of getting an allele for the same trait. When you add up the results of bringing the various alleles together, you are using the sum rule.

✓ Check Your Progress 20.2b

1. A man with a widow's peak has a mother with a straight hairline. Widow's peak (*W*) is dominant over straight hairline (*w*). What is the genotype of the man?

2. Both a man and a woman are heterozygous for freckles. Freckles (*F*) are dominant over no freckles (*f*). What is the chance that their child will have freckles?

3. Both you and your sister or brother have attached earlobes, yet your parents have unattached earlobes. Unattached earlobes (*E*) are dominant over attached earlobes (*e*). What are the genotypes of your parents?

4. A father has dimples, the mother does not have dimples, and all five of their children have dimples. Dimples (*D*) are dominant over no dimples (*d*). Give the probable genotypes of all persons concerned.

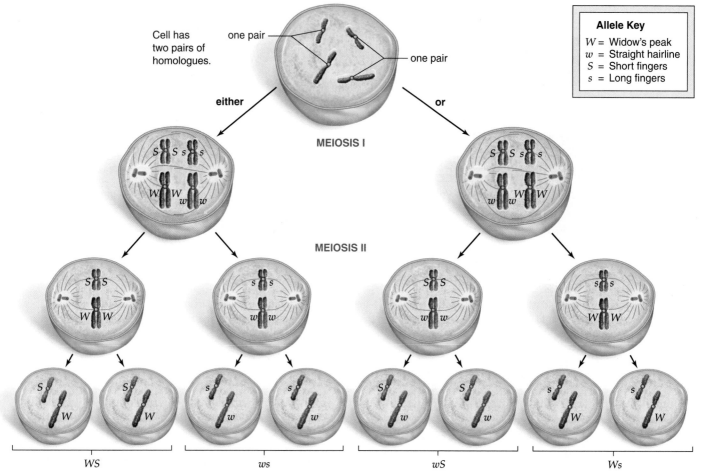

Allele Key

W = Widow's peak
w = Straight hairline
S = Short fingers
s = Long fingers

Figure 20.5   Meiosis and genetic diversity among gametes.
A cell has two pairs of homologous chromosomes (homologues), recognized by length, not color. The long pair of homologues carries alleles for type of hairline and the short pair of homologues carries alleles for finger length. The homologues, and the alleles they carry, align independently during meiosis. Therefore, all possible combinations of chromosomes and alleles occur in the gametes as shown in the last row of cells.

Two-Trait Crosses

Figure 20.5 allows you to relate the events of meiosis to the formation of gametes when a cross involves two traits. In the example given, a cell has two pairs of homologues, recognized by length—one pair of homologues is short, and the other is long. (The color signifies that we inherit chromosomes from our parents; one homologue of each pair is the "paternal" chromosome, and the other is the "maternal" chromosome.)

Because the homologues separate during meiosis I, each gamete receives one member from each pair of homologues. Because the homologues separate independently—it matters not which member of a pair goes into which gamete—all possible combinations of alleles occur in the gametes. In the simplest of terms, a gamete in Figure 20.5 will *receive one short and one long chromosome of either color*. Therefore, all possible combinations of chromosomes and alleles are in the gametes.

Specifically, assume that the alleles for two genes are on these homologues. The alleles S and s are on one pair of homologues, and the alleles W and w are on the other pair of homologues. First notice that because the homologues separate, a gamete will have either an S or an s and either a W or a w and never two of the same letter of the alphabet. Also, because the homologues align independently at the equator, either the paternal or maternal chromosome of each pair can face either pole.

Therefore, there are no restrictions as to which homologue goes into which gamete; a gamete can receive either an S or an s and either a W or a w in any combination. In the end, the gametes will collectively have all possible combinations of alleles. You should be able to transfer this information to any cross that involves two traits. In other words, the process of meiosis explains why a person with the genotype $EeFf$ would produce the gametes EF, ef, Ef, and eF in equal number.

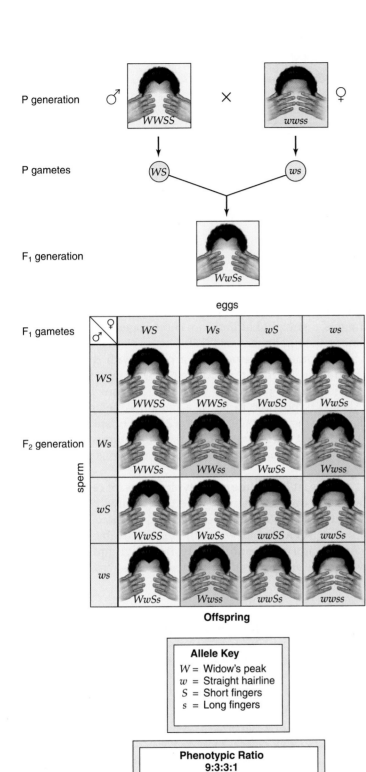

Allele Key

W = Widow's peak
w = Straight hairline
S = Short fingers
s = Long fingers

Phenotypic Ratio
9:3:3:1

9 ☐ Widow's peak, short fingers
3 ☐ Widow's peak, long fingers
3 ☐ Straight hairline, short fingers
1 ☐ Straight hairline, long fingers

Figure 20.6   Dihybrid cross.
Since each dihybrid can form four possible types of gametes, four different phenotypes occur among the offspring in the proportions shown.

The Dihybrid Cross

In the two-trait cross depicted in Figure 20.6, a person homozygous for widow's peak and short fingers (*WWSS*) reproduces with one who has a straight hairline and long fingers (*wwss*). The gametes for the *WWSS* parent must be *WS* and the gametes for the *wwss* parent must be *ws*. Therefore, the offspring will all have the genotype *WwSs* and the same phenotype (widow's peak with short fingers). This genotype is called a **dihybrid** because the individual is heterozygous in two regards: hairline and fingers.

When a dihybrid *WwSs* reproduces with another dihybrid that is *WwSs*, what gametes are possible? Each gamete can have only one letter of each kind in all possible combinations. Therefore, these are the gametes for both dihybrids: *WS*, *Ws*, *wS*, and *ws*.

A Punnett square makes sure that all possible sperm fertilize all possible eggs. If so, these are the expected phenotypic results:

  9 widow's peak and short fingers:
  3 widow's peak and long fingers:
  3 straight hairline and short fingers:
  1 straight hairline and long fingers.

This 9:3:3:1 phenotypic ratio is always expected for a dihybrid cross when simple dominance is present. We can use this expected ratio to predict the chances of each child receiving a certain phenotype. For example, the chance of getting the two dominant phenotypes together is 9 out of 16, and the chance of getting the two recessive phenotypes together is 1 out of 16.

Two-Trait Crosses and Probability

It is also possible to use the rules of probability we discussed on page 425 to predict the results of a dihybrid cross. For example, we know the probable results for two separate monohybrid crosses are as follows:

1. Probability of widow's peak = ¾
   Probability of short fingers = ¾
2. Probability of straight hairline = ¼
   Probability of long fingers = ¼

Using the product rule, we can calculate the probable outcome of a dihybrid cross as follows:

  Probability of widow's peak and short fingers =
    ¾ × ¾ = ⁹⁄₁₆
  Probability of widow's peak and long fingers =
    ¾ × ¼ = ³⁄₁₆
  Probability of straight hairline and short fingers =
    ¼ × ¾ = ³⁄₁₆
  Probability of straight hairline and long fingers =
    ¼ × ¼ = ¹⁄₁₆

In this way, the rules of probability tell us that the expected phenotypic ratio when all possible sperm fertilize all possible eggs is 9:3:3:1.

Other Two-Trait Crosses

It is not possible to tell by inspection whether an individual expressing the dominant alleles for two traits is homozygous dominant or heterozygous. But, if the individual reproduces with the homozygous recessive, it may be possible to tell. For example, if a man homozygous dominant for widow's peak and short fingers reproduces with a female who is homozygous recessive for both traits, then all his children will have the dominant phenotypes. However, if a man is heterozygous for both traits, then each child has a 25% chance of showing either one or both recessive traits. A Punnett square (Fig. 20.7) shows that the expected ratio is 1 wid-

Table 20.1	Phenotypic Ratios of Common Crosses
Genotypes	**Phenotypes**
Monohybrid X monohybrid	3:1 (dominant to recessive)
Monohybrid X recessive	1:1 (dominant to recessive)
Dihybrid X dihybrid	9:3:3:1 (9 both dominant, 3 one dominant, 3 other dominant, 1 both recessive)
Dihybrid X recessive	1:1:1:1 (all possible combinations in equal number)

ow's peak with short fingers: 1 widow's peak with long fingers: 1 straight hairline with short fingers: 1 straight hairline with long fingers, or 1:1:1:1.

For practical purposes, if a parent with the dominant phenotype, in either trait, has an offspring with the recessive phenotype, the parent has to be heterozygous for that trait. Also, it is possible to tell if a person is heterozygous by knowing the parentage. In Figure 20.7, no offspring showing a dominant phenotype is homozygous dominant for either trait. Why? Because the mother is homozygous recessive for that trait.

Table 20.1 gives the phenotypic results for certain crosses we have been studying. These crosses always give these phenotypic results, and therefore, it is not necessary to do a Punnett square to arrive at the results. To facilitate doing crosses, you will want to study Table 20.1 and understand why these are the results expected for these crosses.

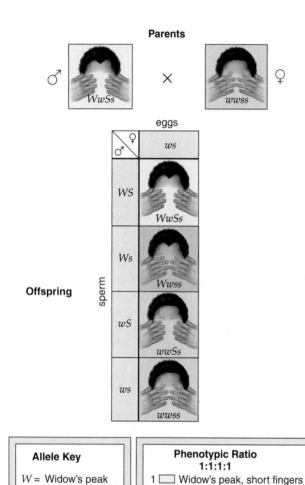

Figure 20.7 Two-traits cross.
The results of this cross indicate that the individual with the dominant phenotypes is heterozygous for both traits because some of the children are homozygous recessive for one or both traits. The chance of receiving any possible phenotype is 25%.

✓ Check Your Progress 20.2c

1. Attached earlobes and straight hairline are recessive. What genotype does a man with unattached earlobes and a widow's peak have if his mother has attached earlobes and a straight hairline?

2. What genotype do children have if one parent is homozygous recessive for earlobes and homozygous dominant for hairline, and the other is homozygous dominant for unattached earlobes and homozygous recessive for hairline?

3. If an individual from this cross reproduces with another of the same genotype, what are the chances that they will have a child with a straight hairline and attached earlobes?

4. A child who does not have dimples or freckles is born to a man who has dimples and freckles (both dominant) and a woman who does not. What are the genotypes of all persons concerned?

Family Pedigrees for Genetic Disorders

When a genetic disorder is autosomal dominant, an individual with the alleles *AA* or *Aa* will have the disorder. When a genetic disorder is autosomal recessive, only individuals with the alleles *aa* will have the disorder. Genetic counselors often construct pedigrees to determine whether a condition that runs in the family is dominant or recessive. A pedigree shows the pattern of inheritance for a particular condition. Consider these two possible patterns of inheritance:

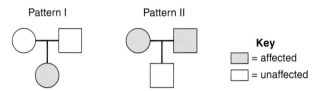

In both patterns, males are designated by squares and females by circles. Shaded circles and squares are affected individuals. A line between a square and a circle represents a union. A vertical line going downward leads, in these patterns, to a single child. (If there are more children, they are placed off a horizontal line.) Which pattern of inheritance do you suppose represents an autosomal dominant characteristic, and which represents an autosomal recessive characteristic?

Autosomal Recessive Disorder

In pattern I, the child is affected, but neither parent is; this can happen if the condition is recessive and the parents are *Aa*. Notice that the parents are **carriers** because they appear to be normal but are capable of having a child with a genetic disorder. Figure 20.8 shows a typical pedigree chart for a recessive genetic disorder. Other ways to recognize an autosomal recessive pattern of inheritance are also listed in the figure. Notice that when both parents are affected, all the children are affected. Why? Because the parents can pass on only recessive alleles for this condition.

It is important to realize that "chance has no memory;" therefore, each child born to heterozygous parents has a 25% chance of having the disorder. In other words, it is possible that if a heterozygous couple has four children, each child might have the condition.

Autosomal Dominant Disorder

In pattern II, the child is unaffected, but the parents are affected. Of the two patterns, this one shows a dominant pattern of inheritance. Because the condition is dominant, the parents can be *Aa* (heterozygous). The child inherited a recessive allele from each parent and, therefore, is unaffected. Figure 20.9 shows a typical pedigree for a dominant disorder. Other ways to recognize an autosomal dominant pattern of inheritance are also listed. When a disorder is dominant, an affected child must have at least one affected parent.

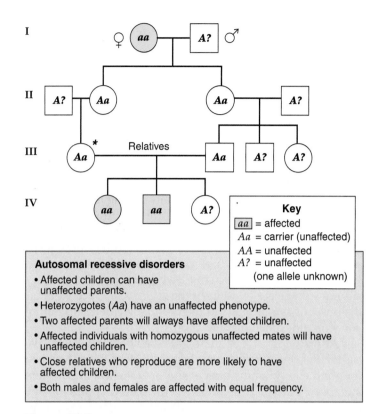

Autosomal recessive disorders
- Affected children can have unaffected parents.
- Heterozygotes (*Aa*) have an unaffected phenotype.
- Two affected parents will always have affected children.
- Affected individuals with homozygous unaffected mates will have unaffected children.
- Close relatives who reproduce are more likely to have affected children.
- Both males and females are affected with equal frequency.

Figure 20.8 **Pedigree for an autosomal recessive.**
The list gives ways to recognize an autosomal recessive disorder. How would you know that the individual at the asterisk is heterozygous?[1]

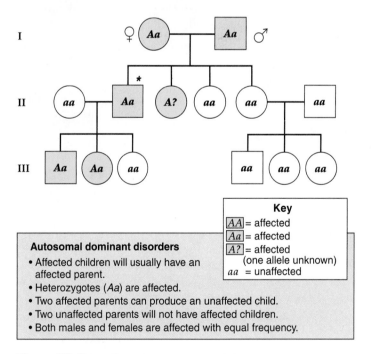

Autosomal dominant disorders
- Affected children will usually have an affected parent.
- Heterozygotes (*Aa*) are affected.
- Two affected parents can produce an unaffected child.
- Two unaffected parents will not have affected children.
- Both males and females are affected with equal frequency.

Figure 20.9 **Pedigree for an autosomal dominant.**
The list gives ways to recognize an autosomal dominant disorder. How would you know that the individual at the asterisk is heterozygous?[1]

[1] See Appendix A for answers.

Genetic Disorders of Interest

Medical genetics has traditionally focused on disorders caused by single gene mutations, and we will discuss a few of the better-known disorders.

Autosomal Recessive Disorders

Inheritance of two recessive alleles is required before an autosomal recessive disorder will appear.

Tay-Sachs Tay-Sachs disease is a well-known autosomal recessive disorder that occurs usually among Jewish people in the United States, most of whom are of central and eastern European descent. Tay-Sachs disease results from a lack of the enzyme hexosaminidase A (Hex A) and the subsequent storage of its substrate, a glycosphingolipid, in lysosomes. Lysosomes build up in many body cells, but the primary sites of storage are the cells of the brain, which accounts for the onset of symptoms and the progressive deterioration of psychomotor functions (Fig. 20.10).

At first, it is not apparent that a baby has Tay-Sachs disease. However, development begins to slow down between four and eight months of age, and neurological impairment and psychomotor difficulties then become apparent. The child gradually becomes blind and helpless, develops uncontrollable seizures, and eventually becomes paralyzed.

Cystic Fibrosis Cystic fibrosis is an autosomal recessive disorder that occurs among all ethnic groups, but it is the most common lethal genetic disorder among Caucasians in the United States. Research has demonstrated that chloride ions (Cl^-) fail to pass through a plasma membrane channel protein in the cells of these patients (Fig. 20.11). Ordinarily, after chloride ions have passed through the membrane, sodium ions (Na^+) and water follow. It is believed that lack of water is the cause of abnormally thick mucus in bronchial tubes and pancreatic ducts. In these children, the mucus in the bronchial tubes and pancreatic ducts is particularly thick and viscous, interfering with the function of the lungs and pancreas. To ease breathing, the thick mucus in the lungs has to be manually loosened periodically, but still the lungs become infected frequently. Clogged pancreatic ducts prevent digestive enzymes from reaching the small intestine, and to improve digestion, patients take digestive enzymes mixed with applesauce before every meal.

Phenylketonuria Phenylketonuria (PKU) is an autosomal recessive metabolic disorder that affects nervous system development. Affected individuals lack an enzyme that is needed for the normal metabolism of the amino acid phenylalanine, and therefore, it appears in the urine and the blood. Newborns are routinely tested in the hospital for elevated levels of phenylalanine in the blood. If elevated levels are detected, newborns will develop normally if they are placed on a diet low in phenylalanine, which must be continued until the brain is fully developed, around the age of seven, or else severe mental retardation develops. Some doctors recommend that the diet continue for life, but in any case, a

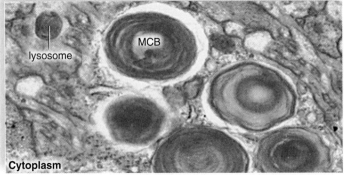

Figure 20.10 Tay-Sachs disease.
A cortical neuron in Tay-Sachs disease contains many membranous cytoplasmic bodies (MCB) stored in the cell because of a deficient lysosomal enzyme. The MCBs are made up of various types of lipids in specific proportions.

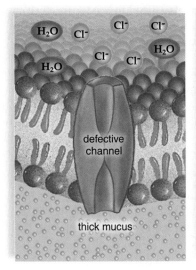

Figure 20.11 Cystic fibrosis.
Cystic fibrosis is due to a faulty protein that is supposed to regulate the flow of chloride ions into and out of cells through a channel protein.

pregnant woman with phenylketonuria must be on the diet in order to protect her unborn child from harm.

Sickle-Cell Disease **Sickle-cell disease** is an autosomal recessive disorder in which the red blood cells are not biconcave disks like normal red blood cells; they are irregular. In fact, many are sickle-shaped. The defect is caused by an abnormal hemoglobin that differs from normal hemoglobin by one amino acid in the protein globin. The single amino acid change causes hemoglobin molecules to stack up and form insoluble rods, and the red blood cells become sickle-shaped.

Because sickle-shaped cells can't pass along narrow capillary passageways as disk-shaped cells can, they clog the vessels and break down. This is why persons with sickle-cell disease suffer from poor circulation, anemia, and low resistance to infection. Internal hemorrhaging leads to further complications, such as jaundice, episodic pain in the abdomen and joints, and damage to internal organs.

Sickle-cell heterozygotes have sickle-cell traits in which the blood cells are normal unless they experience dehydration or mild oxygen deprivation. Still, at present, most experts believe that persons with the sickle-cell trait do not need to restrict their physical activity.

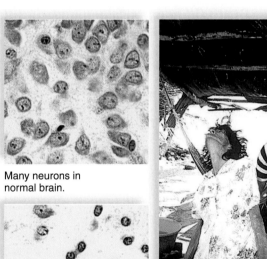

Many neurons in normal brain.

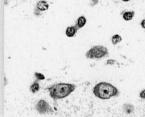

Loss of neurons in Huntington brain.

Figure 20.12 Huntington disease.
Huntington disease is characterized by increasingly serious psychomotor and mental disturbances because of a loss of nerve cells.

Autosomal Dominant Disorders

Inheritance of only one dominant allele is necessary for an autosomal dominant genetic disorder to appear. We discuss only two of the many autosomal dominant disorders here.

Marfan Syndrome **Marfan syndrome,** an autosomal dominant disorder, is caused by a defect in an elastic connective tissue protein, called fibrillin. This protein is normally abundant in the lens of the eye; the bones of limbs, fingers, and ribs; and also in the wall of the aorta. This explains why the affected person often has a dislocated lens, long limbs and fingers, and a caved-in chest. The aorta wall is weak and can possibly burst without warning. A tissue graft can strengthen the aorta, but Marfan patients with aortic symptoms still should not overexert themselves.

Huntington Disease **Huntington disease** is a neurological disorder that leads to progressive degeneration of brain cells (Fig. 20.12). The disease is caused by a mutated copy of the gene for a protein, called huntingtin. Most patients appear normal until they are of middle age and have already had children, who may later also be stricken. Occasionally, the first sign of the disease will appear during the teen years or even earlier. There is no effective treatment, and death comes 10 to 15 years after the onset of symptoms.

Several years ago, researchers found that the gene for Huntington disease was located on chromosome 4. A test was developed for the presence of the gene, but few people want to know if they have inherited the gene because there is no cure. At least now we know that the disease stems from a mutation that causes the huntingtin protein to have too many copies of the amino acid glutamine. The normal version of huntingtin has stretches of between 10 and 25 glutamines. If huntingtin has more than 36 glutamines, it changes shape and forms large clumps inside neurons. Even worse, it attracts and causes other proteins to clump with it. One of these proteins, called CBP, helps nerve cells survive. Researchers hope they may be able to combat the disease by boosting CBP levels.

☑ Check Your Progress 20.2d

1. In a pedigree, all the members of one family are affected. **a.** If the trait is recessive, what are their genotypes? **b.** If the trait is dominant, what are their genotypes?

2. A baby has Tay-Sachs, but the parents are normal. What are the genotypes of the parents and the child?

3. What are the chances that homozygous normal parents for cystic fibrosis will have a child with cystic fibrosis?

4. What causes the sickle-shaped red blood cells in sickle-cell disease?

5. A child has Marfan syndrome. Why would you expect one of the parents to also have Marfan syndrome?

Health Focus

Preimplantation Genetic Diagnosis

If prospective parents are heterozygous for one of the genetic disorders discussed on pages 430–31, they may want the assurance that their offspring will be free of the disorder. Determining the genotype of the embryo will provide this assurance. For example, if both parents are *Aa* for a recessive disorder, the embryo will develop normally if it has the genotype *AA* or *Aa*. On the other hand, if one of the parents is *Aa* for a dominant disorder, the embryo will develop normally only if it has the genotype *aa*.

Following in vitro fertilization (IVF), the zygote (fertilized egg) divides. When the embryo has six to eight cells (Fig. 20A*a*), removal of one of these cells for testing purposes has no effect on normal development. Only embryos that will not have the genetic disorders of interest are placed in the uterus to continue developing.

So far, about 1,000 children with normal genotypes for genetic disorders that run in their families have been born world-wide following embryo testing. In the future, it's possible that embryos who test positive for a disorder could be treated by gene therapy, so that they, too, would be allowed to continue to term.

Testing the egg is possible if the condition of concern is recessive. Recall that meiosis in females results in a single egg and at least two polar bodies (see page 392). Polar bodies later disintegrate and receive very little cytoplasm, but they do receive a haploid number of chromosomes. When a woman is heterozygous for a recessive genetic disorder, about half the first polar bodies will have received the mutated allele, and in these instances, the egg received the normal allele. Therefore, if a polar body tests positive for a recessive mutated allele, the egg received the normal dominant allele. Only normal eggs are then used for IVF. Even if the sperm should happen to carry the mutation, the zygote will, at worst, be heterozygous. But the phenotype will appear normal.

If, in the future, gene therapy becomes routine, it's possible that an egg could be given genes that control traits desired by the parents, such as musical or athletic ability, prior to IVF.

Figure 20A
Preimplantation genetic diagnosis.

a. Following IVF and cleavage, genetic analysis is performed on one cell removed from an eight-celled embryo. If it is found to be free of the genetic defect of concern, the seven-celled embryo is implanted in the uterus and develops into a newborn phenotype.
b. Chromosomal and genetic analysis is performed on a polar body attached to an egg. If the egg is free of a genetic defect, it is used for IVF, and the embryo is implanted in the uterus for further development.

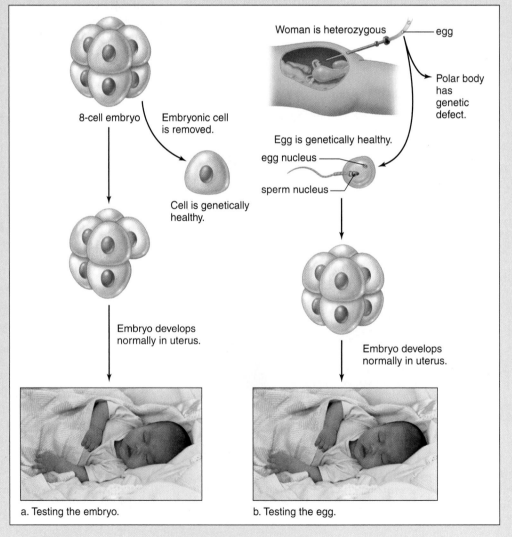

8-cell embryo   Embryonic cell is removed.

Cell is genetically healthy.

Embryo develops normally in uterus.

a. Testing the embryo.

Woman is heterozygous   egg

Polar body has genetic defect.

Egg is genetically healthy.

egg nucleus

sperm nucleus

Embryo develops normally in uterus.

b. Testing the egg.

20.3 Beyond Simple Inheritance Patterns

Certain traits, such as those studied in Section 20.1, are controlled by one set of alleles that follows a simple dominant or recessive inheritance. We now know of many other types of inheritance patterns.

Polygenic Inheritance

Polygenic traits, such skin color and height, are governed by several sets of alleles. The individual has a copy of all allelic pairs, possibly located on many different pairs of chromosomes. Each dominant allele codes for a product, and therefore, the dominant alleles have a quantitative effect on the phenotype, and these effects are additive. The result is a *continuous variation* of phenotypes, resulting in a distribution of these phenotypes that resembles a bell-shaped curve. The more genes involved, the more continuous the variations and distribution of the phenotypes. Also, environmental effects cause many intervening phenotypes; in the case of height, differences in nutrition bring about a bell-shaped curve (Fig. 20.13).

Skin Color

Skin color is an example of a polygenic that is likely controlled by many pairs of alleles. Even so, we will use the simplest model and assume that skin has only two pairs of alleles (*Aa* and *Bb*) and that each capital letter contributes pigment to the skin. When a very dark person reproduces with a very light person, the children have medium-brown skin. When two people with the genotype *AaBb* reproduce with one another, the children may range in skin color from very dark to very light:

Genotypes	Phenotypes
AABB	Very dark
AABb or *AaBB*	Dark
AaBb or *AAbb* or *aaBB*	Medium brown
Aabb or aaBb	Light
aabb	Very light

Notice, again, that a range of phenotypes exists and several possible phenotypes fall between the two extremes. Therefore, the distribution of these phenotypes is expected to follow a bell-shaped curve, meaning that few people have the extreme phenotypes and most people have the phenotype that lies in the middle. Skin color is also a **multifactorial trait**, a polygenic trait that is particularly influenced by the environment. After all, skin color is influenced by sun exposure. Height is also a multifactorial trait because, as mentioned, nutrition can affect height.

Multifactorial Disorders

Many human disorders, such as cleft lip and/or palate, clubfoot, congenital dislocations of the hip, hypertension, diabetes, schizophrenia, and even allergies and cancers, are most likely controlled by polygenes that are subject to environmental influences. The coats of Siamese cats and Himalayan

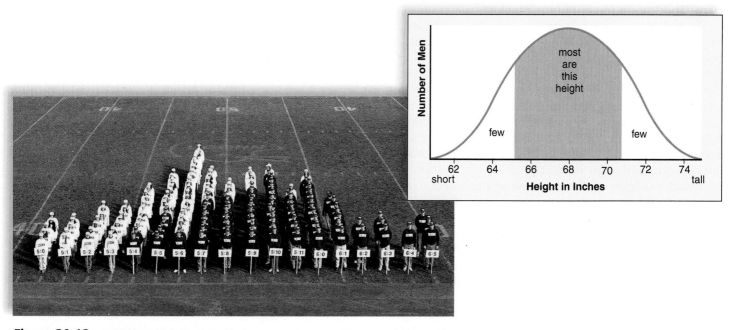

Figure 20.13 Polygenic inheritance.
When you record the heights of a large group of people chosen at random, the values follow a bell-shaped curve. Such a continuous distribution is due to control of a trait by several sets of alleles. Environmental effects are also involved.

Incomplete Dominance and Codominance

Incomplete dominance occurs when the heterozygote is intermediate between the two homozygotes. For example, when a curly-haired individual reproduces with a straight-haired individual, their children have wavy hair. When two wavy-haired persons reproduce, the expected phenotypic ratio among the offspring is 1:2:1—that is, one curly-haired child to two with wavy hair to one with straight hair. We can explain incomplete dominance by assuming that only one allele codes for a product and the single dose of the product gives the intermediate result.

Codominance occurs when alleles are equally expressed in a heterozygote. A familiar example is the human blood type AB, in which the red blood cells have the characteristics of both type A and type B blood. We can explain codominance by assuming that both genes code for a product, and we observe the results of both products being present. Blood type inheritance is said to be an example of multiple alleles.

Incompletely Dominant Disorders

The prognosis in **familial hypercholesterolemia (FH)** parallels the number of LDL-cholesterol receptor proteins in the plasma membrane. A person with two mutated alleles lacks LDL-cholesterol receptors; a person with only one mutated allele has half the normal number of receptors, and a person with two normal alleles has the usual number of receptors.

The presence of excessive cholesterol in the blood causes cardiovascular disease. Therefore, those with no receptors die of cardiovascular disease as children. Individuals with half the number of receptors may die when young or after they have reached middle age. People with the full number of receptors do not have familial hypercholesterolemia (Fig. 20.15).

Figure 20.14 **Coat color in Siamese cats.**
The dark ears, nose, and feet of this cat are believed to be due to a lower body temperature in these areas.

rabbits are darker in color at the ears, nose, paws, and tail (Fig. 20.14). The metabolic cause for this phenomena in Himalayan rabbits is known. Himalayan rabbits are homozygous for the allele *ch*, which is involved in the production of melanin. Experimental evidence suggests that the enzyme coded for by this gene is active only at a low temperature and that, therefore, black fur occurs only at the extremities where body heat is lost to the environment.

In recent years, reports have surfaced that all sorts of behavioral traits, such as alcoholism, phobias, and even suicide, can be associated with particular genes. No doubt behavioral traits are to a degree controlled by genes, but it has not been possible to determine to what degree. Very few scientists would support the idea that these behavioral traits are predetermined by our genes. Therefore, they must be multifactorial traits. Researchers are engaged in trying to determine what percentage of the trait is due to nature (inheritance) and what percentage is due to nurture (the environment). Some studies use identical and fraternal twins separated from birth because, then, it's known that the twins have a different environment. The supposition is that, if identical twins in different environments share the same trait, that trait is most likely inherited. Identical twins are more similar in their intellectual talents, personality traits, and levels of lifelong happiness than are fraternal twins separated from birth. This substantiates the belief that behavioral traits are partly heritable and that genes exert their effects by acting together in complex combinations susceptible to environmental influences.

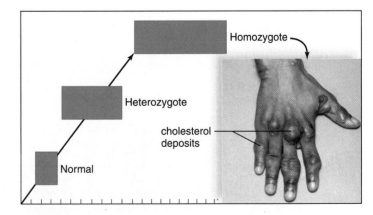

Figure 20.15 **Familial hypercholesterolemia.**
Familial hypercholesterolemia is incompletely dominant. Persons with one mutated allele have an abnormally high level of cholesterol in the blood, and those with two mutated alleles have a higher level still.

Multiple Allele Inheritance

When a trait is controlled by **multiple alleles,** the gene exists in several allelic forms. But each person usually has only two of the possible alleles. In the opening story, Miss Lane asked her students to explain why her mom could have O type blood, while she has A type blood. Let's explore that question now.

ABO Blood Types

Three alleles for the same gene control the inheritance of ABO blood types. These alleles determine the presence or absence of antigens on red blood cells:

$$I^A = \text{A antigen on red blood cells}$$

$$I^B = \text{B antigen on red blood cells}$$

$$i = \text{Neither A nor B antigen on red blood cells}$$

Each person has only two of the three possible alleles, and both I^A and I^B are dominant over i. Therefore, there are two possible genotypes for type A blood and two possible genotypes for type B blood. On the other hand, I^A and I^B are fully expressed in the presence of the other. Therefore, if a person inherits one of each of these alleles, that person will have type AB blood. Type O blood can result only from the inheritance of two i alleles.

The possible genotypes and phenotypes for blood type are as follows:

Phenotype	Genotype
A	$I^A I^A$, $I^A i$
B	$I^B I^B$, $I^B i$
AB	$I^A I^B$
O	ii

So, from this information, we can see that Miss Lane's mother is ii and both I^A and I^B are dominant to an i. Since Miss Lane's blood type is A, what genotypes could her father be? Trying out each genotype in turn, we can see that Miss Lane's father could be $I^A I^A$, $I^A i$, or $I^A I^B$. Any one of these could pass on the allele I^A to Miss Lane.

Blood typing can sometimes aid in paternity suits. However, a blood test of a supposed father can only suggest that he *might* be the father, not that he definitely *is* the father. For example, it is possible, but not definite, that a man with type A blood (genotype $I^A i$) is the father of a child with type O blood. On the other hand, a blood test sometimes can definitely prove that a man is not the father. For example, a man with type AB blood cannot possibly be the father of a child with type O blood. Therefore, blood tests can be used in legal cases only to try to exclude a man from possible paternity.

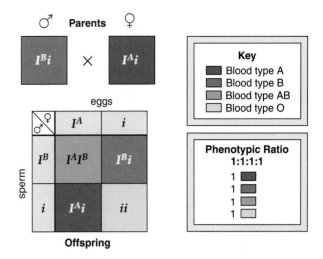

Figure 20.16   Inheritance of blood type.
Blood type exemplifies multiple allele inheritance. The *I* gene has two codominant alleles, designated as I^A and I^B and one recessive allele, designated by i. Therefore, a mating between individuals with type A blood and type B blood can result in any one of the four blood types. Why? Because the parents are $I^A i$ and $I^B i$. If both parents were type AB blood, no child would have what blood type?[1]

Figure 20.16 shows that matings between certain genotypes can have surprising results in terms of blood type.

As a point of interest, the Rh factor is inherited separately from A, B, AB, or O blood types. When you are Rh positive, your red blood cells have a particular antigen, and when you are Rh negative, that antigen is absent. There are multiple recessive alleles for Rh$^-$, but they are all recessive to Rh$^+$.

✅ Check Your Progress 20.3

1. A polygenic trait is controlled by several sets of alleles. Using the example of skin color, what are the two extreme genotypes for this trait?

2. What are some examples of human multifactorial traits?

3. What is the genotype of the lightest child that could result from a mating between two medium-brown individuals?

4. What is an example of incomplete dominance in humans?

5. How is ABO blood type an example of codominance and multiple allele inheritance?

6. A child with type O blood is born to a mother with type A blood. What is the genotype of the child? The mother? What are the possible genotypes of the father?

[1]See Appendix A for answers.

20.4 Sex-Linked Inheritance

Normally, both males and females have 23 pairs of chromosomes; 22 pairs are called **autosomes,** and one pair is the sex chromosomes. These are called the **sex chromosomes** because they differ between the sexes. In humans, males usually have the sex chromosomes X and Y, and females usually have two X chromosomes.

Traits controlled by genes on the sex chromosomes are said to be **sex-linked;** an allele on an X chromosome is **X-linked,** and an allele on the Y chromosome is Y-linked. Most sex-linked genes are only on the X chromosomes, and the Y chromosome is blank for these. Very few Y-linked alleles have been found on the much smaller Y chromosome.

Many of the genes on the X chromosomes, such as those that determine normal as opposed to red-green color blindness, are unrelated to the gender of the individual. It would be logical to suppose that a sex-linked trait is passed from father to son or from mother to daughter, but this is not the case. A male always receives an X-linked allele from his mother, from whom he inherited an X chromosome. *The Y chromosome from the father does not carry an allele for the trait.* Usually, a sex-linked genetic disorder is recessive; therefore, a female must receive two alleles, one from each parent, before she has the condition.

X-Linked Alleles

When considering X-linked traits, the allele on the X chromosome is shown as a letter attached to the X chromosome. For example, this is the key for red-green color blindness, a well-known X-linked recessive disorder:

$$X^B = \text{normal vision}$$

$$X^b = \text{color blindness}$$

The possible genotypes and phenotypes in both males and females are

Genotypes	Phenotypes
$X^B X^B$	Female who has normal color vision
$X^B X^b$	Carrier female who has normal color vision
$X^b X^b$	Female who is color-blind
$X^B Y$	Male who has normal vision
$X^b Y$	Male who is color-blind

The second genotype is a carrier female because, although a female with this genotype appears normal, she is capable of passing on an allele for color blindness. Color-blind females are rare because they must receive the allele from both parents; color-blind males are more common because they need only one recessive allele to be color-blind. The

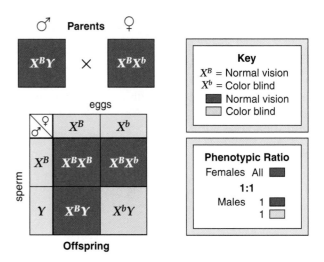

Figure 20.17   Cross involving an X-linked allele.
The male parent is normal, but the female parent is a carrier—an allele for color blindness is located on one of her X chromosomes. Therefore, each son has a 50% chance of being color-blind. The daughters will appear normal, but each one has a 50% chance of being a carrier.

allele for color blindness must be inherited from their mother because it is on the X chromosome; males only inherit the Y chromosome from their father (Fig. 20.17).

Now let us consider a mating between a man with normal vision and a heterozygous woman (Fig. 20.17). What is the chance that this couple will have a color-blind daughter? A color-blind son? All daughters will have normal color vision because they all receive an X^B from their father. The sons, however, have a 50% chance of being color blind, depending on whether they receive an X^B or an X^b from their mother. The inheritance of a Y chromosome from their father cannot offset the inheritance of an X^b from their mother. Because the Y chromosome doesn't have an allele for the trait, it can't possibly prevent color blindness in a son. Notice in Figure 20.17 that the phenotypic results for sex-linked traits are given separately for males and females.

Pedigree for X-Linked Disorders

Like color-blindness, most sex-linked disorders are usually carried on the X chromosome. Figure 20.18 gives a pedigree for an *X-linked recessive disorder.* More males than females have the disorder because recessive alleles on the X chromosome are always expressed in males—the Y chromosome lacks an allele for the disorder. X-linked recessive conditions often pass from grandfather to grandson because the daughters of a male with the disorder are carriers. Recall that Miss Lane in the opening story had a color-blind father, but was not color-blind. Therefore, her pupil, Michael, knew instantly that she was a carrier and

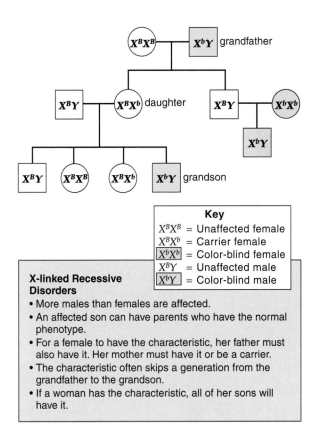

Key

X^BX^B = Unaffected female
X^BX^b = Carrier female
X^bX^b = Color-blind female
X^BY = Unaffected male
X^bY = Color-blind male

X-linked Recessive Disorders

• More males than females are affected.
• An affected son can have parents who have the normal phenotype.
• For a female to have the characteristic, her father must also have it. Her mother must have it or be a carrier.
• The characteristic often skips a generation from the grandfather to the grandson.
• If a woman has the characteristic, all of her sons will have it.

Figure 20.18 Pedigree for an X-linked recessive.

This pedigree for color blindness exemplifies the inheritance pattern of an X-linked recessive disorder. The list gives various ways of recognizing the X-linked recessive pattern of inheritance.

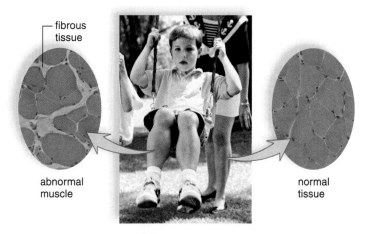

Figure 20.19 Muscular dystrophy

In muscular dystrophy, an X-linked recessive disorder, calves enlarge because fibrous tissue develops as muscles waste away, due to lack of the protein dystrophin.

that her sons would have a 50% chance of being color-blind. Should Miss Lane have told her fiancé she was a carrier for color blindness? Was Miss Lane trying to be tricky when she mentioned that her mom was not color-blind? That information was not necessary to knowing her genotype because she told the class she was not color-blind. Figure 20.18 lists various ways to recognize a recessive X-linked disorder.

Only a few known traits are *X-linked dominant*. If a disorder is X-linked dominant, affected males pass the trait *only* to daughters, who have a 100% chance of having the condition. Females can pass an X-linked dominant allele to both sons and daughters. If a female is heterozygous and her partner is normal, each child has a 50% chance of escaping an X-linked dominant disorder, depending on which maternal X chromosome is inherited.

X-Linked Recessive Disorders of Interest

Color blindness, an X-linked recessive disorder, does not prevent males from leading a normal life. About 8% of Caucasian men have red-green color blindness. Most of these see brighter greens as tans, olive greens as browns, and reds as reddish browns. A few cannot tell reds from greens at all. They see only yellows, blues, blacks, whites, and grays.

Muscular Dystrophy   Duchenne muscular dystrophy is an X-linked recessive disorder characterized by a wasting away of the muscles. Symptoms, such as waddling gait, toe walking, frequent falls, and difficulty in rising, may appear as soon as the child starts to walk. Muscle weakness intensifies until the individual is confined to a wheelchair. Death usually occurs by age 20; therefore, affected males are rarely fathers. The recessive allele remains in the population by passage from carrier mother to carrier daughter.

The absence of a protein, now called dystrophin, is the cause of Duchenne muscular dystrophy. Much investigative work determined that dystrophin is involved in the release of calcium from the sarcoplasmic reticulum in muscle fibers. The lack of dystrophin causes calcium to leak into the cell, which promotes the action of an enzyme that dissolves muscle fibers. When the body attempts to repair the tissue, fibrous tissue forms (Fig. 20.19), and this cuts off the blood supply so that more and more cells die. Immature muscle cells can be injected into muscles, but it takes 100,000 cells for dystrophin production to increase by 30–40%.

Hemophilia   There are two common types of hemophilia, an X-linked recessive disorder. Hemophilia A is due to the absence or minimal presence of a clotting factor known as factor VIII, and hemophilia B is due to the absence of clotting factor IX. Hemophilia is called the bleeder's disease because the affected person's blood either does not clot or clots very slowly. Although hemophiliacs bleed externally after an injury, they also bleed internally, particularly around joints. Hemorrhages can be stopped with transfusions of fresh blood (or plasma) or concentrates of the clotting protein. Also, factors VIII and IX are now available as a biotechnology product.

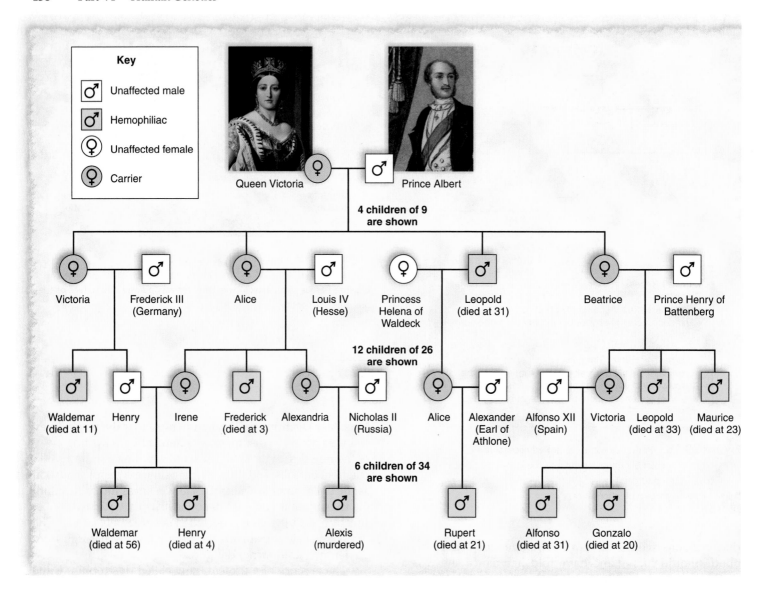

Figure 20.20   **A simplified pedigree showing X-linked inheritance of hemophilia in European royal families.**
Because Queen Victoria was a carrier, each of her sons had a 50% chance of having the disorder, and each of her daughters had a 50% chance of being a carrier. This pedigree shows only the affected descendants. Many others are unaffected, including the members of the present British royal family.

At the turn of the century, hemophilia was prevalent among the royal families of Europe, and all of the affected males could trace their ancestry to Queen Victoria of England. Figure 20.20 shows that, of Queen Victoria's 26 grandchildren, four grandsons had hemophilia and four granddaughters were carriers. Because none of Queen Victoria's relatives were affected, it seems that the faulty allele she carried arose by mutation, either in Victoria or in one of her parents. Her carrier daughters, Alice and Beatrice, introduced the allele into the ruling houses of Russia and Spain, respectively. Alexis, the last heir to the Russian throne before the Russian Revolution, was a hemophiliac (Fig. 20.20). There are no hemophiliacs in the present British royal family because Victoria's eldest son, King Edward VII, did not receive the allele.

☑ **Check Your Progress 20.4**

1. Why are more males than females color-blind?

2. **a.** What phenotypic ratio is expected for a cross in which both parents have one X-linked recessive allele? **b.** in which both parents have one X-linked dominant allele?

3. Both the mother and the father of a son with hemophilia appear to be normal. From whom did the son inherit the allele for hemophilia? What are the genotypes of the mother, the father, and the son?

4. A woman is color-blind. What are the chances that her sons will be color-blind? If she is married to a man with normal vision, what are the chances that her daughters will be color-blind? Will be carriers?

5. Both the husband and wife have normal vision. The wife gives birth to a color-blind daughter. Is it more likely the father had normal vision or was color-blind? What does this lead you to deduce about the girl's parentage?

Genetic Profiling

Various tests are available these days to discover what genetic disorders you now have or may have in the future. For example, knowledge of your genes might indicate your susceptibility to being bipolar, a mental disorder characterized by mood swings. This information would alert a health-care provider that you should be knowledgeable about the disorder and seek professional care if symptoms develop. Are there any reasons not to be in favor of **genetic profiling,** that is knowing your genotype for various illnesses?

Discrimination Concerns

Some people are inclined to believe that insurance companies and employers could use a genetic profile against them. Perhaps employers will not hire, or insurance companies will not insure, those who have a propensity for particular diseases. The federal government, and about 25 states, have passed laws prohibiting genetic discrimination by health insurers, and 11 have passed laws prohibiting genetic discrimination by employers. The legislation states that genetic information cannot be released to anyone without the subject's permission. Is such legislation enough to allay our fears of discrimination? Might an employer not hire you or an insurance company not insure you simply because you will not grant permission to access your genetic profile (Fig. 20B)? Say two women are being considered for the same position, and each meets all the basic requirements for the job. The first denies access to her genetic profile, while the second one grants permission to look at her genetic profile. Thinking that the first woman might have something to hide, the employer hires the second one. The possibility that sick days may be needed by the first woman makes the second woman the more cost-effective choice. People who have genetic profiles proving they are likely to be healthy in the future might even use them in order to have an advantage over those who have profiles showing that they are likely to develop serious illnesses in the future. In this way, we might create a genetic underclass.

Genetic information is sometimes misunderstood, particularly by laypeople. In the past, for example, as an effort to combat sickle-cell disease, many people were screened for it. Unfortunately, those who were found to have the sickle-cell trait, and not the actual disease, experienced discrimination at school or from employers and insurance carriers.

Employer Concerns

On the other hand, employers may fear that the government might use genetic information one day to require them to provide an environment specific to every employee's need, in order to prevent future illness. Would you approve of this, or should individuals be required to leave an area or job that exposes them to an environmental influence that could be detrimental to their health?

Public Benefits

Some researchers believe that free access to genetic profiling data is absolutely essential to developing better preventive care for all. If researchers can match genetic profiles to environmental conditions that bring on illnesses, they could come up with better prevention guidelines for the next generation. Should genetic profiles and health records become public information under these circumstances? It would particularly help in the study of complex diseases, such as mental and cardiovascular disorders, noninsulin-dependent diabetes, and juvenile rheumatoid arthritis. Perhaps there would be some way to protect privacy and still make the information known.

Possible Safeguards

If present legislation to protect privacy is inadequate, what could be done to truly keep such information private? Should the information be coded in some way so that only the medical profession can read it? Should people be responsible for keeping the only copy of their profiles, which would be coded so that even they cannot read it? Or, do you believe that anyone should have access to anyone's profile, for whatever reason?

Decide Your Opinion

1. Should people be encouraged or even required to have a genetic profile so that they can develop programs to possibly prevent future illness?
2. Should employers be encouraged or required to provide an environment suitable to a person's genetic profile? Or should the individual avoid a work environment that could bring on an illness?
3. How can we balance individual rights with the public-health benefit of matching genetic profiles to detrimental environments?

Figure 20B **Interviewing for a position.**
If a prospective employer asked for your genetic profile, would you be inclined to give him permission to view it?

Summarizing the Concepts

20.1 Genotype and Phenotype

Genotype refers to the alleles of the individual, and phenotype refers to the physical characteristics associated with these alleles.

- Homozygous dominant individuals (*EE*) have the dominant phenotype (i.e., unattached earlobes).
- Homozygous recessive individuals (*ee*) have the recessive phenotype (i.e., attached earlobes).
- Heterozygous individuals (*Ee*) have the dominant phenotype (i.e., unattached earlobes).

20.2 One- and Two-Trait Inheritance

One-Trait Crosses

The first step in doing one-trait problems is to determine the genotype and then the gametes.

- An individual has two alleles for every trait, but a gamete has one allele for every trait.

The next step is to combine all possible sperm with all possible eggs. If there is more than one possible sperm and/or egg, a Punnett square is helpful in determining the genotypic and phenotypic ratio among the offspring.

- For a monohybrid × monohybrid cross, a 3:1 ratio is expected among the offspring.
- For a monohybrid × recessive cross, a 1:1 radio is expected among the offspring.
- The expected ratio can be converted to the chance of a particular genotype/phenotype. For example, a 3:1 ratio = a 75% chance of the dominant phenotype and a 25% chance of the recessive phenotype.

Two-Trait Crosses

- If an individual is heterozygous for two traits, 4 gamete types are possible as can be substantiated by knowledge of meiosis.
- For a dihybrid **X** dihybrid cross (*AaBb* X *AaBb*), a 9:3:3:1 ratio is expected among the offspring.
- For a dihybrid **X** recessive cross (*AaBb* X *aabb*), a 1:1:1:1 ratio is expected among the offspring.

Family Pedigrees for Genetic Disorders

A pedigree shows the pattern of inheritance for a trait from generation to generation of a family. This first pattern appears in a family pedigree for a recessive disorder—both parents are carriers. The second pattern appears in a family pedigree for a dominant disorder. Both parents are again heterozygous.

Trait is recessive. Trait is dominant.

Genetic Disorders of Interest

- Tay-Sachs, cystic fibrosis, phenylketonuria, and sickle-cell disease are autosomal recessive disorders.
- Marfan syndrome and Huntington disease are autosomal dominant disorders.

20.3 Beyond Simple Inheritance Patterns

In some patterns of inheritance, the alleles are not just dominant or recessive.

Polygenic Inheritance

Polygenic traits, such as skin color and height, are controlled by more than one set of alleles. The dominant alleles have an additive effect on the phenotype.

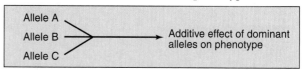

Incomplete Dominance and Codominance

In incomplete dominance (e.g., familial hypercholesterolemia), the heterozygote is intermediate between the two homozygotes. In codominance (e.g., blood type AB), both dominant alleles are expressed equally.

Multiple Allele Inheritance

The multiple allele inheritance pattern is exemplified in humans by blood type inheritance. Every individual has two out of three possible alleles: I^A, I^B, i. Both I^A and I^B are expressed; therefore, this is also a case of codominance.

20.4 Sex-Linked Inheritance

Many genes on the X chromosomes, such as those that determine normal vision as opposed to color blindness, are unrelated to the gender of the individual. Common X-linked genetic crosses are

$X^B X^b$ **x** $X^B Y$ All daughters will be normal, even though they have a 50% chance of being carriers, but sons have a 50% chance of being color-blind.

$X^B X^B$ **x** $X^b Y$ All children are normal (daughters will be carriers).

Pedigree for X-Linked Disorders

- A pedigree for an X-linked recessive disorder shows that the trait often passes from grandfather to grandson by way of a carrier daughter. Also, more males than females have the characteristic.
- Like most X-linked disorders, color blindness, muscular dystrophy, and hemophilia are recessive.

Understanding Key Terms

Match the key terms to these definitions.

a. —————— Gridlike device used to calculate the expected results of simple genetic crosses.

b. —————— Alternate forms of a gene that occur at the same site on homologous chromosomes.

c. —————— Allele that exerts its phenotype effect in the heterozygote; it masks the expression of the recessive allele.

d. —————— Particular site where a gene is found on a chromosome.

e. —————— Alleles of an individual for a particular trait or traits, expressed such as *BB, Aa,* or *BBAa.*

Testing Your Knowledge of the Concepts

1. Parents both have unattached earlobes, but some of their children have attached earlobes. Explain, in terms of phenotype and genotype. (page 422)

2. Explain why the gametes have only one allele for a trait. (page 423)

3. What is the chance of producing a child with the dominant phenotype from each of the following crosses? (pages 424–25)
   a. *AA × AA*
   b. *Aa × AA*
   c. *Aa × Aa*
   d. *aa × aa*

4. Which of the crosses in question 3 can result in an offspring with the recessive phenotype? Explain. (pages 424–25)

5. What are the expected results of the following crosses? (pages 424–28)
   a. monohybrid × monohybrid
   b. monohybrid × recessive
   c. dihybrid × dihybrid
   d. dihybrid × recessive in both traits

6. Which of these crosses would best allow you to determine that an individual with the dominant phenotype is homozygous or heterozygous? (pages 424–28)

7. What is the genotype of a heterozygote with a widow's peak and short fingers? Give all possible gametes for this individual. (page 427)

8. Using a pedigree, show an autosomal recessive disorder pattern and an autosomal dominant disorder pattern. (page 429)

9. What is required for an autosomal recessive disorder to appear? Name four autosomal recessive disorders. What is required for an autosomal dominant disorder to appear? Name two autosomal dominant disorders. (pages 430–31)

10. Define and give an example of each of the following inheritance patterns: polygenic inheritance, multifactorial trait, incomplete dominance, codominance, and multiple alleles. (pages 433–35)

11. What are the phenotypes and genotypes for the ABO blood groups? If both parents are type AB blood, what are the possible genotypes and phenotypes for their children, including the expected percentages? (page 435)

12. How is an X-linked trait different from an autosomal trait? (page 436)

13. A woman with normal vision reproduces with a man with normal vision. Three sons are color-blind, and one son has normal vision. Explain how this occurred, using the genotypes and gametes of all individuals. (pages 436–37)

14. Which of these is a correct statement?
    a. Each gamete contains two alleles for each trait.
    b. Each individual has one allele for each trait.
    c. Fertilization gives each new individual one allele for each trait.
    d. All of these are correct.
    e. None of these is correct.

15. Which of the following indicates a heterozygous individual?
    a. *AB*   c. *Aa*
    b. *AA*   d. *aa*

16. What possible gametes can be produced by *AaBb.*
    a. *Aa, Bb*   c. *AB, ab*
    b. *A, a, B, b*   d. *AB, Ab, aB, ab*

17. In humans, pointed eyebrows (*P*) are dominant over smooth eyebrows (*p*). Mary's father has pointed eyebrows, but she and her mother have smooth. What is the genotype of the father?
    a. *pp*   c. *PPpp*
    b. *Pp*   d. *pPpP*

18. The genotypic ratio from a monohybrid cross is
    a. 1:1.   d. 9:3:3:1.
    b. 3:1.   e. 1:1:1:1.
    c. 1:2:1.

19. A straight hairline is recessive. If two parents with a widow's peak have a child with a straight hairline, then what is the chance that their next child will have a straight hairline?
    a. no chance   d. ½
    b. ¼   e. ¹⁄₁₆
    c. ³⁄₁₆

20. What is the chance that an *Aa* individual will be produced from an *Aa* **x** *Aa* cross?
    a. 50%   d. 25%
    b. 75%   e. 100%
    c. 0%

21. The genotype of an individual with the dominant phenotype can be determined best by reproduction with
    a. the recessive genotype or phenotype.
    b. a heterozygote.
    c. the dominant phenotype.
    d. the homozygous dominant.
    e. Both a and b are correct.

22. Because the homologous chromosomes align independently at the equator during meiosis
    a. all possible combinations of alleles can occur in the gametes.
    b. only the parental combinations of gametes can occur in the gametes.
    c. only the nonparental combinations of gametes can occur in the gametes.

23. What is the chance that a dihybrid cross will produce a homozygous recessive in both traits?
    a. $9/16$
    b. $1/4$
    c. $1/16$
    d. $3/16$
    e. $1/8$

24. Which of the following is not a feature of multifactorial inheritance?
    a. Effects of dominant alleles are additive.
    b. Genes affecting the trait may be on multiple chromosomes.
    c. Environment influences phenotype.
    d. Recessive alleles are harmful.

25. The ABO blood system exhibits
    a. codominance.
    b. multiple alleles.
    c. incomplete dominance.
    d. Both a and b are correct.

26. Assume two normal parents have a color-blind son. Which parent is responsible for color blindness in the son?
    a. the mother
    b. the father
    c. either parent
    d. Neither parent—two normal parents cannot have a color-blind son.

27. If a child has type O blood and the mother is type A, then which of the following could be the blood type of the child's father?
    a. A only
    b. B only
    c. O only
    d. A or O
    e. A, B, or O

28. Under what condition(s) will an autosomal recessive disorder appear?
    a. inheritance of a single recessive allele
    b. inheritance of one dominant and one recessive allele
    c. inheritance of two recessive alleles
    d. Both a and c are correct.

29. Alice and Henry are at the opposite extremes for a multifactorial trait. Their children will
    a. be bell-shaped.
    b. be a phenotype typical of a 3:1 ratio.
    c. have the middle phenotype between their two parents.
    d. look like one parent or the other.

30. Two wavy-haired individuals (neither curly hair nor straight hair are completely dominant) reproduce. What are the chances that their children will have wavy hair?
    a. 0%
    b. 25%
    c. 50%
    d. 100%

Thinking Critically About the Concepts

Remember the nearly identical siblings Nick and Josh from Chapter 18? They looked so much alike, other people didn't realize when they traded places with each other. Their appearance in the opening story in Chapter 18 was to help you visualize how homologous chromosomes separate during meiosis and then reunite at fertilization when the offspring get one of the homologous chromosomes from their mom and one from their dad. The similar physical features of Nick and Josh are the result of inheriting similar genes on the homologous chromosomes from their mom and dad.

1. Could Miss Lane, from the opening story, ever have a full sibling with B or AB blood type? Explain.

2a. Why would Michael be concerned about Miss Lane disclosing her status as a carrier for color blindness to her fiancé?

 b. Would Miss Lane and her fiancé (a normal vision male) ever have a color-blind daughter? Why or why not?

3. Why are the best matches for organ/tissue transplants often a sibling of the person needing the transplant?

DNA Biology and Technology

Will struggled to think of a topic for his biology research paper that was due in a couple of weeks. Then he remembered what his English professor always told his class, "Write what you know about." His uncle recently passed away from liver failure when a donor liver never materialized, so Will decided to do some research on donor organ availability. He remembered hearing about a lab growing a replacement bladder, and wondered if other organs were being grown or not.

He googled "engineered organs for transplant" and was amazed to see a listing of 300,000+ hits for his topic. He noticed several references for the lab-grown bladder he'd heard about on the news. One of the first hits caught his eye because it mentioned the use of pig organs for transplants. He learned that transplants of organs or tissues from one species to another are called xenotransplants, and that newly-created transgenic pigs display human recognition proteins. The display of human proteins minimizes the rejection of a pig organ or tissue that's transplanted into a human.

Will also found a website that discussed the use of transgenic pig livers to support patients with failing livers until human livers could be found for transplant. Both men survived and have not shown any signs of infection by pig viruses. Scientists are trying to assess the risk of pig viruses spreading from human to human before additional transgenic transplants are done. Preparation of engineered organs for transplant is just one aspect of modern DNA technology.

21.1 DNA and RNA Structure and Function

You know that **DNA (deoxyribonucleic acid)** is the genetic material and that DNA is largely found in the chromosomes, located in the nucleus of a cell. This chapter will consider that any genetic material has to be able to do three things: (1) replicate so that it can be transmitted to the next generation, (2) store information, such as how to make the recognition proteins displayed by the cells of the transgenic pigs mentioned in the opening story, and (3) undergo mutations that provide genetic variability.

Structure of DNA

DNA is a **double helix;** it is composed of two strands that spiral about each other (Fig. 21.1a). Each strand is a polynucleotide because it is composed of a series of nucleotides. A nucleotide is a molecule composed of three subunits—phosphoric acid (phosphate), a pentose sugar (deoxyribose), and a nitrogen-con-

taining base. Looking at just one strand of DNA, notice that the phosphate and sugar molecules make up a backbone and the bases project to one side. Put the two strands together, and DNA resembles a ladder (Fig. 21.1b). The phosphate-sugar backbones make up the supports of the ladder, and the rungs of the ladder are the paired bases. The bases are held together by hydrogen bonding: A pairs with T, by forming two hydrogen bonds, and G pairs with C, by forming three hydrogen bonds, or vice versa. This is called **complementary paired bases.**

The bases are important to the functioning of DNA, so it will be helpful to remember that a purine (has two rings) is always paired with a pyrimidine (has one ring), like this:

Purines	Pyrimidines
Adenine (A)	Thymine (T)
Guanine (G)	Cytosine (C)

You might also want to note that two strands of DNA are antiparallel—that is, they run in opposite directions, which you can verify by noticing that in one strand, the sugar

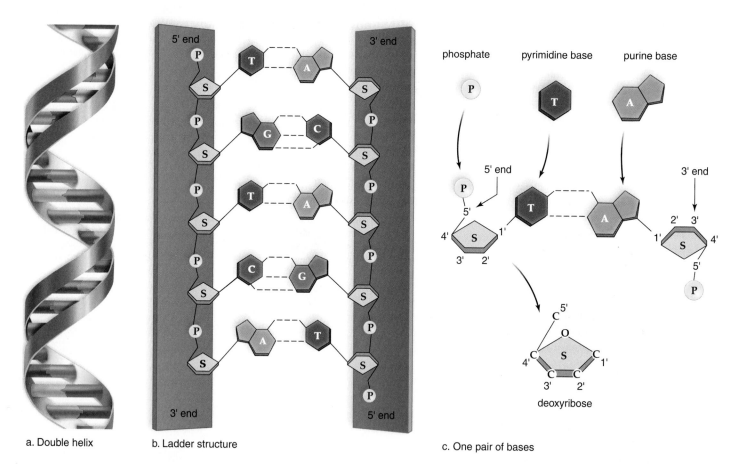

a. Double helix   b. Ladder structure

c. One pair of bases

Figure 21.1   Overview of DNA structure.
a. DNA double helix. **b.** When the helix is unwound, a ladder configuration shows that the supports are composed of sugar (S) and phosphate (P) molecules and the rungs are complementary bases. Notice that the bases in DNA pair in such a way that the phosphate-sugar backbones are oriented in different directions. **c.** The DNA strands are antiparallel, which is apparent by numbering the carbon atoms in deoxyribose.

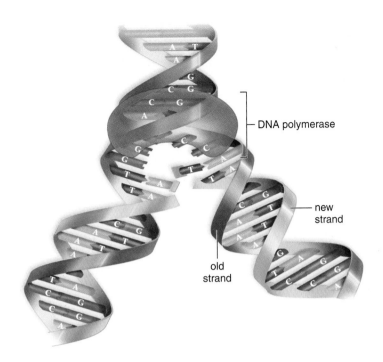

Figure 21.2   **Overview of DNA replication.**
Replication is called semiconservative because each new double helix is composed of an original strand and a new strand.

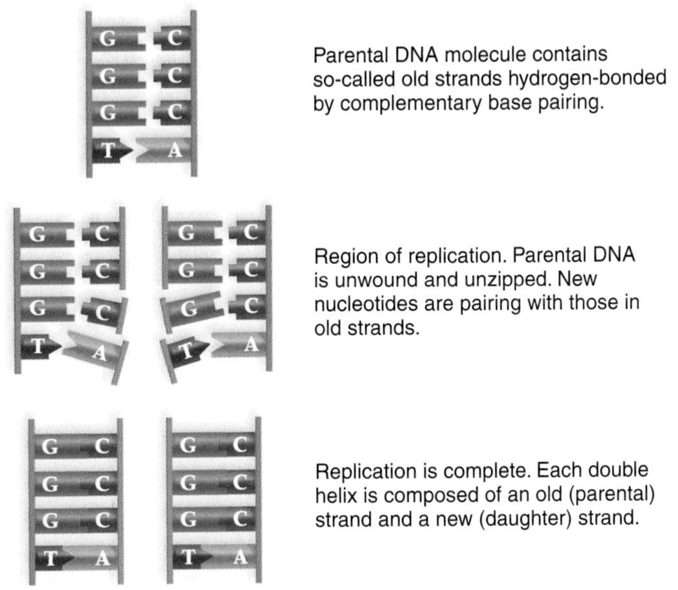

- DNA polymerase
- new strand
- old strand

Parental DNA molecule contains so-called old strands hydrogen-bonded by complementary base pairing.

Region of replication. Parental DNA is unwound and unzipped. New nucleotides are pairing with those in old strands.

Replication is complete. Each double helix is composed of an old (parental) strand and a new (daughter) strand.

Figure 21.3   **Ladder configuration and DNA replication.**
Use of the ladder configuration better illustrates how complementary nucleotides, available in the cell, pair with those of each old strand before they are joined together to form a daughter strand.

molecules appear right-side up, and in the other, they appear upside down. This technicality, while important, need not concern us.

Replication of DNA

When cells divide, each new cell gets an exact copy of DNA. The process of copying a DNA helix is called **DNA replication.** Following up on the opening story, let's consider that in order for all the cells of a pig to display human recognition proteins, the DNA for these genes was inserted into a pig embryo. Then, replication happened as the cell cycle, including mitosis, occurred. In this way, every cell of a genetically modified pig would have the information needed to produce human recognition proteins, making pig organs more acceptable to humans.

During replication, the double-stranded structure of DNA allows each original strand to serve as a **template** (mold) for the formation of a complementary new strand. DNA replication is termed *semiconservative* because each new double helix has one original strand and one new strand. In other words, one of the original strands is conserved, or present, in each new double helix. Because each original strand has produced a new strand through complementary base pairing, there are now two DNA helices identical to each other and to the original molecule (Fig. 21.2).

Figure 21.3 shows how complementary nucleotides pair in the daughter strand.

1. Before replication begins, the two strands that make up parental DNA are hydrogen-bonded to each other.
2. An enzyme unwinds and "unzips" double-stranded DNA (i.e., the weak hydrogen bonds between the paired bases break).
3. New complementary DNA nucleotides, always present in the nucleus, fit into place by the process of complementary base pairing. These are positioned and joined by the enzyme *DNA polymerase.*
4. To complete replication, an enzyme seals any breaks in the sugar-phosphate backbone.
5. The two double-helix molecules are identical to each other and to the original DNA molecule.

Rarely, a replication error occurs making the sequence of the bases in the new strand different from the parental strand. But, if an error does occur, the cell has repair enzymes that usually fix it. A replication error that persists is a **mutation,** a permanent change in the sequence of bases that can possibly cause a change in the phenotype and introduce variability. Such variabilities make you different from your neighbor and humans different from other animals.

The Structure and Function of RNA

RNA (ribonucleic acid) is made up of nucleotides containing the sugar ribose. This sugar accounts for the scientific name of this polynucleotide. The four nucleotides that make up the RNA molecule have the following bases: adenine (A), **uracil (U),** cytosine (C), and guanine (G) (Fig. 21.4). Notice that in RNA, the base uracil replaces the base thymine.

RNA, unlike DNA, is single-stranded (Fig. 21.4), but the single RNA strand sometimes doubles back on itself, and complementary base pairing still occurs. Similarities and differences between these two nucleic acid molecules are listed in Table 21.1.

In general, RNA is a helper to DNA, allowing protein synthesis to occur according to the stored genetic information that DNA provides. There are three types of RNA, each with a specific function in protein synthesis.

Ribosomal RNA

Ribosomal RNA (rRNA) is produced in the nucleolus of a nucleus where a portion of DNA serves as a template for its formation. Ribosomal RNA joins with proteins made in the cytoplasm to form the subunits of ribosomes. The subunits leave the nucleus and come together in the cytoplasm when protein synthesis is about to begin. Proteins are synthesized at the ribosomes, which in low-power electron micrographs

Table 21.1	DNA-RNA Similarities and Differences

DNA-RNA SIMILARITIES

Both are nucleic acids

Both are composed of nucleotides

Both have a sugar-phosphate backbone

Both have four different types of bases

DNA-RNA DIFFERENCES

DNA	RNA
Found in nucleus	Found in nucleus and cytoplasm
The genetic material	Helper to DNA
Sugar is deoxyribose	Sugar is ribose
Bases are A, T, C, G	Bases are A, U, C, G
Double-stranded	Single-stranded
Is transcribed (to give mRNA)	Is translated (to give proteins)

look like granules arranged along the endoplasmic reticulum. The endoplasmic reticulum is a system of tubules and saccules within the cytoplasm. Some ribosomes appear free in the cytoplasm or in clusters called polyribosomes.

Messenger RNA

Messenger RNA (mRNA) is produced in the nucleus where DNA serves as a template for its formation. This type of RNA carries genetic information from DNA to the ribosomes in the cytoplasm where protein synthesis occurs. Messenger RNA is a linear molecule as shown in Figure 21.4.

Transfer RNA

Transfer RNA (tRNA) is produced in the nucleus, and a portion of DNA also serves as a template for its production. Appropriate to its name, tRNA transfers amino acids to the ribosomes, where the amino acids are joined, forming a protein. There are 20 different types of amino acids in proteins; therefore, at least 20 tRNAs must be functioning in the cell. Each type of tRNA carries only one type of amino acid.

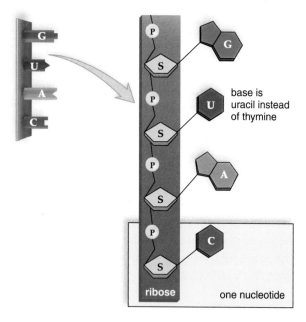

base is uracil instead of thymine

ribose   one nucleotide

Figure 21.4   Structure of RNA.

Like DNA, RNA is a polymer of nucleotides. In an RNA nucleotide, the sugar ribose is attached to a phosphate molecule and to a base: G, U, A, or C. Notice that in RNA, the base uracil replaces thymine as one of the pyrimidine bases. RNA is single-stranded, whereas DNA is double-stranded.

✅ Check Your Progress 21.1

1. How does the structure of DNA allow it to be replicated?

2. How is RNA structure similar to, but also different from, that of DNA?

3. What are the different types of RNA?

21.2 Gene Expression

As we shall see, DNA provides the cell with a blueprint for synthesizing proteins. DNA resides in the nucleus, and protein synthesis occurs in the cytoplasm. First, mRNA carries a copy of DNA's blueprint into the cytoplasm, and second, the other RNA molecules we just discussed are involved in bringing about protein synthesis.

Before discussing the mechanics of gene expression, let's review the structure of proteins.

Structure and Function of Proteins

Proteins are composed of subunits called amino acids (Table 21.2). Twenty different amino acids are commonly found in proteins, which are synthesized at the ribosomes in the cytoplasm of cells. Proteins differ because the number and order of their amino acids differ. Figure 21.5 shows that the sequence of amino acids in a protein leads to its particular shape. Proteins are found in all parts of the body; some are structural proteins, and some are enzymes. The protein hemoglobin is responsible for the red color of red blood cells. Albumins and globulins (antibodies) are well-known plasma proteins. Muscle cells contain the proteins actin and myosin, which give muscles substance and the ability to contract.

Enzymes are organic catalysts that speed reactions in cells. The reactions in cells form metabolic or chemical pathways. A pathway can be represented as follows:

$$E_A \quad E_B \quad E_C \quad E_D$$
$$A \rightarrow B \rightarrow C \rightarrow D \rightarrow E$$

In this pathway, the letters are molecules, and the notations over the arrows are enzymes: Molecule A becomes molecule B, and enzyme E_A speeds the reaction; molecule B becomes molecule C, and enzyme E_B speeds the reaction; and so forth. Enzymes are *specific*: Enzyme E_A can only convert A to B, enzyme E_B can only convert B to C, and so forth.

Proteins determine the structure and function of the various cells in the body, and some of those in the plasma membrane serve as human recognition proteins, as mentioned in the opening story.

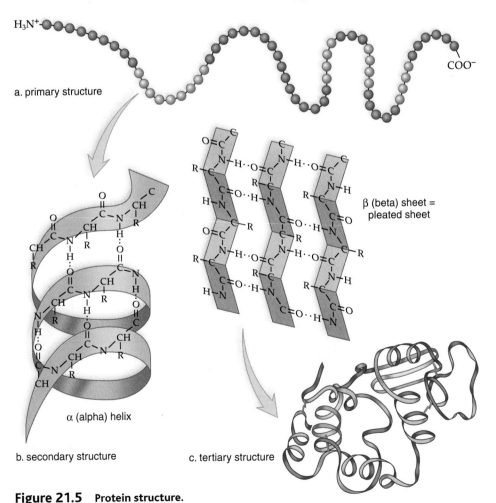

a. primary structure

α (alpha) helix

β (beta) sheet = pleated sheet

b. secondary structure

c. tertiary structure

Figure 21.5   Protein structure.
a. The primary structure of a protein is the sequence of its amino acids. **b.** The secondary structure can be either a helix or pleated sheet. **c.** The tertiary structure is the final three-dimensional shape.

Table 21.2	Amino Acids
Amino Acid	**Abbreviation**
alanine	ala
arginine	arg
asparagine	asn
aspartic acid	asp
cysteine	cys
glutamine	gln
glutamic acid	glu
glycine	gly
histidine	his
isoleucine	ile
leucine	leu
lysine	lys
methionine	met
phenylalanine	phe
proline	pro
serine	ser
threonine	thr
tryptophan	trp
tyrosine	tyr
valine	val

Gene Expression: An Overview

The first step in gene expression is called transcription, and the second step is called translation (Fig. 21.6). During **transcription**, a strand of mRNA forms that is complementary to a portion of DNA. The mRNA molecule that forms is a *transcript* of a gene. Transcription means to make a faithful copy, and in this case, a sequence of nucleotides in DNA is copied to a sequence of nucleotides in mRNA.

We can liken the DNA in the nucleus to a cookbook that contains lots of recipes for making proteins. The original recipe is valuable and must be preserved, so a copy (mRNA) is made, and the copy goes to the production center (ribosomes) to prepare the food (protein).

Protein synthesis requires the process of **translation.** Translation means to put information into a different language. In this case, a sequence of *nucleotides* is translated into the sequence of *amino acids.* This is possible only if the bases in DNA and mRNA code for amino acids. This code is called the genetic code.

The Genetic Code

Recognizing that there must be a genetic code, investigators wanted to know how four bases (A, C, G, U) could provide enough combinations to code for 20 amino acids? If the code were a singlet code (only one base stands for an amino acid), only four amino acids could be encoded. If the code were a doublet (any two bases stand for one amino acid), it would still not be possible to code for 20 amino acids, but if the code were a triplet, then the four bases could supply 64 different triplets, far more than needed to code for 20 different amino acids. It should come as no surprise, then, to learn that the code is a **triplet code.**

Each three-letter (base) unit of an mRNA molecule is called a **codon.** The translation of all 64 mRNA codons has been determined (Fig. 21.7). Sixty-one triplets correspond to a particular amino acid; the remaining three are stop codons, which signal polypeptide termination. The one codon that stands for the amino acid methionine is also a start codon signaling polypeptide initiation. Notice, too, that most amino acids have more than one codon; leucine, serine, and arginine have six different codons, for example. This offers some protection against possibly harmful mutations that change the sequence of the bases.

To crack the code, a cell-free experiment was done: Artificial RNA was added to a medium containing bacterial ribosomes and a mixture of amino acids. Comparison of the bases in the RNA, with the resulting polypeptide, allowed

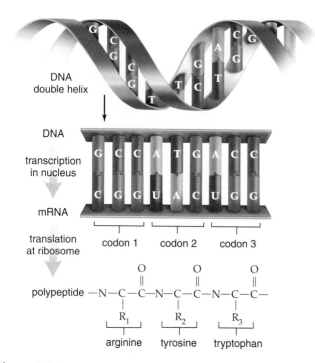

Figure 21.6   Overview of gene expression.
Transcription occurs when DNA acts as a template for RNA (e.g., mRNA) synthesis. (Notice that uracil [U], in RNA, takes the place of thymine [T], in DNA.) Translation occurs when the sequence of codons of mRNA specify the sequence of amino acids in a polypeptide.

First Base	Second Base				Third Base
	U	**C**	**A**	**G**	
U	UUU phenylalanine	UCU serine	UAU tyrosine	UGU cysteine	**U**
	UUC phenylalanine	UCC serine	UAC tyrosine	UGC cysteine	**C**
	UUA leucine	UCA serine	UAA *stop*	UGA *stop*	**A**
	UUG leucine	UCG serine	UAG *stop*	UGG tryptophan	**G**
C	CUU leucine	CCU proline	CAU histidine	CGU arginine	**U**
	CUC leucine	CCC proline	CAC histidine	CGC arginine	**C**
	CUA leucine	CCA proline	CAA glutamine	CGA arginine	**A**
	CUG leucine	CCG proline	CAG glutamine	CGG arginine	**G**
A	AUU isoleucine	ACU threonine	AAU asparagine	AGU serine	**U**
	AUC isoleucine	ACC threonine	AAC asparagine	AGC serine	**C**
	AUA isoleucine	ACA threonine	AAA lysine	AGA arginine	**A**
	AUG *(start)* methionine	ACG threonine	AAG lysine	AGG arginine	**G**
G	GUU valine	GCU alanine	GAU aspartate	GGU glycine	**U**
	GUC valine	GCC alanine	GAC aspartate	GGC glycine	**C**
	GUA valine	GCA alanine	GAA glutamate	GGA glycine	**A**
	GUG valine	GCG alanine	GAG glutamate	GGG glycine	**G**

Figure 21.7   Messenger RNA codons.
Notice that in this chart, each of the codons (white rectangles) are composed of three letters representing the first base, second base, and third base. For example, find the rectangle where C for the first base and A for the second base intersect. You will see that U, C, A, or G can be the third base. CAU and CAC are codons for histidine; CAA and CAG are codons for glutamine.

investigators to decipher the code. For example, an mRNA with a sequence of repeating guanines (GGG'GGG'...) would encode a string of glycine amino acids.

The genetic code is just about universal in living things. This suggests that the code dates back to the very first organisms on Earth and that all living things are related.

Transcription

During transcription, a segment of the DNA serves as a template for the production of an RNA molecule. Although all three classes of RNA are formed by transcription, we will focus on transcription to form mRNA.

Forming mRNA

Transcription begins when the enzyme **RNA polymerase** opens up the DNA helix just in front of it so that complementary base pairing can occur. Then, RNA polymerase joins the RNA nucleotides, and an mRNA molecule results. When mRNA forms, it has a sequence of bases complementary to DNA; wherever A, T, G, or C is present in the DNA template, U, A, C, or G is incorporated into the mRNA molecule (Fig. 21.8). Now, mRNA is a faithful copy of the sequence of bases in DNA.

Processing mRNA

After the mRNA is transcribed in human cells, it must be *processed* before entering the cytoplasm.

The newly synthesized *primary mRNA* molecule becomes a *mature mRNA* molecule after processing (Fig. 21.9). Most genes in humans are interrupted by segments of DNA that are not part of the gene. These portions are called *introns* because they are intragene segments. The other portions of the gene are called *exons* because they are ultimately expressed. Only exons result in a protein product.

Primary mRNA contains bases that are complementary to both exons and introns, but during processing, (1) one end of the mRNA is capped, by the addition of an altered guanine nucleotide, and the other end is given a tail, by the addition of adenosine nucleotides. (2) The introns are removed, and the exons are joined to form a mature mRNA molecule consisting of continuous exons. This *splicing* of mRNA is done by a complex composed of both RNA and protein. Surprisingly, RNA, not the protein, is the enzyme, and so it is called a *ribozyme*.

Ordinarily, processing brings together all the exons of a gene. In some instances, cells use only certain exons rather than all of them to form a mature RNA transcript. The result can be a different protein product in each cell. Alternate mRNA splicing is believed to account for the ability of a single gene to result in two different proteins in a cell.

In order to get pigs to express genes for human recognition proteins, researchers insert DNA for these genes into the embryo of a pig. Replication of DNA occurs prior to mitosis, so every cell of the adult pig contains a copy of this DNA. If all goes well, gene expression occurs. First, transcription and mRNA processing occur. Then translation, which is described next, occurs in each cell. In the end, every cell of the organ to be transplanted bears the human recognition proteins.

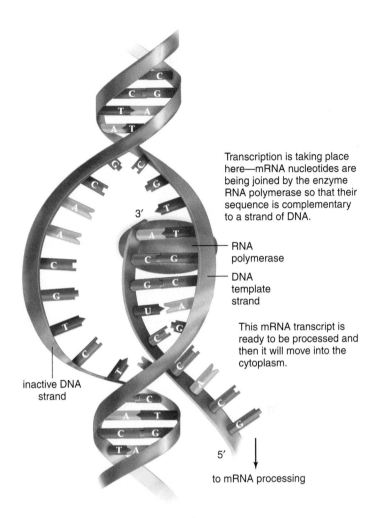

Transcription is taking place here—mRNA nucleotides are being joined by the enzyme RNA polymerase so that their sequence is complementary to a strand of DNA.

3'

— RNA polymerase

— DNA template strand

This mRNA transcript is ready to be processed and then it will move into the cytoplasm.

inactive DNA strand

5'

to mRNA processing

Figure 21.8   Transcription to form mRNA.
During transcription, complementary RNA is made from a DNA template. A portion of DNA unwinds and unzips at the point of attachment of RNA polymerase. A strand of mRNA is produced when complementary bases join in the order dictated by the sequence of bases in template DNA.

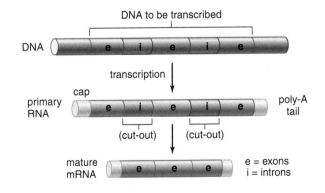

DNA to be transcribed

DNA

transcription

primary RNA   cap   poly-A tail

(cut-out)   (cut-out)

mature mRNA   e = exons
                                                i = introns

Figure 21.9   mRNA processing.
During processing, a cap and tail are added to mRNA, and the introns are removed so that only exons remain.

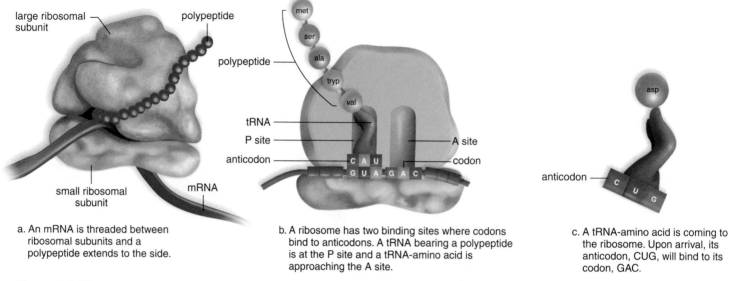

a. An mRNA is threaded between ribosomal subunits and a polypeptide extends to the side.

b. A ribosome has two binding sites where codons bind to anticodons. A tRNA bearing a polypeptide is at the P site and a tRNA-amino acid is approaching the A site.

c. A tRNA-amino acid is coming to the ribosome. Upon arrival, its anticodon, CUG, will bind to its codon, GAC.

Figure 21.10   Ribosome structure and function.

Protein synthesis occurs at a ribosome. **a.** Side view of a ribosome showing mRNA and a growing polypeptide. **b.** Structure and function of a large ribosomal subunit. **c.** tRNA structure and function.

Translation

During translation, transfer RNA (tRNA) molecules bring amino acids to the ribosomes (Fig. 21.10), where polypeptide synthesis occurs. A ribosome consists of a large and a small subunit that join together in the cytoplasm just as protein synthesis begins. Each ribosome has a binding site for mRNA, as well as binding sites for two tRNA molecules at a time. Usually, there is more than one tRNA molecule for each of the 20 amino acids found in proteins. The amino acid binds to one end of the molecule. Therefore, the entire complex is designated as tRNA–amino acid. At the other end of each tRNA is a specific **anticodon,** a group of three bases complementary to an mRNA codon. The tRNA molecules come to the ribosome at the A (for amino acid) site where each anticodon pairs by hydrogen bonding with a codon.

The order in which tRNA–amino acids come to the ribosome is directed by the sequence of the mRNA codons. In this way, the order of codons in mRNA brings about a particular order of amino acids in a protein. The tRNA that is attached to the growing polypeptide is at the P site of a ribosome, as shown in Figure 21.10.

If the codon sequence in a portion of the DNA that codes for a human recognition protein in a pig's cell is ACC, GUA, and AAA, what will be the sequence of amino acids in a portion of the polypeptide? Inspection of Table 21.2 (see page 447) allows us to determine this:

Codon	Anticodon	Amino Acid
ACC	UGG	Threonine
GUA	CAU	Valine
AAA	UUU	Lysine

Polypeptide synthesis requires three steps: initiation, elongation, and termination (Fig. 21.11).

1. During *initiation,* mRNA binds to the smaller of the two ribosomal subunits; then the larger subunit associates with the smaller one.

2. During *elongation,* the polypeptide lengthens, one amino acid at a time. An incoming tRNA–amino acid complex arrives at the A site and then receives the peptide from the outgoing tRNA. The ribosome moves laterally so that again the P site is filled by a tRNA-peptide complex. The A site is now available to receive another incoming tRNA–amino acid complex. In this manner, the peptide grows, and the linear structure of a polypeptide comes about. (The particular shape of a polypeptide is formed later.)

3. Then *termination* of synthesis occurs at a codon that means stop and does not code for an amino acid. The ribosome dissociates into its two subunits and falls off the mRNA molecule.

During elongation, about five amino acids are added to a polypeptide every second. However, many ribosomes are at work forming the same polypeptide. As soon as the initial portion of mRNA has been translated by one ribosome and the ribosome has begun to move down the mRNA, another ribosome attaches to the mRNA. Therefore, several ribosomes, collectively called a **polyribosome,** can move along one mRNA at a time. And several polypeptides of the same type can be synthesized using one mRNA molecule (Fig. 21.12).

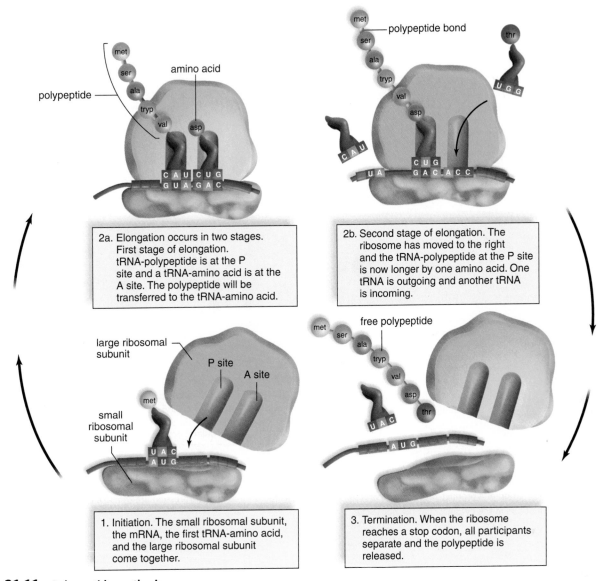

2a. Elongation occurs in two stages. First stage of elongation. tRNA-polypeptide is at the P site and a tRNA-amino acid is at the A site. The polypeptide will be transferred to the tRNA-amino acid.

2b. Second stage of elongation. The ribosome has moved to the right and the tRNA-polypeptide at the P site is now longer by one amino acid. One tRNA is outgoing and another tRNA is incoming.

1. Initiation. The small ribosomal subunit, the mRNA, the first tRNA-amino acid, and the large ribosomal subunit come together.

3. Termination. When the ribosome reaches a stop codon, all participants separate and the polypeptide is released.

Figure 21.11 **Polypeptide synthesis.**
Polypeptide synthesis takes place at a ribosome and has three steps: (1) initiation, (2) elongation, and (3) termination.

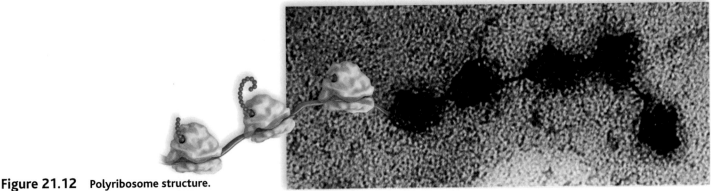

Figure 21.12 **Polyribosome structure.**
Several ribosomes, collectively called a polyribosome, move along an mRNA molecule at one time. They function independently of one another; therefore, several polypeptides can be made simultaneously.

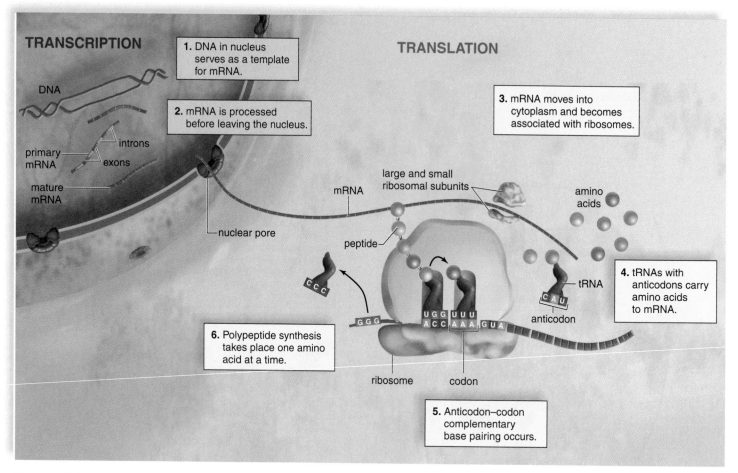

TRANSCRIPTION

TRANSLATION

1. DNA in nucleus serves as a template for mRNA.

DNA

2. mRNA is processed before leaving the nucleus.

3. mRNA moves into cytoplasm and becomes associated with ribosomes.

primary mRNA

introns

exons

mature mRNA

large and small ribosomal subunits

mRNA

amino acids

nuclear pore

peptide

4. tRNAs with anticodons carry amino acids to mRNA.

tRNA

CAU

anticodon

CCC

UGG UUU
ACC AAA GUA

GGG

6. Polypeptide synthesis takes place one amino acid at a time.

ribosome   codon

5. Anticodon–codon complementary base pairing occurs.

Figure 21.13   Gene expression.

Gene expression leads to the formation of a product, most often a protein. The two steps required for gene expression are transcription, which occurs in the nucleus, and translation, which occurs in the cytoplasm at the ribosomes.

Review of Gene Expression

DNA in the nucleus contains a *triplet code.* Each group of three bases stands for a specific amino acid (Fig. 21.13 and Table 21.3). During transcription, a segment of a DNA strand serves as a template for the formation of mRNA. The bases in mRNA are complementary to those in DNA; every three bases is a *codon* for a certain amino acid. Messenger RNA is processed before it leaves the nucleus, during which time the introns are removed and the ends are modified. Messenger RNA carries a sequence of codons to the *ribosomes,* which are composed of rRNA and proteins. A tRNA bonded to a particular amino acid has an *anticodon* that pairs with a codon in mRNA. During translation, tRNAs and their attached amino acids arrive at the ribosomes, where the linear sequence of codons of mRNA determines the order in which amino acids become incorporated into a protein.

Table 21.3	Participants in Gene Expression	
Name of Molecule	**Special Significance**	**Definition**
DNA	Genetic information	Sequence of DNA bases
mRNA	Codons	Sequence of three RNA bases complementary to DNA
tRNA	Anticodon	Sequence of three RNA bases complementary to codon
rRNA	Ribosome	Site of protein synthesis
Amino acid	Building block for protein	Transported to ribosome by tRNA
Protein	Enzyme, structural protein, or secretory product	Amino acids joined in a predetermined order

The Regulation of Gene Expression

All cells receive a copy of all genes; however, cells differ as to which genes are actively expressed. Muscle cells, for example, have a different set of genes that are turned on in the nucleus and different proteins that are active in the cytoplasm than do nerve cells. A variety of mechanisms regulate gene expression, from transcription to protein activity in our cells. These mechanisms can be grouped under four primary levels of control—two that pertain to the nucleus and two that pertain to the cytoplasm:

1. *Transcriptional control* In the nucleus, a number of mechanisms regulate which genes are transcribed and/or the rate at which transcription of genes occurs. These include the organization of chromatin and the use of transcription factors that initiate transcription, the first step in gene expression.

2. *Posttranscriptional control* Posttranscriptional control occurs in the nucleus after DNA is transcribed and mRNA is formed. How mRNA is processed before it leaves the nucleus and also how fast mature mRNA leaves the nucleus can affect the amount of gene expression.

3. *Translational control* Translational control occurs in the cytoplasm after mRNA leaves the nucleus and before there is a protein product. The life expectancy of mRNA molecules (how long they exist in the cytoplasm) can vary, as can their ability to bind ribosomes. It is also possible that some mRNAs may need additional changes before they are translated at all.

4. *Posttranslational control* Posttranslational control, which also occurs in the cytoplasm, occurs after protein synthesis. The polypeptide product may have to undergo additional changes before it is biologically functional. Also, a functional enzyme is subject to feedback control—the binding of an enzyme's product can change its shape so that it is no longer able to carry out its reaction.

Activated Chromatin

For a gene to be transcribed in human cells, the chromosome in that region must first decondense. The chromosomes within the developing egg cells of many vertebrates are called lampbrush chromosomes because they have many loops that appear to be bristles (Fig. 21.14). Here, mRNA is being synthesized in great quantity so that adequate protein synthesis can be carried out.

Transcription Factors

In human cells, **transcription factors** are DNA-binding proteins. Every cell contains many different types of transcription factors, and a specific combination of transcription factors and modifiers is believed to regulate the activity

of any particular gene. After the right combination binds to DNA, an RNA polymerase attaches to DNA and begins the process of transcription.

As cells mature, they differentiate and become specialized. Specialization is determined by which genes are active and, therefore, perhaps, by which transcription factors are active in that cell. Signals received from inside and outside the cell could turn on or off genes that code for certain transcription factors. For example, the gene for fetal hemoglobin ordinarily gets turned off as a newborn matures; therefore, one possible treatment for sickle-cell disease is to turn this gene on again.

Why might the DNA for human recognition proteins not be expressed after the DNA is inserted into cells? It won't be expressed unless the DNA inserts into host DNA at a location scheduled for expression by control elements such as transcription factors.

✅ Check Your Progress 21.2

1. a. What is the structure of a protein, and **(b)** what function do proteins perform in a metabolic pathway?

2. What are the two steps of gene expression, and what has been formed at the end of each step?

3. a. What are the various levels of genetic control in human cells?
   b. How is transcriptional control achieved?

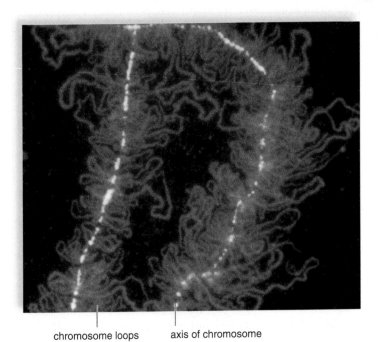

chromosome loops   axis of chromosome

Figure 21.14 Lampbrush chromosomes.
These chromosomes, seen in amphibian egg cells, give evidence that when mRNA is being synthesized, chromosomes decondense. Each chromosome has many loops extending from its axis (white on the micrograph). Many mRNA transcripts are being made off these DNA loops.

Science Focus

The Pig Genome To Be Sequenced

In January 2006, the U.S. Department of Agriculture awarded the University of Illinois $10 million to provide the initial genome sequence of the pig. "After many years of laying the foundation for sequencing the pig genome, it is truly rewarding to see our dreams of a porcine sequence come true," said Jonathan Beever of U of I. "The implications will help tremendously in the swine industry, but will also have biomedical importance as well, such as pig-to-human transplants," said Lawrence Schook, also of U of I who will serve as director of the project. "Now we'll be also be able to see how the various proteins in the genes work together to make, for example, a human toenail as opposed to a pig hoof." Schook explained that although there are remarkable similarities between the pig and the human, they are also very different. "There may be two or three percent of the genome that actually determine whether the organism becomes a pig or a human. This information will show those differences."

Schook explained that the scale of this project is huge. "Picture a million test tubes. We're looking at something that has 2.5 billion pieces and only 500 can be done at time. Each time takes a couple of hours to do. So, if we used the labs at the U of I, it would take us about 10 years of just doing this nonstop to complete. But we are having the Wellcome Trust Sanger Institute in the U.K. do the labor-intensive work. They have the labs and infrastructure to do this kind of work on a big scale."

The pig genome is approximately the same size as the human genome—about 2.5 billion base pairs located on 18 chromosomes, plus the two sex chromosomes. Sequencing of the human genome took almost 10 years and cost approximately $3 billion. "The robotic sequencing technologies and bioinformatics developed from the Human Gene Initiative will permit the same draft sequence for the pig to be concluded in just two years at a cost of $20 million," said Schook. He said the work will be completed five times faster and at 1/150th the cost. This phase of the project is scheduled to be completed in 24 months, coincidentally, just in time for 2007 Chinese year of the pig.

21.3 Genomics

The Human Genome Has Been Sequenced

Genetics in the twenty-first century concerns genomics, the study of genomes—our genes and the genes of other organisms. Due to the Human Genome Project (HGP), a 13-year effort that involved both university and private laboratories around the globe, we now know the order of the 3 billion bases A, T, C, and G in our genome. This biological achievement has been likened to arriving at the periodic table of the elements in chemistry.

How did they do it? First, investigators developed a laboratory procedure that would allow them to decipher a short sequence of base pairs, and then instruments became available that would carry out this procedure automatically. Whose DNA did they use? Sperm DNA was the material of choice because sperm DNA has a much higher ratio of DNA to protein than other type of cells. (Recall that sperm do provide both X and Y chromosomes.) However, white cells from the blood of female donors was also used in order to include female-originated samples. The male and female donors were of European, African American (both North and South), and Asian ancestry.

As you can imagine, many small regions of DNA that vary among individuals—that is, polymorphisms—were identified during the HGP. Most of these are single nucleotide polymorphisms (SNPs). Many SNPs have no physiological effect; others may contribute to the diversity of human beings or possibly increase an individual's susceptibility to disease and response to medical treatments.

A surprising finding has been that genome size is not proportionate to the number of genes and does not correlate to complexity of the organism. For example, the plant *Arabidopsis thaliana,* with a much smaller number of bases, has approximately the same number of genes as a human being. A gene in this context is a sequence of DNA bases that is transcribed into any type of RNA molecule. Some of these RNA molecules are mRNA molecules that direct protein synthesis. Determining that humans have 25,000 genes makes use of a number of techniques, many of which rely on the identification of RNAs in the cell and then working backward to find the DNA that can pair with that RNA. It is not yet known what each of the 25,000 human genes do specifically.

Today, laboratory instruments are commercially available that can automatically analyze up to two million base pairs of DNA in a 24-hour period. Improvements in DNA sequencers are still being designed, however. Researchers are now working on systems that can read as many as 1,700 bases a second! This means that a DNA sequencer might be available by 2014 that could read a genome for only $1,000. The Personal Genome Project is underway, not only to produce faster sequences, but also to determine what are the possible benefits and drawbacks of every person having their own particular genome sequenced. Possible benefits are discussed in the Health Focus on page 457.

Functional and Comparative Genomics

The next step of the HGP is to find out how our 25,000 genes function in cells and how they create a human being. Functional **genomics** also wants to discover the function of gene deserts, 82 regions that comprise 3% of the genome where DNA has no identifiable genes.

Comparing genomes is one way to determine how species have evolved and how genes and noncoding regions of the genome function. In one study, researchers compared our genome to that of chromosome 22 in chimpanzees (Fig. 21.15). They found three types of genes of particular interest: a gene for proper speech development, several for hearing, and several for smell. Genes necessary for speech development are thought to have played an important role in human evolution. You can suppose that changes in hearing may also have facilitated using language for communication between people. To explain differences in smell genes, investigators speculated that the olfaction genes may have affected dietary changes or sexual selection. Or, they may have been involved in other traits, rather than just smell. The researchers who did this study were surprised to find that many of the other genes they located and studied are known to cause human diseases. They wondered if comparing genomes would be a way of finding other genes that are associated with human diseases. Investigators are taking all sorts of avenues to link human base sequence differences to illnesses.

Another surprising discovery has been how very similar the genomes of all vertebrates are. Researchers weren't surprised to learn that the base sequence of chimpanzees and humans is 99% alike, but did not expect to find that our sequence is also 88% like that of a mouse, for example. So, the quest is on to discover how possibly the regulation of genes explains why we have one set of traits and mice have another set, despite the similarity of our base sequences. One possibility is alternative gene splicing, so that we differ from mice by what types of proteins we manufacture and/or by when and where certain proteins are present.

Proteomics and Bioinformatics Are New Endeavors

Proteomics is the study of the structure, function, and interaction of cellular proteins. Many of our genes are translated into proteins at some time, in some of our cells. The translation of all coding genes results in a collection of proteins, called the human proteome. The analysis of proteomes is more challenging than the analysis of genomes for two reasons. Protein concentrations differ widely in cells, and researchers have to be able to identify proteins, no matter whether there is one or thousands of copies of a protein in a cell. Any particular protein differs minute by minute in concentration, interactions, cellular location, and chemical modifications among other features. Yet, to understand a protein, all these features must be analyzed. Computer modeling of the three-dimensional shape of cellular proteins is an important part of proteomics. The study of cellular proteins and how they function is essential to the discovery of better drugs. Most drugs are proteins or molecules that affect the function of proteins.

Bioinformatics is the application of computer technologies to the study of the genome. Genomics and proteomics produce raw data, and these fields depend on computer analysis to find significant patterns in the data. As a result of bioinformatics, scientists are hopeful of finding cause-and-effect relationships between various genetic profiles and genetic disorders caused by multifactorial genes. By correlating any sequence changes with resulting phenotypes, bioinformatics may find that desert regions of the genome do have functions. New computational tools will, most likely, be needed to accomplish these goals.

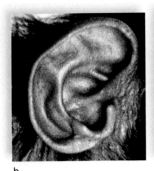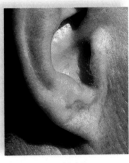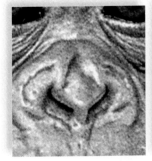

a. b. c.

Figure 21.15 Studying genomic differences between chimpanzees and humans.
Did changes in the genes for (**a**) speech, (**b**) hearing, and (**c**) smell influence the evolution of humans?

A Person's Genome Can Be Modified

Gene therapy is the insertion of genetic material into human cells for the treatment of a disorder. Gene therapy has been used to cure inborn errors of metabolism and also to treat more generalized disorders such as cardiovascular disease and cancer. Both ex vivo (outside the body) and in vivo (inside the body) gene therapy methods are used.

Ex Vivo Gene Therapy   Figure 21.16 describes the methodology for treating children who have SCID (severe combined immunodeficiency). These children lack the enzyme ADA (adenosine deaminase), which is involved in the maturation of T and B cells. To carry out gene therapy, bone marrow stem cells are removed from bone marrow and infected with an RNA retrovirus that carries a normal gene for the enzyme. Then the cells are returned to the patient. Bone marrow stem cells are preferred for this procedure because they divide to produce more cells with the same genes. Patients who have undergone this procedure show significantly improved immune function associated with a sustained rise in the level of ADA enzyme activity in the blood.

Among the many gene therapy trials, one is for the treatment of familial hypercholesterolemia, a condition that develops when liver cells lack a receptor protein for removing cholesterol from the blood. The high levels of blood cholesterol make the patient subject to fatal heart attacks at a young age. A small portion of the liver is surgically excised and then infected with a retrovirus containing a normal gene for the receptor before being returned to the patient. Several patients have experienced lowered serum cholesterol levels following this procedure.

In Vivo Gene Therapy   Cystic fibrosis patients lack a gene that codes for the transmembrane carrier of the chloride ion. They often suffer from numerous and potentially deadly infections of the respiratory tract. In gene therapy trials, the gene needed to cure cystic fibrosis is sprayed into the nose or delivered to the lower respiratory tract by adeno viruses or by the use of liposomes, microscopic vesicles that spontaneously form when lipoproteins are put into a solution. Investigators are trying to improve uptake, and they are also hypothesizing that a combination of all three vectors might be more successful.

Genes are being used to treat medical conditions such as poor coronary circulation. It has been known for some time that VEGF (vascular endothelial growth factor) can cause the growth of new blood vessels. The gene that codes for this growth factor can be injected alone or within a virus into the heart to stimulate branching of coronary blood vessels. Patients report that they have less chest pain and can run longer on a treadmill.

Gene therapy is increasingly used as a part of cancer therapy. Genes are being used to make healthy cells more tolerant of chemotherapy and to make tumors more vulnerable to chemotherapy. The gene *p53* brings about apoptosis, and there is much interest in introducing it into cancer cells and, in that way, killing them off.

☑ Check Your Progress 21.3

1. What are genomics and proteomics?
2. How can a person's genome be modified?

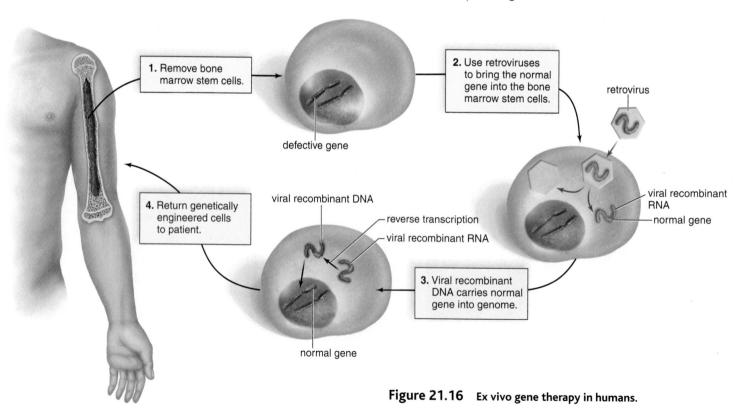

Figure 21.16   **Ex vivo gene therapy in humans.**

Health **Focus**

New Cures on the Horizon

Now that we know the sequence of the bases in the DNA of all the human chromosomes, biologists all over the world believe this knowledge will result in rapid medical advances for ourselves and our children.

First prediction: A longer and healthier life will be yours.

Soon it will be possible to sequence your personal genome and to use it to predict possible illnesses in the future. Many illnesses are caused by genes and also environmental influences. If you have a propensity toward a particular illness, you can avoid any environmental influences that might contribute to the condition.

First prediction:
A longer and healthier life will be yours.

We know that the presence of free radicals cause cellular molecules to become unstable and cells to die. Certain genes are believed to code for antioxidant enzymes that detoxify free radicals. Similarly, some people may have longevity genes that allow them to live longer than others do. Researchers may be able to isolate the antioxidant genes and the longevity genes and make copies available to the general public, so that everyone will have the opportunity to live longer in good health.

Second prediction: Many new medicines tailored to the individual will be available.

Most drugs are proteins or small chemicals that are able to interact with proteins. Today's drugs were usually discovered in a hit-or-miss fashion, but now researchers will be able to take a more systematic approach to finding effective medicines. In a recent search for a medicine that makes wounds heal, researchers cultured skin cells with 14 proteins (found by chance) that can cause skin cells to grow. Only one of these proteins made skin cells grow and did nothing else. They expect this protein to become an effective drug for conditions such as venous ulcers, which are skin lesions that affect many thousands of people in the United States. Tests leading to effective medicines can be carried out with many more proteins that scientists will discover by examining the human genome.

Many drugs, potentially, have unwanted side effects. Why do some people, and not others, have one or more of the side effects? Most likely, this is because people have different genetic mutations. It is expected that physicians will be able to match patients to drugs that are safe for them on the basis of their personal genome.

Second prediction:
Many new medicines will be available.

One study found that various combinations of mutations can lead to the development of asthma. A particular drug, called albuterol, is effective and safe for patients with certain combinations of mutations and not others. This example and others show that many diseases are multifactorial and that only knowledge of the personal genome is able to detect which mutations are causing an individual to have a disease and how it should be properly treated.

Third prediction: You will be able to select your children.

Genome sequence data will be used to identify many more mutant genes that cause genetic disorders than are presently known. In the future, it may be possible to cure genetic disorders before the child is born by adding a normal gene to any egg that carries a mutant gene. Or an artificial chromosome, constructed to carry a large number of corrective genes, could automatically be placed in eggs. In vitro fertilization would have to be utilized

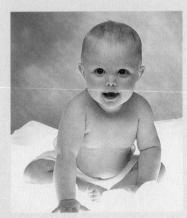

Third prediction:
You will be able to select your children.

in order to take advantage of such measures for curing genetic disorders before conception.

Genome sequence data can also be used to identify polygenic genes for traits such as height, intelligence, or behavioral characteristics. A couple could decide on their own which genes they wish to use to enhance a child's phenotype. In other words, the sequencing of the human genome may bring about a genetically just society, in which all types of genes would be accessible to all parents.

21.4 DNA Technology

Genes Can Be Isolated and Cloned

In biology, **cloning** is the production of genetically identical copies of DNA, cells, or organisms through an asexual means. **Gene cloning** can be done to produce many identical copies of the same gene. **Recombinant DNA,** which contains DNA from two or more different sources, allows genes to be cloned (Fig. 21.17). To create recombinant DNA, a technician needs a **vector,** by which the gene of interest will be introduced into a host cell, such as a bacterium. One common vector is a plasmid. Recall that **plasmids** are small accessory rings of DNA found in bacteria. The ring is not part of the bacterial chromosome and replicates on its own.

① A **restriction enzyme** is used to cleave human DNA and plasmid DNA. Hundreds of restriction enzymes occur naturally in bacteria, where they cut up any viral DNA that enters the cell. They are called restriction enzymes because they restrict the growth of viruses, but they also act as molecular scissors to cleave any piece of DNA at a specific site. For example, the restriction enzyme called *Eco*RI always cuts double-stranded DNA at this sequence of bases and in this manner:

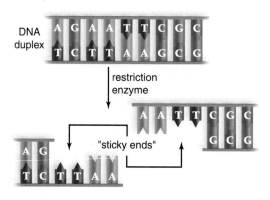

The restriction enzyme creates a gap in plasmid DNA in which foreign DNA (the insulin gene) can be placed if it ends in bases complementary to those exposed by the restriction enzyme. To ensure this, it is only necessary to use the same type of restriction enzyme to cleave both human DNA containing the gene for insulin and plasmid DNA.

② The enzyme called **DNA ligase** is used to seal foreign DNA into the opening created in the plasmid. The single-stranded, but complementary, ends of a cleaved DNA molecule are called "sticky ends" because they can bind a piece of DNA by complementary base pairing. Sticky ends facilitate the sealing of the plasmid DNA with human DNA for the insulin gene. Now the vector is complete, and an rDNA molecule has been prepared.

③ Some of the bacterial cells take up a recombinant plasmid, especially if the bacteria have been treated to make them more permeable.

④a Gene cloning occurs as the plasmid replicates on its own. Scientists clone genes for a number of reasons. They might want to determine the difference in base sequence between a normal gene and a mutated gene. Or, they might use the genes to genetically modify organisms in a beneficial way.

④b The bacterium is also transformed and can make a product (e.g., insulin) that it could not make before. For a human gene to express itself in a bacterium, the gene has to be accompanied by regulatory regions unique to bacteria. Also, the gene should not contain introns because bacteria don't have introns. However, it is possible to make a human gene that lacks introns. The enzyme called reverse transcriptase can be used to make a DNA copy of mRNA. The DNA molecule, called **complementary DNA (cDNA),** does not contain introns.

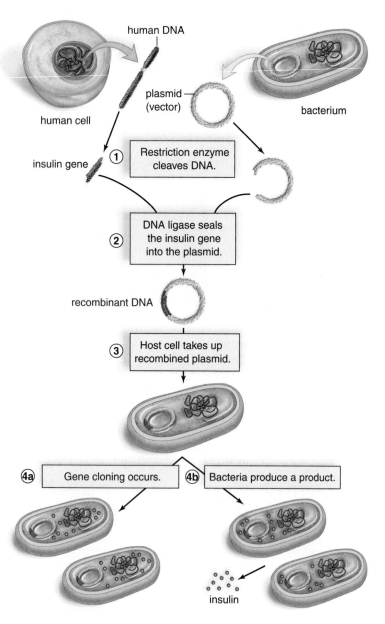

Figure 21.17   Cloning a human gene.

Specific DNA Sequences Can Be Cloned

If only small pieces of identical DNA are needed, the **polymerase chain reaction (PCR),** developed by Kary Mullis in 1985, can create copies of a segment of DNA quickly in a test tube. PCR is very specific—it amplifies (makes copies of) a targeted DNA sequence. The targeted sequence can be less than one part in a million of the total DNA sample!

PCR requires the use of DNA polymerase, the enzyme that carries out DNA replication, and a supply of nucleotides for the new DNA strands. PCR is a chain reaction because the targeted DNA is repeatedly replicated as long as the process continues. The colors in Figure 21.18 distinguish old DNA from new DNA. Notice that the amount of DNA doubles with each replication cycle.

DNA amplified by PCR is often analyzed for various purposes. Mitochondrial DNA base sequences in modern living populations were used to decipher the evolutionary history of human populations. Since so little DNA is required for PCR to be effective, it has even been possible to sequence DNA taken from mummified human brains.

Also, following PCR, DNA can be subjected to DNA fingerprinting. Today, DNA fingerprinting is often carried out by detecting how many times a short sequence (two to five bases) is repeated. Organisms differ by how many repeats they have at particular locations, but how can this be detected? Recall that PCR amplifies only particular portions of the DNA. Therefore, the greater the number of repeats at a location, the greater the amount of DNA that is amplified by PCR. During a process called gel electrophoresis, fragments can be separated according to their charge/size ratios, and the result is a pattern of distinctive bands. If two DNA patterns match, there is a high probability that the DNA came from the same source. It is customary to test for the number of repeats at several locations to further define the source.

DNA fingerprinting has many uses. When the DNA matches that of a virus or mutated gene, it is known that a viral infection, genetic disorder, or cancer is present. Fingerprinting DNA from a single sperm is enough to identify a suspected rapist. DNA fingerprinted from blood or tissues at a crime scene has been successfully used in convicting criminals. Figure 21.19 shows how DNA fingerprinting can be used to identify a criminal. DNA fingerprinting is used to identify the remains of bodies. It was extensively used in identifying the victims of the September 11, 2001, terrorist attacks in the United States.

Applications of PCR and DNA fingerprinting are limited only by our imagination. PCR analysis has been used to identify unknown soldiers and members of the royal Russian family. Paternity suits can be settled, and genetic disorders and illegally poached ivory and whale meat can be identified using these technologies. They have shed new light on evolutionary studies by comparing extracted DNA from certain fossils with that of living organisms.

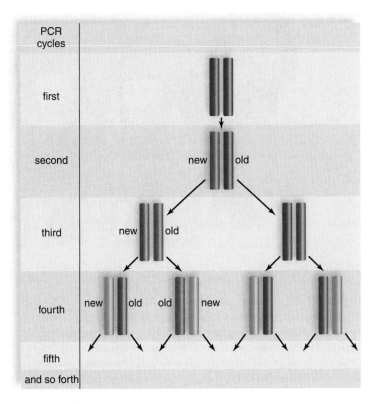

Figure 21.18 Polymerase chain reaction (PCR).
Copies produced per PCR cycle per one original strand.

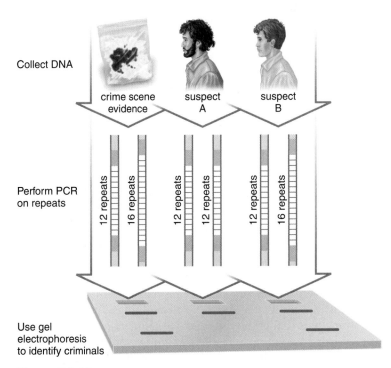

Figure 21.19 DNA fingerprinting.
DNA fingerprinting reveals that suspect B is the criminal.

Biotechnology Products

Today, bacteria, plants, and animals are genetically engineered to produce **biotechnology products**. Organisms that have had a foreign gene inserted into them are called **transgenic organisms.**

From Bacteria

Recombinant DNA technology is used to produce transgenic bacteria, which are grown in huge vats called bioreactors (Fig. 21.20*a*). The gene product is collected from the medium. Biotechnology products on the market that are produced by bacteria include insulin, clotting factor VIII, human growth hormone, t-PA (tissue plasminogen activator), and hepatitis B vaccine. Remember Adam in the opening story for Chapter 15? His doctor gave him growth hormone injections to treat his height deficiency. Transgenic bacteria have many other uses as well. Some have been produced to promote the health of plants. For example, bacteria that normally live on plants and encourage the formation of ice crystals have been changed from frost-plus to frost-minus bacteria. As a result, new crops such as frost-resistant strawberries are being developed.

Bacteria can be selected for their ability to degrade a particular substance, and this ability can then be enhanced by **genetic engineering.** For instance, naturally occurring bacteria that eat oil can be genetically engineered to do an even better job of cleaning up beaches after oil spills (Fig. 21.20*e*). Further, these bacteria were given "suicide" genes that caused them to self-destruct when the job had been accomplished.

From Plants

Corn, potato, soybean, and cotton plants have been engineered to be resistant to either insect predation or herbicides that are widely used (Fig. 21.20*c* and Fig. 21.21). Some corn and cotton plants have been developed that are both insect- and herbicide-resistant. In 2001, transgenic crops were planted on more than 72 million acres worldwide, and the acreage is expected to triple in about five years. If crops are resistant to a broad-spectrum herbicide and weeds are not, then the herbicide can be used to kill the weeds. When herbicide-resistant plants were planted, weeds were easily controlled, less tillage was needed, and soil erosion was minimized.

Crops with other improved agricultural and food quality traits are desirable. A salt-tolerant tomato has already been developed (Fig. 21.22). First, scientists identified a gene coding for a protein that transports Na^+ across the vacuole membrane. Sequestering the Na^+ in a vacuole prevents it from interfering with plant metabolism. Then, the scientists cloned the gene and used it to genetically engineer plants that overproduce the channel protein. The modified plants thrived when watered with a salty solution. Today, crop production is limited by the effects of salinization on about 50% of irrigated lands. Salt-tolerant crops would increase yield on this land. Salt- and also drought- and cold-tolerant crops might help provide enough food for a world population that may nearly double by 2050.

Potato blight is the most serious potato disease in the world. About 150 years ago, it was responsible for the Irish potato famine, which caused the death of millions of people.

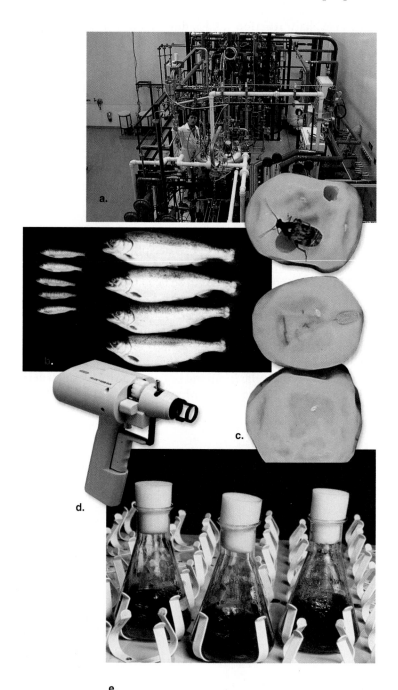

Figure 21.20   Uses of transgenic organisms.

a. Transgenic bacteria, grown industrially, are used to produce medicines. Transgenic animals and plants increase the food supply. **b.** These salmon (*right*) are much heavier and reproduce much sooner because of the growth hormone genes they received as embryos. **c.** Pests didn't consume unblemished peas because the plants that produced them received genes for pest inhibitors. **d.** Gene guns are often used to insert genes in plant embryos while they are in tissue culture. **e.** Transgenic bacteria can also be used for environmental cleanup. These bacteria are able to break down oil.

a. Herbicide-resistant soybean plants   b. Nonresistant potato plant   c. Pest-resistant potato plant

Figure 21.21   Genetically engineered plants.

By placing a gene from a naturally blight-resistant wild potato into a farmed variety, researchers have now made potato plants that are no longer vulnerable to a range of blight strains. Some progress has also been made to increase the food quality of crops. Soybeans have been developed that mainly produce the monounsaturated fatty acid oleic acid, a change that may improve human health.

Other types of genetically engineered plants are also expected to increase productivity. Leaves might be engineered to lose less water and take in more carbon dioxide. To accomplish the latter, the efficiency of the enzyme which captures carbon dioxide in plants, could be improved. A team of Japanese scientists is working on altering the process of photosynthesis in rice so that rice plants can do well in hot, dry weather. These modifications would require a more complete reengineering of plant cells than the single-gene transfers that have been done so far.

Single-gene transfers have allowed plants to produce various products, including human hormones, clotting factors, and antibodies. One type of antibody made by corn can deliver radioisotopes to tumor cells, and another made by soybeans may be developed to treat genital herpes.

Transgenic Crops of the Future	
Improved Agricultural Traits	
Disease-protected	Wheat, corn, potatoes
Herbicide-resistant	Wheat, rice, sugar beets, canola
Salt-tolerant	Cereals, rice, sugarcane
Drought-tolerant	Cereals, rice, sugarcane
Cold-tolerant	Cereals, rice, sugarcane
Improved yield	Cereals, rice, corn, cotton
Modified wood pulp	Trees
Improved Food Quality Traits	
Fatty acid/oil content	Corn, soybeans
Protein/starch content	Cereals, potatoes, soybeans, rice, corn
Amino acid content	Corn, soybeans

a. Desirable traits

b. Salt-intolerant   Salt-tolerant

Figure 21.22   Transgenic crops of the future.

Health Focus

Are Genetically Engineered Foods Safe?

A series of focus groups conducted by the Food and Drug Administration (FDA) in 2000 showed that although most participants believed that genetically engineered foods might offer benefits, they also feared unknown long-term health consequences that might be associated with the technology. In Canada, Conrad G. Brunk, a bioethicist at the University of Waterloo in Ontario, has said, "When it comes to human and environmental safety, there should be clear evidence of the absence of risks. The mere absence of evidence is not enough."

The discovery by activists that a genetically engineered corn called StarLink had inadvertently made it into the food supply triggered the recall of taco shells, tortillas, and many other corn-based foodstuffs from supermarkets. Further, the makers of StarLink were forced to buy back StarLink from farmers and to compensate food producers at an estimated cost of several hundred million dollars. StarLink is a type of "BT" corn. It contains a foreign gene taken from a common soil organism, *Bacillus thuringiensis*, whose insecticidal properties have been long known. About a dozen BT varieties, including corn, potato, and even a tomato, have now been approved for human consumption. These strains contain a gene for an insecticidal protein called CrylA. The makers of StarLink decided to use a gene for a related protein called Cry9C. They thought that using this molecule might slow down the chances of pest resistance to BT corn. In order to get FDA approval for use in foods, the makers of StarLink performed the required tests. Like the other now-approved strains, StarLink wasn't poisonous to rodents, and its biochemical structure is not similar to those of most food allergens. But the Cry9C protein resisted digestion longer than the other BT

proteins when it was put in simulated stomach acid and subjected to heat. Because most food allergens are stable like this, StarLink was not approved for human consumption.

The scientific community is now trying to devise more tests for allergens because it has not been possible to determine conclusively whether Cry9C is or is not an allergen. Also, at this point, it is unclear how resistant to digestion a protein must be in order to be an allergen, and it is also unclear what degree of sequence similarity a potential allergen must have to a known allergen to raise concern. Dean D. Metcalfe, chief of the Laboratory of Allergic Diseases at the National Institute of Allergy and Infectious Diseases, said, "We need to understand thresholds for sensitization to food allergens and thresholds for elicitation of a reaction with food allergens."

Other scientists are concerned about the following potential drawbacks to the planting of BT corn: (1) resistance among populations of the target pest, (2) exchange of genetic material between the transgenic crop and related plant species, and (3) BT crops' impact on nontarget species. They feel that many more studies are needed before it can be said for certain that BT corn has no ecological drawbacks.

Despite controversies, the planting of genetically engineered corn increased in 2001. The USDA reports that U.S. farmers planted genetically engineered corn on 26% of all corn acres, 1% more than in 2000. In all, U.S. farmers planted at least 72 million acres with mostly genetically engineered corn, soybeans, and cotton (Fig. 21A). The public wants all genetically engineered foods to be labeled as such, but this may not be easy to accomplish because, for example, most cornmeal is derived from both conventional and genetically engineered corn. So far, there has been no attempt to sort out one type of food product from the other.

a.

b.

c.

Figure 21A   **Genetically engineered crops.**
Genetically engineered (**a**) corn, (**b**) soybeans, and (**c**) cotton crops are increasingly being planted by today's farmers.

From Animals

Techniques have been developed to insert genes into the eggs of animals. It is possible to microinject foreign genes into eggs by hand, but another method uses vortex mixing. The eggs are placed in an agitator with DNA and silicon-carbide needles, and the needles make tiny holes through which the DNA can enter. When these eggs are fertilized, the resulting offspring are transgenic animals. Using this technique, many types of animal eggs have taken up the gene for bovine growth hormone (bGH). The procedure has been used to produce larger fishes, cows, pigs, rabbits, and sheep. The pigs that are being bred to supply transplant organs for humans can be genetically modified in this manner also.

Gene pharming, the use of transgenic farm animals to produce pharmaceuticals, is being pursued by a number of firms. Genes that code for therapeutic and diagnostic proteins are incorporated into an animal's DNA, and the proteins appear in the animal's milk (Fig. 21.23). Plans are under way to produce drugs for the treatment of cystic fibrosis, cancer, blood diseases, and other disorders by this method. Figure 21.23 outlines the procedure for producing transgenic mammals. The gene of interest is microinjected into donor eggs. Following in vitro fertilization, the zygotes are placed in host females, where they develop. After trans-

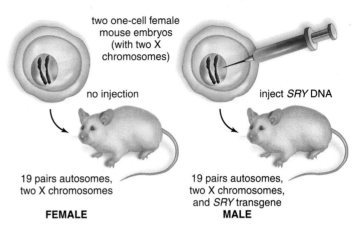

Figure 21.24   Transgenic mice showed that maleness is due to *SRY* **DNA.**

genic female offspring mature, the product is secreted in their milk. Then, cloning can be used to produce many animals that produce the same product. Female clones produce the same product in their milk.

Many researchers are using transgenic mice technology for various research projects in the laboratory. Figure 21.24 shows how this technology demonstrated that a section of DNA called *SRY* (sex-determining region of the Y chromosome) produces a male animal. This *SRY* DNA was cloned, and then one copy was injected into one-celled mouse embryos with two X chromosomes. Injected embryos developed into males, but any that were not injected developed into females. Mouse models have also been created to study human diseases. An allele such as the one that causes cystic fibrosis can be cloned and inserted into mice embryonic stem cells and, occasionally, a mouse embryo homozygous for cystic fibrosis will result. This embryo develops into a mutant mouse that has a phenotype similar to a human with cystic fibrosis. New drugs for the treatment of cystic fibrosis can then be tested in these mice.

As mentioned in the opening story, **xenotransplantation** is the use of animal organs, instead of human organs, in transplant patients. Scientists have chosen to use the pig because animal husbandry has long included the raising of pigs as a meat source, and pigs are prolific. The ability of pigs to express genes for human recognition proteins means that pig organs will be more readily accepted by humans and rejection will not occur.

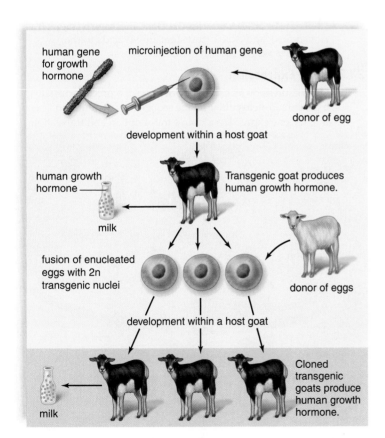

Figure 21.23   Procedure for producing many female clones that produce the same product.

✓ Check Your Progress 21.4

1. What two techniques allow scientists to clone a portion of DNA?

2. a. How and (b) why is DNA fingerprinting done?

3. In what ways does genetic engineering improve agricultural plants and farm animals?

DNA Fingerprinting and the Criminal Justice System

Traditional fingerprinting has been used for years to identify criminals and to exonerate those wrongly accused of crimes. The opportunity now arises to use DNA fingerprinting in the same way. DNA fingerprinting requires only a small DNA sample. This sample can come from blood left at the scene of the crime, semen from a rape case, even a single hair root!

Advocates of DNA fingerprinting claim that identification is "beyond a reasonable doubt." In 2004, Arthur Lee Whitfield was released from prison after DNA fingerprinting proved he had not raped two women in 1981. He had served 22 years of a 63-year sentence, but semen saved all that time proved that he was not the guilty person. Opponents of this technology, however, point out that it is not without its problems. Police or laboratory negligence can invalidate the evidence. For example, during the O. J. Simpson trial, the defense claimed that the DNA evidence was inadmissible because it could not be proven that the police had not "planted" O. J.'s blood at the crime scene. There have also been reported problems with sloppy laboratory procedures and the credibility of forensic experts.

In addition to identifying criminals, DNA fingerprinting can be used to establish paternity and maternity, determine nationality for immigration purposes, and identify victims of a national disaster, such as the terrorist attacks of September 11, 2001.

Considering the usefulness of DNA fingerprints, perhaps everyone should be required to contribute blood to create a national DNA fingerprint databank. Some say, however, this would constitute an unreasonable search, which is unconstitutional.

Decide Your Opinion

1. Would you be willing to provide your DNA for a national DNA databank? Why or why not? What types of privacy restrictions would you want on your DNA?

2. If not everyone, do you think that convicted felons, at least, should be required to provide DNA for a databank?

3. Should all defendants have access to DNA fingerprinting (at government expense) to prove they didn't do a crime? Should this include those already convicted of crimes who want to reopen their cases using new DNA evidence?

Summarizing the Concepts

21.1 DNA and RNA Structure and Function

DNA is the genetic material found in the chromosomes. It replicates, stores information, and mutates for genetic variability.

Structure of DNA

- Double helix composed of two polynucleotide strands. Each nucleotide is composed of a deoxyribose sugar, a phosphate, and a nitrogen-containing base (A, T, C, G).
- The base A is bonded to T, and G is bonded to C.

Replication of DNA

- DNA strands unzip, and a new complementary strand forms opposite each old strand, resulting in two identical DNA molecules:

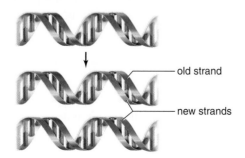

old strand

new strands

The Structure and Function of RNA

- RNA is a single-stranded nucleic acid in which the base U (uracil) occurs instead of T (thymine).
- The three forms of RNA are rRNA (found in the ribosomes); mRNA (carries the DNA message to the ribosomes); and tRNA (transfers amino acids to the ribosomes where protein synthesis occurs).

21.2 Gene Expression

Gene expression leads to the formation of a product, usually a protein. Proteins differ by the sequence of their amino acids. Gene expression requires transcription and translation.

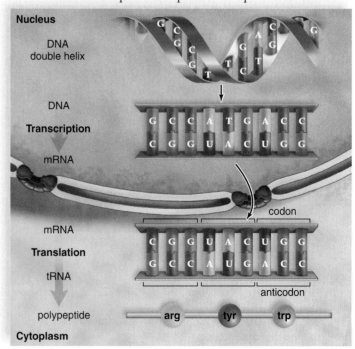

- **Transcription**   Occurs in the nucleus. The DNA triplet code is passed to an mRNA that contains codons. Introns are removed from mRNA during mRNA processing.
- **Translation**   Occurs in the cytoplasm at the ribosomes. tRNA molecules bind to their amino acids, and then their anticodons pair with mRNA codons.

The Regulation of Gene Expression

Regulation of gene expression occurs at four levels in a human cell:

- **Transcriptional Control**   In the nucleus; the degree to which a gene is transcribed into mRNA determines the amount of gene product.
- **Posttranscriptional Control**   In the nucleus; involves mRNA processing and how fast mRNA leaves the nucleus.
- **Translational Control**   In the cytoplasm; afffects when translation begins and how long it continues.
- **Posttranslational Control**   In the cytoplasm; occurs after protein synthesis.

21.3 Genomics

A Person's Genome Can Be Modified

- Gene therapies such as ex vivo gene therapy or in vivo gene therapy can treat various medical conditions.
- The human genome has now been sequenced via the 13-year-long Human Genome Project.
- Genomes of other organisms have also been sequenced.
- The Personal Genome Project is underway, which would allow individuals to have their own personal genome sequenced.

Functional and Comparative Genomics

- Functional genomics is the study of how the 25,000 genes in a human genome function.
- Comparative genomics is a way to determine how species have evolved and how genes and noncoding regions of the genome function.

Proteomics and Bioinformatics Are New Endeavors

- Proteomics is the study of the structure, function, and interaction of cellular proteins.
- Bioinformatics is the application of computer technologies to the study of the genome.

21.4 DNA Technology

- Recombinant DNA contains DNA from two different sources. The foreign gene and vector DNA are cut by the same restriction enzyme and then the foreign gene is sealed into vector DNA. Bacteria take up recombinant plasmids.
- PCR uses DNA polymerase to make multiple copies of a specific piece of DNA. Following PCR, DNA can be subjected to DNA fingerprinting.

Transgenic organisms (bacteria, plants, and animals that have had a foreign gene inserted into them) can produce biotechnology products, such as hormones and vaccines.

- Transgenic bacteria can promote plant health, remove sulfur from coal, clean up toxic waste and oil spills, extract minerals, and produce chemicals.
- Transgenic crops can resist herbicides and pests.
- Transgenic animals can be given growth hormone to produce larger animals; can supply transplant organs; and can produce pharmaceuticals.

Understanding Key Terms

anticodon 450
bioinformatics 455
biotechnology product 460
cloning 458
codon 448
complementary paired bases 444
complementary DNA (cDNA) 458
DNA (deoxyribonucleic acid) 444
DNA ligase 458
DNA replication 445
double helix 444
gene cloning 458
genetic engineering 460
genomics 455
messenger RNA (mRNA) 446
mutation 445
plasmid 458

polymerase chain reaction (PCR) 459
polyribosome 450
proteomics 455
recombinant DNA (rDNA) 458
restriction enzyme 458
ribosomal RNA (rRNA) 446
RNA (ribonucleic acid) 446
RNA polymerase 449
template 445
transcription 448
transcription factor 453
transfer RNA (tRNA) 446
transgenic organism 460
translation 448
triplet code 448
uracil (U) 446
vector 458
xenotransplantation 463

Match the key terms to these definitions.

a. _____ Cluster of ribosomes attached to the same mRNA molecule; each ribosome is producing a copy of the same polypeptide.

b. _____ Free-living organism in the environment that has had a foreign gene inserted into it.

c. _____ Pattern or guide used to make copies.

d. _____ The study of the structure, function, and interactions of cellular proteins.

e. _____ A means to transfer foreign genetic material into a cell.

Testing Your Knowledge of the Concepts

1. Explain why each new DNA double helix is like the parental DNA helix. (pages 444–45)

2. Describe the types of RNA and how they function. (page 446)

3. Explain why the terms transcription and translation are appropriate for these processes. (page 448)

4. If a DNA strand is TAC AAT AAA CGT GTC ATT, what are the codons of mRNA, the anticodons of tRNA, and the amino acid sequence? (pages 448–50)

5. Tell where you would expect each level of genetic control to occur in the cell. Explain. (pages 452–53)

6. Why do the fields of proteomics and bioinformatics depend on genomics? (pages 455)

7. Is ex vivo and in vivo gene therapy harder to perform? Why? (page 456)

8. What is the methodology for producing transgenic bacteria? (page 458)

9. What is the polymerase chain reaction (PCR), and how is it used to produce multiple copies of a DNA segment? (page 459)

10. What are some biotechnology products from bacteria, plants, and animals, and what are they used for? (pages 460–61, 463)

11. The double-helix model of DNA resembles a twisted ladder in which the rungs of the ladder are
    a. complementary base pairs.
    b. A paired with G and C paired with T.
    c. A paired with T and G paired with C.
    d. a sugar-phosphate paired with a sugar-phosphate.
    e. Both a and c are correct.

12. The enzyme responsible for adding new nucleotides to a growing DNA chain during DNA replication is
    a. helicase.
    b. RNA polymerase.
    c. DNA polymerase.
    d. ribozymes.

13. RNA processing
    a. removes the introns, leaving only the exons.
    b. is the same as transcription.
    c. is an event that occurs after RNA is transcribed.
    d. is the rejection of old, worn-out RNA.
    e. Both a and c are correct.

14. During protein synthesis, an anticodon of a tRNA pairs with
    a. amino acids in the polypeptide.
    b. DNA nucleotide bases.
    c. rRNA nucleotide bases.
    d. mRNA nucleotide bases.

15. Which of these associations does not correctly compare DNA and RNA?

DNA	RNA
a. Contains the base thymine.	Contains the base uracil.
b. Is double-stranded.	Is a single-stranded helix.
c. Is found only in the nucleus.	Is found in the nucleus and cytoplasm.
d. The sugar is deoxyribose.	The sugar is ribose.

16. The process of converting the information contained in the nucleotide sequence of RNA into a sequence of amino acids is called
    a. transcription.
    b. translation.
    c. translocation.
    d. replication.

17. Which of the following is involved in controlling gene expression?
    a. the occurrence of transcription
    b. activity of the polypeptide product
    c. life expectancy of the mRNA molecule in the cell
    d. All of these are involved.

18. Complementary base pairing
    a. involves T, A, G, C.
    b. is necessary to replication.
    c. utilizes hydrogen bonds.
    d. All of these are correct.

19. PCR
    a. utilizes RNA polymerase.
    b. takes place in huge bioreactors.
    c. utilizes a temperature-insensitive enzyme.
    d. makes many nonidentical copies of DNA.
    e. All of these are correct.

20. Restriction enzymes found in bacterial cells are ordinarily used
    a. during DNA replication.
    b. to degrade the bacterial cell's DNA.
    c. to degrade viral DNA that enters the cell.
    d. to attach pieces of DNA together.

21. Which of the following is a benefit to having insulin produced by biotechnology?
    a. It is just as effective.
    b. It can be mass-produced.
    c. It is nonallergenic.
    d. It is less expensive.
    e. All of these are correct.

22. Following is a segment of a DNA molecule. (Remember that only one strand is transcribed.) What are (a) the mRNA codons, (b) the tRNA anticodons, and (c) the sequence of amino acids?

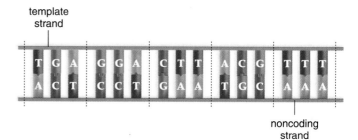

template strand

T G A G G A C T T A C G T T T
A C T C C T G A A T G C A A A

noncoding strand

Thinking Critically About the Concepts

You may recall Adam (page 295), who received growth hormone (GH) treatment. The GH that stimulated additional growth in Adam was produced by transgenic bacteria in a lab. As Will from the opening story learned from his research, prior to the creation of transgenic bacteria that make human GH in labs, GH for treatment was obtained from cadaver pituitary glands. Very little GH was available, and treatment was very expensive. Transgenic bacteria also produce the insulin so many diabetics rely on for blood glucose regulation. Our understanding of DNA may enable us to treat, cure, or prevent any number of diseases or disorders.

1a. Why are transplanted organs so often rejected?

b. Why are the risks of rejection so much higher with xenotransplantation (transplantation from a different species)?

c. Which level of immunity is responsible for the rejection of transplanted organs?

2. Why do you think researchers have focused on using pigs as a potential source of transplant organs/tissues/cells instead of chimpanzees that have DNA more similar to that of humans?

3. Which do you think comes with a higher risk of rejection, an organ transplant or the transplant of some cells, such as brain cells or pancreas cells? Explain.

4. What cellular processes have to occur before human recognition proteins can be displayed by pig cells (how are the proteins produced)?

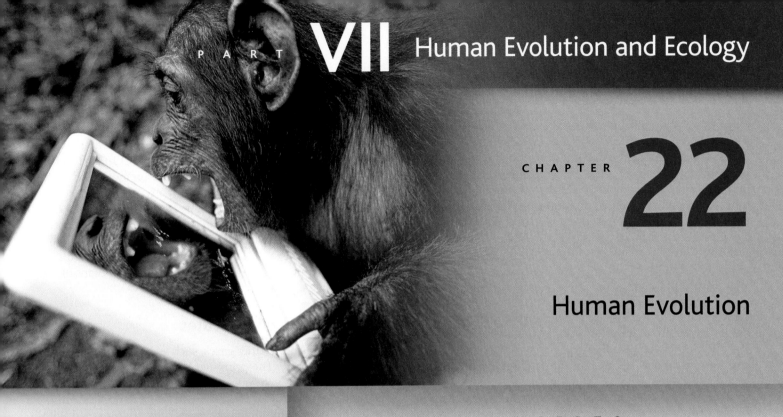

C H A P T E R **22**

Human Evolution

Classmates Ryan and Emily decided to visit the local zoo for their biology class project. They obtained the required worksheet they had to complete, and agreed to meet Saturday morning when the zoo first opened.

They arrived at the zoo in time to watch some of the animals being fed. It was fun watching them open boxes and dig around inside to find their food. "Read that announcement," exclaimed Emily, as they passed the elephant enclosure. "The elephants are going to paint this afternoon. We have to stay and see that!"

Eventually, Ryan and Emily made their way to the chimps' enclosure. A variety of toys had been set up on a ledge outside the chimps' viewing window. Zoo workers and selected visitors manipulated different toys depending on the reaction of the chimps.

When a zoo worker picked up a paper-towel roll and held it to his eye to look at a chimp, the chimp responded by looking back at the worker through the other end of the tube. Another chimp picked up a tube inside the enclosure and looked at the visitors through the tube. A little later, one of the chimps pointed to herself and checked out her teeth in a mirror. Ryan and Emily were amazed by the chimps' participation in play with the zoo workers and visitors and the fact that a chimp showed self-recognition when looking in a mirror.

After studying the concepts of evolution in this chapter, you should understand why many animals, especially chimps, display many humanlike characteristics. Behavior, like appearance, is controlled by genes, which are passed from generation to generation during the evolutionary process.

C H A P T E R   C O N C E P T S

22.1 Origin of Life
Data suggest that a chemical evolution produced the first cell.

22.2 Biological Evolution
Descent from a common ancestor explains the unity of living things—for example, why all living things have a cellular structure and a common chemistry. Adaptation to different environments explains the great diversity of living things.

22.3 Classification of Humans
The classification of humans can be used to trace the ancestry of humans. Humans are hominids in the order primates, a type of mammal in the domain Eukarya.

22.4 Evolution of Hominids
The evolutionary tree of hominids resembles a bush—meaning that there is not a straight line of fossils leading to modern humans.

22.5 Evolution of Humans
Among the members of the genus *Homo*, *Homo habilis* was the first to make tools, *Homo erectus* was the first to have the use of fire, Neandertals were heavily muscled, and Cro-Magnons resembled modern humans, which are believed to have evolved in Africa.

22.1 Origin of Life

Our study of evolution begins with the origin of life. A fundamental principle of biology states that all living things are made of cells and that every cell comes from a preexisting cell (see page 42). But if this is so, how did the first cell come about? Since it was the very first living thing, it had to come from nonliving chemicals. Could there have been a slow increase in the complexity of chemicals—could a **chemical evolution** have produced the first cell(s) on the primitive Earth?

The Primitive Earth

The sun and the planets, including Earth, probably formed over a 10-billion-year period from aggregates of dust particles and debris. At 4.6 billion years ago (BYA), the solar system was in place. Dense silicate minerals became the semiliquid mantle.

The Earth's mass is such that the gravitational field is strong enough to have an atmosphere. If the Earth had had less mass, atmospheric gases would have escaped into outer space. The early Earth's atmosphere was not the same as today's atmosphere. Most likely, the first atmosphere was formed by gases escaping from volcanoes. If so, the primitive atmosphere would have consisted mostly of water vapor (H_2O), nitrogen (N_2), and carbon dioxide (CO_2), with only small amounts of hydrogen (H_2) and carbon monoxide (CO). The primitive atmosphere had little, if any, free oxygen.

At first, the Earth and its atmosphere were extremely hot, and water, existing only as a gas, formed dense, thick clouds. Then, as the Earth cooled, water vapor condensed to liquid water, and rain began to fall. It rained in such enormous quantities over hundreds of millions of years that the oceans of the world were produced.

Small Organic Molecules

The rain washed the other gases, such as N_2 and CO_2, into the oceans (Fig. 22.1a). The primitive Earth had many sources of energy, including volcanoes, meteorites, radioactive isotopes, lightning, and ultraviolet radiation. In the presence of so much available energy, the primitive gases may have reacted with one another and produced small organic compounds, such as

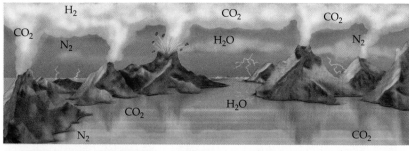

a. The primitive atmosphere contained gases, including H_2O, CO_2, and N_2, that escaped from volcanoes. As the water vapor cooled, some gases were washed into the oceans by rain.

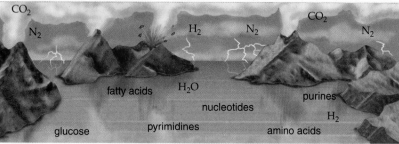

b. The availability of energy from volcanic eruption and lightning allowed gases to form small organic molecules, such as nucleotides and amino acids.

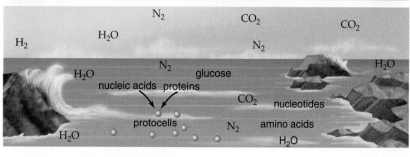

c. Small organic molecules could have joined to form proteins and nucleic acids, which became incorporated into membrane-bound spheres. The spheres became the first cells, called protocells. Later protocells became true cells that could reproduce.

Figure 22.1   Chemical evolution.
A chemical evolution could have produced the protocell, which became a true cell once it had genes composed of DNA and could reproduce.

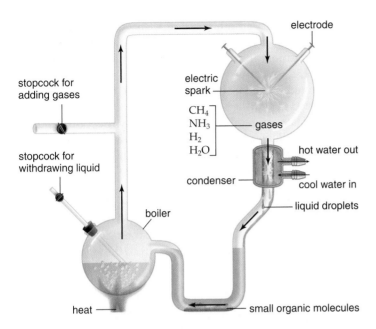

Figure 22.2   Stanley Miller's apparatus and experiment.
Gases that were thought to be present in the early Earth's atmosphere were admitted to the apparatus, circulated past an energy source (electric spark), and cooled to produce a liquid that could be withdrawn. Upon chemical analysis, the liquid was found to contain various small organic molecules.

nucleotides and amino acids (Fig. 22.1*b*). In 1953, Stanley Miller performed an experiment. He placed a mixture of gases resembling the Earth's early atmosphere in a closed system, heated the mixture, and circulated it past an electric spark. After cooling, Miller discovered a variety of small organic molecules in the resulting liquid (Fig. 22.2). Other investigators have achieved the same results using various mixtures of gases.

Macromolecules

The newly formed small organic molecules likely joined to produce organic macromolecules (Fig. 22.1*c*). There are two hypotheses of special interest concerning this stage in the origin of life. One is the **RNA-first hypothesis,** which suggests that only the macromolecule RNA (ribonucleic acid) was needed at this time to progress toward formation of the first cell(s). This hypothesis was formulated after the discovery that RNA can sometimes be both a substrate and an enzyme during RNA processing (see Fig. 21.9). At that time the splicing of mRNA to remove introns was done by a complex composed of both RNA and protein. The RNA and not the protein is the enzyme. RNA enzymes are called ribozymes. Then, too, ribosomes where protein synthesis occurs are composed of rRNA. Perhaps RNA, then, could have carried out the processes of life commonly associated today with DNA (deoxyribonucleic acid) and proteins. Scientists who support this hypothesis are fond of saying that it was

an "RNA world" some 3.5 BYA. DNA, being a double helix, is more stable than RNA, and perhaps this accounts for why DNA became the permanent genetic material and not RNA.

Another hypothesis is termed the **protein-first hypothesis.** Sidney Fox has shown that amino acids join together when exposed to dry heat. He suggests that amino acids collected in shallow puddles along the rocky shore, and the heat of the sun caused them to form proteinoids, small polypeptides that have some catalytic properties. When proteinoids are returned to water, they form microspheres, structures composed only of protein that have many of the properties of a cell.

The Protocell

A cell has a lipid-protein membrane. Fox has shown that if lipids are made available to microspheres, the two tend to become associated, producing a lipid-protein membrane. A **protocell,** which could carry on metabolism but could not reproduce, could have come into existence in this manner.

The protocell would have been able to use the still-abundant small organic molecules in the ocean as food. Therefore, the protocell was, most likely, a **heterotroph,** an organism that takes in preformed food. Further, the protocell would have been a fermenter, because there was no free oxygen.

The True Cell

A true cell can reproduce, and in today's cells, DNA replicates before cell division occurs. Enzymatic proteins carry out the replication process.

How did the first cell acquire both DNA and enzymatic proteins? Scientists who support the RNA-first hypothesis propose a series of steps. According to this hypothesis, the first cell had RNA genes that, like messenger RNA, could have specified protein synthesis. Some of the proteins formed would have been enzymes. Perhaps one of these enzymes, such as reverse transcriptase found in retroviruses, could use RNA as a template to form DNA. Replication of DNA would then proceed normally.

By contrast, supporters of the protein-first hypothesis suggest that some of the proteins in the protocell would have evolved the enzymatic ability to synthesize DNA from nucleotides in the ocean. Then DNA would have gone on to specify protein synthesis, and in this way, the cell could have acquired all its enzymes, even the ones that replicate DNA.

☑ Check Your Progress 22.1

1. What type of evolution produced the first cells?
2. What evidence is there to suggest that the gases of early Earth could have become the first organic molecules?
3. What evidence is there that RNA could have been the original genetic material?

22.2 Biological Evolution

The first true cells were the simplest of life-forms; therefore, they must have been **prokaryotic cells,** which lack a nucleus. Later, **eukaryotic cells** (protists), which have nuclei, evolved. Then, multicellularity and the other kingdoms (fungi, plants, and animals) evolved (see Fig. 1.1). Obviously, all these types of organisms—even prokaryotic cells—are alive today. Each type of organism has its own evolutionary history that is traceable back to the first cell(s).

Biological evolution is the process by which a species changes through time. Biological evolution has two important aspects: descent from a common ancestor and adaptation to the environment. Descent from the original cell(s) explains why all living things have a common chemistry and a cellular structure. An **adaptation** is a characteristic that makes an organism able to survive and reproduce in its environment. Adaptations to different environments help explain the diversity of life—why there are so many different types of living things.

Common Descent

Charles Darwin was an English naturalist who first formulated the theory of evolution that has since been supported by so much independent data. At the age of 22, Darwin sailed around the world as the naturalist on board the HMS *Beagle.* Between 1831 and 1836, the ship sailed in the tropics of the Southern Hemisphere, where life-forms are more abundant and varied than in Darwin's native England.

Even though it was not his original intent, Darwin began to realize and to gather evidence that life-forms change over time and from place to place. The types of evidence that convinced Darwin that common descent occurs were fossil, anatomical, and biogeographical.

Fossil Evidence Supports Evolution

Fossils are the best evidence for evolution because they are the actual remains of species that lived on Earth at least 10,000 years ago and up to billions of years ago. Fossils can be the traces of past life or any other direct evidence that past life existed. Traces include trails, footprints, burrows, worm casts, or even preserved droppings. Fossils can also be such items as pieces of bone, impressions of plants pressed into shale, and even insects trapped in tree resin (which we know as amber). Most fossils, however, are found embedded in or recently eroded from sedimentary rock. Sedimentation, a process that has been going on since the Earth was formed, can take place on land or in bodies of water. Weathering and erosion of rocks produces an accumulation of particles that vary in size and nature and are called sediment. Sediment becomes a stratum (pl., strata), a recognizable layer in a sequence of layers. Any given stratum is older than the one above it and younger than the one immediately below it (Fig. 22.3). This allows fossils to be dated.

Usually when an organism dies, the soft parts are either consumed by scavengers or decomposed by bacteria. This means that most fossils consist only of hard parts such as shells, bones, or teeth, because these are usually not consumed or destroyed. When a fossil is encased by rock, the remains were buried in sediment, the hard parts were preserved by a process called mineralization, and, finally, the surrounding sediment hardened to form rock. Subsequently, the fossil has to be found by a human. Most estimates suggest that less than 1% of past species have been preserved as fossils, and only a small fraction of these have been found.

More and more fossils have been found because researchers, called paleontologists, and their assistants have been out in the field looking for them (Fig. 22.3). Usually, paleontologists remove fossils, from the strata to study them in the laboratory, and then they may decide to exhibit

Figure 22.3 **Paleontologists carefully remove fossils from strata.**

them. The **fossil record** is the history of life recorded by fossils, and paleontology is the science of discovering the fossil record and, from it, making decisions about the history of life, ancient climates, and environments. The fossil record is the most direct evidence we have that evolution has occurred. The species found in ancient sedimentary rock are not the species we see about us today.

Darwin relied on fossils to formulate his theory of evolution, but today we have a far more complete record than was available to Darwin. The record is complete enough to tell us that, in general, life has progressed from the simple to the complex. Unicellular prokaryotes are the first signs of life in the fossil record, followed by unicellular eukaryotes, and then multicellular eukaryotes. Among the latter, fishes evolved before terrestrial plants and animals. On land, nonflowering plants preceded the flowering plants, and amphibians preceded the reptiles, including the dinosaurs. Dinosaurs are directly linked to the birds, but only indirectly linked to the evolution of mammals, including humans.

Transitional fossils are those that have characteristics of two different groups. In particular, they tell us who is related to whom and how evolution occurred. Even in Darwin's day, scientists knew of the *Archaeopteryx* fossils, which are intermediate between reptiles and birds. The dinosaur-like skeleton of these fossils had reptilian features, including jaws with teeth, and a long, jointed tail, but *Archaeopteryx* also had feathers and wings. Figure 22.4 shows not only a fossil of *Archaeopteryx*, it also gives us an artist's representation of the animal based on the fossil remains. Many more semibird fossils have been discovered recently in China.

It had always been thought that whales had terrestrial ancestors. Now, fossils have been discovered that support this hypothesis. *Ambulocetus natans* (meaning the walking whale that swims) was the size of a large sea lion, with broad webbed feet on both fore- and hindlimbs. This animal could both walk and swim. It also had tiny hooves on its toes and the primitive skull and teeth of early whales. Figure 22.5 is an artist's recreation based on fossil remains. It depicts *Ambulocetus* as a predator that patrolled freshwater streams looking for prey.

The origin of mammals is also well documented. The synapsids are mammal-like reptiles whose descendants were wolflike predators and bearlike predators, as well as several types of piglike herbivores. Slowly, mammalian-like fossils acquired features such as a palate, which would have enabled them to breathe and eat at the same time, and a muscular diaphragm and rib cage that would have helped them breathe efficiently, and so forth. The earliest true mammals were shrew-size creatures in fossil beds about 200 million years old.

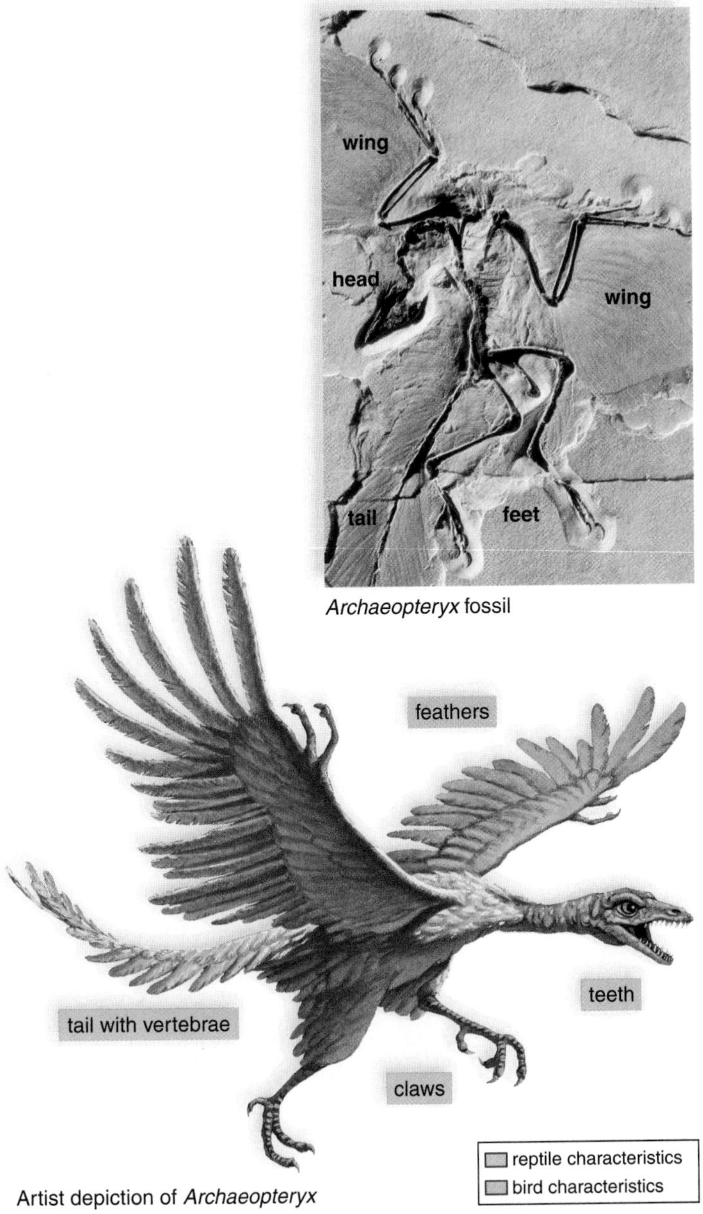

wing

head

wing

tail   feet

Archaeopteryx fossil

feathers

teeth

tail with vertebrae

claws

Artist depiction of *Archaeopteryx*

▢ reptile characteristics
▢ bird characteristics

Figure 22.4 *Archaeopteryx* had a combination of reptilian and bird characteristics.

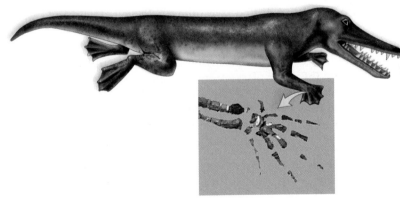

Figure 22.5 Presence of legs in *Ambulocetus* is evidence that terrestrial mammals gave rise to whales.

Other Evidence Supports Evolution

Many different types of evidence support the hypothesis that organisms are related through common descent. Darwin cited much of the evidence to support his theory of evolution. However, he had no knowledge of the genetic and biochemical data that became available after his time.

Biogeographical Evidence **Biogeography** is the study of the distribution of plants and animals in different places throughout the world. Such distributions are consistent with the hypothesis that life-forms evolved in a particular locale and then they may spread out. Therefore, you would expect a different mix of plants and animals whenever geography separates continents, islands, or seas. For example, Darwin noted that South America lacked rabbits, even though the environment was quite suitable for them. He concluded that no rabbits lived in South America because rabbits evolved somewhere else and had no means of reaching South America. Instead, the Patagonian hare lives in South America. The Patagonian hare resembles a rabbit in anatomy and behavior, but it has the face of a guinea pig, from which it probably evolved.

To take another example, both cactuses and euphorbia are plants adapted to a hot, dry environment—both are succulent, spiny, flowering plants. Why do cactuses grow in North American deserts and euphorbia grow in African deserts, when each would do well on the other continent? It seems obvious that they just happened to evolve on their respective continents.

The islands of the world are home to many unique species of animals and plants that are found no place else, even when the soil and climate are the same. Why do so many species of finches live on the Galápagos Islands, when these same species are not on the mainland? The reasonable explanation is that an ancestral finch migrated to all the different islands. Then, geographic isolation allowed the ancestral finch to evolve into a different species on each island.

In the history of the Earth, South America, Antarctica, and Australia were originally connected. Marsupials (pouched mammals) arose at this time and today are found in both South America and Australia. But when Australia separated and drifted away, the marsupials diversified into many different forms suited to various environments of Australia. They were free to do so because there were few, if any, placental mammals in Australia. In South America, where there are placental mammals, marsupials are not as diverse. This supports the hypothesis that evolution is influenced by the mix of plants and animals in a particular continent—that is, by biogeography, not by design.

Anatomical Evidence Darwin was able to show that a common descent hypothesis offers a plausible explanation for anatomical similarities among organisms. Vertebrate forelimbs are used for flight (birds and bats), orientation during swimming (whales and seals), running (horses),

climbing (arboreal lizards), or swinging from tree branches (monkeys). Yet all vertebrate forelimbs contain the same sets of bones organized in similar ways, despite their dissimilar functions (Fig. 22.6). The most plausible explanation for this unity is that the basic forelimb plan belonged to a common ancestor, and then the plan was modified in the succeeding groups as each continued along its own evolutionary pathway. Structures that are anatomically similar because they are inherited from a common ancestor are called **homologous structures**. In contrast, **analogous structures** serve the same function, but they are not constructed similarly, nor do they share a common ancestry. The wings of birds and insects and the jointed appendages of a lobster and humans are analogous structures. The presence of homology, not analogy, is evidence that organisms are related.

Vestigial structures are anatomical features that are fully developed in one group of organisms but that are reduced and may have no function in similar groups. Modern whales have a vestigial pelvic girdle and legs. The ancestors of whales walked on land, but whales are totally aquatic animals today. Most birds have well-developed wings used for flight. Some bird species (e.g., ostrich), however, have greatly reduced wings and do not fly. Similarly, snakes have no use for hindlimbs, and yet some have remnants of a pelvic girdle and legs. Humans have a tailbone but no tail. The presence of vestigial structures can be explained by the common descent. Vestigial structures occur because organisms inherit their anatomy from their ancestors; they are traces of an organism's evolutionary history.

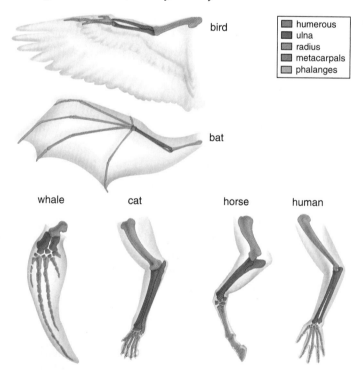

bird

- humerous
- ulna
- radius
- metacarpals
- phalanges

bat

whale   cat   horse   human

Figure 22.6   Despite differences in function, vertebrate forelimbs have the same bones.

Pig embryo

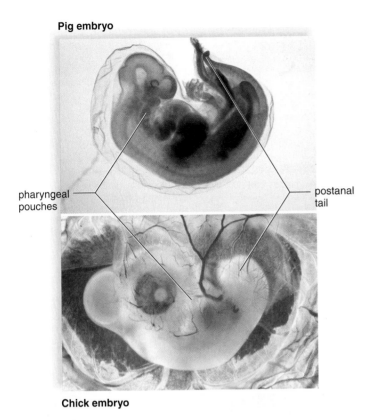

pharyngeal pouches

postanal tail

Chick embryo

Figure 22.7 Vertebrate embryos have features in common, despite different ways of life as adults.

The homology shared by vertebrates extends to their embryological development (Fig. 22.7). At some time during development, all vertebrates have a postanal tail and exhibit paired pharyngeal pouches. In fishes and amphibian larvae, these pouches develop into functioning gills. In humans, the first pair of pouches becomes the cavity of the middle ear and the auditory tube. The second pair becomes the tonsils, while the third and fourth pairs become the thymus and parathyroid glands, respectively. Why should terrestrial vertebrates develop and then modify structures like pharyngeal pouches that have lost their original function? The most likely explanation is that fishes are ancestral to other vertebrate groups.

Biochemical Evidence Almost all living organisms use the same basic biochemical molecules, including DNA (deoxyribonucleic acid), ATP (adenosine triphosphate), and many identical or nearly identical enzymes. Further, organisms use the same DNA triplet code and the same 20 amino acids in their proteins. Since the sequences of DNA bases in the genomes of many organisms are now known, it has become clear that humans share a large number of genes with much simpler organisms. Also of interest, evolutionists who study development have found that many developmental genes are shared in animals ranging from worms to humans. It appears that life's vast diversity has come about by only a slight difference in the regulation of genes. The result has been widely divergent types of bodies.

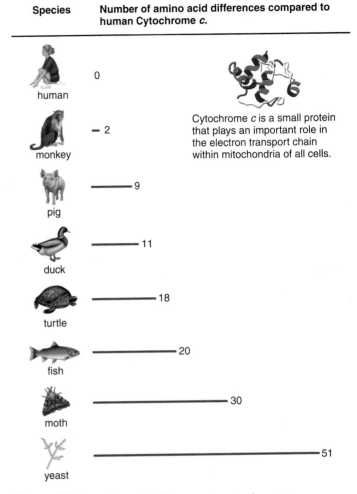

Species	Number of amino acid differences compared to human Cytochrome *c*.
human	0
monkey	2
pig	9
duck	11
turtle	18
fish	20
moth	30
yeast	51

Cytochrome *c* is a small protein that plays an important role in the electron transport chain within mitochondria of all cells.

Figure 22.8 Amino acid differences in cytochrome *c*
The number of amino acid differences between humans and various species are indicated.

When the degree of similarity in DNA base sequences or the degree of similarity in amino acid sequences of proteins is examined, the data are as expected, assuming common descent. Cytochrome *c* is a molecule that is used in the electron transport chain of many organisms. Data regarding differences in the amino acid sequence of cytochrome *c* show that the sequence in a human differs from that in a monkey by only two amino acids, from that in a duck by 11 amino acids, and from that in a yeast by 51 amino acids (Fig. 22.8). These data are consistent with other data regarding the anatomical similarities of these organisms and, therefore, their relatedness.

Intelligent Design

Evolution is a scientific theory, a term reserved for those ideas that scientists have found to be all-encompassing because they are based on evidence (data) collected in a number of different fields, as just discussed. In other words, evolutionary theory has been supported by repeated scientific experiments and observations.

Some persons advocate the teaching of ideas that run contrary to the theory of evolution in schools. The emphasis is usually placed on Intelligent Design, which says that the diversity of life could never have arisen without the involvement of an "intelligent agent." Many scientists, and even religions, argue that intelligent design is faith based and not science because it is not possible to test in a scientific way whether an intelligent agent exists. If it were possible to structure such an experiment, scientists would be the first to do it.

Natural Selection

When Darwin returned home, he spent the next 20 years gathering data to support the principle of biological evolution. Darwin's most significant contribution was to describe a mechanism for adaptation—**natural selection.** During adaptation, a species becomes suited to its environment. On his trip, Darwin visited the Galápagos Islands. He saw a number of finches that resembled one another but had different ways of life. Some were seed-eating ground finches, some cactus-eating ground finches, and some insect-eating tree finches. A warbler-type finch had a beak that could take honey from a flower. A woodpecker-type finch lacked the long tongue of a woodpecker but could use a cactus spine or twig to pull insects from cracks in the bark of a tree. Darwin thought the finches were all descended from a mainland ancestor whose offspring had spread out among the islands and become adapted to different environments.

In order to emphasize the nature of Darwin's natural selection process, it is often contrasted with a process espoused by Jean-Baptiste Lamarck, another nineteenth-century naturalist. Lamarck's explanation for the long neck of the giraffe was based on the assumption that the ancestors of the modern giraffe were trying to reach into the trees to browse on high-growing vegetation (Fig. 22.9). Continual stretching of the neck caused it to become longer, and this acquired characteristic was passed on to the next generation. Lamarck's mechanism will not work because acquired characteristics cannot be inherited (Fig. 22.9).

The critical elements of the natural selection process are variation, competition for limited resources such as food, and adaptation as an end result.

- *Variation.* Individual members of a species vary in physical characteristics. Physical variations can be passed from generation to generation. (Darwin was never aware of genes, but we know today that the inheritance of the genotype determines the phenotype.)
- *Competition for limited resources.* Even though each individual could eventually produce many descendants, the number in each generation usually stays about the same. Why? Because resources are limited and competition for resources results in unequal reproduction among members of a population.
- *Adaptation.* Those members of a population with advantageous traits capture more resources and are more likely to reproduce and pass on these traits. Thus,

Figure 22.9   Mechanism of evolution.
This diagram contrasts Jean-Baptiste Lamarck's process of acquired characteristics with Charles Darwin's process of natural selection.

over time the environment "selects" for the better-adapted traits. Each subsequent generation includes more individuals that are adapted in the same way to the environment.

Natural selection can account for the great diversity of life. Environments differ widely, and, therefore, adaptations are varied. From vampire bats, to sea turtles, to the many finches observed by Darwin, all the different organisms are adapted to their way of life.

✅ Check Your Progress 22.2

1. **a.** What is biological evolution, and **(b)** what are its two most important aspects?

2. What type of evidence convinced Charles Darwin that biological evolution does occur?

3. What process accounts for the great diversity of living things?

22.3 Classification of Humans

To begin a study of human evolution, we can turn to the classification of humans because biologists classify organisms according to their evolutionary relatedness. The **binomial name** of an organism gives its genus and species. Organisms in the same domain have only general characteristics in common; those in the same genus have quite specific characteristics in common. Table 22.1 lists some of the characteristics that help classify humans. The dates in the first column of the table tell us when these groups of animals first appear in the fossil record.

DNA Data and Human Evolution

We are accustomed to using data such as that given in Table 22.1 to determine evolutionary relationships, but research-

Table 22.1	Evolution and Classification of Humans	
BYA/MYA*	**Classification Category**	**Characteristics**
2 bya	Domain Eukarya	Membrane-bound nucleus
600 mya	Kingdom Animalia	Multicellular, motile, heterotrophic
540 mya	Phylum Chordata	Sometime in life history: dorsal tubular nerve cord, notochord, pharyngeal pouches
120 mya	Class Mammalia	Vertebrates with hair, mammary glands
60 mya	Order Primates	Well-developed brain, adapted to live in trees
7 mya	Family Hominidae	Adapted to upright stance and bipedal locomotion
3 mya	Genus *Homo*	Most developed brain, made and used tools
0.1 mya	Species *Homo sapiens*†	Modern humans; speech centers of brain well-developed

*BYA = billions of years ago; mya = millions of years ago.

†To specify an organism, you must use the full binomial name, such as *Homo sapiens*.

ers today are just as apt to use DNA data. Indeed, researchers are depending more and more on DNA data to trace the history of life. DNA data is particularly useful when anatomical differences are unavailable.

For example, in the late 1970s, Carl Woese and his colleagues at the University of Illinois decided to use rRNA sequence data to decide how prokaryotes are related. They knew that the DNA coding for ribosomal RNA (rRNA) changes slowly during evolution, and indeed may change only when there is a major evolutionary event. Woese reported that on the basis of rRNA sequence data, there are three domains of life and the archaea are more closely related to eukaryotes than to bacteria (Fig. 22.10). (See page 6 for a description of these domains.) In other words, major decisions regarding the history of life are now being made on the basis of DNA/rRNA/protein sequencing data. (Figure 22.8 gave an example of the use of protein data to show evolutionary relationships.) Studies of rRNA sequences indicate that among the major groups of eukaryotes, animals are more closely related to fungi than they are to plants, for example.

Later in this chapter, you will learn that DNA sequencing data were used to decide when the last common ancestor for the apes and humans must have existed. As yet, the fossil record has not revealed this ancestor, but comparative DNA data between apes and humans tell us when this ancestor must have existed, about 7 MYA. Mitochondrial DNA (mtDNA) is used to decide the timing of recent evolutionary events because mtDNA changes occur frequently. Mitochondrial DNA data indicate that modern humans arose in Africa and later migrated to Eurasia (see page 484 for a discussion of this study).

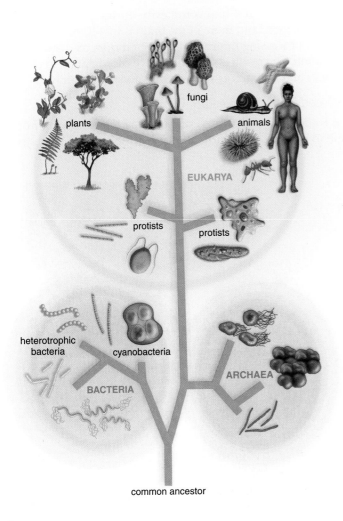

Figure 22.10 The three domain system of classification.
Representatives of each domain are depicted in the ovals. The evolutionary tree of life shows that domain Archaea is more closely related to domain Eukarya than either is to domain Bacteria.

Humans Are Primates

In contrast to the other orders of placental mammals, **primates** are adapted to an arboreal life—that is, for living in trees. Primates have mobile limbs; grasping hands; a flattened face; binocular vision; a large, complex brain; and a reduced reproductive rate. The order Primates has two suborders. The **prosimians** include the lemurs, tarsiers, and lorises. The **anthropoids** include the monkeys, apes (Fig. 22.11), and humans. This classification tells us that humans are more closely related to the monkeys and apes than they are to the prosimians. Which of these traits could Ryan and Emily have noticed when they were observing chimpanzees at the zoo?

Mobile Forelimbs and Hindlimbs

Primate limbs are mobile, and the hands and feet both have five digits each. Many primates, such as chimpanzees, have both an opposable big toe and thumb—that is, the big toe or thumb can touch each of the other toes or fingers. (Humans don't have an opposable big toe, but the thumb is opposable, and this results in a grip that is both powerful and precise.) The opposable thumb allows a primate to easily reach out and bring food, such as fruit, to the mouth. When locomoting, primates grasp and release tree limbs freely because nails have replaced claws. If the enclosure for the chimps included a tree, Ryan and Emily would have certainly observed this behavior in the chimpanzees.

Binocular Vision

In chimps, like other primates, the snout is shortened considerably, allowing the eyes to move to the front of the head. The stereoscopic vision (or depth perception) that results permits primates to make accurate judgments about the distance and position of adjoining tree limbs. Humans and the apes have three different cone cells, which are able to discriminate between greens, blues, and reds. Cone cells require bright light, but the image is sharp and in color. The lens of the eye focuses light directly on the fovea, a region of the retina, where cone cells are concentrated.

Large, Complex Brain

The evolutionary trend among primates is toward a larger and more complex brain—the brain size is smallest in prosimians and largest in modern humans. The cerebral cortex, with many association areas, expands so much that it becomes extensively folded in humans. The portion of the brain devoted to smell gets smaller, and the portions devoted to sight increase in size and complexity during primate evolution. Also, more and more of the brain is involved in controlling and processing information received from the hands and the thumb. The result is good hand-eye coordination in chimpanzees and humans as witnessed by Ryan and Emily.

Reduced Reproductive Rate

It is difficult to care for several offspring while moving from limb to limb, and one birth at a time is the norm in primates.

Asian Apes

African Apes

White-handed gibbon,
Hylobates lar

Orangutan, *Pongo pygmaeus*

Chimpanzee, *Pan troglodytes*

Western lowland gorilla,
Gorilla gorilla

Figure 22.11   Today's apes.
The apes can be divided into the Asian apes (gibbons and orangutans) and the African apes (chimpanzees and gorillas). Molecular data and the location of early hominid fossil remains tell us that we are more closely related to the African than to the Asian apes.

The juvenile period of dependency is extended, and there is an emphasis on learned behavior and complex social interactions. Ryan and Emily may very well have observed interactions between a mother and baby chimpanzee.

Comparing Human Skeleton to the Chimpanzee Skeleton

On page 467, Ryan and Emily were fascinated with the behavior of chimpanzees because they were able to copy our behavior and they had a sense of self. Remarkably, the genomes of humans and chimpanzees are 99% identical; however, this difference of only 1% has apparently resulted in a number of differences in various genes. In one study, investigators discovered differences in genes for speech, hearing, and smell (see Fig. 21.15). Figure 22.12 compares anatomical differences between chimpanzees and humans, which relate to upright stance of humans when they walk compared to the chimpanzees practice of knuckle-walking. No doubt, Ryan and Emily would have observed that when chimpanzees walk, the forearms rest on their knuckles.

These differences in anatomy between chimpanzees and humans include that humans, but not chimps, are adapted for an upright stance: (1) In humans, the spine exits inferior to the center of the skull, and this places the skull in the midline of the body; (2) the longer S-shaped spine of humans places the trunk's center of gravity squarely over the feet; (3) the broader pelvis and hip joint of humans keep them from swaying when they walk; (4) the longer neck of the femur in humans causes the femur to angle inward at the knees; (5) the human knee joint is modified to support the body's weight—the femur is larger at the bottom, and the tibia is larger at the top; and (6) the human toe is not opposable; instead, the foot has an arch. The arch enables humans to walk long distances and run with less chance of injury.

☑ Check Your Progress 22.3

1. How are humans classified—from domain to species?

2. What are the characteristics of primates?

3. What major differences exist between the chimpanzee skeleton and the human skeleton?

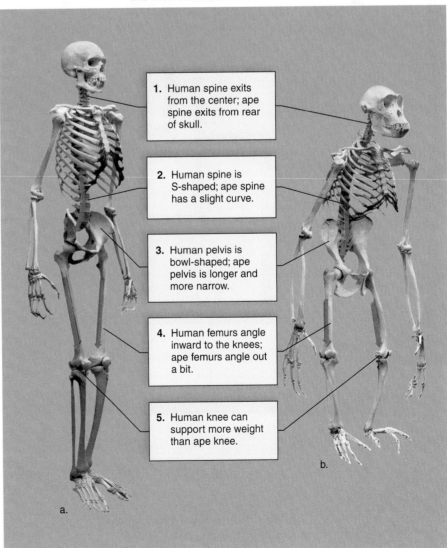

1. Human spine exits from the center; ape spine exits from rear of skull.

2. Human spine is S-shaped; ape spine has a slight curve.

3. Human pelvis is bowl-shaped; ape pelvis is longer and more narrow.

4. Human femurs angle inward to the knees; ape femurs angle out a bit.

5. Human knee can support more weight than ape knee.

Figure 22.12   Adaptations for standing.
a. Human skeleton compared to
(b) chimpanzee skeleton.
©2001 Time Inc. Reprinted by permission.

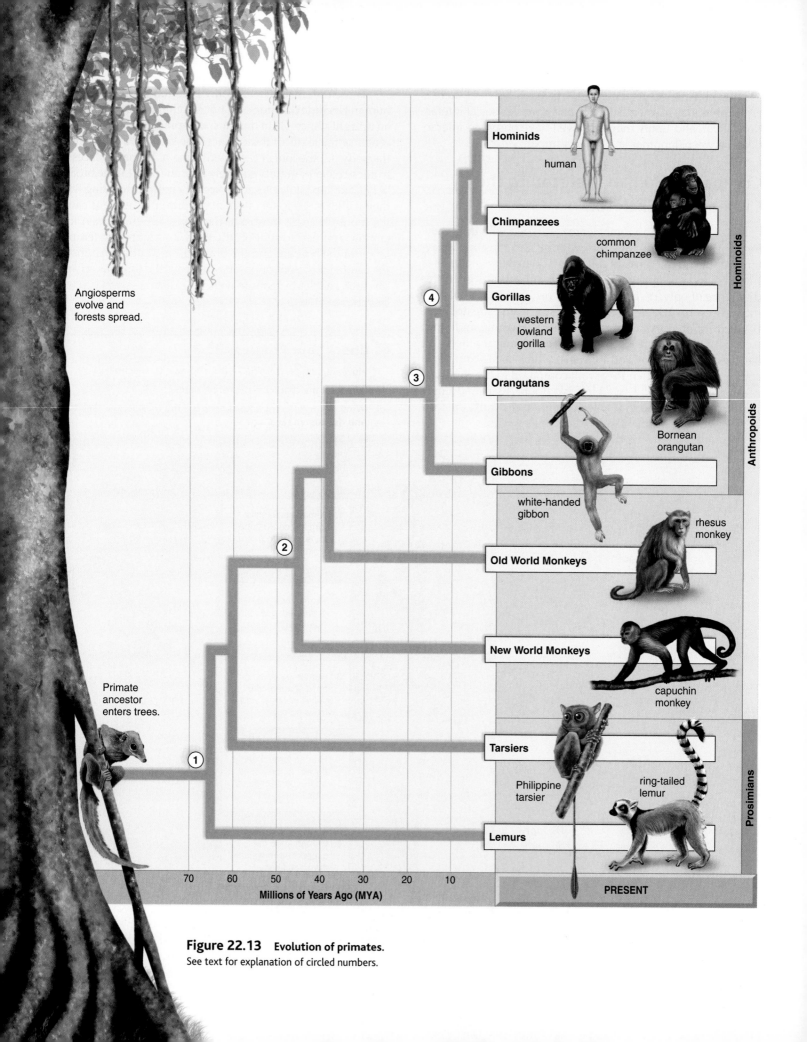

Figure 22.13 **Evolution of primates.**
See text for explanation of circled numbers.

22.4 Evolution of Hominids

Once biologists have studied the characteristics of a group of organisms, they can construct an **evolutionary tree** that is a working hypothesis of their past history. The evolutionary tree in Figure 22.13 shows that all primates share one common ancestor and that the other types of primates diverged from the human line of descent over time. ① The common ancestor for all primates may have resembled a tree shrew. The descendants of this ancestor developed traits such as a shortened snout and nails instead of claws as they adapted to life in trees. The time when each type of primate diverged from the main line of descent is known from the fossil record. A common ancestor was living at each point of divergence; for example, ② there was a common ancestor for monkeys, apes, and hominids about 45 MYA; ③ one for all apes and hominids about 15 MYA; and ④ another for just African apes and hominids about 7 MYA. The split between the ape and human lineage may have occurred about this time.

One of the most unfortunate misconceptions concerning human evolution is the belief that Darwin and others suggested that humans evolved from apes. On the contrary, humans and apes are thought to have shared a common ape-like ancestor. Today's apes are our distant cousins, and we couldn't have evolved from our cousins because we are contemporaries—living on Earth at the same time. Humans and apes have been evolving separately from a common ancestor for about 7 million years. Following the split between humans and apes, different environments selected for the different traits that apes and humans have now.

The First Hominids

Biologists have not been able to agree on which extinct form known only by the fossil record is the first hominid. **Hominid** is a term that refers to our branch of the evolutionary tree. Any fossil placed in the hominid line of descent is closer to us than to one of the African apes.

When any two lines of descent, called a **lineage,** first diverge from a common ancestor, the genes and proteins of the two lineages are nearly identical. As mentioned previously, as time goes by, each lineage accumulates genetic changes, which lead to RNA and protein changes. Many genetic changes are neutral (not tied to adaptation) and accumulate at a fairly constant rate; such changes can be used as a kind of **molecular clock** to indicate the relatedness of two groups and when they diverged from each other. Molecular data also suggest that hominids split from the ape line of descent about 7 MYA.

Hominid Features

Paleontologists use certain anatomical features when they try to determine if a fossil is a hominid. One of these features is **bipedal posture** (walking on two feet). Until recently, many scientists thought that hominids began to walk upright on two feet because of a dramatic change in climate that caused forests to be replaced by grassland. Now, some biologists suggest that the first hominid began to assume a bipedal posture even while it lived in trees. Why? Because they cannot find evidence of a dramatic shift in vegetation about 7 MYA. The first hominid's environment is now thought to have included some forest, some woodland, and some grassland. While still living in trees, the first hominids may have walked upright on large branches as they collected fruit from overhead. Then, when they began to forage on the ground among bushes, it would have been easier to shuffle along on their hindlimbs. Bipedalism would also have prevented them from getting heatstroke because an upright stance exposes more of the body to breezes. Bipedalism may have been an advantage in still another way. Males may have acquired food far afield, and if they could carry it back to females, they would have been more assured of sex.

Two other hominid features of importance are the shape of the face and brain size. Today's humans have a flatter face and a more pronounced chin than do the apes because the human jaw is shorter than that of the apes. Then, too, our teeth are generally smaller and less specialized; we don't have the sharp canines of an ape, for example. Chimpanzees have a brain size of about 400 cm^3, and modern humans have a brain size of about 1,300 cm^3, and this may account for some of the differences in behavior that Ryan and Emily observed when viewing the chimpanzees.

It's hard to decide which fossils are hominids because human features evolved gradually and they didn't evolve at the same rate. Most investigators rely first and foremost on bipedal posture as the hallmark of a hominid, regardless of the size of the brain.

Earliest Fossil Hominids

Fossils, dated at the time the ape and human lineages split, have been found. The oldest of these fossils, called *Sahelanthropus tchadensis*, dated at 7 MYA, was found in Chad, located in central Africa, far from eastern and southern Africa where other hominid fossils were excavated. The only find, a skull, appears to be that of a hominid because it has smaller canines and thicker tooth enamel than an ape. The braincase, however, is very apelike, and it is impossible to tell if this hominid walked upright. Some suggest this fossil is ancestral to the gorilla.

Orrorin tugenensis, dated at 6 MYA and found in eastern Africa, is thought to be another early hominid, especially because the limb anatomy suggests a bipedal posture. However, the canine teeth are large and pointed, and the arm and finger bones retain adaptations for climbing. Some suggest this fossil is ancestral to the chimpanzee.

Ardipithecus kadabba, found in eastern Africa and dated between 5.8 and 5.2 MYA, is closely related to the later-appearing *Ardipithecus ramidus*. This ardipithecine is thought to be closely related to the australopithecines, which are discussed next.

Evolution of Australopithecines

The hominid line of descent begins in earnest with the **australopithecines,** a group of species that evolved and diversified in Africa. Some australopithecines were slight of frame and termed *gracile* (slender) types. Some were *robust* (powerful) and tended to have strong upper bodies and especially massive jaws. The gracile types most likely fed on soft fruits and leaves, while the robust types had a more fibrous diet that may have included hard nuts. In other words, the skull structure of australopithecines was suited to their particular diets.

Southern Africa

The first australopithecine to be discovered was unearthed in southern Africa by Raymond Dart in the 1920s. This hominid, named *Australopithecus africanus,* is a gracile type dated about 2.8 MYA. *A. robustus,* dated from 2 to 1.5 MYA, is a robust type from southern Africa. Both *A. africanus* and *A. robustus* had a brain size of about 500 cm³; their skull differences are essentially due to dental and facial adaptations to different diets.

Limb anatomy suggests these hominids walked upright. Nevertheless, the proportions of the limbs are apelike: The forelimbs are longer than the hindlimbs. Some argue that *A. africanus,* with its relatively large brain, is a possible ancestral candidate for early *Homo,* whose limb proportions are similar to those of this fossil.

Eastern Africa

More than 20 years ago, a team led by Donald Johanson unearthed nearly 250 fossils of a hominid called *A. afarensis.* A now-famous female skeleton dated at 3.18 MYA is known worldwide by its field name, Lucy. (The name derives from the Beatles song "Lucy in the Sky with Diamonds.") Although her brain was quite small (400 cm³), the shapes and relative proportions of her limbs indicate that Lucy stood upright and walked bipedally (Fig. 22.14*a*). Even better evidence of bipedal locomotion comes from a trail of footprints in Laetoli dated about 3.7 MYA. The larger prints are double, as though a smaller-sized being was stepping in the footfalls of another—and there are additional small prints off to the side, within hand-holding distance (Fig. 22.14*b*).

Since the australopithecines were apelike above the waist (small brain) and humanlike below the waist (walked erect), it shows that human characteristics did not evolve all at one time. The term **mosaic evolution** is applied when different body parts change at different rates and, therefore, at different times.

A. afarensis, a gracile type, is most likely ancestral to the robust types found in eastern Africa: *A. aethiopicus* and *A. boisei. A. boisei* had a powerful upper body and the largest molars of any hominid. These robust types died out, and therefore, it is possible that *A. afarensis* is ancestral to both *A. africanus* and early *Homo.*

✓ Check Your Progress 22.4

1. Name three features characteristic of hominids.
2. Name a gracile and a robust australopithecine.

a.

b.

Figure 22.14 *Australopithecus afarensis.*
a. A reconstruction of Lucy on display at the St. Louis Zoo. **b.** These fossilized footprints occur in ash from a volcanic eruption some 3.7 MYA. The larger footprints are double (one followed behind the other), and a third, smaller individual was walking to the side. (A female holding the hand of a youngster may have been walking in the footprints of a male.) The footprints suggest that *A. afarensis* walked bipedally.

22.5 Evolution of Humans

Fossils are assigned to the genus *Homo* if (1) the brain size is 600 cm³ or greater, (2) the jaw and teeth resemble those of humans, and (3) tool use is evident (Fig. 22.15). In this section, we will discuss early *Homo*: *Homo habilis* and *Homo erectus*; and later Homo: the Neandertals and the Cro-Magnons, which are the first modern humans.

Early *Homo*

Homo habilis, dated between 2.0 and 1.9 MYA, may be ancestral to modern humans. Some of these fossils have a brain size as large as 775 cm³, which is about 45% larger than that of *A. afarensis.* The cheek teeth are smaller than even those of the gracile australopithecines. Therefore, it is likely that these early members of the genus *Homo* were omnivores who ate meat in addition to plant material. Bones at their campsites bear cut marks, indicating that they used tools to strip them of meat.

The stone tools made by *H. habilis,* whose name means "handy man," are rather crude. It's possible that these are the cores from which they took flakes sharp enough to scrape away hide, cut tendons, and easily remove meat from bones.

Early *Homo* skulls suggest that the portions of the brain associated with speech areas were enlarged. We can speculate that the ability to speak may have led to hunting cooperatively. Other members of the group may have remained plant gatherers, and if so, both hunters and gatherers most likely ate together and shared their food. In this way, society and culture could have begun.

Culture, which encompasses human behavior and products (e.g., technology and the arts), depends upon the capacity to speak and transmit knowledge. We can further speculate that the advantages of a culture to *H. habilis* may have hastened the extinction of the australopithecines.

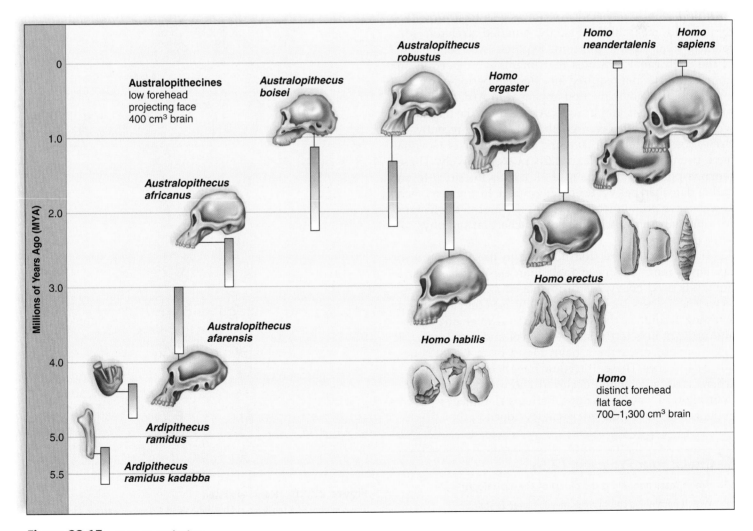

Figure 22.15   Human evolution.
The length of time each species existed is indicated by the vertical gray lines. Notice that there have been times when two or more hominids existed at the same time. Therefore, human evolution resembles a "bush" rather than a single branch. (*Left*) Australopithecines have a low forehead, projecting face, and a brain size of about 400 cm³. (*Right*) Homos have an increasingly high forehead, a flat face, and a brain size of 700 to 1,300 cm³.

Homo erectus

Homo erectus and like fossils are found in Africa, Asia, and Europe and dated between 1.9 and 0.3 MYA. A Dutch anatomist named Eugene Dubois was the first to unearth *H. erectus* bones in Java in 1891, and since that time many other fossils have been found in the same area. Although all fossils assigned the name *H. erectus* are similar in appearance, enough discrepancy exists to suggest that several different species have been included in this group. In particular, some experts suggest that the Asian form is *Homo erectus* and the African form is *Homo ergaster* (Fig. 22.16).

Compared with *H. habilis, H. erectus* had a larger brain (about 1,000 cm³) and a flatter face. The nose projected, however. This type of nose is adaptive for a hot, dry climate because it permits water to be removed before air leaves the body. The recovery of an almost complete skeleton of a ten-year-old boy indicates that *H. ergaster* was much taller than the hominids discussed thus far. Males were 1.8 meters tall (about 6 feet), and females were 1.55 meters (approaching 5 feet). Indeed, these hominids were erect and most likely had a striding gait like ours. The robust and most likely heavily muscled skeleton still retained some australopithecine features. Even so, the size of the birth canal indicates that infants were born in an immature state that required an extended period of care.

H. ergaster may have first appeared in Africa and then migrated into Asia and Europe. At one time, the migration was thought to have occurred about 1 MYA, but recently, *H. erectus* fossil remains in Java and the Republic of Georgia have been dated at 1.9 and 1.6 MYA, respectively. These remains push the evolution of *H. erectus* in Africa to an earlier date than has yet been determined. In any case, such an extensive population movement is a first in the history of humankind and a tribute to the intellectual and physical skills of the species.

H. erectus was the first hominid to use fire and also fashioned more advanced tools than early *Homos*. These hominids used heavy, teardrop-shaped axes and cleavers as well as flakes, which were probably used for cutting and scraping. It could be that *H. ergaster* was a systematic hunter and brought kills to the same site over and over again. In one location, researchers have found over 40,000-bones and 2,647 stones. These sites could have been "home bases," where social interaction occurred and a prolonged childhood allowed time for learning. Perhaps a language evolved and a culture more like our own developed.

☑ Check Your Progress 22.5a

1. Which fossil hominid is the oldest of the australopithecines?
2. Which hominid is the first to have culture? Explain.
3. Which hominid was the first to migrate out of Africa?

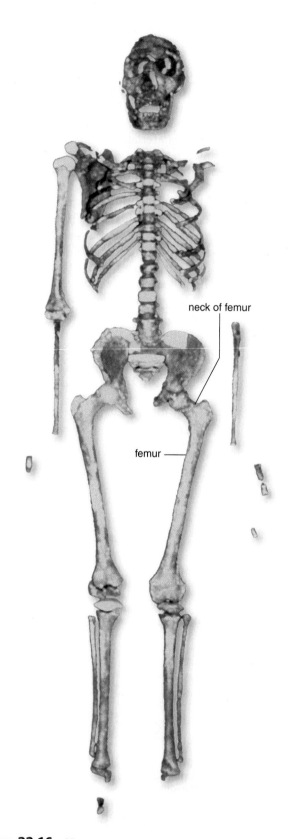

neck of femur

femur

Figure 22.16   Homo ergaster.
This skeleton of a ten-year-old boy who lived 1.6 MYA in eastern Africa shows femurs that are angled because the femur neck is quite long.

Science Focus

Homo floresiensis

In 2003, scientists made one of the most spectacular discoveries in evolutionary history. Nine skeletons of a hobbit-like species were discovered in a cave on the isle of Flores. Flores is an Indonesian island east of Bali, located midway between Asia and Australia. This new species of humans grew no taller than a modern three-year-old child (Fig. 22A) One skeleton was that of an adult female who died when she was approximately 30 years old. She stood 1 meter tall (3.3 feet) and weighed approximately 25 kilograms (55 pounds). Scientists estimate that she died around 18,000 years ago.

After examining the first skeleton, the research team concluded that they had discovered a new human species. They named the species *Homo floresiensis* after the island where it was found. The workers at the excavation site nicknamed the tiny creatures "hobbits" after the fictional creatures in the *Lord of the Rings* books.

Classification of *Homo floresiensis*

Homo floresiensis have skulls the size of a grapefruit and a brain size of approximately 417 cc. The teeth are humanlike, the eyebrow ridges are thick, the forehead slopes sharply, and the face lacks a chin. Even though the hobbits stand only 3 feet tall, they are not classified as pygmies. Despite the small body size, small brain size and a mixture of primitive and advanced anatomical features, *H. floresiensis* is distinctly a member of genus *Homo*. The researchers believe that *H. floresiensis* possibly evolved from a population of *Homo erectus* that reached Flores approximately 840,000 years ago.

Culture of *Homo floresiensis*

Many of the habits exhibited by *H. floresiensis* are remarkably similar to those of other *Homo* species. Archaeological evidence indicates that *H. floresiensis* had the use of fire. The skeletons discovered on Flores were found in sediment deposits that also contained stone tools and the bones of dwarf elephants, giant rodents, and Komodo dragons. The dwarf elephants, or stegodons, weighed about 1,000 kilograms (2,200 pounds) and would have posed a serious challenge to men who were only 1 meter tall. Successful hunting would have required communication among members of the hunting party. The Flores diet also included fish, frogs, birds, rodents, snakes, and tortoises.

The hobbits produced sophisticated stone tools, hunted successfully in groups, and crossed at least two bodies of water to reach Flores from mainland Asia. And yet, their brain was about one-third the size of modern humans. *H. floresiensis* is the smallest species of human ever discovered. Pound for pound, they outcompete every other member of genus *Homo*.

a. *Homo floresiensis*,
artist's impression

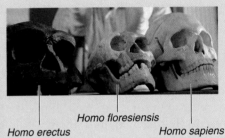

Homo erectus   Homo floresiensis   Homo sapiens

b. Comparison of skulls

Figure 22A   *Homo floresiensis*
a. Artist's recreation of *H. floresiensis*.
b. The *H. floresiensis* is smaller than that of *H. erectus* and that of *H. sapiens*.

Further Research Needed

Researchers are interested in determining why the hobbits were so small. The first-discovered skeleton was believed to be that of a small child. There is no evidence of any other 1 meter tall adults in genus *Homo*. Modern pygmies are 1.4 to 1.5 meters (4.6 to nearly 5 feet) tall. Over thousands of years, it is possible that a population of *Homo erectus* evolved into *H. floresiensis*. If so, a smaller body size would have been favored by natural selection. The members of each generation could have been smaller than the previous generation. Dwarfing of mammals on islands is a well-known process that can be seen worldwide. Islands generally have a limited food supply, few predators, and at least a few species competing for the same ecological niche. It behooves species living on islands to minimize their daily energy requirements. The smaller the body size, the fewer the calories per day required for survival. At this point, there is no substantiation for this hypothesis, but continued research may produce an answer as to why hobbits are so small.

The Extinction of *Homo floresiensis*

It appears that many of Flores' inhabitants became extinct approximately 12,000 years ago due to a major volcanic eruption. Researchers found *Homo floresiensis* and pygmy stegodon remains below a 12,000-year-old volcanic ash layer. Hobbits reached the island approximately 11,000 years ago and possibly intermingled with modern humans. Rumors, myths, and legends among the indigenous tribes of Flores about "the tiny people who lived in the forest" have persisted.

Evolution of Modern Humans

Most researchers accept the idea that ***Homo sapiens*** (modern humans) evolved from *H. erectus,* but they differ as to the details. Perhaps *Homo sapiens* evolved from *H. erectus* separately in Asia, Africa, and Europe. The hypothesis that *Homo sapiens* evolved in several different locations is called the **multiregional continuity hypothesis** (Fig. 22.17*a*). This hypothesis proposes that evolution to modern humans was essentially similar in several different places. If so, each region should show a continuity of its own anatomical characteristics from the time when *H. erectus* first arrived in Europe and Asia.

Opponents argue that it seems highly unlikely that evolution would have produced essentially the same result in these different places. They suggest, instead, the **out-of-Africa hypothesis,** which proposes that *H. sapiens* evolved from *H. erectus* only in Africa, and thereafter *H. sapiens* migrated to Europe and Asia about 100,000 years BP (before present) (Fig. 22.17*b*). If so, there would be no continuity of characteristics between fossils dated 200,000 years BP and 100,000 years BP in Europe and Asia.

According to which hypothesis would modern humans be most genetically alike? The multiregional continuity hypothesis states that human populations have been evolving separately for a long time, and therefore, genetic differences are expected. The out-of-Africa hypothesis states that we are all descended from a few individuals from about 100,000-years BP. Therefore, the out-of-Africa hypothesis suggests that we are more genetically similar.

A few years ago, a study attempted to show that all the people of Europe (and the world, for that matter) have essentially the same mitochondrial DNA. Called the "mitochondrial Eve" hypothesis by the press (note that this is a misnomer because no single ancestor is proposed), the statistics that calculated the date of the African migration were found to be flawed. Still, the raw data—which indicate a close genetic relationship among all Europeans—support the out-of-Africa hypothesis.

These opposing hypotheses have sparked many other innovative studies to test them. Most scientists have come to the conclusion that the out-of-Africa hypothesis is supported.

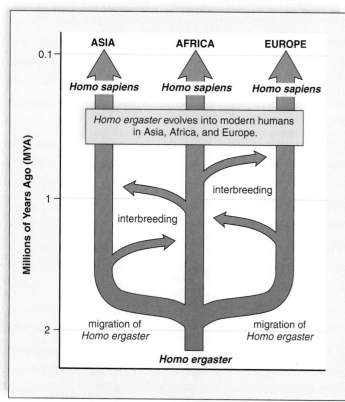

a. Multiregional continuity

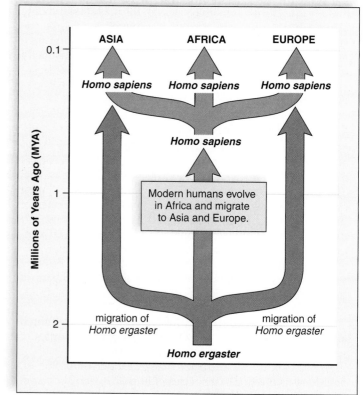

b. Out of Africa

Figure 22.17 **Evolution of modern humans.**
a. The multiregional continuity hypothesis proposes that *Homo sapiens* evolved separately in at least three different places: Asia, Africa, and Europe. Therefore, continuity of genotypes and phenotypes is expected in each region but not between regions. **b.** The out-of-Africa hypothesis proposes that *Homo sapiens* evolved only in Africa and then migrated out of Africa, as *H. ergaster* did many years before. *H. sapiens* would have supplanted populations of *Homo* in Asia and Europe about 100,000 years BP.

Neandertals

Neandertals *(H. neandertalensis)* take their name from Germany's Neander Valley, where one of the first Neandertal skeletons, dated some 200,000 years BP, was discovered. The Neandertals had massive brow ridges, and their nose, jaws, and teeth protruded far forward. The forehead was low and sloping, and the lower jaw lacked a chin. New fossils show that the pubic bone was long compared with ours.

According to the out-of-Africa hypothesis, Neandertals were eventually supplanted by modern humans. Surprisingly, however, the Neandertal brain was, on the average, slightly larger than that of *Homo sapiens* (1,400 cm³, compared with 1,360 cm³ in most modern humans). The Neandertals were heavily muscled, especially in the shoulders and neck (Fig. 22.18). The bones of the limbs were shorter and thicker than those of modern humans. It is hypothesized that a larger brain than that of modern humans was required to control the extra musculature. The Neandertals lived in Europe and Asia during the last ice age, and their sturdy build could have helped conserve heat.

The Neandertals give evidence of being culturally advanced. Most lived in caves, but those living in the open may have built houses. They manufactured a variety of stone tools, including spear points, which could have been used for hunting, and scrapers and knives, which would have helped in food preparation. They most likely successfully hunted bears, woolly mammoths, rhinoceroses, reindeer, and other contemporary animals. They used and could control fire, which probably helped them cook meat and keep themselves warm. They even buried their dead with flowers and tools and may have had a religion. Perhaps they believed in life after death, and if so, they were capable of thinking symbolically.

Cro-Magnons

Cro-Magnons are the oldest fossils to be designated *Homo sapiens*. In keeping with the out-of-Africa hypothesis, the Cro-Magnons, who are named after a fossil location in France, were the modern humans who entered Asia and Europe from Africa 100,000 years BP or even earlier. Cro-Magnons had a thoroughly modern appearance (Fig. 22.19). Recent analysis of Neandertal DNA indicates that it is so different from Cro-Magnon DNA, these two groups of people did not interbreed. Instead Cro-Magnons seem to have replaced the Neandertals in the Middle East and then spread to Europe 40,000 years ago where they lived side by side with the Neandertals for several thousand years. If so, the Neandertals are cousins and not ancestors to us.

Cro-Magnons made advanced stone tools, including compound tools, as when stone flakes were fitted to a wooden handle. They may have been the first to throw spears, enabling them to kill animals from a distance, and to make knifelike blades. They were such accomplished hunters that some researchers suggest they were responsible for the extinction of many larger mammals, such as the giant sloth, the mammoth, the saber-toothed tiger, and the giant ox, during the late Pleistocene epoch.

Cro-Magnons hunted cooperatively, and perhaps they were the first to have a language. Most likely, they lived in small groups, with the men hunting by day, while the women remained at home with the children. It's quite possible that this hunting way of life among prehistoric people influences our behavior even today. The Cro-Magnon culture included art. They sculpted small figurines out of reindeer bones and antlers. They also painted beautiful drawings of animals on cave walls in Spain and France (Fig. 22.19).

Figure 22.18 Neandertals.
This drawing shows that the nose and the mouth of the Neandertals protruded from their faces, and their muscles were massive. They made stone tools and were most likely excellent hunters.

Figure 22.19 Cro-Magnons.
Cro-Magnon people are one of the earliest groups to be designated *Homo sapiens*. Their tool-making ability and other cultural attributes, including artistic talents, are legendary.

Human Variation

Human beings have been widely distributed about the globe ever since they evolved. As with any other species that has a wide geographical distribution, phenotypic and genotypic variations are noticeable between populations. Today, we say that people have different ethnicities (Fig. 22.20*a*).

It has been hypothesized that human variations evolved as adaptations to local environmental conditions. One obvious difference among people is skin color. A darker skin is protective against the high ultraviolet (UV) intensity of bright sunlight. On the other hand, a white skin ensures vitamin D production in the skin when the UV intensity is low. Harvard University geneticist Richard Lewontin points out, however, that this hypothesis concerning the survival value of dark and light skin has never been tested.

Two correlations between body shape and environmental conditions have been noted since the nineteenth century. The first, known as Bergmann's rule, states that animals in colder regions of their range have a bulkier body build. The second, known as Allen's rule, states that animals in colder regions of their range have shorter limbs, digits, and ears. Both of these effects help regulate body temperature by increasing the surface-area-to-volume ratio in hot climates and decreasing the ratio in cold climates. For example, Figure 22.20*b,c* shows that the Massai of East Africa tend to be slightly built with elongated limbs, while the Eskimos, who live in northern regions, are bulky and have short limbs.

Other anatomical differences among ethnic groups, such as hair texture, a fold on the upper eyelid (common in Asian peoples), or the shape of lips, cannot be explained as adaptations to the environment. Perhaps these features became fixed in different populations due simply to genetic drift. As far as intelligence is concerned, no significant disparities have been found among different ethnic groups.

Genetic Evidence for a Common Ancestry

The two hypotheses regarding the evolution of humans, discussed on page 484, pertain to the origin of ethnic groups. The multiregional hypothesis suggests that different human populations came into existence as long as a million years ago, giving time for significant ethnic differences to accumulate despite some gene flow. The out-of-Africa hypothesis, on the other hand, proposes that all modern humans have a relatively recent common ancestor who evolved in Africa and then spread into other regions. Paleontologists tell us that the variation among modern populations is considerably less than among human populations some 250,000 years ago. This would mean that all ethnic groups evolved from the same single, ancestral population.

A comparative study of mitochondrial DNA shows that the differences among human populations are consistent with their having a common ancestor no more than a million years ago. Lewontin, mentioned above, found that the genotypes of different modern populations are extremely similar.

Figure 22.20   Ethnic groups.
a. Some of the differences between the three prevalent ethnic groups in the United States may be due to adaptations to the original environment. **b.** The Massai live in East Africa. **c.** Eskimos live near the Arctic circle.

He examined variations in 17 genes, including blood groups and various enzymes, among seven major geographic groups: Caucasians, black Africans, mongoloids, south Asian Aborigines, Amerinds, Oceanians, and Australian Aborigines. He found that the great majority of genetic variation—85%—occurs within ethnic groups, not among them. In other words, the amount of genetic variation between individuals of the same ethnic group is greater than the variation between ethnic groups.

✔ Check Your Progress 22.5b

1. How might *Homo habilis*, living about 2 MYA, have differed from the australopithecines?

2. How might *Homo erectus*, living about 1.9–1.6 MYA, have differed from *Homo habilis*?

3. a. What is the out-of-Africa hypothesis, and (b) what bearing does it have on the evolution of humans?

4. How do Cro-Magnons, the first of the modern humans, differ from the other species in the genus Homo?

Effects of Population Growth

Human beings today undergo **biocultural evolution** because culture has developed to the point that adaptation to the environment is not dependent on genes but on the passage of culture from one generation to the next.

Tool Use and Language Began

The first step toward biocultural evolution began when *Homo habilis* made his primitive stone tools. *Homo erectus* continued the tradition and most likely was a hunter of sorts. It's possible that the campsites of *H. erectus* were "home bases," where the women stayed behind with the children while the men went out to hunt. Hunting was an important event in the development of culture, especially because it encourages the development of language. If *Homo erectus* didn't have the use of language, certainly Cro-Magnon did. People who have the ability to speak a language would have been able to cooperate better as they hunted and even as they sought places to gather plants. Among animals, only humans have a complex language that allows them to communicate their experiences. Words are not objects and events, they stand for objects and events that can be pictured in the mind.

Agriculture Began

About 10,000 years ago, people gave up being full-time hunter-gatherers and became at least part-time farmers. What accounts for the rise of agriculture? The answer is not known, but several explanations have been put forth. About 12,000 years ago, a warming trend occurred as the ice age came to a close. A variety of big game animals became extinct, including the saber-toothed cats, mammoths, and mastodons; and this may have made hunting less productive. As the weather warmed, the glaciers retreated and left fertile valleys, where rivers and streams were full of fish and the soil was good. The Fertile Crescent in Mesopotamia is one such example. Here, fishing villages may have sprung up, causing people to settle down.

The people were probably already knowledgeable about what crops to plant. Most likely, as hunter-gatherers, people had already selected seeds with desirable characteristics for propagation. Then, a chance mutation may have made these plants particularly suitable as a source of food. So now they began to till the good soil where they had settled, and they began to systematically plant crops (Fig. 22B).

As people became more sedentary, they may have had more children, especially since the men were home more often. Population increases may have tipped the scales and caused them to adopt agriculture full time, especially if agriculture could be counted on to provide food for hungry mouths. The availability of agricultural tools must have contributed to making agriculture worthwhile. The digging stick, the hoe, the sickle, and the plow were improved when iron tools replaced bronze in the stone-bronze-iron sequence of ancient tools. Irrigation began as a way to control water supply, especially in semiarid areas and regions of periodic rainfall.

If evolutionary success is judged by population size, agriculture was extremely beneficial to our success because it caused a rapid increase of human numbers all over the Earth. Also, agriculture ushered in civilization as we know it. When crops became bountiful, some people were freed from raising their own food, and they began to specialize for other ways of life in towns and then cities. Some people became traders, shopkeepers, bakers, and teachers, to name a few possibilities. Others became the nobility, priests, and soldiers. Today, farming is highly mechanized and cities are extremely large. However, today we are on a treadmill. As the human population increases, we need new innovations in order to produce greater amounts of food. As soon as food production increases, populations grow once again, and the demand for food becomes still greater. Will there be a point when the population is greater than the food capacity? Perhaps that time is already upon us?

Industrial Revolution Began

The industrial revolution began in England during the eighteenth century and with it a demand for energy in the form of coal and oil that today seems unlimited. Our ability to construct any number of tools, including high-tech computers, is not stored in our genes; we learn it from the previous generation. Our modern civilization that began due to the advent of agriculture is now altering the global environment in a way that affects the evolution of other species. Species are becoming extinct, unless they are able to adapt to the presence of our civilization. It could be that biocultural evolution will be so harmful to the biosphere that the human species will eventually be driven to extinction also.

Decide Your Opinion

1. Can technology be used to help us not pollute the environment? How?
2. Should the extinction of other species be prevented? How can it be prevented?
3. Should the human population be reduced in size? How could this occur?

Figure 22B   Primitive agriculture.
A primitive form of agriculture began in several locations on Earth about 10,000 years ago.

Summarizing the Concepts

22.1 Origin of Life

A chemical evolution could have produced the protocell.

- Using an outside energy source, small organic molecules were produced by reactions between early Earth's atmospheric gases.
- Macromolecules evolved and interacted.
- The RNA-first hypothesis—only macromolecule RNA was needed for the first cell(s).
- The protein-first hypothesis—amino acids join to form polypeptides when exposed to dry heat.
- The protocell, a heterotrophic fermenter, lived on preformed organic molecules in the ocean.

The protocell eventually became a true cell once it had genes composed of DNA and could reproduce.

22.2 Biological Evolution

Biological evolution explains both the unity and diversity of life.

- Descent from a common ancestor explains the unity (sameness) of living things.
- Adaptation to different environments explains the great diversity of living things.
- **Fossil evidence supports evolution**   The fossil record gives us the history of life in general and allows us to trace the descent of a particular group.

Darwin discovered much evidence for common descent.

- **Biogeographical evidence**   The distribution of organisms on Earth is explainable by assuming that organisms evolved in one locale.
- **Anatomical evidence**   The common anatomies and development of a group of organisms are explainable by descent from a common ancestor.
- **Biochemical evidence**   All organisms have similar biochemical molecules.

Darwin developed a mechanism for adaptation known as natural selection:

	Observation	Result	Conclusion
1	a. Organisms have variations.	b. New adaptations to the environment arise.	
2	a. Organisms struggle to exist.	b. More organisms are present than can survive.	Organisms become more adapted with each generation.
3	a. Organisms differ in fitness.	b. Organisms best suited to the environment survive and reproduce.	

- The result of natural selection is a population adapted to its local environment.

22.3 Classification of Humans

The classification of humans can be used to trace their ancestry.

- Humans are primates.
- A primate evolutionary tree shows that humans share a common ancestor with African apes.

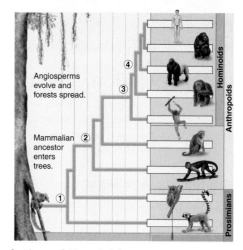

22.4 Evolution of Hominids

- The first hominid (includes humans) most likely lived about 6–7 MYA.
- Certain features (bipedal posture, flat face, and brain) identify fossil hominids.
- Ardipithecines were most likely hominids.

Evolution of Australopithecines

The evolutionary tree of hominids resembles a bush (not a straight line of fossils leading to modern humans).

- Australopithecines (a hominid) lived about 3 MYA.
- They could walk erect, but they had a small brain.
- This testifies to a mosaic evolution for humans (not all advanced features evolved at the same time).

22.5 Evolution of Humans

Fossils are classified as *Homo* with regard to brain size (over 600 cm³), jaws and teeth (resemble modern humans'), and evidence of tool use.

- *H. habilis* made and used tools.
- *H. erectus* was the first *Homo* to have a brain size of more than 1,000 cm³.
- *H. erectus* migrated from Africa into Europe and Asia.
- *H. erectus* used fire and may have been big-game hunters.

Evolution of Modern Humans

Two hypotheses of modern human evolution are being tested.

- The multiregional continuity hypothesis suggests that modern humans evolved separately in Europe, Africa, and Asia.
- The out-of-Africa hypothesis says that *H. sapiens* evolved in Africa but then migrated to Asia and Europe.

Neandertals and Cro-Magnons

- The Neandertals were already living in Europe and Asia before modern humans arrived.
- They had a culture, but did not have the physical traits of modern humans.
- Cro-Magnons are the oldest fossil to be designated *H. sapiens*. Their tools were sophisticated, and they had a culture.

Understanding Key Terms

adaptation 470
analogous structure 472
anthropoid 476
australopithecine 480
binomial name 475
biocultural evolution 487
biogeography 472
biological evolution 470
bipedal posture 479
chemical evolution 468
Cro-Magnon 485
culture 481
eukaryotic cell 470
evolutionary tree 479
fossil 470
fossil record 471
heterotroph 469
hominid 479
Homo erectus 482

Homo habilis 481
homologous structure 472
Homo sapiens 484
lineage 479
molecular clock 479
mosaic evolution 480
multiregional continuity
   hypothesis 484
natural selection 474
Neandertal 485
out-of-Africa hypothesis 484
primate 476
prokaryotic cell 470
prosimian 476
protein-first hypothesis 469
protocell 469
RNA-first hypothesis 469
vestigial structure 472

Match the key terms to these definitions.

a. _____ Process by which populations become adapted to their environment.

b. _____ Organism's modification in structure, function, or behavior suitable to the environment.

c. _____ Structure that is similar in two or more species because of common ancestry.

d. _____ Increase in the complexity of chemicals over time that could have led to the first cells.

e. _____ Any remains of an organism that have been preserved in the Earth's crust.

Testing Your Knowledge of the Concepts

1. List and discuss the steps by which a chemical evolution could have produced a protocell. (pages 468–69)

2. You are studying finches on the Galápagos Islands. What evidence do you need to show common descent and adaptation to the environment? (pages 470–73)

3. Describe the evidences that support descent from a common ancestor. (pages 470–73)

4. Describe the critical elements of Darwin's natural selection process. (page 474)

5. List a characteristic of each category of human classification. (page 475)

6. What type of life are primates adapted to? Discuss several primate characteristics that aid them in their adaptation. (pages 476–77)

7. How is the human skeleton like that of a chimpanzee skeleton? What's the major difference? (page 477)

8. Why do most investigators use bipedal locomotion, and not size of brain, to designate a hominid? (page 479)

9. Describe the characteristics of australopithecines, early *Homo*, *Homo erectus*, Neandertals, and Cro-Magnons. (pages 480–85)

10. Contrast the multiregional continuity hypothesis with the out-of-Africa hypothesis for the evolution of *Homo sapiens*. (page 484)

11. Which of these did Stanley Miller place in his experimental system to show that organic molecules could have arisen from inorganic molecules on the early Earth?
    a. microspheres
    b. purines and pyrimidines
    c. gases in the atmosphere of early Earth
    d. only RNA
    e. All of these are correct.

12. Which of these is the chief reason the protocell was probably a fermenter?
    a. The protocell didn't have any enzymes.
    b. The atmosphere didn't have any oxygen.
    c. Fermentation provides the most energy.
    d. There was no ATP yet.
    e. All of these are correct.

13. Evolution of the DNA → RNA → protein system was a milestone because the protocell could now
    a. be a heterotrophic fermenter.
    b. pass on genetic information.
    c. use energy to grow.
    d. take in preformed molecules.
    e. All of these are correct.

14. According to Darwin,
    a. the adapted individual is the one who survives and passes on its genes to offspring.
    b. changes in phenotype are passed on by way of the genotype to the next generation.
    c. organisms are able to bring about a change in their phenotype.
    d. evolution is striving toward particular traits.
    e. All of these are correct.

15. Organisms
    a. compete with other members of their species.
    b. vary in physical characteristics.
    c. are adapted to their environment.
    d. are related by descent from common ancestors.
    e. All of these are correct.

16. If evolution occurs, we would expect different biogeographical regions with similar environments to
    a. all contain the same mix of plants and animals.
    b. each have its own specific mix of plants and animals.
    c. have plants and animals with similar adaptations.
    d. have plants and animals with different adaptations.
    e. Both b and c are correct.

17. The fossil record offers direct evidence for evolution because you can
    a. see that the types of fossils change over time.
    b. sometimes find common ancestors.
    c. trace the ancestry of a particular group.
    d. trace the biological history of living things.
    e. All of these are correct.

18. Organisms such as whales and sea turtles that are adapted to an aquatic way of life
    a. will probably have homologous structures.
    b. will have similar adaptations but not necessarily homologous structures.
    c. may very well have analogous structures.
    d. will have the same degree of fitness.
    e. Both b and c are correct.

19. Which of these gives the correct order of divergence from the main line of descent leading to humans?
    a. prosimians, monkeys, Asian apes, African apes, humans
    b. gibbons, baboons, prosimians, monkeys, African apes, humans
    c. monkeys, gibbons, prosimians, African apes, baboons, humans
    d. African apes, gibbons, monkeys, baboons, prosimians, humans
    e. *H. habilis, H. erectus, H. neanderthalensis,* Cro-Magnon

20. Lucy is a member of what species?
    a. *Homo erectus*
    b. *Australopithecus afarensis*
    c. *H. habilis*
    d. *A. robustus*
    e. *A. anamensis* and *A. afarensis* are alternative forms of Lucy.

21. What possibly may have influenced the evolution of bipedalism?
    a. A larger brain developed. d. Both b and c are correct.
    b. Food gathering was easier. e. Both a and c are correct.
    c. The climate became colder.

22. *H. ergaster* could have been the first to
    a. use and control fire.
    b. migrate out of Africa.
    c. make tools.
    d. Both a and b are correct, but c is not.
    e. a, b, and c are correct.

23. Which of these characteristics is not consistent with the others?
    a. opposable thumb   d. well-developed brain
    b. learned behavior   e. stereoscopic vision
    c. multiple births

24. The last common ancestor for African apes and hominids
    a. has been found, and it resembles a gibbon.
    b. has not yet been identified, but it is expected to be dated from about 6–7 MYA.
    c. has been found, and it has been dated at 30 MYA.
    d. is not expected to be found because there was no such common ancestor.
    e. most likely lived in Asia, not Africa.

25. If the multiregional continuity hypothesis is correct, then
    a. hominid fossils in China after 100,000 BP are not expected to resemble earlier fossils.
    b. hominid fossils in China after 100,000 BP are expected to resemble earlier fossils.
    c. the mitochondrial Eve study must be invalid.
    d. Both a and c are correct.
    e. Both b and c are correct.

26. A primate evolutionary tree
    a. exists only for humans.
    b. shows the evolutionary relationship among the different types of primates.
    c. should not include extinct forms.
    d. indicates that the ape lineage and human lineage are still joined.
    e. All of these are correct.

27. Classify humans by filling in the missing lines.

Domain	Eukarya
Kingdom	Animalia
Phylum	a. _____
b. _____	Mammalia
c. _____	Primates
Family	d. _____
e. _____	*Homo*
Species	f. _____

In questions 28–31, match each description to a type of evolutionary evidence in the key.

Key:
    a. biogeography   c. comparative biochemistry
    b. fossil record   d. comparative anatomy

28. Species change over time.

29. Forms of life are variously distributed.

30. A group of related species have homologous structures.

31. The same types of molecules are found in all living things.

Thinking Critically About the Concepts

Emily, from the opening story, was watching her day-care children play in exactly the same fashion as the chimps at the zoo. One child started out peeking at another through a plastic tube, which prompted the other child to go to the other end of the tube to peer back. A couple of little girls were laughing as they checked out their "dress-up" fashions in a mirror. Aside from the physical differences in the appearance of the children, she thought "I could be watching the chimps playing right now."

1. More and more zoo animals are offered enrichments that are stimulating and engaging. What did Ryan and Emily observe at the zoo that might be classified as an enrichment?

2. Since the goal of offering enrichments for zoo animals is to stimulate natural species-specific behaviors, what kind of enrichments do you think might be offered to lions and/or tigers?

3. Many zoo animals are part of a worldwide species survival plan designed to help ensure the survival of various species in the future. Part of the plan includes a planned breeding program. What would a species gain from a planned breeding program that selects which animals are bred to each other?

4. Many animals have been observed playing with each other and with objects in their environment, and some are known to use tools to obtain food and do other tasks. Explain why this kind of behavior (playing and/or tool use) might be common to a variety of animals.

Global Ecology and Human Interferences

T he title of an online article about "green" burials caught Tyler's eye. His grandfather, "Pops," unexpectedly passed away a month ago, and no one knew what his burial wishes were. Tyler's mom argued that Pops should be cremated because he'd grumbled about the expenses associated with a casket burial when his children tried to get him to preplan his funeral. Her siblings all wanted to have Pops embalmed, entombed in a gasketed casket, and buried next to their mother. Since Tyler's mom was outnumbered by her siblings, Pops had been embalmed, placed in an airtight and watertight casket, and buried next to his wife.

The article Tyler stumbled upon described the increasing popularity of green burials and the natural cemeteries that offer them. Most green cemeteries prohibit embalming of bodies before burial, and they recommend caskets made of cardboard, wicker, or the use of a cloth shroud. Green burials tend to be less expensive ($1,000–$2,000) than conventional burials ($5,000 and higher) and have the added benefit of sustaining the environment and conserving the land used for green cemeteries. This return to a more natural form of burial appealed to Tyler, since he loved observing nature and spending time outdoors hiking and kayaking. It sounded more spiritual to him than the hoopla that accompanied the conventional funeral of his grandfather.

As you study the material in this chapter, it should become evident how green burials benefit the environment. It's a "dust-to-dust" way to ensure the cycle of life continues.

CHAPTER CONCEPTS

23.1 The Nature of Ecosystems
The biosphere encompasses that part of Earth where living things live in ecosystems. Populations interact among themselves and with the physical environment in ecosystems. Ecosystems are characterized by energy flow and chemical cycling.

23.2 Energy Flow
Ecosystems contain food webs, in which the various populations are connected by who eats whom. Both energy and chemicals are passed from one population to the next, as one population feeds on another. Food chains have a limited length because, as demonstrated by food pyramids, only about 10% of energy is passed from one feeding level to the other. Eventually all the energy dissipates but the chemicals cycle back to the photosynthesizers.

23.3 Global Biogeochemical Cycles
Biogeochemical cycles contain reservoirs, which retain nutrients; exchange pools, where nutrients are readily available; and the biotic community, which passes nutrients from one population to the next. The water, carbon, and nitrogen cycles are gaseous because the exchange pool is the atmosphere. The phosphorus cycle is a sedimentary cycle. Particular ecological problems arise because human activities alter the normal transfer rates within each cycle.

Figure 23.1 **The major terrestrial ecosystems.**
Temperate forests have moderate temperatures and occur where rainfall is moderate, yet sufficient to support trees. Deserts have changeable temperatures with minimal rainfall. Tropical rain forests, which generally occur near the equator, have a high average temperature and the greatest amount of rainfall of all the terrestrial ecosystems. A tropical grassland (savanna) has high temperatures and moderate/seasonal rainfall. A temperate grassland (prairie) has low to high temperatures, with low annual rainfall. The taiga, a coniferous forest that encircles the globe, has a low average temperature, but moderate rainfall. The tundra is the northernmost terrestrial ecosystem and has the lowest average temperature of all the terrestrial ecosystems, with minimal to moderate rainfall.

23.1 The Nature of Ecosystems

The **biosphere** is where organisms are found on planet Earth, from the atmosphere above to the depths of the oceans below and everything in between. Taking the global view, the entire biosphere is one giant **ecosystem**, a place where organisms interact among themselves and with the physical and chemical environment. These interactions help maintain ecosystems and, in turn, the biosphere. Human activities can alter the interactions between organisms and their environments in ways that reduce the abundance and diversity of life in an ecosystem. It is important to understand how ecosystems function so that we can repair past damage and predict how human activities might change normal conditions.

Ecosystems

Scientists recognize several distinctive major types of terrestrial ecosystems, also called biomes (Fig. 23.1). Temperature and rainfall define the biomes, which contain communities of organisms adapted to the regional climate The tropical rain forest, which occurs at the equator, is dominated by large evergreen, broad-leaved trees. The savanna is a tropical grassland that supports many types of grazing animals. Temperate grasslands receive less rainfall than temperate forests (in which trees lose their leaves during the winter) and more water than deserts, which lack trees. The taiga is a very cold northern coniferous forest, and the tundra, which borders the North Pole, is also very cold, with long winters and a short growing season. A permafrost persists even during the summer in the tundra and prevents large plants from becoming established.

Aquatic ecosystems are divided into those composed of freshwater and those composed of salt water (marine ecosystems). The ocean is a marine ecosystem that covers 70% of the Earth's surface. Two types of freshwater ecosystems are those with standing water, such as lakes and ponds, and those with running water, such as rivers and streams. The richest marine ecosystems lie near the coasts. Coral reefs are located offshore, while marshes occur where rivers meet the sea (Fig. 23.2).

Figure 23.2   The major aquatic ecosystems.
Aquatic ecosystems are divided into those that have salt water, such as the ocean (**a**) and those that have freshwater, such as a river (**b**). Saltwater, or marine, ecosystems also include coral reefs (**c**) and marshes (**d**).

Biotic Components of an Ecosystem

The abiotic components of an ecosystem are the nonliving components. The biotic components are living things that can be categorized according to their food source (Fig. 23.3). Some populations (species) are autotrophs, and some are heterotrophs.

Autotrophs

Autotrophs require only inorganic nutrients and an outside energy source to produce organic nutrients for their own use and for all the other members of a community. Therefore, they are called **producers**—they produce food (Fig. 23.3*a*). Photosynthetic organisms produce most of the organic nutrients for the biosphere. Algae of all types possess chlorophyll and carry on photosynthesis in freshwater and marine habitats. Algae photosynthesize in aquatic ecosystems, and green plants are the dominant photosynthesizers on land.

Heterotrophs

Heterotrophs need a source of organic nutrients. They are the **consumers**—they consume food. **Herbivores** are animals that graze directly on plants or algae (Fig. 23.3*b*). In terrestrial habitats, insects are small herbivores, while in aquatic habitats, protists, such as protozoans, play that role. **Carnivores** feed on other animals; birds that feed on insects are carnivores, and so are hawks that feed on birds (Fig. 23.3*c*). This example allows us to mention that there are primary consumers (e.g., insects), secondary consumers (e.g., insect-eating birds), and tertiary consumers (e.g., hawks). Sometimes tertiary consumers are called top predators. **Omnivores** are animals that feed both on plants and animals. As you most likely know, humans are omnivores.

   Detritus feeders are organisms that feed on detritus, which is decomposing particles of organic matter. Marine fan worms take detritus from the water, while clams take it from the substratum. Earthworms and some beetles, termites, and ants are all terrestrial detritus feeders. Bacteria and fungi, including mushrooms are decomposers; they acquire nutrients by breaking down dead organic matter, including animal wastes. Decomposers perform a valuable service because they release inorganic substances that are taken up by plants once more (Fig. 23.3*d*). Otherwise, plants would be completely dependent only on physical processes, such as the release of minerals from rocks, to supply them with inorganic nutrients.

Niche

A **niche** is the role of an organism in an ecosystem: how it gets its food and what eats it, and how it interacts with other populations in the same community. Now that we have studied the niche of organisms in an ecosystem, we can better understand the concept of a green burial and natural cemeteries discussed in the opening story. It is a way for a human being to be a part of the cycle of life and for the chemicals making up the body to return to nature.

a. Producers

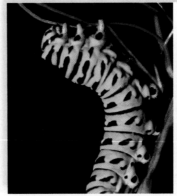

b. Herbivores

c. Carnivores

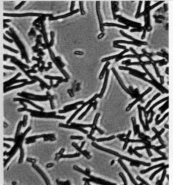

d. Decomposers

Figure 23.3   Biotic components.
a. Diatoms and green plants are autotrophs. **b.** Caterpillars and rabbits are herbivores. **c.** Spiders and osprey are carnivores. **d.** Bacteria and some mushrooms are decomposers.

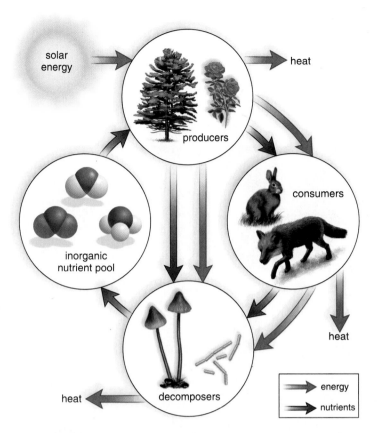

Figure 23.4   **Energy flow and chemical cycling.**
Chemicals cycle, but energy flows through an ecosystem. As energy transformations repeatedly occur, all the energy derived from the sun eventually dissipates as heat.

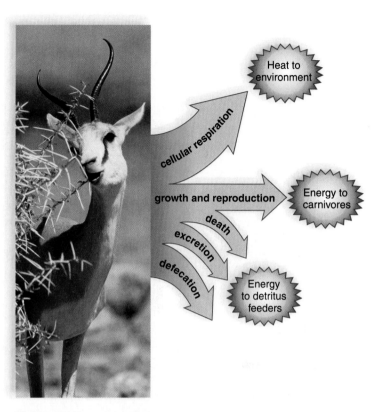

Figure 23.5   **Energy balances.**
Only about 10% of the food energy taken in by a herbivore is passed on to carnivores. A large portion goes to detritus feeders via defecation, excretion, and death, and another large portion is used for cellular respiration.

Energy Flow and Chemical Cycling

When we diagram the interactions of all the populations of an ecosystem, it is possible to illustrate that every ecosystem is characterized by two phenomena: energy flow and chemical cycling. Energy flow begins when producers absorb solar energy, and chemical cycling begins when producers take in inorganic nutrients from the physical environment. Thereafter, via photosynthesis, producers make organic nutrients (food) directly for themselves and indirectly for the other populations of the ecosystem. Energy flow occurs because as nutrients pass from one population to another, all the energy content is eventually converted to heat, which dissipates in the environment. Therefore, most ecosystems cannot exist without a continual supply of solar energy. Chemicals cycle when inorganic nutrients are returned to the producers from the atmosphere or soil (Fig. 23.4).

Only a portion of the organic nutrients made by autotrophs is passed on to heterotrophs because plants use organic molecules to fuel their own cellular respiration. Similarly, only a small percentage of nutrients taken in by heterotrophs is available to higher-level consumers. Figure 23.5 shows why. Some of the food eaten by a herbivore is never digested and is eliminated as feces. Metabolic wastes are excreted as urine. Of the assimilated energy, a large portion is utilized during cellular respiration and thereafter becomes heat. Only the remaining food, converted into increased body weight (or additional offspring), becomes available to carnivores. Because plants carry on cellular respiration only about 55% of the original energy absorbed by plants is available to an ecosystem. And, as organisms feed on one another, less and less of this 55% is available in a usable form.

The elimination of feces and urine by a heterotroph, and indeed the death of all organisms, does not mean that substances are lost to an ecosystem. Instead they are nutrients made available to decomposers. Decomposers convert the organic nutrients, such as glucose, back into inorganic chemicals, such as carbon dioxide and water, and release them to the soil or atmosphere. Chemicals complete their cycle within an ecosystem when inorganic chemicals are absorbed by the producers from the atmosphere or from soil.

✅ Check Your Progress 23.1

1. Why could it be said that the biosphere is a giant ecosystem?
2. a. Why are autotrophs called producers and (b) heterotrophs called consumers?
3. What are the different types of consumers in an ecosystem?
4. What two processes characterize an ecosystem? Explain.

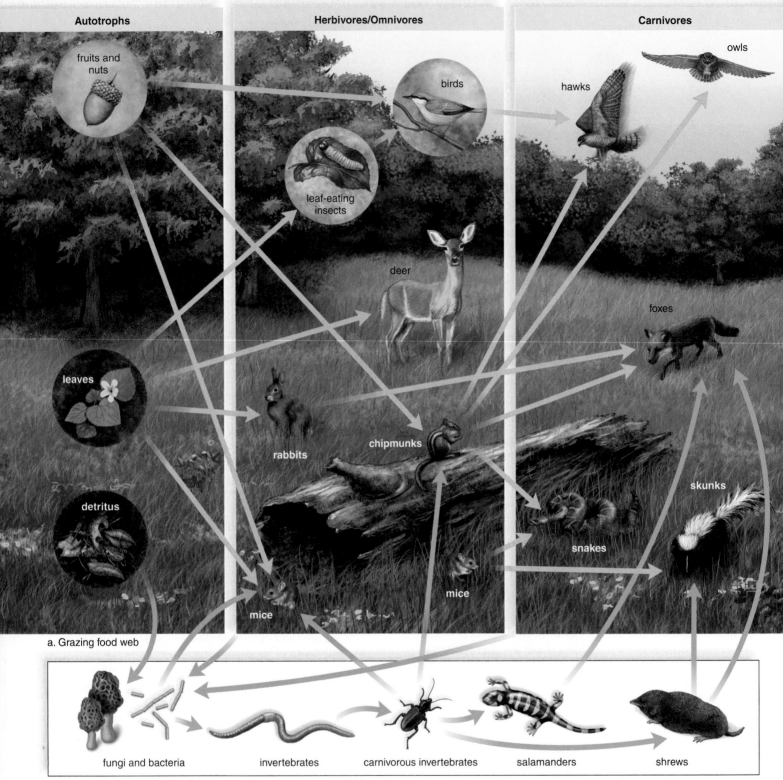

Autotrophs

fruits and nuts

leaves

detritus

a. Grazing food web

Herbivores/Omnivores

birds

leaf-eating insects

deer

rabbits

chipmunks

mice

mice

Carnivores

owls

hawks

foxes

skunks

snakes

fungi and bacteria   invertebrates   carnivorous invertebrates   salamanders   shrews

b. Detrital food web

Figure 23.6   Grazing and detrital food webs.

Food webs are descriptions of who eats whom. **a.** Tan arrows illustrate possible grazing food webs. For example, birds, which feed on nuts, may be eaten by a hawk. Autotrophs such as the tree are producers (first trophic, or feeding, level), the first series of animals are primary consumers (second trophic level), and the next group of animals are secondary consumers (third trophic level). **b.** Green arrows illustrate possible detrital food webs, which begin with detritus—the bacteria and fungi of decay and the remains of dead organisms. A large portion of these remains are from the grazing food web illustrated in (**a**). The organisms in the detrital food web are sometimes fed on by animals in the grazing food web, as when chipmunks feed on bugs. Thus, the grazing food web and the detrital food web are connected to one another.

23.2 Energy Flow

The principles we have been discussing can now be applied to an actual ecosystem—a forest. The various interconnecting paths of energy flow are represented by a **food web,** a diagram that describes **trophic (feeding) relationships.** Figure 23.6*a* is a **grazing food web** because it begins with an oak tree and grass. Caterpillars feed on oak leaves, while mice, rabbits, and deer feed on leaves and grass at or near the ground. Birds, chipmunks, and mice feed on seeds and nuts, but they are in fact omnivores because they also feed on caterpillars. These herbivores and omnivores are then food for a number of different carnivores.

Figure 23.6*b* is a **detrital food web,** which begins with detritus and the decomposers found within detritus. Detritus is food for soil organisms such as earthworms. Earthworms are, in turn, fed on by carnivorous invertebrates that may be eaten by shrews or salamanders. Because the members of detrital food webs may become food for aboveground carnivores, the detrital and grazing food webs are joined.

We naturally tend to think that aboveground plants, such as trees, are the largest storage form of organic matter and energy, but this is not necessarily the case. In this particular forest, the organic matter lying on the forest floor and mixed into the soil contains over twice as much energy as the leaves of living trees. Therefore, more energy in a forest may be funneling through the detrital food web than through the grazing food web.

Trophic Levels

The arrangement of the species in Figure 23.6 suggests that organisms are linked to one another in a straight line, according to feeding relationships or who eats whom. Diagrams that show a single path of energy flow are called **food chains.** For example, in the grazing food web, we could find this **grazing food chain:**

Leaves → caterpillars → birds → hawks

And in the detrital food web (Fig. 23.6*b*), we could find this **detrital food chain:**

detritus → earthworms → shrews

A **trophic level** is composed of all the organisms that feed at a particular link in a food chain. In the grazing food web in Figure 23.6*a,* going from left to right, the trees are producers (first trophic level), the first series of animals are primary consumers (second trophic level), and the next group of animals are secondary consumers (third trophic level).

Ecological Pyramids

The shortness of food chains can be attributed to the loss of energy between trophic levels. As mentioned, only about 10% of the energy of one trophic level is available to the next trophic level. Therefore, if a herbivore population consumes 1,000 kg of plant material, only about 100 kg is converted to herbivore tissue, 10 kg to first-level carnivores, and 1 kg to second-level carnivores. The so-called 10% rule of thumb explains why few carnivores can be supported in a food web. The flow of energy with large losses between successive trophic levels is sometimes depicted as an **ecological pyramid** (Fig. 23.7).

Energy losses between trophic levels also result in pyramids based on the number of organisms in each trophic level. When constructing a pyramid based on number of organisms, problems arise, however. For example, in Figure 23.6*a,* each tree would contain numerous caterpillars; therefore, there would be more herbivores than autotrophs. The explanation, of course, has to do with size. An autotroph can be as tiny as a microscopic alga or as big as a beech tree; similarly, an herbivore can be as small as a caterpillar or as large as an elephant.

Pyramids of biomass eliminate size as a factor because **biomass** is the number of organisms multiplied by the weight of organic matter within one organism. You would certainly expect the biomass of the producers to be greater than the biomass of the herbivores and that of the herbivores to be greater than that of the carnivores. In aquatic ecosystems, such as lakes and open seas where algae are the only producers, the herbivores may have a greater biomass than

Figure 23.7   Ecological pyramid.
The biomass, or dry weight (g/m²), for trophic levels in a grazing food web in a bog at Silver Springs, Florida. There is a sharp drop in biomass between the producer level and herbivore level, which is consistent with the common knowledge that the detrital food web plays a significant role in bogs.

the producers when you take their measurements. Why? The reason is that, over time, the algae reproduce rapidly, but they are also consumed at a high rate.

These kinds of problems are making some ecologists hesitant about using pyramids to describe ecological relationships. One more problem is what to do with the decomposers, which are rarely included in pyramids, even though a large portion of energy becomes detritus in many ecosystems.

☑ Check Your Progress 23.2

1. What type of diagram represents the various paths of energy flow in an ecosystem?

2. What is the difference between a grazing food web and a detrital food web?

3. What type of diagram represents a single path of energy flow in an ecosystem?

4. What type of diagram illustrates that usable energy is lost in ecosystems?

23.3 Global Biogeochemical Cycles

In this section, we will examine in more detail how chemicals cycle through ecosystems. All organisms require a variety of organic and/or inorganic nutrients. For example, carbon dioxide and water are necessary nutrients for photosynthesizers. Nitrogen is a component of all the structural and functional proteins and nucleic acids that sustain living tissues. Phosphorus is essential for ATP and nucleotide production.

The pathways by which chemicals circulate through ecosystems involve both living (biotic) and nonliving (geological) components; therefore, they are known as **biogeochemical cycles.** A biogeochemical cycle can be gaseous or sedimentary. In a gaseous cycle, such as the carbon and nitrogen cycles, the element returns to and is withdrawn from the atmosphere as a gas. The phosphorus cycle is a sedimentary cycle: The chemical is absorbed from the soil by plant roots, passed to heterotrophs, and eventually returned to the soil by decomposers.

Chemical cycling involves the components of ecosystems shown in Figure 23.8. A *reservoir* is a source normally unavailable to producers, such as carbon in calcium carbonate shells on ocean bottoms. An *exchange pool* is a source from which organisms do generally take chemicals, such as the atmosphere or soil. Chemicals move along food chains in a *biotic community*, perhaps never entering an exchange pool.

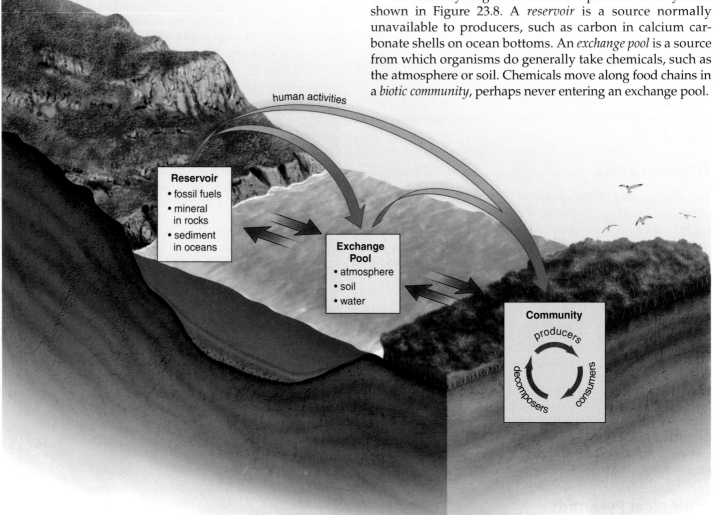

Figure 23.8   Model for chemical cycling.
Reservoirs, such as fossil fuels, minerals in rocks, and sediments in oceans, are normally relatively unavailable sources of nutrients for the biotic community. Nutrients in exchange pools, such the atmosphere, soil, and water, are available sources of chemicals for the biotic community. When human activities (purple arrows) remove chemicals from a reservoir or an exchange pool and make them available to the biotic community, pollution can result because not all the nutrients are utilized. For example, when humans burn fossil fuels, CO_2 (a nutrient for plants) rises in the atmosphere and contributes to global warming (see page 501).

Human activities (purple arrows) remove chemicals from reservoirs and exchange pools and make them available to the biotic community. In this way, human activities result in pollution because it upsets the normal balance of nutrients for producers in the environment.

The Water Cycle

The **water (hydrologic) cycle** is described in Figure 23.9. The width of the arrows in this and the other cycles that will be examined indicate the transfer rate of water between components of an ecosystem

① During **evaporation** in the water cycle, the sun's rays cause freshwater to evaporate from seawater, and the salts are left behind. Net condensation occurs. During condensation, a gas is changed into a liquid. ② Vaporized freshwater rises into the atmosphere, condenses, and then falls as **precipitation** (e.g., rain, snow, sleet, hail, and fog) over the oceans and the land.

③ Water also evaporates from land and from plants (evaporation from plants is called transpiration). ④ Because land lies above sea level, gravity eventually returns all freshwater to the sea. In the meantime, much water is contained within standing waters (lakes and ponds), flowing water (streams and rivers), and groundwater. ⑤ **Runoff** is water that flows directly into nearby streams, lakes, wetlands or the ocean.

Instead of running off, some precipitation sinks, or percolates, into the ground and saturates the earth to a certain level. The top of the saturation zone is called the ground-water table, or simply, the water table. ⑥ Sometimes, groundwater is also located in **aquifers,** rock layers that contain water and release it in appreciable quantities to wells or springs. Aquifers are recharged when rainfall and melted snow percolate into the soil.

Human Activities

Humans interfere with the water cycle in three ways. First, they withdraw water from aquifers; second, they clear vegetation from land and build roads and buildings that prevent percolation and increase runoff; and third, they interfere with the natural processes that purify water and instead add pollutants like sewage and chemicals to water.

In some parts of the United States, especially the arid West and southern Florida, withdrawals from aquifers exceed any possibility of recharge. This is called "groundwater mining." In these locations, the groundwater is dropping, and residents may run out of groundwater, at least for irrigation purposes, within a few short years. Freshwater, which makes up only about 3% of the world's supply of water, is called a renewable resource because a new supply is always being produced. But it is possible to run out of freshwater when the available supply runs off instead of entering bodies of freshwater and aquifers or has become so polluted that it is not usable.

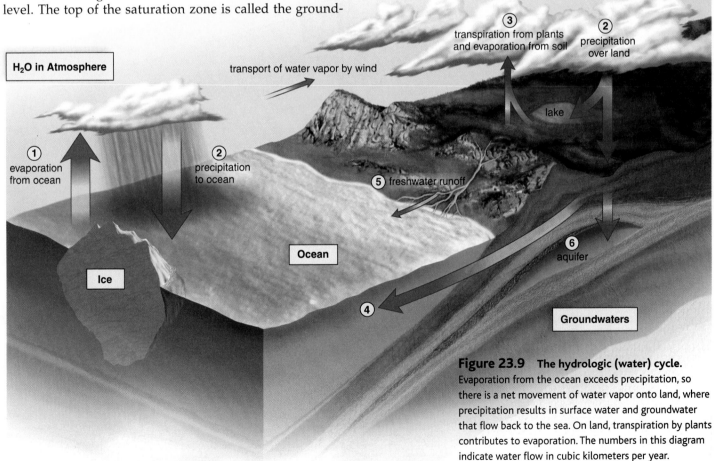

Figure 23.9   The hydrologic (water) cycle.
Evaporation from the ocean exceeds precipitation, so there is a net movement of water vapor onto land, where precipitation results in surface water and groundwater that flow back to the sea. On land, transpiration by plants contributes to evaporation. The numbers in this diagram indicate water flow in cubic kilometers per year.

The Carbon Cycle

The carbon dioxide (CO_2) in the atmosphere is the exchange pool for the carbon cycle. In this cycle, organisms in both terrestrial and aquatic ecosystems exchange carbon dioxide with the atmosphere (Fig. 23.10). ① On land, plants take up carbon dioxide from the air, and through photosynthesis, they incorporate carbon into nutrients that are used by autotrophs and heterotrophs alike. ② When organisms, including plants, respire, carbon is returned to the atmosphere as carbon dioxide. Therefore, carbon dioxide recycles to plants by way of the atmosphere.

In aquatic ecosystems, the exchange of carbon dioxide with the atmosphere is indirect. ③ Carbon dioxide from the air combines with water to produce bicarbonate ion (HCO_3^-), a source of carbon for algae that produce food for themselves and for heterotrophs. Similarly, when aquatic organisms respire, the carbon dioxide they give off becomes bicarbonate ion. ④ The amount of bicarbonate in the water is in equilibrium with the amount of carbon dioxide in the air.

Figure 23.10   The carbon cycle.
The carbon cycle is a gaseous biogeochemical cycle. Producers take in carbon dioxide from the atmosphere and convert it to organic molecules that feed all organisms. The transfer rate of carbon into the atmosphere due to respiration approximately matches the rate due to withdrawal by plants for photosynthesis. Fossil fuels arise when organisms die but do not decompose. When humans burn fossil fuels and destroy vegetation (purple arrows), more carbon dioxide is added to the atmosphere than is withdrawn. This causes environmental pollution.

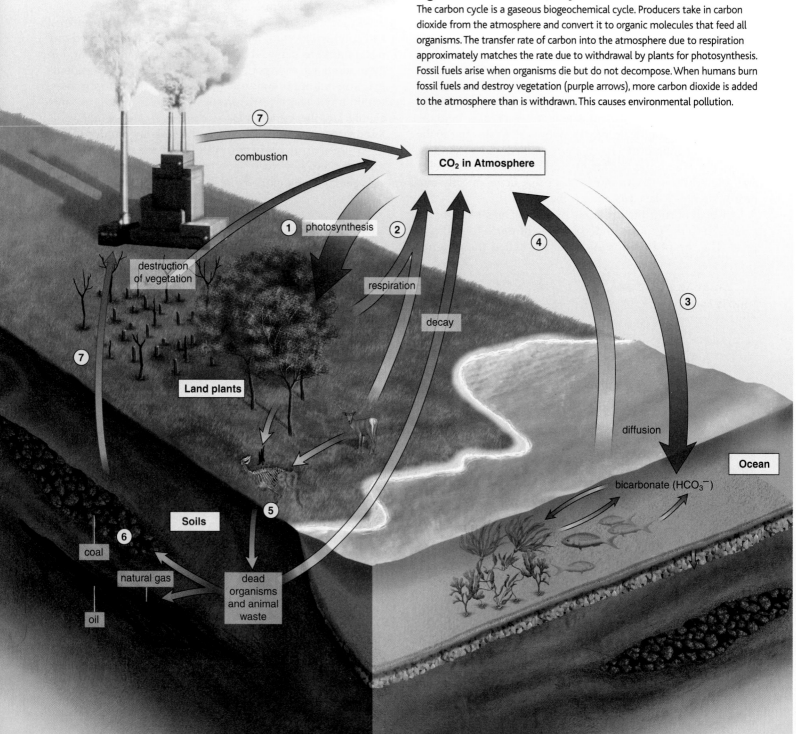

Reservoirs Hold Carbon

⑤ Living and dead organisms contain organic carbon and serve as one of the reservoirs for the carbon cycle. The world's biotic components, particularly trees, contain 800-billion tons of organic carbon, and an additional 1,000–3,000 billion metric tons are estimated to be held in the remains of plants and animals in the soil. Ordinarily, decomposition of animals, as, for example, when humans elect to have a green burial, discussed in the opening story, returns CO_2 to the atmosphere.

⑥ In the history of the Earth, some 300 MYA, plant and animal remains were transformed into coal, oil, and natural gas. We call these materials the **fossil fuels.** Another reservoir for carbon is the inorganic carbonate that accumulates in limestone and in calcium carbonate shells. Many marine organisms have calcium carbonate shells that remain in bottom sediments long after the organisms have died. Geological forces change these sediments into limestone.

Human Activities

The transfer rates of carbon dioxide due to photosynthesis and cellular respiration, which includes the work of decomposers, are just about even. However, more carbon dioxide is being deposited in the atmosphere than is being removed. ⑦ This increase is largely due to the burning of fossil fuels and the destruction of forests to make way for farmland and pasture. When we do away with forests, we reduce a reservoir and also the very organisms that take up excess carbon dioxide. Today, the amount of carbon dioxide released into the atmosphere is about twice the amount that remains in the atmosphere. It's believed that much of this has been dissolving into the ocean.

CO₂ and Global Warming Carbon dioxide and also other gases are being emitted due to human activities. The other gases include nitrous oxide (N_2O) from fertilizers and animal wastes and methane (CH_4) from bacterial decomposition, particularly in the guts of animals, in sediments, and in flooded rice paddies. These gases are known as **greenhouse gases** because, just like the panes of a greenhouse, they allow solar radiation to pass through but hinder the escape of infrared rays (heat) back into space. The greenhouse gases are contributing significantly to an overall rise in the Earth's ambient temperature, a phenomenon called **global warming,** because of their **greenhouse effect.**

Figure 23.11 shows the Earth's radiation balances. One thing to be learned from this diagram is that water vapor is a greenhouse gas, so clouds (which are composed of water vapor) also reradiate heat back to Earth. If the Earth's temperature rises, more water will evaporate, forming more clouds and setting up a positive feedback effect that could increase global warming still more. The global climate has already warmed about 0.6°C since the Industrial Revolution. Computer models are unable to consider all possible variables, but the Earth's temperature may rise 1.5–4.5°C by 2100 if greenhouse emissions continue at the current rates.

Global warming will bring about other effects, which computer models attempt to forecast. It is predicted that, as the oceans warm, temperatures in the polar regions will rise to a greater degree than in other regions. If so, glaciers will melt, and sea levels will rise, not only due to this melting but also because water expands as it warms. Water evaporation will increase, and most likely there will be increased rainfall along the coasts and dryer conditions inland. The occurrence of droughts will reduce agricultural yields and also cause trees to die off. Expansion of forests into arctic areas might not offset the loss of forests in the temperate zones. Coastal agricultural lands, such as the deltas of Bangladesh and China, will be inundated, and billions of dollars will have to be spent to keep coastal cities such as New Orleans, New York, Boston, Miami, and Galveston from disappearing into the sea.

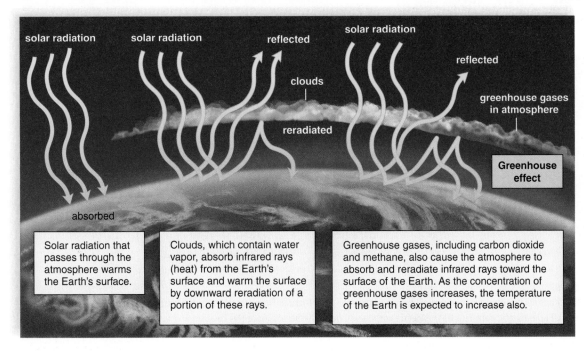

solar radiation · solar radiation · reflected · solar radiation · reflected · clouds · greenhouse gases in atmosphere · reradiated · **Greenhouse effect** · absorbed

Solar radiation that passes through the atmosphere warms the Earth's surface.

Clouds, which contain water vapor, absorb infrared rays (heat) from the Earth's surface and warm the surface by downward reradiation of a portion of these rays.

Greenhouse gases, including carbon dioxide and methane, also cause the atmosphere to absorb and reradiate infrared rays toward the surface of the Earth. As the concentration of greenhouse gases increases, the temperature of the Earth is expected to increase also.

Figure 23.11 Earth's radiation balances.
The effect of the greenhouse gases (*far right*) is contributing to global warming. A positive feedback cycle is predicted. As the Earth's temperature rises, more water will evaporate, causing clouds to thicken and absorb more solar radiation. Water vapor in clouds and the other greenhouse gases prevent the escape of heat (infrared rays), and it is reradiated back to the Earth, causing even more evaporation, and so forth.

The Nitrogen Cycle

Nitrogen gas (N_2) makes up about 78% of the atmosphere, but plants cannot make use of nitrogen gas. Therefore, nitrogen can be a nutrient that limits the amount of growth in an ecosystem.

Ammonium (NH_4^+) Formation and Use

① In the nitrogen cycle, **nitrogen fixation** occurs when nitrogen gas (N_2) is converted to ammonium (NH_4^+), a form plants can use (Fig. 23.12). Some cyanobacteria in aquatic ecosystems and some free-living bacteria in soil are able to fix atmospheric nitrogen in this way. Other nitrogen-fixing bacteria live in nodules on the roots of legumes, such as beans, peas, and clover. They make organic compounds containing nitrogen available to the host plants so that the plant can form proteins and nucleic acids.

Nitrate (NO_3^-) Formation and Use

Plants can also use nitrates (NO_3^-) as a source of nitrogen. The production of nitrates during the nitrogen cycle is called

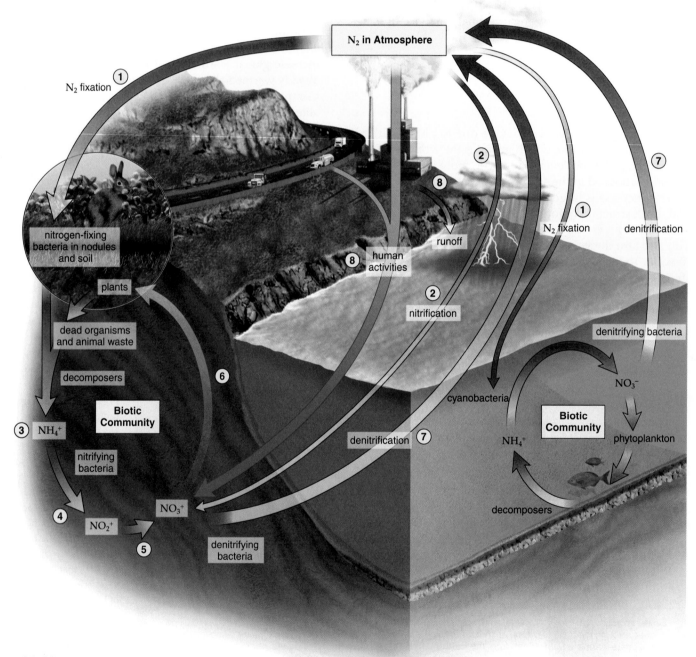

Figure 23.12 The nitrogen cycle.
Nitrogen is primarily made available to biotic communities by internal cycling of the element. Without human activities, the amount of nitrogen returned to the atmosphere (denitrification in terrestrial and aquatic communities) exceeds withdrawal from the atmosphere (N_2 fixation and nitrification). Human activities (dark purple arrow) result in an increased amount of NO_3^- in terrestrial communities, with resultant runoff to aquatic biotic communities.

nitrification. Nitrification can occur in these ways: ② Nitrogen gas (N_2) is converted to nitrate (NO_3^-) in the atmosphere when cosmic radiation, meteor trails, and lightning provide the high energy needed for nitrogen to react with oxygen. ③ Ammonium (NH_4^+) in the soil from various sources, including decomposition of organisms and animal wastes, is converted to nitrate by soil bacteria: ④ nitrite-producing bacteria convert ammonium to nitrite (NO_2^-), and then ⑤ nitrate-producing bacteria convert nitrite to nitrate. ⑥ During the process of **assimilation**, plants take up ammonia and nitrate from the soil and use these ions to produce proteins and nucleic acids.

Notice that the subcycle involving the biotic community, which occurs on land and in the ocean, need not depend on the presence of nitrogen gas at all.

Formation of Nitrogen Gas from Nitrate

⑦ **Denitrification** is the conversion of nitrate back to nitrogen gas, which enters the atmosphere. Denitrifying bacteria living in the anaerobic mud of lakes, bogs, and estuaries carry out this process as a part of their own metabolism. In the nitrogen cycle, denitrification would counterbalance nitrogen fixation except for human activities.

Human Activities

⑧ Human activities significantly alter the transfer rates in the nitrogen cycle by producing fertilizers from N_2—in fact, they nearly double the fixation rate. Fertilizer, which also contains phosphate, runs off into lakes and rivers and results in an overgrowth of algae and rooted aquatic plants. As discussed on page 505, the end result is cultural eutrophication (over-enrichment), which can lead to an algal boom. When the algae die off, enlarged decomposer populations use up all the oxygen in the water, and the result is a massive fish kill.

Acid deposition occurs because nitrogen oxides (NO_x) and sulfur dioxide (SO_2) enter the atmosphere from the burning of fossil fuels (Fig. 23.13). Both these gases combine with water vapor to form acids that eventually return to the earth. Acid deposition has drastically affected forest and lakes in northern Europe, Canada, and northeastern United States because their soils are naturally acidic and their surface waters are only mildly alkaline (basic). Acid deposition reduces agricultural yields and corrodes marble, metal, and stonework.

Nitrogen oxides and hydrocarbons (HC) from the burning of fossil fuels react with one another in the presence of sunlight to produce smog which contains dangerous pollutants. Warm air near the Earth usually escapes into the atmosphere, taking pollutants with it. However, during a thermal inversion, pollutants are trapped near the Earth beneath a layer of warm, stagnant air. Because the air does not circulate, pollutants can build up to dangerous levels. Areas surrounded by hills are particularly susceptible to the effects of a thermal inversion because the air tends to stagnate and little turbulent mixing can occur (Fig. 23.14).

a.

b.

Figure 23.13 Acid deposition.

a. Many forests in higher elevations of northeastern North America and northern Europe are dying due to acid deposition. **b.** Air pollution due to fossil-fuel burning in factories and modes of transportation is the major cause of acid deposition.

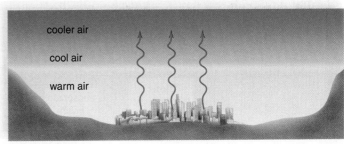

a. Normal pattern

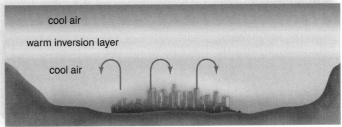

b. Thermal inversion

Figure 23.14 Thermal inversion.

a. Normally, pollutants escape into the atmosphere when warm air rises.
b. During a thermal inversion, a layer of warm air (warm inversion layer) overlies and traps pollutants in cool air below.

The Phosphorus Cycle

In the phosphorus cycle (Fig. 23.15), ① phosphorus trapped in oceanic sediments moves onto land after a geological upheaval. ② On land, the very slow weathering of rocks places phosphate ions (PO_4^{3-} and HPO_4^{2-}) in the soil. ③ Some of this becomes available to plants, which use phosphate in a variety of molecules including phospholipids, ATP, and the nucleotides that become a part of DNA and RNA. ④ Animals eat producers and incorporate some of the phosphate into teeth, bones, and shells, which take many years to decompose. ⑤ Death and decay of all organisms and also decomposition of animal wastes do, however, make phosphate ions available to producers once again. Because the available amount of phosphate is already being used within food chains, phosphate is usually a limiting inorganic nutrients for plants—that is, the lack of it limits the size of populations in ecosystems.

⑥ Some phosphate naturally runs off into aquatic ecosystems, where algae acquire phosphate from the water before it becomes trapped in sediments. ⑦ Phosphate in marine sediments does not become available to producers

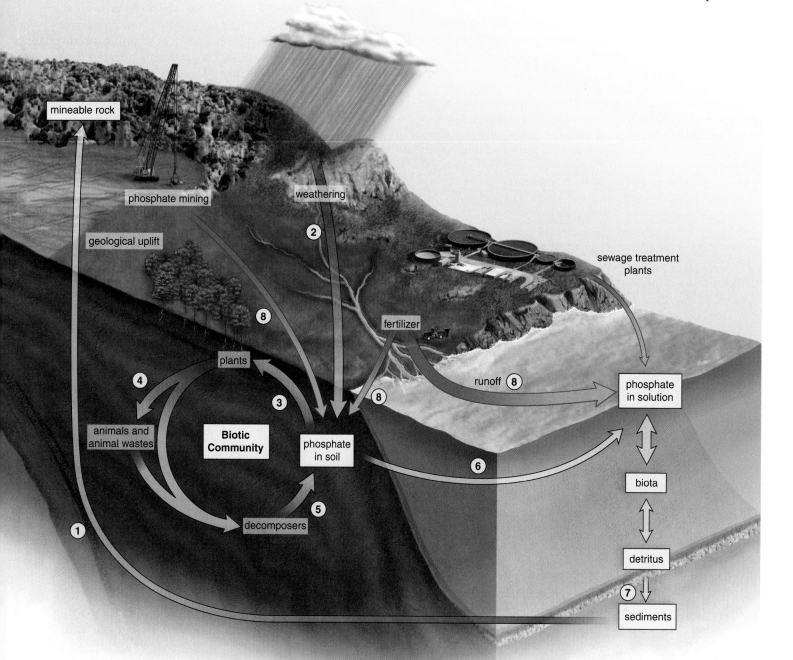

Figure 23.15 The phosphorus cycle.
The weathering of rocks provides phosphorus, which cycles locally in both terrestrial and aquatic biota. Human activities (purple arrows) produce fertilizers, which add to the amount of phosphorus available to biotic communities—eventually, fertilizers become a part of the runoff that enriches waters. Sewage treatment plants directly add phosphorus to local waters. When phosphorus becomes a part of oceanic sediments, it is lost to biotic communities for many years.

on land again until a geological upheaval exposes sedimentary rocks on land. Now, the cycle begins again. Phosphorus does not enter the atmosphere; therefore, the phosphorus cycle is called a sedimentary cycle.

Phosphorus and Water Pollution

⑧ Human beings boost the supply of phosphate by mining phosphate ores for fertilizer and detergent production. As mentioned previously, runoff of phosphate and nitrogen due to fertilizer use, animal wastes from livestock feedlots, and discharge from sewage treatment plants results in **cultural eutrophication** (overenrichment) of waterways.

Figure 23.16 lists the various sources of water pollution. Point sources of pollution are specific, and nonpoint sources are those caused by runoff from the land. Industrial wastes can include heavy metals and organochlorides, such as those in some pesticides. These materials are not readily degraded under natural conditions or in conventional sewage treatment plants. **Biological magnification** occurs as they pass along a food chain and become more and more concentrated because they remain in the body and are not excreted. Biological magnification occurs more readily in aquatic food chains, which have more links than terrestrial food chains. In any case, humans are the final consumers in food chains, and in some areas, human milk contains detectable amounts of DDT and PCBs, which are organochlorides.

Coastal regions are the immediate receptors for local pollutants and the final receptors for pollutants carried by rivers that empty at a coast. Waste dumping occurs at sea, but ocean currents sometimes transport both trash and pollutants back to shore. Offshore mining and shipping add pollutants to the oceans. Some 5 million metric tons of oil a year—or more than 1 gram per 100 square meters of the oceans' surfaces—end up in the oceans. Large oil spills kill plankton, fish, and shellfishes, as well as birds and marine mammals.

In the last 50 years, humans have polluted the seas and exploited their resources to the point that many species are at the brink of extinction. Fisheries once rich and diverse, such as George's Bank off the coast of New England, are in severe decline. Haddock was once the most abundant species in this fishery, but now it accounts for less than 2% of the total catch. Cod and bluefin tuna have suffered a 90% reduction in population size. In warm, tropical regions, many areas of coral reefs are now overgrown with algae because the fish that normally keep the algae under control have been killed off.

✅ Check Your Progress 23.3

1. Chemical cycling in the biosphere may involve what three components?
2. **a.** What are two examples of gaseous biogeochemical cycles?
   b. What is an example of a sedimentary biogeochemical cycle?
3. What roles do bacteria play in the nitrogen cycle?
4. What ecological problems are associated with water, carbon, nitrogen, and phosphorus cycles?

Sources of Water Pollution	
Leading to Cultural Eutrophication	
oxygen-demanding waste	Biodegradable organic compounds (e.g., sewage, wastes from food-processing plants, paper mills, and tanneries)
plant nutrients	Nitrates and phosphates from detergents, fertilizers, and sewage treatment plants
sediments	Enriched soil in water due to soil erosion
thermal discharges	Heated water from power plants
Health Hazards	
disease-causing agents	Bacteria and viruses from sewage and barnyard waste (causing, for example, cholera, food poisoning, and hepatitis)
synthetic organic compounds	Pesticides, industrial chemicals (e.g., PCBs)
inorganic chemicals and minerals	Acids from mines and air pollution; dissolved salts; heavy metals (e.g., mercury) from industry
radiation	Radioactive substances from nuclear power plants, medical and research facilities

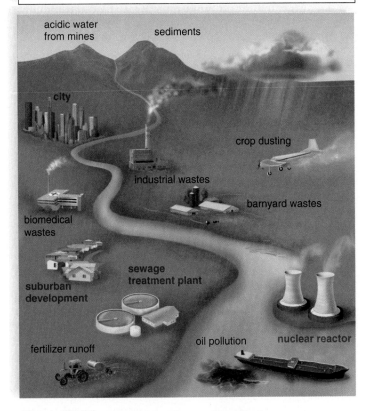

Figure 23.16 Sources of surface water pollution.
Many bodies of water are dying due to the introduction of pollutants from point sources, which are easily identifiable, and nonpoint sources, which cannot be specifically identified.

Science Focus

Ozone Shield Depletion

In the stratosphere, some 50 kilometers above the Earth, ozone forms the **ozone shield,** a layer of ozone that absorbs most of the ultraviolet (UV) rays of the sun so that fewer rays strike the Earth. Ozone forms when ultraviolet radiation from the sun splits oxygen molecules (O_2), and then the oxygen atoms (O) combine with other oxygen molecules to produce ozone (O_3).

Cause of Depletion

The absorption of UV radiation by the ozone shield is critical for living things. In humans, UV radiation causes mutations that can lead to skin cancer and can make the lens of the eye develop cataracts. In addition, it adversely affects the immune system and our ability to resist infectious diseases. UV radiation also impairs crop and tree growth and kills off algae and tiny shrimplike animals (krill) that sustain oceanic life. Without an adequate ozone shield, therefore, our health and food sources are threatened.

It became apparent in the 1980s that depletion of ozone had occurred worldwide and that the depletion was most severe above the Antarctic every spring. There, ozone depletion became so great that it covered an area two and a half times the size of Europe, and exposed not only Antarctica but also the southern tip of South America and vast areas of the Pacific and Atlantic oceans to harmful ultraviolet rays. In the popular press, severe depletions of the ozone layer are called **ozone holes** (Fig. 23A). Of even greater concern, an ozone hole has now appeared above the Arctic as well, and ozone holes were also detected within northern and southern latitudes, where many people live. Whether or not these holes develop in the spring depends on prevailing winds, weather conditions, and the type of particles in the atmosphere. A United Nations Environmental Program report predicts a 26% rise in cataracts and nonmelanoma skin cancers for every 10% drop in the ozone level. A 26% increase translates into 1.75 million additional cases of cataracts and 300,000 more skin cancers every year, worldwide.

The seriousness of the situation caused scientists around the globe to begin studying the cause of ozone depletion. The cause was found to be chlorine atoms (Cl), which can destroy up to 100,000 molecules of ozone before settling to the Earth's surface as chloride years later.

Control of CFCs

The chlorine atoms that enter the troposphere and eventually reach the stratosphere come primarily from the breakdown of **chlorofluorocarbons (CFCs),** chemicals much in use by humans. The best-known CFC is Freon, a coolant found in refrigerators and air conditioners. CFCs are also used as cleaning agents and as foaming agents during the production of Styrofoam coffee cups, egg cartons, insulation, and paddings. Formerly, CFCs were used as propellants in spray cans, but this application is now banned in the United States

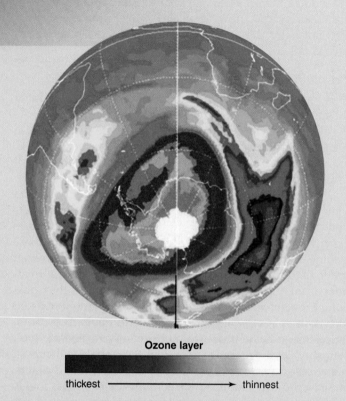

Ozone layer

thickest ————————→ thinnest

Figure 23A   Ozone shield depletion.
Map of ozone levels in the atmosphere of the Southern Hemisphere, September 2000. The ozone depletion, often called an ozone hole, is larger than the size of Europe.

and several European countries. Other molecules, such as the cleaning solvent methyl chloroform, are also sources of harmful chlorine atoms.

Most of the countries of the world have stopped using CFCs, and the United States halted production in 1995. Since that time, satellite measurements indicate that the amount of harmful chlorine pollution in the stratosphere has started to decline. It is clear, however, that recovery of the ozone shield may take several more years and involve other pollution-fighting approaches, aside from lowering chlorine pollution. Researchers report that during the winter of 2000, there were more and longer-lasting polar clouds than previously. Why might that be? As the Earth's surface warms due to global warming, less heat reradiates into the stratosphere. Mathematical modeling suggests that stratospheric clouds could last twice as long over the Arctic before the year 2010, when the coldest winter ever is expected.

Cloud cover contributes to the breakdown of the ozone shield by chlorine pollution. It is speculated that once polar stratospheric clouds become twice as persistent, there could still be an ozone loss of 30%.

Oil Drilling in the Arctic

In the aftermath of hurricanes Katrina and Rita in 2005, disruptions in oil drilling in the Gulf of Mexico sent shockwaves through the American economy. Oil prices topped $74 a barrel, more than triple the cost of the same amount barely five years prior. Continued oil production disruptions, political unrest, and animosity toward the United States in many oil-producing countries also conspire to keep oil prices at or near record highs. Amid this turmoil, a decades-old controversy blazes with a renewed fury—should the Arctic National Wildlife Refuge (ANWR) be opened to oil exploration to help alleviate our dependence on foreign oil and help lower prices?

Established by an act of Congress in 1980, ANWR covers a total of 19 million acres of northernmost Alaska far above the Arctic Circle. ANWR is home to a variety of wildlife, such as caribou, migratory birds, grizzly and polar bears, wolves, and musk oxen (Fig. 23B). In the summer, the adjacent coastal waters host a variety of marine mammals, such as bowhead and gray whales. ANWR also contains substantial oil reserves. These aspects of ANWR have become the focus of much debate, in which the value of pristine wilderness is weighed against the value of its energy resources.

Benefits of Drilling

When ANWR was initially established, its coastal plain area, which covers 1.5 million acres, was designated off-limits to oil drilling without Congressional authorization. No oil production has yet been permitted in the coastal plain. However, it has been proposed that a portion of the coastal plain be developed for oil drilling. This action could, according to one estimate, yield up to a million barrels of oil per day over the next 20 years. Supporters of this plan note that it would affect only a small portion of the total area of ANWR—approximately 2,000 acres of an area that covers over 19.5 million acres, about the size of the state of South Carolina. By using the newest drilling methods, removing the oil and disposing of wastes would cause much less disruption to the surface environment compared with techniques used in the past. If drilling were permitted, Alaskans would benefit by the creation of new jobs and a boost to the state economy. Advocates of ANWR drilling say that this approach would reduce our dependence on foreign oil, and help to buffer us against oil price spikes and supply shocks.

Risks of Drilling

Those who are opposed to drilling in the ANWR believe that we should instead focus on reducing our need for energy by making better use of existing resources. For example, it has been suggested that a modest increase in vehicle fuel efficiency would save even more oil than would be gained by drilling in the ANWR coastal plain. Opponents point to the fact that the United States uses approximately 20 million barrels of oil per day, so even if the yield from coastal plain drilling met with advocates' expectations, its contribution to our oil needs would be relatively minor. The argument is also made that despite claims that only 2,000 acres would be developed, the land is located on scattered sites that would need to be connected with roads, increasing the amount of land that would be subject to disruption. And finally, those who are against ANWR development because of environmental concerns point to a large oil spill from a pipeline in nearby Prudhoe Bay as evidence that even modern drilling techniques are not immune to pollution events.

Decide Your Opinion

1. In the United States, Republicans generally favor oil drilling and Democrats generally oppose. Should party politics play a role in our decision making?
2. On what basis would you go about making a decision whether to drill or not to drill?
3. Compared to a tropical rain forest, ANWR has a limited number of species. Does this influence your decision-making process? Why or why not?

Figure 23B   The terrain and wildlife of the Arctic National Wildlife Refuge.

Summarizing the Concepts

23.1 The Nature of Ecosystems

Ecology is the study of the interactions of organisms with each other and with the physical environment.

- Organisms interact with the physical and chemical environment, and the result is an ecosystem.
- Terrestrial ecosystems are forests (tropical rain forests, coniferous, temperate deciduous), grasslands (savanna and prairie), and deserts, which includes the tundra.
- Aquatic ecosystems are either salt water (i.e., seashores, oceans, coral reefs, estuaries) or freshwater (i.e., lakes, ponds, rivers, and streams).

Biotic Components of an Ecosystem

- In a community, each population has a habitat (residence) and a niche (its role in the community).
- Autotrophs (producers) produce organic nutrients for themselves and others from inorganic nutrients and an outside energy source.
- Heterotrophs (consumers) consume organic nutrients.
- Consumers are herbivores (eat plants/algae), carnivores (eat other animals), and omnivores (eat both plants/algae and animals).
- Decomposers feed on detritus, releasing inorganic substances back into the ecosystem.

Energy Flow and Chemical Cycling

Ecosystems are characterized by energy flow and chemical cycling.

- Energy flows through the populations of an ecosystem.
- Chemicals cycle within and among ecosystems.

23.2 Energy Flow

Various interconnecting paths of energy flow are called a food web.

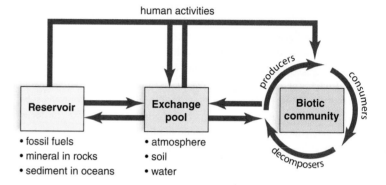

- A food web is a diagram showing how various organisms are connected by eating relationships.
- Grazing food webs begin with vegetation eaten by a herbivore that becomes food for a carnivore.
- Detrital food webs begin with detritus, food for decomposers and for detritivores.
- Members of detrital food webs can be eaten by aboveground carnivores, joining the two food webs.

Trophic Levels

A trophic level is all the organisms that feed at a particular link in a food chain.

- Ecological pyramids illustrate that biomass and energy content decrease from one trophic level to the next because of energy loss.

23.3 Global Biogeochemical Cycles

Chemicals circulate through ecosystems via biogeochemical cycles, pathways involving both biotic and geological components. Biogeochemical cycles

- can be gaseous or sedimentary.
- have reservoirs (e.g., ocean sediments, the atmosphere, and organic matter) that contain inorganic nutrients available to living things on a limited basis.

Exchange pools are sources of inorganic nutrients.

- Nutrients cycle among the biotic communities (producers, consumers, decomposers) of an ecosystem.

The Water Cycle

- The reservoir of the water cycle is freshwater that evaporates from the ocean.
- Water that falls on land enters the ground, surface waters, or aquifers and evaporates again.
- All water returns to the ocean.

The Carbon Cycle

- The reservoirs of the carbon cycle are organic matter (e.g., forests and dead organisms for fossil fuels), limestone, and the ocean (e.g., calcium carbonate shells).
- The exchange pool is the atmosphere.
- Photosynthesis removes carbon dioxide from the atmosphere.
- Respiration and combustion add carbon dioxide to the atmosphere.

The Nitrogen Cycle

- The reservoir of the nitrogen cycle is the atmosphere.
- Nitrogen gas must be converted to a form usable by plants (producers).
- Nitrogen-fixing bacteria (in root nodules) convert nitrogen gas to ammonium, a form producers can use.
- Nitrifying bacteria convert ammonium to nitrate.
- Denitrifying bacteria convert nitrate back to nitrogen gas.

The Phosphorus Cycle

- The reservoir of the phosphorus cycle is ocean sediments.
- Phosphate in ocean sediments becomes available through geological upheaval, which exposes sedimentary rocks to weathering.
- Weathering slowly makes phosphate available to the biotic community.
- Phosphate is a limiting nutrient in ecosystems.

Understanding Key Terms

acid deposition 503
aquifer 499
assimilation 503
autotroph 494
biogeochemical cycle 498
biological magnification 505
biomass 497
biosphere 493
carnivore 494
chlorofluorocarbon (CFC) 506
consumer 494
cultural eutrophication 505
denitrification 503
detrital food chain 497
detrital food web 497
detritus feeder 494
ecological pyramid 497
ecosystem 493
evaporation 499
food chain 497
food web 497

fossil fuel 501
global warming 501
grazing food chain 497
grazing food web 497
greenhouse effect 501
greenhouse gases 501
herbivore 494
heterotroph 494
niche 494
nitrification 503
nitrogen fixation 502
omnivore 494
ozone hole 506
ozone shield 506
precipitation 499
producer 494
runoff 499
trophic level 497
trophic relationship 497
water (hydrologic) cycle 499

Match the key terms to these definitions.

a. —————— Animals that feed on both plants and animals.

b. —————— The organisms that feed at a particular link in a food chain.

c. —————— Remains of once-living organisms that are burned to release energy, such as coal, oil, and natural gas.

d. —————— Process by which atmospheric nitrogen gas is changed to forms that plants can use.

e. —————— Photosynthetic organism at the start of a grazing food chain that makes its own food.

Testing Your Knowledge of the Concepts

1. What are the major types of terrestrial ecosystems? Describe each with its temperature, rainfall, and type of vegetation. (pages 492–93)

2. What are the major types of aquatic ecosystems? (page 493)

3. What is a niche? (page 494)

4. Name the four different types of consumers (heterotrophs) found in ecosystems. (page 494)

5. Explain why energy flows but chemicals cycle through an ecosystem. (page 495)

6. Describe the two types of food webs and two types of food chains in terrestrial ecosystems. Which of these typically moves more energy through an ecosystem? (pages 496–97)

7. What is a trophic level? An ecological pyramid? (page 497)

8. What is a biochemical cycle? What is a reservoir and an exchange pool? Give an example of each. (page 498)

9. Draw a diagram to illustrate the water, carbon, nitrogen, and phosphorus biochemical cycles. Include how human activities can change a particular cycle's balance. (pages 499–505)

In questions 10–13, match each description to a population in the key.

Key:

a. producer   c. decomposer
b. consumer   d. herbivore

10. Heterotroph that feeds on plant material.

11. Autotroph that manufactures organic nutrients.

12. Any type of heterotroph that feeds on plant material or on other animals.

13. Heterotroph that breaks down detritus as a source of nutrients.

14. Of the total amount of energy that passes from one trophic level to another, about 10% is
    a. respired and becomes heat.
    b. passed out as feces or urine.
    c. stored as body tissue.
    d. recycled to autotrophs.
    e. All of these are correct.

15. Compare this food chain:
    algae → water fleas → fish→ green herons
    with this food chain:
    trees → tent caterpillars → red-eyed vireos → hawks

    Both water fleas and tent caterpillars are
    a. carnivores.
    b. primary consumers.
    c. detritus feeders.
    d. present in grazing and detrital food webs.
    e. Both a and b are correct.

16. In what way are decomposers like producers?
    a. Either may be the first member of a grazing or a detrital food chain.
    b. Both produce oxygen for other forms of life.
    c. Both require nutrient molecules and energy.
    d. Both are present only on land.
    e. Both produce organic nutrients for other members of ecosystems.

17. Why are ecosystems dependent on a continual supply of solar energy?
    a. Carnivores have a greater biomass than producers.
    b. Decomposers process the greatest amount of energy in an ecosystem.
    c. Energy transformation results in a loss of usable energy to the environment.
    d. Energy cycles within and between ecosystems.

18. Nutrient cycles always involve
    a. rocks as a reservoir.
    b. movement of nutrients through the biotic community.
    c. the atmosphere as an exchange pool.
    d. loss of the nutrients from the biosphere.

19. Which of the following contribute(s) to the carbon cycle?
    a. respiration
    b. photosynthesis
    c. fossil fuel combustion
    d. decomposition of dead organisms
    e. All of these are correct.

20. How do plants contribute to the carbon cycle?
    a. When plants respire, they release CO_2 into the atmosphere.
    b. When plants photosynthesize, they consume CO_2 from the atmosphere.
    c. When plants photosynthesize, they provide oxygen to heterotrophs.
    d. When plants emigrate, they transport carbon molecules between ecosystems.
    e. Both a and b are correct.

21. How do nitrogen-fixing bacteria in the soil contribute to the nitrogen cycle?
    a. They return nitrogen to the atmosphere.
    b. They change ammonium to nitrate.
    c. They change nitrogen to ammonium.
    d. They withdraw nitrate from the soil.
    e. They decompose and return nitrogen to autotrophs.

22. What is the reservoir in the phosphorus cycle?
    a. oceans   c. plants
    b. marine sediments   d. animals

For questions 23–26, match each human activity with one or more of the cycles listed in the key. More than one answer can be used, and answers can be used more than once.

Key:
    a. water cycle   d. phosphorus cycle
    b. carbon cycle   e. none of these
    c. nitrogen cycle   f. all of these

23. Drive cars

24. Use fertilizers

25. Take showers

26. Grow crops

For questions 27–31, match each characteristic to the cycles listed in the key for questions 23–26. More than one answer can be used, and answers can be used more than once.

27. Occurs on land but not in the water

28. Can occur without the participation of humans

29. Always involves the participation of decomposers

30. The atmosphere is involved

31. Rocks are the reservoir in this cycle

32. Label the following diagram of an ecosystem.

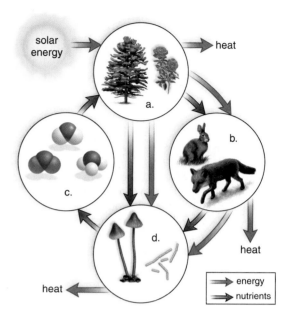

Thinking Critically About the Concepts

You may remember that crime-scene investigators in the Chapter 12 opening story were interested in the extent of rigor mortis so they might determine the time of death of a victim. The rate and extent of decomposition of a corpse are also of interest to crime-scene investigators who would like to identify a body and determine the time and cause of death. Dr. William Bass, Professor Emeritus, at the University of Tennessee in Knoxville, has spent considerable time investigating the rate of decomposition of human corpses at the UT Anthropologic Research Facility, sometimes called "The Body Farm." The research facility also plays an important role in training law-enforcement individuals and testing new forensic techniques.

1. How do you think the nutrient cycles would be affected by the use of fungicides and pesticides?

2. What environmental conditions influence the rate of decomposition?

3. What kinds of things could you do at home to facilitate the cycling of nutrients?

4. Worldwide desertification is increasing rapidly. Of what concern is the loss of other kinds of terrestrial ecosystems to desertification?

24

Human Population, Planetary Resources, and Conservation

The four friends Jacob, Megan, Eva, and Andrew volunteered to assist with the university's Earth Day celebration. Since meeting three years ago through the Outdoors Club, the foursome had been active in environmental activities on campus. Now, with graduation less than a month away, they were contemplating their futures and how to help make the world a better place.

The day of the Earth Day festivities, the friends reported to the booths to which they had been assigned. Jacob assisted representatives from Population Connection, who showed a video of the rate of growth of the human population from 0 AD projected through 2050. Eva helped people determine the size of their ecological footprint on the planet. Andrew, who volunteered at the local zoo, presented information about biodiversity and the zoo's role in maintaining biodiversity with their species survival programs. He also distributed fact sheets provided by the regional aquarium about overfishing and a list of the best fish to eat. Megan, an avid gardener herself, helped a gardening club present information about recycling and composting.

At the end of a long day, the friends were presented with an added bonus of working at the Earth Day celebration: T-shirts and native tree species to plant. They commented on the fun they'd had and the impact they may have had on other students. This chapter should also increase your awareness of the connection between resource consumption and the occurrence of pollution, with an eye to conserving all the Earth's species.

CHAPTER CONCEPTS

24.1 Human Population Growth
The present growth rate for the world's population has decreased to 1.2%, but still 78 million more people are expected within a year because the human population is so large.

24.2 Human Use of Resources and Pollution
Human beings use land, water, food, energy, and minerals to meet their basic needs, such as a place to live and food to eat, and to make products for their daily lives. Use of these resources leads to pollution.

24.3 Biodiversity
A biodiversity crises is upon us because of habitat loss followed by introduction of alien species, pollution, overexploitation, and disease. Yet, wildlife has both a direct value and an indirect value for us.

24.4 Working Toward a Sustainable Society
Our present day society is not sustainable, and ways are given to make it more sustainable.

511

24.1 Human Population Growth

In the opening story, Jacob helped man the Population Connection booth, where people learned that the world's population has risen steadily to a present size of about 7 billion people (Fig. 24.1). Prior to 1750, the growth of the human population was relatively slow, but as more reproducing individuals were added, growth increased, until the curve began to slope steeply upward, indicating that the population was undergoing **exponential growth.** The number of people added annually to the world population peaked at about 87 million around 1990, and currently it is a little over 78 million per year. This is roughly equal to the collective current population of Argentina, Chile, and Peru.

The **growth rate** of a population is determined by considering the difference between the number of persons born per year (birthrate, or natality) and the number who die per year (death rate, or mortality). It is customary to record these rates per 1,000 persons. For example, the world at the present time has a birthrate of 21 per 1,000 per year, but it has a death rate of 9 per 1,000 per year. This means that the world's population growth, or simply its growth rate, is

$$\frac{21-9}{1,000} = \frac{1.2}{1,000} = 0.012 \times 100 = 1.2\%$$

(Notice that while the birthrate and death rate are expressed in terms of 1,000 persons, the growth rate is expressed per 100 persons, or as a percentage.) After 1750, the world population growth rate steadily increased, until it peaked at 2% in 1965. It has since fallen to its present 1.2%. Yet, the world population is still steadily growing because of its past exponential growth.

In the wild, exponential growth indicates that a population is enjoying its **biotic potential**—that is, the maximum growth rate under ideal conditions. Growth begins to decline because of limiting factors such as food and space. Finally, the population levels off at the carrying capacity. The **carrying capacity** is the maximum population that the environment can support for an indefinite period. The carrying capacity of the Earth for humans has not been determined. Some authorities think the Earth is potentially capable of supporting 50–100 billion people. Others think we already have more humans than the Earth can adequately support.

The MDCs Versus the LDCs

The countries of the world today can be divided into two groups. The more-developed countries (MDCs), typified by countries in North America and Europe, are those in which population growth is modest and the people enjoy a good standard of living. The less-developed countries (LDCs), typified by some countries in Asia, Africa, and Latin America, are those in which population growth is dramatic and the majority of people live in poverty.

The MDCs

The MDCs did not always have low population increases. Between 1850 and 1950, they doubled their populations, largely because of a decline in the death rate due to development of modern medicine and improvements in public health and socioeconomic conditions. The decline in the death rate was followed shortly thereafter by a decline in the birthrate, so that populations in the MDCs have experienced only modest growth since 1950 (Fig. 24.1).

The growth rate for the MDCs as a whole is now about 0.1%, but several are not growing at all or are actually decreasing in size. The MDCs are expected to increase by 52 million

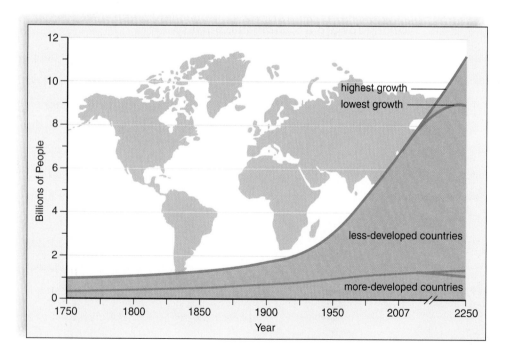

Figure 24.1   Human population growth.
The world's population size is now about 7 billion. It is predicted that the world's population size may level off at 9 billion or increase to more than 11 billion by 2250, depending on the speed with which the growth rate declines.

between 2002 and 2050, but this amount will still keep their total population at just about 1.2 billion. In contrast to other MDCs, there is no leveling off and no end in sight to U.S. population growth. The United States has a growth rate of 0.6%, and many people immigrate to the U.S. each year. In addition, a baby boom between 1947 and 1964 means that a large number of U.S. women are still of reproductive age.

The LDCs

The death rate began to decline steeply in the LDCs following World War II with the introduction of modern medicine, but the birthrate remained high. The growth rate of the LDCs peaked at 2.5% between 1960 and 1965. Since that time, the collective growth rate for the LDCs has declined to 1.6%, but some 46 countries have not participated in this decline. Thirty-five of these countries are in sub-Saharan Africa, where women on the average are presently having more than five children each.

Between 2002 and 2050, the population of the LDCs may jump from 5 billion to at least 8 billion. Some of this increase will occur in Africa, but most will occur in Asia because many deaths from AIDS are slowing the growth of the African population. Asia already has 56% of the world's population living on 31% of its arable (farmable) land. Therefore, Asia is expected to experience acute water scarcity, a significant loss of biodiversity, and more urban pollution. Twelve of the world's 15 most polluted cities are in Asia.

Comparing Age Structure

The LDCs are experiencing a population momentum because they have more women entering the reproductive years than older women leaving them. Populations have three age groups: prereproductive, reproductive, and postreproductive. This is best visualized by plotting the proportion of individuals in each group on a bar graph, thereby producing an age-structure diagram (Fig. 24.2).

Laypeople are sometimes under the impression that if each couple has two children, zero population growth will take place immediately. However, **replacement reproduction,** as this practice is called, will still cause most LDCs today to have a positive growth rate due to the age structure of the population. Because there are more young women entering the reproductive years than older women leaving them, the population continues to increase.

Most MDCs—not including the United States—have a stabilized age-structure diagram. Therefore, their populations are expected to remain just about the same or decline if couples are having fewer than two children each.

✓ Check Your Progress 24.1

1. Are the MDCs or the LDCs undergoing a larger population growth?

2. The growth rate of the world's population is decreasing. Why, then, is the population increasing so wildly?

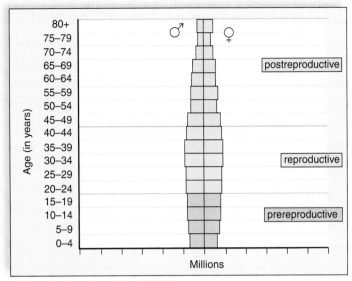
a. More-developed countries (MDCs)

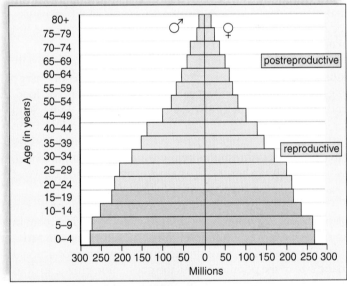
b. Less-developed countries (LDCs)

c.

Figure 24.2   Age-structure diagrams (2002).
The shape of these age-structure diagrams allows us to predict that (**a**) the populations of MDCs are approaching stabilization, and (**b**) the populations of LDCs will continue to increase for some time. **c.** Improved women's rights and increasing contraceptive use could change this scenario. Here a community health worker is instructing women in Bangladesh about the use of contraceptives.

24.2 Human Use of Resources and Pollution

Human beings have certain basic needs, and a resource is anything from the biotic or abiotic environment that helps meet these needs. Land, water, food, energy, and minerals are the maximally used resources that will be discussed in this chapter (Fig. 24.3). In the opening story on page 511, Eva was at the booth that helped individuals determine their ecological footprint: the amount of resources a person consumes and the amount of waste a person produces. A person can make their ecological footprint smaller by driving an energy efficient car, living in a smaller house, owning fewer possessions, eating vegetables as opposed to meat, and so forth.

Some resources are nonrenewable, and some are renewable. **Nonrenewable resources** are limited in supply. For example, the amount of land, fossil fuels, and minerals is finite and can be exhausted. Better extraction methods can make more fossil fuels and minerals available. Efficient use, recycling, or substitution can make the supply last longer, but eventually these resources will run out.

Renewable resources are capable of being naturally replenished. We can use water and certain forms of energy (e.g., solar energy) or harvest plants and animals for food, and more supply will always be forthcoming. Even with renewable resources, though, we have to be careful not to squander them. Consider, for example, that most species have population thresholds below which they cannot recover, as when the huge herds of buffalo that once roamed the West disappeared after being overexploited.

Unfortunately, a side effect of resource consumption can be pollution. **Pollution** is any alteration of the environment in an undesirable way. Pollution is often caused by human activities. The effect of humans on the environment is proportional to the size of the population. As the population grows, so does the need for resources and the amount of pollution caused by using these resources. Consider that seven people adding waste to the ocean may not be alarming, but seven billion people doing so would certainly affect its cleanliness. Actually, in modern times, the consumption of mineral and energy resources has grown faster than population size, most likely because people in the LDCs have increased their use of them.

Human population

Figure 24.3 **Resources.**
Human beings use land, water, food, energy, and minerals to meet their basic needs, such as a place to live, food to eat, and products that make their lives easier.

Land

People need a place to live. Worldwide, there are currently more than 32 persons for each square kilometer (83 persons per square mile) of all available land, including Antarctica, mountain ranges, jungles, and deserts. Naturally, land is also needed for a variety of uses aside from homes, such as agriculture, electric power plants, manufacturing plants, highways, hospitals, schools, and so on.

Beaches and Human Habitation

At least 40% of the world population lives within 100 km (60 mi) of a coastline, and this number is expected to increase. In the United States today, over half of the population lives within 80 km (50 mi) of the coasts (including the Great Lakes). Living right on the coast is an unfortunate choice because it leads to beach erosion and loss of habitat for marine organisms and loss of a buffer zone for storms. An estimated 70% of the world's beaches are eroding; Figure 24.4 shows how severe the problem can be in the United States. Humans fill in coastal wetlands, such as saltwater marshes in the northern and mangrove swamps in the southern United States. One reason to protect coastal wetlands is

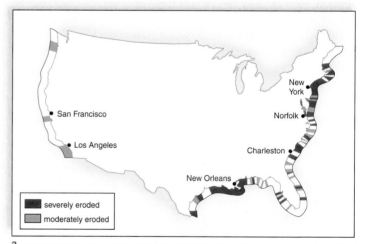

severely eroded
moderately eroded

a.

b.

Figure 24.4 **Beach erosion.**
a. Most of the U.S. coastline is subject to beach erosion. **b.** Therefore, people who choose to live near the coast may eventually lose their homes.

that they are spawning areas for fish and other forms of marine life. They are also habitats for certain terrestrial species, including many types of birds. Wetlands also protect coastal areas from storms. The loss of wetlands contributed to the devastation suffered by New Orleans from hurricane Katrina in 2006, for example. The coast is particularly subject to pollution because toxic substances placed in freshwater lakes, rivers, and streams may eventually find their way to the coast.

Semiarid Lands and Human Habitation

Forty percent of the Earth's lands are already deserts, and land adjacent to a desert is in danger of becoming unable to support human life if it is improperly managed by humans (Fig. 24.5). **Desertification** is the conversion of semiarid land to desertlike conditions.

Quite often, desertification begins when humans allow animals to overgraze the land. The soil can no longer hold rainwater, and it runs off instead of keeping the remaining plants alive or replenishing wells. Humans then remove whatever vegetation they can find to use as fuel or fodder for their animals. The end result is a lifeless desert, which is

then abandoned as people move on to continue the process someplace else. Some estimate that nearly three-quarters of all rangelands worldwide are in danger of desertification. Many famines in Africa and elsewhere are due, at least in part, to degradation of the land to the point that it can no longer support human beings and their livestock.

Tropical Rain Forest and Human Habitation

Deforestation, the removal of trees, has long allowed humans to live in areas where forests once covered the land. The concern of late has been that people are settling in tropical rain forests, such as the Amazon, following the building of roads (Fig. 24.6). This land, too, is subject to desertification. Soil in the tropics is often thin and nutrient-poor because all the nutrients are tied up in the trees and other vegetation. When the trees are felled and the land is used for agriculture or grazing, it quickly loses its fertility and becomes subject to desertification.

Loss of biodiversity also sometimes results from the destruction of rain forests. For example, logging in the Congo Republic has contributed to the growth of a city called Pokola, whose residents routinely eat wild bush animals.

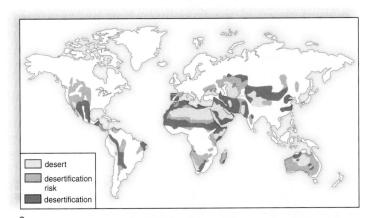

a.

b.

Figure 24.5   Desertification.
a. Desertification is a worldwide occurrence that (**b**) reduces the amount of land suitable for human habitation.

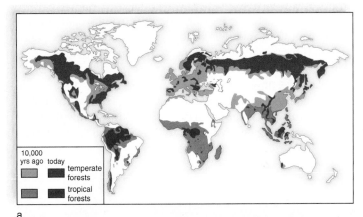

a.

b.

Figure 24.6   Deforestation.
a. Nearly half of the world's forest lands have been cleared for farming, logging, and urbanization. **b.** The soil of tropical rain forests is not suitable for long-term farming.

Water

In the water-poor areas of the world, people may not have ready access to drinking water, and if they do, the water may be impure. It's considered a human right for people to have clean drinking water, but actually, most freshwater is utilized by industry and agriculture (Fig. 24.7). Worldwide, 70% of all freshwater is used to irrigate crops! Much of a recent surge in demand for water stems from increased industrial activity and irrigation-intensive agriculture, the type of agriculture that now supplies about 40% of the world's food crops. Domestically, in the MDCs, more water is usually used for bathing, flushing toilets, and watering lawns than for drinking and cooking.

Increasing Water Supplies

Although the needs of the human population overall do not exceed the renewable supply, this is not the case in certain regions of the United States and the world. About 40% of the world's land is desert, and deserts are bordered by semiarid land. When needed, humans increase the supply of freshwater by damming rivers and withdrawing water from aquifers.

Dams   The world's 45,000 large dams catch 14% of all precipitation runoff, provide water for up to 40% of irrigated land, and give some 65 countries more than half their electricity. Damming of certain rivers has been so extensive that they no longer flow as they once did. The Yellow River in China fails to reach the sea most years; the Colorado River barely makes it to the Gulf of California, and even the Rio Grande dries up before it can merge with the Gulf of Mexico. The Nile in Egypt and the Ganges in India are also so overexploited that at some times of the year, they hardly make it to the ocean.

Dams have other drawbacks: (1) They lose water due to evaporation and seepage into underlying rock beds. The amount of water lost sometimes equals the amount made available! (2) The salt left behind by evaporation and agricultural runoff increases salinity and can make a river's water unusable farther downstream. (3) Dams hold back less water with time because of sediment buildup. Sometimes a reservoir becomes so full of silt that it is no longer useful for storing water.

Aquifers   To meet their freshwater needs, people are pumping vast amounts of water from **aquifers,** which are reservoirs found just below or as much as 1 km below the surface. Aquifers hold about 1,000 times the amount of water that falls on land as precipitation each year. This water accumulates from rain that fell in far-off regions even hundreds of thousands of years ago. In the past 50 years, groundwater depletion has become a problem in many areas of the world. In substantial portions of the High Plains Aquifer, which stretches from South Dakota to the Texas panhandle, more than half of the water has been pumped out. In the 1950s, India had 100,000 motorized pumps in operation; today, India has 20 million pumps, a huge increase in groundwater pumping.

Consequences of Groundwater Depletion   Removal of water is causing land **subsidence,** a settling of the soil as it dries out. In California's San Joaquin valley, an area of more than 13,000 km_2 has subsided at least 30 cm due to groundwater depletion, and in the worst spot, the surface of the ground has dropped more than 9 meters! In some parts of Gujarat,

a. Agriculture uses most of the fresh water consumed.

b. Industrial use of water is about half that of agricultural use.

c. Domestic use of water is about half that of industrial use.

Figure 24.7   Global water use.
a. Agriculture primarily uses water for irrigation. **b.** Industry uses water variously. **c.** Households use water to drink, shower, flush toilets, and water lawns.

India, the water table has dropped as much as 7 meters. Subsidence damages canals, buildings, and underground pipes. Withdrawal of groundwater can cause **sinkholes,** in which an underground cavern collapses when water no longer holds up its roof (Fig. 24.8).

Saltwater intrusion is another consequence of aquifer depletion. The flow of water from streams and aquifers usually keeps them fairly free of seawater. But as water is withdrawn, the water table can lower to the point that seawater backs up into streams and aquifers. Saltwater intrusion reduces the supply of freshwater along the coast.

Conservation of Water

By 2025, two-thirds of the world's population may be living in countries that are facing serious water shortages. Some solutions for expanding water supplies have been suggested. Planting drought- and salt-tolerant crops would help a lot. Using drip irrigation delivers more water to crops and saves about 50% over traditional methods, while increasing crop yields as well (Fig. 24.9). Although the first drip systems were developed in 1960, they're used on only less than 1% of irrigated land. Most governments subsidize irrigation so heavily that farmers have little incentive to invest in drip systems or other water-saving methods. Reusing water and adopting conservation measures could help the world's industries cut their water demands by more than half.

Figure 24.8 **Sinkholes.**
Sinkhole caused by heavy rain in Richmond, Virginia.

Food

In 1950, the human population numbered 2.5 billion, and there was only enough food to provide less than 2,000 calories per person per day; now, with 7 billion people on Earth, the world food supply provides more calories per person per day. Generally speaking, food comes from three activities: growing crops, raising animals, and fishing the seas. The increase in the food supply has largely been possible because

a.

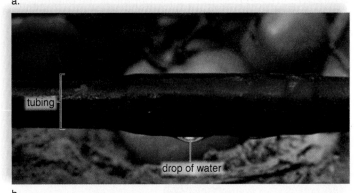

b.

c.

Figure 24.9 **Conservation measures to save water.**
a. Planting drought-resistant crops in the field and drought-resistant plants in parks and gardens cuts down on the need to irrigate. **b.** When irrigation is necessary, drip irrigation is preferable to using sprinklers. **c.** Wastewater can be treated and reused instead of withdrawing more water from a river or aquifer.

of modern farming methods, which, unfortunately, include some harmful practices such as these:

1. Planting of a few genetic varieties. The majority of farmers practice monoculture. Wheat farmers plant the same type of wheat, and corn farmers plant the same type of corn. Unfortunately, monoculture means that a single type of parasite can cause much devastation.

2. Heavy use of fertilizers, pesticides, and herbicides. Fertilizer production is energy intensive, and fertilizer runoff contributes to water pollution. Pesticides reduce soil fertility because they kill off beneficial soil organisms as well as pests, and some pesticides and herbicides are linked to the development of cancer. **Agricultural runoff** places these chemicals in our water supply.

3. Generous irrigation. As already discussed, water is sometimes taken from aquifers whose water content may, in the future, become so reduced that it could be too expensive to pump out any more.

4. Excessive fuel consumption. Irrigation pumps remove water from aquifers, and large farming machines are used to spread fertilizers, pesticides, and herbicides, as well as to sow and harvest the crops. In effect, modern farming methods transform fossil fuel energy into food energy.

Figure 24.10 shows ways to minimize the harmful effects of modern farming practices. Intercropping is the planting of two or more different crops in the same area. In Figure 24.10a, a farmer has planted alfalfa in between strips of corn. The alfalfa replenishes the nitrogen content of the soil so that fertilizer doesn't have to be added. In Figure 24.10b, contour farm-

ing with no-till conserves topsoil because it reduces agricultural runoff. Contour farming is planting and plowing according to the slope of the land, and no-till is the absence of tillage, during which the previous crop is removed so that the land is laid bare. Biological control (Fig. 24.10c) is discussed on page 462.

Soil Loss and Degradation

Land suitable for farming and grazing animals is being degraded worldwide. Topsoil, the topmost portion of the soil, is the richest in organic matter and the most capable of supporting grass and crops. When bare soil is acted on by water and wind, soil erosion occurs and topsoil is lost. As a result, marginal rangeland becomes desertized, and farmland loses its productivity.

The custom of planting the same crop in straight rows that facilitates the use of large farming machines has caused the United States and Canada to have one of the highest rates of soil erosion in the world. Conserving the nutrients now being lost could save farmers $20 billion annually in fertilizer costs. Much of the eroded sediment ends up in lakes and streams, where it reduces the ability of aquatic species to survive.

Between 25% and 35% of the irrigated western croplands are thought to have undergone **salinization,** an accumulation of mineral salts due to the evaporation of excess irrigation water. Salinization makes the land unsuitable for growing crops.

Green Revolutions

About 50 years ago, researchers began to breed tropical wheat and rice varieties specifically for farmers in the

a. Polyculture

b. Contour farming

c. Biological pest control

Figure 24.10   Conservation methods.
a. Polyculture reduces the ability of one parasite to wipe out an entire crop and reduces the need to use a herbicide to kill weeds. This farmer has planted alfalfa in between strips of corn, which also replenishes the nitrogen content of the soil (instead of adding fertilizers). Alfalfa, a legume, has root nodules that contain nitrogen-fixing bacteria. **b.** Contour farming with no-till conserves topsoil because water has less tendency to run off. **c.** Instead of pesticides, it is possible to use a natural predator. Here, ladybugs are feeding on cottony-cushion scale insects on citrus trees.

LDCs. The dramatic increase in yield due to the introduction of these new varieties around the world was called "the green revolution." These plants helped the world food supply keep pace with the rapid increase in world population. Most green revolution plants are called "high responders" because they need high levels of fertilizer, water, and pesticides in order to produce a high yield. In other words, they require the same subsidies and create the same ecological problems as do modern farming methods.

Genetic Engineering   As we discussed in Chapter 21 (see page 460), genetic engineering can produce transgenic plants with new and different traits, among them, resistance to both insects and herbicides. When herbicide-resistant crops are planted, weeds are easily controlled, less tillage is needed, and soil erosion is minimized. Researchers also want to produce crops that tolerate salt, drought, and cold. Some progress has also been made in increasing the food quality of crops so that they will supply more of the proteins, vitamins, and minerals people need. Genetically engineered crops could result in still another green revolution.

Nevertheless, people are opposed to the use of genetically engineered crops, fearing that they will damage the environment and lead to health problems in humans. The Health Focus on page 462 discusses these problems.

Domestic Livestock

A low-protein, high-carbohydrate diet consisting only of grains such as wheat, rice, or corn can lead to malnutrition. In the LDCs, kwashiorkor, caused by a severe protein deficiency, is seen in infants and children ages 1–3, usually after a new arrival in the family and the older children are no longer fed milk but starches. Such children are lethargic, irritable, and have bloated abdomens. Mental retardation is expected.

In the MDCs, many people tend to have more than enough protein in their diet. Almost two-thirds of U. S. cropland is devoted to producing livestock feed. This means that a large percentage of the fossil fuel, fertilizer, water, herbicides, and pesticides we use are actually for the purpose of raising livestock. Typically, cattle are range-fed for about four months, and then they are brought to crowded feedlots where they receive growth hormone and antibiotics, while they feed on grain or corn. Most pigs and chickens spend their entire lives cooped up in crowded pens and cages (Fig. 24.11).

If livestock eat a large proportion of the crops in the United States, then raising livestock accounts for much of the pollution associated with farming. Consider, also, that presently, fossil fuel energy is needed not just to produce herbicides and pesticides and to grow food, but also to make the food available to the livestock. Raising livestock is extremely energy-intensive in the MDCs. In addition, water is used to wash livestock wastes into nearby bodies of water, where they add significantly to water pollution. Whereas human wastes are sent to sewage treatment plants, raw animal wastes are not.

For these reasons, it is prudent to recall the ecological energy pyramid (see Fig. 23.7), which shows that as you move up the food chain, energy is lost. As a rule of thumb, for every 10 calories of energy from a plant, only 1 calorie is available for the production of animal tissue in a herbivore. In other words, it is extremely wasteful for the human diet to contain more protein than is needed to maintain good health. It is possible to feed ten times as many people on grain as on meat.

☑ Check Your Progress 24.2a

1. What five resources are maximally utilized by humans?
2. Which ecosystems suffer the most due to human habitation?
3. a. How do humans increase the supply of freshwater, and (b) what are the consequences of doing so?
4. a. What farming methods are in use today, and (b) what are the possible environmental consequences?

Figure 24.11   Crowding of livestock.
Hogs are packed into a feedlot pen.

Energy

Modern society runs on various sources of energy. Some of these energy sources are nonrenewable, and others are renewable. The consumption of nonrenewable energy supplies results in environmental degradation. Renewable energy is expected to be utilized more in the future, and a solar-hydrogen revolution is expected.

Nonrenewable Sources

Presently, about 6% of the world's energy supply comes from nuclear power, and 75% comes from fossil fuels; both of these are finite, nonrenewable sources. Although it was once predicted that the nuclear power industry would fulfill a significant portion of the world's energy needs, this has not happened for two reasons: (1) People are very concerned about nuclear power dangers, such as the meltdown that occurred in 1986 at the Chernobyl nuclear power plant in Russia. (2) Radioactive wastes from nuclear power plants remain a threat to the environment for thousands of years, and we still have not decided how best to safely store them.

Fossil fuels (oil, natural gas, and coal) are so named because they are derived from the compressed remains of plants and animals that died many thousands of years ago. Although the U.S. population makes up about 5% of the world's population, it uses more than half of the fossil fuel energy supply. Comparatively speaking, each person in the MDCs uses approximately as much energy in one day as a person in an LDC does in one year.

Among the fossil fuels, oil burns more cleanly than coal, which may contain a considerable amount of sulfur. So despite the fact that the United States has a goodly supply of coal, imported oil is our preferred fossil fuel today. Even so, the burning of any fossil fuel causes environmental problems because as it burns, pollutants are emitted into the air. Acid deposition is discussed on page 503.

Fossil Fuels and Global Climate Change In 1850, the level of carbon dioxide in the atmosphere was about 280 parts per million (ppm), and today it is about 350 ppm. This increase is largely due to the burning of fossil fuels and the burning and clearing of forests to make way for farmland and pasture. Human activities are causing the emission of other gases as well. For example, the amount of methane given off by oil and gas wells, rice paddies, and all sorts of organisms, including domesticated cows, is increasing by about 1% a year. These gases are known as **greenhouse gases** because, just like the panes of a greenhouse, they allow solar radiation to pass through but hinder the escape of infrared heat back into space.

Computer models predict the Earth may warm to temperatures never before experienced by living things. The global climate has already warmed about 0.6°C, and it may rise as much as 1.5—4.5°C by 2100. If so, sea levels will rise as glaciers melt and warm water expands. Major coastal cities of the United States, such as New York, Boston, Miami, and Galveston, could eventually be threatened. The present wetlands will be inundated and the loss of aquatic habitat will be great, wherever wetlands cannot move inward because of coastal development and levees. Coral reefs, which prefer shallow waters, will most likely "drown" as the waters rise.

On land, regions of suitable climate for various species will shift toward the poles and higher elevations. Plants migrate when seeds disperse and growth occurs in a new locale. Even so, the present assemblages of species in ecosystems will be disrupted as some species migrate northward faster than others. Trees, for example, cannot migrate as fast as nonwoody plants. Also, too many species of organisms are confined to relatively small habitat patches that are surrounded by agricultural or urban areas they would not be able to cross. And even if they have the capacity to disperse to new sites, suitable habitats may not be available.

Renewable Energy Sources

Renewable types of energy include hydropower, geothermal, wind, and solar.

Hydropower Hydroelectric plants convert the energy of falling water into electricity (Fig. 24.12a). Hydropower accounts for about 10% of the electric power generated in the United States and almost 98% of the total renewable energy used. Brazil, New Zealand, and Switzerland produce at least 75% of their electricity with water power, but Canada is the world's leading hydropower producer. Worldwide, hydropower presently generates 19% of all electricity utilized, but this percentage is expected to rise because of increased use in certain countries. For example, Iceland has an ambitious hydropower project under way because presently it uses only 10% of its potential capacity.

Much of the hydropower development in recent years has been due to the construction of enormous dams, which are known to have detrimental environmental effects (see page 516). The better choice is believed to be small-scale dams that generate less power per dam but do not have the same environmental impact.

Geothermal Energy Elements such as uranium, thorium, radium, and plutonium undergo radioactive decay below the Earth's surface and then heat the surrounding rocks to hundreds of degrees Celsius. When the rocks are in contact with underground streams or lakes, huge amounts of steam and hot water are produced. This steam can be piped up to the surface to supply hot water for home heating or to run steam-driven turbogenerators. The California's Geysers project is the world's largest geothermal electricity-generating complex.

Wind Power Wind power is expected to account for a significant percentage of our energy needs in the future. Despite the common belief that a huge amount of land is required for the "wind farms" that produce commercial electricity, the actual amount of space for a wind farm compares favorably with the amount of land required by a coal-fired power plant or a solar thermal energy system (Fig. 24.12b).

A community that generates its own electricity by using wind power can solve the problem of uneven energy production by selling electricity to a local public utility when an excess is available and buying electricity from the same facility when wind power is in short supply.

Energy and the Solar-Hydrogen Revolution   Solar energy is diffuse energy that must be (1) collected, (2) converted to another form, and (3) stored if it is to compete with other

a.

b.

c.

d.

Figure 24.12 Other renewable energy sources.

a. Hydropower dams provide a clean form of energy but can be ecologically disastrous in other ways. **b.** Wind power requires land on which to place enough windmills to generate energy. **c.** Photovoltaic cells on rooftops and **(d)** sun-tracking mirrors on land can collect diffuse solar energy more cheaply than could be done formerly.

available forms of energy. Passive solar heating of a house is successful when the windows of the house face the sun, the building is well insulated; and heat can be stored in water tanks, rocks, bricks, or some other suitable material.

In a **photovoltaic (solar) cell,** a wafer of the electron-emitting metal is in contact with another metal that collects the electrons and passes them along into wires in a steady stream. Spurred by the oil shocks of the 1970s, the U. S. government has been supporting the development of photovoltaics ever since. As a result, the price of buying one has dropped from about $100 per watt to around $4. The photovoltaic cells placed on roofs, for example, generate electricity that can be used inside a building and/or sold back to a power company (Fig. 24.12c).

Several types of solar power plants are now operational in California. In one type, huge reflectors focus sunlight on a pipe containing oil. The heated pipes boil water, generating steam that drives a conventional turbogenerator. In another type, 1,800 sun-tracking mirrors focus sunlight onto a molten salt receiver mounted on a tower. The hot salt generates steam that drives a turbogenerator (Fig. 24.12d).

Scientists are working on the possibility of using solar energy to extract hydrogen from water via electrolysis. The hydrogen can then be used as a clean-burning fuel; when it burns, water is produced. Presently, cars have internal combustion engines that run on gasoline. In the future,

vehicles are expected to be powered by fuel cells, which use hydrogen to produce electricity (Fig. 24.13). The electricity runs a motor that propels the vehicle. Fuel cells are now powering buses in Vancouver and Chicago, and more buses are planned.

Hydrogen fuel can be produced locally or in central locations, using energy from photovoltaic cells. If in central locations, hydrogen can be piped to filling stations using the natural gas pipes already plentiful in the United States. Two major hurdles are storage and production. The advantages of a solar-hydrogen revolution are at least twofold: (1) The world would no longer be dependent on the Middle East region for oil, and (2) environmental problems, such as global warming, acid deposition, and smog, would begin to lessen.

Minerals

Minerals are nonrenewable raw materials in the Earth's crust that can be mined (extracted) and used by humans. Nonrenewable minerals include fossil fuels; nonmetallic raw materials, such as sand, gravel, and phosphate; and metals, such as aluminum, copper, iron, lead, and gold.

The most dangerous metals to human health are the heavy metals: lead, mercury, arsenic, cadmium, tin, chromium, zinc, and copper. They are used to produce batteries, electronics, pesticides, medicines, paints, inks, and dyes. In

a.

b.

c.

Figure 24.13   Solar-hydrogen revolution.
a. Hydrogen fuel cells. **b.** This bus is powered by hydrogen fuel. **c.** The use of fuel-cell hybrid vehicles, such as this prototype, will reduce air pollution and dependence on fossil fuels.

the ionic form, they enter the body and inhibit vital enzymes. That's why these items should be discarded carefully and taken to hazardous waste sites.

One of the greatest threats to the maintenance of ecosystems and biodiversity is surface mining, called strip mining. In the United States, huge machines can go as far as removing mountaintops in order to reach a mineral. The land devoid of vegetation takes on a surreal appearance, and rain washes toxic waste deposits into nearby streams and rivers.

Hazardous Wastes

The consumption of minerals contributes to the buildup of hazardous wastes, including synthetic organic chemicals, in the environment. Every year, countries of the world discard billions of tons of solid waste, some on land and some in fresh- and marine waters. Using an allocation of monies called the Superfund, the Environmental Protection Agency oversees the cleanup of hazardous waste disposal sites in the United States. The nine most commonly found contaminants of the environment are heavy metals (lead, arsenic, cadmium, chromium) and synthetic organic compounds (trichloroethylene, toluene, benzene, polychlorinated biphe-

nyls [PC
disrupting

Synthet
tion of plastics
solvents, wood
ucts. Synthetic o
hydrocarbons, in wl
rine) have replaced ce
comprises the **chloroflu**
genated hydrocarbon in
atoms replace some of the
brought about a thinning of tl
protects terrestrial life from the
violet radiation. Now that MDCs
the ozone shield is predicted to re

Other synthetic organic chemicals
ous threat to the health of living thing
Rachel Carson's book *Silent Spring*, publi
the public aware of the deleterious effects of
times, they accumulate in the mud of deltas a
highly polluted rivers and cause environmenta
disturbed. These wastes enter bodies of water and
to **biological magnification** (Fig. 24.14). Decomp
unable to break down these wastes. They enter and re
the bodies of organisms because they accumulate in f
are not excreted. Therefore, they become more concentr
as they pass along a food chain. Biological magnification
most apt to occur in aquatic food chains, which have more
links than terrestrial food chains. Humans are the final consumers in both types of food chains, and over the past 25–30 years, a number of toxic chemicals have found their way into breast milk—PCBs, DDT, solvents, and heavy metals.

Raw sewage causes oxygen depletion in lakes and rivers. As the oxygen level decreases, the diversity of life is greatly reduced. Also, human feces can contain pathogenic microorganisms that cause cholera, typhoid fever, and dysentery. In regions of the LDCs where sewage treatment is practically nonexistent, many children die each year from these diseases. Typically, sewage treatment plants use bacteria to break down organic matter to inorganic nutrients, such as nitrates and phosphates, which then enter surface waters. The end result can be cultural eutrophication discussed on page 505.

Figure 24.14 Biological magnification.
Various synthetic organic chemicals, such as DDT, accumulate in animal fat. Therefore, the chemical becomes increasingly concentrated at higher tropic levels. By the time DDT was banned in the United States, it had interfered with predatory bird reproduction by causing egg-shell thinning.

DDT Concentration

25 ppm in predatory birds

2 ppm in large fish

0.5 ppm in small fish

0.04 ppm in zooplankton

0.000003 ppm in water

✓ Check Your Progress 24.2b

1. Why is it better for humans to use renewable rather than nonrenewable sources of energy?

2. The consumption of mineral resources has what environmental consequences?

3. Name some hazardous wastes present in today's environment.

agriculture, and accidental transport. For example, the pilgrims brought the dandelion to the United States as a familiar
salad green. Kudzu is a vine from Japan that the U. S. Department of Agriculture thought would help prevent soil erosion.
The plant now covers much landscape in the South, including
walnut, magnolia, and sweet gum trees. The zebra mussel
from the Caspian Sea was accidentally introduced into the
Great Lakes in 1988. It now forms dense beds that squeeze out
native mussels. Alien species that crowd out native species are
termed **invasive**. One way to counteract alien species is to
replant native (original to the area) species. Recall that Jacob,
Eva, Andrew and Megan were given native tree species to
plant because they had demonstrated an interest in keeping the
environment healthy by participating in Earth Day festivities.

Pollution

Pollution brings about environmental change that adversely
affects the lives and health of living things. Biodiversity is
particularly threatened by the following types of environmental pollution.

Acid deposition Acid deposition decimates forests because it causes trees to weaken and increases their susceptibility to disease and insects.

Global warming Global warming, an increase in the Earth's
temperature due to the presence of greenhouse gases in the
atmosphere, is expected to have many detrimental effects,

24.3 Biodi

Figure 24.15   Loss of biodiversity.
a. Habitat loss, alien species, pollution,
overexploitation, and disease have been
identified as causes of extinction of organisms.
b. Macaws that reside in South American
tropical rain forests are endangered for the
reasons listed in the graph.

such as destruction of coastal wetlands due to a rise in sea levels; loss of habitat due to temperature shifts; and the death of coral reefs if the temperature increases by 4°C.

Ozone depletion   The release of CFCs into the atmosphere causes the shield to break down, leading to impairment of crop and tree growth and death of plankton that sustains oceanic life.

Synthetic organic chemicals   Many of the organic chemicals released into the environment are endocrine-disrupting contaminants that can possibly affect the endocrine system and reproductive potential of food species and humans.

Overexploitation

Overexploitation occurs when the number of individuals taken from a wild population is so great that the population becomes severely reduced in numbers. A positive feedback cycle explains overexploitation: the smaller the population, the more valuable its members, and the greater the incentive to exploit the few remaining organisms.

Markets for decorative plants and exotic pets support both legal and illegal trade in wild species. Rustlers dig up rare cacti, such as the single-crested saguaro, and sell them to gardeners. Parakeets and macaws are among the birds taken from the wild for sale to pet owners. For every bird delivered alive, many more have died in the process. The same holds true for tropical fish, which often come from the coral reefs of Indonesia and the Philippines. Divers dynamite reefs or use plastic squeeze-bottles of cyanide to stun them; in the process, many fish die (see the Bioethical Focus on page 528).

Declining species of mammals are still hunted for their hides, tusks, horns, or bones. Because of its rarity, a single Siberian tiger is now worth more than $500,000—its bones are

pulverized and used as a medicinal powder. The horns of rhinoceroses become ornate carved daggers, and their bones are ground up to sell as a medicine. The ivory of an elephant's tusk is used to make art objects, jewelry, or piano keys. The fur of a Bengal tiger sells for as much as $100,000 in Tokyo.

Fish are a renewable resource if harvesting does not exceed the ability of the fish to reproduce. Today, larger and more efficient fishing fleets decimate fishing stocks (Fig. 24.16). Tuna and like fishes are captured by purse-seine fishing, in which a very large net surrounds a school of fish, and then the net is closed in the same manner as a drawstring purse. Dolphins that accompany the tuna are killed by this type of net. Other fishing boats drag huge trawling nets, large enough to accommodate 12 jumbo jets, along the sea-floor to capture bottom-dwelling fish. Trawling has been called the marine equivalent of clear-cutting trees because after the net goes by, the sea bottom is devastated. Only large fish are kept; undesirable small fish and sea turtles are discarded, dying, back into the ocean. Cod and haddock, once the most abundant bottom-dwelling fish along the northeast coast, are now often outnumbered by dogfish and skate.

A marine ecosystem can be disrupted by overfishing, as exemplified on the U.S. west coast. When sea otters began to decline in numbers, investigators found that they were being eaten by orcas (killer whales). Usually, orcas prefer seals and sea lions to sea otters, but they began eating sea otters when few seals and sea lions could be found. What caused a decline in seals and sea lions? Their preferred food sources—perch and herring—were no longer plentiful due to overfishing. Ordinarily, sea otters keep the population of sea urchins, which feed on kelp, under control. But with fewer sea otters around, the sea urchin population exploded and decimated

a.

b.

Figure 24.16   Fishing.
a. The world fish catch has declined in recent years (insert) because **(b)** modern fishing methods are overexploiting fisheries.

the kelp beds. Thus, overfishing set in motion a chain of events that detrimentally altered the food web of an ecosystem.

Disease

Wildlife is subject to emerging diseases just as humans are. Exposure to domestic animals and their pathogens occurs due to the encroachment of humans on wildlife habitats. Wildlife can also be infected by animals not ordinarily encountered. For example, African elephants carry a strain of herpes virus that is fatal to Asian elephants, and deaths can result if the two types of elephants are housed together.

The significant effect of diseases on biodiversity is underscored by a National Wildlife Health Center study that found that almost half of sea otter deaths along the coast of California are due to infectious diseases. Scientists tell us that the number of pathogens that cause disease are on the rise, and just as human health is threatened, so is that of wildlife. Extinctions due simply to disease may occur.

Direct Value of Biodiversity

Various individual species perform useful services for human beings and contribute greatly to the value we should place on biodiversity. The direct value of wildlife is related to their medicinal value, agricultural value, and consumptive use value. These are the most obvious values that are discussed here and illustrated in Figure 24.17.

Medicinal Value

Most of the prescription drugs used in the United States were originally derived from living organisms. The rosy periwinkle from Madagascar is an excellent example of a tropical plant that has provided us with useful medicines. Potent chemicals from this plant are now used to treat two forms of cancer: leukemia and Hodgkin disease. Because of these drugs, the survival rate for childhood leukemia has gone from 10% to 90%, and Hodgkin disease is usually curable. Although the value of saving a life cannot be calculated, it is still sometimes easier for us to appreciate the worth of a resource if it is explained in monetary terms. Thus, researchers tell us that, judging from the success rate in the past, an additional 328 types of drugs are yet to be found in tropical rain forests, and the value of this resource to society is probably $147 billion.

You may already know that the antibiotic penicillin is derived from a fungus and that certain species of bacteria

Figure 24.17 Direct value of wildlife.
The direct services of wild species benefit human beings immensely, and it is sometimes possible to calculate the monetary value, which is always surprisingly large.

Wild species, like the rosy periwinkle, are sources of many medicines.

Wild species, like many marine species, provide us with food.

Wild species, like the nine-banded armadillo, play a role in medical research.

produce the antibiotics tetracycline and streptomycin. These drugs have proven to be indispensable in the treatment of diseases, including certain sexually transmitted diseases.

Leprosy is among those diseases for which there is, as yet, no cure. The bacterium that causes leprosy will not grow in the laboratory, but scientists discovered that it grows naturally in the nine-banded armadillo. Having a source for the bacterium may make it possible to find a cure for leprosy. The blood of horseshoe crabs contains a substance called limulus amoebocyte lysate, which is used to ensure that medical devices, such as pacemakers, surgical implants, and prosthetic devices, are free of bacteria. Blood is taken from 250,000 crabs a year, and then they are returned to the sea unharmed.

Agricultural Value

Crops, such as wheat, corn, and rice, are derived from wild plants that have been modified to be high producers. The same high-yield, genetically similar strains tend to be grown worldwide. When rice crops in Africa were being devastated by a virus, researchers grew wild rice plants from thousands of seed samples until they found one that contained a gene for resistance to the virus. These wild plants were then used in a breeding program to transfer the gene into high-yield rice plants. If this variety of wild rice had become extinct before it could be discovered, rice cultivation in Africa might have collapsed.

Biological pest controls—natural predators and parasites—are often preferable to using chemical pesticides. When a rice pest, called the brown planthopper, became resistant to pesticides, farmers began to use natural brown planthopper enemies instead. The economic savings were calculated at well over $1 billion. Similarly, cotton growers in Cañete Valley, Peru, found that pesticides were no longer working against the cotton aphid because of resistance. Research identified natural predators that are now being used to an ever greater degree by cotton farmers. Again, savings have been enormous.

Most flowering plants are pollinated by animals, such as bees, wasps, butterflies, beetles, birds, and bats. The honeybee, *Apis mellifera*, has been domesticated, and it pollinates almost $10 billion worth of food crops annually in the United States. The danger of this dependency on a single species is exemplified by mites that have now wiped out more than 20% of the commercial honeybee population in the United States. Where can we get resistant bees? From the wild, of course. The value of wild pollinators to the U.S. agricultural economy has been

Wild species, like the lesser long-nosed bat, are pollinators of agricultural and other plants.

Wild species, like ladybugs, play a role in biological control of agricultural pests.

Wild species, like rubber trees, can provide a product indefinitely if the forest is not destroyed.

Cyanide Fishing and Coral Reefs

Coral reefs are areas of biological abundance found in shallow, warm, tropical waters just below the surface of the water. They occur in the Caribbean and off the coasts of the Southern Hemisphere continents. The Great Barrier Reef off the coast of Australia is the largest coral reef in the world.

Coral Reefs: Areas of Biological Abundance

The chief constituents of a coral reef are stony corals, animals that exist as small polyps with a calcium carbonate (limestone) exterior (Fig. 24A). Corals do not usually occur individually; rather they form colonies derived from just one individual. When the corals die off, they leave behind a hard, stony, branching limestone structure. The large number of crevices and caves of a reef provide shelter for many animals, including sponges, nudibranchs, small fish (groupers, clown fish, parrotfish, snapper, and scorpion fish), jellyfish, anemones, sea stars (including the destructive Crown of Thorns), crustaceans (like crabs, shrimp, and lobsters), turtles, sea snakes, snails, and mollusks (like octopuses, nautilus, and clams). The barracuda, moray eel, and sharks are top predators in coral reefs.

Coral reefs are such incredibly diverse communities, they are often referred to as the "rain forests of the sea." They are home to thousands of known species, with perhaps millions more yet to be documented. Coral reef degradation is a worldwide problem that involves overfishing for the food and aquarium trades, careless tourist divers, impacts from boat anchors and propellers, oil spills, nutrient pollution, global climate change, and an increase in coral diseases. It has been estimated that 58% of all coral reefs are being harmed by human activities.

Coral Reefs: Sources of Aquarium Fish

Many aquarium hobbyists enjoy collecting and caring for the beautiful fish, star fish, sea anemones, and other organisms that call the reef home. One of the most damaging methods of collecting aquarium fish from coral reefs is known as cyanide fishing. Cyanide is a poison that interferes with the mitochondrial electron transport system and effectively shuts down aerobic respiration. Reef fish can be challenging for divers to capture with handheld nets alone, since they are able to seek refuge in hard-to-reach nooks and crannies. However, spewing a cyanide solution over the fish stuns them and makes them much easier to catch. Unfortunately, cyanide dosing in open water is a tricky proposition. Many fish die immediately, and even more expire later from aftereffects of the poison. To make matters worse, the coral itself, which takes thousands of years to grow and form a reef, can succumb to cyanide. Even the divers do not always escape unscathed; they can inadvertently poison themselves.

The United States imports almost half of all marine aquarium organisms. Approximately two-thirds of them come from the Philippines, where cyanide fishing began, and Indonesia, throughout which the practice has spread. Even though cyanide fishing is officially illegal, many of the people in this region live in poverty. As a result, numerous individuals feel that they must assign a lower priority to protection of the coral reefs than the effort to earn a living.

There is hope that cyanide fishing may become a less popular way of obtaining organisms for the aquarium trade, as hobbyists and retailers become better educated about the harmful effects of this practice. The Marine Aquarium Council (MAC) is a nonprofit, international organization that certifies marine organisms as having been harvested using environmentally responsible methods. However, some scientists contend that removing fish by any means may be detrimental to the reef community, since some of the most popular aquarium fish are herbivores that keep the plants of the reef from growing too much and overwhelming the coral.

Decide Your Opinion

1. How far should we go to preserve natural coral reefs? Should aquarium fish be grown in tanks?
2. Can you think of alternatives to fishing the coral reefs that would help people who live near reefs make a living?
3. At this time, some nations, such as the United States and Australia, have designated small "no-take" zones within their reef systems, where harvesting of reef organisms is strictly prohibited. Do you think this is enforceable?

Figure 24A   Coral reef ecosystem.
Gathering tropical fish in coral reefs with cyanide.

calculated at $4.1 to $6.7 billion a year. And yet, modern agriculture often kills bees by spraying fields with pesticides.

The pesticides and also the fertilizer applied to fields ends up in standing bodies of freshwater because of agricultural runoff.

Consumptive Use Value

We have had much success cultivating crops, keeping domesticated animals, growing trees in plantations, and so forth. But so far, aquaculture, the growing of fish and shellfish for human consumption, has contributed only minimally to human welfare—instead, most freshwater and marine harvests depend on the catching of wild animals, such as fishes (e.g., trout, cod, tuna, and flounder), crustaceans (e.g., lobsters, shrimps, and crabs), and mammals (e.g., whales). Obviously, these aquatic organisms are an invaluable biodiversity resource.

The environment provides all sorts of other products that are sold in the marketplace worldwide, including wild fruits and vegetables, skins, fibers, beeswax, and seaweed. Also, some people obtain their meat directly from the environment. In one study, researchers calculated that the economic value of wild pig in the diet of native hunters in Sarawak, East Malaysia, was approximately $40 million per year.

Similarly, many trees are still felled in the natural environment for their wood. Researchers have calculated that a species-rich forest in the Peruvian Amazon is worth far more if the forest is used for fruit and rubber production than for timber production. Fruit and the latex needed to produce rubber can be brought to market for an unlimited number of years, whereas once the trees are gone, no more timber can be harvested.

Indirect Value of Biodiversity

To bring about the preservation of wildlife, it is necessary to make all people aware that biodiversity is a resource of immense value. If we want to preserve wildlife, it is more economical to save ecosystems than individual species. Ecosystems perform many useful services for modern humans, who increasingly live in cities. These services are said to be indirect because they are pervasive and not easily discernible. Even so, our very survival depends on the functions that ecosystems perform for us, and they tell us the indirect value of wildlife. Therefore, the indirect value of biodiversity can be associated with the following services.

Waste Disposal

Decomposers break down dead organic matter and other types of wastes to inorganic nutrients that are used by the producers within ecosystems. This function aids humans immensely because we dump millions of tons of waste material into natural ecosystems each year. If it were not for decomposition, waste would soon cover the entire surface of our planet. We can build sewage treatment plants, but they are expensive, and few of them break down solid wastes completely to inorganic nutrients. It is less expensive and

more efficient to water plants and trees with partially treated wastewater and let soil bacteria cleanse it completely.

Biological communities are also capable of breaking down and immobilizing pollutants, such as heavy metals and pesticides, that humans release into the environment. A review of wetland functions in Canada assigned a value of $50,000 per hectare (2.471 acres, or 10,000 m2) per year to the ability of natural areas to purify water and take up pollutants.

Provision of Freshwater

Few terrestrial organisms are adapted to living in a salty environment—they need freshwater. The water cycle continually supplies freshwater to terrestrial ecosystems. Humans use freshwater in innumerable ways, including drinking it and irrigating their crops. Freshwater ecosystems, such as rivers and lakes, also provide us with fish and other types of organisms for food.

Unlike other commodities, there is no substitute for freshwater. We can remove salt from seawater to obtain freshwater, but the cost of desalination is about four to eight times the average cost of freshwater acquired via the water cycle.

Forests and other natural ecosystems exert a "sponge effect." They soak up water and then release it at a regular rate. When rain falls in a natural area, plant foliage and dead leaves lessen its impact, and the soil slowly absorbs it, especially if the soil has been aerated by organisms. The water-holding capacity of forests reduces the possibility of flooding. The value of a marshland outside Boston, Massachusetts, has been estimated at $72,000 per hectare per year solely on its ability to reduce floods. Forests release water slowly for days or weeks after the rains have ceased. Rivers flowing through forests in West Africa release twice as much water halfway through the dry season and between three and five times as much at the end of the dry season, as do rivers from coffee plantations.

Prevention of Soil Erosion

Intact ecosystems naturally retain soil and prevent soil erosion. The importance of this ecosystem attribute is especially observed following deforestation. In Pakistan, the world's largest dam, the Tarbela Dam, is losing its storage capacity of 12 billion cubic meters many years sooner than expected because silt is building up behind the dam, due to deforestation. At one time, the Philippines was exporting $100 million worth of oysters, mussels, clams, and cockles each year. Now, silt carried down rivers following deforestation is smothering the mangrove ecosystem that serves as a nursery for these shellfish. Most coastal ecosystems are not as bountiful as they once were because of deforestation and a myriad of other assaults.

Biogeochemical Cycles

You'll recall from Chapter 23 that ecosystems are characterized by energy flow and chemical cycling. The biodiversity within ecosystems contributes to the workings of the water, phosphorus, nitrogen, carbon, and other biogeochemical cycles. We depend on these cycles for freshwater, provision of phosphate, uptake of excess soil nitrogen, and removal of carbon dioxide from the atmosphere. When human activities upset the usual workings of biogeochemical cycles, the dire environmental consequences include the release of excess pollutants that are harmful to us. Technology is unable to substitute for any of the biogeochemical cycles.

Regulation of Climate

At the local level, trees provide shade and reduce the need for fans and air conditioners during the summer.

Globally, forests ameliorate the climate because they take up carbon dioxide. The leaves of trees use carbon dioxide when they photosynthesize, and the bodies of the trees store carbon. When trees are cut and burned, carbon dioxide is released into the atmosphere. Carbon dioxide makes a significant contribution to global warming, which is expected to be stressful for many plants and animals. Only a small percentage of wildlife will be able to move northward, where the weather will be suitable for them.

Ecotourism

Almost everyone prefers to vacation in the natural beauty of an ecosystem. In the United States, nearly 100 million people enjoy vacationing in a natural setting. To do so, they spend $4 billion each year on fees, travel, lodging, and food. Many tourists want to go sport fishing, whale watching, boat riding, hiking, birdwatching, and the like.

☑ Check Your Progress 24.3

1. What are the chief causes of species extinctions?
2. What are the direct and indirect benefits of preserving wildlife?

24.4 Working Toward a Sustainable Society

A **sustainable** society would always be able to provide the same amount of goods and services for future generations, as it does at present. At the same time, biodiversity would be preserved.

In order to achieve a sustainable society, resources can not be depleted and must be preserved. In particular, future generations need clean air, water, an adequate amount of food, and enough space in which to live. This goal is not possible unless we carefully regulate our consumption of resources today, taking into consideration that the human population is still increasing.

Today's Unsustainable Society

While we are quick to realize that population growth in the LDCs creates an environmental burden, we need to consider that the excessive resource consumption of the MDCs also stresses the environment. Sustainability is, more than likely, incompatible with the level of consumption plus the generation of wastes practiced by the MDCs today. Both overpopulation of the LDCs and overconsumption by the MDCs accounts for the increasing amount of worldwide pollution and the extinction of wildlife observable today (Fig. 24.18).

At present, a considerable proportion of land is being used for human purposes (homes, agriculture, factories, etc.). Agriculture uses large inputs of fossil fuels, fertilizer, and

deforestation
and desertification

erosion of topsoil
and groundwater
depletion

overfishing and loss
of terrestrial habitats

urban sprawl
and wildlife
habitat loss

pesticide, fertilizer,
and hazardous waste
accumulation

air and water pollution
and global warming

Figure 24.18   Unsustainable activities.
Arrows point outward to signify that these types of
activities reduce the carrying capacity of the Earth.

pesticides, which create much pollution. More freshwater is used for agriculture than in homes. Almost half of the agricultural yield in the United States goes toward feeding animals. According to the ten-to-one rule of thumb, it takes 10 lb of grain to grow 1 lb of meat. Therefore, it is wasteful for citizens in MDCs to eat as much meat as they do.

Farm animals and crops require freshwater from surface water and groundwater, and so do humans. Available supplies are dwindling, and what groundwater remains is in danger of being contaminated. Sewage and animal wastes wash into bodies of surface water and cause overenrichment, which robs aquatic animals of the oxygen they need to survive.

Our society primarily utilizes nonrenewable fossil fuel energy, which leads to global warming, acid deposition, and smog. The result is weakened ecosystems. The demand for goods has increased to the point that facilities to meet the demand are strained, and construction of improved infrastructure to support increased transportation needs only increases the utilization of nonrenewable energy resources. LDCs soon will have increased needs for energy, and, therefore, it is imperative for the MDCs to develop renewable energy sources.

Because the human population is expanding into all regions on the face of the planet, habitats for other species are being lost, and a wildlife extinction crisis is expected.

Characteristic of a Sustainable Society

A natural ecosystem can offer clues about how to make today's society sustainable. A natural ecosystem makes use of only renewable solar energy, and its materials cycle through the various populations back to the producer once again. For example, coral reefs have been sustaining themselves for millions of years and, at the same time, offering sustenance to the LDCs. The value of coral reefs has been assessed at over $300 billion a year. Their aesthetic value is not measurable.

It is clear that if we want to develop a sustainable society, we too should use renewable energy sources and recycle materials, and we should protect natural ecosystems that help sustain our modern society. At least a quarter of the coral reefs exist close to the shores of an MDC country, and the chances are good that these coral reefs will be protected. Unfortunately, the other coral reefs are threatened by unsustainable practices. Even so, reefs are remarkably regenerative and will return to their former condition if left alone. The message of today's environmentalists is not gloom and doom but what can be done to improve matters and make the environment sustainable (Fig. 24.19).

Sustainability should be practiced in various areas of human endeavor, from agriculture to business enterprises. Efficiency is the key to sustainability. For example, an efficient car would be ultralight and gas thrifty. Efficient cars could be just as durable and speedy as the inefficient ones of today. Only through efficiency can we meet the challenges of limited resources and finances in the future.

People generally live in either the country or the city, but the two regions depend on one another. The city needs goods and services provided by the country, and the country needs goods and services provided by the city. Achieving sustainability will require that we understand how the two regions are interdependent. It would be impossible to have one sustainable and not the

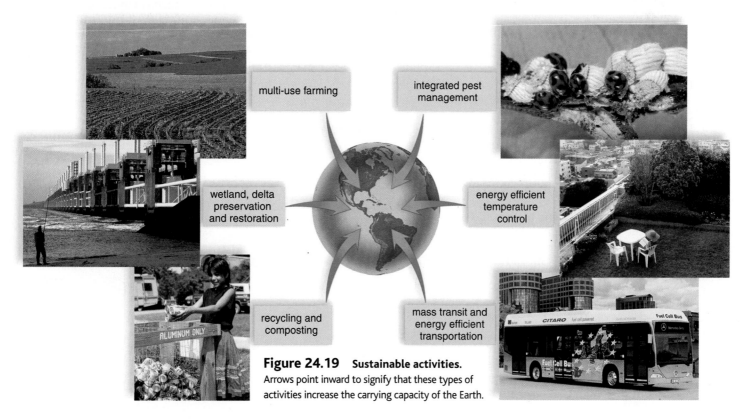

multi-use farming

integrated pest management

wetland, delta preservation and restoration

energy efficient temperature control

recycling and composting

mass transit and energy efficient transportation

Figure 24.19 Sustainable activities.
Arrows point inward to signify that these types of activities increase the carrying capacity of the Earth.

Figure 24.20   Multipurpose trees.
Trees planted by a farmer to break the wind and prevent soil erosion can also have other purposes, such as supplying nuts and fruits.

other because the two regions are linked, and what happens within one will ultimately affect the other. Let's consider, therefore, the importance of both rural and urban sustainability.

Rural Sustainability

In rural areas, we should put the emphasis on preservation. We should preserve ecosystems, including terrestrial ecosystems—such as forests and prairies—and aquatic ecosystems—both freshwater and brackish ones along the coast. We should also preserve agricultural land, groves of fruit trees, and other areas that provide us with renewable resources.

It is imperative that we take all possible steps to preserve what remains of our topsoil and replant areas with native grasses, as necessary. Native grasses stabilize the soil and rebuild soil nutrients, while also serving as a source of renewable biofuel. Trees can be planted to break the wind and protect the soil from erosion, while providing a product as well. Creative solutions to today's ecological problems are very much needed.

Here are some other possible ways to help make rural areas sustainable:

- Plant *cover crops,* which often are a mixture of legumes and grasses, to stabilize the soil between rows of cash crops or between seasonal plantings of cash crops.
- Use *multiuse farming* by planting a variety of crops and use a variety of farming techniques to increase the amount of organic matter in the soil.
- Replenish soil nutrients through composting, organic gardening, or other self-renewable methods.
- Use low flow or trickle irrigation, retention ponds, and other water-conserving methods.
- Increase the planting of *cultivars* (plants propagated vegetatively), which are resistant to blight, rust, insect damage, salt, drought, and encroachment by noxious weeds.
- Use *precision farming (PF)* techniques that rely on accumulated knowledge to reduce habitat destruction, while improving crop yields.
- Use *integrated pest management (IPM),* which encourages the growth of competitive beneficial insects and uses biological controls to reduce the abundance of a pest.
- Plant a variety of species, including native plants, to reduce our dependence on traditional crops.

- Plant *multipurpose trees*—trees with the ability to provide numerous products and perform a variety of functions, in addition to serving as windbreakers (Fig. 24.20). Remember that mature trees can provide many different types of products: Mature rubber trees provide us with rubber, and tagua nuts are an excellent substitute for ivory, for example.
- Maintain and restore wetlands, especially in hurricane or tsunami-prone areas. Protect deltas from storm damage. By protecting wetlands, we protect the spawning ground for many valuable fish nurseries.
- Use renewable forms of energy, such as wind and biofuel.
- Support local farmers, fishermen, and feed stores by buying food products produced close to home.

Urban Sustainability

More and more people are moving to a city. Much thought needs to be given about how to serve the needs of new arrivals, without overexpansion of the city. Resources need to be shared in a way that will allow urban sustainability.

Here are some other possible ways to help make a city sustainable:

- Design an energy-efficient transportation system to rapidly move people about.
- Use solar or geothermal energy to heat buildings; cool them with an air-conditioning system that uses seawater; and, in general, use conservation methods to regulate the temperature of buildings.
- Utilize *green roofs*—a wild garden of grasses, herbs, and vegetables on the tops of buildings—to assist temperature control, supply food, reduce the amount of rainwater runoff, and be visually appealing (Fig. 24.21).
- Improve storm-water management by using sediment traps for storm drains, artificial wetlands, and holding ponds. Increase use of porous surfaces for walking paths, parking lots, and roads. These surfaces reflect less heat, while soaking up rainwater runoff.

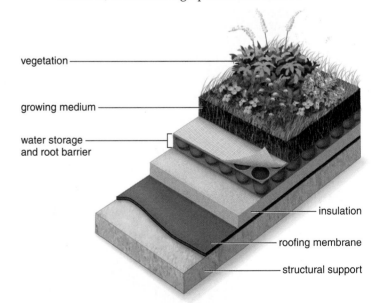

Figure 24.21   Green roofs are ecologically sound and can be visually beautiful.

- Instead of traditional grasses, plant native species that attract bees and butterflies and require less water and fertilizers.
- Create *greenbelts* that suit the particular urban setting. Include plentiful walking and bicycle paths.
- Revitalize old sections of a city, before developing new sections.
- Use lighting fixtures that hug the walls or ground and send light down; control noise levels by designing quiet motors.
- Promote sustainability by encouraging recycling of business equipment; use low-maintenance building materials, rather than wood.

Assessing Economic Well-Being and Quality of Life

The gross national product (GNP) is a measure of the flow of money from consumers to businesses, in the form of goods and services purchased. It can also be considered the total costs of all manufacturing, production, and services in the form of salaries and wages, mortgage and rent, interest and loans, taxes, and profit within and outside the country. In other words, GNP pertains solely to economic activities.

When calculating GNP, economists do not necessarily consider whether an activity is environmentally or socially harmful. For example, destruction of forests due to clear-cutting, strip mining, or land development is not a part of the GNP. In the same way, the cost of medical services does not include the pain or suffering caused by illness, for example.

Measures that include noneconomic indicators are most likely better at revealing our quality of life than is the GNP. The Index of Sustainable Economic Welfare (ISEW) includes real per capita income, distributional equity, natural resources depletion, environmental damage, and the value of unpaid labor. Notice, then, that the ISEW *does* take into account other

forms of value, beyond the purely monetary value of goods and services. Another such index is called the Genuine Progress Indicator (GPI). This indicator attempts to consider the quality of life, an attribute that does not necessarily depend on worldly goods. For example, the quality of life might depend on how much respect we give other human beings. The Grameen Bank in Bangladesh decided that if women were loaned small amounts of money, they would pay it back after starting up small businesses. The loans give women the opportunity to make choices that can improve the quality of their lives. For these women, a loan is a way to sustain their lives while, in part, fulfilling their dreams. It is difficult to assign a value to well-being or happiness, but economists are searching for a way to measure these values such as the following:

Use value: actual price we pay to use or consume a resource, such as the entrance fees into national parks.

Option value: preserving options for the future, such as saving a wetland or a forest.

Existence value: saving things we might not realize exist yet, such as flora and fauna in a tropical rain forest that, one day, could be the source of new drugs.

Aesthetic value: appreciating an area or creature for its beauty and/or contribution to biodiversity (Fig. 24.22).

Cultural value: factors such as language, mythology, and history that are important for cultural identity.

Scientific and educational value: valuing the knowledge of naturalists, or even an experience of nature, as types of rational facts.

Development of the environment will always continue, but perhaps we can use these values to help us direct future development. Growth creates increases in demand, but development includes the direction of growth. If we permit unbridled growth, resources will become depleted. However, if development restrains resource consumption, while still promoting economic growth, perhaps a balance can be reached that will allow us to retain sources for future generations.

Each person has a particular comfort level, and human beings do not like to make sacrifices that reduce their particular comfort level. So, despite our knowledge of the need to protect fisheries and forests, we continue to exploit them. People from LDCs directly depend on these resources to survive, and so have much to lose. Even so, it is difficult for them to sacrifice today for the sake of the future. Yet, there is still hope because nature is incredibly resilient. One solution to deforestation is reforestation. Costa Rica has been successfully reforesting since the early 1980s. Also, declining fisheries can be restocked and then managed for sustainability. What will it take to move toward sustainability? It will take an informed citizenry, creativity, and a desire to bring about change for the better.

Figure 24.22 **Learning to appreciate the value of ecosystems.**

✔ Check Your Progress 24.4

1. What are the characteristics of today's unsustainable society?
2. How can we change today's society to one that is sustainable?
3. How can we assess how well we are doing?

Summarizing the Concepts

24.1 Human Population Growth
- Populations have a biotic potential for increase in size.
- Biotic potential is normally held in check by environmental resistance.
- Population size usually levels off at carrying capacity.

The MDCs Versus the LDCs
- The MDCs have a 0.1% growth rate since 1950.
- The LDC growth rate is presently 1.6% after peaking at 2.5% in the 1960s.

  Age-structure diagrams can be used to predict population growth.
- MDCs are approaching a stable population size.
- LDC populations will continue to increase in size.

24.2 Human Use of Resources and Pollution
Five resources are maximally used by humans:

Human population

land   water   food   energy   minerals

Resources are either nonrenewable or renewable.
- Nonrenewable resources are not replenished and are limited in quantity (e.g., land, fossil fuels, minerals).
- Renewable resources are replenished but still are limited in quantity (e.g., water, solar energy, food).

Land
Human activities, such as habitation, farming, and mining, contribute to erosion, pollution, desertification, deforestation, and loss of biodiversity.

Water
Industry and agriculture use most of the freshwater supply. Water supplies are increased by damming rivers and drawing from aquifers. As aquifers are depleted, subsidence, sinkhole formation, and saltwater intrusion can occur. If used by industries, water conservation methods could cut world water consumption by half.

Food
Food comes from growing crops, raising animals, and fishing.
- Modern farming methods increase the food supply, but some methods harm the land, pollute water, and consume fossil fuels excessively.
- Genetically engineered plants increase the food supply and reduce the need for chemicals.
- Raising livestock contributes to water pollution and uses fossil fuel energy.
- The increased number and high efficiency of fishing boats have caused the world fish catch to decline.

Energy
Fossil fuels (oil, natural gas, coal) are nonrenewable sources. Burning fossil fuels and burning to clear land for farming cause pollutants and gases to enter the air.

- Greenhouse gases include CO_2 and other gases. Greenhouse gases cause global warming because solar radiation can pass through, but infrared heat cannot escape back into space.
- Renewable resources include hydropower, geothermal, wind, and solar power.

Minerals
Minerals are nonrenewable resources that can be mined. These raw materials include sand, gravel, phosphate, and metals. Mining causes destruction of the land by erosion, loss of vegetation, and toxic runoff into bodies of water. Some metals are dangerous to health. Land ruined by mining can take years to recover.

Hazardous Wastes Billions of tons of solid waste are discarded on land and in water.
- Heavy metals (lead, arsenic, cadmium, chromium).
- Synthetic organic chemicals include chlorofluorocarbons (CFCs), which are involved in the production of plastics, pesticides, herbicides, and other products.
- Ozone shield destruction is associated with CFCs.
- Other synthetic organic chemicals enter the aquatic food chain, where the toxins become more concentrated (biological magnification).

24.3 Biodiversity
Biodiversity is the variety of life on Earth. The five major causes of biodiversity loss and extinction are
- habitat loss,
- introduction of alien species,
- pollution,
- overexploitation of plant and animals, and
- disease.

Direct Value of Biodiversity
Direct values of biodiversity are
- medicinal value (medicines derived from living organisms),
- agricultural value (crops derived from wild plants; biological pest controls and animals pollinators), and
- consumptive use values (food production).

Indirect Value of Biodiversity
Biodiversity in ecosystems contributes to
- waste disposal (through the action of decomposers and the ability of natural communities to purify water and take up pollutants),
- freshwater provision through the water biogeochemical cycle,
- prevention of soil erosion, which occurs naturally in intact ecosystems,
- function of biogeochemical cycles,
- climate regulation (plants take up carbon dioxide), and
- ecotourism (human enjoyment of a beautiful ecosystem).

24.4 Working Toward a Sustainable Society

A sustainable society would use only renewable energy sources, would reuse heat and waste materials, and would recycle

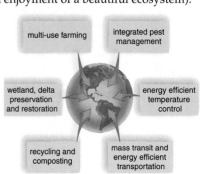

multi-use farming

integrated pest management

wetland, delta preservation and restoration

energy efficient temperature control

recycling and composting

mass transit and energy efficient transportation

almost everything. It would also provide the same goods and services presently provided and would preserve biodiversity.

Understanding Key Terms

agricultural runoff 518
alien species 524
aquifer 516
biodiversity 524
biological magnification 523
biotic potential 512
carrying capacity 512
chlorofluorocarbon (CFC) 523
deforestation 515
desertification 515
exponential growth 512
fossil fuel 520
greenhouse gases 520

growth rate 512
invasive 524
mineral 522
nonrenewable resource 514
photovoltaic (solar) cell 522
pollution 514
renewable resource 514
replacement reproduction 513
salinization 518
saltwater intrusion 517
sinkhole 517
subsidence 516
sustainable 530

Match the key terms to these definitions.

a. ———— The ability of a society or ecosystem to maintain itself while also providing services to human beings.

b. ———— Largest number of organisms of a particular species that can be maintained indefinitely in an ecosystem.

c. ———— Gases such as carbon dioxide and methane in the atmosphere that trap heat.

d. ———— Concentration of a synthetic organic chemical as it passes along a food chain.

e. ———— Water reservoir below the Earth's surface.

Testing Your Knowledge of the Concepts

1. Define carrying capacity and discuss the relevancy of this concept to the size of today's human population. (page 512)

2. Distinguish between MDCs and LDCs. Why are most LDCs, but not most MDCs, increasing in size? (pages 512–13)

3. Name three locales where humans have settled with unfortunate environmental consequences. What are those consequences? (pages 514–15)

4. What steps can be taken to conserve water? (page 517)

5. What are the environmental benefits of reducing the amount of meat in the American diet? (pages 517–18)

6. What are the types of fossil fuels, and what environmental problems are associated with burning fossil fuels? (page 520)

7. What renewable energy sources are available? What are the benefits of the solar-hydrogen revolution? (pages 520–22)

8. What are minerals, and what are the drawbacks of mining them? (pages 522–23)

9. Exponential growth is best described by
   a. steep unrestricted growth.
   b. an S-shaped growth curve.
   c. a constant rate of growth.
   d. growth that levels off after rapid growth.
   e. Both b and d are correct.

10. When the carrying capacity of the environment is exceeded, the population will typically
    a. increase, but at a slower rate.
    b. stabilize at the highest level reached.
    c. decrease.
    d. die off entirely.

11. Decreased death rate followed by decreased birthrate has occurred in
    a. MDCs.   c. MDCs and LDCs.
    b. LDCs.   d. neither MDCs nor LDCs.

12. Renewable resources
    a. are in limited supply compared to nonrenewable resources.
    b. are always forthcoming but still may be inadequate for human needs.
    c. include such energy sources as wind, solar, and biomass.
    d. All of these are correct.

13. Desertification is often caused by
    a. overuse of aquifers.   c. air pollution.
    b. urban sprawl.   d. overgrazing.

14. Soil in the tropics is often nutrient-poor because
    a. nutrients are tied up in plants.
    b. it is mostly sand.
    c. it has a high pH.
    d. rainfall leaches out minerals.

15. Most freshwater is used for
    a. domestic purposes, such as bathing, flushing toilets, and watering lawns.
    b. domestic purposes, such as cooking and drinking.
    c. agriculture.
    d. industry.

16. Which of the following is not a major problem with the damming of rivers?
    a. increase in salinity
    b. loss of water through evaporation
    c. change in the level of the water table
    d. buildup of sediment

17. Removal of groundwater from aquifers may cause
    a. pollution.   d. soil erosion.
    b. subsidence.   e. All of the above are correct.
    c. mineral depletion.

18. Which of the following is not a component of modern agriculture?
    a. dependency on chemical inputs
    b. frequent irrigation
    c. high fuel consumption
    d. high diversity of cultivars planted

19. The raising of domestic livestock
    a. consumes large amounts of fossil fuels.
    b. leads to water pollution.
    c. is energetically wasteful.
    d. All of the above are correct.

20. The best way to maintain fish supplies is to
    a. limit harvesting to the ability of fish to reproduce.
    b. do away with all the other animals that feed on fish.
    c. use larger and better kinds of nets
    d. All of these are good ways to maintain fish supplies.

For questions 21–25, match the description to the type of fuel in the key. Each answer can be used more than once. Each question may have more than one answer.

Key:
    a. nuclear power   d. solar power
    b. fossil fuels   e. wind power
    c. hydropower   f. geothermal power

21. Renewable source of power

22. Waste products are harmful

23. Detrimental environmental effects

24. More available in certain geographic locations

25. Contribute(s) to global warming

26. Heavy metals are dangerous to humans because they
    a. inhibit important enzymes.
    b. cause apoptosis.
    c. reduce the body's ability to carry oxygen.
    d. break down mitochondria.

27. The Earth's ozone shield has been damaged by
    a. heavy metals.
    b. strip mining.
    c. chlorofluorocarbons.
    d. fossil fuels.

28. The preservation of ecosystems indirectly provides freshwater because
    a. trees produce water as a result of photosynthesis.
    b. animals excrete water-based products.
    c. forests soak up water and release it slowly.
    d. ecosystems promote the growth of bacteria that release water into the environment.

29. Which of these are indirect values of species?
    a. participation in biogeochemical cycles
    b. participation in waste disposal
    c. provision of freshwater
    d. prevention of soil erosion
    e. All of these are indirect values.

30. Which of the following is not a function that ecosystems can perform for humans?
    a. purification of water
    b. immobilization of pollutants
    c. reduction of soil erosion
    d. removal of excess soil nitrogen
    e. breakdown of heavy metals

31. A transition to hydrogen fuel technology will
    a. be long in coming and not likely to be of major significance.
    b. lessen many current environmental problems.
    c. not be likely since it will always be expensive and as polluting as natural gas.
    d. be of major consequence, but resource limitations for obtaining hydrogen will hinder its progress.

32. In which of the following is biological magnification most pronounced?
    a. aquatic food chains   d. energy pyramids
    b. terrestrial food chains e. Both a and c are correct.
    c. long food chains

33. Complete the following graph by labeling each bar with a cause of extinction, from the most influential to the least.

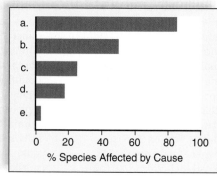

% Species Affected by Cause

Thinking Critically About the Concepts

After graduation, the four friends from the opening story found themselves in various parts of the world. Andrew joined the Peace Corps and was assigned to a nutritional center in a Togo village. Megan joined church members on a mission trip to Mexico. Eva went to China to teach English, and Jacob backpacked around Europe for the summer.

The friends kept in touch with each other via e-mail and shared their amazement at how differently environmental concerns were dealt with in the respective countries. Jacob commented on how convenient it was to travel using public transportation and recycle throughout Europe. He described a number of energy-saving techniques, like an escalator that moves only when someone steps on it, he'd seen. The number of childhood illnesses and deaths associated with malnutrition stunned Andrew, who was not accustomed to food shortages. Eva sent pictures of numerous huge construction projects in China and shared that she'd heard China was using 50% of the world's concrete. Megan's pictures of her trip into Mexico City all looked hazy thanks to the terrible air pollution there. The foursome all agreed that their world experiences were eye-opening with regard to environmental concerns of the planet and vowed to continue to make a difference.

1. The Vatican has been urging Italian married couples to have more children.
   a. Would Italy be considered an MDC or an LDC?
   b. Why might Italy's low birthrate be a concern?

2. What environmental reasons would you give a friend for becoming a vegetarian?

3a. What size footprint do you currently have on Earth? Visit http://www.earthday.net/footprint/index.asp and determine the size of your footprint.
   b. What kinds of things could you do to decrease the size of your footprint on Earth?

4. How would failure to recycle items that can be recycled (they end up in a landfill) affect the nutrient cycles, such as the carbon cycle covered in Chapter 23?

5a. What kinds of environmental activities/initiatives (Earth Day festivities, recycling program, etc.) are in place at your school?
   b. How could you increase awareness of environmental issues on your campus?

Appendix

This appendix contains the answers to the Understanding Key Terms, Testing Your Knowledge of the Concepts, and Thinking Critically About the Concepts questions, which appear at the end of each chapter, and the Check Your Progress questions, which appear within each chapter.

Chapter 1
Check Your Progress
(1.1) 1. Organization, acquiring materials and energy, reproducing, growing and developing, being homeostatic, responding to stimuli, evolution; **2.** Ancestry can be traced to the first cells. **(1.2) 1.** Presence of vertebrae tells us humans are vertebrates; hair and mammary glands tell us humans are mammals; **2.** Brains, upright stance, language, usage of tools; **3.** Continuation of our species. **(1.3) 1.** Observation, hypothesis, experiment, conclusion; **2.** A control group and one or more tests groups; **3.** Review process and the investigator usually is the author. **(1.4) 1a.** Not reliable because not based on studies involving large numbers of subjects; **1b.** Not all correlations are causations; **2a.** Conclusion is an interpretation of data (results); **2b.** conclusion; **3.** Graphs summarize data, statistics evaluate data. **(1.5). 1.** Nuclear physics (cancer therapy, nuclear bomb), biotechnology (GM plants/bacteria, possibly endangering biosphere), gene technology (stem cells/destroying embryos); **2.** All members; **3.** Yes.

Understanding Key Terms
a. biosphere; **b.** cell; **c.** scientific theory; **d.** homeostasis; **e.** experiment

Testing Your Knowledge of the Concepts
9. c; **10.** b; **11.** e; **12.** a; **13.** a; **14.** c; **15.** b; **16.** c; **17.** a; **18.** c; **19.** a; **20.** d; **21.** a; **22.** c; **23.** b; **24.** e; **25.** d

Thinking Critically About the Concepts
1. The sympathetic NS increases heart rate and the respiratory rate. Digestive activities are inhibited. Blood flow to the muscles of the legs and arms increases in preparation for fighting or running away. Body temperature increases and stimulates perspiration to get rid of the excess heat; **2.** Prey animals share a common ancestor with humans; **3.** Exams; test anxiety is the result of sympathetic NS responses; **4.** Heart rate, blood pressure, and respiratory rate stay at normal levels; digestive activities are stimulated and blood flow to the limbs' skeletal muscles decreases.

Chapter 2
Check Your Progress
(2.1) 1. Central nucleus containing protons and neutrons, with electrons orbiting about the nucleus in shells; **2a.** Isotopes that release various types of energy in the form of rays and subatomic particles; **2b.** as tracers (imaging); **3.** ionic, covalent. **(2.2) 1.** liquid at room temperature, temperature of liquid water rises and falls slowly, has a high heat of vaporation, frozen water is less dense than liquid water, water molecules are cohesive, solvent for polar molecules; **2.** Water contains an intermediate concentration of H^+ and has a neutral pH (pH 7), acidic solutions contain high concentrations of H^+ and have a pH below 7, basic solutions contain low concentrations of H^+ and have a pH above 7; **3.** Buffers. **(2.3) 1.** carbohydrates, lipids, proteins, and nucleic acids; **2.** dehydration reaction. **(2.4) 1.** quick and short term energy; **2.** Simple carbohydrates contain low numbers of carbon atoms (3–7), complex carbohydrates contain high numbers of carbon atoms in

chains of sugar (glucose) units; **3.** Insoluble fiber stimulates movements of the large intestine, soluble fiber prevents the absorption of cholesterol. **(2.5) 1a.** energy storage; **1b.** one glycerol molecule and three fatty acids; **2.** phospholipids—cellular membranes, steroids—sex hormones. **(2.6) 1.** support, enzymes, transport, defense, hormones, motion; **2.** —NH_2 (amino group) and the —COOH (acid group); **3.** Once a protein loses its normal shape, it is no longer able to perform its usual function. **(2.7) 1a.** DNA contains genes, RNA conveys DNA's instructions regarding the amino acid sequence in a protein; **1b.** DNA contains deoxyribose and the bases A, T, C, and G; RNA contains ribose, and the bases A, U, C, G (see Table 2.1); **2.** hydrogen; **3.** adenosine and three phosphate groups, high energy molecule.

Understanding Key Terms
a. emulsification; **b.** ion; **c.** covalent bond; **d.** hydrophilic; **e.** hydrogen bond

Testing Your Knowledge of the Concepts
15. b; **16.** c; **17.** b; **18.** c; **19.** d; **20.** a; **21.** b; **22.** d; **23.** c; **24.** b; **25.** a; **26.** d; **27.** b; **28.** d; **29.** a; **30.** c; **31.** a; **32.** c; **33.** d; **34.** b; **35. a.** subunits; **b.** dehydration reaction; **c.** macromolecule; **d.** hydrolysis reaction

Thinking Critically About the Concepts
1. Rita's iron deficiency anemia resulted in less oxygen being delivered to her cells. Cells use oxygen to make ATP. less oxygen means less ATP made and the feeling of tiredness; **2.** Whenever energy is transformed, some is lost as heat. When less energy is made, less energy will be lost as heat. The heat lost during metabolism is often used to maintain body temperature; **3.** Organ meats, such as liver, leafy greens, such as spinach and kale, and some cereals are sources of iron; **4.** Sickle cell anemia; **5.** The word lyse means rupture or burst so hemolytic anemia occurs when the red blood cells rupture/burst; **6.** Acidic conditions like those in the tissues would cause hemolglobin to release oxygen.

Chapter 3
Check Your Progress
(3.1) 1. A cell is the basic unit of life, all living things are made up of cells, new cells arise only from preexisting cells; **2.** Small cells have a greater surface-area-to-volume ratio, thus a greater ability to get material in and out of the cell; **3.** Light microscopes use light rays to magnify objects and can be use to view living specimens; electron microscopes use a stream of electrons to magnify objects and have a higher resolving power than light microscopes. **(3.2) 1a.** nucleus, plasma membrane, cytoplasm; **1b.** plasma membrane, cytoplasm; **1c.** nuclues; **2.** Nucleus and endomembrane system arose from invagination of the plasma membrane; mitochondria and chloroplast evolved by engulfing prokaryotic cells. **(3.3) 1.** Structure is a phospholipid bilayer with attached or embedded proteins (fluid-mosaic model), function is to keep cells intact and selectively allow passage of molecules and ions; **2a.** diffusion, osmosis, facilitated diffusion, active transport; **2b.** passive—diffusion, osmosis, facilitated diffusion; active—active transport; **3a.** isotonic solutions have the same solute concentration as in the cell, hypotonic solutions have a lower solute concentration than in the cell, hypertonic solutions have a higher solute concentration than in the cell; **3b.** isotonic solution—cell's shape stays the same, hypotonic solution—cell gains water, hypertonic solution—cell loses water. **(3.4) 1.** nucleus—stores genetic information that specifies the proteins in cells, ribosomes—site of protein synthesis, rough endoplasmic reticulum—contains ribosomes, which are the sites of protein synthesis;

2. endoplasmic reticulum—processing, modification and formation of transport vesicles, Golgi apparatus—processing, packaging, and secretion; **3.** smooth endoplasmic reticulum—synthesizes the phospholipids that occur in membranes, lysosomes—fuse with vesicles and digest their contents; **4.** move molecules from the ER to the Golgi. **(3.5) 1.** microtubules, actin filaments, intermediate filaments; **2.** microtubules; **3.** The cytoskeleton helps maintain a cell's shape and either anchors organelles or assists in the movement of organelles, cilia are involved in movement (sweeping of debris in respiratory tract, moving of egg along oviduct). **(3.6) 1a.** citric acid cycle; **1b.** electron transport chain; **2.** So the energy within a glucose molecule can be released slowly and ATP can be produced gradually; **3a.** Enzymes are protein molecules with specific active sites that speed up metabolic reactions, coenzymes are nonprotein molecules that assist the activity of an enzyme; **3b.** carries hydrogen atoms; **4a.** to receive electrons at the end of the electron transport chain; **4b.** glucose; **5a.** burst of energy for a short time; **5b.** only produces two ATP per glucose molecule.

Understanding Key Terms
a. fluid-mosaic model; **b.** osmosis; **c.** selectively permeable; **d.** fermentation; **e.** cellular respiration

Testing Your Knowledge of the Concepts
13. c; **14.** b; **15.** e; **16.** a; **17.** c; **18.** c; **19.** a; **20.** d; **21.** b; **22.** c; **23.** c; **24.** a; **25.** e; **26.** c; **27.** b; **28. a.** carbohydrate chain; **b.** glycolipid; **c.** glycoprotein; **d.** hydrophilic heads; **e.** phospholipid bilayer; **f.** hydrophobic tails; **g.** filaments of cytoskeleton; **h.** membrane protein; **i.** cholesterol; **j.** membrane protein; **29. a.** active site; **b.** substrate; **c.** product; **d.** enzyme; **e.** enzyme-substrate complex; **f.** enzyme.

Thinking Critically About the Concepts
1a. Diffusion; **1b.** Facilitated diffusion; **2a.** Testosterone is a steroid, which is a lipid, and there will be a lot of smooth endoplasmic reticulum, the organelle associated with lipid synthesis; **2b.** Growth hormone is a peptide/protein hormone, so a cell-making growth hormone will have lots of ribosomes and extensive rough endoplasmic reticulum, the organelles associated with protein synthesis.

Chapter 4
Check Your Progress
(4.1) 1. connective tissue, muscular tissue, nervous tissue, and epithelial tissue. **(4.2) 1.** fibrous connective tissue—loose fibrous tissue, adipose tissue, dense fibrous connective tissue; supportive connective tissue—cartilage, bone; fluid connective tissue—blood, lymph; **2.** Loose fibrous tissue supports epithelium and forms a protective covering enclosing many internal organs allowing them to expand, dense fibrous tissue contains many collagen fibers that are packed together and has more specific functions than loose connective tissue does; **3.** Compact bone makes up the shaft of a long bone and consists of osteons composed of rings of hard matrix, spongy bone is contained in the ends of a long bone and has an open bony latticework; **4.** Blood is located in the blood vessels and consists of formed elements (Figure 4.3) and plasma, lymph is carried by lymphatic vessels and is derived from tissue fluid and contains white blood cells. **(4.3) 1.** skeletal muscle—voluntary movement of body, striated cells with multiple nuclei; smooth muscle—movement of substances in lumens of body, spindle-shaped cells each with a single nuclei; cardiac muscle—pumping blood, branching striated cells with a single nucleus; **2.** skeletal muscles—attached to the skeleton, smooth muscle—blood vessels and walls of the digestive tract, cardiac muscle—walls of the heart. **(4.4) 1a,b.** Dendrites—receive signals from sensory receptors or other neurons; cell body—contains most of the cell's cytoplasm and the nucleus; axon—conducts the nerve impulses; **2.** support and nourish neurons, engulf bacterial and cellular debris, produce hormones, form myelin sheaths. **(4.5) 1.** Squamous epithelium is composed of flattened cells—lining the air sacs of lungs and walls of blood vessels; cuboidal epithelium is composed of cube-shaped cells—glands, covering ovaries and lining kidney tubules; columnar epithelium is composed of rectangular cells—lining

the digestive tract; **2a.** Appears to be layered, irregularly placed nuclei; **2b.** upward motion of the cilia carries mucus to the back of the throat; **3a.** Stratified epithelia have layers of cells piled one on top of the other; **3b.** and lines the nose, mouth, esophagus, anal canal, the outer portion of the cervix, and the vagina. **(4.6) 1a,b.** tight junctions—hold epithelial cells close together creating an impermeable barrier; adhesion junctions—firmly attach cytoskeletal fibers of one cell to that of another cell and allow the layer of cells to bend; gap junctions—occur when adjacent plasma membranes come together and leave a tiny channel between them allowing the passage of small molecules; **2.** Epithelial cells must create a layer that prevents leakage from one side of the sheet to the other because they cover the surface of organs and line the body cavities; **3a.** gap junctions; **b.** they allow the heart to beat as a coordinated whole. **(4.7) 1.** The epidermis is made up of stratified epithelium, forms a protective covering over the entire body, contains Langerhans cells and melanocytes, and produces vitamin D; the dermis (located beneath the epidermis) consists of dense fibrous connective tissue, which prevents the skin from being torn, contains blood vessels and sensory receptors; **2a.** melanin, carotene, and hemoglobin; **2b.** broad-spectrum sunscreen, protective clothing, sunglasses, and avoid tanning machines; **3a.** Sweat glands are present in all regions of the skin and play a role in modifying body temperature, oil glands are located in hair follicles and secrete sebum; **3b.** Secretions are acidic and retard bacterial growth; **4a.** beneath the dermis; **4b.** stores fat. **(4.8) 1.** integumentary system—homeostatic functions, cardiovascular system—circulates blood, lymphatic and immune systems—transports lymph and defends against disease, digestive system—digests food and eliminates waste, respiratory system—brings oxygen into the body and removes carbon dioxide, urinary system—rids the body of metabolic wastes, skeletal system—protects body parts, muscular system—maintains posture and accounts for movement, nervous system—conducts nerve impulses, endocrine system—secretes hormones, reproductive system—produces sex cells; **2a,b.** Ventral cavity—thoracic and abdominal cavities, Dorsal cavity—cranial cavity and vertebral canal; **3.** mucous, serous, synovial, and meninges. **(4.9) 1a.** The body's ability to maintain a relative constancy of its internal environment by adjusting its physiological processes; **1b.** Fluctuation in internal conditions can result in illness; **2.** Issue electrochemical signals, release hormones, supply oxygen, maintain body temperature, remove waste, maintain adequate nutrient levels, adjust the water-salt and acid-base balance of the blood (see Figure 4.15); **3.** negative feedback keeps a variable close to a particular value, positive feedback brings about an ever greater change in the same direction, which assists the body in completing a process.

Understanding Key Terms
a. ligament; **b.** epidermis; **c.** carcinoma; **d.** homeostasis; **e.** spongy bone

Testing Your Knowledge of the Concepts
12. a; **13.** d; **14.** d; **15.** d; **16.** c; **17.** c; **18.** b; **19.** b; **20.** d; **21.** c; **22.** d; **23.** b; **24.** d; **25.** d; **26.** b; **27.** a; **28.** b; **29.** b; **30. a.** cranial cavity; **b.** vertebral cavity; **c.** dorsal cavity; **d.** diaphragm; **e.** pelvic cavity; **f.** abdominal cavity; **g.** thoracic cavity; **h.** ventral cavity

Thinking Critically About the Concepts
1. No, the large intestine bladder lacks sphincter muscles and the transisitonal epithelium. She will have to empty the new bladder frequently until it stretches to accommodate an increased volume of urine. **2a.** Nervous and endocrine; **2b.** The nervous system uses neurotransmitters and the endocrine system uses hormones.; **3a.** Cartilaginous rings, mucus producing cells, and ciliated epithelium; **3b.** The bladder needs to increase in size as the volume of urine being stored increases, which transitional epithelial allows it to do.; **4a.** Enzymes; **4b.** Blood pH; **4c.** Respiratory and urinary systems; **5a.** Shivering involves muscle contractions, causing energy loss, which can increase body temperature.; **5b.** No, as a component of a negative feedback mechanism, shivering would cease.

Chapter 5
Check Your Progress

(5.1) 1. heart and blood vessels; 2. generate blood pressure, transport blood, create exchange at the capillaries, and regulate blood flow; 3. lymphatic vessels collect excess tissue fluid and return it to the cardiovascular system. (5.2) 1. arteries, capillaries, veins; 2. Arteries carry blood away from the heart and have strong walls; capillaries have thin walls, where exchange can occur; veins carry blood to the heart, have thinner walls than arteries, and contain valves. (5.3) 1. The right ventricle of the heart sends blood through the lungs, and the left ventricle sends blood throughout the body; 2. "Lub" occurs when increasing pressure of blood inside a ventricle forces the cusps of the AV valve to slam shut, "dup" occurs when the ventricles relax and blood in the arteries pushes back, causing the semilunar valves to close; 3. cardiac control center of the medulla oblongata. (5.4) 1. heart rate; 2. blood pressure; 3. skeletal muscle pump, respiratory pump, and valves. (5.5) 1. O_2-poor blood in pulmonary artery is on its way to the lungs and O_2-rich blood in the pulmonary vein is coming from the lungs. 2. Right ventricle→pulmonary trunk→pulmonary arteries→pulmonary capillaries→pulmonary veins→left atrium. 3. left ventricle→aorta→mesenteric arteries→digestive tract capillary bed→hepatic portal vein→liver capillary bed→hepatic vein→inferior vena cava→right atrium; (5.6) 1. Blood pressure causes water to exit at the arterial end and osmotic pressure causes water to enter at the venous end. Solutes diffuse according to their concentration gradient; 2. It is collected by the lymphatic capillaries and eventually returned to the systemic venous blood. (5.7) 1a. Hypertension, stroke, heart attack, aneurysm, heart failure; 1b. drugs to lower blood pressure, nitroglycerin given at onset of heart attack, replace diseased/damaged portion of the vessel, open clogged arteries, dissolve clots, surgery, heart transplant.

Figure Question
Fig. 5.11: Heart

Understanding Key Terms
a. diastole; b. inferior vena cava; c. pulse; d. venule; e. systemic circuit

Testing Your Knowledge of the Concepts
15. b; 16. a; 17. c; 18. b; 19. a; 20. e; 21. b; 22. d; 23. c; 24. c; 25. c; 26. e; 27. b; 28. d; 29. e; 30. d; 31. c; 32. a. jugular vein; b. pulmonary artery; c. superior vena cava; d. inferior vena cava; e. hepatic vein; f. hepatic portal vein; g. renal vein; h. iliac vein; i. carotid artery; j. pulmonary vein; k. aorta; l. mesenteric arteries; m. renal artery; n. iliac artery. 33. a. blood pressure; b. osmotic pressure; c. blood pressure; d. osmotic pressure

Thinking Critically About the Concepts
1a. Arteries have thick walls of smooth muscle and connective tissue; 1b. The aorta carries such a large volume of blood at such high pressure that an aneurysm typically results in vessel rupture and internal bleeding; 2a. They are thicker than normal leaving little room in the chamber for blood to accumulate; 2b. Much less blood is pumped out of the heart; 3. Yes, it is a one-way system; 4. Smoking and diets high in fat impact artery structure, minimizing the vessels ability to transport blood; 5. For example, capillaries are like small side streets. Side streets do not carry a lot of traffic, which is similar to the small volume of blood that travels through a capillary.

Chapter 6
Check Your Progress

(6.1) 1a. transport, defense and regulation; 1b. formed elements and plasma; 2a. 92% water, other 8% consists of various salts and organic molecules; 2b. help maintain homeostasis. (6.2) 1. hemoglobin; 2. they lose a nucleus during maturation; 3. anemia, sickle-cell disease, hemolytic disease of the newborn. (6.3) 1. granular leukocytes (neutrophils, eosinophils, basophils), agranular leukocytes (lymphocytes, monocytes); 2. neutrophils—granular with a multilobed nucleus, first responders in bacterial infection; eosinophils—granular with a bilobed nucleus, increase in number during a parasitic worm infection or al-

lergic reaction; basophils—granular with a U-shaped nucleus, release histamine; lymphocytes—do not have granules and have nonlobular nuclei, responsible for specific immunity to particular pathogens and their toxins; monocytes—do not have granules and are the largest of the white blood cells, phagocytize pathogens and stimulate other white blood cells; 3. severe combined immunodeficiency—stem cells of white blood cells lack adenosine deaminase; leukemia—uncontrolled white blood cell proliferation; infectious mononucleosis—EBV infection of lymphocytes. (6.4) 1. platelets—clump at the site of puncture; thrombin—activates fibrinogen, forming long threads of fibrin; fibrin threads—wind around platelets, plasmin—destroys the fibrin network; serum (released after blood clots)—contains all the components of plasma. (6.5) 1a. A, B, AB, and O; 1b. based on the presence or absence of type A antigen and type B antigen; 2. Type O is the universal donor because it has neither type A or type B antigens on the red blood cells, type AB is the universal acceptor because it has neither anti-A nor anti-B antibodies in plasma; 3. when an RH^- mother is pregnant with her second RH^+ baby. (6.6) 1. delivers oxygen from the lungs and nutrients from the digestive system, and removes metabolic wastes; 2. digestive system—provides molecules for plasma protein and blood cell formation, urinary system—helps maintain water-salt balance of blood, muscular system—moves blood, nervous system—regulates heart contraction and constriction/dilation of blood vessels, endocrine system—produces hormones that regulate blood pressure and volume, respiratory system—gas exchange, lymphatic system—helps maintain blood volume, skeletal system—protects heart and red bone marrow; 3. Glycogen stored in the liver can be broken down to make glucose available to the blood.

Understanding Key Terms
a. hemoglobin; b. formed element; c. plasma; d. leukemia; e. prothrombin

Testing Your Knowledge of the Concepts
10. a; 11. a; 12. b; 13. d; 14. c; 15. c; 16. a; 17. e; 18. d; 19. b; 20. c; 21. a; 22. d; 23. d; 24. e; 25. d; 26. b; 27. d; 28. e; 29. d; 30. e

Thinking Critically About the Concepts
1. A car with a damaged exhaust system, barbeque grilling; 2a. Protein; 2b. Will complain of tiredness and feeling cold.; 3. They will appear normal, skin will have a pink/red appearance; if suffering from hypoxia, skin will appear blue due to low levels of O_2 in blood.; 4a. Too many red blood cells may clog capillaries and decrease the rate of gas exchange.; 4b. At night, heartbeat rate decreases and red blood cells move more slowly through the capillaries and gas exchange rate is even lower.; 5. The respiratory system receives oxygen that the cardiovascular system transports; the nervous system can increase the respiratory rate, which increases the amount of oxygen transported by the cardiovascular system.

Chapter 7
Check Your Progress

(7.1) 1. capsule, flagella, fimbriae, pilus, plasmids, and toxins; 2. Viruses must replicate inside a living cell. (7.2) 1a. red bone marrow, thymus; 1b. lymph nodes, spleen; 2. Primary lymphatic organs are the sites where white blood cells are produced and mature, secondary lymphatic organs are the sites where white blood cells react to pathogens and blood and lymph are purified. 3. red bone marrow, spleen, lymph nodes, thymus gland. (7.3) 1. skin and mucous membranes, chemical barriers, resident bacteria, inflammatory response, and protective proteins; 2. neutrophils and macrophages engulf pathogens by phagocytosis; 3. They "complement" certain immune responses; (7.4) 1. Nonspecific defenses act indiscriminately against all pathogens, specific defenses respond to antigens; 2. Antibody-mediated immunity, cell-mediated immunity; 3. B cells produce plasma and memory cells, plasma cells produce antibodies, memory cells produce antibodies in the future, T cells regulate immune responses and produce cytotoxic T cells and helper T cells, cytotoxic T cells kill virus infected cells and cancer cells, helper T cells regulate immunity, and memory T cells kill in the future. (7.5) **1a.** Immunity that occurs naturally through infection or is brought about artificially by medical intervention; **1b, c.** active immunity—infection with a pathogen, immunization; passive immunity—transfer of IgG antibodies across the placenta, breast feeding, gamma globulin injection; **2.** monoclonal antibodies, cytokine therapy. (7.6) **1.** allergies—hypersensitivities to substances that ordinarily would do no harm to the body, tissue rejection—the recipient's immune system recognizes that the transplanted tissue is not self, autoimmune disease—cytotoxic T cells or antibodies mistakenly attack the body's own cells, as if they bear foreign antigens.

Understanding Key Terms
a. vaccine; **b.** complement system; **c.** antigen; **d.** apoptosis; **e.** T cell

Testing Your Knowledge of the Concepts
17. b; **18.** d; **19.** e; **20.** b; **21.** a; **22.** c; **23.** d; **24.** d; **25.** b; **26.** c; **27.** c; **28.** c; **29.** a; **30.** e; **31.** d; **32.** d; **33.** c

Thinking Critically About the Concepts
1. Ice; **2.** Applying heat inceases blood flow to the injured area, which facilitates the healing process; **3.** They have a greater volume of blood, more blood results in more interstitial fluid developing.; **4.** Veins; **5.** Barrier defenses are supposed to prevent the entrance of pathogens to someone's body, similar as a fence keeping intruders off your property; **6.** They should get another shot of anti-venom, since the first shot gave them passive immunity.

Chapter 8
Check Your Progress
(8.1) 1. ingestion—the mouth takes in food, digestion—divides food into pieces and hydrolyzes food to molecular nutrients, movement—food is passed along from one organ to the next and indigestible remains are expelled, absorption—unit molecules produced by digestion cross the wall of the GI tract and enter the blood for delivery to cells, elimination—removal of indigestible wastes through the anus; **2.** mucosa—diverticulitis, submucosa—inflammatory bowel disease, muscularis—irritable bowel syndrome, serosa—appendicitis. **(8.2) 1.** mechanical digestion—teeth chew food into pieces convenient for swallowing and the tongue moves food around the mouth; chemical digestion—salivary amylase begins the process of digesting starch; **2.** The soft palate moves back to close off the nasal passages, and the trachea moves up under the epiglottis to cover the glottis. **(8.3) 1a.** store food, initiate the digestion of protein, and control the movement of chyme into the small intestine; **1b.** The muscularis contains an oblique layer that allows the stomach to stretch and mechanically break food down, the mucosa has millions of gastric pits, which lead into gastric glands that produce gastric juice; **2a.** Complete digestion using enzymes, which digest all types of food and absorb the products of the digestive process; **2b.** It contains villi that have an outer layer of columnar epithelial cells, each containing thousands of microvilli. **(8.4) 1.** pancreas, liver, gallbladder; **2.** Pancreas produces pancreatic juice, liver produces bile, gallbladder stores bile; **3.** nervous system and hormones. **(8.5) 1.** cecum, colon, rectum, and anal canal; **2.** absorb water and vitamins, form feces, defecation. **(8.6) 1a.** Gives an idea of how much weight is due to adipose tissue; **1b.** premature death, diabetes type 2, hypertension, cardiovascular disease, stroke, gall bladder disease, respiratory disfunction, osteoarthritis, and certain cancers; **2.** High intake of refined carbohydrates and fructose sweeteners as well as a high intake of fat leads to increased deposition of fat; **3a.** saturated fats and trans-fatty acids; **3b.** unsaturated, particularly omega-3 fatty acids; **4.** Brain cells require glucose and excess nitrogen excretion stresses the kidneys; **5.** Which foods should be eaten often and which foods should not be eaten on a regular basis; **6.** anorexia nervosa—severe psychological disorder characterized by an irrational fear of getting fat, bulimia nervosa—binge eating and then purging to avoid gaining weight, binge eating

disorder—episodes of overeating that are not followed by purging, muscle dysmorphia—thinking ones body is underdeveloped.

Understanding Key Terms
a. vitamin; **b.** lipase; **c.** lacteal; **d.** esophagus; **e.** gallbladder

Testing Your Knowledge of the Concepts
9. d; **10.** a; **11.** b; **12.** e; **13.** c; **14.** a; **15.** b; **16.** e; **17.** d; **18.** d; **19.** d; **20.** e; **21.** c; **22.** e; **23.** b; **24.** c; **25.** d; **26.** a; **27.** a; **28.** b; **29.** c; **30.** b; **31.** a; **32.** d; **33.** a; **34.** c; **35.** c; **36.** d; **37.** a; **38.** e; **39.** b, c; **40.** a; **41.** d; **42.** e; **43.** a; **44.** c; **45.** b

Thinking Critically About the Concepts
1a. To prepare the food for enzymes to perform chemical digestion, such as chewing does. The smaller stomach from bariatric surgery does not no longer performs a significant amount of mechanical digestion; **1b.** Because the stomach now holds only a few ounces at a time; **2.** If stomach is overfilled, stomach contents would flow into the esophagus causing heartburn; if chronic, serious problems can result; **3a.** It would have been digested chemically; **3b.** It would not have been digested chemically, carbohydrates arre digested in the mouth and small intestine.

Chapter 9
Check Your Progress
(9.1) 1. nasal cavity→pharynx→glottis→larynx→trachea→bronchus→bronchioles→lungs; **2.** ensure that oxygen enters the body and carbon dioxide leaves the body. **(9.2) 1.** nose—filter, warm, and moisten the air; pharynx—connect nasal and oral cavities to the larynx; larynx—sound production; **2.** nasal cavity, oral cavity, and larynx; **3a.** Air passage and food passage; **3b.** The larynx, which receives air, is in front of the esophagus, which receives food. **(9.3) 1.** trachea—keeps lungs clean by sweeping mucus upwards and connects the larynx to the primary bronchi; bronchial tree—passage of air to the lungs; lungs—site of gas exchange between air in the alveoli and blood in the capillaries; **2.** alveoli. **(9.4) 1a.** During inspiration, the diaphragm contracts expanding the lungs, a partial vacuum is created, and air moves into the lungs. During expiration, the diaphragm relaxes, the lungs recoil and abdominal organs press up against the diaphragm, forcing air out of the lungs; **1b.** expiration; **2a.** A free flow of air from the nose or mouth to the lungs and from the lungs to the nose or mouth is vitally important; **2b.** tidal volume, vital capacity, inspiratory and expiratory reserve volume, and residual volume. **(9.5) 1.** The rhythm of ventilation is controlled by a respiratory control center, located in the medulla oblongata; **2.** Olivia's working cells produced carbon dioxide, which lowered the pH of the blood, stimulating chemoreceptors and the respiratory center that overrode Olivia's voluntary inhibition of respiration. **(9.6) 1.** External respiration refers to the exchange of gases between air in the alveoli and blood in the pulmonary capillaries; internal respiration refers to the exchange of gases between the blood in systemic capillaries and the tissue fluid; **2a.** hemoglobin; **2b.** Takes up oxygen and becomes oxyhemoglobin, takes up carbon dioxide and becomes carbaminohemoglobin. **(9.7) 1a.** strep throat, sinusitis, otitis media, tonsillitis, laryngitis; **1b.** lower respiratory infections, restrictive pulmonary disorders, obstructive pulmonary disorders; **2.** chronic bronchitis, emphysema, lung cancer.

Understanding Key Terms
a. pharynx; **b.** surfactant; **c.** vocal cord; **d.** bicarbonate ion; **e.** expiration

Testing Your Knowledge of the Concepts
10. d; **11.** d; **12.** c; **13.** f; **14.** d; **15.** a; **16.** b; **17.** e; **18.** d; **19.** b; **20.** d; **21.** a; **22.** c; **23.** b; **24.** d; **25.** e; **26.** d; **27.** b; **28. a.** nasal cavity; **b.** nostril; **c.** pharynx; **d.** epiglottis; **e.** glottis; **f.** larynx; **g.** trachea; **h.** bronchus; **i.** bronchiole

Thinking Critically About the Concepts
1. The expression describes the movement of food particles past the epiglottis, through the glottis, and into the trachea; **2.** Holding her breath increased the amount of CO_2 in the blood. Some of which combines with water and forms carbonic acid, which dissociates and froms H^+ and the bicarbonate ion that influence blood pH. This triggers involuntary exhalation; **3a.** Since ciliated cells are damaged from smoking, coughing prevents dust, bacteria, and other airborne contaminants from reaching the

lungs; **3b.** Their ciliated epithelium is fully functioning; **4.** diaphragm and abdominal muscles; **5.** Hyperventilation lowers the level of CO_2 in the blood, which affects the blood pH. Breathing from a bag causes the individual to rebreathe the CO_2 and increases the CO_2, in the blood.

Chapter 10
Check Your Progress

(10.1) 1. kidneys—primary organs of excretion, ureters—conduct urine from the kidneys to the bladder, urinary bladder—stores urine until it is expelled, urethra—small tube that extends from the urinary bladder to an external opening; **2.** excretion of metabolic wastes, maintenance of water-salt balance, maintenance of acid-base balance, and secretion of hormones. **(10.2) 1.** renal cortex, renal medulla, and renal pelvis; **2.** nephrons; **3.** The specific structure of each part is especially suited to a particular function. **(10.3) 1.** glomerular filtration, tubular reabsorption, and tubular secretion; **2.** Due to glomerular blood pressure, water and small blood molecules move from the glomerulus to the inside of the glomerular capsule; tubular reabsorption occurs as molecules and ions are both passively and actively reabsorbed from the nephron into the blood of the peritubular capillary network; certain molecules are then actively secreted from the peritubular capillary network into the convoluted tubules. **(10.4) 1a.** loop of the nephron and the collecting duct; **1b.** Salt is reabsorbed, the solute gradient is established, and water leaves the descending limb of the nephron and the collecting duct because of the osmotic gradient within the renal medulla; **2a.** aldosterone, atrial natriuretic hormone, and antidiuretic hormone; **2b.** Aldosterone promotes the excretion of potassium ions and the reabsorption of sodium ions, leading to a reabsorption of water; atrial natriuretic hormone promotes the excretion of sodium, leading to a decrease in water reabsorption; antidiuretic hormone increases water reabsorption; **3.** If the blood is acidic, hydrogen ions are excreted and bicarbonate ions are reabsorbed; if the blood is basic, hydrogen ions are not excreted and bicarbonate ions are not reabsorbed. **(10.5) 1.** infections, diabetes, hypertension, and inherited conditions cause renal disease, which can be treated using hemodialysis or replacing the kidney. **(10.6) 1.** excrete waste molecules—cleanse blood of nitrogenous wastes and excrete them in urine, water-salt balance—increase or decrease Na^+ and water absorption, acid-base balance—remove a wide range of acidic and basic substances. **2a.** Only kidneys excrete nitrogenous waste, maintain water-salt balance, and sufficiently keep pH constant. **2b.** A single kidney is sufficient to perform the required functions.

Understanding Key Terms
a. diuretic; **b.** excretion; **c.** urethra; **d.** renal pelvis; **e.** glomerular filtrate

Testing Your Knowledge of the Concepts
11. d; **12.** d **13.** c; **14.** a; **15.** c; **16.** b; **17.** c; **18.** c; **19.** a; **20.** c; **21.** d; **22.** c; **23.** a; **24.** b; **25.** b; **26.** c; **27.** d; **28.** b; **29.** a; **30.** d; **31.** f; **32.** a; **33.** c; **34.** d; **35. a.** glomerular capsule; **b.** efferent arteriole; **c.** afferent arteriole; **d.** proximal convoluted tubule; **e.** loop of the nephron; **f.** descending limb; **g.** ascending limb; **h.** peritubular capillaries; **i.** Distal convoluted tubule; **j.** renal vein; **k.** renal artery; **l.** collecting duct

Thinking Critically About the Concepts
1. Tired from low red blood count—medication for chemotherapy patients is a synthetic eythropoetin, hormone—endocrine system, produced by the kidneys—urinary system, that stimulate red blood count synthesis—cardiovascular system, by the bones—skeletal system. Trouble urinating—urinary system, enlarged prostate—reproductive system; **2.** If there are more red blood cells, more ozygen will be delivered to the cells that will use the oxygen to make more ATP; **3.** It calms the smooth muscle of the bladder, which the large intestine also has. When the large intestine's contractions are lessened, constipation may result; **4.** The man is likely to be incontinent; **5a.** The normal number of red blood cells is known, so a higher number would indicate EPO use; **5b.** Too many red blood cells increase the thickness of the blood, which increases the risk of blood clots, heart attack, or stroke.

Chapter 11
Check Your Progress

(11.1) 1a. compact bone is highly organized and composed of tubular units called osteons; **1b.** spongy bone has an unorganized appearance and is composed of trabeculae; **2.** hyaline cartilage—at the ends of long bones, in the nose, at the ends of the ribs, and in the larynx and trachea; fibrocartilage—the disks located between the vertebrae and also in the cartilage of the knee; elastic cartilage—the ear flaps and the epiglottis; **3.** fibrous connective tissue; **4.** The main portion of the bone (diaphysis) is composed of compact bone, the expanded region at each end of a long bone (epiphysis) is composed of spongy bone, and the long bone is covered by a layer of fibrous connective tissue (periosteum). **(11.2) 1.** osteoblasts, osteocytes, and osteoclasts; **2.** Through intramembranous ossification, in which bone develops between sheets of fibrous connective tissue, and endochondral ossification, in which bone replaces a cartilage model; **3.** A vitamin D converted hormone, growth hormone, and thyroid hormone; **4.** allows the body to regulate the amount of calcium in the blood; **5.** A hematoma is formed, next tissue repair begins, and a fibrocartilaginous callus is formed between the ends of the broken bone, then the fibrocartilaginous callus is converted into a bony callus and remodeled. **(11.3) 1.** skull, hyoid bone, vertebral column, rib cage, and the ear ossicles; **2.** The frontal bone forms the forehead, the parietal bones extend to the sides, and the occipital bone curves to form the base of the skull, each temporal bone is located below the parietal bones, the sphenoid bone extends across the floor of the cranium from one side to the other, and the ethmoid bone lies in front of the sphenoid; **3.** mandible forms the lower jaw and chin, maxillae forms the upper jaw and the anterior portion of the hard palate, zygomatic bones are the cheekbone prominences, and the nasal bones form the bridge of the nose; **4a.** cervical vertebrae—located in the neck and allow movement of the head; thoracic vertebrae—form the thoracic curvature and have long, thin, spinous processes and articular facets for the attachment of the ribs; lumbar vertebrae—form the lumbar curvature and have a large body and thick processes; sacral vertebrae—fused together, forming the pelvic curvature; coccyx—fused vertebrae that form the tailbone; **4b.** Structure consists of thoracic vertebrae, the ribs, and the sternum; function is to protect the heart and lungs and to move during inspiration and expiration. **(11.4) 1a.** scapula, clavicle; **1b.** humerus, radius, ulna, carpals, metacarpals, phalanges. **2a.** two coxal bones; **2b.** femur, patella, tibia, fibula, tarsals, metatarsals, phalanges. **(11.5) 1a.** fibrous, cartilaginous, synovial; **1b.** flexion and extension, adduction and abduction, rotation and circumduction, inversion and eversion.

Understanding Key Terms
a. ligament; **b.** compact bone; **c.** sinus; **d.** fontanel; **e.** osteocyte

Testing Your Knowledge of the Concepts
18. c; **19.** d; **20.** a; **21.** g; **22.** f; **23.** e; **24.** e; **25.** a; **26.** b; **27.** c; **28.** f; **29.** d; **30.** b; **31.** b; **32.** a; **33.** b; **34.** c; **35.** e; **36.** c; **37.** c; **38.** b; **39.** d; **40.** c; **41.** c; **42.** F; **43.** T; **44.** F; **45.** T; **46.** T; **47.** b; **48.** a; **49.** e; **50.** c; **51.** d; **52. a.** frontal bone; **b.** zygomatic bone; **c.** maxilla; **d.** mandible; **e.** clavicle; **f.** scapula; **g.** sternum; **h.** ribs; **i.** costal cartilages; **j.** coxal bones; **k.** sacrum; **l.** coccyx; **m.** patella; **n.** metatarsals; **o.** phalanges; **p.** temporal bone; **q.** vertebral column; **r.** humerus; **s.** ulna; **t.** radius; **u.** carpals; **v.** metacarpals; **w.** femur; **x.** fibula; **y.** tibia; **z.** tarsals

Thinking Critically About the Concepts
1. A calcium-rich diet. Also, being a non-smoker contributed to her fast vertebral fusion; **2.** Calcium absorption and osteoblast activity decrease with age, so it is harder for bone cells to get the calcium needed for bone repair; **3a.** low blood calcium; **3b.** Kidney stones and osteoporosis are possible diseases; **4.** Very little space is between the spinal cord and the vertebral foramen, which means the spinal cord is easily crushed or damaged with a cervical vertebrae break; **5a.** Use of EPO increases red blood cell synthesis, which results in an increase in blood oxygen; **5b.** Greater blood oxygen means a greater ability of synthesize ATP, which could give an athelete an advantage to win a competition.

Chapter 12
Check Your Progress
(12.1) 1. smooth, cardiac and skeletal; **2.** make bones move, help maintain a constant body temperature, assist movement in cardiovascular and lymphatic vessels, protect internal organs and stabilize joints; **3.** In opposite pairs, for example, one flexes and the other extends; **4.** See Figure 12.4. **(12.2) 1.** Muscle fiber consists of many myofibrils, myofibril has many sarcomeres, sarcomeres contain two types of protein myofilaments; **2.** Muscle fibers are stimulated to contract by motor neurons whose axons are in nerves.; **3.** ATP supplies the energy for muscle contraction. **(12.3) 1a.** A nerve fiber together with all of the muscle fibers it innervates; **1b.** Because all the muscle fibers in a motor unit are stimulated at once; **2.** Not all motor units contract at the same time; **3a.** creatine phosphate pathway, fermentation, and cellular respiration; **3b.** fast-twitch. **(12.4) 1.** Tendinitis occurs when the smooth gliding motion of a tendon is impaired; bursitis is an inflammation of the bursa.; **2.** Myalgia and fibromyalgia, muscular dystrophy, myasthenia gravis, amyotrophic lateral sclerosis. **(12.5) 1.** body movement and protection of body parts; **2a.** store and release calcium, and production of blood cells; **2b.** provides heat.

Understanding Key Terms
a. sarcomere; **b.** insertion; **c.** tetanus; **d.** rigor mortis; **e.** strain

Testing Your Knowledge of the Concepts
10. c; **11.** b; **12.** a; **13.** c; **14.** d; **15.** e; **16** a; **17.** d; **18.** d; **19.** b; **20.** c; **21.** a; **22.** e; **23.** c; **24.** e; **25.** d; **26.** a; **27.** b; **28.** c; **29.** c; **30.** a; **31.** d; **32.** b; **33.** c; **34.** a; **35.** b; **36.** a; **37.** c; **38.** b; **39.** b; **40.** a; **41.** c; **42. a.** T tubule; **b.** sarcoplasmic reticulum; **c.** myofibril **d.** Z line; **e.** sarcomere **f.** sarcolemma

Thinking Critically About the Concepts
1a. Rigor mortis would take longer to occur; **1b.** Metabolism slows down in the cold, so the reactions that cause rigor mortis will occur more slowly; **2.** Decay begins to occur and causes the body to lose its stiffness; **3a.** body location (i.e., submerged, in a car, in the shade/sun, shallow garve/deep grave), insect activity, and effects of temperature and humidity; **3b.** Decomposition can occur differently depending on the geographical area; **4a.** Meat chilled too quickly or before the onset of rigor mortis is tougher; **4b.** Aged beef becomes more tender; **4c.** They are generally young animals that do not have tough muscle tissue yet.

Chapter 13
Check Your Progress
(13.1) 1. central nervous system and peripheral nervous system; **2a.** sensory neurons, interneurons, and motor neurons; **2b.** cell body, dendrites, and axon; **3a.** action potential traveling along a neuron; **3b.** Exchange of ions generated along the length of an axon; **4.** neurotransmitters. **(13.2) 1.** spinal cord and brain; **2.** provide a means of communication between the brain and the peripheral nerves, the center for reflex actions; **3.** cerebrum—main part of the brain, communicates and coordinates the activities of the other parts of the brain; diencephalon—contains the hypothalamus and thalamus, maintains homeostasis, receives sensory input; cerebellum—sends out motor impulses by way of the brain stem to the skeletal muscles, produces smooth, coordinated voluntary movements; brain stem—contains the midbrain, pons, and medulla oblongata; relay station and medulla had reflex centers. **(13.3) 1.** blend primitive emotions and higher mental functions into a united whole; **2.** amygdala and hippocampus; **3a.** Wernicke's area and Broca's area; **3b.** left hemisphere. **(13.4) 1a.** 12 pairs; **1b.** 31 pairs; **2.** reflex; **3a.** A system that regulates the activity of cardiac and smooth muscles and glands; **3b.** Functions automatically, has two systems that generally cause opposite responses. **(13.5) 1.** Affects the limbic system and either promotes or decreases the action of a particular neurotransmitter.

Understanding Key Terms
a. reflex; **b.** neurotransmitter; **c.** autonomic system; **d.** ganglion; **e.** acetylcholine

Testing Your Knowledge of the Concepts
18. b; **19.** d; **20.** c; **21.** a; **22.** a; **23.** b; **24.** c; **25.** b; **26.** c; **27.** a; **28.** c; **29.** c; **30.** b; **31.** d; **32.** c; **33.a.** central canal, **b.** dorsal horn; **c.** white matter; **d.** dorsal root; **e.** dorsal root ganglion; **f.** spinal nerve; **g.** cell body of motor neuron; **h.** cell body of interneuron.

Thinking Critically About the Concepts
1. Myelin enables the impulse to jump from node to node quickly because the depolarization process occurs only at the nodes of Ranvier. **2a.** triglyceride (fat or oil) **2b.** Unsaturated fatty acids are characterized by one or more double bonds between carbons, while saturated fatty acids are all single bonds. **2c.** Animal fat, such as butter and fatty cuts of meat. **3.** Myelination enables impulses to travel quickly down an axon, which helps motor skills. **4.** By inheriting a recessive allele for the disease from each parent.

Chapter 14
Check Your Progress
(14.1) 1. To detect certain types of stimuli; **2.** chemoreceptors, photoreceptors, mechanoreceptors, and thermoreceptors; **3.** Sensation is the conscious perception of stimuli that occurs after sensory receptors generate a nerve impulse that arrives at the cerebral cortex. **(14.2) 1.** Assist the brain in knowing the position of the limbs in space; **2.** To make the skin sensitive to touch, pressure, pain, and temperature. **(14.3) 1a.** chemoreceptors; **1b.** sweet, sour, salty, bitter, and perhaps umami; **2a.** olfactory cells; **2b.** within olfactory epithelium, high in the roof of the nasal cavity; **3.** Because it activates a different combination of receptor proteins on olfactory cells, thus stimulating different neurons. **(14.4) 1.** cornea, assisted by the lens and the humors; **2.** rod cells, cone cells; **3.** Rods and cones synapse with bipolar cells, which synapse with ganglion cells whose axons become the optic nerve. **(14.5) 1.** tympanic membrane, malleus, incus, stapes; **2a.** mechanoreceptors; **2b.** inner ear; **2c.** sensitive to mechanical stimulation. **(14.6) 1a.** semicircular canals; **1b.** Ampullae of the semicircular canals contain hair cells with stereocilia embedded in a cupula; when the head rotates, the cupula is displaced, bending the stereocilia.; **2a.** utricle and saccule; **2b.** Utricle and saccule contain hair cells with stereocilia embedded in an otolithic membrane; when the head bends, otoliths are displaced causing the stereocilia to bend.

Understanding Key Terms
a. sensory receptor; **b.** retina; **c.** sclera; **d.** chemoreceptor; **e.** spiral organ

Testing Your Knowledge of the Concepts
14. d; **15.** a; **16.** c; **17.** b; **18.** c; **19.** a; **20.** d; **21.** c; **22.** d; **23.** b; **24.** e; **25.** b; **26.** c; **27.** b; **28.** d; **29.** d; **30.** c; **31.** e; **32.** a; **33.** e; **34.** d; **35. a.** retina; **b.** choroid; **c.** sclera; **d.** optic nerve; **e.** fovea centralis; **f.** ciliary body; **g.** lens; **h.** iris; **i.** pupil; **j.** cornea

Thinking Critically About the Concepts
1. Pressoreceptors in certain arteries recognize changes in blood pressure; **2.** changes in glucose levels, blood calcium, and blood oxygen; **3a.** Chemoreceptors (smells and tastes), photoreceptors (tracks on roller coaster), and mechanoreceptors (sounds and spinning of some of the rides, and pressure from seat harnesses); **3b.** cerebrum (interpretation of all stimuli), temporal lobe (smells and sounds), parietal lobe (taste and touch), and occipital lobe (vision); **4.** hearing loss, because loud noises damage the hair cells in the spiral organ.

Chapter 15
Check Your Progress
(15.1) 1. To respond to stimuli, the nervous system uses nerve impulses to produce a quick response; the endocrine system uses hormones deposited in the bloodstream for a response that is slower but lasts longer; **2.** A chemical signal, a means of communication between cells, between body parts, and even between individuals.; **3.** Peptide hormones cause the formation of cyclic AMP, which results in a series of enzymatic reactions, steroid hormones promote the synthesis of a specific enzyme.

(15.2) **1.** Controls the glandular secretions of the pituitary gland; **2.** thyroid-stimulating hormone, adrenocorticotropic hormone, gonadotropic hormone, prolactin hormone, melanocyte-stimulating hormone, and growth hormone. **(15.3) 1a.** triidothyronine and thyroxine; **1b.** increase the metabolic rate; **2a.** parathyroid hormone; **2b.** Causes the blood Ca^{2+} level to increase; **2c.** calcitonin. **(15.4) 1.** mineralocorticoids—regulate salt and water balance, glucocorticoids—regulate carbohydrate, protein and fat metabolism. **(15.5) 1.** insulin—lowers the blood glucose level, glucagon—raises the blood glucose level; **2.** A hormonal disease in which liver cells, and most body cells, are unable to take up glucose as they should. **(15.6) 1a.** testes, ovaries, thymus gland, and pineal gland; **1b.** testes and ovaries; **2.** Kidneys secrete renin and erythropoietin, heart produces atrial natriuretic hormone, adipose tissue produces leptin, tissue cells secrete prostaglandins; **3a.** A hormone that acts where it is produced; **3b.** prostaglandins. **(15.7) 1.** The hormone aldosterone from the adrenal cortex will act on the kidney tubules to conserve Na^+ and water reabsorption will follow; **2.** Hypothalamus acts directly through the nerves of the autonomic system to control *internal* organs and produces hormones released by the posterior pituitary, and produces hormones that control the anterior pituitary.

Understanding Key Terms
a. thyroid gland; **b.** diabetes mellitus; **c.** adrenocorticotropic hormone (ACTH); **d.** peptide hormone; **e.** oxytocin

Testing Your Knowledge of the Concepts
14. d; **15.** b; **16.** b; **17.** d; **18.** c; **19.** c; **20.** b; **21.** e; **22.** e; **23.** d; **24.** d; **25.** d; **26.** b; **27.** c; **28.** d; **29.** a; **30.** d; **31.** e; **32.** c; **33.** b; **34.** f; **35.** b; **36.** c; **37.** a; **38.** e; **39.** d; **40. a.** inhibits , **b.** inhibits, **c.** releasing hormone, **d.** stimulating hormone, **e.** target gland hormone

Thinking Critically About the Concepts
1. Follicle stimulating hormone and growth hormone care both peptide hormones; both will bond to membrane receptors and activate the second messenger, cAMP; **2.** Noses, ears, and feet continue to grow slightly larger as people grow older; **3a.** Could possibly increase muscle mass, improve memory, increase sleep quality, and minimize other effects of aging; **3b.** increase in blood glucose; **4a.** athletes and children of below average height or with a desire of above average heighth; **4b.** Answer may vary; **4c.** Answer may vary; **5.** Spermatogenesis and oogenesis may be affected, possibility of thyroid diseases, or adrenal cortex may be affected.

Chapter 16
Check Your Progress
(16.1) 1. Mitosis is duplication division (number of chromosomes stays the same), meiosis is reduction division (number of chromosomes is reduced); **2a.** 23; **2b.** 23; **3a.** testes; **3b.** ovaries. **(16.2) 1a.** testes; **1b.** seminal vesicles, prostate gland, and bulbourethral gland; **1c.** urethra; **2a.** hypothalamus, anterior pituitary, testes; **2b.** gonadotropin-releasing hormone, follicle-stimulating hormone, luteinizing hormone, testosterone. **(16.3) 1a.** ovary; **1b.** oviducts; **1c.** uterus; **1d.** vagina; **2.** glans clitoris. **(16.4) 1a.** follicle-stimulating hormone, luteinizing hormone; **1b.** estrogen, progesterone; **2a.** Corpeus luteum is maintained in the ovary and produces increasing concentrations of progesterone; progesterone shuts down the hypothalamus and anterior pituitary so that no new follicles begin in the ovary; **2b.** The hormones in birth control pills feedback to inhibit the hypothalamus and the anterior pituitary; therefore, no new follicles begin in the ovary. **(16.5) 1.** abstinence, birth control pills, intrauterine devices, diaphragm, condom, contraceptive implants, contraceptive injections, contraceptive vaccines, vasectomy, tubal ligation, morning-after pills; **2.** artificial insemination by donor, in vitro fertilization, gamete intrafallopian transfer, surrogate mothers, intracytoplasmic sperm injection. **(16.6) 1.** pelvic inflammatory disease; **2.** cancer of the cervix.

Understanding Key Terms
a. ovulation; **b.** progesterone; **c.** semen; **d.** cervix; **e.** acrosome

Testing Your Knowledge of the Concepts
12a. seminal vesicle; **b.** ejaculatory duct; **c.** prostate gland; **d.** bulbourethral gland; **e.** anus; **f.** vas deferens; **g.** epididymis; **h.** testis; **i.** scrotum; **j.** foreskin; **k.** glans penis; **l.** penis; **m.** urethra; **n.** vas deferens; **o.** urinary bladder. Path of sperm: h, g, f, n, b, m; **13.** c; **14.** d; **15.** c; **16.** b; **17.** b; **18.** d; **19.** d; **20.** c; **21.** c; **22.** b; **23.** c; **24.** d; **25.** c; **26.** d; **27.** a; **28.** d; **29.** e; **30.** c; **31.** d; **32.** d; **33.** c

Thinking Critically About the Concepts
1a. Drawing should show the introduction of progesterone early in the cycle, inhibiting FSh and LH; **1b.** Drawing should show that HCH prevents the corpus luteum from regressing, which means progeserone remains high and the uterine lining remains thick; **2.** Birth control pills inhibit FSH, which in turn inhibit follicle development. Fewer follicles may be the cause of a lower risk of ovarian cancer. Fertility drugs increase the number of follicles, which may be the cause of the higher risk of ovarian cancer; **3.** A low body fat composition may prevent the female althete from producing estrogen and progesterone, which prevents the occurrence of a menstrual cycle; **4.** Increases testosterone inhibits GNRH, which inhibits FSH, thereby inhibiting sperm production.

AIDS Supplement
Check Your Progress
(S.1) 1. sub-Saharan Africa; **2.** Russia has a high incidence of HIV infections due to injection drug users and the Caribbean epidemic is fueled by the sex industry. **(S.2) 1.** body produces enough new CD4 T cells to keep pace with the destruction of these cells by HIV; **2.** category A individual—above 500/mm^3, category B individual—between 499/mm^3 and 200 mm^3, category C individual—below 200 mm^3; **3.** The immune system is not able to adequately fight infections when the level of CD4 T cells decreases. **(S.3) 1.** reverse transcriptase—catalyzes reverse transcription, integrase—catalyzes the integration of viral DNA into the DNA of the host cell, protease—catalyzes the breakdown of the newly synthesized viral polypeptides into functional viral proteins; **2.** reverse transcriptase inhibitors (AZT) interfere with reverse transcriptase, integrase inhibitors interfere with integrase and prevent HIV from inserting its genetic material, protease inhibitors prevent protease from cutting up newly created polypeptides; **3.** Because the vaccine would be a preventive measure that helps people who are not yet infected escape infection or slow down the progression of the disease.

Chapter 17
Check Your Progress
(17.1) 1. A sperm makes its way through the corona radiata, acrosome releases digestive enzymes to digest zona pellucida, sperm binds to the egg, their plasma membranes fuse, sperm enters the egg, egg and sperm nuclei fuse. **(17.2) 1.** cleavage, growth, morphogenesis, differentiation; **2a.** Membranes that are outside of the embryo, not part of the embryo and fetus; **2b.** chorion—develops into fetal half of the placenta, allantois—forms umbilical blood vessels, yolk sac—produces blood cells, amnion—contains fluid to cushion and protect embyo; **3a.** Zygote divides repeatedly, develops into a morula, and then a blastocyst; **3b.** Embryo implants itself in the uterine wall, gastrulation occurs and primary germ layers are formed, organ systems appear and develop; **(17.3) 1a.** *chorionic villi of placenta→umbilical vein→venous ducts→inferior vena cava→heart→blood is shunted into the left atrium by way of the oval opening and some enters the aorta by way of the arterial duct→aorta→ umbilical arteries*; **1b.** structures unique to fetus in italics; **2.** skeleton becomes ossified, sex of fetus is distinguishable, fetus grows and gains weight; **3.** The presence of an *SRY* gene, on the Y chromosome, leads to the development of testes and male genitals; otherwise, ovaries and female genitals develop; **(17.4) 1.** placental hormones; **2.** creates a concentration gradient favorable to the flow of carbon dioxide from fetal blood to maternal blood at the placenta; **3.** stage 1—cervix dilates; stage

2—baby is born; stage 3—placenta/afterbirth is delivered. **(17.5) 1.** genetic in origin, whole-body process, extrinsic factors; **2.** skin—becomes thinner and less elastic, homeostatic adjustment to heat is limited; processing and transporting—heart shrinks because of a reduction in cardiac muscle size, arteries become rigid, cardiovascular problems are often accompanied by respiratory disorders, blood supply to kidney and liver is reduced; integration and coordination—reaction time slows, greater stimulation is needed for sense receptors to function, decline in bone density, loss of skeletal muscle mass; reproductive system—females undergo menopause, males undergo andropause; **3.** good health habits.

Understanding Key Terms
a. lanugo; **b.** afterbirth; **c.** gerontology; **d.** fertilization; **e.** cleavage

Testing Your Knowledge of the Concepts
12. c; **13.** e; **14.** b; **15.** c; **16.** b; **17.** a; **18.** b; **19.** d; **20.** d; **21.** b; **22.** c; **23.** d; **24.** d; **25.** c; **26.** a; **27.** e; **28.** e; **29.** d; **30.** e; **31.** e; **32.** d; **33.** b; **34. a.** chorion—fetal portion of the placenta, for nutrient, waste, and gas exchange; **b.** amnion—contains fluid to protect and cushion the embryo; **c.** embryo; **d.** allantois—becomes the umbilical vessels; **e.** yolk sac—first site of blood cell formation; **f.** fetal portion of placenta where fetal blood makes exchanges with maternal blood; **g.** maternal portion of placenta; **h.** umbilical cord—connects the embryo to the placenta; **35. a.** artrial duct; **b.** oval opening; **c.** venus duct; **d.** umbilical vein; **e.** umbilical arteries

Thinking Critically About the Concepts
1. HCG is filtered from the blood by the nephron during glomerular filtration. It is not reabsorbed during tubular reabsorption, so it remains in the tubule and is a urine component; **2a.** HCG is a hormone that is distributed by the cardiovascular system; **2b.** It binds to a membrane receptor and activates the second messenger, cAMP; **3a.** Consume calcium-rich foods and participate in bone stressing exercise; **3b.** Osteoblast activity and calcium absoprtion decline with age, making the building of new bone more difficult; **4a.** FSH; **4b.** decreased; **4c.** FSH stimulates follicle development and estrogen production by the follicle. Without FSH, follicle development doesn't occur and estrogen secretion does not cause the development of the uterine lining that is shed during menstruation.

Chapter 18
Check Your Progress
(18.1) 1. G_1, S, G_2; **2.** G_1:organelles are doubled; S:DNA replicates; G_2:synthesis of proteins needed for cell division. **(18.2) 1.** They are the same; **2.** prophase—chromosomes attach to the spindle, metaphase—chromosomes align at the equator, anaphase—chromatids separate and chromosomes move toward poles, telophase—nuclear envelopes form around chromosomes; **3.** During cytokinesis, a contractile ring pinches the cell in two. **(18.3) 1.** Each daughter cell contains half as many chromosomes as the parent cell; **2.** In prophase I, crossing-over occurs, in metaphase I, independent alignment occurs; **3.** meiosis I—homologous chromosomes pair and then separate, meiosis II—sister chromatids separate, resulting in four cells with a haploid number of chromosomes. **(18.4) 1.** b; **2.** a; **3.** a; **4a.** a; **4b.** b; **5.** a; **6.** b; **7.** d; **8.** b; **9.** a. **(18.5) 1.** nondisjunction; **2.** trisomy 21; **3.** Turner syndrome, Klinefelter syndrome, poly-X females, Jacobs syndrome; **4.** changes in chromosome structure, including deletions, duplications, inversions and translocations.

Understanding Key Terms
a. Barr body; **b.** homologous chromosome; **c.** monosomy; **d.** inversion; **e.** haploid

Testing Your Knowledge of the Concepts
12. b; **13. a.** G_1 phase—cells grow, organelles double; **b.** S phase—DNA synthesis; **c.** G_2 phase—cell prepares to divide; **d.** mitosis—nuclear division and cytokinesis; **14.** d; **15.** b; **16.** a; **17.** d; **18.** e; **19.** c; **20.** c; **21.** a; **22.** d; **23.** a; **24.** c; **25.** c; **26.** d; **27.** c; **28.** b; **29.** e; **30.** a; **31.** d; **32.** b; **33.** c; **34.** d; **35.** c; **36.** d; **37.** a; **38.** b

Thinking Critically About the Concepts
1. Neutrophils and macrophages; **2.** Homologous chromosomes separate randomly. Siblings that look similar inherited similar homologous chromosomes; siblings that are dissimilar inherited different combinations of homologous chromosomes; **3a.** They do not undergo the phases of mitosis to form new cells that would replace the damaged cells; **3b.** They are not likley to recover fully.

Chapter 19
Check Your Progress
(19.1) 1. do not undergo apoptosis, have unlimited replicative potential due to telomerase, do not exhibit contact inhibition, have no need for growth factors, undergo angiogenesis and metastasis; **2.** proto-oncogenes, tumor-suppressor genes; **3.** lung cancer, colorectal cancer, prostate cancer, breast cancer. **(19.2) 1.** DNA-linkage studies have revealed breast cancer genes (*BRCA1* and *BRCA2*), a tumor suppressor gene has been associated with retinoblastoma, and an abnormal RET gene predisposes and individual to thyroid cancer; **2.** ionizing radiation, tobacco smoke, pollutants, certain viruses; **3.** good nutrition, and, for example, not smoking. **(19.3) 1.** self-examination, Pap test, mammography, CAT scan, MRI, radioactive scan, ultrasound, biopsy; **2.** tumor marker tests and genetic tests. **(19.4) 1.** surgery, radiation therapy, chemotherapy; **2a.** cancer vaccines and immune cells genetically engineered to bear the tumor's antigen; **2b.** monoclonal antibodies; **3.** raise the level of *p53* in all cells and use an adenovirus to kill cells that lack a *p53* gene; **4.** Antiangiogenic drugs confine and reduce tumors by breaking up the network of new capillaries in the vicinity of a tumor.

Understanding Key Terms
a. telomere; **b.** carcinogen; **c.** proto-oncogene; **d.** angiogenesis; **e.** metastasis

Testing Your Knowledge of the Concepts
10. d; **11.** d; **12.** c; **13.** c; **14.** c; **15.** b; **16.** d; **17.** a; **18.** c; **19.** e; **20.** b; **21.** b; **22.** c; **23.** b; **24.** c; **25.** d; **26.** d; **27.** d; **28.** a; **29.** d; **30.** a; **31.** b

Thinking Critically About the Concepts
1a. A number of chemicals in tobacco smoke are know mutagens. They may cause protooncogenes to become oncogenes that cause cancer; **1b.** Many pollutants and viral infections cause cancer. Her dietary choices may have contributed to the occurrence of cancer, as well; **2.** The use of a condom during sexual intercourse, which helps prevent several other sexually transmitted diseases; **3a.** Women already sexually active may have exposed to HPV and won't gain anything from vaccination, young girls would; **3b.** Active immunity, since the vaccine contains an HPV antigen that stimulates the immune system response; **4.** The lymphatic system is a transport system, similar to the cardiovascular system; cancerous cells in the lymphatic system have a chance to travel throughout the body and begin developing in other organs or tissue; **5.** Don't smoke and avoid second-hand smoke; minimize your exposure to sun and wear sunscreen; eat high-fiber foods, foods rich in vitamins A and C, and vegetables form the cabbage family; Avoid radiation; test for cancer, such as self-breast exams and mammagrams for women and colonoscopies for older men and women.

Chapter 20
Check Your Progress
(20.1) 1. Genotype refers to the genes of an individual; phenotype is a characteristic of an individual.; **2.** homozygous dominant (*EE*) and heterozygous (*Ee*) give the dominant phenotype, homozygous recessive (*ee*) gives the recessive phenotype. **(20.2a) 1a.** W; **1b.** WS, Ws; **1c.** T, t; **1d.** Tg, tg; **1e.** AB, Ab, aB, ab; **2a.** gamete; **2b.** genotype; **2c.** gamete; **2d.** genotype; **3a.** EeSs; **3b.** eeSS. **(20.2b) 1.** Ww; **2.** 75% or 3:1; **3.** Ee; **4.** father-DD, mother-dd, children-Dd. **(20.2c) 1.** EeWw; **2.** EeWw; **3.** ¹⁄₁₆; **4.** father-DdFf, mother-ddff, child-ddff. **(20.2d) 1a.** aa; **1b.** A?, the questions mark means that each could be *AA* or *Aa*; **2.** parents are

heterozygous-*Aa*, child is homozygous recessive-*aa*; **3.** 0%; **4.** abnormal hemoglobin stacks up; **5.** Marfan syndrome is dominant. **(20.3)** **1.** *AABB* (very dark), *aabb* (very light); **2.** cleft lip, club-foot, congenital hip dislocation, hypertension, diabetes, schizophrenia, allergies, cancer, and behavioral traits; **3.** *Aabb* or *aaBb* (light); **4.** familial hypercholesterolemia; **5.** codominance—I^A and I^B are fully expressed in the presence of the other; multiple allele inheritance—three alleles for the same gene exist; **6.** child-*ii*, mother-*I^Ai*, father-*I^Ai*, *I^Bi*, *ii*. **(20.4)** **1.** Color-blindness is a recessive sex-linked genetic disorder, a female must receive two recessive alleles before she receives the condition, the male must only receive the X-linked recessive allele; **2a.** girls: 50% recessive phenotype, 50% dominant phenotype, boys: 50% dominant phenotype, 50% recessive phenotype; **2b.** girls: 100% dominant phenotype, boys: 50% dominant phenotype, 50% recessive phenotype; **3.** mother, mother-$X^H X^h$, father-$X^H Y$, son-$X^h Y$; **4.** 100%, 0%, 100%; **5.** color blind, husband is not the father.

Figure Questions
Fig. 20.8, page 429: The individual does not have the disorder but has a child with the disorder. Therefore, the individual has a recessive allele; **Fig. 20.9,** page 429: The individual has the disorder and has a child without the disoder. Therefore, the individual has a recessive allele; **Fig. 20.16,** page 435: Type O.

Understanding Key Terms
a. Punnett square; **b.** allele; **c.** dominant allele; **d.** locus; **e.** genotype

Testing Your Knowledge of the Concepts
14. e; **15.** c; **16.** d; **17.** b; **18.** c; **19.** b; **20.** a; **21.** a; **22.** a; **23.** c; **24.** d; **25.** d; **26.** a; **27.** e; **28.** c; **29.** c; **30.** c

Thinking Critically About the Concepts
1. No, her mother has type O blood (homozygous recessive for blood type); her father has type A blood (homozygous dominate for A type or heterozygous). Neither parent has an allele for B type blood, so together they are unable to have a child with B or AB blood; **2a.** Since she is a carrier for color blindness, she could contribute; the recessive allele to her male offspring; **2b.** No, since females inherit two X chromosomes, at most, a daughter would inherit only one recessive allele from the mother. The father would contribute the dominant allele (normal vision); **3.** Full siblings are most likely to have similar alleles to the person needing a transplant, since the same two parents contributed alleles to the siblings.

Chapter 21
Check Your Progress
(21.1) 1. The double-stranded structure of DNA allows each original strand to serve as a template for the formation of a complementary new strand; **2.** Both RNA and DNA are polynucleotides, RNA contains the sugar ribose, the base U instead of T, and is single stranded; **3.** ribosomal RNA, messenger RNA, transfer RNA. **(21.2) 1a.** chain of amino acid subunits with a specific three-dimensional shape; **1b.** enzymes; **2.** transcription forms an mRNA, translation synthesizes a polypeptide; **3a.** transcriptional control, posttranscriptional control, translational control, posttranslational control; **3b.** The organization of the chromatin and the use of transcription factors. **(21.3) 1.** genomics-the study of genomes, proteomics-the study of the structure, function, and interaction of cellular proteins; **2.** ex vivo gene therapy and in vivo gene therapy. **(21.4) 1.** recombinant DNA technology or polymerase chain reaction; **2a.** By detecting how many times a short sequence is repeated. **2b.** DNA fingerprinting can be used to identify a virus, a mutated gene, criminals, or remains of bodies; **3.** Can provide resistance to insect predation and herbicides, allow plants to be salt tolerant, improve the food quality of crops, increase productivity, larger animals, production of therapeutic and diagnostic proteins in animal milk, research projects.

Understanding Key Terms
a. polyribosome; **b.** transgenic organism; **c.** template; **d.** proteomics; **e.** vector

Testing Your Knowledge of the Concepts
11. e; **12.** c; **13.** e; **14.** d; **15.** b; **16.** b; **17.** d; **18.** d; **19.** c; **20.** c; **21.** e; **22.** a. ACU'CCU'GAA'UGC'AAA; **b.** UGA'GGA'CUU'ACG'UUU; **c.** thr-pro-glu-cys-lys

Thinking Critically About the Concepts
1a. The immune system responds to foreign recognition proteins (antigens) by attacking them. Organ transplant patients take immunosuppressive drugs to prevent organ rejection; **1b.** Different species have unique recognition proteins. All human cells have recognition proteins that identify them as human cells; **1c.** Specific immunity; **2.** Pigs are prolific reproducers; **3.** Organ transplant, since an organ is composed of multiple tissues and cells with foreign recognition proteins, risk for rejection is higher; **4.** The DNA that codes for the human recognition proteins must be transcribed and translated by the pigs' cells.

Chapter 22
Check Your Progress
(22.1) 1. chemical; **2.** Stanley Miller experiment; **3.** RNA can sometimes be both a substrate and an enzyme during RNA processing. **(22.2) 1a.** change in a population or species over time; **1b.** descent from a common ancestor and adaptation to the environment; **2.** fossil, biogeographical, and anatomical; **3.** natural selecion. **(22.3) 1.** Domain Eukarya, Kingdom Animalia, Phylum Chordata, Class Mammalia, Order Primates, Family Hominidae, Genus *Homo*, Species *Homo sapiens*; **2.** mobile limbs, grasping hands, flattened face, binocular vision, large complex brain, reduced reproductive rate; **3.** human skull is in the midline of the body, human spine is S-shaped, human pelvis is broader, human femur has a longer neck, human toe is not opposable. **(22.4) 1.** bipedal posture, shape of face, and size of brain; **2.** gracile-*A. africanus, A. afarensis*; robust-*A. robustus, A. aethiopicus, A. boisei*. **(22.5a) 1.** *Australopithecus afarensis* (Lucy) **2.** *Homo habilis* because he made tools. **3.** *Homo ergaster*. **(22.5b) 1.** larger brain size, smaller teeth, culture; **2.** larger brain, flatter face, used fire, fashioned advanced tools; **3a.** *H. sapiens* evolved from *H. ergaster* only in Africa, and thereafter *H. sapiens* migrated to Europe and Asia; **3b.** suggests that we all descended from a few individuals and are more genetically similar; **4.** a thoroughly modern appearance, advanced stone tools, possibly first with a language, culture included art.

Understanding Key Terms
a. natural selection; **b.** adaptation; **c.** homologous structure; **d.** chemical evolution; **e.** fossil

Testing Your Knowledge of the Concepts
11. c; **12.** b; **13.** b; **14.** a; **15.** e; **16.** e; **17.** e; **18.** e; **19.** a; **20.** b; **21.** b; **22.** d; **23.** c; **24.** b; **25.** e; **26.** b; **27. a.** chordata; **b.** class; **c.** order; **d.** hominidae; **e.** genus; **f.** *Homo sapiens*; **28.** b; **29.** a; **30.** d; **31.** c

Thinking Critically About the Concepts
1. Food "hidden" in boxes; challenges and games for the chimps; elephant painting; **2.** Since lions and tigers are hunters naturally, they might be given the chance to hunt for their food; **3.** A species with limited genetic diversity is much more vulnerable to extinction. Planned breeding should increase the genetic diversity of a species and prevent inbreeding, which tends to cause the expression of undesirable genes; **4.** Behvaior is controlled, in part, by genes. Animals are related to each other evolutionarily, so they chare common genes. Some of the common genes may stimulate game playing and/or tool use.

Chapter 23
Check Your Progress
(23.1) 1. Because it is a place where organisms interact among themselves and with the physical and chemical environment; **2a.** Autotrophs require only inorganic nutrients and energy to produce food; **2b.** Heterotrophs need a source of organic nutrients and they must

consume food; **3.** herbivores, carnivores, omnivores, detritus feeders; **4.** energy flow—occurs because, as nutrients pass from one population to another, all the energy is eventually converted into heat; chemical cycling—inorganic nutrients are returned to the producers from the atmosphere or soil. **(23.2) 1.** food web; **2.** Grazing food web begins with trees and grass, detrital food web begins with detritus, more energy may be found funneling through a detrital food web; **3.** food chain; **4.** ecological pyramid. **(23.3) 1.** reservoir, exchange pool, biotic community; **2a.** carbon, nitrogen; **2b.** phosphorus cycle; **3.** convert nitrogen gas to ammonium; **4.** ground water shortage, global warming, acid deposition, culturual eutrophication.

Understanding Key Terms
a. omnivore; **b.** trophic level; **c.** fossil fuel; **d.** nitrogen fixation; **e.** producer

Testing Your Knowledge of the Concepts
10. d; **11.** a; **12.** b; **13.** c; **14.** c; **15.** b; **16.** c; **17.** c; **18.** b; **19.** e; **20.** e; **21.** c; **22.** b; **23.** b, c; **24.** c, d; **25.** a; **26.** f; **27.** e; **28.** f; **29.** b, c, d; **30.** a, b, c; **31.** d; **32. a.** producers; **b.** consumers; **c.** inorganic nutrient pool; **d.** decomposers

Thinking Critically About the Concepts
1. Fungicides kill fungi that decompose dead/discarded tissues and pesticides kill many insects that are detritivores that recycle dead/discarded tissues. The nutrients in the dead/discarded tissues will not be released without the activities of the fungi and insects causing adverse effects to the cycles; **2.** temperature and humidity; **3.** Answers will vary; may include composting organic waste and recycling newspapers, plastic, and aluminum; **4.** Deserts have very low primary productivity—number and variety of species able to survive in a desert ecosystem are few. It is likely that more species will become endangered and/or extinct as more ecosystems are converted into deserts.

Chapter 24
Check Your Progress
(24.1) 1. LDCs; **2.** Because of its past exponential growth. **(24.2a) 1.** land, water, food, energy, and minerals; **2.** beaches, semiarid lands, tropical rain forest; **3a.** damming rivers and withdrawing water from aquifers; **3b.** land subsidence and saltwater intrusion; **4a.** planting few genetic varieties, heavy use of fertilizers, generous irrigation, excessive fuel consumption; **4b.** A single parasite can cause devastation of a crop, agricultural runoff, water shortage, soil loss, salinization. **(24.2b) 1.** The consumption of nonrenewable energy supplies results in environmental degradation; **2.** Strip

mining minerals results in land devoid of vegetation, allowing rain to wash toxic waste deposits into nearby streams and rivers; **3.** heavy metals, synthetic organic compounds, and raw sewage. **(24.3) 1.** habitat loss, alien species, pollution, overexploitation, and disease; **2.** direct value—medicinal value, agricultural value, consumptive use value; indirect value—waste disposal, provision of fresh water, prevention of soil erosion, biogeochemical cycles, regulation of climate, ecotourism. **(24.4) 1.** large portion of land used for human purposes, agriculture uses large amounts of nonrenewable energy and creates pollution, more freshwater is used in agriculture than used in homes, almost half of the agriculture yield goes toward feeding animals, decrease in surface water, use of nonrenewable fossil energy, expansion of the population into all regions of the planet; **2.** To make rural areas sustainable: plant cover crops, plant multiuse crops, use low flow irrigation, use precision farming to reduce habitat destruction, use integrated pest management, plant multipurpose trees, restore wetlands, use renewable forms of energy. To make urban areas sustainable: use energy-efficient modes of transport, use solar or geothermal energy to heat buildings, utilize green roofs, improve storm-water management, plant native grasses for lawns, create greenbelts, revitalize old sections of cities, use more efficient light fixtures, recycle business equipment, use low maintenance building materials. **3.** ISEW, GPI, and additional ways ecological economists are in the process of developing.

Understanding Key Terms
a. sustainable; **b.** carrying capacity; **c.** greenhouse gases; **d.** biological magnification; **e.** aquifer

Testing Your Knowledge of the Concepts
9. a; **10.** c; **11.** a; **12.** d; **13.** d; **14.** a; **15.** c; **16.** c; **17.** b; **18.** d; **19.** d; **20.** a; **21.** c,d,e,f; **22.** a,b; **23.** a,b,c; **24.** c,d,e,f; **25.** b; **26.** a; **27.** c; **28.** c; **29.** e; **30.** e; **31.** b; **32.** e; **33. a.** habitat loss; **b.** alien species; **c.** pollution; **d.** overexploitation; **e.** disease

Thinking Critically About the Concepts
1a. MDC; **1b.** There may not be enough young people to support the aging population; **2.** Less energy is needed to grow plant material than animals for human consumption; **3a.** Answers will vary; **3b.** Answers will vary. Example: drive less or drive an energy efficient vehicle, use alternative means of heating/cooling a home, purchase locally grown food; **4.** Most items in a landfill are not exposed to decomposers, so the nutrients to not cycle back; **5a.** Anwers will vary; **5b.** Anwers will vary. Examples: recycling programs, ride sharing/bus use, water/electricity use, events associated with Earth Day.

Glossary

A

absorption Taking in of substances by cells or membranes. 145

acetylcholine (ACh) (uh-seet-ul-koh-leen) Neurotransmitter active in both the peripheral and central nervous systems. 252

acetylcholinesterase (AChE) (uh-seet-ul-koh-luh-nes-tuh-rays) Enzyme that breaks down acetylcholine bound to postsynaptic receptors within a synapse. 252

acid Molecules tending to raise the hydrogen ion concentration in a solution and to lower its pH numerically. 26

acidosis Excessive accumulation of acids in body fluids. 199

acid deposition The return to Earth in rain or snow of sulfate or nitrate salts of acids produced by commercial and industrial activities. 503

acne Inflammation of sebaceous glands. 73

acquired immunodeficiency syndrome (AIDS) (im-yuh-noh-dih-fish-un-see) Disease caused by HIV and transmitted via body fluids; characterized by failure of the immune system. 344

acromegaly (ak-roh-meg-uh-lee) Condition resulting from an increase in growth hormone production after adult height has been achieved. 302

acrosome (ak-ruh-sohm) Cap at the anterior end of a sperm that partially covers the nucleus and contains enzymes that help the sperm penetrate the egg. 323

actin (ak-tin) One of two major proteins of muscle; makes up thin filaments in myofibrils of muscle fibers. See myosin. 232

actin filament Cytoskeletal filaments of eukaryotic cells composed of the protein actin; also refers to the thin filaments of muscle cells. 51

action potential Electrochemical changes that take place across the axomembrane; the nerve impulse. 250

active immunity Reistance to disease due to the immune system's response to a microorganism or a vaccine. 136

active site Region on the surface of an enzyme where the substrate binds and where the reaction occurs. 53

active transport Use of a plasma membrane carrier protein and energy to move a substance into or out of a cell from lower to higher concentration. 48

acute bronchitis (brahn-ky-tis) Infection of the primary and secondary bronchi. 180

adaptation Organism's modification in structure, function, or behavior suitable to the environment. 5, 470

Addison disease Condition resulting from a deficiency of adrenal cortex hormones; characterized by low blood glucose, weight loss, and weakness. 307

adenine (A) (ad-uh-neen) One of four nitrogen bases in nucleotides composing the structure of DNA and RNA. 35

adhesion junction Junction between cells in which the adjacent plasma membranes do not touch but are held together by intercellular filaments attached to buttonlike thickenings. 70

adipose tissue (ah-duh-pohs) Connective tissue in which fat is stored. 63

ADP (adenosine diphosphate) (ah-den-ah-seen dy-fahs-fayt) Nucleotide with two phosphate groups that can accept another phosphate group and become ATP. 36

adrenal cortex (uh-dree-nul kor-teks) Outer portion of the adrenal gland; secretes mineralocorticoids, such as aldosterone, and glucocorticoids, such as cortisol. 305

adrenal gland (uh-dree-nul) An endocrine gland that lies atop a kidney; consisting of the inner adrenal medulla and the outer adrenal cortex. 305

adrenal medulla (uh-dree-nul muh-dul-uh) Inner portion of the adrenal gland; secretes the hormones epinephrine and norepinephrine. 305

adrenocorticotropic hormone (ACTH) (uh-dree-noh-kawrt-ih-koh-troh-pik) Hormone secreted by the anterior lobe of the pituitary gland that stimulates activity in the adrenal cortex. 300

aerobic Requiring oxygen. 55

afterbirth Placenta and the extraembryonic membranes, which are delivered (expelled) during the third stage of parturition. 369

agglutination (uh-gloot-un-ay-shun) Clumping of red blood cells due to a reaction between antigens on red blood cell plasma membranes and antibodies in the plasma. 114

aging Progressive changes over time, leading to loss of physiologic function and eventual death. 370

agranular leukocyte White blood cell that does not contain distinctive granules. 110

agricultural runoff Water from precipitation and irrigation that flows over fields into bodies of water or aquifers. 240

albumin (al-byoo-mun) Plasma protein of the blood having transport and osmotic functions. 107

aldosterone (al-dahs-tuh-rohn) Hormone secreted by the adrenal cortex that decreases sodium and increases potassium excretion; raises blood volume and pressure. 197, 306

alien species Nonnative species that migrate or are introduced by humans into a new ecosystem; also called exotics. 524

alkalosis Excessive accumulation of bases in body fluids. 199

allantois (uh-lan-toh-is) Extraembryonic membrane that contributes to the formation of umbilical blood vessels in humans. 355

allele (uh-leel) Alternative form of a gene; alleles occur at the same locus on homologous chromosomes. 422

allergen (al-ur-jun) Foreign substance capable of stimulating an allergic response. 138

allergy Immune response to substances that usually are not recognized as foreign. 138

all-or-none law Law that states that muscle fibers either contract maximally or not at all, and that neurons either conduct a nerve impulse completely or not at all. 236

alveolus (pl., alveoli) (al-vee-uh-lus) Air sac of a lung. 173

Alzheimer disease (AD) Brain disorder characterized by a general loss of mental abilities. 266

amino acid Organic molecule having an amino group and an acid group, which covalently bonds to produce peptide molecules. 33

amnion (am-nee-ahn) Extraembryonic membrane that forms an enclosing, fluid-filled sac. 355

ampulla (am-pool-uh, -pul-uh) Base of a semicircular canal in the inner ear. 291

amygdala (uh-mig-duh-luh) Portion of the limbic system that functions to add emotional overtones to memories. 260

amylotrophic lateral sclerosis (ALS) Chronic, progressive motor neuron disease characterized by the gradual degeneration of the nerve cells resulting in death. 518

anabolic steroid (a-nuh-bahl-ik) Synthetic steroid that mimics the effect of testosterone. 311

anaerobic Growing or metabolizing in the absence of oxygen. 54

analogous structure Structure that has a similar function in separate lineages but differs in anatomy and ancestry. 472

anaphase Mitotic phase during which daughter chromosomes move toward the poles of the spindle. 382

anaphylactic shock Severe systemic form of allergic reaction involving bronchiolar contriction, impaired breathing, vasodilation, and a rapid drop in blood pressure with a threat of circulatory failure. 138

androgen (an-druh-jun) Male sex hormone (e.g., testosterone). 311

anemia (uh-nee-mee-uh) Inefficiency in the oxygen-carrying ability of blood due to a shortage of hemoglobin. 109

aneurysm Saclike expansion of a blood vessel wall. 97

angina pectoris (an-jy-nuh pek-tuh-ris) Condition characterized by thoracic pain resulting from occluded coronary arteries; precedes a heart attack. 97

angiogenesis (an-jee-oh-jen-uh-sis) Formation of new blood vessels; one mechanism by which cancer spreads. 405

angioplasty (an-jee-uh-plas-tee) Surgical procedure for treating clogged arteries, in which a plastic tube is threaded through a major blood vessel toward the heart and then a balloon at the end of the tube is inflated, forcing open the vessel. A stent is then placed in the vessel. 100

anorexia nervosa (a-nuh-rek-see-uh nur-voh-suh) Eating disorder characterized by a morbid fear of gaining weight. 165

anterior pituitary (pih-too-ih-tair-ee) Portion of the pituitary gland that is controlled by the hypothalamus and produces six types of hormones, some of which control other endocrine glands. 300

anthropoid Group of primates that includes monkeys, apes, and humans. 476

antibody (an-tih-bahd-ee) Protein produced in response to the presence of an antigen; each antibody combines with a specific antigen. 110

antibody-mediated immunity Specific mechanism of defense in which plasma cells derived from B cells produce antibodies that combine with antigens. 131

antibody titer Amount of antibody present in a sample of blood serum. 136

anticodon (an-tih-koh-dahn) Three-base sequence in a tRNA molecule base that pairs with a complementary codon in mRNA. 450

antidiuretic hormone (ADH) (an-tih-dy-uh-ret-ik) Hormone secreted by the posterior pituitary that increases the permeability of the collecting ducts in a kidney. 198, 300

antigen (an-tih-jun) Foreign substance, usually a protein or a polysaccharide, that stimulates the immune system to produce antibodies. 110, 130

antigen-presenting cell (APC) Cell that displays the antigen to the cells of the immune system so they can defend the body against that particular antigen. 134

anus Outlet of the digestive tract. 154

aorta (ay-or-tuh) Major systemic artery that receives blood from the left ventricle. 94

apoptosis (ap-uh-toh-sis, -ahp-) Programmed cell death involving a cascade of specific cellular events leading to death and destruction of the cell. 131, 379, 404

appendicular skeleton (ap-un-dik-yuh-lur) Portion of the skeleton forming the pectoral girdles and upper extremities and the pelvic girdle and lower extremities. 220

appendix In humans, small, tubular appendage that extends outward from the cecum of the large intestine. 145

aquaporin Protein membrane channel through which water can diffuse. 197

aqueous humor (ay-kwee-us, ak-wee-) Clear, watery fluid between the cornea and lens of the eye. 280

aquifer (ahk-wuh-fur) Rock layers that contain water that is released in appreciable quantities to wells or springs. 499, 516

arteriole (ar-teer-ee-ohl) Vessel that takes blood from an artery to capillaries. 87

articular cartilage (ar-tik-yuh-lur) Hyaline cartilaginous covering over the articulating surface of the bones of synovial joints. 208

assimilation Action of chemically changing absorbed substances. 503

association area One of several regions of the cerebral cortex related to memory, reasoning, judgment, and emotional feelings. 257

aster Short, radiating fibers about the centrioles at the poles of a spindle. 381

asthma (az-muh, as-) Condition in which bronchioles constrict and cause difficulty in breathing. 138, 182

astigmatism (uh-stig-muh-tiz-um) Blurred vision due to an irregular curvature of the cornea or the lens. 284

atherosclerosis (ath-uh-roh-skluh-roh-sis) Condition in which fatty substances accumulate abnormally beneath the inner linings of the arteries. 97

atom Smallest particle of an element that displays the properties of the element. 2, 20

atomic mass Mass of an atom equal to the number of protons plus the number of neutrons with the nucleus. 21

atomic number Number of protons within the nucleus of an atom. 20

ATP (adenosine triphosphate) (uh-den-uh-seen try-fahs-fayt) Nucleotide with three phosphate groups. The breakdown of ATP into ADP + ⓟ makes energy available for energy-requiring processes in cells. 36

atrial natriuretic hormone (ANH) (ay-tree-ul nay-tree-yoo-ret-ik) Hormone secreted by the heart that increases sodium excretion and, therefore, lowers blood volume and pressure. 198, 306

atrioventricular (AV) bundle (ay-tree-oh-ven-trik-yuh-lur) Group of specialized fibers that conduct impulses from the atrioventricular node to the ventricles of the heart; also called AV bundle. 90

atrioventricular (AV) valve Valve located between the atrium and the ventricle. 88

atrium (ay-tree-um) One of the upper chambers of the heart, either the left atrium or the right atrium, that receives blood. 88

auditory canal Curved tube extending from the pinna to the tympanic membrane. 286

auditory (Eustachian) tube Extension from the middle ear to the nasopharynx that equalizes air pressure on the eardrum. 171, 286

australopithecine (aw-stray-loh-pith-uh-syn) Any of the first evolved hominids; classified into several species of *Australopithecus*. 480

autoimmune disease Disease that results when the immune system mistakenly attacks the body's own tissues. 139

autonomic system (aw-tuh-nahm-ik) Branch of the peripheral nervous system that has control over the internal organs; consists of the sympathetic and parasympathetic systems. 265

autosome (aw-tuh-sohm) Any chromosome other than the sex chromosomes. 436

autotroph Organism that can capture energy and synthesize organic nutrients from inorganic nutrients. 494

AV (atrioventricular) node Small region of neuromuscular tissue that transmits impulses received from the sinoatrial node to the ventricles. 90

axial skeleton (ak-see-ul) Portion of the skeleton that supports and protects the organs of the head, the neck, and the trunk. 216

axon (ak-sahn) Elongated portion of a neuron that conducts nerve impulses typically from the cell body to the synapse. 249

axon terminal Small swelling at the tip of one of many endings of the axon. 252

B

bacteria One of three domains of life; prokaryotic cells other than archaea with unique genetic, biochemical, and physiological characteristics. 122

Barr body Dark-staining body (discovered by M. Barr) in the nuclei of female mammals that contains a condensed, inactive X chromosome. 393

basal body Cytoplasmic structure that is located at the base of—and may organize—cilia or flagella. 51

base Molecules tending to lower the hydrogen ion concentration in a solution and raise the pH numerically. 26

basement membrane Layer of nonliving material that anchors epithelial tissue to underlying connective tissue. 68

basophil (bay-suh-fil) White blood cell with a granular cytoplasm; able to be stained with a basic dye. 110

B cell (B lymphocyte) Lymphocyte that matures in the bone marrow and, when stimulated by the presence of a specific antigen, gives rise to antibody-producing plasma cells. 130

bicarbonate ion Ion that participates in buffering the blood; the form in which carbon dioxide is transported in the bloodstream.   178

bile Secretion of the liver that is temporarily stored and concentrated in the gallbladder before being released into the small intestine, where it emulsifies fat.   150, 152

binge-eating disorder Condition characterized by overeating episodes that are not followed by purging.   165

binomial name Two-part scientific name of an organism. The first part designates the genus, the second part the specific epithet.   475

biodiversity Total number of species, the variability of their genes, and the communities in which they live.   7, 524

biogeochemical cycle (by-oh-jee-oh-kem-ih-kul) Circulating pathway of elements such as carbon and nitrogen, involving exchange pools, storage areas, and biotic communities.   498

biogeography Study of the geographical distribution of organisms.   472

bioinformatics Computer technologies used to study the genome.   455

biological evolution Change in life-forms that has taken place in the past and will take place in the future; includes descent from a common ancestor and adaptation to the environment.   470

biological magnification Process by which substances become more concentrated in organisms in the higher trophic levels of a food web.   505, 523

biology Scientific study of life.   2

biomass The number of organisms multiplied by their weight.   497

biosphere (by-oh-sfeer) Zone of air, land, and water at the surface of the Earth in which living organisms are found.   2, 493

biotechnology product Product created by using biotechnology techniques.   487

biotic potential Maximum reproductive rate of an organism, given unlimited resources and ideal environmental conditions. Compare with environmental resistance.   512

bipedal posture Ability to walk upright on two feet.   479

birth control method Prevents either fertilization or implantation of an embryo in the uterine lining.   331

birth control pill Oral contraceptive containing estrogen and progesterone.   331

blastocyst (blas-tuh-sist) Early stage of human embryonic development that consists of a hollow, fluid-filled ball of cells.   356

blind spot Region of the retina lacking rods or cones where the optic nerve leaves the eye.   283

blood Type of connective tissue in which cells are separated by a liquid called plasma.   64

blood doping Practice of boosting the number of red blood cells in the blood in order to enhance athletic performance.   109

blood pressure Force of blood pushing against the inside wall of a vessel.   92

blood transfusion Introduction of whole blood or a blood component directly into the bloodstream.   114

body mass index (BMI) Calculation used to determine whether or not a person is overweight or obese.   156

bolus Small lump of food that has been chewed and swallowed.   147

bone Connective tissue having protein fibers and a hard matrix of inorganic salts, notably calcium salts.   63

bone marrow transplant A cancer patient's stem cells are harvested and stored before chemotherapy beings. Then, the stored cells are returned to the patient by injection.   415

bone remodeling Ongoing mineral deposits and withdrawals from bone that adjust bone strength and maintain levels of calcium and phophorus in blood.   212

brain Enlarged superior portion of the central nervous system located in the cranial cavity of the skull.   256

brain stem Portion of the brain consisting of the medulla oblongata, pons, and midbrain.   259

Braxton Hicks contraction Strong, late-term uterine contractions prior to cervical dilation; also called false labor.   368

breech birth Birth in which the baby is positioned rump first.   363

Broca's area Region of the frontal lobe that coordinates complex muscular actions of the mouth, tongue, and larynx, making speech possible.   258

bronchiole (brahng-kee-ohl) Smaller air passages in the lungs that begin at the bronchi and terminate in alveoli.   173

bronchus (pl., bronchi) (brahng-kus) One of two major divisions of the trachea leading to the lungs.   173

buffer Substance or group of substances that tend to resist pH changes of a solution, thus stabilizing its relative acidity and basicity.   26, 199

bulbourethral gland (bul-boh-yoo-ree-thrul) Either of two small structures located below the prostate gland in males; each adds secretions to semen.   321

bulimia nervosa (byoo-lee-mee-uh, -lim-ee-, nur-voh-suh) Eating disorder characterized by binge eating followed by purging via self-induced vomiting or use of a laxative.   165

bursa (bur-suh) Saclike, fluid-filled structure, lined with synovial membrane, that occurs near a joint.   240

bursitis (bur-sy-tis) Inflammation of any of the friction-easing sacs called bursae within the knee joint.   240

C

calcitonin (kal-sih-toh-nin) Hormone secreted by the thyroid gland that increases the blood calcium level.   304

calorie Amount of heat energy required to raise the temperature of 1 g of water 1°C.   24

cancer Malignant tumor whose nondifferentiated cells exhibit loss of contact inhibition, uncontrolled growth, and the ability to invade tissue and metastasize.   404

capsule Gelatinous layer surrounding the cells of blue-green algae and certain bacteria.   122

carbaminohemoglobin Hemoglobin carrying carbon dioxide.   178

carbohydrate Class of organic compounds that includes monosaccharides, disaccharides, and polysaccharides.   28

carbonic anhydrase (kar-bahn-ik an-hy-drays, -drayz) Enzyme in red blood cells that speeds the formation of carbonic acid from the reactants water and carbon dioxide.   178

carcinogen (kar-sin-uh-jun) Environmental agent that causes mutations leading to the development of cancer.   408

carcinogenesis (kar-suh-nuh-jen-uh-sis) Development of cancer.   405

carcinoma (kar-suh-noh-muh) Cancer arising in epithelial tissue.   407

cardiac cycle One complete cycle of systole and diastole for all heart chambers.   90

cardiac muscle Striated, involuntary muscle found only in the heart.   65, 228

cardiovascular system (kar-dee-oh-vas-kyuh-lur) Organ system in which blood vessels distribute blood powered by the pumping action of the heart.   74

carnivore (kar-nuh-vor) Consumer in a food chain that eats other animals.   494

carrier Heterozygous individual who has no apparent abnormality but can pass on an allele for a recessively inherited genetic disorder.   429

carrying capacity Maximum number of individuals of any species that can be supported by a particular ecosystem on a long-term basis.   512

cartilage (kar-tul-ij, kart-lij) Connective tissue in which the cells lie within lacunae separated by a flexible proteinaceous matrix.   63, 208

cecum (see-kum) Small pouch that lies below the entrance of the small intestine and is the blind end of the large intestine.   154

cell Smallest unit that displays the properties of life; always contains cytoplasm surrounded by a plasma membrane.   2

cell body Portion of a neuron that contains a nucleus and from which dendrites and an axon extend. 249

cell cycle Repeating sequence of cellular events that consists of interphase, mitosis, and cytokinesis. 379

cell-mediated immunity Specific mechanism of defense in which T cells destroy antigen-bearing cells. 135

cell theory One of the major theories of biology; states that all organisms are made up of cells and cells come only from preexisting cells. 42

cellular respiration Metabolic reactions that use the energy primarily from carbohydrates but also from fatty acid or amino acid breakdown to produce ATP molecules. 52

cellulose (sel-yuh-lohs, -lohz) Polysaccharide that is the major complex carbohydrate in plant cell walls. 28

central nervous system (CNS) Portion of the nervous system consisting of the brain and spinal cord. 248

centriole (sen-tree-ohl) Cellular structure, existing in pairs, that possibly organizes the mitotic spindle for chromosomal movement during mitosis and meiosis. 51, 381

centromere (sen-truh-meer) Constriction where sister chromatids of a chromosome are held together. 378

centrosome Central microtubule organizing center of cells. In animal cells, it contains two centrioles. 51, 381

cerebellum (ser-uh-bel-um) Part of the brain located posterior to the medulla oblongata and pons that coordinates skeletal muscles to produce smooth, graceful motions. 259

cerebral cortex (suh-ree-brul, ser-uh-brul kor-teks) Outer layer of cerebral hemispheres; receives sensory information and controls motor activities. 257

cerebral hemisphere One of the large, paired structures that together constitute the cerebrum of the brain. 256

cerebrospinal fluid (sair-uh-broh-spy-nul, suh-ree-broh-) Fluid found in the ventricles of the brain, in the central canal of the spinal cord, and in association with the meninges. 254

cerebrum (sair-uh-brum, suh-ree-brum) Main part of the brain consisting of two large masses, or cerebral hemispheres; the largest part of the brain in mammals. 256

cervix (sur-viks) Narrow end of the uterus, which projects into the vagina. 325

cesarean section Birth by surgical incision of the abdomen and uterus. 363

chemical evolution Increase in the complexity of chemicals over time that could have led to the first cells. 468

chemical signal Molecule that brings about a change in a cell, tissue, organ, or individual when it binds to a specific receptor. 298

chemoreceptor (kee-moh-rih-sep-tur) Sensory receptor that is sensitive to chemical stimuli—for example, receptors for taste and smell. 177, 274

chancre Sore that appears on the skin; first sign of syphilis. 338

chlamydia (kluh-mid-ee-uh) Sexually transmitted disease, caused by the bacterium *Chlamydia trachomatis;* can lead to pelvic inflammatory disease. 337

chlorofluorocarbons (CFCs) (klor-oh-floor-oh-kar-buns) Organic compounds containing carbon, chlorine, and fluorine atoms. CFCs, such as Freon, can deplete the ozone shield by releasing chlorine atoms in the upper atmosphere. 506, 523

chondrocyte Type of cell found in the lacunae of cartilage. 208

chordae tendineae (kor-dee ten-din-ee-ee) Tough bands of connective tissue that attach the papillary muscles to the atrioventricular valves within the heart. 88

chorion (kor-ee-ahn) Extraembryonic membrane that contributes to placenta formation. 356

chorionic villi (kor-ee-ahn-ik vil-eye) Treelike extensions of the chorion that project into the maternal tissues at the placenta. 359

choroid (kor-oyd) Vascular, pigmented middle layer of the eyeball. 280

chromatin (kroh-muh-tin) Network of fine threads in the nucleus that are composed of DNA and proteins. 49

chromosome (kroh-muh-som) Chromatin condensed into a compact structure. 49

chronic bronchitis Obstructive pulmonary disorder that tends to recur; marked by inflamed airways filled with mucus and degenerative changes in the bronchi, including loss of cilia. 181

chyme (kym) Thick, semiliquid food material that passes from the stomach to the small intestine. 149

ciliary body (sil-ee-air-ee) Structure associated with the choroid layer that contains ciliary muscle and controls the shape of the lens of the eye. 280

cilium (pl., cilia) (sil-ee-um) Short, hair-like projection from the plasma membrane, occurring usually in large numbers. 51

circadian rhythm (sur-kay-dee-un) Biological rhythm with a 24-hour cycle. 312

circumcision Removal of the prepuce (foreskin) of the penis. 322

cirrhosis (sih-roh-sis) Chronic, irreversible injury to liver tissue; commonly caused by frequent alcohol consumption. 153

citric acid cycle Cycle of reactions in mitochondria that begins with citric acid; it breaks down an acetyl group as CO_2, ATP, NADH, and $FADH_2$ are given off; also called the Krebs cycle. 54

cleavage Cell division without cytoplasmic addition or enlargement; occurs during the first stage of animal development. 355

cleavage furrow Indentation that begins the process of cleavage, by which human cells undergo cytokinesis. 383

clonal selection model Concept that an antigen selects which lymphocyte will undergo clonal expansion and produce more lymphocytes bearing the same type of antigen receptor. 131

cloning Production of identical copies; can be either the production of identical individuals or, in genetic engineering, the production of identical copies of a gene. 458

clotting Process of blood coagulation, usually when injury occurs. 113

cochlea (kohk-lee-uh, koh-klee-uh) Portion of the inner ear that resembles a snail's shell and contains the spiral organ, the sense organ for hearing. 286

cochlear nerve Either of two cranial nerves that carry nerve impulses from the spiral organ to the brain; also called the auditory nerve. 287

codominance Inheritance pattern in which both alleles of a gene are equally expressed. 434

codon Three-base sequence in mRNA that causes the insertion of a particular amino acid into a protein or termination of translation. 448

coenzyme (koh-en-zym) Nonprotein organic molecule that aids the action of the enzyme to which it is loosely bound. 53

collagen fiber (kahl-uh-jun) White fiber in the matrix of connective tissue; gives flexibility and strength. 62

collecting duct Duct within the kidney that receives fluid from several nephrons; the reabsorption of water occurs here. 193, 198

colon (koh-lun) The major portion of the large intestine, consisting of the ascending colon, the transverse colon, and the descending colon. 154

colony-stimulating factor (CSF) Protein that stimulates differentiation and maturation of white blood cells. 110

color blindness Deficiency in one or more of the three kinds of cone cells responsible for color vision. 437

color vision Ability to detect the color of an object; dependent on three kinds of cone cells. 282

columnar epithelium (kuh-lum-nur ep-uh-thee-lee-um) Type of epithelial tissue with cylindrical cells. 68

community Assemblage of populations interacting with one another within the same environment. 2

compact bone Type of bone that contains osteons consisting of concentric layers of matrix and osteocytes in lacunae. 63, 208

the ionic form, they enter the body and inhibit vital enzymes. That's why these items should be discarded carefully and taken to hazardous waste sites.

One of the greatest threats to the maintenance of ecosystems and biodiversity is surface mining, called strip mining. In the United States, huge machines can go as far as removing mountaintops in order to reach a mineral. The land devoid of vegetation takes on a surreal appearance, and rain washes toxic waste deposits into nearby streams and rivers.

Hazardous Wastes

The consumption of minerals contributes to the buildup of hazardous wastes, including synthetic organic chemicals, in the environment. Every year, countries of the world discard billions of tons of solid waste, some on land and some in fresh- and marine waters. Using an allocation of monies called the Superfund, the Environmental Protection Agency oversees the cleanup of hazardous waste disposal sites in the United States. The nine most commonly found contaminants of the environment are heavy metals (lead, arsenic, cadmium, chromium) and synthetic organic compounds (trichloroethylene, toluene, benzene, polychlorinated biphe-

nyls [PCBs], and chloroform). Some of these are endocrine-disrupting contaminants.

Synthetic organic chemicals play a role in the production of plastics, pesticides, herbicides, cosmetics, coatings, solvents, wood preservatives, and hundreds of other products. Synthetic organic chemicals include halogenated hydrocarbons, in which halogens (chlorine, bromine, fluorine) have replaced certain hydrogens. One such molecule comprises the **chlorofluorocarbons (CFCs),** a type of halogenated hydrocarbon in which both chlorine and fluorine atoms replace some of the hydrogen atoms. CFCs have brought about a thinning of the Earth's ozone shield, which protects terrestrial life from the dangerous effects of ultraviolet radiation. Now that MDCs are no longer using CFCs, the ozone shield is predicted to recover by 2050.

Other synthetic organic chemicals pose a direct and serious threat to the health of living things, including humans. Rachel Carson's book *Silent Spring*, published in 1962, made the public aware of the deleterious effects of pesticides. Sometimes, they accumulate in the mud of deltas and estuaries of highly polluted rivers and cause environmental problems if disturbed. These wastes enter bodies of water and are subject to **biological magnification** (Fig. 24.14). Decomposers are unable to break down these wastes. They enter and remain in the bodies of organisms because they accumulate in fat and are not excreted. Therefore, they become more concentrated as they pass along a food chain. Biological magnification is most apt to occur in aquatic food chains, which have more links than terrestrial food chains. Humans are the final consumers in both types of food chains, and over the past 25–30 years, a number of toxic chemicals have found their way into breast milk—PCBs, DDT, solvents, and heavy metals.

Raw sewage causes oxygen depletion in lakes and rivers. As the oxygen level decreases, the diversity of life is greatly reduced. Also, human feces can contain pathogenic microorganisms that cause cholera, typhoid fever, and dysentery. In regions of the LDCs where sewage treatment is practically nonexistent, many children die each year from these diseases. Typically, sewage treatment plants use bacteria to break down organic matter to inorganic nutrients, such as nitrates and phosphates, which then enter surface waters. The end result can be cultural eutrophication discussed on page 505.

Figure 24.14 Biological magnification.
Various synthetic organic chemicals, such as DDT, accumulate in animal fat. Therefore, the chemical becomes increasingly concentrated at higher tropic levels. By the time DDT was banned in the United States, it had interfered with predatory bird reproduction by causing egg-shell thinning.

DDT Concentration

25 ppm in predatory birds

2 ppm in large fish

0.5 ppm in small fish

0.04 ppm in zooplankton

0.000003 ppm in water

✓ Check Your Progress 24.2b

1. Why is it better for humans to use renewable rather than nonrenewable sources of energy?

2. The consumption of mineral resources has what environmental consequences?

3. Name some hazardous wastes present in today's environment.

24.3 Biodiversity

Biodiversity can be defined as the variety of life on Earth, described in terms of the number of different species. We are presently in a biodiversity crisis—the number of extinctions (loss of species) expected to occur in the near future is unparalleled in the history of the Earth.

Loss of Biodiversity

Figure 24.15 identifies the major causes of extinction.

Habitat Loss

Human occupation of the coastline, semiarid lands, tropical rain forests (see pages 514–15), and other areas have contributed to the loss of biodiversity. Scientists are especially concerned about the tropical rain forests and coral reefs because they are particularly rich in species. Already, tropical rain forests have been reduced from their original 14% of landmass to the present 6%. Also, 60% of coral reefs have been destroyed or are on the verge of destruction; it's possible that all coral reefs may disappear during the next 40 years.

Alien Species

Alien species, sometimes called exotics, are nonnative members of an ecosystem. Humans have introduced alien species into new ecosystems chiefly due to colonization, horticulture and agriculture, and accidental transport. For example, the pilgrims brought the dandelion to the United States as a familiar salad green. Kudzu is a vine from Japan that the U. S. Department of Agriculture thought would help prevent soil erosion. The plant now covers much landscape in the South, including even walnut, magnolia, and sweet gum trees. The zebra mussel from the Caspian Sea was accidentally introduced into the Great Lakes in 1988. It now forms dense beds that squeeze out native mussels. Alien species that crowd out native species are termed **invasive**. One way to counteract alien species is to replant native (original to the area) species. Recall that Jacob, Eva, Andrew and Megan were given native tree species to plant because they had demonstrated an interest in keeping the environment healthy by participating in Earth Day festivities.

Pollution

Pollution brings about environmental change that adversely affects the lives and health of living things. Biodiversity is particularly threatened by the following types of environmental pollution.

Acid deposition   Acid deposition decimates forests because it causes trees to weaken and increases their susceptibility to disease and insects.

Global warming   Global warming, an increase in the Earth's temperature due to the presence of greenhouse gases in the atmosphere, is expected to have many detrimental effects,

Figure 24.15   Loss of biodiversity.
a. Habitat loss, alien species, pollution, overexploitation, and disease have been identified as causes of extinction of organisms. **b.** Macaws that reside in South American tropical rain forests are endangered for the reasons listed in the graph.

complementary DNA (cDNA) DNA that has been synthesized from mRNA by the action of reverse transcriptase.   458

complementary paired bases Hydrogen bonding between particular bases; in DNA thymine (T) pairs with adenine (A), and guanine (G) pairs with cytosine (C); in RNA, uracil (U) paits with A, and G pairs with C.   36, 444

complement system Series of proteins in plasma that form a nonspecific defense mechanism against a microbe invasion; it complements the antigen-antibody reaction.   130

compound Substance having two or more different elements united chemically in a fixed ratio.   22

conclusion Statement made following an experiment as to whether the results support the hypothesis.   8

cone cell Photoreceptor in retina of eye that responds to bright light; detects color and provides visual acuity.   282

congenital hypothyroidism Condition resulting from improper development of the thyroid in an infant; characterized by stunted growth and mental retardation.   303

connective tissue Type of tissue that binds structures together, provides support and protection, fills spaces, stores fat, and forms blood cells; adipose tissue, cartilage, bone, and blood are types of connective tissue.   62

constipation (kahn-stuh-pay-shun) Delayed and difficult defecation caused by insufficient water in the feces.   154

consumer Organism that feeds on another organism in a food chain; primary consumers eat plants, and secondary consumers eat animals.   494

contraceptive (kahn-truh-sep-tiv) Medication or device used to reduce the chance of pregnancy.   331

contraceptive implant Birth control method utilizing synthetic progesterone; prevents ovulation by disrupting the ovarian cycle.   332

contraceptive injection Birth control method utilizing progesterone or estrogen and progesterone together; prevents ovulation by disrupting the ovarian cycle.   332

contraceptive vaccine Under development, this birth control method immunizes against the hormone HCG, crucial to maintaining implantation of the embryo.   332

control group Sample that goes through all the steps of an experiment but lacks the factor or is not exposed to the factor being tested; a standard against which results of an experiment are checked.   10

convulsion Sudden attack characterized by a loss of consciousness and severe, sustained, rhythmic contractions of some or all voluntary muscles.   240

cornea (kor-nee-uh) Transparent, anterior portion of the outer layer of the eyeball.   280

coronary artery (kor-uh-nair-ee) Artery that supplies blood to the wall of the heart.   95

coronary bypass operation Therapy for blocked coronary arteries in which part of a blood vessel from another part of the body is grafted around the obstructed artery.   99

corpus luteum (kor-pus loot-ee-um) Yellow body that forms in the ovary from a follicle that has discharged its secondary oocyte; it secretes progesterone and some estrogen.   327

cortisol (kor-tuh-sawl) Glucocorticoid secreted by the adrenal cortex that responds to stress on a long-term basis; reduces inflammation and promotes protein and fat metabolism.   306

cough Sudden expulsion of air from the lungs that clears the air passages; a common symptom of upper respiratory infections.   171

covalent bond (coh-vay-lent) Chemical bond in which atoms share one pair of electrons.   23

cramp Muscle contraction that causes pain.   240

cranial nerve Nerve that arises from the brain.   262

creatinine (kree-ah-tuhn-een) Nitrogenous waste; the end product of creatine phosphate metabolism.   189

Cro-Magnon (kroh-mag-nun) Common name for first fossils to be designated *Homo sapiens.*   485

crossing-over Exchange of segments between nonsister chromatids of a tetrad during meiosis.   388

cuboidal epithelium (kyoo-boyd-ul) Type of epithelial tissue with cube-shaped cells.   68

cultural eutrophication Enrichment of water by inorganic nutrients used by phytoplankton. Often, overenrichment caused by human activities leads to excessive bacterial growth and oxygen depletion.   505

culture Total pattern of human behavior; includes technology and the arts, and depends upon the capacity to speak and transmit knowledge.   7, 481

Cushing syndrome (koosh-ing) Condition resulting from hypersecretion of glucocorticoids; characterized by thin arms and legs and a "moon face," and accompanied by high blood glucose and sodium levels.   307

cutaneous receptor Sensory receptors for pressure and touch found in the dermis of the skin.   276

cyclic adenosine monophosphate (cAMP) (sy-klik, sih-klik) ATP-related compound that acts as the second messenger in peptide hormone transduction; it initiates activity of the metabolic machinery.   299

cyclin Protein that regularly increases and decreases in concentration during the cell cycle.   406

cystic fibrosis A generalized, autosomal recessive disorder of infants and children in which there is widespread dysfunction of the exocrine glands.   430

cystitis Inflammation of the urinary bladder.   200

cytokine (sy-tuh-kyn) Type of protein secreted by a T cell that stimulates cells of the immune system to perform their various functions.   129, 135, 138

cytokinesis (sy-tuh-kyn-ee-sus) Division of the cytoplasm following mitosis and meiosis.   379

cytoplasm (sy-tuh-plaz-um) Contents of a cell between the nucleus and the plasma membrane that contains the organelles.   44

cytosine (C) (sy-tuh-seen) One of four nitrogen bases in nucleotides composing the structure of DNA and RNA.   35

cytoskeleton Internal framework of the cell, consisting of microtubules, actin filaments, and intermediate filaments.   51

cytotoxic T cell (sy-tuh-tahk-sik) T cell that attacks and kills antigen-bearing cells.   135

D

data Facts or pieces of information collected through observation and/or experimentation.   8

daughter cell Cell that arises from a parent cell by mitosis or meiosis.   381

dead air space Volume of inspired air that cannot be exchanged with blood.   176

defecation (def-ih-kay-shun) Discharge of feces from the rectum through the anus.   154

deforestation (dee-for-eh-stay-shun) Removal of trees from a forest in a way that ever reduces the size of the forest.   515

dehydration reaction Chemical reaction resulting in a covalent bond with the accompanying loss of a water molecule.   27

dehydrogenase Coenzyme that removes hydrogen atoms (electrons plus hydrogen ions) and carries electrons to the electron transport chain in mitochondria.   53

delayed allergic response Allergic response initiated at the site of the allergen by sensitized T cells, involving macrophages and regulated by cytokines.   139

deletion Change in chromosome structure in which the end of a chromosome breaks off, or two simultaneous breaks lead to the loss of an internal segment; often causes abnormalities (e.g., cri du chat syndrome).   397

denaturation (dee-nay-chuh-ray-shun) Loss of normal shape by an enzyme so that it no longer functions; caused by a less than optimal pH or temperature.   33

dendrite (den-dryt) Branched ending of a neuron that conducts signals toward the cell body.   249

denitrification Conversion of nitrate or nitrite to nitrogen gas by bacteria in soil.   5503

dense fibrous connective tissue Type of connective tissue containing many collagen fibers packed together; found in tendons and ligaments, for example.   63

dental caries (kar-eez) Tooth decay that occurs when bacteria within the mouth metabolize sugar and give off acids that erode teeth; a cavity.   146

deoxyhemoglogin Hemoglobin not carrying oxygen.   108

dermis (dur-mus) Region of skin that lies beneath the epidermis.   72

desertification Denuding and degrading a once-fertile land, initiating a desert-producing cycle that feeds on itself and causes long-term changes in the soil, climate, and biota of an area.   515

detrital food chain (dih-tryt-ul) Straight-line linking of organisms according to who eats whom, beginning with detritus.   497

detrital food web (dih-tryt-ul) Complex pattern of interlocking and crisscrossing food chains, beginning with detritus.   497

detritus feeder Any organism that obtains most of its nutrients from the detritus in an ecosystem.   494

development Group of stages by which a zygote becomes an organism or by which an organism changes during its life span; includes puberty and aging, for example.   4

diabetes mellitus (dy-uh-bee-teez mel-ih-tus, muh-ly-tus) Condition characterized by a high blood glucose level and the appearance of glucose in the urine, due to a deficiency of insulin production and failure of cells to take up glucose.   195, 310

dialysate Material that passes through the membrane in dialysis.   200

diaphragm (dy-uh-fram) Dome-shaped horizontal sheet of muscle and connective tissue that divides the thoracic cavity from the abdominal cavity.   77, 147 Also, a birth control device consisting of a soft rubber or latex cup that fits over the cervix.   332

diarrhea (dy-uh-ree-uh) Excessively frequent bowel movements.   154

diastole (dy-as-tuh-lee) Relaxation period of a heart chamber during the cardiac cycle.   90

diastolic pressure (dy-uh-stahl-ik) Arterial blood pressure during the diastolic phase of the cardiac cycle.   92

diencephalon (dy-en-sef-uh-lahn) Portion of the brain in the region of the third ventricle that includes the thalamus and hypothalamus.   258

differentiation Cell specialization.   355

diffusion (dih-fyoo-zhun) Movement of molecules or ions from a region of higher to lower concentration; it requires no energy and stops when the distribution is equal.   47

digestion Breaking down of large nutrient molecules into smaller molecules that can be obsorbed.   144

digestive system Organ system including the mouth, esophagus, stomach, small intestine, and large intestine (colon) that receives food and digests it into nutrient molecules. Also has associated organs: teeth, tongue, salivary glands, liver, gallbladder, and pancreas.   74

dihybrid Individual that is heterozygous for two traits; shows the phenotype governed by the dominant alleles but carries the recessive alleles.   427

diploid (2n) Cell condition in which two of each type of chromosome are present in the nucleus.   381

disaccharide (dy-sak-uh-ryd) Sugar that contains two units of a monosaccharide (e.g., maltose).   28

distal convoluted tubule Final portion of a nephron that joins with a collecting duct; associated with tubular secretion.   193

diuretic (dy-uh-ret-ik) Drug used to counteract hypertension by causing the excretion of water.   198

DNA (deoxyribonucleic acid) Nucleic acid polymer produced from covalent bonding of nucleotide monomers that contain the sugar deoxyribose; the genetic material of nearly all organisms.   35, 444

DNA ligase (ly-gays) Enzyme that links DNA fragments; used during production of rDNA to join foreign DNA to vector DNA.   458

DNA replication Synthesis of a new DNA double helix prior to mitosis and meiosis in eukaryotic cells, and during prokaryotic fission in prokaryotic cells.   445

domain The primary taxonomic group above the kingdom level; all living organisms may be placed in one of three domains.   6

dominant allele (uh-leel) Allele that exerts its phenotypic effect in the heterozygote; it masks the expression of the recessive allele.   422

dopamine Neurotransmitter in the central nervous system.   252

dorsal-root ganglion (gang-glee-un) Mass of sensory neuron cell bodies located in the dorsal root of a spinal nerve.   262

double helix Double spiral; describes the three-dimensional shape of DNA.   444

drug abuse Dependence on a drug, which assumes an "essential" biochemical role in the body following habituation and tolerance.   267

Duchenne muscular dystrophy Chronic progressive disease affecting the shoulder and pelvic girdles, commencing in early childhood. Characterized by increasing weakness of the muscles, followed by atrophy and a peculiar swaying gait with the legs kept wide apart. Transmitted as an X-linked trait, and affected individuals, predominantly males, rarely survive to maturity. Death is usually due to respiratory weakness or heart failure.   240, 437

duodenum (doo-uh-dee-num) First part of the small intestine where chyme enters from the stomach.   150

duplication Change in chromosome structure in which a particular segment is present more than once in the same chromosome.   397

E

ecological pyramid Pictorial graph based on the biomass, number of organisms, or energy content of various trophic levels in a food web—from the producer to the final consumer populations.   497

ecosystem (ek-oh-sis-tum, ee-koh-) Biological community together with the associated abiotic environment; characterized by energy flow and chemical cycling.   2, 493

ectopic pregnancy Implantation of the embryo in a location other than the uterus, most often in an oviduct.   357

edema (ih-dee-muh) Swelling due to tissue fluid accumulation in the intercellular spaces.   116

effector Muscle or gland that responds to stimulation.   249

egg Female gamete having the haploid number of chromosomes that is fertilized by a sperm, the male gamete.   324

elastic cartilage Type of cartilage composed of elastic fibers, allowing greater flexibility.   63

elastic fiber Yellow fiber in the matrix of connective tissue, providing flexibility.   62

electrocardiogram (ECG) (ih-lek-troh-kar-dee-uh-gram) Recording of the electrical activity associated with the heartbeat.   91

electron Negative subatomic particle, moving about in an energy level around the nucleus of an atom.   20

electron transport chain Passage of electrons along a series of membrane-bound carrier molecules from a higher to lower energy level; the energy released is used for the synthesis of ATP.   55

element Substance that cannot be broken down into substances with different properties; composed of only one type of atom.   20

elimination Process of expelling substances from the body.   145

embolus (em-buh-lus) Moving blood clot that is carried through the bloodstream.   97

embryo (em-bree-oh) Immature developmental stage that is not recognizable as a human being.   357

embryonic development Period of development from the second through eighth weeks.   357

embryonic disk Stage of embryonic development following the blastocyst stage that has two layers; one layer will be endoderm, and the other will be ectoderm.   357

emphysema (em-fih-see-muh) Degenerative lung disorder in which the bursting of alveolar walls reduces the total surface area for gas exchange.   181

emulsification (ih-mul-suh-fuh-kay-shun) Breaking up of fat globules into smaller droplets by the action of bile salts or any other emulsifier.   30

endochondral ossification Ossification that begins as hyaline cartilage that is subsequently replaced by bone tissue.   211

endocrine gland (en-duh-krin) Ductless organ that secretes (a) hormone(s) into the bloodstream.   69, 296

endocrine system Organ system involved in the coordination of body activities; uses hormones as chemical signals secreted into the bloodstream.   75

endomembrane system A collection of membranous structures involved in transport within the cell.   44, 50

endometrium Mucous membrane lining the interior surface of the uterus.   325

endoplasmic reticulum (ER) (en-duh-plaz-mik reh-tik-yuh-lum) System of membranous saccules and channels in the cytoplasm, often with attached ribosomes.   49

eosinophil (ee-oh-sin-oh-fill) White blood cell containing cytoplasmic granules that stain with acidic dye.   110

epidermis (ep-uh-dur-mus) Region of skin that lies above the dermis.   71

epididymis (ep-uh-did-uh-mus) Coiled tubule next to the testes where sperm mature and may be stored for a short time.   321

epiglottis (ep-uh-glaht-us) Structure that covers the glottis during the process of swallowing.   147, 172

epinephrine (ep-uh-nef-rin) Hormone secreted by the adrenal medulla in times of stress; adrenaline.   305

episiotomy (ih-pee-zee-aht-uh-mee) Surgical procedure performed during childbirth in which the opening of the vagina is enlarged to avoid tearing.   369

episodic memory Capacity of brain to store and retrieve information with regard to persons and events.   260

epithelial tissue (ep-uh-thee-lee-ul) Type of tissue that lines hollow organs and covers surfaces; also called epithelium.   68

erectile dysfunction Failure of the penis to achieve or maintain erection.   322

erythropoietin (EPO) (ih-rith-roh-poy-ee-tin) Hormone, produced by the kidneys, that speeds red blood cell formation.   109, 189

esophagus (ih-sahf-uh-gus) Muscular tube for moving swallowed food from the pharynx to the stomach.   147

essential amino acids Amino acids required in the human diet because the body cannot make them.   158

essential fatty acid Fatty acid required in the human diet because the body cannot make them.   159

estrogen (es-truh-jun) Female sex hormone that helps maintain sex organs and secondary sex characteristics.   311, 328

eukaryotic cell Type of cell that has a membrane-bound nucleus and membranous organelles.   44, 470

evaporation Conversion of a liquid or a solid into a gas.   499

evolution Descent of organisms from common ancestors with the development of genetic and phenotypic changes over time that make them more suited to the environment.   5

evolutionary tree Diagram that describes the evolutionary relationship of groups of organisms; a common ancestor is presumed to have been present at points of divergence.   479

excretion Removal of metabolic wastes from the body.   188

exocrine gland Gland that secrets its product to an epithelial surface directly or through ducts.   69

exophthalmic goiter (ek-sahf-thal-mik) Enlargement of the thyroid gland accompanied by an abnormal protrusion of the eyes.   303

experiment Artificial situation devised to test a hypothesis.   10

experimental variable Value that is expected to change as a result of an experiment; represents the factor that is being tested by the experiment.   10

expiration (ek-spuh-ray-shun) Act of expelling air from the lungs; also called exhalation.   170, 174

expiratory reserve volume (ik-spy-ruh-tor-ee) Volume of air that can be forcibly exhaled after normal exhalation.   176

exponential growth Growth at a constant rate of increase per unit of time; can be expressed as a constant fraction or exponent.   512

external respiration Exchange of oxygen and carbon dioxide between alveoli and blood.   178

exteroceptor Sensory receptor that detects stimuli from outside the body (e.g., taste, smell, vision, hearing, and equilibrium).   274

extinction Total disappearance of a species or higher group.   7

extraembryonic membrane (ek-struh-em-bree-ahn-ik) Membrane that is not a part of the embryo but is necessary to the continued existence and health of the embryo.   355

F

facial tic Involuntary muscle movement of the face.   240

facilitated transport Use of a plasma membrane carrier to move a substance into or out of a cell from higher to lower concentration; no energy required.   47

familial hypercholesterolemia (FH) Inability to remove cholesterol from the bloodstream; predisposes individual to heart attack.   434

farsighted Vision abnormality due to a shortened eyeball from front to back; light rays focus in back of retina when viewing close objects.   284

fat Organic molecule that contains glycerol and fatty acids; found in adipose tissue.   30

fatty acid Molecule that contains a hydrocarbon chain and ends with an acid group.   30

female condom Large polyurethane tube with a flexible ring that fits onto the cervix. Functions as a contraceptive and helps minimize the risk of transmitting infection.   332

fermentation Anaerobic breakdown of glucose that results in a gain of two ATP and end products, such as alcohol and lactate.   56

fertilization Union of a sperm nucleus and an egg nucleus, which creates a zygote.   354, 385

fetal development Period of development from the ninth week through birth.   363

fiber Structure resembling a thread; also, plant material that is nondigestible.   154

fibrin (fy-brun) Insoluble protein threads formed from fibrinogen during blood clotting.   113

fibrinogen (fy-brin-uh-jun) Plasma protein that is converted into fibrin threads during blood clotting.   107

fibroblast (fy-bruh-blast) Cell in connective tissues that produces fibers and other substances.   62

fibrocartilage (fy-broh-kar-tul-ij, -kart-lij) Cartilage with a matrix of strong collagenous fibers.   63

fibromyalgia Chronic, widespread pain in muscles and soft tissues surrounding joints.   240

fibrous connective tissue Tissue composed mainly of closely packed collagenous fibers and found in tendons and ligaments.   208

fimbriae Small bristlelike fiber on the surface of a bacterial cell, which attaches bacteria to a surface.   122, 324

first messenger Chemical signal, such as a peptide hormone, that binds to a plasma membrane receptor protein and alters the metabolism of a cell because a second messenger is activated.   299

flagellum (pl., flagella) (fluh-jel-um) Slender, long extension that propels a cell through a fluid medium.   51

floating kidney Kidney that has been dislodged from its normal position.   188

fluid-mosaic model Model for the plasma membrane based on the changing location and pattern of protein molecules in a fluid phospholipid bilayer.   46

focus Bending of light rays by the cornea, lens, and humors so that they converge and create an image on the retina.   281

follicle (fahl-ih-kul) Structure in the ovary that produces a secondary oocyte and the hormones estrogen and progesterone.   326

follicle-stimulating hormone (FSH) Hormone secreted by the anterior pituitary gland that stimulates the development of an ovarian follicle in a female or the production of sperm in a male.   324

fontanel (fahn-tun-el) Membranous region located between certain cranial bones in the skull of a fetus or infant.   216, 363

food chain Order in which one population feeds on another in an ecosystem, from detritus (detrital food chain) or producer (grazing food chain) to final consumer.   497

food web In ecosystems, complex pattern of interlocking and crisscrossing food chains.   497

foramen magnum (fuh-ray-mun mag-num) Opening in the occipital bone of the vertebrate skull through which the spinal cord passes.   216

formed element Constituent of blood that is either cellular (red blood cells and white blood cells) or at least cellular in origin (platelets).   107

fossil Any past evidence of an organism that has been preserved in the Earth's crust.   470

fossil fuel Fuels, such as oil, coal, and natural gas, that are the result of partial decomposition of plants and animals coupled with exposure to heat and pressure for millions of years.   501, 520

fossil record History of life recorded from remains from the past.   471

fovea centralis Region of the retina consisting of densely packed cones; responsible for the greatest visual acuity.   281

G

GABA (gamma aminobutyric acid) Major inhibitory neurotransmitter in the CNS. 252

gallbladder Organ attached to the liver that serves to store and concentrate bile.   152

gallstone Crystalline bodies formed by from concentration of normal and abnormal bile components within the gallbladder.   152

gamete (ga-meet, guh-meet) Haploid sex cell; the egg or a sperm, which join in fertilization to form a zygote.   334, 385

gamma globulin Large proteins found in the blood plasma and on the surface of immune cells, functioning as antibodies (IgG).   137

ganglion Collection or bundle of neuron cell bodies usually outside the central nervous system.   262

gap junction Junction between cells formed by the joining of two adjacent plasma membranes; it lends strength and allows ions, sugars, and small molecules to pass between cells.   70

gastric Pertaining to the stomach.   149

gastric gland Gland within the stomach wall that secretes gastric juice.   149

gastrulation Stage of animal development during which germ layers form, at least in part, by invagination.   358

gene Unit of heredity existing as alleles on the chromosomes; in diploid organisms, typically two alleles are inherited—one from each parent.   4

gene cloning Production of one or more copies of the same gene.   458

genetic engineering Alteration of DNA for medical or industrial purposes.   460

genomics Study of all the nucleotide sequences, including structural genes, regulatory sequences, and noncoding DNA segments, in the chromosomes of an organism.   455

genotype (jee-nuh-typ) Genes of an individual for a particular trait or traits; often designated by letters, for example, *BB* or *Aa*.   422

gerontology (jer-un-tahl-uh-jee) Study of aging.   370

gestation Period of development, from the start of the last menstrual cycle until birth; in humans, typically 280 days.   355

gland Epithelial cell or group of epithelial cells that are specialized to secrete a substance.   69

glaucoma (glow-koh-muh, glaw-koh-muh) Increasing loss of field of vision; caused by blockage of the ducts that drain the aqueous humor, creating pressure buildup and nerve damage.   280

global warming Predicted increase in the Earth's temperature, due to human activities that promote the greenhouse effect.   501

globulin Type of protein in blood plasma. There are alpha, beta, and gamma globulines.   107

glomerular capsule (gluh-mair-yuh-lur) Double-walled cup that surrounds the glomerulus at the beginning of the nephron.   193

glomerular filtrate Filtered portion of blood contained within the glomerular capsule.   195

glomerular filtration Movement of small molecules from the glomerulus into the glomerular capsule due to the action of blood pressure.   195

glomerulus (gluh-mair-uh-lus, gloh-mair-yuh-lus) Cluster; for example, the cluster of capillaries surrounded by the glomerular capsule in a nephron, where glomerular filtration takes place.   192

glottis (glaht-us) Opening for airflow in the larynx.   147, 172

glucagon (gloo-kuh-gahn) Hormone secreted by the pancreas that causes the liver to break down glycogen and raises the blood glucose level.   308

glucocorticoid (gloo-koh-kor-tih-koyd) Type of hormone secreted by the adrenal cortex that influences carbohydrate, fat, and protein metabolism; see cortisol.   305

glucose (gloo-kohs) Six-carbon sugar that organisms degrade as a source of energy during cellular respiration.   28

glutamate Major excitatory CNS neurotransmitter.   252

glycemic index (GI) Blood glucose response of a given food.   158

glycogen (gly-koh-jun) Storage polysaccharide that is composed of glucose molecules joined in a linear fashion but having numerous branches.   28

glycolysis Anaerobic breakdown of glucose that results in a gain of two ATP molecules.   54

Golgi apparatus (gohl-jee) Organelle, consisting of saccules and vesicles, that processes, packages, and distributes molecules about or from the cell.   50

gonad (goh-nad) Organ that produces gametes; the ovary produces eggs, and the testis produces sperm.   311

gonadotropic hormone (goh-nad-uh-trahp-ic, -troh-pic) Chemical signal secreted by the anterior pituitary that regulates the activity of the ovaries and testes; principally, follicle-stimulating hormone (FSH) and luteinizing hormone (LH).   300

gonadotropin-releasing hormone (GnRH) Hormone secreted by the hypothalamus that stimulates the anterior pituitary to secrete follicle-stimulating hormone (FSH) and luteinizing hormone (LH).   324

gout Joint inflammation caused by accumulation of uric acid.   189

granular leukocyte (gran-yuh-lur loo-kuh-syt) White blood cell with prominent granules in the cytoplasm.   110

gravitational equilibrium Maintenance of balance when the head and body are motionless.   291

gray matter Nonmyelinated axons and cell bodies in the central nervous system.   254

grazing food chain Straight-line linking of organisms according to who eats whom, beginning with a producer.   497

grazing food web Complex pattern of interlocking and crisscrossing food chains that begins with populations of autotrophs serving as producers.   497

greenhouse effect Reradiation of solar heat toward the Earth because gases, such as carbon dioxide, methane, nitrous oxide, and water vapor, allow solar energy to pass through toward the Earth but block the escape of heat back into space.   501

greenhouse gases Gases that are involved in the greenhouse effect.   501, 520

growth Increase in the number of cells and/or the size of these cells.   355

growth factor Chemical signal that regulates mitosis and differentiation of cells that have receptors for it; important in such processes as fetal development, tissue maintenance and repair, and hematopoiesis; sometimes a contributing factor in cancer.   406

growth hormone (GH) Substance secreted by the anterior pituitary; controls size of individual by promoting cell division, protein synthesis, and bone growth.   300

growth plate Cartilaginous layer within an epiphysis of a long bone that permits growth of bone to occur.   211

growth rate A percentage that reflects the difference between the number of persons in a population who are born and the number who die each year.   512

guanine (G) (gwah-neen) One of four nitrogen-containing bases in nucleotides composing the structure of DNA and RNA; pairs with cytosine.   35

H

hair cell Cell with stereocilia (long microvilli) that is sensitive to mechanical stimulation; mechanoreceptor for hearing and equilibrium in the inner ear.   286

hair follicle Tubelike depression in the skin in which a hair develops.   73

haploid (n) (hap-loyd) The n number of chromosomes—half the diploid number; the number characteristic of gametes, which contain only one set of chromosomes.   385

hard palate (pal-it) Bony, anterior portion of the roof of the mouth.   146

hay fever Seasonal variety of allergic reaction to a specific allergen. Characterized by sudden attacks of sneezing, swelling of nasal mucosa, and often asthmatic symptoms.   138

heart Muscular organ located in the thoracic cavity whose rhythmic contractions maintain blood circulation.   88

heart attack Damage to the myocardium due to blocked circulation in the coronary arteries; also called a myocardial infarction (MI).   97

heartburn Burning pain in the chest that occurs when part of the stomach contents escape into the esophagus.   147

heart failure Syndrome characterized by distinctive symptoms and signs resulting from disturbances in cardiac output or from increased pressure in the veins.   100

helper T cell T cell that secretes cytokines that stimulate all kinds of immune system cells.   135

hemodialysis (he-moh-dy-al-uh-sus) Cleansing of blood by using an artificial membrane that causes substances to diffuse from blood into a dialysis fluid.   200

hemoglobin (Hb) (hee-muh-gloh-bun) Iron-containing pigment in red blood cells that combines with and transports oxygen.   32, 108

hemolysis (he-mahl-uh-sus) Rupture of red blood cells accompanied by the release of hemoglobin.   109

hemolytic desease of the newborn Destruction of a fetus's red blood cells by the mother's immune system, caused by differing Rh factors between mother and fetus.   109

hemophilia (he-moh-fil-ee-uh) Genetic disorder in which the affected individual is subject to uncontrollable bleeding.   113

hemorrhoid (hem-uh-royd, hem-royd) Abnormally dilated blood vessels of the rectum.   155

hepatic portal vein Vein leading to the liver and formed by the merging blood vessels leaving the small intestine.   95

hepatic vein Vein that runs between the liver and the inferior vena cava.   95

hepatitis (hep-uh-ty-tis) Inflammation of the liver. Viral hepatitis occurs in several forms.   153

herbivore (hur-buh-vor) Primary consumer in a grazing food chain; a plant eater.   494

heterotroph Organism that cannot synthesize organic molecules from inorganic nutrients, and therefore must take in organic nutrients (food).   469, 494

heterozygous Possessing unlike alleles for a particular trait.   422

hexose Six-carbon sugar.   28

hippocampus (hip-uh-kam-pus) Portion of the limbic system where memories are stored.   260

histamine (his-tuh-meen, -mun) Substance, produced by basophils in blood and mast cells in connective tissue, that causes capillaries to dilate.   129

HLA (human leukocyte antigen) Protein in a plasma membrane that identifies the cell as belonging to a particular individual and acts as an antigen in other organisms.   134

homeostasis (hoh-mee-oh-stay-sis) Maintenance of normal internal conditions in a cell or an organism by means of self-regulating mechanisms.   4, 78

hominid (hahm-uh-nid) Member of the family Hominidae, which contains australopithecines and humans.   479

Homo erectus (hoh-moh ih-rek-tus) Hominid who used fire and migrated out of Africa to Europe and Asia.   482

Homo habilis (hoh-moh hab-uh-lus) Hominid of 2 MYA who is believed to have been the first tool user.   481

homologous chromosome (hoh-mahl-uh-gus, huh-mahl-uh-gus) Member of a pair of chromosomes that are alike and come together in synapsis during prophase of the first meiotic division.   385

homologous structure Structure that is similar in two or more species because of common ancestry.   472

homologue Member of a homologous pair of chromosomes.   385

Homo sapiens (hoh-moh say-pe-nz) Modern humans.   482

homozygous dominant Possessing two identical alleles, such as *AA,* for a particular trait.   422

homozygous recessive Possessing two identical alleles, such as *aa,* for a particular trait.   422

hormone (hor-mohn) Chemical signal produced by one set of cells that affects a different set of cells.   152, 212, 296

human chorionic gonadotropin (HCG) (kor-ee-ahn-ik, goh-nad-uh-trahp-in, -troh-pin) Hormone produced by the chorion that functions to maintain the uterine lining.   330, 357

human immunodeficiency virus (HIV) Virus responsible for AIDS.   344

Huntington disease Genetic disease marked by progressive deterioration of the nervous system due to deficiency of a neurotransmitter.   431

hyaline cartilage (hy-uh-lin) Cartilage whose cells lie in lacunae separated by a white, translucent matrix containing very fine collagen fibers.   63

hydrogen bond Weak bond that arises between a slightly positive hydrogen atom of one molecule and a slightly negative atom of another, or between parts of the same molecule.   24

hydrolysis reaction (hy-drahl-ih-sis re-ak-shun) Splitting of a compound by the addition of water, with the H^+ being incorporated in one fragment and the OH^- in the other.   27

hydrophilic (hy-druh-fil-ik) Type of molecule that interacts with water by dissolving in water and/or forming hydrogen bonds with water molecules.   25

hydrophobic (hy-druh-foh-bik) Type of molecule that does not interact with water because it is nonpolar.   25

hypertension Elevated blood pressure, particularly the diastolic pressure.   97

hypothalamic-inhibiting hormone (hy-poh-thuh-lah-mik) One of many hormones produced by the hypothalamus that inhibits the secretion of an anterior pituitary hormone.   300

hypothalamic-releasing hormone One of many hormones produced by the hypothalamus that stimulates the secretion of an anterior pituitary hormone.   300

hypothalamus (hy-poh-thal-uh-mus) Part of the brain located below the thalamus that helps regulate the internal environment of the body and produces releasing factors that control the anterior pituitary.   258, 300

hypothesis (hy-pahth-ih-sis) Supposition that is formulated after making an observation; it can be tested by obtaining more data, often by experimentation.   8

I

immediate allergic response Allergic response that occurs within seconds of contact with an allergen, caused by the attachment of the allergen to IgE antibodies.   138

immune system White blood cells and lymphatic organs that protect the body against foreign organisms and substances and also cancerous cells.   74, 110

immunity Ability of the body to protect itself from foreign substances and cells, including disease-causing agents.   128

immunization (im-yuh-nuh-zay-shun) Use of a vaccine to protect the body against specific disease-causing agents.   136

immunosuppressive Inactivating the immune system to prevent organ rejection, usually via a drug.   139

implantation Attachment and penetration of the embryo into the lining of the uterus (endometrium).   325, 357

incomplete dominance Inheritance pattern in which the offspring has an intermediate phenotype, as when a red-flowered plant and a white-flowered plant produce pink-flowered offspring.   434

incus (ing-kus) The middle of three ossicles of the ear that serve to conduct vibrations from the tympanic membrane to the oval window of the inner ear.   286

infant respiratory distress syndrome Condition in newborns, especially premature ones, in which the lungs collapse because of a lack of surfactant lining the alveoli.   174

infectious mononucleosis Acute, self-limited infectious disease of the lymphatic system caused by the Epstein-Barr virus and characterized by fever, sore throat, lymph node and spleen swelling, and the proliferation of moncytes and abnormal lymphocytes.   111

inferior vena cava (vee-nuh kay-vuh) Large vein that enters the right atrium from below and carries blood from the trunk and lower extremities.   94

infertility Inability to have as many children as desired.   334

inflammatory response Tissue response to injury that is characterized by redness, swelling, pain, and heat.   128

ingestion The taking of food or liquid into the body by way of the mouth.   144

inner cell mass An aggregation of cells at one pole of the blastocyte, which is destined to form the embryo proper.   356

inner ear Portion of the ear consisting of a vestibule, semicircular canals, and the cochlea, where equilibrium is maintained and sound is transmitted.   286

insertion End of a muscle that is attached to a movable bone.   229

inspiration (in-spuh-ray-shun) Act of taking air into the lungs; also called inhalation.   170, 174

inspiratory reserve volume (in-spy-ruh-tohr-ee) Volume of air that can be forcibly inhaled after normal inhalation.   176

insulin (in-suh-lin) Hormone secreted by the pancreas that lowers the blood glucose level by promoting the uptake of glucose by cells, and the conversion of glucose to glycogen by the liver and skeletal muscles.   308

integrase Viral enzyme that enables the integration of viral genetic material into a host cell's DNA.   348

integration Summing up of excitatory and inhibitory signals by a neuron or by some part of the brain.   253, 275

integumentary system (in-teg-yoo-men-tuh-ree, -men-tree) Organ system consisting of skin and various organs, such as hair, that are found in skin.   71

intercalated disk (in-tur-kuh-lay-tud) Region that holds adjacent cardiac muscle cells together; disks appear as dense bands at right angles to the muscle striations.   65, 228

interferon (in-tur-feer-ahn) Antiviral agent produced by an infected cell that blocks the infection of another cell.   130

interkinesis Period of time between meiosis I and meiosis II, during which no DNA replication takes place.   385

interleukin (in-tur-loo-kun) Cytokine produced by macrophages and T cells that functions as a metabolic regulator of the immune response.   138

intermediate filament Ropelike assemblies of fibrous polypeptides in the cytoskeleton that provide support and strength to cells; so called because they are intermediate in size between actin filaments and microtubules.   51

internal respiration Exchange of oxygen and carbon dioxide between blood and tissue fluid.   178

interneuron Neuron located within the central nervous system that conveys messages between parts of the central nervous system.   249

interoceptor Sensory receptor that detects stimuli from inside the body (e.g., pressoreceptors, osmoreceptors, and chemoreceptors).   274

interphase Cell cycle stage during which growth and DNA synthesis occur when the nucleus is not actively dividing.   379

interstitial cell (in-tur-stish-ul) Hormone-secreting cell located between the seminiferous tubules of the testes.   323

intervertebral disk (in-tur-vur-tuh-brul) Layer of cartilage located between adjacent vertebrae.   218

intramembranous ossification Ossification that forms from membranelike layers of primitive connective tissue.   210

intrauterine device (IUD) (in-truh-yoo-tur-in) Birth control device consisting of a small piece of molded plastic inserted into the uterus; believed to alter the uterine environment so that fertilization does not occur.   332

invasive Cells, such as tumor cells, that invade normal cells.   524

inversion Change in chromosome structure in which a segment of a chromosome is turned around 180°; this reversed sequence of genes can lead to altered gene activity and abnormalities.   397

ion (eye-un, -ahn) Charged particle that carries a negative or positive charge.   22

ionic bond (eye-ahn-ik) Chemical bond in which ions are attracted to one another by opposite charges.   22

iris (eye-ris) Muscular ring that surrounds the pupil and regulates the passage of light through this opening.   280

isotope (eye-suh-tohp) One of two or more atoms with the same atomic number but a different atomic mass due to the number of neutrons.   21

J

jaundice (jawn-dis) Yellowish tint to the skin caused by an abnormal amount of bilirubin (bile pigment) in the blood, indicating liver malfunction.   153

joint Articulation between two bones of a skeleton.   208

juxtaglomerular apparatus (juk-stuh-gluh-mer-yuh-lur) Structure located in the walls of arterioles near the glomerulus; regulates renal blood flow.   197

K

kidney Organ in the urinary system that produces and excretes urine.   188

kingdom One of the categories used to classify organisms; the category above phylum.   6

L

lacteal (lak-tee-ul) Lymphatic vessel in an intestinal villus; it aids in the absorption of lipids.   150

lactose intolerance (lak-tohs) Inability to digest lactose because of an enzyme deficiency.   150

lacuna (pl., lacunae) (luh-koo-nuh, -kyoo-nuh) Small pit or hollow cavity, as in bone or cartilage, where a cell or cells are located.   63

Langerhans cell Specialized epidermal cells that assist the immune system.   72

lanugo (luh-noo-goh) Short, fine hair that is present during the later portion of fetal development.   363

large intestine Last major portion of the digestive tract, extending from the small intestine to the anus and consisting of the cecum, the colon, the rectum, and the anal canal.   154

laryngitis (lar-un-jy-tis) Infection of the larynx with accompanying hoarseness.   180

larynx (lar-ingks) Cartilaginous organ located between the pharynx and the trachea that contains the vocal cords; also called the voice box.   172

learning Relatively permanent change in behavior that results from practice and experience.   260

lens Clear, membranelike structure found in the eye behind the iris; brings objects into focus.   280

leptin Hormone produced by adipose tissue that acts on the hypothalamus to signal satiety.   312

leukemia (loo-kee-mee-uh) Cancer of the blood-forming tissues leading to the overproduction of abnormal white blood cells.   111, 407

ligament (lig-uh-munt) Tough cord or band of dense fibrous connective tissue that joins bone to bone at a joint.   63, 208

limbic system Association of various brain centers, including the amygdala and hippocampus; governs learning and memory and various emotions, such as pleasure, fear, and happiness.   260

lineage Evolutionary line of descent.   479

lipase (ly-pays, ly-payz) Fat-digesting enzyme secreted by the pancreas.   150

lipid (lip-id, ly-pid) Class of organic compounds that tends to be soluble only in nonpolar solvents, such as alcohol; includes fats and oils.   30

liver Large, dark red internal organ that produces urea and bile, detoxifies the blood, stores glycogen, and produces the plasma proteins, among other functions.   152

locus (pl., loci) Particular site where a gene is found on a chromosome. Homologous chromosomes have corresponding gene loci.   422

long-term memory Retention of information that lasts longer than a few minutes.   260

long-term potentiation (LTP) (puh-ten-shee-ay-shun) Enhanced response at synapses within the hippocampus; likely essential to memory storage.   261

loop of the nephron (nef-rahn) Portion of the nephron lying between the proximal convoluted tubule and the distal convoluted tubule that functions in water reabsorption.   193, 198

loose fibrous connective tissue Tissue composed mainly of fibroblasts widely separated by a matrix containing collagen and elastic fibers.   62

lumen (loo-mun) Cavity inside any tubular structure, such as the lumen of the digestive tract.   145

lung cancer Malignant growth that often begins in the bronchi.   182

lungs Paired, cone-shaped organs within the thoracic cavity; function in internal respiration and contain moist surfaces for gas exchange.   173

luteinizing hormone (LH) Hormone that controls the production of testosterone by interstitial cells in males and promotes the development of the corpus luteum in females.   324

lymph (limf) Fluid, derived from tissue fluid, that is carried in lymphatic vessels.   64, 126

lymphatic nodule Mass of lymphatic tissue not surrounded by a capsule.   128

lymphatic organ Organ other than a lymphatic vessel that is part of the lymphatic system; includes lymph nodes, tonsils, spleen, thymus gland, and bone marrow.   127

lymphatic system (lim-fat-ik) Organ system consisting of lymphatic vessels and lymphatic organs that transport lymph and lipids, and aids the immune system.   74, 86

lymph node Mass of lymphatic tissue located along the course of a lymphatic vessel.   128

lymphocyte (lim-fuh-syt) Specialized white blood cell that functions in specific defense; occurs in two forms—T cell and B cell.   111

lymphoma Cancer of lymphatic tissue (reticular connective tissue).   407

lysosome (ly-suh-sohm) Membrane-bound vesicle that contains hydrolytic enzymes for digesting macromolecules.   50

lysozyme Enzyme found in tears, milk, saliva, mucus, and other body fluids that destroys bacteria by digesting their cell walls.   128

M

macromolecule Extremely large biological molecule; refers specifically to proteins, nucleic acids, polysaccharides, lipids, and complexes of these.   27

macrophage (mak-ruh-fayj) Large phagocytic cell derived from a monocyte that ingests microbes and debris.   129

male condom Sheath used to cover the penis during sexual intercourse; used as a contraceptive and, if latex, to minimize the risk of transmitting infection.   332

malleus (mal-ee-us) The first of three ossicles of the ear that serve to conduct vibrations from the tympanic membrane to the oval window of the inner ear.   286

Marfan syndrome Congenital disorder of connective tissue characterized by abnormal length of the extremities.   431

mass Sum of the number of protons and neutrons in an atom's nucleus.   21

mast cell Cell to which antibodies, formed in response to allergens, attach, causing it to release histamine, thus producing allergic symptoms.   111, 129

mastoiditis (mas-toyd-eye-tis) Inflammation of the mastoid sinuses of the skull.   216

matter Anything that takes up space and has mass.   20

matrix (may-triks) Unstructured semifluid substance that fills the space between cells in connective tissues or inside organelles.   62

mechanoreceptor (mek-uh-noh-rih-sep-tur) Sensory receptor that responds to mechanical stimuli, such as that from pressure, sound waves, and gravity.   274

medulla oblongata (muh-dul-uh ahb-lawng-gah-tuh) Part of the brain stem that is continuous with the spinal cord; controls heartbeat, blood pressure, breathing, and other vital functions.   259

medullary cavity (muh-dul-uh-ree) Cavity within the diaphysis of a long bone containing marrow.   208

megakaryocyte (meg-uh-kar-ee-oh-syt,-uh-syt) Large cell that gives rise to blood platelets.   113

meiosis (my-oh-sis) Type of nuclear division that occurs as part of sexual reproduction in which the daughter cells receive the haploid number of chromosomes in varied combinations.   385

melanocyte Melanin-producing cell found in skin.   72

melanocyte-stimulating hormone (MSH) Substance that causes melanocytes to secrete melanin in lower vertebrates.   300

melatonin (mel-uh-toh-nun) Hormone, secreted by the pineal gland, that is involved in biorhythms.   311

membrane attack complex Group of complement proteins that form channels in a microbe's surface, thereby destroying it.   130

memory Capacity of the brain to store and retrieve information about past sensations and perceptions; essential to learning.   260

memory T cell T cell that differentiates during an initial infection and responds rapidly during subsequent exposure to the same anitgen. 135

meninges (sing., meninx) (muh-nin-jeez) Protective membranous coverings about the central nervous system. 77, 254

meningitis Condition that refers to inflammation of the brain or spinal cord meninges (membranes). 77

menopause (men-uh-pawz) Termination of the ovarian and uterine cycles in older women. 328

menstruation (men-stroo-ay-shun) Loss of blood and tissue from the uterus at the end of a uterine cycle. 328

messenger RNA (mRNA) Type of RNA formed from a DNA template that bears coded information for the amino acid sequence of a polypeptide. 446

metabolism All of the chemical reactions that occur in a cell. 4, 53

metaphase Mitotic phase during which chromosomes are aligned at the equator of the mitotic spindle. 382

metastasis (muh-tas-tuh-sis) Spread of cancer from the place of origin throughout the body; caused by the ability of cancer cells to migrate and invade tissues. 405

microtubule (my-kro-too-byool) Small cylindrical structure that contains 13 rows of the protein tubulin around an empty central core; present in the cytoplasm, centrioles, cilia, and flagella. 51

micturition Emptying of the bladder; urination. 188

midbrain Part of the brain located below the thalamus and above the pons; contains reflex centers and tracts. 259

middle ear Portion of the ear consisting of the tympanic membrane, the oval and round windows, and the ossicles; where sound is amplified. 286

mineral Naturally occurring inorganic substance containing two or more elements; certain minerals are needed in the diet. 160, 522

mineralocorticoid (min-ur-uh-loh-kor-tih-koyd) Type of hormone secreted by the adrenal cortex that regulates water-salt balance, leading to increases in blood volume and blood pressure. 305

mitochondrion (my-tuh-kahn-dree-un) Membrane-bound organelle in which ATP molecules are produced during the process of cellular respiration. 52

mitosis (my-toh-sis) Type of cell division in which daughter cells receive the exact chromosomal and genetic makeup of the parent cell; occurs during growth and repair. 378, 379, 381

mitotic spindle Microtubule structure that brings about chromosomal movement during nuclear division. 381

molecular clock Mutational changes that accumulate at a presumed constant rate in regions of DNA not involved in adaptation to the environment. 479

molecule Union of two or more atoms of the same element; also, the smallest part of a compound that retains the properties of the compound. 2, 22

monoclonal antibody One of many antibodies produced by a clone of hybridoma cells that all bind to the same antigen. 137

monocyte (mahn-uh-syt) Type of agranular white blood cell that functions as a phagocyte and an antigen-presenting cell. 111

monohybrid Individual that is heterozygous for one trait; shows the phenotype of the dominant allele but carries the recessive allele. 424

monosaccharide (mahn-uh-sak-uh-ryd) Simple sugar; a carbohydrate that cannot be decomposed by hydrolysis (e.g., glucose). 28

monosomy One less chromosome than usual. 393

morphogenesis Emergence of shape in tissues, organs, or entire embryo during development. 355

morula Spherical mass of cells resulting from cleavage during animal development, prior to the blastula stage. 356

mosaic evolution Concept that human characteristics did not evolve at the same rate; for example, some body parts are more humanlike than others in early hominids. 480

motor neuron Nerve cell that conducts nerve impulses away from the central nervous system and innervates effectors (muscles and glands). 249

motor unit Motor neuron and all the muscle fibers it innervates. 236

movement Motion. 144

mucosa Membrane that lines tubes and body cavities that open to the outside of the body; mucous membrane. 145

mucous membrane (myoo-kus) Membrane lining a cavity or tube that opens to the outside of the body; also called mucosa. 77

multicellular Organism composed of many cells; usually has organized tissues, organs, and organ systems. 2

multifactorial trait Controlled by several allelic pairs; each dominant allele contributes to the phenotype in an additive and like manner. 433

multiple allele (uh-leelz) Inheritance pattern in which there are more than two alleles for a particular trait; each individual has only two of all possible alleles. 435

multiple sclerosis (MS) Disease in which the outer myelin layer of nerve fiber insulation becomes scarred, interfering with normal conduction of nerve impulses. 139

multiregional continuity hypothesis Proposal that modern humans evolved independently in at least three different places: Asia, Africa, and Europe. 484

muscle dysmorphia Mental state where a person thinks his or her body is underdeveloped, and becomes preoccupied with body-building and diet; affects more men than women. 165

muscle fiber Muscle cell. 228

muscle tone Continuous, partial contraction of muscle.

muscle twitch Contraction of a whole muscle in response to a single stimulus. 236

muscular dystrophy (mus-ku-lar dis-tro-fee) Progressive muscle weakness and atrophy caused by deficient dystrophin protein. 240

muscularis Two layers of muscle in the gastrointestinal tract. 145

muscular system System of muscles that produces movement, both within the body and of its limbs; principal components are skeletal, smooth, and cardiac muscle. 75

muscular (contractile) tissue Type of tissue composed of fibers that can shorten and thicken. 65

mutagen (myoo-tuh-jun) Agent, such as radiation or a chemical, that brings about a mutation. 408

mutation Alteration in chromosome structure or number and also an alteration in a gene due to a change in DNA composition. 445

myalgia Muscular pain. 240

myasthenia gravis (mi-as-thee-ne-ah grav-is) Chronic disease characterized by muscles that are weak and easily fatigued. It results from the immune system's attack on neuromuscular junctions so that stimuli are not transmitted from motor neurons to muscle fibers. 139, 240

myelin sheath (my-uh-lin) White, fatty material, derived from the membrane of Schwann cells, that forms a covering for nerve fibers. 249

myocardium (my-oh-kar-dee-um) Cardiac muscle in the wall of the heart. 88

myofibril (my-uh-fy-brul) Contractile portion of muscle cells that contains a linear arrangement of sarcomeres and shortens to produce muscle contraction. 232

myoglobin Pigmented molecule in muscle tissue that stores oxygen. 237

myosin (my-uh-sin) One of two major proteins of muscle; makes up thick filaments in myofibrils of muscle fibers. See actin. 232

myxedema (mik-sih-dee-muh) Condition resulting from a deficiency of thyroid hormone in an adult. 303

N

NAD (nicotinamide adenine dinucleotide) Coenzyme that functions as a carrier of electrons and hydrogen ions, especially in cellular respiration. 53

nail Protective covering of the distal part of fingers and toes.　73

nasal cavity One of two canals in the nose, separated by a septum.　171

natural selection Mechanism resulting in adaptation to the environment.　474

Neandertal (nee-an-dur-thahl) Hominid with a sturdy build who lived during the last ice age in Europe and the Middle East; hunted large game and has left evidence of being culturally advanced.　485

nearsighted Vision abnormality due to an elongated eyeball from front to back; light rays focus in front of retina when viewing distant objects.　284

negative feedback Mechanism of homeostatic response in which a stimulus initiates reactions that reduce the stimulus.　80

nephron (nef-rahn) Microscopic kidney unit that regulates blood composition by glomerular filtration, tubular reabsorption, and tubular secretion.　191

nerve Bundle of long axons outside the central nervous system.　66

nerve impulse Action potential (electrochemical change) traveling along a neuron.　250

nervous system Organ system consisting of the brain, spinal cord, and associated nerves that coordinates the other organ systems of the body.　75

nervous tissue Tissue that contains nerve cells (neurons), which conduct impulses, and neuroglia, which support, protect, and provide nutrients to neurons.　66

neuroglia (noo-rahg-lee-uh, noo-rohg-lee-uh) Nonconducting nerve cells that are intimately associated with neurons and function in a supportive capacity.　66, 248

neuromuscular junction Region where an axon terminal approaches a muscle fiber; the synaptic cleft separates the axon terminal from the sarcolemma of a muscle fibre.　234

neuron (noor-ahn, nyoor-) Nerve cell that characteristically has three parts: dendrites, cell body, and axon.　66, 248

neurotransmitter Chemical stored at the ends of axons that is responsible for transmission across a synapse.　252

neutron (noo-trahn) Neutral subatomic particle, located in the nucleus and having a weight of approximately one atomic mass unit.　20

neutrophil (noo-truh-fil) Granular leukocyte that is the most abundant of the white blood cells; first to respond to infection.　110

niche (nich) Role an organism plays in its community, including its habitat and its interactions with other organisms.　494

nitrification Process by which nitrogen in ammonia and organic molecules is oxidized to nitrites and nitrates by soil bacteria.　503

nitrogen fixation Process whereby free atmospheric nitrogen is converted into compounds, such as ammonium and nitrates, usually by bacteria.　502

node of Ranvier (rahn-vee-ay) Gap in the myelin sheath around a nerve fiber.　249

nondisjunction Failure of homologous chromosomes or daughter chromosomes to separate during meiosis I and meiosis II, respectively.　393

nonrenewable resource Minerals, fossil fuels, and other materials present in essentially fixed amounts (within human time scales) in our environment.　514

norepinephrine (NE) (nor-ep-uh-nef-rin) Neurotransmitter of the postganglionic fibers in the sympathetic division of the autonomic system; also, a hormone produced by the adrenal medulla.　252, 305

nuclear envelope Double membrane that surrounds the nucleus and is connected to the endoplasmic reticulum; has pores that allow substances to pass between the nucleus and the cytoplasm.　49

nuclear pore Opening in the nuclear envelope that permits the passage of proteins into the nucleus and ribosomal subunits out of the nucleus.　49

nucleolus (noo-klee-uh-lus, nyoo-) Dark-staining, spherical body in the cell nucleus that produces ribosomal subunits.　49

nucleoplasm (noo-klee-uh-plaz-um) Semifluid medium of the nucleus, containing chromatin.　49

nucleotide Monomer of DNA and RNA consisting of a 5-carbon sugar bonded to a nitrogen-containing base and a phosphate group.　35

nucleus (noo-klee-us, nyoo-) (pl., nuclei) Membrane-bounded organelle that contains chromosomes and controls the structure and function of the cell.　44, 49, 259

nutrient Chemical substances in foods that are essential to the diet and contribute to good health.　158

O

obesity (oh-bee-sih-tee) Excess adipose tissue; exceeding ideal weight by more than 20%.　156

oil Substance, usually of plant origin and liquid at room temperature, formed when a glycerol molecule reacts with three fatty acid molecules.　30

oil gland Gland of the skin associated with hair follicle; secretes sebum; also called sebaceous gland.　73

olfactory cell (ahl-fak-tuh-ree, -tree, ohl-) Modified neuron that is a sensory receptor for the sense of smell.　279

omnivore (ahm-nuh-vor) Organism in a food chain that feeds on both plants and animals.　494

oncogene (ahng-koh-jeen) Cancer-causing gene.　406

oncologist Physician who specializes in one or more types of cancers.　407

oncology The study of cancer.　406

oogenesis (oh-uh-jen-uh-sis) Production of an egg in females by the process of meiosis and maturation.　326, 392

opportunistic infection Infection that has an opportunity to occur because the immune system has been weakened.　344

optic chiasma X-shaped structure on the underside of the brain formed by a partial crossing-over of optic nerve fibers.　283

optic nerve Either of two cranial nerves that carry nerve impulses from the retina of the eye to the brain, thereby contributing to the sense of sight.　281

organ Combination of two or more different tissues performing a common function.　2, 71

organelle (or-guh-nel) Small membranous structure in the cytoplasm having a specific structure and function.　44

organic Molecule that always contains carbon and hydrogen, and often contains ozygen as well; organic molecules are associated with living things.　27

organic molecule Type of molecule that contains carbon and hydrogen—and often contains oxygen also.　27

organism Individual living thing.　2

organ system Group of related organs working together.　2, 71

origin End of a muscle that is attached to a relatively immovable bone.　229

osmosis (ahz-moh-sis, ahs-) Diffusion of water through a selectively permeable membrane.　47

osmotic pressure Measure of the tendency of water to move across a selectively permeable membrane; visible as an increase in liquid on the side of the membrane with higher solute concentration.　47, 107

ossicle (ahs-ih-kul) One of the small bones of the middle ear—malleus, incus, and stapes.　286

ossification Formation of bone tissue.　210

osteoblast (ahs-tee-uh-blast) Bone-forming cell.　210

osteoclast (ahs-tee-uh-klast) Cell that causes erosion of bone.　210

osteocyte (ahs-tee-uh-syt) Mature bone cell located within the lacunae of bone.　208, 210

osteoporosis Condition in which bones break easily because calcium is removed from them faster than it is replaced.　160

otitis media (oh-ty-tis mee-dee-uh) Infection of the middle ear, characterized by pain and possibly by a sense of fullness, hearing loss, vertigo, and fever.　180

otolith (oh-tuh-lith) Calcium carbonate granule associated with ciliated cells in the utricle and the saccule.　291

outer ear Portion of ear consisting of the pinna and auditory canal.　286

out-of-Africa hypothesis Proposal that modern humans originated only in Africa; then migrated out of Africa and supplanted populations of early *Homo* in Asia and Europe about 100,000 years ago.   484

oval window Membrane-covered opening between the stapes and the inner ear.   286

ovarian cycle (oh-vair-ee-un) Monthly follicle changes occurring in the ovary that control the level of sex hormones in the blood and the uterine cycle.   326

ovary Female gonad that produces eggs and the female sex hormones.   311, 324

oviduct (oh-vuh-dukt) Tube that transports eggs to the uterus; also called uterine tube.   324

ovulation (ahv-yuh-lay-shun, ohv-) Release of a secondary oocyte from the ovary; if fertilization occurs, the secondary oocyte becomes an egg.   326

oxygen debt Amount of oxygen needed to metabolize lactate, a compound that accumulates during vigorous exercise.   237

oxyhemoglobin (ahk-see-hee-muh-gloh-bin) Compound formed when oxygen combines with hemoglobin.   108, 178

oxytocin (ahk-sih-toh-sin) Hormone released by the posterior pituitary that causes contraction of uterus and milk letdown.   300

ozone hole Seasonal thinning of the ozone shield in the lower stratosphere at the North and South Poles.   506

ozone shield Accumulation of O_3, formed from oxygen in the upper atmosphere; a filtering layer that protects the Earth from ultraviolet radiation.   506

P

pacemaker See sinoatrial (SA) node.   90

pain receptor Sensory receptor that is sensitive to chemicals released by damaged tissues or excess stimuli of heat or pressure.   274

pancreas (pang-kree-us, pan-) Internal organ that produces digestive enzymes and the hormones insulin and glucagon.   152, 308

pancreatic amylase (pang-kree-at-ik am-uh-lays, -layz) Enzyme in the pancreas that digests starch to maltose.   152

pancreatic islets (islets of Langerhans) Masses of cells that constitute the endocrine portion of the pancreas.   308

pandemic An increase in the occurrence of a disease within a large and geographically widespread population (often refers to a worldwide epidemic).   344

Pap test Analysis done on cervical cells for detection of cancer.   325

parasympathetic division That part of the autonomic system that is active under normal conditions; uses acetylcholine as a neurotransmitter.   265

parathyroid gland (par-uh-thy-royd) Gland embedded in the posterior surface of the thyroid gland; it produces parathyroid hormone.   304

parathyroid hormone (PTH) Hormone secreted by the four parathyroid glands that increases the blood calcium level and decreases the blood phosphate level.   304

parent cell Cell that divides so as to form daughter cells.   381

Parkinson disease Progressive deterioration of the central nervous system due to a deficiency in the neurotransmitter dopamine.   266

parturition (par-tyoo-rish-un, par-chuh-) Processes that lead to and include birth and the expulsion of the afterbirth.   369

passive immunity Protection against infection acquired by transfer of antibodies to a susceptible individual.   136

pathogen (path-uh-jun) Disease-causing agent.   64, 122

pectoral girdle (pek-tur-ul) Portion of the skeleton that provides support and attachment for an arm; consists of a scapula and a clavicle.   220

pelvic girdle Portion of the skeleton to which the legs are attached; consists of the coxal bones.   221

pelvis Bony ring formed by the sacrum and coxae.   221

penis External organ in males through which the urethra passes; also serves as the organ of sexual intercourse.   322

pentose (pen-tohs, -tohz) Five-carbon sugar. Deoxyribose is the pentose sugar found in DNA; ribose is a pentose sugar found in RNA.   28

pepsin (pep-sin) Enzyme secreted by gastric glands that digests proteins to peptides.   149

peptide bond Type of covalent bond that joins two amino acids.   33

peptide hormone Type of hormone that is a protein, a peptide, or derived from an amino acid.   298

pericardium (pair-ih-kar-dee-um) Protective serous membrane that surrounds the heart.   88

periodontitis Inflammation of the periodontal membrane that lines tooth sockets, causing loss of bone and loosening of teeth.   146

periosteum (pair-ee-ahs-tee-um) Fibrous connective tissue covering the surface of bone.   208

peripheral nervous system (PNS) (puh-rif-ur-ul) Nerves and ganglia that lie outside the central nervous system.   248

peristalsis (pair-ih-stawl-sis) Wavelike contractions that propel substances along a tubular structure, such as the esophagus.   147

peritonitis (pair-ih-tuh-ny-tis) Generalized infection of the lining of the abdominal cavity.   77, 145

peritubular capillary network (pair-ih-too-byuh-lur) Capillary network that surrounds a nephron and functions in reabsorption during urine formation.   192

Peyer's patches Lymphatic organs located in the small intestine.   128

phagocytosis (fag-uh-sy-toh-sis) Process by which amoeboid-type cells engulf large substances, forming an intracellular vacuole.   48

pharynx (far-ings) Portion of the digestive tract between the mouth and the esophagus that serves as a passageway for food and also for air on its way to the trachea.   147, 171

phenotype (fee-nuh-typ) Visible expression of a genotype—for example, brown eyes or attached earlobes.   422

phenylketonuria (PKU) Result of accumulation of phenylalanine, characterized by mental retardation, light pigmentation, eczema, and neurologic manifestations unless treated by a diet low in phenylalanine.   430

pheromone Chemical signal released by an organism that affects the metabolism or influences the behavior of another individual of the same species.   298

phospholipid (fahs-foh-lip-id) Molecule that forms the bilayer of the cell's membranes; has a polar, hydrophilic head bonded to two nonpolar, hydrophobic tails.   31

photoreceptor Sensory receptor in retina that responds to light stimuli.   274

photovoltaic (solar) cell An energy-conversion device that captures solar energy and directly converts it to electrical current.   522

pH scale Measurement scale for hydrogen ion concentration.   26

pilus Elongated, hollow appendage on bacteria used to transfer DNA from one cell to another.   122

pineal gland (pin-ee-ul, py-nee-ul) Endocrine gland located in the third ventricle of the brain; produces melatonin.   311

pinna Part of the ear that projects on the outside of the head.   286

pituitary dwarfism (pih-too-ih-tair-ee, -tyoo-) Condition in which an affected individual has normal proportions but small stature; caused by inadequate growth hormone.   302

pituitary gland Endocrine gland that lies just inferior to the hypothalamus; consists of the anterior pituitary and posterior pituitary.   300

placebo Treatment that is an inactive substance (pill, liquid, etc.), which is administered as if it were a therapy in an experiment, but which has no therapeutic value.   10

placenta (pluh-sen-tuh) Structure that forms from the chorion and the uterine wall and allows the embryo, and then the fetus, to acquire nutrients and rid itself of wastes.   330, 355

plaque (plak) Accumulation of soft masses of fatty material, particularly cholesterol, beneath the inner linings of the arteries.   97, 159

plasma (plaz-muh) Liquid portion of blood; contains nutrients, wastes, salts, and proteins.   107

plasma cell Cell derived from a B lymphocyte that is specialized to mass-produce antibodies.   131

plasma membrane Membrane surrounding the cytoplasm that consists of a phospholipid bilayer with embedded proteins; functions to regulate the entrance and exit of molecules from the cell.   44

plasma protein Protein dissolved in blood plasma.   107

plasmid (plaz-mid) Self-replicating ring of accessory DNA in the cytoplasm of bacteria.   122, 458

platelet (thrombocyte) (playt-lit) Component of blood that is necessary to blood clotting; also called a thrombocyte.   64, 113

pleura Serous membrane that encloses the lungs.   77

pneumonectomy (noo-muh-nek-tuh-mee, nyoo-) Surgical removal of all or part of a lung.   182

pneumonia (noo-mohn-yuh, nyoo-) Infection of the lungs that causes alveoli to fill with mucus and pus.   180

polar Combination of atoms in which the electrical charge is not distributed symmetrically.   24

polar body In oogenesis, a nonfunctional product; two to three meiotic products are of this type.   392

pollution Any environmental change that adversely affects the lives and health of living things.   514

polygenic trait Trait is controlled by several allelic paris; each dominant allele contributes to the phenotype in an additive and like manner.   433

polymerase chain reaction (PCR) (pahl-uh-muh-rays, -rayz) Technique that uses the enzyme DNA polymerase to produce millions of copies of a particular piece of DNA.   459

polyp (pahl-ip) Small, abnormal growth that arises from the epithelial lining.   155

polypeptide Polymer of many amino acids linked by peptide bonds.   34

polyribosome (pahl-ih-ry-buh-sohm) String of ribosomes simultaneously translating regions of the same mRNA strand during protein synthesis.   49, 450

polysaccharide (pahl-ee-sak-uh-ryd) Polymer made from sugar monomers; the polysaccharides starch and glycogen are polymers of glucose monomers.   28

pons (pahnz) Portion of the brain stem above the medulla oblongata and below the midbrain; assists the medulla oblongata in regulating the breathing rate.   259

population Organisms of the same species occupying a certain area.   2

positive feedback Mechanism in which the stimulus initiates reactions that lead to an increase in the stimulus.   81, 300

posterior pituitary Portion of the pituitary gland that stores and secretes oxytocin and antidiuretic hormone, which are produced by the hypothalamus.   300

precipitation   Water deposited on the Earth in the form of rain, snow, sleet, hail, or fog.   499

pre-embryonic development Development of the zygote in the first week, including fertilization, the beginning of cell division, and the appearance of the chorion.   356

prefrontal area Association area in the frontal lobe that receives information from other association areas and uses it to reason and plan actions.   258

primary germ layer Three layers (ectoderm, mesoderm, and endoderm) of embryonic cells that develop into specific tissues and organs.   358

primary motor area Area in the frontal lobe where voluntary commands begin; each section controls a part of the body.   257

primary somatosensory area (soh-mat-uh-sens-ree, -suh-ree) Area dorsal to the central sulcus where sensory information arrives from skin and skeletal muscles.   257

primate (pry-mayt) Animal that belongs to the order Primate; includes prosimians, monkeys, apes, and humans, all of whom have adaptations for living in trees.   476

principle Theory that is generally accepted by an overwhelming number of scientists; a law.   8

prion An infectious particle that is the cause of diseases, such as scrapie in sheep, mad cow disease, and Creutzfeldt-Jakob disease in humans; it has a protein component, but no nucleic acid has yet been detected.   124

producer Photosynthetic organism at the start of a grazing food chain that makes its own food (e.g., green plants on land and algae in water).   494

product Substance that forms as a result of a reaction.   53

progesterone (proh-jes-tuh-rohn) Female sex hormone that helps maintain sex organs and secondary sex characteristics.   311, 328

prokaryotic cell Type of cell that lacks a membrane-bounded nucleus and organelles.   44, 470

prolactin (PRL) (proh-lak-tin) Hormone secreted by the anterior pituitary that stimulates the production of milk from the mammary glands.   300

prophase (proh-fayz) Mitotic phase during which chromatin condenses so that chromosomes appear; chromosomes are scattered.   382

proprioceptor (proh-pree-oh-sep-tur) Sensory receptor in skeletal muscles and joints that assists the brain in knowing the position of the limbs.   276

prosimian Group of primates that includes lemur and tarsiers and may resemble the first primates to have evolved.   476

prostaglandin (prahs-tuh-glan-din) Hormone that has various and powerful local effects.   312

prostate gland (prahs-tayt) Gland located around the male urethra below the urinary bladder; adds secretions to semen.   321

protease   Enzyme capable of breaking peptide bonds in a protein.   348

protein Molecule consisting of one or more polypeptides.   33

protein-first hypothesis In chemical evolution, the proposal that protein originated before other macromolecules and allowed the formation of protocells.   469

proteomics The study of all proteins in an organism.   455

prothrombin (proh-thrahm-bin) Plasma protein that is converted to thrombin during the steps of blood clotting.   113

prothrombin activator Enzyme that catalyzes the transformation of the precursor prothrombin to the active enzyme thrombin.   113

protocell In biological evolution, a possible cell forerunner that became a cell once it could reproduce.   469

proton Positive subatomic particle, located in the nucleus and having a weight of approximately one atomic mass unit.   20

proto-oncogene (proh-toh-ahng-koh-jeen) Normal gene that can become an oncogene through mutation.   406

provirus   Latent form of a virus in which the viral DNA is incorporated into the chromosome of the host.   348

proximal convoluted tubule Highly coiled region of a nephron near the glomerular capsule, where tubular reabsorption takes place.   193

pseudostratified columnar epithelium Appearance of layering in some epithelial cells when, actually, each cell touches a baseline and true layers do not exist.   68

pulmonary artery (pool-muh-nair-ee, puul-) Blood vessel that takes blood away from the heart to the lungs.   89

pulmonary circuit Circulatory pathway that consists of the pulmonary trunk, the pulmonary arteries, and the pulmonary veins; takes O_2-poor blood from the heart to the lungs and O_2-rich blood from the lungs to the heart.   94

pulmonary fibrosis (fy-broh-sis) Accumulation of fibrous connective tissue in the lungs; caused by inhaling irritating particles, such as silica, coal dust, or asbestos.   181

pulmonary tuberculosis Tuberculosis of the lungs, caused by the bacillus *Mycobacterium tuberculosis*.   180

pulmonary vein Blood vessel that takes blood from the lungs to the heart.   89

pulse Vibration felt in arterial walls due to expansion of the aorta following ventricle contraction.   92

Punnett square (pun-ut) Gridlike device used to calculate the expected results of simple genetic crosses.   424

pupil (pyoo-pul) Opening in the center of the iris of the eye.   280

Purkinje fibers (pur-kin-jee) Specialized muscle fibers that conduct the cardiac impulse from the AV bundle into the ventricles.   90

pus Thick, yellowish fluid composed of dead phagocytes, dead tissue, and bacteria.   110, 129

pyelonephritis Inflammation of the kidney due to bacterial infection.   200

R

radio isotope Unstable form of an atom that spontaneously emits radiation in the form of radioactive particles or radiant energy.   21

reactant (re-ak-tunt) Substance that participates in a reaction.   53

recessive allele (uh-leel) Allele that exerts its phenotypic effect only in the homozygote; its expression is masked by a dominant allele.   422

recombinant DNA (rDNA) DNA that contains genes from more than one source.   458

rectum (rek-tum) Terminal end of the digestive tube between the sigmoid colon and the anus.   154

red blood cell (erythrocyte) Formed element that contains hemoglobin and carries oxygen from the lungs to the tissues; also called erythrocyte.   64, 108

red bone marrow Blood-cell-forming tissue located in the spaces within spongy bone.   127, 208

reduced hemoglobin (hee-muh-gloh-bun) Hemoglobin that is carrying hydrogen ions.   178

referred pain Pain perceived as having come from a site other than that of its actual origin.   277

reflex Automatic, involuntary response of an organism to a stimulus.   263

refractory period (rih-frak-tuh-ree) Time following an action potential when a neuron is unable to conduct another nerve impulse.   251

renal artery (ree-nul) Vessel that originates from the aorta and delivers blood to the kidney.   188

renal cortex (ree-nul kor-teks) Outer portion of the kidney that appears granular.   191

renal medulla (ree-nul muh-dul-uh) Inner portion of the kidney that consists of renal pyramids.   191

renal pelvis Hollow chamber in the kidney that lies inside the renal medulla and receives freshly prepared urine from the collecting ducts.   191

renal vein (ree-nul) Vessel that takes blood from the kidney to the inferior vena cava.   188

renewable resource Resources normally replaced or replenished by natural processes; resources not depleted by moderate use. Examples include solar energy, biological resources such as forests and fisheries, biological organisms, and some biogeochemical cycles.   514

renin (ren-in) Enzyme released by kidneys that leads to the secretion of aldosterone and a rise in blood pressure.   197, 306

replacement reproduction Population in which each person is replaced by only one child.   513

reproduce To produce a new individual of the same kind.   4

reproductive system Organ system that contains male or female organs and specializes in the production of offspring.   75

reproductive cloning Genetically identical to the original individual.   362

residual volume Amount of air remaining in the lungs after a forceful expiration.   176

respiratory control center Group of nerve cells in the medulla oblongata that sends out nerve impulses on a rhythmic basis, resulting in involuntary inspiration on an ongoing basis.   177

respiratory pump Mechanism whereby reductions in thoracic pressure during the breathing cycle tend to aid the return of blood to the heart from peripheral veins.   93

respiratory system Organ system consisting of the lungs and tubes that bring oxygen into the lungs and take carbon dioxide out.   74

resting potential Polarity across the plasma membrane of a resting neuron due to an unequal distribution of ions.   250

restriction enzyme Bacterial enzyme that stops viral reproduction by cleaving viral DNA; used to cut DNA at specific points during production of recombinant DNA.   458

reticular fiber (rih-tik-yuh-lur) Very thin collagen fibers in the matrix of connective tissue, highly branched and forming delicate supporting networks.   62

reticular formation (rih-tik-yuh-lur) Complex network of nerve fibers within the central nervous system that arouses the cerebrum.   259

retina (ret-n-uh, ret-nuh) Innermost layer of the eyeball that contains the rod cells and the cone cells.   281

retinal (ret-n-al, -awl) Light-absorbing molecule that is a derivative of vitamin A and a component of rhodopsin.   282

retrovirus RNA virus containing the enzyme reverse transcriptase that carries out RNA to DNA transcription.   348

reverse transcriptase Enzyme that speeds the conversion of viral RNA to viral DNA.   348

rheumatoid arthritis Persistent inflammation of synovial joints, often causing cartilage destruction, bone erosion, and joint deformities.   139

rhodopsin (roh-dahp-sun) Light-absorbing molecule in rod cells and cone cells that contains a pigment and the protein opsin.   282

ribosomal RNA (rRNA) (ry-buh-soh-mul) Type of RNA found in ribosomes where protein synthesis occurs.   446

ribosome (ry-buh-sohm) RNA and protein in two subunits; site of protein synthesis in the cytoplasm.   49

rigor mortis Contraction of muscles at death due to lack of ATP.   228

RNA (ribonucleic acid) (ry-boh-noo-klee-ik) Nucleic acid produced from covalent bonding of nucleotide monomers that contain the sugar ribose; occurs in three forms: messenger RNA, ribosomal RNA, and transfer RNA.   35, 446

RNA-first hypothesis In chemical evolution, the proposal that RNA originated before other macromolecules and allowed the formation of the first cell(s).   469

RNA polymerase (pahl-uh-muh-rays) During transcription, an enzyme that joins nucleotides complementary to a DNA template.   449

rod cell Photoreceptor in retina of eyes that responds to dim light.   282

rotational equilibrium Maintenance of balance when the head and body are suddenly moved or rotated.   291

rotator cuff Tendons that encircle and help form a socket for the humerus, and also help reinforce the shoulder joint.   220

round window Membrane-covered opening between the inner ear and the middle ear.   286

rugae Deep folds, as in the wall of the stomach.   149

runoff Water, from rain, snowmelt, or other sources, that flows over the land surface, adding to the water cycle.   499

S

SA (sinoatrial) node (sy-noh-ay-tree-ul) Small region of neuromuscular tissue that initiates the heartbeat; also called the pacemaker.   90

saccule (sak-yool) Saclike cavity in the vestibule of the inner ear; contains sensory receptors for gravitational equilibrium. 291

salinization Process in which mineral salts accumulate in the soil, killing plants; occurs when soils in dry climates are irrigated profusely. 518

salivary amylase (sal-uh-vair-ee am-uh-lays, -layz) Secreted from the salivary glands; the first enzyme to act on starch. 146

salivary gland Gland associated with the mouth that secretes saliva. 146

saltatory conduction Movement of nerve impulses from one neurofibral node to another along a myelinated axon. 251

saltwater intrusion Movement of salt water into freshwater aquifers in coastal areas where groundwater is withdrawn faster than it is replenished. 517

sarcolemma (sar-kuh-lem-uh) Plasma membrane of a muscle fiber; also forms the tubules of the T system involved in muscular contraction. 232

sarcoma Cancer that arises in muscles and connective tissues. 407

sarcomere (sar-kuh-mir) One of many units, arranged linearly within a myofibril, whose contraction produces muscle contraction. 232

sarcoplasmic reticulum (sar-kuh-plaz-mik rih-tik-yuh-lum) Smooth endoplasmic reticulum of skeletal muscle cells; surrounds the myofibrils and stores calcium ions. 232

saturated fatty acid Fatty-acid molecule that lacks double bonds between the atoms of its carbon chain. 30

scavenger Animal that specializes in the consumption of dead animals. 494

schistosomiasis Infection by a blood fluke (*Schistosoma*), often as a result of contact with contaminated water in rivers and streams. Symptoms include fever, chills, diarrhea, and cough. Infection may be chronic. 449

Schwann cell Cell that surrounds a fiber of a peripheral nerve and forms the myelin sheath. 249

science Development of concepts about the natural world, often by utilizing the scientific method. 8

scientific method Process of attaining knowledge by making observations, testing hypotheses, and coming to conclusions. 8

scientific theory Concept supported by a broad range of observations, experiments, and conclusions. 8

sclera (skleer-uh) White, fibrous, outer layer of the eyeball. 280

scoliosis Abnormal laterial (side-to-side) curvature of the vertebral column. 218

scrotum (skroh-tum) Pouch of skin that encloses the testes. 321

secondary oocyte In oogenesis, the functional product of meiosis I; becomes the egg. 392

second messenger Chemical signal such as cyclic AMP that causes the cell to respond to the first messenger—a hormone bound to a receptor protein in the plasma membrane. 299

selectively permeable Having degrees of permeability; the cell is impermeable to some substances and allows others to pass through at varying rates. 46

self (protein) Proteins in plasma membranes that contribute to the specificity of a tissue. 134

semantic memory Capacity of the brain to store and retrieve information with regard to words or numbers. 260

semen (see-mun) Thick, whitish fluid consisting of sperm and secretions from several glands of the male reproductive tract. 321

semicircular canal (sem-ih-sur-kyuh-lur) One of three tubular structures within the inner ear that contain sensory receptors responsible for the sense of rotational equilibrium. 286, 291

semilunar valve (sem-ee-loo-nur) Valve resembling a half moon located between the ventricles and their attached vessels. 88

seminal vesicle (sem-uh-nul) Convoluted structure attached to the vas deferens near the base of the urinary bladder in males; adds secretions to semen. 321

seminiferous tubule (sem-uh-nif-ur-us) Long, coiled structure contained within chambers of the testis; where sperm are produced. 323

sensation Conscious awareness of a stimulus due to nerve impulses sent to the brain from a sensory receptor by way of sensory neurons. 275

sensory adaptation Phenomenon of a sensation becoming less noticeable once it has been recognized by constant repeated stimulation. 275

sensory neuron Nerve cell that transmits nerve impulses to the central nervous system after a sensory receptor has been stimulated. 249

sensory receptor Structure that receives either external or internal environmental stimuli and is a part of a sensory neuron or transmits signals to a sensory neuron. 249, 274

septum Wall between two cavities; in the human heart, a septum separates the right side from the left side. 88

serosa Membrane that covers internal organs and lines cavities without an opening to the outside of the body. 145

serotonin A neurotransmitter. 252

serous membrane (seer-us) Membrane that covers internal organs and lines cavities without an opening to the outside of the body; also called serosa. 77

Sertoli cell Cell associated with developing germ cells in seminiferous tubule; secretes

fluid into seminiferous tubule and mediates hormonal effects on tubule. 323

serum (seer-um) Light yellow liquid left after clotting of blood. 113

severe combined immunodeficiency disease (SCID) Congenital illness in which both antibody- and cell-mediated immunity are lacking or inadequate. 111, 139

sex chromosome Chromosome that determines the sex of an individual; in humans, females have two X chromosomes, and males have an X and Y chromosome. 436

sex-linked Allele that occurs on the sex chromosomes but may control a trait that has nothing to do with the sex characteristics of an individual. 436

short-term memory Retention of information for only a few minutes, such as remembering a telephone number. 260

sickle-cell disease Genetic disorder in which the affected individual has sickle-shaped red blood cells that are subject to hemolysis. 109, 431

simple goiter (goy-tur) Condition in which an enlarged thyroid produces low levels of thyroxine. 303

sinkhole Large surface crater caused by the collapse of an underground channel or cavern; often triggered by groundwater withdrawal. 517

sinus (sy-nus) Cavity or hollow space in an organ such as the skull. 216

sinusitis (sy-nuh-sy-tis) Infection of the sinuses, caused by blockage of the openings to the sinuses and characterized by postnasal discharge and facial pain. 180

sister chromatids One of two genetically identical chromosomal units that are the result of DNA replication and are attached to each other at the centromere. 378

skeletal muscle Striated, voluntary muscle tissue found in muscles that move the bones. 65, 228

skeletal muscle pump Pumping effect of contracting skeletal muscles on blood flow through underlying vessels. 93

skeletal system System of bones, cartilage, and ligaments that works with the muscular system to protect the body and provide support for locomotion and movement. 75

skill memory Capacity of the brain to store and retrieve information necessary to perform motor activities, such as riding a bike. 260

skin Outer covering of the body; can be called the integumentary system because it contains organs such as sense organs. 71

skull Bony framework of the head, composed of cranial bones and the bones of the face. 216

sliding filament model An explanation for muscle contraction based on the movement of actin filaments in relation to myosin filaments. 232

small intestine Long, tubelike chamber of the digestive tract between the stomach and large intestine.   150

smooth (visceral) muscle Nonstriated, involuntary muscle tissue found in the walls of internal organs.   65, 228

sodium-potassium pump Carrier protein in the plasma membrane that moves sodium ions out of and potassium ions into cells; important in nerve and muscle cells.   250

soft palate (pal-it) Entirely muscular posterior portion of the roof of the mouth.   146

somatic system That portion of the peripheral nervous system containing motor neurons that control skeletal muscles.   263

spasm Sudden, involuntary contraction of one or more muscles.   240

species Group of similarly constructed organims capable of interbreeding and producing fetile offspring; organisms that share a common gene pool.   2

sperm Male gamete having a haploid number of chromosomes and the ability to fertilize an egg, the female gamete.   323

spermatogenesis (spur-mat-uh-jen-ih-sis) Production of sperm in males by the process of meiosis and maturation.   323, 392

sphincter (sfingk-tur) Muscle that surrounds a tube and closes or opens the tube by contracting and relaxing.   147

spinal cord Part of the central nervous system; the nerve cord that is continuous with the base of the brain plus the vertebral column that protects the nerve cord.   254

spinal nerve Nerve that arises from the spinal cord.   262

spiral organ Organ in the cochlear duct of the inner ear that is responsible for hearing; also called the organ of Corti.   287

spleen Large, glandular organ located in the upper left region of the abdomen; stores and purifies blood.   128

spongy bone Porous bone found at the ends of long bones where red bone marrow is sometimes located.   63, 208

sprain Injury to a ligament caused by abnormal force applied to a joint.   240

squamous epithelium (skway-mus, skwah-) Type of epithelial tissue that contains flat cells.   68

standard error Number used in evaluating statistical data to show the range of error in the data.   12

stapes (stay-peez) The last of three ossicles of the ear that serve to conduct vibrations from the tympanic membrane to the oval window of the inner ear.   286

starch Storage polysaccharide found in plants that is composed of glucose molecules joined in a linear fashion with few side chains.   28

steroid (steer-oyd) Type of lipid molecule having a complex of four carbon rings; examples are cholesterol, progesterone, and testosterone.   32

stimulus Change in the internal or external environment that a sensory receptor can detect, leading to nerve impulses in sensory neurons.   274

stomach Muscular sac that mixes food with gastric juices to form chyme, which enters the small intestine.   148

strain Injury to a muscle resulting from overuse or improper use.   240

striae gravidarum Linear, depressed, scarlike lesions occurring on the abdomen, breasts, buttocks, and thighs due to the weakening of the elastic tissues during pregnancy.   368

striated (stry-ayt-ud) Having bands; in cardiac and skeletal muscle, alternating light and dark crossbands produced by the distribution of contractile proteins.   65

stroke Condition resulting when an arteriole in the brain bursts or becomes blocked by an embolism; also called cerebrovascular accident.   97

subcutaneous layer (sub-kyoo-tay-nee-us) Tissue layer that lies just beneath the skin and contains adipose tissue.   71

submucosa Layer of connective tissue underneath a mucous membrane.   145

subsidence Occurs when a portion of the Earth's surface gradually settles downward.   516

substrate Reactant in a reaction controlled by an enzyme.   53

sudden infant death syndrome (SIDS) Any sudden and unexplained death of an apparently healthy infant aged one month to one year.   177

superior vena cava (vee-nuh kay-vuh) Large vein that enters the right atrium from above and carries blood from the head, thorax, and upper limbs to the heart.   94

surfactant Agent that reduces the surface tension of water; in the lungs, a surfactant prevents the alveoli from collapsing.   174

sustainable Ability of a society or ecosystem to maintain itself while also providing services to human beings.   530

suture (soo-chur) Type of immovable joint articulation found between bones of the skull.   222

sweat gland Skin gland that secretes a fluid substance for evaporative cooling; also called sudoriferous gland.   73

sympathetic division The part of the autonomic system that usually promotes activities associated with emergency (fight-or-flight) situations; uses norepinephrine as a neurotransmitter.   265

synapse (sin-aps, si-naps) Junction between neurons consisting of the presynaptic (axon) membrane, the synaptic cleft, and the postsynaptic (usually dendrite) membrane.   252

synapsis (sih-nap-sis) Pairing of homologous chromosomes during prophase I of meiosis I.   385

synaptic cleft (sih-nap-tik) Small gap between presynaptic and postsynaptic membranes of a synapse.   252

syndrome Group of symptoms that appear together and tend to indicate the presence of a particular disorder.   380

synovial membrane Membrane that forms the inner lining of the capsule of a freely movable joint.   77

systemic circuit Blood vessels that transport blood from the left ventricle and back to the right atrium of the heart.   94

systemic lupus erythematosus (SLE) Syndrome involving the connective tissues and various organs, including kidney.   139

systole (sis-tuh-lee) Contraction period of the heart during the cardiac cycle.   90

systolic pressure (sis-tahl-ik) Arterial blood pressure during the systolic phase of the cardiac cycle.   92

T

T cell (T lymphocyte) Lymphocyte that matures in the thymus. Cytotoxic T cells kill antigen-bearing cells outright; helper T cells release cytokines that stimulate other immune system cells.   130

T (transverse) tubule Membranous channel that extends inward.   232

taste bud Sense organ containing the receptors associated with the sense of taste.   278

Tay-Sachs disease Lethal genetic disease in which the newborn has a faulty lysosomal digestive enzyme.   430

technology The science or study of the practical or industrial arts.   14

tectorial membrane (tek-tor-ee-ul) Membrane that lies above and makes contact with the hair cells in the spiral organ.   287

telomere (tel-uh-meer) Tip of the end of a chromosome.   404

telophase (tel-uh-fayz) Mitotic phase during which daughter chromosomes are located at each pole.   383

template (tem-plit) Pattern or guide used to make copies; parental strand of DNA serves as a guide for the production of daughter DNA strands, and DNA also serves as a guide for the production of messenger RNA.   445

tendinitis (ten-din-eye-tis) An inflammation of muscle tendons and their attachments.   240

tendon (ten-dun) Strap of fibrous connective tissue that connects skeletal muscle to bone.   63, 208, 229

testes (sing., testis) (tes-teez, tes-tus) Male gonads that produce sperm and the male sex hormones.   311, 321

test group Group exposed to the experimental variable in an experiment, rather than the control group.   10

testosterone (tes-tahs-tuh-rohn) Male sex hormone that helps maintain sexual organs and secondary sex characteristics.   311, 324, 364

tetanus (tet-n-us) Sustained muscle contraction without relaxation.   236

tetany (tet-n-ee) Severe twitching caused by involuntary contraction of the skeletal muscles due to a calcium imbalance.   304

thalamus (thal-uh-mus) Part of the brain located in the lateral walls of the third ventricle that serves as the integrating center for sensory input; it plays a role in arousing the cerebral cortex.   258

therapeutic cloning Used to create mature cells of various cell types. Also, used to learn about specialization of cells and provice cells and tissue to treat human illnesses.   362

thermoreceptor Sensory receptor that is sensitive to changes in temperature.   274

threshold Electrical potential level (voltage) at which an action potential or nerve impulse is produced.   250

thrombin (thrahm-bin) Enzyme that converts fibrinogen to fibrin threads during blood clotting.   113

thrombocytopenia Insufficient number of platelets in the blood.   113

thromboembolism (thrahm-boh-em-buh-liz-um) Obstruction of a blood vessel by a thrombus that has dislodged from the site of its formation.   97, 113

thrombus (thrahm-bus) Blood clot that remains in the blood vessel where it formed.   97

thymine (T) (thy-meen) One of four nitrogen-containing bases in nucleotides composing the structure of DNA; pairs with adenine.   35

thymosin Group of peptides secreted by the thymus gland that increases production of certain types of white blood cells.   311

thymus gland Lymphatic organ, located along the trachea behind the sternum, involved in the maturation of T lymphocytes in the thymus gland. Secretes hormones called thymosins, which aid the maturation of T cells and perhaps stimulate immune cells in general.   127, 311

thyroid gland Endocrine gland in the neck that produces several important hormones, including thyroxine, triiodothyronine, and calcitonin.   303

thyroid-stimulating hormone (TSH) Substance produced by the anterior pituitary that causes the thyroid to secrete thyroxine and triiodothyronine.   300

thyroxine (T_4) (thy-rahk-sin) Hormone secreted from the thyroid gland that promotes growth and development; in general, it increases the metabolic rate in cells.   303

tidal volume Amount of air normally moved in the human body during an inspiration or expiration.   176

tight junction Junction between cells when adjacent plasma membrane proteins join to form an impermeable barrier.   70

tissue Group of similar cells that perform a common function.   2, 62, 176

tissue fluid Fluid that surrounds the body's cells; consists of dissolved substances that leave the blood capillaries by filtration and diffusion.   64, 86

tonicity (toh-nis-ih-tee) Osmolarity of a solution compared with that of a cell. If the solution is isotonic to the cell, there is no net movement of water; if the solution is hypotonic, the cell gains water; and if the solution is hypertonic, the cell loses water.   47

tonsillectomy (tahn-suh-lek-tuh-mee) Surgical removal of the tonsils.   180

tonsillitis Infection of the tonsils that causes inflammation and can spread to the middle ears.   180

tonsils Partially encapsulated lymph nodules located in the pharynx.   128, 180

total artificial heart (TAH) A mechanical replacement for the heart, as opposed to a partial replacement.   100

toxin Poisonous substance produced by living cells or organisms. Toxins are nearly always proteins that are capable of causing disease on contact or absorption with body tissues.   123

tracer Substance having an attached radioisotope that allows a researcher to track its whereabouts in a biological system.   21

trachea (tray-kee-uh) Passageway that conveys air from the larynx to the bronchi; also called the windpipe.   173

tracheostomy (tray-kee-ahs-tuh-mee) Creation of an artificial airway by incision of the trachea and insertion of a tube.   173

tract Bundle of myelinated axons in the central nervous system.   254

transcription Process whereby a DNA strand serves as a template for the formation of mRNA.   448

transcription factor In eukaryotes, protein required for the initiation of transcription by RNA polymerase.   453

trans fat Fats, which occur naturally in meat and dairy products of ruminants, that are also industrially created through partial hydrogenation of plant oils and animal fats.   30

transfer RNA (tRNA) Type of RNA that transfers a particular amino acid to a ribosome during protein synthesis; at one end, it binds to the amino acid, and at the other end it has an anticodon that binds to an mRNA codon.   446

transgenic organism Free-living organism in the environment that has a foreign gene in its cells.   460

translation Process whereby ribosomes use the sequence of codons in mRNA to produce a polypeptide with a particular sequence of amino acids.   448

translocation Movement of a chromosomal segment from one chromosome to another nonhomologous chromosome, leading to abnormalities; e.g, Down syndrome.   397

triglyceride (trih-glis-uh-ryd) Neutral fat composed of glycerol and three fatty acids.   30

triplet code Each sequence of three nucleotide bases in the DNA of genes stands for a particular amino acid.   448

trisomy One more chromosome than usual.   393

trophic level Feeding level of one or more populations in a food web.   497

trophic relationship In ecosystems, feeding relationships such as grazing food webs or detrital food webs.   497

tropomyosin (trahp-uh-my-uh-sin, trohp-) Protein that functions with troponin to block muscle contraction until calcium ions are present.   235

troponin (troh-puh-nin) Protein that functions with tropomyosin to block muscle contraction until calcium ions are present.   235

trypsin (trip-sin) Protein-digesting enzyme secreted by the pancreas.   152

tubal ligation Method for preventing pregnancy in which the uterine tubes are cut and sealed.   332

tubular reabsorption Movement of primarily nutrient molecules and water from the contents of the nephron into blood at the proximal convoluted tubule.   195

tubular secretion Movement of certain molecules from blood into the distal convoluted tubule of a nephron so that they are added to urine.   195

tumor (too-mur) Cells derived from a single mutated cell that has repeatedly undergone cell division; benign tumors remain at the site of origin, and malignant tumors metastasize.   404

tumor marker test Blood test for a substance, such as a tumor antigen, that indicates a patient has cancer.   414

tumor-suppressor gene Gene that codes for a protein that ordinarily suppresses cell division; inactivity can lead to a tumor.   406

tympanic membrane (tim-pan-ik) Located between the outer and middle ear where it receives sound waves; also called the eardrum.   286

U

umbilical cord Cord connecting the fetus to the placenta through which blood vessels pass.   359

unsaturated fatty acid Fatty-acid molecule that has one or more double bonds between the atoms of its carbon chain. 30

uracil (U) (yoor-uh-sil) The base in RNA that replaces thymine found in DNA; pairs with adenine. 35, 446

urea (yoo-ree-uh) Primary nitrogenous waste of humans derived from amino acid breakdown. 189

uremia High level of urea nitrogen in the blood. 200

ureter (yoor-uh-tur) One of two tubes that take urine from the kidneys to the urinary bladder. 188

urethra (yoo-ree-thruh) Tubular structure that receives urine from the bladder and carries it to the outside of the body. 189, 321

urethritis Inflammation of the urethra. 200

uric acid (yoor-ik) Waste product of nucleotide metabolism. 189

urinary bladder Organ where urine is stored before being discharged by way of the urethra. 188

urinary system Organ system consisting of the kidneys and urinary bladder; rids the body of nitrogenous wastes and helps regulate the water-salt balance of the blood. 75

uterine cycle (yoo-tur-in, -tuh-ryn) Monthly occurring changes in the characteristics of the uterine lining (endometrium). 328

uterus (yoo-tur-us) Organ located in the female pelvis where the fetus develops; also called the womb. 325

utricle (yoo-trih-kul) Saclike cavity in the vestibule of the inner ear that contains sensory receptors for gravitational equilibrium. 291

V

vaccine Antigens prepared in such a way that they can promote active immunity without causing disease. 136

vagina Organ that leads from the uterus to the vestibule and serves as the birth canal and organ of sexual intercourse in females. 325

valve Membranous extension of a vessel of the heart wall that opens and closes, ensuring one-way flow. 87

vas deferens (vas def-ur-unz, -uh-renz) Tube that leads from the epididymis to the urethra in males. 321

vasectomy Method for preventing pregnancy in which the vasa deferentia are cut and sealed. 333

vector (vek-tur) In genetic engineering, a means to transfer foreign genetic material into a cell (e.g., a plasmid). 458

ventilation Process of moving air into and out of the lungs; also called breathing. 170

ventricle (ven-trih-kul) Cavity in an organ, such as a lower chamber of the heart or the ventricles of the brain. 88, 254

venule (ven-yool, veen-) Vessel that takes blood from capillaries to a vein. 87

vermiform appendix Small, tubular appendage that extends outward from the cecum of the large intestine. 154

vernix caseosa (vur-niks kay-see-oh-suh) Cheeselike substance covering the skin of the fetus. 363

vertebral column (vur-tuh-brul) Series of joined vertebrae that extends from the skull to the pelvis. 218

vertebrate (vur-tuh-brit, -brayt) An animal with a vertebral column. 6

vesicle Small, membrane-bounded sac that stores substances within a cell. 50

vestibule (ves-tuh-byool) Space or cavity at the entrance of a canal, such as the cavity that lies between the semicircular canals and the cochlea. 286

vestigial structure Remains of a structure that was functional in some ancestor but is no longer functional in the organism in question. 472

villus (pl., villi) (vil-us) Small, finger-like projection of the inner small intestinal wall. 150

virus Noncellular, parasitic agent consisting of an outer capside and an inner core of nucleic acid. 123

visual accommodation Ability of the eye to focus at different distances by changing the curvature of the lens. 281

vital capacity Maximum amount of air moved in or out of the human body with each breathing cycle. 176

vitamin Essential requirement in the diet, needed in small amounts. They are often part of coenzymes. 162

vitamin D Required for proper bone growth. 72

vitreous humor (vit-ree-us) Clear, gelatinous material between the lens of the eye and the retina. 281

vocal cord Fold of tissue within the larynx; creates vocal sounds when it vibrates. 172

vulva External genitals of the female that surround the opening of the vagina. 326

W

water (hydrologic) cycle Interdependent and continuous circulation of water from the ocean to the atmosphere, to the land, and back to the ocean. 499

Wernicke's area Brain area involved in language comprehension. 258

white blood cell (leukocyte) Type of blood cell that is transparent without staining and protects the body from invasion by foreign substances and organisms; also called a leukocyte. 64, 110

white matter Myelinated axons in the central nervous system. 254

X

xenotransplantation Use of animal organs, instead of human organs, in human transplant patients. 139, 463

X-linked Allele located on an X chromosome, but may control a trait that has nothing to do with the sex characteristics of an individual. 436

Y

yolk sac Extraembryonic membrane that encloses the yolk of birds; in humans, it is the first site of blood cell formation. 355

Z

zygote (zy-goht) Diploid cell formed by the union of sperm and egg; the product of fertilization. 324, 354, 385

Credits

Photographs

Chapter 1
Opener: © image100/Alamy; 1.1(giraffes): © Dr. Sylvia S. Mader; 1.1(monarch): © Dave Cavagnaro/Peter Arnold, Inc.; 1.1(sequoia): © Vol. 74/PhotoDisc/Getty Images; 1.1(mushrooms): © Hans Pfletschinger/Peter Arnold, Inc.; 1.1(family): © David Young/PhotoEdit; 1.1(angelfish): © Tom Stack/Tom Stack & Associates; 1.3a: © John Cancalosi/Peter Arnold, Inc.; 1.3b: © Vol. 124/Corbis; 1.4a(seedling): © Herman Eisenbeiss/Photo Researchers, Inc.; 1.4a(sapling): Courtesy Paul Wray, Iowa State University; 1.4b(2 cell stage): Lennart Nilsson/Albert Bonniers Forlag AB, *A Child is Born*, Delacorte Press; 1.4b(fetus): Lennart Nilsson, *A Child is Born*, 1990 Delacorte Press; 1.6a: © William M. Smithy/Getty Images; 1.6b: © Don and Pat Valenti/DRK; 1.7: Courtesy Leica Microsystems Inc.; 1.8 main: © Tony McDonough/epa/Corbis; 1.8 inset: © Eye of Science/Photo Researchers, Inc.; 1.9(people): © image100 Ltd.; 1.9(endoscopy): © Phanie/Photo Researchers, Inc.; 1A(left): © Royalty-Free/Corbis; 1A(right): © Jim Craigmyle/Corbis; 1B: © Daniel Bedell/AnimalsAnimals; 1.11a: © Edgar Bernstein/Peter Arnold, Inc.; 1.11b: © Luiz C. Marigo/Peter Arnold, Inc.

Chapter 2
Opener: © Larry Bray/Getty Images; 2.3a: © Biomed Commun/Custom Medical Stock Photo; 2.3b(left): © Mazzlota et al./Photo Researchers, Inc; 2.3b(right): © Hank Morgan/Rainbow; 2.4: (radiographers): © SPL/Photo Researchers, Inc.; 2.5b(crystals): © Charles M. Falco/Photo Researchers, Inc. 2.5b(shaker): © The McGraw-Hill Companies, Inc./Evelyn Jo Hebert, photographer; 2.8a: © Martin Dohrn/SPL/Photo Researchers, Inc.; 2.8b: © Amanda Langford/SuperStock RF; 2.10a: © Ray Pfortner/Peter Arnold, Inc.; 2.10b: © Frederica Georgia/Photo Researchers, Inc.; 2.12: © Jeremy Burgess/SPL/Photo Researchers, Inc.; 2.13: © Don W. Fawcett/Photo Researchers, Inc.; p. 29(teacup): © Royalty-Free/Corbis, 2.17(man): © Royalty-Free/Corbis, 2.17(woman): © Royalty-Free/Corbis; p. 32(Nicole Kidman): © Reuters/Corbis; p. 32(red blood cells): © P. Motta & S. Correr/Photo Researchers, Inc.; p. 32(woman running): © Duomo/Corbis; 2A: © John Nuhn.

Chapter 3
Opener: © Dennis MacDonald/PhotoEdit; 3.1a(RBCs): © Prof. P. Motta, Dept. of Anatomy, Univ. LaSapienza Rome/SPL/Photo Researchers, Inc.; 3.1a(nerve cells): © Dr. Dennis Kunkel/Visuals Unlimited; 3.1a(hyaline cartilage): © Ed Reschke, 3.2a: © David M. Phillips/Visuals Unlimited; 3.2b: © Alfred Pasieka/Photo Researchers, Inc.; 3.2c: © Warren Rosenberg/BPS; 3.2d: © Inga Spence/Visuals Unlimited; 3.4: © Alfred Pasieka/Photo Researchers, Inc.; 3.8(all): © Dennis Kunkel/Phototake; 3.11(left): Courtesy E.G. Pollock; 3.11(right): © R. Bolender & D. Fawcett/Visuals Unlimited; 3.13b: © Y. Nikas/Photo Researchers, Inc.; 3.13c: © David M. Phillips/Photo Researchers, Inc.; 3.14: © Dr. Don W. Fawcett/Visuals Unlimited; 3.18: © Creasource/Corbis; 3A: © Jerry Cooke/Photo Researchers, Inc.

Chapter 4
Opener: © Jose Luis Pelaez, Inc./Corbis; 4.2a(all), 4.5a: © Ed Reschke; 4.5b: © The McGraw-Hill Companies, Inc./Dennis Strete, Photographer; 4.5c, 4.6: © Ed Reschke; 4Ba: © AP Photo/Ellcot D. Novak; 4Bb: © Diana De Rosa/Press Link; 4.7(all): © Ed Reschke; 4.10a: © John D. Cunningham/Visuals Unlimited; 4.10b: © Ken Greer/Visuals Unlimited; 4.10c: © James Stevenson/SPL/Photo Researchers, Inc.; 4.14: © AP Photo/Obed Zilwa.

Chapter 5
Opener: © Warren Morgan/Corbis; 5.2(left): © Ed Reschke; 5.2(right): © Biophoto Associates/Photo Researchers, Inc.; 5.3b: © SIU/Visuals Unlimited; 5.4b: © Dr. Don W. Fawcett/Visuals Unlimited; 5.5d: © Biophoto Associates/Photo Researchers, Inc.; 5.6(left): © Ed Reschke; 5.6(right): © David Joel/MacNeal Hospital/Getty Images; 5.7: © Sheila Terry/SPL/Photo Researchers, Inc.; 5.9c: © Gondelon/Photo Researchers, Inc.; 5.14(left): © Ed Reschke; 5.14(right): © Biophoto

Associates/Photo Researchers, Inc.; 5A: Ryan McVay/Getty Images; 5.15: © Pascal Goethgheluck/SPL/Photo Researchers, Inc.; 5.16b: ABIOMED, Inc.; 5B: © Paul Hardy/Corbis.

Chapter 6
Opener: © Medical-on-Line/Alamy; 6.1: © Doug Menuez/Getty Images; 6.2: © Michael Keller/Corbis; 6.3a: © Lennart Nilsson, *Behold Man*, Little Brown and Company, Boston; 6.3b: © Andrew Syred/Photo Researchers, Inc.; p.109(sickle cell): © Phototake, Inc./Alamy; 6.5(all): © Dr. Fred Hossler/Visuals Unlimited; 6.7: © Bettman/Corbis; 6A: © Michael Del Guercio/Photo Researchers, Inc.; 6.8: © Eye of Science/Photo Researchers, Inc.; 6.12: © Royalty-Free/Corbis.

Chapter 7
Opener: © Joel Benard/Masterfile; 7.1b: © Dr. David M. Phillips/Visuals Unlimited; 7.1c: © Dr. Dennis Kunkel/Visuals Unlimited; 7.1d: © Dr. Gary D. Gaugler/Phototake; 7.5: © AP Photo/Xinhua, Huang Benquiang; 7A: © Vassil Donev/epa/Corbis; 7.8(marrow): © R. Valentine/Visuals Unlimited; 7.8(thymus): © Ed Reschke/Peter Arnold, Inc.; 7.8(lymph): © Fred E. Hossler/Visuals Unlimited; 7.8(spleen): © Ed Reschke/Peter Arnold, Inc.; 7.12b: Courtesy Dr. Arthur J. Olson, Scripps Institute; 7Ba: © AP/Wide World Photo; 7.14b: © Bohringer Ingelheim International, photo by Lennart Nilsson; 7.15a: © Michael Newman/PhotoEdit; 7.16b: © Digital Vision/Getty Images; 7.16c: © Aaron Haupt/Photo Researchers, Inc.; 7.18(pollen): © David Scharf/SPL/Photo Researchers, Inc.; 7.18(girl): © Damien Lovegrove/SPL/Photo Researchers, Inc.; 7.19: © Richard Anderson.

Chapter 8
Opener: © Roger Ressmeyer/Corbis; 8A(bike): © Royalty-Free/Corbis; 8A(man): © Stockbyte/Getty Images; 8.5c: © Dr. Fred Hossler/Visuals Unlimited; 8.6(left): © Manfred Kage/Peter Arnold, Inc.; 8.6(right): Photo by Susumu Ito, from Charles Flickinger *Medical Cellular Biology*, W.B. Saunders, 1979; 8B: © Stockdisc/PunchStock; p.157(vegetables): © Photolink/Getty Images; p.157(couple): © Ronnie Kaufman/Corbis; p. 157(scale): © Dennis D. Potokar/Photo Researchers, Inc.; 8.12: Courtesy ® Quaker Oats; 8.13: © Volume 20/Photo Disc/Getty Images; 8.15a: © Biophoto Associates/Photo Researchers, Inc.; 8.15b: © Ken Greer/Visuals Unlimited; 8.15c(top): *A Colour Atlas and Text of Nutritional Disorders* by Dr. Donald D. McLaren/Mosby-Wolfe Europe Lt.; 8.15c(bottom): © ISM/Phototake; 8.17a: © Tony Freeman/PhotoEdit; 8.17b: © Donna Day/Stone/Getty Images; 8.17c: © Royalty-Free/Corbis.

Chapter 9
Opener: © H. Benser/zefa/Corbis; 9.4: © CNRI/Phototake; 9.5: © Dr. Kessel & Dr. Kardon/Tissues & Organs/Visuals Unlimited; 9.9: © Steinmark/Custom Medical Stock Photo; 9.13(both): © Martin Rotker/Martin Rotker Photography.

Chapter 10
Opener: © Sie Productions/zefa/Corbis; 10.4(top left): © Prof. P.M. Motta & M. Castellucci/SPL/Photo Researchers, Inc.; 10.4(top right, lower left): © 1966 Academic Press, from A.B. Maunsbach, *Journal of Ultrastructure and Molecular Structure Research*, Vol. 15, pp 242-282. Reprinted by permission of Elsevier; 10.5a: © Joseph F. Gennaro Jr./Photo Researchers, Inc.; 10B: © Ian Hooton/Photo Researchers, Inc.; 10.10: © AJPhoto/Photo Researchers, Inc.; 10.12: © James Cavallini/Photo Researchers, Inc.; 10C: © APPhoto/Steven Senne.

Chapter 11
Opener: © Royalty-Free/Corbis; 11.1(hyaline cartilage, compact bone): © Ed Reschke; 11.1(osteocyte): © Biophoto Associates/Photo Researchers, Inc.; 11A: © Jose Luis Pelaez, Inc./Corbis; 11Aa, b: © Michael Klein/Peter Arnold, Inc.; 11Ac: © Bill Aaron/PhotoEdit; 11.5b: © Tony Freeman/PhotoEdit; 11.8b: © Royalty-Free/Corbis; 11.13a: © Gerard Vandystadt/Photo Researchers, Inc.

Chapter 12
Opener: © Steve Helber/AP Photo; 12.1(all): © Ed Reschke; 12.2c: © David Young-Wolff/PhotoEdit; 12.3:

© David R. Frazier Photolibrary, Inc./Alamy; 12.5(left): © Royalty-Free/Corbis; 12.5(right): © Biology Media/Photo Researchers, Inc.; 12.6a: © Victor B. Eichler; 12.11(left): © Lawrence Manning/Corbis; 12.11(center): © G. W. Willis/Visuals Unlimited; 12.11(right): © Royalty-Free/Corbis; 12A: © Ben Margot/AP Images; 12.13: © Julie Lemberger/Corbis.

Chapter 13
Opener: © Scott T. Baxter/Getty Images; 13.2(top): © M.B. Bunge/Biological Photo Service; 13.2(bottom): © Manfred Kage/Peter Arnold, Inc.; 13.5a: Courtesy Dr. E.R. Lewis, University of California Berkeley; 13.7a: © Karl E. Deckart/Phototake; 13.7d: Posterior aspect of the cervical segment of the spinal cord (right side) from *The Photographic Atlas of the Human Body* by Branislav Vidic and Faustino R. Suarez. The C.V. Mosby Co., St. Louis, 1984; 13.8b: © Colin Chumbley/Science Source/Photo Researchers, Inc.; 13.13(all): © Marcus Raichle; 13.14: © Dr. Richard Kessel & Dr. Randy Kardon/Tissues & Organs/Visuals Unlimited; 13.17: © Betts Anderson Loman/PhotoEdit; 13.18: © Vol. 94 PhotoDisc/Getty Images.

Chapter 14
Opener: © Tom & Dee Ann McCarthy/Corbis; 14.4a: © Coneyl Jay/Photo Researchers, Inc.; 14.4c, d: © Omikron/SPL/Photo Researchers, Inc.; 14.7: © Dennis McDonald/PhotoEdit; 14.9: © Lennart Nilsson, from *The Incredible Machine*; 14.10: © Biophoto Associates/Photo Researchers, Inc.; 14A: © Pascal Goetgheluck/Photo Researchers, Inc.; 14.14: © P. Motta/SPL/Photo Researchers, Inc.; 14B(both): Robert S. Preston and Joseph E. Hawkins, Kresge Hearing Research Institute, University of Michigan; 14.16: © Myrleen Ferguson Cate/PhotoEdit.

Chapter 15
Opener: © Kevin Fleming/Corbis; 15.7a: © AP/Wide World Photos; 15.7b: © Ewing Galloway/Corbis; 15.8(all): From Clinical Pathological Conference, "Acromegaly, Diabetes, Hypermetabolism, Proteinuria and Heart Failure", *American Journal of Medicine* 20 (1956) 133, with permission from Excerpta Medica, Inc.; 15.9a: © Bruce Coleman, Inc./Alamy; 15.9b: © Medical-on-Line/Alamy; 15.9c: © Dr. P. Marazzi/Photo Researchers, Inc.; 15.13a: © Custom Medical Stock Photos; 15.13b: © NMSB/Custom Medical Stock Photos; 15.14(both): *Atlas of Pediatric Physical Diagnosis*, Second Edition by Zitelli & Davis, 1992. Mosby-Wolfe Europe Limited, London, UK; 15.15: © Peter Arnold, Inc./Alamy; 15A: © Vol. 83/PhotoDisc/Getty Images; 15.19: © The McGraw-Hill Companies, Inc./Evelyn Jo Hebert, photographer; 15B: © Royalty-Free/Corbis; 15.21: © Royalty-Free/Corbis.

Chapter 16
Opener: © Chris Rout/Alamy; 16.3: © Anatomical Travelogue/Photo Researchers, Inc.; 16.4b: © Ed Reschke; 16.4c: © Dr. Dennis Kunkel/Visuals Unlimited; 16.8: © Ed Reschke/Peter Arnold, Inc.; 16.13a: © Saturn Stills/Photo Researchers, Inc.; 16.13b: © Michael Keller/Corbis; 16.13c: © LADA/Photo Researchers, Inc.; 16.13d: © SIU/Visuals Unlimited; 16.13e: © Keith Brofsky/Getty Images; 16.13f: © The McGraw-Hill Companies, Inc./Lars A. Niki, photographer; 16.13g: © Phanie/Photo Researchers, Inc.; 16.13h: © Getty Images; 16.15: © CC Studio/SPL/Photo Researchers; 16A: © Ilene MacDonald/Alamy; 16.16: © Scott Camazine/Photo Researchers; 16.17(both): Courtesy of the Centers for Disease Control, Atlanta, GA. (Crooks, R. and Baur, K. *Our Sexuality*, 8/e, Wadsworth 2000 17.4, pg. 489; 16.18a: © Science VU/CDC/Visuals Unlimited; 16.18b(left, both): © Carroll H. Weiss/Camera M.D. Studios; 16.18b(right): © Science VU/Visuals Unlimited; 16B(left): © Vol. 161/Corbis; 16B(right): © Vol. 178/Corbis; 16C: © Argus Fotoarchiv/Peter Arnold, Inc.

AIDS Supplement
Opener: © Lennart Nilsson/Boehringer Ingelheim International GmbH; S.3(all): © Nicholas Nixon; S.A: © 2000–2006 Custom Medical Stock Photo.

Chapter 17
Opener: © 2000–2006 Custom Medical Stock Photo; 17.1: © David M. Phillips/Visuals Unlimited; 17.3(all): © Photo Lennart Nilsson/Albert Bonniers Forlag AB, *A Child is*

Born, Delacorte Press; 17.6a: © Lennart Nilsson, *A Child is Born*, Dell Publishing; p. 362(kittens): © Ben Margot/AP Images; p. 362(monkeys): © Jack Smith/AP Images; p. 362(twins): © Norbert Schaefer/Corbis; p. 362(Dolly): © Associated Press; 17.8: © James Stevenson/SPL/Photo Researchers, Inc.; 17.9(both): © Photo Lennart Nilsson/Albert Bonniers Forlag AB, *A Child is Born*, Delacorte Press; 17.10: Courtesy Dr. Howard Jones, Eastern VA Medical School; 17A: Courtesy Dr. Ann Streissguth, University of Washington School of Medicine, Fetal Alcohol and Drug Unit; 17B: © PhotoDisc Vol. 170/Getty Images; 17.11c: © Karen Kasmauski/Corbis; 17.12(top): © David Young-Wolff/PhotoEdit; 17.12(bottom): © Laura Dwight/PhotoEdit; 17.13: © Ronnie Kaufman/Corbis; 17C: © David Young-Wolff/PhotoEdit.

Chapter 18

Opener: © Joseph Sohm; ChromoSohmInc/Corbis; 18.1: © CNRI/SPL/Photo Researchers, Inc.; 18Ac: © Yoav Levy/Phototake; 18.5(early prophase, prophase): © Ed Reschke; 18.5(early metaphase): © Michael Abbey/Photo Researchers, Inc.; 18.5(metaphase, anaphase, telophase): © Ed Reschke; 18.6a(left): © Royalty-Free/Corbis; 18.6a(top right): © Scott Camazine/Photo Researchers, Inc.; 18.6a(bottom right): © Edward Kinsman/Photo Researchers, Inc.; 18.14a: Courtesy Chris Burke; 18.14b: © CNRI/SPL/Photo Researchers, Inc.; 18.15a(top): UNC Medical Illustration and Photography; 18.15a(bottom): © CNRI/SPL/Photo Researchers, Inc.; 18.15b(top): Courtesy Robert H. Shelton/Klinefelter Syndrome & Associates; 18.15b(bottom): © CNRI/SPL/Photo Researchers, Inc.; 18.18b: Courtesy The Williams Syndrome Association; 18.19b(both): From N.B. Spinner et al., *American Journal of Human Genetics* 55 (1994):p. 239. The University of Chicago Press; 18B: © Corbis/Sygma.

Chapter 19

Opener: © 2000–2006 Custom Medical Stock Photo; 19.1a: © Biology Media/Photo Researchers, Inc.; 19.1b: © SPL/Photo Researchers, Inc.; 19.1c: © Martin M. Rotker/Phototake; 19.6: © Dr. M.A. Ansary/SPL/Photo Researchers, Inc.; 19.8A, D: © Custom Medical Stock Photo, Inc.; 19.8B: © ISM/Phototake; 19.8C: © Dr. P. Marazzi/Photo Researchers, Inc.; 19.9: © Chris Bjornberg/Photo Researchers, Inc.; 19.10: © 2000–2006 Custom Medical Stock Photo; © Simon Fraser/Royal Victoria Infirmary/Photo Researchers, Inc.; 19D(man): © S. Terry/Photo Researchers, Inc.; 19D(lighter): © The McGraw Hill Companies, Inc./Evelyn Jo Hebert, photographer.

Chapter 20

Opener: © David Grossman/Photo Researchers, Inc.; 20.2a: © Superstock; 20.2b: © Michael Grecco/Stock Boston; 20.2c–f, h: © The McGraw-Hill Companies, Inc./Bob Coyle, photographer; 20.2g: © Royalty-Free/Corbis; 20.10(baby): © PM Images/Getty Images; 20.10(lysosomes): Courtesy Robert D. Terry/Univ. of California, San Diego, School of Medicine; 20.11: © Pat Pendarvis; 20.12(normal brain): Courtesy Dr. Hemachandra Reddy, The Neurological Science Institute, Oregon Health & Science University; 20.12(Huntington brain): Courtesy Dr. Hemachandra Reddy, The Neurological Science Institute, Oregon Health & Science University; 20.12(woman): © Steve Uzzell; 20A(both): © Brand X/SuperStock RF; 20.13a: Courtesy University of Connecticut, Peter Morenus, photographer; 20.14: © Royalty-Free/Corbis; 20.15: © Medican/Medical-On-Line; 20.19(abnormal, normal): Courtesy Dr. Rabi Tawil, Director, Neuromuscular Pathology Laboratory, University of Rochester Medical Center; 20.19(boy): Courtesy Muscular Dystrophy Association; 20.20(queen): © Stapleton Collection/Corbis; 20.20(prince): © Huton Archive/Getty Images; 20B: © Bob Daemmrich/PhotoEdit.

Chapter 21

Opener: © Ted Horowitz/Corbis; 21.12: Courtesy Alexander Rich; 21.14: From M.B. Roth and J.G. Gall, *Cell Biology* 105:1047-1054, 1987. © Rockefeller University Press; 21.15(chimp mouth, chimp ear, chimp nose): © Digital Vision/Getty Images; 21.15(human mouth, human nose): © Royalty-Free/Getty Images; 21.15(human ear): © The McGraw Hill Companies, Inc; p. 457(woman): © Paul Markow/FPG International/Getty Images; p. 457(couple): © Ron Chapple/FPG International/Getty

Images; p. 457(baby): Steven Jones/FPG International/Getty Images; 21.20a: © Nita Winter; 21.20b: Courtesy Robert H. Devlin, Fisheries and Oceans Canada; 21.20c: © Richard Shade; 21.20d: Courtesy Bio-Rad; 21.20e: © Jerry Mason/Photo Researchers, Inc; 21.21(all): Courtesy Monsanto; 21.22b(both): Courtesy Eduardo Blumwald; 21Aa: © Larry Lefever/Grant Heilman Photography, Inc.; 21Ab: © James Shaffer/PhotoEdit; 21Ac: Courtesy Carolyn Merchant/UC Berkeley.

Chapter 22

Opener: © BIOS Ruoso Cyril/Peter Arnold, Inc.; 22.3: © Richard T. Nowitz/Corbis; 22.4(fossil): © Jean-Claude Carton/Bruce Coleman Inc.; 22.4(art): © Joe Tucciarone; 22.5(fossil): © J.G.M. Thewissen, www.neoucom.edu/DEPTS/ANAT/Thewissen/index.html; 22.7(both): © Carolina Biological Supply/Phototake; 22.10(gibbon): © Hans & Judy Beste/Animals Animals/Earth Scenes; 22.11(orangutans, chimpanzees, gorillas): © Creatas/PunchStock; 22.12: © 2001 Time Inc. Reprinted by Permission; 22.14a: © Dan Dreyfus and Associates; 22.14b: © John Reader/Photo Researchers, Inc.; 22.16: © National Museum of Kenya; 22A(art): © Mick Tsikas/epa/Corbis; 22A(skulls): © Richard Lewis/Associated Press; 22.17: Courtesy The Field Museum # A102513c; 22.18: Transp. #608 Courtesy Department of Library Services, American Museum of Natural History; 22.19a: © PhotoDisc/Getty Images; 22.19b: © Dr. Sylvia S. Mader; 22.19c: © B & C Alexander/Photo Researchers, Inc.; 22B: © Peter Bowater/Photo Researchers, Inc.

Chapter 23

Opener: © Michael Newman/PhotoEdit; 23.1(temperate): © Vol. 6/Corbis; 23.1(desert, savanna): © Gregory Dimijan/Photo Researchers; 23.1(rain forest): © Vol. 121/Corbis; 23.1 (prairie): © Vol. 86/Corbis; 23.1(taiga): © Vol. 63/Corbis; 23.1(tundra): © Vol. 121/Corbis; 23.2a: © Vol. 121/Corbis; 23.2b: © The McGraw-Hill Companies, Inc./Carlyn Iverson, photographer; 23.2c: © David Hall/Photo Researchers; 23.2d: © Vol. 121/Corbis; 23.3a(diatom): © Ed Reschke/Peter Arnold, Inc.; 23.3a (tree): © Herman Eisenbeiss/Photo Researchers, Inc.; 23.3b(caterpillar): © Royalty-Free/Corbis; 23.3b(rabbit): © Gerald C. Kelley/Photo Researchers, Inc.; 23.3c(spider): © Bill Beatty/Visuals Unlimited; 23.3c(osprey): © Joe McDonald/Visuals Unlimited; 23.3d(bacteria): © SciMAT/Photo Researchers, Inc.; 23.3d(mushrooms): © Michael Beug; 23.5: © George D. Lepp/Photo Researchers, Inc.; 23.13a: © John Shaw/Tom Stack & Assoc.; 23.13b: © Thomas Kitchin/Tom Stack & Assoc.; 23A: Courtesy GSFC/NASA; 23B(geese): © Emi Allen/SuperStock RF; 23B(brown bear): © Royalty-Free/Corbis; 23B(ox): © Arthur Morris/Visuals Unlimited; 23B(caribou, polar bear): © Royalty-Free/Corbis; 23B(wolf): © Digital Vision/Getty Images; 23B(tundra): © Royalty-Free/Corbis.

Chapter 24

Opener: © Christi Carter/Grant Heilman Photography; 24.2c: © Still Pictures/Peter Arnold, Inc.; 24.3(land): © Vol. 39/PhotoDisc/Getty Images; 24.3(ocean): © The McGraw-Hill Companies, Inc./ Evelyn Jo Hebert, photographer; 24.3(food): © The McGraw-Hill Companies, Inc./John Thoeming, photographer; 24.3(solar power): © Gerald and Buff Corsi/Visuals Unlimited; 24.3(mining): © National Geographic; 24.4: © Melvin Zucker/Visuals Unlimited; 24.5b: © Carlos Dominguez/Photo Researchers, Inc.; 24.6b: Courtesy Carla Montgomery; 24.7a, c: © Gregg Vaughn/Tom Stack & Associates; 24.7b: © Jeremy Samuelson/FoodPix/Getty Images; 24.8: © Mark Gormus/Associated Press; 24.9a: © Jodi Jacobson/Peter Arnold, Inc.; 24.9b, c: © Peter Essick/Aurora; 24.10a: © Laish Bristol/Visuals Unlimited; 24.10b: © Inga Spence/Visuals Unlimited; 24.10c: V. Jane Windsor, Division of Plant Industry, Florida Department of Agriculture & Consumer Services; 24.11: © Jim Richardson/Corbis; 24.12a: © David L. Pearson/Visuals Unlimited; 24.12b: © S.K. Patrick/Visuals Unlimited; 24.12c: © Argus Fotoarchiv/Peter Arnold, Inc.; 24.12d: © Gerald & Buff Corsi/Visuals Unlimited; 24.13a: © Matt Stiveson/NREL/DOE; 24.13b: Courtesy DaimlerChrysler; 24.13c: © Warren Gretz/NREL/DOE; 24.15b: © Gunter Ziesler/Peter Arnold, Inc.; 24.16a: © Shane Moore/Animals Animals/Earth Scenes; 24.16b: © PhotoLink/Getty Images RF; 24.17(periwinkle): © Kevin Schafer/Peter Arnold, Inc.; 24.17(armadillo):

© PhotoDisc/Getty Images; 24.17(fishing): © Herve Donnezan/Photo Researchers, Inc.; 24.17(bat): © Merlin D. Tuttle/Bat Conservation International; 24.17(lady bug): © Anthony Mercieca/Photo Researchers, Inc.; 27.17(rubber tree): © Bryn Campbell/Stone/Getty Images; 24A: Courtesy Marine Aquarium Council; 24.18(soil erosion): © USDA/Nature Source/Photo Researchers, Inc.; 24.18(desertification): © Carlos Dominguez/Photo Researchers, Inc.; 24.18(fish): © Shane Moore/Animals Animals/Earth Scenes; 24.18(eutrophic lake): © Michael Gadomski/Animals Animals/Earth Scenes; 24.18(pollution): © Thomas Kitchin/Tom Stack & Assoc.; 24.18(cougar): © Bill Zimmer/Associated Press; 24.19(farming): © Inga Spence/Visuals Unlimited; 24.19(storm barriers): © Peter DeJong/Associated Press; 24.19(recycling): © Jeffrey Greenberg/Photo Researchers, Inc.; 24.19(fuel cell): © Matt Stiveson/NREL/DOE; 24.19(rooftop): © Reuters/Corbis; 24.19(scale insect): V. Jane Windsor, Division of Plant Industry, Florida Department of Agriculture & Consumer Services; 24.20: © Dan McCoy/Rainbow; 24.22: © Bob Rowan/Progressive Image/Corbis.

Text and Line Art

Inside front cover

ScienCentral news stories are supported in part by the National Science Foundation (http://www.nsf.gov) <http://www.nsf.gov)>, Division of Informal Science Education, under Grants No. ESI0206184, ESI021155, and ESI9904457. Any opinions, findings, and conclusions or recommendations expressed in this material are those of the author(s) and do not necessarily reflect the views of the National Science Foundation.

Chapter 7

7.14a & b: Approved by the Advisory Committee on Immunization Practices (ACIP), the American Academy of Pediatrics (AAP), and the American Academy of Family Physicians.

Chapter 8

8.11: Evidence Report of Clinical Guidelines on the Identification, Evaluation, and Treatment of Overweight and Obesity in Adults, 1998. NIH/National Heart, Lung, and Blood Institute; and U.S. Department of Agriculture, Report of the Dietary Guidelines Advisory Committee on the Dietary Guidelines for Americans, 2000; 8.16: Data from the U.S. Department of Agriculture.

Chapter 9

Health Focus: Adapted from The Most Often Asked Questions About Smoking Tobacco and the Answers, Revised July 1993, The American Cancer Society, Inc.

Chapter 14

Table 14A: National Institute on Deafness and Other Communication Disorders, January 1990, National Institute of Health.

Chapter 17

Table 17.2: Reprinted from Gestation Encyclopedia of Human Biology, 7e, vol. 7, Thomas R. Moore. © 1997 with permission of Elsevier.

Chapter 18

Health Focus: Reprinted with permission from Stefan Schwarz. For more information regarding Klinefelter Syndrome, please visit http://klinefeltersyndrome.org.

Chapter 21

Science Focus: Debra Levey Larson is a writer for the College of Agricultural, Consumer and Environmental Sciences at the University of Illinois (217-244-2880; dlarson@uiuc.edu; www.aces.uiuc.edu/news).

Chapter 24

24.1: Population Reference Bureau and United Nations, World Population Projections to 2100 (1998); Figure 24.2: United Nations Population Division, 1998; 24.5: Data from A. Goudie and J. Wilkinson, The Warm Desert Environment. Copyright 1977 by Cambridge University Press, New York; 24.6: United Nations Environment Program, World Resources Institute; 24.16: Data from U.S. Marine Fisheries Service.

Index